C0-AOE-277

Analysis of electric circuits

1 81

McGRAW-HILL ELECTRICAL AND ELECTRONIC ENGINEERING SERIES

FREDERICK EMMONS TERMAN, *Consulting Editor*
W. W. HARMAN AND J. G. TRUXAL,
Associate Consulting Editors

AHRENDT AND SAVANT · Servomechanism Practice
ANGELO · Electronic Circuits
ASELTINE · Transform Method in Linear System Analysis
ATWATER · Introduction to Microwave Theory
BAILEY AND GAULT · Alternating-current Machinery
BERANEK · Acoustics
BRACEWELL · The Fourier Transform and Its Applications
BRENNER AND JAVID · Analysis of Electric Circuits
BROWN · Analysis of Linear Time-invariant Systems
BRUNS AND SAUNDERS · Analysis of Feedback Control Systems
CAGE · Theory and Application of Industrial Electronics
CAUER · Synthesis of Linear Communication Networks
CHEN · The Analysis of Linear Systems
CHEN · Linear Network Design and Synthesis
CHIRLIAN · Analysis and Design of Electronic Circuits
CHIRLIAN AND ZEMANIAN · Electronics
CLEMENT AND JOHNSON · Electrical Engineering Science
COTE AND OAKES · Linear Vacuum-tube and Transistor Circuits
CUCCIA · Harmonics, Sidebands, and Transients in Communication Engineering
CUNNINGHAM · Introduction to Nonlinear Analysis
D'AZZO AND HOUPIS · Feedback Control System Analysis and Synthesis
EASTMAN · Fundamentals of Vacuum Tubes
ELGERD · Control Systems Theory
FEINSTEIN · Foundations of Information Theory
FITZGERALD AND HIGGINBOTHAM · Basic Electrical Engineering
FITZGERALD AND KINGSLEY · Electric Machinery
FRANK · Electrical Measurement Analysis
FRIEDLAND, WING, AND ASH · Principles of Linear Networks
GHAUSI · Principles and Design of Linear Active Circuits
GHOSE · Microwave Circuit Theory and Analysis
GREINER · Semiconductor Devices and Applications
HAMMOND · Electrical Engineering
HANCOCK · An Introduction to the Principles of Communication Theory
HAPPELL AND HESSELBERTH · Engineering Electronics
HARMAN · Fundamentals of Electronic Motion
HARMAN · Principles of the Statistical Theory of Communication
HARMAN AND LYTLE · Electrical and Mechanical Networks
HARRINGTON · Introduction to Electromagnetic Engineering
HARRINGTON · Time-harmonic Electromagnetic Fields
HAYASHI · Nonlinear Oscillations in Physical Systems
HAYT · Engineering Electromagnetics
HAYT AND KEMMERLY · Engineering Circuit Analysis
HILL · Electronics in Engineering
JAVID AND BRENNER · Analysis, Transmission, and Filtering of Signals
JAVID AND BROWN · Field Analysis and Electromagnetics
JOHNSON · Transmission Lines and Networks
KOENIG AND BLACKWELL · Electromechanical System Theory
KOENIG, TOKAD, AND KESAVAN · Analysis of Discrete Physical Systems
KRAUS · Antennas
KRAUS · Electromagnetics
KUH AND PEDERSON · Principles of Circuit Synthesis
KUO · Linear Networks and Systems
LEDLEY · Digital Computer and Control Engineering
LEPAGE · Analysis of Alternating-current Circuits

LePage · Complex Variables and the Laplace Transform for Engineering
LePage and Seely · General Network Analysis
Levi and Panzer · Electromechanical Power Conversion
Ley, Lutz, and Rehberg · Linear Circuit Analysis
Linvill and Gibbons · Transistors and Active Circuits
Littauer · Pulse Electronics
Lynch and Truxal · Introductory System Analysis
Lynch and Truxal · Principles of Electronic Instrumentation
Lynch and Truxal · Signals and Systems in Electrical Engineering
Manning · Electrical Circuits
McCluskey · Introduction to the Theory of Switching Circuits
Meisel · Principles of Electromechanical-energy Conversion
Millman · Vacuum-tube and Semiconductor Electronics
Millman and Seely · Electronics
Millman and Taub · Pulse and Digital Circuits
Millman and Taub · Pulse, Digital, and Switching Waveforms
Mishkin and Braun · Adaptive Control Systems
Moore · Traveling-wave Engineering
Nanavati · An Introduction to Semiconductor Electronics
Pettit · Electronic Switching, Timing, and Pulse Circuits
Pettit and McWhorter · Electronic Amplifier Circuits
Pfeiffer · Concepts of Probability Theory
Pfeiffer · Linear Systems Analysis
Reza · An Introduction to Information Theory
Reza and Seely · Modern Network Analysis
Rogers · Introduction to Electric Fields
Ruston and Bordogna · Electric Networks: Functions, Filters, Analysis
Ryder · Engineering Electronics
Schwartz · Information Transmission, Modulation, and Noise
Schwarz and Friedland · Linear Systems
Seely · Electromechanical Energy Conversion
Seely · Electron-tube Circuits
Seely · Electronic Engineering
Seely · Introduction to Electromagnetic Fields
Seely · Radio Electronics
Seifert and Steeg · Control Systems Engineering
Siskind · Direct-current Machinery
Skilling · Electric Transmission Lines
Skilling · Transient Electric Currents
Spangenberg · Fundamentals of Electron Devices
Spangenberg · Vacuum Tubes
Stevenson · Elements of Power System Analysis
Stewart · Fundamentals of Signal Theory
Storer · Passive Network Synthesis
Strauss · Wave Generation and Shaping
Su · Active Network Synthesis
Terman · Electronic and Radio Engineering
Terman and Pettit · Electronic Measurements
Thaler · Elements of Servomechanism Theory
Thaler and Brown · Analysis and Design of Feedback Control Systems
Thaler and Pastel · Analysis and Design of Nonlinear Feedback Control Systems
Thompson · Alternating-current and Transient Circuit Analysis
Tou · Digital and Sampled-data Control Systems
Tou · Modern Control Theory
Truxal · Automatic Feedback Control System Synthesis
Tuttle · Electric Networks: Analysis and Synthesis
Valdes · The Physical Theory of Transistors
Van Bladel · Electromagnetic Fields
Weinberg · Network Analysis and Synthesis
Williams and Young · Electrical Engineering Problems

Analysis of electric circuits

SECOND EDITION

EGON BRENNER
Professor of Electrical Engineering
The City College of the City University
of New York

MANSOUR JAVID
Professor of Electrical Engineering
The City College of the City University
of New York

McGraw-Hill Book Company
NEW YORK, ST. LOUIS, SAN FRANCISCO
TORONTO, LONDON, SYDNEY

Analysis of electric circuits

Library of Congress Catalog Card Number 67-12958

ISBN 07-007630-8
67890 HDHD 7987

Preface

The subject of linear electric circuit analysis has always formed a part of the electrical engineering curriculum. In recent years, however, the need to broaden the study of circuit analysis and thus provide the student with the necessary background of fundamental concepts and techniques has become evident. The study of such concepts and techniques is aimed at preparing the student for various areas of more recent interest, including *feedback control systems, spectral analysis and signal theory, noise and communication theory, network synthesis, computer sciences, electromagnetic field theory,* and *pulse and digital circuits,* as well as older disciplines such as *electronic circuits, radio communications, transmission lines,* and *energy-conversion devices.*

We believe this book serves as a suitable text for introducing the subject of engineering analysis via circuit theory at the earliest possible level, that is, the sophomore-junior level.

We have in this new edition retained the emphasis on the dynamical behavior of electrical circuits, one of the basic features of the first edition. The emphasis on waveforms and complete response is preferred as a pedagogical approach to the compartmentalization of circuit analysis into d-c, a-c, and transients.

To adjust the students' changed needs we have, in this new edition, introduced additional analytical techniques, namely, *state analysis, signal-flow analysis,* and a study of the relationship between *pole-zero analysis* and the *Laplace transform.* In order to include this new material we have somewhat deemphasized the details of calculations as they apply to circuits with many elements. Further economy has been achieved by introducing the symbolism of the differential operator early in the text. The Laplace transform is introduced through the concept of superposition of exponential functions after a thorough treatment of time-domain analysis and response to exponential sources. In this way the student is led to understand transform techniques as opposed to simply utilizing them as a mechanical means of "cranking out" answers. This understanding is enhanced through the study of the relationship between the poles of transform network function and the time-domain properties of source-free response of the network. Solution of a given problem by time-domain techniques as well as by Laplace transform method provides the student with additional insight into the merits of each technique.

The book is divided into four parts. Chapters 1 to 6 constitute the first part and treat the following areas: *waveforms, Kirchhoff's laws,* the powerful *differential-operator method* for formulating equilibrium equations including the *operational-network or system-function concept, Thévenin's theorem* and the fundamental concepts of *time constant, natural frequency,* and *complex frequency* as they are related to *exponential waveforms.* Chapters 7 to 10 form the second stage of the development. They deal with *pole-zero analysis, Laplace transform, sinusoidal steady-state* and *frequency response.* The general development is completed in Chapters 11 to 13 where the techniques of *state theory, node* and *loop equations,* and *signal flow graphs* are treated. The remaining chapters, 14 to 16, include the additional topics of *magnetically coupled circuits, three-phase circuits,* and *Fourier series.* Appendixes deal with *rms values, complex numbers, phasors,* the elements of *matrix algebra,* and the *computer solution* of network problems.

At all points in the text, care has been taken to motivate the analysis by the discussion of meaningful applications that can be understood at the students' level. In addition all subjects, including topics such as state theory and signal flow graphs that are new to this edition, have been approached at a fundamental level, and the conventional freshman mathematics courses form a completely adequate background.

The amount of material in the book exceeds what can be covered in a conventional one-year course by about 15 per cent. The classroom teacher will have no difficulty in the selection process as there is sufficient overlap between chapters so that continuity is not lost; moreover, various patterns of emphasis can be developed to suit students' needs. Thus, for example, the sections dealing with locus diagrams can be omitted or treated very briefly; the subject of tuned transformer circuits can be postponed to the next course; three-phase circuits can be omitted and taught as part of a machinery course; Fourier series and harmonic analysis can be postponed to a course in signal theory; details of solution of problems in several of the chapters can be omitted. To relieve the teacher of the burden of using valuable class time discussing computational details we have again included a large number of worked-out examples. These examples, as well as the detail with which the fundamental proofs and theorems are presented, account in part for the bulk of this volume. General reading lists have been omitted: the circuits literature is particularly extensive, and we feel the classroom teacher is best able to choose references that suit his students.

A large number of problems, many newly designed for this edition, is provided at the end of each chapter. The exercises range in difficulty from simple application of text material to theorems (or similar problems) intended to permit the interested student to carry forward the theory independently.

For their many helpful comments and criticisms we thank our colleagues who have taught from the first edition as well as from the classroom notes that were used at The City College of New York prior to that edition. In this respect the contributions of Professors L. Bergstein (now at the Polytechnic Institute of Brooklyn), G. J. Clemens, D. Eitzer, L. Echtman, W. T. Hunt, Jr., S. R. Parker (now at the U. S. Naval Postgraduate School), L. Rosenthal (now at Stevens Institute of Technology), A. G. Schillinger (now at the Polytechnic Institute of Brooklyn), C. Shulman, R. Stein, Doctor D. Politis, and Messrs. A. Van Gelder, R. D. Klafter and P. M. Brown deserve special mention. The generous help given by Dr. Peter Weiner (now at Princeton University) in the preparation of Chapter 13 is particularly appreciated.

Finally, the authors once again acknowledge with gratitude the invaluable assistance of Miss Sadie Silverstein in the preparation of the manuscript.

Egon Brenner
Mansour Javid

Contents

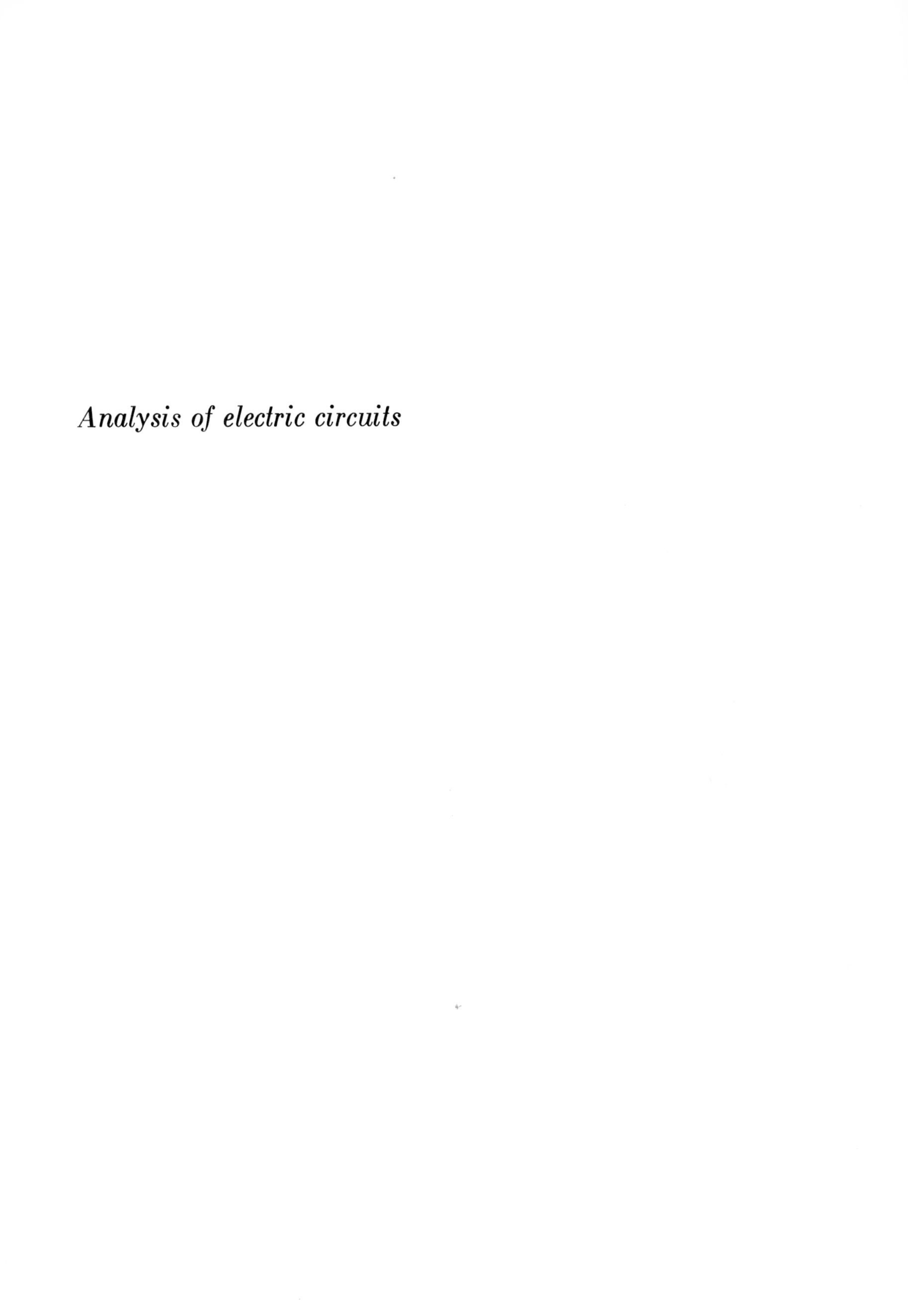

Analysis of electric circuits

Chapter 1 Circuits and circuit elements

Electrical engineering deals with the design, construction, utilization, and maintenance of electrical apparatus. The electrical engineer must not only understand the new devices which are continually being introduced, he must also conceive and develop novel equipment. Complete understanding of each device requires specialized knowledge of the branches of electrical engineering. For example, understanding the operation of a radio transmitter requires a knowledge of electronics, and understanding the behavior of electric motors requires a background in the theory of electrical machinery. Yet the same basic tool is employed in the study of all these divers specializations within electrical engineering. This tool is called *circuit theory*.

The reader probably recalls that basic electrical phenomena in physics are described by the laws of electromagnetism as formulated by Coulomb, Oersted, Ampère, Faraday, and Maxwell. None of these laws deals with electric circuits; they all deal with fields. Although a knowledge of field theory is *not* essential for understanding circuit theory, it is the authors' belief that, at the outset, the reader must be made conscious of the relation between the field and circuit concepts, as well as the aims and limitations of circuit theory.

To serve this purpose a brief qualitative exposition of basic electromagnetic field concepts is given in this chapter. Following this, the elementary, or "primitive," circuit concepts and circuit elements are defined. The remaining chapters of the book deal exclusively with circuit concepts and require no knowledge of electromagnetic theory.

1-1 Field concepts

It is assumed that the reader is familiar with the concept of energy and power as introduced in the problems of mechanics. To illustrate the significance of terms used in connection with electricity and magnetism, we shall use examples from mechanics.

FORCE FIELD The concepts of force and mass are closely related. For example, gravitational force presupposes the existence of mass. Similarly, observation of the attraction which a magnet possesses for a piece of iron depends on the existence of the mass of iron to be attracted. Nonetheless, through a process of abstraction, we can think of force as a quality of space without explicit reference to the ultimate cause of the force or to the mass upon which the force will be exerted. *With this understanding, we define a force field as a region of space where a force is associated with its every point.* The gravitational field is a familiar example.

The field intensity at a point in a field is defined as the force applied to some unit quantity placed at that point. For example, in a gravitational field, the intensity at a given point is the force applied to a unit mass placed at that point.

ENERGY STORED IN SPACE Consider the motion of a mass in the gravitational field of the earth. Let a mass M, which was at rest at point A, at a height h_1, fall to a point B at the height of h_2. At point B the mass M has acquired a kinetic energy which it did not possess at point A. Where did the energy come from? This question is answered by introducing the concept of potential energy. The potential energy of the mass at point A with respect to point B is $Mg(h_1 - h_2)$, where g is the gravitational constant. Thus, in falling from point A to point B, the mass has lost $Mg(h_1 - h_2)$ of its potential energy and acquired its equivalent in kinetic energy. This argument is a familiar one. However, we may ask another question. At point A the mass had a certain potential energy; where was this energy located? This question is not meaningful unless we are convinced that energy must be located somewhere, and such a conviction can be achieved only if we know what energy is. The questions about the nature of fundamental concepts such as mass, time, length, energy, etc., are outside the domain of natural science and belong to the study of metaphysics. In the field theory of gravitation we *assume* that the potential energy is stored in the space (field) where the gravitational force exists. As a mass "falls" in this field, it "draws" from the potential energy stored in the field and transforms it into other forms of energy. In order to "raise" a mass in a gravitational field, work must be done. The energy spent in raising the mass is stored in the gravitational field and is available in the form of potential energy. Thus a mass "rising" in a gravitational field "delivers" energy to the field.

So far we have discussed only the familiar gravitational field. From our observation of nature we come to the conclusion that there are force fields other than the gravitational field. Two such fields are the *electric* and *magnetic* fields. The arguments given for the storage of energy in a gravitational field can be extended to apply to electric and magnetic fields. This is discussed in Sec. 1-2.

ENERGY DENSITY As soon as we accept the concept of energy stored in space, it is logical to inquire about the quantity of energy stored in a unit volume. If an energy distribution is uniform in space, the energy stored in any unit volume will be the same as that stored in any other, regardless of the position of the unit volume. However, most often this distribution is not uniform. For example, in the gravitational field of the earth, the force applied to a mass decreases as the mass moves away from the earth. Since the storage of energy is related to the force, it is plausible to assume that the gravitational energy stored in a unit volume located "far" away from the earth will be smaller than the energy stored in a unit volume "closer" to the earth. Moreover, within a given unit volume, the energy distribution is not uniform. As the dimensions of a given volume decrease (approach zero), however, the variation of energy distribution within that volume becomes smaller (approaches zero). In the limit, as the volume approaches a point, the energy stored in that volume approaches zero, but the ratio of the energy stored to the volume remains finite. Thus we define energy density *at a point* as the ratio of the energy stored in an infinitesimal volume located at that point to the volume.

FIELD QUANTITIES With each point of a force field we associate a field intensity (a vector quantity) and an energy density (a scalar quantity). These quantities are called field quantities, and may vary from point to point, as well as with time. For a given field we may define various other field quantities which would best describe that field.

STATIC FIELD A field whose field quantities are independent of time is called a static field. In a static field the field quantities may vary from point to point, but at a given point their value is constant for all time.

DYNAMIC FIELD A field whose field quantities vary with time is called a dynamic field.

1-2 Electromagnetic fields

In this section we shall attempt to present a theory which will explain the phenomena of electricity and magnetism. This presentation is qualitative because we wish to show only the relationship between the fundamental concepts of field and circuit theories without presenting involved analytical derivations.

Electric charge The concepts of electric charge and the associated electric and magnetic fields are introduced in physics to explain and analyze certain observed phenomena. The electric charge is a fundamental quantity, and is described by its attributes as follows:

1. The electric charge is always associated with mass.
2. It exists in two forms, called the positive and the negative charge, respectively.
3. Two force fields are associated with a moving charge. These fields are called the electric and magnetic fields. The intensity of these fields at a given point of space will depend on the magnitude of the charge; its velocity; its acceleration; the distance of the point under discussion from the charge; and three qualities of the medium, ϵ, μ, and ρ, called the permittivity, permeability, and resistivity of the medium, respectively. The (vector) intensities of the electric and magnetic fields are designated by the letters $\mathcal{E}$ and $\mathcal{H}$, respectively.

Electromagnetic field quantities The field associated with the field intensities $\mathcal{E}$ and $\mathcal{H}$ is called the electromagnetic field. Electromagnetic theory relates the values of $\mathcal{E}$ and $\mathcal{H}$ to each other and to the distribution and motion of the charges in space. The fields associated with charges moving with nonuniform acceleration are dynamic fields, and the relationships between these dynamic field quantities are given by a set of equations called *Maxwell's equations.* Three important assumptions which are postulated in conjunction with Maxwell's equations are of great interest to us and will be given here:

1. The density of energy stored in an electric field at a given point in space[1] is $W_E = \epsilon E^2/2$, where ϵ and E are the permittivity of the space and the magnitude of the electric field intensity, respectively, at the point in question. We shall call the energy stored in an electric field the electric energy.
2. The density of the energy stored in a magnetic field at a given point in space[1] is $W_M = \mu H^2/2$, where μ is the permeability of the space and H the magnitude of the magnetic field intensity at that point in space. We shall call the energy stored in a magnetic field the magnetic energy.
3. At a point in space where the magnitude of the electric field intensity is E, electric energy will be transformed into heat at a *rate* equal to E^2/ρ per unit volume per second, where ρ is a property of the space at the point under discussion and is called the resistivity of the space at that point. The quantities ϵ, μ, and ρ describe the medium in which the electromagnetic field is established.

[1] Details of the conditions under which this statement is valid can be found in texts on electromagnetic theory. See, for example, M. Javid and P. M. Brown, "Field Analysis and Electromagnetics," secs. 6-12 to 6-14 and 8-7 to 8-9, McGraw-Hill Book Company, New York, 1963.

1-3 Exchange of energy between charge and fields

ELECTROMAGNETIC ENERGY We shall use the term electromagnetic energy whenever we wish to refer to either electric or magnetic energy or both. A charge moving in an electromagnetic field will exchange energy with the field in the same manner as a mass moving in a gravitational field exchanges energy with that field. If the charge is accelerated by the field, it will abstract energy from the field and transform it into kinetic energy. If it is decelerated, it gives up its kinetic energy, which will be stored in the electromagnetic field. If the medium has finite conductivity, the motion of the charge will be affected by the material in the medium, and electromagnetic energy will be converted into heat. The fundamental problem of the electromagnetic theory is the determination of the field quantities and computation of energy distribution and transformation in the space.

1-4 Circuit concepts

The solution of a field problem requires the determination of the field quantities $\mathcal{E}$ and $\mathcal{H}$ at every point in space as a function of time, when the initial distribution of charges, their velocities, and accelerations are given. Since these field quantities are functions of three space variables (for example, x, y, z) and time, the solution of field problems is generally complex.

Any electrical device entails the movement of charges. An exact analysis of the behavior of electrical devices should therefore be made by the use of the field theory. Since such an analysis can be exceedingly complex, it has been found necessary to develop an approximate, but sufficiently accurate, method of dealing with many electrical engineering problems. This method is called *circuit analysis*. The basic problem of circuit analysis is the determination of energy distribution and transformation in *different parts* of a device as a function of time *only*.[1] When an electrical device is functioning, there will be charges distributed in different parts of the device, and most often these charges move with nonuniform velocities. Hence, an electromagnetic field is established over all space; energy is dissipated into heat, stored or withdrawn from this field, as the charges move in the device. It follows that, as a result of accelerating or decelerating charges in a device, energy will be transferred from one region of space to other regions. This phenomenon is known as the radiation of energy, and in electromagnetic theory it is shown that the energy is transmitted (radiated) with the velocity of light. (In fact, light itself is an electromagnetic phenomenon.)

When we discuss an electrical device, we assume that charges exist and move in that device. Therefore the phenomenon stated above, which also entails radiation of energy, will take place when the charges in the

[1] It is possible to extend the meaning of the term circuit analysis to apply to systems with distributed parameters, such as transmission lines. In such cases space is also an independent variable. In this book we do not deal with such systems.

device are accelerated or decelerated. However, the ratio of the energy associated with the rest of space will depend on the dimensions of the device and the magnitude of the acceleration of the charges. It can be shown that, if the product of the maximum value of the acceleration of the charges in a device and the largest dimension of the device is much smaller than the square of the velocity of light, then the greater part of the energy storage and its transformations will be confined to the device and its immediate vicinity, and the distribution and transformation of energy at distant points can be neglected. For devices with dimensions of the order of tens of feet, the acceleration of the charges must be of the order of 10^{16} ft/sec^2 before any appreciable radiation will take place. When radiation can be neglected, all energy storage and transformation take place in the immediate neighborhood of the device. In addition, we assume that the fields due to the motion of charges in one part of the device are established instantaneously in other parts instead of propagating with the velocity of light.

In the solution of the field problem the energy density at a given point of space was given in terms of the electric field intensity $\mathcal{E}$ and the magnetic field intensity $\mathcal{H}$. We have said that the electromagnetic energy stored per unit volume of space is $W_{EM} = (\epsilon E^2 + \mu H^2)/2$ and the rate of transformation of energy into heat is E^2/ρ per unit volume.

In circuit theory, instead of computing the values of $\mathcal{E}$ and $\mathcal{H}$ for every point in space as a function of time and evaluating from them the energy density and the energy transformation (into heat), we define two other variables, namely, *voltage* and *current*, associated with a pair of terminals in the device, and relate these to the energy delivered to that region of the device which is identified by the terminals in question. Thus any electrical device is divided in regions or volumes, symbolically shown by rectangles or other symbols identified by two terminals and called a *terminal pair* (Fig. 1-1). In this way a device is represented by an interconnection of terminal

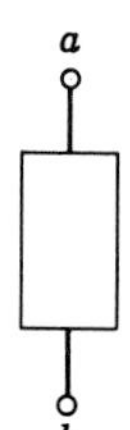

***Fig. 1-1** Graphical symbol for a terminal pair representing a part of a device between two terminals.*

pairs, connected at their accessible terminals and called a terminal-pair network, as shown in Fig. 1-2. Such a terminal-pair network is said to be the equivalent network of the device. Instead of discussing energy storage and transformation at all points of the device and computing them from $\mathcal{E}$ and $\mathcal{H}$, we focus our attention on the total energy storage and transformation within each terminal pair. These processes are computed from two variables, voltage and current, which are associated with the terminals of the terminal pair. Thus field problems, whose variables ($\mathcal{E}$ and $\mathcal{H}$) were functions of space (x,y,z) and time (t), have been reduced to circuit problems, whose variables, voltage and current, will be functions of time only and refer to the terminals of a given terminal pair.

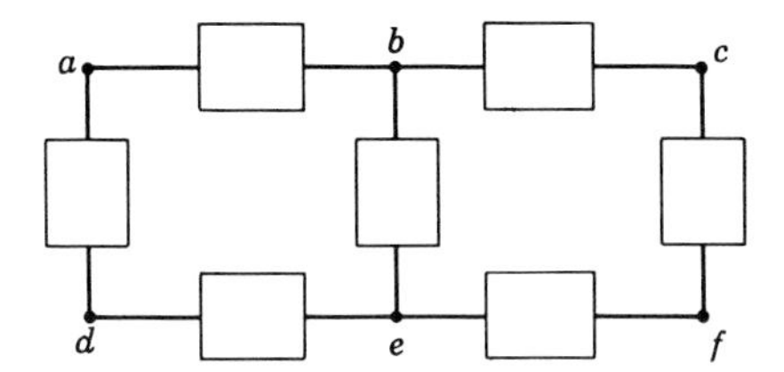

Fig. 1-2 A terminal-pair network which may represent a device.

1-5 Electric circuits

We assume that the electric charges which enter one terminal of a terminal pair will come out of the other terminal. Since we have assumed that charges move inside the device, they must be confined to motion through terminal pairs which represent different parts of the device, and therefore must move through a closed path formed by terminal pairs connected together. Any such closed path is called a *circuit.* It must be noted that the term circuit may be used to refer to a connection of terminal pairs which is not closed, but it is always implied that such connections will be closed eventually.

Before defining the principal variables of circuit theory, we digress for a moment to explain the notation used.

NOTATION In this text, variables which are or may be time-dependent are denoted by small *italic* letters. Thus, for example, a voltage v which is a function of time will be denoted by $v(t)$ or v, but never by V or $V(t)$. The reader is cautioned to note that the small letter v stands for the time function $v(t)$ even when the argument (t) is omitted. Consistent with the above, constants or parameters which are time-invariant are denoted by capital letters. Thus V stands for a constant value of the variable v, and the equation $v = V$ is meaningful; i.e., it reads: The function v, which is in general time-dependent, is (in this example) constant. Similarly, the use of capital letters for circuit resistance R, etc., implies time invariance of these parameters.

We use the term *waveform* of $v(t)$ to mean the graphical representation of v as a function of t. Occasionally, we shall use the term *waveform* as a synonym for *time function.*

We shall now proceed to define the two principal variables of circuit theory.

ELECTRIC CURRENT An orderly motion of charges as defined by the (time) rate of flow of positive charges from the terminal a to the terminal b of a terminal pair a-b is called the current flowing from a to b and is shown by the symbol $i_{ab}(t)$. The order of the subscripts ab indicates that the function $i_{ab}(t)$ is the rate of flow of positive charges from a to b inside the terminal pair a-b. The definition of current implies that, in the terminal pair a-b, $i_{ab}(t) = -i_{ba}(t)$. Thus if, at one instant of time $t = t_1$, $i_{ab}(t_1)$ is a positive number, then at that instant positive charges are flowing from a to b through the terminal pair. If, at an instant $t = t_2$, $i_{ab}(t_2)$ is a negative number, then at that instant positive charges are flowing in the terminal pair, from b to a. We emphasize that the current i_{ab} is associated with the terminals a-b, although we may have no information about what is the exact motion of charges inside that particular part of the device which is represented by the terminal pair.

VOLTAGE FUNCTION With the assumption made in the foregoing paragraph, it can be shown that the rate of delivery of energy to a terminal pair can always be given as the product of the current $i_{ab}(t)$ and another function of time associated with the terminals a and b. This function is called the voltage between the two terminals of the terminal pair and is indicated by $v_{ab}(t)$. The order of the subscripts ab indicates that the function $v_{ab}(t)$ is the voltage of point a with respect to point b. Thus, if $p_{ab}(t)$ is the rate of delivery of the energy to the terminal pair a-b, we can always find a function $v_{ab}(t)$ such that $p_{ab}(t) \equiv v_{ab}(t)i_{ab}(t)$. This equation may be considered to be the defining equation of the voltage function $v_{ab}(t)$. Since $v_{ab}(t)$ and $i_{ab}(t)$ are both functions of time and at any given instant of time one may have a positive value and the other a negative value (resulting, at that instant, in a negative value of p_{ab}), we have to interpret the significance of the positive and negative values of p_{ab}. If, at a given instant of time, $p_{ab}(t)$ is a positive value, by this we understand that energy is being delivered to the terminal pair a-b at the rate given by the magnitude of p_{ab} at that instant. On the other hand, if $p_{ab}(t)$ is negative at another instant of time, by this we understand that the terminal pair is delivering energy to the rest of the circuit at a rate given by the magnitude of p_{ab} at that instant. Consider a terminal pair a-b. By definition, the rate of the delivery of energy to this terminal pair at time t is

$$p_{ab}(t) = v_{ab}(t)i_{ab}(t)$$

and also

$$p_{ba}(t) = v_{ba}(t)i_{ba}(t)$$

Since $p_{ba}(t)$ and $p_{ab}(t)$ refer to the rate of the delivery of energy to the same terminal pair, it follows that

$$p_{ab}(t) = p_{ba}(t)$$

and

$$v_{ab}(t)i_{ab}(t) = v_{ba}(t)i_{ba}(t)$$

Since from the definition of current

$$i_{ba}(t) = -i_{ab}(t)$$

it follows that

$$v_{ba}(t) = -v_{ab}(t)$$

It can be shown that the methods of circuit analysis may be applied to the study of other systems such as mechanical systems. In such cases $i_{ab}(t)$ may represent the velocity of a (point) mass a with respect to some reference (point) b, and $v_{ab}(t)$ may represent the force applied to a (point) mass a in a reference system designated by b. In this case also, the product of $v_{ab}i_{ab}$ will be power (work done on a per unit time), but the interpretations of the variables are different.

1-6 Idealized lumped circuit elements

In the field problem we discuss the energy density at a given point of space and compute it in terms of the $\mathcal{E}$ and $\mathcal{H}$ variables. In circuit theory we discuss the rate of delivery of energy to a terminal pair or combination of terminal pairs and compute it in terms of the voltage functions and the currents associated with the terminal pairs. In field theory energy stored in a unit volume of the magnetic field is $\mu H^2/2$, and $\epsilon E^2/2$ is the energy stored in the unit volume of the electric field, E^2/ρ being the rate of transformation of electromagnetic energy into heat per second per unit volume. The factors μ, ϵ, and ρ are the parameters of the medium in which the electric and magnetic fields exist. We should be able to find analogous parameters for terminal pairs which would correspond to the permeability, permittivity, and resistivity of the medium in which the fields exist. The latter parameters are called distributed parameters of the medium since their value may change continually from one point to another in the medium. In circuit theory we are not concerned with values which change from point to point, but deal with quantities which are defined in connection with two terminals. When energy is delivered to a terminal pair a-b at the rate $p_{ab} = v_{ab}i_{ab}$, the total energy delivered to the terminal pair between time t_1 and t_2 is $W(t_1,t_2) = \int_{t_1}^{t_2} v_{ab}i_{ab}\,dt$. Part of this energy will be stored in the electric field associated with the terminal pair; another part will be stored in its magnetic field; and the rest of the energy will be transformed into heat. This partition of energy within a terminal pair is shown symbolically in a diagram called a circuit diagram. Such a diagram consists of interconnected symbols called *circuit elements*.[1] The number of such elements and the manner in which they are interconnected are determined by the physical properties of the device or part of the device which the terminal pair represents. Five elements are necessary to represent the energy-storage and energy-conversion processes in an electrical device. These basic elements describe the

[1] Some authors use the term "parameter" instead of "element."

following processes:

1. Storage of energy in magnetic form; such energy storage is accounted for by the element *inductance.*
2. Storage of energy in electrical form; accounted for by the element *capacitance.*
3. Conversion of electromagnetic energy into heat; accounted for by the element *resistance.*
4. Transfer of energy from one part of a device to another part through a magnetic field. The circuit model for this phenomenon is *mutual inductance.*
5. Conversion of other forms of energy into electromagnetic energy for delivery to parts of a device, or reception of electromagnetic energy from parts of a device and its transformation into other forms of energy. *Ideal sources* are introduced in circuit analysis to account for these processes.

In contrast to the "distributed" parameters of the field theory, these elements are called *idealized lumped circuit elements.* The values attributed to these elements depend on the geometry of the path of the current between the terminals of the terminal pair, and are computed from the dimensions and field properties (ϵ, μ, ρ) of the path. The computation of the values of circuit elements corresponding to a given path of current is not studied in circuit analysis. Instead, we define circuit elements in terms of energy-storage and energy-conversion processes which they represent. It should be noted that this is not the only possible way of defining circuit elements. For example, the voltage-current relationship at the terminals of the elements can be used to define such elements. Although the computation of the values of circuit elements is based on field theory, it is possible to determine, by experiment, the values of the elements of a network representing a device.

1-7 Inductance

In circuit theory, the characteristic of a part of a device (circuit) which accounts for the storage of energy in a magnetic field associated with that part is termed the *inductance* of that part. The two-terminal element represented by the symbol shown in Fig. 1-3, whose value is designated by

***Fig. 1-3** Graphical symbol for the element inductance.*

the letter L, represents an inductance. Quantitatively, inductance L may be defined by analogy with the field expression for energy density $\frac{1}{2}\mu H^2$ as follows: In Fig. 1-3, if the current in the inductance is i_{ab}, the energy stored in the inductance (in the form of magnetic energy) is related to the current by the equation

$$w_M = \tfrac{1}{2}Li_{ab}^2 \tag{1-1}$$

From this definition one can deduce the *voltage-current* relationship at the terminals of the inductance as follows: The rate at which energy is delivered to a terminal pair *a-b* is given by the power $v_{ab}(t)i_{ab}(t)$. Therefore

$$v_{ab}i_{ab} = \frac{d}{dt}\left(\frac{1}{2}Li_{ab}^2\right)$$

Since L is defined as independent both of t and of i_{ab}, the result of the differentiation is

$$v_{ab}i_{ab} = \left(L\frac{di_{ab}}{dt}\right)i_{ab}$$

Thus, for an inductance L, connected between terminals *a-b*, the voltage v_{ab} is related to the current i_{ab} by the basic equation

$$v_{ab} = L\frac{di_{ab}}{dt} \tag{1-2}$$

If, from Eq. (1-2), we express the current i_{ab} as a function of the voltage v_{ab}, the result reads

$$i_{ab} = \frac{1}{L}\int v_{ab}\,dt + \text{const} \tag{1-3}$$

The constant can be made explicit if we recognize that the energy stored in the inductance at time t has been delivered over the period of time for which $v_{ab}(t)$ has existed. The current and energy are related by Eq. (1-1), and it is seen that the value of the current at any time t, like the value of the stored energy, will depend on the past history of the voltage across the inductance. This is taken care of by writing the integral expression (1-3) with the lower limit at minus infinity and allowing the integral to be a function of an upper limit t.

$$i_{ab}(t) = \frac{1}{L}\int_{-\infty}^{t} v_{ab}(\tau)\,d\tau \tag{1-4}$$

It frequently happens that interest is focused on the function i_{ab} beginning at some arbitrary instant of time, usually $t = 0$. In such cases it is convenient to write (1-4) in the form

$$i_{ab}(t) = \frac{1}{L}\int_{-\infty}^{0} v_{ab}\,dt + \frac{1}{L}\int_{0}^{t} v_{ab}(\tau)\,d\tau \tag{1-5}$$

Since the first of the integrals in (1-5) has numerical limits, it represents a number. This number is the value of the current at $t = 0$. Denoting this value by $i(0)$, we have, for an inductance,

$$i_{ab}(t) = i_{ab}(0) + \frac{1}{L}\int_0^t v_{ab}(\tau)\,d\tau \tag{1-5a}$$

Equation (1-5*a*) reads, in words,

$$\begin{pmatrix}\text{Current in inductance}\\ \text{for all } t > 0\end{pmatrix} = \begin{pmatrix}\text{current at}\\ t = 0\end{pmatrix} + \begin{pmatrix}\text{current due to } v_{ab}\\ \text{from } t = 0 \text{ on}\end{pmatrix}$$

The value $i_{ab}(0)$ is usually called the initial value of the current i_{ab}. The reciprocal of inductance is called *inverse self-inductance* and is represented by the symbol $\Gamma = 1/L$.

From Eq. (1-2) we observe that the voltage-current relationship for an inductance is *linear;* that is, it is a *first-degree* differential relationship. We note that, if i_{ab} is a function of time represented, for example, by the graph shown in Fig. 1-4*a*, then v_{ab} will be a function of time proportional to the *slope*

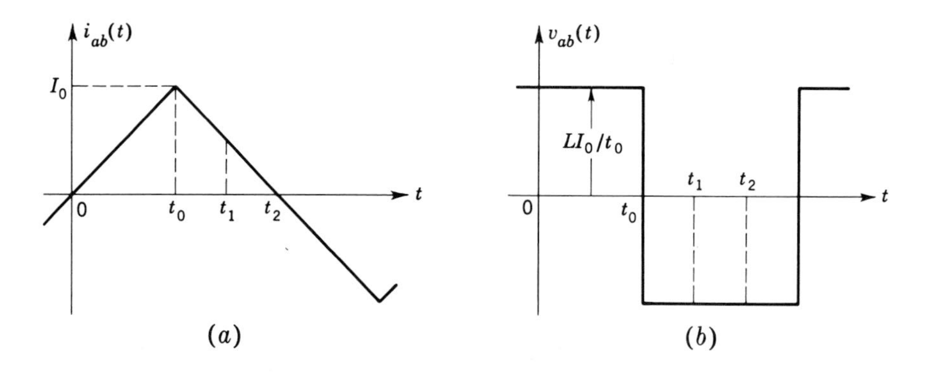

***Fig. 1-4** (a) Example of a current waveform i_{ab}. (b) Voltage v_{ab} across an inductance L when the current i_{ab} has the waveform of (a).*

of the i_{ab} graph as shown in Fig. 1-4*b*. We note also that the algebraic *sign* of v_{ab} depends on the *sign of the slope of* i_{ab}, *not on* i_{ab} *itself;* for example, i_{ab} is positive at t_1 whereas v_{ab} is negative. In connection with Fig. 1-4*a*, we observe that, at $t = t_0$, the slope of the graph i_{ab} changes abruptly, resulting in an abrupt change in v_{ab}.

Another interesting observation concerning inductance is the following: If we specify in addition to L the current and its slope at some instant, for example, $t = t_1$, we can calculate the stored energy in the inductance at that instant as $w_M(t_1) = \frac{1}{2}Li_{ab}^2(t_1)$ and the voltage v_{ab} at t_1 as

$$v_{ab}(t_1) = L(di_{ab}/dt)_{t=t_1}$$

In contrast, if we specify v or dv/dt at some instant, we cannot calculate i_{ab} at that moment because [from Eq. (1-4)] $i_{ab}(t_1)$ does not depend only on $v_{ab}(t_1)$, but also on how v_{ab} varied up to the time t_1. This dependence of one circuit variable on the "history" of the other is characteristic of energy-storing devices.

1-8 Capacitance

In circuit theory, the characteristic of a part of a device which accounts for storage of energy in an electric field associated with that part is termed *capacitance* of that part. It is represented by the symbol shown in Fig. 1-5

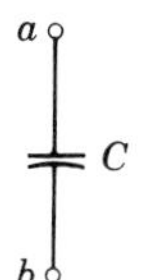

Fig. 1-5 Graphical symbol for the element capacitance.

and quantitatively denoted by the letter C. By analogy with the expression for electric field energy density $\frac{1}{2}\epsilon E^2$, the defining equation for energy stored in a capacitance owing to a voltage v_{ab} is given by the expression

$$w_E = \tfrac{1}{2}Cv_{ab}^2 \qquad (1\text{-}6)$$

The rate of delivery of energy to a capacitance is given by

$$v_{ab}i_{ab} = \frac{dw_E}{dt} = \frac{d}{dt}\left(\frac{1}{2}Cv_{ab}^2\right)$$

If C is independent of t and of v_{ab}, then

$$v_{ab}i_{ab} = \left(C\frac{dv_{ab}}{dt}\right)v_{ab}$$

Thus, for a capacitance connected between terminals a-b,

$$i_{ab} = C\frac{dv_{ab}}{dt} \qquad (1\text{-}7)$$

or in integral form,

$$v_{ab} = \frac{1}{C}\int_{-\infty}^{t} i_{ab}(\tau)\,d\tau = v_{ab}(0) + \frac{1}{C}\int_0^t i_{ab}(\tau)\,d\tau \qquad (1\text{-}8)$$

where $v_{ab}(0)$ is the value of the voltage across the terminals of the capacitance at $t = 0$. This value is called the initial value of the voltage across the capacitance. The reciprocal of capacitance is termed *elastance* and is denoted by $S = 1/C$.

We observe that the voltage-current equations for capacitance are analogous to those of inductance. Comparison of Eq. (1-7) with (1-2) shows that voltage and current have exchanged roles. Thus, in a capacitance, the current depends on the instantaneous rate of change (slope) of the voltage, whereas the voltage depends on the "history" of the current as indicated by Eq. (1-8).

1-9 Charge and flux linkages We stated at the outset that current is the orderly motion of charge such that i_{ab} represents, when positive, the motion of positive charge from a to b. Thus one can define at a terminal pair a-b

$$i_{ab} = \frac{dq_a}{dt} \tag{1-9}$$

where q_a represents, when positive, positive charge entering the a terminal. Using the inverse relationship

$$q_a = \int_{-\infty}^{t} i_{ab}(\tau)\, d\tau \tag{1-10}$$

in Eq. (1-8), we obtain the familiar charge-voltage relationship for capacitance,

$$q_a = Cv_{ab} \tag{1-11}$$

In the integral form of the voltage-current relation for an inductance [Eq. (1-4)], we define the flux linkage Ψ_a as the integral of the voltage, i.e.,

$$\Psi_a(t) = \int_{-\infty}^{t} v_{ab}(\tau)\, d\tau \tag{1-12}$$

or

$$i_{ab} = \frac{1}{L}\Psi_a \tag{1-13}$$

Although we have based our field concepts on the assumption of the existence of electric charge and are familiar with its significance, the significance of flux linkage can be discussed only in terms of concepts based on (the surface integral of) magnetic flux density associated with a current. The reader may remember Faraday's law which relates voltage and flux linkages in a circuit through the expression $|v| = |d\Psi/dt| = N|d\phi/dt|$. In circuit analysis we do not need to deal with the concept of "flux," and we consider Eq. (1-12) to define a quantity which is given the name "flux linkage."

1-10 Voltage-current relation in a resistance

A resistance is defined as that circuit element which accounts for the transformation of electric energy into heat *at the rate* given by

$$p_R = \frac{v_{ab}^2}{R} = v_{ab}i_{ab} \tag{1-14}$$

Thus, for a resistance R connected between terminals a-b, represented by the

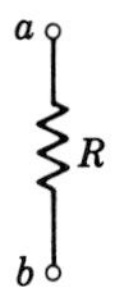

Fig. 1-6 Graphical symbol for the element resistance.

symbol shown in Fig. 1-6,

$$i_{ab} = \frac{v_{ab}}{R} \tag{1-15}$$

or

$$v_{ab} = Ri_{ab} \tag{1-15a}$$

The reciprocal of resistance is termed *conductance*, denoted by the symbol $G = 1/R$.

1-11 Mutual inductance In circuit theory, the characteristic of part of a device which accounts for the transfer of energy from one region to another through a magnetic field is called *mutual inductance*. It is designated by the letter M. It is always related to two inductances as shown in Fig. 1-7. Quantitatively, M is defined as follows: If

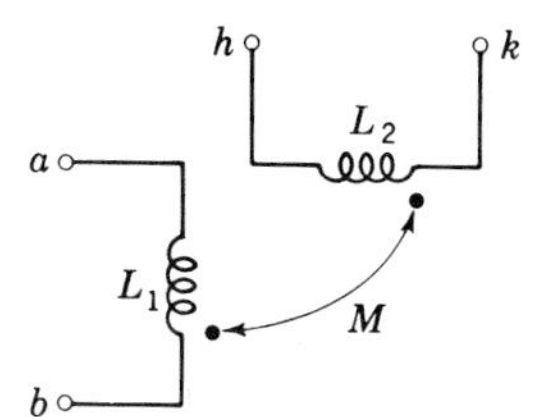

Fig. 1-7 Graphical symbol used in connection with mutual inductance.

L_1 and L_2 (Fig. 1-7) are coupled through mutual inductance, total magnetic energy stored is

$$w_M = \tfrac{1}{2}L_1 i_{ab}^2 + \tfrac{1}{2}L_2 i_{hk}^2 + M i_{ab} i_{hk} \tag{1-16}$$

We note that the effect of mutual inductance is to add an energy-storage term which is proportional to the product of the currents i_{ab} and i_{hk}. The rate of transfer of energy into magnetic form $p_M = dw_M/dt$ is given by

$$p_M = \left(L_1 \frac{di_{ab}}{dt} + M \frac{di_{hk}}{dt}\right) i_{ab} + \left(L_2 \frac{di_{hk}}{dt} + M \frac{di_{ab}}{dt}\right) i_{hk} \tag{1-17}$$

In Chap. 15 it is shown how Eq. (1-17) is interpreted as a two-terminal-pair relationship. For the present we consider circuits without mutual inductance.

1-12 Ideal sources

In circuit theory, the characteristic of part of a device which accounts for the conversion of other forms of energy into electromagnetic energy or the conversion of electromagnetic energy into other forms (excluding heat, magnetic, or electric) is called a source. An attribute of an *ideal* source is its capability of delivering (or absorbing) energy without limit. Unlike the case of R, L, and C elements, the voltage across an ideal source is *not* restricted (by power or energy relations) if the current is prescribed, nor is the current restricted if the voltage function is defined.

For this reason two types of ideal sources are defined: (1) an ideal voltage source, where the waveform of the voltage at its terminals is specified but the waveform of the current through it will depend on the nature of the terminal pair connected to the ideal voltage source; (2) an ideal current source, where the waveform of the current at its terminals is specified but the waveform of the voltage across it will depend on the nature of the terminal pair connected to the ideal current source.

IDEAL VOLTAGE SOURCE As implied above, an ideal voltage source exists between terminals a-b if the waveform v_{ab} is independent of the nature or value of the circuit elements connected to these terminals. In most cases

v_{ab} will vary with time, but by the above definition the waveform $v_{ab}(t)$ will be independent of the circuit elements connected to it.

The symbolic representation of an ideal voltage source is shown in Fig. 1-8*a*; it is a circle with appropriate symbol or sign beside it, designating the function $v_{ab}(t)$. Occasionally, the waveform symbol appears within the circle as in Fig. 1-8*b*, where a sinusoidal time function is indicated.

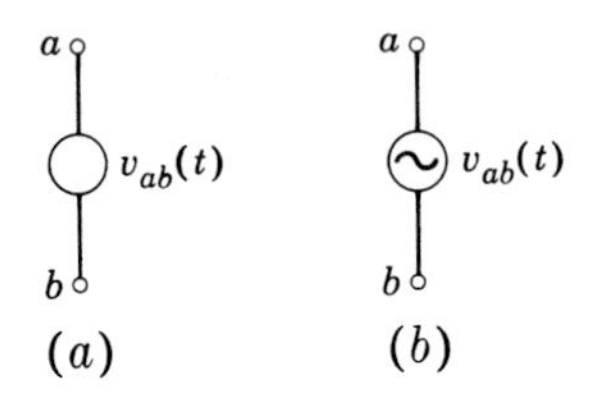

***Fig. 1-8** (a) Graphical symbol for an ideal voltage source. (b) Graphical symbol for an ideal voltage source furnishing sinusoidal waveform.*

Ideal voltage sources are often called "voltage sources" for brevity. They are idealized concepts, and do not exist in practice. The voltage between the terminals of all practical sources of voltage will depend on the current flowing through them. In circuit analysis, however, the concept of an ideal voltage source is used to advantage. In the literature the term "constant voltage source" is used to designate what we have called "ideal voltage source." The former term is appropriate to sources which do not vary with time. Since we are dealing with time-varying sources, we prefer the use of the term "ideal" rather than "constant."

The exact equivalent circuit of an actual source will depend on the nature of the source. It is the subject of study in courses dealing with power supplies, alternators, electronic oscillators, and other sources. It suffices to say that often practical sources can be represented to a good degree of approximation by a combination of ideal sources and additional circuit elements (R, L, and C).

IDEAL CURRENT SOURCES If the waveform of the current which flows between two terminals *a-b* is independent of the circuit elements connected to those terminals, an ideal current source is said to be connected between terminals *a-b*. The symbol for an ideal current source is an arrow inside a circle, as shown in Fig. 1-9. The direction of the arrow has the meaning

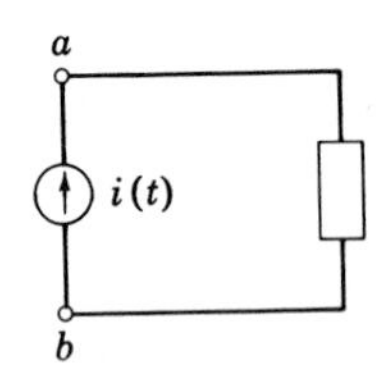

***Fig. 1-9** Graphical symbol for an ideal current source connected to a terminal pair.*

explained in Sec. 1-14. Certain electronic components and circuits provide approximations for ideal current sources. Practical sources may be represented by combinations of ideal current sources and other (R, L, C) circuit elements.

1-13 Additional discussion of circuit elements

The circuit elements R, L, and C (as well as M) are termed *passive* elements because they can only store or dissipate electromagnetic energy. Voltages and currents in them can be produced only by sources. In contrast, ideal voltage and current sources are *active* elements since they can make electromagnetic energy available to a circuit. It is, of course, understood that the ideal sources are not physically realizable, but since practical situations can be represented by a combination of ideal sources with passive elements, they are of great utility in analysis.

The elements R, L, C, M and *sources* are termed *lumped idealized* circuit elements. They describe the basic energy-transfer processes in circuits. Thus the term resistance represents the degradation of electromagnetic energy into heat, and inductance represents storage of energy in the magnetic field. It is true that in elementary courses in physics the symbol used for resistance was defined in terms of a metallic wire, inductance in connection with a coil, and capacitance in terms of parallel metallic plates. Nonetheless, many parts of an electrical device can be represented by a combination of R, L, and C elements whose actual construction does not contain wires, coils, or parallel plates. A good example of such a device is a crystal used in connection with electronic oscillators. The behavior of a crystal can be predicted from its equivalent circuit, shown in Fig. 1-10. An actual

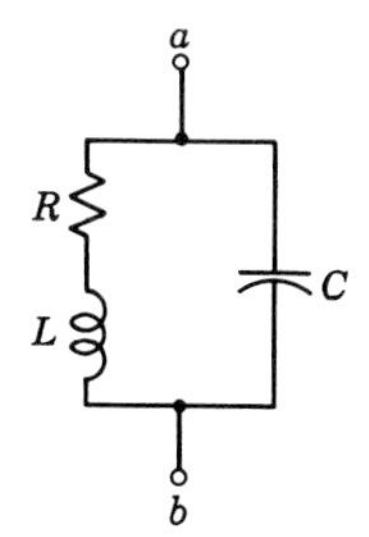

***Fig. 1-10** An equivalent circuit of a crystal.*

crystal does not contain a coil or a metallic resistor, but its equivalent circuit will contain resistive and inductive elements.

In effect, the behavior of many mechanical devices may be predicted from their analogous electric circuits. For example, the problem of the stability of an airplane in flight or the oscillations of a suspension bridge may be solved by the study of their respective analogous circuits.

Any part of an electrical apparatus which carries a current will have all the three characteristics of resistance, inductance, and capacitance. It will be seen that the relative importance of these characteristics will depend on the nature of the variations of the current with time. For example, a part of an electrical device known as a resistor can be represented by a resistance R if the rate of change of the current with time is sufficiently small. In such cases the effect of the inductance and capacitance of the part may be neglected compared with its resistance. However, if the time rate of change of current in this part is sufficiently high, the inductive or capacitive character of the resistor may become comparable with the resistance of the part, and, in some instances, the dominant characteristic. In such a case a good approximation of the equivalent circuit of a resistor may be a capacitance or an inductance or a combination of the two. The dominant characteristic of some other part of a device may be the inductive characteristic. In this case the part is represented by an inductance, and the resistive and the capacitive effects are neglected.

LINEARITY In the foregoing discussion, we have defined the three passive elements such that their voltage-current relationships are linear; i.e.,

For inductance: $$v_{ab} = L \frac{di_{ab}}{dt} \qquad i_{ab} = \frac{1}{L} \int_{-\infty}^{t} v_{ab}\, d\tau \tag{1-18a}$$

For capacitance: $$i_{ab} = C \frac{dv_{ab}}{dt} \qquad v_{ab} = \frac{1}{C} \int_{-\infty}^{t} i_{ab}\, d\tau \tag{1-18b}$$

For resistance: $$v_{ab} = Ri_{ab} \qquad i_{ab} = \frac{1}{R} v_{ab} \tag{1-18c}$$

We recall that the power and energy relationships are not linear. For this reason, the voltage-current relationships (1-18) are taken as fundamental in analysis since their application leads to linear equations. It is of course understood that the actual physical device which is represented by a circuit diagram consisting of linear elements may actually not be linear, so that the analysis is an approximation. However, many portions of practical devices are, in fact, practically linear, so that linear analysis agrees with performance within measurable precision.

NOTE ON AN OVERSIMPLIFICATION At this point the authors wish to draw the attention of the reader to a very common oversimplification about the current through a capacitance. In some elementary books on electricity it is stated that current cannot flow through a capacitance. This statement refers to conduction current. In circuit analysis we do not distinguish between conduction and displacement current, but use the terminal relationship $i_{ab} = C\, dv_{ab}/dt$. Hence, in accordance with the circuit concept of terminal pairs, it is seen that a current does flow through a capacitance, and its value at any instant is proportional to the rate of change of the voltage across its terminals. This current is identically zero when the voltage across the capacitance does not change with time. It must be remembered that i_{ab} in a capacitance depends on the *rate of change* of v_{ab}, and not on v_{ab} itself. Thus it is possible to have very large voltages across a capacitance, without any current through it, if the voltage does not change with time. Alternatively, it is possible that at any instant the voltage across a capacitance will be zero while current flows through the element at that instant. Similarly, the voltage v_{ab} across an inductance does not depend on i_{ab} through it, but on the *rate of change* of i_{ab}. Hence, if i_{ab} in an inductance does not change with time, then, regardless of the value of the current, the voltage v_{ab} across the inductance will be zero. Alternatively, at a given instant, the current in an inductance may be zero, but its rate of change may be nonzero. At such instants there will be a voltage across the inductance without any current through the element.

1-14 Arrow notation for currents

The "double-subscript" notation for current is not convenient when several elements are connected between a pair of terminals as in Fig. 1-11. In this case the symbol i_{ab} may mean i_{ab} in R or i_{ab} in L or i_{ab} in the R-L combination. To avoid such ambiguity, the element currents in R and L are

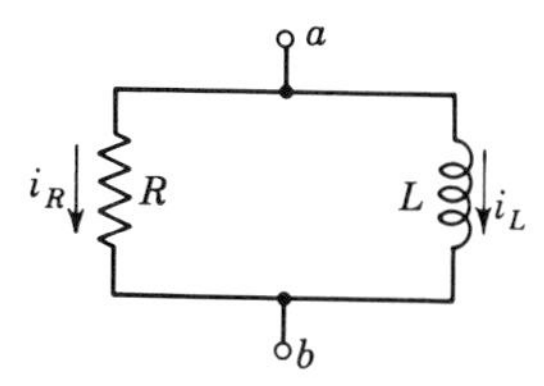

Fig. 1-11 Arrow notation for currents.

shown as i_R and i_L, respectively, and an arrow is shown with each of the symbols i_R and i_L. These arrows represent *reference directions* in accordance with the following rule:

> If an arrow is associated with a current $i(t)$, then at an instant when $i(t)$ is positive, positive charges move[1] in the direction of the arrow.

Thus the arrow for i_R which points from a to b does not mean that positive charges move from a to b; it does mean that if $i(t)$ is a positive number, then positive charges move from a to b. It follows that, from the reference arrow of a current *alone,* we cannot say in which direction the positive charges are flowing. The waveform of the current associated with the arrow must be known before the direction of the flow of the positive charges can be ascertained at a given instant of time. Similarly, if the waveform of i_L is given but a reference arrow is not shown on the circuit, we cannot tell in which direction the positive charges flow at a given instant, although the magnitude of the current at that instant can be found from the waveform.

1-15 Notation for voltages

Although the double-subscript notation is quite often the best for indicating which voltage is under discussion, other notations are commonly used. These notations accomplish what the arrow notation accomplishes for the current. The information linking the waveform with the voltage across an element is transferred from the double subscripts to *marks* on the circuit diagram, so that the symbol representing the voltage is interpreted in connection with the circuit diagram. With double-subscript notation the

[1] In circuit theory we discuss current as the motion of charges without regard to the actual physical charge carriers, which may be electrons, holes, ions, etc. Thus, when we say "positive charges move," a purist might say "it is as if positive charges move."

function $v_{ab}(t)$ represented the variation of the voltage of a with respect to b. If we show this function as $v_1(t)$ (Fig. 1-12a), we must mark the points a and b *on the circuit diagram* so that it is clear that the waveform $v_1(t)$ represents the voltage of point a with respect to point b. We may do this by putting an X mark near a and a 0 mark near b, stating the rule that whenever a waveform $v(t)$ is associated with the terminal pair a-b, marked as above, the waveform represents the voltage of the point marked X with respect to the point marked 0. The conventional marks used are + and − signs. From Fig. 1-12b, with the waveform and the signs shown, we understand that the waveform $v_1(t)$ represents the voltage of point a (marked +) with respect to point b (marked −). Thus, for $0 < t < t_1$, a is negative with

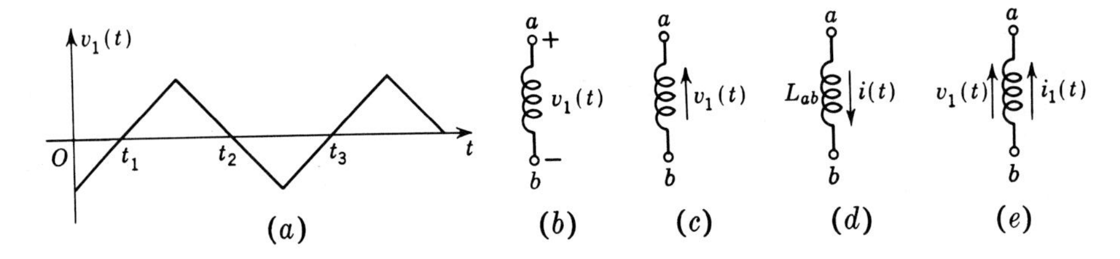

Fig. 1-12 Illustration of polarity marks, and arrow notation for voltages.

respect to b, since during this interval the waveform of $v_1(t)$ has a negative value. For $t_1 < t < t_2$, a is positive with respect to b, since during this interval the waveform $v_1(t)$ has a positive value. Thus it is seen that from the + and − signs alone we cannot say whether a is positive or negative with respect to b at a given instant, and we need the waveform of $v_1(t)$ to ascertain this fact.

An alternative notation for voltages is the arrow notation. In Fig. 1-12c the arrow shown with $v_1(t)$ indicates that the waveform $v_1(t)$ represents the voltage of the tip of the arrow with respect to the tail. In summary, the plus, minus, or arrow notation means that:

> If the voltage v has ± marks (or an arrow) associated with it, then if at an instant v is a positive number, at that instant the terminal marked + (tip of arrow) is positive with respect to the terminal marked − (tail of arrow).

To illustrate the arrow notation we note that, for Fig. 1-12d, $v_{ab} = L\, di_{ab}/dt$ and $i = i_{ab}$; hence $v_{ab} = +L\, di/dt$. In Fig. 1-12e, v_1 is defined with the tip of the arrow at a; hence $v_1 = v_{ab}$. The arrow of i_1 points toward a; hence $i_1 = i_{ba} = -i_{ab}$ and $v_1 = -L\, di_1/dt$.

1-16 Potential difference, voltage drop, and electromotive force The terms potential difference, voltage drop, and electromotive force (usually abbreviated emf) are used interchangeably in many introductions to electric-circuit analysis. In such introductions it has become customary to treat only circuits in which the charges are moving at constant velocity, i.e., circuits in which the currents are constant. The term potential difference is applicable only to such circuits. The reader has undoubtedly noticed that we have avoided using these terms. The reason for this is our interest in time-varying voltages and currents. It is the purpose of this section to show how the above-mentioned terms relate to the term "voltage" as voltage is defined above.

The term *potential difference* has been carried over into circuit analysis from electrostatics. When an electric field in space is a static field, the potential difference between two points is defined as the work done to carry a (unit) charge from one point to the other. In such static fields the work done in moving the charge is independent of the path through which the charge is moved. When we deal with problems in which the charges are moving with variable velocities (or, in circuit terminology, when the voltages and currents are time-dependent), the work done in carrying a charge from one point in a region of a dynamic field to another point is no longer independent of the path through which the charge is carried, and the original definition of potential difference becomes meaningless. When the variations of voltages and currents with time approach zero, the term voltage, as defined in Sec. 1-5, becomes synonymous with the term potential difference. In electrical engineering it has become customary to ignore the original definition of potential difference (which deals with static fields) and use the term as a synonym for voltage even when the voltage is time-dependent. Thus a voltage between terminals a-b, $v_{ab}(t)$, may be referred to as the potential of point a with respect to point b.

As in the case of potential difference, the term *voltage drop* is very descriptive when we deal with time-independent currents. In such a case, if, for example, a current $I_{ab} = 3$ amp flows through a resistance $R_{ab} = 2$ ohms, the voltage V_{ab} will be 6 volts, and point a is positive with respect to point b. Hence, one may say that in going from a to b the voltage has "dropped" by 6 volts. If, however, through the same resistance, the current i_{ab} is a function of time so that it may be positive or negative at different instants of time, then $v_{ab}(t) = R_{ab}i_{ab}(t)$ and, although at one instant of time a may be positive with respect to b, at another instant of time a may be negative with respect to b. Now it is perfectly permissible to talk about a negative voltage drop between terminals a-b. If we were to say that the voltage drop from a to b is -6 volts, we should mean that a is 6 volts below b. The authors feel, however, that the term voltage drop from a to b tends to mislead the reader by the excessive use of double negatives. The term voltage drop as commonly used has, however, the same meaning as the term voltage; i.e., the voltage drop or voltage v_{ab} is related to the power flowing into terminals a-b through the product $p_{ab}(t) = v_{ab}(t)i_{ab}(t)$.

The term *electromotive force* is introduced in field theory to describe a particular effect which occurs under dynamic field conditions. It is found that, if a physical circuit (i.e., materials such as metals) is placed in a region of space in which there are dynamic fields, currents may flow (i.e., charges are set in motion) in this circuit. This effect is attributed to a cause called electromotive force. We shall now qualitatively describe how emf is related to the field intensity and how this phenomenon is represented in circuit analysis.

If we start with a region in space in which there are dynamic fields, the placing of material into that space will result in a motion of charge (flow of current) which modifies the field distribution in the space. Let us imagine that the material which constitutes the physical circuit is removed and a closed geometrical line replaces the path. In this closed geometrical line there is no longer any current, but an electric field intensity will still exist at any point of the line. At a given instant the line integral of the electric field intensity around the closed line (with the physical circuit removed) will have a value which is called the emf associated with that closed path at that instant.

In the circuit representation this emf (which is associated with a given closed path) is represented by an ideal voltage source. We have said that, when the material circuit is placed back in the path, charges in the material will be set in motion. The distribution of these charges disturbs the original electric field. In the circuit representation the effect of this charge distribution is represented by the circuit element capacitance. Capacitance does not take into account the effect of the motion of the charges. If the charges move with variable velocity, aside from the effect of their distribution they will cause another disturbance in the electric field. The sum total of this disturbance in the *closed* path is represented by the circuit element inductance, and the effect is called the "counter emf of the inductance." Finally, the effect of the dissipation of electromagnetic energy into heat along the path of the current is represented by the circuit element resistance. In the light of the above explanation the term lumped circuit element becomes more meaningful. The various effects which take place around a closed path are represented in "lumps." For example, the integral of the electric field intensity around a closed loop is shown as a terminal pair (voltage source for emf, inductance for counter emf); the effect of the existence of charge around a closed path is represented by the terminal-pair capacitance. In this way these *field* phenomena are represented by circuit elements. When the voltage source and R, L, and C are connected properly[1] to make a closed circuit, the current flowing in this closed circuit will, under certain conditions, approximate the actual current flowing in the physical circuit.

In circuit analysis it is common to use the term emf interchangeably with voltage in connection with ideal voltage sources, and counter emf interchangeably with voltage across an inductance.

1-17 *Circuit elements and circuit components* In elementary physics the idea of resistance is introduced in connection with the properties of metallic materials, inductance is associated with coils, and capacitance is associated with dielectric regions. In this book we have introduced the elements R, L, and C without reference to materials, simply stating the energy process which the elements represent and the voltage-current relationship which is associated with each type of element. Now, when one builds a circuit out of material components (wire, coils, etc.), the representation of such a circuit in terms of the symbols R, L, and C may require more than one such element for each physical circuit component. Thus a circuit component such as a coil may be represented by several elements, that is, R-L, or a combination of several R-L-C elements. When a coil is placed into a circuit, often the intention is to impart inductance to that circuit. We shall distinguish between the property inductance (as defined by the voltage-current relationship $v_{ab} \equiv L_{ab}\, di_{ab}/dt$) and the component which is placed into the circuit with the intent of imparting inductance to the circuit. This is done by referring to the component in the circuit as an inductor, with the understanding that the component may also introduce other circuit properties. In general, the ending "-or" will be used to describe circuit components, and the ending "-ance" will be used to describe the "pure" elements. Thus a resistor is a circuit component whose "dominant" circuit property may be resistance. In every case an "ideal" circuit component is identical with a circuit element. Thus an ideal capacitor is represented on a circuit diagram by a capacitance. The above discussion may be summarized by observing that circuit diagrams using (ideal) R-L-C elements are *models* of physical circuits.

VARIABLES IN CIRCUIT PROBLEMS In field problems the field quantities depend on the distribution, velocities, and acceleration of the charges in space. Analogously, one can consider the main variables in circuit problems to be the charge at any point, say, point a, the current flowing from a to b, $i_{ab} = dq_a/dt$, and the second time derivative of the charge, $di_{ab}/dt = d^2q_a/dt^2$.

Voltage and current are the most commonly used circuit variables, although their derivatives and integrals may also be used.

1-18 *Units*

All relationships which have been so far presented in the various equations hold, of course, in any consistent system of units. In this book we

[1] A detailed discussion of this subject is given in Javid and Brown, *op. cit.*, chap. 11.

shall use the mks (meter-kilogram-second-ampere) system of units as shown in Table 1-1. In Table 1-2 common prefixes for multiples and decimal fractions of units are shown. The unit hertz is employed in place of the traditional cycles per second (cps).

Table 1-1 Units

Quantity	Units	Abbreviation or symbol†
Energy	Joules	W
Power	Watts = joules/second	w
Time	Seconds	sec
Voltage	Volts	v
Current	Amperes = coulombs/second	amp or a
Charge	Coulombs	q
Flux linkages	Webers = volt-seconds	Ψ
Resistance	Ohms	Ω
Inductance	Henrys	h
Capacitance	Farads	f

† In this book the abbreviations w, v, h, f are used only in conjunction with a prefix (see Table 1-2) and on diagrams. The symbol Ω for ohms is used on diagrams only.

Table 1-2

Factor	Prefix	Abbreviation	Example
10^{12}	tera-	T	5.88 teramiles/light year = 5.88×10^{12} miles/light year
10^{9}	giga-	G	3 gigahertz = 3×10^{9} hertz
10^{6}	meg-	M	5 megohms = 5×10^{6} ohms
10^{3}	kilo-	k	15 kw = 15×10^{3} watts
10^{-3}	milli-	m	2.5 ma = 2.5×10^{-3} amp
10^{-6}	micro-	μ	5 μf = 5×10^{-6} farad
10^{-9}	nano-	n	2 nf = 2×10^{-9} farad
10^{-12}	pico- or micromicro-	p or $\mu\mu$	5 pf = 5×10^{-12} farad 5 pf = 5 $\mu\mu$f

PROBLEMS

1-1 A certain physical element stores 20 joules. Calculate the value of the element if the element is (*a*) an inductance carrying a current of 2 amp (find L); (*b*) a capacitance with 500 volts across its terminals (find C); (*c*) a spring stretched 2 cm (find elastance); (*d*) a mass moving at 2 m/sec (find M); (*e*) a disk revolving at 10 rad/sec (find moment of inertia).

1-2 The voltage across an inductance of 3 henrys is constant and equal to 12 volts in the time intervals discussed below. (*a*) If at $t = 0$ the current in the inductance is zero $[i(0) = 0]$, calculate the energy stored in the inductance at $t = 0.5$ sec. (*b*) If at $t = 0.5$ the energy stored in the inductance is zero, calculate the energy stored in L at $t = 0$. (c) If at $t = 0.5$ sec the energy stored in the inductance is 15 joules, calculate the two possible values that the magnitude of the current may have at $t = 0$.

1-3 An inductance L carries a current of 10 amp. To what value should the current be increased if the stored energy is to be (a) doubled; (b) tripled; (c) halved?

1-4 The current i_{ab} in a 0.6-henry inductance is defined through the graph of Fig. P1-4. (a) Sketch to scale $v_{ab}(t)$ in the interval $-0.002 < t < 0.042$. (b) Sketch to scale the power $p_{ab}(t)$ and the stored energy $w_L(t)$ in the interval $0 < t < 0.04$. (c) Calculate the time-average stored energy in the interval $0 < t < 0.04$.

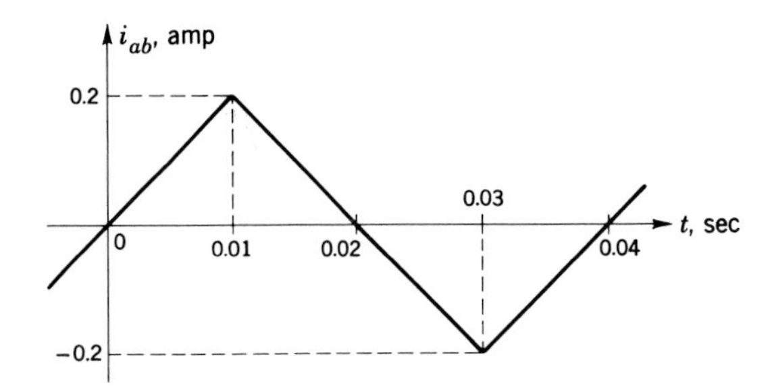

Fig. P1-4

1-5 A capacitance has 100 volts across its terminals. To what value should the voltage be changed if the stored energy is to be (a) doubled; (b) tripled; (c) halved?

1-6 A constant current $I = 0.2$ ma is passed through a capacitance in the time interval $t = 0$ to $t = 2$ sec. The capacitance stores zero energy at $t = 0$. (a) Calculate the value C of the capacitance if the voltage across C is to be 80 volts at $t = 2$. (b) What is the value of the energy stored in C at $t = 2$ if C has the value found in part a?

1-7 In the time interval $t = 1$ to $t = 5$, the current in a 4-μf capacitance is given by $i_{ab} = 2 \times 10^{-5}t$. The voltage across the capacitance at the instant $t = 3$ is given as $v_{ab}(3) = 15$. Calculate the voltage v_{ab} at the instant (a) $t = 1$; (b) $t = 5$.

1-8 An ideal voltage source $v(t)$ is connected across a passive element. Calculate the energy delivered to the element by the source in the time interval $t = 0$ to $t = 2$ sec in the following cases: (a) The element is a $\frac{1}{3}$-farad capacitance and $v(t)$ is given by (1) $v(t) = 10t$, (2) $v(t) = 10t^2$; (b) The element is a 2-ohm resistance and (1) $v(t) = 10t$, (2) $v(t) = 10t^2$; (c) The element is a $\frac{1}{2}$-henry inductance with $i(0) = 0$ and (1) $v(t) = 10t$, (2) $v(t) = 10t^2$.

1-9 The current from an ideal source is passed through a certain element. Its value is 1 amp at $t = 1$ and 2 amp at $t = 2$. Calculate the energy delivered to the element in the interval $t = 1$ to $t = 2$ for the following cases: (a) The element is a 2-henry inductance. (b) The element is a 2-ohm resistance, and the current increases linearly from 1 to 2 amp in the given time interval. (c) The element is a 2-ohm resistance, and the current is given by $i(t) = 1$ for $1 < t < 1.5$ and $i(t) = 2$ for $1.5 < t < 2$.

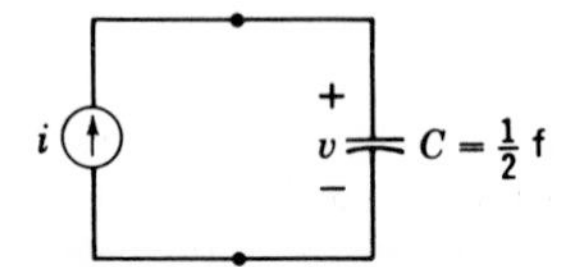

Fig. P1-10

1-10 An ideal current source $i(t) = 5t$ is connected across a capacitance as shown in Fig. P1-10 at $t = 0$. The value of v at $t = 0$ is -100 volts. (*a*) Calculate the value v at $t = 2$. (*b*) How much energy is delivered to the source in the time interval $t = 0$ to $t = 2$? (*c*) At what instant $t = t_1$ will $v(t_1) = 0$?

1-11 The current in a resistance R is given by $i(t) = I_m \cos(2\pi t/T)$. (*a*) Show that the energy delivered to the resistance in the time interval $t = 0$ to $t = T$ is given by $W(0,T) = \frac{1}{2}I_m{}^2RT$. (*b*) A constant current I replaces $i(t)$ in R. Find the value of I in terms of I_m so that the energy delivered to R by I in an interval T is also $\frac{1}{2}I_m{}^2RT$.

1-12 The voltage v in Fig. P1-10 is given by $v = 10 \sin 2\pi t$. Find i at the instants (*a*) $t = 0$; (*b*) $t = \frac{1}{2}$; (*c*) $t = \frac{3}{4}$; (*d*) $t = \frac{7}{8}$.

1-13 The current i_{ab} in an inductance L is given by $i_{ab}(t) = (1/L) \cos 2\pi t$. Find the voltage v_{ab} at the instants (*a*) $t = 0$; (*b*) $t = \frac{1}{4}$, (*c*) $t = \frac{3}{8}$; (*d*) $t = \frac{3}{2}$.

1-14 The element in the terminal pair of Fig. P1-14 is either an inductance or a capacitance. If $i = 2 \cos 3t$ and $v = \sin 3t$, what is the value of the element if (*a*) $v = v_{ab}$; (*b*) $v = v_{ba}$?

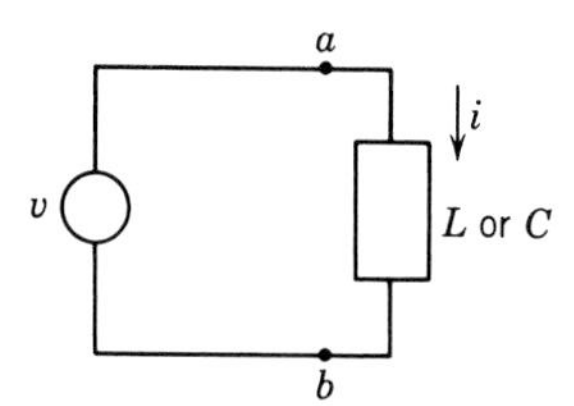

Fig. P1-14

1-15 In Fig. P1-15 it is suspected that the terminal pair within the square is an ideal source. When $R = 1$ ohm, $v_{ab} = 7$ volts. (*a*) If the source is an ideal current source, what will be the value of v_{ab} when (1) $R = 2$ ohms; (2) $R = \frac{1}{2}$ ohm? (*b*) If the source is an ideal voltage source, what will be the value of v_{ab} when (1) $R = 2$ ohms; (2) $R = \frac{1}{2}$ ohm?

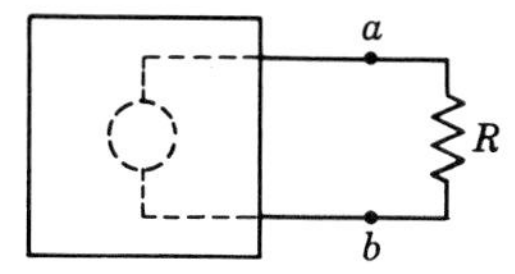

Fig. P1-15

1-16 Express the terminal relationships for resistance, capacitance, and inductance, using as variables (*a*) charge and voltage; (*b*) current and flux linkages; (*c*) charge and flux linkages.

Chapter 2 Fundamental waveforms and element response

The fundamental problem of circuit analysis can be stated as follows: Given an interconnection of passive elements and sources that form a network, and given further the waveforms of all the sources in the network, find the waveforms of all the currents in and voltages across the elements and study the process of energy transfer.

In a network the sources can be considered as cause, and the voltages and currents associated with the elements as effect. The network structure and composition will determine the relationship between the cause and effect. Analogous with the terminology used in connection with the nervous system, the sources of the network may be called the excitation, or stimuli, and the currents and voltages associated with the passive elements, the responses.

The source and response waveforms in a network can be specified by formula or by graphical means as in the case of the waveforms shown in Fig. 2-1. In this chapter we define certain basic waveforms and obtain the responses of single elements to these waveforms.

2-1 Basic waveforms

In the study of circuits certain simple waveforms are of special interest. The *ramp-function* waveform is illustrated in Fig. 2-2*a*. This waveform is described analytically by the equations

$$\begin{aligned} f_a(t) &= 0 \qquad t < 0 \\ f_a(t) &= At \qquad t \geq 0 \end{aligned} \tag{2-1}$$

(The symbol f is used to indicate that the waveform may represent any time variable such as a voltage or a current.)

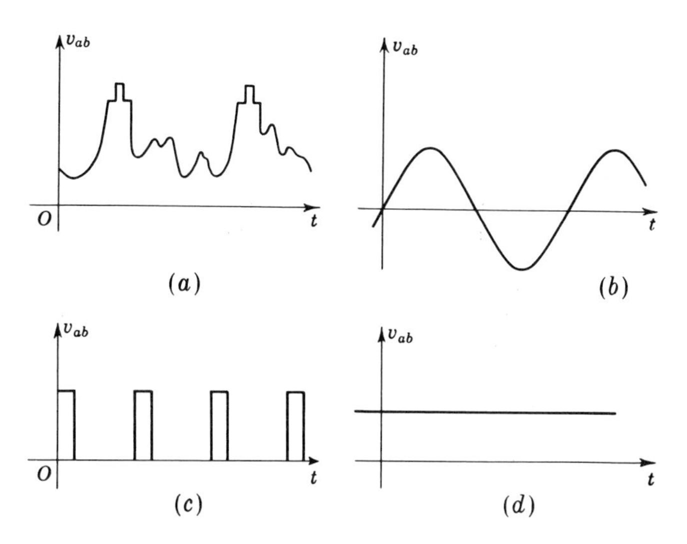

***Fig. 2-1** Voltage waveforms. (a) Waveform of a television signal. (b) Waveform of household electricity (sinusoidal). (c) Waveform of a train of rectangular pulses such as may be used in a radar modulator. (d) Waveform of a constant voltage (as from a battery).*

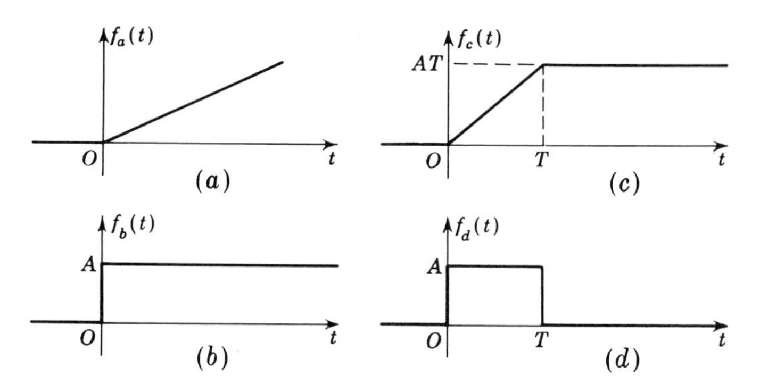

***Fig. 2-2** Basic waveforms. (a) Ramp function. (b) Step function. (c) Modified ramp function. (d) Rectangular pulse.*

In Fig. 2-2*b* the "step function" is illustrated. This function is defined by

$$\begin{aligned} f_b(t) &= 0 \qquad t < 0 \\ f_b(t) &= A \qquad t > 0 \end{aligned} \tag{2-2}$$

Note that the step function is proportional to the time derivative (slope) of a ramp function.

The step which is illustrated in Fig. 2-2*b* is a step of value A. If $A = 1$, the function is called the *unit step* and denoted by the symbol $u(t)$ and defined by

$$u(t) = \begin{cases} 0 & t < 0 \\ 1 & t > 0 \end{cases} \tag{2-2a}$$

The ramp-function waveform of Fig. 2-2*a* represents a time function which increases continuously with increasing time. The *modified ramp function* shown in Fig. 2-2*c* is often useful because it does not increase indefinitely. The modified ramp function is defined through the equations

$$\begin{aligned} f_c(t) &= 0 \qquad t \le 0 \\ f_c(t) &= At \qquad 0 \le t \le T \\ f_c(t) &= AT \qquad t \ge T \end{aligned} \tag{2-3}$$

In Eq. (2-3) T is a constant, as shown in Fig. 2-2*c*.

A *rectangular pulse* of duration T shown in Fig. 2-2*d* is another waveform which will be used frequently. It may be described by the equations

$$\begin{aligned} f_d(t) &= 0 \qquad t < 0 \text{ and } t > T \\ f_d(t) &= A \qquad 0 < t < T \end{aligned} \tag{2-4}$$

Note that the rectangular pulse is proportional to the time derivative (slope) of the modified ramp (Fig. 2-2*c*).

2-2 Continuous and discontinuous waveforms

In dealing with discontinuous waveforms [such as $u(t)$], it is necessary to define the values of a finite waveform at a point of discontinuity. The following notation is used in this connection: Let ϵ be a positive value, then the instants t_1^+ and t_1^- are defined by

$$\lim_{\epsilon \to 0} (t_1 + \epsilon) = t_1^+ \qquad \lim_{\epsilon \to 0} (t_1 - \epsilon) = t_1^-$$

Thus $f(t_1^+)$ is the value of f at t_1 if t_1 is approached from the "right" and $f(t_1^-)$ is the value of f at t_1 if t_1 is approached from the "left." With reference to Fig. 2-3, $f(t_1^+) = f(t_1^-)$ but $f(t_3^+) = C$ and $f(t_3^-) = B$.

The difference $f(t_1^+) - f(t_1^-)$ is termed the *jump* of the function at $t = t_1$. A function is *continuous* at $t = t_1$ if the value of the jump is zero. Thus, for example, with reference to Fig. 2-3, the function f is continuous at all instants shown, including $t = t_1$, t_2, t_4, and t_5, except at t_3, when it has a jump of value $C - B$ (a negative value).

MULTIPLICATION BY $u(t)$ In circuit analysis one often deals with waveforms which vary with time but which are zero prior to some arbitrary instant of time. If a function[1] $g(t) \equiv 0$ for all $t < 0$ and is given by $f(t)$

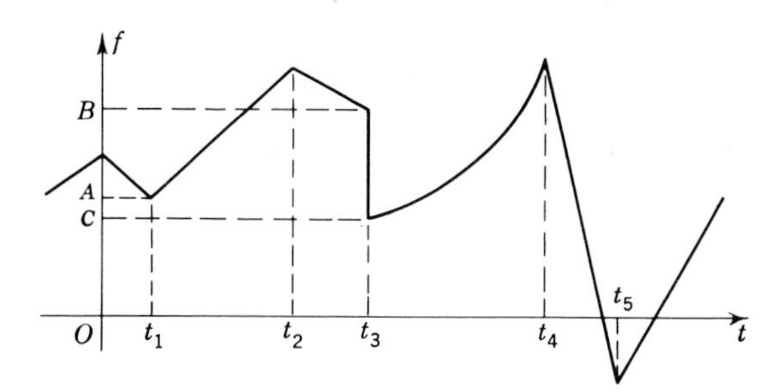

***Fig. 2-3** A waveform with discontinuities and corners.*

for $t > 0$, we can represent the function analytically through the two equations

$$\begin{aligned} g(t) &\equiv 0 \qquad t < 0 \\ g(t) &= f(t) \qquad t > 0 \end{aligned} \tag{2-5}$$

It is not necessary to use two equations if we take advantage of the properties of the unit step function as previously defined by Eqs. (2-2*a*). Since $u(t)$ is zero for all negative values of t, multiplying any function $f(t)$ by $u(t)$ makes the product zero for all negative values of t. Similarly, since $u(t)$ is unity for positive values of t, the product of any function $f(t)$ and $u(t)$ will be given by $f(t)$ for positive values of t. Hence the function $g(t)$, which was defined by the two equations shown in (2-5), can be defined through the single equation

$$g(t) = f(t)u(t) \tag{2-6}$$

As an example, consider first the straight line defined by

$$f_1(t) = At \tag{2-7}$$

This function is represented by the waveform shown in Fig. 2-4*a*. Note that $f_1(t)$ has nonzero values for both positive and negative instants of time. The ramp function

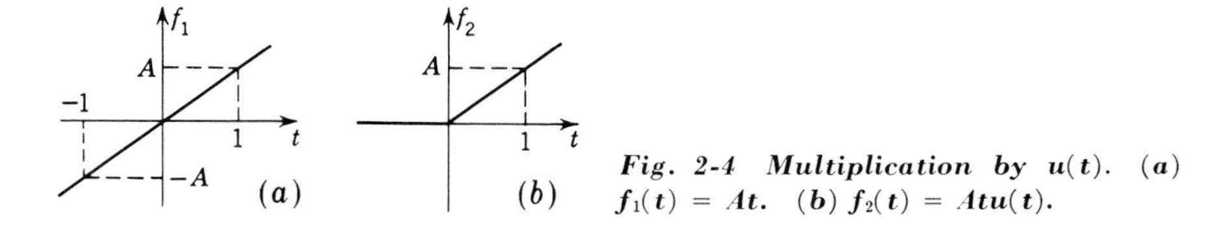

***Fig. 2-4** Multiplication by $u(t)$. (a) $f_1(t) = At$. (b) $f_2(t) = Atu(t)$.*

[1] The symbol $\equiv$, to be read "is identically," is different from the symbol $=$, "equals," inasmuch as, when we write $f(t) = 0$, this means that there exists a value of t for which $f(t)$ is equal to zero; whereas $g(t) \equiv 0$ signifies that $g(t)$ is zero for all values of t (or for a specified range of values of t).

shown in Fig. 2-4*b* can be defined through the equation

$$f_2(t) = f_1(t)u(t) = Atu(t) \tag{2-8}$$

In general, the unit step function can be used to represent a function which has one discontinuity by using the product of a continuous function and a unit step function. Consider the general functions $g_1(t)$ and $g_2(t)$, and assume that both of these functions are continuous. Then the combination $g(t) = g_1(t) + g_2(t)u(t)$ is discontinuous at $t = 0$ if $g_2(0)$ is not zero. This follows from the definition of $t = 0^-$ and $t = 0^+$ as stated in the previous section. Since both g_1 and g_2 were defined as continuous functions,

$$g(0^+) = g_1(0^+) + g_2(0^+)u(0^+) = g_1(0) + g_2(0)$$

and

$$g(0^-) = g_1(0^-) + g_2(0^-)u(0^-) = g_1(0)$$

This representation is illustrated in Fig. 2-5 for the special functions $g_1(t) = 4t$ and $g_2(t) = 3$.

In general, we observe that, if $f(t)$ is continuous, $f(t)u(t)$ has a jump of value $f(0)$ at $t = 0$. Below, we show that a function which has a jump J can be represented as the sum of a continuous function and J times a displaced step function.

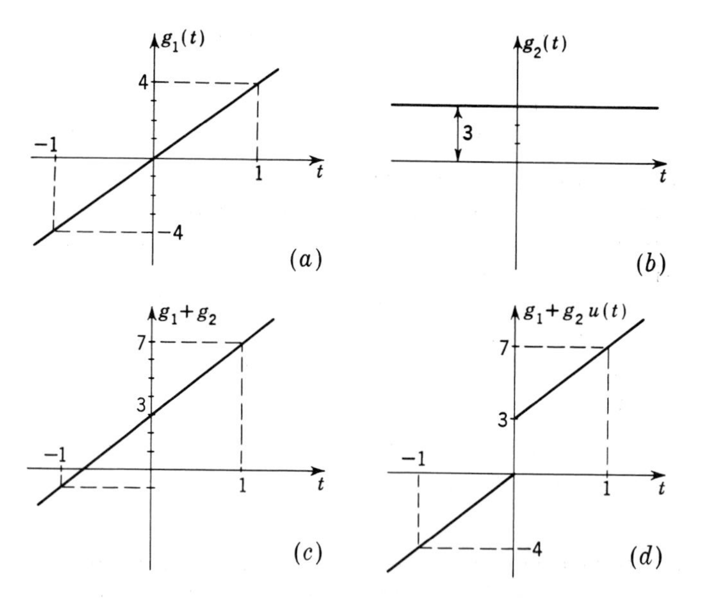

***Fig. 2-5** Creation of a discontinuity by multiplication with a step function.* (*a*) $g_1(t) = 4t$. (*b*) $g_2(t) = 3$. (*c*) $g_1(t) + g_2(t)$. (*d*) $g_1(t) + g_2(t)u(t)$.

TIME DISPLACEMENT The reader will recall from analytic geometry that a translation of the axis leaves the shape of the graph of a function unchanged. Thus, for example, the parabola $y = x^2$ has the vertex at the origin, the curve $y = (x - 1)^2$ represents a parabola of identical shape, but with its vertex at $y = 0$ and $x = 1$.

Applying these ideas to a time function $f(t)$, we note that if, at the instant $t = t_1$, $f(t_1)$ has the value A, then the function $f(t - t_d)$ has the value A at

$t = t_1 + t_d$. If t_d is a positive number, the function $f(t - t_d)$ is delayed in time by t_d as illustrated in Fig. 2-6.

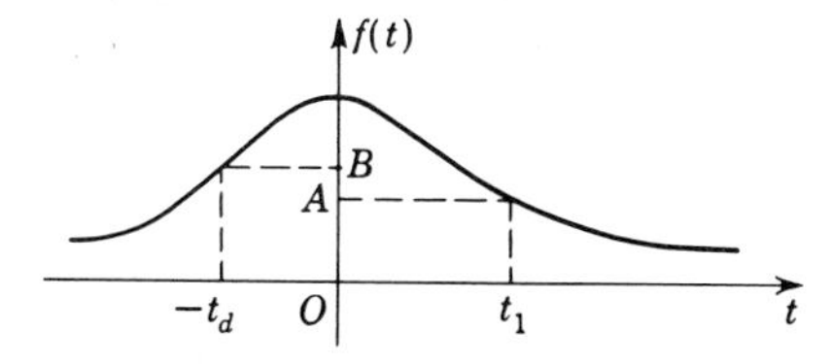

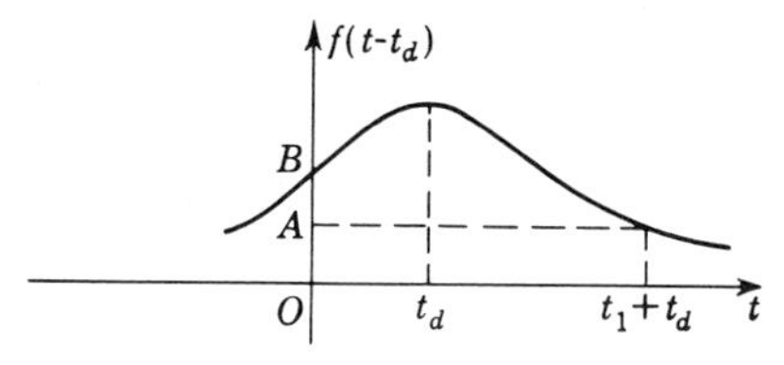

***Fig. 2-6** Time displacement. The waveform $f(t - t_d)$ is delayed by t_d with respect to $f(t)$.*

Since the unit step is defined as the function

$$u(t) = \begin{cases} 0 & t < 0 \\ 1 & t > 0 \end{cases}$$

the function $u(t - t_d)$ represents a step occurring at $t = t_d$ since

$$u(t - t_d) = 0 \qquad t - t_d < 0$$
$$u(t - t_d) = 1 \qquad t - t_d > 0$$

As an example, consider the ramp function

$$f_1(t) = Atu(t)$$

which "starts" at $t = 0$ as illustrated in Fig. 2-7*b*. The function

$$f_2(t) = A(t - t_d)u(t - t_d)$$

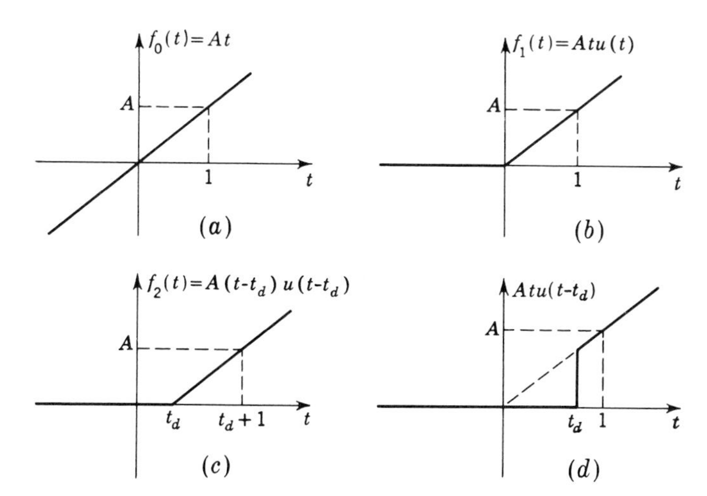

***Fig. 2-7** The straight line of (a) becomes the ramp of (b) when multiplied by $u(t)$. In (c) this ramp is delayed by t_d. In (d) the straight line of (a) has been multiplied by a delayed step function.*

also represents a ramp function, but it starts at $t = t_d$ (Fig. 2-7*c*). Note, however, that the function $Atu(t - t_d)$, illustrated in Fig. 2-7*d*, is no longer a ramp function. Given the equation of a waveform, the equation of the same waveform delayed in time by t_d is obtained by replacing t with $t - t_d$ *everywhere* in the equation of the original waveform.

2-3 Superposition of waveforms

We now show how the unit step function can be used to represent complex waveforms by addition (superposition) of relatively simple waveforms. Such representations are of particular importance in the analysis of *linear* systems because, as we shall show below, the response of a linear system to a complex waveform can be analyzed as the *sum* of the responses to the component waveforms.

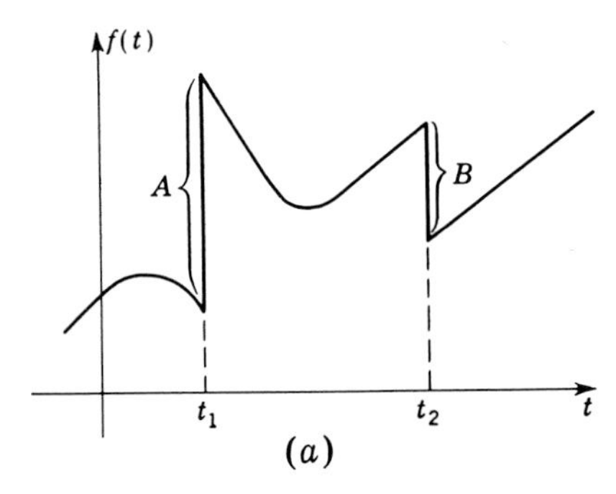

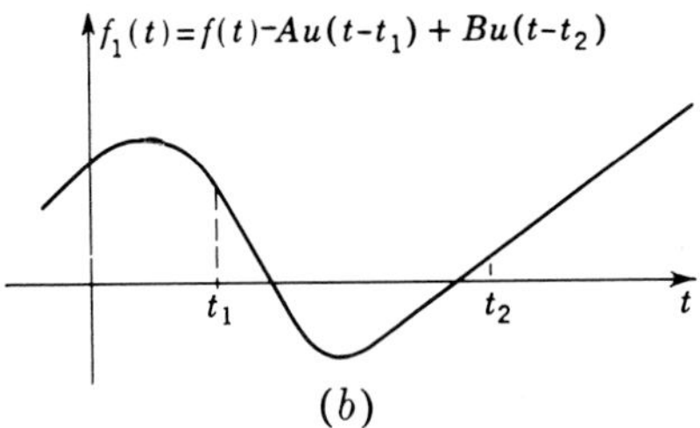

Fig. 2-8 Relating a discontinuous to a continuous waveform.

First we show that a waveform with jumps can be represented as the sum of a continuous function and step functions. Consider the waveforms shown in Fig. 2-8*a* and *b*. The function $f(t)$ is related to $f_1(t)$ by

$$\begin{aligned} f_1(t) &= f(t) && t < t_1 \\ f_1(t) &= f(t) - A && t_1 < t < t_2 \\ f_1(t) &= f(t) - A + B && t_2 < t \end{aligned}$$

From the definition of $u(t)$ in Eq. (2-2*a*), we can write

$$f_1(t) = f(t) - Au(t - t_1) + Bu(t - t_2)$$

or

$$f(t) = f_1(t) + Au(t - t_1) - Bu(t - t_2) \tag{2-9}$$

so that we have represented $f(t)$ as the sum of the continuous function $f_1(t)$ and two (time-displaced) step functions.

As a second example, we observe that the addition of the positive step $Au(t)$ and the negative step $-Au(t - T)$ gives the rectangular pulse shown in Fig. 2-2*d*.

The addition of waveforms can, of course, also be used when the waveforms are continuous, but defined sectionally, as in the case of the modified ramp function. If we add the ramp function $Atu(t)$ to the delayed ramp function $-A(t - T)u(t - T)$, the modified ramp illustrated in Fig. 2-2*c* results. A triangular pulse is obtained by the addition of two ramps and a step function, i.e., the addition of a negative step $-ATu(t - T)$ to the modified ramp.

SUPERPOSITION The representation of waveforms as a sum of step functions can be generalized. We shall discuss this generalization for functions which are zero prior to $t = 0$; an example of such a function is shown in Fig. 2-9*a*. In Fig. 2-9*b* is shown a (staircase) waveform consisting of a sum

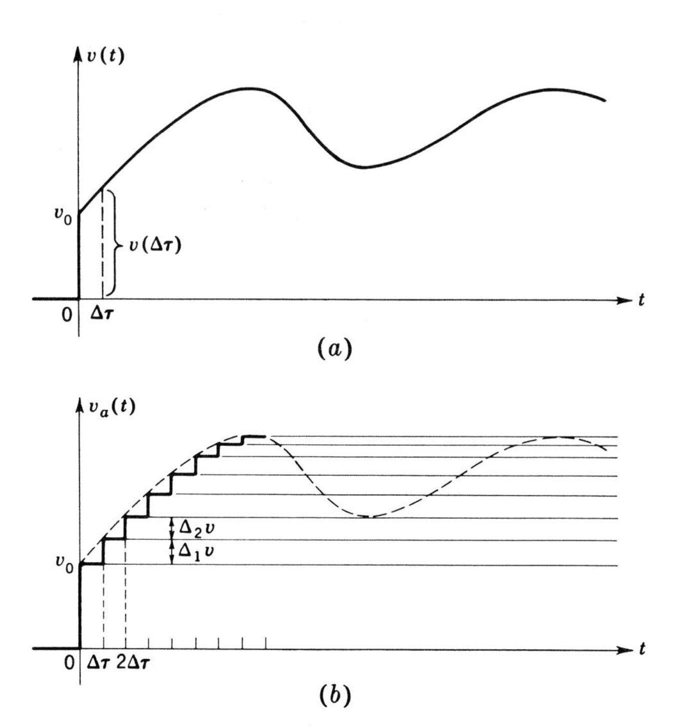

***Fig. 2-9** Approximation of the waveform v(t) shown in (a) by a sequence of step functions.*

of step functions. We observe that the steps in the function $v_a(t)$ (Fig. 2-9*b*) occur at interval $\Delta\tau$ and that $v(k\,\Delta\tau) = v_a(k\,\Delta\tau^+)$. Thus the function $v_a(t)$ is an approximation for the function $v(t)$. The smaller the intervals $\Delta\tau$, the better the approximation; in the limit as $\Delta\tau \to 0$, $v_a(t) \to v(t)$.

The approximation $v_a(t)$ can be written

$$v_a(t) = v_0 u(t) + \sum_{k=1}^{\infty} \Delta_k v u(t - k\,\Delta\tau) \tag{2-10}$$

The increment $\Delta_k v$ can be approximated by the differential:

$$\Delta_k v \approx \left(\frac{dv}{dt}\right)_{t=k\Delta\tau} \Delta\tau$$

Hence

$$v(t) \approx v_0 u(t) + \sum_{k=1}^{\infty} \left(\frac{dv}{dt}\right)_{t=k\Delta\tau} u(t - k\,\Delta\tau)\,\Delta\tau \tag{2-11}$$

If the limit is taken in Eq. (2-11) as $\Delta\tau \to 0$, one obtains a general formula for $v(t)$ as the sum of an infinite number of infinitesimal steps.[1]

The importance of Eq. (2-10) or (2-11) [or the exact representation of $v(t)$] as a sum of step functions stems from the superposition property of linear systems: The response of a linear system to a sum of sources is equal to the sum of the responses of the system to the component sources. Thus, if the response of a system to $u(t)$ is $a(t)$, the response to $v_0 u(t)$ is $v_0 a(t)$, and the response to $\Delta_k v u(t - k\,\Delta\tau)$ is $\Delta_k v a(t - k\,\Delta\tau)$. Thus the response of the system to $v_a(t)$ is

$$\text{Response to } v_a(t) = v_0 a(t) + \sum_{k=1}^{\infty} (\Delta_k v) a(t - k\,\Delta\tau) \tag{2-12}$$

or approximately,

$$\text{Response to } v(t) \approx v_0 a(t) + \sum_{k=1}^{\infty} \left(\frac{dv}{dt}\right)_{t=k\Delta\tau} a(t - k\,\Delta\tau)\,\Delta\tau \tag{2-13}$$

If in Eq. (2-13) the limit $\Delta\tau \to 0$ is approached, a general superposition formula results.[2] The function $a(t)$ [response of system to $u(t)$] is termed *indicial response.*

[1] See, for example, M. Javid and E. Brenner, "Analysis, Transmission, and Filtering of Signals," chap. 6, p. 148, McGraw-Hill Book Company, New York, 1963, where it is shown that

$$v(t) = v(0)u(t) + \int_{\tau=0}^{\tau=t} \frac{dv(\tau)}{d\tau} u(t - \tau)\,d\tau$$

[2] The result is

$$\text{Response to } v(t) = v_0 a(t) + \int_{\tau=0}^{\tau=t} \frac{dv}{d\tau} a(t - \tau)\,d\tau$$

See also Appendix E for computer solutions of network problems.

Example 2-1 In Fig. 2-10a is shown $a(t)$, the response of a linear system to $u(t)$. Construct graphically the response of the system to $v(t)$ as shown in Fig. 2-10b, using the step-approximation technique.

Solution In Fig. 2-10c a four-step approximation is shown:

$$v_a(t) = 10[u(t) - \tfrac{1}{4}u(t - \tfrac{1}{2}) - \tfrac{1}{4}u(t - 1) - \tfrac{1}{4}u(t - \tfrac{3}{2}) - \tfrac{1}{4}u(t - 2)]$$

The response to $v_a(t)$ is therefore given by

$$\text{Response} = 10[a(t) - \tfrac{1}{4}a(t - \tfrac{1}{2}) - \tfrac{1}{4}a(t - 1) - \tfrac{1}{4}a(t - \tfrac{3}{2}) - \tfrac{1}{4}a(t - 2)]$$

as illustrated in Fig. 2-10d.

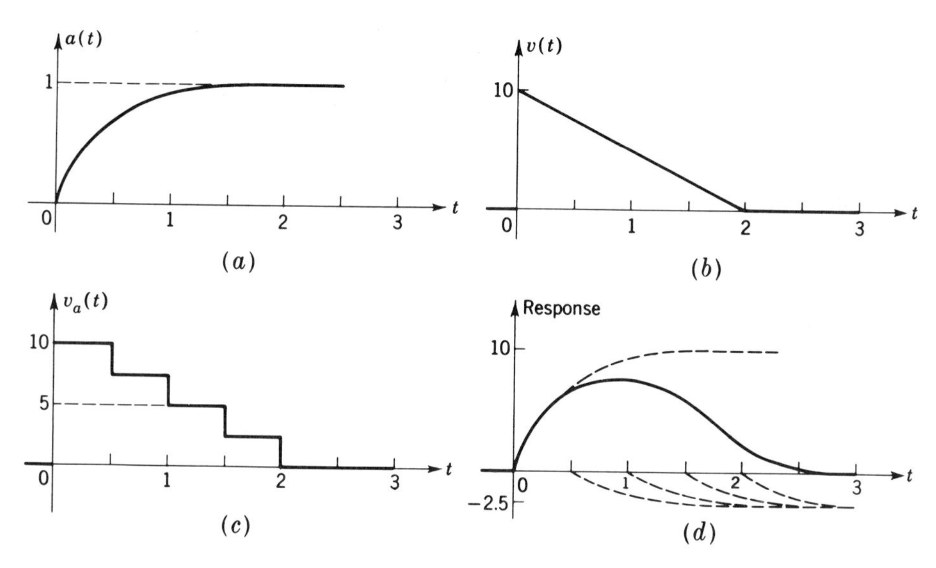

Fig. 2-10 **(*a*) *Response of a system to u*(*t*). (*b*) *A triangular pulse.* (*c*) *Five-step approximation for the pulse of* (*b*). (*d*) *Response of the system to the approximand shown in* (*c*) *obtained by superposition.***

2-4 *Integration of waveforms*

Recalling that the voltage-current relations for inductance and capacitance [Eqs. (1-2), (1-4), (1-7), and (1-8)] are integral or differential relationships, we recognize that waveforms that are related through integration and differentiation are of importance in circuit analysis. We discuss first the integral of a waveform; that is, if the graph of a function $f_1(t)$ is given, the waveform of its integral $f_2(t)$ is defined to be a new graph whose ordinate at any point t_1 is equal to the total area under the graph of $f_1(t)$ between $t = -\infty$ and $t = t_1$. The analytical expression for this interpretation of an integral[1] reads

$$f_2(t) = \int_{-\infty}^{t} f_1(\tau)\, d\tau \tag{2-14}$$

This interpretation differs from that of an indefinite integral, since, as in Eq. (2-14), the "constant of integration" is determined.

If $f_1(t)$ is zero prior to some arbitrary time, say, $t = 0$, as, for example, in the waveform $f_1(t) = f_3(t)u(t)$, then its integral $f_2(t)$ is

$$f_2(t) = \int_{-\infty}^{t} f_3(\tau)u(\tau)\, d\tau = \int_{-\infty}^{0} f_3(\tau)u(\tau)\, d\tau + \int_{0}^{t} f_3(\tau)u(\tau)\, d\tau$$

Since $u(\tau) = 0$ for $\tau < 0$, we have $\int_{-\infty}^{0} f_3(\tau)u(\tau)\, d\tau \equiv 0$. Then

$$f_2(t) = \int_{0}^{t} f_3(\tau)u(\tau)\, d\tau = \left[\int_{0}^{t} f_3(\tau)\, d\tau\right] u(t) \qquad t > 0$$

In Fig. 2-11*a* and *b*, a waveform and its integral are shown. Note that $f(t)$ has discontinuities at $t = -1$, 1, 3, and 5 and that the integral has no

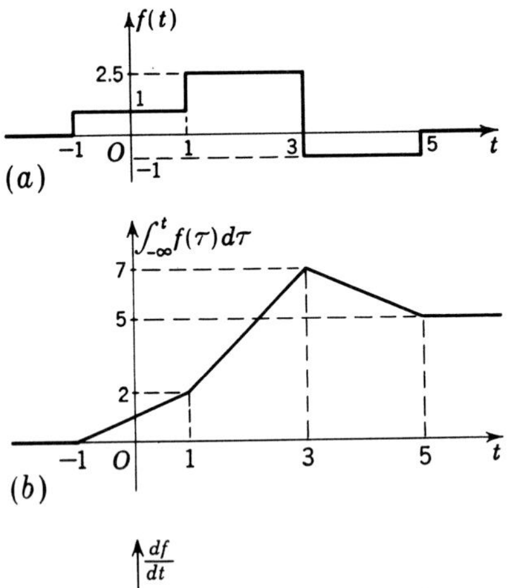

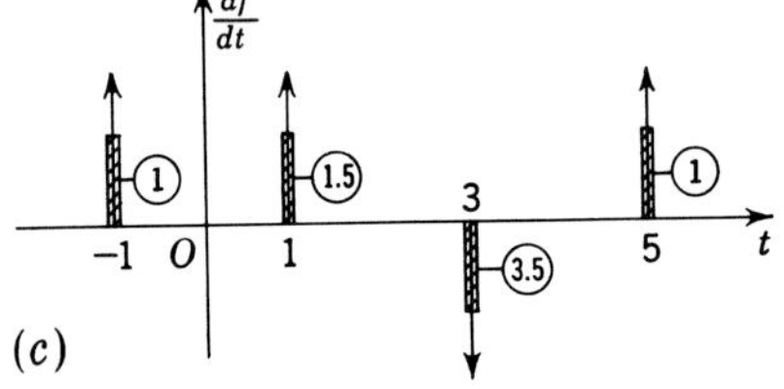

Fig. 2-11 (a) A waveform. (b) Integral of the waveform of (a). (c) Derivative of the waveform of (a). The number in each small circle is the strength of the corresponding impulse.

discontinuities. Since, in general, the integral of most discontinuous waveforms with which we deal has no discontinuities, we ascribe to the process of integration a *smoothing action* on the waveform. Conversely, differentiation results in discontinuities in the derivative of the waveform at its corners.

[1] The integrand is written with τ rather than t as the variable of integration because $f_2(t)$ is a function of the upper limit t, and not a function of the variable of integration.

2-5 *Differentiation of waveforms; impulse function*

When a waveform $f_2(t)$ is related to $f_1(t)$ by differentiation, that is, when

$$f_2(t) = \frac{df_1}{dt}$$

the value of f_2 at every point is equal to the value of the *slope* of f_1 at that point. We observe that, for example, the step function $Au(t)$ shown in Fig. 2-2*b* is proportional to the derivative of the ramp function shown in Fig. 2-2*a*. We have already implied that jumps in waveforms play an important role in analysis. For this reason we need to introduce certain definitions in connection with the derivative of $u(t)$. From Fig. 2-12*a* we see that the derivative of a step function is zero except when t is zero; i.e.,

$$\frac{du(t)}{dt} = 0 \qquad \text{for } t < 0 \text{ and } t > 0$$

At $t = 0$ the value of the step function is not defined. We do, however, call the slope at the point of discontinuity "infinite." In this sense we say that the derivative of the step is zero except at $t = 0$, when it is infinite. The derivative of a unit step function is called the unit impulse function

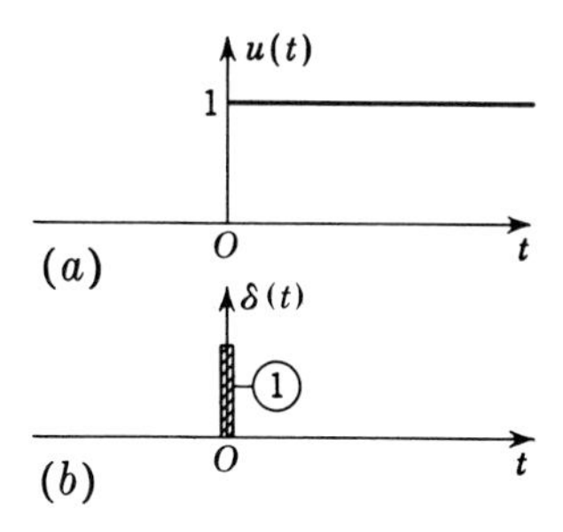

***Fig. 2-12** A unit step function and the symbol for its derivative, the unit impulse function ("delta function"). The number 1 in the small circle indicates that the strength of (area under) the impulse is unity.*

and denoted by the symbol $\delta(t)$, shown graphically by the symbol indicated in Fig. 2-12*b*. Thus

$$\frac{d}{dt}u(t) \equiv \delta(t) \tag{2-15}$$

so that

$$\begin{aligned} \delta(t) &= 0 && t < 0 \\ \delta(t) &= 0 && t > 0 \\ \delta(t) &= \text{“}\infty\text{”} && t = 0 \end{aligned} \tag{2-15a}$$

Now, if we were merely to call the derivative of the step at $t = 0$ infinite, the impulse function would be very poorly defined and quite useless because the same "value" would be assigned to the derivatives of all steps, regardless of the value of the step. In order to have a consistent definition, we note that, *if the unit impulse is the derivative of the unit step, then the unit step must be the integral of the unit impulse.* We therefore supplement the definition of the unit impulse which was stated in (2-15) by the relationship

$$\int_{-a}^{+b} \delta(t)\, dt = 1 \qquad \text{or} \qquad \int_{-\infty}^{t} \delta(\tau)\, d\tau = u(t) \tag{2-16}$$

where a and b are any positive numbers.

In general, the derivative of a waveform at a point of discontinuity is an impulse. The *strength* of the impulse is given by the value of the jump at the point of discontinuity of the waveform being differentiated, that is, it is the *area under the impulse function.* Applying this definition to the derivative of the waveform shown in Fig. 2-11a, the derivative shown in Fig. 2-11c results, and is given by the formula

$$\frac{df}{dt} = \delta(t+1) + 1.5\delta(t-1) - 3.5\delta(t-3) + \delta(t-5)$$

Since a step function can be considered to be the limiting case of a modified ramp function, the study of the derivative of a modified ramp function is useful in illuminating the notion of impulses. Consider the modified ramp function shown in Fig. 2-13a. The derivative of this function is shown in

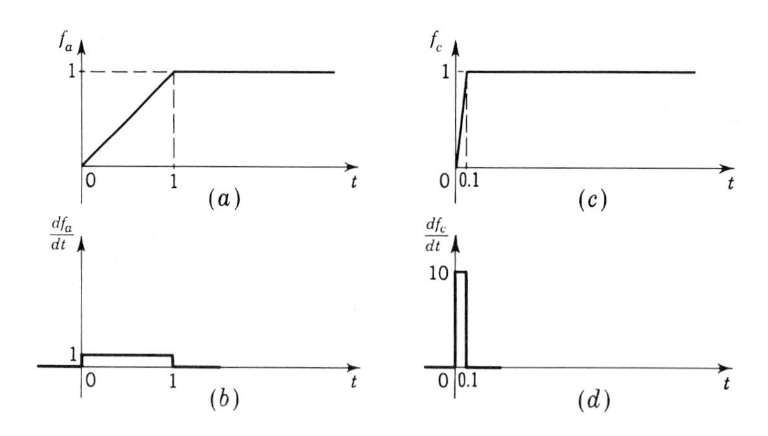

Fig. 2-13 Modified ramp functions and their derivatives.

Fig. 2-13b and is seen to be a rectangular pulse of unit area and unit duration. If the slope of the ramp portion of the modified ramp function is now increased so that the rise takes place in one-tenth the original time as shown in Fig. 2-13c, the derivative of the new modified ramp function is the rectangular pulse shown in Fig. 2-13d. Note that this rectangular pulse has ten times the height and one-tenth the duration of the pulse in Fig. 2-13b. In the limit, as the duration of the rise in the modified ramp function is reduced to zero, the modified ramp function becomes the step function, and the derivative of the modified ramp function becomes the impulse. Note, however, that the area under the derivative of the modified ramp function remains constant as the limit is approached. *This area is the quantity which we have defined as the strength of the impulse.*

SCANNING INTEGRAL The concept of $\delta(t)$ as the "derivative" of $u(t)$ is a plausible but inexact concept. More precisely, we may define $\delta(t)$ through an integral termed scanning integral. This definition was used by the physicist Dirac,[1] who introduced the impulse function into functional analysis, and states that, if $f(t)$ is continuous at $t = t_1$, then

$$\int_{-\infty}^{+\infty} f(t)\delta(t - t_1)\, dt = f(t_1) \tag{2-17}$$

Equation (2-17) is the *scanning integral*. We note that

$$\delta(t - t_1) = 0 \qquad \text{when } t \neq t_1$$

Hence

$$\int_{-\infty}^{+\infty} f(t)\delta(t - t_1)\, dt = \int_{t_1^-}^{t_1^+} f(t)\delta(t - t_1)\, dt$$

If now $f(t)$ is continuous at t_1, it is constant in the infinitesimal interval t_1^- to t_1^+. Thus

$$\int_{t_1^-}^{t_1^+} f(t)\delta(t - t_1)\, dt = f(t_1) \int_{t_1^-}^{t_1^+} \delta(t - t_1)\, dt$$

and we see from Eq. (2-17) that the area under $\delta(t - t_1)$ must have unit value and occurs in a zero (t_1^- to t_1^+) interval. Thus our discussion of $\delta(t)$ in terms of Eqs. (2-15) and (2-16) is consistent with Dirac's definition (2-17). The significance of the term scanning becomes clear when we observe that the value of the integral is the result of scanning the waveform of $f(t)$ and taking its value at the time t_1 when $\delta(t - t_1)$ is nonzero.

2-6 *Response of single elements*

When an ideal source is applied to a single R or L or C element, the response is either proportional to the source or proportional to the derivative or integral of the source. In this connection an ideal voltage source which is zero prior to $t = 0$ may be shown, as in Fig. 2-14*a*, by means of an ideal

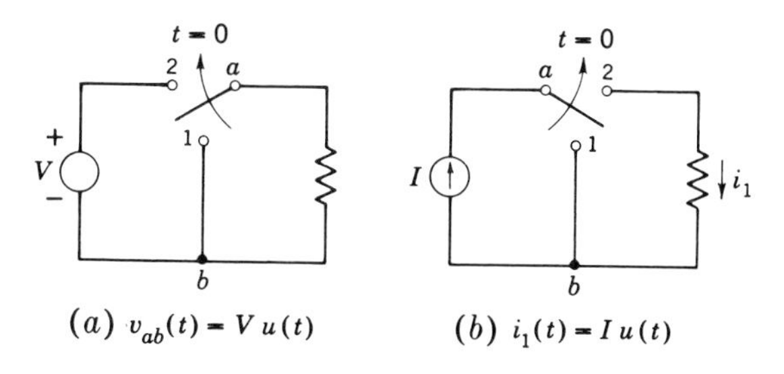

Fig. 2-14 ***Switching arrangements that may be used to visualize (a) a step-function voltage source applied to a resistance; (b) a step-function current source applied to a resistance.***

switch, and the short circuit from 1 to b. Note that the short circuit (ideal connection) between 1 and b assures that $v_{ab} = 0$ for $t < 0$. Similarly, the arrangement shown in Fig. 2-14*b* can be used to visualize the source $i(t)u(t)$.

[1] Paul Adrian Maurice Dirac, 1902–1963, Nobel laureate, 1933.

Example 2-2 Given a voltage source $v_{ab}(t) = 10tu(t)$ as shown in Fig. 2-15*a*, calculate and sketch the current i_{ab} through an element connected to v_{ab} if the element is (*a*) a 2-ohm resistance; (*b*) a 2-henry inductance with $i(0) = 3$; (*c*) a 2-farad capacitance.

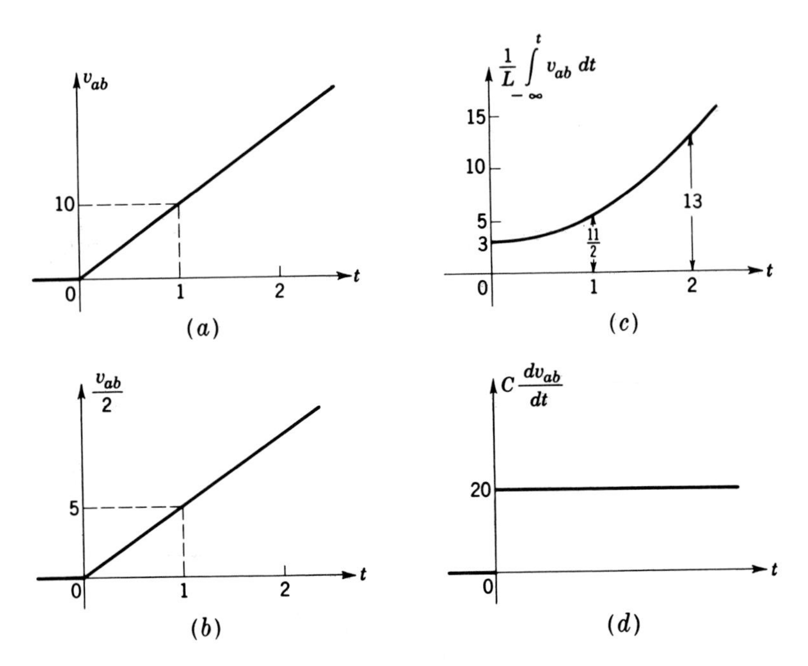

***Fig. 2-15** (**a**) Waveform of the ramp-function voltage source $v_{ab}(t) = 10tu(t)$. (**b**) Current i_{ab} in a 2-ohm resistance with v_{ab} impressed. (**c**) Current i_{ab} in a 2-henry inductance with $i(0) = 3$ and v_{ab} impressed. (**d**) Current i_{ab} in a 2-farad capacitance with v_{ab} impressed.*

Solution (*a*) Since $i_{ab} = v_{ab}/R = 5tu(t)$, the response waveform is the ramp function shown in Fig. 2-15*b*.

(*b*) Since $i_{ab} = i(0) + \frac{1}{L}\int_0^t v_{ab}\,d\tau$, $i_{ab} = 3 + \frac{5}{2}t^2u(t)$as illustrated in Fig. 2-15*c*.

(*c*) Since $i_{ab} = C\,dv_{ab}/dt$, $i_{ab} = 20u(t)$ as shown in Fig. 2-15*d*.

DUALITY We note that, since $i = C\,dv/dt$ and $v = L\,di/dt$ for capacitance and inductance, respectively, the two types of elements are similar. We note further that, if the waveform of the current in a capacitance has the same shape as the waveform of the voltage across an inductance, the waveform of the voltage across the capacitance will have the same shape as the waveform of the current in the inductance. This exchange of roles between voltage and current is termed *duality*, and is further explored in the next chapter.

SUPERPOSITION We note that

$$\frac{d}{dt}(f_1 + f_2) = \frac{df_1}{dt} + \frac{df_2}{dt}$$

Thus, if a voltage $v_{ab}(t) = v_1 + v_2$ is applied to a capacitance, the current i_{ab} in C will be given by

$$i_{ab} = C\frac{dv_1}{dt} + C\frac{dv_2}{dt}$$

Thus the response to a sum of voltages is the sum of the responses to the components in the sum. Similar statements can be deduced for other sources and R, L, or C elements.

2-7 Continuity

We observe that in an inductance a jump in current waveform results in an impulse of voltage. Conversely, a jump in current waveform can be produced only by voltage impulses. Such voltage-impulse sources are sources of infinite power and cannot occur in practical circuits. They can occur in the solution of problems which represent *idealized* circuits. Thus we conclude that, *in the absence of impulse voltage sources, the current in every inductance must be a continuous function of time.* Similarly, *in the absence of impulse current sources, the voltage across every capacitance must be a continuous function of time.* These statements are termed *continuity conditions* for inductance and capacitance. A little thought will show that these conditions are equivalent to stating that, in the absence of impulse sources, the stored energy of each energy-storing element must be continuous with time.

PROBLEMS

2-1 Sketch the waveform $u(t)u(2 - t)$. *Hint:* To sketch $u(2 - t)$, first let $x = 2 - t$.

2-2 Given the function $f(t) = t + 1$. Sketch to scale, in the interval $t = -2$ to $t = +2$: (*a*) $f(t)$; (*b*) $f(t)u(t)$; (*c*) $f(t)u(-t)$; (*d*) $f(t)u(t - 1)$; (*e*) $f(t - 1)u(t)$; (*f*) $f(t - 1)u(t - 1)$; (*g*) $f(t) + u(t)$; (*h*) $f(t)u(t + 1)$.

2-3 Sketch the waveforms of these functions: (*a*) $2tu(t - 1)$; (*b*) $2(t - 1)u(t - 1)$.

2-4 Show that $t^2u(t - 1) = (t - 1)^2u(t - 1) + 2(t - 1)u(t - 1) + u(t - 1)$.

2-5 If $f(t)$ is given by $f(t) = t^2u(t) - tu(t - 2) + u(t - 4)$, evaluate (*a*) $f(-3)$; (*b*) $f(1)$; (*c*) $f(2^-)$; (*d*) $f(2^+)$; (*e*) $f(3)$; (*f*) $f(5)$.

2-6 Write the equation of the waveform of Fig. P2-6 as a sum of three ramp functions.

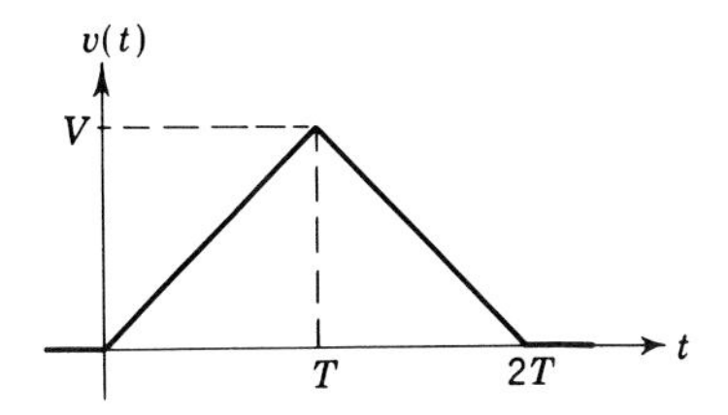

Fig. P2-6

2-7 Write the equations for the waveforms shown in Fig. P2-7 as sums of step and ramp functions.

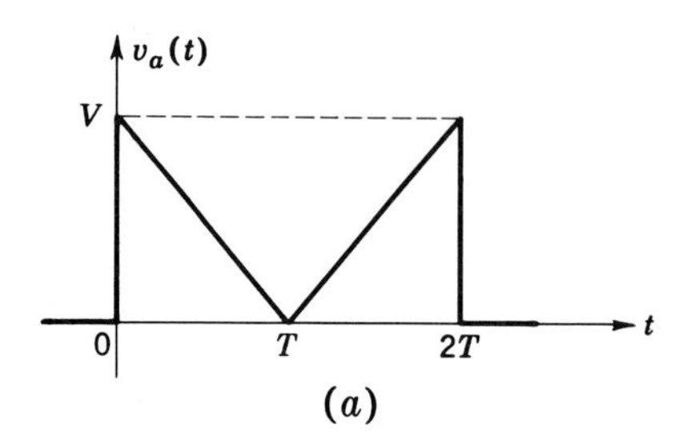

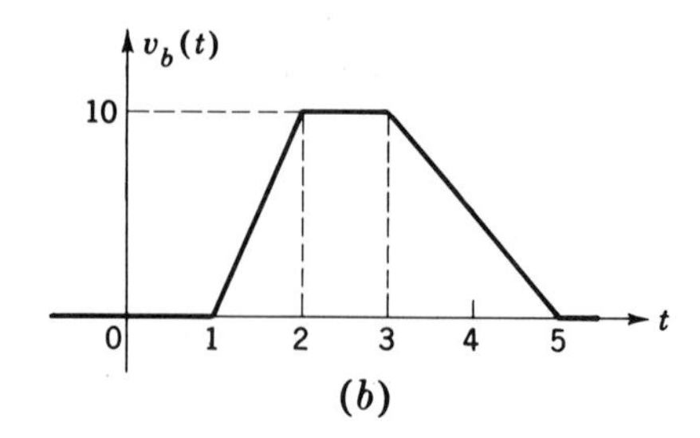

Fig. P2-7

2-8 (*a*) For the waveform shown in Fig. P2-8, state the points of discontinuity and the value of the jump at each of these points. (*b*) Show that the equation of this waveform may be written

$$f(t) = tu(t) - (1 + t)u(t - 1) + (9 - 2t)u(t - 3) + 2(t - 4)u(t - 4)$$

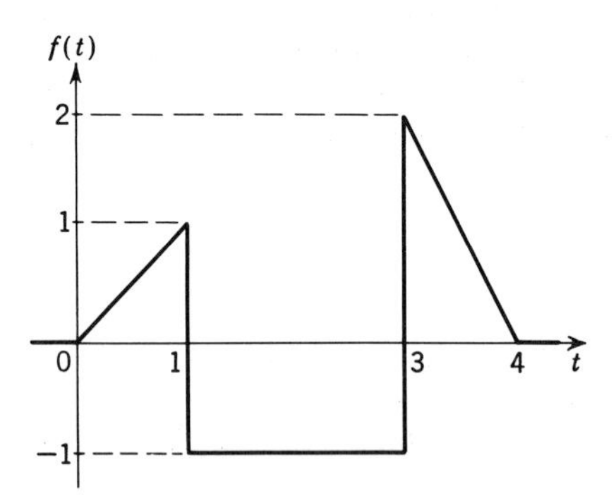

Fig. P2-8

2-9 Assume that the response of a linear system to the function $u(t)$ is given (approximately) by the modified ramp function $a(t) = tu(t) - (t - 1)u(t - 1)$. Use the superposition principle to sketch to scale the response of the system to the rectangular pulse $v(t) = u(t) - u(t - T)$ in the cases (*a*) $T = 10$; (*b*) $T = 2$; (*c*) $T = 1$; (*d*) $T = \frac{1}{2}$.

2-10 The response of a linear system to the step-function source $v(t) = u(t)$ is given by $a(t) = e^{-t}u(t)$. Sketch the response of the system to the source $v(t) = u(t) - u(t - 1)$, and give the numerical value of the response at the following instants (*a*) $t = 0^+$; (*b*) $t = \frac{1}{2}$; (*c*) $t = 1^-$; (*d*) $t = 1^+$; (*e*) $t = 2$.

2-11 Prove that, if $\int_{-\infty}^{t} f(\tau)u(\tau)\, d\tau = g(t)u(t)$, then $\int_{-\infty}^{t} f(\tau - a)u(\tau - a)\, d\tau = g(t - a)u(t - a)$.

2-12 The function $f_1(t)$ is given by the waveform shown in Fig. P2-12. Sketch the waveform of $f_2(t) = \int_{-\infty}^{t} f_1(\tau)\, d\tau$ for the cases (*a*) $a = 1$, $b = 2$; (*b*) $a = 2$, $b = 3$; (*c*) $a = 1$, $b = 3$.

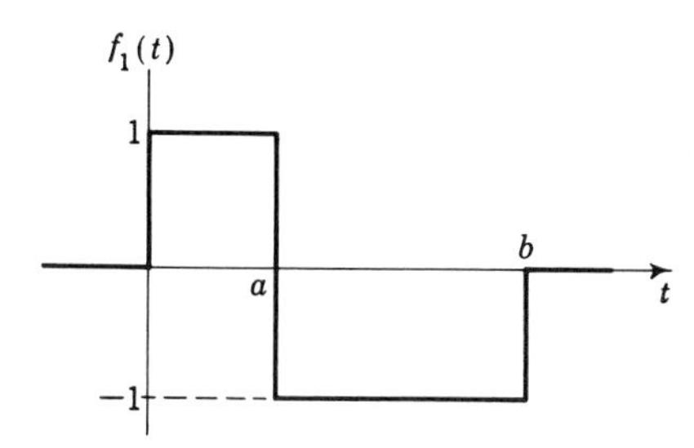

Fig. P2-12

2-13 Integrate graphically $f(t)$ as specified by Fig. P2-13, and verify the result analytically [$f(t) \equiv 0$, $t > 3$].

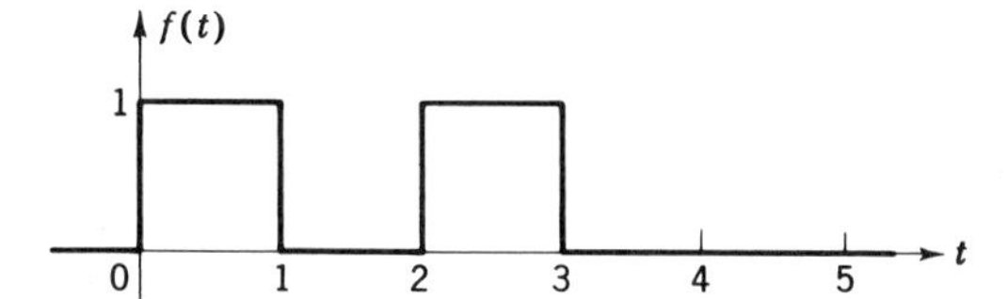

Fig. P2-13

2-14 For the triangular pulse $v(t)$ as given in Fig. P2-6, sketch the waveform of (*a*) the integral of $v(t)$; (*b*) the derivative dv/dt.

2-15 Integrate and differentiate graphically the function shown in (*a*) Fig. P2-15*a*; (*b*) Fig. P2-15*b*.

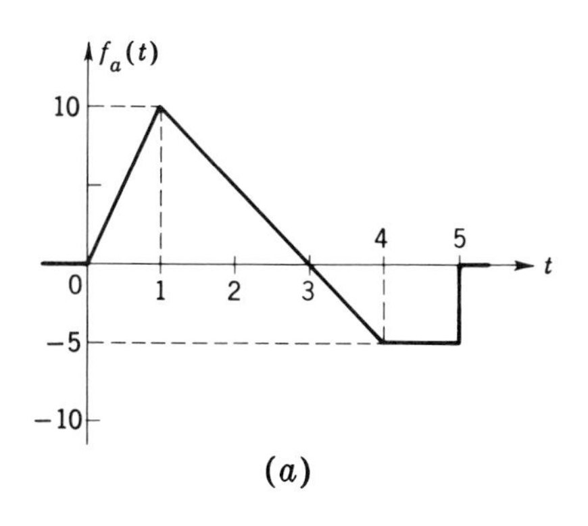

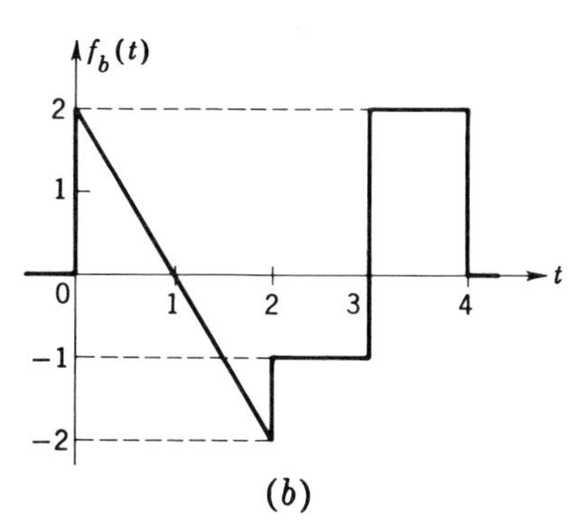

Fig. P2-15

2-16 Sketch the functions f and df/dt if (*a*) $f(t) = (\cos t)u(t) + [\sin (t - \pi/2)]u(t - \pi/2)$; (*b*) $f(t) = e^{-t}u(t)$.

2-17 (*a*) Prove that the area under the graph $f(t) = \alpha e^{-\alpha t}u(t)$ is equal to unity independently of the value α. (*b*) Show that $\lim_{\alpha \to \infty} f(t)$ has the properties of the unit impulse as given by Eqs. (2-15*a*) and (2-16).

2-18 Evaluate (*a*) $\int_{-\infty}^{+\infty} x^2\delta(x - 3)\, dx$; (*b*) $\int_{-\infty}^{+\infty} e^{-x}\delta(x - 1)\, dx$.

2-19 For the triangular pulse shown in Fig. P2-19: (*a*) Show that the area under the pulse is independent of T. (*b*) Show that, as T approaches zero, the triangular pulse shown becomes the unit impulse. (*c*) Sketch the derivative df/dt. (*d*) The derivative of the unit impulse, termed *unit doublet*, may be defined as the limit of df/dt as T approaches zero. Obtain the properties of the unit doublet by use of this definition.

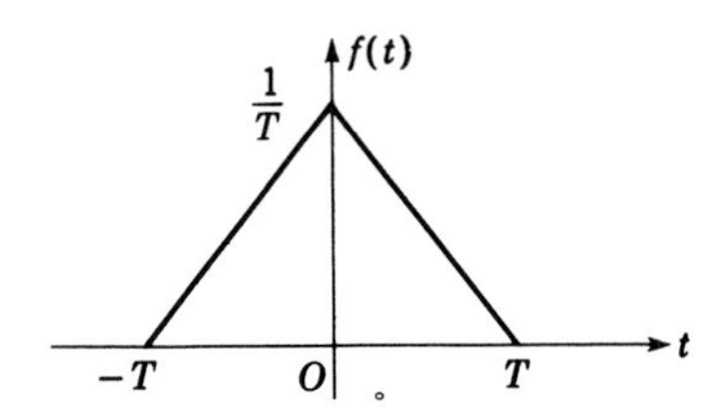

Fig. P2-19

2-20 The derivative of a certain waveform is given by $df/dt = \delta(t) - u(t) + u(t - 1)$. Find and sketch $f(t)$ if $f(t) = 0$ for $t < 0$.

2-21 A parabolic source function $f(t) = 6t^2u(t)$ is applied to an element. Calculate the response and sketch the result if (*a*) the source is a voltage source and the element is (1) an inductance $L = 2$ henrys with $i(0) = 0$; (2) a capacitance $C = 2$ farads; (*b*) the source is a current source and the element is (1) an inductance $L = 2$ henrys, (2) a capacitance $C = 2$ farads with $v(0) = 0$; (3) a 2-ohm resistance.

2-22 A source function, given by $f(t) = (a_1 + a_2t)u(t)$, is applied to a single element. Calculate the response if $a_1 = a_2 = 10$ and the source and element are given as in Prob. 2-21*a* and *b*.

2-23 Do Prob. 2-22, with $a_1 = 10$ and $a_2 = -5$.

2-24 A voltage $v_{ab} = 10u(t) - 25u(t - 2)$ is applied to a $\frac{1}{2}$-henry inductance. If, initially, the current $i_{ab}(0) = I_0$, calculate graphically the instant $t = t_1 > 0$ at which i_{ab} has the largest positive value and the instant $t = t_2 > 0$ when $i_{ab}(t_2) = 0$ for the cases (*a*) $I_0 = 0$; (*b*) $I_0 = 10$; (*c*) $I_0 = -10$.

2-25 A 3-μf capacitance connected across terminals *a*-*b* has 50 volts across it at $t = 0$ [$v_{ab}(0) = 50$]. A current source produces the current $i_{ab} = 20u(t) - 50u(t - 3)$ μa in the capacitance. Calculate (*a*) the instant $t = t_1$ when v_{ab} has its largest value; (*b*) the value $v_{ab}(t_1)$; (*c*) the instant $t = t_2$ when $v_{ab}(t_2) = 0$; (*d*) the instant $t = t_3$ when $v_{ab}(t_3) = -50$; (*e*) the instant $t = t_4$ when $v_{ab}(t_4) = -v_{ab}(t_1)$.

2-26 A modified-ramp-function voltage source is applied to a capacitance at $t = 0$. Calculate the response, and sketch the result.

2-27 Calculate the response of a capacitance C to a unit-impulse (*a*) voltage source; (*b*) current source.

2-28 Calculate the response of an inductance to a unit-impulse voltage source.

2-29 Let the triangular pulse shown in Fig. P2-19 represent the waveform of a voltage source $f = v_{ab}$, and let $T = 1$. (*a*) Calculate the response i_{ab} if this source is applied to an inductance of 0.5 henry [$i_{ab}(-1) = 0$]. (*b*) Calculate the energy stored in the inductance at $t = -1$, $+1$, and ∞. (*c*) Plot the power $p_{ab}(t)$. (*d*) Plot the stored energy $w_L(t)$. (*e*) What should be the duration of a rectangular voltage pulse which has unit amplitude to store the same final energy?

2-30 Assume that the waveform specified in Prob. 2-29 (Fig. P2-19 with $T = 1$) is produced by a current source $f = i_{ab}$. (*a*) Calculate the response if an inductance of 0.5 henry is connected to the source terminals *a-b*. (*b*) Calculate the energy stored in the passive element at $t = 1$ and $t = \infty$. (*c*) Plot the power $p(t)$. (*d*) Plot the stored energy $w(t)$. (*e*) Answer (*a*) to (*d*) if the element is a capacitance C with $v_{ab}(-1) = 0$.

2-31 The waveform specified in Prob. 2-29 (Fig. P2-19 with $T = 1$) represents the current in a 3-ohm resistance. Plot the power delivered to, and the energy dissipated by, the resistance as a function of time.

2-32 (*a*) Sketch to scale the current i_{ab} in a 1-henry inductance with the voltage waveform shown in Fig. P2-32 applied. Use $i(0) = 0$. (*b*) Sketch to scale the stored energy as a function of time.

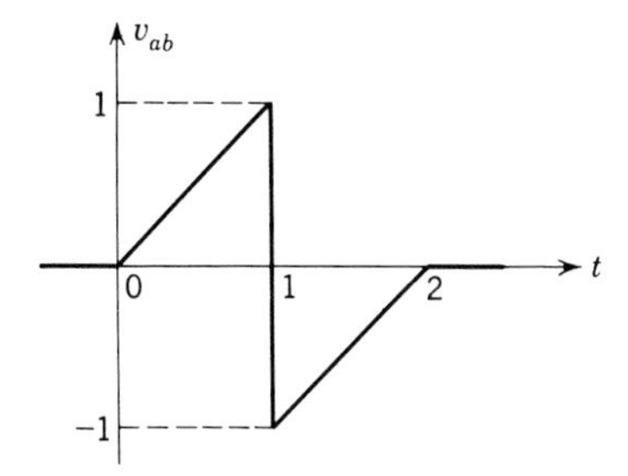

Fig. P2-32

2-33 In the circuit shown in Fig. P2-33, the switch is in position 1 from $t = 0$ to $t = 1$, in position 2 from $t = 1$ to $t = 2$, and in position 3 from $t = 2$ on. The source function is given by $i(t) = 4t$. The initial value of the stored energy in C is zero. (*a*) Calculate and plot $v_{1b}(t)$, $v_{2b}(t)$, $v_{3b}(t)$, and $v_{ab}(t)$. (*b*) Calculate the energy delivered by the source to the elements in the time interval (1) $t = 0$ to $t = 2^-$, (2) $t = 0$ to $t = 2^+$, (3) $t = 0$ to $t = 3$. (*c*) The source is modified to have the form $i(t) = A + 4t$. Find the value of A so that p_{ab} is always finite, and do the calculations required in (*a*) and (*b*) for this value of A.

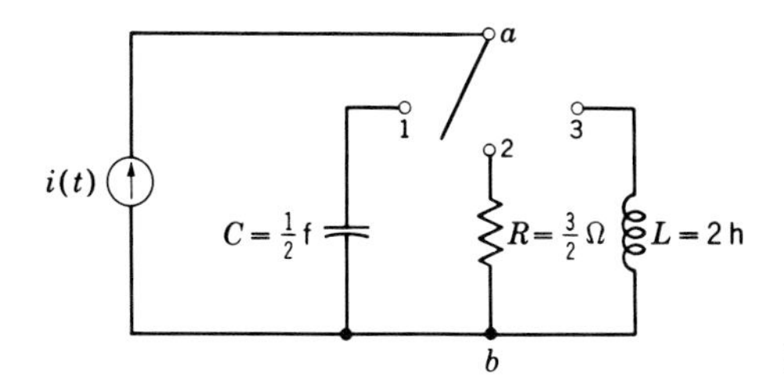

Fig. P2-33

2-34 A sinusoidal voltage source $v_{ab} = V_m \cos \omega t$ is applied to the terminals *a-b* of a passive element. Sketch the waveform of v_{ab} and of i_{ab} in the element if the element is (*a*) a resistance R; (*b*) a capacitance C; (*c*) an inductance L [$i_{ab}(0) = 0$].

2-35 Do Prob. 2-34, changing v_{ab} to $v_{ab} = V_m \sin \omega t$ [in part *c* use $i_{ab}(0) = I_0$].

2-36 Do Prob. 2-34, changing v_{ab} to Ve^{st}.

Chapter 3 Kirchhoff's laws, network equations, and introduction to network functions

By combining circuit elements in a proper manner, a circuit can be constructed to represent the behavior of an actual device. For example, under certain conditions, the field coil of a generator can be represented as a combination of R and L elements placed end to end as shown in Fig. 3-1.

Referring to the field coil of a generator and its equivalent circuit as shown in Fig. 3-1, the two terminals a and b represent actual terminals

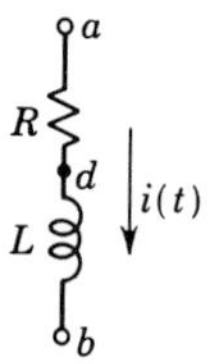

Fig. 3-1 Series connection of two elements.

of the field coil, but the junction between the two elements (point d) does not correspond to any particular point in the field coil. This is not surprising since, according to our definitions of the circuit elements, R accounts for the dissipation of energy into heat, and L accounts for the storage of energy in the magnetic field, whenever a current flows between terminals a-b of the actual field coil. In the actual device the "resistance" and the "inductance" are not separate "parts," but in the equivalent circuit we choose to represent them as such.

The only true similarity between the response of the actual device and its equivalent circuit is in the correspondence between their voltage-current characteristic at the two terminals a-b. What goes on inside the equivalent circuit may have no counterpart in the actual device. For example, as a result of the flow of current from a to b, a voltage will develop across R (Fig. 3-1). This voltage does not correspond to any voltage which can be *measured* in the actual field coil. On the other hand, voltage and current at terminals a-b of the field coil and of its equivalent circuit will be identical. We may consider the circuit shown in Fig. 3-1 to be a terminal pair a-b representing the actual device. In such a case the terminals a and b are called *accessible* terminals, and the terminal d is called an *inaccessible* terminal.

3-1 Series and parallel connections of elements

Elements connected end to end, such that they carry the *same current*, are said to be in *series*. In Fig. 3-1, R and L are connected in series. If at any instant of time there is a current $i_{ad}(t)$ in R, there will also be a current $i_{db}(t)$ such that $i_{ad}(t) \equiv i_{db}(t) \equiv i$. This result is arrived at from the fundamental assumption that in a terminal pair the charges which enter one terminal must come out from the other terminal. This is called the "assumption of continuity of current."

Elements connected between a pair of terminals are said to be connected in "parallel." Elements connected in parallel will have the *same voltage* across them.

The connection of the elements R and C in Fig. 3-2 is an example of *parallel connection* of two elements. In this circuit the two elements are

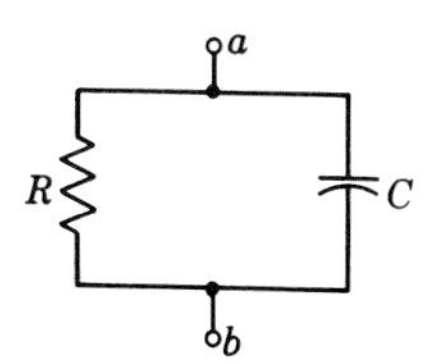

Fig. 3-2 Parallel connection of two elements.

placed so that the same voltage will always exist across them. Although this circuit may correspond to the actual connection of two physical devices, it may also be the equivalent circuit for many practical devices. In this circuit terminals a-b may represent the input terminals of an electronic amplifier. In that case the capacitance C may not be placed in the circuit intentionally, but may represent the combination of certain unavoidable features of the electronic device. As another example, the same circuit may, with respect to the voltage-current characteristic at terminals a-b, represent the input to a telephone cable.

The circuit shown in Fig. 3-3*a* is an example of a "series-parallel" connection of circuit elements. The series connection of R and L is connected in parallel with C. It is interesting to note that this circuit might represent a coil of wire. The resistance and inductance represent the same types of energy conversion as explained in connection with Fig. 3-1. The capacitance C would account for the electric-energy storage in the electric field between the windings of the coil.

The circuit of Fig. 3-3*b* is another example of a series-parallel circuit. In this circuit the resistance R_1 is in series with the parallel combination

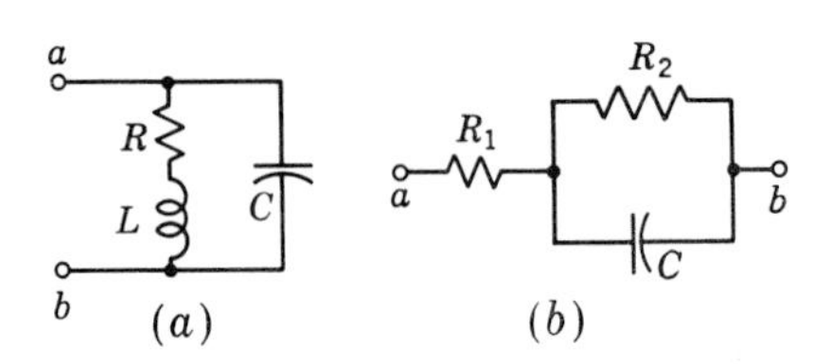

Fig. 3-3 Series-parallel connection of elements.

of R_2 and C. This circuit, like all the others, may represent the actual combination of several nearly "pure" elements or may be the equivalent circuit of an actual device or a portion of such a device. (Figure 3-3*b* may represent the input circuit of an oscilloscope probe.)

3-2 Network terminology

The representation of electrical systems frequently requires the combination of many sources and elements connected in a more complicated manner than the series or parallel arrangement described above. To facilitate discussion and analysis of networks the special terminology given below is used.

NETWORK An interconnection of circuit elements is termed a network, or circuit. A network may contain both active and passive elements or may consist of passive elements only. In the former case it is termed an *active* network, and in the latter case the term *passive,* or *source-free,* applies. If the passive elements in a network are all of the same type, i.e., all resistances or all inductances or all capacitances, the network is termed a resistive, inductive, or capacitive network, respectively.

RESPONSES OF A NETWORK The waveform, or function, representing the current in or voltage across a passive element or a combination of elements, as well as the voltage across an ideal current source and the current in ideal voltage sources, is termed a *response of a network.* Thus all voltage and current waveforms which are not specified through the ideal sources in a network are responses of the network.

NODES We shall call a point in the network common to two or more elements a node. For example, in Fig. 3-4, the points a, b, d, f, g, h, k, m, and n are nodes.

JUNCTIONS We shall call a node common to three or more elements a junction. In Fig. 3-4 the nodes a, b, d, and m are junctions, whereas the nodes g, h, k, and n are not junctions, since each of them is common to two elements only.

BRANCH We shall call a single element or a series connection of elements between any two junctions a branch. In Fig. 3-4 the following branches are identified: branch ab, consisting of the current source $i(t)$; the resistive branch ab, consisting of R_1; branch ad, consisting of C_1; branch db, consisting

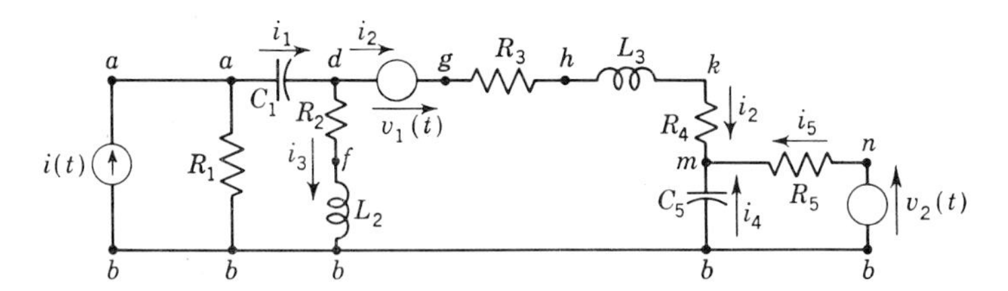

Fig. 3-4 Diagram of a network.

of R_2 and L_2; branch dm, consisting of $v_1(t)$, R_3, L_3, and R_4; branch mb, consisting of C_5; branch mnb, consisting of R_5 and $v_2(t)$. Note that, in accordance with our definition, in Fig. 3-4 km, hkm, and df are not branches since they do not connect two junctions.

PASSIVE AND ACTIVE BRANCHES If a branch contains no sources, it is called a passive branch; otherwise it is an active branch. In Fig. 3-4 the branch mnb is an active branch, and the branch dfb is a passive branch. The current source $i(t)$ in this figure constitutes a branch by itself.

BRANCH CURRENT A current flowing in a branch is called a branch current. The same branch current flows through all the elements in a given branch, since the elements in a branch are in series.

LOOP We shall call any closed path through the circuit elements of a network a loop. In Fig. 3-4 the following loops are identified: loop a-b-a, consisting of R_1 and $i(t)$; loop a-d-f-b-a, consisting of C_1, R_2, L_2, and R_1; loop a-b-f-d-a, consisting of $i(t)$, L_2, R_2, and C_1; loop d-g-h-k-m-n-b-f-d, consisting of $v_1(t)$, R_3, L_3, R_4, R_5, $v_2(t)$, L_2, and R_2. The reader may trace a few more loops as an exercise. If a loop contains one element of a branch, it will contain all the elements of that branch. In Fig. 3-4 the loop a-b-f-d-a, which contains R_2, also contains L_2, which with R_2 forms the branch dfb. In Fig. 3-4 loop a-b-a has two branches, that is, $i(t)$ and R_1; and loop a-b-f-d-a has three branches, R_1, C_1, and dfb. A branch may be common to more than two loops. In Fig. 3-4 the branch R_1 is common to loops a-d-f-b-a, a-d-g-h-k-m-b-a, and others.

MESH In a given diagram of a network a loop which does not encircle or enclose another loop and cannot be divided into other loops is called a mesh. In Fig. 3-4 the loop *m-n-b-m* is a mesh, whereas the loop *a-d-g-h-k-m-b-a* is not a mesh, since it encloses the loop *a-d-f-b-a*. A branch may not be common to more than two meshes. The reasons for defining meshes as well as loops are discussed in Chap. 11.

3-3 Restrictions on variables of networks: Kirchhoff's laws

As a result of the interconnection of elements in a network, certain restrictions (*constraints*) are placed on the currents and voltages associated with these elements. For example, in the series connection of the R and L elements in Fig. 3-1, the principle of continuity of current requires that $i_{ad} \equiv i_{db}$. In the parallel connection of the R and C elements in Fig. 3-2, the definition of the voltage requires that the same voltages exist across R and C. Therefore, if the current i_{ad} in Fig. 3-1 or the voltage across R in Fig. 3-2 is specified, the current i_{db} in Fig. 3-1 or the voltage across C in Fig. 3-2, respectively, will also be specified. *Thus, when we connect elements in series, we place a restriction on the current through them. When we connect elements in parallel, a restriction is established on the voltage across them.*

The application of the principles of continuity of current and the law of conservation of energy establishes certain additional restrictions on the currents and voltages associated with elements in a network. We shall first state these restrictions and then deduce them from the above principles. The treatment of circuit problems can be approached by stating these restrictions as "laws" of circuits. These laws are called, after the physicist Gustav Robert Kirchhoff (1824–1887), Kirchhoff's laws.

> *Kirchhoff's voltage law* At any instant the sum of the voltages around any loop is identically zero.

As an illustration, we apply this law to the loop *a-d-f-b-a* in Fig. 3-4:

$$v_{ad} + v_{df} + v_{fb} + v_{ba} = 0$$

Similarly, for the loop *a-d-g-h-k-m-n-b-a*,

$$v_{ad} + v_{dg} + v_{gh} + v_{hk} + v_{km} + v_{mn} + v_{nb} + v_{ba} = 0$$

To state Kirchhoff's current law in compact form, we shall give two definitions in connection with reference arrows for currents.

If the head of the reference arrow of a current points toward (or away from) a node, we say that the current is entering (or leaving) that node. If a current i enters a node through an element, a current $-i$ leaves the node through that element.

Using these definitions, Kirchhoff's current law is stated as follows:

> *Kirchhoff's current law* At any instant the algebraic sum of the currents *entering* a node is identically zero, or at any instant the sum of the currents *leaving* a node is identically zero.

As an illustration of this law, consider the node m in Fig. 3-4, where $i_2 + i_4 + i_5 = 0$. The reference arrows of the three currents point toward the junction m, and therefore i_2, i_4, and i_5 enter this junction, and their sum is zero. At the junction d, i_2 and i_3 both leave d, but i_1 enters d. If i_1 enters d, then $-i_1$ leaves d, and the application of Kirchhoff's current law gives $-i_1 + i_2 + i_3 = 0$. The latter equation may be written as

$$i_1 = i_2 + i_3$$

which can be interpreted as: The sum of currents whose reference arrows enter a node (i_1 entering d) is equal to the sum of currents whose reference arrows leave that node (i_2 and i_3 leaving d).

Notice that, in the branch $dghkm$ of Fig. 3-4, i_2 is the branch current and flows through all the elements of that branch. Thus

$$i_{dg} = i_{gh} = i_{hk} = i_{km} = i_2$$

The equation $i_{dg} = i_{gh}$ can also be considered an expression of Kirchhoff's current law since i_{dg} is the current entering node g, and i_{gh} is the current leaving node g.

In discussing the series connection of elements, we referred to the assumption of continuity of current. The application of this principle to the junction of elements results in Kirchhoff's current law. If the sum of the currents entering a junction were not equal to the sum of the currents leaving that junction, there would be an accumulation of charge at that junction. The effect of accumulation of charge is represented by the circuit element capacitance, and accumulation of charge at a terminal has no meaning in terms of the concepts already defined.

Example 3-1 Let i_1 and i_2 in Fig. 3-5 be given by $i_1 = 8u(t)$ and $i_2 = 3tu(t)$. With these two currents specified, we are no longer at liberty to specify $i_3(t)$ since $i_3(t) = i_1 + i_2 = (8 + 3t)u(t)$.

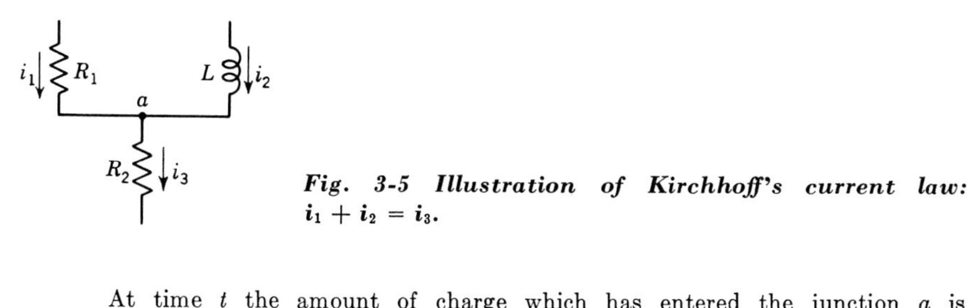

Fig. 3-5 Illustration of Kirchhoff's current law: $i_1 + i_2 = i_3$.

At time t the amount of charge which has entered the junction a is $\int_{-\infty}^{t} (i_1 + i_2)\, d\tau$, and the amount of charge which has left a is $\int_{-\infty}^{t} i_3\, d\tau$.

If $i_3 \neq i_1 + i_2$, charge equal to $\int_{-\infty}^{t} (i_1 + i_2 - i_3)\, d\tau$ has accumulated at a. As mentioned above, this is contrary to our concept of terminal pairs, and therefore $i_1 + i_2 = i_3$.

3-4 Derivation of Kirchhoff's voltage law from the current law and the law of conservation of energy

In Fig. 3-6*a* a closed path is shown. Let us assume that the terminal pairs shown in this figure are passive (do not include sources) and that the only source in the path is $v(t)$.

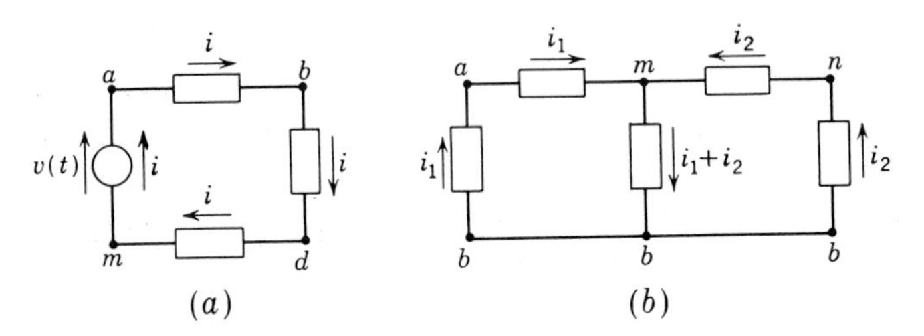

Fig. 3-6 (a) Network consisting of one loop. (b) A two-mesh network.

At any instant, the power input to the terminal pairs *a*-*b*, *b*-*d*, and *d*-*m* is $v_{ab}i_{ab}$, $v_{bd}i_{bd}$, and $v_{dm}i_{dm}$, respectively. Since $i_{ab} = i_{bd} = i_{dm} = i$, the rate of delivery of energy to the passive elements is

$$v_{ab}i_{ab} + v_{bd}i_{bd} + v_{dm}i_{dm} = i(v_{ab} + v_{bd} + v_{dm})$$

The source delivers energy at the rate iv_{am}, and therefore

$$iv_{am} = i(v_{ab} + v_{bd} + v_{dm})$$

so that

$$v_{am} = v_{ab} + v_{bd} + v_{dm} = -v_{ma} \tag{3-1}$$

or

$$v_{ab} + v_{bd} + v_{dm} + v_{ma} = 0 \tag{3-2}$$

which is the statement of Kirchhoff's voltage law for this circuit.

Most often, when the circuit contains a source in series with passive terminal pairs, it is more convenient to express Kirchhoff's voltage law in the form of Eq. (3-1), namely, with the source function on one side of the equation, rather than in the form of Eq. (3-2).

The above derivation was made in the special case of a network consisting of one loop. The argument can be extended to apply to any network, regardless of its geometry. In Fig. 3-6*b* a two-mesh network, without any source, is shown. In conformity with Kirchhoff's current law, the current in branch *mb* of Fig. 3-6*b* is $i_1 + i_2$.

By the law of conservation of energy the total power received by all the terminal pairs is zero. (This means that at any given instant some of the terminal pairs receive and others deliver energy.)

$$-v_{ab}i_1 + v_{am}i_1 + v_{mb}(i_1 + i_2) - v_{mn}i_2 - v_{nb}i_2 = 0$$
$$i_1(v_{am} + v_{mb} - v_{ab}) + i_2(v_{mb} - v_{mn} - v_{nb}) = 0 \tag{3-2a}$$

In the network of Fig. 3-6*b* the waveform of i_2 can be changed without any change in the waveform of i_1. This can be achieved by changing the elements in the terminal pairs of the network (for example, terminal pairs *m-n*, *n-b*, and *m-b*). In other words, the only restriction placed by the network on the waveform of the current is that given by $i_{mb} = i_1 + i_2$. This is a restriction on the waveform of i_{mb}. One of the currents i_1 and i_2 can be specified independently of the other.

Returning now to Eq. (3-2*a*), we note that the form of this equation is

$$i_1(t)f_1(t) + i_2(t)f_2(t) = 0 \tag{3-2b}$$

Now, since $i_1(t)$ or $i_2(t)$ can be specified arbitrarily, it follows that $f_1(t) \equiv 0$ and $f_2(t) \equiv 0$. To clarify this argument, let us assume that $i_2(t) \equiv 0$ and $i_1(t) \not\equiv 0$. Then $f_1(t) \equiv 0$. If, on the other hand, we let $i_1(t) \equiv 0$ and choose $i_2(t) \not\equiv 0$, then $f_2(t) \equiv 0$. Hence, in Eq. (3-2*a*), $v_{am} + v_{mb} - v_{ab} = 0$, and $v_{mb} - v_{mn} - v_{nb} = 0$.

These equations correspond to Kirchhoff's voltage law for each of the two meshes *a-m-b-a* and *m-n-b-m*.

Although the derivation of Kirchhoff's voltage law has dealt only with particular examples (Fig. 3-6*a* and *b*), the result is general and can be proved generally by application of a similar procedure.

3-5 The application of Kirchhoff's laws

We shall now show how the application of Kirchhoff's laws leads to integro-differential or differential equations for network responses. Before generalizing, two examples are appropriate.

Example 3-2 Apply Kirchhoff's voltage law to the circuit shown in Fig. 3-7, and show that the equation for $i(t)$ is an integrodifferential equation.

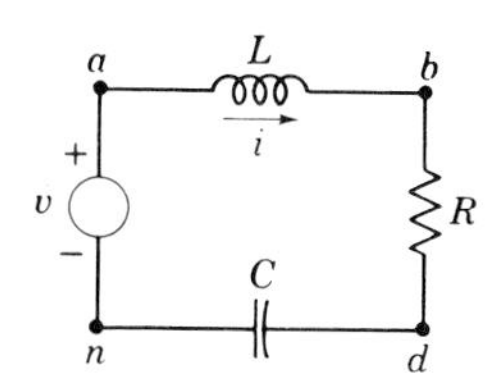

***Fig. 3-7** A series R-L-C circuit excited by an ideal voltage source.*

Solution Kirchhoff's voltage law applied to the single-loop circuit of Fig. 3-7 yields the equation

$$v_{ab} + v_{bd} + v_{dn} + v_{na} = 0$$

or since $v = v_{an} = -v_{na}$,

$$v_{ab} + v_{bd} + v_{dn} = v \tag{3-3}$$

We observe that the three voltage variables v_{ab}, v_{bd}, and v_{dn} can all be expressed in terms of the current i through the voltage-current relations for the L, R, and C elements; that is,

$$v_{ab} = L\frac{di}{dt} \qquad v_{bd} = Ri \qquad v_{dn} = \frac{1}{C}\int_{-\infty}^{t} i\,d\tau$$

Hence, using the above relations in Eq. (3-3),

$$L\frac{di}{dt} + Ri + \frac{1}{C}\int_{-\infty}^{t} i\,d\tau = v \tag{3-4}$$

Equation (3-4) is the required integrodifferential equation.

Example 3-3 Apply Kirchhoff's current law to the circuit shown in Fig. 3-8, and deduce the integrodifferential equation for $v(t)$.

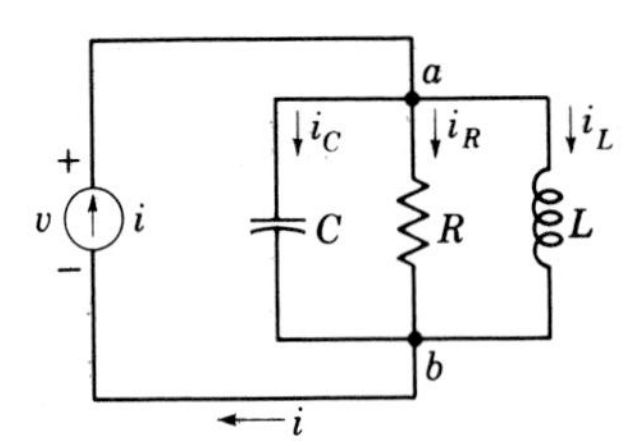

Fig. 3-8** **A parallel R-L-C circuit excited by an ideal current source.

Solution Kirchhoff's current laws applied at junction b of Fig. 3-8 yields the equation

$$i_C + i_R + i_L - i = 0$$

or

$$i_C + i_R + i_L = i \tag{3-5}$$

We observe that the three current variables i_C, i_R, and i_L can all be expressed in terms of the same voltage $v = v_{ab}$ through the current-voltage relations for the elements C, R, and L; that is,

$$i_C = C\frac{dv}{dt} \qquad i_R = \frac{1}{R}v \qquad i_L = \frac{1}{L}\int_{-\infty}^{t} v\,d\tau$$

Hence, using the above relations in Eq. (3-5),

$$C\frac{dv}{dt} + \frac{1}{R}v + \frac{1}{L}\int_{-\infty}^{t} v\,d\tau = i \tag{3-6}$$

Equation (3-6) is the required integrodifferential equation.

The solution of integrodifferential equations such as Eqs. (3-4) and (3-6) is the subject of other chapters. Examples 3-2 and 3-3 illustrate the technique for arriving at the equation relating a response to the source of a network and point out several general properties of Kirchhoff-law equations as follows:

1. Application of Kirchhoff's laws to network loops and junctions will generally result in integrodifferential equations.
2. Application of the voltage law to series elements is most conveniently done by expressing each element voltage in the series connection in terms of the common current.
3. Application of the current law to parallel elements is most conveniently carried out by expressing each element current of the parallel combination in terms of the common voltage.
4. Different networks can yield integrodifferential equations of the same mathematical form [compare Eqs. (3-4) and (3-6)]. Such analogies are useful, and are discussed further in Chap. 4.

At this point in the development we note that in *resistive networks* every voltage-current relationship is algebraic; hence the application of Kirchhoff's laws will result in algebraic rather than integrodifferential equations. Below we show that *inductive* and *capacitive* networks can also be described by algebraic equations. Finally, in Sec. 3-10, we show that the use of operational notation allows the treatment of general networks in algebraic form.

3-6 Resistive, inductive, and capacitive ladder networks

When the passive elements in a network are all of one type, the relationship between response and the sources can be obtained by algebraic means. In this section we show how such equilibrium relations can be obtained in ladder networks. Ladder networks consist alternately of series and parallel elements as shown in Fig. 3-9.

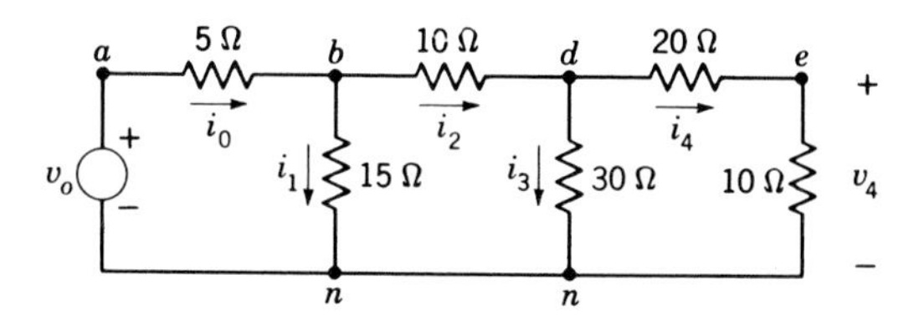

***Fig. 3-9** Resistive ladder network.*

Example 3-4 In the resistive ladder network of Fig. 3-9, deduce the relationship between v_4 and v_o, between i_4 and v_o, and between i_0 and v_o.

Solution In Fig. 3-9, a current with a reference arrow has been assigned to every branch. We now *begin* at the *output* v_4 and apply, in turn, the resistance voltage-current relationship, the voltage law, and the current law. At each step the variables are expressed in terms of v_4.

Thus, from the resistance voltage-current relation,

$$i_4 = \tfrac{1}{10}v_4 \qquad \text{and} \qquad v_{de} = 20i_4 = 2v_4 \tag{3-7}$$

Hence, from the voltage law,

$$v_{dn} = v_{de} + v_4 = 3v_4 \tag{3-8}$$

and from the resistance relation for the 30-ohm resistance,

$$i_3 = \frac{v_{dn}}{30} = \frac{v_4}{10} \tag{3-9}$$

From the current law,

$$i_2 = i_3 + i_4 = \frac{v_4}{10} + \frac{v_4}{10} = \frac{v_4}{5} \tag{3-10}$$

and from the resistance relation for R_{bd},

$$v_{bd} = 10i_2 = 2v_4 \tag{3-11}$$

From the voltage law,

$$v_{bn} = v_{bd} + v_{dn} = 5v_4 \tag{3-12}$$

From the resistance relation for the 15-ohm resistance,

$$i_1 = \tfrac{1}{15}v_{bn} = \tfrac{1}{3}v_4 \tag{3-13}$$

From the current law,

$$i_0 = i_1 + i_2 = \tfrac{1}{3}v_4 + \tfrac{1}{5}v_4 = \tfrac{8}{15}v_4 \tag{3-14}$$

From the resistance relation for R_{ab},

$$v_{ab} = 5i_0 = \tfrac{8}{3}v_4 \tag{3-15}$$

and from the voltage law,

$$v_o = v_{ab} + v_{bn}$$

$$v_o = \tfrac{8}{3}v_4 + 5v_4 = \tfrac{23}{3}v_4 \tag{3-16}$$

Equation (3-16) is one required result. Since $i_4 = v_4/10$, the relationship between i_4 and v_o is given by $v_o = \frac{23}{3}(10i_4) = 230i_4/3$. Finally, from Eq. (3-14),

$$v_4 = \tfrac{15}{8}i_0$$

so that

$$v_o = \tfrac{23}{3} \times \tfrac{15}{8}i_0 = \tfrac{115}{8}i_0 \tag{3-17}$$

which is also a required result.

Example 3-5 In the capacitance network of Fig. 3-10, deduce the relationship between i_3 and i, v_4 and i, and v_o and i.

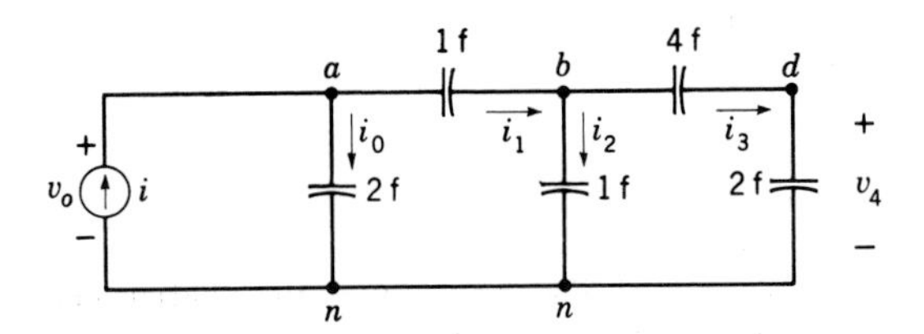

Fig. 3-10 Capacitive ladder network.

Solution The procedure which may be followed is completely analogous to that used in the resistive case if one uses

$$q_3 = \int_{-\infty}^{t} i_3\,d\tau \qquad q_2 = \int_{-\infty}^{t} i_2\,d\tau \qquad \cdots$$

We shall however use voltages and currents as variables and begin with

$$v_4 = \tfrac{1}{2}\int_{-\infty}^{t} i_3\,d\tau = \tfrac{1}{2}q_3$$

$$v_{bd} = \tfrac{1}{4}\int_{-\infty}^{t} i_3\,d\tau = \tfrac{1}{4}q_3$$

Hence, from the voltage law, $v_{bn} = v_3 + v_4$, or

$$v_{bn} = \tfrac{3}{4}\int_{-\infty}^{t} i_3\,dt = \tfrac{3}{4}q_3$$

From the capacitance voltage-current relation,

$$i_2 = \frac{dv_{bn}}{dt} = \frac{3}{4}i_3$$

From the current law,

$$i_1 = i_2 + i_3 = \tfrac{7}{4}i_3 \tag{3-18}$$

The procedure should now be clear since it is analogous to the resistive case treated in Example 3-4. It is left as an excercise for the reader to show that the results are

$$i = \tfrac{27}{4} i_3 \tag{3-19}$$

$$i = \frac{27}{2}\frac{dv_4}{dt} \tag{3-20}$$

and

$$i = 2.7\,\frac{dv_0}{dt} \tag{3-21}$$

The calculation of an inductive ladder is left as an exercise for the reader (Prob. 3-13).

3-7 *Equivalent elements*

We observe that in Example 3-4 the voltage-current relationship at the terminals of the source is as given by Eq. (3-17),

$$v_o = \tfrac{115}{8} i_0$$

so that at the source terminal the voltage-current relationship is the same as if a resistance of $\tfrac{115}{8}$ ohms were connected. Similarly, in Example 3-5,

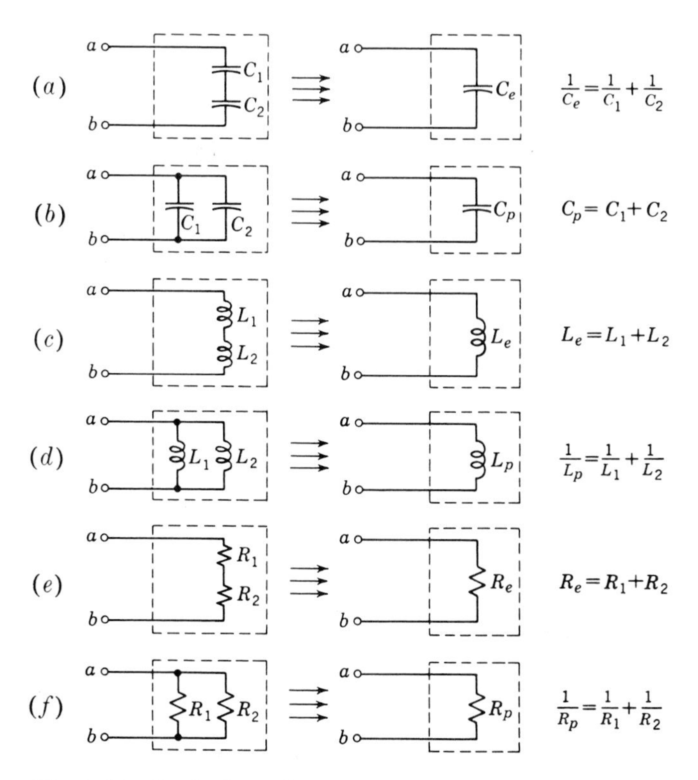

Fig. 3-11 ***Equivalent elements.***

the relationship at the source terminals is

$$i_0 = 2.7 \frac{dv_o}{dt}$$

Hence it appears that the effect of the entire ladder at the source terminals is the same as if a 2.7-farad capacitance were connected in place of the ladder.

Series and parallel connection of similar elements It is left as an exercise for the reader to show that the equivalences shown in Fig. 3-11 are valid provided that, in those given by Fig. 3-11*a* and *d*, the proper initial conditions exist.

3-8 Voltage division and current division for similar elements

Consider the series connection of the two inductances shown in Fig. 3-12. (Although we are formally dealing with only two elements in series, the

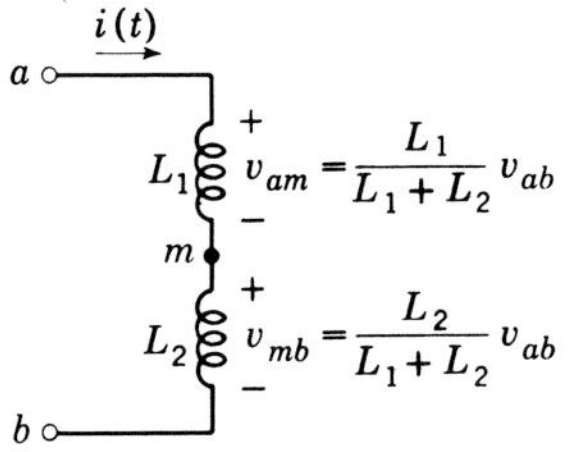

Fig. 3-12 Voltage division for series inductances.

reader will certainly recognize that each of these two elements may be the equivalent element for a combination of elements.) We desire to formulate the relationship between the voltage across one inductance and the voltage across both inductances.

The voltage across L_1 is, at any instant of time,

$$v_{am} = L_1 \frac{di}{dt}$$

and the voltage across L_2 is

$$v_{mb} = L_2 \frac{di}{dt}$$

Hence the ratio v_{am}/v_{mb} is equal to the ratio of the inductances,

$$\frac{v_{am}}{v_{mb}} = \frac{L_1}{L_2} \tag{3-22}$$

In words, when two inductances are connected in series, the voltage across each inductance is proportional to the value of that inductance.

Since the equivalent inductance of the two elements in series is

$$L_e = L_1 + L_2$$

we may also write

$$\frac{v_{am}}{v_{ab}} = \frac{L_1}{L_1 + L_2} \tag{3-23}$$

Equation (3-23) is the *voltage-division formula* for inductances. In words, this formula states that the ratio of the voltage across one inductance in a series connection of inductances is to the voltage across all the series-connected inductances as the one inductance is to the sum of the inductances.

For the series connection of several resistances, the same statement applies if the word inductance is replaced by resistance everywhere. This result is illustrated in Fig. 3-13. The proof is left as an exercise for the reader.

a
b
m
R_1
R_2
$v_{am} = \frac{R_1}{R_1 + R_2} v_{ab}$
$v_{mb} = \frac{R_2}{R_1 + R_2} v_{ab}$

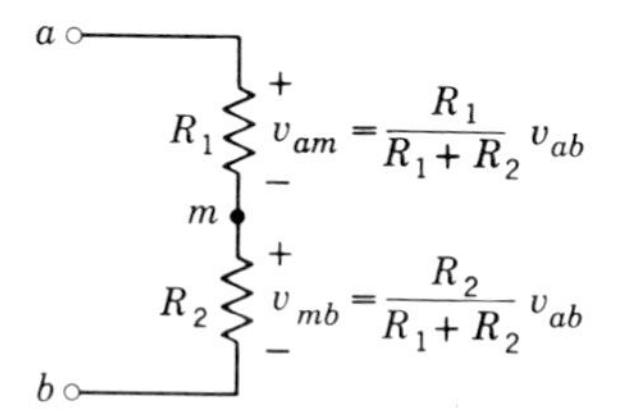

***Fig. 3-13** Voltage division across series resistances.*

In the case of capacitances, we can show that, if the same current has been flowing through the series-connected capacitances for all time, the voltages will divide as the reciprocals of the capacitances, i.e., proportionally to the elastances, as illustrated in Fig. 3-14. The proof is left an an exercise for the reader (Prob. 3-15).

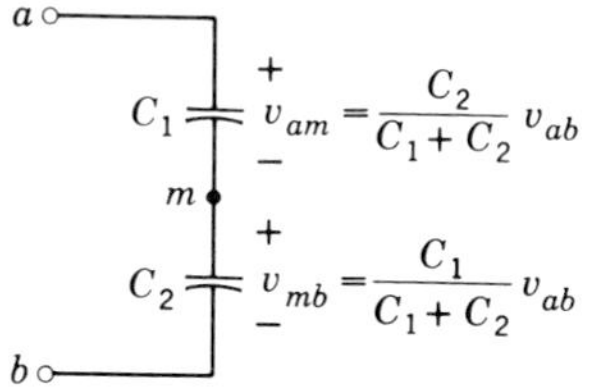

a
b
m
C_1
C_2
$v_{am} = \frac{C_2}{C_1 + C_2} v_{ab}$
$v_{mb} = \frac{C_1}{C_1 + C_2} v_{ab}$

***Fig. 3-14** Voltage division across series capacitances.*

Consider now the parallel inductances L_1 and L_2 shown in Fig. 3-15. The current in each inductance is given, for all $t > 0$, by

$$i_1(t) = i_1(0) + \frac{1}{L_1}\int_0^t v_{ab}\, d\tau$$

and

$$i_2(t) = i_2(0) + \frac{1}{L_2}\int_0^t v_{ab}\, d\tau$$

If we now assume that the ratio $i_1(0)/i_2(0) = L_2/L_1$ [or that both $i_1(0)$ and $i_2(0)$ are zero], then

$$\frac{i_1(t)}{i_2(t)} = \frac{L_2}{L_1} \tag{3-24}$$

so that (under the assumed conditions) the current in two parallel inductances divides proportionally to the reciprocals of the inductances.

a
i(t)
i1(t) = L2/(L1 + L2) i(t)
L1
L2
i2(t) = L1/(L1 + L2) i(t)
b

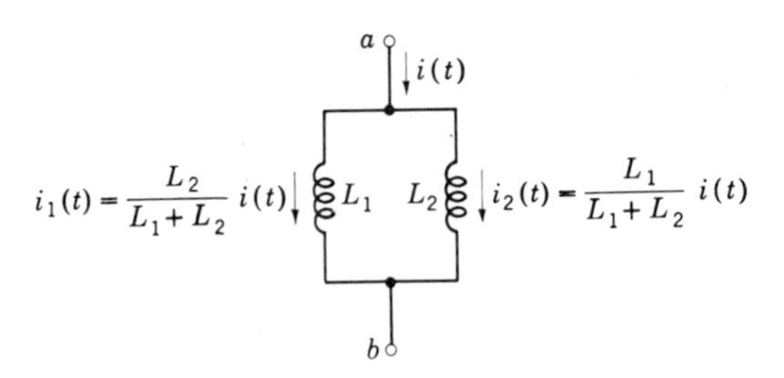

***Fig. 3-15** Current division in parallel inductances.*

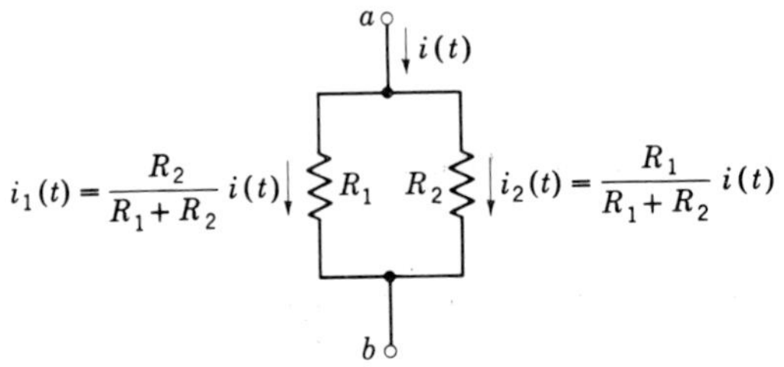

a
i(t)
i1(t) = R2/(R1 + R2) i(t)
R1
R2
i2(t) = R1/(R1 + R2) i(t)
b

***Fig. 3-16** Current division in parallel resistances.*

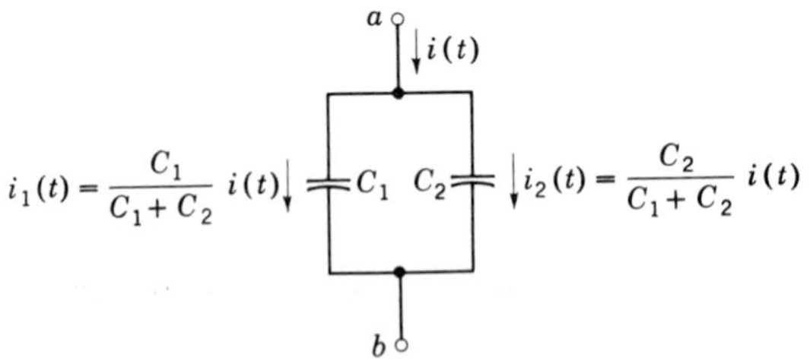

a
i(t)
i1(t) = C1/(C1 + C2) i(t)
C1
C2
i2(t) = C2/(C1 + C2) i(t)
b

***Fig. 3-17** Current division in parallel capacitances.*

In the case of two parallel resistances (as shown in Fig. 3-16) the current divides proportionally to the conductances (reciprocal resistances), and for parallel capacitances the current divides proportionally to the capacitances as shown in Fig. 3-17. The proof of these statements is left as an exercise for the reader.

3-9 Equilibrium equations

We recall that, in general, any voltage or current which is not generated by an ideal source can be a network response; thus many responses are associated with a network. In most practical examples one is not interested in *every* response, but only in certain responses, which may constitute the "output." Thus the output of a network may be a voltage across several elements rather than the voltage across a single element. In such a case the desired response may not occur as a variable in the Kirchhoff-law equations (for example, it may be the voltage v_{ad} in Fig. 3-18, as discussed below,

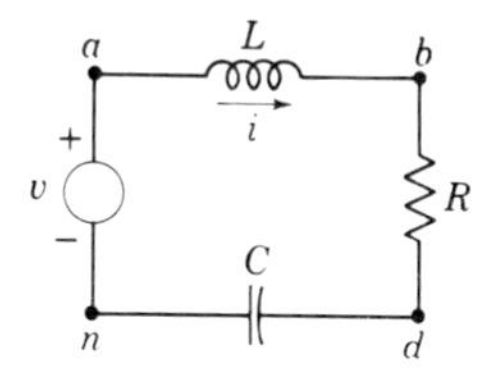

***Fig. 3-18** A series R-L-C circuit excited by an ideal voltage source.*

in Example 3-6). In other cases it may happen that the application of Kirchhoff's laws leads to simultaneous equations for several variables, only one of which is the desired response variable (Example 3-7). In either event one can manipulate the Kirchhoff-law equations to *derive an equation which relates the desired response function to the source function.* Such a derived equation is termed an *equilibrium equation;* the remainder of this chapter deals with efficient formulation of such an equation for series-parallel circuits; additional general techniques are given in Chaps. 11, 12, and 13.

Example 3-6 In the circuit of Fig. 3-18 obtain the equilibrium equation relating the response v_{ad} to the source v.

Solution From Eq. (3-4),

$$L\frac{di}{dt} + Ri + \frac{1}{C}\int_{-\infty}^{t} i\,d\tau = v \tag{3-25}$$

and from Fig. 3-18,

$$v_{ad} = L\frac{di}{dt} + Ri \tag{3-26}$$

Equations (3-25) and (3-26) are simultaneous equations for the variables v_{ad} and i. To eliminate i from the equations, substitute (3-26) into (3-25).

$$v_{ad} + \frac{1}{C}\int_{-\infty}^{t} i\,d\tau = v \tag{3-27}$$

Now differentiate each term in Eq. (3-27) with respect to t and solve for i:

$$i = C\frac{dv}{dt} - C\frac{dv_{ad}}{dt} \tag{3-28}$$

Substituting (3-28) in (3-25) yields

$$LC\frac{d^2v}{dt^2} - LC\frac{d^2v_{ad}}{dt^2} + RC\frac{dv}{dt} - RC\frac{dv_{ad}}{dt} + v - v_{ad} = v$$

or

$$LC\frac{d^2v_{ad}}{dt^2} + RC\frac{dv_{ad}}{dt} + v_{ad} = LC\frac{d^2v}{dt^2} + RC\frac{dv}{dt} \tag{3-29}$$

Equation (3-29) is the desired equilibrium equation. We observe that it is not a Kirchhoff-law equation since sums of the individual terms do not represent voltages added up around a loop or currents entering or leaving a junction.

Example 3-7 In the circuit of Fig. 3-19 apply Kirchhoff's current law to obtain simultaneous equations for v_{an} and v_{bn} and obtain the equilibrium equation for v_{bn}.

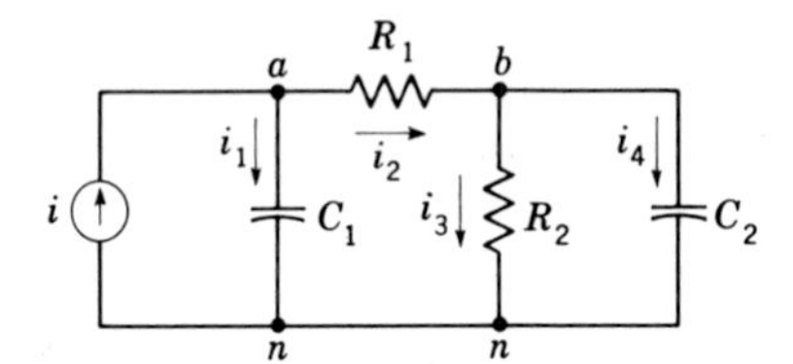

Fig. 3-19 Circuit for Example 3-7.

Solution Application of Kirchhoff's current law at junction a yields $i_1 + i_2 = i$, or

$$C_1\frac{dv_{an}}{dt} + \frac{1}{R_1}(v_{an} - v_{bn}) = i \tag{3-30}$$

At junction b the result is $i_2 = i_3 + i_4$, or

$$\frac{1}{R_1}(v_{an} - v_{bn}) = \frac{1}{R_2}v_{bn} + C_2\frac{dv_{bn}}{dt} \tag{3-31}$$

Equations (3-30) and (3-31) are the simultaneous differential equations for v_{an} and v_{bn}. From Eq. (3-31),

$$v_{an} = v_{bn} + \frac{R_1}{R_2}v_{bn} + R_1C_2\frac{dv_{bn}}{dt} \tag{3-32}$$

Substituting for v_{an} from Eq. (3-32) into (3-30) gives

$$C_1\left(1 + \frac{R_1}{R_2}\right)\frac{dv_{bn}}{dt} + C_1R_1C_2\frac{d^2v_{bn}}{dt^2} + \frac{1}{R_2}v_{bn} + C_2\frac{dv_{bn}}{dt} = i$$

or collecting terms,

$$R_1C_1C_2\frac{d^2v_{bn}}{dt^2} + \frac{(R_1 + R_2)C_1 + R_2C_2}{R_2}\frac{dv_{bn}}{dt} + \frac{1}{R_2}v_{bn} = i \tag{3-33}$$

Equation (3-33) is the required equilibrium equation.

At this point in the discussion we emphasize that the preceding examples are meant only to illustrate the idea of equilibrium equation. Efficient techniques for formulating such equations are discussed in the remainder of this chapter.

3-10 Operational notation

When a circuit consists of several types of passive elements in addition to sources, the formulation of equilibrium relations is no longer algebraic, but involves the manipulation of *simultaneous integrodifferential equations.* To facilitate such manipulations we use the *operational notation* of the calculus. The use of this notation allows the *manipulation* of integrodifferential equations in algebraic form.

We shall denote the operation "differentiate with respect to time" by the symbol p thus:

$$p \equiv \frac{d}{dt} \qquad \text{so that} \qquad p(f) \equiv \frac{df}{dt} \tag{3-34}$$

The symbol p is termed differential operator. Using this operator, the second derivative is written as $p(pf) = d^2f/dt^2$. We write pp as p^2, exactly as if p were an algebraic quantity, thus:

$$p^n \equiv \frac{d^n}{dt^n} \tag{3-35}$$

It can be shown[1] that simultaneous differential equations can be manipulated in the same way as simultaneous algebraic equations by writing them in operational form and then treating p as if it were an algebraic entity.

In the same manner that the symbol p is defined to mean differentiation with respect to time, we define division with p to mean integration with respect to time. We further define the operation p followed by $1/p$ *or* $1/p$ followed by p, operating on a function f, to mean

$$p \cdot \frac{1}{p} f = \frac{1}{p} \cdot pf = f \tag{3-36}$$

Since integration involves an integration constant, Eq. (3-36) does not apply unless this constant is set to zero. In general, for inductance and capacitance, respectively,

$$i_{ab} = i_{ab}(0) + \frac{1}{L_{ab}} \int_0^t v_{ab}\, d\tau \qquad \text{or} \qquad v_{ab} = v_{ab}(0) + \frac{1}{C} \int_0^t i_{ab}\, d\tau \tag{3-37}$$

we define

$$\frac{1}{p}(\quad) = \int_0^t (\quad)\, d\tau \tag{3-38}$$

so that for inductance L_{ab},

$$i_{ab} = i_{ab}(0) + \frac{1}{pL_{ab}} v_{ab} \tag{3-39}$$

and for capacitance C_{ab},

$$v_{ab} = v_{ab}(0) + \frac{1}{pC_{ab}} i_{ab} \tag{3-40}$$

It is shown in Chap. 6 that the initial values $i_{ab}(0)$ or $v_{ab}(0)$ can be represented by ideal sources and that, as a consequence, the formulation of *input-output* (source-response) equations is correctly carried out if the initial-condition terms in Eqs. (3-39) and (3-40) are set to zero. This statement also means that in the formulation of equilibrium equations (and net-

[1] See Chap. 8.

work functions; see Sec. 3-11) we *define a passive network as one consisting of R, L, and C elements, where the L's and C's store zero energy at* $t = 0$, so that the definition of $1/p$ given by Eq. (3-38) applies. In operational form, the voltage-current relationships for the elements R, L, and C are therefore

$$v = Ri \qquad v = pLi \text{ or } i = \frac{1}{pL} v \qquad i = pCv \text{ or } v = \frac{1}{pC} i \tag{3-41}$$

The *form* of these relationships is algebraic; the operator[1] pL plays the same role for inductance that R plays for resistance and $1/pC$ for capacitance. We now illustrate the derivation of equilibrium equations using operational notation.

We shall derive the equilibrium equations for $v_{ad}(t)$ in the resistive circuit of Fig. 3-20*a* and in the *R-L-C* circuit of Fig. 3-20*b* so as to show the similarity of technique when operational notation is used. We observe that each of the circuits consists of three passive elements in series with an ideal voltage source. To show the similarity of technique, we shall work out the solutions for both circuits side by side.

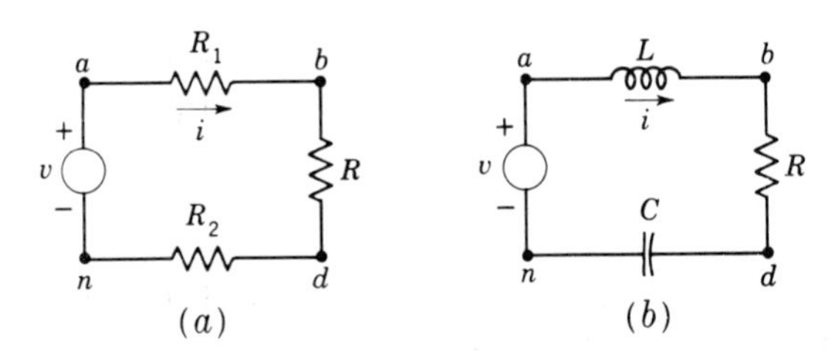

***Fig. 3-20** (a) A resistive circuit. (b) An R-L-C circuit with the same number of elements as the circuit of (a).*

For both circuits,

$$v_{ab} + v_{bd} + v_{dn} = v \tag{3-42}$$

For the circuit of Fig. 3-20a:

$$v_{ab} = R_1 i$$
$$v_{bd} = Ri$$
$$v_{dn} = R_2 i$$

Hence

$$(R_1 + R + R_2)i = v$$

$$i = \frac{1}{R_1 + R + R_2} v$$

$$v_{ad} = (R_1 + R)i$$

$$v_{ad} = \frac{R_1 + R}{R_1 + R + R_2} v \tag{3-43a}$$

For the circuit of Fig. 3-20b:

$$v_{ab} = pLi$$
$$v_{bd} = Ri$$
$$v_{dn} = \frac{1}{pC} i$$

Hence

$$\left(pL + R + \frac{1}{pC}\right) i = v$$

$$i = \frac{1}{pL + R + 1/pC} v$$

$$v_{ad} = (pL + R)i$$

$$v_{ad} = \frac{pL + R}{pL + R + 1/pC} v \tag{3-43b}$$

[1] For time-invariant elements the operator pL is identical with the operator Lp; that is, $pLi = d(Li)/dt = L\, di/dt = Lpi$. This is not true if L is a function of time.

Now, in the case of the resistive circuit, we can arrive at the result Eq. (3-43*a*) by voltage division:

$$\frac{v_{ad}}{v_{an}} = \frac{R_{ad}}{R_{an}} = \frac{R_1 + R}{R_1 + R + R_2} \tag{3-44}$$

The result for the analogous *R-L-C* circuit [Eq. (3-43*b*)] is seen to be related to the voltage-division method for resistive circuits as follows: Replace R_1 by pL and replace R_2 by $1/pC$ in Eq. (3-44). This gives

$$\frac{v_{ad}}{v} = \frac{pL + R}{pL + R + 1/pC} \tag{3-45}$$

Equation (3-45) is seen to be identical with Eq. (3-43*b*).

It is necessary now to emphasize the *difference* between Eq. (3-44) and Eq. (3-45). Equation (3-44) represents the *solution* of the problem: Find the response v_{ad} in terms of the source v. In the resistive circuit the *solution* is that v_{ad} is the fraction $(R_1 + R)/(R_1 + R + R_2)$ of v. On the other hand, Eq. (3-45) does not give v_{ad} numerically in terms of v, but rather, we have the equation

$$v_{ad} = \frac{pL + R}{pL + R + 1/pC}\, v \tag{3-46}$$

which, upon clearing the fraction, is seen to be

$$\left(pL + R + \frac{1}{pC}\right) v_{ad} = (pL + R)v$$

or, replacing p by d/dt and $1/p$ by the integral operation, we obtain the integrodifferential equation

$$L\frac{dv_{ad}}{dt} + Rv_{ad} + \frac{1}{C}\int_0^t v_{ad}\, d\tau = L\frac{dv}{dt} + Rv \tag{3-47a}$$

Differentiating both sides of this equation, we have

$$L\frac{d^2v_{ad}}{dt^2} + R\frac{dv_{ad}}{dt} + \frac{1}{C}v_{ad} = L\frac{d^2v}{dt^2} + R\frac{dv}{dt} \tag{3-47b}$$

which is identical with Eq. (3-29) as derived without the use of operational notation. The differentiation involved in going from Eq. (3-47*a*) to Eq. (3-47*b*) is identical with the "multiplication" with p of the numerator and denominator of the fraction in Eq. (3-46). Such multiplication (by Cp) results in

$$v_{ad}(t) = \frac{RCp + LCp^2}{RCp + LCp^2 + 1}\, v(t) \tag{3-48a}$$

This equation is interpreted as stating

$$(RCp + LCp^2 + 1)v_{ad}(t) = (RCp + LCp^2)v(t) \tag{3-48b}$$

which again is identical with Eq. (3-47*b*).

In summary, we conclude that the operational voltage-current relationships for inductance and capacitance, $v = pLi$ and $i = pCv$, respectively, allow the manipulation of integrodifferential relationships in algebraic form. *It is important to remember that the manipulations of integrodifferential relationships in algebraic form result in a (differential) equilibrium equation which must be solved to obtain the response in terms of the source.* In contrast, the analogous manipulations for the resistive network [as in the derivation of Eq. (3-43*a*)] result in the numerical relationship between response and source.

3-11 Introduction to operational network function

In Eq. (3-43*a*) the response v_{ad} of the resistive network of Fig. 3-20*a* is related to the source v by the formula

$$v_{ad} = \frac{R_1 + R}{R_1 + R + R_2} v$$

so that response v_{ad} is related to the source by the fraction $(R_1 + R)/(R_1 + R + R_2)$. In Eq. (3-48*a*) the analogous fraction $(LCp^2 + RCp)/(LCp^2 + RCp + 1)$ relates the response v_{ad} of the *R-L-C* circuit of Fig. 3-20*b* to the source $v(t)$. As described above, this fraction is a *differential operator* since Eq. (3-47) *must always* be interpreted as the differential equation (3-47*b*) [or its integrodifferential version (3-47*a*)]. The fraction $(LCp^2 + RCp)/(LCp^2 + RCp + 1)$ may be considered a function of the (operator) p. *When such a function of the operator p relates a response function in a network to a source function, the function of p is termed operational network function.* Denoting, in general, an operational network function by $H(p)$, we can write formally

$$\text{Response function} = [H(p)] \text{ source function}$$

In the example of Fig. 3-20*b* the operational network function $H(p)$ which relates the response v_{ad} to the source $v(t)$ is given by

$$H(p) = \frac{LCp^2 + RCp}{LCp^2 + RCp + 1} \qquad v_{ad} = [H(p)]v$$

We can regard the resistive network as a special case where H is independent of p and is an algebraic, rather than a differential, operator.

Example 3-8 In the circuit of Fig. 3-21, find the network functions which relate i_4, v_{bn}, and v_{ab} to the source current i_s.

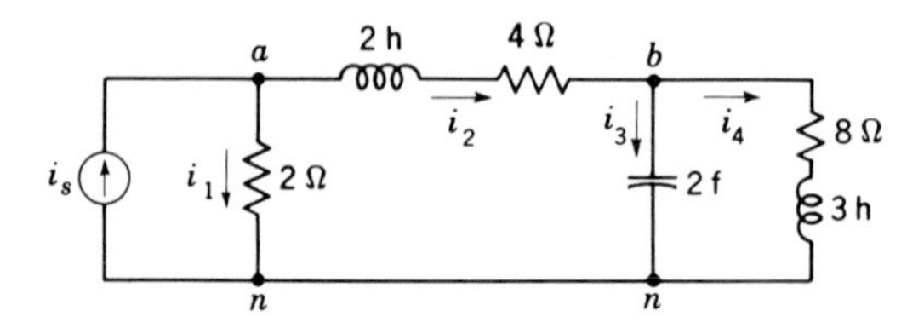

Fig. 3-21 Ladder network analyzed in Example 3-8.

Solution As in resistive ladder networks, we begin with the relationship between i_4 and v_{bn}:

$$(8 + 3p)i_4 = v_{bn} \qquad i_4 = \frac{1}{8 + 3p} v_{bn} \tag{3-49}$$

$$i_3 = 2pv_{bn}$$

Hence, since $i_2 = i_3 + i_4$,

$$i_2 = \left(2p + \frac{1}{8 + 3p}\right) v_{bn}$$
$$= \frac{6p^2 + 16p + 1}{8 + 3p} v_{bn}$$

Further,

$$v_{ab} = (2p + 4)i_2$$
$$= \frac{(2p + 4)(6p^2 + 16p + 1)}{8 + 3p} v_{bn}$$
$$= \frac{12p^3 + 56p^2 + 66p + 4}{8 + 3p} v_{bn} \tag{3-50}$$

Since

$$v_{an} = v_{ab} + v_{bn}$$

we have

$$v_{an} = \left(\frac{12p^3 + 56p^2 + 66p + 4}{8 + 3p} + 1\right) v_{bn}$$
$$= \frac{12p^3 + 56p^2 + 69p + 12}{8 + 3p} v_{bn}$$

Finally,

$$i_1 = \frac{v_{an}}{2} = \frac{12p^3 + 56p^2 + 69p + 12}{16 + 6p} v_{bn}$$

so that

$$i_s = i_2 + i_1$$
$$= \left(\frac{6p^2 + 16p + 1}{8 + 3p} + \frac{12p^3 + 56p^2 + 69p + 12}{16 + 6p}\right) v_{bn}$$
$$= \frac{12p^3 + 68p^2 + 101p + 14}{16 + 6p} v_{bn}$$

or

$$v_{bn} = \frac{16 + 6p}{14 + 101p + 68p^2 + 12p^3} i_s$$

From Eq. (3-49),

$$i_4 = \frac{2}{14 + 101p + 68p^2 + 12p^3} i_s$$

and from Eq. (3-50),

$$v_{ab} = \frac{8 + 132p + 112p^2 + 24p^3}{14 + 101p + 68p^2 + 12p^3} i_s$$

so that the required network functions are H_i, H_a, and H_b, where

$$i_4 = [H_i(p)]i_s \qquad H_i(p) = \frac{2}{14 + 101p + 68p^2 + 12p^3}$$

$$v_{ab} = [H_a(p)]i_s \qquad H_a(p) = \frac{8 + 132p + 112p^2 + 24p^3}{14 + 101p + 68p^2 + 12p^3}$$

and

$$v_{bn} = [H_b(p)]i_s \qquad H_b(p) = \frac{16 + 6p}{14 + 101p + 68p^2 + 12p^3}$$

We recall again that the above results correspond to the differential equations

$$12\frac{d^3i_4}{dt^3} + 68\frac{d^2i_4}{dt^2} + 101\frac{di_4}{dt} + 14i_4 = 2i_s$$

$$12\frac{d^3v_{ab}}{dt^3} + 68\frac{d^2v_{ab}}{dt^2} + 101\frac{dv_{ab}}{dt} + 14v_{ab} = 8i_s + 132\frac{di_s}{dt} + 112\frac{d^2i_s}{dt^2} + 24\frac{d^3i_s}{dt^3}$$

$$12\frac{d^3v_{bn}}{dt^3} + 68\frac{d^2v_{bn}}{dt^2} + 101\frac{dv_{bn}}{dt} + 14v_{bn} = 16i_s + 6\frac{di_s}{dt}$$

3-12 Driving point and transfer functions: immittance and gain

The operational voltage-current relations at the terminals of a terminal-pair network are often of special interest. A network function which relates a *voltage* source to a *current* response, or vice-versa, is termed *operational immittance.*[1] When the voltage and current in question are those at the terminals of a terminal-pair network, the term operational *driving-point* immittance is used. When the source and response are *not* associated with the *same* pair of terminals, the network function is termed a *transfer function.* Thus, in Fig. 3-22*a*, the network function which relates the source at *m-b* to

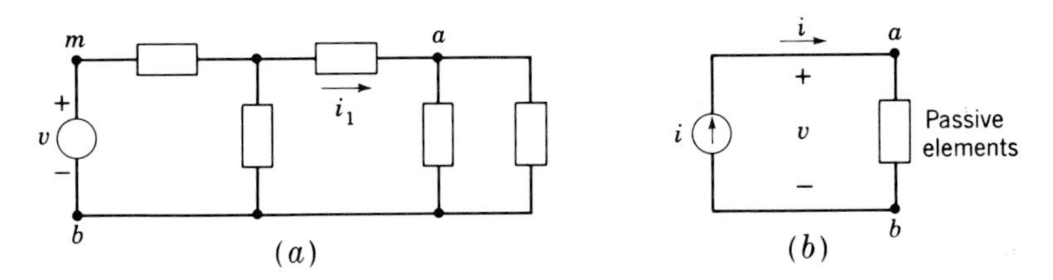

***Fig. 3-22** (a) The network functions that relate v_{ab} or i_1 to v_{mb} is a transfer function. (b) Driving-point immittance relates v and i.*

the response at *a-b* is a transfer function. This is in contrast to the situation described by Fig. 3-22*b*, where the source and response are associated with the same terminal pair *a-b*, and the combination of passive, initially unenergized elements which form the terminal pair *a-b* can be described by a driving-point immittance. If, in Fig. 3-22*b*, the source is a current source, the response v is written, symbolically,

$$v = [Z(p)]i \tag{3-51}$$

where $Z(p)$ is termed *operational driving-point impedance.* When $v(t)$ is the source, the response is $i(t)$, and one writes

$$i(t) = [Y(p)]v \qquad Y(p) = \frac{1}{Z(p)} \tag{3-51a}$$

where $Y(p)$ *is termed operational driving-point admittance.*

[1] Immittance is a synthetic word meaning *either* impedance or admittance.

For the three basic elements R, L, and C, we have

$$v = Ri \qquad v = pLi \qquad v = \frac{1}{pC}\,i$$

Hence

$$\text{For a resistance:} \qquad Z = R \qquad Y = G = \frac{1}{R}$$

$$\text{For an inductance:} \qquad Z = pL \qquad Y = \frac{1}{pL} \tag{3-52}$$

$$\text{For a capacitance:} \qquad Z = \frac{1}{pC} \qquad Y = pC$$

Since Eq. (3-51) is analogous to $v = Ri$ and Eq. (3-51a) is analogous to

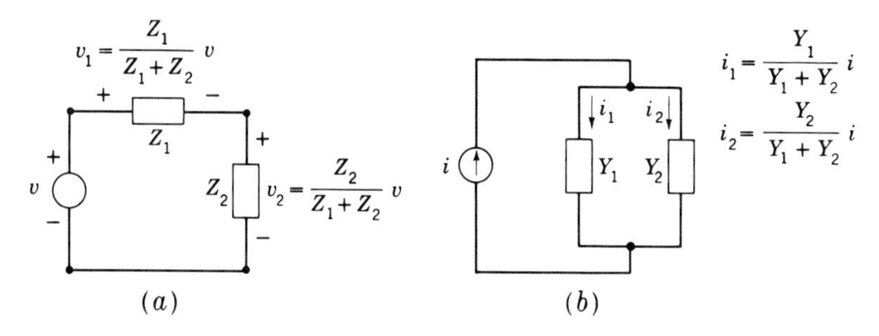

***Fig. 3-23** (a) Voltage division for operational impedances. (b) Current division for operational admittances.*

$i = Gv$, we conclude that operational impedances in series add and admittances in parallel add. Similarly, we can derive the operational voltage and current-division formulas by analogy: In the series circuit of Fig. 3-23a,

$$v_1 = \frac{Z_1(p)}{Z_1(p) + Z_2(p)}\,v(t) \tag{3-53}$$

and in Fig. 3-23b,

$$i_1(t) = \frac{Y_1(p)}{Y_1(p) + Y_2(p)}\,i \tag{3-54}$$

Gain functions A transfer function may be an impedance or admittance function. For example, in Fig. 3-22a, the network function which relates the response $i_1(t)$ to the source $v(t)$ is a transfer admittance. If a transfer function relates a voltage (current) response to a voltage (current) source, it is referred to as a voltage (current) gain function. The reader may easily show that, in Fig. 3-24, $i_1(t) = \frac{1}{4}i(t)$. In this expression the network func-

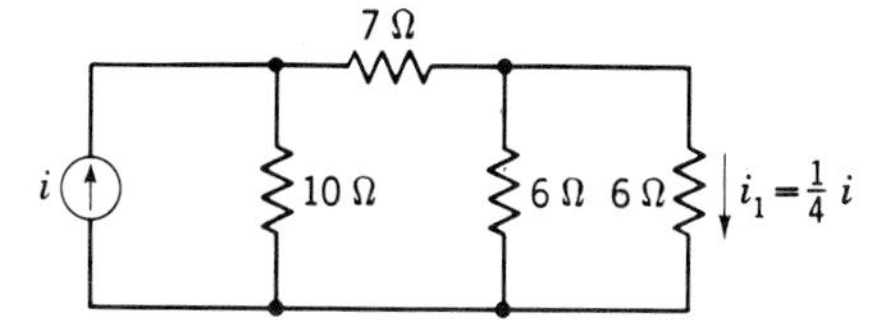

***Fig. 3-24** Current gain in a resistive network.*

tion $\frac{1}{4}$ relates the current response $i_1(t)$ to the source $i(t)$ and is therefore a current gain[1] function. Similarly, in Eq. (3-53), the transfer function $Z_1(p)/[Z_1(p) + Z_2(p)]$ is a voltage gain function relating the voltage response $v_1(t)$ to the voltage source $v(t)$.

We note that, in general, the reciprocal of a driving-point immittance is also a driving-point immittance, but the reciprocal of a transfer function has no significance as a network function.

Example 3-9 In the circuit of Fig. 3-25 derive the network function relating v_1 to v by operational voltage division.

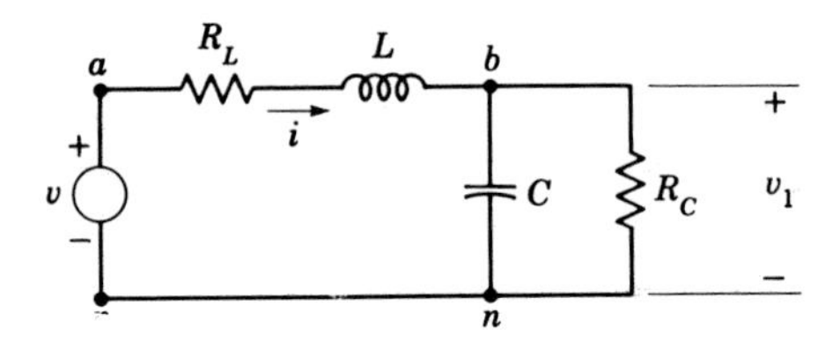

Fig. 3-25 Circuit for Example 3-9.

Solution From voltage division

$$v_1 = \frac{Z_{bn}(p)}{Z_{an}(p)} v$$

we formulate Z_{bn} and Z_{an}. The admittance Y_{bn} is given by

$$Y_{bn}(p) = \frac{1}{R_C} + pC = \frac{1 + pCR_C}{R_C}$$

Since the driving-point impedance Z_{bn} is the reciprocal of the driving-point admittance Y_{bn}, we have

$$Z_{bn}(p) = \frac{R_C}{1 + pR_C C}$$

Since

$$Z_{ab}(p) = R_L + pL$$

$$Z_{an}(p) = Z_{ab}(p) + Z_{bn}(p) = R_L + pL + \frac{R_C}{1 + pR_C C}$$

$$= \frac{p^2 R_C LC + p(R_L R_C C + L) + R_L + R_C}{1 + pR_C C}$$

Hence

$$v_1 = \frac{R_C}{p^2 R_C LC + p(R_L R_C C + L) + R_L + R_C} v = H(p)v$$

which is the desired result. Note that in the above expression $H(p)$ is a transfer (voltage-gain) function, and in contrast with the relationship $Z_{bn} = 1/Y_{bn}$ used above, the reciprocal of transfer function $H(p)$ is not a transfer function of the network.

[1] Loss function would be a more appropriate term in this case. In general, however, gain functions are differential operators, and not numbers, and distinction between gain and loss is not relevant.

3-13 Introduction to the solution of equilibrium equations

The foregoing discussion and examples illustrate procedures for the formulation of network functions or the corresponding equilibrium equations. We shall now discuss some of the general properties of such functions and equations.

FORM OF EQUILIBRIUM EQUATION The equilibrium equation which relates a response function to source functions in a linear network shows how *a certain linear combination of response function with its derivatives is related to another linear combination of the source function(s) with its (their) derivatives.* Thus, for example, in the circuit of Fig. 3-20*b*, from Eq. (3-47*b*), we have that the linear combination of the response v_{ad} with its derivatives,

$$\left(Lp^2 + Rp + \frac{1}{C}\right) v_{ad}$$

is equal to

$$(Lp^2 + Rp)v$$

which is seen to be a linear combination of the source function and its derivatives. In Chap. 11 it is shown that this general form always arises; without going through details of the proof, we may anticipate this result from the fact that all voltages and currents in the network are linearly related by algebraic or differential operations. The reader can verify the form for all the preceding examples. Before proceeding, however, we present an example involving more than one source function.

Example 3-10 In the circuit of Fig. 3-26 establish the equilibrium equation relating the response $v_2(t)$ to the source functions $i(t)$, $v_a(t)$, and $v_b(t)$.

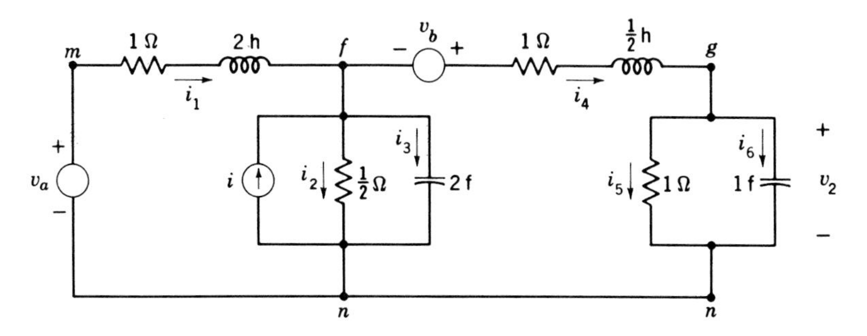

Fig. 3-26 Circuit with three source functions used in Example 3-10.

Solution Since $i_5 = v_2/1$ and $i_6 = pv_2$, we have, from the current law,

$$i_4 = (1 + p)v_2$$

and

$$\begin{aligned} v_{fg} &= -v_b + (1 + \tfrac{1}{2}p)(1 + p)v_2 \\ &= -v_b + (\tfrac{1}{2}p^2 + \tfrac{3}{2}p + 1)v_2 \end{aligned}$$

But

$$v_{fn} = v_{fg} + v_2$$
$$= -v_b + (\tfrac{1}{2}p^2 + \tfrac{3}{2}p + 2)v_2$$

and

$$i_2 = 2v_{fn} \qquad i_3 = 2pv_{fn} \qquad i_2 + i_3 = (2 + 2p)v_{fn}$$

and also

$$\begin{aligned} i_1 &= -i + i_2 + i_3 + i_4 \\ &= -i + (2 + 2p)[-v_b + (\tfrac{1}{2}p^2 + \tfrac{3}{2}p + 2)v_2] + (1 + p)v_2 \\ &= -i - (2 + 2p)v_b + (p^3 + 4p^2 + 8p + 5)v_2 \\ v_{mf} &= (1 + 2p)i_1 \\ &= -(1 + 2p)i - (4p^2 + 6p + 2)v_b + (2p^4 + 9p^3 + 20p^2 + 18p + 5)v_2 \end{aligned}$$

so that

$$\begin{aligned} v_{mn} &= v_{mf} + v_{fn} \\ &= -(1 + 2p)i - (4p^2 + 6p + 3)v_b + (2p^4 + 9p^3 + \tfrac{41}{2}p^2 + \tfrac{39}{2}p + 7)v_2 \end{aligned}$$

Since $v_{mn} = v_a$, we have, finally,

$$(2p^4 + 9p^3 + \tfrac{41}{2}p^2 + \tfrac{39}{2}p + 7)v_2 = v_a + (1 + 2p)i + (4p^2 + 6p + 3)v_b \quad (3\text{-}55)$$

We observe that the left side of Eq. (3-55) is a linear combination of v_2 with its derivatives and the right side is a linear combination of the source functions with their derivatives.

FORM OF NETWORK FUNCTION A network function which relates a selected network response variable (such as v_2 of the foregoing example) to a single source function (such as v_a in the example) can be written as the ratio of two polynomials in p. Thus, in general,

$$H(p) = \frac{N(p)}{D(p)} \quad (3\text{-}56)$$

where both $N(p)$ and $D(p)$ are polynomials in p, that is,

$$N(p) = b_0 + b_1p + \cdots + b_np^m$$

$D(p) = a_0 + a_1p + \cdots + a_np^n$, so that

$$H(p) = \frac{b_np^m + \cdots + b_1p + b_0}{a_np^n + \cdots + a_1p + a_0}$$

The above observation follows directly from the nature of the equilibrium equation. Denoting a response by $y(t)$ and a source by $x(t)$,

$D(p)y(t)$ = linear combination of response function y with its derivatives
$N(p)x(t)$ = linear combination of source function x with its derivatives

Since

$$\begin{aligned} D(p)y(t) &= N(p)x(t) \\ y(t) &= \frac{N(p)}{D(p)}x(t) \end{aligned}$$

as stated above.

FORCING FUNCTION In the equation

$$[D(p)]y(t) = [N(p)]x(t)$$

the terms $[N(p)]x(t)$ describe how the source function x influences the response y. Thus if, as in Eq. (3-55), we assume $v_a \equiv 0$, $i \equiv 0$, $x = v_b$, and $y = v_2$, we have the equation

$$(2p^4 + 9p^3 + \tfrac{41}{2}p^2 + \tfrac{39}{2}p + 7)v_2 = (4p^2 + 6p + 3)v_b$$

where the terms $(4p^2 + 6p + 3)v_b$ describe the relative weights with which v_b and its first two derivatives influence v_2. This linear combination of the source function with its derivatives is termed a *forcing function* of the equilibrium equation. Thus, in general, the sum of terms in an equilibrium equation which represent linear combination of the source functions with their derivatives is termed the *forcing function* of the equilibrium equation. Denoting the forcing function by $f(t)$, we have, for example, in Eq. (3-55),

$$(2p^4 + 9p^3 + \tfrac{41}{2}p^2 + \tfrac{39}{3}p + 7)v_2 = f(t) \tag{3-57a}$$

where, for the response $v_2(t)$ of the network of Fig. 3-26,

$$f(t) = v_a + (1 + 2p)i + (4p^2 + 6p + 3)v_b \tag{3-57b}$$

We note that *if the source functions are all identically zero, the forcing function must be zero.* In passing, it is pointed out that the converse of the above is *not true.* If in Eq. (3-57*b*) $v_a \equiv 0$ and $v_b \equiv 0$ but $i \not\equiv 0$, we can still have $f(t) \equiv 0$ if $i(t) = Ae^{-t/2}$, where A is any constant. If $f(t) \equiv 0$ when all the source functions are not zero, a *resonance* is said to exist (Chap. 10).

Superposition If a forcing function $f(t)$ consists of, or can be represented as, a sum of forcing functions, the response to $f(t)$ can be represented as the sum of the responses to the components of f. That is, if

$$(a_0 + a_1p + \cdots + a_np^n)v_2(t) = f_a(t) + f_b(t) = f(t)$$

and if

$$(a_0 + a_1p + \cdots + a_np^n)v_{2a}(t) = f_a(t)$$

and

$$(a_0 + a_1p + \cdots + a_np^n)v_{2b}(t) = f_b(t)$$

then since

$$a_kp^kv_{2a} + a_kp^kv_{2b} = a_kp^k(v_{2a} + v_{2b})$$

$$(a_0 + a_1p + \cdots + a_np^n)(v_{2a} + v_{2b}) = f(t)$$

and the response is the sum $v_2 = v_{2a} + v_{2b}$.

PARTICULAR AND COMPLEMENTARY SOLUTIONS The ensuing sections and chapters deal with the development of efficient techniques for the solution of equilibrium equations. The remaining discussion in this section is intended to be a general introduction to this subject.

Since in the equation

$$(a_0 + a_1p + a_2p^2 + \cdots + a_np^n)v_2(t) = f(t) \tag{3-58}$$

one can *always* interpret $f(t)$ as the sum $f(t) + 0$, the solution of Eq. (3-58), if it is to be complete, must include the solution of the homogeneous equation

$$(a_0 + a_1 p + \cdots + a_n p^n)v_1(t) = 0$$

Since we associate zero forcing function with *source-free* networks, we term the solution of the homogeneous equation the *source-free component of the response* and often use the subscript f to denote this component. Thus, for

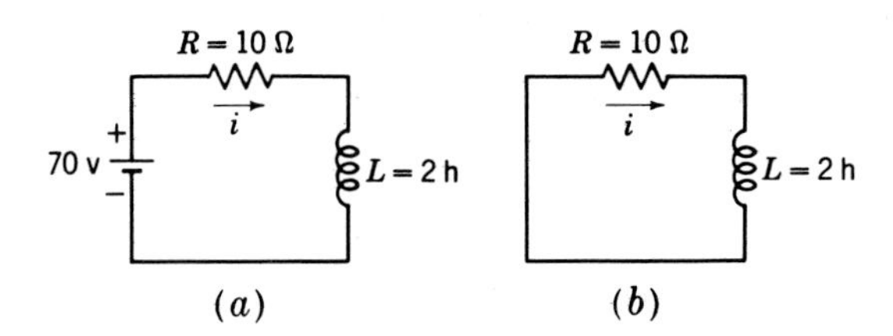

***Fig. 3-27** (a) An R-L circuit with a source. (b) Source-free circuit for (a).*

example, the equilibrium equation for the current of Fig. 3-27*a* is

$$2\frac{di}{dt} + 10i = 70 \tag{3-59}$$

In this equation $i(t)$ has as a component the solution $i_f = Ke^{-5t}$, where

$$2\frac{di_f}{dt} + 10i_f = 0$$

which is the solution to the source-free circuit of Fig. 3-27*b*. We note further that Eq. (3-59) has a solution $i_s = 7$, which is the current in the circuit of Fig. 3-27*a* when the time variations of $i_f = Ke^{-5t}$ have died out as t approaches infinity. Furthermore, the linear combination

$$i = 7 + Ke^{-5t}$$

satisfies Eq. (3-59). In the theory of differential equations, the solution $i = 7$ is referred to as a *particular* solution of Eq. (3-59). In circuit analysis, this term is referred to as *response due to the source.* We note that $i = 7 + 14e^{-5t}$ is also a particular solution, whereas $i = 7 + Ke^{-5t}$ with K an unspecified (arbitrary) constant represents the *totality of solutions.* Generalization of the concepts illustrated in the above example leads to the following mathematical definitions:

Particular solution: Any solution of the inhomogeneous equation.

Complementary solution: The general solution of the homogeneous equation.

General solution, or "totality of solutions": A linear combination of the particular and complementary solutions.

In network theory there are two particular solutions of special interest, the component of the response due to the source and the complete response.

The component of the response due to the source is that particular solution which, *independently of initial conditions*, vanishes when the forcing function vanishes. Thus, in the equation

$$2\frac{di}{dt} + 10i = 70$$

the solution $i = 7 = i_s$ is the component of the response due to the source. We shall often distinguish this part of the response by the subscript s, as in i_s or $[i(t)]_s$ or $[v(t)]_s$.

COMPLETE RESPONSE OF A NETWORK We recall that, in the absence of an impulsive voltage source in series with an inductance, the current through the inductance must be a continuous function of time. Therefore, for the example of Fig. 3-27, the value of K in the solution $i(t) = 7 + Ke^{-5t}$ must be evaluated so that the current (in the inductance) is continuous. Thus, if we are given $i(0^-) = I_0$, it follows that

$$i(0^+) = i(0^-) = I_0 = 7 + Ke^{-0} = 7 + K$$

$$K = I_0 - 7$$

and

$$i(t) = 7 + (I_0 - 7)e^{-5t}$$

The above expression for $i(t)$ has the following properties:

1 It satisfies the equilibrium equation of the circuit given in Eq. (3-59).
2 It satisfies the specified initial condition $i(0^+) = I_0$.

A solution of an equilibrium equation which satisfies the above two conditions is referred to as a *complete response.*

3-14 Response due to constant sources

In the example of the circuit of Fig. 3-27a we noted that the equilibrium equation for $i(t)$ is

$$10i + 2\frac{di}{dt} = 70 \tag{3-59}$$

and the complete response for $i(0) = I_0$ is

$$i(t) = i_s + i_f = 7 + (I_0 - 7)e^{-5t} \tag{3-60}$$

In the above solution the term $i_s = 7$ is obtained by reasoning that since the source (70 volts) is constant, the response due to the source, i_s, will also be constant, and $di_s/dt \equiv 0$. Substituting zero for the term di/dt in Eq. (3-58a), we have $10i_s = 70$. We now observe that Eq. (3-59) may be written

$$i(t) = \frac{1}{10 + 2p}70 \tag{3-61}$$

Substitution of $di/dt \equiv 0$ (for constant source) in Eq. (3-59) is equivalent to setting p equal to zero in Eq. (3-61). This results in

$$i_s = \frac{1}{10 + 2p}\bigg|_{p=0} 70 = 7$$

We now generalize the result of this example by stating that, if the source $x = X_0$ in a network is constant (time-invariant) and a response $y(t)$ is given by

$$y(t) = [H(p)]x(t) = [H(p)]X_0$$

then the component of the response due to the constant source is given as

$$y_s = Y_s = H(0)X_0 \qquad \text{if } \frac{1}{H(0)} \neq 0$$

As an example for the circuit of Fig. 3-26, the response $v_2(t)$ due to sources v_a, v_b, and i is given by Eq. (3-55). Using network-function symbolism, this equation may be rewritten

$$v_2(t) = H_a(p)v_a + H_b(p)v_b + H_i(p)i \tag{3-62}$$

Designating $D(p)$ by

$$D(p) = 2p^4 + 9p^3 + 20.5p^2 + 19.5p + 7 \tag{3-63}$$

from Eq. (3-55) we have

$$v_2(t) = \frac{1}{D(p)} v_a + \frac{4p^2 + 6p + 3}{D(p)} v_b + \frac{1 + 2p}{D(p)} i \tag{3-64}$$

If the sources v_a, v_b, and i are constant, say, $v_a = V_a = 70$ volts, $v_b = V_b = 14$ volts, $i = I = 21$ amp, then, in accordance with the above discussion the component of response due to these *constant* sources is

$$v_{2s} = H_a(0)v_a + H_b(0)v_b + H(0)i$$

Replacing p with zero in expressions for $H_a(p)$, $H_b(p)$, and $H(p)$, we have

$$v_{2s} = \tfrac{1}{7} \times 70 + \tfrac{3}{7} \times 14 + \tfrac{1}{7} \times 21 = 19 \text{ volts}$$

The foregoing procedure for calculation of response due to constant sources, illustrated by this example, is based on the theory of differential equations where it is known that a particular solution of an inhomogeneous linear differential equation with constant coefficients can generally be constructed as a *linear combination of the forcing function with all its derivatives*. When the forcing function $f(t)$ is constant, F_0, the equation

$$(a_n p^n + a_{n-1}p^{n-1} + \cdots + a_1 p + a_0)v(t) = F_0$$

is seen to have as solution the constant

$$v(t) = V = \frac{F_0}{a_0}$$

provided that $a_0 \neq 0$. This statement is verified by direct substitution.

In a network with a constant source $v_1(t) = V_1$, if the forcing function is given as $N(p)v_1(t)$, we have

$$\begin{aligned} f(t) &= (b_0 + b_1 p + \cdots + b_m p^m)v_1(t) \\ &= b_0 V_1 = F_0 \end{aligned}$$

and the response due to the source v (denoted by the subscript s) is given by

$$v_{2s} = V_2 = \frac{b_0}{a_0} V_1$$

As mentioned above, if in

$$H(p) = \frac{b_0 + b_1 p + \cdots + b_m p^m}{a_0 + a_1 p + \cdots + a_n p^n}$$

we set $p = 0$, $H(0) = b_0/a_0$, and

$$v_{2s} = \frac{b_0}{a_0} V_1 = H(0) V_1 \qquad a_0 \neq 0$$

As mentioned before, replacement of p with zero, when the source is constant, is due to the fact that all derivatives of constants are zero. In the above discussion it is assumed that $a_0 \neq 0$. When $a_0 = 0$ and $a_1 \neq 0$, one may define $pv = y$ and solve for the constant $dv/dt = F_0/a_1$.

3-15 Response due to exponential sources: the special role of exponential functions in linear analysis

The exponential function is unique in that it is the only function whose derivative is proportional to itself. Thus, if

$$f(t) = Ke^{st}$$
$$pf(t) = sKe^{st} = sf(t) \tag{3-65}$$

Equation (3-65), $pf = sf$, should not be read as $p = s$. Rather, it states that, when f is exponential, the operation of differentiation with respect to time is identical with multiplication by s.

Below we show that,[1] if the source in a network is exponential and of the form

$$x(t) = X_g e^{s_g t} \tag{3-66}$$

where the subscript g refers to "generator" and both V_g and s_g are constants, then if the response $y(t)$ is related to this source by $H(p)$,

$$y(t) = H(p)x(t) = H(p)X_g e^{s_g t}$$

it follows that the response due to the source is

$$y_s(t) = H(s_g)X_g e^{s_g t} \qquad \frac{1}{H(s_g)} \neq 0$$

For example, if in the circuit of Fig. 3-26 the source $v_b(t) = V_g e^{s_g t} = 20e^{-2t}$, $v_a(t) = 70$ (a constant source), and $i(t) \equiv 0$, then from

$$v_2(t) = H_a(p)v_a + H_b(p)v_b + H(p)i$$

and the corresponding expressions for H_a, H_b, and H given in Eq. (3-64), we have

$$\begin{aligned} v_2(t)_s &= H_a(0)v_a + H_b(s_g)v_b + 0 \\ &= \left.\frac{1}{2p^4 + 9p^3 + 20.5p^2 + 19.5p + 7}\right|_{p=0} \times 70 \\ &\qquad + \left.\frac{4p^2 + 69p + 3}{2p^4 + 9p^3 + 20.5p^2 + 19.5p + 7}\right|_{p=s_g=-2} \times 20e^{-2t} \\ &= \tfrac{1}{7} \times 70 + \tfrac{7}{10} \times 20e^{-2t} = 10 + 14e^{-2t} \end{aligned}$$

[1] With certain exceptional cases discussed in Chap. 8.

In this result we note that such a constant source $v_a = 70$ may be considered to be exponential with $s_g = 0$; that is,

$$70 = 70e^{(0)t}$$

The case of response due to a constant source is a special case of the response due to exponential sources.

To show that the procedure used above is general, we note that, when a source $v_1(t)$ has the form $v_1(t) = V_1e^{s_gt}$, the forcing function is given by

$$f(t) = [N(p)]v_1(t) = (b_0 + b_1p + b_2p^2 + \cdots + b_mp^m)v_1(t)$$

But since $pv_1 = s_gv_1$, for exponential v_1,

$$\begin{aligned} f(t) &= (b_0 + b_1s_g + b_2s_g^2 + \cdots + b_ms_g^m)V_1e^{s_gt} \\ &= N(s_g)V_1e^{s_gt} \end{aligned}$$

so that

$$f(t) = F_1e^{s_gt} \qquad F_1 = N(s_g)V_1$$

Thus, for example, if $(2p^2 + 8p + 58)v_2 = (4p + 7)e^{-4t}$, then

$$f(t) = [4 \times (-4) + 7]e^{-4t} = -9e^{-4t}$$

If

$$D(p)v_2 = (a_0 + a_1p + a_2p^2 + \cdots + a_np^n)v_2 = F_1e^{s_gt} \tag{3-67}$$

is to have a solution

$$v_2 = V_2e^{s_gt} \tag{3-68}$$

then

$$D(p)v_2 = (a_0 + a_1p + \cdots + a_np^n)V_2e^{s_gt} = F_1e^{s_gt}$$

but

$$p^kV_2e^{s_gt} = s_g{}^kV_2e^{s_gt}$$

so that (3-68) is a solution of (3-67) if

$$D(s_g)V_2e^{s_gt} = (a_0 + a_1s_y + a_1s_g{}^2 + \cdots + a_ns_g{}^n)V_1e^{s_gt} = F_1e^{s_gt}$$

or if

$$D(s_g)V_2 = F_1$$

which is possible wherever $D(s_g) \neq 0$ by letting

$$V_2 = \frac{F_1}{D(s_g)} \qquad D(s_g) \neq 0$$

We now recall from the relations above that if $v_1 = V_1e^{s_gt}$,

$$F_1 = N(s_g)V_1$$

so that

$$V_2 = \frac{N(s_g)}{D(s_g)}V_1 = H(s_g)V_1$$

Thus, in general, if

$$v_2(t) = H(p)v_1(t)$$

and when $v_1(t)$ is the exponential source function,

$$v_1(t) = V_1e^{s_gt}$$

and if $1/[H(s_g)] \neq 0$, then[1] the response due to the source will be proportional to the source, and the proportionality constant will be the network

[1] This is the exceptional case mentioned above.

function with the operator p replaced by the number s_g:

$$v_{2s} = H(s_g)V_1e^{s_gt} \qquad D(s_g) \neq 0$$

[The special case $D(s_g) = 0$ is treated in Chap. 8.]

Thus, for example, if

$$(2p^2 + 8p + 58)v_2 = (4p + 7)e^{-4t}$$

$$v_{2s} = \frac{4p + 7}{2p^2 + 8p + 58} e^{-4t}$$

and the response component due to the source is

$$v_{2s} = \frac{(4)(-4) + 7}{2 \times 16 + 8(-4) + 58} e^{-4t} = -\tfrac{9}{58}e^{-4t}$$

PROBLEMS

3-1 A 10-ohm resistance is connected in series with a 2-henry inductance. Current in this combination is given by $i(t) = 2t$. Calculate (*a*) the magnitude of voltage across the combination at $t = 0$ and at $t = 1$; (*b*) the energy transferred to the series combination in the time interval $t = 0$ to $t = 1$.

3-2 In the series *R-L-C* circuit shown in Fig. 3-7, let $R = 2$ ohms, $L = 2$ henrys, and $C = \frac{1}{2}$ farad. Calculate and sketch $v_{an}(t)$ if (*a*) $i(t) = 2u(t)$; (*b*) $i(t) = 2tu(t)$; (*c*) $i(t) = e^{-2t}u(t)$. The capacitance is initially uncharged.

3-3 In the parallel *R-L-C* circuit shown in Fig. 3-8, let $R = 1$ ohm, $L = 1$ henry, and $C = 1$ farad. Calculate and sketch $i(t)$ if (*a*) $v(t) = u(t)$; (*b*) $v(t) = 3tu(t)$; (*c*) $v(t) = (2 + 3t)u(t)$; (*d*) $e^{-2t}u(t)$. The inductance carries zero current at $t = 0$.

3-4 (*a*) In the circuit of Fig. P3-4, apply Kirchhoff's current law at junction b, using v as the variable, and obtain the differential equation that relates v to v_o. *Hint:* Note that $i_{ab} = (v_o - v)/12$. (*b*) Sketch the waveform of v_o if v is given by (1) $v = u(t)$; (2) $v = tu(t)$; (3) $v = (1 + t)u(t)$; (4) $e^{-2t/3}$; (5) $e^{-2t/3}u(t)$.

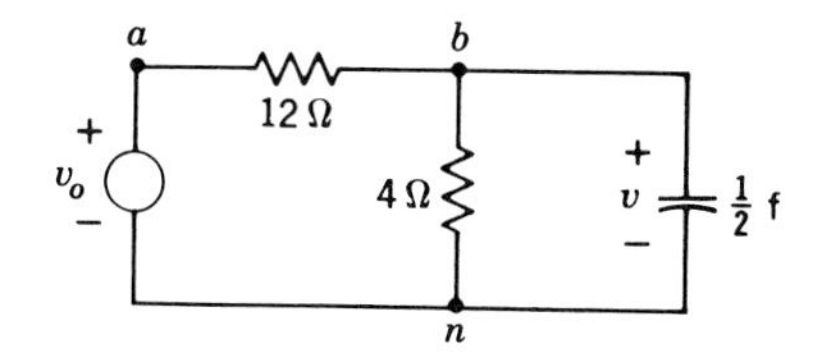

Fig. P3-4

3-5 In the circuit of Fig. P3-5: (*a*) Express i_2 in terms of i_1. (*b*) Express i in terms of i_1. (*c*) Express v_{ab} in terms of i_1. (*d*) Find the integrodifferential equation that relates i_1 to v. (*e*) Find the waveform of v if $i_1(t) = 2tu(t)$.

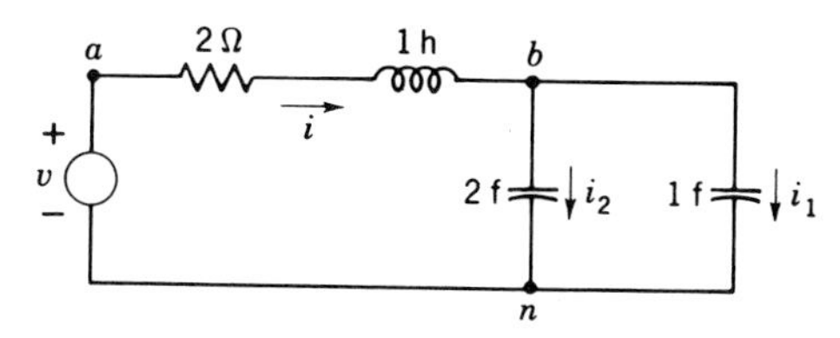

Fig. P3-5

3-6 (*a*) In the circuit of Fig. P3-6, apply Kirchhoff's voltage law on loop *a-b-n-a* and obtain the differential equation that relates i to i_s. *Hint:* Express the current in the 4-ohm resistance in terms of i and i_s. (*b*) Find the waveform of i_s if i is given by (1) $i = (2 + t^2)u(t)$; (2) $i = e^{-2t}$; (3) $i = e^{-2t}u(t)$; (4) $(e^{-2t} - 1)u(t)$.

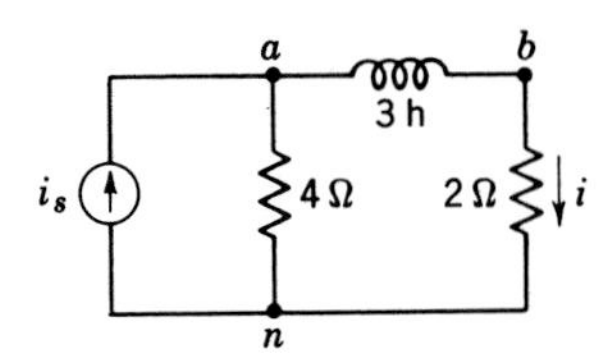

Fig. P3-6

3-7 (*a*) In the circuit of Fig. P3-7 show by use of Kirchhoff's current law at junction *a* that $i = 4v_1 + 3v_2 - 7v_{ab}$. (*b*) Show that $i = 2v_{ab} + 2dv_{ab}/dt$. (*c*) Use the results of parts *a* and *b* to find the differential equation that relates v_{ab} to the sources v_1 and v_2.

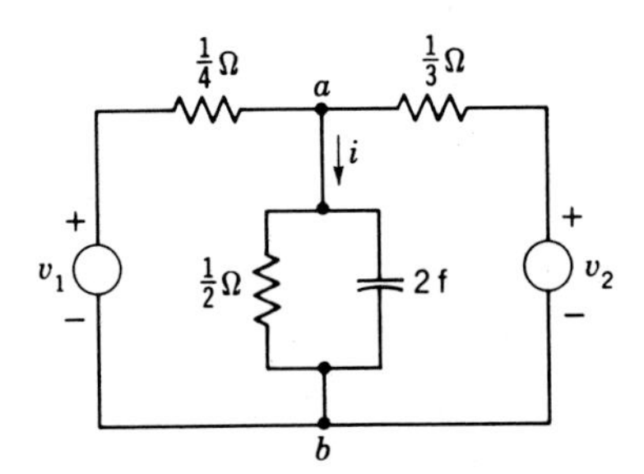

Fig. P3-7

3-8 In Fig. P3-6 replace the 4-ohm resistance by a 4-henry inductance and show that $7di/dt + 2i = 4di_s/dt$.

3-9 In the circuit of Fig. P3-9, let $R_1 = R_2 = 5$ ohms, $R_3 = 60$ ohms, $R_4 = R_5 = 10$ ohms, and replace R_6 and R_7 with open circuit ($R_{6,7} \to \infty$). Calculate (*a*) the ratio i_3/i_s; (*b*) the ratio v_{dn}/i_s; (*c*) the ratio v_{an}/i_s.

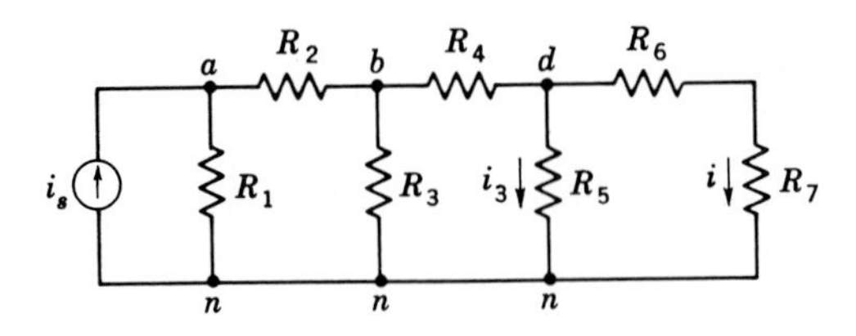

Fig. P3-9

3-10 In the circuit of Fig. P3-9 it is known that $i(t) = I = 2$ when $i_s = I_s = 15$. Find $i(t)$ if i_s is given by (*a*) $i_s = 3t$; (*b*) $i_s = 10 \cos 5t$.

3-11 In the circuit of Fig. P3-9 prove that the equivalent resistance of the seven elements at the source terminals is less than R_1 and greater than $R_1R_2/(R_1 + R_2)$.

3-12 In the capacitive ladder network shown in Fig. P3-12: (*a*) Calculate v/v_o. (*b*) Obtain the relationship between i and v_o.

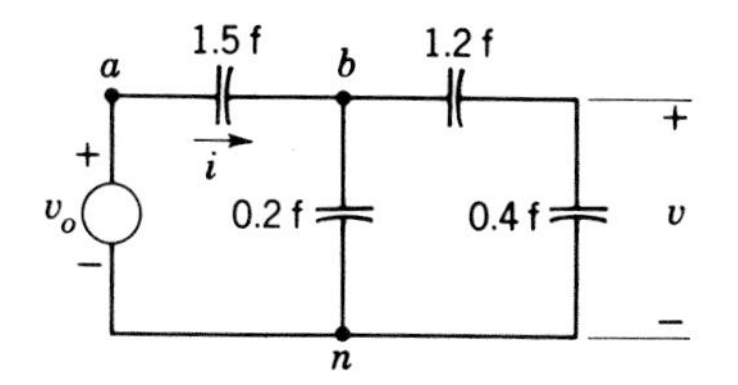

Fig. P3-12

3-13 In the inductive ladder circuit shown in Fig. P3-13, calculate (*a*) v_2/v_o; (*b*) v_{an}/v_o; (*c*) the relationship between v_o and i.

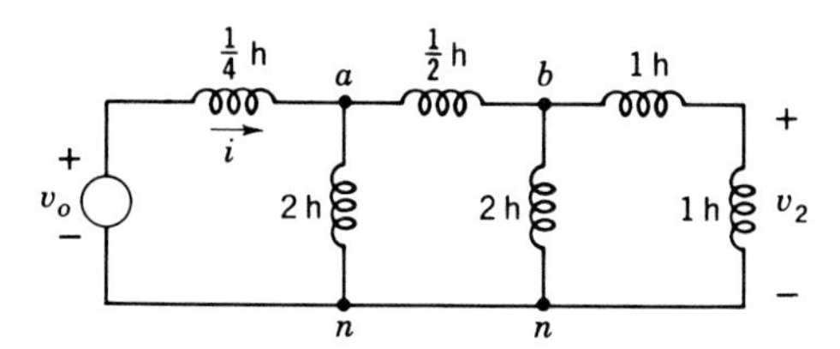

Fig. P3-13

3-14 In the resistive ladder circuit shown in Fig. P3-14, obtain a formula relating the sources v_A, v_B, and i_C to v_2.

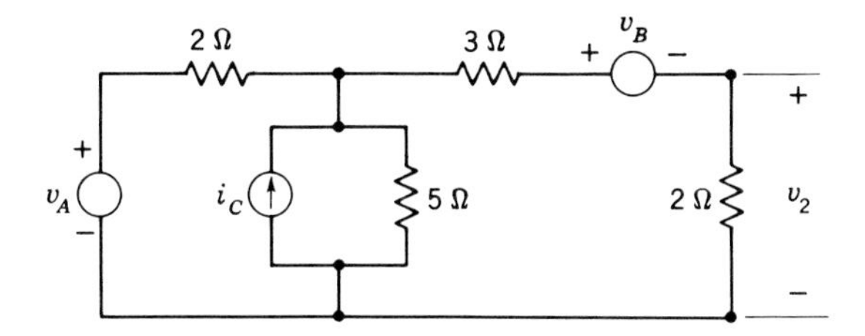

Fig. P3-14

3-15 (*a*) Prove the relationships indicated in Fig. 3-11. (*b*) Prove the voltage-division formula for series resistance and for series capacitance indicated in Figs. 3-13 and 3-14, respectively. (*c*) Prove the current-division formulas for parallel resistances and capacitances indicated in Figs. 3-16 and 3-17, respectively.

3-16 Calculate the equivalent inductance L_{ab} of Fig. P3-16 if $L = \frac{1}{4}$ henry, $L_1 = \frac{1}{2}$ henry, and L_2 is variable. Calculate also the largest and smallest value L_{ab} can have as L_2 is varied.

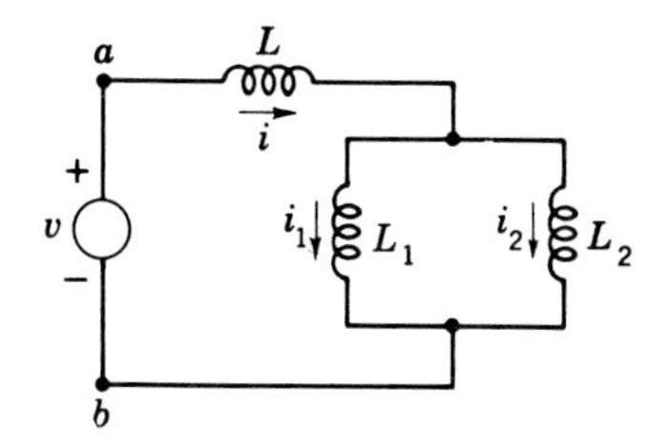

Fig. P3-16

3-17 In the circuit of Fig. P3-16, let $L = 0$ and set $i_1(0) = i_2(0) = 0$. The source waveform is the sawtooth pulse $v(t) = tu(t)u(T - t)$. It is specified that neither i_1 nor i_2 may exceed 8 amp. Calculate the largest allowable value of T and the maximum allowable value of i if $L_1 = 4L_2 = \frac{1}{2}$ henry.

3-18 The equivalent resistance of two resistances in series is 10 ohms. When the two resistances are placed in parallel, the equivalent resistance is 2.4 ohms. Calculate the values of the resistances.

3-19 The equivalent capacitance of two capacitances in series is 10 μf. When the same capacitances are connected in parallel, the equivalent capacitance is 50 μf. Calculate the values of the individual capacitances.

3-20 In the two circuits shown in Fig. P3-20 the same current I_0 flows for the same voltage V_{ab}. It is also known that the ratio I_1/I_0 is the same in both circuits. Calculate R_a and R_b.

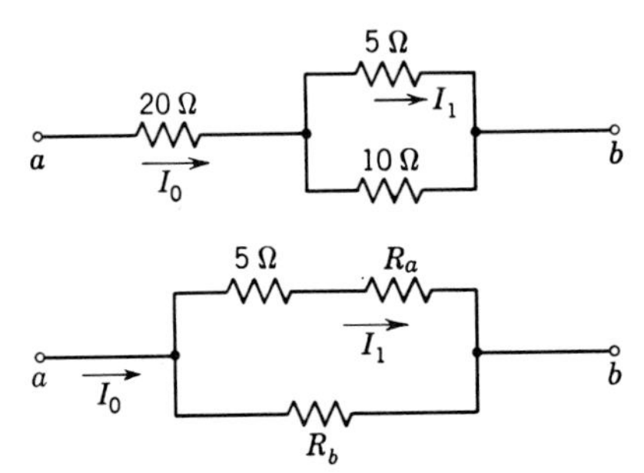

Fig. P3-20

3-21 In the circuit shown in Fig. P3-21, $v_{db} = 50$ mv when $v_{ab} = 150$ volts. Calculate R (voltmeter "multiplier").

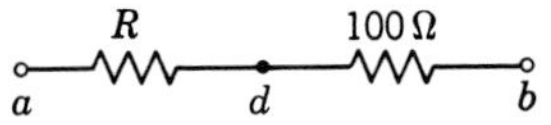

Fig. P3-21

3-22 In the circuit shown in Fig. P3-22, $I_1 = 100$ μa when $I = 5$ amp. Calculate R (ammeter "shunt").

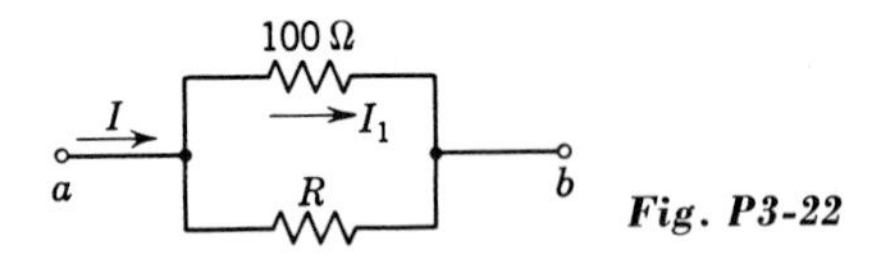

Fig. P3-22

3-23 (*a*) Two uncharged capacitances C_1 and C_2 are connected in series. If $C_1 = 2$ μf and C_2 is adjustable from 0.01 to 0.1 μf, calculate the maximum and the minimum values of the capacitance of the series combination. (*b*) Calculate the maximum and minimum values of the equivalent capacitance if the two capacitances are placed in parallel.

3-24 In the circuit shown in Fig. P3-24, the voltage value indicated below each capacitance value specifies the highest permissible voltage across the capacitance. Calculate the maximum permissible value of v_{ab}.

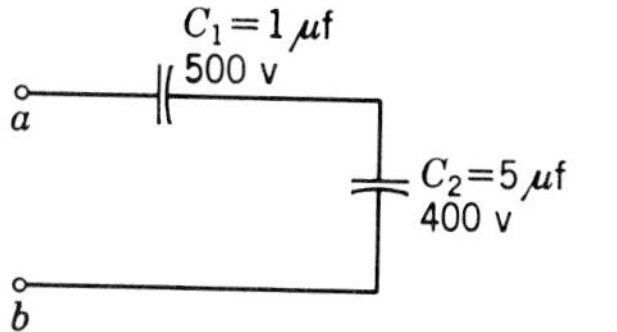

Fig. P3-24

3-25 In the circuit of Fig. P3-25, calculate V_{ab}.

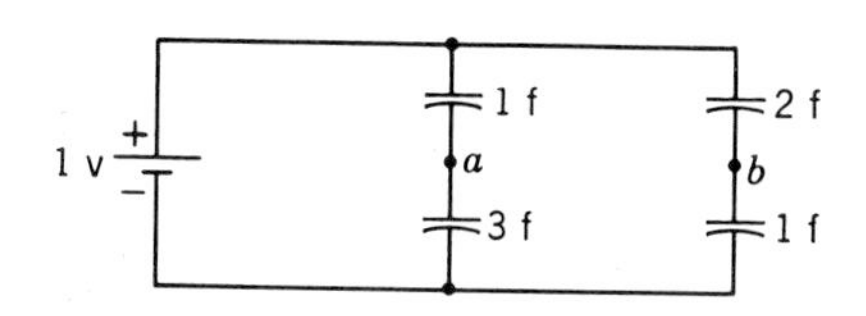

Fig. P3-25

3-26 In the circuit of Fig. P3-26, calculate I if (*a*) $R = 3$ ohms; (*b*) $R = 1.5$ ohms.

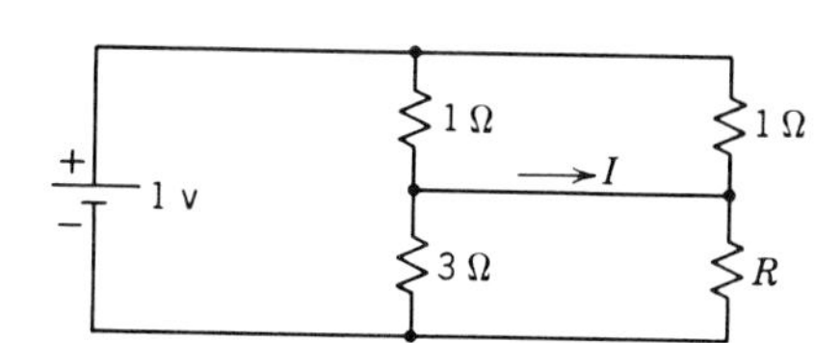

Fig. P3-26

3-27 In the circuit of Fig. P3-27, let $R = 0$ and calculate the ratio v_{ab}/v.

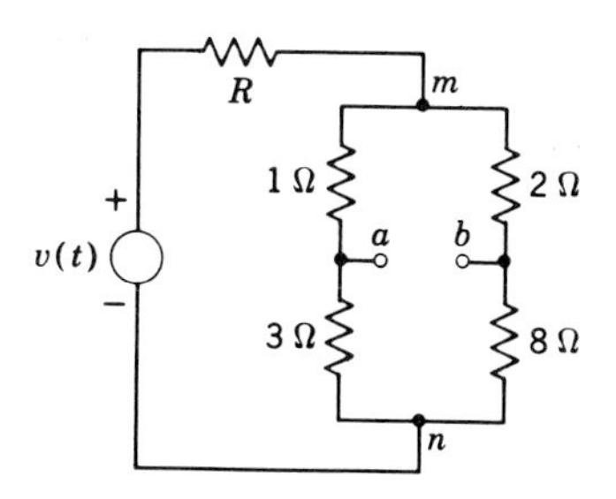

Fig. P3-27

3-28 In the circuit of Fig. P3-27, let $R = 1$ ohm and calculate the ratio v_{ab}/v.

3-29 In the circuit of Fig. P3-27, let $R = 1$ ohm and place a short circuit (zero resistance) between terminals *a-b*. Calculate the current in the short circuit I_{ab} if the source is constant and given by $v(t) = V = 100$ volts.

3-30 In the circuit of Fig. P3-30, $v_{an}/v_o = \frac{1}{2}$ and $v_{bn}/v_o = \frac{1}{5}$. Find R_1 and R_2.

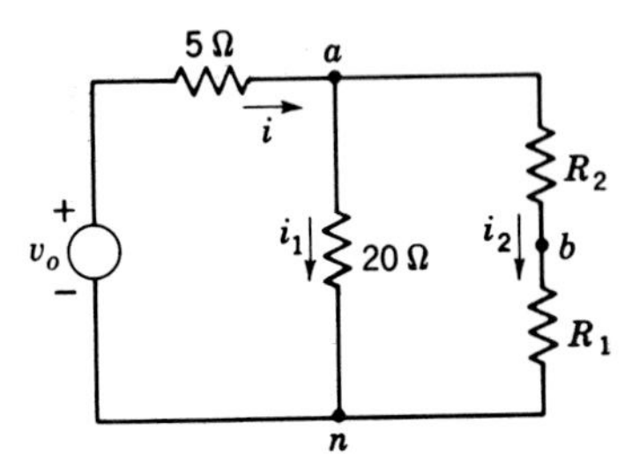

Fig. P3-30

3-31 In the circuit of Fig. P3-30, $i_1/i = \frac{3}{4}$ and $v_{bn}/v_o = \frac{1}{2}$. Find R_1 and R_2.

3-32 If $f(t) = (2p^3 + 4p^2 + 3p + 4)g(t)$, find $f(t)$ for (*a*) $g(t) = e^{-t}$; (*b*) $g(t) = te^{-t}$; (*c*) $e^{-t} + te^{-t}$; (*d*) $t^2 + 4t$.

3-33 Show that, if $f(t)$ is a forcing function for a response, then $Kf(t)$ is also a forcing function if $K = \text{const.}$

3-34 In the circuit of Fig. P3-34, show that (*a*) $i_1 = (2p + 2)v_2$; (*b*) $v_{an} = (4p^2 + 6p + 3)v_2$; (*c*) $i_0 = (4p^3 + 6p^2 + 5p + 2)v_2$; (*d*) $v_o = (8p^3 + 16p^2 + 16p + 7)v_2$; (*e*) the driving-point impedance $Z_{a'n}(p)$ is given by $Z_{a'n}(p) = (8p^3 + 16p^2 + 16p + 7)/(4p^3 + 6p^2 + 5p + 2)$.

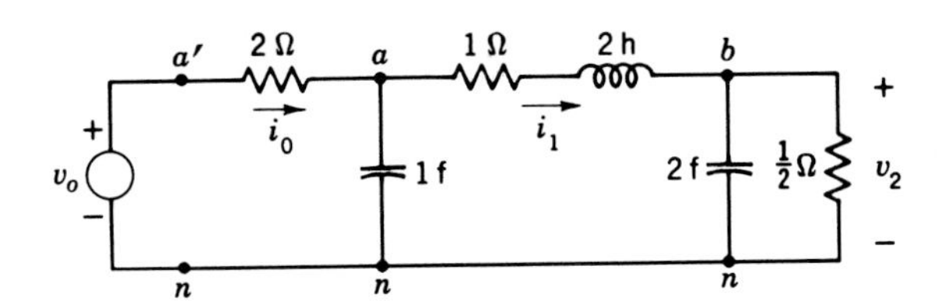

Fig. P3-34

3-35 In the circuit of Fig. P3-35: (*a*) Show that the differential equation that relates the response v_2 to the source function v_o is $(p^3R^3C^3 + 5p^2R^2C^2 + 6pRC + 1)v_2 = v_o$. (*b*) Use the result of part *a* to deduce the differential equation relating v_2 to v_o if each resistance R is replaced by inductance L.

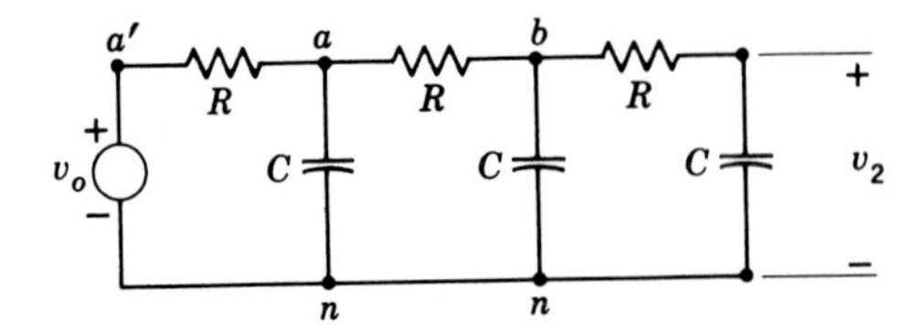

Fig. P3-35

3-36 In the circuit of Fig. P3-35, find $v_o(t)$ if $v_2(t) = e^{-t/RC}$.

3-37 In the circuit of Fig. P3-37, obtain the equilibrium equations relating i_2 and v_{ab} to the source function i_s.

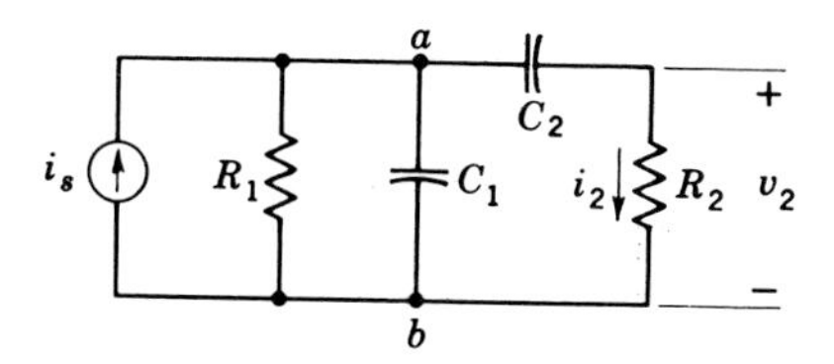

Fig. P3-37

3-38 In the circuit of Fig. 3-25, let $C = 1$ farad, $L = 2$ henrys, $R_L = 1$ ohm, and $R_C = \frac{1}{2}$ ohm. Obtain the equilibrium equation that relates v_{ab} to v.

3-39 In the circuit of Fig. P3-39, obtain the equilibrium equation that relates v_{ab} to $v_o(0 < a < 1)$.

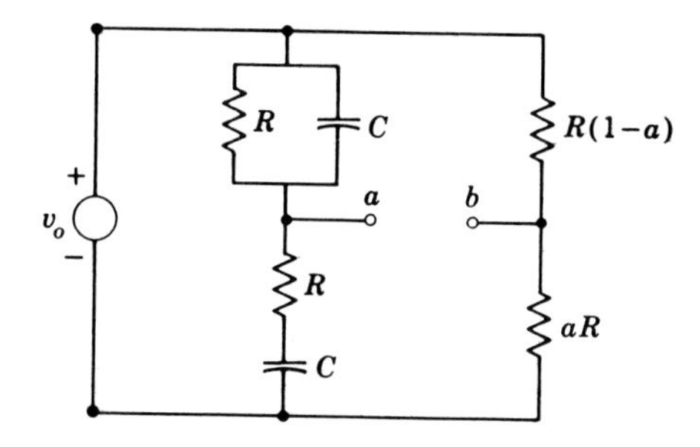

Fig. P3-39

3-40 In the circuit of Fig. P3-34 (whose driving-point impedance is as given in Prob. 3-34*e*), find the forcing functions for the response $i_0(t)$ if (*a*) $v_o(t) = 1$; (*b*) $v_o(t) = t$; (*c*) $v_o(t) = 1 + t + t^2$; (*d*) $v_o(t) = e^t$; (*e*) $v_o(t) = e^{-2t}$.

3-41 For the circuit of Fig. P3-37, let $R_2 = \frac{1}{9}R_1 = 1$ and $C_1 = \frac{1}{9}C_2 = 1$ and use the result to find the forcing function for the response v_2 if the source function is given by (*a*) $i_s = 12$; (*b*) $i_s = 12t$; (*c*) $i_s = 5e^{-2t}$.

3-42 In the circuit of Prob. 3-38, find the forcing function for the response $v_1(t)$ for the cases (*a*) $v(t) = 5 + 5t$; (*b*) $v(t) = e^{-t}$; (*c*) $v(t) = e^{-t/2}$.

3-43 Find the component of the response v_2 due to a constant source of value $v_o = V_o = 10$ in the circuit of (*a*) Fig. P3-34; (*b*) Fig. P3-35; (*c*) Fig. P3-39.

3-44 Find the component of the response due to the source $v = v_o(t) = 10e^{-2t}$ in the circuit of (*a*) Fig. P3-34 with v_2 as the response; (*b*) Fig. P3-35 with v_2 as the response and $RC = 2$; (*c*) Prob. 3-38 with v_1 as the response.

3-45 In a certain circuit the response v_2 is related to the source function v_o by the $v_2 = [(\frac{1}{2}p + 2)/(p + 5)]v_o$. Find the component of the response due to the source (*a*) $v_o = 10$; (*b*) $v_o = 5e^{-t}$; (*c*) $v_o = 10e^{-4t}$; (*d*) $v_o = 10e^{-t/4}$.

Chapter 4 Natural modes of networks

In Chap. 3 we showed how a network function, or its corresponding differential equation, relates a desired response of a network to a source function. We recall that the complete response can be calculated as the sum of the source-free component and the component of the response due to the source. In this chapter we deal in detail with the source-free component of the response. The following are two of the important properties of the source-free component of response which will be discussed in detail:

1. We shall see that the source-free component of response for a given network will have the same mathematical form regardless of the specific response function selected.
2. The source-free response component shows how the network elements interact when a finite amount of energy is initially put into the circuit. In this sense the source-free condition represents the response in the absence of sources, and is termed the "natural response" of the network.

4-1 The concept of source-free network

The terminology used in connection with network theory implies that in a network the application of a source "stimulus" leads to a "response." Consequently, the term "source-free network" may appear to be trivial: How can we obtain responses without sources? To answer this point, consider the circuit of Fig. 3-25, discussed in Example 3-9, redrawn as Fig. 4-1a for convenience. In Example 3-9 we found that the response v_1 is related to the source v by

$$v_1 = \frac{R_C}{p^2R_CLC + p(R_LR_CC + L) + R_C + R_L} v \tag{4-1}$$

If in this circuit we let

$$R_L = R_C = 1 \qquad L = 1 \qquad C = 1$$

we have

$$v_1 = \frac{1}{p^2 + 2p + 2} v \tag{4-2}$$

or

$$(p^2 + 2p + 2)v_1 = v \tag{4-3}$$

If we set the source v to zero, then the new circuit is as shown in Fig. 4-1*b*,

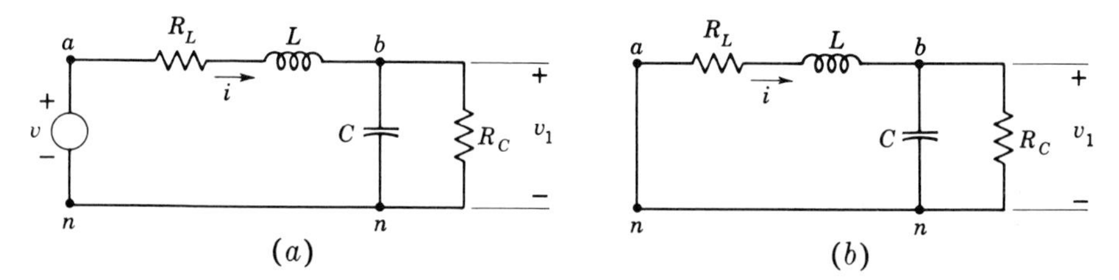

Fig. 4-1 ***(a) Circuit with ideal source.*** ***(b) Source-free circuit corresponding to (a).***

and its equilibrium equation, obtained by setting $v \equiv 0$ in Eq. (4-3), is

$$(p^2 + 2p + 2)v_1 = 0 \qquad \text{or} \qquad \frac{d^2v_1}{dt^2} + 2\frac{dv_1}{dt} + 2v_1 = 0$$

The source-free response v_1 in the circuit of Fig. 4-1*b* is due to energy stored in the circuit at $t = 0$ when the source in the circuit of Fig. 4-1*a* is set to zero. This initial stored energy is equal to

$$W_0 = \tfrac{1}{2}L[i(0)]^2 + \tfrac{1}{2}C[v_1(0)]^2$$

and will cause a response v_1 after $t = 0$.

In general, a response $y(t)$ is related to a source $x(t)$ by the relationship

$$y(t) = \frac{N(p)}{D(p)} x(t) \qquad \text{or} \qquad D(p)y(t) = N(p)x(t)$$

and when the source is set to zero, the response must satisfy the equation

$$D(p)y(t) = 0 \tag{4-4}$$

In words, the denominator polynomial of a network function determines the source-free equilibrium equation of the network.[1] This is one of the impor-

[1] It is recalled from Sec. 3-13 that a network with special source waveform may, in certain cases, lead to equilibrium equations that are homogeneous, as is Eq. (4-4). This situation has been illustrated in Sec. 3-13. Although the formalism for solving such special resonant cases is the same as in source-free circuits, we did not emphasize this point in that chapter. The matter is discussed in Chaps. 7 and 8.

tant properties of a network function and may always be used to find the equilibrium equation for a source-free response for any network. In the rest of this chapter we concern ourselves with free response (natural modes) of simple circuits. The study of these circuits reveals many fundamental facts about natural modes in general.

4-2 The source-free R-C circuit

To gain insight into the details of circuit behavior it is useful to study simple source-free circuits. We shall see that, in addition to demonstrating the general properties of source-free response, the analysis of such simple two- and three-element combinations is useful in itself, because such simple circuits find extensive practical application.

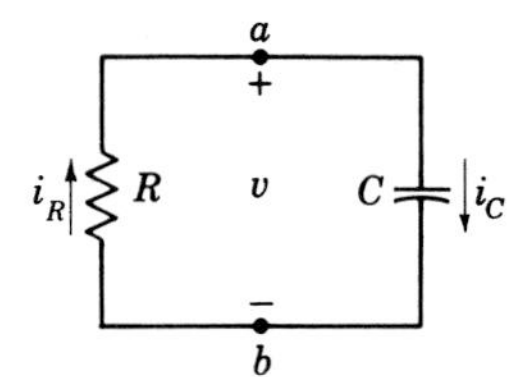

Fig. 4-2 Source-free R-C circuit.

Consider now the source-free *R-C* circuit shown in Fig. 4-2, and let it be specified that the value of v_{ab} *at the instant* $t = 0^+$ is V_0; that is,

$$v_{ab}(0^+) = v(0^+) = V_0$$

It is desired to find the formula for $v(t)$ for all $t \geq 0^+$. Before proceeding to quantitative analysis of the problem, a short qualitative study is useful. We observe that the statement $v(0^+) = V_0$ implies that, at $t = 0^+$, the stored energy in the capacitance is $\frac{1}{2}CV_0^2$; further, at $t = 0^+$, the current i_{ab} in the resistance R is V_0/R; hence, at the instant $t = 0^+$, the resistance dissipates electric energy at the *rate* V_0^2/R. It follows that the resistance absorbs energy stored in the capacitance and that the voltage across the capacitance must continually decrease until all the initial energy $(\frac{1}{2}CV_0^2)$ has been used up.

To find the quantitative description for the decay of the voltage, we apply Kirchhoff's current law at node *a*, where $i_C = i_R$, or in terms of v, we have the homogeneous equation

$$Cpv = C\frac{dv}{dt} = -\frac{v}{R} \tag{4-5}$$

or

$$pv = \frac{dv}{dt} = -\frac{1}{RC}v \tag{4-5a}$$

Equation (4-5*a*) shows that $v(t)$ is a function whose derivative is proportional to the function itself. From differential calculus it is recalled that this property defines the *exponential function*. The solution of Eq. (4-5*a*) may therefore be written in the form

$$v(t) = Ke^{st} \tag{4-6}$$

To find s we note that, for $v(t)$ as given by Eq. (4-6), the derivative pv is given by

$$pv = sKe^{st} = sv \tag{4-7}$$

and substitution of (4-6) and (4-7) in (4-5*a*) gives

$$sv = -\frac{1}{RC}v \tag{4-8}$$

If Eq. (4-8) is to hold for all values of v, it is necessary that

$$s = -\frac{1}{RC} \tag{4-9}$$

Hence the general solution for $v(t)$ is given by

$$v(t) = Ke^{-t/RC} \qquad t \geq 0^+ \tag{4-10}$$

Introducing the initial condition $v(0^+) = V_0$ into (4-10) gives

$$V_0 = Ke^{-0/RC} \qquad \text{or} \qquad K = V_0$$

It follows that the solution which satisfies both the equilibrium equation and the given initial condition is

$$v(t) = V_0e^{-t/RC} \qquad t \geq 0^+ \tag{4-10a}$$

We observe that the exponent $st = -t/RC$ is negative, indicating that, as anticipated, the voltage v decays. The decay process is also referred to as the "transient" process. The exact energy balance can be verified if the energy dissipated in the resistance is calculated in the interval $t = 0$ to $t = \infty$:

$$p_R(t) = \frac{v^2}{R} = \frac{V_0^2}{R}e^{-2t/RC}$$

$$W = \int_0^\infty \frac{V_0^2}{R}e^{-2t/RC}\,dt$$

or

$$W = \frac{V_0^2e^{-2t/RC}}{R(-2/RC)}\Bigg|_0^\infty = \tfrac{1}{2}CV_0^2$$

so that when the voltage v has decayed to zero (at $t = \infty$), all the energy which had initially been stored in the capacitance has been dissipated by the resistance.

4-3 Time constant

To study the waveform of the solution we first examine the exponent $st = -t/RC$. The exponent must be dimensionless; otherwise the exponential function is meaningless. This is easily verified since R has the dimensions of (volts/ampere), and the capacitance C is dimensionally (ampere-sec/volt); therefore $1/RC$ has the dimensions (1/sec), and the product RC, the dimensions of time. This product is termed the *time constant* of the R-C circuit, and will be denoted by the capital T:

$$RC = T \tag{4-11}$$

Introducing Eq. (4-11) in the solution [Eq. (4-10*a*)]

$$v(t) = V_0 e^{-t/T} \tag{4-12}$$

From inspection of Eq. (4-12) we note that all R-C circuits with the same time constant will give the same waveshape of voltage when deenergized. This universality of the parameter T is a matter of considerable importance, which is to be explored more fully.

Consider the table of exponentials shown as Table 4-1. Using this table

Table 4-1

x	0	1	2	3	4	5	6
e^{-x}	1	0.37	0.14	0.05	0.018	0.0067	0.0025

in Eq. (4-12), we obtain the values shown in Table 4-2. We note that the

Table 4-2

t	0	T	$2T$	$3T$	$4T$
v	V_0	$0.37V_0$	$0.14V_0$	$0.05V_0$	$0.018V_0$

voltage decays to 37 per cent of its initial value in a time-interval equation to one time constant. After three time constants, the voltage is 5 per cent of its initial value. Since at $t = 3T$ the stored energy is then less than one-fourth of 1 per cent of its initial value (why?), it is common practice ("rule of thumb") to say that (in the absence of other specifications) the deenergizing process, although theoretically lasting to $t = \infty$, is completed for practical purposes in a time equal to three time constants.

To study the waveshape of the voltage $v(t)$ as given by Eq. (4-12), it is convenient to rewrite the equation in dimensionless form:

$$\frac{v(t)}{V_0} = e^{-t/T} \qquad t \geq 0^+ \tag{4-12a}$$

Equation (4-12*a*) is a dimensionless, or "normalized," equation; that is, v/V_0 represents the fraction of the initial value which the voltage has at any time. Moreover, it is convenient to regard t/T as a time variable in which time is measured, not in seconds, but in time constants. Introducing this notion explicitly, we may set

$$\frac{v(t)}{V_0} = y = \text{normalized voltage}$$

and

$$\frac{t}{T} = x = \text{normalized time}$$

Then the equation

$$y = e^{-x} \tag{4-13}$$

represents the decay of voltage in *any* R-C circuit which is deenergized. This relationship, showing the waveshape of such a decaying voltage, is plotted to scale in Fig. 4-3. Note that the graph of Fig. 4-3 can be sketched

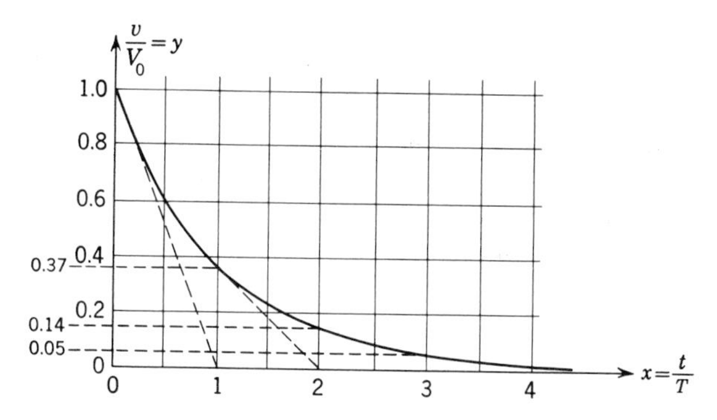

Fig. 4-3 Normalized exponential decay.

quickly if the numbers $e^{-1} = 0.37$, $e^{-2} = 0.14$, $e^{-3} = 0.05$ are remembered. Since exponential decays are so frequently encountered and since the response waveforms are of such importance, the reader should practice sketching responses involving exponential decays by using the values at $t = T, 2T$, and $3T$ (this method of sketching exponentials may be termed the "three-time-constant" method).

Another interpretation of the time constant is possible by considering the slope of y

$$y = e^{-x} \qquad x \geq 0^+$$

$$\frac{dy}{dx} = -e^{-x} \qquad x \geq 0^+$$

Hence, at $x = 0^+$ $(t = 0^+)$,

$$\left.\frac{dy}{dx}\right|_{x=0^+} = -1 \qquad \text{or} \qquad \left.\frac{dv}{dt}\right|_{t=0^+} = -\frac{1}{T}V_0$$

so that the slope of the normalized response curve is -1 at $x = 0^+$, and the slope of $v(t)$ is $-V_0/T$ at $t = 0^+$. Consequently, *if the voltage were to decay at its initial rate,* the value zero *would* be reached in a time equal to one time constant (Prob. 4-3).

4-4 Initial derivatives; examples

The foregoing analysis deals with the voltage across the capacitance C in the source-free R-C circuit. We now note that the current in C (or in R), *as well as any other linear combination of v with any of its derivatives, also has exponential waveform.* This follows by observing that if

$$v = Ke^{st}$$

then

$$pv = sv, \quad \ldots, \quad p^n v = s^n v$$

so that the linear combination

$$\begin{aligned} f(t) &= (a_0 + a_1 p + a_1 p^2 + \cdots + a_n p^n)v = (a_0 + a_1 s + \cdots + a_n s^n)v \\ &= K_1 v \end{aligned}$$

In general, any source-free response of a network (for example, i_C in Fig. 4-2) is obtained from a given source-free response (for example, v in Fig. 4-2) through linear combination of derivatives of the given response (for example, $i_C = Cpv$). Therefore the form $Ke^{-t/T}$ represents the waveform of any network variable in a source-free R-C circuit. For this reason this form is termed the *natural* (i.e., source-free) response of such a circuit. Thus one network variable such as i_R is distinguished from any other such as p^3v only in the value (and dimensions) of the scale factor K.

Since for any variable $f(t)$,

$$f(t) = Ke^{-t/T}$$

if follows that $K = f(0^+)$, the initial value of f. Now, in the discussion of Secs. 4-2 and 4-3, the function f is the voltage across the capacitance v, and it was stated that $v(0^+) = V_0$. This given initial condition can be used to derive the initial value for any function f by use of the equilibrium equation (4-5a):

$$pv = -\frac{1}{RC}v = -\frac{1}{T}v \qquad t \geq 0^+ \tag{4-5a}$$

Equation (4-5a) applies at $t = 0^+$; hence

$$(pv)_{0^+} = \left(\frac{dv}{dt}\right)_{0^+} = -\frac{1}{T}V_0 \tag{4-14}$$

If Eq. (4-5a) is differentiated n times with respect to t, we obtain

$$p^n v = -\frac{1}{T}p^{n-1}v = \left(-\frac{1}{T}\right)^n v \tag{4-15}$$

so that

$$(p^n v)_{0^+} = \left(\frac{d^n v}{dt^n}\right)_{0^+} = \left(-\frac{1}{T}\right)^n V_0$$

The foregoing procedure illustrates how initial derivatives can be evaluated *without solving the equilibrium equation* (see also Prob. 4-4).

Example 4-1 In the circuit of Fig. 4-4, $v_C(0^+) = 10$. Find v_{ab} for all $t \geq 0^+$.

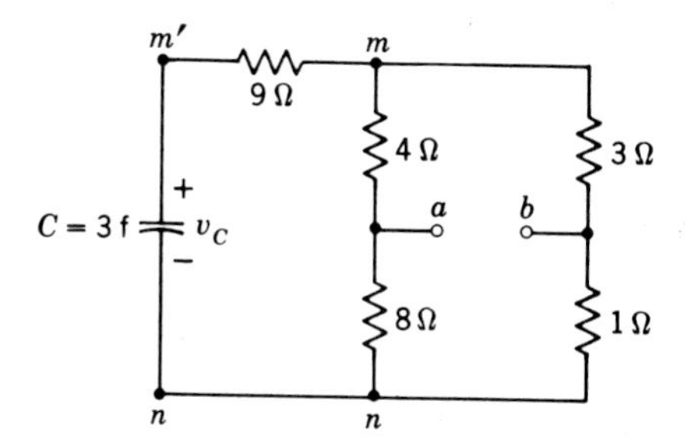

Fig. 4-4 Circuit for Example 4-1.

Solution From voltage division,

$$v_{ab} = v_C \frac{R_{mn}}{R_{m'n}} \left(\frac{R_{an}}{R_{man}} - \frac{R_{bn}}{R_{mbn}} \right)$$

Since

$$R_{mn} = \frac{12 \times 4}{12 + 4} = 3 \text{ ohms}$$

so that

$$R_{m'n} = 12 \text{ ohms}$$

and

$$v_{ab} = v_C \times \tfrac{3}{12}(\tfrac{8}{12} - \tfrac{1}{4}) = v_C \times \tfrac{5}{48}$$

The time constant of this circuit is

$$T = R_{m'n}C = 12 \times 3 = 36 \text{ sec}$$

Hence

$$v_{ab} = 10e^{-t/36} \times \tfrac{5}{48} = \tfrac{25}{24}e^{-t/36} \qquad t \geq 0^+$$

Ans.

Example 4-2 In the circuit of Fig. 4-5, $v(0^+) = 10$. Solve for $i(t)$ directly, without finding $v(t)$.

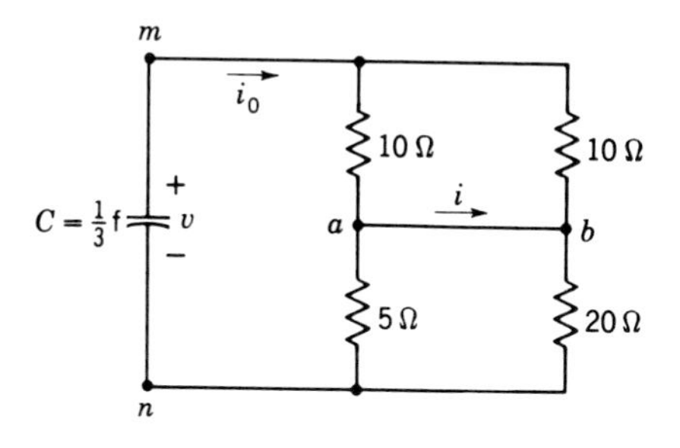

Fig. 4-5 Circuit for Example 4-2.

Solution The resistance connected across C, R_{mn}, is given by

$$R_{mn} = \frac{10 \times 10}{20} + \frac{20 \times 5}{25} = 9 \text{ ohms}$$

Hence

$$T = R_{mn}C = 3 \text{ sec}$$

and

$$i(t) = i(0^+)e^{-t/3}$$

To determine $i(0^+)$ we note that, generally,

$$i = i_0(\tfrac{1}{2} - \tfrac{20}{25}) = -\tfrac{3}{10}i_0 \tag{4-16}$$

Further,

$$i_0 = \frac{v_{mn}}{R_{mn}}$$

Hence

$$i_0(0^+) = \frac{v_{mn}(0^+)}{9} = \frac{10}{9}$$

and using this result in Eq. (4-16),

$$i(0^+) = -\tfrac{3}{10} \times \tfrac{10}{9} = -\tfrac{1}{3} \text{ amp}$$

so that the final result is

$$i(t) = -\tfrac{1}{3}e^{-t/3} \qquad t \geq 0^+ \qquad\qquad \textit{Ans.}$$

CONTINUITY In many practical problems circuit elements are added or removed from circuits by switching operations at various times. In such cases one can often calculate the waveforms of the network variables by use of the *continuity condition*, namely, that in the *absence of impulses*, the energy stored in an inductance or capacitance (and therefore the corresponding current or voltage, respectively) must be continuous. As is illustrated below, the fact that the *voltage* across a *capacitance* is *continuous* usually implies that the *current* through it has *jumps*. Analogously, continuity of current in an inductance often implies discontinuity for its voltage.

Example 4-3 In the circuit of Fig. 4-6 the switch is open for $0 < t < t_1$ and closed for $t > t_1$. If $v_{ab}(0^+) = V_0$, formulate $v_{ab}(t)$ and $i(t)$ for all $t \geq 0^+$.

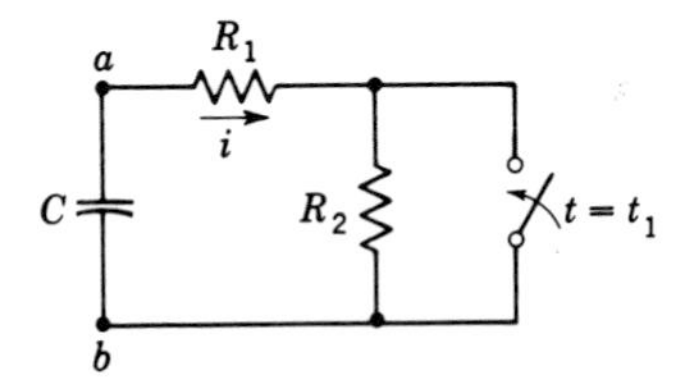

Fig. 4-6 Circuit for Example 4-3.

Solution For the interval $t = 0$ to $t = t_1$, we have the solution, as before,

$$v_{ab} = V_0 e^{-[t/(R_1+R_2)C]} \qquad 0^+ \leq t \leq t_1$$

and

$$i = \frac{v_{ab}}{R_{ab}} = \frac{V_0}{R_1 + R_2} e^{-[t/(R_1+R_2)C]} \qquad 0 < t < t_1$$

Now, we note that, at $t = t_1^-$,

$$i(t_1^-) = \frac{v_{ab}(t_1^-)}{R_1 + R_2}$$

but at $t = t_1^+$, $R_{ab} = R_1$, so that

$$i(t_1^+) = \frac{v_{ab}(t_1^+)}{R_1}$$

It follows that $i(t_1^-)$ cannot be the same as $i(t_1^+)$ if v_{ab} is continuous; rather, since $v_{ab}(t_1^-) = v_{ab}(t_1^+)$,

$$i(t_1^+) = \frac{R_1 + R_2}{R_1} i(t_1^-)$$

To show that v_{ab} is continuous (i.e., to show that there are no impulse currents), we need only observe that, for finite R_{ab}, the equation

$$C \frac{dv_{ab}}{dt} = -\frac{1}{R_{ab}} v_{ab}$$

requires finite dv_{ab}/dt for all finite v_{ab}. Hence, no impulses occur.

The solution is therefore

$$v_{ab}(t) = v_{ab}(t_1^-)e^{-(t-t_1)/R_1C} \qquad t \geq t_1$$

and

$$i(t) = \frac{v_{ab}(t_1^-)}{R_1} e^{-(t-t_1)/R_1C} \qquad t > t_1$$

where

$$v_{ab}(t_1^-) = V_0 e^{-[t_1/(R_1+R_2)C]}$$

The result is illustrated for $R_2 = 2R_1$ at $t_1 = (R_1 + R_2)C$ in Fig. 4-7.

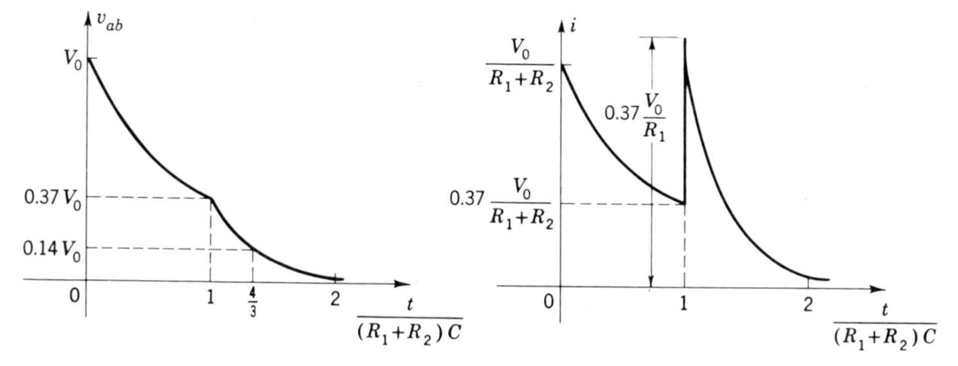

***Fig. 4-7** Voltage and current waveforms for Example 4-3. The graphs are drawn for the numerical case $R_2 = 2R_1$.*

4-5 *The source-free R-L circuit* The equation of the source-free R-C circuit, Eq. (4-5*a*), has the form

$$\frac{dy}{dt} + \frac{1}{T} y = 0 \qquad t \geq 0^+ \tag{4-17}$$

and the solution

$$y = y(0^+)e^{-t/T} \qquad t \geq 0^+ \tag{4-18}$$

It is evident that Eq. (4-17) and its solution, Eq. (4-18), need not be restricted to R-C circuits, but rather represent the equilibrium equation and its general solution for all source-free systems that are represented by a first-order linear homogeneous differential equation with constant coefficients. Thus, for the R-L circuit shown in Fig. 4-8, we have, for $t \geq 0^+$,

$$L \frac{di}{dt} + Ri = 0 \qquad i(0^+) = I_0 \tag{4-19}$$

or

$$\frac{di}{dt} + \frac{1}{L/R} i = 0 \qquad t \geq 0^+ \tag{4-19a}$$

and the solution is

$$i = I_0 e^{-t/T} \tag{4-20}$$

where the time constant T for the R-L circuit is given by

$$T = \frac{L}{R} \tag{4-21}$$

The waveform of $i(t)$ is seen to be given by the graph of Fig. 4-3 if $y = i/I_0$ and $T = L/R$.

Further development of the R-L circuit is reserved for reader's exercises (Probs. 4-9 to 4-15).

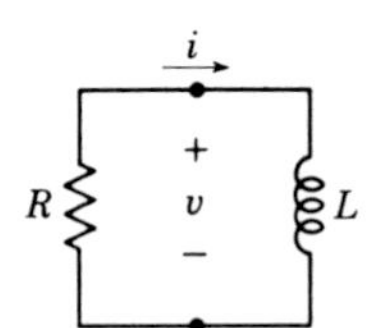

Fig. 4-8 ***Source-free R-L circuit.***

4-6 Some analogies In general, systems described by the same type of equilibrium equation are termed *analogous systems*. Thus, since both the source-free R-L and the source-free R-C circuits are described by the equation

$$\left(p + \frac{1}{T}\right) y = 0 \qquad y(0^+) = Y_0 \tag{4-22}$$

these two circuits are *analogous*. There are several other (nonelectrical) physical systems which are described by Eq. (4-22) and are therefore analogous to R-C and R-L circuits.

An example of a mechanical system described by Eq. (4-22) consists of a mass M moving through a viscous fluid. Denoting the velocity of the mass by u, its acceleration is $du/dt = pu$, so that the inertial force is pMu. Assuming the viscous fluid to offer linear damping friction,[1] the frictional force is Du (where D is the coefficient of *viscous* friction). Thus the application of Newton's laws of motion to the mass-friction system results in the equation

$$Mpu + Du = \text{applied force}$$

and when the applied force is identically zero, the equation has the form of Eq. (4-22):

$$(pM + D)u = 0$$

or

$$\left(p + \frac{D}{M}\right) u = 0$$

The time constant of this mechanical system is seen to be M/D.

Another example of a physical system analogous to the R-C circuit is a d-c motor with constant field excitation (Fig. 4-9). In such a motor an electromotive force is induced in the armature which is proportional to angular velocity ω. Representing the effect of this emf by the voltage e_a, $e_a = K\omega$, and hence, from Fig. 4-9,

$$i = \frac{v - K\omega}{R_a} \tag{4-23}$$

[1] The reader will recall that, in mechanics, one defines two types of friction, namely, the speed-independent *Coulomb friction* (proportional to "normal" force, or torque) and *viscous friction*, which is velocity-dependent. When the viscous friction is *linear*, the frictional force (or torque) is *proportional* to velocity.

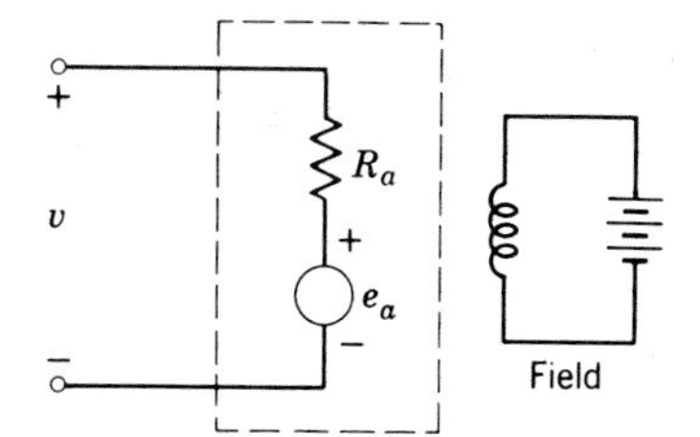

***Fig. 4-9** Schematic of a d-c motor. The source e_a represents the induced counter emf, and R_a represents the armature circuit resistance.*

The electromagnetic torque developed by the motor is proportional to the armature current i. It is used to accelerate the armature, whose moment of inertia is J, and to overcome viscous friction. Assuming linear viscous friction,

$$J p\omega + D\omega = \text{developed torque} = K'i$$

Thus, using Eq. (4-23),

$$(pJ + D)\omega = K'\left(\frac{v}{R_a} - \frac{K\omega}{R_a}\right)$$

or

$$\left[pJ + \left(D + \frac{KK'}{R_a}\right)\right]\omega = K''v$$

The externally applied voltage source v causes the motor to increase its angular velocity from zero to a value determined by v. If, now, the applied voltage v is made identically zero, the equation for the angular velocity is

$$\left(p + \frac{D + KK'/R_a}{J}\right)\omega = 0$$

This equation has the form of Eq. (4-22) with the (electromechanical) time constant J/D_e, where D_e represents the equivalent damping (due to both electrical and mechanical causes) as is given by

$$D_e = D + \frac{KK'}{R_a}$$

Evidently, analogous systems have the same type of response (since they are described by the same type of equation). Thus, for systems of the type

$$\left(p + \frac{1}{T}\right)y = 0 \qquad t \geq 0^+$$

the response is

$$y = y(0^+)e^{-t/T}$$

It should be clear that systems can be mathematically analogous even if they do not deal with behavior as a function of time. Thus the system

$$\frac{dy}{dx} + kx = 0 \tag{4-24}$$

has the solution

$$y = Ce^{-kx}$$

independently of the physical interpretation given to either the dependent variable y or the independent variable x, and the system given by Eq. (4-24) is analogous to that of Eq. (4-22).

An interesting aspect of the subject of linear first-order systems is that they represent, on the average, certain models of human perception as embodied in the Weber-Fechner[1]

[1] Ernst Heinrich Weber (1745–1878) and Gustav Theodor Fechner (1801–1887).

law. This law deals with the response sensation of the human body to sense stimuli (such as sound, light, odor, and even pain) and states that the minimum perceptible increment of stimulus needed to produce a perceptible response is proportional to the amount of stimulus already present. In mathematical form, let

y = stimulus present
Δy = increment in stimulus to produce perceptible response
Δx = perceptible response

Then the Weber-Fechner law states that

$$\Delta y = k(\Delta x)y$$

or

$$\frac{\Delta y}{\Delta x} = ky$$

taking limits as $\Delta x \to 0$, $\Delta y/\Delta x \to dy/dx$, and

$$\frac{dy}{dx} - ky = 0$$

so that the law relating total stimulus y to total response x is

$$e^{kx} = Cy$$

where C is a constant. In logarithmic form we note that

$$x = K \log y \tag{4-25}$$

that is, the total response varies with the logarithm of the stimulus.

Equation (4-25) has become the basis for certain *volume units*, specifically, the *bel*,[1] the *decibel*,[1] and the *neper*.[2] These units, originally used as measures of relative sound power level, are now generally used in connection with relative power levels, as well as voltage and current levels.

THE BEL *Relative power level in bels* is defined as the logarithm *to the base 10* of the corresponding power ratio:

$$\text{bels} = \log_{10} \frac{P}{P_0} \tag{4-26}$$

The quantity P_0 is often referred to as "reference power."

THE DECIBEL A decibel is defined as one-tenth of a bel:

$$\text{decibel} = \text{db} = 10 \log_{10} \frac{P}{P_0} \tag{4-27}$$

If, for example, P_0 = 1 mw, the power level P = 2 mw is $10 \log_{10} 2$ = 3.03 db above 1 mw. It is convenient to remember that doubling the power corresponds (approximately) to a 3-db increase in decibel level.

Very often, relative voltage and current level are expressed in decibels. To make such definition meaningful, one compares the voltages (or the currents) by considering the power they would produce in a certain resistance. Thus, if V_0 is applied to a resistance R, the power is V_0^2/R; if V is applied to the *same* resistance, the power is V^2/R. Evidently, the power *ratio* is the square of the voltage ratio and is independent of R. One

[1] Named after Alexander Graham Bell (1847–1922), inventor of the telephone.
[2] John Napier (or Nepa) (1550–1617), inventor of logarithms.

therefore defines the relative decibel level of voltages or of currents as follows:

$$\text{Relative voltage decibel level} = 10 \log_{10} \frac{V^2}{V_0^2} = 20 \log_{10} \frac{V}{V_0} \tag{4-28}$$

or

$$\text{Relative current decibel level} = 20 \log_{10} \frac{I}{I_0} \tag{4-29}$$

THE NEPER The relative *voltage* (or *current*) level of two voltages (or currents) in *nepers* is defined as the *natural* logarithm of their quotient:

$$\text{Relative current level in nepers} = n_i = \ln \frac{I}{I_0} \tag{4-30}$$

$$\text{Relative voltage level in neper} = n_v = \ln \frac{V}{V_0} \tag{4-30a}$$

Since, in general,

$$\ln x \approx 2.3 \log_{10} x$$

we can convert relative voltage (or current) level from decibel to nepers, and vice versa, by noting that

$$\text{Relative voltage level in db} = 8.69 \times \text{relative voltage level in nepers}$$

As an example we observe that, in a source-free R-C circuit, the voltage level is 1 neper, or 8.69 db, *below* its initial value after one time constant.

4-7 The source-free L-C circuit[1]

Since in all electrical devices a certain amount of energy is transformed into heat, a circuit containing no resistive element does not correspond to any physical device. Nevertheless, the study of circuits containing only L and C is instructive because this combination represents the simplest type of circuit in which both types of energy storage are present.

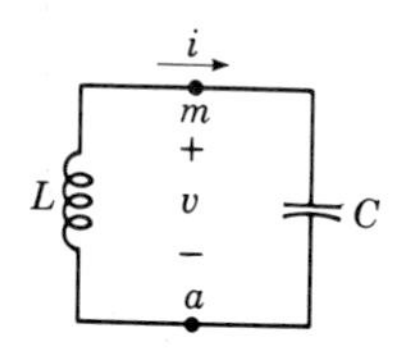

Fig. 4-10 Source-free L-C circuit.

A source-free L-C circuit is shown in Fig. 4-10. The continuity conditions in this circuit are

$$v(0^-) = v(0^+) = V_0$$
$$i(0^-) = i(0^+) = I_0$$

and from the definition of L and C we have the relationships

$$i = pCv \qquad v = -pLi$$

[1] *The reader who is not familiar with complex numbers should read Appendix B before proceeding with the remainder of this chapter.*

so that $pi = p^2Cv$ and

$$v = -p^2LCv$$

or

$$\left(p^2 + \frac{1}{LC}\right)v = 0 \tag{4-31}$$

We observe that $1/LC$ has the dimensions (time)$^{-2}$. We define ω_0 as an abbreviation for $1/\sqrt{LC}$:

$$\omega_0{}^2 = \frac{1}{LC} \tag{4-32}$$

The differential equation of the *L-C* circuit is then written

$$(p^2 + \omega_0{}^2)v = 0 \tag{4-33}$$

Before proceeding to a formal solution of this differential equation, a physical examination of the energy processes is instructive. Suppose that at $t = 0$ there is zero current in the circuit but that there is a finite voltage across the capacitance: $v(0) = V_0$. Since, from $t = 0$ on, the inductance is also connected between terminals *a-m*, the voltage V_0 will be across the inductance at $t = 0$. Hence the current through the inductance will change because the voltage across an inductance is proportional to the rate of change of the current through it. Now, as the current through the inductance grows, the stored energy in the inductance grows so that the stored energy in the capacitance must decrease, since the total stored energy in the circuit cannot exceed the initial energy in the capacitance. Eventually, a time is reached at which all the energy which was stored in the capacitance at $t = 0$ has been transferred and is stored in the inductance. At that instant v will be zero (since w_C will be zero), but the current $i(t)$ will have a value. Hence this current will again carry charges to the capacitance, and the stored energy will return from L to C. Now, in this energy-transfer process, no energy is lost since the circuit contains no resistance (resistance is the only passive element which can make energy unavailable to the rest of the circuit elements). We anticipate, therefore, on physical grounds, that the source-free response of the *L-C* circuit will involve a periodic interchange of stored energy between inductance and capacitance for all time after $t = 0$. (See also Probs. 4-19 and 4-20.)

Returning now to Eq. (4-33), we observe that we could solve the equation by using sine and cosine functions directly. As it is, however, since one of the objectives of this text is to illustrate analytical techniques that are generally useful, we shall use the exponential form of solution. We begin with Eq. (4-33), i.e.,

$$(p^2 + \omega_0{}^2)v = 0 \qquad t \geq 0^+ \tag{4-33}$$

and let

$$v = Ke^{st} \tag{4-34}$$

so that $pv = sv$ and $p^2v = s^2v$. Substitution in Eq. (4-33) gives the relationship

$$(s^2 + \omega_0{}^2)v = 0 \qquad t \geq 0^+ \tag{4-34a}$$

Since v cannot be identically zero for *all* $t \geq 0^+$ (except in the trivial case), we have the condition

$$s^2 + \omega_0^2 = 0 \tag{4-35}$$

as the characteristic equation of the *L-C* circuit. The two values of s which satisfy Eq. (4-35) are the imaginary values[1]

$$\mathbf{s}_{1,2} = \pm j\omega_0$$

where j denotes the unit imaginary quantity

$$j = \sqrt{-1}$$

The solution of the *L-C* circuit consists of a linear combination of the two modes $e^{j\omega_0 t}$ and $e^{-j\omega_0 t}$:

$$v = \mathbf{K}_1 e^{j\omega_0 t} + \mathbf{K}_2 e^{-j\omega_0 t} \tag{4-36}$$

where the constants $\mathbf{K}_1$ and $\mathbf{K}_2$ are (generally) complex.

We shall now show that $\mathbf{K}_1$ and $\mathbf{K}_2$ are in general *conjugate complex* quantities. This proof can be done in several ways; we shall show it by recalling that $v(t)$ is a real function of a real variable since it is the solution of a real physical problem. Since $v(t)$ is real, we have, from Eq. (4-36),

$$\text{Im}\,(\mathbf{K}_1 e^{j\omega_0 t} + \mathbf{K}_2 e^{-j\omega_0 t}) \equiv 0 \tag{4-37}$$

when the operation Im denotes the imaginary part.[2] Denoting $\mathbf{K}_1$ and $\mathbf{K}_2$ in exponential form,

$$\mathbf{K}_1 = K_1 e^{j\gamma_1} \qquad \mathbf{K}_2 = K_2 e^{j\gamma_2}$$

we have, from Eq. (4-37),

$$\text{Im}\,(K_1 e^{j(\omega_0 t+\gamma_1)} + K_2 e^{-j(\omega_0 t-\gamma_2)}) = 0$$

or

$$\text{Im}\,(K_1 e^{j(\omega_0 t+\gamma_1)}) = -\,\text{Im}\,(K_2 e^{-(j\omega_0 t-\gamma_2)})$$

Hence

$$K_1 \sin\,(\omega_0 t + \gamma_1) = -[-K_2 \sin\,(\omega_0 t - \gamma_2)]$$

or

$$K_1 \sin\,(\omega_0 t + \gamma_1) = K_2 \sin\,(\omega_0 t - \gamma_2) \tag{4-38}$$

A little thought shows that Eq. (4-38) can be true for *all* $t \geq 0^+$ *only* if

$$K_2 = K_1 \qquad \text{and} \qquad \gamma_2 = -\gamma_1$$

so that

$$\mathbf{K}_2 = \mathbf{K}_1^*$$

where the asterisk (*) denotes the complex conjugate.

The solution of the *L-C* circuit [Eq. (4-36)] can therefore be written

$$v = \mathbf{K}_1 e^{j\omega_0 t} + \mathbf{K}_1^* (e^{j\omega_0 t})^*$$

or [see Eq. (B-6)]

$$v = \text{Re}\,(2\mathbf{K}_1 e^{j\omega_0 t})$$

We define

$$\mathbf{K} = 2\mathbf{K}_1$$

and therefore

$$v = \text{Re}\,(\mathbf{K} e^{j\omega_0 t}) \tag{4-39}$$

[1] In this book complex quantities are denoted by **boldface** type.

[2] The reader is referred to Appendix B for notation conventions in connection with complex numbers.

It is pointed out that Eq. (4-39) is the general solution of Eq. (4-33) and that the complex constant **K** is determined by the initial conditions. Thus, for example, if

$$v(0^+) = V_0 \qquad \text{and} \qquad C\left(\frac{dv}{dt}\right)_{0^+} = I_0$$

then, since

$$\begin{aligned} v &= \text{Re}\,(\mathbf{K}e^{j\omega_0 t}) \\ \text{Re}\,\mathbf{K} &= V_0 \end{aligned} \tag{4-40}$$

and[1]

$$\begin{aligned} \frac{dv}{dt} &= \frac{d}{dt}[\text{Re}\,(\mathbf{K}e^{j\omega_0 t})] \\ &= \text{Re}\left[\frac{d}{dt}(\mathbf{K}e^{j\omega_0 t})\right] \\ &= \text{Re}\,(j\omega_0\mathbf{K}e^{j\omega_0 t}) \end{aligned}$$

Letting $t = 0^+$, we have

$$\left(\frac{dv}{dt}\right)_{0^+} = \frac{I_0}{C} = \text{Re}\,(j\omega_0\mathbf{K}) \tag{4-41}$$

Equations (4-40) and (4-41) can be solved for **K**, so that, if **K** is expressed in polar form,

$$\begin{aligned} \mathbf{K} &= K\underline{/\gamma} \\ v(t) &= K\cos(\omega_0 t + \gamma) \end{aligned}$$

Example 4-4 In the circuit of Fig. 4-10, $i(0^+) = 5$ and $v(0^+) = 50$. Find $v(t)$ for all $t \geq 0^+$ for $L = \frac{1}{2}$ henry and $C = 0.005$ farad.

Solution $1/LC = 400$, $\omega_0 = 20$, since $i = +C\,dv/dt$, $dv/dt = i/C = 200i$, $(dv/dt)_{0^+} = 1{,}000$. Starting now with the general solution

$$v = \text{Re}\,(\mathbf{K}e^{j20t})$$

we have

$$\left(\frac{dv}{dt}\right) = \text{Re}\,(j20\mathbf{K}e^{j20t})$$

Introducing the initial condition $v(0^+) = 50$ gives

$$\text{Re}\,\mathbf{K} = 50$$

so that

$$\mathbf{K} = 50 + jk'$$

where $k' = \text{Im}\,\mathbf{K}$. Using the initial value of dv/dt,

$$1{,}000 = \left(\frac{dv}{dt}\right)_{0^+} = \text{Re}\,[j20(50 + jk')]$$

[1] The identity

$$\frac{d}{dt}(\text{Re}\,e^{j\omega_0 t}) = \text{Re}\left[\frac{d}{dt}(e^{j\omega_0 t})\right]$$

is of great importance in this and other developments. It is discussed generally in Appendix C, Sec. C-6.

or
$$1{,}000 = -20k'$$
$$-50 = k'$$

Hence
$$\mathbf{K} = 50 - j50 = 70.7\underline{/-45°}$$

so that
$$v = 70.7 \cos (20t - 45°) \qquad \textit{Ans.}$$

4-8 Waveform of L-C circuit response

The form of the solution for voltage and current is seen to be sinusoidal. To study a typical waveform it is sufficient to consider the special example

$$I_0 = 0 \qquad V_0 \neq 0 \tag{4-42}$$

Physically, this case can be visualized as the situation which arises when an initially nonenergized inductance is connected across a charged capacitance at $t = 0$. For this case

$$v(t) = V_0 \cos \omega_0 t \tag{4-43}$$
$$i(t) = -\omega_0 C V_0 \sin \omega_0 t \tag{4-44}$$

The waveforms corresponding to Eqs. (4-43) and (4-44) are shown in Fig. 4-11a and b, respectively. These diagrams show oscillations at the radian frequency ω_0 (or at the frequency $\omega_0/2\pi$ hertz). In the sketches V_0 has been

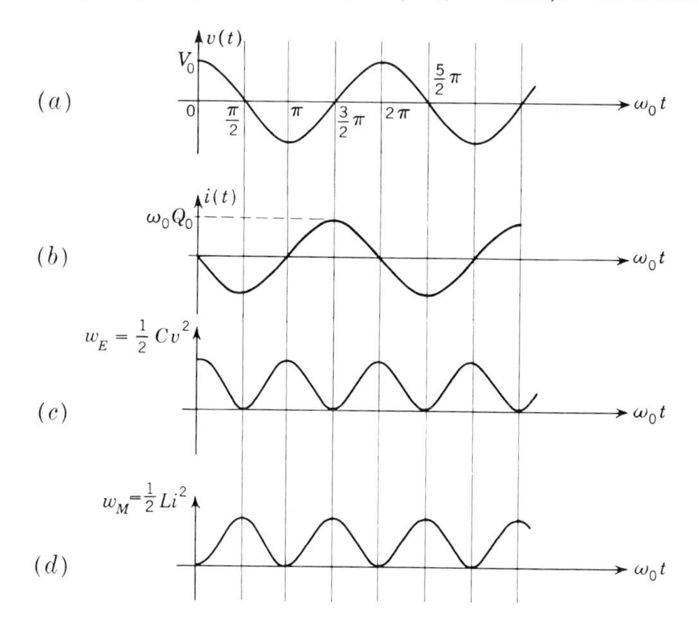

Fig. 4-11 ***Waveforms in the L-C circuit for $I_0 = 0$, $V_0 > 0$.***

taken as a positive number, so that the initial time rate of change of the current is negative.

$$\left(\frac{di}{dt}\right)_{t=0^+} = -\frac{V_0}{L} \tag{4-45}$$

Since there is no resistive element in the L-C circuit, there is no time constant associated with such a circuit. Instead, the circuit is characterized by the natural radian frequency $\omega_0 = 1/\sqrt{LC}$ (rad/sec). The oscillations of the current and voltage in the circuit are at this *natural angular or radian frequency of the circuit.*

4-9 The source-free R-L-C circuit

A circuit consisting of a single resistance, a single inductance, and a single capacitance is the simplest circuit in which energy dissipation as well as electric- and magnetic-energy storage takes place. For this reason the source-free R-L-C circuit does, to a first order of approximation, represent the equivalent circuit of many electrical devices. For example, a receiving or transmitting antenna, cavity resonators and matching devices used in microwave techniques, piezoelectric crystals used in audio- and radio-frequency work, automatic control systems, as well as electromechanical and purely mechanical devices such as galvanometers and vibrating systems, can be represented by series or parallel connections of R, L, and C elements. The technique of such representations is briefly discussed in Sec. 4-19, and the reader will find more extended discussions in textbooks dealing with particular devices. In this section we study the behavior of the variables (voltages and currents) associated with R-L-C circuits. We show that, depending on the relative values of the three elements, any variable $y(t)$ associated with a source-free R-L-C circuit will be of one of the following three forms:

$$y(t) = Ke^{-\alpha t}\cos(\omega t + \gamma)$$
$$y(t) = K_1e^{-\alpha_1 t} + K_2e^{-\alpha_2 t}$$
$$y(t) = (K_3 + K_4t)e^{-\alpha t}$$

where α and ω are positive values, independent of time, and are determined by the values of R, L, and C; K, K_1, K_2, K_3, K_4, and γ are also independent of time, and their values are determined by the initial conditions of the circuit as well as the values of R, L, and C. The three possible waveforms associated with a source-free R-L-C circuit are shown in Figs. 4-13*c*, 4-14, and 4-15, respectively.

A study of these figures will show that any variable associated with a source-free R-L-C circuit will "decay" to zero, although, as in the case of Fig. 4-13*c*, this decay may be associated with oscillations. Thus the voltage or current associated with an element in an R-L-C circuit (or, in mechanical analogues, the velocity or displacement) will vary with time in accordance with the waveform of one of these figures. Furthermore, in Chap. 11, it is shown that the variables associated with any connection of passive elements, no matter how complicated, will vary in accordance with a combination of the above-mentioned waveforms.

For this reason a knowledge of the behavior of an *R-L-C* circuit enables us to understand the behavior of any "complicated" source-free network.

In Fig. 4-12*a* is shown an *R-L-C* circuit connected to a source prior to $t = 0$, and by switching from d to a', the circuit is source-free for $t \geq 0^+$ as shown in Fig. 4-12*b*.

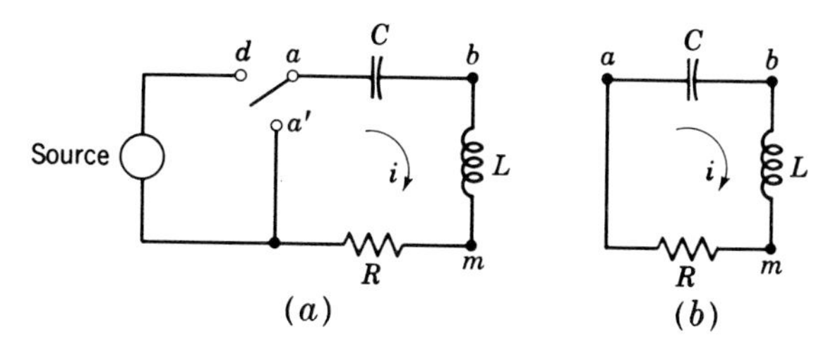

***Fig. 4-12** (a) R-L-C circuit with source. (b) Source-free series R-L-C circuit.*

For this circuit the equilibrium equation with current i as the variable is

$$\left(pL + R + \frac{1}{pC}\right) i = 0 \tag{4-46}$$

or, operating on the equation with p (i.e., differentiating with respect to t) and dividing by L,

$$\left(p^2 + \frac{R}{L} p + \frac{1}{LC}\right) i = 0 \tag{4-47}$$

If instead of current we choose v_{ab}, the voltage across the capacitance, as the variable, then substitution of the relation

$$i = pCv_{ab}$$

(4-46) gives the same equilibrium equation for v_{ab} as for i, that is,

$$(p^2LC + pRC + 1)v_{ab} = 0$$

or

$$\left(p^2 + \frac{R}{L} p + \frac{1}{LC}\right) v_{ab} = 0$$

It can, of course, be shown (Prob. 4-28) that the equilibrium equation for any linear combination of network variables with their derivatives has the form

$$\left(p^2 + \frac{R}{L} p + \frac{1}{LC}\right) y = 0 \qquad t \geq 0^+ \tag{4-48}$$

where y represents the combination of variables with their derivatives referred to above.

INITIAL CONDITIONS In the circuit of Fig. 4-12*b* if we assume that at $t = 0^-$ there was energy stored in the inductance and in the capacitance, then the solution of the equilibrium equations is subject to the continuity conditions

Because of the inductance:

$$i(0^-) = i(0^+) \equiv I_0$$

and

Because of the capacitance:

$$v_{ab}(0^-) = v_{ab}(0^+) \equiv V_0$$

since, exactly as in the cases of the *R-C*, *R-L*, and *L-C* circuits, no impulses can occur.

In addition to the continuity conditions it is possible to obtain initial conditions for any of the variables of the circuit or their derivatives by using the continuity conditions in the equilibrium equations. Several examples of this procedure follow.

The initial value of di/dt can be obtained from Fig. 4-12*b* by observing that

$$v_{bm} = L\frac{di}{dt} = v_{ba} + v_{am}$$

and

$$v_{ba} + v_{am} = -v_{ab} - Ri$$

Hence

$$\frac{di}{dt} = -\frac{1}{L}(v_{ab} + Ri) \qquad (4\text{-}49)$$

and setting $t = 0^+$ gives the initial value of di/dt:

$$\left(\frac{di}{dt}\right)_{0^+} = -\left(\frac{R}{L}I_0 + \frac{1}{L}V_0\right) \qquad (4\text{-}50)$$

The condition (4-50) gives the initial value of the first derivative of the current. This condition is *not* a continuity condition; in fact, the derivative of the current is generally not continuous at $t = 0$. Hence Eq. (4-50) gives only $(di/dt)_{0^+}$ and gives *no* information about $(di/dt)_{0^-}$. We recall from Chap. 3 that an initial condition which is *not* a continuity condition is referred to as a *derived* initial condition. We can obtain as many derived initial conditions as we choose by using the continuity conditions in the various forms of the equilibrium equation. If, for example, Eq. (4-49) is differentiated once with respect to time, and using $i = pCv_{ab}$ we have

$$\frac{d^2i}{dt^2} = -\left(\frac{R}{L}\frac{di}{dt} + \frac{1}{LC}i\right) \qquad t \geq 0^+ \qquad (4\text{-}51)$$

Hence, substituting $t = 0^+$, we have the initial condition for the second derivative of the current,

$$\left(\frac{d^2i}{dt^2}\right)_{0^+} = -\left[\frac{R}{L}\left(\frac{di}{dt}\right)_{0^+} + \frac{1}{LC}i(0^+)\right] \qquad (4\text{-}52)$$

The numerical value of $(d^2i/dt^2)_{0^+}$ is obtained by substitution of the result (4-50) and the continuity condition for i in (4-52).

The process of differentiating the equilibrium equation and substituting $t = 0^+$ can be repeated indefinitely, and as many initial values for the higher derivatives can be deduced as are desired. Although some of these derived initial conditions are subject to immediate physical interpretation [for example, $L(di/dt)_{0^+} = v_{bm}(0^+)$], the higher derivatives should generally be looked upon as describing the slopes and curvatures (and their rates of change) of the waveform for the variable.

4-10 The characteristic equation

The response of the source-free *R-L-C* circuit, using the voltage across the capacitance, v_{ab}, as the variable, is determined by the solution of the equilibrium equation

$$\frac{d^2v_{ab}}{dt^2} + \frac{R}{L}\frac{dv_{ab}}{dt} + \frac{1}{LC}v_{ab} = 0 \qquad t \geq 0^+ \tag{4-53}$$

subject to the initial conditions

$$v_{ab}(0^+) = V_0 \qquad \text{and} \qquad \left(\frac{dv_{ab}}{dt}\right)_{0^+} = \frac{1}{C}I_0 \tag{4-54}$$

The equilibrium equation (4-53) can be satisfied at every instant of time only if the waveshapes of v_{ab}, its first derivative dv_{ab}/dt, and its second derivative d^2v_{ab}/dt^2 are all similar to each other. In addition, we note that, at any instant, at least one of these terms must have a sign opposite to the others, so that the sum of the three terms can be identically zero (for all values of t). This property which is required of the function $v_{ab}(t)$ is also a property of exponential functions. It is recalled that a similar argument was used in obtaining the solution of the first-order homogeneous equation which is associated with a source-free *R-L* or *R-C* circuit. In fact, the same argument applies to any homogeneous linear differential equation with constant coefficients. The solution of such an equation will therefore be of the form e^{st}, where s is independent of t. We therefore introduce the solution

$$v_{ab} = Ke^{st} \tag{4-55}$$

into the equilibrium equation (4-53) and obtain

$$\left(s^2 + \frac{R}{L}s + \frac{1}{LC}\right)v_{ab} = 0 \tag{4-56}$$

Since v_{ab} cannot be identically zero for all t, the condition

$$\left(s^2 + \frac{R}{L}s + \frac{1}{LC}\right) = 0 \tag{4-57}$$

must hold, and the assumed form (4-55) is a solution.

Equation (4-57) is the characteristic equation associated with the equilibrium equation (4-53). This equation resulted from the equilibrium equation with $v_{ab}(t)$ as its variable. We recall now that the equilibrium equation in terms of any other variable of this source-free circuit has exactly the form of Eq. (4-53). The solution for any one of the variables, therefore, also has the same form, i.e., the form Ke^{st}. Since the same equation, (4-57), must be satisfied for any of the variables, this equation is called the *characteristic* equation. The word "characteristic" is used to imply that the relationship applies to *any variable of the source-free circuit.*

The algebraic equation (4-57) has two roots, called the *characteristic roots* of the equation (or of the circuit). Applying the quadratic formula to Eq. (4-57), the two characteristic roots are

$$\begin{aligned} s_1 &= -\frac{R}{2L} + \sqrt{\left(\frac{R}{2L}\right)^2 - \frac{1}{LC}} \\ s_2 &= -\frac{R}{2L} - \sqrt{\left(\frac{R}{2L}\right)^2 - \frac{1}{LC}} \end{aligned} \tag{4-58}$$

Either of the terms $K_1e^{s_1t}$ and $K_2e^{s_2t}$ can therefore satisfy the equilibrium equation. Since the circuit contains two independent energy-storing elements, the solution must contain two, as yet arbitrary, constants to fit the two arbitrarily chosen continuity conditions. If $s_1 \neq s_2$ the solution is written in the form

$$v_{ab}(t) = K_1e^{s_1t} + K_2e^{s_2t} \tag{4-59}$$

We note again that the solutions for any other variable can also be written in the same form, for example,

$$\begin{aligned} i(t) &= K_3e^{s_1t} + K_4e^{s_2t} \\ \text{or} \qquad v_{bm}(t) &= K_5e^{s_1t} + K_6e^{s_2t} \end{aligned} \tag{4-60}$$

MODES OF RESPONSE Each of the time functions $K_3e^{s_1t}$ and $K_4e^{s_2t}$ in Eq. (4-60) is called a *natural mode* (mode for brevity) of the source-free series *R-L-C* circuit. In general, the number of modes of a source-free circuit is identical with the number of characteristic roots of the characteristic equation associated with the circuit. In the case of an *R-C* circuit which contains one energy-storing element, the characteristic equation was $s + 1/RC = 0$, resulting in the single characteristic root $s = -1/RC$ and its corresponding mode $Ke^{-t/RC}$. The *R-L-C* circuit which contains two energy-storing elements has two characteristic roots, given in Eq. (4-58), and two corresponding modes, $K_1e^{s_1t}$ and $K_2e^{s_2t}$. Later we shall see that the number of natural modes of a source-free network is the same as the number of its "independent" energy-storing elements.

SIMPLE AND MULTIPLE CHARACTERISTIC ROOTS If the discriminant of the characteristic equation (4-57), namely, $(R/2L)^2 - 1/LC$, is *not* zero, the two roots s_1 and s_2 are not equal to each other; they are "distinct," and are called simple roots. If $(R/2L)^2 = 1/LC$, the two roots are identical, since then $s_1 = s_2 = -R/2L = -1/\sqrt{LC}$. This type of root is called a multiple root. With multiple roots the complete solution cannot be of the form (4-60) since, for $s_1 = s_2$,

$$K_1e^{s_1t} + K_2e^{s_2t} = K_1e^{s_1t} + K_2e^{s_1t} = (K_1 + K_2)e^{s_1t} = Me^{s_1t}$$

This form has only one arbitrary constant, M, and therefore cannot be made to fit two (arbitrarily specified) initial conditions. The case for which the discriminant of the characteristic equation is zero is therefore a special case. The special form of the solution in this case is taken up in Sec. 4-16.

USE OF THE INITIAL CONDITIONS Excluding the case $s_1 = s_2$, the solution for the voltage $v_{ab}(t)$ can be written in the form

$$v_{ab}(t) = K_1e^{s_1t} + K_2e^{s_2t} \qquad s_1 \neq s_2,\ t \geq 0^+ \tag{4-61}$$

and its first derivative is

$$\frac{dv_{ab}}{dt} = s_1K_1e^{s_1t} + s_2K_2e^{s_2t} \tag{4-62}$$

The constants K_1 and K_2 can now be evaluated by introducing the initial conditions (4-54).

$$v_{ab}(0^+) = V_0 = K_1e^{s_10} + K_2e^{s_20} = K_1 + K_2$$

$$\left(\frac{dv_{ab}}{dt}\right)_{0^+} = \frac{I_0}{C} = s_1K_1e^{s_10} + s_2K_2e^{s_20} = s_1K_1 + s_2K_2$$

Hence K_1 and K_2 are the solutions of the two simultaneous algebraic equations

$$K_1 + K_2 = V_0 \tag{4-63}$$

$$s_1K_1 + s_2K_2 = \frac{I_0}{C} \tag{4-64}$$

Solving

$$K_1 = \frac{I_0/C - s_2V_0}{s_1 - s_2} \tag{4-65a}$$

$$K_2 = \frac{s_1V_0 - I_0/C}{s_1 - s_2} \tag{4-65b}$$

[At this point we note again that the case $s_1 = s_2$ is excluded. If $s_1 = s_2$, then Eqs. (4-65) are meaningless because of the division by $s_1 - s_2$.] The solution for v_{ab} can now be written by substituting the values of K_1 and K_2 in (4-61).

$$v_{ab}(t) = \frac{1}{s_1 - s_2}\left[\left(\frac{I_0}{C} - s_2V_0\right)e^{s_1t} + \left(s_1V_0 - \frac{I_0}{C}\right)e^{s_2t}\right] \qquad s_1 \neq s_2,\ t > 0^+ \tag{4-66}$$

If the current $i(t)$ is desired, we may use the relationship $i(t) = C\, dv_{ab}/dt$.

$$i(t) = \frac{C}{s_1 - s_2}\left[\left(\frac{s_1 I_0}{C} - s_1 s_2 V_0\right)e^{s_1 t} + \left(s_1 s_2 V_0 - \frac{I_0 s_2}{C}\right)e^{s_2 t}\right] \qquad s_1 \neq s_2,\ t \geq 0^+$$

We again point out that the mathematical form of $v_{ab}(t)$ is identical with the form of $i(t)$.

4-11 Types of response

In order to study the waveshapes of the solution and to include the case heretofore excluded, three cases, depending on the nature of the characteristic roots, may be distinguished, as defined in Table 4-3. The reasons for

Table 4-3

Condition	Name of case	Nature of the roots
$\left(\frac{R}{2L}\right)^2 < \frac{1}{LC}$	Underdamped or oscillatory	Complex conjugates with negative real parts
$\left(\frac{R}{2L}\right)^2 = \frac{1}{LC}$	Critically damped	Negative, real, and equal ($s_1 = s_2$)
$\left(\frac{R}{2L}\right)^2 > \frac{1}{LC}$	Overdamped	Negative, real, and unequal ($s_1 \neq s_2$)

the names assigned to these conditions will become clear as each case is discussed in detail. The conditions which lead to these three cases depend on the relative values of two numbers: $R/2L$ and $1/\sqrt{LC}$. In order to show this more explicitly, the following abbreviations are introduced:

$$\omega_0 = \frac{1}{\sqrt{LC}} \qquad \alpha = \frac{R}{2L} \tag{4-67}$$

The meaning of ω_0 is already clear: it is the radian frequency with which the response *would* oscillate *if* the resistance were zero. The term α is interpreted as a "damping" constant since it is the term due to the resistance. With the symbols defined in Eqs. (4-67), the roots of the characteristic equation are

$$s_{1,2} = -\alpha \pm \sqrt{\alpha^2 - \omega_0{}^2}$$

The critically damped case occurs when $\alpha = \omega_0$, so that in this case

$$s_1 = s_2 = -\omega_0 = -\alpha$$

It is convenient to define a single number which, for this circuit, indicates the type of response. Since, for critical damping $s_1 = s_2$, $(R/2L)^2 = 1/LC$, it follows that

For critical damping:

$$\frac{R}{2\sqrt{L/C}} = 1$$

It is often convenient to define the *damping ratio* of the circuit (see also Prob. 4-42), denoted by ζ (Greek zeta), as

$$\frac{R}{2\sqrt{L/C}} = \frac{\alpha}{\omega_0} = \zeta \qquad (4\text{-}68)$$

Thus the circuit is

Overdamped	for $\zeta > 1$
Critically damped	for $\zeta = 1$
Underdamped	for $\zeta < 1$

In some applications it is more convenient to define a number which *increases* when the damping ratio decreases. For this purpose one defines a number called the Q ("the cue") of the circuit. In the case of the series R-L-C circuit, one defines

$$Q_0 = \frac{\sqrt{L/C}}{R} = \frac{\omega_0 L}{R} = \frac{1}{2\zeta} \qquad (4\text{-}69)$$

and therefore the three cases of damping are related to the Q by

$Q_0 > \frac{1}{2}$	underdamped
$Q_0 = \frac{1}{2}$	critically damped
$Q_0 < \frac{1}{2}$	overdamped

The symbol Q_0 is called the quality factor of the circuit and should not be confused with the symbol for charge (or initial charge). The Q of a circuit is a basic quantity whose value depends on the circuit elements only. The relationship given in Eq. (4-69) holds only for the series R-L-C circuit. A Q is also defined for other simple R-L-C circuits; the expressions for other such cases will be deduced as these cases are discussed.

4-12 The underdamped (oscillatory) case

When the Q of the circuit is more than $\frac{1}{2}$, then $\omega_0 > \alpha$ and the characteristic roots are the conjugate complex values

$$\mathbf{s}_{1,2} = -\alpha \pm \sqrt{(-1)(\omega_0^2 - \alpha^2)}$$

In this case it is convenient to define

$$\omega_d = +\sqrt{\omega_0^2 - \alpha^2}$$

so that the roots can be written

$$\begin{aligned} \mathbf{s}_1 &= -\alpha + j\omega_d \quad \text{and} \quad \mathbf{s}_2 = -\alpha - j\omega_d \\ \mathbf{s}_1 - \mathbf{s}_2 &= j2\omega_d \quad \text{and} \quad \mathbf{s}_1\mathbf{s}_2 = \alpha^2 + \omega_d^2 = \omega_0^2 \end{aligned} \qquad (4\text{-}70)$$

The response of the circuit as given by the equation for $v_{ab}(t)$ has the form [see Eq. (4-67)]

$$v_{ab}(t) = \mathbf{K}_1 e^{\mathbf{s}_1 t} + \mathbf{K}_2 e^{\mathbf{s}_2 t} \tag{4-71}$$

where

$$\mathbf{K}_1 = \frac{I_0/C - \mathbf{s}_2 V_0}{\mathbf{s}_1 - \mathbf{s}_2} \qquad \mathbf{K}_2 = \frac{\mathbf{s}_1 V_0 - I_0/C}{\mathbf{s}_1 - \mathbf{s}_2} \tag{4-72}$$

Boldface type is used wherever the quantity *is or may be* complex. The form of Eq. (4-71) shows that $v_{ab}(t)$ is the sum of two complex terms; yet we know that $v_{ab}(t)$ is a real function of time, because it is the solution of a differential equation which represents a real physical problem. We shall now show that $v_{ab}(t)$ can be written in real form and that, *once this general form has been established*, we need not use the complex form (4-71). Since $\mathbf{s}_1$ and $\mathbf{s}_2$ are conjugate complex quantities, $\mathbf{K}_1$ and $\mathbf{K}_2$ are also conjugate complex quantities. This can be shown by substituting the expressions for $\mathbf{s}_1$ and $\mathbf{s}_2$ from Eqs. (4-70) into Eqs. (4-72).

$$\mathbf{K}_1 = \frac{I_0/C - (-\alpha - j\omega_d)V_0}{2j\omega_d} = \frac{I_0/C + \alpha V_0}{2j\omega_d} + \frac{j\omega_d V_0}{2j\omega_d} = \tfrac{1}{2}V_0 - j\,\frac{I_0/C + \alpha V_0}{2\omega_d} \tag{4-73a}$$

and

$$\mathbf{K}_2 = \frac{(-\alpha + j\omega_d)V_0 - I_0/C}{2j\omega_d} = \frac{j\omega_d V_0}{2j\omega_d} + \frac{-\alpha V_0 - I_0/C}{2j\omega_d} = \tfrac{1}{2}V_0 + j\,\frac{I_0/C + \alpha V_0}{2\omega_d} \tag{4-73b}$$

In Eqs. (4-73*a*) and (4-73*b*) we note that

$$\text{Re } \mathbf{K}_1 = \tfrac{1}{2}V_0 = \text{Re } \mathbf{K}_2 \qquad \text{Im } \mathbf{K}_1 = -\frac{I_0/C + \alpha V_0}{2\omega_d} = -\text{Im } \mathbf{K}_2 \tag{4-73c}$$

Hence $\mathbf{K}_1$ and $\mathbf{K}_2$ are conjugate complex numbers,

$$\mathbf{K}_1^* = \mathbf{K}_2$$

Since $\mathbf{K}_1$ and $\mathbf{K}_2$ are complex numbers, we can write them in polar or exponential form: Let

$$\mathbf{K}_1 = K_1 e^{j\gamma} = K_1\underline{/\gamma}$$

Then

$$\mathbf{K}_2 = \mathbf{K}_1^* = K_1 e^{-j\gamma} = K_1\underline{/-\gamma}$$

Noting that $\mathbf{s}_1$ and $\mathbf{s}_2$ are also conjugate complex numbers, Eq. (4-71) can be written as follows:

$$v_{ab}(t) = \mathbf{K}_1 e^{\mathbf{s}_1 t} + \mathbf{K}_1^* e^{\mathbf{s}_1^* t} \tag{4-71a}$$

Now

$$\mathbf{K}_1 e^{\mathbf{s}_1 t} = K_1 e^{-\alpha t} e^{j(\omega_d t + \gamma)}$$

and

$$\mathbf{K}_1^* e^{\mathbf{s}_1^* t} = K_1 e^{-\alpha t} e^{-j(\omega_d t + \gamma)} = (\mathbf{K}_1 e^{\mathbf{s}_1 t})^* = \mathbf{K}_2 e^{\mathbf{s}_2 t}$$

Hence Eq. (4-71*a*) can be written as the sum of the above complex conjugate terms. We now recall that, for any complex quantity **z**,

$$\mathbf{z} + \mathbf{z}^* = 2 \operatorname{Re} \mathbf{z}$$

Hence the general form of the solution in the underdamped case can be written

$$v_{ab}(t) = 2 \operatorname{Re} (\mathbf{K}_1 e^{\mathbf{s}_1 t})$$

or, letting $\mathbf{K} = 2\mathbf{K}_1$,

$$v_{ab}(t) = \operatorname{Re} (\mathbf{K} e^{\mathbf{s}_1 t}) \tag{4-74}$$

4-13 Different forms of the solution in the underdamped case; concept of complex frequency

In Eq. (4-74) the value of the complex number $\mathbf{K} = 2\mathbf{K}_1$ will depend on the initial conditions as well as the circuit elements. In Example 4-5, below, a detailed numerical example is presented. At any rate, **K** can be written in the exponential form $\mathbf{K} = Ke^{j\gamma}$, where K and γ will be evaluated from Eq. (4-73*c*). By convention, K is always a positive number, and the value of γ, depending on the initial conditions, may be anywhere between zero and 2π. In contrast $\mathbf{s}_1$ is a complex number whose value is determined by the circuit elements only and is independent of the initial conditions.

$$\mathbf{s}_1 = -\frac{R}{2L} + j\sqrt{\omega_0^2 - \alpha^2} = -\alpha + j\omega_d$$

or

$$\mathbf{s}_1 = \sqrt{\alpha^2 + \omega_d^2} \,\Big/ \tan^{-1} \frac{\omega_d}{-\alpha}$$

Since α and ω_d are both positive numbers, $s_1 = -\alpha + j\omega_d$ will be in the second quadrant of the complex plane. With reference to Fig. 4-16*a*, let

$$\delta = \tan^{-1} \frac{\omega_d}{-\alpha} \qquad \frac{\pi}{2} < \delta < \pi$$

From $\omega_0^2 = \alpha^2 + \omega_d^2$ and the above equations, we have the polar form

$$\mathbf{s}_1 = \omega_0 \underline{/\delta}$$

where ω_0 and δ are both independent of the initial conditions, depending only on the values of the circuit elements.

Using the exponential form of **K** in Eq. (4-74), we have

$$\begin{aligned} v_{ab}(t) &= \operatorname{Re} [Ke^{j\gamma} e^{(-\alpha + j\omega_d)t}] \\ &= \operatorname{Re} [Ke^{-\alpha t} e^{j(\omega_d t + \gamma)}] \\ &= Ke^{-\alpha t} \cos (\omega_d t + \gamma) \end{aligned} \tag{4-74a}$$

Expanding the cosine term in Eq. (4-74*a*), we have

$$v_{ab}(t) = e^{-\alpha t}(K_3 \cos \omega_d t + K_4 \sin \omega_d t) \tag{4-74b}$$

where $\quad K_3 = K \cos \gamma \qquad \text{and} \qquad K_4 = -K \sin \gamma$

Depending on the type of the problem and the required result, one or another of the above forms will be found to be more convenient.

We have already stated that the form of the response for the variables or for any of the derivatives of a circuit variable will be of the same form as indicated for $v_{ab}(t)$. Thus the waveshape of these variables and their derivatives will be the same (damped oscillatory), and the difference in their waveforms will be in the value of the constant K and the phase angle of the cosine term.

COMPLEX FREQUENCY The derivatives of circuit variables (v_{ab} and i_{ab}) are of particular interest in circuit analysis. When the mode of response of a circuit is underdamped, the most convenient form for differentiation is the complex form of Eq. (4-74), discussed in Appendix C and repeated here:

$$v_{ab}(t) = \text{Re}\,(\mathbf{K}e^{\mathbf{s}_1 t})$$

Differentiating with respect to time,

$$\frac{d}{dt}\,v_{ab}(t) = \frac{d}{dt}\,\text{Re}\,(\mathbf{K}e^{\mathbf{s}_1 t}) = \text{Re}\,(\mathbf{s}_1\mathbf{K}_1 e^{\mathbf{s}_1 t}) \tag{4-75}$$

Substituting the exponential forms of $\mathbf{s}_1$ and $\mathbf{K}$ in Eq. (4-75), we have

$$\begin{aligned} \frac{d}{dt}\,v_{ab}(t) &= \text{Re}\,[\omega_0 e^{j\delta} K e^{j\gamma} e^{(-\alpha+j\omega_d)t}] \\ &= \omega_0 K e^{-\alpha t} \cos\,(\omega_d t + \gamma + \delta) \end{aligned} \tag{4-76}$$

A comparison of Eqs. (4-74*a*) and (4-76) shows that the operation of differentiation on the former has resulted in multiplying Eq. (4-74*a*) by ω_0 and advancing the phase angle of its cosine term by δ.

If $v_{ab} = \text{Re}\,(\mathbf{K}e^{\mathbf{s}_1 t})$ is differentiated twice, we have

$$\begin{aligned} \frac{d^2}{dt^2}\,v_{ab}(t) &= \text{Re}\,(\mathbf{K}\mathbf{s}_1^2 e^{\mathbf{s}_1 t}) \\ &= \text{Re}\,[K e^{j\gamma} \omega_0^2 e^{j2\delta} e^{(-\alpha+j\omega_d)t}] \\ &= \omega_0^2 K e^{-\alpha t} \cos\,(\omega_d t + \gamma + 2\delta) \end{aligned}$$

Thus, to obtain the second derivative of $v_{ab}(t)$, we have only to multiply it by ω_0^2 and advance the phase of its cosine term by 2δ. Similarly, the nth derivative of a damped oscillatory waveform is obtained by multiplying the waveform by ω_0^n and advancing the phase of its oscillatory term by $n\delta$.

$$\frac{d^n}{dt^n}\,\text{Re}\,(\mathbf{K}e^{\mathbf{s}_1 t}) = \omega_0^n K e^{-\alpha t} \cos\,(\omega_d t + \gamma + n\delta)$$

From the study of sinusoidal functions (Appendix C) we recall that

$$\begin{aligned} \text{Re}\,e^{j\omega t} &= \cos\,\omega t \\ \frac{d}{dt}\cos\,\omega t &= \frac{d}{dt}\,\text{Re}\,e^{j\omega t} = \text{Re}\,(j\omega e^{j\omega t}) \end{aligned}$$

Since

$$j = e^{j\pi/2}$$

$$\frac{d}{dt}\cos\omega t = \text{Re}\,(\omega e^{j\pi/2}e^{j\omega t}) = \omega\cos\left(\omega t + \frac{\pi}{2}\right)$$

Thus, in differentiating $\cos\omega t$, its amplitude is multiplied by ω and its phase is advanced by $\pi/2$. Similarly,

$$\frac{d^2}{dt^2}\cos\omega t = \omega^2\cos\left(\omega t + 2\frac{\pi}{2}\right)$$

$$\frac{d^n}{dt^n}\cos\omega t = \omega^n\cos\left(\omega t + n\frac{\pi}{2}\right)$$

Thus $e^{j\omega t}$ is recognized to be a special case of $e^{\mathbf{s}_1 t}$, where the value of α is zero. Whereas ω is called the (real) radian frequency of the function $\cos\omega t$, the complex number $\mathbf{s}_1 = -\alpha + j\omega_d$ is called the complex radian frequency of the function $e^{\mathbf{s}_1 t}$. Complex frequencies are characteristic of damped oscillations and play an important part in the study of circuits. The real part of the complex frequency is the attenuation factor of the natural mode of a source-free circuit, and its imaginary part, ω_d, is the damped radian frequency of the oscillatory term of the mode.

It is noted that $\int e^{\mathbf{s}t}\,dt = (1/\mathbf{s})e^{\mathbf{s}t}$, and by analogy with differentiation,

$$\int e^{-\alpha t}\cos\omega_d t\,dt = \frac{1}{\omega_0}e^{-\alpha t}\cos(\omega_d t - \delta)$$

In words, integration of a mode with complex frequency results in division of its amplitude by ω_0 and delay of the phase of its oscillatory component of δ.

Example 4-5 In the circuit of Fig. 4-12*a* the switch is thrown from position d to position a' at $t = 0$. It is known that, at $t = 0^-$, $v_{ab}(0^-) = 12.5$ volts and $i(0^-) = -0.6$ amp. The numerical values of the elements are

$$R = 3 \text{ ohms} \qquad L = 0.5 \text{ henry} \qquad C = 0.08 \text{ farad}$$

Solve for $v_{ab}(t)$ and $i(t)$ for all $t \geq 0^+$.

Solution The continuity conditions require that

$$v_{ab}(0^+) = v_{ab}(0^-) = 12.5 \text{ volts}$$

and

$$i(0^+) = i(0^-) = C\left(\frac{dv_{ab}}{dt}\right)_{0^+} = -0.6 \text{ amp}$$

The equilibrium equation for the voltage v_{ab}

$$0.5\frac{d^2}{dt^2}v_{ab} + 3\frac{d}{dt}v_{ab} + \tfrac{100}{8}v_{ab} = 0$$

so that the characteristic equation is

$$0.5s^2 + 3s + 12.5 = 0$$

or

$$s^2 + 6s + 25 = 0$$

The characteristic roots are

$$\mathbf{s}_{1,2} = -3 \pm \sqrt{9 - 25}$$
$$\mathbf{s}_{1,2} = -3 \pm j4$$

The circuit is therefore underdamped. We immediately write the general form of the solution as

$$v_{ab}(t) = \text{Re}\,[\mathbf{K}e^{(-3+j4)t}]$$

and

$$\frac{dv_{ab}}{dt} = \text{Re}\,[(-3 + j4)\mathbf{K}e^{(-3+j4)t}]$$

We now introduce the initial conditions

$$v_{ab}(0^+) = 12.5 = \text{Re}\,\mathbf{K}$$
$$\left(\frac{dv_{ab}}{dt}\right)_{0^+} = \frac{i(0^+)}{C} = -\frac{0.6}{8 \times 10^{-2}} = \text{Re}\,[(-3 + j4)\mathbf{K}]$$

Now we write **K** in *rectangular* form:

$$\mathbf{K} = k + jk'$$

Then

$$12.5 = \text{Re}\,(k + jk') = k$$

Now

$$\mathbf{K} = 12.5 + jk'$$

Since

$$\text{Re}\,[(-3 + j4)(k + jk')] = -3k - 4k'$$

we have

$$\frac{-0.6}{8 \times 10^{-2}} = (-3 \times 12.5) - 4k'$$

Then

$$k' = -7.5$$

so that

$$\mathbf{K} = 12.5 - j7.5$$

Now we write **K** in *polar* form:

$$\mathbf{K} = 14.6\underline{/-31^\circ}$$

Hence

$$v_{ab}(t) = \text{Re}\,(14.6\underline{/-31^\circ}\,e^{-3t}e^{j4t})$$

or

$$v_{ab}(t) = 14.6e^{-3t}\cos\,(4t - 31^\circ) \qquad \textit{Ans.}$$

where $4t$ is in radians and 31 is in degrees.

The root $\mathbf{s}_1$ in polar form is

$$\mathbf{s}_1 = -3 + j4 = 5\underline{/126.9^\circ}$$

Hence

$$\frac{d}{dt}v_{ab} = 5 \times 14.6e^{-3t}\cos\,(4t - 31^\circ + 126.9^\circ)$$

or

$$\frac{d}{dt}v_{ab} = 73e^{-3t}\cos\,(4t + 95.9^\circ)$$

Since

$$i = C\frac{dv_{ab}}{dt} = 8 \times 10^{-2}\frac{dv_{ab}}{dt}$$

$$i(t) = 5.84e^{-3t}\cos\,(4t + 95.9^\circ) \qquad \textit{Ans.}$$

4·14 Sketching of waveforms in the underdamped case

The general forms of solution in the underdamped case are proportional to

$$f(t) = e^{-\alpha t} \cos (\omega_d t + \gamma) \tag{4-77}$$

We shall discuss in this section a reasonably accurate method for sketching $f(t)$ so that elaborate point-by-point computation of values can often be avoided. Such "reasonably accurate" sketches of functions can usually be obtained by locating the instants at which $f(t)$ is zero, has a maximum, or has a minimum. While the maxima and minima of Eq. (4-77) will be discussed, the form of the equation suggests a slightly different method. Equation (4-77) gives $f(t)$ as the *product* of two functions, both of which were discussed in detail in earlier chapters,

$$f(t) = f_1(t)f_2(t)$$

where

$$f_1(t) = e^{-\alpha t}$$
$$f_2(t) = \cos (\omega_d t + \gamma)$$

These functions are sketched in Figs. 4-13*a* and *b*, respectively. (In Fig. 4-13*b* it is assumed that $0 < \gamma < \pi/2$.) We note now that the following points on the product $f(t)$ are evident:

When $\cos (\omega_d t + \gamma) = 0$, then $f(t) = 0$. These instants are denoted by $t = t_z$ in Fig. 4-13.

When $\cos (\omega_d t + \gamma) = \pm 1$, then $f(t) = \pm e^{-\alpha t}$. These instants are denoted by $t = t_e$ in Fig. 4-13.

To sketch the product $f(t)$, the instants $t = t_z$ and $t = t_e$ are located on the t axis, and the curves $+e^{-\alpha t}$ and $-e^{-\alpha t}$ are sketched so that both positive and negative values at $t = t_e$ are readily located. [The curves $+e^{-\alpha t}$ and $-e^{-\alpha t}$ are called the envelopes of $f(t)$.] The curve is now sketched in through the t_e and t_z points. Before discussing the location of maxima and minima of the curve, the location of the values of t_e and t_z will be discussed in the following example.

Example 4-6 Find the values of t_z and t_e for $f(t) = e^{-\alpha t} \cos (\omega_d t + \gamma)$ when $\alpha = 3$, $\omega_d = 4$, and $\gamma = \pi/3$.

Solution By definition of t_z,

$$\cos \left(4t_z + \frac{\pi}{3}\right) = 0$$

Hence

$$4t_z + \frac{\pi}{3} = \frac{\pi}{2}, \frac{3\pi}{2}, \ldots, (2n - 1)\frac{\pi}{2}$$

so that the first zero occurs at

$$4t_{z_1} = \frac{\pi}{2} - \frac{\pi}{3} = \frac{\pi}{6} \qquad t_{z_1} = \frac{\pi}{24}$$

The second zero occurs at

$$4t_{z_2} = \frac{3\pi}{2} - \frac{\pi}{3} = \frac{7\pi}{6} \qquad t_{z_2} = \frac{7\pi}{24}$$

so that the zeros are separated by $\pi/\omega_d = \pi/4$,

$$t_{z_{n+1}} - t_{z_n} = \frac{\pi}{\omega_d} = \frac{\pi}{4}$$

All the t_z points can therefore be found if any one t_z point is located. The others are then found by spacing them at the t axis π/ω_d units from each other.

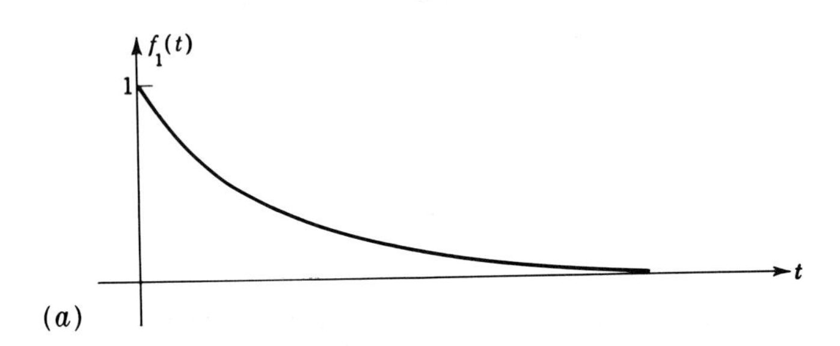

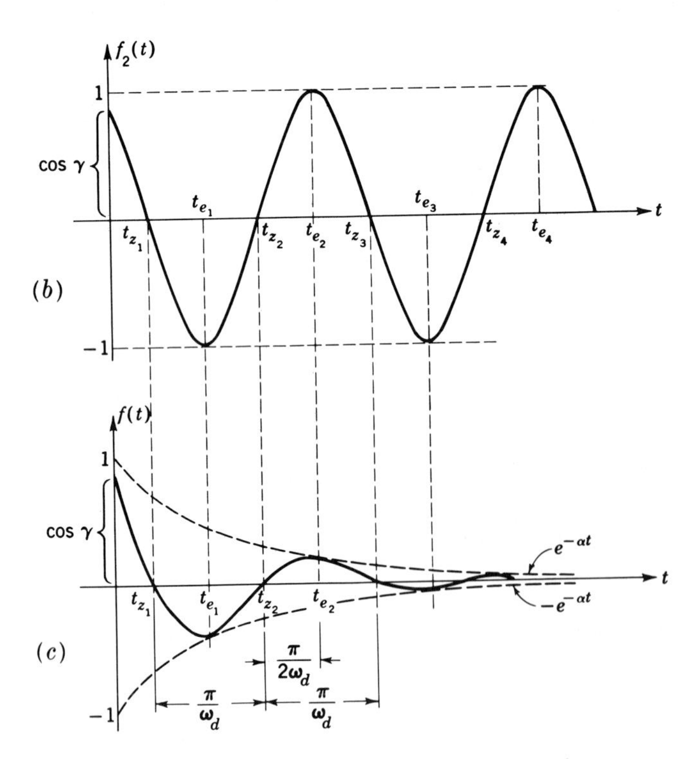

Fig. 4-13 Construction of a waveform in the underdamped case.

The values of t_e are defined by

$$\cos\left(4t_e + \frac{\pi}{3}\right) = \pm 1$$

Hence

$$4t_e + \frac{\pi}{3} = \pi,\ 2\pi,\ 3\pi,\ \ldots,\ m\pi$$

so that the first value of t_e is given by $4t_{e_1} = \pi - \pi/3$; $t_{e_1} = \pi/6$, and

$$t_{e_2} - t_{e_1} = \frac{\pi}{\omega_d} = \frac{\pi}{4}$$

$$t_{e_{n+1}} - t_{e_n} = \frac{\pi}{\omega_d} = \frac{\pi}{4}$$

Note also that a value of t_e is separated from an adjacent value of t_z by $\pi/2\omega_d$. Hence the location of any t_z point or any t_e point, together with the number $\pi/2\omega_d$, is sufficient for locating all these points. Figure 4-13 illustrates the example.

Although the location of the maxima and minima of the damped oscillations is not, in general, necessary for the sketching of the waveforms, these values are of interest and will therefore be examined. There is a tendency to think that the maxima and minima of $f(t)$ will occur at the position of the maxima and minima of the cosine function. This, of course, is not true, since the maxima of a product $f_1(t)f_2(t)$ are not generally at the same points as the maxima of either factor. The maxima and minima of $f(t)$ occur at the zeros of df/dt. We have seen that, if

$$f(t) = e^{-\alpha t}\cos(\omega_d t + \gamma)$$

then

$$\frac{df}{dt} = \omega_0 e^{-\alpha t}\cos(\omega_d t + \gamma + \delta)$$

where ω_0 and δ are connected with the characteristic roots through the relationship

$$-\alpha + j\omega_d = \omega_0\underline{/\delta}$$

The maxima and minima therefore occur when $\cos(\omega_d t + \gamma + \delta) = 0$.

4-15 The overdamped case

If the characteristic roots are real and unequal, the response of the source-free circuit is given as the sum of two real exponential functions as shown by Eq. (4-59).

An example suffices to illustrate the technique of solution.

Example 4-7 In the series *R-L-C* circuit of Fig. 4-12*b*,

$$R = 8 \text{ ohms} \qquad L = 0.5 \text{ henry} \qquad C = 8 \times 10^{-2} \text{ farad}$$

It is known that $v_{ab}(0^+) = 12.5$ volts and $i(0^+) = -0.6$ amp. Find $v_{ab}(t)$ and $i(t)$ for all $t \geq 0^+$.

Solution The characteristic equation is

$$0.5s^2 + 8s + \frac{10^2}{8} = 0$$

or

$$s^2 + 16s + 25 = 0$$

so that

$$s_{1,2} = -8 \pm \sqrt{64 - 25}$$

and

$$s_1 = -1.75 \qquad s_2 = -14.25$$

Since the roots are real and unequal, we write

$$v_{ab}(t) = K_1 e^{-1.75t} + K_2 e^{-14.25t}$$

$$\frac{dv_{ab}}{dt} = -1.75K_1 e^{-1.75t} - 14.25K_2 e^{-14.25t}$$

The constants K_1 and K_2 are related to the initial conditions through the equations

$$v_{ab}(0^+) = K_1 + K_2$$

$$\left(\frac{dv_{ab}}{dt}\right)_{0^+} = -1.75K_1 - 14.25K_2$$

$v_{ab}(0^+)$ is given; $(dv_{ab}/dt)_{0^+}$ is found from $i(0^+)$ by noting that

$$i = C\frac{dv_{ab}}{dt}$$

and hence

$$i(0^+) = C\left(\frac{dv_{ab}}{dt}\right)_{0^+}$$

We therefore write

$$12.5 = K_1 + K_2$$

$$\frac{-0.6}{8 \times 10^{-2}} = -1.75K_1 - 14.25K_2$$

Solving these two simultaneous equations, we have

$$K_1 = 13.65 \qquad \text{and} \qquad K_2 = -1.15$$

The solution for $v_{ab}(t)$ is

$$v_{ab}(t) = 13.65e^{-1.75t} - 1.15e^{-14.25t}$$

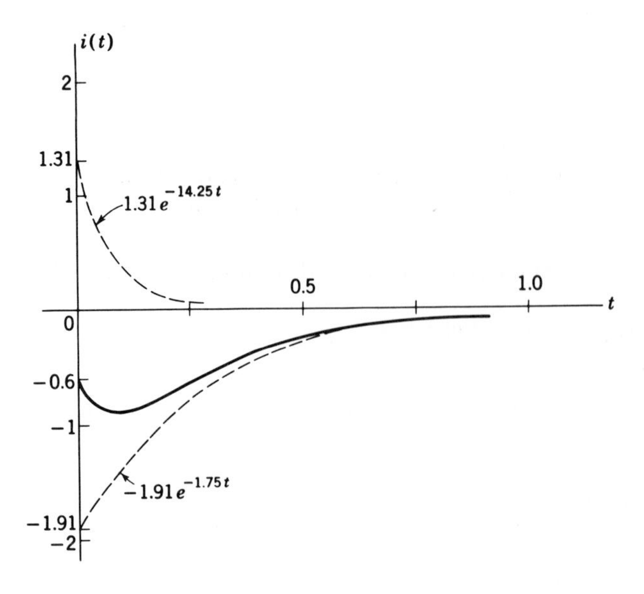

Fig. 4-14 Response waveform in the overdamped circuit of Example 4-7.

The equation for the current is obtained from the relationship $i = C(dv_{ab}/dt)$,

$$i(t) = 8 \times 10^{-2}[-1.75(13.65e^{-1.75t}) + 14.25(1.15e^{-14.25t})]$$

or

$$i(t) = -1.91e^{-1.75t} + 1.31e^{-14.25t}$$

The reader should sketch the above waveforms for v_{ab}. The graph of $i(t)$ is shown in Fig. 4-14.

4-16 The critically damped case (multiple roots)

It is pointed out in Sec. 4-10 that the case

$$s_1 = s_2 = -\alpha = -\omega_0$$

is a special case because the exponential function gives only one independent mode of the circuit. This case was explicitly excluded from the "general" solution (4-66).

To obtain the complete solution in this case, several procedures may be used. One may, for instance, start with the solution (4-66) and take the limit as s_1 approaches s_2. It is also possible to start with the oscillatory case and take the limit as ω_d approaches zero. The last-named method is used here. The solution for the underdamped case is, for our present purpose, most convenient in the form shown below. This form is obtained by expanding Eq. (4-74a) and using the values for γ and K in terms of the initial conditions.

$$v_{ab}(t) = e^{-\alpha t}\left[V_0 \cos \omega_d t + \left(\alpha V_0 + \frac{I_0}{C}\right)\frac{\sin \omega_d t}{\omega_d}\right]$$

We desire to obtain

$$\lim_{\omega_d \to 0} v_{ab}(t) = \lim_{\omega_d \to 0}\left\{e^{-\alpha t}\left[V_0 \cos \omega_d t + \left(\alpha V_0 + \frac{I_0}{C}\right)\frac{\sin \omega_d t}{\omega_d}\right]\right\}$$

Now

$$\lim_{\omega_d \to 0} \cos \omega_d t = 1$$

and

$$\lim_{\omega_d \to 0} \frac{\sin \omega_d t}{\omega_d} = \lim_{\omega_d \to 0}\left(t\,\frac{\sin \omega_d t}{\omega_d t}\right) = \lim_{x \to 0}\left(t\,\frac{\sin x}{x}\right) = t$$

so that

$$\lim_{\omega_d \to 0} v_{ab}(t) = e^{-\alpha t}\left[V_0 + \left(\alpha V_0 + \frac{I_0}{C}\right)t\right]$$

Since $\alpha = \omega_0$, the solution in the critically damped case reads

$$v_{ab}(t) = e^{-\omega_0 t}\left[V_0 + \left(\omega_0 V_0 + \frac{I_0}{C}\right)t\right] \tag{4-78}$$

It is, of course, not necessary to remember solution (4-78). If, in a numerical problem, the solution of the characteristic equation leads to the critically damped case, the general form of the solution is

$$v_{ab}(t) = (K_1 + K_2 t)e^{-\omega_0 t} \tag{4-79}$$

where the constants K_1 and K_2 are evaluated from the initial conditions.

WAVEFORM IN THE CRITICALLY DAMPED CASE Although the graph $(K_1 + K_2t)e^{-\omega_0 t}$ can be constructed as the product of the straight line $K_1 + K_2t$ with the exponential decay $e^{-\omega_0 t}$, it is helpful to do further analysis.

Given that

$$f(t) = (K_1 + K_2t)e^{-\omega_0 t}$$

we note that $f(t)$ is zero at $t = t_0$, which is given by

$$t_0 = -\frac{K_1}{K_2}$$

Next we calculate the derivative df/dt. The result is

$$\frac{df}{dt} = [(1 - \omega_0 t)K_2 - \omega_0 K_1]e^{-\omega_0 t}$$

Hence $f(t)$ has a maximum or a minimum at $t = t_M$, where

$$\begin{aligned} t_M &= -\frac{K_1}{K_2} + \frac{1}{\omega_0} \\ &= t_0 + \frac{1}{\omega_0} \end{aligned} \tag{4-80}$$

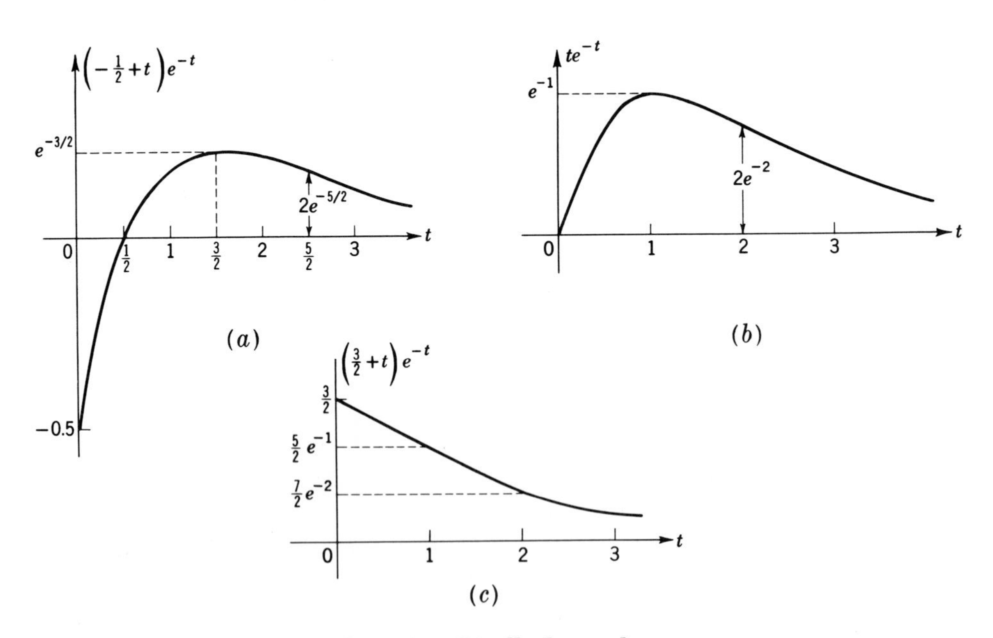

***Fig. 4-15** Response waveforms in critically damped circuits. (a) Zero crossing and maximum both occur for $t > 0$. (b) Zero crossing is at $t = 0$. (c) Zero crossing and maxima are to the left of $t = 0$. Note that this decay is slower than that of a pure exponential, i.e., $(\frac{3}{2} + t)e^{-t} > \frac{3}{2}e^{-t}$.*

Thus we conclude that the extremum is reached one time constant ($1/\omega_0$) after the graph crosses the zero axis. This result is useful in construction even if t_0 or t_M are negative values. Three examples are illustrated in Fig. 4-15.

4-17 Comparison of the three cases. The complex-frequency plane, or s plane

The underdamped, overdamped, and critically damped cases give different response waveshapes as determined by the nature of the characteristic roots. It is recalled that the roots of the characteristic equation are

$$s_{1,2} = -\frac{R}{2L} \pm \sqrt{\left(\frac{R}{2L}\right)^2 - \frac{1}{LC}}$$

or

$$s_{1,2} = -\alpha \pm \sqrt{\alpha^2 - \omega_0{}^2}$$

An interesting relationship exists between the roots of the characteristic equation. We note that, in any case, the geometric mean of the roots is the undamped natural radian frequency ω_0,

$$s_1 s_2 = (-\alpha + \sqrt{\alpha^2 - \omega_0{}^2})(-\alpha - \sqrt{\alpha^2 - \omega_0{}^2})$$

or

$$s_1 s_2 = \omega_0{}^2$$

Let us now study the series *R-L-C* circuit with *fixed values of L and C* but for all possible values of R. Since L and C are fixed, ω_0 remains constant. Since the roots of the characteristic equation are complex in the underdamped case, these roots may be identified with points in the complex plane. In Fig. 4-16*a* we have shown a set of rectangular coordinates; the real axis on that diagram corresponds to the real part of **s**, and the imaginary axis shows the imaginary part of **s**. The plane defined by these axes is then called the **s** plane. When the roots are complex, they can be written in polar form as

$$\mathbf{s}_{1,2} = \omega_0 \underline{/\pm\delta} = \omega_0 \underline{\bigg/ \pm \tan^{-1} \frac{\omega_d}{-\alpha}}$$

Hence the locus of the roots for all possible values of R which result in the underdamped case forms the semicircle with the center at the origin and radius ω_0. When the roots are real, they will be located on the negative real axis in the complex plane so that their geometric mean is ω_0. In the critically damped case the two roots are identical and will be designated as a "double" point in the **s** plane. The relationships are illustrated in Fig. 4-16*a*.

Now the study of these relationships leads to the following conclusions: When the resistance is zero, the circuit is undamped (*L-C*) and the roots are imaginary. In that case the response is undamped; i.e., it consists of periodic sinusoidal oscillations of radian frequency ω_0. Hence *imaginary* roots shown in Fig. 4-16*b* correspond to sinusoidal oscillations. On the other

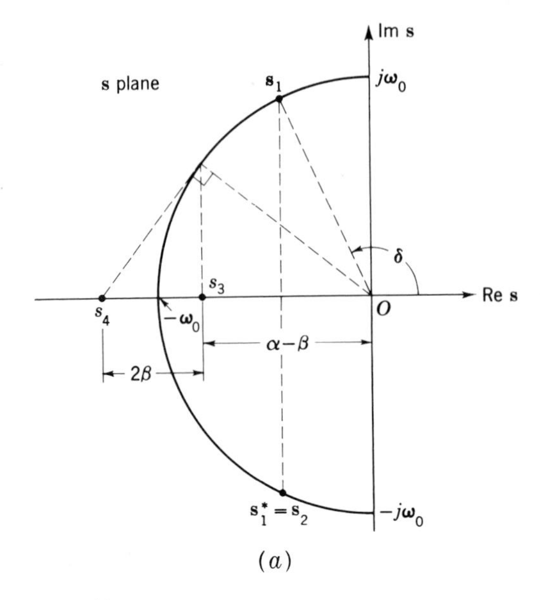

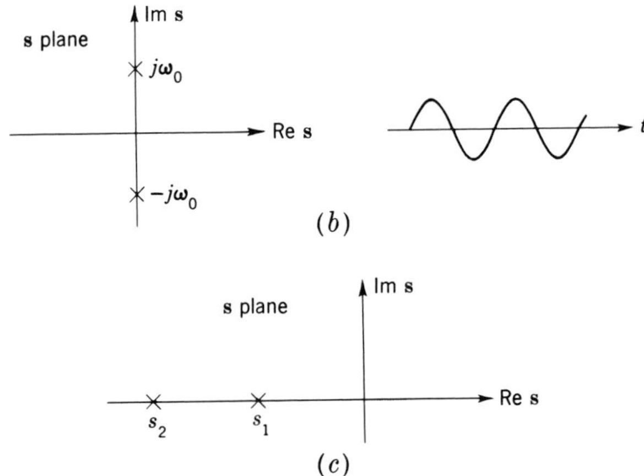

***Fig. 4-16** (a) Root locus in the s plane. Points s_1 and s_2 are examples of complex roots; s_3 and s_4 are a pair of real roots. The locus, consisting of the semicircle and the negative real axis, represents all possible characteristic roots that can be obtained by varying α with fixed ω_0. (b) A pair of conjugate imaginary roots corresponds to sinusoidal oscillations. (c) Illustration of real roots in the overdamped case.*

hand, real roots as shown in Fig. 4-16*c* correspond to decaying exponentials. We now note that

$$e^{-\alpha t} = e^{j(j\alpha)t}$$

corresponds to exponential damping. We therefore use the term imaginary frequency to describe exponential functions with real exponents. It then follows that the exponentially damped oscillations which occur in the underdamped case may be termed oscillations with complex frequency. It will be recalled that, in Sec. 4-13, we illustrated some properties of the complex frequency $\mathbf{s}$. The real part of the complex frequency ($\text{Re}\,\mathbf{s} = -\alpha$) describes the damping envelope of the response curve, and the imaginary part of the complex frequency ($\text{Im}\,\mathbf{s} = \omega_d$) describes the radian frequency of the oscillations. In Fig. 4-16*a* a multiple root, corresponding to the critically damped case, occurs where the semicircle intersects the real axis at $s = -\alpha = -\omega_0$. We shall further discuss the concept of the complex frequencies and its use in the study of circuits in Chaps. 7 and 8.

4-18 The parallel and other R-L-C circuits The techniques of solution and waveforms discussed above in connection with the source-free series R-L-C circuit can evidently be applied to the analysis of any system described by a second-order homogeneous differential equation. This section deals with such applications.

For the parallel R-L-C circuit shown in Fig. 4-17, the equilibrium equation in terms of v is

$$\left(\frac{1}{pL} + \frac{1}{R} + pC\right)v = 0$$

or

$$\left(p^2 + \frac{1}{RC}p + \frac{1}{LC}\right)v = 0 \tag{4-81}$$

By comparison of Eq. (4-81) with Eq. (4-47) we conclude that the analysis of the series

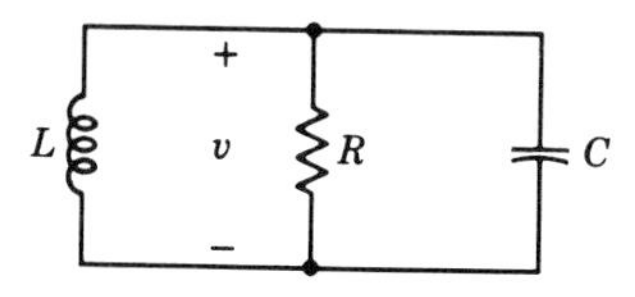

Fig. 4-17 Parallel R-L-C circuit.

circuit can be applied to the parallel circuit of Fig. 4-17 *provided* that

$$\alpha = \frac{1}{2RC}$$

is used in place of $\alpha = R/2L$. Thus, for the parallel circuit, the damping ratio is

$$\zeta_{\text{parallel}} = \frac{\sqrt{L/C}}{2R} = \frac{G}{2\sqrt{C/L}} = \frac{\alpha}{\omega_0} \tag{4-82}$$

and the quality factor Q_0 is given by

$$Q_0 = \frac{\omega_0 C}{G} = \frac{R}{\omega_0 L} \tag{4-83}$$

As a second example we observe that, for the circuit of Fig. 4-18, the equilibrium equation for the voltage v is given by

$$\left(\frac{1}{R_1 + pL} + pC + \frac{1}{R}\right) v = 0$$

or simplifying,

$$\left[p^2 + \left(\frac{1}{RC} + \frac{R_1}{L}\right) p + \frac{1 + R_1/R}{LC}\right] v = 0$$

Comparison with the above results indicates that, for this circuit, α and ω_0 are

$$\alpha = \frac{1}{2}\left(\frac{1}{RC} + \frac{R_1}{L}\right)$$

$$\omega_0 = \sqrt{\frac{1 + R_1/R}{LC}}$$

Further circuit examples are given as reader exercises (Probs. 4-58 to 4-60).

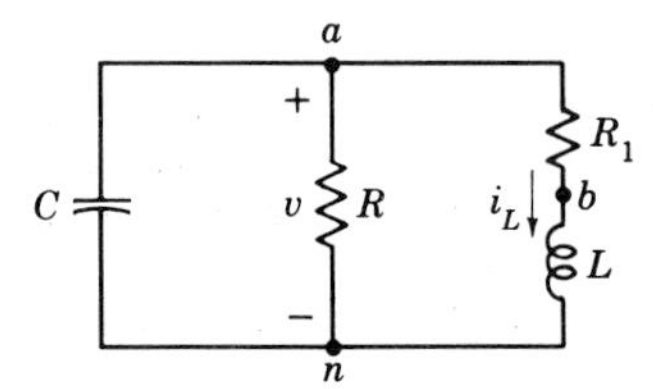

Fig. 4-18 Series-parallel source-free R-L-C circuit.

We conclude the discussion of source-free second-order systems by citing an electromechanical analogy.

4-19 Galvanometer dynamics

It is assumed that the reader is familiar with the principle of operation of moving coil meters. In Fig. 4-19*a* and *b* the essential parts of such a meter are shown. Assuming that the permanent magnet (Fig. 4-19*b*) provides a radial as well as constant magnetic flux density, a current-carrying conductor in each side of the movable coil will experience a force in accordance with Ampère's law (force = $Bi_c l$), and as a result a torque proportional to the current is generated.

$$\text{Torque on coil} = K_m i_c$$

where $K_m = Bnld$, and B is the magnetic flux density, n is the number of turns on the coil, l is the length of the coil frame, and d is the width of the coil frame. As a result of the electromagnetic torque, the coil will rotate around its axis of suspension so that the electromagnetic torque is in equilibrium with the total mechanical counter torque; that is,

Applied torque = (inertial torque) + (frictional torque) + (restoring torque due to spring)

$$K_m i_c = (Jp^2 + fp + k)\theta \qquad (4\text{-}84)$$

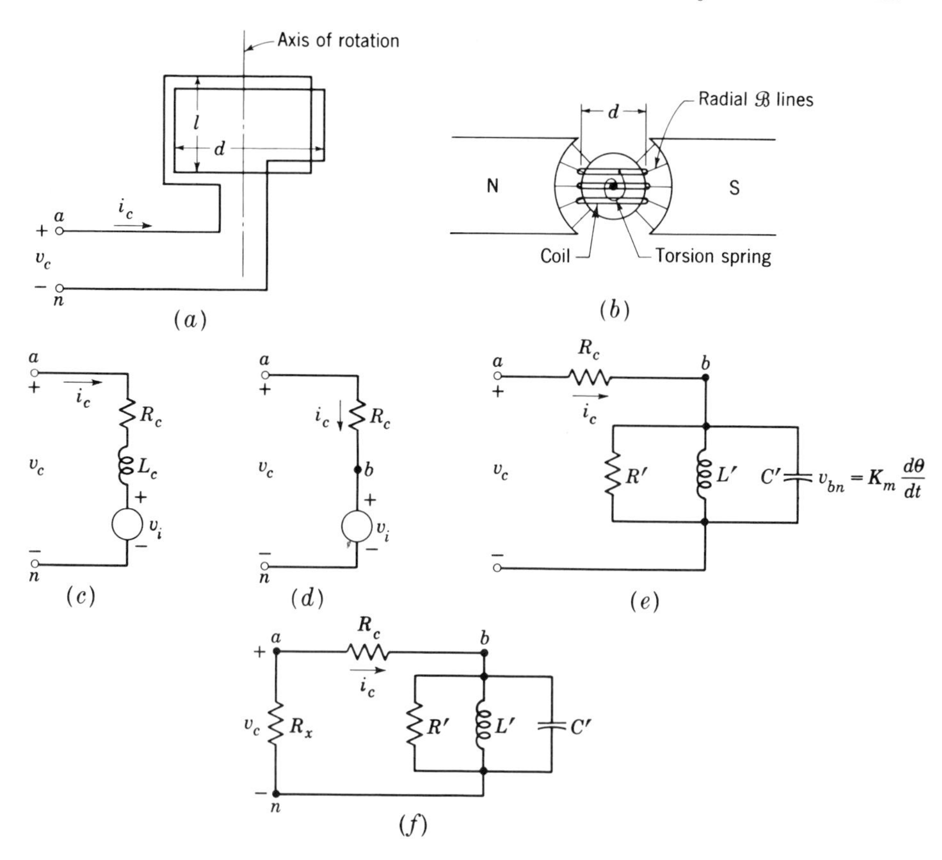

Fig. 4-19 Sketch of a galvanometer and equivalent circuits.

where J is the coil moment of inertia, f is the frictional torque per unit angular velocity, k is the spring elastance, and θ is the angle of rotation.

We observe that, from Eq. (4-84), if the current in the coil is constant, $i_c = I_c$, the steady-state solution for θ is the constant $K_m I_c/k$, so that the steady-state coil deflection is proportional to the coil current. Below, we analyze briefly the dynamics of coil motion.

Since the moving coil is in a magnetic field, there will be an induced voltage v_i (counter emf) in the coil, proportional to the coil velocity $p\theta$, in accordance with Faraday's law,

$$v_i = K_m p\theta$$

Thus, taking into account coil resistance R_c and inductance L_c, the terminal pair shown in Fig. 4-19c is electrically equivalent to the meter movement. If the inductance L_c is neglected (which is an allowable approximation), the terminal pair shown in Fig. 4-19d can be used to represent the meter electrically. From this circuit we observe that

$$v_c = R_c i_c + v_i$$
$$= R_c i_c + K_m p \theta \qquad (4\text{-}85)$$

To study the electromechanical interaction, we deduce the equilibrium equation relating v_c and i_c; that is, we derive the expression for the operational driving-point impedance $Z_c(p)$ of the terminal pair shown in Fig. 4-19d by eliminating θ from Eqs. (4-84) and (4-85). From Eq. (4-85),

$$\theta = \frac{v_c - R_c i_c}{K_m p}$$

Substituting this in Eq. (4-84) gives

$$K_m i_c = \frac{Ip^2 + fp + k}{K_m p}(v_c - R_c i_c)$$

or

$$[K_m{}^2 p + R_c(Jp^2 + fp + k)]i_c = (Jp^2 + fp + k)v_c$$

Since $v_c = Z_c i_c$,

$$Z_c(p) = \frac{R_c(Jp^2 + fp + k) + K_m{}^2 p}{Jp^2 + fp + k}$$
$$= R_c + \cfrac{1}{\cfrac{J}{K_m{}^2}p + \cfrac{f}{K_m{}^2} + \cfrac{k}{K_m{}^2 p}} \qquad (4\text{-}86)$$

The form of Eq. (4-86) suggests an equivalent circuit, namely, the series combination of R_c with Z_c', where

$$Z_c' = \cfrac{1}{\cfrac{Jp}{K_m{}^2} + \cfrac{f}{K_m{}^2} + \cfrac{k}{K_m{}^2 p}}$$

Hence the admittance Y_c' is given by

$$Y_c' = \frac{Jp}{K_m{}^2} + \frac{f}{K_m{}^2} + \frac{k}{K_m{}^2 p} \qquad (4\text{-}87)$$

From Eq. (4-87) we define an equivalent parallel R-L-C circuit,

$$Y_c' = pC' + \frac{1}{R'} + \frac{1}{pL'}$$

where $C' = J/K_m{}^2$, $R' = K_m{}^2/f$, $L' = K_m{}^2/k$. The equivalent circuit of the meter is shown in Fig. 4-19e.

From Fig. 4-19e we observe that, if the terminals of the meter are open-circuited, than R_c has no effect and the characteristic roots are given by $s_{1,2} = -\alpha \pm \sqrt{\alpha^2 - \omega_0{}^2}$, where $\alpha = 1/2R'C'$, $\omega_0{}^2 = 1/L'C' = k/J$. In the construction of ballistic galvanometers, the mechanical constants f, J, and k are generally chosen so that the circuit is highly oscillatory. The movement of the galvanometer will therefore also be highly oscillatory. This is seen from Eq. (4-85) as follows: The voltage $v_i = K_m p\theta$ is proportional to the angular velocity of the coil. *Hence v_{bn} in Fig. 4-19e is proportional to coil velocity.* It follows that the mechanical clamping can be controlled by electric elements; specifically, if a resistance R_x is placed across terminals a-n, then, for the electric circuits, $\omega_0^2 = 1/L'C'$, $\alpha = 1/2R''C'$, where

$$\frac{1}{R''} = \frac{1}{R_c + R_x} + \frac{f}{K_m^2} \tag{4-88}$$

If R_x can be adjusted for a desired damping, the free response v_{bn}, and therefore the motion of the galvanometer, will have the desired damping.

PROBLEMS

4-1 (*a*) For a source-free *R-C* circuit, sketch on the same set of axes $v(t)/v(0^+)$ and $w_C(t)/w_C(0^+)$ as a function of t/RC (w_C is the stored energy in the capacitance). (*b*) In a source-free *R-C* circuit, what fraction of the initial energy is stored in the capacitance when the voltage across it is (1) 50 per cent of its initial value; (2) 5 per cent of its initial value?

4-2 In a source-free *R-C* circuit, $R = 10$, $C = 2$. At $t = 20$ sec, the current in the resistance is 3 amp. Calculate the value of the energy stored in the capacitance at $t = 0$.

4-3 Show that a straight line tangent to the line $y = Ke^{-t/T}$ at $t = t_1$ will intersect the t axis at $t = t_1 + T$.

4-4 (*a*) In the circuit of Fig. P4-4, $v(0^+) = 100$. Find, without solving for $v(t)$, $i(t)$, or $i_1(t)$, the following initial values: (1) $i(0^+)$; (2) $i_1(0^+)$; (3) $(di/dt)_{0^+}$; (4) $(di_1/dt)_{0^+}$. (*b*) Sketch $v_{ab}(t)$ using $v(0^+) = 100$.

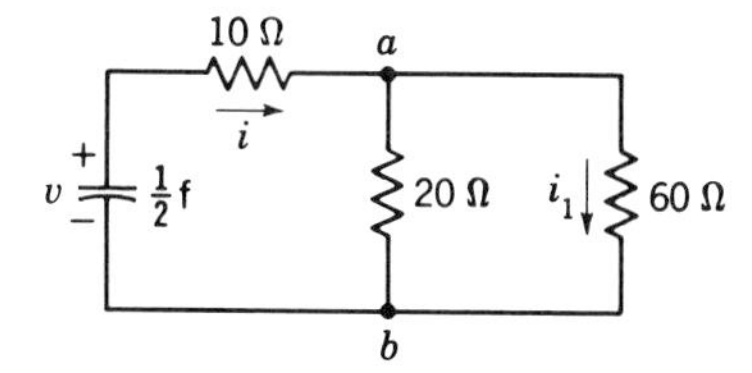

Fig. P4-4

4-5 In the circuit of Fig. P4-4, $i_1(0^+) = \frac{1}{3}$ amp. Calculate $v(0^+)$ and find $v(t)$ for all $t \geq 0^+$.

4-6 In the circuit of Fig. P4-6, the switch is in position 1 from $t = 0^-$ to $t = 1$; at $t = 1$ it is moved to position 2; it remains in position 2 for all $t > 1$. For $v(0^+) = 10$, perform all calculations necessary to sketch $v(t)$ and $i(t)$ for all $t \geq 0^+$ and make such sketches.

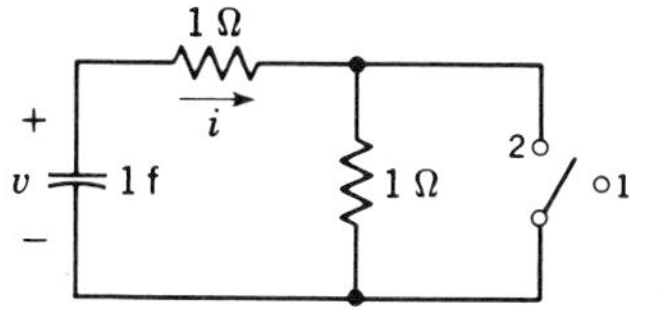

Fig. P4-6

4-7 In the circuit of Fig. P4-6, the switch is in position 2 from $t = 0$ until $t = 1$; for $t > 1$ it is in position 1. Given that $v(0^+) = 10$, perform all calculations necessary to sketch $v(t)$ and $i(t)$ for all $t > 0^+$ and make such sketches.

4-8 In the circuit of Fig. P4-8, $v(0^+) = 100$. The switch occupies the following positions in the time interval given: position 1 for $0 < t < 1$, position 2 for $1 < t < 2$, position 1 for $2 < t < 3$, and position 2 for $t > 3$. Sketch $v(t)$ and $i(t)$ in the interval $t = 0^-$ to $t = 5$.

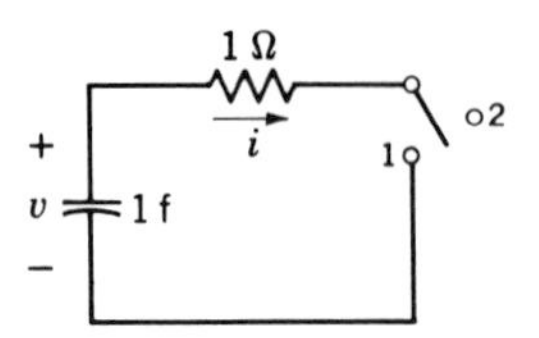

Fig. P4-8

4-9 Show that, if the resistance in a source-free *R-L* circuit [with $i(0) \neq 0$] is changed from R_1 to R_2 at $t = t_1$, the voltage across the inductance at t_1^+ will be related to its value at t_1^- by $v(t_1^+) = v(t_1^-)R_2/R_1$.

4-10 For a source-free *R-L* circuit [as described by Eq. (4-19)] find the initial value $(d^n i/dt^n)_{0^+}$ in terms of $T = L/R$ and I_0.

4-11 In the circuit of Fig. 4-4, the 3-farad capacitance is replaced by a $\frac{1}{3}$-henry inductance. It is known that $v_{m'm}(0^+) = 18$ volts. Find $v_{ab}(t)$ for all $t > 0^+$.

4-12 In the circuit of Fig. 4-5 the $\frac{1}{3}$-farad capacitance is replaced by a 3-henry inductance and $i_0(0^+) = 2$ amp. Find $i(t)$ for all $t \geq 0^+$.

4-13 In the circuit of Fig. P4-6, the 1-farad capacitance is replaced by a 2-henry inductance and $i(0^+) = 5$. Sketch to scale $i(t)$ and $v(t)$ for all $t \geq 0^+$ if (*a*) the switch is in position 1 from $t = 0^-$ to $t = 1$ and in position 2 for $t > 1$; (*b*) the switch is in position 2 from $t = 0$ to $t = 1$ and in position 1 for $t > 1$.

4-14 When a physical device represented by an *R-L* circuit is disconnected from a source at $t = t_1$ so that no closed path exists for $t > t_1$ as indicated in Fig. P4-14, an arc will occur across the switch. Explain.

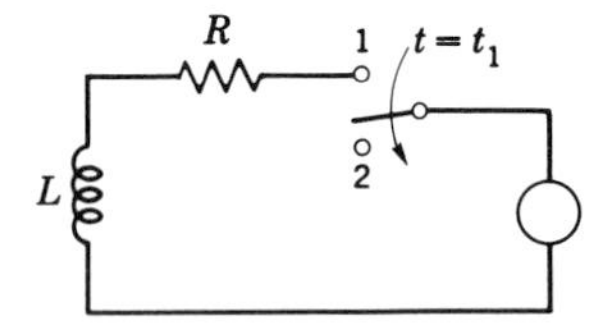

Fig. P4-14

4-15 (*a*) In the circuit of Fig. P4-15 the switch is thrown from position 1 to position 2 at $t = 0$. Calculate $v_{ab}(t)$ for all $t > 0^+$ if $i(0^+) = 2$ amp. (*b*) The 20,000-ohm resistance represents a voltmeter rated at 0–300 volts that was used to measure v_{ab} for $t < 0$. Comment on the suitability of the switching operation and suggest an alternative method for connecting the voltmeter.

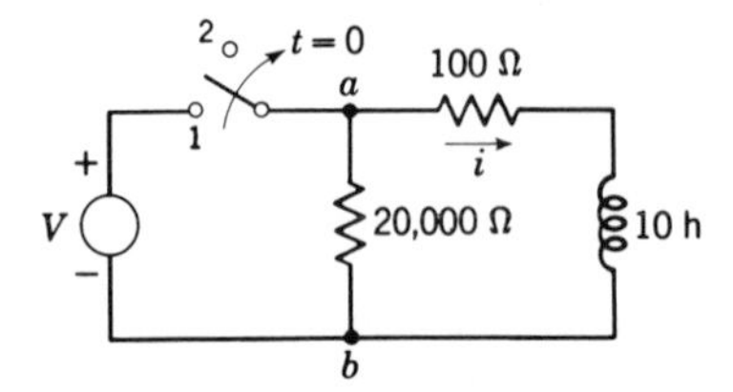

Fig. P4-15

4-16 In the circuit of Fig. P4-16 the switch is moved from position b to a' at $t = 0$. Find $v_{ba}(t)$ for all $t \geq 0^+$ and sketch the result if the element marked X is (*a*) a 2-farad capacitance with $v(0^+) = 20$ volts; (*b*) a 4-henry inductance with $i(0^+) = 2$ amp.

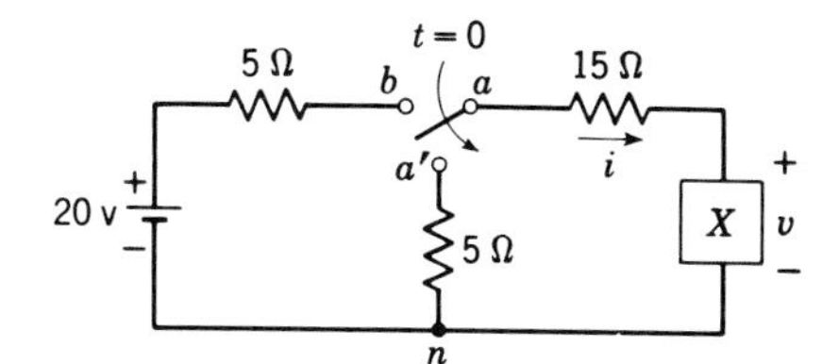

Fig. P4-16

4-17 (*a*) Noting that the equation in mechanics, relating force f and velocity u for a mass M, namely, $f = M\, du/dt$, is analogous to the equation relating voltage to current in an inductance, $v = L\, di/dt$, complete the following tabulation of electromechanical analogies:

Voltage—force	Inductance—mass
Current—velocity	Capacitance—
Flux linkages—	Resistance—
Charge—	

(*b*) The equation relating force to mass, $f = M\, du/dt$, is also analogous to the current-voltage relationship for capacitance, $i = C\, dv/dt$. Use this analogy to complete the following tabulation of electromechanical analogies:

Current—force	Inductance—
Voltage—velocity	Capacitance—mass
Charge—	Resistance—
Flux linkages—	

4-18 (*a*) Assuming that the plane on which the mass M slides in Fig. P4-18 is frictionless, write the differential equation of motion for the mass M (the spring is linear, with elastance K). (*b*) Show that the mechanical system of Fig. P4-18 is analogous to an L-C circuit excited by either a voltage source or a current source, and state analogous quantities for each case. (*c*) Show that the electrical analogue of a pendulum is, for small displacements, a linear L-C circuit.

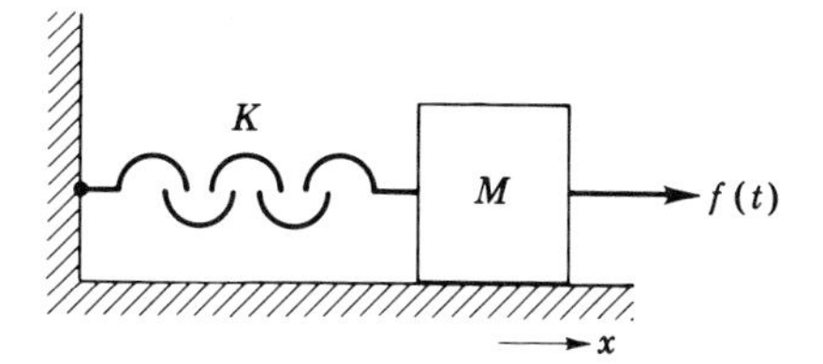

Fig. P4-18

4-19 A capacitance C is discharged through a resistance R beginning at $t = 0$. What is the decibel level of the voltage across the capacitance with respect to the initial value of voltage at (*a*) $t = T = RC$; (*b*) $t = \frac{3}{2}RC$; (*c*) $t = 2RC$; (*d*) $t = 3RC$?

4-20 (*a*) In a source-free L-C circuit the initial value of the energy stored in the inductance $w_L(0^+) = W_L$, and the initial value of the stored energy in the capacitance $w_C(0^+) = W_C$. Use the law of conservation of energy to show that the maximum value of the current is $i_{\max} = [2(W_L + W_C)/L]^{\frac{1}{2}}$. (*b*) In a source-free *L-C* circuit the initial value of the voltage across the capacitance is 100 volts, and the initial value of the current is zero. It is desired that $LC = 0.0004$ and that at no instant after $t = 0$ the current exceed 0.2 amp. Calculate the range of values which L and C can have. (*c*) In the circuit of (*b*), $i(0^+) = 0.5$, and the initial value of the voltage across the capacitance is zero. Determine the range of values which L and C can have so that at no instant after $t = 0$ the voltage across the capacitance exceeds 100 volts ($LC = 0.0004$).

4-21 In a source-free *L-C* circuit $L = 3$, $C = 0.03$. At $t = 0^+$, $i(0^+) = 0$ and $(di/dt)_{0^+} = 10$. At a certain instant of time the energy stored in the capacitance is 10 joules. Calculate the magnitude of the current at that instant of time.

4-22 From the defining equations for L and C, verify the dimensions of $1/LC$.

4-23 Sketch the quantities shown in Fig. 4-11 for the initial conditions $V_0 = 0$, $I_0 \neq 0$.

4-24 In the source-free *L-C* circuit shown in Fig 4-10, it is known that $C = 0.4$ farad and $v_{ma} = 10 \cos (5t - 30°)$. Calculate (*a*) $v_{ma}(0^+)$; (*b*) $i(0^+)$; (*c*) L.

4-25 In a certain source-free *L-C* circuit $\omega_0 = 2{,}000$, $L = 0.4$ henry. The initial value of the voltage across the capacitance is $v(0^+) = 12$ volts and $i(0^+) = C(dv/dt)_{0^+} = 0.02$ amp. (*a*) Find the maximum value of v and that of i. (*b*) Find the smallest positive value t_1 so that $v(t_1) = 0$. (*c*) Find the smallest positive value t_2 so that $|v(t_2)|$ has its maximum value at t_2. (*d*) Find the energy stored in each of the two elements at (1) $t = 0$; (2) $t = 0.001$ sec; (3) $t = 0.002$ sec; (4) $t = 0.004$ sec.

4-26 In the circuit of Fig. P4-26, $v(0^+) = 10$ volts. The switch is in position 1 from $t = 0$ to $t = 1$; it is in position 2 for $t > 1$. The current i_L is zero for $t < 1$. Find the sketch $v(t)$ for all $t \geq 0^+$.

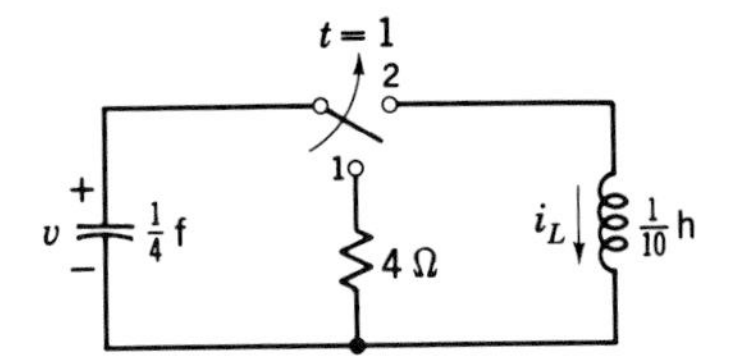

Fig. P4-26

4-27 In the circuit of Fig. P4-27, $v_{an}(0^+) = 0$ and $i(t) = 5tu(t)$. The switch is in position 1 from $t = 0$ to $t = 4$ [$v_{1n}(0^+) = 0$] and in position 2 for $t > 4$. Find and sketch $v_{an}(t)$ if the element marked X is (*a*) a resistance $R = 4$ ohms, (*b*) an inductance $L = \frac{1}{8}$ henry, storing zero energy at $t = 0$.

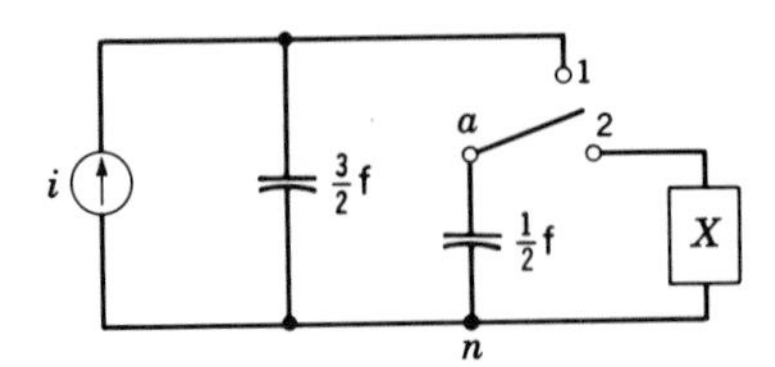

Fig. P4-27

4-28 Recall that the equilibrium equation for a variable y (such as v_C) in a source-free *R-L-C* circuit is given by $(p^2 + 2\alpha p + \omega_0^2)y = 0$. (*a*) Show that, if a new variable $x = py$ is defined, the equilibrium equation for x is $(p^2 + 2\alpha p + \omega_0^2)x = 0$. (*b*) Show that, if a new variable x is defined as a linear combination of y with N of its derivatives, $x = \sum_0^N a_n p^n y$, with the a_n's constant, then the equilibrium equation for x is the same as that given in (*a*).

4-29 In the circuit of Fig. P4-29 obtain the equilibrium equation for the variable (a) v_{mb}; (b) v_{mn}.

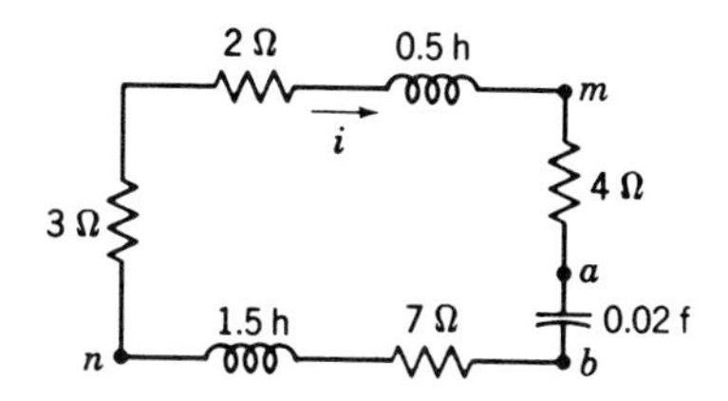

Fig. P4-29

4-30 In the circuit of Fig. P4-29, $i(0^+) = 3$ amp and $v_{ab}(0^+) = 5$ volts. Calculate the following initial values: (a) $v_{ma}(0^+)$; (b) $v_{mb}(0^+)$; (c) $(di/dt)_{0^+}$; (d) $v_{bn}(0^+)$; (e) $v_{mn}(0^+)$; (f) $(dv_{ab}/dt)_{0^+}$; (g) $(dv_{mb}/dt)_{0^+}$; (h) $(di^2/dt^2)_{0^+}$; (i) $(d^2v_{mn}/dt^2)_{0^+}$.

4-31 In the circuit of Fig. P4-29, it is known that, at the instant $t = t_1$, $i(t_1) = 4$ amp and $v_{mn}(t_1) = -10$ volts. Calculate (a) $(di/dt)_{t_1}$; (b) $v_{ab}(t_1)$; (c) $(dv_{mn}/dt)_{t_1}$.

4-32 (a) Determine the characteristic roots of the circuit shown in Fig. P4-29. (b) The value of the capacitance in the circuit of Fig. P4-29 is to be changed to a new value, C, so that the circuit is critically damped. Find C.

4-33 In the circuit shown in Fig. P4-33, $i(0^+) = 0$ and $v_{an}(0^+) = 100$ volts. Calculate $v_{an}(t)$ and $i(t)$ for all $t \geq 0^+$ if (a) $R = 0$; (b) $R = 0.4$ ohm; (c) $R = 2.4$ ohms; (d) $R = 4$ ohms; (e) $R = 5.66$ ohms; (f) $R = 10.4$ ohms.

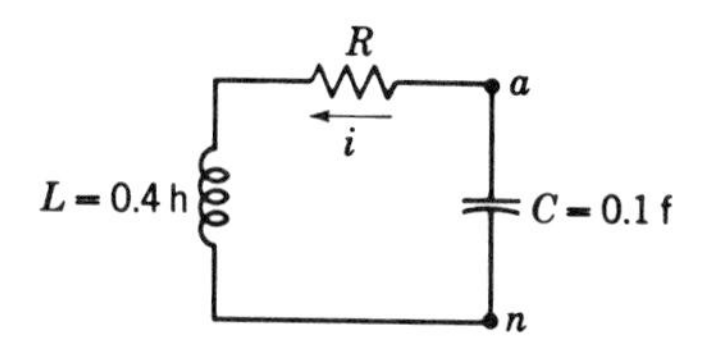

Fig. P4-33

4-34 In each of the cases given in Prob. 4-33, calculate the first maximum value of $i(t)$. In each case state the instant at which the maximum occurs.

4-35 Sketch the waveforms of $i(t)$ and $v_{an}(t)$ for the various cases given in Prob. 4-33.

4-36 For the circuit of Fig. P4-29 and the initial conditions given in Prob. 4-30, use the results of Prob. 4-30b and g to solve for $v_{mb}(t)$. Do not solve for $v_{ab}(t)$ or $i(t)$.

4-37 In the circuit of Fig. P4-33, change L to 0.2 henry and C to 0.05 μf and let $R =$ 400 ohms. For $v_{an}(0^+) = 100$ and $i(0^+) = 0$, find and sketch $v_{an}(t)$ for all $t \geq 0^+$.

4-38 (a) Show that the form $Kte^{-\omega_0 t}$, $\omega_0^2 = 1/LC$, is a solution of the source-free series R-L-C circuit if $R/2L = \omega_0$, by direct substitution of this form in the equilibrium equation. (b) Start with the general solution of the source-free series R-L-C circuit as given by Eq. (4-66), and show that Eq. (4-79) is the solution for the critically damped case by taking the limit as s_2 approaches s_1 in Eq. (4-66).

4-39 The characteristic roots of a certain R-L-C circuit are $s_{1,2} = -5 \pm j200$. What will be the characteristic roots if the value of both L and C are multiplied by 10^{-4}?

4-40 Sketch the following waveforms: (a) $f_a(t) = e^{-t} \cos 2\pi t$; (b) $f_b(t) = e^{-t} \cos (2\pi t - 120°)$; (c) $f_c(t) = e^{-2t} \cos (30t + 120°)$; (d) $f_d(t) = 10e^{-t} - 5e^{-3t}$; (e) $f_e(t) = 10e^{-t} - 20e^{-3t}$; (f) $f_f(t) = 10e^{-t} + 20e^{-3t}$; (g) $f_g(t) = (4 + 2t)e^{-t}$; (h) $f(t) = (4 - 2t)e^{-t}$.

4-41 In the circuit of Fig. P4-14, a capacitance C is placed across the terminals of the switch, terminals 1-2. With reference to your answer in Prob. 4-14, explain how this capacitance functions as an "arc suppressor."

4-42 (*a*) Using the definition of damping ratio, ζ, as given by Eq. (4-68), show that the solution for the voltage across the capacitance in a series *R-L-C* circuit for $t \geq 0^+$ and for $v(0^+) = V_0$, $(dv/dt)_{0^+} = \pm i(0^+)/C = 0$, is given, for the underdamped case, by

$$v_c(t) = \frac{V_0}{\sqrt{1-\zeta^2}} e^{-\omega_0 \zeta t} \cos\left(\omega_0 \sqrt{1-\zeta^2}\, t - \tan^{-1} \frac{\zeta}{\sqrt{1-\zeta^2}}\right)$$

(*b*) Show that the maxima and minima of v_c in (*a*) are independent of ω_0 and are given by

$$|v|_{\text{extreme}} = V_0 e^{-n\pi\zeta/\sqrt{1-\zeta^2}} \qquad n = 0, 1, 2, \ldots$$

4-43 (*a*) For a highly oscillatory ($\alpha \ll \omega_0$) source-free *R-L-C* circuit, show that (1) $\omega_d \approx \omega_0$ and (2) the value of $\mathbf{K}$ in the solution $v = \text{Re}(\mathbf{K}e^{st})$ can be found approximately by satisfying the initial conditions for $v = \text{Re}(\mathbf{K}e^{j\omega_0 t})$. (*b*) Illustrate this approximation in the circuit of Fig. P4-43, with the switch closed (position 1), using the initial conditions $v_{an}(0^+) = 10$, $i(0^+) = 0$.

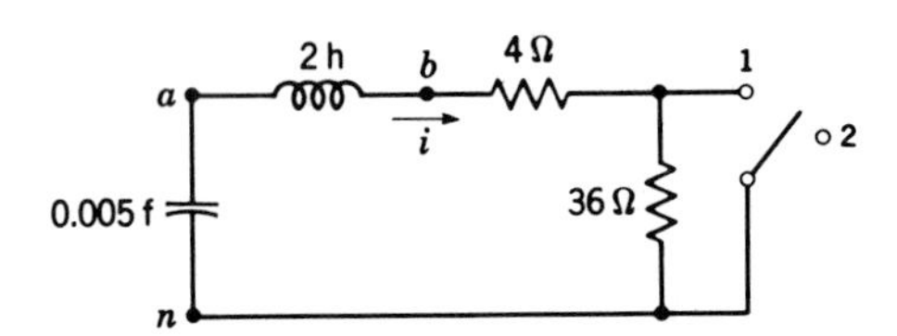

Fig. P4-43

4-44 In the circuit of Fig. P4-43, the switch is closed (position 1) from $t = 0^-$ to $t = 0.6$; it is open (position 2) for $t > 0.6$. The initial conditions are $v_{an}(0^+) = 10$, $i(0^+) = 0$. (*a*) Find and sketch $v_{an}(t)$ for all $t \geq 0^+$. (*b*) What are the values $v_{ab}(0.6^-)$ and $v_{ab}(0.6^+)$?

4-45 In the circuit of Fig. P4-43, the switch is open (position 2) from $t = 0^-$ to $t = 0.1$; it is closed (position 1) for $t > 0.1$. The initial conditions are $v_{an}(0^+) = 10$, $i(0^+) = 0$. (*a*) Find and sketch $v_{an}(t)$ for all $t \geq 0^+$. (*b*) What are the values $v_{ab}(0.1^-)$ and $v_{ab}(0.1^+)$?

4-46 In the circuit of Fig. P4-46, $v_{ab}(0^+) = 50$ and $i_L = 0$ for $t < \frac{1}{4}$. The switch is closed (position 1) from $t = 0$ to $t = \frac{1}{4}$; it is open (position 2) for $t > \frac{1}{4}$. (*a*) Find $v_{ab}(t)$ and $i(t)$ in the interval $0^+ < t < \frac{1}{4}$. (*b*) Find $v_{ab}(\frac{1}{4}^-)$, $v_{ab}(\frac{1}{4}^+)$, $i(\frac{1}{4}^-)$, and $i(\frac{1}{4}^+)$. (*c*) Find $v_{mb}(\frac{1}{4}^+)$. (*d*) Find $i(t)$ for $t > \frac{1}{4}$ and sketch $i(t)$ in the interval $t = 0$ to $t = 1$.

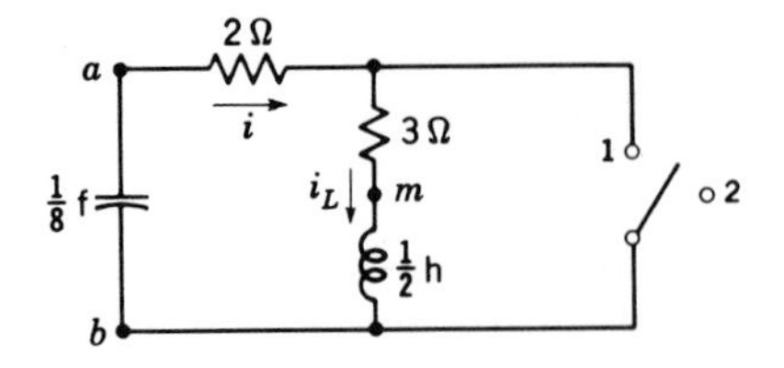

Fig. P4-46

4-47 In the circuit of Fig. P4-46, the switch is open for $t < t_1$ and closed for $t > t_1$. It is known that $v_{ab}(t_1) = 10$ volts and $i(t_1^-) = 2$ amp. (*a*) Find $i(t_1^+)$ and $i_L(t_1^+)$. (*b*) Find $i(t)$ and $i_L(t)$ for $t > t_1$.

4-48 When the resistance in a source-free series *R-L-C* circuit is 1,000 ohms, a response has the form $K_1e^{-3{,}730t} + K_2e^{-270t}$. (*a*) Calculate ω_0. (*b*) The values of L and

C remain fixed, and R is changed so that the circuit is critically damped. Find the new value of R. (c) The values of L and C remain fixed, and R is changed so that the circuit is underdamped and two cycles of transient oscillations occur in a time equal to the envelope time constant. Find R.

4-49 For a certain source-free R-L-C circuit, $\omega_0 = 100$ and $\alpha = 60$. Find Q_0 and ζ.

4-50 In a source-free R-L-C circuit the characteristic roots are $s_{1,2} = -3 \pm j6$, when $R = 5$. Calculate the value to which R should be changed if the circuit is to be (a) critically damped; (b) overdamped with one characteristic root $s_1 = -3$ if the circuit is (1) a series circuit, (2) a parallel R-L-C circuit.

4-51 In the circuit shown in Fig. P4-51, the switch is in position 1. Obtain the equilibrium equation with (a) i_L as the variable; (b) v_{an} as the variable.

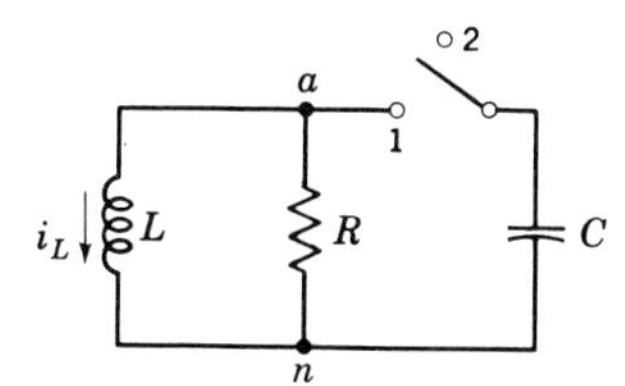

Fig. P4-51

4-52 In the circuit of Fig. P4-51, the switch is in position 1 and $i_L(0^+) = I_0$ and $v_{an}(0^+) = V_0$. Express in terms of R, L, C, I_0, and V_0 the initial values (a) $(dv_{an}/dt)_{0^+}$; (b) $(d^2v_{an}/dt^2)_{0^+}$.

4-53 In the circuit of Fig. P4-51, the switch is in position 1 and $L = 1$, $C = 1$, and $R = 0.5$. Find and sketch $v_{an}(t)$ for all $t \geq 0^+$ if the initial conditions are (a) $v_{an}(0^+) = 10$, $i_L(0^+) = 0$; (b) $v_{an}(0^+) = 0$, $i_L(0^+) = 2$; (c) $v_{an}(0^+) = 10$, $i_L(0^+) = 2$.

4-54 In the circuit of Fig. P4-51, the switch is open (position 2) for $t < 1$ and closed (position 1) for $t > 1$. (a) If $v_{2n} = V_0$ for $t \leq 1^+$, express (1) $(di_L/dt)_{t=1^+}$ in terms of V_0; (2) $(dv_{an}/dt)_{t=1^+}$ in terms of V_0 and $i_L(1^-)$. (b) Solve for $v_{an}(t)$ for $t > 0^+$ if $i_L(0^+) = 3$ and $v_{an}(t \leq 1^+)$ is equal to (1) zero; (2) 2 volts. Let $R = \frac{1}{2}$, $L = 1$, $C = 1$.

4-55 In the circuit of Fig. P4-51, the switch is closed (position 1) for $t < 1$ and open (position 2) for $t > 1$. For $v_{an}(0^+) = 10$ volts and $i_L(0^+) = 7.5$ amp find and sketch for all $t \geq 0^+$: (a) $v_{2n}(t)$; (b) $v_{an}(t)$. Let $R = \frac{1}{2}$, $L = 1$, $C = 1$.

4-56 Three elements R, L, and C are connected in series as shown in Fig. P4-56a; the same elements, connected in parallel, are shown in Fig. P4-56b. In both circuits $R = 2.50$ ohms, $v_c(0^+) = 0$, and $i_L(0^+) = 1$ amp. Calculate $v_c(t)$ for each circuit for all $t \geq 0^+$.

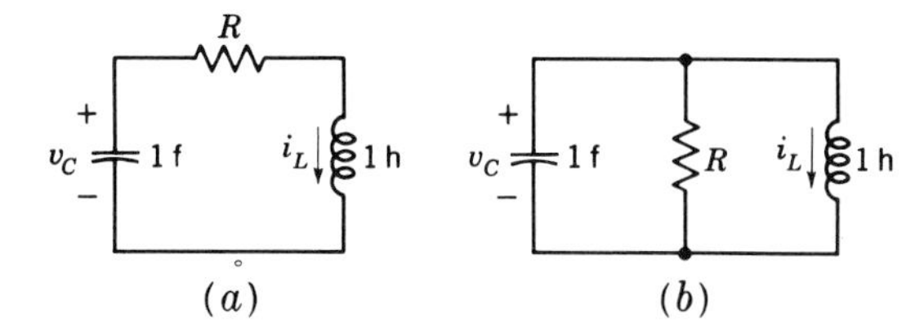

Fig. P4-56

4-57 Given the three elements R, $L = 1$ henry, $C = 1$ farad, these elements are either connected in series as shown in Fig. P4-56*a* or in parallel as shown in Fig. P4-56*b*. (*a*) Calculate the value of R for which both circuits have the same characteristic roots. (*b*) There exists a range of values for R, $R_{\max} > R > R_{\min}$, so that both circuits have the same type of response (overdamped or underdamped). Calculate $R_{\max}$ and $R_{\min}$.

4-58 In a critically damped source-free parallel R-L-C circuit, the initial energy stored in the inductance is zero. If $L = 1$ and $C = 1$ and if the energy stored in the capacitance at $t = 2$ is 2 joules, calculate (*a*) the energy stored in the capacitance at $t = 0$; (*b*) the energy stored in the inductance at $t = 2$.

4-59 In the circuit of Fig. 4-18, (*a*) express $v_{bn}(0^+)$ and $(dv_{bn}/dt)_{0^+}$ in terms of $v(0^+) = V_0$ and $i_L(0^+) = I_0$. (*b*) Find the relationship between R, R_1, L, and C for critical damping in the circuit. (*c*) Solve for $v_{bn}(t)$ for all $t \geq 0^+$ if $V_0 = 10$, $I_0 = 1$, and all elements have unit value.

4-60 For the circuit shown in Fig. P4-60, $v_{ab}(0^+) = 10$ and $i_L(0^+) = 2$. (*a*) Find the following initial values: (1) $v_{bn}(0^+)$; (2) $v_{an}(0^+)$; (3) $(di_L/dt)_{0^+}$; (4) $(dv_{ab}/dt)_{0^+}$; (5) $i_1(0^+)$; (6) $(di_1/dt)_{0^+}$; (7) $(d^2v_{ab}/dt^2)_{0^+}$. (*b*) Find the equilibrium equation for v_{an}. (*c*) Find the characteristic roots of the circuit.

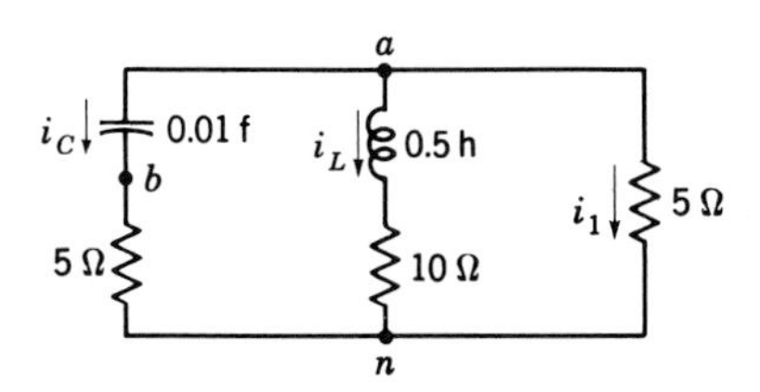

Fig. P4-60

4-61 In the circuit of Fig. P4-61, $v_{ab}(0^+) = 10$ and $v_{am}(0^+) = 10$. (*a*) Find the equilibrium equation for v_{ab}. (*b*) Find the characteristic roots. (*c*) Find $(dv_{ab}/dt)_{0^+}$. (*d*) Solve for $v_{ab}(t)$ for all $t \geq 0^+$.

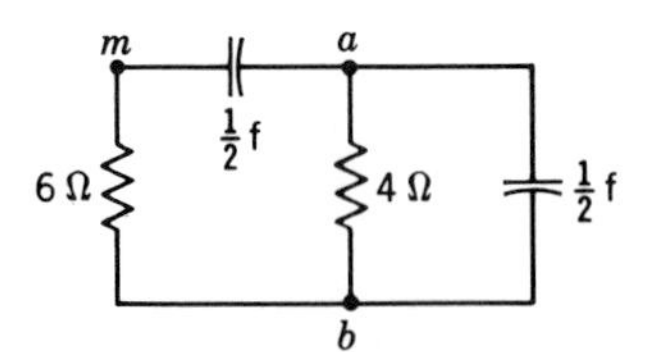

Fig. P4-61

4-62 The movable coil of a ballistic galvanometer that is not connected to a circuit (Fig. 4-19*e*) is given an initial velocity, and the motion observed is described by $\theta = 10e^{-t/24} \sin 0.25t$ deg. The coil resistance R_c is known to be 100 ohms. When a resistance $R_x = 9{,}900$ ohms is placed across the galvanometer terminals (Fig. 4-19*f*), it is observed that $\theta = Ke^{-t/12} \sin \omega_d t$, $\omega_d \approx 0.25$. (*a*) Show that the natural undamped radian frequency of the galvanometer is approximately $\omega_0 = 0.25$. (*b*) Show that $R' = 10{,}000$ ohms. (*c*) Find R_x so that the galvanometer circuit is critically damped.

Chapter 5 Complete response of simple circuits

In the preceding chapter circuits with initial energy storage but without other sources are analyzed. As pointed out in Chap. 3, the complete response of circuits with sources includes, as a component, the source-free response (complementary solution) in addition to the component of response produced by the source. In this chapter we discuss the response calculations of *R-C*, *R-L*, and *R-L-C* circuits with two goals in mind. First, we wish to continue development of the general theory of linear analysis, and second, we shall, for certain simple cases, develop results which are of extensive practical value.

5-1 Review and general discussion

The application of Kirchhoff's laws to a network results in differential equations that relate the network response functions to the source functions. We recall (from Chap. 3) that these equations are termed *equilibrium equations* and are often conveniently obtained by manipulating the differential relationships in operational form. We recall further that an equilibrium equation, in general, equates a certain linear combination, consisting of a response function and its derivatives, to a linear combination of the source functions with their derivatives. Thus, if a certain network response is denoted by v_2, the general form of its equilibrium equation is

$$\sum_{n=0}^{N} a_n p^n v_2 = f(t) \tag{5-1}$$

or

$$[D(p)]v_2 = f(t) \tag{5-1a}$$

where p^n is the nth derivative operator, $p^n = d^n/dt^n$, and $f(t)$ is a linear combination of the source function(s) with its (their) derivatives. The function $D(p)$ is a polynomial in p.

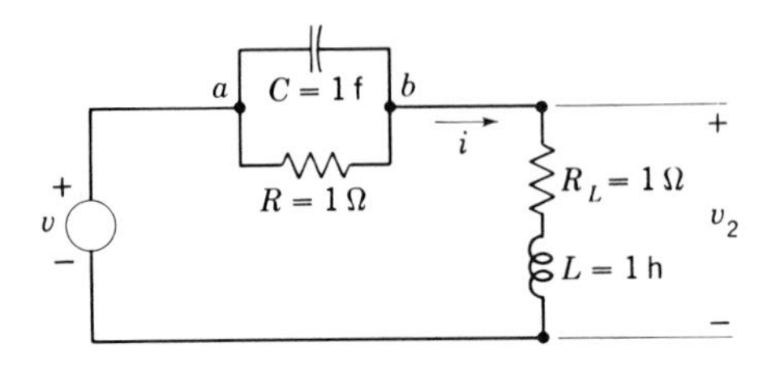

***Fig. 5-1** Circuit used to illustrate operational voltage division.*

In the circuit of Fig. 5-1, for example, we have, from operational voltage division,

$$v_2 = \frac{1 + p}{1 + p + 1/(1 + p)} v$$

$$= \frac{p^2 + 2p + 1}{p^2 + 2p + 2} v$$

or

$$(p^2 + 2p + 2)v_2 = (p^2 + 2p + 1)v \qquad (5\text{-}2)$$

Equation (5-2) is seen to be a special case of Eq. (5-1), where $a_n = 0$ for $n > 2$, $a_2 = 1$, $a_1 = 2$, $a_0 = 2$, and where the forcing function is given by

$$f(t) = (p^2 + 2p + 1)v = \frac{d^2v}{dt^2} + 2\frac{dv}{dt} + v$$

COMPLETE RESPONSE We have defined a *complete response* of a network as that particular solution of the equilibrium equation which satisfies the specified initial conditions. As mentioned before, since the equation

$$[D(p)]v_2 = f(t)$$

can be interpreted as

$$[D(p)]v_2 = f(t) + 0$$

we conclude from the superposition property (Sec. 3-13) that the solution of the source-free equation

$$[D(p)]v_2 = 0$$

will be a response component (termed "free component of the response"). Thus the complete response can be analyzed as the "(source-) free component of the response" and "component of the response due to the source," as discussed further below.

With reference to the example of Fig. 5-1 and its equilibrium equation (5-2), we observe that the function

$$v_2 = 5t + 5 + 14.1e^{-t} \cos (t - 45°) \qquad (5\text{-}3)$$

satisfies the equilibrium equation for $v(t) = 10t$ with the initial conditions $v_2(0^+) = 15$ and $(dv_2/dt)_{0^+} = 5$. The reader can verify the preceding statement by direct substitution. In Eq. (5-3) the term

$$(v_2)_f = 14.1e^{-t} \cos (t - 45°) = \text{Re}\,[(10 - j10)e^{(-1+j)t}]$$

is seen to be a solution of the source-free equation

$$(p^2 + 2p + 2)v_2 = 0$$

and the term

$$(v_2)_s = 5t + 5$$

satisfies the equilibrium equation

$$(p^2 + 2p + 2)(5t + 5) = (p^2 + 2p + 1)10t$$

5-2 *The source-free modes*

Let $y(t)$ be a (current or voltage) response in a network and $x(t)$ a source (current or voltage) of that network. In Chap. 3 we have illustrated, and in Chap. 11 we shall show generally, that in a linear system a response is related to a source by

$$y(t) = H(p)x(t)$$

where $H(p) = N(p)/D(p)$ and $N(p)$ and $D(p)$ are polynomials in p.

For example, in the circuit of Fig. 5-1, $y(t) = v_2(t)$, $x(t) = v(t)$, and from Eq. (5-2),

$$v_2(t) = \frac{p^2 + 2p + 1}{p^2 + 2p + 2} v(t) = \frac{N(p)}{D(p)} v(t)$$

which is another way of writing the differential equation

$$(p^2 + 2p + 2)v_2(t) = (p^2 + 2p + 1)v(t) \tag{5-4}$$

If we set the source $v(t)$ to zero, from Eq. (5-4) we have the source-free equilibrium equation $(p^2 + 2p + 2)v_2(t) = 0$ corresponding to the circuit of Fig. 5-1 when the source is set to zero.

In general, the denominator of a network function is a polynomial; i.e., it is of the form

$$D(p) = a_n p^n + a_{n-1} p^{n-1} + \cdots + a_1 p + a_0 = \sum_{k=0}^{n} a_k p^k$$

and the source-free equilibrium equation of the circuit has the corresponding form

$$(a_n p^n + a_{n-1} p^{n-1} + \cdots + a_1 p + a_0)y(t) = 0 \tag{5-4a}$$

where n is the number of independent energy-storing elements in the circuit.

The general solution of the above nth-order source-free equilibrium equation consists of n linearly independent functions, termed *source-free modes* of the response. These modes are in general derived by introducing into the source-free equilibrium equation the exponential form

$$(y)_f = Ke^{st} \tag{5-5}$$

Introducing (5-5) into (5-4), we note that

$$p(y)_f = sKe^{st} = s(y)_f$$

or

$$p^n(y)_f = s^n(y)_f$$

so that the exponential form (5-5) satisfies the homogeneous (source-free) equation (5-4*a*) if

$$(a_n s^n + a_{n-1}s^{n-1} + \cdots + a_1 s + a_0)(y)_f = 0$$

that is, if

$$[D(s)](y)_f = 0 \tag{5-6}$$

As $(y)_f$ need not be identically zero for all t, it follows from Eq. (5-6) that the form (5-5), i.e.,

$$(y)_f = Ke^{st}$$

is a solution of Eq. (5-4) if the value of s is chosen from the equation

$$D(s) = \sum_{k=0}^{n} a_k s^k = 0 \tag{5-7}$$

Equation (5-7) is termed the *characteristic equation* of the equilibrium equation (5-4), and the values of s that satisfy the characteristic equation (5-7) are called the *characteristic roots* of the equilibrium equation. Thus the characteristic equation can be obtained by replacing p with s in the denominator polynomial of the network function and setting the resulting polynomial to zero.

If the n characteristic roots are all distinct, the linear combination

$$(v_2)_f = \sum_{k=1}^{n} K_k e^{s_k t} \tag{5-8}$$

consists of a sum of n linearly independent functions and therefore forms the general solution. (It is recalled that the general solution of a linear differential equation always consists of a linear combination of a sum of linearly independent functions equal in number to the order of the equation.)

If any of the characteristic roots are equal to each other (as in the case of the critically damped *R-L-C* circuit), the solution of the source-free equation contains modes of the form $te^{s_k t}$ or $t^m e^{s_k t}$. Thus, for example, the characteristic equation

$$(s + 2)^2(s + 3) = 0$$

corresponds to the modes te^{-2t}, e^{-2t}, and e^{-3t}, and the form

$$y_f = (K_1 + K_2 t)e^{-2t} + K_3 e^{-3t}$$

is the general solution of the equation

$$(p + 2)^2(p + 3)y = 0$$

that is, the equation

$$\frac{d^3y}{dt^3} + 7\frac{d^2y}{dt^2} + 16\frac{dy}{dt} + 12y = 0$$

We observe, incidentally, that complex characteristic roots always occur in conjugate pairs (why?) and that the two corresponding complex modes $\mathbf{K}_1 e^{s_1 t} + \mathbf{K}_1^* e^{s_1^* t}$ can always be written as the real function $2 \operatorname{Re}(\mathbf{K}_1 e^{s_1 t})$.

Except in lossless (resistanceless) networks, the source-free (component of the) response always "dies out," i.e., approaches zero, as t becomes infinite. For this reason this response component is often termed the "transient."

5-3 Component of response due to source

In the relationship

$$y(t) = \frac{N(p)}{D(p)} x(t) \tag{5-9}$$

we may replace $N(p)x(t)$ by the forcing function $f(t)$ to write the equilibrium equation in the form

$$D(p)y(t) = f(t) \tag{5-10}$$

This equation has a solution that vanishes when the forcing function is identically zero. As mentioned before, this solution is termed the *component of the response due to the source* and is denoted by the subscript s; that is, $(v_2)_s$ denotes the component of the (complete) response v_2 due to the source. This response component can be obtained by various procedures. Except in certain (resonant) cases, when the forcing function is continuous, *the component of the response due to the sources can be found as a linear combination of the forcing function with all its derivatives.* The special resonant cases referred to above are discussed in Chaps. 7 and 8.

When the source function is constant or periodic, the component of the response due to the source will, in general, also be constant or periodic. Such constant or periodic response components are also termed *steady-state* response (components). Another term used as a synonym for "component of the response due to the source" is the term *forced response* (component), to emphasize that it is caused by the forcing function, and not by the source-free interaction of the energy stored in passive elements.

Example 5-1 In the circuit of Fig. 5-1, where v_2 is related to $v(t)$ by the equilibrium equation

$$(p^2 + 2p + 2)v_2 = (p^2 + 2p + 1)v$$

find the component of the response for $t \geq 0^+$ due to the sources (*a*) $v = 10\,u(t)$; (*b*) $v = (10t)u(t)$; (*c*) $v = (10 + 10t)u(t)$; (*d*) $v = 5e^{-3t}u(t)$; (*e*) $v = (10t + 5e^{-3t})u(t)$.

Solution (*a*) From the discussion of Sec. 3-14 we know that, when the source is $x(t) = V_0u(t)$, the response due to this source is

$$y_0(t) = \frac{N(0)}{D(0)} V_0u(t) \qquad \text{if } D(0) \neq 0$$

Applying this to the problem at hand, we have

$$v_{2s}(t) = \left.\frac{p^2 + 2p + 1}{p^2 + 2p + 2}\right|_{p=0} 10u(t) = 5u(t)$$

At this point it is instructive to consider the solution of this problem in more detail. The function $10u(t)$ has a jump at $t = 0$, so that $(p^2 + 2p + 1)u(t) = d\delta(t)/dt + 2\delta(t) + u(t)$. For $t \geq 0^+$, $\delta(t)$ and its derivative are zero and $u(t) = 1$. Thus

$$f(t) = (p^2 + 2p + 1)10 = 10$$

Hence

$$(p^2 + 2p + 2)v_{2s} = 10 \qquad t \geq 0^+$$

Since f = constant, $v_{2s} = V_{2s}$, $(p^2 + 2p)V_{2s} = 0$, and exactly as before,

$$V_{2s} = 5$$

(*b*) For $v = 10t$, the forcing function is given by

$$f(t) = (p^2 + 2p + 1)(10t) \qquad t > 0$$
$$= 0 + 20 + 10t$$

and

$$(p^2 + 2p + 2)(v_2)_s = 20 + 10t \qquad t \geq 0^+$$

We now note that the form

$$(v_2)_s = A + Bt \tag{5-11}$$

is a linear combination of $f(t)$ with all its derivatives. Using Eq. (5-11),

$$(p^2 + 2p + 2)(A + Bt) = 0 + 2B + 2A + 2Bt$$

which must equal $20 + 10t$, or

$$2(A + B) + 2Bt \equiv 20 + 10t \qquad t \geq 0^+$$

Hence, by coefficient comparison,

$$B = 5 \qquad A = 5$$

or

$$(v_2)_s = 5 + 5t$$

(*c*) Using the results of parts *a* and *b* and the principle of superposition,

$$(v_2)_s = 10 + 5t \qquad t \geq 0^+$$

(*d*) For $v = 5e^{-3t}$,

$$f(t) \doteq (p^2 + 2p + 1)5e^{-3t} = [(-3)^2 + 2(-3) + 1] \times 5e^{-3t} = 20e^{-3t}$$

All linear combinations of $f(t)$ with its derivatives are, because of the unique property of the exponential function, proportional to $f(t)$. Hence we let

$$[v_2(t)]_s = Ae^{-3t}$$

Then

$$(p^2 + 2p + 2)v_{2s} = [(-3)^2 + 2(-3) + 2]Ae^{-3t} = 5Ae^{-3t}$$

and

$$5Ae^{-3t} = 20e^{-3t}$$
$$A = 4$$

and the solution is

$$(v_2)_s = 4e^{-3t} \qquad t \geq 0^+$$

This same answer can be obtained more rapidly by

$$v_{2s}(t) = \left.\frac{p^2 + 2p + 1}{p^2 + 2p + 2}\right|_{p=-3} \times 5e^{-3t} = 4e^{-3t}$$

(*e*) Using the results of parts *b* and *d*, the solution is, by superposition,

$$(v_2)_s = 5 + 5t + 4e^{-3t} \qquad t \geq 0^+$$

NETWORKS WITH D-C SOURCES[1] Constant sources are of considerable importance in circuit analysis. Since the analysis to such sources represents an important special case, certain shortcuts can be deduced for it.

We have seen above that a network with d-c sources approaches its steady state, and the voltages and currents in the network are then independent of time. Since the voltage across an inductance is proportional to the *rate of change* of the current in the inductance [$v_{ab} = L(d/dt)i_{ab}$], the flow of a d-c current in an inductance will result in zero voltage across the inductance. Hence, in the d-c steady state, we have a circuit element L_{ab}, wherein there is a current I_{ab} but with voltage $V_{ab} = 0$. As far as the d-c steady state is concerned, the inductance L_{ab} can be replaced by a line (short circuit) as shown in Fig. 5-2*a*. The d-c voltages and currents at the terminals *a-b* of the

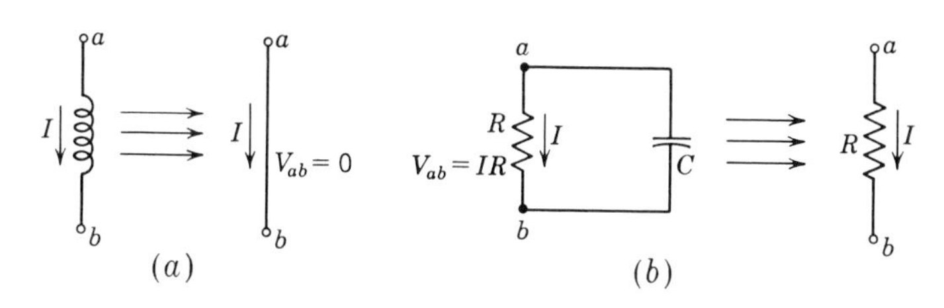

Fig. 5-2 Direct-current equivalents for the elements L and C.

L_{ab} and the short circuit *a-b* are identical when the current I_{ab} is independent of time. Similarly, the current through a capacitance is proportional to the rate of change of the voltage across the capacitance [$i_{ab} = C(dv_{ab}/dt)$]. If v_{ab} is independent of time, the current i_{ab} is equal to zero. Since in the d-c steady state there is no current through C_{ab}, it can be taken out of the circuit (open-circuited) without any change in the currents or voltages of the network. In Fig. 5-2*b* this equivalence is shown. Therefore, when the steady-state response of a network with d-c sources is required, all the capacitances in the network can be open-circuited and all inductances short-circuited. This reduction of form of the network produces results identical with that obtained when we set $p \equiv 0$ in $H(p)$ to obtain the steady-state response due to a constant source, but in cases where only the d-c steady-state response is derived, formulation of $H(p)$ is superfluous.

5-4 Related sources theorem

For a given equilibrium equation of the form

$$[D(p)]v(t) = f(t)$$

[1] The abbreviation d-c refers to direct current. Throughout the literature this abbreviation is used to indicate a source which is independent of time (for all values of time between $-\infty$ and $+\infty$).

the component of the response due to the source is denoted by v_s. If the forcing function $f(t)$ is replaced by $g(t)$, a new forcing function related to $f(t)$ by

$$g(t) = A \frac{df}{dt}$$

where A is a constant, then the new response component due to the source is given by $A\ dv_s/dt$, that is,

$$[D(p)]pv_s = g(t) = pf(t) \tag{5-12}$$

A little thought shows that the proof of the statement is implicit in Eq. (5-12). The relationship discussed above is summarized as the related sources theorem:

> The response to the derivative of a given forcing function is the same as the derivative of the response to the given forcing function.

The same relationship holds for integration; in that case an integration constant must be evaluated as in the example below. The related sources theorem as stated above applies not to complete response, but merely to the component of response due to the source. It can be shown that it can be applied to the complete response when the initial-energy storage in all L's and C's of the network is zero.

Example 5-2 Recall that the response component due to the source $v = 10tu(t)$ in the equilibrium equation

$$(p^2 + 2p + 2)v_2 = (p^2 + 2p + 1)v$$

is given by

$$[v_2(t)]_s = 5 + 5t \qquad t \geq 0^+$$

Find the component of the response due to (*a*) $v(t) = 20u(t)$; (*b*) $v(t) = 20t^2u(t)$.

Solution (*a*) Since $20u(t) = 2p[10tu(t)]$, the desired response component will be

$$\begin{aligned}[v_2(t)]_s &= 2p(5 + 5t) \qquad t \geq 0^+ \\ &= 10\end{aligned}$$

(*b*) Since $20t^2 = 4\int 10t\ dt + \text{const}$, the new response component due to the source is given by

$$\begin{aligned}(v_2)_s &= 4\int(5 + 5t)\ dt \\ &= 20t + 10t^2 + B\end{aligned}$$

where B is the constant of integration. It is determined by substitution in the equilibrium equation:

$$(p^2 + 2p + 2)v_2 = f(t) = 20t^2 + 80t + 40$$

Since

$$(p^2 + 2p + 2)(10t^2 + 20t + B) = 20t^2 + 80t + 60 + 2B$$

we have $60 + 2B = 40$, $B = -10$, and

$$[v_2(t)]_s = 10t^2 + 20t - 10 \qquad t \geq 0^+$$

5-5 *Initial conditions*

We recall again that if a network response $y(t)$ is related to a source $x(t)$ by the linear differential equation

$$y(t) = [H(p)]x \qquad H(p) = \frac{N(p)}{D(p)}$$

the complete response $y(t)$ has the form

$$y(t) = y_s(t) + y_f(t)$$

We further recall that the free component of the response, $y_f(t)$, consists of the sum of free modes and that each mode is multiplied by a constant (the "mode amplitude"). If, for example, all the characteristic roots are distinct, the form of the complete response is given by

$$y(t) = y_s(t) + \sum_{k=1}^{n} K_k e^{s_k t} \tag{5-13}$$

where n is the order of the equilibrium equation and is equal to the number of independent energy-storing elements of the network. From Eq. (5-13) we observe that once the component of the response due to the source is known, the constants K_k (free-mode amplitudes) must be evaluated. These values are found from the initial conditions in the network. Most conveniently, the n constants can be found if the initial value of y and the initial values of the first $n - 1$ derivatives of y are known. Then

$$\begin{aligned}
y(0^+) &= y_s(0^+) + K_1 + K_2 + \cdots + K_n \\
\left(\frac{dy}{dt}\right)_{0^+} &= \left(\frac{dy_s}{dt}\right)_{0^+} + s_1K_1 + s_2K_2 + \cdots + s_nK_n \\
&\vdots \\
\left(\frac{d^{n-1}y}{dt^{n-1}}\right)_{0^+} &= \left(\frac{d^{n-1}y_s}{dt^{n-1}}\right)_{0^+} + s_1{}^{n-1}K_1 + s_2{}^{n-2}K_2 + \cdots + s_n{}^{n-1}K_n
\end{aligned} \tag{5-14}$$

and the constants $K_1, \ldots, K_n$ are found by solving the n simultaneous algebraic equations (5-14).

For example, in the circuit of Fig. 5-1, if the source $v(t)$ is given as $10u(t)$, we have $(v_2)_s = 5$. The characteristic roots are given by $s_{1,2} = -1 \pm j$; hence the complete response is given by

$$v_2 = \begin{cases} 5 + \mathbf{K}_1 e^{(-1+j)t} + \mathbf{K}_1^* e^{(-1-j)t} & t \geq 0^+ \\ 5 + \operatorname{Re}\,(\mathbf{K}e^{(-1+j)t}) & t \geq 0^+ \end{cases}$$

The two constants Re $\mathbf{K}$ and Im $\mathbf{K}$ can be found from the relations

$$\begin{aligned}
\operatorname{Re}\mathbf{K} &= v_2(0^+) - 5 \\
\operatorname{Re}\,[(-1 + j)\mathbf{K}] &= \left(\frac{dv_2}{dt}\right)_{0^+} - 0
\end{aligned}$$

If the initial values $v_2(0^+)$ and $(dv_2/dt)_{0^+}$ in the preceding example [or the values $y(0^+)$, $\left(\frac{dy}{dt}\right)_{0^+}, \ldots, \left(\frac{d^{n-1}y}{dt^{n-1}}\right)_{0^+}$, in the general case] are given, formal solution is complete. In practical cases, however, the initial values and initial derivatives are *not* generally known, but must be found as part of the solution of the problem. What is known in practical cases is the *continuity conditions,* that is, the initial values for voltages across capacitances and currents in inductances. The desired initial conditions are then generally derived from the continuity conditions, the circuit configuration, and the source function.

It is always possible to derive the desired initial conditions by expressing them in terms of the continuity conditions with the aid of the Kirchhoff-law equations. The following example illustrates this procedure.

Example 5-3 In the circuit of Fig. 5-3 (which is identical with Fig. 5-1) derive the initial values $v_2(0^+)$ and $(dv_2/dt)_{0+}$ in terms of the continuity conditions $v_{ab}(0^+) = V_0$, $i(0^+) = I_0$, and the source function $v(t)$.

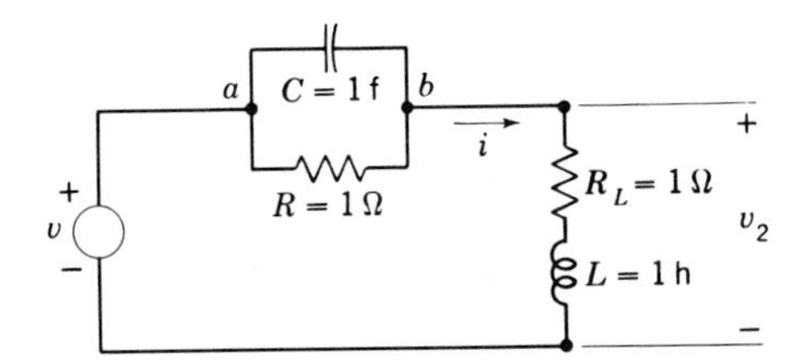

Fig. 5-3 Circuit used in Example 5-3.

Solution From the Kirchhoff voltage law we observe that

$$v_2 = v(t) - v_{ab} \tag{5-15}$$

Hence, setting $t = 0^+$ in Eq. (5-15),

$$\begin{aligned} v_2(0^+) &= v(0^+) - v_{ab}(0^+) \\ &= v(0^+) - V_0 \end{aligned} \tag{5-15a}$$

Differentiation of Eq. (5-15) with respect to t gives

$$\frac{dv_2}{dt} = \frac{dv}{dt} - \frac{dv_{ab}}{dt} \tag{5-16}$$

To find $(dv_2/dt)_{0+}$ from Eq. (5-16) it is necessary to find $(dv_{ab}/dt)_{0+}$. Note, however, that $(dv/dt)_{0+}$, the initial value of the derivative of the source function, will be known once the source is specified. To find $(dv_{ab}/dt)_{0+}$, we apply Kirchhoff's current law at junction b:

$$C\frac{dv_{ab}}{dt} + \frac{v_{ab}}{R} = i$$

Using $R = 1$ and $C = 1$,

$$\frac{dv_{ab}}{dt} = i - v_{ab} \tag{5-17}$$

Equation (5-17) is substituted in Eq. (5-16) to give the result

$$\frac{dv_2}{dt} = \frac{dv}{dt} - i + v_{ab} \qquad t \geq 0^+ \tag{5-18}$$

Letting $t = 0^+$ in Eq. (5-18) gives the desired result

$$\left(\frac{dv_2}{dt}\right)_{0+} = \left(\frac{dv}{dt}\right)_{0+} - I_0 + V_0 \tag{5-19}$$

Equations (5-15*a*) and (5-19) are the required initial conditions.

5-6 Introduction to state equations and state variables

We have seen in the course of the preceding discussion about derived initial conditions that the variables associated with stored energy (voltage across capacitance and current in inductance) play a special role in analysis because they are (in the absence of impulses) continuous functions of time. Below, we illustrate (and in Chaps. 11 and 12 we discuss more fully) how a network which includes n independent energy-storing elements can be described in terms of n simultaneous first-order differential equations for n unknowns, where each unknown is either a voltage across a capacitance or a current through an inductance. These "unknowns" in this context are called *state variables* of the network (because they describe the stored-energy states), and the corresponding equations are termed *state equations.*[1] State variables and state equations are introduced in this chapter because they are often useful for finding derived initial conditions. Their main use is discussed in Chaps. 11 and 12.

To illustrate the procedure, we refer to the example of Fig. 5-3, i.e., the network described by the equilibrium equation

$$(p^2 + 2p + 2)v_2(t) = f(t) \qquad f(t) = (p^2 + 2p + 1)v \tag{5-20}$$

Any differential equation of nth order such as Eq. (5-20) may be expressed as n simultaneous first-order differential equations by introducing $n - 1$ new variables. These variables may be derivatives of the original variables. For example, in Eq. (5-20), if we let $pv_2 \equiv x$, we have the equations

$$\begin{aligned} \frac{dx}{dt} + 2x + 2v_2 &= f(t) \\ \frac{dv_2}{dt} &= x \end{aligned} \tag{5-21}$$

and Eqs. (5-21) are the two simultaneous first-order equations describing the same network as the original second-order equation (5-20).

The variable $x(t) \equiv dv_2/dt$ used in Eq. (5-21) does not have any special physical significance, and is chosen to demonstrate that a second-order equation may be expressed as two simultaneous first-order equations. However,

[1] Variables other than voltages across capacitances and currents in inductances can be chosen as state variables.

if we choose the voltages across capacitances and currents through inductances of a network as the variables, we describe a network containing n energy-storing elements by n first-order differential equations which are especially meaningful. We shall now illustrate the state equations for the network of Fig. 5-4 (which is identical with Fig. 5-3 in form, but no numerical

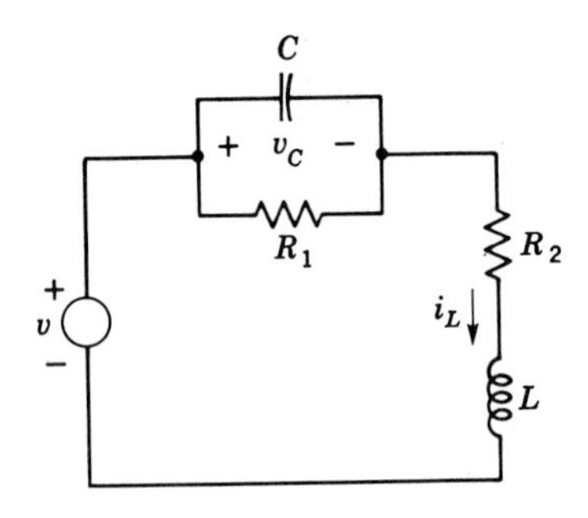

Fig. 5-4 Circuit used to illustrate equations involving state variables.

values are used). For the purpose of this example we derive the equations by direct application of Kirchhoff's laws; below we describe an alternative procedure.

In the circuit of Fig. 5-4, the state variables are v_C and i_L. We observe from the Kirchhoff voltage law that

$$v = v_C + R_2 i_L + L\frac{di_L}{dt}$$

or

$$\frac{di_L}{dt} = -\frac{R_2}{L}i_L - \frac{1}{L}v_C + \frac{1}{L}v \tag{5-22}$$

Equation (5-22) is one of the "state equations" because it relates the first derivative of a state variable, i_L, to an algebraic combination of state variables, i_L and v_C, and source functions. The second state equation can be obtained by application of Kirchhoff's current law at junction b:

$$\frac{v_C}{R_1} + C\frac{dv_C}{dt} = i_L$$

or

$$\frac{dv_C}{dt} = \frac{1}{C}i_L - \frac{v_C}{R_1 C} \tag{5-23}$$

The two state equations are therefore

$$\begin{aligned} \frac{di_L}{dt} &= -\frac{R_2}{L}i_L - \frac{1}{L}v_C + \frac{1}{L}v \\ \frac{dv_C}{dt} &= \frac{1}{C}i_L - \frac{1}{R_1 C}v_C + 0 \end{aligned} \tag{5-24}$$

For purpose of application to derived initial conditions, we observe that we can obtain $(di_L/dt)_{0+}$ and $(dv_C/dt)_{0+}$ directly from Eqs. (5-24) by setting $t = 0^+$.

$$\begin{aligned} \left(\frac{di_L}{dt}\right)_{0+} &= -\frac{R_2}{L}i_L(0^+) - \frac{1}{L}v_C(0^+) + \frac{1}{L}v(0^+) \\ \left(\frac{dv_C}{dt}\right)_{0+} &= \frac{1}{C}i_L(0^+) - \frac{1}{R_1 C}v_C(0^+) \end{aligned}$$

Higher initial derivatives can be derived by successive differentiation and substitution of previously derived initial conditions. Once the derived initial derivatives of the state variables are known, desired initial conditions for the variables are easily found; in the circuit of Fig. 5-4, for example,

$$\frac{dv_2}{dt} = R_2 \frac{di_L}{dt} + L \frac{d^2 i_L}{dt^2}$$

so that $(di_L/dt)_{0+}$ and $(d^2 i_L/dt^2)_{0+}$ determine the value of the desired initial derivative.

5-7 General form of state equations

Inspection of Eqs. (5-24) shows that the general form of a state equation is the following: *The derivative of a state variable is equal to a linear combination of the state variables and the source functions.* Thus, if there are two state variables in a circuit (two independent energy-storing elements) and if the state variables are denoted as $y_1(t)$ and $y_2(t)$, the state equations have the form

$$\begin{aligned} \frac{dy_1}{dt} &= a_{11}y_1 + a_{12}y_2 + g_1(t) \\ \frac{dy_2}{dt} &= a_{21}y_1 + a_{22}y_2 + g_2(t) \end{aligned} \qquad (5\text{-}25)$$

where a_{11}, a_{12}, a_{21}, and a_{22} depend only on the values of the circuit elements, and the functions $g_1(t)$ and $g_2(t)$ are linear combinations of the source functions. If a network has three independent energy-storing elements, the state equations are

$$\begin{aligned} \frac{dy_1}{dt} &= a_{11}y_1 + a_{12}y_2 + a_{13}y_3 + g_1(t) \\ \frac{dy_2}{dt} &= a_{21}y_1 + a_{22}y_2 + a_{23}y_3 + g_2(t) \\ \frac{dy_3}{dt} &= a_{31}y_1 + a_{32}y_2 + a_{33}y_3 + g_3(t) \end{aligned} \qquad (5\text{-}26)$$

where y_1, y_2, y_3 are the state variables, g_1, g_2, g_3 depend on the sources, and the a's are time-invariant and depend only on the circuit elements.

The form of a typical state equation suggests a simple method for finding the a's and the g functions as follows: Consider the typical equation

$$\frac{dy_1}{dt} = a_{11}y_1 + a_{12}y_2 + a_{13}y_3 + g_1(t)$$

In this equation we may consider the term on the left-hand side of the equation, dy_1/dt, as an "effect," and the state variables on the right-hand side of the equation as a number of individual "causes." Thus y_1 may be considered a cause and $a_{11}y_1$ its "contribution" to the effect dy_1/dt. Similarly, y_2 is another cause, and $a_{12}y_2$ is its contribution to the effect dy_1/dt. From this point of view it is evident that a_{11} may be obtained by setting

y_2, y_3, and g_1 to zero to give

$$\left(\frac{dy_1}{dt}\right)_{\substack{y_2=0\\y_3=0\\g_1=0}} = a_{11}y_1$$

Similarly,

$$\left(\frac{dy_1}{dt}\right)_{\substack{y_1=0\\y_3=0\\g_1=0}} = a_{12}y_2 \qquad \left(\frac{dy_1}{dt}\right)_{\substack{y_1=0\\y_2=0\\g_1=0}} = a_{13}y_3 \qquad \left(\frac{dy_1}{dt}\right)_{\substack{y_1=0\\y_2=0\\y_3=0}} = g_1(t)$$

The above results are arrived at without concern as to the nature of y_1, y_2, y_3, or $g_1(t)$, and *are a direct result of the linearity of the system;* that is, *the effect due to a number of causes is equal to the sum of the effects due to individual causes.*

In an electric circuit, setting a state variable equal to zero has the following significance:

1 If the state variable is the voltage across a capacitance, setting this voltage to zero means *replacing the capacitance* by a *short circuit.* This ensures that the given state variable (voltage across capacitance) is zero, *while permitting flow of current as required by another state variable.*
2 If the state variable is the current through an inductance, setting the current to zero means replacing the inductance with an open circuit or opening the branch at the terminals of the inductance. This ensures zero current in that branch, *but allows establishment of voltages across the open circuit required by other state variables.*
3 When the term set to zero in the right side of a state equation is the term depending on an ideal source (or sources), the source (or sources) involved must be set to zero.

Thus the coefficients involved in state equations can be found by successive application of Kirchhoff's laws to the circuit in which all but one state variable has been set to zero. The general procedure is illustrated using the circuit of Fig. 5-4, which is shown on page 151 as Fig. 5-5*a*. One of the state equations of this circuit is $di_L/dt = a_1 i_L + a_2 v_C + a_3 v$. To find a_1, we have $(di_L/dt)_{v_C=0,v=0} = a_1 i_L$. Setting v_C and v to zero means that, in the circuit of Fig. 5-5*a*, the capacitance C and the source $v(t)$ are replaced by a short circuit, resulting in Fig. 5-5*b*. From this circuit it is seen that

$$v_{en} = \left(L\frac{di_L}{dt}\right)_{\substack{v=0\\v_C=0}} = -R_2 i_L \tag{5-27}$$

$$\left(\frac{di_L}{dt}\right)_{\substack{v=0\\v_C=0}} = -\frac{R_2}{L}i_L = a_1 i_L$$

which gives the result

$$a_1 = -\frac{R_2}{L}$$

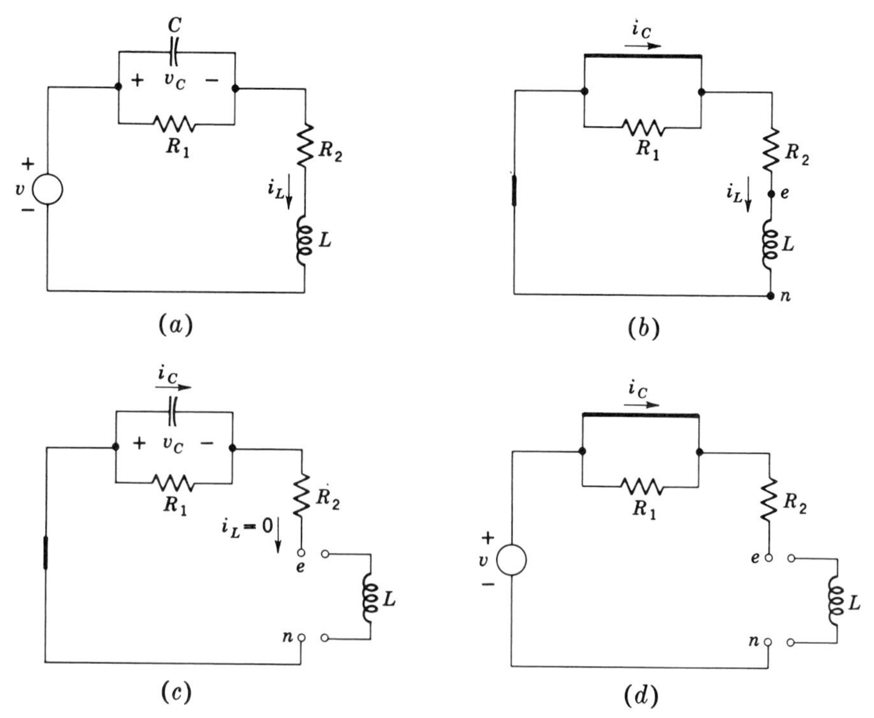

***Fig. 5-5** Illustration of setting up state equations. (**a**) Complete circuit. (**b**) Circuit with v_C and v set to zero. (**c**) Circuit with v and i_L set to zero. (**d**) Circuit with v_C and i_L set to zero.*

To set i_L and v to zero in Fig. 5-5*a*, we replace i_L with an open circuit and v with a short circuit, resulting in Fig. 5-5*c*. The voltage $L(di_L/dt)$ which would appear across L, that is,

$$v_{en} = \left(L\frac{di_L}{dt}\right)_{\substack{v=0\\i_L=0}} = -v_C \tag{5-28}$$

$$\left(\frac{di_L}{dt}\right)_{\substack{v=0\\i_L=0}} = -\frac{1}{L}v_C = a_2v_C$$

gives the result

$$a_2 = -\frac{1}{L}$$

Finally, setting v_C and i_L to zero, we have the circuit of Fig. 5-5*d*, in which

$$v_{en} = \left(L\frac{di_L}{dt}\right)_{\substack{v_C=0\\i_L=0}} = v$$

$$\left(\frac{di_L}{dt}\right)_{\substack{v=0\\i_L=0}} = \frac{1}{L}v = a_3v$$

and

$$a_3 = \frac{1}{L}$$

It follows that

$$\frac{di_L}{dt} = a_1 i_L + a_2 v_C + a_3 v = -\frac{R_2}{L} i_L - \frac{1}{L} v_C + \frac{1}{L} v \tag{5-29}$$

which is identical with the result in the first of Eqs. (5-24). Obtaining the second of Eqs. (5-24) is reserved as a student exercise.

Example 5-4 In the circuit of Fig. 5-6: (*a*) Obtain the state equations. (*b*) Find $v_2(0^+)$, $(dv_2/dt)_{0^+}$, and $(d^2v_2/dt^2)_{0^+}$ if $i_1(0^+) = 2$, $i_2(0^+) = -1$, and $v_C(0^+) = 3$.

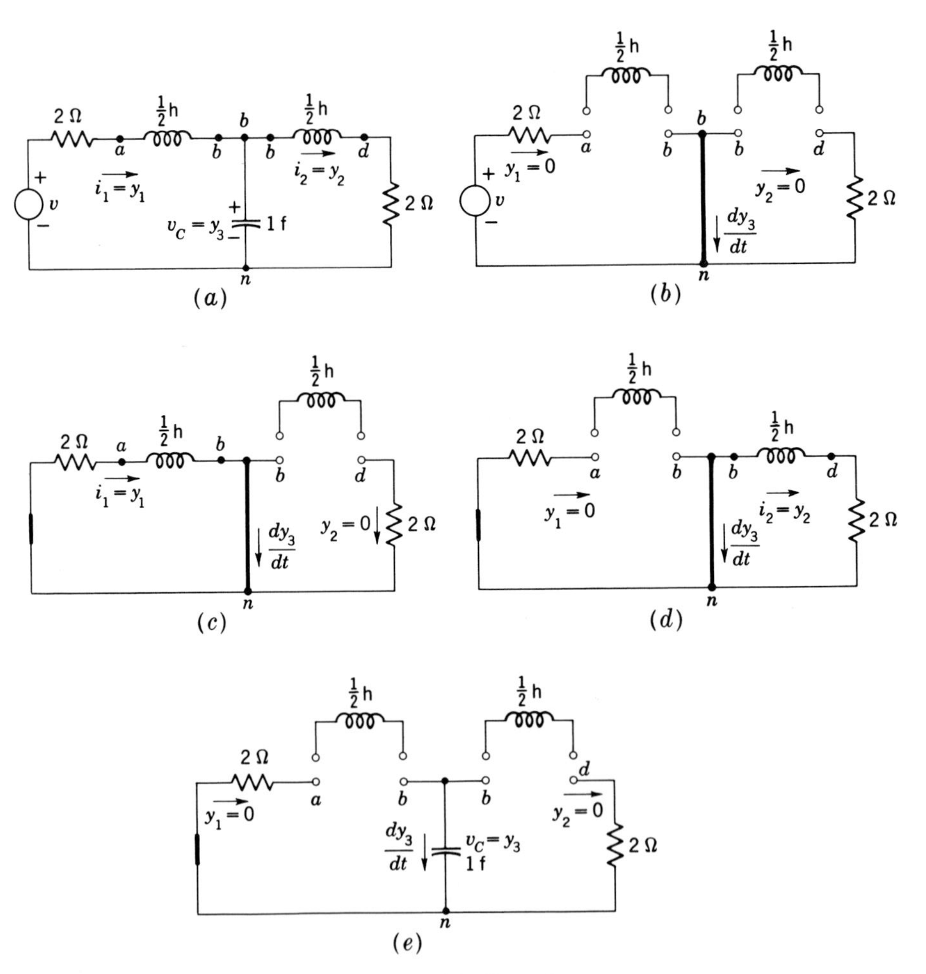

Fig. 5-6 Circuits for Example 5-4. ***(a) Complete circuit. (b) Circuit with all state variables set to zero. (c) Circuit with source set to zero and*** $y_2 = y_3 = 0$. ***(d) Circuit with source set to zero and*** $y_1 = y_3 = 0$. ***(e) Circuit with source set to zero and*** $y_1 = y_2 = 0$.

Solution (*a*) Let $y_1 \equiv i_1$, $y_2 \equiv i_2$, and $y_3 \equiv v_C$. The state equations have the form

$$\frac{dy_1}{dt} = a_{11}y_1 + a_{12}y_2 + a_{13}y_3 + g_1 \tag{5-30a}$$

$$\frac{dy_2}{dt} = a_{21}y_1 + a_{22}y_2 + a_{23}y_3 + g_2 \tag{5-30b}$$

$$\frac{dy_3}{dt} = a_{31}y_1 + a_{32}y_2 + a_{33}y_3 + g_3 \tag{5-30c}$$

In Fig. 5-6*b* the network is drawn with all state variables set to zero. For this circuit we can write, from Eqs. (5-30),

$$\frac{dy_1}{dt} = g_1 \qquad \frac{dy_2}{dt} = g_2 \qquad \frac{dy_3}{dt} = g_3 \qquad y_1 = y_2 = y_3 = 0$$

From Fig. 5-6*b* we have

$$\frac{1}{2}\frac{dy_1}{dt} = v_{ab} = v \qquad \frac{dy_1}{dt} = 2v$$

Hence $g_1(t) = 2v$. Since $v_{bd} = v_{bn} = 0$, $\frac{1}{2}(dy_2/dt) = 0$ and $g_2 \equiv 0$. Similarly, $i_C = 1(dy_3/dt) = 0$; hence $g_3 = 0$. Thus

$$g_1(t) = 2v \qquad g_2(t) = 0 \qquad g_3(t) = 0$$

To obtain a_{11}, a_{21}, and a_{31}, we set $y_2 = y_3 = 0$, and set the source v to zero, so that the g functions are zero. This is shown in Fig. 5-6*c*. In this circuit[1] we have

$$v_{bn} = 0 = -2y_1 - \frac{1}{2}\frac{dy_1}{dt}$$

Hence

$$\left(\frac{dy_1}{dt}\right)_{\substack{y_2=0\\y_3=0\\v=0}} = -4y_1$$

so that $a_{11} = -4$. Since $i_C = i_1 = 1 \times dy_3/dt$, we have

$$\left(\frac{dy_3}{dt}\right)_{\substack{y_2=0\\y_3=0\\v=0}} = y_1$$

and $a_{31} = 1$. As $v_{bd} = v_{bn} = 0$, $\frac{1}{2}dy_2/dt = 0$ and $a_{21} = 0$. Thus $a_{11} = -4$, $a_{21} = 0$, and $a_{31} = 1$.

To obtain a_{12}, a_{22}, and a_{32}, we set $y_1 = y_3 = v = 0$ as shown in Fig. 5-6*d*. In this circuit $v_{ab} = v_{nb} = 0$. Hence $dy_1/dt = 0$, so that $a_{12} = 0$. Since

$$\frac{1}{2}\frac{dy_2}{dt} + 2y_2 = 0$$

$$\frac{dy_2}{dt} = -4y_2$$

and $a_{22} = -4$. Also, $y_2 = -i_C = -dy_3/dt$; hence $a_{32} = -1$. Thus $a_{12} = 0$, $a_{22} = -4$, and $a_{32} = -1$.

[1] The reader who may find it strange to have a current without a source in the circuit can use several interpretations to overcome this difficulty. He can visualize, for example, a moment when the source is zero yet there are currents in some of the passive elements. The use of compensating sources, as explained in Sec. 6-8, is also convenient in visualizing this situation.

Finally, a_{13}, a_{23}, and a_{33} are obtained from Fig. 5-6*e*. In this circuit

$$y_3 = v_{bn} = v_{ba} = -\frac{1}{2}\frac{dy_1}{dt} \qquad a_{13} = -2$$

$$y_3 = v_{bn} = v_{bd} = \frac{1}{2}\frac{dy_2}{dt} \qquad a_{23} = 2$$

$$i_C = \frac{dy_3}{dt} = 0 \qquad a_{33} = 0$$

Thus we have the final result, the state equations for the circuit of Fig. 5-6*a*:

$$\begin{aligned} \frac{dy_1}{dt} &= -4y_1 + 0 - 2y_3 + 2v \\ \frac{dy_2}{dt} &= 0 - 4y_2 + 2y_3 + 0 \\ \frac{dy_3}{dt} &= y_1 - y_2 + 0 + 0 \end{aligned} \tag{5-31}$$

(*b*) Since $v_2 = 2i_2 = 2y_2$, we have

$$v_2(0^+) = 2 \times 2 = 4 \text{ volts}$$

Since

$$\frac{dv_2}{dt} = 2\frac{dy_2}{dt} \tag{5-32}$$

we have (from the second state equation)

$$\left(\frac{dy_2}{dt}\right)_{0^+} = -4y_2(0^+) + 2y_3(0^+) = 10$$

Hence

$$\left(\frac{dv_2}{dt}\right)_{0^+} = 20 \text{ volts/sec}$$

Differentiating Eq. (5-32) gives

$$\frac{d^2v_2}{dt^2} = 2\frac{d^2y_2}{dt^2}$$

and from the derivative of the second state equation,

$$\frac{d^2y_2}{dt^2} = -4\frac{dy_2}{dt} + 2\frac{dy_3}{dt} \tag{5-33}$$

Hence

$$\left(\frac{d^2v_2}{dt^2}\right)_{0^+} = -8\left(\frac{dy_2}{dt}\right)_{0^+} + 4\left(\frac{dy_3}{dt}\right)_{0^+}$$

The value $(dy_2/dt)_{0^+}$ has already been found to be 10. Since

$$\frac{dy_3}{dt} = y_1 - y_2$$

$$\left(\frac{dy_3}{dt}\right)_{0^+} = 2 - (-1) = 3$$

and

$$\left(\frac{d^2v_2}{dt^2}\right)_{0^+} = -8(10) + 4(3) = -68 \text{ volts/sec}^2$$

The subject of state variables and state equations is discussed further in Chaps. 6, 11, and 12. It is clear that state equations describe networks in terms of *simultaneous differential equations* of a special form. A little thought shows that one can use these equations for calculation of the network functions and for complete response calculations directly, not merely for the derivation of initial conditions. These matters are discussed in Chaps. 11 and 12.

5-8 Initial-condition generators

In this section we show how initial conditions can be represented by *ideal* sources. The purpose of this representation is fourfold:

1 To gain further understanding of the continuity conditions.
2 To be able to evaluate separately the effect of initial-energy storage on the complete response.
3 By the use of superposition, to be able to prove certain theorems generally, although in the proof it will be assumed that the initial-energy storage is zero.
4 To be able to solve certain problems involving impulses.

We begin the discussion by recalling that the continuity conditions require that, in the absence of impulses, the current in an inductance and the voltage across a capacitance must be continuous functions of time. Based on these relationships, we shall see that an initially charged capacitance as shown in Fig. 5-7 can be represented, with respect to its terminals, as a series combination of a step-function voltage source and an initially uncharged capacitance. Similarly, an inductance with initial current can be represented as the parallel combination of a step-function current source and an inductance with no initial current as indicated in Fig. 5-7.

INITIALLY CHARGED CAPACITANCE Consider first the capacitance C connected between terminals a-b as shown in Fig. 5-7a on the left side. The voltage-current relationship for a capacitance is given by

$$v_{ab} = \frac{1}{C}\int_{-\infty}^{t} i_{ab}\,d\tau \qquad (5\text{-}34)$$

where the limits, from $-\infty$ to t, mean that we wish to define q_a as *all* the charge carried to the a side of the capacitance up to the instant of time t. Now, in view of the linearity of the capacitance, we may rewrite Eq. (5-34) as

$$q_a(t) = q_a(0+) + q'(t) \qquad (5\text{-}35)$$

where
$$q_a(0+) = \int_{-\infty}^{0+} i_{ab}\,dt$$

and
$$q'(t) = \int_{0+}^{t} i_{ab}\,d\tau$$

The quantities $q_a(0^+)$ and $q'(t)$ are easily interpreted. By $q_a(0^+)$ we mean the charge on the a side of the capacitance at the instant $t = 0^+$. The symbol $q'(t)$ stands for the additional charge which the current $i_{ab}(t)$ carries to the a side of the capacitance after $t = 0^+$ until time t. Upon substituting Eq. (5-35) in Eq. (5-34), the voltage-current relationship for the capacitance reads: For all time $t \geq 0^+$,

$$v_{ab} = \frac{1}{C} q_a(0^+) + \frac{1}{C} q'(t)$$

or since

$$\frac{q_a(0^+)}{C} = v_{ab}(0^+) = V_0$$

$$v_{ab} = v_{ab}(0^+) + \frac{1}{C} q'(t) \tag{5-36}$$

Since we specify that the problem starts at $t = 0^+$, we may write Eq. (5-36), employing the step function $u(t)$, as

$$v_{ab} = V_0 u(t) + \frac{q'(t)}{C} \tag{5-36a}$$

Equation (5-36*a*) can be interpreted as representing the circuit shown on the right side in Fig. 5-7*a*, if we set $q_{a'}(t)$ equal to the charge on the a' side of the capacitance C in that circuit. The equation relating v_{ab} and i_{ab} in the circuit on the right side of Fig. 5-7*a* reads

$$v_{ab} = V_0 u(t) + \frac{1}{C} \int_{0^+}^{t} i_{ab}\, d\tau$$

where

$$\int_{0^+}^{t} i_{ab}\, d\tau = q_{a'}(t) = q'(t)$$

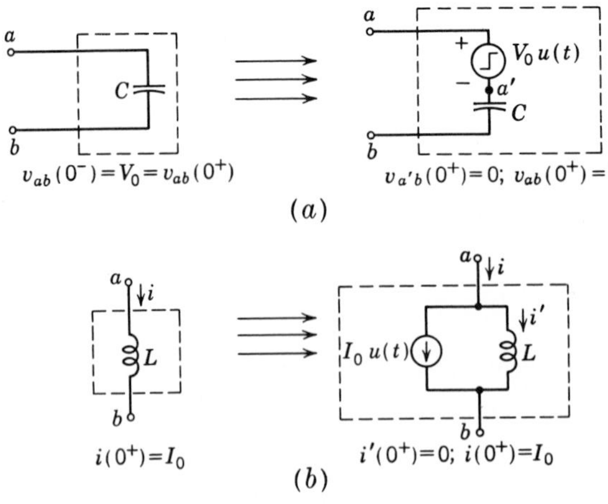

Fig. 5-7 Initial-condition generators. (a) An initially charged capacitance and its voltage-source equivalent. (b) An initially energized (charged) inductance and its current-source equivalent.

It follows that the two circuits shown in Fig. 5-7a are equivalent with respect to the terminals a-b. The charge on the capacitance in the circuit on the right, $q_{a'}(t)$, represents the additional charge which the current i_{ab} carries to either capacitance after $t = 0^+$. The step-function voltage source $V_0u(t)$ represents the initial voltage across the original capacitance.

It is concluded that the series combination of a step-function voltage source and a series capacitance which is uncharged at $t = 0^+$ can be used to represent a capacitance with initial charge for all $t \geq 0^+$. Note that the correspondence has been established only for the voltage-current relationship[1] after $t = 0$ since $v_{ab}(0^-)$ is zero in the circuit on the right side of Fig. 5-7a.

If we wish to calculate the complete response of a circuit in which there are initially charged capacitances, we may therefore represent each of the capacitances as the series combination of source and initially uncharged capacitance and calculate the complete response by superposing the responses due to the individual sources, treating a source which represents an initial condition exactly as any other ideal source.

INDUCTANCE WITH INITIAL CURRENT We now consider an inductance L connected between terminals a-b as shown on the left side of Fig. 5-7b. The voltage-current relationship for an inductance may be written

$$i_{ab} = \frac{1}{L} \int_{-\infty}^{t} v_{ab}\, d\tau$$

Exactly as in the case of the analogous integral for a capacitance, we may attach meaning to $\int_{-\infty}^{t} v_{ab}\, d\tau$. When we write $\int_{-\infty}^{t} v_{ab}\, d\tau$, we mean the total flux linkages which link the inductance,[2] $\Psi(t)$. Since, in a practical circuit, the current in an inductance must satisfy the continuity condition

$$i_{ab}(0^-) = i_{ab}(0^+) \qquad \text{or} \qquad Li_{ab}(0^-) = Li_{ab}(0^+)$$

we may write

$$i_{ab}(t) = \frac{1}{L} \int_{-\infty}^{0^+} v_{ab}\, dt + \frac{1}{L} \int_{0^+}^{t} v_{ab}\, d\tau \qquad t \geq 0^+ \tag{5-37}$$

We now note that the term $\int_{-\infty}^{0^+} v_{ab}\, dt$ represents the flux linkages at the time $t = 0^+$. Hence

$$\frac{1}{L} \int_{-\infty}^{0^+} v_{ab}\, dt = i(0^+)$$

[1] Energy relations must be calculated by using the total charge $q_a(t)$ because the energy stored in a capacitance is not linearly related to the charge.

[2] Since $v_{ab} = L(di_{ab}/dt)$, in a linear inductance where $|v| = N(d\phi/dt)$ we define the flux linkages $N\phi = \Psi = Li$; hence $\Psi = \int_{-\infty}^{t} v\, d\tau$.

Using this result in Eq. (5-37), we have the following voltage-current relationship for an inductance with initial current:

$$i_{ab}(t) = i_{ab}(0^+) + \frac{1}{L}\int_{0^+}^{t} v_{ab}\, d\tau \qquad t \geq 0^+ \tag{5-38}$$

If we define

$$i'(t) = \frac{1}{L}\int_{0^+}^{t} v_{ab}\, d\tau \qquad t \geq 0^+$$

then Eq. (5-38) becomes

$$i_{ab}(t) = i_{ab}(0^+)u(t) + i'(t) \tag{5-38a}$$

But Eqs. (5-38) and (5-38*a*) represent the relationships for the circuit on the right side of Fig. 5-8*b*. Thus, with respect to its terminals, an inductance which carries an initial current can be represented as the parallel combination of a step-function current source and an inductance which carries no initial current. The current source which represents the initial condition may be treated exactly as any other ideal source in the analysis of a circuit.

5-9 Impulse sources for initial conditions

We shall now show that the step-function voltage source in series with the capacitance C (Fig. 5-8) can be represented with respect to its terminal

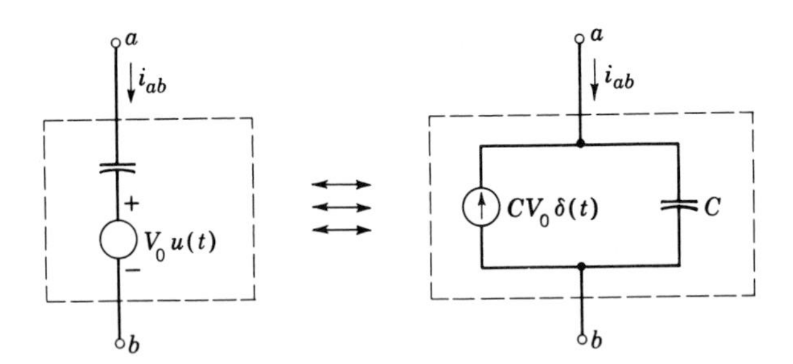

***Fig. 5-8** Representation of initial condition for capacitance by impulse-function sources.*

characteristics by the impulse current sources shown in parallel with C as on the right side of Fig. 5-8. Similarly, the step-function current source in parallel with L (Fig. 5-9) can be represented as an impulse-function voltage source as shown in the illustration.

To prove the equivalences given above, we observe that, in Fig. 5-8, on the left,

$$v_{ab} - V_0 u(t) = \frac{1}{C}\int_{0}^{t} i_{ab}\, d\tau \tag{5-39}$$

Differentiation of both sides of Eq. (5-39) gives

$$\frac{dv_{ab}}{dt} - V_0\delta(t) = \frac{1}{C} i_{ab}$$

or

$$i_{ab} = C\frac{dv_{ab}}{dt} - CV_0\delta(t)$$

Hence i_{ab} consists of the sum of two terms represented by the two parallel branches on the right side of Fig. 5-8. Proof of the dual, the equivalences of Fig. 5-9, is reserved for the reader's exercise. The relationships derived

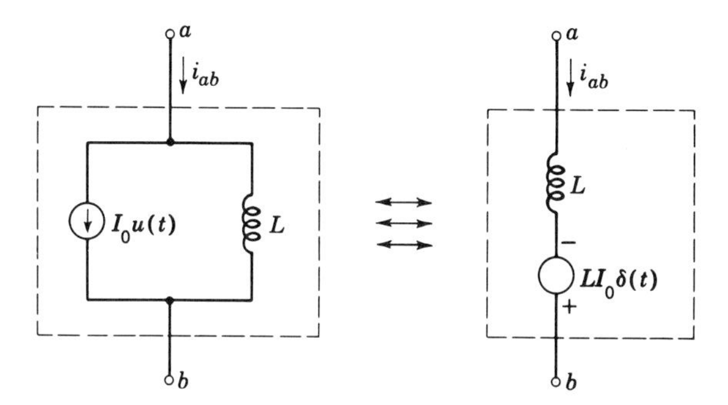

Fig. 5-9 Impulse-function initial-condition generator for inductance.

in this section are special cases of the source-conversion theorems discussed in Chap. 6.

We conclude that the initial-energy storage in an inductance or a capacitance may be correctly represented by either step- or impulse-function sources as shown in Figs. 5-8 and 5-9.

5-10 Response to impulse sources

The equivalences demonstrated in the two preceding sections are useful not only for the purpose of calculating the contribution (to a complete response) of the initial conditions, but also for *interpretation and study* of the response to step and impulse sources. If, for example, a step-function voltage $Vu(t)$ in series with a capacitance C, as shown at the right in Fig. 5-10,

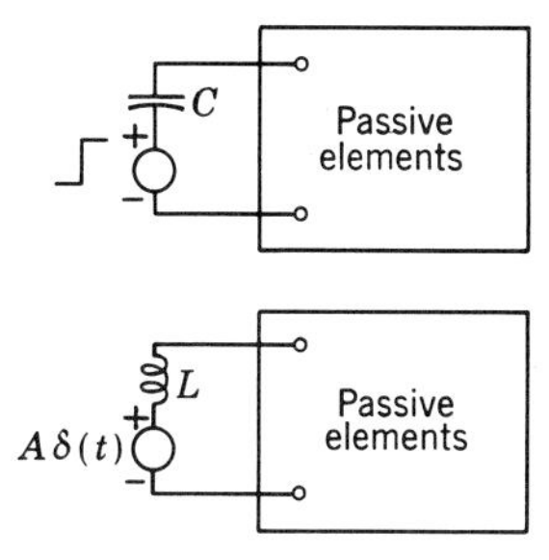

Fig. 5-10 The ideal sources shown can be considered initial-condition generators.

or an impulse-function voltage source $A\delta(t)$ in series with L is *applied* to a passive terminal pair, the effect of such a source is exactly the same as the effect of connecting the element C or L, with the respective initial conditions, to the terminal pair; i.e., the response will be identical with the response of a source-free circuit! The same statement applies if the equivalent current sources are impressed.

A very simple physical argument may serve to illustrate the mathematical conclusions which have been reached. An impulse function, although having infinite amplitude, lasts for zero time. Hence it represents a theoretical mechanism for transferring finite energy in zero time. Moreover, if the impulse occurs at $t = 0$, it has passed (i.e., its value is zero) at $t = 0^+$. Any response observed from $t = 0^+$ onward must therefore be the source-free response since the circuit, for $t \geq 0^+$, is indeed source-free.

In Sec. 2-3 we show that if $a(t)$ is the response of a network to a unit step function $u(t)$, the response of that network to an arbitrary source $v(t)$ can be expressed in terms of $a(t)$ in accordance with Eq. (2-13). In analogy with this result, in Appendix E we show that the response of a network to an arbitrary source may be calculated from the network's response to an impulsive source. Since we have already demonstrated that the impulsive response of a network is the source-free response of that network, we conclude that, if the source-free responses of a network are known, its response to any arbitrary source can be calculated. In Appendix E we use this fact to write a computer program for evaluation of response of a network to different types of sources.

Example 5-5 Calculate the current at $t = 0^+$ in the circuit of Fig. 5-11. The current at $t = 0^-$ is zero.

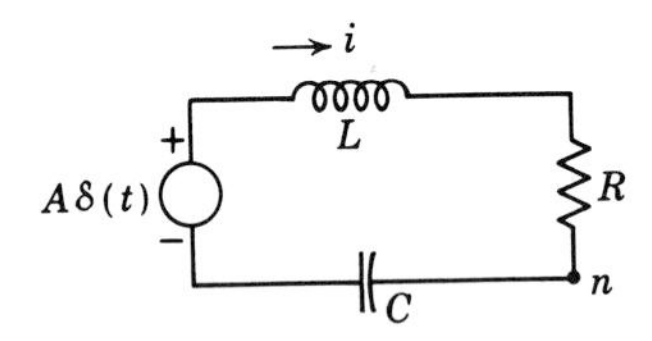

Fig. 5-11 Series R-L-C circuit with impulse-function source.

Solution *Method* 1 (*by initial-condition generator*) The series combination of the impulse source $A\delta(t)$ and the inductance L can be interpreted as an inductance with initial-energy storage. The inductance with initial current I_0 is represented as the series combination of an impulse-function voltage source $LI_0\delta(t)$ and L, as in Fig. 5-9. Therefore the current at $t = 0^+$ in the circuit of Fig. 5-11 is A/L.

Method 2 If we do not interpret the impulse source $A\delta(t)$ as an initial-condition generator, the result for $i(0^+)$ can be obtained by use of the equilibrium equation

$$L\frac{di}{dt} + Ri + \frac{1}{C}q_n = A\delta(t)$$

Now let us integrate this equation from $t = 0^-$ to $t = 0^+$.

$$\int_{0^-}^{0^+} L\frac{di}{dt}\,dt + R\int_{0^-}^{0^+} i\,dt + \frac{1}{C}\int_{0^-}^{0^+} q_n\,dt = \int_{0^-}^{0^+} A\delta(t)$$

By definition of the impulse function, $\int_{0^-}^{0^+} A\,\delta(t) = A$; hence, integrating,

$$L[i(0^+) - i(0^-)] + R[q_n(0^+) - q_n(0^-)] + \frac{1}{C}\int_{0^-}^{0^+} q_n\,dt = A$$

Since the current in the circuit is at all times *finite*, it follows that a finite current cannot transfer nonzero charge to the capacitance in zero time.

$$q_n(0^+) = q_n(0^-) \qquad \int_{0^-}^{0^+} q_n\,dt = 0$$

If the equilibrium equation is to be satisfied at $t = 0$, the current i must experience a jump. Hence

$$L[i(0^+) - i(0^-)] = A$$

and since $i(0^-)$ was given as zero, $i(0^+) = A/L$.

5-11 Detailed examples of complete response calculations. The R-C circuit

In the preceding sections of this chapter general principles of complete response calculations are discussed and the various steps are illustrated individually. In this and the remaining sections of the chapter we review these steps and show solution of complete problems with applications.

Consider as the first example the series R-C circuit shown in Fig. 5-12. Assume that $v(t)$ is the step-function voltage source

$$v(t) = Vu(t)$$

and let it be assumed that $v_{mb}(0^+) = V_0$ is the given continuity condition.

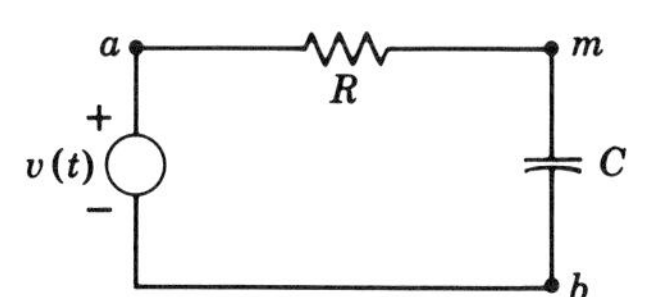

Fig. 5-12 Series R-C circuit with voltage source.

The equilibrium equation for v_{mb} is given by

$$v_{mb} = \frac{1/pC}{R + 1/pC}\,v$$

or, clearing fractions and defining $T = RC$,

$$pTv_{mb} + v_{mb} = V \qquad t \geq 0^+ \tag{5-40}$$

In order to obtain the complete response, we need to add the two components $(v_{mb})_f$, the source-free component, and $(v_{mb})_s$, the component due to the source. The source-free component is the solution of the homogeneous equation

$$T\,\frac{d(v_{mb})_f}{dt} + (v_{mb})_f = 0 \tag{5-41}$$

which is given by the exponential form

$$(v_{mb})_f = Ke^{st} \tag{5-42}$$

Substituting (5-42) in (5-41) gives the now familiar characteristic equation for this circuit,

$$sT + 1 = 0 \qquad \text{or} \qquad s = -\frac{1}{T}$$

so that

$$(v_{mb})_f = Ke^{-t/T}$$

In order to obtain the component of the response due to the source, we set $p = 0$ in Eq. (5-40) so that

$$(v_{mb})_s = V$$

The complete response is found by adding the components

$$v_{mb}(t) = (v_{mb})_f + (v_{mb})_s$$

namely,

$$v_{mb}(t) = V + Ke^{-t/T} \qquad t \geq 0^+ \tag{5-43}$$

The as yet undetermined constant K in Eq. (5-43) is evaluated from the initial condition

$$V_0 = V + K \qquad K = V_0 - V$$

The complete response which satisfies both the initial condition and the equilibrium equation is therefore

$$v_{mb}(t) = V + (V_0 - V)e^{-t/T} \tag{5-44}$$

The form of Eq. (5-44) is the sum of two terms, the component due to the source (in this case a steady-state value V) and the free component $(V_0 - V)e^{-t/T}$. It is sometimes convenient to regroup the terms as follows:

$$v_{mb}(t) = V(1 - e^{-t/T}) + V_0e^{-t/T}$$

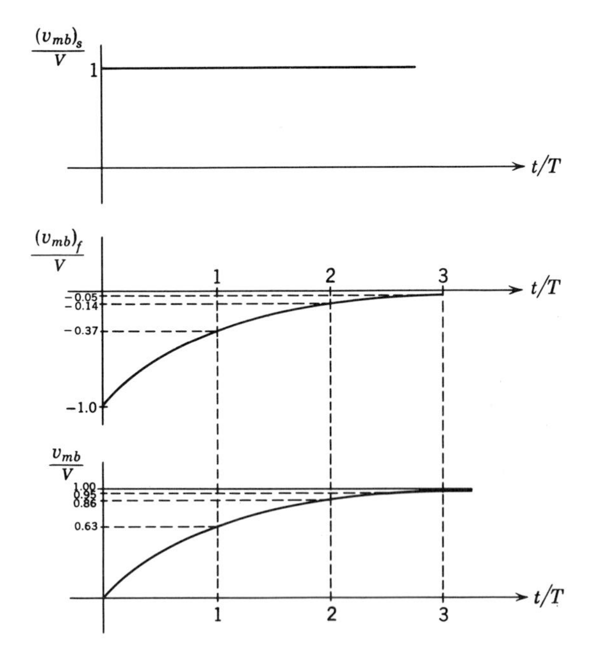

Fig. 5-13 Response of an initially unenergized R-C circuit to step-function voltage source obtained by superposition of forced and free-response components.

In this form the two terms may be interpreted differently:

$$v_{mb}(t) = \begin{pmatrix}\text{complete response if} \\ \text{capacitance is initially} \\ \text{uncharged, } V_0 = 0\end{pmatrix} + \begin{pmatrix}\text{component of response} \\ \text{due to initial charge} \\ \text{on capacitance}\end{pmatrix}$$

This is the form in which the solution would have been found if the initial voltage across the capacitance had been represented by an initial-condition generator. The reader should solve the problem using such a generator and verify this result. To study the waveforms of the response, it is convenient to work with the form of Eq. (5-44), normalizing $v_{mb}(t)$ with respect to the value of the source, V,

$$\frac{v_{mb}(t)}{V} = 1 + \left(\frac{V_0}{V} - 1\right)e^{-t/T}$$

As an example, consider the case $V_0 = 0$. (The case $V_0 = V$ is trivial because in this case no change in stored energy, and therefore no transient, occurs.) The construction of the waveform for the case $V_0 = 0$ is illustrated in Fig. 5-13, where the two components $(v_{mb})_s$ and $(v_{mb})_f$ are added.

Note that the value V is approached exponentially, so that the response is within 5 per cent of this value after a time equal to three time constants has elapsed. The complete response for two nonzero values of V_0 is illustrated in Fig. 5-14. In each

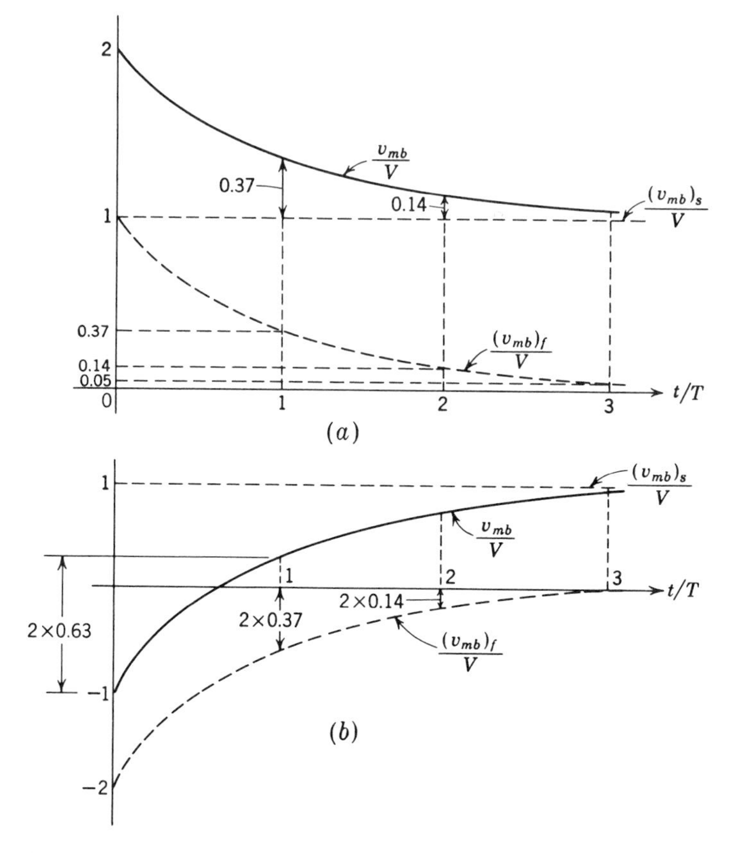

***Fig. 5-14** Response of R-C circuit to step-function voltage source for (a) $V_0 = 2V$; (b) $V_0 = -V$.*

case the resultant starts from its initial value and approaches the final value in such a way that, after an interval equal to one time constant, the difference between the resultant and the final value is reduced by 37 per cent.

$$\begin{aligned} V - v_{mb}(0^+) &= V - V_0 \\ V - v_{mb}(T) &= (V - V_0) \times 0.37 \\ V - v_{mb}(2T) &= (V - V_0) \times 0.14 \\ V - v_{mb}(3T) &= (V - V_0) \times 0.05 \end{aligned}$$

Since the output voltage (across the capacitance) will approach (after three time constants) the step input, another way of describing the response of the *R-C* circuit is to say that it furnishes a *time delay*.

5-12 Integrating property of the R-C circuit

An interesting and exceedingly useful property of the *R-C* circuit is noticed if the response is examined in the time interval $0 < t < T$, that is, at the beginning of the transient. Consider the case in which the capacitance of Fig. 5-12 is initially deenergized so that, for a step voltage source, the response is given by

$$v_{mb}(t) = V(1 - e^{-t/T}) \qquad t \geq 0^+$$

Let the voltage $v_{ab}(t)$ be called the input voltage and $v_{mb}(t)$ the output voltage. If the *beginning* of the transient is studied, $t/T \ll 1$, then the exponential function may be approximated by a finite number of terms in the power series

$$e^{-t/T} = 1 + \left(-\frac{t}{T}\right) + \frac{(-t/T)^2}{2!} + \cdots$$

Using the first two terms only,

$$v_{mb}(t) \approx V\left(1 - 1 + \frac{t}{T}\right)$$

so that for sufficiently "small" values of t the response (voltage across the capacitance) may be approximated by the *ramp* function

$$v_{mb}(t) \approx V\frac{t}{T}u(t) \qquad t \ll T$$

Since the input to the circuit is the step

$$v(t) = Vu(t)$$

it follows that, within the limits of the approximation $(t^2/2T^2 \ll t/T)$, the waveform of the voltage across the capacitance is proportional to the integral of the waveform of the source of voltage.

The special case of the foregoing example can be generalized as follows: In the *R-C* circuit of Fig. 5-12, let $v(t)$ be the input and define $v_C = v_{mb}$ as the response. For this circuit the differential equation is given by

$$T\frac{dv_C}{dt} + v_C = v(t) \tag{5-45}$$

If at any time t

$$\left| T \frac{dv_C}{dt} \right| \gg v_C \tag{5-46}$$

then

$$v_C \approx \frac{1}{T} \int_0^t v \, d\tau + V_0 \tag{5-47}$$

The inequality (5-46) is most often satisfied for $t/T \ll 1$. Under such conditions, and in the interval of time $t \ll T$, Eq. (5-47) is a good approximation. Furthermore, if the initial condition $v_{mb}(0^+) = 0 = V_0$, then

$$v_{mb}(t) \approx \frac{1}{T} \int_0^t v \, d\tau \tag{5-48}$$

These relationships state that, in a series R-C circuit for values of $t \ll T$, and to a first order of approximation, the waveform of the voltage across the capacitance is the integral of the waveform of the voltage across the R and C elements plus the initial value of the voltage across the capacitance.

Thus the R-C circuit can be used to produce, approximately, the integral of a waveform. The inequality (5-46) is the necessary and sufficient condition for an R-C circuit to have the above-mentioned integrating property. The form of this inequality is not very helpful since it refers to the voltage across the capacitance, which is not usually given but is to be found. When the voltage source and R-C circuit are specified and a judgment is required as to whether the voltage across the capacitance is the integral of the applied voltage, we can say that in most cases this is true for values of $t \ll T$. No further generalizations are made beyond this if the source is not specified.

In Fig. 5-15*a* the actual voltage across the capacitance of Fig. 5-15*b* due to a step-function input voltage is shown with the integral of the input voltage, i.e., a ramp function. Note that in this case the difference between the output voltage and the integral of the input voltage is 37 per cent at $t = T$, but only 0.5 per cent at $t = 0.1T$. The integrating property of the R-C circuit will be illustrated again in a later section of this chapter for other specific input waveforms.

5-13 Differentiating property of the R-C circuit

Suppose that in the circuit of Fig. 5-15*b* with the step voltage applied, it is required to find the current in the circuit, $i(t)$, or the voltage across the resistance, $v_{am}(t)$. Since we have the solution for $v_{mb}(t)$, we could apply the equation $q_m = Cv_{mb}$ and find i from dq_m/dt. Alternatively, we could solve for $i(t)$ directly by use of a derived initial condition for $i(0^+)$. Here it is convenient to apply Kirchhoff's voltage law.

$$v_{am} + v_{mb} = v = Vu(t) \qquad v_{mb}(0^+) = V_0$$

Using the solution for the voltage across the capacitance, we obtain

$$v_{am}(t) = V - [V + (V_0 - V)e^{-t/T}] \qquad t \geq 0^+$$

or

$$v_{am}(t) = (V - V_0)e^{-t/T} \qquad t \geq 0^+ \tag{5-49}$$

Also,

$$i(t) = \frac{V - V_0}{R} e^{-t/T} \qquad t \geq 0^+$$

The solution is seen to be the familiar exponential decay shown in Fig. 5-15*c*. We note that the shape of this response could have been predicted from the curves shown in Figs. 5-13 and 5-14; since $i = dq/dt$, Fig. 5-15*c* is proportional *to the slope* of the curves shown in Figs. 5-13 and 5-14. Referring to

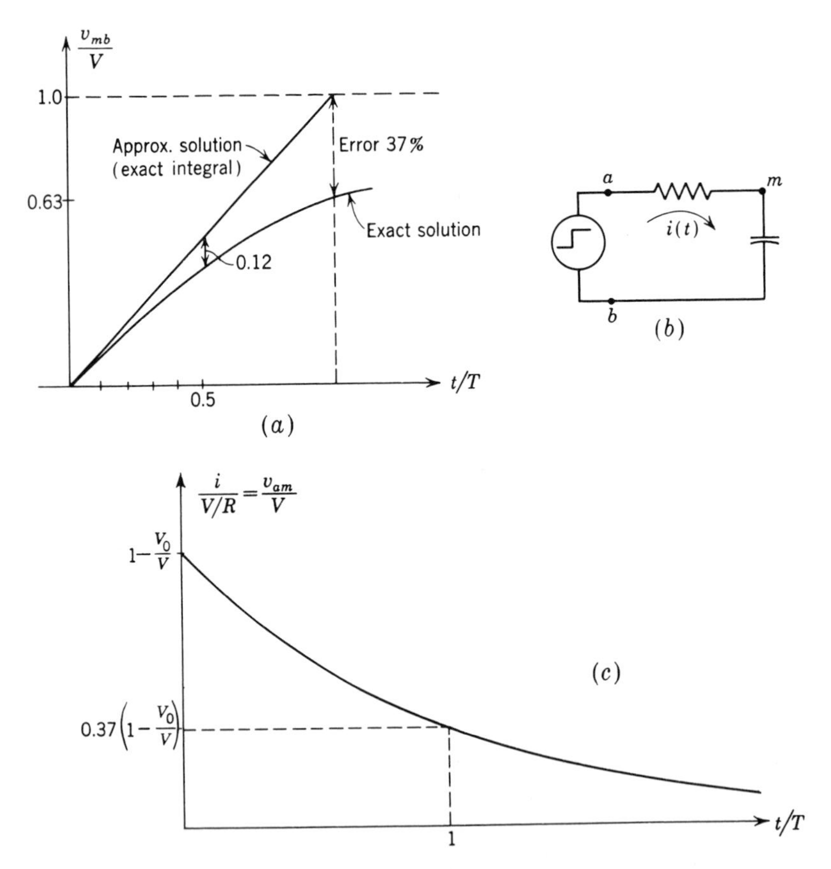

***Fig. 5-15** (a) Illustration of integration property of the R-C circuit for step-function source. (b) Series R-C circuit with step-function voltage source. (c) Illustration of differentiation property of the R-C circuit: The voltage across the resistance for step input approaches zero as t increases.*

Fig. 5-15*c*, we note that the voltage across R is less than 5 per cent of its initial value after three time constants; i.e., it is "practically" zero or equal to the *slope* of the input voltage. For this reason, the *R-C* circuit is called a *differentiating* circuit when the "output" is considered to be the voltage across the resistance, and for $t > 3T$.

The results illustrated in this example can be generalized by noting that, in the R-C circuit of Fig. 5-15b,

$$\begin{aligned} v_{am} &= Ri = RC \frac{d}{dt} v_{mb}(t) \\ &= T \frac{d}{dt} [v(t) - v_{am}(t)] \end{aligned} \tag{5-50}$$

If for all values of t under consideration

$$\left| \frac{dv_{am}}{dt} \right| \ll \left| \frac{dv}{dt} \right| \tag{5-51}$$

then

$$v_{am} \approx T \frac{dv}{dt} \tag{5-52}$$

The condition (5-51) requires that the rate of change of the source voltage be much larger than that of the voltage across the resistance. This condition, which is necessary, is misleading when stated alone, because it appears that the inequality will best be fulfilled when the source function has a jump ("infinite" rate of change). We must recall, however, that because of the continuity of the voltage across the capacitance, a discontinuity in $v(t)$ will result in an equal discontinuity in the voltage across the resistance. It is therefore useful to examine the equation of the R-C circuit in a slightly different form.

We start with $v_{mb} = v(t) - v_{am}$. Since

$$v_{mb} = \frac{1}{C} \int_{-\infty}^{t} i \, d\tau = \frac{1}{T} \int_{-\infty}^{t} v_{am} \, d\tau$$

we observe that

$$|v(t)| \gg |v_{am}| \tag{5-53}$$

then

$$v_{mb} = \frac{1}{T} \int_{-\infty}^{t} v_{am} \, d\tau \approx v(t)$$

or, differentiating both sides,

$$v_{am} \approx T \frac{dv}{dt} \tag{5-52}$$

We therefore conclude that both conditions, (5-51) and (5-53), must be fulfilled for the approximation (5-52) to hold. Thus the source function must be larger than the voltage across the resistance, and the rate of change of the source function must be much larger than the rate of change of the voltage across the resistance to obtain (approximate) differentiating action in an R-C circuit. These conditions are often fulfilled more than three time constants after the source has had a jump. The result of this paragraph and the preceding paragraphs may be summarized as follows:

Under certain conditions the series R-C circuit can furnish an output voltage which is *approximately* the *integral* of the impressed voltage if the voltage across the capacitance is considered the "output" and if the beginning of the time response ($t < T$) is considered only (Fig. 5-16*a*).

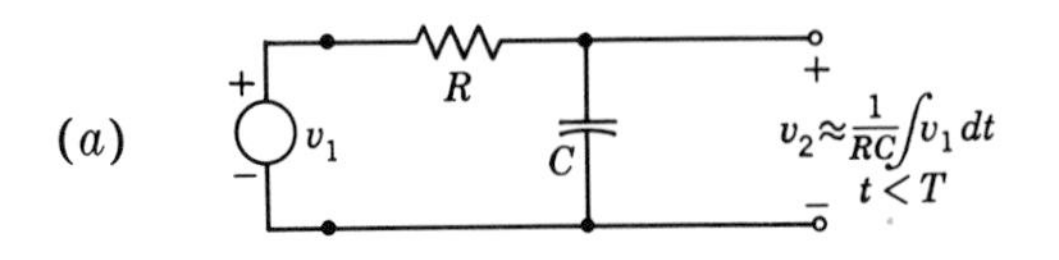

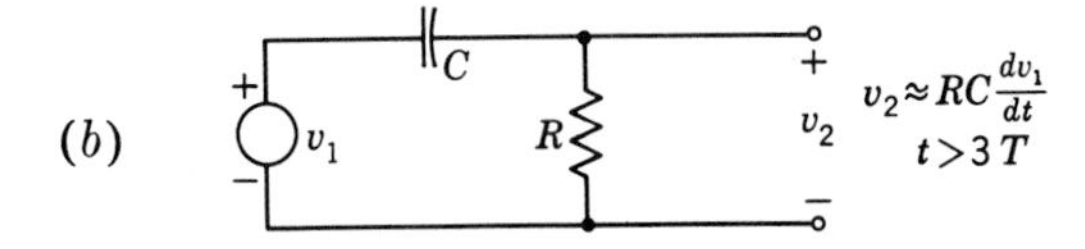

Fig. 5-16 The R-C circuit as an integrating and as a differentiating circuit. In addition to the condition stated on each diagram, inequality (5-46) must be satisfied for (a), and inequalities (5-51) and (5-53) apply for (b).

Under certain conditions the series R-C circuit can furnish an output voltage which is *approximately* the *derivative* of the impressed voltage if the voltage across the resistance is considered the "output" voltage (Fig. 5-16*b*).

5-14 Response to impulse Let us consider again the equation of the current in the R-C circuit of Fig. 5-12 for step input of value $v_1(t) = Vu(t)$.

$$i(t) = \frac{V - V_0}{R} e^{-t/T} \qquad t \geq 0^+ \tag{5-54}$$

The *area* under the curve of the current [Eq. (5-54)] until any time t represents the charge carried to the m side of the capacitance in the time interval $t = 0^+$ to the arbitrary time chosen,

$$q(t) - Q_0 = \int_{\tau=0}^{\tau=t} i\,d\tau = \frac{V - V_0}{R}(-T)e^{-\tau/T}\Big|_{\tau=0}^{\tau=t} = C(V - V_0)(1 - e^{-t/T})$$

where $T = RC$. Let us assume now that $Q_0 = 0$ ($V_0 = 0$); then the area under the curve represented by Eq. (5-54) is $Q = CV$ if we consider the time interval from $t = 0$ to $t = \infty$. This result is independent of the resistance R. Consider now the four examples of response curves for $i(t)$ shown in Fig. 5-17 with *non*normalized axes. If curve 1 represents the current response for a certain R-C circuit when C is fixed and R is adjusted to a value R_1, then curve 2 represents the response of the same circuit if $R = \frac{1}{2}R_1$, curve 3 represents the response when $R = \frac{1}{3}R_1$, and curve 4 represents the response when $R = \frac{1}{5}R_1$. As R is decreased, the value of the current $i(t)$ at $t = 0^+$ *increases, but the area under the curve remains constant and equal to* CV. In the *limit,* as R is reduced to zero, the time constant becomes zero, the initial value of the current $i(0^+) = I_0$ becomes infinite, *but the area under the curve which is independent of* R *remains* CV. We recall that this type of infinite pulse, lasting zero time but with finite area, defines an impulse function

$$\lim_{R \to 0} \frac{V}{R} e^{-t/RC} u(t) = VC\delta(t)$$

$$\lim_{T \to 0} \frac{e^{-t/T}}{T} u(t) = \delta(t)$$

Therefore, if the resistance is zero, the capacitance will charge to the value of the impressed voltage in zero time, and an impulse of current will occur to transfer this finite charge. In practice, such a situation is impossible because of the ever-present "residual" resistance. The present discussion is nevertheless of great value since it emphasizes the possibility of obtaining large-amplitude current pulses with this circuit in spite of the fact that no constant (direct) current can flow. This property may be useful (as in pulse-forming networks) or undesirable (as in gas-tube rectifiers).

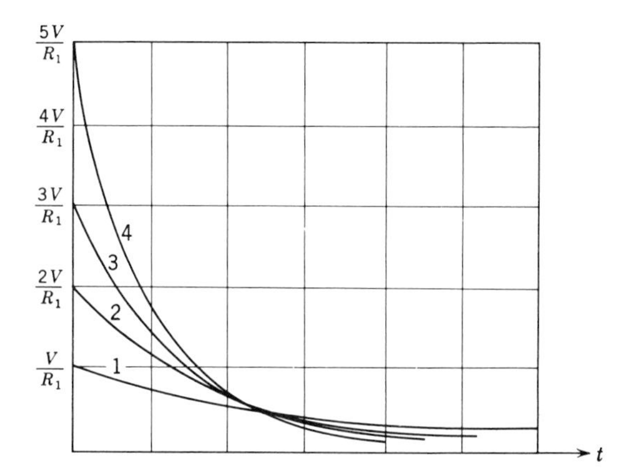

Fig. 5-17 Current response of step-excited R-C circuit with different resistance values.

5-15 *Rectangular pulse applied to R-C circuit* Consider now the rectangular pulse

$$v(t) = 0 \qquad t < 0$$
$$v(t) = V \qquad 0 < t < T_1$$
$$v(t) = 0 \qquad t > T_1$$

applied to an initially deenergized R-C circuit as shown in Fig. 5-18. The response may be found either by solving the problem first in the time interval from $t = 0$ to $t = T_1$,

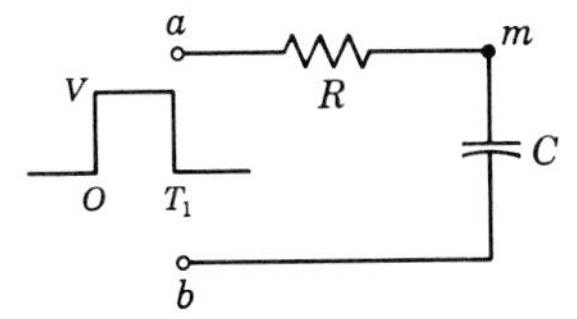

Fig. 5-18 Rectangular pulse applied to R-C circuit.

then to $t = \infty$, or by superposition of the responses to two step functions. Both methods will be illustrated.

If we consider the time interval $t = 0$ to $t = T_1$, the response is given by the step response [Eq. (5-44) or (5-49)] with $V_0 = 0$. (There is no loss in generality in taking this case since the case $V_0 \neq 0$ can be included easily by using an initial-condition generator.)

$$v_{mb}(t) = V(1 - e^{-t/T}) \qquad 0^+ \leq t \leq T_1 \tag{5-55}$$

or

$$v_{am}(t) = Ve^{-t/T} \qquad 0^+ \leq t < T_1$$

At $t = T_1^-$ the voltage across the capacitance is $v_{mb}(T_1^-) = V(1 - e^{-T_1/T})$. Since the voltage across the capacitance cannot change in zero time, $v_{mb}(T_1^-) = v_{mb}(T_1^+)$. Now consider the new time scale $t' = t - T_1$. When $t' = 0$, the terminals a-b become short-circuited and the voltage across the capacitance will decay exponentially to zero,

$$v_{mb}(t') = V(1 - e^{-T_1/T})e^{-t'/T} \qquad t' > 0$$

since, for $t' > 0$, $v_{mb} + v_{am} = 0$,

$$v_{am}(t') = -V(1 - e^{-T_1/T})e^{-t'/T} \qquad t' > 0$$

Note that the voltage across the resistance just before $v_{ab} = 0$, at $t = T_1^-$, is

$$v_{am}(T_1^-) = Ve^{-T_1/T}$$

and just after v_{ab} becomes zero, at $t = T_1^+$,

$$v_{am}\Big|_{\substack{t=T_1^+ \\ t'=0^+}} = -V + Ve^{-T_1/T}$$

The voltage across the resistance is seen to change discontinuously at $t = T$; in fact, the voltage v_{am} decreases by the value V; $v_{am}(T_1^-) - v_{am}(T_1^+) = V$. This can be reasoned immediately from the following considerations: At $t = T_1$ the impressed voltage decreases abruptly. Since the voltage across C cannot change discontinuously, the impressed voltage jump appears across the resistance at $t = T_1$. These results are illustrated in Fig. 5-19 for the case $T_1 = T$.

It is left as an exercise for the reader to show that, when the time constant of the circuit is short *compared* with the duration of the pulse, the voltage across the capacitance is a "good" reproduction of the input pulse, and the voltage across the resistance (except

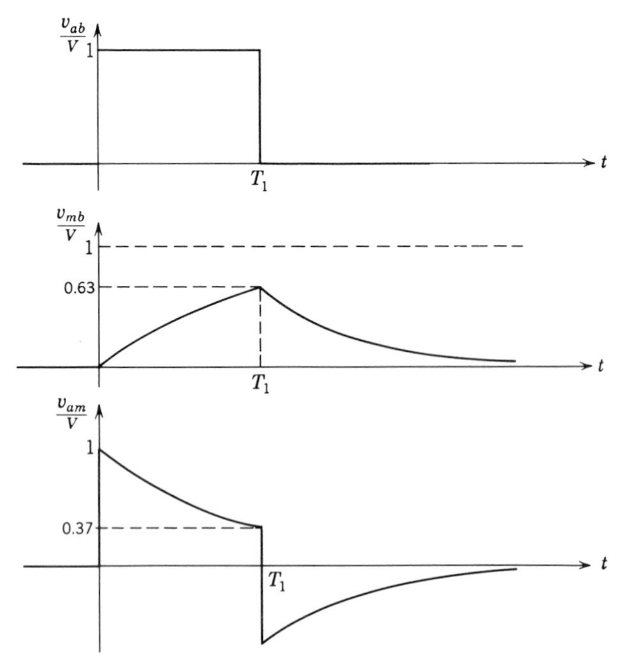

Fig. 5-19 Responses of R-C circuit to rectangular pulse of duration $T_1 = T = RC$.

near $t = 0$ and near $t = T_1$) is approximately proportional to the derivative of the rectangular pulse (i.e., zero). Furthermore, when the time constant of the R-C circuit is large *compared* with the duration of the pulse, the voltage across the resistance is a "good" reproduction of the input waveform, and the voltage across the capacitance is proportional (approximately) to the integral of the rectangular pulse (i.e., modified ramp function).

As was stated above, the response to the rectangular pulse may also be obtained by superposition. The input pulse is represented as the sum of two steps,

$$v_{ab}(t) = Vu(t) - Vu(t - T)$$

so that the responses are

$$v_{mb}(t) = V(1 - e^{-t/T})u(t) - V(1 - e^{-(t-T_1)/T})u(t - T_1)$$

and

$$v_{am}(t) = Ve^{-t/T}u(t) - Ve^{-(t-T_1)/T}u(t - T_1)$$

This procedure is illustrated for the case $T = T_1$ in Fig. 5-20.

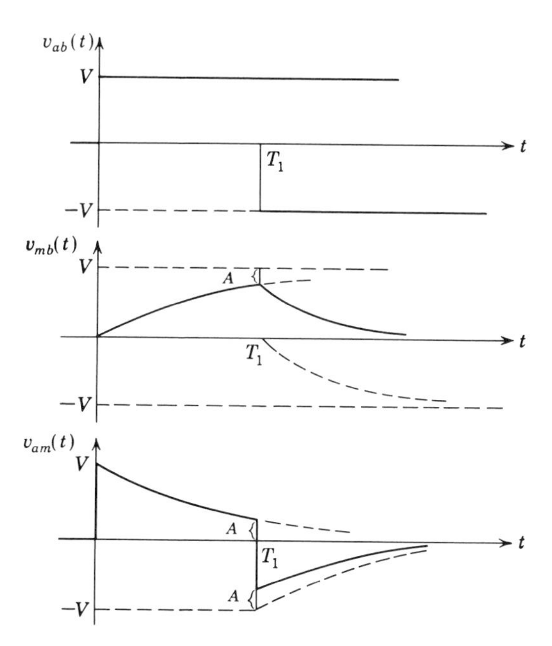

Fig. 5-20 Response of R-C circuit to rectangular pulse obtained by superposition.

An interesting result concerning the analytical utility of impulses may be obtained if a rectangular pulse of very short duration compared with the time constant of the circuit is studied. Consider a rectangular pulse of amplitude V and duration $T_1 \ll T$, shown in Fig. 5-21a. As shown above [(Eq. 5-55)], the voltage across the capacitance follows the equation

$$v_{mb}(t) = V(1 - e^{-t/T}) \qquad 0^+ \leq t \leq T_1$$

Thus, at the time $t = T_1^-$, $v_{mb}(T_1^-)$ is

$$v_{mb}(T_1^-) = V(1 - e^{-T_1/T})$$

Since $T_1/T \ll 1$,

$$e^{-T_1/T} \approx 1 - \frac{T_1}{T}$$

or

$$v_{mb}(T_1^-) \approx V\frac{T_1}{T}$$

For $t > T_1$ the capacitance will discharge exponentially through the resistance

$$v_{mb}(t) = v_{mb}(T_1^+)e^{-(t-T_1)/T} \qquad t \geq T_1^+$$

Since $v_{mb}(T_1^+) = v_{mb}(T_1^-)$, the equation for the discharge of the capacitance is given approximately by

$$v_{mb}(t) \approx V\frac{T_1}{T}e^{-(t-T_1)/T} \qquad t \geq T_1^+$$

This result is illustrated in Fig. 5-21*b*.

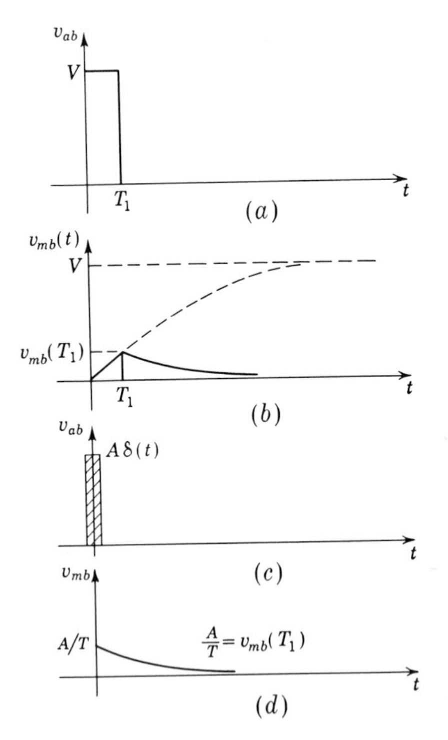

Fig. 5-21 Response of R-C circuit to pulse and impulse sources.

Consider now an impulse of voltage applied to the circuit as indicated in Fig. 5-21*c*.

$$v(t) = A\,\delta(t)$$

Because of the impulse, the voltage across the capacitance will be changed discontinuously, and for $t \geq 0^+$ there will be a voltage across it although no energy was stored in it at $t = 0^-$. The change in voltage across the capacitance is found from the equation

$$T\frac{dv_{mb}}{dt} + v_{mb} = A\,\delta(t)$$

Since at $t = 0$ the impulse is infinite, v_{mb} will be insignificant compared with its time derivative. Hence

$$v_{mb}(t)\Big|_{0^-}^{0^+} = \frac{A}{T}\int_{0^-}^{0^+} \delta(t)\,dt = \frac{A}{T}$$

and the capacitance will charge to the voltage A/T and then (for $t > 0^+$) discharge exponentially through the resistance as indicated in Fig. 5-21*c*.

On comparing Fig. 5-21*b* with Fig. 5-21*d*, it is noted that if we adjust the area of the rectangular pulse (Fig. 5-21*a*) so that $VT_1 = A$, the response for $t > T_1$ (Fig. 5-21*b*) will differ only slightly from the response to an impulse (Fig. 5-21*d*).

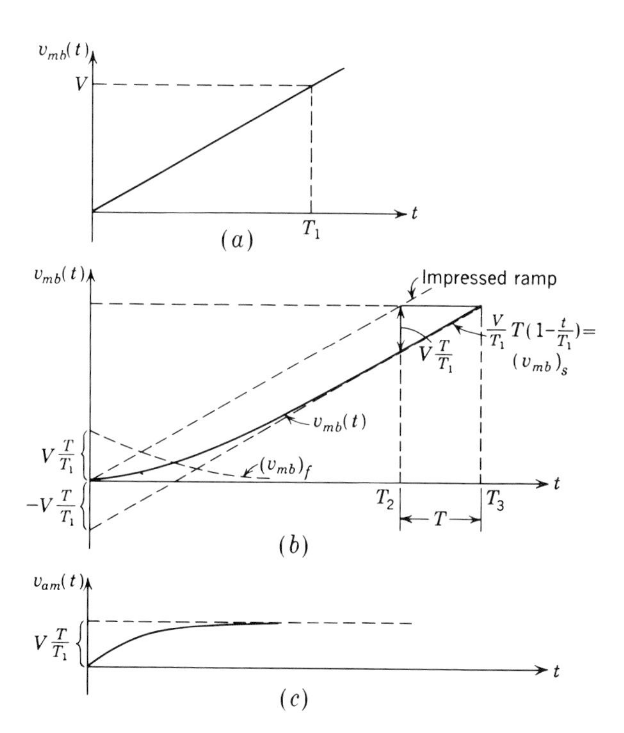

Fig. 5-22 *Response of initially unenergized R-C circuit to ramp-function source.*

5-16 *Ramp-function voltage source applied to the series R-C circuit* If the ramp function (Fig. 5-22*a*)

$$v(t) = \frac{V}{T_1}\,tu(t)$$

is applied to the *R-C* circuit of Fig. 5-18, the equation for the voltage across the capacitance is

$$pTv_{mb} + v_{mb} = \frac{V}{T_1}\,tu(t) \qquad t \geq 0^+$$

For the purpose of this illustration, we shall assume that the capacitance is initially uncharged so that $v_{mb}(0^-) = 0$; hence $v_{mb}(0^+) = 0$.

The complete response is of the form

$$v_{mb}(t) = [v_{mb}(t)]_s + Ke^{-t/T} \qquad T = RC_1 t \geq 0^+ \tag{5-56}$$

In Eq. (5-56) the free component of the response is written with the undetermined constant K. The component of the response due to the source $[v_{mb}(t)]_s$ must now be found. We know this solution to have the form

$$[v_{mb}(t)]_s = A_1 + A_2 t \qquad \text{and} \qquad \frac{d(v_{mb})_s}{dt} = A_2 \tag{5-57}$$

Substituting the solution (5-57) in the equilibrium equation, we obtain

$$TA_2 + A_1 + A_2 t = \frac{V}{T_1} t$$

or

$$TA_2 + A_1 + t\left(A_2 - \frac{V}{T_1}\right) = 0 \tag{5-58}$$

Since Eq. (5-58) must apply at all instants of time after $t = 0$, it follows that

$$A_2 = \frac{V}{T_1} \qquad \text{and} \qquad A_1 = -TA_2 = -\frac{VT}{T_1}$$

so that the complete response at this point is

$$v_{mb}(t) = \frac{VT}{T_1}\left(-1 + \frac{t}{T}\right) + Ke^{-t/T} \qquad t \geq 0^+ \tag{5-59}$$

The constant K in Eq. (5-59) is evaluated from the initial condition.

$$v_{mb}(0^+) = v_{mb}(0^-) = 0 = \frac{VT}{T_1}(-1 + 0) + Ke^0$$

or

$$K = \frac{VT}{T_1}$$

The complete response is therefore

$$v_{mb}(t) = \frac{VT}{T_1}\left(-1 + \frac{t}{T} + e^{-t/T}\right) \qquad t \geq 0^+ \tag{5-60}$$

This solution is illustrated in Fig. 5-22*b*. We note that after the free component has decayed ($t > 3T$), the voltage across the capacitance has the same shape as the input voltage but is delayed in time. The difference between the input and the voltage across the capacitance is seen to be VT/T_1. The time delay is calculated as follows: At some time $t = T_2 > 3T$, the input is VT_2/T_1. Voltage across the capacitance is $VT_2/T_1 - VT/T_1$. The voltage across the capacitance will have the value VT_2/T_1 at $t = T_3$.

$$V\frac{T_2}{T_1} = \frac{VT}{T_1}\left(-1 + \frac{T_3}{T}\right)$$

or

$$T_2 + T = T_3$$

It is therefore seen that the time delay introduced by the R-C circuit is (after the exponential has decayed) equal to the time constant of the R-C circuit.

The voltage across the resistance may be found by subtraction.

$$v_{am} = v_{ab} - v_{mb}$$

so that

$$v_{am}(t) = \frac{VT}{T_1}(1 - e^{-t/T}) \qquad t \geq 0^+$$

The result is illustrated in Fig. 5-22*c*. Note that the voltage across the resistance after a time equal to three time constants will be within 5 per cent of $T(V/T_1)$, so that the voltage across the resistance (for $t > 3T$) is T times the slope of the input ramp.

The integrating property of the circuit for this impressed waveform can be seen by studying the voltage across the capacitance [Eq. (5-60)] for $t < T$. The expansion of the exponential function

$$e^{-t/T} = 1 - \frac{t}{T} + \frac{(t/T)^2}{2!} + \cdots$$

is substituted in Eq. (5-60)

$$v_{mb}(t) \approx \frac{VT}{T_1}\left(-1 + \frac{t}{T} + 1 - \frac{t}{T} + \frac{t^2}{2T^2} - \cdots\right)$$

Using the first nonzero term in the result,

$$v_{mb}(t) \approx \frac{V}{2T_1T}t^2 \qquad 0^+ < t < T \tag{5-61}$$

Equation (5-61) is the equation of a parabola, the integral of the ramp function.

5-17 *Exponentially decaying voltage applied to the series R-C circuit* As a final example in the study of the *R-C* circuit, consider the voltage

$$v(t) = Ve^{-t/T_1}u(t)$$

applied to the series *R-C* circuit of Fig. 5-12. The equilibrium relationship for v_{mb} is given by

$$v_{mb} = \frac{1}{1 + pT}Ve^{-t/T_1}$$

so that (Sec. 3-15), if $T_1 \neq T$, the component of the response due to the source is given by

$$v_{mb}(t) = \frac{1}{1 - T/T_1}Ve^{-t/T_1} \qquad T \neq T_1$$

The complete response has the form

$$v_{mb} = \frac{1}{1 - T/T_1}Ve^{-t/T_1} + Ke^{-t/T} \tag{5-62}$$

and if $v_{mb}(0^+) = V_0$, the constant K is given by

$$K = V_0 - \frac{V}{1 - T/T_1} \qquad T \neq T_1 \tag{5-63}$$

so that the solution (for $T \neq T_1$) is

$$v_{mb}(t) = \frac{V}{1 - T/T_1}(e^{-t/T_1} - e^{-t/T}) + V_0e^{-t/T} \qquad T \neq T_1,\ t \geq 0^+ \tag{5-64}$$

The case $T = T_1$ is discussed in Chap. 8.

5-18 The parallel R-C circuit; other single-energy circuits The equilibrium equation of the parallel R-C circuit shown in Fig. 5-23 with the current source $i(t)$ is

$$\frac{1}{R}v_{ab} + C\frac{dv_{ab}}{dt} = i(t) \qquad v_{ab}(0^+) = v_{ab}(0^-) \tag{5-65}$$

This equation is identical in form with the equation of the series circuit, and may be treated exactly as in the preceding examples. Details are reserved for reader exercises.

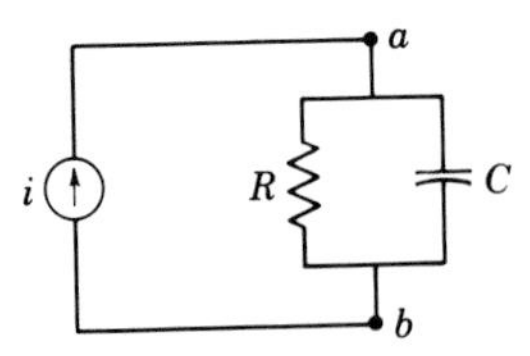

Fig. 5-23 Parallel R-C circuit with source.

R-L CIRCUIT The equilibrium equations of the two circuits of Fig. 5-24*a* and *b* are given by

Fig. 5-24*a*:
$$Ri + L\frac{di}{dt} = v(t) \qquad i(0^+) = i(0^-) \tag{5-66}$$

Fig. 5-24*b*:
$$\frac{v_{ab}}{R} + \frac{1}{L}\int_{-\infty}^{t} v_{ab}\, d\tau = i(t)$$

or
$$\frac{1}{R}\left(L\frac{di_L}{dt}\right) + i_L = i(t) \qquad i_L(0^+) = i_L(0^-) \tag{5-67}$$

Equations (5-66) and (5-67) are again first-order linear differential equations, so that the procedures for solution are identical with those for the R-C circuit. Attention is

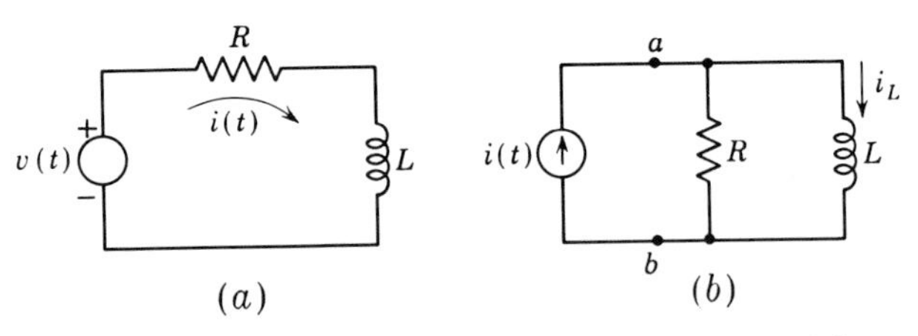

Fig. 5-24 Series and parallel R-L circuits with sources.

called to the fact that, in the inductance, the current cannot change abruptly (unless impulse voltages are impressed), whereas for a capacitance, the voltage cannot change abruptly (unless impulse currents are impressed).

A SERIES-PARALLEL CIRCUIT In the circuit of Fig. 5-25, let $v_C(0^+) = 4$ and let the source be given by the ramp:

$$v(t) = 10tu(t)$$

Let it be required to formulate $i(t)$ for all $t \geq 0^+$. We first formulate the equilibrium equation for $i(t)$:

$$i(t) = \frac{v_C}{R_1} - \frac{v_2}{R_2}$$

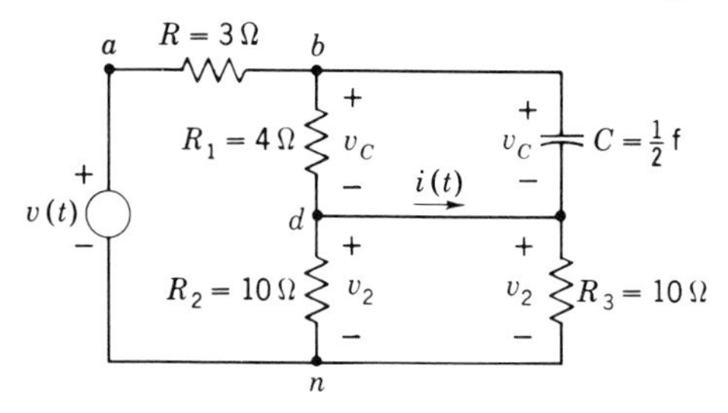

Fig. 5-25 Series-parallel circuit analyzed in Sec. 5-18.

and from voltage division,

$$v_C = \frac{\dfrac{R_1/pC}{R_1 + 1/pC}}{R + \dfrac{R_1/pC}{R_1 + 1/pC} + \dfrac{R_2R_3}{R_2 + R_3}} v$$

$$v_2 = \frac{R_2R_3/(R_2 + R_3)}{R + \dfrac{R_1/pC}{R_1 + 1/pC} + \dfrac{R_2R_3}{R_2 + R_3}} v$$

Clearing fractions and using the numerical values given,

$$v_C = \frac{1}{4p + 3} v \qquad \frac{v_C}{R_1} = \frac{1}{16p + 12}$$

$$v_2 = \frac{10p + 5}{16p + 12} v \qquad \frac{v_2}{R_2} = \frac{p + \frac{1}{2}}{16p + 12}$$

Hence the equilibrium equation for $i(t)$ is given by

$$i(t) = \frac{\frac{1}{2} - p}{16p + 12} v$$

or

$$(16p + 12)i = (\tfrac{1}{2} - p)10t \qquad t \geq 0^+$$

or

$$(16p + 12)i = 5t - 10 \qquad t \geq 0^+$$

The component of the response due to the source has the form

$$i_s = A + Bt$$

Hence

$$(16p + 12)(A + Bt) = 16B + 12A + 12Bt$$

and

$$16B + 12A + 12Bt = 5t - 10$$

so that $12B = 5$, $B = \frac{5}{12}$, and $A = -\frac{25}{18}$ and

$$i_s = \tfrac{5}{6}(\tfrac{1}{2}t - \tfrac{5}{3})$$

Since the characteristic root is $s = -\frac{3}{4}$ [$(16s + 12) = 0$], the complete response has the form

$$i(t) = \tfrac{5}{6}(\tfrac{1}{2}t - \tfrac{5}{3}) + Ke^{-3t/4}$$

To evaluate K we need $i(0^+)$. Since $v_C(0^+) = 4$, $(v_C/R_1)_{0^+} = \frac{4}{4} = 1$ amp. Because

$$v_{ab}(0^+) + v_C(0^+) + v_2(0^+) = v(0^+) = 0$$
$$(v_{ab} + v_2)_{0^+} = -4$$

and because $R_{dn} = 5$ and $R_{ab} = 3$, the initial value of v_2 is

$$v_2(0^+) = \tfrac{5}{8} \times (-4) = -\tfrac{5}{2}$$

Hence

$$\frac{v_2(0^+)}{R_2} = -\tfrac{1}{4}$$

and

$$i(0^+) = 1 - (-\tfrac{1}{4}) = \tfrac{5}{4}$$

Thus

$$-\tfrac{25}{18} + K = \tfrac{5}{4}$$

and

$$K = +\tfrac{95}{36}$$

so that the complete response is given by

$$i(t) = 0.417t - 1.39 + 2.64e^{-3t/4} \qquad t \geq 0^+$$

5-19 Double energy and other circuits The techniques described above can be applied to the solution of second- and higher-order equilibrium equations describing the behavior of circuits with two or more independent energy-storing elements. As the number of energy-storing elements increases, the complexity of the problem increases not only because two or more initial conditions must be satisfied but also because these initial conditions may have to be derived from the continuity conditions and the network equations. The techniques described in Chap. 8 (Laplace Transform) are suited for the solution of such problems.

PROBLEMS

5-1 The source-free response of networks is given as follows; in each case write the characteristic equation in factored form. (*a*) $v_f = 5e^{-3t} + 2e^{-2t} - 7e^{-2t}\cos 5t$; (*b*) $v_f = 10te^{-t} + e^{-2t}$; (*c*) $v_f = 2e^{-t} + 3te^{-t}\cos(t - 30°) - 4e^{-t}\cos(t + 60°)$; (*d*) $v_f = 5t^3e^{-t} + 3e^{-t}$.

5-2 In the circuit of Fig. 5-4, let $R_1 = 1$ ohm, $C = 1$ farad, $L = 1$ henry, and find the value of R_2 so that the circuit is critically damped.

5-3 A series *R-L* circuit is defined by the equilibrium equation $(p + 1)i = v$. In each of the following cases the complete response is given; in each case find the component of the response due to the source. (*a*) $5e^{-t} + t + 2e^{-4t}$; (*b*) $t + 5e^{-t}$; (*c*) $t - 1 + 6e^{-t}$; (*d*) $te^{-t} + 5$.

5-4 In a certain circuit the response v_2 is related to the source i_s by the equation $(p^2 + 4p + 3)v_2 = (p + 2)i_s$. (*a*) What are the characteristic roots? (*b*) Identify the source function if the component of the response due to the source is (1) $(v_2)_s = V_2 = 10$; (2) $(v_2)_s = 10t$; (3) $(v_2)_s = 10 + 10t$; (4) $(v_2)_s = 5 + 6t$; (5) $(v_2)_s = 5e^{-5t}$. (*c*) Find the component of the response due to the source (1) $i_s = I_s = 10$; (2) $i_s = 3t$; (3) $i_s = 6t - 5$; (4) $i_s = 5e^{-2t}$; (5) $i_s = 5e^{-6t}$. (*d*) Show that the source function $i_s = 10e^{-t}$ produces the response component Ate^{-t}, and find A by substitution in the given equilibrium equation.

5-5 In the circuit shown in Fig. P5-5, calculate the steady-state value V_{ab}.

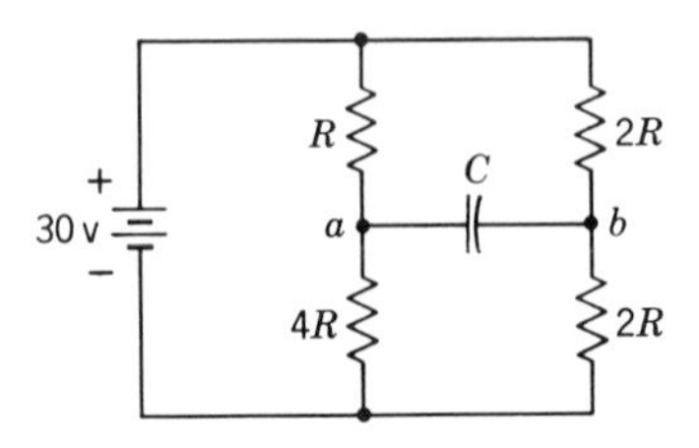

Fig. P5-5

5-6 In the circuit shown in Fig. P5-6, calculate the steady-state value I.

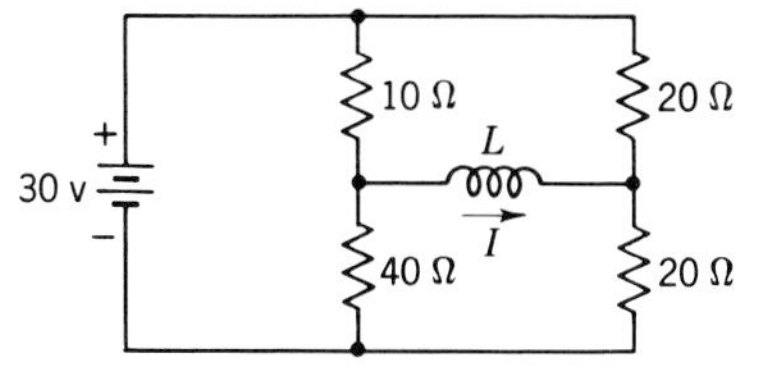

Fig. P5-6

5-7 In the circuit shown in Fig. P5-6, calculate the value of L if it stores 2 joules in the steady state.

5-8 Calculate the steady-state value of the voltage across each capacitance in the circuit of Fig. P5-8.

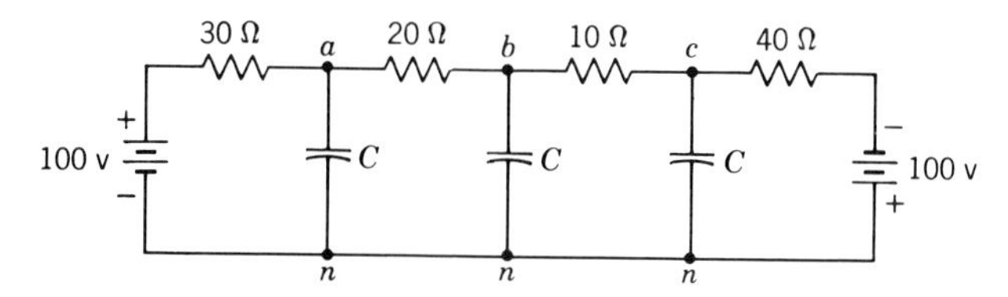

Fig. P5-8

5-9 Calculate the steady-state values V_{an} and I in the circuit shown in (*a*) Fig. P5-9*a*; (*b*) Fig. P5-9*b*.

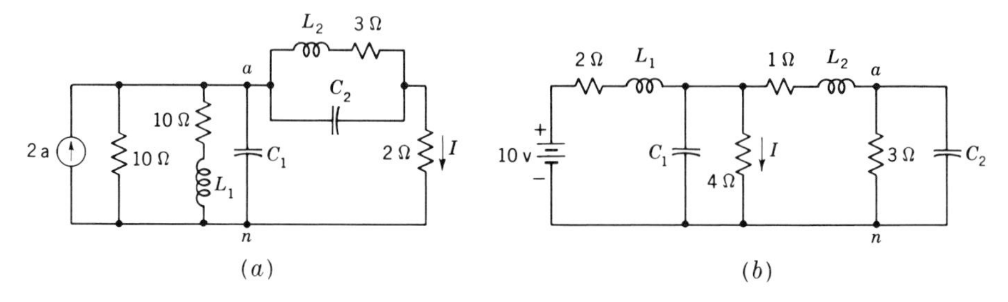

Fig. P5-9

5-10 In the circuit shown in Fig. P5-10, the power dissipated by R_1 in the steady state is 4,500 watts. The steady-state value of the voltage V_{dn} is 90 volts. Calculate the following steady-state values: (*a*) V_{ab}; (*b*) I; (*c*) the power delivered to R_2; (*d*) the power delivered by the 100-volt source.

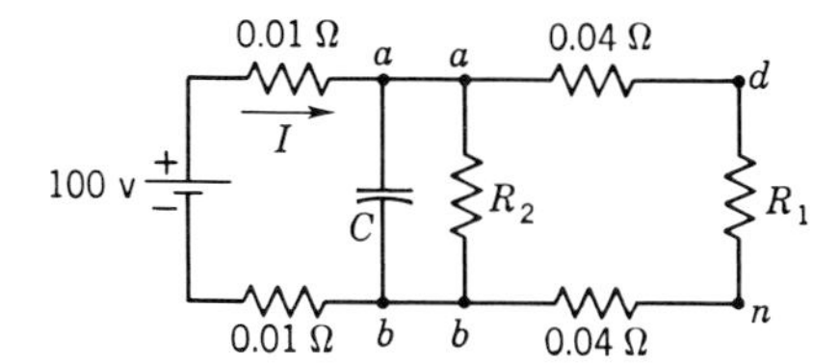

Fig. P5-10

5-11 In the circuit shown in Fig. P5-11, E_1, E_2, and E_3 have certain fixed values. When $E_4 = 0$, $V_{ab} = 30$ volts. Calculate E_4 so that $V_{ab} = 20$ volts.

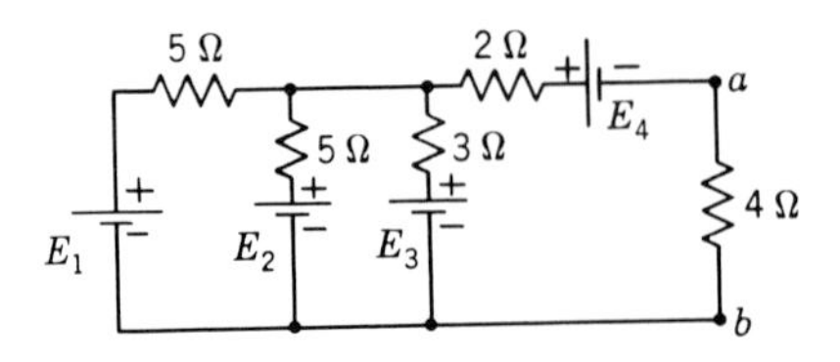

Fig. P5-11

5-12 Use the principle of superposition to find the equilibrium equation that relates v_{an} to the source functions in the circuit of Fig. P5-12.

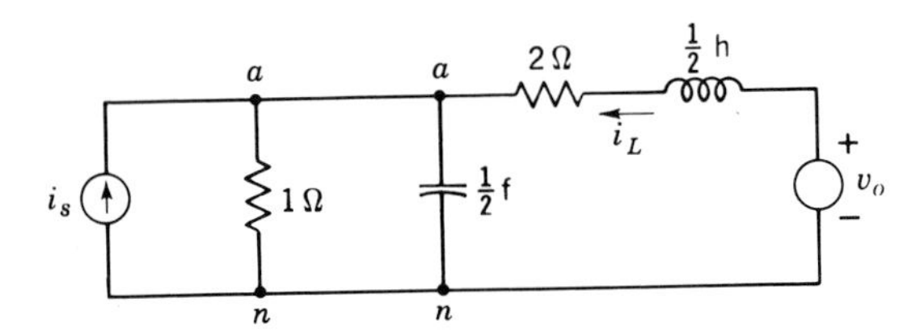

Fig. P5-12

5-13 In a certain linear circuit the source $v(t) = t^2$ produces the response component $[v_2(t)]_s = 0.5t^2 + 0.4t$. What is the component of the response $(v_2)_s$ due to the source (*a*) $v(t) = 10t^2 + 5t$; (*b*) $10t + 10$?

5-14 In the circuit of Fig. P5-14: (*a*) Show that the initial value $(dv_{an}/dt)_{0^+}$ is given by $(dv_{an}/dt)_{0^+} = \frac{1}{2}i_s(0^+) - \frac{1}{20}v_{an}(0^+)$. (*b*) Evaluate $(dv_{an}/dt)_{0^+}$ if $v_{an}(0^+) = 50$ volts and $i_s(t)$ is given by (1) $4u(t)$, (2) $8u(t)$, (3) $12u(t)$, (4) $8tu(t)$, (5) $16 \cos 5t$, (6) $16 \cos (5t + 60°)$. (*c*) Evaluate $(d^2v_{an}/dt^2)_{0^+}$ if $v_{an}(0^+) = 50$ volts for each of the functions $i_s(t)$ given in (*b*).

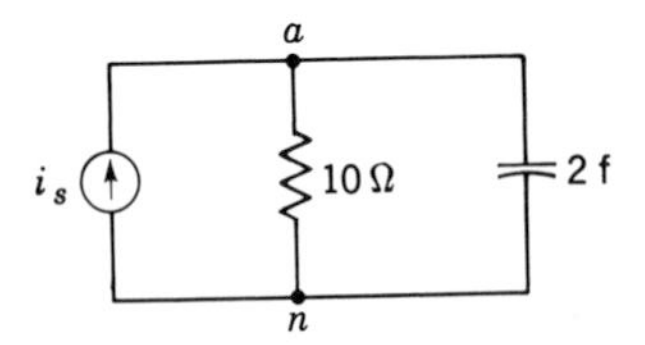

Fig. P5-14

5-15 (*a*) Show that the state equations for the circuit of Fig. P5-15 are $di/dt = (-Ri - v_C + v)/L$, $dv_C/dt = i/C$. (*b*) Express d^2i/dt^2 in terms of v_C, i, R, L, and C.

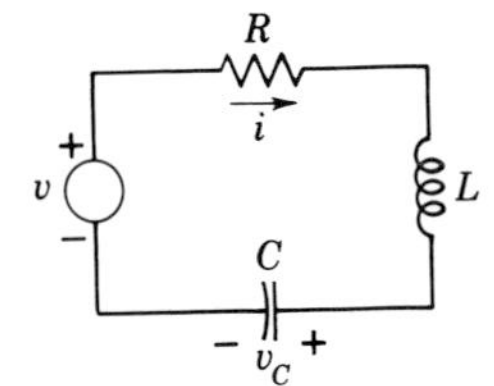

Fig. P5-15

5-16 In the series *R-L-C* circuit of Fig. P5-15, $R = 10$ ohms, $L = 0.1$ henry, and $C = 0.025$ farad. The continuity conditions are $v_C(0^+) = 10$ volts, $i(0^+) = 0.5$ amp. Use the result of Prob. 5-15 to find the initial values $(di/dt)_{0^+}$ and $(d^2i/dt^2)_{0^+}$ if the source function $v(t)$ is given by (*a*) $15u(t)$; (*b*) $20u(t)$; (*c*) $(15 + 30t)u(t)$; (*d*) $25 \cos (8t - 60°)$.

5-17 (*a*) Using v_1 and v_2 as the variables, find the state equations for the circuit shown in Fig. P5-17. (*b*) Use the result of part *a* to evaluate $(dv_1/dt)_{0^+}$, $(dv_2/dt)_{0^+}$, and $(d^2v_2/dt^2)_{0^+}$ if $v_1(0^+) = 5$, $v_2(0^+) = 0$, and if the source function is given by (1) $v(t) = 25 \cos 3t$, (2) $v(t) = 25 \cos (3t - 30°)$, (3) $35u(t)$.

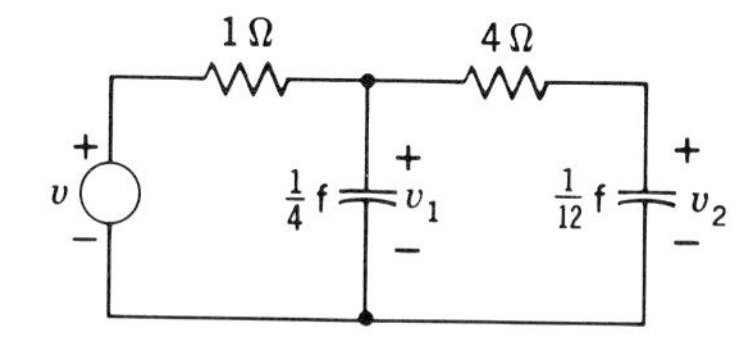

Fig. P5-17

5-18 (*a*) In the circuit of Fig. P5-17, find the initial values $v_1(0^+)$ and $v_2(0^+)$ that will give the initial values $(dv_1/dt)_{0^+} = 0$ and $(dv_2/dt)_{0^+} = 0$ when the source function is given by $v(t) = 10 \cos 4t$. (*b*) Using the initial value for $v_1(0^+)$ and $v_2(0^+)$ of part *a*, find $(dv_1/dt)_{0^+}$ if the source function is changed to $v(t) = 10 \cos (4t - 60°)$.

5-19 (*a*) Show that the state equations for the circuit of Fig. P5-12 are $pi_L = -4i_L - 2v_{an} + 2v_o$; $pv_{an} = 2i_L - 2v_{an} + 2i_s$. (*b*) Use the result of part *a* to find the initial values $(dv_{an}/dt)_{0^+}$ and $(d^2v_{an}/dt^2)_{0^+}$ if $i_L(0^+) = 0.5$ amp, $v_{an}(0^+) = 3$ volts, $i_s(t) = (4 + 6t)u(t)$, and $v_o(t) = 5e^{-t}u(t)$.

5-20 In the circuit of Fig. P5-15, $L = 3$ henrys, $i(0^-) = 6$ amp, and $v(t) = 12\delta(t)$. Find $i(0^+)$.

5-21 In Fig. P5-21 the capacitance C_1 is initially charged to 100 volts; C_2 is initially uncharged. The switch is closed at $t = 0$. (*a*) Calculate $v_{ab}(t)$ for all $t \geq 0^+$. (Use an initial-condition generator.) (*b*) Calculate the energy stored on C_1 at $t = 0$. (*c*) Calculate the total energy stored on C_1 and C_2 at $t = \infty$. (*d*) Account for the difference between total initial stored energy and total final stored energy by calculating the total energy dissipated in the 3-ohm resistance between $t = 0$ and $t = \infty$. (*e*) The 3-ohm resistance is replaced by an arbitrary resistance *R*. Show that the energy dissipated by *R* between $t = 0$ and $t = \infty$ is independent of the value of *R*.

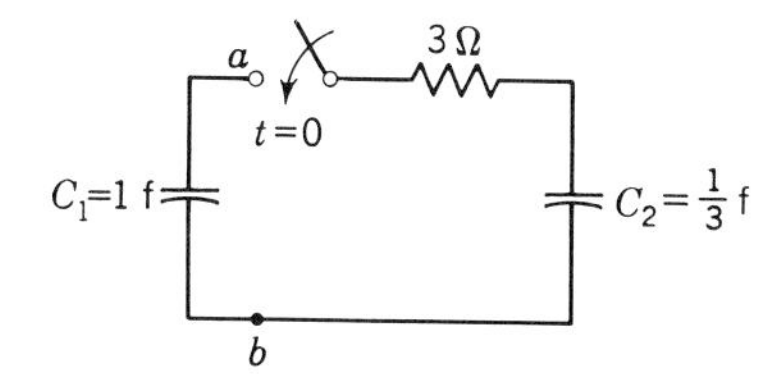

Fig. P5-21

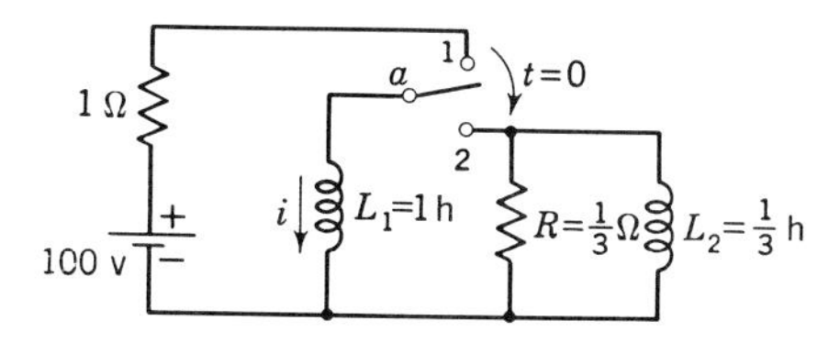

Fig. P5-22

5-22 The circuit shown in Fig. P5-22 is in the steady state until $t = 0$ with the switch in position 1. At $t = 0$ the switch is thrown to position 2. (*a*) Calculate $i(t)$ for all $t \geq 0^+$. The initial value of the current in L_2 is zero. *Hint:* Represent the initial energy stored in L_1 by an ideal current source. (*b*) Calculate the energy stored in L_1 and L_2 at $t = 0$ and at $t = \infty$. (*c*) Calculate the total energy delivered to *R* in the interval $t = 0$ to $t = \infty$. (*d*) Compare the results of this problem with the results of Prob. 5-21.

5-23 The circuit shown in Fig. P5-23 is in the steady state before $t = 0$ with $V = 50$ volts. At $t = 0$ the value of V is abruptly changed to 80 volts. Calculate (*a*) $i(0^-)$; (*b*) $v_{ab}(0^-)$; (*c*) $v_{bn}(0^-)$; (*d*) $v_{bn}(0^+)$; (*e*) $v_{ab}(0^+)$; (*f*) $i(0^+)$; (*g*) the new steady-state values of v_{ab} and v_{bn}.

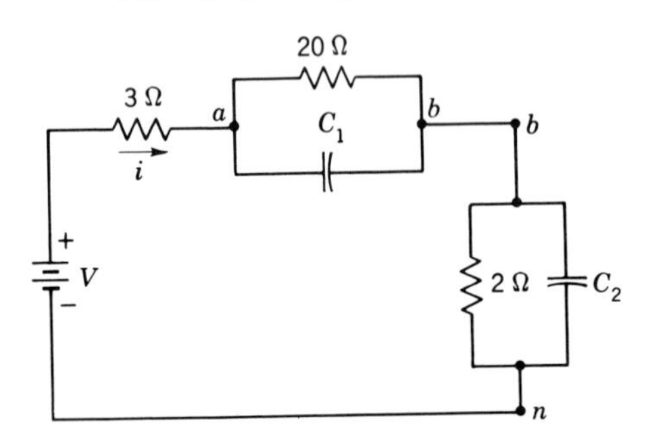

Fig. P5-23

5-24 In the circuit of Fig. P5-23, the 3-ohm resistance is replaced by a short circuit, $C_1 = 0.5$ farad, and $C_2 = 1$ farad. The circuit is in the steady state before $t = 0$ with $V = 44$ volts. At $t = 0$ the value of V is abruptly changed to 66 volts. Find (*a*) $v_{bn}(0^-)$; (*b*) $v_{bn}(0^+)$; (*c*) the new steady-state value of v_{bn}.

5-25 In the circuit of Fig. P5-14, $v_{an}(0^-) = 10$ and $v_{an}(0^+) = 15$. The source function is an impulse, $i_s(t) = Q\delta(t)$. Find Q.

5-26 The circuit of Fig. P5-26 is in the steady state for $t < 0$ with the switch in position 1. At $t = 0$ the switch is thrown to position 2. Calculate $v_{2n}(0^+)$, $v_{an}(0^+)$, and the new steady-state value of v_{an}.

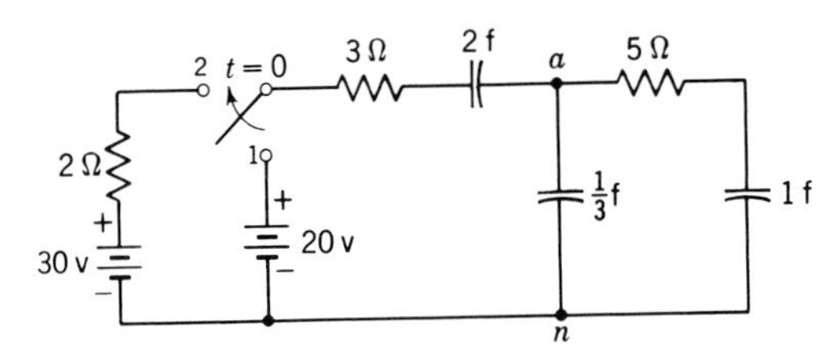

Fig. P5-26

5-27 In the circuit of Fig. P5-14, $v_{an}(0^-) = 20$ volts. Formulate and sketch to scale $v_{an}(t)$ for all $t \geq 0^+$ if $i_s(t)$ is given by (*a*) $2u(t)$; (*b*) $4u(t)$; (*c*) $2u(t) + 8\delta(t)$; (*d*) $2tu(t)$.

5-28 A ramp-function voltage source $v(t) = Vt/T_1$ is applied to a series *R-C* circuit. The initial value of the voltage across the capacitance is V_0. Calculate the value of V_0 (and show the reference direction on a circuit diagram) if the free component of the response is zero.

5-29 The rectangular voltage pulse shown in Fig. P5-29 is impressed on the series *R-C* circuit. Calculate and sketch $v_{ab}(t)$ and $i(t)$ for all $t \geq 0^+$ if (*a*) $v_{ab}(0^+) = 0$; (*b*) $v_{ab}(0^+) = 5$ volts; (*c*) $v_{ab}(0^-) = -5$ volts. *Hint:* It is convenient to represent the initial-energy storage by an equivalent ideal voltage source and to use the principle of superposition.

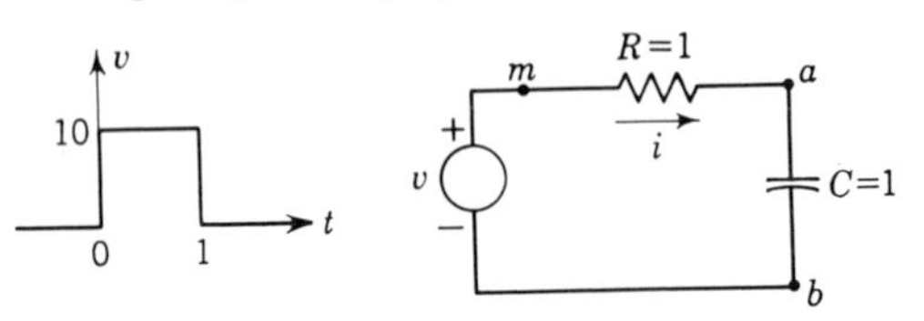

Fig. P5-29

5-30 The triangular pulse shown in Fig. P5-30 represents the voltage waveform v of the source in the circuit of Fig. P5-29. Calculate and sketch carefully to scale $v_{ab}(t)$ and $v_{ma}(t)$ if $v_{ab}(0^+) = 0$ and (*a*) $T_1 = 1$; (*b*) $T_1 = 10$; (*c*) $T_1 = 0.1$.

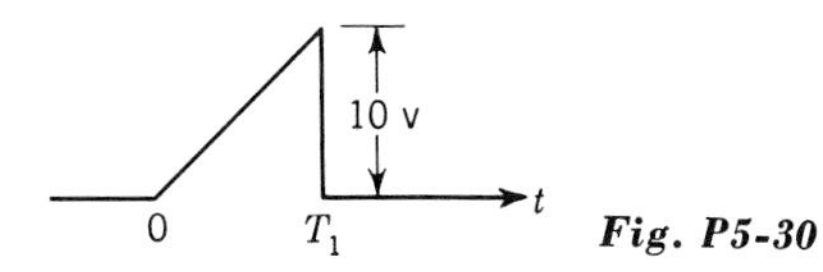

Fig. P5-30

5-31 Analyze the results of Prob. 5-30 with the integrating and differentiating property of *R-C* circuits in mind; i.e., compare the waveforms obtained with those obtained from an ideal integrator and differentiator.

5-32 Show that an *R-L* circuit has (approximate) integrating and differentiating properties by deducing the appropriate inequalities, and illustrate your result, using a ramp-function source with zero initial-energy storage.

5-33 Calculate the current response of a series *R-L* circuit to a voltage source which is a rectangular pulse of duration T_1, for the following cases: (*a*) $T_1 = L/R$; (*b*) $T_1 = 10L/R$; (*c*) $T_1 = L/5R$. Assume that the initial-energy storage is zero.

5-34 In the circuit of Fig. P5-34, $i(0^+) = 0$. Calculate $v_{ab}(t)$ for all $t \geq 0^+$ if (*a*) $v(t) = 10u(t)$; (*b*) $v(t) = 10tu(t)$; (*c*) $v(t) = (5 + 3t)u(t)$; (*d*) $v(t) = e^{t/3}u(t)$.

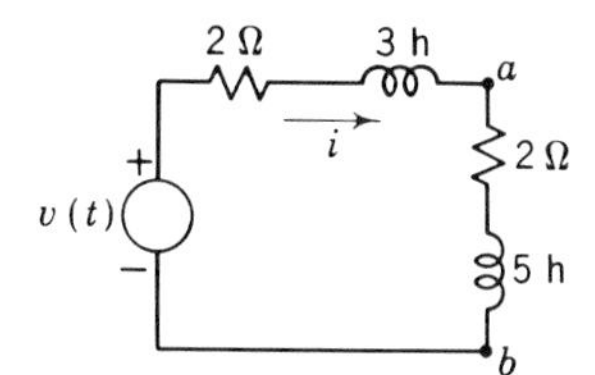

Fig. P5-34

5-35 In the circuit of Fig. P5-34, the source function is given by $v(t) = 20u(t)$. Find $i(0^-)$ so that the steady-state response $(v_{ab})_s$ is also the complete response.

5-36 In the circuit of Fig. P5-34, $i(0^-) = 0$. Find $i(t)$ and $v_{ab}(t)$ if $v(t) = 20u(t) + 16\delta(t)$.

5-37 In the circuit of Fig. P5-37 the source waveform $v(t)$ is a rectangular pulse of 0.5 sec duration and unit amplitude. Calculate and sketch $v_{an}(t)$ for all $t \geq 0^+$ in the following cases: (*a*) $v_{bn}(0^+) = 0$, $R_1 = R_2 = 5$ ohms; (*b*) $v_{bn}(0^+) = 0$, $R_1 = 4R_2 = 8$ ohms; (*c*) $v_{bn}(0^+) = 2$, $R_1 = R_2 = 5$ ohms.

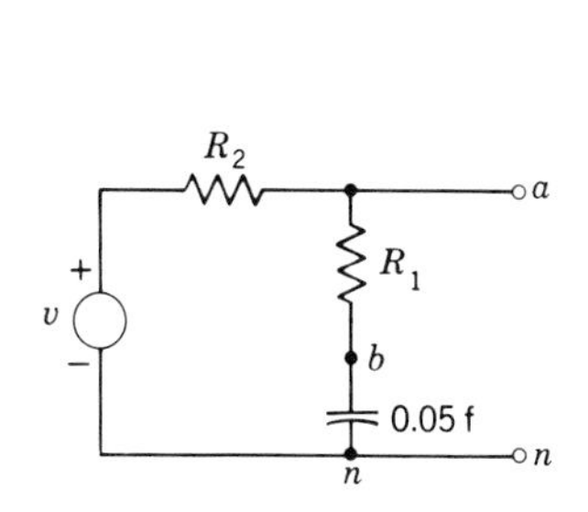

Fig. P5-37

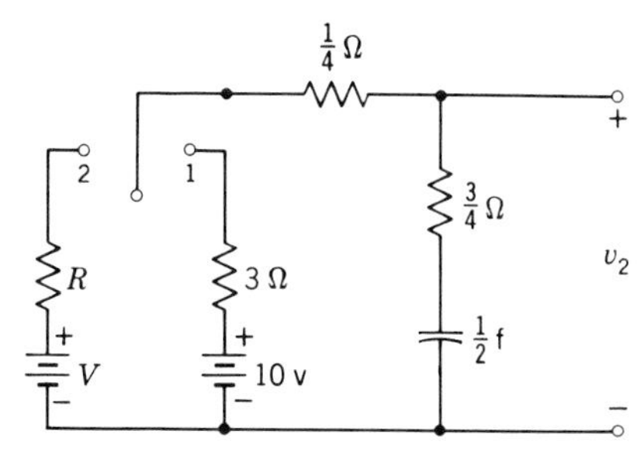

Fig. P5-38

5-38 The circuit shown in Fig. P5-38 is in the steady state for $t < 0$ (switch in position 1). The switch is moved to position 2 at $t = 0$. It remains in position 2 until $t = 2$; it is in position 1 for $t > 2$. Sketch to scale $v_2(t)$ for $-1 < t < 5$ in each of the following cases: (*a*) $V = 0$, $R = 1$ ohm; (*b*) $V = 20$ volts, $R = 1$ ohm; (*c*) $V = -10$ volts, $R = 1$ ohm.

5-39 In the circuit of Fig. P5-38, $V = 0$, $R = 7$ ohms, and the switch is in position 2 for $t < 0$. The capacitance is uncharged at $t = 0^-$. The switch is moved to position 1 at $t = 0$, and it remains in that position until $t = 2$; it is in position 2 for $t > 2$. Sketch to scale $v_2(t)$ for all $t \geq 0^+$.

5-40 In the series *R-L-C* circuit of Fig. P5-15, $v_C(0^+) = 0$ and $i(0^+) = 0$. The element values are $L = 0.4$ henry, $C = 0.1$ farad, and R is as given below. (*a*) Find and sketch $v_C(t)$ for $t \geq 0^+$ if $v(t) = 10u(t)$ and (1) $R = 0.4$ ohm; (2) $R = 2.4$ ohms; (3) $R = 5.66$ ohms. (*b*) Find and sketch $v_C(t)$ and $i(t)$ if $v(t) = 10tu(t)$ for each of the three values of R given in part *a*.

5-41 In the series *R-L-C* circuit of Fig. P5-15, $v_C(0^+) = 0$ and $i(0^+) = 0$; the element values are $L = 2$ henrys, $C = 0.125$ farad, $R = 8$ ohms. Find the complete response $v_C(t)$ if $v(t)$ is (*a*) $30e^{-5t}$; (*b*) $v(t) = 10(e^{j2t} + e^{-j2t}) = 20 \cos 2t$.

5-42 (*a*) Show that the equilibrium equation (of a critically damped series *R-L-C* circuit), $(p^2 + 2p + 1)i = e^{-t}$, has the solution $i = (At^2 + K_1t + K_2)e^{-t}$, where K_1 and K_2 depend on the initial conditions, and find A. (*b*) Show that an *L-C* circuit, $L = \frac{1}{2}$, $C = \frac{1}{8}$, with the source $v(t) = \cos 4t$ impressed in series, has the solution $v_C = At \sin 4t$ for the voltage across the capacitance, and find A.

5-43 In the circuit of Prob. 5-4, $v_2(0^+) = 2$ and $(dv_2/dt)_{0+} = -2$. Find the complete response $v_2(t)$ for each of the source functions given in Prob. 5-4*c*.

5-44 In the circuit of Fig. P5-12, $i_s(t) = (1 + 3t)u(t)$ and $v_0(t) = e^{-3t}$. If $v_{an}(0^+) = 2$ volts and $i_L(0^+) = 0$, solve for $v_{an}(t)$ for all $t \geq 0^+$.

5-45 In the circuit of Fig. P5-17, $v_1(0^+) = 0$ and $v_2(0^+) = 0$. Solve for $v_1(t)$ and $v_2(t)$ if the source function is (*a*) $v(t) = 10u(t)$; (*b*) $v(t) = 10tu(t)$; (*c*) $v(t) = 10e^{-8t}$; (*d*) $v(t) = 10e^{-3t}$.

5-46 In the circuit of Fig. P5-5, let $R = 1$ ohm and $C = \frac{1}{3}$ farad. (*a*) Show that, when C is replaced by a short circuit, then $i_{ab} = 5$. (*b*) Show that, when the 30-volt source is set to zero and a voltage $v_{ab} \neq 0$ exists across C, then $i_{ab} = -5v_{ab}/9$. (*c*) Use the result of parts *a* and *b* to show that the state equation for the given circuit is $pv_{ab} = -5v_{ab}/3 + 15$. (*d*) What is the time constant of the circuit?

5-47 (*a*) Find the state equation, $pi = ai + \text{const}$, i = current in L, for the circuit of Fig. P5-6. (*b*) What is the time constant of the circuit?

5-48 In the circuit of Fig. P5-48, $v_C(0^+) = 4$ volts and $i(0^+) = 0.1$ amp. (*a*) Find the initial conditions $v_{ab}(0^+)$ and $(dv_{ab}/dt)_{0^+}$ if (1) $v = 12u(t)$; (2) $v = 12tu(t)$; (3) $v = 24e^{-10t}u(t)$. (*b*) For each of the source functions given in part *a*, formulate the complete response $v_{ab}(t)$.

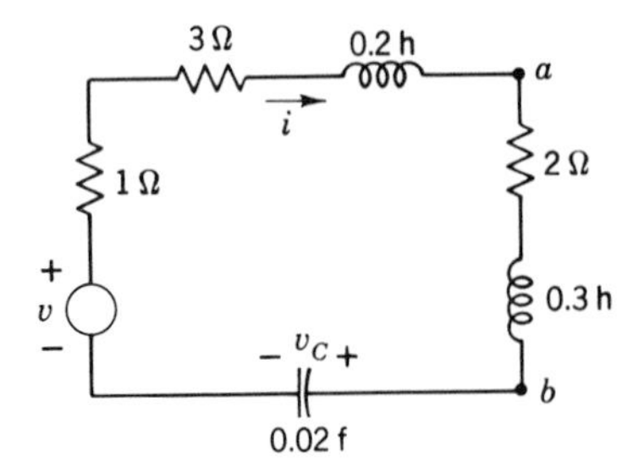

Fig. P5-48

5-49 For the *R-C* circuit shown in Fig. P5-49*a*, the time constant $RC = 10$ sec and the capacitance is uncharged prior to application of the source. The purpose of this problem is to illustrate that the response of the circuit to a pulse whose duration

T_d is much smaller than RC is similar to the response to an impulse whose strength is the same as the area under the pulse. (*a*) Show that, if the source $v(t) = A\,\delta(t)$, then $v_C(0^+) = A/RC = A/10$. (*b*) For each of the pulses shown in Fig. P5-49, find $v_C(0^+)$ if $v_C(-1) = 0$. (*c*) Show that the response v_C is approximately the same for the pulses of (*b*) as for the impulse for $t > 0^+$.

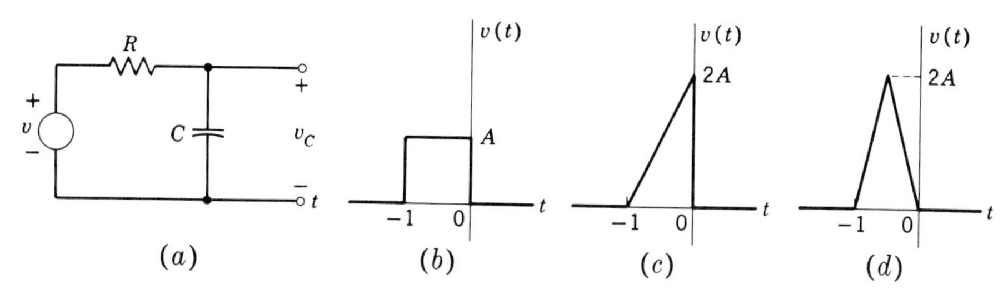

Fig. P5-49

5-50 The purpose of this problem is to illustrate that a current pulse of duration T_d applied to a ballistic galvanometer will cause a peak deflection proportional to the area under the pulse. The equivalent circuit of the galvanometer is shown in Fig. P5-50, the resistance R is the parallel combination of R' and $R_c + R_x$ (Fig. 4-19*f*), and the circuit is critically damped with $\omega_0^2 = (1/2RC')^2 = 1/L'C' = 10^{-2}$. (*a*) For the source $i = Q\delta(t)$, show that, if $i_L(0^-) = 0$ and $v_{an}(0^-) = 0$, then $i_L(0^+) = 0$ and $v_{an}(0^+) = Q_0/C'$. (*b*) Show that, for the conditions of part *a*, $i_L(t) = \omega_0^2 Q_0 t e^{-\omega_0 t}$, $t \geq 0^+$. (*c*) Show that the peak value of i_L (which is proportional to peak galvanometer deflection θ) is proportional to Q_0. (*d*) The current pulse $i(t) = Q_0[u(t+1) - u(t)]$ is applied with $i_L(-1) = 0$ and $v_{an}(-1) = 0$. (1) Find $i_L(0^+)$ and $v_{an}(0^+)$. (2) Find $i_L(t)$ for $t \geq 0^+$, and show that its peak value is approximately proportional to Q_0.

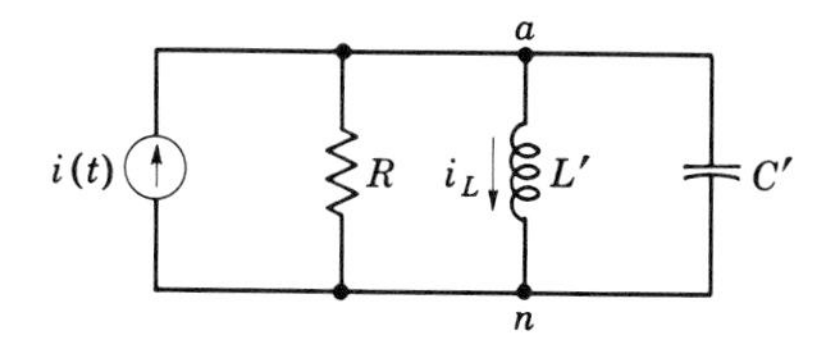

Fig. P5-50

Chapter 6 Sources, Thévenin's theorem, and related topics

In Chap. 5 we showed how to obtain the complete response of simple circuits which are excited by ideal sources. In this chapter we show, first of all, how practical sources can be represented as a combination of ideal sources and passive elements. We also consider the matter of equivalent representations for sources, which is generalized in Thévenin's theorem.

6-1 Ideal and nonideal sources

The reader will recall that, in Chap. 1, two types of *ideal* sources of electric energy were defined. The *ideal voltage source* (represented by the symbol shown in Fig. 6-1a) has the property of maintaining across its terminals

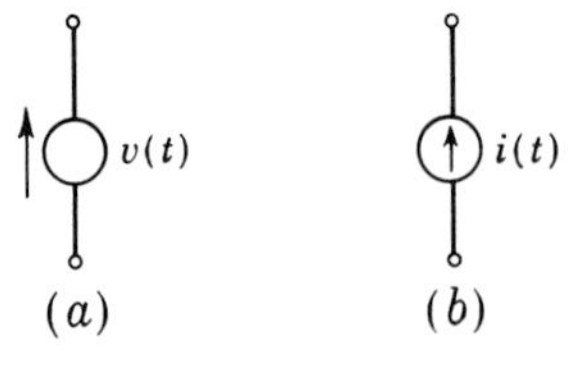

$v(t)$ (a) $i(t)$ (b)

***Fig. 6-1** Symbols for ideal sources. (a) Ideal voltage source. (b) Ideal current source.*

a waveform of voltage $v(t)$ which is independent of any elements connected to its terminals. The *ideal current source* (represented by the symbol shown in Fig. 6-1b) delivers a current $i(t)$ whose waveform is independent of any element connected to its terminals. Practical sources such as generators, batteries, power supplies, electronic oscillators, etc., differ from ideal sources in that the waveforms at their accessible terminals are influenced by the circuit components connected to those terminals. This happens because, in the process of transferring energy from the source to the circuit components connected to it, energy is either dissipated or stored in elements representing certain internal characteristics of the source.

Under certain conditions, which need not be discussed here, practical sources may be represented by a combination of ideal sources and passive elements, i.e., equivalent circuits. For example, the series combination of an ideal constant voltage source and resistance shown in Fig. 6-2*a* may

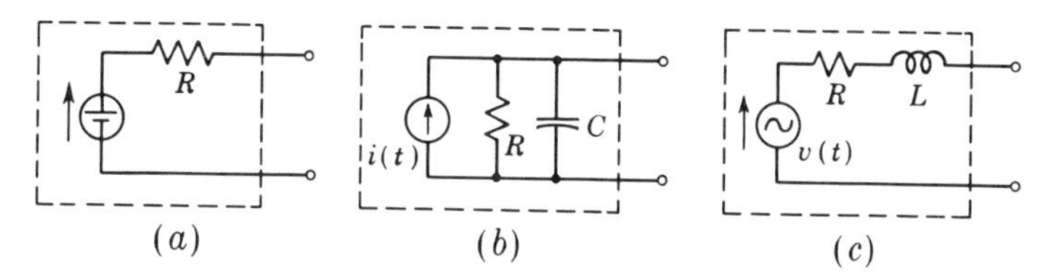

***Fig. 6-2** Examples of equivalent circuits for practical circuits. (a) Battery. (b) Electronic oscillator. (c) Alternator.*

represent a battery. The connection of current source, resistance, and capacitance shown in Fig. 6-2*b* may represent an electronic source. The series connection of voltage source, resistance, and inductance shown in Fig. 6-2*c* may represent a generator.

6-2 Resistive voltage source

Because, under certain conditions, many (but not all) sources may be represented as the combination of an ideal source and a resistance, this case is treated first. We shall call this type of source a *resistive source*. Consider the series combination of an ideal voltage source $v_o(t)$ and a resistance R_s, representing a practical source, delivering energy to a load resistance R_L as shown in Fig. 6-3*a*. We note that, when $R_s = 0$, the source is ideal and the terminal voltage $v_{ab}(t)$ is independent of R_L, as indicated by the dotted line in Fig. 6-3*b*. To obtain the dependence of terminal voltage on load resistance for finite values of R_s, apply voltage division to the circuit of Fig. 6-3*a*:

$$v_{ab}(t) = v_o(t)\,\frac{R_L}{R_L + R_s} \tag{6-1}$$

This expression may be written in normalized form,

$$\frac{v_{ab}(t)}{v_o(t)} = \frac{1}{1 + R_s/R_L} \tag{6-2}$$

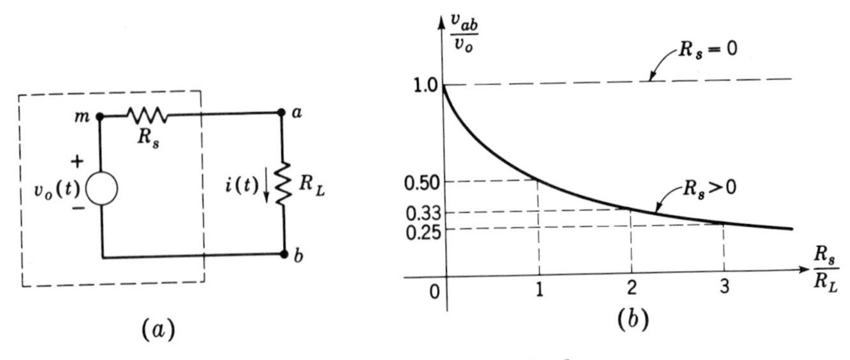

Fig. 6-3 A resistive voltage source and its terminal characteristics with resistive load.

and is illustrated in Fig. 6-3*b*. Note that the terminal characteristics of the *resistive*-source–*resistive*-load combination are independent of the waveform of the source.

The power delivered by the source to the load resistance R_L, at any instant of time, is given by

$$p_L(t) = \frac{v_{ab}^2}{R_L} \tag{6-3}$$

so that

$$p_L(t) = [v_o(t)]^2 \frac{R_L}{(R_L + R_s)^2} \tag{6-4}$$

Consider now an instant of time at which $v_o(t)$ is not zero. We note then that no power is delivered to the load resistance if its value is zero (short circuit) or infinite (open circuit). Between these extremes the power is always finite and positive. Consequently, we expect that, for a particular value of R_L, the power will be a maximum. This value of R_L is found by solving the equation $\partial p_L/\partial R_L = 0$. The partial derivative is used because $v_o(t)$ and R_s are kept fixed. Therefore the results apply to the case of a given v_o and a given nonzero R_s.

From Eq. (6-4),

$$\frac{\partial p_L}{\partial R_L} = [v_o(t)]^2 \frac{(R_L + R_s)^2 - 2(R_L + R_s)R_L}{(R_L + R_s)^4}$$

Setting this derivative to zero, the load receives maximum power when

$$R_L = R_s \tag{6-5}$$

The result obtained in Eq. (6-5) is called the *maximum-power-transfer theorem for resistive sources.* (It will be shown in the next section that the same result applies to both voltage and current sources.) This theorem states: For an adjustable load and fixed source resistances, at all instants of time, maximum power is transferred to a load resistance by a resistive source if the load resistance is made equal to the source resistance.[1] It is noted that, for this condition ($R_L = R_s$), the power which is dissipated in the load (R_L) is equal to the power dissipated within the source (R_s). In many applications where the total amount of power or energy involved is sufficiently small, the rather low (50 per cent) efficiency is tolerated for the sake of obtaining the maximum power which the resistive source is *capable* of delivering; i.e., when $R_L = R_s$, one draws from the source all *available* power.

It is instructive to calculate the ratio of the power delivered for any value of R_L to the power delivered when $R_L = R_s$. Substituting Eq. (6-5) in (6-4),

$$[p_L(t)]_{R_L=R_s} = [v_o(t)]^2 \frac{1}{4R_s} \tag{6-6}$$

Hence

$$\frac{p_L(t)}{[p_L(t)]_{R_L=R_s}} = \frac{4R_sR_L}{(R_s + R_L)^2} = \frac{4(R_L/R_s)}{(1 + R_L/R_s)^2} \tag{6-7}$$

The result of Eq. (6-7) is shown in Fig. 6-4.

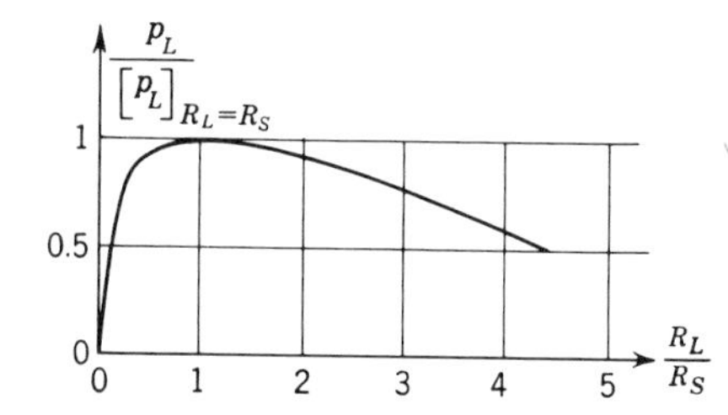

Fig. 6-4 Relative-power-transfer curve for resistive voltage source and resistive load.

We now note that every practical voltage source must contain some series "internal" resistance R_s in its equivalent circuit because no transfer of energy from the source with 100 per cent efficiency is possible. Consequently, every practical voltage source can deliver only *finite* maximum power. *We conclude, therefore, that no impulse currents can be produced by such a source,* since impulse currents are associated with infinite power.

Resistive current source A nonideal current source may frequently be represented as the parallel combination of an ideal current source and a resistance. Such a current source is shown connected to a load resistance R_L in Fig. 6-5*a*. We note that in this case the source is ideal when R_s is infinite; $i_s(t) = i_L(t)$, independently of R_L. This ideal case is illustrated by the dashed line of Fig. 6-5*b*.

[1] It is again pointed out that the condition $R_L = R_s$ deals with maximum power transfer to R_L for any value of $v_o(t)$. This condition should not be confused with the maximum instantaneous power, which occurs for any value of R_L at those instants of time at which $v_o(t)$ is a maximum. See Prob. 6-5.

To obtain the dependence of load current on load resistance for the case of the practical source, apply the current-division formula, giving the result

$$i_L(t) = i_s(t)\,\frac{1}{1 + R_L/R_s} \tag{6-8}$$

or

$$\frac{i_L(t)}{i_s(t)} = \frac{1}{1 + R_L/R_s} \tag{6-9}$$

The result (6-9) is illustrated in Fig. 6-5*b*.

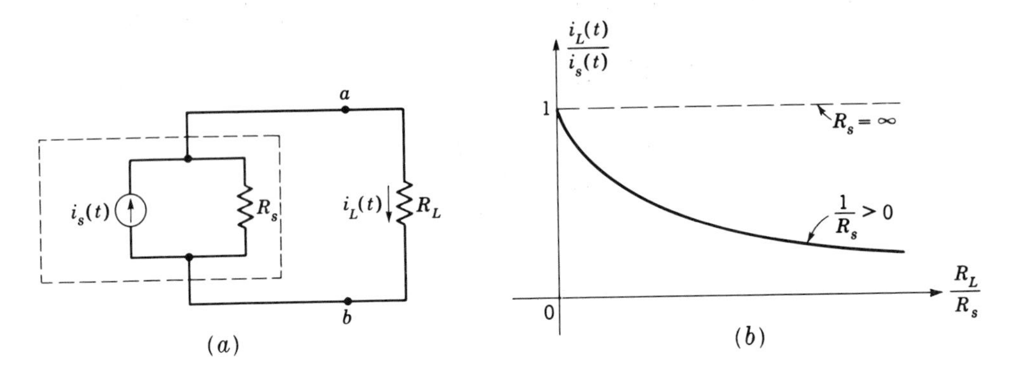

***Fig. 6-5** Resistive current source and its terminal characteristic with resistive load.*

The power delivered to the load resistance R_L is

$$p_L(t) = [i_L(t)]^2 R_L = v_{ab}(t)\,i_L(t) \tag{6-10}$$

or

$$p_L(t) = [i_s(t)R_s]^2\,\frac{R_L}{(R_L + R_s)^2} \tag{6-11}$$

Comparing Eq. (6-11) with (6-4), we note that the result of solving $\partial p_L/\partial R_L$ gives again

$$R_L = R_s \tag{6-12}$$

as the condition for maximum power transfer. For this condition the power is

$$[p_L(t)]_{R_L=R_s} = [i_s(t)]^2\,\frac{R_s}{4} \tag{6-13}$$

This result is identical with that obtained for the voltage source [Eq. (6-6)] as illustrated in Fig. 6-4. The remarks above concerning the need for an internal resistance, which limits the power of a practical source, are applicable to current sources.

6-3 Equivalent resistive sources

The similarity between the characteristics of practical voltage and current sources, however, leads one to suspect that there is a general equivalence between these two *types* of sources [compare Figs. 6-3*b* and 6-5*b* and Eqs. (6-1) and (6-9)]. Such an equivalence does indeed exist, and will be discussed now. We shall first state this general equivalence:

A voltage source consisting of an ideal voltage source $v_o(t)$ and a *series* resistance R_s may, *with respect to its output terminals only*, be represented as the *parallel* combination of an ideal current source $i_s(t) = v_o(t)/R_s$ and the resistance R_s. Conversely, a current source consisting of the *parallel* combination of an ideal source $i_s(t)$ and the internal resistance R_s may, *with respect to its output terminals only*, be represented as the *series* combination of an ideal voltage source $v_o(t) = i_s(t)R_s$ and the resistance R_s. The reader may ask himself why one type of source should be converted into another type. The answer is to be found in the simplification which can be achieved in some network calculations when a suitable conversion is made. This point will become clear as we progress. The equivalences, illustrated in Fig. 6-6, will now be proved.

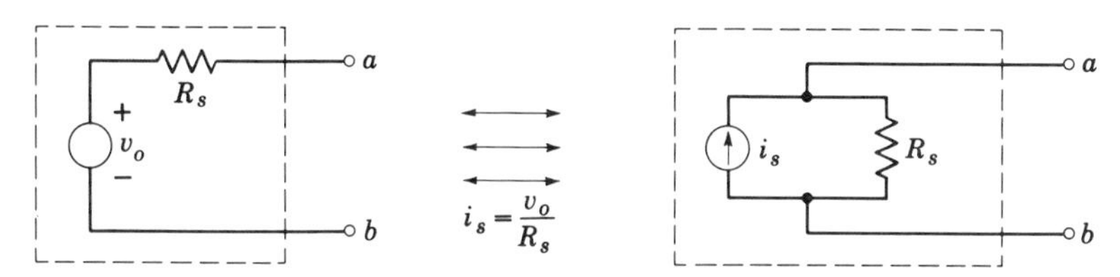

***Fig. 6-6** Equivalent resistive voltage and current sources.*

If a load resistance R_L is connected to the terminals *a-b* of the voltage source of Fig. 6-7, the current in this resistance is given by Eq. (6-2):

$$i_L = \frac{v_o(t)}{R_L + R_s} \tag{6-14}$$

If the same resistance R_L is connected to the current source, the current in R_L is given by

$$i_L(t) = i_s(t)\frac{1}{1 + R_L/R_s}$$

which may be written

$$i_L(t) = i_s(t)R_s\frac{1}{R_L + R_s} \tag{6-15}$$

Comparing Eqs. (6-14) and (6-15), we note that the same current flows *in the load resistance* if

$$i_s(t)R_s = v_o(t) \tag{6-16}$$

thus establishing the equivalence indicated in Fig. 6-6 for a resistance connected to terminals *a-b*. We note here that R_L may be the equivalent resistance of a series-parallel combination, so that the proof above applies to resistive loading generally. Note also that the equivalence applies only to the elements connected to terminals *a-b* external to the sources. *Within* the sources *no* equivalence exists. If, for example, $R_L = \infty$, no power is dissipated in R_s of the voltage source but $i_s(t)^2 R_s$ is the power dissipated within the current source.

We shall now show that the equivalence indicated in Fig. 6-6 applies for *any* combination of elements connected to terminals *a-b*. Such an arrangement is indicated in Fig. 6-7.

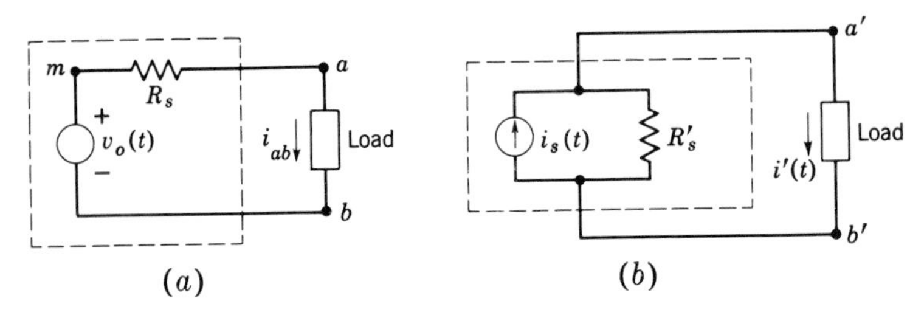

Fig. 6-7 ***Circuits used to establish the equivalence indicated in Fig. 6-6. The terminal pairs marked "load" are identical.***

Let the terminal pair shown in Fig. 6-7*a* connected across the terminals *a-b* of the resistive voltage source contain an arbitrary combination of circuit elements. The *same* terminal pair is connected across the terminals *a′-b′* of the resistive current source shown in Fig. 6-7*b*. If the response of the elements represented by the terminal pair to the two sources is to be identical (so that the sources are equivalent with respect to their output terminals), then the following two (equivalent) conditions must be fulfilled:

$$v_{a'b'}(t) = v_{ab}(t) \tag{6-17}$$

and

$$i'(t) = i_{ab}(t) \tag{6-18}$$

Given these conditions, we are required to find the relation between $v_o(t)$, $i_s(t)$, R_s, and R'_s. We cannot evaluate i_{ab} or i' flowing in the loads since the nature and the value of the elements in the load are not known and are arbitrary.

In Fig. 6-7*a*, independently of the nature of the load, application of Kirchhoff's voltage law gives

$$i_{ab}(t) = \frac{v_o(t)}{R_s} - \frac{v_{ab}(t)}{R_s} \tag{6-19}$$

In Fig. 6-7*b* application of Kirchhoff's current law gives

$$i'(t) = i_s(t) - \frac{v_{a'b'}(t)}{R'_s} \tag{6-20}$$

The conditions (6-17) and (6-18) must hold for all *arbitrary* time functions. Using (6-19) and (6-20),

$$i_s(t) - \frac{v_{a'b'}(t)}{R'_s} = \frac{v_o(t)}{R_s} - \frac{v_{ab}(t)}{R_s} \tag{6-21}$$

If the equivalence is to be valid, $v_{a'b'} = v_{ab}$, and Eq. (6-21) may be written

$$v_{ab}(t)\left(\frac{1}{R_s} - \frac{1}{R'_s}\right) = \frac{v_o(t)}{R_s} - i_s(t) \tag{6-21a}$$

Now $v_o(t)$ is an unspecified time function; $v_{ab}(t)$ is arbitrary in the sense that it will depend on the elements in the load as well as on $v_o(t)$. Equation (6-21a) states that the arbitrary time function $v_o(t)/R_s - i_s(t)$ must equal the arbitrary time function $v_{ab}(t)(1/R_s - 1/R'_s)$. This is possible only if each side of Eq. (6-21a) is equal to zero.

The conditions for equivalence (with respect to their output terminals) of the two sources shown in Fig. 6-7 are therefore

$$R_s = R'_s$$

and

$$\frac{v_o(t)}{R_s} = i_s(t) \tag{6-22}$$

We note now that

$$i_s(t)R_s = v_o(t)$$

is the voltage across the terminals of either source if *no* elements are connected to those terminals (open circuit) and

$$i_s(t) = \frac{v_o(t)}{R_s}$$

is the current which flows from terminal a to b in either circuit if a short circuit is connected between those terminals.

We observe that, in converting a resistive voltage source, the corresponding current source $i_s(t)$ has the same waveform as the current which will flow in R_s (Fig. 6-7) if the terminals of the voltage source are short-circuited together. This is called the short-circuited current of the practical sources. Similarly, if a current source is converted to a voltage source, the waveform of the latter is identical with that of the voltage appearing across R'_s (Fig. 6-7) if no external element is connected across the terminals of the practical current source. This is called the open-circuited voltage of the resistive sources.

Thus we conclude that two resistive sources (voltage or current sources) are equivalent *with respect to their terminal characteristics* if they furnish both the same *open-circuit* voltage and the same *short-circuit* current. Below we show that the case of resistive sources is a special case and that, in fact, all two-terminal active networks which supply the same open-circuit voltage and the same short-circuit current are equivalent with respect to their terminal characteristics.

6-4 Inductive and capacitive sources

Under certain conditions it is possible to identify the combination of an ideal source and an inductance or a capacitance as a practical source. For such combinations one can also represent the combination of passive element (L or C) and ideal source with respect to the terminal characteristic as a voltage source or as a current source.

This equivalence for capacitive sources is illustrated in Fig. 6-8. To prove the equivalence we connect the same terminal pair (load) across each source

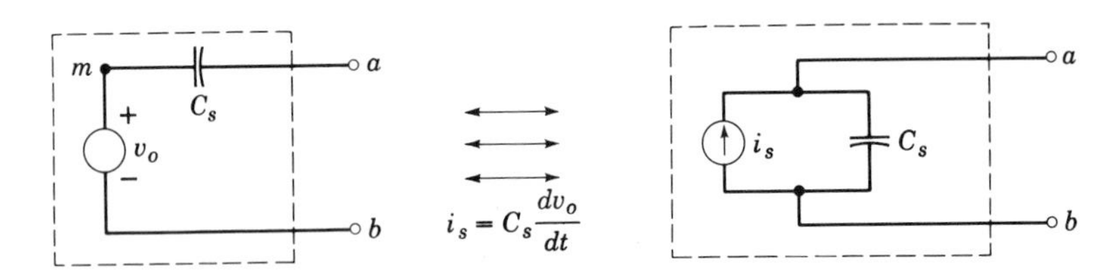

Fig. 6-8 Equivalent capacitive sources.

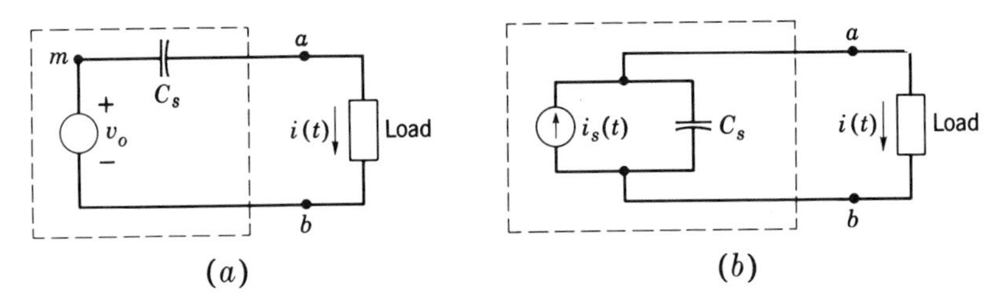

Fig. 6-9 Circuits used to establish the equivalence indicated in Fig. 6-8.

as shown in Fig. 6-9. Since the load is the same in both circuits, the function i will also be the same when v_{ab} is the same. Thus we have from the voltage-source circuit

$$i = C_s \frac{d}{dt} (v_o - v_{ab})$$

or

$$i = pC_s v_o - pC_s v_{ab} \tag{6-23}$$

and in the circuit with the current source,

$$i = i_s - pC_s v_{ab} \tag{6-24}$$

Evidently, the result in Eq. (6-23) is identical with that of Eq. (6-24) if

$$i_s = pC_s v_o = C_s \frac{dv_o}{dt} \tag{6-25}$$

as indicated in Fig. 6-8.

Proof of the dual result for inductive sources as shown in Fig. 6-10 is left as an exercise for the reader.

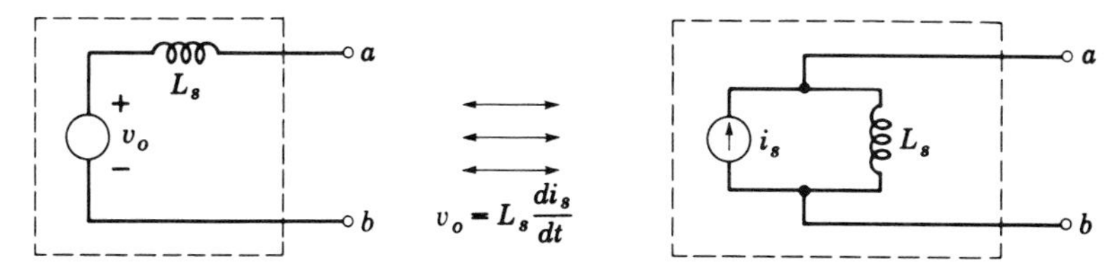

Fig. 6-10 Equivalent inductive sources.

INITIAL CONDITIONS When source conversion is applied in a problem involving complete response, it is necessary to realize that the initial-energy storage in a circuit in which an inductive or capacitive source has been converted is not the same as in the circuit with the original source. Thus, if in the circuit of Fig. 6-9a the voltage across C_s at $t = 0^+$ is given by $v_{ma}(0^+)$, the voltage across C_s at $t = 0^+$ in the circuit of Fig. 6-9b will not be $v_{ma}(0^+)$, but will have to be given by $v_{ab}(0^+) = v_o(0^+) + v_{ma}(0^+)$ if the two sources are to be equivalent.

An awareness of the changed initial conditions contributes insight into the nature of equivalent sources by bringing out again the limitation of the equivalences to terminal characteristics. Frequently, however, source conversion is used, not in solution of complete response problems, but only in connection with the component of the response due to the source (e.g., in steady-state problems), so that initial conditions do not influence the solution.

6-5 Source conversion with internal operational driving-point immittances

We recall that, at the terminals of an interconnection of passive elements, the network function that relates voltage to current is the operational driving-point immittance. Thus, in connection with the representation of Fig. 6-11, we write, symbolically,

$$v_{ab} = [Z_{ab}(p)]i(t) \qquad \text{or} \qquad i(t) = [Y_{ab}(p)]v_{ab}(t)$$

where $Z_{ab}(p)$ is the operational driving-point *impedance* and

$$Y_{ab}(p) = \frac{1}{Z_{ab}(p)}$$

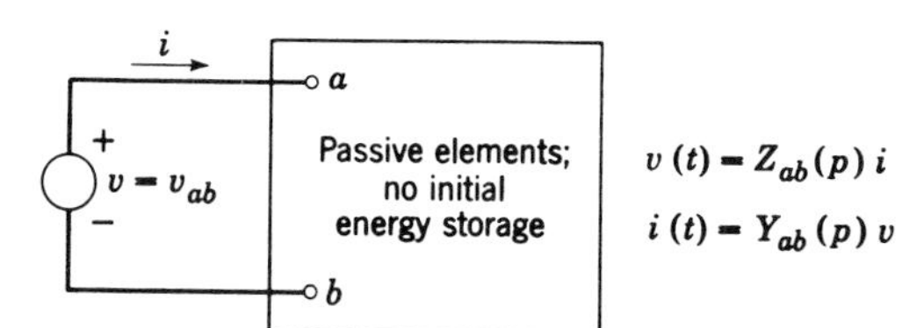

$v(t) = Z_{ab}(p)\, i$

$i(t) = Y_{ab}(p)\, v$

Fig. 6-11 Illustration of driving-point immittance.

is the operational driving-point *admittance.* We recall, further, that $Z_{ab}(p)$ and $Y_{ab}(p)$ are ratios of polynomials in p; that is,

$$Z_{ab}(p) = \frac{a_0 + a_1p + a_2p^2 + \cdots + a_np^n}{b_0 + b_1p + b_2p^2 + \cdots + b_mp^m}$$

so that the voltage-current relationship at the terminals is given by the *differential equation*

$$(b_0 + b_1p + \cdots + b_mp^m)v_{ab}(t) = (a_0 + a_1p + \cdots + a_np^n)i(t)$$

that is, in words, a linear combination of the voltage function with certain of its derivatives is in equilibrium with linear combination of the current function with certain of its derivatives.

When operational notation is used to analyze a network, we shall also use the term *operational element,* or *p element,* to describe a passive terminal-pair network. In contrast, we shall refer to the elements R, L, C as *primitive elements.* Since operational driving-point impedances (admittances) in series (parallel) add, we can consider a p element as a "building block" of a network, exactly as the primitive elements R-L-C are the building blocks of a p element.

We shall now use the idea of a terminal-pair network represented by a driving-point immittance to deduce some useful network theorems.

In Sec. 6-3 we saw how a voltage source in series with an element (R, L, or C) can be converted into a current source in parallel with an element (R, L, or C). If the element in series with the voltage source is a p element rather than R, L, or C, we can make the conversion in a similar manner to that described above. In the following discussion we shall refer to the immittance in series with a voltage source which is being converted (or in parallel with a current source which is being converted) as the internal immittance of the source.

In Fig. 6-12, $Z_1(p)$ is the internal impedance of the voltage source between terminals a-b. The admittance $Y_2(p)$ is the internal admittance of the current source between terminals h-k.

If the two sources in Fig. 6-12a and b are to be equivalent between their respective terminals a-b and h-k, then, if we connect a p element given by $Z(p) = 1/Y(p)$ at either of these terminals, the current through this p element must be identical in both cases. In Fig. 6-12c,

$$i_1(t) = \frac{1}{Z(p) + Z_1(p)} v_o(t) \tag{6-26}$$

From Fig. 6-12d,

$$i_2(t) = \frac{Y(p)}{Y(p) + Y_2(p)} i_s(t) \tag{6-27}$$

Assuming that the two practical sources are equivalent at their terminals [i.e., that they deliver the same current to the same load immittance

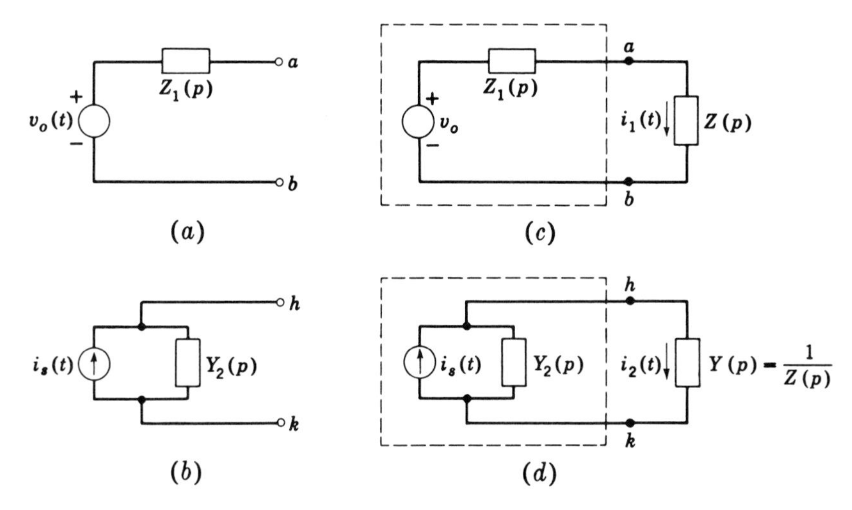

Fig. 6-12 Illustrations pertinent to source conversion with internal p elements.

$Z(p) = 1/Y(p)$], we set $i_1 = i_2$:

$$\frac{1}{Z(p) + Z_1(p)} v_o(t) = \frac{Y(p)}{Y(p) + Y_2(p)} i_s(t) \tag{6-28}$$

or

$$i_s(t) = \frac{Y(p) + Y_2(p)}{Z(p)Y(p) + Z_1(p)Y(p)} v_o(t) \tag{6-29}$$

Since $Z(p)Y(p) = 1 = Z_1(p)Y_1(p)$, we can reduce Eq. (6-29) to

$$i_s(t) = Y_1(p) \frac{Y(p) + Y_2(p)}{Y(p) + Y_1(p)} v_o(t) \tag{6-30}$$

The expression (6-30) is the condition for equivalence. However, the two sources must be equivalent *independently* of the load admittance $Y(p)$. This condition can be fulfilled only if $Y_2(p) = Y_1(p)$. Then

$$i_s(t) = \frac{1}{Z_1(p)} v_o(t) \quad \text{or} \quad i_s(t) = Y_1(p)v_o(t) \quad \text{or} \quad v_o(t) = Z_1(p)i_s(t) \tag{6-31}$$

Equations (6-31) can then be used to convert a voltage source with a series p-element immittance to a current source with a parallel p-element immittance, and vice versa. It is emphasized that, exactly as in the case of source conversion with single internal elements, the two sources are equivalent only at their "accessible" terminals.

Example 6-1 Convert the voltage-source–series-impedance combination of Fig. 6-13a to the left of terminals a-b to a current source. Use the result to obtain the transfer impedance which relates $i_1(t)$ to $v_o(t)$.

Solution The converted source is shown in Fig. 6-13*b*. In this circuit $i_s(t) = [1/(1 + p)]v_o(t)$. Using the current-division formula,

$$i_1(t) = \frac{2}{(2+p) + 1/(1+p)} i_s(t) \tag{6-32}$$

or

$$i_1(t) = \frac{2}{(2+p) + 1/(1+p)} \frac{1}{1+p} v_o(t) \tag{6-33}$$

Hence

$$i_1(t) = \frac{2}{p^2 + 3p + 3} v_0(t) \tag{6-34}$$

As illustrated by the example, the source-conversion formulas

$$i_s(t) = Y(p)v_o(t) \qquad \text{and} \qquad v_o(t) = Z(p)i_s(t) \tag{6-35}$$

do not explicitly give $i_s(t)$ in terms of $v_o(t)$, or vice versa; rather, Eqs. (6-35) are differential equations that relate the new source waveform, given the original

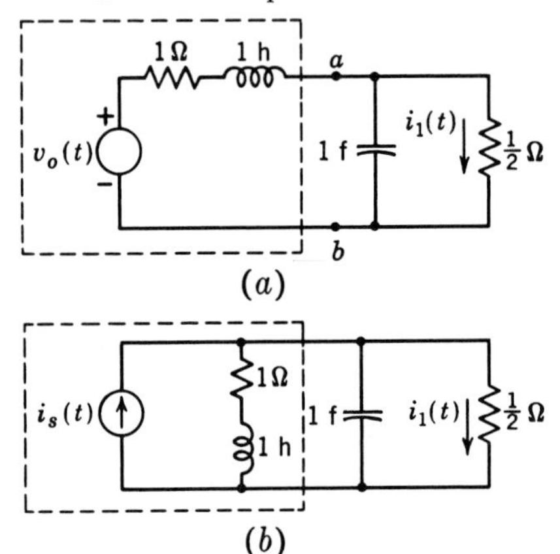

Fig. 6-13 Circuits for Example 6-1.

source waveform. We observe that, in Example 6-1, the waveform $i_s(t)$ was not found, but was rather represented by the operation

$$i_s(t) = \frac{1}{1+p} v_o(t)$$

This example is a typical application of the generalized source-conversion formulas: The conversion is not made explicitly, but only in operational form; thus only the final equilibrium equation is solved.

Another important observation must be made. A differential equation which describes the source conversion, e.g.,

$$i_s(t) = Y(p)v_o(t) \tag{6-36}$$

involves only the elements of the source. Thus, for example, in the case of Fig. 6-13,

$$i_s(t) = \frac{1}{1+p} v_o$$

or

$$(1 + p)i_s(t) = v_o(t) \tag{6-37}$$

involves only the series branch $Z = 1 + p$. Consequently, *complete* solution of Eq. (6-37) is not necessary since the free response component [Ke^{-t} in the case of

Eq. (6-37)] represents the free response of the open-circuited source *alone* and will not be part of the solution when the source is connected to a load. In the mathematics above (Example 6-1) this is brought out in going from Eqs. (6-33) to (6-34) where the factor $1 + p$ has been canceled.

INITIAL CONDITIONS In discussing the conversion of sources we have already mentioned the initial energy which may be stored in the source immittance. If this initial energy is represented by equivalent ideal sources (initial-condition generators), the source conversion is carried out with several sources (some of them being initial-condition generators). It is, however, not necessary to consider the initial-condition generators explicitly, provided that we use the correct initial conditions at the terminals of the source. Thus, using source conversion, we obtain the differential equation which relates an impressed source to the response. The differential equation is independent of the initial conditions, and can be solved completely if the initial value of the response function and a sufficient number of its initial derivatives (derived initial conditions) are known. These derived initial conditions are obtained from the circuit before source conversion.

6-6 Thévenin's theorem

We have seen that many passive terminal pairs can be represented by an operational immittance function $Z(p)$ or $Y(p)$. In Chap. 11 we shall show that all such terminal pairs can be represented by such an operator. Further, we have seen that any ideal voltage source in series with an operational element can be converted to a current source in parallel with that element, and vice versa. We shall now discuss an important theorem in circuit analysis. This theorem, named after its discoverer, M. L. Thévenin, deals with active terminal pairs. Thévenin's theorem may be stated in two parts as follows:

> *Thévenin's theorem* (1) Any active terminal pair *a-b* composed of combinations of active and passive elements can, with respect to its terminals, be represented as the series connection of an ideal voltage source $v_o(t)$ and an operational element $Z_s(p)$ between terminals *a-b*. (2) The voltage source $v_o(t)$, referred to in (1) of the theorem, is the voltage function found at the terminals *a-b* of the active terminal pair (due to the sources within it), with no external elements connected to these terminals. The impedance function $Z_s(p)$ is the driving-point impedance function at the terminals *a-b* of the terminal pair when all the ideal sources within the terminal pair are set to zero.

In the above statement of Thévenin's theorem the voltage v_o is the component of the response due to all the internal sources (representing initial conditions by initial-condition generators). It also applies when the component of the response due to the internal sources is studied, for example in steady-state calculations. *Frequently the theorem is used for the latter purpose.*

Thus Thévenin's theorem states that the active terminal pair shown in Fig. 6-14*a* can be represented with respect to terminals *a-b* as shown in Fig. 6-14*b*. The function $v_o(t)$ is given by $v_{ab}(t)$ in Fig. 6-14*a*, when no external elements are connected to the "box," which represents the active terminal pair. We refer to the series combination of Fig. 6-14*b* as "Thévenin's equivalent source."

PROOF OF THEOREM We shall not prove the first part of this theorem. We shall, however, show that, for every network, there is a function $v_o(t)$ and an impedance $Z_s(p)$ which satisfies the condition of equivalence with respect to the selected terminal pair.

Consider the active network shown in Fig. 6-14*d*, connected to the series combination of an ideal voltage source $v(t)$ and an impedance $Z_T(p)$. The current $i(t)$ which flows in the impedance Z_T can be calculated by superposition as follows:

$$i(t) = [i(t) \text{ due to } v(t)] + [i(t) \text{ due to sources within network}] \qquad (6\text{-}38)$$

We shall now consider the waveform of source $v(t)$ as adjustable and adjust $v(t)$ until the total current $i(t) \equiv 0$. The special function $v(t)$ which makes $i(t) \equiv 0$ in Eq. (6-38) will be denoted by $v_1(t)$, as indicated in Fig. 6-14*e*. Hence

$$0 \equiv [i(t) \text{ due to } v_1(t)] + [i(t) \text{ due to sources within network}] \qquad (6\text{-}39)$$

If we now set all the "independent"[1] sources within the network equal to zero, the network is no longer active, and can be represented by the driving-point impedance Z_s at terminals a'-b', as indicated in Fig. 6-14*f*. The current $i(t)$ due to $v_1(t)$ is then given by

$$[i(t) \text{ due to } v_1(t)] = -\frac{1}{Z_s + Z_T} v_1(t) \qquad (6\text{-}40)$$

Substituting (6-40) in (6-39) gives

$$0 \equiv \frac{-1}{Z_s + Z_T} v_1(t) + [i(t) \text{ due to sources within network}] \qquad (6\text{-}41)$$

Now, since the total current $i(t)$ is zero, the voltage across terminals a'-b' is exactly equal to the voltage at those terminals if they are open-circuited. It follows that $v_1(t) = v_o(t) =$ voltage v_{ab} in Fig. 6-14*a* where terminals *a-b* are open-circuited. Using this result in Eq. (6-41),

$$[i(t) \text{ due to sources within network}] = \frac{1}{Z_s + Z_T} v_o(t) \qquad (6\text{-}42)$$

This result proves Thévenin's theorem since Eq. (6-42) can be represented by the circuit shown in Fig. 6-14*g*.

Before giving some examples of the application of this theorem, several statements need to be made. First, the reader should recognize that the

[1] Independent sources are ideal sources. Dependent sources are defined in Sec. 6-7.

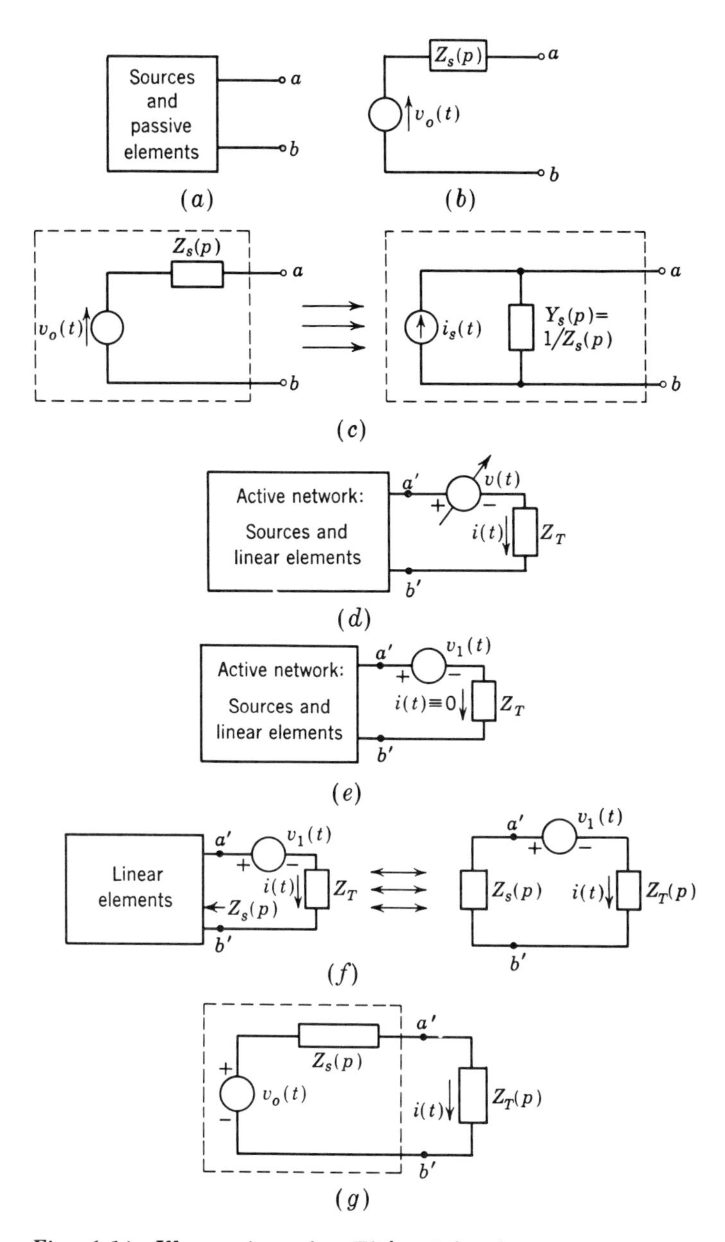

Fig. 6-14 Illustrations for Thévenin's theorem. (a) Active-terminal-pair network. (b) Thévenin's equivalent. (c) Conversion to Norton's equivalent. (d) Active terminal pair of (a) with "active load" consisting of $v(t)$ and $Z_T(p)$. (e) The voltage $v(t)$ is adjusted to be the special function $v_1(t)$ which gives the result $i(t) \equiv 0$. Under this condition $v_{a'b'}$ is the open-circuit voltage, that is, $v_{a'b'} = v_{ab}$ in (a). (f) When the sources in the active terminal pair are set to zero, the terminal pair is represented by a driving-point immittance. (g) Thévenin's equivalent circuit with load $Z_T(p)$.

conversion of voltage to current sources, and vice versa, which has already been discussed, are special cases of the application of Thévenin's theorem. Second, one of the advantages of the general theorem over the source-conversion procedure lies in the fact that the active network may have several sources acting simultaneously and such a network may be represented by a Thévenin equivalent source.

NORTON'S THEOREM Since every series combination of a voltage source and an impedance can be represented as the parallel combination of a current source and an admittance as shown in Fig. 6-14*c* every active network can be represented (with respect to two terminals) by such a current source. Note that the current source $i_s(t)$ in Fig. 6-14*c* is the current which flows out of the active network when its two terminals are shorted. It is therefore possible to calculate this short-circuit current and obtain *Norton's equivalent circuit* without first obtaining Thévenin's voltage-source circuit. We also note that the impedance Z_s is related to the two circuits by the equation

$$Z_s(p)i_s(t) = v_o(t) \tag{6-43}$$

so that either Thévenin's or Norton's equivalent circuit can be obtained from a knowledge of the open-circuit voltage and the short-circuit current.

OUTPUT IMPEDANCE In electronic circuits the term output impedance is used as a synonym for the internal impedance of Thévenin's equivalent circuit of a device such as an oscillator.

The following examples illustrate the application of Thévenin's and Norton's theorems.

Example 6-2 In Fig. 6-15 obtain Thévenin's equivalent circuit with respect to terminals *a-b*, and use the result to determine (*a*) the value of a resistance R to be connected between terminals *a-b* so that R receives maximum power from the active terminal pair; (*b*) the value of the maximum power delivered to R.

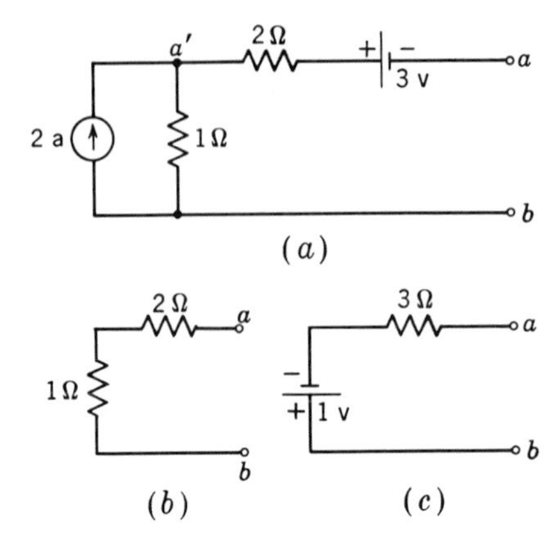

Fig. 6-15 Circuits for Example 6-2. (a) Given active terminal pair. (b) Terminal pair with ideal sources set to zero. (c) Thévenin's equivalent circuit for (a).

Solution In the circuit of Fig. 6-15*a*, $v_{a'b} = 2$ and $v_{aa'} = -3$; hence $v_{ab} = v_o = -1$ volt. If the sources are set to zero as in Fig. 6-15*b*, $Z_{ab} = R_{ab} = 3$ ohms. Hence Thévenin's equivalent circuit is the circuit shown in Fig. 6-15*c*. For maximum power transfer $R = 3$. The maximum power is $(\frac{1}{6})^2 \times 3 = \frac{3}{36} = 0.0833$ watt.

As in the case of source conversion with p-element internal impedances, application of Thévenin's theorem does not require solution for the complete open-circuit voltage response, nor does Norton's theorem require finding the complete response for the short-circuit current (although finding these responses is not incorrect, see Prob. 6-19). In applications of these theorems it is sufficient to find the open-circuit-voltage or short-circuit-current components due to the sources.

Example 6-3 (*a*) Replace the circuit of Fig. 6-16*a* by a Thévenin equivalent (the capacitance is initially uncharged). (*b*) Use the result of (*a*) to calculate the complete response if a load is connected across terminals *a-b* and this load is (1) a 1-ohm resistance or (2) the series impedance $(1 + p)$ (initially deenergized).

Solution (*a*) Applying Thévenin's theorem to the left of terminals *a′-b′* (Fig. 6-16*b*), we have from Kirchhoff's current law

$$-\frac{v_{a'b'}}{2} + 6 + \frac{4 - v_{a'b'}}{2} = 0$$

or

$$v_{a'b'} = 8 \text{ volts}$$

and $Z_{a'b'} = 1$ ohm when the sources are set to zero. We now replace the 1-farad capacitance and apply Thévenin's theorem at terminals *a-b* on the left side of Fig. 6-16*c*. In that circuit $Y_s(p) = 1 + p$ and Thévenin's equivalent circuit is shown on the right side of Fig. 6-16*c*. The complete response v_{ab} for the circuit of Fig. 6-16*c* is not needed; we need only the component of the response due to the source $(v_o)_s = V_o = 8$ volts.

(*b*) When a 1-ohm resistance is placed across terminals *a-b*, as shown in Fig. 6-16*d*, we have, from voltage division,

$$v_{ab}(t) = \frac{1}{1 + 1/(1 + p)} 8 \qquad t \geq 0^+$$

or

$$(2 + p)v_{ab}(t) = (1 + p)(8) \qquad t \geq 0^+$$

and

$$(2 + p)v_{ab}(t) = 8$$

Hence

$$v_{ab} = 4 + Ke^{-2t}$$

The initial value of v_{ab} is equal to zero *because* $v_{ab}(0^+) = 0$ *in Fig. 6-16a is zero;* hence

$$v_{ab}(t) = 4(1 - e^{-2t}) \qquad t \geq 0^+$$

When the deenergized impedance $1 + p$ is placed across terminals *a-b*, the circuit of Fig. 6-16*e* results. In this circuit

$$v_{ab} = \frac{1 + p}{1 + p + 1/(1 + p)} v_0$$

or

$$(p^2 + 2p + 2)v_{ab} = (1 + p)^2(8) \qquad t \geq 0^+$$

Now $v_{ab} = (v_{ab})_f + (v_{ab})_s$. By voltage division across the resistances $(v_{ab})_s = 4$. $(v_{ab})_f$ is the solution of

$$(p^2 + 2p + 2)(v_{ab})_f = 0$$

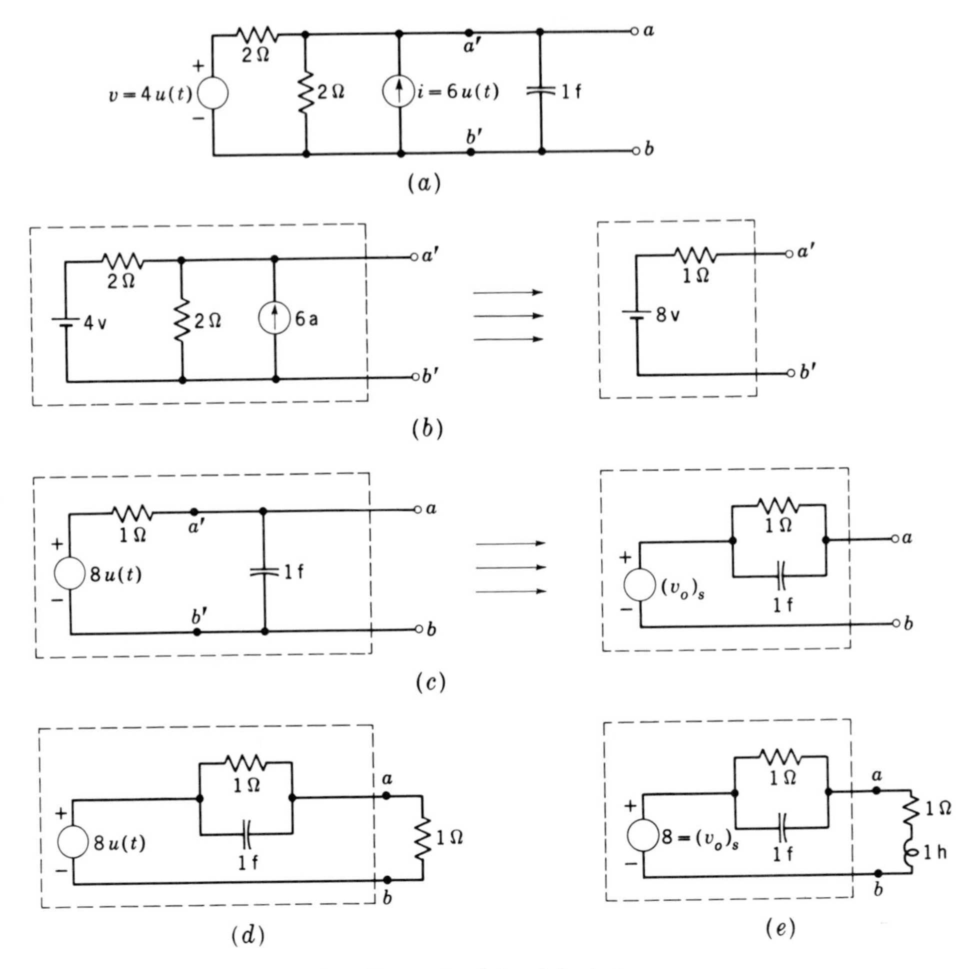

***Fig. 6-16** Circuits for Example 6-3. (a) Active terminal pair. (b) Portion of terminal pair to the left of a′-b′ and its Thévenin equivalent. (c) Representation of the terminal pair of (a) by Thévenin's theorem. (d) Thévenin's equivalent circuit with 1-ohm load. (e) Thévenin's equivalent circuit with load impedance $Z_{ab}(p) = 1 + p$.*

Hence the characteristic equation is $s^2 + 2s + 2 = 0$. The roots are $\mathbf{s} = -1 \pm j$. Now

$$v_{ab} = 4 + \text{Re}\,[\mathbf{K}e^{(-1+j)t}] \qquad t \geq 0^+$$

and

$$\frac{dv_{ab}}{dt} = \text{Re}\,[\mathbf{K}(-1 + j)e^{(-1+j)t}]$$

We have $v_{ab}(0^+) = 0$ and $i_{ab}(0^+) = 0$. At junction a we observe that i_{ab} can be expressed as $(1 + p)(v_o - v_{ab}) = i_{ab}$; hence

$$v_o(0^+) + (dv_o/dt)_{0^+} - v_{ab}(0^+) - (dv_{ab}/dt)_{0^+} = i_{ab}(0^+)$$

and $(dv_{ab}/dt)_{0^+} = (dv_o/dt)_{0^+} + v_o(0^+)$. Since $[v_o(0^+)]_s = 8$ and $[(dv_o/dt)_{0^+}]_s = 0$, it follows that $(dv_{ab}/dt)_{0^+} = 8$. Hence Re $\mathbf{K} = -4$, Re $[\mathbf{K}(-1 + j)] = 8$, and $\mathbf{K} = -4 - j4$, so that $v_{ab}(t) = 4[1 + \sqrt{2}\, e^{-t} \cos (t - 135°)]$ for $t \geq 0^+$.

This example illustrates the convenience of using Thévenin's theorem when a given active terminal-pair network is applied to several loads in turn.

Example 6-4 For the circuit shown in Fig. 6-17*a*, obtain Norton's equivalent circuit with respect to terminals *a-b*.

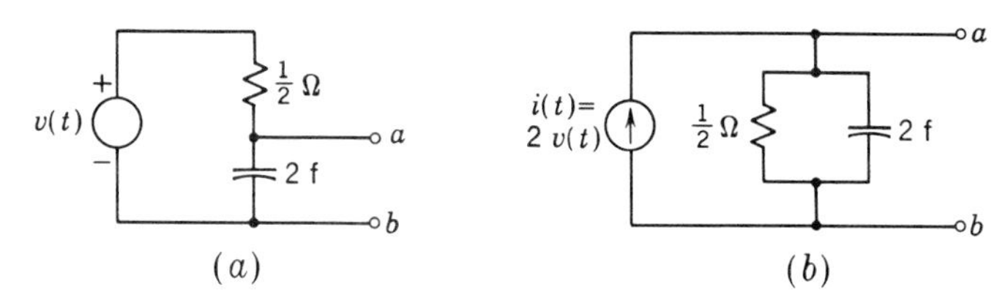

Fig. 6-17 Illustration of Norton's theorem.

Solution When $v(t)$ is set to zero, $Y_{ab}(p) = 2 + 2p$. When terminals *a-b* are short-circuited, the current through the short circuit is $i_s = v(t)/\frac{1}{2} = 2v(t)$. Norton's circuit is shown in Fig. 6-17*b*. The same result can of course be obtained by converting the series connection of the $\frac{1}{2}$-ohm resistance and $v(t)$ to a current source.

In the foregoing examples the Thévenin and Norton equivalent circuits were found by solving for either the open-circuit voltage and the output impedance or for the short-circuit current and the output impedance. We cannot conclude this introduction to Thévenin's theorem without pointing out that very often the output impedance cannot be found conveniently, but is determined from the relationship between open-circuit voltage and short-circuit current (Prob. 6-25).

6-7 Dependent, or controlled, sources

In many practical electronic circuits the sources are not ideal, but are controlled by a circuit variable. Thus, for example, in an amplifier, the output voltage may be h_0 times the input voltage when h_0 is the "gain" of the amplifier. Under these circumstances the amplifier output voltage is *controlled* by, or *dependent* upon, the input and must be represented by a corresponding source. Such dependent sources are best understood by first studying the compensation theorem.

The compensation theorem (also called "substitution theorem") is so deceptively simple that it is often neglected as trivial. There are, however, several types of applications, such as electronic circuits and circuits with mutual inductance (Chap. 15) in which application of the compensation theorem is helpful. This theorem may be stated as follows:

Compensation theorem If a passive element or branch in a circuit is defined by the voltage-current relationship $v_{ab} = Z_{ab}(p)i_{ab}$ or

$i_{ab} = Y_{ab}(p)v_{ab}$, then that element or branch may be replaced by a compensating voltage source whose waveform is given by $Z_{ab}(p)i_{ab}$ or a current source given by $Y_{ab}(p)v_{ab}$.

This theorem follows from Kirchhoff's laws. That is, it makes no difference which symbol is used to describe the voltage-current relationship for an element or branch. If the application of Kirchhoff's laws results in the correct equilibrium equations, the circuit is correctly formulated.

In order to apply the theorem fruitfully, we must understand the difference (Fig. 6-18) between an ideal source such as $v(t)$ and a source such as

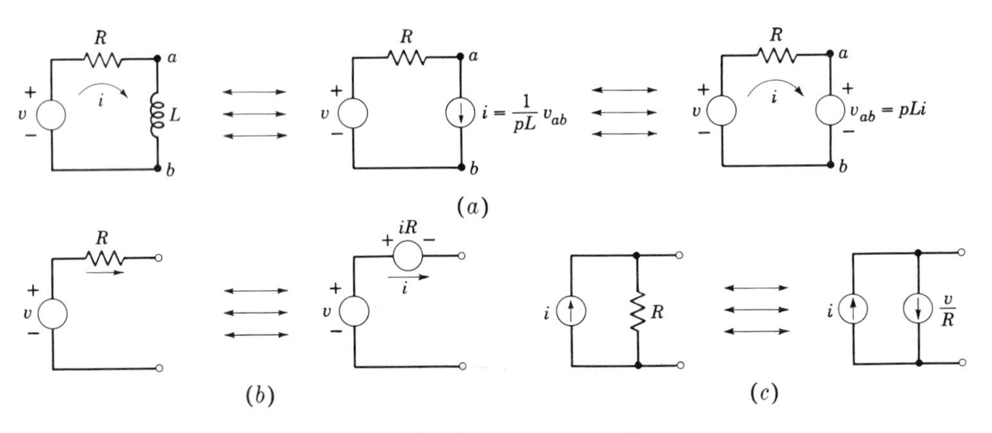

***Fig. 6-18** Illustrations of the compensation theorem. (a) Representation of inductance as a dependent voltage or current source. (b) Representation of resistance as dependent voltage source. (c) Representation of resistance as dependent current source.*

$v_{ab}(t)$. We recall that an ideal source is a source that generates a function $v(t)$ or $i(t)$ which is *independent* of the elements connected to the terminals of the source. Clearly, the waveform of the compensating voltage source pLi discussed above is dependent on the circuit elements. Hence a source which is introduced into the circuit diagram as a result of applying the compensation theorem is not an ideal source. We shall call such a source a *compensating, dependent,* or *controlled,* source, as distinguished from an *impressed, independent,* or *ideal* source. We note, incidentally, that a resistive source may be represented as the combination of an independent and a dependent source as shown in Fig. 6-18*b* and *c*.

Application of the compensation theorem is illustrated for passive networks in Chap. 15, where circuits with mutual inductance are analyzed with the aid of dependent sources. Such sources also occur in the analysis of active networks such as electronic amplifiers. The following example illustrates the use of dependent sources in such a circuit.

Example 6-5 Shown in Fig. 6-19*a* is a linear equivalent circuit of a transistor circuit. The source v is ideal, but the sources $h_{21}i_1$ and $h_{12}v_2$ are controlled (dependent) sources. Determine Thévenin's equivalent circuit with respect to terminals *a-b*.

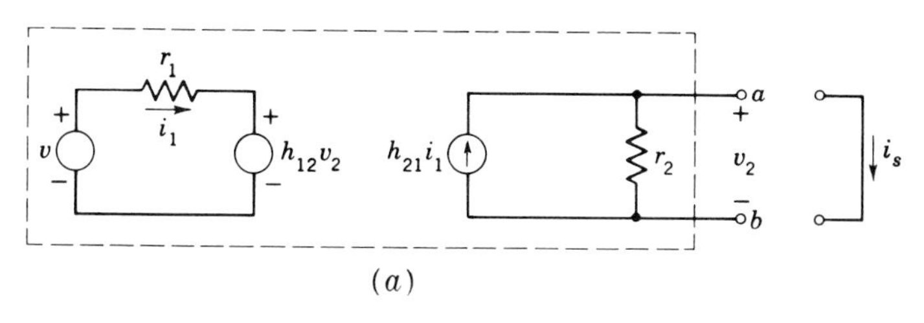

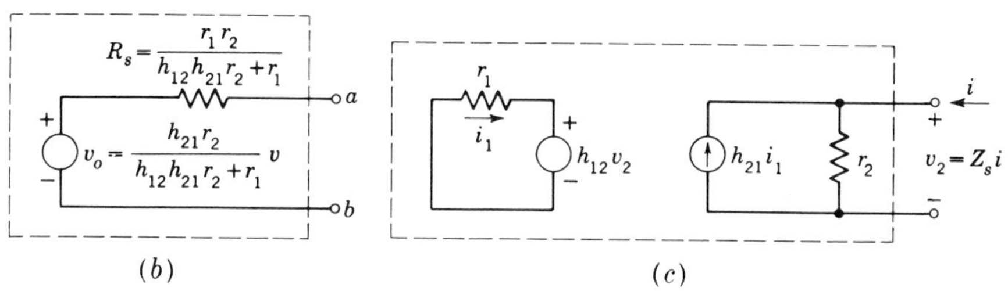

Fig. 6-19 ***Illustration of use of Thévenin's theorem with dependent sources. (a) Linear equivalent circuit of transistor. (b) Thévenin equivalent. (c) Circuit of (a) with ideal sources set to zero.***

Solution The open-circuit voltage $(v_2)_{oc}$ is found as v_2 in Fig. 6-19*a* as follows: Since

$$r_1 i_1 + h_{12}v_2 = v$$

and

$$h_{21}i_1 r_2 = v_2$$

$$[r_1 + h_{12}h_{21}r_2]i_1 = v$$

$$i_1 = \frac{v}{r_1 + h_{12}h_{21}r_2}$$

and

$$(v_2)_{oc} = h_{21}i_1 r_2 = h_{21}\frac{r_2 v}{h_{12}h_{21}r_2 + r_1} \tag{6-44}$$

If terminals *a-b* are short-circuited, $v_2 \equiv 0$ and $i_1 = v/r_1$ and $i_s = h_{21}i_1 = h_{21}v/r_1$. Thus

$$i_s = \frac{h_{21}v}{r_1}$$

and the equivalent Thévenin impedance is

$$Z_s = \frac{(v_2)_{oc}}{i_s}$$

so that

$$Z_s = \frac{r_1 r_2}{h_{12}h_{21}r_2 + r_1} \tag{6-45}$$

Thévenin's equivalent is shown in Fig. 6-19*b*. We observe that use of the open-circuit voltage–short-circuit current method avoided the step of setting ideal sources to zero, so that no formal distinction between ideal and other sources was needed. One can, of course, calculate Z_s from the circuit of Fig. 6-19*c*, where the ideal source v has been set to zero but the controlled sources are not zero.

In the circuit of Fig. 6-19*c*,

$$Z_s = \frac{v_2}{i}$$

but
$$i = \frac{v_2}{r_2} - h_{21}i_1$$

and
$$i_1 = -\frac{h_{12}v_2}{r_1}$$

so that
$$i = \left(\frac{1}{r_2} + \frac{h_{12}h_{21}}{r_1}\right)v_2$$

and
$$\frac{v_2}{i} = \frac{r_1r_2}{r_1 + h_{12}h_{21}r_2} = Z_s = R_s \tag{6-46}$$

which is the same result as obtained before [Eq. (6-45)].

6-8 Use of dependent sources to formulate state equations

Compensating sources can be used conveniently to derive state equations. If every *inductance* in a circuit is represented as a compensating *current* source and every *capacitance* as a compensating *voltage* source, the procedure of setting a state variable to zero is interpreted in the circuit as setting one of the compensating sources to zero. If all but one state variable is made zero, the compensating source which represents this variable can be treated as the source which produces all the voltages and currents in the circuit. An illustration will help to clarify the procedure.

Consider the circuit of Fig. 6-20*a*. In Fig. 6-20*b* is shown the corresponding diagram using dependent sources. The state equations for this circuit are of the form

$$\frac{di_L}{dt} = a_{11}i_L + a_{12}v_C + g_1(t)$$

$$\frac{dv_C}{dt} = a_{21}i_L + a_{22}v_C + g_2(t)$$

To find g_1 and g_2 we set $i_L = v_C = 0$, giving the circuit of Fig. 6-20*c*. In this circuit $i_C = Cpv_C = 0$.

$$\left(\frac{dv_C}{dt}\right)_{\substack{v_C=0\\ i_L=0}} = \frac{1}{C}(i_C)_{\substack{v_C=0\\ i_L=0}} = 0 = g_2(t)$$

since voltage across R_2 is zero. Similarly,

$$\left(\frac{di_L}{dt}\right)_{\substack{v_C=0\\ i_L=0}} = \frac{1}{L}(v_{ab})_{\substack{v_C=0\\ i_L=0}} = \frac{1}{L}v(t) = g_1(t)$$

To find a_{11} and a_{21}, we allow i_L only to "act" as shown in Fig. 6-20*d*. We observe that

$$\left(\frac{dv_C}{dt}\right)_{\substack{v=0\\ v_C=0}} = \frac{1}{C}(i_C)_{\substack{v=0\\ v_C=0}} = \frac{1}{C}i_L = a_{21}i_L$$

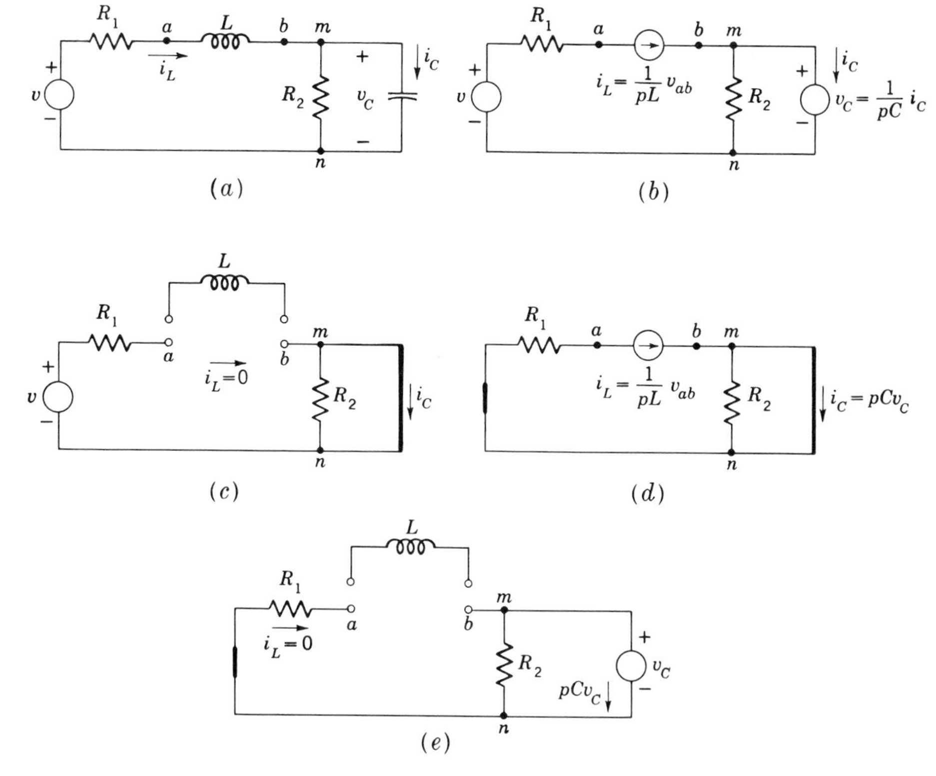

***Fig. 6-20** Use of dependent sources to formulate state equations of a network. (a) Given network. (b) Circuit of (a) with energy-storing elements replaced by controlled sources. (c) Circuit with both state-variable sources set to zero. (d) Circuit with $v = 0$ and $v_C = 0$. (e) Circuit with $v = 0$ and $i_L = 0$.*

and $a_{21} = 1/C$; also,

$$\left(\frac{di_L}{dt}\right)_{\substack{v=0 \\ v_C=0}} = \frac{1}{L}(v_{ab})_{\substack{v=0 \\ v_C=0}} = -R_1 i_L = a_{11} i_L \qquad \text{and} \qquad a_{11} = -\frac{R_1}{L}$$

To find a_{12} and a_{22}, we use Fig. 6-20*e*. Here $v_C = v_{ba} = -pLi_L$, so that $a_{12} = -1/L$. Since $i_C = -v_C/R_2 = pCv_C$, $pv_C = (-1/R_2C)v_C$ and

$$a_{12} = -\frac{1}{R_2C}$$

Thus the state equations are

$$\frac{di_L}{dt} = \left(-\frac{R_1}{L}\right) i_L + \left(-\frac{1}{L}\right) v_C + \left(\frac{1}{L}\right) v$$

$$\frac{dv_C}{dt} = \left(\frac{1}{C}\right) i_L + \left(-\frac{1}{R_2C}\right) v_C$$

It is pointed out that this method of dependent sources for the derivation of state equations is identical in procedure with the method described in Chap. 5. The difference is only one of interpretation. We look upon the state variables as "causes" and thus represent them as sources. Since, however, the state variables are associated with passive elements, these sources must be *dependent* rather than ideal. The use of dependent sources helps visualize the influence of the individual state variables in the circuit diagram.

PROBLEMS

6-1 Show that, for a resistive current or voltage source with load resistance R_L, the power delivered to the load R_{L1} is the same as that delivered to R_{L2} when $R_{L1}R_{L2} = R_s^2$.

6-2 A resistive voltage source has a source resistance of 20 ohms; it supplies a 10-ohm resistive load. (*a*) What fraction of the available power is transferred to the load? (*b*) The load is changed from 10 ohms to another value R_L. Find this value of R_L so that it receives the same power as the 10-ohm resistance did.

6-3 (*a*) Show that the resistive voltage source with resistive load (Fig. 6-3*a*) approximates an ideal current source for $R_L \ll R_S$. (*b*) Show that the resistive current source with resistive load (Fig. 6-5*a*) approximates an ideal voltage source for $R_L \gg R_S$. (*c*) In each case give the inequality relating R_S and R_L, so that the resistive voltage or current source acts as an approximately ideal voltage or current source, respectively.

6-4 In the circuit of Fig. P6-4: (*a*) Which switch position will result in maximum power transfer to R_L? (*b*) For the switch position as selected in part *a*, find the power delivered to R_L.

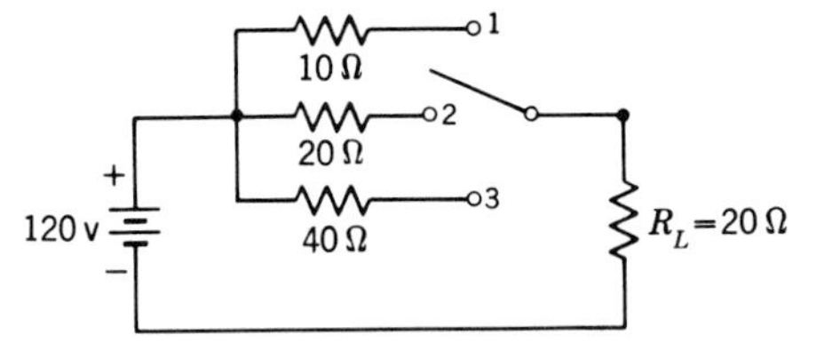

Fig. P6-4

6-5 In the circuit of Fig. 6-3*a*, $R_S = 5$ ohms and $v_o(t) = 20 \cos 2\pi t$. Calculate the maximum instantaneous power delivered to R_L if (*a*) $R_L = 2$ ohms; (*b*) $R_L = 5$ ohms; (*c*) $R_L = 12.5$ ohms.

6-6 (*a*) Calculate the time-average power delivered to R_L in each of the cases given in Prob. 6-5. (*b*) Defining per cent efficiency as 100 times the ratio of average power delivered to R_L to average power supplied by the source, find the efficiency of the circuit for each of the given values of R_L.

6-7 For a resistive source with resistive load, per cent efficiency may be defined as 100 times the ratio of power delivered to load to power supplied by the ideal source in the circuit. (*a*) For a resistive voltage source calculate the efficiency if (1) $R_L = 3R_S$; (2) $R_L = \frac{1}{3}R_S$; (3) $R_L = R_S$. (*b*) For a resistive current source calculate the efficiency if (1) $R_L = 3R_S$; (2) $R_L = \frac{1}{3}R_S$; (3) $R_L = R_S$.

6-8 For a resistive source with source resistance R_S, define the reference power p_0 as the power delivered to a load resistance $R_L = R_S$. What is the power decibel level referred to p_0 when (*a*) $R_L = \frac{1}{2}R_S$; (*b*) $R_L = 2R_S$; (*c*) $R_L = 0.1R_S$?

6-9 For each of the resistive sources shown in Fig. P6-9, find the value of R_L for maximum (available) power transfer to R_L.

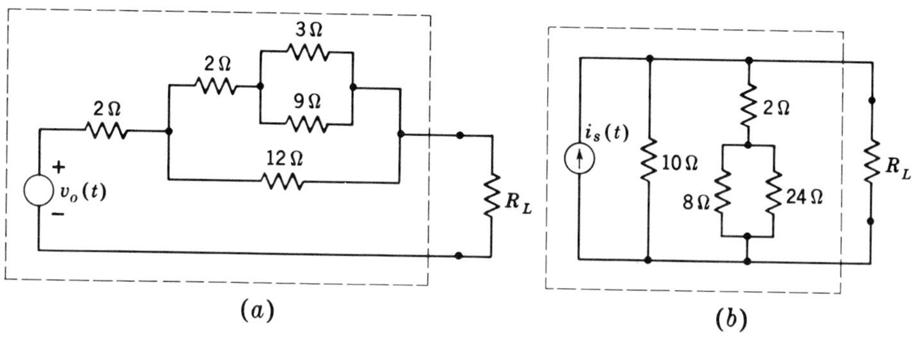

Fig. P6-9

6-10 (*a*) In each of the two circuits shown in Fig. P6-10, show that the power delivered to R_L is independent of R', and find the value of R_L for maximum (available) power transfer. (*b*) In each of the circuits referred to in part *a*, show that the power supplied by the ideal source depends on R' and find that power in Fig. P6-10*a*, using $v_o(t) = V_o = 30$ volts, $R' = 3$ ohms, $R_L = 3$ ohms, and in Fig. P6-10*b*, $i_S = I_S = 2$ amp, $R' = 3$ ohms, $R_L = 6$ ohms.

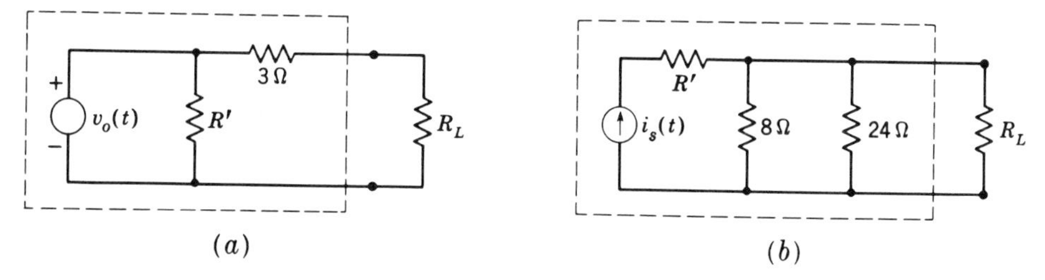

Fig. P6-10

6-11 In the circuit of Fig. P6-11: (*a*) Find the value of R_L for maximum power transfer to R_L. (*b*) Find the power delivered to R_L if it has the value you found in part *a*. (*c*) Calculate the power delivered to the circuit by each of the ideal constant sources.

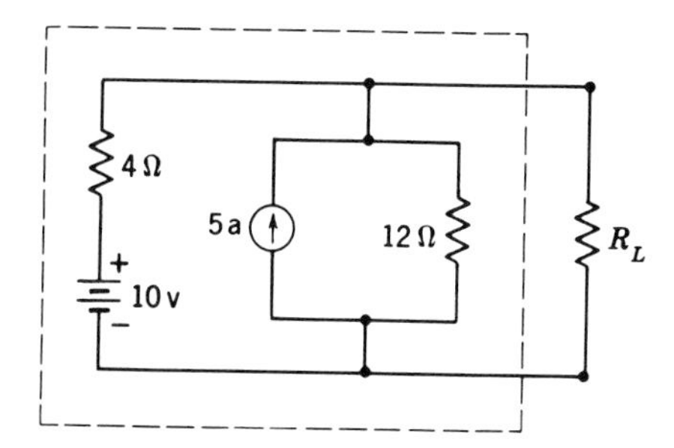

Fig. P6-11

6-12 In the circuit of Fig. P6-12: (*a*) Represent all voltage sources as current source. (*b*) Calculate V_{ab}. (*c*) Calculate the power delivered by each ideal source.

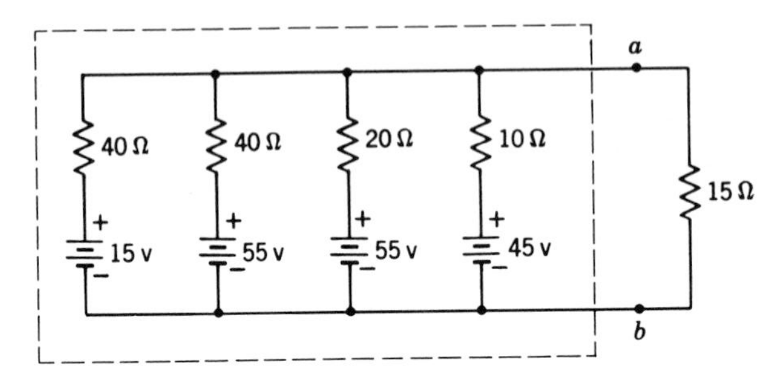

Fig. P6-12

6-13 (*a*) Represent the circuit to the left of terminals *a*-*b* in Fig. P6-13 as (1) a resistive voltage source; (2) a resistive current source. (*b*) Calculate the power delivered to R_L if R_L is connected to terminals *a*-*b* and if its value is chosen for maximum power transfer.

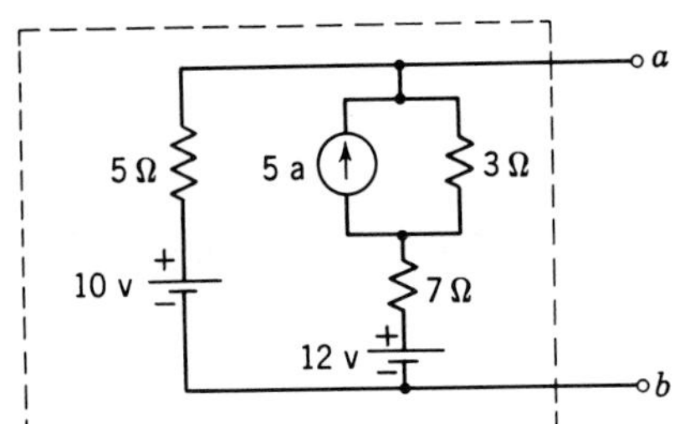

Fig. P6-13

6-14 Prove the source conversion for inductive sources indicated in Fig. 6-10.

6-15 In the circuit of Fig. P6-15 let $i(t) = 10^{-3}u(t)$. (*a*) Relate $v_{ab}(t)$ to the current i using voltage division. (*b*) Find v_{ab} by representing the circuit to the left of terminals *a*-*b* as a capacitive voltage source.

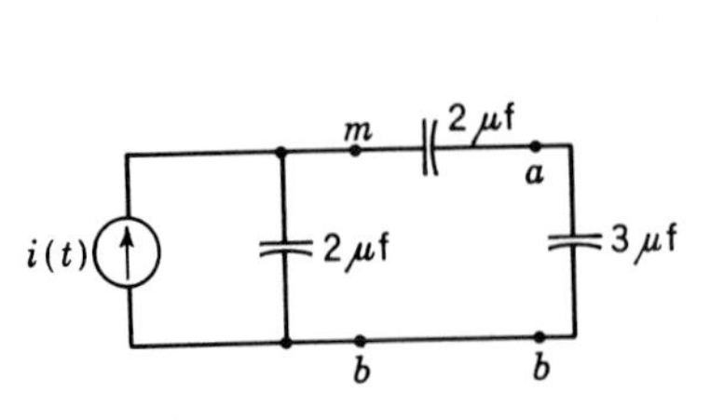

Fig. P6-15

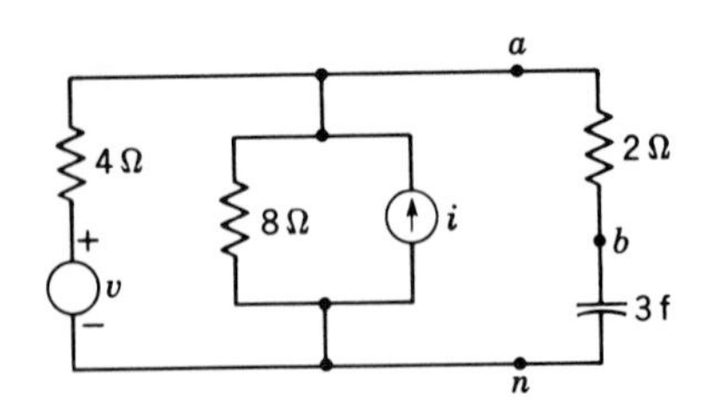

Fig. P6-16

6-16 In the circuit of Fig. P6-16, $v_{bn}(0^+) = 0$. Find $v_{ab}(t)$ and $v_{an}(t)$ for all $t \geq 0^+$ if (*a*) $v(t) = 14u(t)$, $i(t) = 7tu(t)$; (*b*) $v(t) = 14u(t)$, $i(t) = 7e^{-t/10}$.

6-17 (a) In the circuit of Fig. P6-17 the energy-storing elements store zero energy at $t = 0$. Find $v_{an}(t)$ for all $t \geq 0^+$. (b) The series combination of the 4-ohm resistance, the $\frac{1}{2}$-henry inductance, and the 0.01-farad capacitance is replaced by the parallel combination of the same three elements. If the initial-energy storage is zero, find $v_{an}(t)$ for all $t \geq 0^+$.

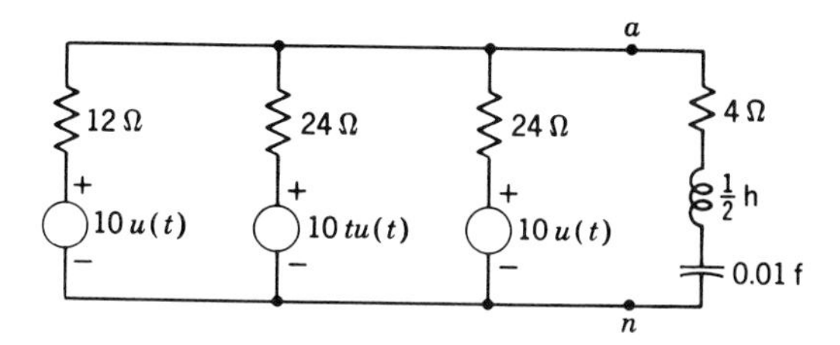

Fig. P6-17

6-18 For the circuit shown in Fig. P6-18: (a) Find the operational driving-point impedance $Z_{an}(p)$. (b) An ideal ramp-function voltage source, $v_{an}(t) = 6tu(t)$, is impressed at terminals a-n; show that, if $v_{bn}(0^+) = v_{mn}(0^+) = 0$, then $i(0^+) = 0$, $(di/dt)_{0^+} = 6$ amp/sec, and find the complete response $i(t)$. (c) An ideal current source $i_s(t) = i(t)$ is impressed at terminals a-n. What are the characteristic roots of the response v_{an}?

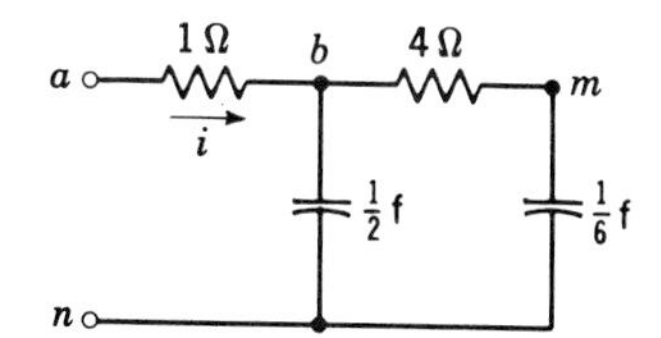

Fig. P6-18

6-19 In the circuit of Fig. 6-12c a series circuit with a voltage source is given; in Fig. 6-12d is shown its current source equivalent with respect to terminals a-b, so that $Z_1 = 1/Y_2 = Z_S$; hence, $i_s(t) = [Y_S(p)]v_o(t)$. Let $Z_S(p) = N_S(p)/D_S(p)$ and $Z(p) = N(p)/D(p)$, where N_S, N, D_S, and D are polynomials in p. (a) Show that the equilibrium equation relating i_s to v_o has the source-free response component whose characteristic roots are the zeros of $N_S(s)$. (b) Show that, in the circuit of Fig. 6-12c, the equilibrium equation relating v_{ab} to v_o has characteristic roots at the zeros of $N(s)D_S(s) + N_S(s)D(s) = 0$. (c) Show that the characteristic roots of the equation relating v_{hk} to $i_s = Y_S v_o$ in Fig. 6-12d are the same as those for v_{ab} in part b. (d) Use the preceding result to prove generally that, in the application of the source-conversion equation, only the forced response component need be found when solving for v_o or i_S from Eq. (6-31).

6-20 In the circuit of Fig. 6-12c and d, $v_o(t) = A + Bt$, $Z_1(p) = (p + 1)/p$, and $Z(p) = p + 2$. (a) Show that the complete solution for $i_s(t)$, using Eq. (6-31), is given by $i_s(t) = B + Ke^{-t}$. (b) Show that the equilibrium equation for v_{ab} in the circuit of Fig. 6-12c and d is given by $(p^2 + 3p + 1)v_{ab} = (p + 1)(p + 2)i_s = f(t)$. (c) Use the result of part a to show that the forcing function of the equilibrium equation for v_{ab} in part b, $f(t)$, independently of the value K, depends only on $v_o(t)$.

6-21 In Fig. P6-21 are shown two circuits that are equivalent with respect to terminals *a-b*. Given that $v_o(t) = 10 + 5t$, $v_{am}(0^+) = 10$ volts, and $i_L(0^+) = 2$ amp. (*a*) What is $v_C(0^+)$ in the circuit of Fig. P6-21*b*? (*b*) What is the value $(dv_{ab}/dt)_{0^+}$ in either circuit? (*c*) Find $i_s(t)$, using only the component due to v_o.

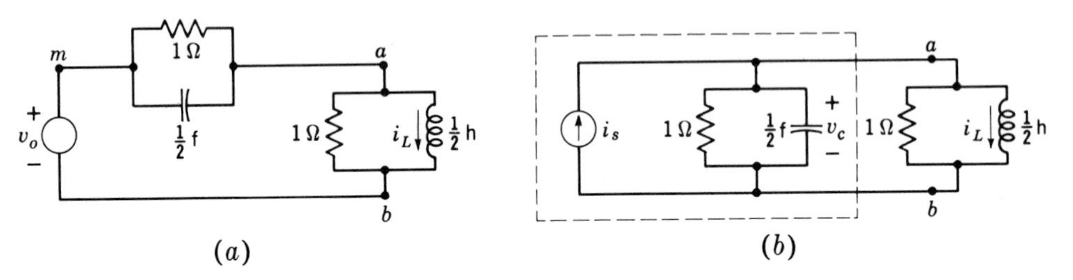

Fig. P6-21

6-22 Explain how the proof of Prob. 6-19 applies to Thévenin's theorem.

6-23 (*a*) Deduce Thévenin's equivalent circuit with respect to terminals *a-b* for the terminal pair shown in Fig. P6-23. (*b*) The parallel combination of R and C is placed across terminals *a-b*. If R is chosen for maximum power transfer in the steady state and if the time constant of the circuit is to be 0.2 sec, find R and C.

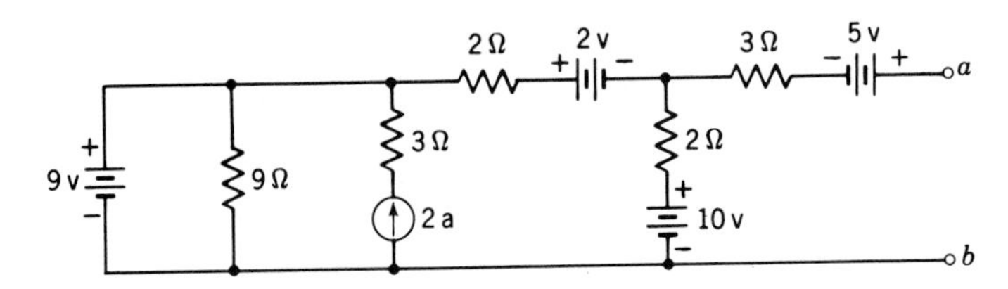

Fig. P6-23

6-24 In the circuit of Fig. P6-24 find Thévenin's equivalent circuit with respect to terminals *a-b* by the following procedure: (*a*) Calculate the voltage $V_{ab} = V_o$. (*b*) Short-circuit terminals *a-b*, and calculate the short-circuit current $I_{ab} = I_s$. Deduce R_S. (*c*) Replace the 18-ohm resistance by a $\frac{1}{2}$-farad capacitance, and connect a 2-henry inductance in series with the 6-ohm resistance between terminals *n-b*. Deduce Thévenin's equivalent circuit.

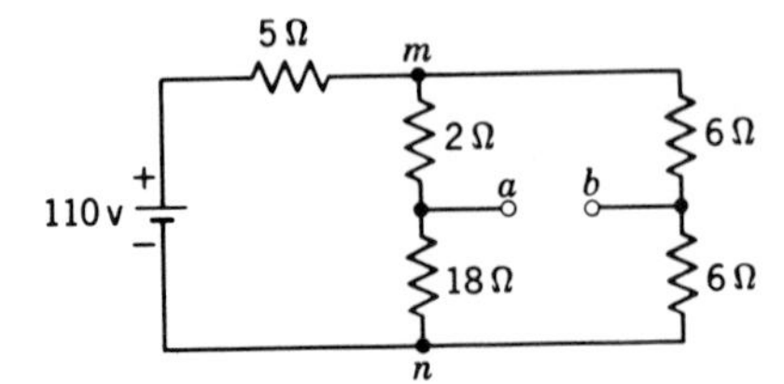

Fig. P6-24

6-25 Use the result of Prob. 6-24*a* to find v_{ab} for all $t \geq 0^+$ if (*a*) an initially uncharged capacitance of value 0.2 farad is placed across terminals *a-b*; (*b*) an initial deenergized series combination of 0.2 farad and 4 ohms is placed across terminals a-b; (c) the elements of part b are placed in parallel across terminals a-b, $v_{ab}(0^+) = 0$; (d) an initially deenergized inductance of 0.2 henry is placed across terminals a-b.

6-26 In the circuit of Fig. P6-26, use Thévenin's theorem to deduce the relationship between the elements R and C if the circuit is to be critically damped for L = 3 henrys.

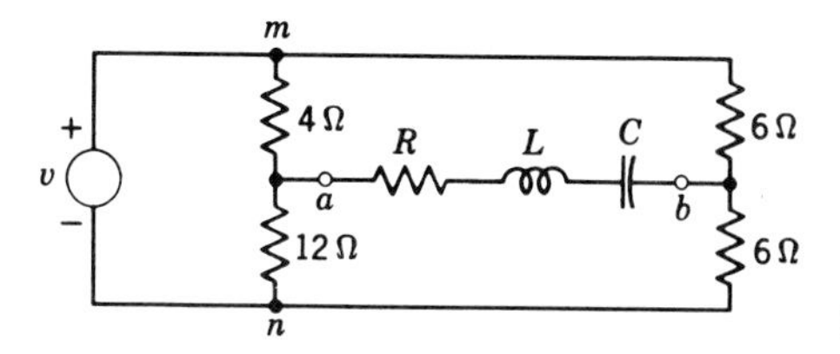

Fig. P6-26

6-27 The Wheatstone-bridge circuit shown in Fig. P6-27 is balanced for $x = 0$. (a) Obtain Thévenin's equivalent circuit with respect to terminals a-b (with R_g

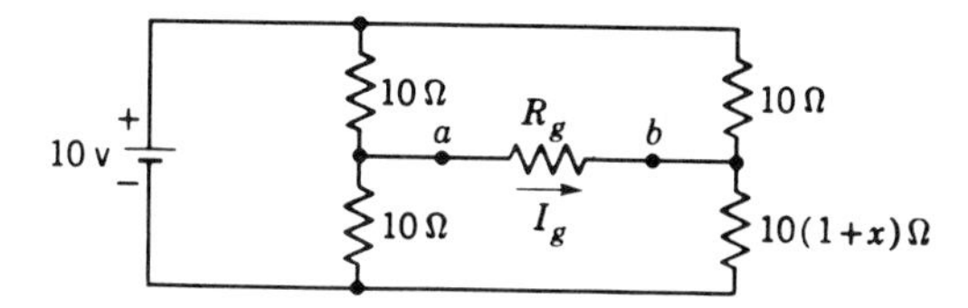

Fig. P6-27

removed). (b) Find the formula for I_g in terms of R_g and x. (c) Show that, when x is sufficiently small, then I_g is proportional to x, and explain how this result may be used to make incremental resistance measurements with an unbalanced bridge.

6-28 The circuit shown in Fig. P6-28 is a potentiometer with load resistance aR. (a) Show that the open-circuit voltage ($a \to \infty$) v_{ab} is given by $v_{ab} = v_o = xv$. (b) Show that the circuit to the left of terminals a-b may be represented by the series combination of v_o and $R_S = (x - x^2)R$. (c) Show that the potentiometer approximates an ideal voltage source worst when $x = \frac{1}{2}$ (Prob. 6-3c). (d) Show that $v_{ab}/v = ax/(a + x - x^2)$. ($e$) When a is infinite, v_{ab} varies linearly with x, $v_o/v = x$; study the effect of loading (finite a) by plotting v_{ab}/v as a function of x on the same set of axes for $a = 1$ and $a = 5$, and compare the graph with the straight line $v_o/v = x$. (f) Define the per cent deviation from linearity as $d = (v_{ab}/v_o - 1) \times 100$ per cent, and plot d as a function of x for $a = 1$ and $a = 5$.

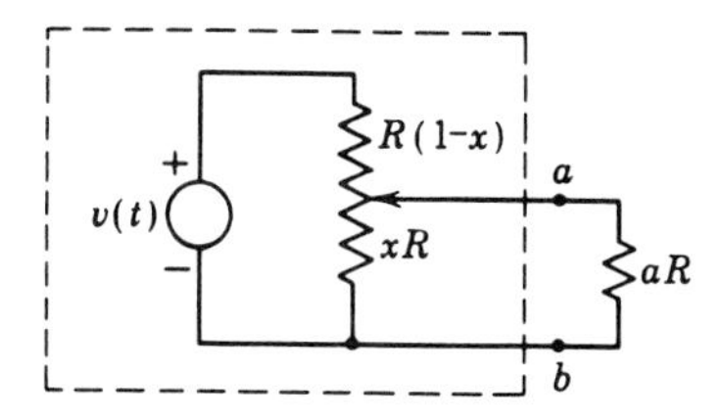

Fig. P6-28

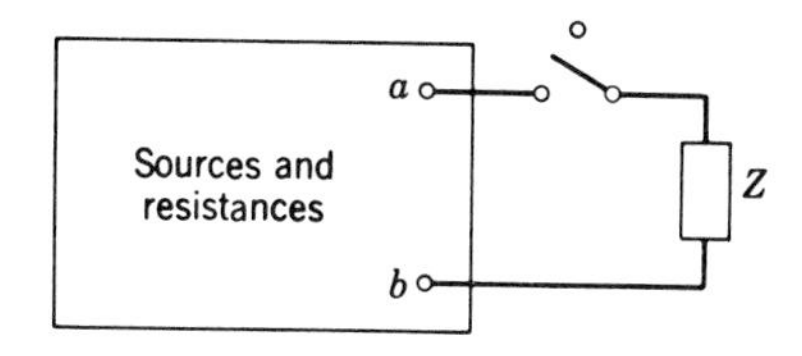

Fig. P6-29

6-29 The terminal pair shown in Fig. P6-29 consists of ideal *constant* sources and resistances only. The switch is closed, and $Z = R$. When $R = 15$ ohms, $V_{ab} = 45$ volts, and when $R = 10$ ohms, $V_{ab} = 40$ volts. Find V_{ab} when (a) $R = 5$ ohms; (b) $R = 1$ ohm.

6-30 In the active terminal pair shown in Fig. P6-29 it is known that all sources are constant and the switch is closed for all t. The following data are obtained: When Z consists of a capacitance, the steady-state value of V_{ab} is 60 volts. When Z consists of an inductance, the magnitude of the steady-state current in the inductance is 12 amp. Use this information to deduce Thévenin's equivalent circuit for the terminal-pair network.

6-31 In the network indicated in Fig. P6-29 the active terminal pair consists of ideal constant sources and resistances only. The load Z consists of the series combination of a resistance $R = 2$ ohms and a capacitance C. The capacitance stores zero energy before the switch is closed. The switch is closed at $t = 0$, and it is observed that $v_{ab} = 15 - 9e^{-t/2}$ for $t \geq 0^+$. (*a*) Find Thévenin's equivalent circuit for the terminal pair. (*b*) Find the value C.

6-32 The active terminal pair shown in Fig. P6-29 consists of an ideal source with exponential waveform and resistances as indicated. The load Z consists of the series combination of a 2-henry inductance and an 8-ohm resistance. The switch is closed at $t = 0$, and it is observed that $v_{ab} = 20e^{-6t} + 10e^{-3t}$ for $t \geq 0^+$. (*a*) Identify the time constant of the circuit. (*b*) Find Thévenin's equivalent circuit for the active-terminal-pair network.

6-33 The active terminal pair shown in Fig. P6-29 consists of resistive step- and ramp-function sources. The load Z consists of the parallel combination of a 12-ohm resistance and a capacitance C. The switch is closed at $t = 0$, and it is observed that $v_{ab} = 18 - 6t - 18e^{-t/2}$. (*a*) Deduce Norton's equivalent circuit for the active-terminal-pair network. (*b*) Use the answer to part *a* to find Thévenin's equivalent for the active-terminal-pair network.

6-34 In the circuit of a transistor, Fig. 6-19*a*, let $r_1 = 1{,}000$ ohms, $r_2 = 40$ k-ohms, $h_{21} = -50$, and $h_{12} = 2.5 \times 10^{-4}$. A load resistance R_L is connected across terminals *a-b*. (*a*) Calculate the driving-point impedance v/i_1 for $R_L = 40$ k-ohms and $R_L = 120$ k-ohms. (*b*) Calculate the ratio v_2/v for each of the values of R_L given in part *a*.

6-35 In the circuit of Fig. 6-19*a* let $r_1 = 20$ ohms, $r_2 = 500$ k-ohms, $h_{12} = 3 \times 10^{-4}$, and $h_{21} = -0.98$. A load resistance R_L is connected across terminals *a-b*. (*a*) Calculate the driving-point impedance v/i_1 for $R_L = 100$ k-ohms and $R_L = 500$ k-ohms. (*b*) Calculate the ratio v_2/v for each of the values of R_L given in part *a*.

6-36 In the circuit of Fig. P6-36 the source μv_{ab} is partially an ideal source and partially a controlled source ($v_{ab} = v - iR_k$). Obtain Thévenin's equivalent circuit with respect to terminals *m-n* using the following procedure: (*a*) Show that $v_o = -\mu v$. (*b*) Show that if terminals *m-n* are short-circuited, then i_{mn}, the current in the short circuit, is given by $i_s = -\mu v/[r_p + (\mu + 1)R_k]$. (*c*) Use the result of part *a* to find the output resistance R_s and draw Thévenin's equivalent circuit. (*d*) Verify the result for R_s by calculation of the driving-point impedance at terminals *m-n* with $v \equiv 0$. (*e*) For $\mu = 30$, $r_p = 10$ k-ohms and a load $R_L = 40$ k-ohms connected between terminals *m-n*, find the ratio v_{mn}/v for (1) $R_k = 400$ ohms; (2) $R_k = 0$.

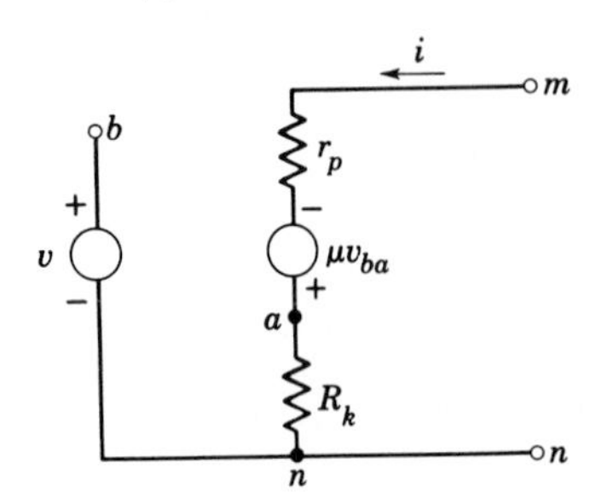

Fig. P6-36

6-37 In the circuit of Fig. P6-36 terminals m-n are permanently connected together (short-circuited). Use the open-circuit voltage–short-circuit current method to find Thévenin's equivalent circuit with respect to terminals a-n.

6-38 In the circuit of Fig. P6-18, an ideal voltage source $v_o(t)$ is connected between terminals a-n. (*a*) Use the compensating-source method to deduce the state equations of the circuit. (*b*) The $\frac{1}{6}$-farad capacitance is replaced by a $\frac{1}{6}$-henry inductance. Use v_{bn} and $i_{bmn} = i_L$ as state variables and find the state equations.

6-39 Deduce the state equations for the circuit of Fig. P6-21*a*, using i_L and v_{ma} as variables.

6-40 In the circuit of Fig. P6-24, the 18-ohm resistance has a $\frac{1}{2}$-μf capacitance connected in parallel. Obtain the state equation for v_{an} and find the time constant of the circuit.

6-41 In the circuit of Fig. P6-41, there are two dependent sources, $v_a = pi_1$ and $v_2 = pi_2$. Obtain an expression for the operational driving-point impedance $Z_{ab}(p)$.

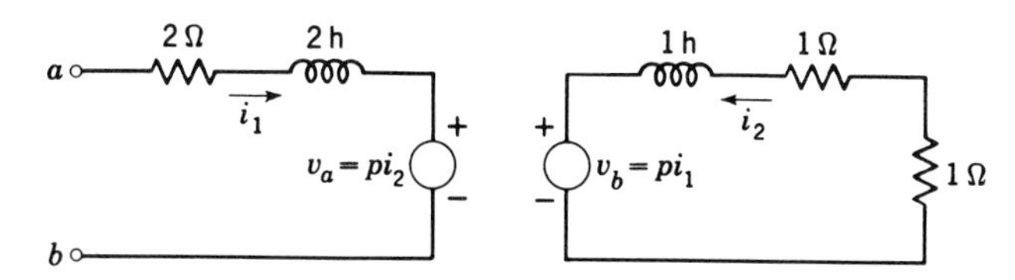

Fig. P6-41

Chapter 7 Introduction to pole-zero analysis

In the preceding chapters it was shown how the operational notation allows the formulation of equilibrium equations by algebraic means. In this chapter we explore further the algebraic properties of network functions. These properties are of special importance when the source waveform is an exponential function of time or when it can be represented as a sum of such exponential functions.

7-1 Review and classification of network functions

We recall that an operator which relates a response of a network to a source function is termed an *operational network function* of that network. In the discussion of network functions it was found convenient to define different types of network functions, depending on the type of source and response involved. In this section we summarize these definitions.

OPERATIONAL IMMITTANCE An operational immittance function is a network function that relates a voltage response to a current source or a current response to a voltage source. An operational *driving-point* immittance is a network function that relates voltage to current at the *same* pair of terminals.

TRANSFER FUNCTION A transfer function is a network function which relates the response at a pair of terminals (or in a branch) to a source at

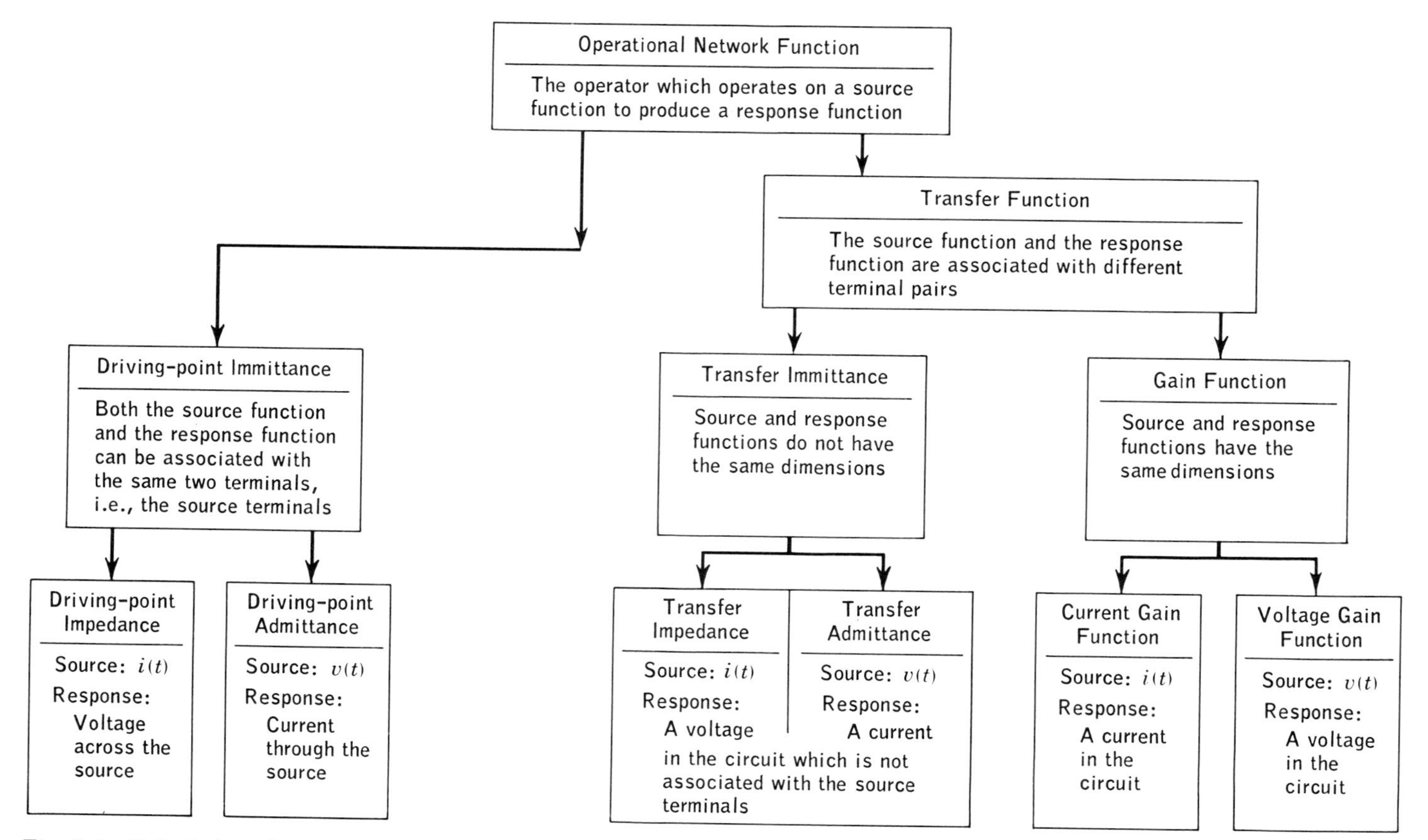

Fig. 7-1 Tabulation showing definitions of operational network functions.

another pair of terminals (or in another branch) of the network. A *gain function* is a transfer function which relates a voltage response to a voltage source or a current response to a current source.

IMPEDANCE AND ADMITTANCE An immittance function can be a driving point or a transfer function. If an immittance function (driving-point or transfer) relates a voltage response to a current source, it is called an operational impedance function (driving-point or transfer) and is denoted by the symbol $Z(p)$ with appropriate subscripts. If the source is a voltage source, the immittance function is called an admittance and is denoted by $Y(p)$.

These definitions are summarized in the tabulation of Fig. 7-1 and are illustrated in the following examples.

Example 7-1 In the circuit of Fig. 7-2, obtain the following network functions: (*a*) the driving-point admittance at the source terminals; (*b*) the voltage gain relating v_2 to v_o; (*c*) the transfer admittance relating i_2 to v_o.

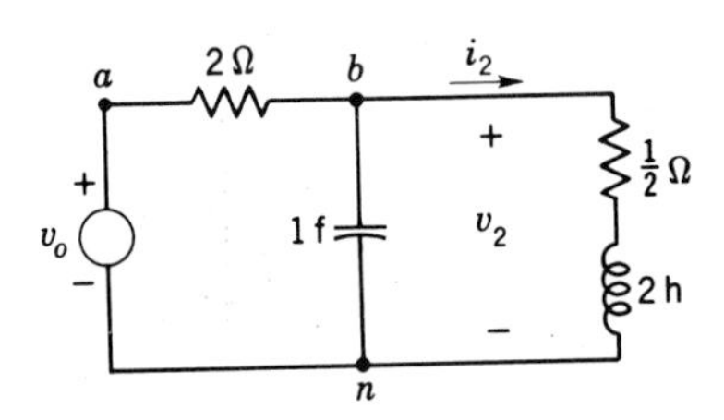

Fig. 7-2 Circuit for Example 7-1.

Solution (*a*) The driving-point impedance is found as the sum

$$Z_{an}(p) = Z_{ab} + Z_{bn}$$

where

$$Z_{ab} = 2 \qquad Y_{bn} = p + \frac{1}{2p + \frac{1}{2}} = \frac{4p^2 + p + 2}{4p + 1}$$

Hence

$$Z_{an} = 2 + \frac{4p + 1}{4p^2 + p + 2} = \frac{8p^2 + 6p + 5}{4p^2 + p + 2}$$

and

$$Y_{an} = \frac{4p^2 + p + 2}{8p^2 + 6p + 5} \tag{7-1}$$

(*b*) The voltage gain function relating v_2 to v_o can be found by voltage division.

$$\begin{aligned} v_2 &= \frac{Z_{bn}}{Z_{bn} + Z_{ab}} v_o \\ &= \frac{(4p + 1)/(4p^2 + p + 2)}{(4p + 1)/(4p^2 + p + 2) + 2} v_o \\ &= \frac{4p + 1}{8p^2 + 6p + 5} v_o \end{aligned} \tag{7-2}$$

Denoting the result by $H_v(p)$ [that is, $v_2 = H_v(p)v_o$], we have

$$H_v(p) = \frac{4p + 1}{8p^2 + 6p + 5} \tag{7-3}$$

(*c*) The transfer admittance relating i_2 to v_o can be found by observing that

$$i_2 = \frac{1}{2p + \frac{1}{2}} v_2 \tag{7-4}$$

and substituting for v_2 in Eq. (7-4) from Eq. (7-2). Hence

$$i_2 = \frac{1}{2p + \frac{1}{2}} \frac{4p + 1}{8p^2 + 6p + 5} v_o$$

or

$$i_2 = \frac{2}{8p^2 + 6p + 5} v_o$$

Denoting the required admittance function by $Y_t(p)$, we have

$$Y_t(p) = \frac{2}{8p^2 + 6p + 5} \tag{7-5}$$

Example 7-2 In the circuit shown in Fig. 7-3, obtain (*a*) the transfer impedance relating v_2 to i_s; (*b*) the driving-point impedance at the source terminals.

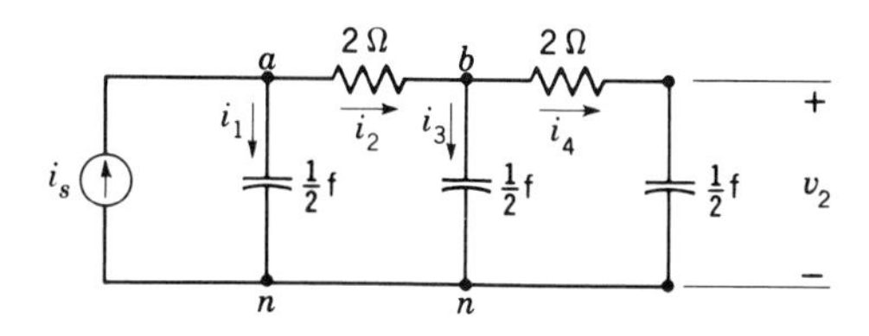

Fig. 7-3 Circuit for Example 7-2.

Solution To express v_2 in terms of i_s we begin by writing

$$i_4 = \tfrac{1}{2}pv_2$$

so that

$$v_{bn} = 2 \times \tfrac{1}{2}pv_2 + v_2 = (p + 1)v_2$$

Hence

$$i_3 = \tfrac{1}{2}p(p + 1)v_2 = (\tfrac{1}{2}p^2 + \tfrac{1}{2}p)v_2$$

and

$$\begin{aligned} i_2 &= i_3 + i_4 \\ &= (\tfrac{1}{2}p^2 + p)v_2 \end{aligned}$$

Then

$$v_{ab} = (p^2 + 2p)v_2$$

and

$$\begin{aligned} v_{an} &= v_{ab} + v_{bn} \\ &= (p^2 + 3p + 1)v_2 \end{aligned} \tag{7-6}$$

We can now write

$$\begin{aligned} i_1 &= \tfrac{1}{2}pv_{an} \\ &= (\tfrac{1}{2}p^3 + \tfrac{3}{2}p^2 + \tfrac{1}{2}p)v_2 \end{aligned}$$

and

$$\begin{aligned} i_s &= i_1 + i_2 \\ &= (\tfrac{1}{2}p^3 + 2p^2 + \tfrac{3}{2}p)v_2 \end{aligned} \tag{7-7}$$

From Eq. (7-7) we have

$$v_2 = \frac{2}{p^3 + 4p^2 + 3p} i_s = Z_t(p)i_s \tag{7-8}$$

so that the required transfer impedance is given by

$$Z_t(p) = \frac{2}{p^3 + 4p^2 + 3p}$$

The driving-point impedance at terminals a-n can be obtained by substitution from Eq. (7-8) into Eq. (7-6):

$$v_{an} = (p^2 + 3p + 1)\frac{2}{p^3 + 4p^2 + 3p}i_s$$

$$= \frac{2p^2 + 6p + 2}{p^3 + 4p^2 + 3p}i_s$$

Hence

$$Z_{an}(p) = \frac{2p^2 + 6p + 2}{p^3 + 4p^2 + 3p} \tag{7-9}$$

7-2 *Nature of the denominator of network functions*

In every linear network a response function $y(t)$ is related to a source function $\phi(t)$ through an equilibrium equation of the form

$$(a_0 + a_1p + a_2p^2 + \cdots + a_Mp^M)y(t) = (b_0 + b_1p + b_2p^2 + \cdots + b_Np^N)\phi(t)$$

or, defining the polynomial functions in p, $N(p)$ and $D(p)$, as

$$N(p) = b_0 + b_1p + b_2p^2 + \cdots + b_Np^N$$

and

$$D(p) = a_0 + a_1p + a_2p^2 + \cdots + a_Mp^M$$

we have defined the operational network function $H(p)$ as

$$H(p) = \frac{N(p)}{D(p)}$$

or

$$y(t) = [H(p)]\phi(t)$$

We now recall from the preceding chapters, as well as from Examples 7-1 and 7-2, that for a given network with a single source, all network functions have the same denominator polynomial $D(p)$ but the numerator polynomial depends on the particular response function which has been selected.[1] In the equation

$$D(p)y(t) = N(p)\phi(t)$$

we recall that the function $N(p)\phi(t)$ is the *forcing function* for the selected response and that the equilibrium equation which relates one response of a network to a given source function differs from that which relates another response to the source in the nature of the forcing function or, equivalently, in the numerator of the network function. The foregoing observations are the result of examining special examples. A formal proof is given in Chap. 11.

FREE COMPONENT OF THE RESPONSE If the source function is set to zero, the equilibrium equation for any network variable $y(t)$ has the same form: $D(p)y(t) = 0$. This linear homogeneous equation with constant coefficients

[1] This statement assumes that common factors of N and D have not been canceled. The matter of common factors is discussed in Chap. 12.

can be solved by introducing the modes Ke^{st} so that the characteristic roots are given by solution of the algebraic equation $D(s) = 0$. The polynomial $D(s)$ can be factored. If the roots are $s = s_1, \ldots, s_M$, then $D(s)$ can be written in the form

$$a_M s^M + a_{M-1}s^{M-1} + \cdots + a_1 s + a_0 = a_M(s - s_1)(s - s_2) \cdots (s - s_M)$$

In this connection the symbolic use of the product symbol Π is useful. We write

$$\prod_{k=1}^{M} (s - s_k) \equiv (s - s_1)(s - s_2) \cdots (s - s_M)$$

The characteristic equation can therefore be written in the form

$$\prod_{k=1}^{M} (s - s_k) = 0$$

If no two roots of this equation are equal, the free component of the response is

$$y(t)_f = \sum_{k=1}^{M} K_k e^{s_k t}$$

If m' of the M roots are equal and denoted by s_1, the characteristic equation has the form

$$(s - s_1)^{m'}(s - s_{m'+1})(s - s_{m'+2}) \cdots (s - s_M) = 0$$

or

$$(s - s_1)^{m'} \prod_{m'+1}^{M} (s - s_k) = 0$$

In this case the free component of the response can be shown to have the form[1]

$$y(t)_f = \sum_{k=1}^{m'} K_k t^{k-1} e^{s_1 t} + \sum_{k=m'+1}^{M} K_k e^{s_k t} \tag{7-10}$$

or

$$y(t)_f = (K_1 + K_2 t + K_3 t^2 + \cdots + K_{m'} t^{m'-1})e^{s_1 t} + K_{m'+1} e^{s_{m'+1}} t + \cdots + K_M e^{s_M t}$$

7-3 *Exponential sources and transform network functions*

As was shown in Chap. 3, the use of operational network functions enables us to compute the component of the response due to an exponential source by a very simple algebraic procedure. We recall that if a source function has the form

$$\phi(t) = \phi_0 e^{s_0 t}$$

[1] See Sec. 8-10.

(where the subscript g indicates that the value of s is related to the source or generators), then, provided that s_g is *not* a characteristic root (that is, a zero of the denominator of the network function), the component of the response due to the source, y_s, is given by

$$y_s(t) = H(s_g)\phi_0 e^{s_g t} \qquad \frac{1}{H(s_g)} \neq 0 \tag{7-11}$$

Because this result, Eq. (7-11), is of fundamental importance, its proof is given again in this section. To prove Eq. (7-11) we first note that if

$$\phi(t) = \phi_0 e^{s_g t} \tag{7-12}$$

then

$$\begin{aligned} p\phi &= s_g\phi \\ p^2\phi &= s_g{}^2\phi \\ &\vdots \\ p^n\phi &= s_g{}^n\phi \end{aligned}$$

and therefore

$$(b_0 + b_1 p + b_2 p^2 + \cdots + b_M p^M)\phi = (b_0 + b_1 s_g + b_2 s_g{}^2 + \cdots + b_M s_g{}^M)\phi$$

so that if

$$N(p) = b_0 + b_1 p + \cdots + b_M p^M$$

the forcing function corresponding to the source function $\phi(t)$ is given by

$$N(p)\phi(t) = N(s_g)\phi(t) \tag{7-13}$$

Equation (7-13) states, in words, that for source functions that vary exponentially with time, the forcing function will be *proportional* to the source function and the proportionality factor can be obtained by replacing the operator p in the numerator polynomial of the network function by s_g. It follows that, for $\phi(t)$ given by Eq. (7-12), the equilibrium equation has the form

$$D(p)y(t) = N(s_g)\phi_0 e^{s_g t}$$

If we let $y = Ae^{s_g t}$, then $D(p)Ae^{s_g t} = D(s_g)Ae^{s_g t}$, so that the formula $y = Ae^{s_g t}$ is a solution if

$$D(s_g)Ae^{s_g t} = N(s_g)\phi_0 e^{s_g t} \tag{7-14}$$

or

$$A = \frac{N(s_g)}{D(s_g)}\phi_0 \qquad D(s_g) \neq 0 \tag{7-15}$$

We observe that

$$A = H(s_g)\phi_0 \tag{7-16}$$

provided that $D(s_g) \neq 0$. This condition corresponds to the requirement that s_g not be a characteristic root. The above proof shows that, for exponential source functions of the form $\phi_0 e^{s_g t}$, when s_g is not a characteristic root, the component of the response due to the source is *proportional* to the source function, and the proportionality factor is obtained by replacing the operator p in the network function $H(p)$ by s_g.

If we regard s as an algebraic variable rather than as a particular value such as s_g, then, for a source

$$\phi(t) = \phi_0 e^{st}$$

$$y_s = H(s)\phi \qquad \frac{1}{H(s)} \neq 0 \tag{7-17}$$

The algebraic function $H(s)$ is termed the *transform* network function. This term is used because, for exponential source functions, the differential equation $y = H(p)\phi$ is transformed into the algebraic relationship (7-17). Moreover, we shall see (in Chap. 8) that $H(s)$ is also the ratio of the Laplace transform of the response to that of the source.

RESONANCE Two special cases of exponential source functions are of interest; both of these result in "resonant" situations. One such case results when the value of s_g in the source exponential results in $N(s_g) = 0$. Thus, for example, if

$$y(t) = \frac{p+2}{p^2+p+5}\,\phi(t)$$

the transform network function is given by

$$H(s) = \frac{s+2}{s^2+s+5}$$

and is zero for $s = -2$ $[H(-2) = 0]$. This means that the component of the response y due to a source of waveform $\phi_0 e^{-2t}$ will be zero. To describe the special values of s which result in resonance, the following terms are used:

The zeros of the *numerator* of a transform network function are called the *zeros* of that network function. The reason for this terminology should be clear from the foregoing discussion.

The zeros of the *denominator* of a transform network function are called the *poles* of that network function. The reason for this terminology is probably the fact that, for exponential sources where s_g is a zero of the denominator of the network function, the value of the network function is infinity (goes up the pole).

The second special case mentioned above deals with the source $\phi_0 e^{st}$, where s is a pole of the network function. Below (Sec. 8-8), we show that, if s_1 is an nth-order pole of $H(s)$, that is, if

$$H(s) = \frac{N(s)}{(s-s_1)^n D_1(s)} \tag{7-18}$$

where $N(s_1) \neq 0$ and $D_1(s_1) \neq 0$, then the component of the response y_s due to a source

$$\phi = \phi_0 e^{s_1 t}$$

is

$$y_s = At^n e^{s_1 t} \tag{7-19}$$

7-4 Exponential source functions with complex exponent

We observe that sources of the form

$$\phi(t) = \Phi_m \cos(\omega t + \gamma) \tag{7-20a}$$

or

$$\phi(t) = \Phi_0 e^{\sigma t} \cos(\omega t + \gamma) \tag{7-20b}$$

are special cases of the source function

$$\phi(t) = \text{Re}\,(\boldsymbol{\phi}_0 e^{\mathbf{s}_g t}) \tag{7-21}$$

where in Eq. (7-20a) $\boldsymbol{\phi}_0 = \phi_m\underline{/\gamma}$ and $\mathbf{s}_g = j\omega$ and in Eq. (7-20b) $\boldsymbol{\phi}_0 = \phi\underline{/\gamma}$ and $\mathbf{s}_g = \sigma + j\omega$. We shall now show that if

$$\phi(t) = \text{Re}\,(\boldsymbol{\phi}_0 e^{\mathbf{s}_g t})$$

then, if $\mathbf{s}_g$ is *not* a pole of network function, the component of the response due to such a source is given by

$$y_s(t) = \text{Re}\,[H(\mathbf{s}_g)\boldsymbol{\phi}_0 e^{\mathbf{s}_g t}] \tag{7-22}$$

Note: The beginning student has some tendency to write Eq. (7-22) incorrectly as $[H(\mathbf{s}_g)] \times \text{Re}\,(\boldsymbol{\phi}_0 e^{\mathbf{s}_g t})$. This form leads to complex solutions which are not only wrong, but are meaningless as well.

To prove Eq. (7-22), we begin by observing that the proof of Eq. (7-17) is valid for complex values of s; that is, if

$$\boldsymbol{\phi}(t) = \tfrac{1}{2}\boldsymbol{\phi}_0 e^{\mathbf{s}_g t}$$

then if $D(\mathbf{s}_g) \neq 0$,

$$\mathbf{y}_s(t) = \frac{N(\mathbf{s}_g)}{D(\mathbf{s}_g)}\frac{1}{2}\boldsymbol{\phi}_0 e^{\mathbf{s}_g t}$$

that is,

$$\mathbf{y}_s(t) = H(\mathbf{s}_g) \cdot \tfrac{1}{2}\boldsymbol{\phi}_0 e^{\mathbf{s}_g t} \tag{7-23}$$

Similarly, if

$$\boldsymbol{\phi}(t) = \tfrac{1}{2}\boldsymbol{\phi}_0^* e^{\mathbf{s}_g^* t}$$

where the asterisk denotes the complex conjugate, the response will have the form of Eq. (7-22), with $\mathbf{s}_g$ replaced by $\mathbf{s}_g^*$. Since $H(\mathbf{s}_g)$ is a real function (of a complex variable),

$$\tfrac{1}{2}H(\mathbf{s}_g^*)\boldsymbol{\phi}_0^* e^{\mathbf{s}_g^* t} = [\tfrac{1}{2}H(\mathbf{s}_g)\boldsymbol{\phi}_0 e^{\mathbf{s}_g t}]^* \tag{7-24}$$

then if the source has the form

$$\begin{aligned}\phi(t) &= \tfrac{1}{2}\boldsymbol{\phi}_0 e^{\mathbf{s}_g t} + \tfrac{1}{2}\boldsymbol{\phi}_0^* e^{\mathbf{s}_g^* t} \\ &= \text{Re}\,(\boldsymbol{\phi}_0 e^{\mathbf{s}_g t})\end{aligned}$$

it follows from the foregoing discussion, together with the principle of superposition, that

$$y_s(t) = \text{Re}\,[H(\mathbf{s}_g)\boldsymbol{\phi}_0 e^{\mathbf{s}_g t}] \tag{7-25}$$

as stated above.

Example 7-3 In a certain network the response $y(t)$ is related to the source by the equation

$$y(t) = \frac{p+2}{(p+1)(p^2+2p+5)}\phi(t)$$

Calculate the component of the response due to each of the following sources:

(*a*) $\phi(t) = 10u(t)$ (*b*) $\phi(t) = 10tu(t)$ (*c*) $\phi(t) = 10e^{-2t}u(t)$
(*d*) $\phi(t) = 10e^{-4t}u(t)$ (*e*) $\phi(t) = e^{-t}u(t)$ (*f*) $\phi(t) = 10\cos(2t - 17°)$
(*g*) $\phi(t) = 10e^{-2t}\cos(2t - 10°)$

Solution (*a*) The source $\phi(t) = 10u(t) = 10e^{0t}u(t)$; hence we replace p with zero and obtain

$$y_s(t) = \tfrac{2}{5} \times 10 = 4$$

(*b*) Multiplying out $D(p)$, we get

$$\begin{aligned}(p^3 + 3p^2 + 7p + 5)y &= (p+2)10t \\ &= 10 + 20t\end{aligned}$$

Letting

$$y_s = A + Bt \qquad py_s = B$$

we have

$$7B + 5A + 5Bt = 10 + 20t$$

so that

$$5B = 20 \qquad B = 4$$

and

$$\begin{aligned}7B + 5A &= 10 \\ 5A &= -18 \\ A &= -3.6\end{aligned}$$

and

$$y_s = -3.6 + 4t$$

(*c*) When $\phi = 10e^{-2t}$, $H(-2) = 0$ and $y_s \equiv 0$.
(*d*) For $\phi = 10e^{-4t}$,

$$y_s = \frac{-4+2}{(-4+1)(16-8+5)} \times 10e^{-4t}$$

or

$$y_s = \tfrac{20}{39}e^{-4t}$$

(*e*) For $\phi(t) = e^{-t}$, $s_g = -1$, which is a pole of the network function. Hence the solution has the form

$$y_s = Ate^{-t}$$

To find A we substitute in the equilibrium equation

$$\begin{aligned}(p^3 + 3p^2 + 7p + 5)y &= (p+2)e^{-t} \\ &= e^{-t}\end{aligned} \tag{7-26}$$

If

$$y_s = Ate^{-t}$$

then

$$py_s = A(-t+1)e^{-t} \qquad p^2y_s = A(t-2)e^{-t} \qquad p^3y_s = A(-t+3)e^{-t}$$

Substituting in Eq. (7-26) gives

$$A(-t+3) + 3A(t-2) + 7A(-t+1) + 5At = 1$$

or

$$t(-A + 3A - 7A + 5A) + A(3 - 6 + 7) = 1$$

or

$$A = \tfrac{1}{4}$$

so that

$$y_s = \tfrac{1}{4}te^{-t}$$

(*f*) When $\phi(t) = 10 \cos (2t - 17°)$, we write

$$\phi(t) = \text{Re}\,(10\underline{/-17°}\, e^{j2t})$$

so that

$$y_s = \text{Re}\left[\frac{j2 + 2}{(j2 + 1)(-4 + j4 + 5)}\, 10\underline{/-17°}\, e^{j2t}\right]$$

Since

$$\frac{j2 + 2}{(1 + j2)(1 + j4)} = 0.305\underline{/-94.4°}$$

$$y_s = 3.05 \cos (2t - 111.4°)$$

(*g*) We write

$$\phi(t) = \text{Re}\,(10\underline{/-10°}\, e^{(-2+j2)t})$$

Hence

$$y_s = \text{Re}\left[10\underline{/-10°}\, \frac{j2}{(-1 + j2)(-j8 - 4 + j4 + 5)}\, e^{(-2+j2)t}\right]$$

Since

$$\frac{j2}{(-1 + j2)(1 - j4)} = 0.216\underline{/49.4°}$$

$$y_s = 2.16e^{-2t} \cos (2t + 39.4°)$$

RELATION BETWEEN POLES AND ZEROS OF THE NETWORK FUNCTION AND RESPONSE DUE TO AN EXPONENTIAL SOURCE Operational network functions are the ratios of polynomials in p. These polynomials can be written in factored form:

$$\frac{N(p)}{D(p)} = \frac{b_N p^N + b_{N-1}p^{N-1} + \cdots + b_1 p + b_0}{a_M p^M + a_{M-1}p^{M-1} + \cdots + a_1 p + a_0}$$
$$= H\,\frac{(p - s_A)(p - s_B)\cdots(p - s_N)}{(p - s_1)(p - s_2)\cdots(p - s_M)}$$

where $H = b_N/a_M$ is a real number, and the poles $s_1, \ldots, s_M$ and zeros $s_A, \ldots, s_N$ are in general complex. The transform network function is obtained by replacing p by[1] s.

$$\frac{N(s)}{D(s)} = H\,\frac{(s - s_A)(s - s_B)\cdots(s - s_N)}{(s - s_1)(s - s_2)\cdots(s - s_M)} \tag{7-27}$$

We recall that the zeros of the denominator, $s_1, s_2, \ldots, s_M$, which determine the source-free modes, are called the *poles* of the network function. Similarly, the zeros of the numerator, $s_A, s_B, \ldots, s_N$, are called the *zeros* of the network function. We now come to a very important conclusion: For any source function e^{st} the component of the response due to the source is determined by the *differences* between s and the zero and pole values of the network function. Below, in Sec. 7-5, it is shown how this fact is used to simplify the evaluation of response components due to exponential sources. We also recall that when the value of s in the source function e^{st} is identical

[1] Although s may be real or complex, boldface type will *not* be used to indicate general values of s in the remainder of this chapter.

with a zero of the network function, the *component* of the response due to the source as well as the network function is zero.

7-5 Graphical interpretation of zeros and poles

The general form of a transform network function $H(s)$ is given by Eq. (7-27),

$$H(s) = H\frac{(s - s_A)(s - s_B) \cdots (s - s_N)}{(s - s_1)(s - s_2) \cdots (s - s_M)}$$

where the zeros and poles may be real, imaginary, or complex. In any event the value of $H(s)$ for any given value of s is completely specified if all zero and pole locations are specified and if the (real) numerical constant H is known. Since the zero and pole values are generally complex, they can be made to correspond to points in the complex plane. Thus, if the zeros and poles of a (transform) network function are given in the complex-frequency plane, the function is specified except for the multiplying factor H. We shall follow the practice of marking the poles by a cross (**x**) and the zeros by a small circle (**o**) in the s plane as shown in Fig. 7-4. The location of zeros and poles in Fig. 7-4 gives the following information:

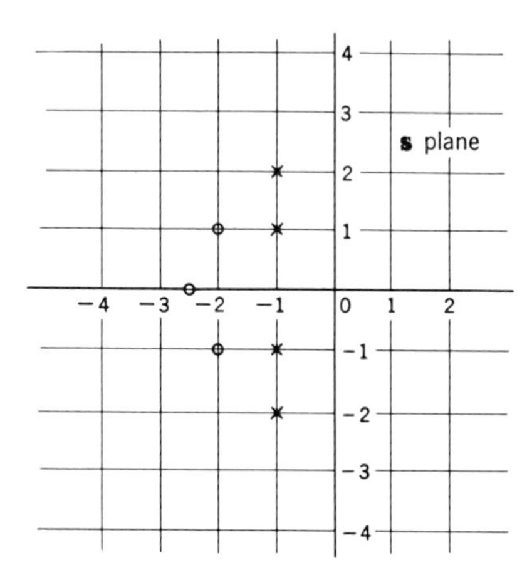

Fig. 7-4 Pole-zero locations for a network function.

$$H(s) = H\frac{(s + 2.5)(s + 2 - j)(s + 2 + j)}{(s + 1 - j)(s + 1 + j)(s + 1 - j2)(s + 1 + j2)}$$

It is seen that, from the location of the poles alone, we know that the free component of the response to any source function is

$$y(t)_f = \mathbf{K}_1e^{(-1+j)t} + \mathbf{K}_2e^{(-1-jt)} + \mathbf{K}_3e^{(-1+j2)t} + \mathbf{K}_4e^{(-1-j2)t}$$

or

$$y(t)_f = \text{Re}\ (2\mathbf{K}_1e^{-t}e^{jt}) + \text{Re}\ (2\mathbf{K}_3e^{-t}e^{j2t})$$

We can correlate the location of poles and the free modes as follows:

1 Pole on the real axis (Fig. 7-5). The free mode is an exponential function with real exponent.

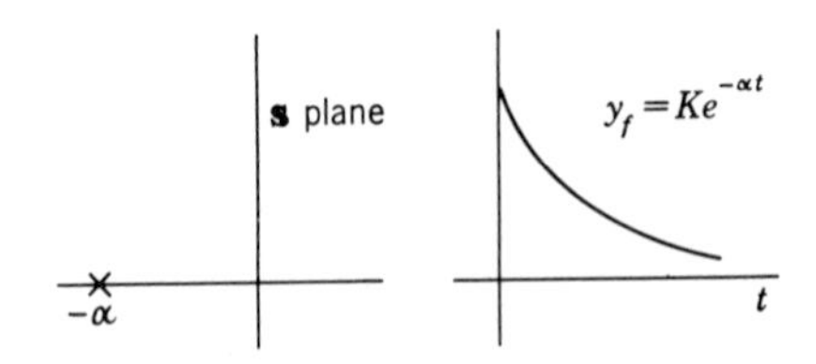

Fig. 7-5 A pole on the negative portion of the real axis corresponds to an exponential decay.

2 A pair of conjugate poles on the imaginary axis at $s = \pm j\omega$ (Fig. 7-6). The free modes form one sinusoid.

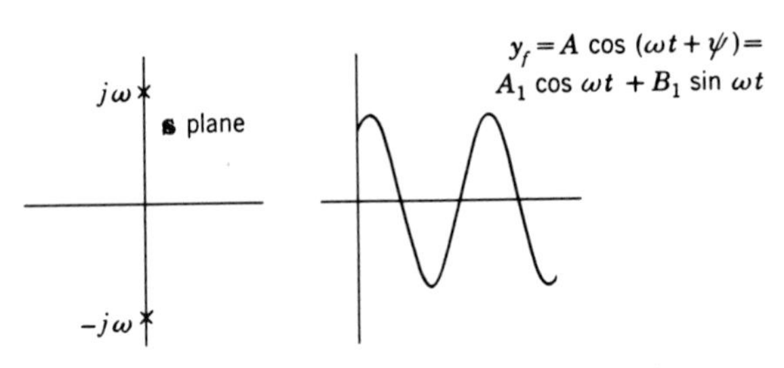

Fig. 7-6 A pair of poles on the imaginary axis corresponding to a sinusoid.

3 A pair of conjugate poles in the left hand plane with conjugate complex values (Fig. 7-7) of $s = -\alpha \pm j\omega_d$. The two free modes form one exponentially damped sinusoid.

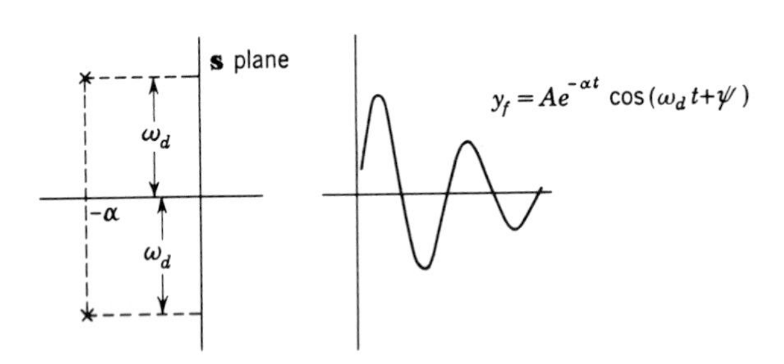

Fig. 7-7 A pair of complex conjugate poles in the left half s plane corresponds to an exponentially damped sinusoid.

MULTIPLICITY So far we have assumed that all the poles arise from simple roots of the characteristic equation; i.e., in the equation

$$(s - s_1)(s - s_2) \cdots (s - s_M) = 0$$

$s_j \neq s_k$; that is, no two factors are identical. We know already (e.g., from the critically damped *R-L-C* circuit) that it is possible for two or more poles to coincide. Such poles are called *multiple poles*. Graphically we shall represent such multiple poles by a cross, together with a roman numeral to indicate the multiplicity of the pole. We can classify multiple poles as follows:

1 Double pole on the negative real axis (Fig. 7-8*a*) at $s = -\alpha$. This corresponds to a factor $(s + \alpha)^2$ in the characteristic equation. Hence the free modes corresponding to this are $y_f = (K_1 + K_2t)e^{-\alpha t}$.

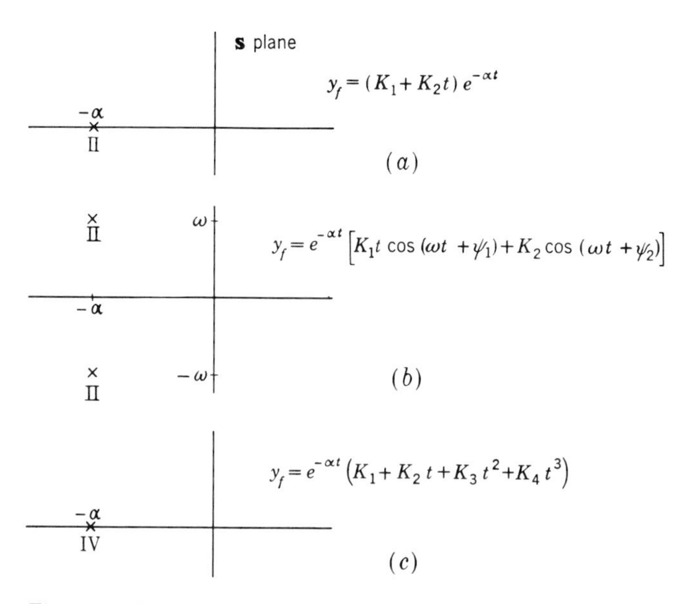

***Fig. 7-8** Multiple poles and their significance.*

2 Double poles at complex-conjugate values of s as shown in Fig. 7-8*b*. Since a double pole at $s = -\alpha + j\omega$ corresponds to the modes $(\mathbf{K}_a + \mathbf{K}_bt)e^{(-\alpha+j\omega)t}$, the two complex-conjugate double poles correspond to the four modes $\mathbf{K}_ae^{(-\alpha+j\omega)t} + \mathbf{K}_a^*e^{(-\alpha-j\omega)t} + \mathbf{K}_bte^{(-\alpha+j\omega)t} + \mathbf{K}_b^*te^{(-\alpha-j\omega)t}$ or

$$y_f = e^{-\alpha t}[K_1t \cos(\omega t + \psi_1) + K_2 \cos(\omega t + \psi_2)]$$

3 Higher multiplicity. The interpretation for multiplicity of orders higher than second should now be clear. For example, the fourth-order pole shown in Fig. 7-8*c* corresponds to the response component

$$y_f(t) = (K_1 + K_2t + K_3t^2 + K_4t^3)e^{-\alpha t}$$

7-6 The origin in the s plane

A mode $e^{-\alpha t}$ corresponds to a pole $s = -\alpha$ in the s plane. If $\alpha \to 0$, the pole moves to the origin. Hence a simple pole at the origin corresponds to a constant term in the free response. Similarly, a double pole at the origin corresponds to the $\lim_{\alpha\to 0} (K_1 + K_2t)e^{-\alpha t} = K_1 + K_2t$.

To summarize our conclusions so far, we observe that the location (and multiplicity) of the poles of a transform network function in the s plane tell us the nature of the free modes, and if the zeros also are known, the network function is known to within a factor H.

7-7 Representation of source functions by poles

In this section we shall show that exponential source functions can also be represented by poles in the s plane. We are led to this consideration by means of the following reasoning: Suppose that we know that the *complete* response of a circuit is given by $f(t) = 3e^{-t} + 4e^{-2t}$. Can we tell the nature of the circuit? The answer is that we *cannot*, since the following are some of the possibilities which apply:

1 The circuit consists of resistances only, and the source has the form $Ae^{-t} + Be^{-2t}$.
2 The circuit is an R-L circuit *or* an R-C circuit with time constant either 1 or $\frac{1}{2}$, and the source is either Be^{-2t} or Ae^{-t}.
3 The circuit is a source-free R-L-C circuit.

It is therefore clear that in the complete response we cannot tell, without additional data, which terms are free modes and which terms form the component of the response due to an exponential source (the component of the response due to the source is also called the sum of the *forced* modes). To show how this situation arises mathematically, consider the equation

$$D(p)y(t) = (p - s_1)(p - s_2) \cdots (p - s_n)y(t) = e^{s_g t} \qquad (7\text{-}28)$$

where the subscript g refers to generator. If Eq. (7-28) is differentiated, we have

$$p(p - s_1)(p - s_2) \cdots (p - s_n)y(t) = s_g e^{s_g t} \qquad (7\text{-}29)$$

Now we multiply Eq. (7-28) by $-s_g$ and add to (7-29).

$$(p - s_g)D(p)y(t) = 0 \qquad (7\text{-}30)$$

Consequently, the inhomogeneous equation (7-28) is equivalent to the homogeneous equation (7-30). The forced mode $Ae^{s_g t}$, which forms the component of the response due to the source in Eq. (7-28), is a free mode of Eq. (7-30).

This discussion indicates that we can represent the exponential sources in the s plane by poles in the same manner as a free mode is characterized by a pole. To distinguish these poles from the poles of the network function, the letter G will appear near the pole symbol. Thus a d-c source is represented by the pole at the origin as shown in Fig. 7-9*a*. A ramp- plus a step-function source is illustrated in Fig. 7-9*b*. A sinusoidal source is represented by the two poles on the axis of imaginaries as shown in Fig. 7-9*c*.

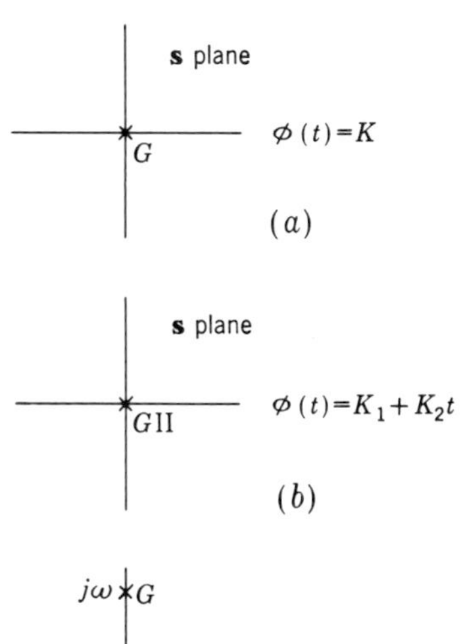

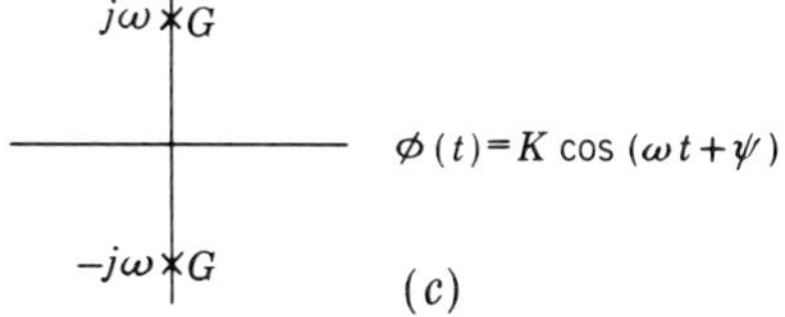

***Fig. 7-9** Poles marked with the letter G represent modes of the source function.*

In order to avoid confusion, the poles of the network function are sometimes marked; we shall use the letter F (for free) to mark poles of the network function.

At this point it is clear what happens if a pole of the network function coincides with a pole of the source function: the complete response will have a term due to a double pole. Thus, in Fig. 7-10*a*, the complete response is $K_1e^{-\alpha t} + Ae^{s_gt}$, where $K_1e^{-\alpha t}$ is the free mode and Ae^{s_gt} is the forced mode. In Fig. 7-10*b* the source was changed so that $s_g = -\alpha$. In

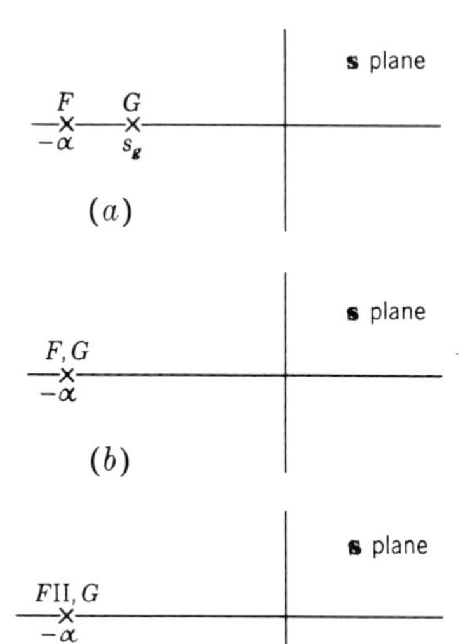

***Fig. 7-10** Poles representing the complete response. (a) Simple network pole and simple, distinct source pole. (b) Coincidence of source pole and network pole. (c) A source pole coincides with a double pole of the network.*

this case the complete response is $(K_1 + At)e^{-\alpha t}$, where $A \equiv 0$ if the source is set to zero. *Hence $Ate^{-\alpha t}$ is the component of the response due to the source.* If a source $\Phi_0 e^{-\alpha t}$ is impressed on a network which has a double pole at $s = -\alpha$ (Fig. 7-10*c*), a triple pole results; *the forced mode (i.e., the component of the response due to the source) is $At^2e^{-\alpha t}$.* These are examples of resonance mentioned in Sec. 7-3 and discussed further in Sec. 8-10.

7-8 Graphical determination of forced modes

We have seen that, if a source function $\phi(t) = Ae^{s_g t}$ is impressed on a network, then, if s_g is not a pole of the network function, the response component due to the source (forced mode) is

$$y_s(t) = \frac{N(s_g)}{D(s_g)} Ae^{s_g t} = H(s_g)Ae^{s_g t}$$

or

$$y_s(t) = H \frac{(s_g - s_A)(s_g - s_B) \cdots (s_g - s_N)}{(s_g - s_1)(s_g - s_2) \cdots (s_g - s_M)} Ae^{s_g t}$$

Hence the evaluation of the complete response depends on the evaluation of products of the form $\prod_k (s_g - s_k)$. Since each term $s_g - s_k$ may be a complex number, it is convenient to carry out the multiplication and division in polar form. The values of the factors $s_g - s_k$ can be obtained with ease, in polar form, from the s-plane diagram.

Suppose we wish to find the factor $s_g - s_1$, where s_g and s_1 are as shown in Fig. 7-11; that is, we wish to subtract two complex numbers. Since the

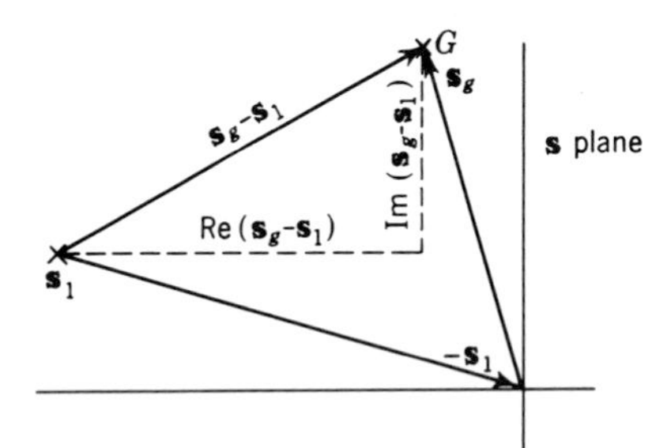

Fig. 7-11 The line from s_1 to s_g represents the factor $s_g - s_1$.

subtraction of complex numbers is identical with vector subtraction, the vector drawn from the origin to s_g represents s_g. The vector *from* s_1 *to* the origin represents $-s_1$. Hence the line *from* s_1 *to* s_g represents the complex number $s_g - s_1$. The transform network function can therefore be evaluated as follows:

1 Join all zero points to the s_g point, and determine the factors in the numerator ($s_g - s_A$, $s_g - s_B$, . . .).
2 Join all pole points to the s_g point, and determine the factors in the denominator ($s_g - s_1$, $s_g - s_2$, . . .).

3 Multiply the factors of (1), and divide by the product of the factors in (2). Multiply the result by H.

This method applies only when s_g is not a pole of the network function.

Example 7-4 In a network, current source $i(t)$ produces response $v(t)$ such that

$$v(t) = \frac{p}{p^2 + p + 1}\, i(t)$$

Use the graphical method to determine the component of the response due to (*a*) a sinusoidal source $i(t) = 10 \cos (t + 30°)$; (*b*) a source $i(t) = 10e^{-t} \cos t$.

Solution The network function is

$$H(s) = \frac{s}{s^2 + s + 1} = \frac{s}{\left[s + \dfrac{(1 - j\sqrt{3})}{2}\right]\left[s + \dfrac{(1 + j\sqrt{3})}{2}\right]}$$

The poles and zeros are located in Fig. 7-12*a* at $s_A = 0$ a zero, $s_1 = -\frac{1}{2} + j(\sqrt{3}/2)$ a pole, $s_2 = -\frac{1}{2} - j(\sqrt{3}/2)$ a pole.

(*a*) The sinusoidal source $10 \cos (t + 30°)$ can be written as $10 \operatorname{Re} e^{j(t+\pi/6)}$; hence we consider the source $e^{j(t+\pi/6)}$. This source is represented by a pole at $s = j$ as marked in Fig. 7-12*a*. The numerator is $s_g - 0 = s_g = j$. The denominator is

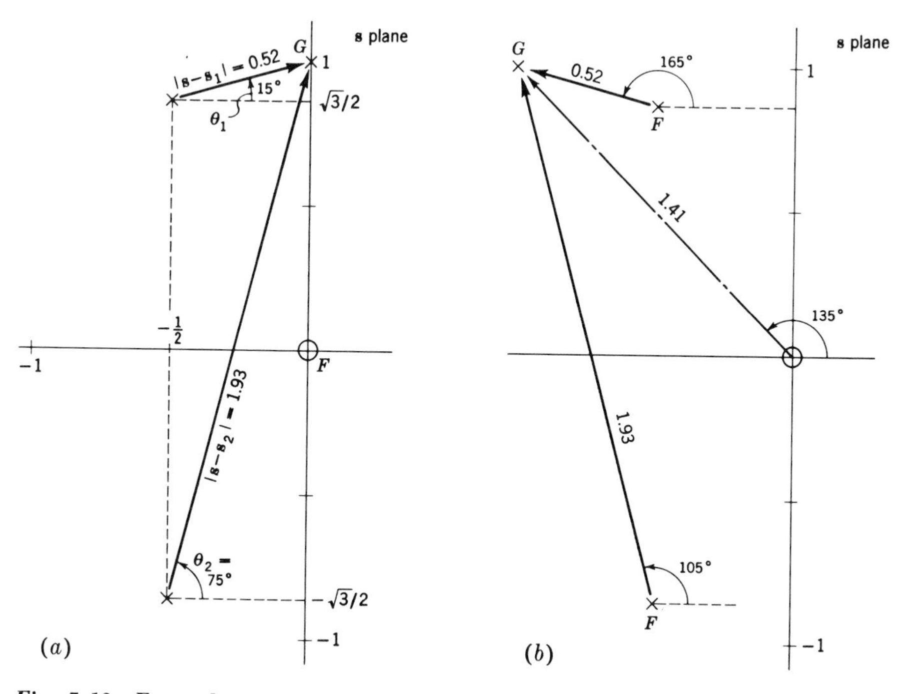

Fig. 7-12 ***Examples of graphical construction to obtain the factors of $H(s_g)$.***

$(s - s_1)(s - s_2)$. Measuring, $|s_g - s_1| = 0.52$, $\theta_1 = 15.0°$, $|s_g - s_2| = 1.93$, $\theta_2 = 75°$. Hence

$$\frac{N(s_g)}{D(s_g)} = \frac{1\angle 90°}{(0.52)(1.93)\angle 90°} = 1\angle 0$$

and $v_s = 10 \cos (t + 30°)$.

(*b*) When the source is $i(t) = 10e^{-t} \cos t$, it is represented by a pole at $s = -1 + j$ as shown in Fig. 7-12*b*. Now $s_g - s_1 = 0.52\angle 165°$, $s_g - s_2 = 1.93\angle 105°$, $D(s_g) = 1\angle 270°$, $N(s_g) = 1.41\angle 135°$, $H(s_g) = 1.41\angle -135°$, $v_s(t) = 14.1e^{-t} \cos (t - 135°)$.

It is clear that the graphical procedure is completely equivalent to the analytical procedure but can result in a considerable saving of effort when the transfer function has many factors. We observe, however, that the graphical procedure makes it *necessary* to factor $N(s)$ and $D(s)$; this algebraic procedure is not necessary with the analytical method.

7-9 An application of the pole-zero diagram

A-C STEADY-STATE RESPONSE We have seen that specification of the pole-zero locations (and the scale factor H) of a network function is equivalent to specifying the differential equation which relates a response of that network to a source function. As a result, in Example 7-4 we observed that components of response due to exponential sources can be calculated directly if the poles and zeros (as well as the scale factor H) are specified. In the following sections and chapters there are numerous occasions to return to these concepts. This section deals with an application of pole-zero diagrams, namely, the evaluation of the component of response due to a sinusoidal source. This response component is termed a-c steady-state response.

A sinusoidal source can be considered the superposition of two sources of the form $e^{j\omega t}$ and $e^{-j\omega t}$, that is, two sources of the form $e^{s_g t}$, where s_g is imaginary. Hence, if a source

$$\phi(t) = \Phi_m \cos (\omega t + \gamma) = \text{Re} (\Phi_m e^{j\gamma} e^{j\omega t}) = \tfrac{1}{2}\Phi_m(e^{j\gamma}e^{j\omega t} + e^{-j\gamma}e^{-j\omega t})$$

is applied to a network, and if the network function which relates a response $y(t)$ to the source function is $H(p)$, then the component of the response due to the source is the sinusoidal function,

$$y_s(t) = \text{Re} [H(j\omega)\Phi_m e^{j\gamma}e^{j\omega t}] = \tfrac{1}{2}[H(j\omega)\Phi_m e^{j\gamma}e^{j\omega t} + H(-j\omega)\Phi_m e^{-j\gamma}e^{-j\omega t}]$$

or, if the complex number $H(j\omega)$ is written in polar form,

$$\begin{aligned} H(j\omega) &= |H(j\omega)|\angle\theta \\ y_s(t) &= \text{Re} [\Phi_m|H(j\omega)|e^{j(\omega t+\gamma+\theta)}] \\ &= |H(j\omega)|\Phi_m \cos (\omega t + \gamma + \theta) \end{aligned} \tag{7-31}$$

From the discussion above it is clear that the sinusoidal steady-state condition in a circuit is merely a special case of excitation by means of exponential sources. Yet this is such an important special case that many definitions and short-cut procedures have been devised for the solution of problems. These short cuts are discussed in subsequent chapters. However, the problem of "frequency response" can best be understood from a study of the poles and zeros. To explain what is meant by frequency response, consider Eq. (7-31). The source function Re $(\Phi_m e^{j(\omega t+\gamma)})$ has an amplitude Φ_m, and the response Re $[H(j\omega)\Phi_m e^{(j\omega t+\gamma)}]$ has an amplitude $|H(j\omega)|\Phi_m$. Therefore

$$\frac{\text{Response amplitude}}{\text{Source amplitude}} = |H(j\omega)| = H \left| \frac{(j\omega - s_A)(j\omega - s_B) \cdots}{(j\omega - s_1)(j\omega - s_2) \cdots} \right|$$

The response due to the source has the same frequency as the source. Hence, if we know the ratio of the amplitudes of these two functions (as well as their phase angle), we have complete information about the response. To study the effect of changing the frequency on this "response ratio," we note that, in the above expression, the absolute value of products is the same as the product of the absolute values. The study of the effect of variation of frequency can be accomplished by examining each of the factors $|j\omega - s_k|$. If such a factor is in the numerator and increases, the response ratio will increase; if it is in the denominator, the response ratio will decrease. Because the magnitude of each factor $j\omega - s_k$ is represented in the s plane by the line joining the point s_k to the point $j\omega$, the variation of this magnitude can often easily be seen from an inspection of the pole-zero diagram. Similarly, if the radian frequency of the source, ω, remains fixed, the effect of changing the pole-zero locations of the network can be predicted. As we have stated, subsequent chapters will deal with many general problems of this type; in this section two simple examples will serve to illustrate the basic idea.

Example 7-5 In the circuit of Fig. 7-13, $v(t)$ is a variable-frequency source. Discuss the ratio of the response amplitude of v_{ab} to the source amplitude.

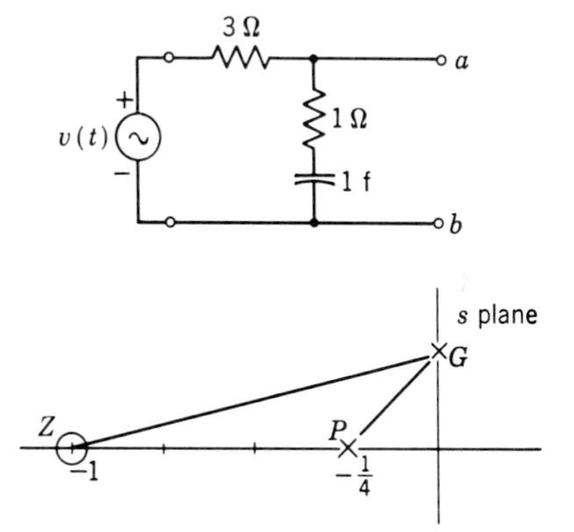

***Fig. 7-13** (a) Circuit diagram and (b) the transfer-function pole-zero diagram for the relationship between the amplitudes of v_{ab} and v in the sinusoidal steady state.*

Solution From voltage division,

$$v_{ab}(t) = \frac{1 + 1/p}{4 + 1/p} v(t)$$

Hence the transform network function under consideration is

$$H(s) = \frac{1}{4}\frac{s+1}{s+\frac{1}{4}}$$

When $s_g = 0$ (d-c), the transform network function is 1. (This can also be deduced by observing that the capacitance is an open circuit for the d-c steady-state condition.) As the radian frequency of the source increases to G, with reference to Fig. 7-13, we observe that the distance PG increases more rapidly than the distance ZG. It follows that the amplitude ratio decreases with increasing frequency.

Example 7-6 A certain transform network function is given as

$$H(s) = \frac{s+1}{(s+1-j10)(s+1+j10)}$$

Discuss the amplitude response ratio in the sinusoidal steady state.

Solution Referring to the pole-zero diagram of Fig. 7-14*a*, we observe that the distance from Z to G is a minimum when $\omega = 0$. As ω increases, the G point moves upward on the $j\omega$ axis; hence P_1G decreases, and P_2G and ZG increase. The situation is

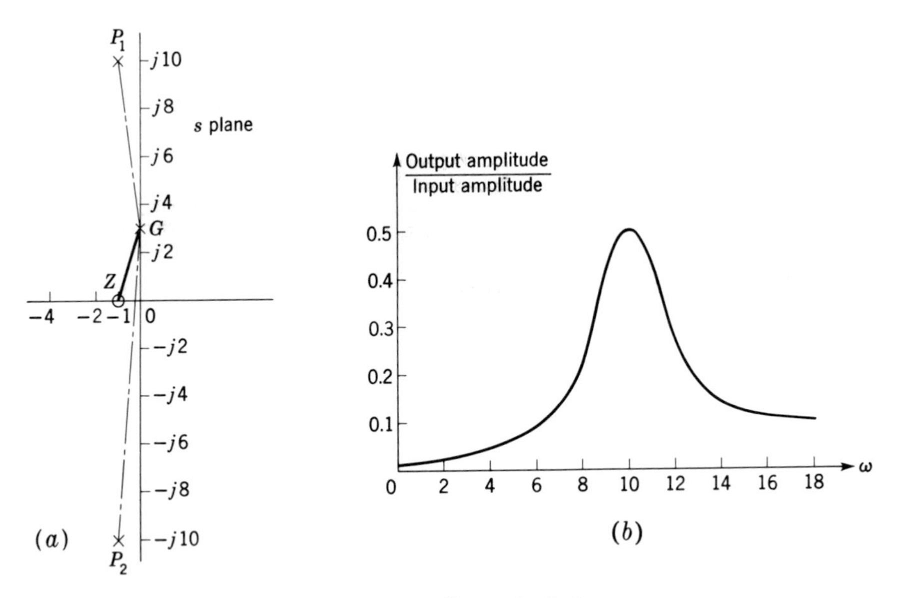

Fig. 7-14** **(a) Pole-zero diagram for Example 7-6 and (b) corresponding ratio of output to input amplitude as a function of frequency in the sinusoidal steady state.

evidently more complex than in Example 7-5. Nevertheless, near $\omega = 10$, P_1G decreases more rapidly than the other two factors increase. We therefore expect that the response ratio will reach a maximum as the G point moves close to P_1. By taking the limit as $s \to \infty$, we observe that $H(s)$ approaches zero for frequencies far beyond $\omega = 10$. A curve of $|H(j\omega)|$ can easily be constructed by *measuring* the distances P_1G, P_2G, ZG for several values of ω. Then, for each value of ω, the response ratio is the quotient $ZG/(P_1G)(P_2G)$ as shown in Fig. 7-14*b*.

In the examples above we have seen how the pole-zero diagram gives information about the network behavior under various conditions. The following are several conclusions which we can draw from these examples. First, we note that those zeros and poles which are closest to the pole which represents the source function have the largest influence on response changes because the corresponding distances will vary most "rapidly." This useful fact can be developed into an approximation scheme when only "dominant" poles and zeros are considered. This is discussed in Chap. 10.

PROBLEMS

7-1 (*a*) For the network shown in Fig. P7-1, obtain the equilibrium equation relating v_{ab} to the source functions. (*b*) Use the result of part *a* to find (1) the operational transfer impedance relating $v_{ab}(t)$ to i; (2) the operational current gain relating

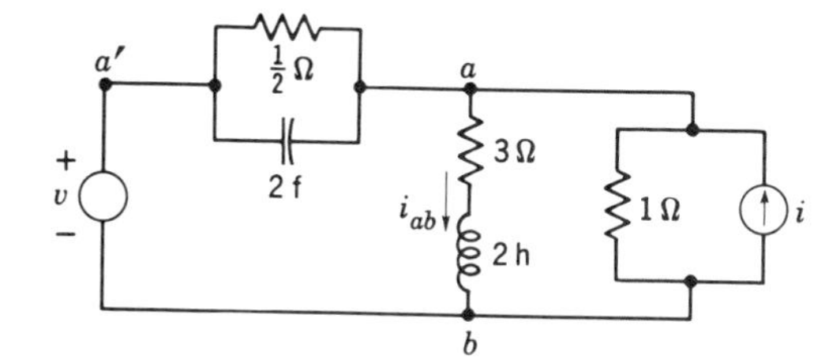

Fig. P7-1

i_{ab} to i; (3) the operational transfer admittance relating i_{ab} to v; (4) the operational voltage gain relating v_{ab} to v. (*c*) What are the natural modes of the circuit? (*d*) What are the forcing functions for v_{ab} and for i_{ab} if the source functions are given by (1) $v(t) = 10 + 10t$, $i(t) = 5t$; (2) $v(t) = 10e^{-t}$, $i(t) = 5e^{-t}$; (3) $v(t) = 10e^{-2t}$, $i(t) = 10e^{-3t/2}$? (*e*) Find the component of the responses v_{ab} and i_{ab} due to the source functions for each of the cases given in part *d*.

7-2 Substitute the "trial solution" $t^n e^{-t}$ in the equation $(p + 1)^3 y = 0$, and show that only the integral values $n = 0, 1, 2$ satisfy the equation.

7-3 Locate the poles of the network function in the s plane if the free response component is given by (*a*) $y_f = e^{-2t} - te^{-2t} + te^{-3t} + e^{-3t}\cos 4t$; (*b*) $y_f = t^2e^{-3t}$; (*c*) $y_f = te^{-t}\cos(2t + 30°) + e^{-2t}$.

7-4 When a source voltage $v(t) = 5e^{-2t}$ is impressed on a network, the complete response v_2 is given by $v_2(t) = 3te^{-2t} + 2e^{-3t}\sin 6t$. Locate in the s plane the poles of the voltage gain function relating v_2 to v.

7-5 It is known that a network function has two zeros and two poles, all at real (and negative) values of s. A source $\phi(t) = e^{st}$ is impressed, and the complete response $y(t)$ is analyzed. The following results are noted: When $s = -1$, $y(t) = \frac{1}{2}(e^{-2t} + e^{-4t})$. When $s = -2$, $y(t) = \frac{1}{4}(e^{-2t} - 2te^{-2t} + 3e^{-4t})$. When $s = -3$, $y(t) = \frac{1}{2}(-e^{-2t} + 3e^{-4t})$. When $s = -4$, $y(t) = \frac{1}{4}(-e^{-2t} - 6te^{-4t} + 5e^{-4t})$. (*a*) Deduce the form of the network function; specify numerically the location of all zeros and poles. (*b*) Write the differential equation which relates $y(t)$ to $\phi(t)$. (*c*) Calculate the complete response $y(t)$ if the source is given as $\phi(t) = e^{-5t}$. Use $H = 1$, $y(0^+) = 1$, $(dy/dt)_{0^+} = 4$.

7-6 If $f(t) = (p^3 + 4p^2 + 36p + 16)\phi(t)$, find $f(t)$ [using the relationship $p \operatorname{Re} e^{\mathbf{s}t} = \operatorname{Re}(\mathbf{s}e^{\mathbf{s}t})$ when possible] if $\phi(t) = \operatorname{Re}[(10/30°)e^{\mathbf{s}_g t}]$ with (*a*) $\mathbf{s}_g = -2$; (*b*) $\mathbf{s}_g = j2$; (*c*) $\mathbf{s}_g = j5$; (*d*) $\mathbf{s}_g = j6$; (*e*) $\mathbf{s}_g = -2 + j2$.

7-7 In a certain network the response v_2 is related to the source function v by the equation $v_2 = [(p + 2)/(p + 5)]v$. (*a*) Find $(v_2)_s$, the component of the response due to the source for the following cases: (1) $v = 10$; (2) $v(t) = 10t$; (3) $v = (5 + 4t)$; (4) $v = 10e^{-2t}$; (5) $v = 10e^{-3t}$; (6) $v = 10e^{-5t}$; (7) $v = 10 \cos (5t - 20°) = \text{Re}\,[(10\underline{/-20°})e^{j5t}]$; (8) $v = 15e^{-2t} \cos (4t - 13°) = \text{Re}\,[(15\underline{/-13°})e^{(-2+j4)t}]$. (*b*) For $v_2(0^+) = 5$, find the complete response $v_2(t)$ if $v(t) = 10e^{-3t} + 5 \cos (5t - 30°)$.

7-8 In a certain network the response $v_2(t)$ is related to the source $v(t)$ by the equation $v_2(t) = [1/(p^2 + 2p + 2)]v$. (*a*) Find $(v_2)_s$, the component of the response due to each of the following source functions: (1) $v(t) = 8e^{-4t}$; (2) $v(t) = 5 \cos (t + 36.9°)$; (3) $v(t) = 10e^{-t} \cos (2t + 50°)$. (*b*) A source $v(t) = A \cos \omega t$ is used. It is desired that $[v_2(t)]_s$ be given by $(v_2)_s = B \sin \omega t$. Find ω.

7-9 In a certain network the response $v_2(t)$ is related to the source $i(t)$ by $v_2(t) = [p/(p^2 + 6p + 16)]i$. A source function $i(t) = A \cos (\omega t + \theta)$ is used. (*a*) It is desired that the component of the response due to the source be in phase with the source function; that is, $(v_2)_s = B \cos (\omega t + \theta)$. Find ω. (*b*) It is desired that the phase angle of $(v_2)_s$ differ from θ by 45°. Find the two values of ω that satisfy this condition.

7-10 When a voltage source $v(t)$ is applied to a series *R-L* circuit as shown in Fig. P7-10, the component of $i(t)$ due to the source is given by $i_s(t) = 2$ when $v(t) = 5u(t)$ and $i_s(t) = A \cos (2t - 45°)$ when $v(t) = 5 \cos 2t$. Calculate (*a*) R; (*b*) L; (*c*) A; (*d*) the component i_s when $v(t) = 5e^{-t}$.

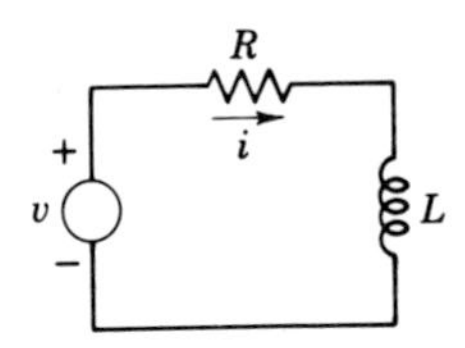

Fig. P7-10

7-11 For a certain response-source relationship the transform network function is $H(s) = (s + 1)/[(s + 2)(s^2 + 2s + 10)]$. Calculate *graphically* the component of the response due to each of the source functions (*a*) $20e^{-3t}$; (*b*) $20 \cos 3t$; (*c*) $20 \cos (3t - 30°)$; (*d*) $20e^{-3t} \cos t$.

7-12 A source $v(t) = \text{Re}\, e^{j\omega t} = \cos \omega t$ is applied to a critically damped series *R-L-C* circuit. Use the pole-zero diagram of the transform driving-point admittance to show that $i_s(t)$ is a sinusoid in phase with $v(t)$ when $\omega = \omega_0 = 1/\sqrt{LC}$.

7-13 In the circuit of Fig. P7-13, $v(t) = V_m \cos \omega t$ and in the steady state $v_{ab} = (V_{ab})_m \cos (\omega t - \theta)$. (*a*) Find the transform network function that relates v_{ab} to v. (*b*) Plot $(V_{ab})_m/V_m$ as a function of ω. (*c*) Plot θ as a function of ω.

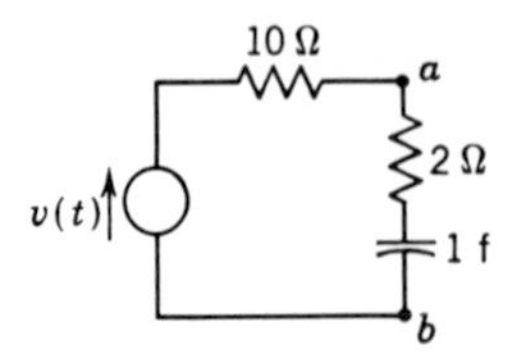

Fig. P7-13

Chapter 8 Introduction to Laplace transform

In the preceding chapters we showed how network responses are analyzed by study and solution of the differential equations that relate responses to sources. Since we dealt with time functions, the method explained was termed *time-domain* method of analysis. In this chapter we deal with another method of analysis, termed *Laplace transform* method. The technique described below is a generalization of the use of exponential functions e^{st}. Since (as discussed in Chap. 4) s is interpreted as generalized, or complex, frequency, the Laplace transform method is also referred to as the (complex) *frequency-domain* method.

8-1 Introduction to Laplace transform

We have seen that the response component of a network to exponential source functions is proportional to the source function; that is, if $v_2(t) = H(p)v(t)$ and if the source $v(t) = V_g e^{s_g t}$, then (except where s_g is a pole of H), $v_{2s}(t) = H(s_g)V_g e^{s_g t}$. This relationship can be generalized by means of superposition ideas.

Suppose that a source function consists of the two components

$$v(t) = V_1 e^{s_1 t} + V_2 e^{s_2 t}$$

The response due this source is, by superposition (assuming that neither s_1 nor s_2 is a pole of H),

$$v_{2s}(t) = H(s_1)V_1 e^{s_1 t} + H(s_2)V_2 e^{s_2 t}$$

Extending the reasoning to $2N + 1$ exponential functions, we consider the source

$$v(t) = \sum_{n=-N}^{n=+N} V_n e^{s_n t} \tag{8-1}$$

and the response due to the source is

$$v_{2s} = \sum_{-N}^{+N} H(s_n) V_n e^{s_n t} \tag{8-2}$$

To generalize the representation of Eq. (8-1), we envisage a source function with continuous range of variations of s, thus shifting from the summation to an integration. Equation (8-2) would then be of the form

$$v(t) = \int V(s) e^{st}\, ds \tag{8-3}$$

Using the idea represented by Eq. (8-3) for the response equation (8-2), the result would be

$$v_2(t) = \int H(s) V(s) e^{st}\, ds \tag{8-4}$$

In going from Eqs. (8-1) and (8-2) to Eqs. (8-3) and (8-4), we have assumed that the integrals involved in the latter equations exist (are not infinite), as the limit of the sums involved in Eqs. (8-1) and (8-2), as $N \to \infty$ and $V_n \to V(s)\, ds$. Furthermore, in the integral representations we have not shown the limits, since s in e^{st} may be complex, $s = \sigma + j\omega$ representing real functions of the form $e^{\sigma t} \cos \omega t$, and the integration involved must be carried out in the complex s plane. Nonetheless, the basic idea of representing a source $v(t)$ as the integral (sum) of a continuous range of exponential functions, and finding the response due to $v(t)$ by superposition as given in Eq. (8-3), is mathematically sound, and is embodied in the theory of transforms, a special case of which are the Laplace transforms discussed in this section.

In Eq. (8-3), $V(s)\, ds$ may be considered the "amplitude" of the exponential source e^{st}, and the source $v(t)$ as the superposition of these sources. In an introductory discussion of Laplace transform, it is convenient to start with a time function $f(t)$ and define its Laplace transform denoted by the symbol[1] $\mathcal{L}[f(t)]$ or $F(s)$ as

$$\mathcal{L}[f(t)] \triangleq \int_{t=0}^{\infty} f(t) e^{-st}\, dt \triangleq F(s) \tag{8-5}$$

where the symbol $\triangleq$ indicates a definition, and it is assumed that s is a complex variable whose real part is positive. It can be shown that, given the Laplace transform $F(s)$ of the function $f(t)$, we can find the latter as

$$f(t) = \frac{1}{2\pi j} \int_{s=-\sigma-j\infty}^{-\sigma+j\infty} F(s) e^{st}\, ds \tag{8-6}$$

[1] We shall use the notation of capital letters, as in $F(s)$, for the Laplace transform of a function of t denoted by lower-case letter as in $f(t)$. The symbol $\mathcal{L}$ denotes the operation indicated by Eq. (8-5). If $F(s)$ is given, the symbol $\mathcal{L}^{-1}$ represents the operation of finding the corresponding $f(t)$.

where the significance of the limits of integration in Eq. (8-6) is beyond our scope, and is discussed in texts on theory of functions of a complex variable. Except for the factor $1/2\pi j$ and limits of integration, the similarity between Eqs. (8-3) and (8-6) is clear, and the implication is that, if $v_2(t) = H(p)v(t)$ is a response and source relationship and the source is given by

$$v(t) = \frac{1}{2\pi j} \int_{-\sigma - j\infty}^{-\sigma + j\infty} V(s)e^{st}\, ds \tag{8-7}$$

the response is

$$v_2(t) = \frac{1}{2\pi j} \int_{-\sigma - j\infty}^{-\sigma + j\infty} H(s)V(s)e^{st}\, ds \tag{8-8}$$

In this chapter we study techniques other than Eq. (8-8) for evaluation of $v_2(t)$ when $v(t)$ or $V(s)$ is known.

8-2 *Properties of Laplace transform*

The defining equation for the Laplace transform of a time function $f(t)$ has been given as

$$\mathcal{L}[f(t)] = F(s) = \int_0^\infty f(t)e^{-st}\, dt \tag{8-9}$$

In the application of Eq. (8-9) it can be shown that, although s is generally complex, it can, for purpose of integration, be treated as if it were a positive real number. Before discussing some of the specific techniques associated with the use of Laplace transform in the solution of linear-system problems, it is appropriate to discuss the ideas that motivate these techniques.

To a given waveform $f(t) = f(t)u(t)$ (where f is continuous) there corresponds a function $F(s)$ defined through Eq. (8-9), and it can be shown that this correspondence is unique; i.e., there exists one and only one $F(s)$ for a given $f(t)u(t)$, and vice versa. Since the lower limit of integration in Eq. (8-9) is $t = 0$ (or, as we shall discuss below, $t = 0^-$), the form $f(t)u(t)$ is used to indicate that the waveform is zero for $t \leq 0^-$.

To relate a response waveform to a source waveform a differential equation involving time functions must be solved. We shall see that the relationship between Laplace transform of a source and that of its response is an *algebraic* equation with s, the complex frequency, as its variable. Thus, by dealing with a source transform, we can use algebraic techniques to find response transforms. Naturally, we shall have to find the response waveform by finding the "inverse" transform. In the problems dealt with in this text this can also be accomplished by algebraic means. The idea outlined above is similar in *concept* to the use of logarithms. If we want to *multiply* two numbers, we may *represent* them by their logs, add the logs, and then find the product as the inverse logarithm of the sum of the logs. We shall see that the use of Laplace transforms systematizes the procedure of network analysis by giving a general procedure that allows the computation of complete response with arbitrary initial conditions for a large class of source waveforms. In addition, the representation of waveforms by poles in the s plane is clarified.

This point can be illustrated by observing that

$$\mathcal{L}[u(t)] = \int_0^\infty e^{-st}\,dt = \frac{1}{s}$$

so that the unit step which we have represented as a pole at $s = 0$ has a Laplace transform that has this pole at $s = 0$. Similarly,

$$\mathcal{L}[e^{-\alpha t}u(t)] = \int_0^\infty e^{-\alpha t}e^{-st}\,dt = \frac{1}{s+\alpha}$$

so that the transform of the exponential decay $e^{-\alpha t}u(t)$ has a pole at $s = -\alpha$. We now discuss some basic properties of the Laplace transform.

SUPERPOSITION From the definition of the Laplace transform

$$\mathcal{L}[f(t)] = F(s) = \int_0^\infty f(t)e^{-st}\,dt \tag{8-10}$$

it follows that, if

$$\mathcal{L}[f_1(t)] = F_1(s) \qquad \text{and} \qquad \mathcal{L}[f_2(t)] = F_2(s)$$

then

$$\mathcal{L}[f_1(t) + f_2(t)] = F_1(s) + F_2(s)$$

DIFFERENTIATION AND INTEGRATION WITH RESPECT TO TIME It is our intention to apply Laplace transform to the solution of linear integrodifferential equations. Consider, for example, a series R-L-C circuit with a known voltage source impressed. The equilibrium equation is

$$L\frac{di}{dt} + Ri + \frac{1}{C}\int_{-\infty}^{t} i\,d\tau = v(t) \tag{8-11}$$

We shall now show that the application of Laplace transform to such a linear equation results in an algebraic equation relating the transform of $i(t)$ (the response) to the transform of $v(t)$ (the source). Taking the Laplace transform of Eq. (8-11),

$$\mathcal{L}\left[L\frac{di}{dt} + Ri + \frac{1}{C}\int_{\infty}^{t} i\,d\tau\right] = \mathcal{L}[v(t)] \equiv V(s)$$

and using the superposition property,

$$L\mathcal{L}\left[\frac{di}{dt}\right] + R\mathcal{L}[i(t)] + \frac{1}{C}\mathcal{L}\left[\int_{-\infty}^{t} i\,d\tau\right] = V(s) \tag{8-12}$$

In Eq. (8-12) we assume that $V(s)$ is "known" as $v(t)$ is known. We denote

$$\mathcal{L}[i(t)] = I(s)$$

and have

$$L\mathcal{L}\left[\frac{di}{dt}\right] + RI(s) + \frac{1}{C}\mathcal{L}\left[\int_{-\infty}^{t} i\,d\tau\right] = V(s) \tag{8-13}$$

From Eq. (8-13) it is evident that if we wish to solve for $I(s)$, we need the relationship between the Laplace transform of a function and the Laplace transform of its derivatives and integrals. Consider first the following problem: If

$$\mathcal{L}[f(t)] = F(s)$$

express $\mathcal{L}[df/dt]$ in terms of $F(s)$. Applying the definition of Laplace transform,

$$\mathcal{L}\left[\frac{df}{dt}\right] = \int_0^\infty \frac{df}{dt} e^{-st}\, dt$$

and integrating by parts,

$$\mathcal{L}\left[\frac{df}{dt}\right] = [f(t)e^{-st}]_0^\infty + s\int_0^\infty f(t)e^{-st}\, dt \tag{8-14}$$

Since s can be treated as a positive real variable, $\lim_{t\to\infty} [f(t)e^{-st}] = 0$ and, recognizing the integral in (8-14) as $\mathcal{L}[f(t)] = F(s)$, we have

$$\mathcal{L}\left[\frac{df}{dt}\right] = sF(s) - f(0) \tag{8-15}$$

The result, Eq. (8-15), shows that the relation between the Laplace transform of a function and that of its derivative is algebraic. As stated, however, Eq. (8-15) is vague when $f(t)$ is not continuous at $t = 0$, when $f(0)$ is not defined, or when $f(0^-)$ differs from $f(0^+)$. We shall now show that in Eq. (8-15), as well as in the lower limit of the defining integral for $F(s)$, the lower limit is most conveniently interpreted as 0^-.

TRANSFORM OF UNIT IMPULSE One of several possible definitions[1] of the unit impulse $\delta(t)$ is as the limit of a rectangular pulse whose area remains unity as its duration shrinks to zero; that is,

$$\delta(t) = \lim_{\tau\to\infty} \frac{u(t+\tau/2) - u(t-\tau/2)}{\tau}$$

This definition is the same as defining $\delta(t)$ as the derivative of the unit step:

$$\delta(t) \equiv \frac{du(t)}{dt}$$

Using this definition together with Eq. (8-15),

$$\begin{aligned}\mathcal{L}[\delta(t)] &= \mathcal{L}\left[\frac{d}{dt}u(t)\right] \\ &= s\mathcal{L}[u(t)] - u(0) \\ &= s\int_0^\infty e^{-st}\, dt - u(0) \\ &= 1 - u(0)\end{aligned} \tag{8-16}$$

[1] For a rigorous treatment of delta functions, see, for example, A. H. Zemanian, "Distribution Theory and Transform Analysis," McGraw-Hill Book Company, New York, 1965.

We notice here the vagueness mentioned above; $u(0)$ is not defined. We observe that if we define zero as 0^+, then $\mathcal{L}[\delta(t)] = 0$, but if we use 0^-, then

$$\mathcal{L}[\delta(t)] = 1$$

and
$$\mathcal{L}\left[\frac{df}{dt}\right] = sF(s) - f(0^-) \tag{8-17}$$

We shall find it convenient to have $\mathcal{L}[\delta(t)] = 1$. For this reason we choose 0^- as the lower limit in the defining integral for the Laplace transform; i.e.,

$$F(s) = \mathcal{L}[f(t)] = \int_{0^-}^{\infty} f(t)e^{-st}\,dt \tag{8-17a}$$

HIGHER DERIVATIVES It is left as an exercise for the reader to show that, as a result of Eq. (8-17),

$$\mathcal{L}\left[\frac{d^2f}{dt^2}\right] = s^2F(s) - sf(0^-) - \left(\frac{df}{dt}\right)_{0^-} \tag{8-18}$$

and in general,

$$\mathcal{L}\left[\frac{d^nf}{dt^n}\right] = s^nF(s) - s^{n-1}f(0^-) - s^{n-2}\left(\frac{df}{dt}\right)_{0^-} - \cdots - \left(\frac{d^{n-1}f}{dt^{n-1}}\right)_{0^-} \tag{8-18a}$$

INTEGRATION We now relate the Laplace transform of the integral of a function to the transform of the function. Considering

$$g(t) = \int_{-\infty}^{t} f(\tau)\,d\tau \qquad t \geq 0$$

we have
$$\frac{dg}{dt} = f(t)$$

so that, using Eq. (8-17),

$$sG(s) - g(0^-) = F(s)$$

and
$$G(s) = \frac{1}{s}F(s) + \frac{1}{s}g(0^-)$$

Hence we have the result

$$\mathcal{L}\left[\int_{-\infty}^{t} f(\tau)\,d\tau\right] = \frac{1}{s}F(s) + \frac{1}{s}\int_{-\infty}^{0^-} f(\tau)\,d\tau \qquad t \geq 0 \tag{8-19}$$

CONCLUSIONS We now return to the Laplace-transformed equation of the series *R-L-C* circuit.

$$L\mathcal{L}\left[\frac{di}{dt}\right] + R\mathcal{L}[i(t)] + \frac{1}{C}\mathcal{L}\left[\int_{-\infty}^{t} i\,d\tau\right] = V(s)$$

Denoting $\mathcal{L}[i(t)]$ by $I(s)$ and using (8-17) as well as (8-19),

$$L[sI(s) - i(0^-)] + RI(s) + \frac{1}{C}\left[\frac{I(s)}{s} + \frac{1}{s}\int_{-\infty}^{0^-} i\,d\tau\right] = V(s)$$

and hence
$$\left(sL + R + \frac{1}{sC}\right)I(s) = V(s) + Li(0^-) - \frac{1}{sC}\int_{-\infty}^{0^-} i\,d\tau \tag{8-20}$$

Equation (8-20) relates the Laplace transform of the current to the Laplace transform of the source voltage and to the initial values $i(0^-)$ and

$$q(0^-) = \int_{-\infty}^{0^-} i \, d\tau$$

Below, we show that these "initial-condition" terms may be represented by initial-condition generators.

At this point we note that the terms $sLI(s)$ correspond to the time-domain terms Li [if $i(0^-) = 0$]. Hence multiplication with s in the complex-frequency (Laplace transform) domain corresponds to differentiation with respect to time in the time domain. Similarly, division by s represents integration with respect to time. This result is exactly parallel to the multiplication and division by s that represent differentiation and integration, respectively, in the time domain when the waveforms are exponentials of the form e^{st}.

Defining,

$$V_i(s) = Li(0^-) - \frac{1}{sC} \int_{-\infty}^{0^-} i \, d\tau$$

and

$$V_T(s) = V(s) + V_i(s)$$

(where the subscripts T and i refer to total and initial); Eq. (8-20) reads

$$Z(s)I(s) = V_T(s)$$

and

$$I(s) = \frac{V_T(s)}{Z(s)}$$

Thus the Laplace transform $I(s)$ of the response $i(t)$ is obtained by algebraic means. If we perform the inversion $i(t) = \mathcal{L}^{-1}[I(s)]$, the response obtained is the complete response, satisfying all initial conditions, because $V_T(s)$ includes the effect of initial-energy storage.

8-3 Initial-condition generators

In linear-system analysis we generally deal with sources that are assumed zero prior to $t = 0$. If $f_c(t)$ denotes a function which at $t = 0$ is continuous and has continuous derivatives, the function $f(t) = f_c(t)u(t)$ may represent a source function applied at $t = 0$. We note that the assumption $f(t) = f_c(t)u(t)$ leads to

$$f(0^-) = 0$$

$$\left.\frac{d^n f}{dt^n}\right|_{0^-} = 0 \qquad n = 1, 2, \ldots$$

As far as response functions are concerned, we know that, because of initial-energy storage, responses at $t = 0^-$ need not be zero. If, however, the initially stored energy is replaced by initial-condition generators, the response functions can also be taken as zero at $t = 0^-$. As a consequence, Eq.

(8-18a) reduces to

$$\mathcal{L}\left[\frac{d^n f(t)}{dt^n}\right] = s^n F(s) \tag{8-21a}$$

and

$$\mathcal{L}\left[\int_{-\infty}^{t} f(t)\,dt\right] = \frac{F(s)}{s} \tag{8-21b}$$

when $f(t) = f_c(t)u(t)$.

These simple forms, compared with Eqs. (8-18a) and (8-19), are further reasons for the use of 0^- as the lower limit in the defining integral for $F(s)$. Thus, *once initial-condition generators are used, initial values of responses or of their derivatives are never needed for sources of the form* $f_c(t)u(t)$.

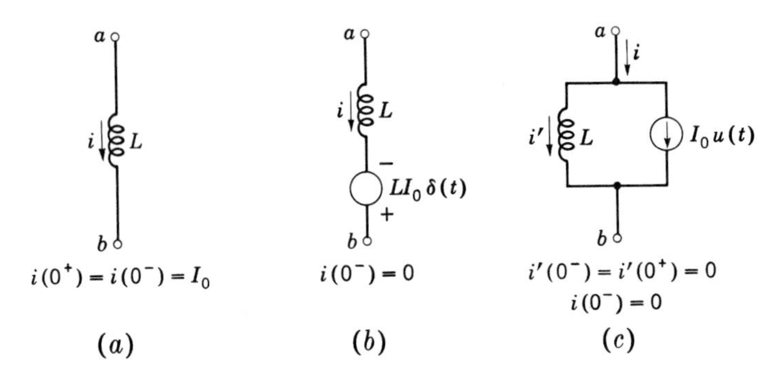

***Fig. 8-1** (a) Inductance with $i(0^-) = I_0$. (b) Inductance with $i(0^-) = 0$ and initial-condition (impulse) generator. (c) Current-source initial-condition generator for inductance.*

Consider now an inductance L carrying an initial current $i(0^-) = I_0$ with voltage v_{ab} across the terminals as shown in Fig. 8-1a. For the inductance the voltage-current relationship is $v_{ab} = L\,di/dt$; hence

$$V_{ab}(s) = sLI(s) - LI_0$$

or

$$V_{ab}(s) + LI_0 = sLI(s)$$

If we now consider the same inductance L in series with the source

$$v_i(t) = LI_0\delta(t)$$

but with initial current $i(0^-) = 0$, as shown in Fig. 8-1b, the voltage-current relationship is

$$v_{ab} = L\,di/dt - LI_0\delta(t)$$

and since $\mathcal{L}[\delta(t)] = 1$,

$$V_{ab}(s) = sLI(s) - Li(0^-) - LI_0$$

In Fig. 8-1b the initial current $i(0^-)$ is assumed to be zero; therefore

$$V_{ab}(s) = sLI(s) - LI_0 \tag{8-22}$$

We observe that the term $Li(0^-)$ in $\mathcal{L}[L\,di/dt] = sLI(s) - Li(0^-)$, corresponding to the circuit of Fig. 8-1*a*, is interpreted as the transform of the initial-condition generator $v_i = LI_0\delta(t)$ placed in series with inductance L as shown in Fig. 8-1*b*.

We may also represent the initial current $i(0^-) = I_0$ by a parallel current source as shown in Fig. 8-1*c*. To find the waveform of this source, from Eq. (8-22) we have

$$I(s) = \frac{1}{sL} V_{ab}(s) + \frac{1}{s} I_0 \tag{8-23}$$

Recalling that $\mathcal{L}[u(t)] = 1/s$, we interpret this equation as the transform of $i(t) = (1/L)\int_0^t v_{ab}\,dt + I_0 u(t)$, which is the voltage-current relationship of the circuit of Fig. 8-1*c*. Therefore Fig. 8-1*c* can also be used to represent the effect of initial-energy storage in L. From duality, the analogous initial-condition generators for capacitances are deduced as in Fig. 8-2.

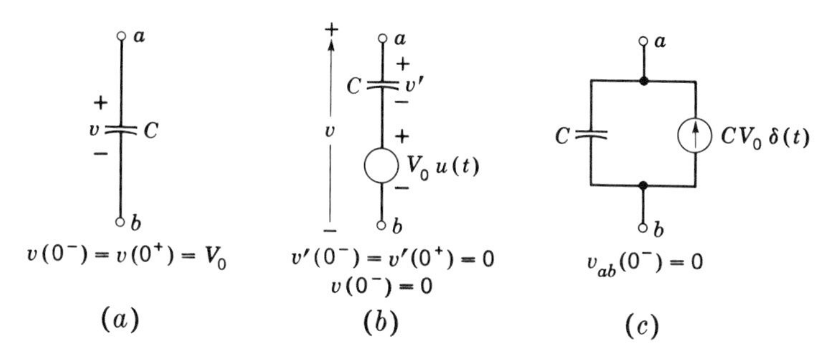

***Fig. 8-2** (a) Capacitance with $v(0^-) = V_0$. (b) Step-function initial-condition voltage generator for capacitance. (c) Impulse current-source initial-condition generator for capacitance.*

We conclude that, with the use of Laplace transform, the initial-condition generators introduced in Chap. 5 "automatically" arise whenever the Laplace transform of derivatives or integrals of currents and voltages in a circuit is obtained. We shall see below that it is generally convenient to include the initial-condition generators in *circuit diagrams* before developing the equilibrium equations of the circuit and their transforms. These initial-condition generators represent the "residue" of the past history of the response of the system.

8-4 Additional theorems; a short table of transforms

As indicated above, the use of Laplace transform in circuit problems requires the formulation of the transform of the source function at the beginning of the problem. Moreover, once the Laplace transform of a response is known, its inverse must be found in order to obtain the corresponding response function in the time domain. Although both $F(s)$ and $f(t)$ can be evaluated by use of the integrals (8-5) and (8-6), such formal integration is often inconvenient and unnecessary. Tables of Laplace transforms can be used for both direct and inverse transformations. Such tables are actually compendiums of the evaluations of the corresponding integrals. Complete and extensive tables are available, but we shall see that, for many purposes, knowledge of a few transform pairs suffices, particularly when coupled with several theorems pertinent to transforms. Some of these basic transforms will now be deduced.

We know already that

$$\mathcal{L}[\delta(t)] = 1 \qquad \text{and} \qquad \mathcal{L}[u(t)] = \frac{1}{s}$$

Since

$$tu(t) = \int_{-\infty}^{t} u(\tau)\, d\tau$$

$$\mathcal{L}[tu(t)] = \frac{1}{s}\left(\frac{1}{s}\right) - 0 = \frac{1}{s^2}$$

and similarly,

$$\tfrac{1}{2}t^2u(t) = \int_{-\infty}^{t} \tau u(\tau)\, d\tau$$

Hence

$$\mathcal{L}[\tfrac{1}{2}t^2u(t)] = \frac{1}{s^3} \qquad \mathcal{L}[t^2u(t)] = \frac{2}{s^3}$$

It is left as an exercise for the reader to show that

$$\mathcal{L}\left[\frac{1}{n!}t^n u(t)\right] = \frac{1}{s^{n+1}} \qquad \mathcal{L}[t^n u(t)] = \frac{n!}{s^{n+1}} \qquad n = 0, 1, 2, \ldots$$

COMPLEX FREQUENCY DISPLACEMENT There exists a simple relationship between the transform of a function $f(t)$ and the transform of the product of that function with an exponential function of time $f(t)e^{s_1 t}$. Specifically, if

$$\mathcal{L}[f(t)] = F(s)$$

then

$$\mathcal{L}[f(t)e^{s_1 t}] = F(s - s_1) \tag{8-24}$$

where s_1 is any (real or complex) *number*. The proof consists of recognizing that

$$\mathcal{L}[f(t)e^{s_1 t}] = \int_0^\infty f(t)e^{s_1 t}e^{-st}\, dt = \int_0^\infty f(t)e^{-(s-s_1)t}\, dt = F(s - s_1)$$

This theorem may be used to deduce several commonly useful transform pairs: Since

$$\mathcal{L}[u(t)] = \frac{1}{s}$$

$$\mathcal{L}[e^{-\alpha t}u(t)] = \frac{1}{s + \alpha}$$

and

$$\mathcal{L}[e^{\pm j\omega t}u(t)] = \frac{1}{s \mp j\omega}$$

Table 8-1 Seven basic Laplace transform pairs

Waveform	$f(t)$	$F(s)$	Pole-zero diagram
1, t	$\delta(t)$	1	s plane
1, t	$u(t)$	$\frac{1}{s}$	s plane
1, 1, t	$tu(t)$	$\frac{1}{s^2}$	s plane, II
t	$t^n u(t)$	$\frac{n!}{s^{n+1}}$ $n = 0, 1, 2, \ldots$	s plane, (N)
t	$e^{-\alpha t} u(t)$	$\frac{1}{s+\alpha}$	s plane
1, t	$[\sin(\omega t)]\, u(t)$	$\frac{\omega}{s^2+\omega^2}$	s plane
1, t	$[\cos(\omega t)]\, u(t)$	$\frac{s}{s^2+\omega^2}$	s plane

Since $\cos \omega t = \frac{1}{2}e^{j\omega t} + \frac{1}{2}e^{-j\omega t}$ and $\sin \omega t = \frac{1}{2j}e^{j\omega t} - \frac{1}{2j}e^{-j\omega t}$,

$$\mathfrak{L}[\cos \omega t u(t)] = \frac{1}{2}\left(\frac{1}{s - j\omega} + \frac{1}{s + j\omega}\right) = \frac{s}{s^2 + \omega^2}$$

and $$\mathfrak{L}[\sin \omega t u(t)] = \frac{1}{2j}\left(\frac{1}{s - j\omega} - \frac{1}{s + j\omega}\right) = \frac{\omega}{s^2 + \omega^2}$$

The seven basic transform pairs are shown in Table 8-1.

Example 8-1 Use Table 8-1 to find

$$F(s) = \mathfrak{L}[e^{-4t} \cos (10t - 30°)u(t)]$$

Solution Since $\cos (10t - 30°) = 0.866 \cos 10t + 0.500 \sin 10t$,

$$\mathfrak{L}[\cos (10t - 30°)u(t)] = \frac{0.866s + 5}{s^2 + 100}$$

and, using frequency displacement,

$$\mathfrak{L}[e^{-4t} \cos (10t - 30°)u(t)] = \frac{0.866(s + 4) + 5}{(s + 4)^2 + 100}$$

$$F(s) = \frac{0.866s + 8.46}{s^2 + 8s + 116}$$

Example 8-2 Evaluate the inverse Laplace transform of $F(s)$ if

$$F(s) = \frac{2s + 4}{s^2 + 2s + 5}$$

Solution Since $s^2 + 2s + 5 = (s + 1)^2 + 4$,

$$F(s) = \frac{2s + 4}{(s + 1)^2 + 4} = \frac{2(s + 1) + 2}{(s + 1)^2 + 4}$$

and $$F(s) = F_1(s) + F_2(s)$$

where $$F_1(s) = \frac{2(s + 1)}{(s + 1)^2 + 4} \qquad F_2(s) = \frac{+2}{(s + 1)^2 + 4}$$

Since

$$\mathfrak{L}^{-1}\left[\frac{2s}{s^2 + 4}\right] = 2 \cos 2tu(t)$$

and $$\mathfrak{L}^{-1}\left[\frac{2}{s^2 + 4}\right] = \sin 2tu(t)$$

by frequency displacement and superposition,

$$\begin{aligned} f(t) &= e^{-t}(2 \cos 2t + \sin 2t)u(t) \\ &= [2.24e^{-t} \cos (2t - 26.6°)]u(t) \end{aligned}$$

TIME DISPLACEMENT We know that many waveforms can be represented as the superposition of simple waveforms that are delayed with respect to each other. In Fig. 8-3 several examples are shown. Although direct integration can, of course, be used to find the Laplace transform of such func-

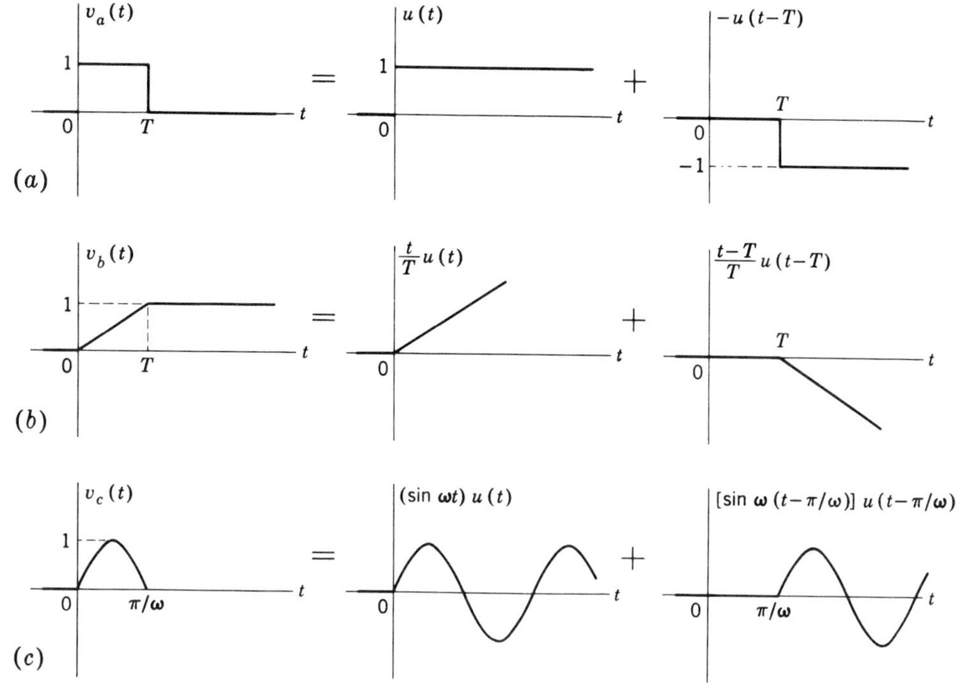

***Fig. 8-3** Examples of pulse-type waveforms analyzed into sums of simple waveforms.* (a) $v_a(t) = u(t) - u(t-T)$; (b) $v_b(t) = tu(t)/T - [(t-T)/T]u(t-T)$; (c) $v_c(t) = (\sin \omega t)u(t) + \sin[\omega(t - \pi/\omega)]\,u(t-\pi/\omega)$.

tions, the use of the time-displacement theorem makes it possible to dispense with formal integration. Moreover, we shall see that the same theorem is useful in identifying inverse Laplace transforms.

The time-displacement theorem states that if

$$\mathcal{L}[f(t)u(t)] = F(s)$$

then

$$\mathcal{L}[f(t-T)u(t-T)] = F(s)e^{-sT} \tag{8-25}$$

In this equation it is implied that T is positive because $f(t-T)u(t-T)$ must be zero for $t < 0$.

Proof of time displacement Applying the definition of Laplace transform,

$$\mathcal{L}[f(t-T)u(t-T)] = \int_0^\infty f(t-T)u(t-T)e^{-st}\,dt$$

Since $u(t-T) = 0$ for $t < T$, the integral can be written

$$\mathcal{L}[f(t-T)u(t-T)] = \int_0^T 0\,dt + \int_T^\infty f(t-T)u(t-T)e^{-st}\,dt$$
$$= \int_T^\infty f(t-T)e^{-st}\,dt$$

or, letting $\tau = t - T$,

$$\int_T^\infty f(t-T)e^{-st}\,dt = \int_0^\infty f(\tau)e^{-s(\tau+T)}\,d\tau = \left[\int_0^\infty f(\tau)e^{-st}\,d\tau\right]e^{-sT}$$

Hence

$$\mathcal{L}[f(t-T)u(t-T)] = F(s)e^{-sT}$$

Example 8-3 Obtain the Laplace transform of the modified ramp function $v_b(t)$ shown in Fig. 8-3*b*, and use the result to find the Laplace transform of the trapezoidal pulse shown in Fig. 8-4.

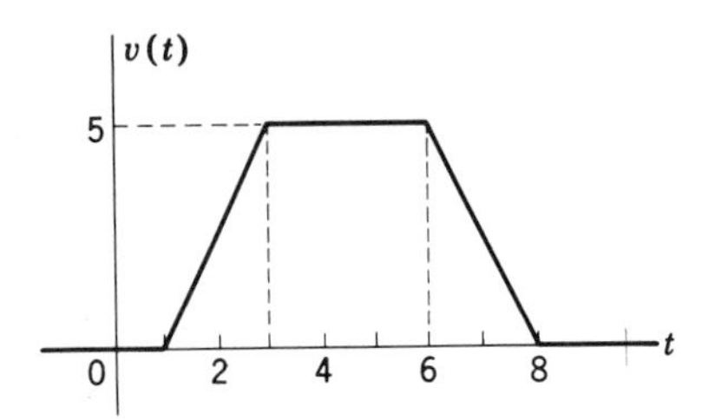

Fig. 8-4 Trapezoidal pulse.

Solution Since

$$v_b(t) = \frac{t}{T}u(t) - \frac{t-T}{T}u(t-T)$$

$$\mathcal{L}[v_b(t)] = \frac{1}{T}\left(\frac{1}{s^2} - \frac{1}{s^2}e^{-sT}\right)$$

The trapezoidal pulse $v(t)$ can be represented as the superposition

$$v(t) = 5v_b(t-1)u(t-1) - 5v_b(t-6)u(t-6) \qquad T = 2$$

Hence

$$\mathcal{L}[v(t)] = V(s) = \frac{5}{2s^2}(e^{-s} - e^{-3s} - e^{-6s} + e^{-8s})$$

Example 8-4 In the circuit of Fig. 8-5*a*, $v_{ab}(0^-) = 10$ and $v_1(t) = 20[u(t) - u(t-1)]$. Use Laplace transform to obtain the response $v_2(t)$.

Solution Using an initial-condition generator and representing $v_1(t)$ as the sum of two source functions, the circuit shown in Fig. 8-5*b* is obtained. Hence the Laplace-

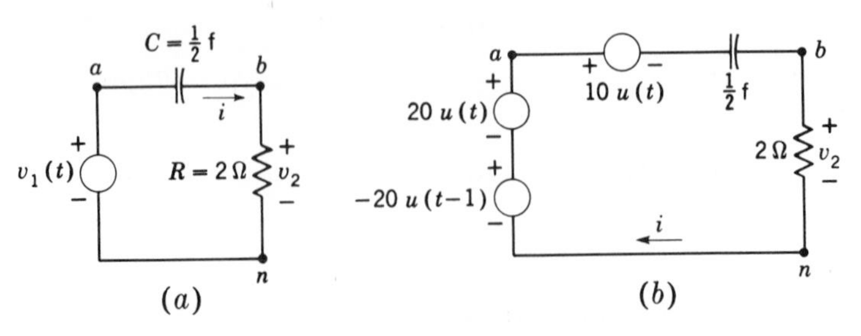

***Fig. 8-5** (a) R-C circuit for Example 8-4. (b) Circuit of (a) showing the initial-condition generator and with the pulse source represented as the sum of two step sources.*

transformed equilibrium equation is

$$\left(\frac{2}{s} + 2\right) I(s) = \frac{10}{s} - \frac{20}{s} e^{-s}$$

$$2I(s)\left(\frac{s+1}{s}\right) = \frac{10}{s}(1 - 2e^{-s})$$

But $2I(s) = V_2(s)$, so that

$$V_2(s) = \frac{10(1 - 2e^{-s})}{s+1} = \frac{10}{s+1} - \frac{20e^{-s}}{s+1}$$

We recall that $\mathcal{L}[e^{-t}u(t)] = 1/(s+1)$; hence

$$v_2(t) = 10[e^{-t}u(t) - 2e^{-(t-1)}u(t-1)]$$

8-5 Formulation of response transforms

Once the term $f(0^-)$ in

$$\mathcal{L}\left[\frac{df}{dt}\right] = sF(s) - f(0^-)$$

and the term $\frac{1}{s}\int_{-\infty}^{0^-} f(\tau)\,d\tau$ in

$$\mathcal{L}\left[\int_{-\infty}^{t} f(\tau)\,d\tau\right] = \frac{1}{s}F(s) + \frac{1}{s}\int_{-\infty}^{0^-} f(\tau)\,d\tau$$

are identified as initial-condition generators, the Laplace-transformed equilibrium equations can be set up directly, without writing the corresponding differential equations. Treating the initial-condition generators exactly as any other ideal source, one may use for each element R, L, or C its transform impedance R, sL, and $1/sC$, respectively, and apply all the methods that are developed in connection with operational or complex immittance: Series impedances add, parallel admittances add; voltage and current division, superposition, Thévenin's and Norton's theorems, as well as other methods discussed in Chap. 11, can then all be used to formulate algebraically the Laplace transform of any desired response. In this section some of these are illustrated. In the succeeding sections we deal with the problem of formulating the inverse Laplace transform.

Example 8-5 The circuit shown in Fig. 8-6*a* is in the steady state at $t = 0^-$ with the switch in position 1. At $t = 0$ the switch is thrown to position 2. Formulate $V_{ab}(s)$, the Laplace transform of $v_{ab}(t)$.

Solution The circuit with initial-condition generators is shown in Fig. 8-6*b* [the symbol V_0 and I_0 are used for $v_{ab}(0^-)$ and $i_L(0^-)$]; in Fig. 8-6*c* Laplace transform variables are shown. Applying Kirchhoff's current law at junction a,

$$\frac{V(s) + LI_0 - V_{ab}(s)}{1+s} = \tfrac{5}{3}V_{ab}(s) + \tfrac{1}{3}s\left[V_{ab}(s) - \frac{V_0}{s}\right]$$

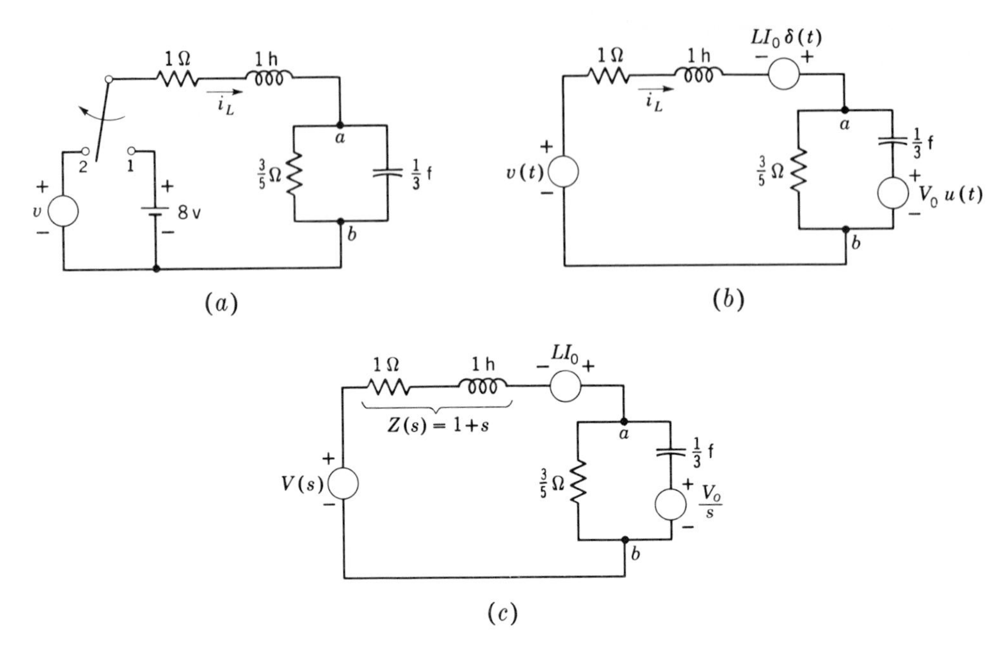

***Fig. 8-6** (a) Circuit for Example 8-5. (b) Circuit of (a) for $t \geq 0^+$ showing initial-condition generators. (c) Circuit of (b) showing transform sources.*

Solving,

$$V_{ab}(s) = \frac{3}{s^2 + 6s + 8} V(s) + \frac{3}{s^2 + 6s + 8} LI_0 + \frac{1 + s}{s^2 + 6s + 8} V_0 \quad (8\text{-}26)$$

where $L = 1$.

Referring to Fig. 8-6*a*, it is seen that $i_L(0^-) = 8/(1 + \frac{3}{5}) = 5$ amp, and $v_{ab}(0^-) = V_0 = 3$ volts. This results in

$$V_{ab}(s) = \frac{3V(s) + 3s + 18}{s^2 + 6s + 8} \quad (8\text{-}26a)$$

For any source waveform $v(t)$, $V(s)$ is found and substituted in Eq. (8-26*a*).

8-6 Initial-value theorems

In some systems we may be interested only in the initial or final values of a response waveform or its derivatives; we may not need to know the waveform for all time. In electronic switching circuits, for example, we may wish to know only if a voltage is large enough at $t = 0$ to cause or prevent conduction of a transistor, or we may wish to know only whether, at $t = 0$, a certain current in a branch of the network is increasing (or decreasing) at some specified rate. Similarly, in mechanical systems such as projectile motion, we may be interested in initial derivatives such as velocity and acceleration. We now discuss a family of initial-value formulas which, by use of Laplace transform, permit computations of values at $t = 0^+$ in a convenient fashion.

INITIAL-VALUE FORMULAS If $f(t)$ is given by $f(t) = f_c(t)u(t)$, where $f_c(t)$ and its derivatives are continuous at $t = 0$, then

$$f(0^+) = \lim_{s \to \infty} [sF(s)] \tag{8-27a}$$

$$\left(\frac{df}{dt}\right)_{0^+} = \lim_{s \to \infty} [s^2F(s) - sf(0^+)] \tag{8-27b}$$

$$\left(\frac{d^2f}{dt^2}\right)_{0^+} = \lim_{s \to \infty} \left[s^3F(s) - s^2f(0^+) - s\left(\frac{df}{dt}\right)_{0^+}\right] \tag{8-27c}$$

and in general,

$$\left(\frac{d^nf}{dt^n}\right)_{0^+} = \lim_{s \to \infty} \left[s^{n+1}F(s) - \sum_{m=0}^{n-1} s^{n-m}\left(\frac{d^mf}{dt^m}\right)_{0^+}\right] \tag{8-27d}$$

Proof Since $f(t) = f_c(t)u(t)$, $f(0^-) = 0$, and $f(0^+) = f_c(0^+)$, the derivative df/dt has an impulse of strength $f(0^+)$ at $t = 0$. Consider now the Laplace transform of df/dt given by Eq. (8-17); that is,

$$\int_{0^-}^{\infty} \frac{df}{dt} e^{-st}\, dt = sF(s) - f(0^-) \tag{8-28}$$

In order to take into account the impulse of df/dt at $t = 0$, the integral above is written as the sum

$$\int_{0^-}^{\infty} \frac{df}{dt} e^{-st}\, dt = \int_{0^-}^{0^+} \frac{df}{dt} e^{-(s)(0)}\, dt + \int_{0^+}^{\infty} \frac{df}{dt} e^{-st}\, dt$$

$$= f(0^+) - f(0^-) + \int_{0^+}^{\infty} \frac{df}{dt} e^{-st}\, dt \tag{8-29}$$

Using Eqs. (8-29) and (8-28), we have

$$sF(s) - f(0^+) = \int_{0^+}^{\infty} \frac{df}{dt} e^{-st}\, dt \tag{8-30}$$

We now take the limit of both sides of Eq. (8-30) as s becomes infinity:

$$\lim_{s \to \infty} [sF(s) - f(0^+)] = \lim_{s \to \infty} \int_{0^+}^{\infty} \frac{df}{dt} e^{-st}\, dt \tag{8-31}$$

The limit of the right side of Eq. (8-31) approaches zero even if df/dt has impulses in the interval 0^+ to ∞ because $\delta(t - t_1)e^{-st} = \delta(t - t_1)e^{-st_1}$; that is, it is an impulse of strength e^{-st_1} which goes to zero as $s \to \infty$ if $t_1 > 0$ as it is in the given interval. Hence Eq. (8-27a), the initial-value theorem, is proved. By treating in a similar fashion the second derivative as the derivative of the function whose initial value we seek, the initial-slope theorem [Eq. (8-27)] can be proved. The same procedure is used to prove the general case [Eq. (8-27d)].

Example 8-6 For $f(t) = e^{-2t}u(t)$, find the initial value and slope both by letting $t = 0^+$ and from the initial-value theorem.

Solution Substitution of $t = 0^+$ in $f(t)$ and in df/dt results in $f(0^+) = 1$ and $(df/dt)_{0+} = -2$, since $(de^{-2t}/dt)_{0+} = -2$. Since $F(s) = 1/(s+2)$, $sF(s) = s/(s+2)$ and

$$f(0^+) = \lim_{s\to\infty} \frac{s}{s+2} = 1$$

Since $s^2F(s) = s^2/(s+2)$,

$$\left(\frac{df}{dt}\right)_{0+} = \lim_{s\to\infty} \left[\frac{s^2}{s+2} - s(1)\right]$$

$$= \lim_{s\to\infty} \frac{-2s}{s+2} = -2$$

Example 8-7 Find $v(0^+)$ and $(dv/dt)_{0+}$ if

$$V(s) = \frac{-36s^2 - 24s + 2}{12s^3 + 17s^2 + 6s}$$

Solution

$$sV(s) = \frac{-36s^2 - 24s + 2}{12s^2 + 17s + 6}$$

Hence

$$v(0^+) = \lim_{s\to\infty} [sV(s)] = -3$$

Since

$$sV(s) - v(0^+) = \frac{-36s^2 - 24s + 2}{12s^2 + 17s + 6} + 3$$

$$= \frac{27s + 20}{12s^2 + 17s + 6}$$

we have
$$s[sV(s) - v(0^+)] = \frac{27s^2 + 20s}{12s^2 + 12s + 6}$$

so that
$$\left(\frac{dv}{dt}\right)_{0+} = \lim_{s\to\infty} [s^2V(s) - sv(0^+)] = 2.25$$

8-7 Transform network functions

We recall that, in a network, a response $v_2(t)$ is related to a source $v_1(t)$ through the operational network function $H(p)$; that is,

$$v_2(t) = H(p)v_1(t)$$

where $H(p)$ is a ratio of polynomials in p with real coefficients. When the initial conditions for voltages across capacitances and currents in inductances are zero, the Laplace transform relations $sLI_L(s) = V_L(s)$ and $sCV_C(s) = I_C(s)$ are exactly of the same form as the operational relations between time functions $pLi(t) = v_L(t)$ and $pCv_C(t) = i_C(t)$, respectively. It follows that the algebraic manipulations that lead to $H(p)$ from the Kirchhoff-law equations are the same as those that lead to the relationship between source and response *transforms* when Laplace transform is used; that is,

$$V_2(s) = H(s)V_1(s)$$

when
$$v_2(t) = [H(p)]v_1(t)$$

subject to the condition that the initial-energy storage in the network is zero.

When the initial-energy storage is not zero, but is represented by initial-condition generators, superposition is used to find the complete response. In general, the external ideal source will not be in the same branch of the network (or between the same nodes) as any initial-condition generators. Therefore the network function that gives the response due to the initial-condition generators will not be the same as that which relates the response to other sources. This is illustrated in the results found in Example 8-5 and Eq. (8-26). We observe that, in that example,

$$V_{ab}(s) = \frac{3}{s^2 + 6s + 8} V(s) + \frac{3}{s^2 + 6s + 8} LI_0 + \frac{1 + s}{s^2 + 6s + 8} V_0$$

Since $V(s)$ and LI_0 *are* in the same branch, the same network function $3/(s^2 + 6s + 8)$ appears for those sources. The initial-condition generator V_0/s is in a *different* branch, and therefore the corresponding network function is different; it is $s(s + 1)/(s^2 + 6s + 8)$.

An important result of the above discussion is obtained by choosing an impulse-function input $v_1(t) = \delta(t)$. *Then* $V_1(s) = 1$ *and*

$$V_2(s) = H(s)$$

Hence $h(t) = \mathcal{L}^{-1}[H(s)]$ *is the response of the initially unenergized system to an impulse source.*

8-8 Inversion of Laplace transforms by partial-fractions expansion

In the section we discuss the problem of finding $f(t)$ if $F(s)$ is a ratio of polynomials in s; that is,

$$F(s) = \frac{N(s)}{D(s)} = \frac{F_0(s^m + b_1 s^{m-1} + b_2 s^{m-2} + \cdots + b_{m-1}s + b_m)}{s^n + a_1 s^{n-1} + a_2 s^{n-2} + \cdots + a_{n-1}s + a_n} \tag{8-32}$$

The generality of this form for lumped-parameter systems is understood from the following reasoning: The Laplace transform of a response $V_2(s)$ to a source $V_1(s)$ is given by

$$V_2(s) = H(s)V_1(s) \tag{8-32a}$$

where $H(s)$ is the transform network function and is therefore the ratio of polynomials in s. The source function $v_1(t)$ has a Laplace transform $V_1(s)$ which most often is either a ratio of polynomials in s (Table 8-1) or it may consist of a sum

$$V_1(s) = V_a(s) + V_b(s)e^{-a_1 s}$$

where $V_b(s)$ is a ratio of polynomials.

Then

$$V_2(s) = H(s)V_a(s) + H(s)V_b(s)e^{-a_1 s}$$

and

$$v_2(t) = \mathcal{L}^{-1}[H(s)V_a(s)] + v_2'(t)$$

where

$$v_2'(t) = v_2''(t - a_1)u(t - a_1)$$

and

$$v_2''(t) = \mathcal{L}^{-1}[H(s)V_b(s)]$$

We now consider a Laplace transform $F(s)$ of the form given in Eq. (8-32) with the proviso[1] that the degree of the numerator polynomial be less than the degree of the denominator polynomial, that is, $m < n$. To find $f(t) = \mathcal{L}^{-1}[F(s)]$, the denominator polynomial is now factored (and it is assumed that common factors in numerator and denominator have been canceled). We have now

$$F(s) = \frac{N(s)}{D(s)} = \frac{N(s)}{(s - s_1)(s - s_2)(s - s_3) \cdots (s - s_n)} \tag{8-33}$$

One now distinguishes the following two cases.

CASE 1: SIMPLE ROOTS ONLY Suppose that no two roots of the polynomial equation $D(s) = 0$ are identical; that is,

$$s_1 \neq s_2 \neq s_3 \neq s_4 \neq \cdots \neq s_n$$

In that case one may expand $F(s)$ in partial fractions:

$$F(s) = \frac{K_1}{s - s_1} + \frac{K_2}{s - s_2} + \cdots + \frac{K_n}{s - s_n} \tag{8-34}$$

and once $K_1, \ldots, K_n$ are found, we have

$$f(t) = (K_1 e^{s_1 t} + K_2 e^{s_2 t} + \cdots + K_n e^{s_n t})u(t) \tag{8-35}$$

Thus each term in the partial-fractions expansion corresponds to a response mode. To find the scale factor $K_1, \ldots, K_n$, we note that if

$$F(s) = \frac{N(s)}{(s - s_1)(s - s_2) \cdots (s - s_n)} = \frac{K_1}{s - s_1} + \frac{K_2}{s - s_2} + \cdots + \frac{K_n}{s - s_n}$$

then

$$(s - s_1)F(s) = \frac{N(s)}{(s - s_2)(s - s_3) \cdots (s - s_n)}$$
$$= K_1 + \frac{s - s_1}{s - s_2} K_2 + \cdots + \frac{s - s_1}{s - s_n} K_n$$

[1] If this condition is not fulfilled, long division can be used to arrive at the proper form. If, for example, $m = n + 2$,

$$F(s) = F_0 \frac{s^{n+2} + b_1 s^{n-1} + b_2 s^n + \cdots}{s^n + a_1 s^{n-1} + \cdots} = F_0 \left[s^2 + As + B + \frac{N_1(s)}{D(s)} \right]$$

where the degree of N_1 is less than the degree of D. This situation does not occur frequently, because if

$$F(s) = F_0(s^2 + As + B) + F_1(s)$$

then

$$f(t) = F_0 \left[\frac{d^2}{dt^2} \delta(t) + a \frac{d}{dt} \delta(t) + B\delta(t) \right] + f_1(t)u(t)$$

and derivatives of impulses are generally not system response functions (unless the system is "overidealized" or they are assumed sources).

and if we let $s = s_1$, then

$$[(s - s_1)F(s)]_{s=s_1} = \frac{N(s_1)}{[D(s)/(s - s_1)]_{s=s_1}} = K_1$$

or in general,

$$[(s - s_k)F(s)]_{s=s_k} = \frac{N(s_k)}{[D(s)/(s - s_k)]_{s=s_k}} = K_k \tag{8-36}$$

When the zeros of the denominator $D(s)$ (the poles of the function) are known, it is not necessary to factor $D(s)$, and the following formula may be used: Let

$$D_k(s) = \frac{D(s)}{s - s_k}$$

so that $\qquad (s - s_k)D_k(s) = D(s)$

and $\qquad \frac{d}{ds}[(s - s_k)D_k(s)] = (s - s_k)\frac{d}{ds}D_k(s) + D_k(s) = \frac{dD(s)}{ds}$

so that if we set $s = s_k$,

$$D_k(s_k) = \left[\frac{dD(s)}{ds}\right]_{s=s_k} = \left[\frac{D(s)}{(s - s_k)}\right]_{s=s_k}$$

Hence an alternative formula for Eq. (8-36) is

$$K_k = \frac{N(s_k)}{[(d/ds)D(s)]_{s=s_k}} \tag{8-37}$$

Note that, for (8-36) and (8-37) to apply, the only requirements are that $m < n$ and that the zeros of $D(s)$ be *simple*.

COMPLEX ROOTS Since $s_1, \ldots, s_n$ are the solutions of a polynomial equation with real coefficients, complex roots, when they exist, occur in conjugate pairs. If $\mathbf{s}_1$ and $\mathbf{s}_2$ are such a conjugate pair ($\mathbf{s}_2 = \mathbf{s}_1^*$), the partial-fractions expansion will include the sum

$$\frac{\mathbf{K}_1}{s - \mathbf{s}_1} + \frac{\mathbf{K}_2}{s - \mathbf{s}_1^*}$$

where $\mathbf{K}_1$ and $\mathbf{K}_2$ are also in general complex. Moreover, since the coefficients $b_1, \ldots, b_m$ and $a_1, \ldots, a_n$ are all real, $\mathbf{K}_2$ will be the conjugate of $\mathbf{K}_1$. It follows that

$$\mathcal{L}^{-1}\left[\frac{\mathbf{K}_1}{s - \mathbf{s}_1} + \frac{\mathbf{K}_1^*}{s - \mathbf{s}_1^*}\right] = \mathbf{K}_1 e^{\mathbf{s}_1 t} + \mathbf{K}_1^* e^{\mathbf{s}_1^* t} = 2\,\mathrm{Re}\,(\mathbf{K}_1 e^{\mathbf{s}_1 t})$$

so that if $\mathbf{s}_1 = -\alpha + j\omega_d$ and if $\mathbf{K}_1 = K_1\underline{/\gamma}$,

$$\mathcal{L}^{-1}\left[\frac{\mathbf{K}_1}{s - \mathbf{s}_1} + \frac{\mathbf{K}_1^*}{s - \mathbf{s}_1^*}\right] = 2K_1 e^{-\alpha t}\cos(\omega_d t + \gamma)$$

CASE 2: MULTIPLE ROOTS In the *R-C* circuit of Fig. 8-5*a*, let $v_1(t) = e^{-t}u(t)$.

Since

$$v_2(t) = H(p)v_1(t) = \frac{p}{p+1} v_1(t)$$

and assuming that the initial condition for v_{ab} is zero, we have

$$V_2(s) = H(s)V_1(s) = \frac{s}{s+1}\frac{1}{s+1} = \frac{s}{(s+1)^2} \tag{8-38}$$

where $1/(s+1) = \mathcal{L}[e^{-t}u(t)]$.

In Eq. (8-38), the pole of the network function $H(s) = s/(s+1)$ is at $s = -1$, which coincides with the value of s_g in the source function $v_1(t) = e^{s_g t}u(t) = e^{-t}u(t)$. As a result of this coincidence, the transform $V_2(s) = s/(s+1)^2$ contains a double pole at $s = -1$. The multiplicity of poles of a transform function may arise from other causes. For example, in $V_2(s) = H(s)V_1(s)$, the transform $V_1(s) = 1/(s+1)^2$ if $v_1(t) = te^{-t}u(t)$ (Table 8-1) and $V_2(s) = H(s)V_1(s)$ will contain a double pole at $s = -1$ due to $V_1(s) = 1/(s+1)^2$. If, in addition, $H(s) = s/(s+1)$, then $V_2(s) = s/(s+1)^3$ contains a triple pole at $s = -1$. The multiplicity of poles may also be due to multiple poles in $H(s)$. For example, in a critically damped series *R-L-C* circuit with a voltage source, the admittance is

$$Y(s) = \frac{1}{R + sL + 1/sC} = \frac{1}{L}\frac{s}{(s+\omega_0)^2} \tag{8-39}$$

where $\omega_0 = 1/\sqrt{LC} = \alpha = R/2L$.

The relationship between voltage and current is then $I(s) = Y(s)V(s)$, and $I(s)$ contains a double pole at $s = -\alpha$. In the foregoing discussion, for the sake of simplicity, the examples given were for multiple poles at $s = -1$ or $s = -\omega_0$. Simple as well as multiple poles may also occur at imaginary or complex values of s. For example, if the voltage source is of the form $e^{-at}\cos \omega t$ and is applied to a series *R-L-C* circuit, such that $a = \alpha = R/2L$ and $\omega^2 = \omega_d{}^2 = 1/LC - a^2$, then the resulting current transform is

$$I(s) = Y(s)V(s) = \frac{1}{L}\frac{s}{s^2 + (R/L)s + 1/LC}\frac{s+a}{(s+a)^2+\omega^2}$$

$$= \frac{1}{L}\frac{s(s+a)}{[(s+a)^2+\omega^2]^2}$$

and has double poles at $s = -a \pm j\omega = -\alpha \pm j\omega_d$. In general, a transform $F(s)$ may have multiple poles at more than one value of s; for example, $F(s) = (s+1)(s+2)/(s+3)^2(s+4)^3(s+5)$ has multiple poles at $s = -3$ and $s = -4$. It will be sufficient however, to consider a function $F(s)$ which has one zero of the denominator repeated P times, for example,

$$F(s) = \frac{N(s)}{D(s)} = \frac{N(s)}{(s-s_1)^P D_1(s)} \tag{8-40}$$

where $D_1(s) = (s - s_2)(s - s_3) \cdots (s - s_n)$, $D_1(s_1) \neq 0$. Assuming again that the degree requirement is fulfilled, Eq. (8-40) can be expanded into partial fractions. The terms arising from the factor $(s - s_1)^P$ are of interest. Thus we write

$$F(s) = \frac{M_0}{(s - s_1)^P} + \frac{M_1}{(s - s_1)^{P-1}} + \frac{M_2}{(s - s_1)^{P-2}} + \frac{M_3}{(s - s_1)^{P-3}} + \cdots + \frac{M_{P-1}}{s - s_1} + F_1(s) \quad (8\text{-}41)$$

where $[1/F_1(s)]_{s=s_1} \neq 0$. Corresponding to Eq. (8-41), the inverse Laplace transform is

$$f(t) = \left[\left(\frac{M_0}{(P-1)!}t^{P-1} + \frac{M_1 t^{P-2}}{(P-2)!} + \cdots + \frac{M_{P-2}t}{1!} + M_{P-1}\right)e^{s_1 t}\right]u(t) + f_1(t)$$

where $f_1(t) = \mathcal{L}^{-1}[F_1(s)]$. The scale factors $M_0, \ldots, M_{P-1}$ are found by multiplying (8-41) by $(s - s_1)^P$:

$$(s - s_1)^P F(s) = M_0 + M_1(s - s_1) + M_2(s - s_1)^2 + M_3(s - s_1)^3 + \cdots + M_{P-1}(s - s_1)^{P-1} + (s - s_1)^P F_1(s) \quad (8\text{-}42a)$$

and $$\frac{d}{ds}[(s - s_1)^P F(s)] = M_1 + 2M_2(s - s_1) + 3M_3(s - s_1)^2 + \cdots + (s - s_1)^P F_1'(s) + P(s - s_1)^{P-1}F_1(s) \quad (8\text{-}42b)$$

also $$\frac{d^2}{ds^2}[(s - s_1)^P F(s)] = 2M_2 + 3 \times 2(s - s_1)M_3 + \cdots \quad (8\text{-}42c)$$

From (8-42a) we have, setting $s = s_1$,

$$M_0 = [(s - s_1)^P F(s)]_{s=s_1} = \frac{d^0}{ds^0}[(s - s_1)^P F(s)]_{s=s_0} \quad (8\text{-}43a)$$

$$M_1 = \left\{\frac{d}{ds}[(s - s_1)^P F(s)]\right\}_{s=s_1} \quad (8\text{-}43b)$$

$$M_2 = \tfrac{1}{2}\left\{\frac{d^2}{ds^2}[(s - s_1)^P F(s)]\right\}_{s=s_1} \quad (8\text{-}43c)$$

and in general,

$$M_k = \frac{1}{k!}\left\{\frac{d^k}{ds^k}[(s - s_1)^P F(s)]\right\}_{s=s_1} \quad (8\text{-}44)$$

"FINAL-VALUE" THEOREM We observe that, if $F(s)$ has an nth-order pole at $s = 0$, such that

$$F(s) = \frac{N(s)}{s^n D_1(s)} \qquad D_1(0) \neq 0$$

then the partial-fractions expansion includes the terms $(K_1 t^{n-1} + K_2 t^{n-2} + \cdots + K_{n-1})u(t)$. It follows that, if $n = 1$, then $[sF(s)]_{s=0} = K_0$ and $f(t)$ includes $K_0 u(t)$. If $n > 1$, $f(t)$ includes $tu(t)$ and powers of t. If now $D_1(s)$ has zeros only in the left-half s plane, the corresponding exponential modes decay to zero. From this we conclude that the behavior of $f(t)$, as t becomes infinite, depends on n. In particular, when the poles of $F(s)$ are in the left-half s plane and

$$\lim_{s\to 0} [sF(s)] = \text{finite}$$

then

$$\lim_{s\to 0} [sF(s)] = f(\infty)$$

This expression is referred to as the final-value theorem. A formal proof of this theorem consists in considering the expression

$$\mathcal{L}\left[\frac{df}{dt}\right] = \int_{0^-}^{\infty} \frac{df}{dt} e^{-st}\, dt = sF(s) - f(0^-)$$

If in the expression $\int_{0^-}^{\infty} [(df/dt)e^{-st}\, dt]$ we set $s = 0$, we have $\int_{t=0^-}^{t=\infty} df$, and if the integral is convergent [this is the same requirement as the location of the poles of $F(s)$ in the left-half s plane], then

$$\int_{t=0^-}^{t=\infty} df = f(\infty) - f(0^-) = \lim_{s\to 0} [sF(s)] - f(0^-)$$

or

$$\lim_{s\to 0} sF(s) = \lim_{t\to\infty} [f(t)]$$

For instance, in Example 8-7,

$$sV(s) = \frac{-36s^2 - 24s + 2}{12s^2 + 17s + 6}$$

Therefore

$$\lim_{t\to\infty} v(t) = \lim_{s\to 0} sV(s) = \tfrac{1}{3}$$

8-9 Examples

Example 8-8 Evaluate $f(t)$ if

$$F(s) = \frac{2s^2 + 13s + 17}{(s+1)(s+3)}$$

Solution Since the degree of the numerator is equal to (rather than smaller than) the degree of the denominator, we first divide

$$F(s) = 2 + \frac{N_1(s)}{(s+1)(s+3)} = 2 + \frac{K_1}{s+1} + \frac{K_2}{s+3}$$

where

$$K_1 = \left(\frac{2s^2 + 13s + 17}{s+3}\right)_{s=-1} = \frac{2 - 13 + 17}{-1 + 3} = 3$$

$$K_2 = \left(\frac{2s^2 + 13s + 17}{s+1}\right)_{s=-3} = \frac{18 - 39 + 17}{-3 + 1} = 2$$

Hence

$$F(s) = 2 + \frac{3}{s+1} + \frac{2}{s+3}$$

and

$$f(t) = 2\delta(t) + (3e^{-t} + 2e^{-3t})u(t)$$

Example 8-9 A response transform $V_{ab}(s)$ is related to a source transform $V(s)$ by

$$V_{ab}(s) = \frac{3V(s) + 3s + 27}{s^2 + 6s + 8}$$

Find $v_{ab}(t)$ if $v(t) = 16e^{-3t}u(t)$.

Solution We have $V(s) = 16/(s + 3)$, so that

$$V_{ab}(s) = \frac{48/(s+3) + 3s + 27}{s^2 + 6s + 8}$$

$$= \frac{3s^2 + 36s + 129}{(s+4)(s+2)(s+3)} = \frac{K_1}{s+4} + \frac{K_2}{s+2} + \frac{K_3}{s+3}$$

and

$$K_1 = \left[\frac{3s^2 + 36s + 129}{(s+2)(s+3)}\right]_{s=-4} = \frac{48 - 144 + 129}{(-4+2)(-4+3)} = \frac{33}{2}$$

$$K_2 = \left[\frac{3s^2 + 36s + 129}{(s+4)(s+3)}\right]_{s=-2} = \frac{12 - 72 + 129}{(-2+4)(-2+3)} = \frac{69}{2}$$

$$K_3 = \left[\frac{3s^2 + 36s + 129}{(s+4)(s+2)}\right]_{s=-3} = \frac{27 - 108 + 129}{(-3+4)(-3+2)} = -48$$

Hence

$$v_{ab}(t) = \tfrac{1}{2}[33e^{-4t} + 69e^{-2t} - 96e^{-3t}]u(t)$$

Example 8-10 If $v(t) = 10e^{-4t}u(t)$ and $V_{ab}(s)$ is related to $V(s)$ as in Example 8-9, formulate $v_{ab}(t)$.

Solution Substituting $V(s) = \mathcal{L}[v(t)] = 10/(s + 4)$ gives

$$V_{ab}(s) = \frac{[30/(s+4)] + 3s + 27}{(s+2)(s+4)} = \frac{3s^2 + 39s + 138}{(s+4)^2(s+2)}$$

Hence

$$V_{ab}(s) = \frac{M_0}{(s+4)^2} + \frac{M_1}{s+4} + \frac{K}{s+2}$$

where

$$M_0 = \left(\frac{3s^2 + 39s + 138}{s+2}\right)_{s=-4} = -15$$

$$M_1 = \left(\frac{d}{ds}\frac{3s^2 + 39s + 138}{s+2}\right)_{s=-4}$$

$$= \left[\frac{(s+2)(6s+39) - (3s^2 + 39s + 138)}{(s+2)^2}\right]_{s=-4} = -15$$

$$K = \left(\frac{3s^2 + 39s + 138}{(s+4)^2}\right)_{s=-2} = \frac{72}{4}$$

and hence $v_{ab}(t) = \frac{1}{4}[(-60t - 60)e^{-4t} + 72e^{-2t}]u(t)$.

Example 8-11 Evaluate $f(t)$ if

$$F(s) = \frac{3s^2 + 17s + 47}{(s+2)(s^2 + 4s + 29)}$$

Solution Factoring the denominator, the partial-fractions expansion is

$$F(s) = \frac{3s^2 + 17s + 47}{(s+2)(s+2-j5)(s+2+j5)} = \frac{K_1}{s+2} + \frac{\mathbf{K}_2}{s+2-j5} + \frac{\mathbf{K}_2^*}{s+2+j5}$$

where $K_1 = \left(\dfrac{3s^2 + 17s + 47}{s^2 + 4s + 29}\right)_{s=-2} = 1$

$$\mathbf{K}_2 = \left[\frac{3s^2 + 17s + 47}{(s+2)(s+2+j5)}\right]_{s=-2+j5} = \frac{-50 + j25}{(j5)(j10)} = 1.12\underline{/-26.6^\circ}$$

Hence

$$f(t) = [e^{-2t} + 2.24e^{-2t}\cos(5t - 26.6^\circ)]u(t)$$

Example 8-12 A voltage $V_m \cos t$ is impressed on a series *L-C* circuit. Calculate $i(t)$ if $L = 1$ and $C = 1$.

Solution $Z(s) = s + 1/s$; hence

$$I(s)\left(s + \frac{1}{s}\right) = \frac{sV_m}{s^2 + 1}$$

$$I(s) = \frac{V_m s^2}{(s^2+1)^2} = \frac{V_m s^2}{(s-j)^2(s+j)^2}$$

Expanding,

$$I(s) = V_m\left[\frac{\mathbf{M}_0}{(s-j)^2} + \frac{\mathbf{M}_0^*}{(s+j)^2} + \frac{\mathbf{M}_1}{s-j} + \frac{\mathbf{M}_1^*}{s+j}\right]$$

where $\mathbf{M}_0 = \left[\dfrac{s^2}{(s+j)^2}\right]_{s=j} = \dfrac{j^2}{(2j)^2} = \dfrac{1}{4}$

$$\mathbf{M}_1 = \left[\frac{d}{ds}\frac{s^2}{(s+j)^2}\right]_{s=j} = \left[\frac{(s+j)^2(2s) - s^2(2)(s+j)}{(s+j)^4}\right]_{s=j}$$

$$= -\frac{j}{4} = \frac{1}{4}\underline{/-90^\circ}$$

Hence

$$i(t) = \left[\frac{V_m}{2}t\cos t + \frac{V_m}{2}\cos(t - \pi/2)\right]u(t)$$

8-10 Comparison of time-domain and Laplace transform methods for complete-response calculation

In Chap. 5 it is shown how a complete response may be formulated by superposition of the source-free and forced components. The method described in Chap. 5 requires that the constants associated with the free modes (source-free response component) be evaluated from continuity or derived initial conditions after the formula for complete response has been obtained.

We observe that use of the Laplace transform method allows formulation of the Laplace transform with the continuity conditions accounted for at the outset through the initial-condition generators. Thus the inverse Laplace transform contains no unknown constants. For example, Eq.

(8-26*a*),

$$V_{ab}(s) = \frac{3V(s) + 3s + 18}{s^2 + 6s + 8}$$

is the Laplace transform of the complete response $v_{ab}(t)$, subject to the conditions $i_L(0^-) = 5$ amp, $v_{ab}(0^-) = 3$ volts, in the circuit of Fig. 8-6*a*. Equation (8-26),

$$V_{ab}(s) = \frac{3}{s^2 + 6s + 8} V(s) + \frac{3}{s^2 + 6s + 8} LI_0 + \frac{1 + s}{s^2 + 6s + 8} V_0$$

shows how Laplace transform resolves the response into a component due to a source (complete response with zero initial-energy storage) and a component due to the initial conditions (energy storage).

The partial-fractions expansion yields a further advantage of Laplace transform; namely, any one of the forced or free modes of a complete response can be found independently of the others. This is done by evaluating only the coefficients K_i [in Eq. (8-35)] or M_i [in Eq. (8-41)] which correspond to the desired mode. Specifically, for example, if the component of the response due to the source is desired, one need only find those terms in the fractions expansion that arise from source poles. In Eq. (8-26*a*), for example, if the source is a unit ramp $v(t) = tu(t)$, then $V(s) = 1/s^2$ and

$$\begin{aligned} V_{ab}(s) &= \frac{3s^3 + 18s^2 + 3}{s^2(s^2 + 6s + 8)} \\ &= \frac{M_0}{s^2} + \frac{M_1}{s} + [V_{ab}(s)]_f \end{aligned} \tag{8-45}$$

where $[V_{ab}(s)]_f$ represents the remaining terms (transform source-free response component) in the expansion. From Eqs. (8-43*a*) and (8-43*b*),

$$\begin{aligned} M_0 &= \left(\frac{3s^3 + 18s^2 + 3}{s^2 + 6s + 8}\right)_{s=0} = \frac{3}{8} \\ M_1 &= \left[\frac{d}{ds}\left(\frac{3s^3 + 18s^2 + 3}{s^2 + 6s + 8}\right)\right]_{s=0} = -\frac{9}{32} \end{aligned}$$

and therefore

$$[v_{ab}(t)]_s = \tfrac{1}{32}(12t - 9)$$

If, on the other hand, in Eq. (8-45), we desire only the mode of the form K_1e^{-2t}, we have, from Eq. (8-36),

$$\begin{aligned} K_1 &= \left[\frac{3s^3 + 18s^2 + 3}{s^2(s^2 + 6s + 8)}(s + 2)\right]_{s=-2} \\ &= \left(\frac{3s^3 + 18s^2 + 3}{s^2(s + 4)}\right)_{s=-2} = \frac{51}{8} \end{aligned}$$

If a simple source pole coincides with a simple network pole, as, for example, when $v(t) = e^{-t}$ is applied to $Y(s) = s/(s + 1)$ and

$$I(s) = \frac{s}{(s + 1)^2} = \frac{M_0}{(s + 1)^2} + \frac{M_1}{s + 1}$$

The term $M_1/(s + 1)$ is identified as the Laplace transform of the source-free component because the network has a simple pole at $s = -1$. Therefore the term $M_0/(s + 1)^2$ is the Laplace transform of the component of the response due to the source.

Similar reasoning will show that, when a source $v(t) = t^n e^{-\alpha t} u(t)$ whose transform is $n!/(s + \alpha)^{n+1}$ is applied to a network

$$H(s) = \frac{N(s)}{(s + \alpha)^m D_1(s)} \qquad N(\alpha) \neq 0,\ D_1(\alpha) \neq 0$$

so that the response $V_2(s)$ is

$$V_2(s) = \frac{N(s)}{(s + \alpha)^{n+m+1} D_1(s)}$$

the component of the response due to the source has the transform

$$[V_2(s)]_s = \frac{M_0}{(s + \alpha)^{n+m+1}} + \frac{M_1}{(s + \alpha)^{m+n}} + \frac{M_2}{(s + \alpha)^{m+n-1}} + \cdots + \frac{M_n}{(s + \alpha)^{m+1}}$$

so that

$$[v_2(t)]_s = \left[\frac{M_0 t^{m+n}}{(m + n)!} + \frac{M_1 t^{m+n-1}}{(m + n - 1)!} + \cdots + \frac{M_n t^n}{n!} \right] e^{-\alpha t} u(t)$$

In Example 8-12 the source has two conjugate imaginary poles, each coinciding with a pole of $Y(s)$ (at $s = \pm j$), and the above reasoning shows that the component of the response $i(t)$ due to the source is $i_s(t) = (\frac{1}{2} V_m t \cos t) u(t)$.

8-11 *Periodicity* We now formulate the Laplace transform of semiperiodic functions, that is, functions that are periodic for $t > 0$ and zero for $t < 0$. A semiperiodic function

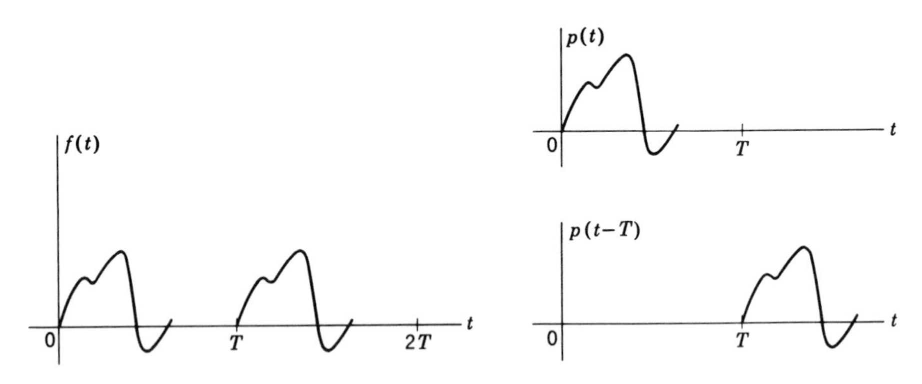

Fig. 8-7 ***A semiperiodic function shown as the (infinite) sum of successive pulses.***

can be represented as the infinite sum of pulses that are delayed relative to each other as shown in Fig. 8-7. Denoting the first pulse by $p(t)$, we have

$$f(t) = p(t) + p(t - T) + \cdots + p(t - nT) + \cdots$$

and if the Laplace transform of the *first* pulse is $P(s)$,

$$\mathcal{L}[p(t)] = P(s)$$

Then the infinite series

$$F(s) = P(s) + P(s)e^{-sT} + P(s)e^{-2sT} + \cdots + P(s)e^{-snT} + \cdots \tag{8-46}$$

represents the Laplace transform of the semiperiodic function. Rewriting Eq. (8-46), factoring $P(s)$,

$$F(s) = P(s)(1 + e^{-sT} + e^{-s2T} + \cdots + e^{-snT} + \cdots)$$

In the series

$$1 + e^{-sT} + e^{-s2T} + \cdots + e^{-snT} + \cdots \tag{8-47}$$

$|e^{-sT}| < 1$ because Re $s = \sigma > 0$, and therefore the geometric series (8-47) is convergent and its sum is

$$1 + e^{-sT} + e^{-s2T} + \cdots + e^{-snT} + \cdots = \frac{1}{1 - e^{-sT}}$$

The Laplace transform of a semiperiodic function is then

$$F(s) = \frac{P(s)}{1 - e^{-sT}} \tag{8-48}$$

Equation (8-48) is useful, not only in formulating the transform of semiperiodic functions, but also in identifying inverse transforms.

Example 8-13 Find the Laplace transform of the output of an ideal full-wave rectifier with input $V_m \sin \omega t u(t)$.

Solution The first pulse $p(t)$ is given by (Fig. 8-3*c*)

$$p(t) = V_m \sin \omega t u(t) + V_m \sin \omega \left(t - \frac{\pi}{\omega}\right) u \left(t - \frac{\pi}{\omega}\right)$$

Hence

$$P(s) = \frac{V_m \omega}{s^2 + \omega^2} (1 + e^{-s\pi/\omega})$$

and for the semiperiodic function the period is π/ω. Hence

$$F(s) = \frac{V_m \omega}{s^2 + \omega^2} \frac{1 + e^{-s\pi/\omega}}{1 - e^{-s\pi/\omega}} = \frac{V_m \omega}{s^2 + \omega^2} \coth \frac{\pi s}{2\omega}$$

Example 8-14 Find the inverse Laplace transform of

$$F(s) = \frac{1 - 2e^{-s} + e^{-2s}}{s^2(1 - e^{-2s})}$$

Solution We write

$$F(s) = \frac{(1 - 2e^{-s} + e^{-2s})/s^2}{1 - e^{-2s}}$$

so that $f(t)$ is a semiperiodic function with period 2, *provided that* $(1 - 2e^{-s} + e^{-2s})/s^2$ *represents a pulse.* Since

$$\mathcal{L}^{-1}\left[\frac{1 - 2e^{-s} + e^{-2s}}{s^2}\right] = tu(t) - 2(t - 1)u(t - 1) + (t - 2)u(t - 2) = p(t)$$

represents triangular pulse shown in Fig. 8-8, $f(t)$ is the semiperiodic function shown in Fig. 8-8.

We observe that, when a Laplace transform $F(s)$ has the denominator $1 - e^{-sT}$, the result is a periodic function if the numerator represents *a pulse,* that is, if the inverse transform of the numerator is zero for $t > T$. When this is *not* the case, $f(t)$ is *periodically modified,* as illustrated in the next example.

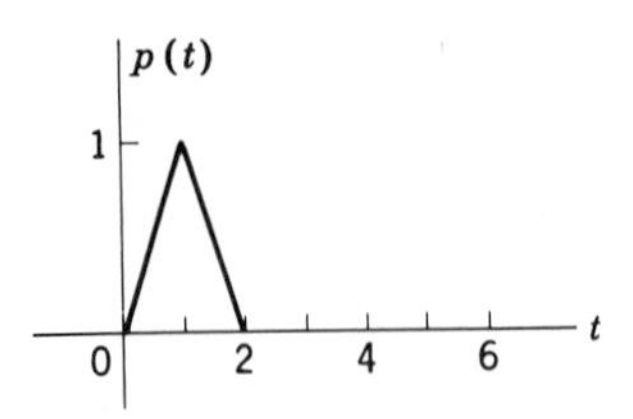

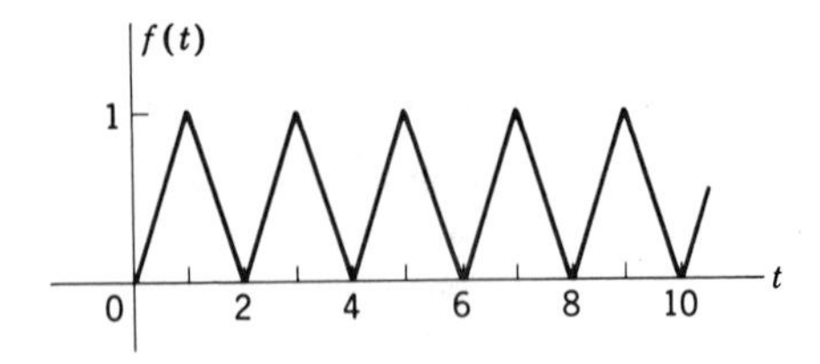

Fig. 8-8 Solution of Example 8-14.

Example 8-15 Sketch $f(t)$ if $F(s) = (1/s)/(1 - e^{-s})$.

Solution In this case the numerator is $\mathcal{L}[u(t)]$. Hence

$$f(t) = u(t) + u(t - 1) + \cdots + u(t - n) + \cdots$$

As shown in Fig. 8-9, $f(t)$ is "modified" periodically.

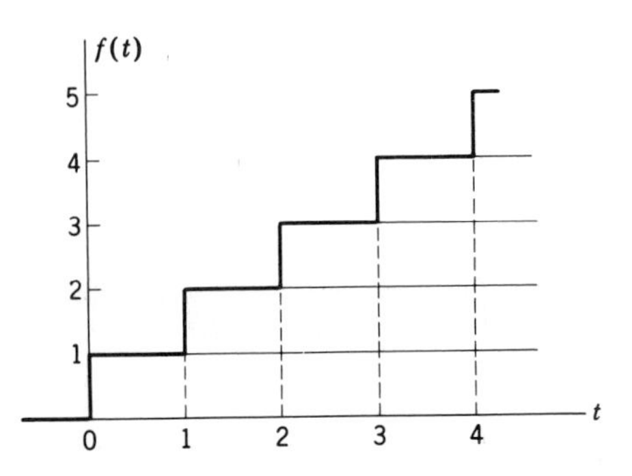

Fig. 8-9 Solution of Example 8-15.

8-12 Convolution We now pose the following question: If $\mathcal{L}^{-1}[F_1(s)] = f_1(t)$ and if $\mathcal{L}^{-1}[F_2(s)] = f_2(t)$, what is the inverse Laplace transform of the product $F_1(s)F_2(s)$? We shall show that

$$\mathcal{L}^{-1}[F_1(s)F_2(s)] = \int_{\tau=0}^{\tau=t} f_1(\tau)f_2(t - \tau)\, d\tau \tag{8-49}$$

where the right-hand side of Eq. (8-49) is referred to as the convolution integral. For proof we start with

$$\int_{\tau=0}^{\tau=t} f_1(\tau)f_2(t - \tau)\, d\tau = \int_0^\infty f_1(\tau)f_2(t - \tau)u(t - \tau)\, d\tau$$

and taking the Laplace transform,

$$\begin{aligned}\mathcal{L}\left[\int_0^\tau f_1(\tau)f_2(t - \tau)\, d\tau\right] &= \int_{t=0}^{t=\infty}\left[\int_{\tau=0}^{\tau=\infty} f_1(\tau)f_2(t - \tau)u(t - \tau)\, d\tau\right] e^{-st}\, dt \\ &= \int_{\tau=0}^{\tau=\infty} f_1(\tau)\left[\int_{t=0}^{t=\infty} f_2(t - \tau)u(t - \tau)e^{-st}\, dt\right] d\tau\end{aligned}$$

Since $u(t - \tau) = 0$ for $t < \tau$,

$$\int_{t=0}^{t=\infty} f_2(t - \tau)u(t - \tau)e^{-st}\, dt = \int_{t=\tau}^{t=\infty} f_2(t - \tau)e^{-st}\, dt$$

Letting $x = t - \tau$,

$$\int_{t=\tau}^{t=\infty} f_2(t - \tau)e^{-st}\, dt = \int_{x=0}^{x=\infty} f_2(x)e^{-s(x+\tau)}\, dx$$

Hence

$$\mathcal{L}\left[\int_0^T f_1(\tau)f_2(t-\tau)\,d\tau\right] = \int_{\tau=0}^{\tau=\infty} f_1(\tau)\left[\int_{x=0}^{x=\infty} f_2(x)e^{-sx}\,dx\right]e^{-s\tau}\,d\tau$$
$$= F_1(s)F_2(s)$$

Example 8-16 Solve the equation

$$\sin t = \int_0^t f(\tau)e^{-(t-\tau)}\,d\tau$$

for $f(t)$.

Solution

$$\mathcal{L}[\sin t] = \mathcal{L}\left[\int_0^t f(\tau)e^{-(t-\tau)}\,d\tau\right]$$

Hence

$$\frac{1}{s^2+1} = F(s)\,\frac{1}{s+1}$$

$$\frac{s+1}{s^2+1} = F(s)$$

$$f(t) = (\cos t + \sin t) = \sqrt{2}\cos\left(t - \frac{\pi}{4}\right)$$

The convolution integral can, of course, be used to find inverse Laplace transforms when $F(s)$ has the form of a product.

This product form is of particular interest in system analysis since

$$V_2(s) = H(s)V_1(s)$$

then by convolution

$$v_2(t) = \int_0^t v_1(\tau)h(t-\tau)\,d\tau$$

the response $v_2(t)$ to the source $v_1(t)$ is related to the response $h(t)$ which is the output for unit-impulse input. An application of this result is given in Appendix E.

8-13 Introduction to physical realizability of networks[1]

We have observed, and shall prove generally in Chap. 11, that every transform network function is a ratio of polynomials in s. We now ask the question regarding the converse: Is every ratio of polynomials in s the network function of a linear lumped parameter network? The answer is that it is not, and this section deals with the restrictions to which such ratios are subject in order to be network functions.

First of all, we observe that the coefficients a_k and b_k in the ratio

$$H(s) = \frac{b_0 + b_1 s + \cdots + b_m s^m}{a_0 + a_1 s + \cdots + a_n s^n} = \frac{B(s)}{A(s)} \tag{8-50}$$

must be real numbers because they are the result of multiplying, adding, subtracting, and dividing the R, L, and C values of the network. Consequently (from algebra), we conclude that poles and zeros, if complex, must occur in conjugate pairs.

[1] A somewhat more comprehensive, but still elementary, discussion of this topic is found in M. Javid and E. Brenner, "Analysis, Transmission, and Filtering of Signals," chap. 13, McGraw-Hill Book Company, New York, 1963.

Next we recall that the characteristic roots of the source-free network are the solutions of the equation $A(s) = 0$. Now, if any solution of this equation $s = s_k$ is positive or complex with a positive real part, the source-free response could have a component of the form $e^{\alpha t}$ or $e^{\alpha t} \cos(\omega t + \varphi)$, $\alpha > 0$. Such a component increases indefinitely with time. Such increasing source-free response is impossible since the stored energy in a source-free network cannot increase with time. It follows that *a physically realizable network function cannot have any poles in the right-half s plane.*

If the function $H(s)$ is a *driving-point* immittance [$Z_{ab}(s)$ or $Y_{ab}(s)$], then its reciprocal is also a *driving-point* immittance, and the poles of the reciprocal are the zeros of the original function. It follows that, for *driving-point immittances,* there may be *neither zeros nor poles* in the right-half s plane. Transfer functions may (but need not) have zeros in the right-half s plane.

With respect to poles for imaginary values of s, we recall that a pair of simple poles at $s = \pm j\omega$ corresponds to free responses of sinusoidal form. This occurs when there is no dissipation of energy. However, a double pole at $s = \pm j\omega$ corresponds to source-free modes of the form $t \cos \omega t$. Such a mode increases with time, and is not permissible as a free mode. Therefore, when the poles of a network function are located on the imaginary axis, they must be simple conjugate poles.[1]

Another realizability condition for a driving-point immittance function is that the degree of the numerator and denominator polynomial cannot differ by more than 1. Thus, in Eq. (8-50), this means that either $m = n$ or they differ by 1.[2]

The conditions for physical realizability for driving-point immittances dis-

[1] If function is a *driving-point* immittance, there is an additional restriction on the poles of Z or Y, namely, the residues at these poles must be positive; that is, if Z or Y is expanded in partial fractions, the constant K of the term $K/(s - j\omega)$ must be positive and real.

[2] To prove this statement we start with

$$Z_{ab}(s) = \frac{a_n s^n + a_{n-1}s^{n-1} + \cdots + a_1 s + a_0}{b_n s^n + b_{n-1}s^{n-1} + \cdots + b_1 s + b_0}$$

so that either $m = n$ or $m \neq n$. Now, if $m = n$, then

$$Z_{ab}(s) = \frac{a_m s^m + a_{m-1}s^{m-1} + \cdots + a_1 s + a_0}{b_n s^n + b_{n-1}s^{n-1} + \cdots + b_1 s + b_0} = \frac{a_n + \dfrac{a_{n-1}}{s} + \dfrac{a_{n-2}}{s^2} + \cdots + \dfrac{a_1}{s^{n-1}} + \dfrac{a_0}{s^n}}{b_n + \dfrac{b_{n-1}}{s} + \cdots + \dfrac{b_1}{s^{n-1}} + \dfrac{b_0}{s^n}}$$

so that $\lim_{s \to \infty} Z_{ab}(s) = a_n/b_n$ if $n = m$. Hence the point $s = \infty$ is neither a zero nor a pole of $Z_{ab}(s)$. Now, if $n \neq m$, then $\lim_{s \to \infty} Z_{ab}(s) = 0$ or approaches infinity. Hence we have either a zero or a pole at infinity. Now it is an axiom that infinity in the s plane is a single point; i.e., if we imagine the s plane to be on the surface of a large (infinite) sphere with the origin at the "south pole," then infinity is at the "north pole." The imaginary axis which passes through this south pole also passes through the north pole. But zeros and poles on this axis must be simple. Hence the degree of the polynomial $A(s)$ cannot differ from the degree of the polynomial $B(s)$ by more than unity.

cussed in the paragraphs above are all deduced from the properties of the source-free response, in particular from the fact that the total initially stored energy, which is finite, cannot produce a response which increases without bound. These properties show that a passive lumped-element driving-point immittance function must satisfy the following requirements:

1. It is the ratio of polynomials with real positive coefficients.
2. The degree of the numerator polynomial may be the same as the degree of the denominator polynomial or may differ by unity.
3. No poles or zeros may be in the right-half s plane.
4. Poles and zeros on the imaginary axis must be simple, with positive real residues.
5. Complex poles and zeros must occur in conjugate pairs.

"POSITIVE REAL" PROPERTY The conditions enumerated above are necessary, but not sufficient. For example,

$$F(s) = \frac{s+4}{s^2+2s+5} = \frac{s+4}{[s-(-1+j2)][s-(-1-j2)]}$$

satisfies the conditions stated above, but it is not a transform driving-point immittance. This is seen from the fact that, at a radian frequency $\omega = 5$, $F(j\omega) = (j5+4)/(-5^2+j10+5)$, or

$$F(j\omega) = \frac{\sqrt{41}\,\underline{/51.4^\circ}}{10\sqrt{5}\,\underline{/153.4^\circ}} = 0.285\underline{/-102^\circ} = -0.06 - j0.28$$

If $F(s)$ were a driving-point immittance, then at $s = j5$ its negative real part would represent *negative* resistance, since, in the sinusoidal steady state, the average power input to a terminal-pair network has the same sign as the real part of the driving-point immittance (Prob. 8-25). Hence, if this function were an immittance function, the network would deliver electromagnetic energy to the source, without bound, instead of converting the energy supplied by the source into heat or storing it. Since a passive network is not capable of delivering energy without bound, it is evident that $F(s) = (s+4)/(s^2+2s+5)$ cannot be a driving-point function.

It can be shown[1] that a necessary condition that $H(s) = A(s)/B(s)$ be a driving-point immittance function is the requirement $\text{Re}\, H(s) \geq 0$ when $\text{Re}\, s \geq 0$. A function which satisfies this condition is referred to as a "positive real function."

Example 8-17 Show that $Z(s) = (s+1)/(s+5)$ is a positive real function.

Solution. Let $s = \sigma + j\omega$, where $\sigma = \text{Re}\, s > 0$; then $Z(s) = (\sigma+1+j\omega)/(\sigma+5+j\omega)$. Rationalizing,

$$Z(\sigma + j\omega) = \frac{(\sigma+1)(\sigma+5)+\omega^2+j4\omega}{(\sigma+5)^2+\omega^2}$$

$$\text{Re}\,[Z(\sigma+j\omega)] = \frac{(\sigma+1)(\sigma+5)+\omega^2}{(\sigma+5)^2+\omega^2} > 0 \qquad \text{for } \sigma \geq 0$$

[1] Javid and Brenner, *op. cit.*

Therefore $Z(s) = (s + 1)/(s + 5)$ is a positive real function. From the definition of positive real functions it is seen that the condition $\text{Re}\,[H(s)] \geq 0$ for $\text{Re}\,s = 0$, that is, $\text{Re}\,[H(j\omega)] \geq 0$, is a necessary condition. It is noted that the function $F(s) = (s + 4)/(s^2 + 2s + 5)$ is such that $\text{Re}\,[F(s)] < 0$ for certain imaginary values of s ($\text{Re}\,s = 0$).

REALIZABILITY OF TRANSFER FUNCTIONS Transfer functions differ from driving-point immittance functions in that a transfer function which, for example, relates a response to a voltage source does not apply when a current source is substituted for the voltage source. In contrast, a driving-point immittance and its reciprocal relate either type of source to response. Transfer functions are therefore restricted less severely. However, it can be shown that for transfer functions, the restrictions for physical realizability are:

1 They are ratios of polynomials in s.
2 No poles may be in the right-half s plane.
3 Poles if on the imaginary axis must be simple.
4 The degree of the numerator polynomial cannot exceed the degree of the denominator polynomial by more than unity.

There is no restriction in the location of the zeros of the transfer functions. In Prob. 8-29 a network which has a zero in the right-half s plane is shown. Note also that the restriction on the numerator polynomial states only that the degree of this polynomial cannot *exceed* the degree of the denominator by more than unity.

Example 8-18 In the circuit of Fig. 8-10 find the transform gain function which relates v_2 to v.

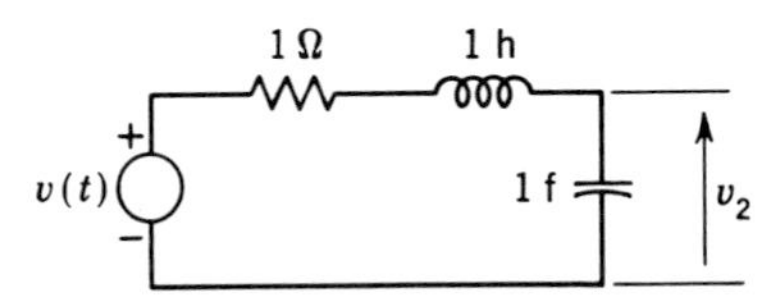

Fig. 8-10 ***In the transfer function that relates v_2 to v, the degree of the denominator polynomial exceeds that of the numerator by 2.***

Solution From voltage division,

$$v_2(t) = \frac{1/p}{1 + p + 1/p}\,v(t)$$

Hence

$$v_2 = \frac{1}{p^2 + p + 1}\,v(t)$$

The transform gain function is $1/(s^2 + s + 1)$. $N(s) = 1$ $D(s) = s^2 + s + 1$. Hence the degree of $N(s)$ is 0, and the degree of $D(s)$ is 2.

PROBLEMS

8-1 By application of the defining integral for Laplace transform, find $\mathcal{L}[f(t)]$ if (*a*) $f(t) = e^{-t}[u(t) - u(t - 1)]$; (*b*) $t[u(t) - u(t - 1)]$.

8-2 Prove that, if $\mathcal{L}[f(t)] = F(s)$, then (*a*) $\mathcal{L}[f(t/a)] = aF(as)$, where a is positive. (*b*) $\mathcal{L}[tf(t)] = -dF/ds$.

8-3 Find $\mathcal{L}[v(t)]$ if $(p^2 + 2p + 10)v = te^{-2t}$, $v(0^-) = 0$, $(dv/dt)_{0^-} = -5$.

8-4 Find $\mathcal{L}[v(t)]$ if $v(t) = 0$, $t < 0$; $v(t) = \cos 2\pi t$, $0 < t < 5$; $v(t) = 0$, $t > 5$.

8-5 For the (pulse-type) functions given below, $f(t) = 0$ for $t < 0$, $f(t) = 0$ for $t > 1$, and $f(t)$ is defined by the formula given below in the interval $t = 0$ to $t = 1$. Sketch the pulse and find the Laplace transform for each of the pulses given. (*a*) $f(t) = 1 - t$; (*b*) $f(t) = \cos \pi t/2$; (*c*) $f(t) = \cos \pi t$.

8-6 Sketch $f(t)$ if (*a*) $F(s) = (1 - e^{-4s})/s^2$; (*b*) $F(s) = (s - 1 + e^{-2s} + se^{-4s})/s^2$.

8-7 Find $F(s)$ if $f(t)$ is given as (*a*) $[\cos (10t + 30°)]u(t)$; (*b*) $[e^{-5t} \cos (10t + 30°)]u(t)$; (*c*) $\cosh tu(t)$; (*d*) $(t \cos 10t)u(t)$.

8-8 In the circuit of Fig. P8-8, $R = 1$ ohm and $L = 1$ henry. Find $I(s)$ and $i(t)$ if (*a*) $v(t) = (1 - e^{-t})u(t)$, $i(0^-) = 0$; (*b*) $v(t) = u(t)$, $i(0^-) = 1$; (*c*) $v(t) = (e^{-t} \cos t)u(t)$, $i(0^-) = 0$.

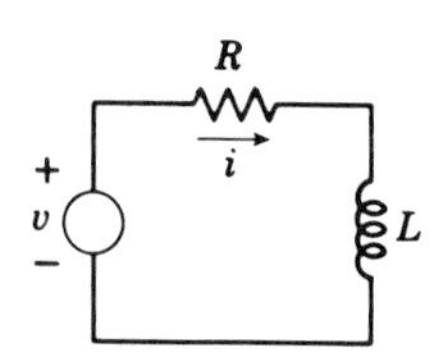

Fig. P8-8

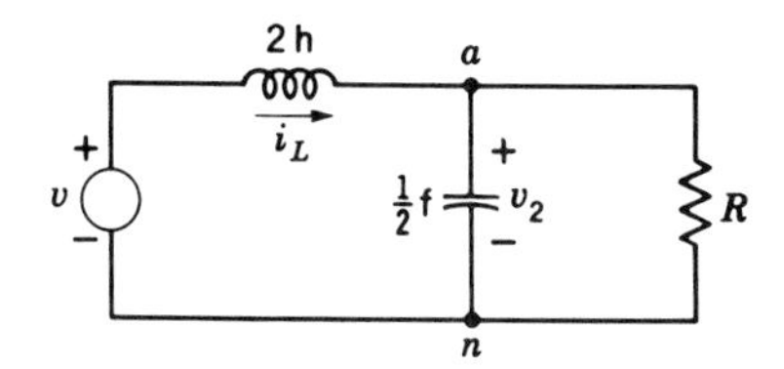

Fig. P8-9

8-9 In the circuit of Fig. P8-9, $v(t) = 5e^{-t}u(t)$, $i_L(0^-) = I_0$, $v_2(0^-) = V_0$. Formulate $\mathcal{L}[v_2(t)]$ in terms of R, V_0, and I_0.

8-10 In the circuit of Fig. P7-1, let $\mathcal{L}[v] = V(s)$, $\mathcal{L}[i] = I(s)$, $v_{a'a}(0^-) = V_0$, $i_{ab}(0^-) = I_0$, and formulate $\mathcal{L}[v_{ab}(t)]$.

8-11 In the circuit shown in Fig. P8-11, $v_{an}(0^-) = v_{bn}(0^-) = 10$. Formulate $V_{an}(s)$.

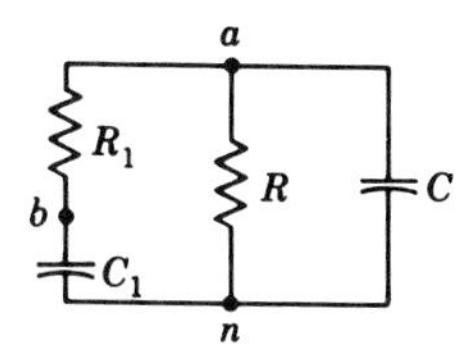

Fig. P8-11

8-12 In the circuit of Fig. P8-11, find $V_{an}(s)$ and $v_{an}(t)$ if (*a*) $v_{an}(0^-) = 10$, $v_{bn}(0^-) = -20$; (*b*) $v_{an}(0^-) = 10$, $v_{bn}(0^-) = +20$. Use $R_1 = 2R = 1$; $C = 2C_1 = 1$.

8-13 A passive network consists of a capacitance of 1 farad in parallel with the series combination of R ohms and 1 henry. A voltage source $v(t)$ is applied to the terminals of the capacitance. The current from the source is $\delta(t)$. Find $v(t)$ if (*a*) $R = 0$, (*b*) $R = 1.2$ ohms.

8-14 Find $v(0^+)$, $(dv/dt)_{0^+}$, and $(d^2v/dt^2)_{0^+}$ in each of the following cases:

(*a*) $V(s) = \dfrac{2s^2 + 10s + 3}{s(s^2 + 4s + 5)}$ (*b*) $V(s) = \dfrac{2s^2 + 10s + 3}{s^2(s^2 + 4s + 5)}$

8-15 Find the inverse Laplace transform of each of the following functions:

(a) $\dfrac{s+2}{s^2+8s+12}$ (b) $\dfrac{s^2+2s}{s^2+8s+12}$ (c) $\dfrac{1-e^{-s}}{s(s+1)}$ (d) $\dfrac{s+2}{s^2+4}$

(e) $\dfrac{3(s^2-4)}{(s^2+4)^2}$ (f) $\dfrac{s+2}{[(s+1)^2+1]^2}$ (g) $\dfrac{s^2+7s+22}{(s+3)^2(s+5)}$ (h) $\dfrac{8}{s^4+1}$

8-16 In the circuit of Fig. P8-9, $R = 1$ and the source function is given by $v(t) = 5e^{-t}u(t)$. Find $v_2(t)$ if (a) $i_L(0^-) = 0$, $v_2(0^-) = 0$; (b) $i_L(0^-) = 1$, $v_2(0^-) = 2$; (c) $i_L(0^-) = 2$, $v_2(0^-) = 2$.

8-17 A unit ramp-function source $v_1(t) = tu(t)$ is applied to a network, and the Laplace transform of the response $v_2(t)$ is given by

$$V_2(s) = H(s)V_1(s) \qquad H(s) = \frac{s+1}{s^2+2s}$$

Find the component of the response due to the source.

8-18 Find the steady-state response component in each of the following response transforms [$H(s)$ has no poles on the $j\omega$ axis]:

(a) $V(s) = \dfrac{8s^3+2s+40}{s(s^3+8s^2+12s+10)}$

(b) $V(s) = \dfrac{10^5}{(s^2+9)(s+3)^2(s^2+6s+18)}$

(c) $V(s) = \dfrac{30s^3+90}{s(s^2+9)(s+2)(s+4)}$

8-19 In a certain network the response $r(t)$ is related to a source $\phi(t)$ by $r(t) = [p/(p+2)]\phi(t)$. Find $r_s(t)$, the component of the response due to the source if (a) $\phi(t) = tu(t)$; (b) $\phi(t) = t^2u(t)$; (c) $\phi(t) = t^3u(t)$; (d) $\phi(t) = te^{-t}u(t)$; (e) $\phi(t) = t\cos 2tu(t)$; (f) $\phi(t) = e^{-2t}u(t)$; (g) $\phi(t) = te^{-2t}u(t)$.

8-20 Solve the following problems by use of Laplace transform: (a) Prob. 5-28; (b) Prob. 5-30; (c) Prob. 5-34; (d) Prob. 5-37; (e) Prob. 5-42.

8-21 Sketch to scale $f(t)$ if

(a) $F(s) = \dfrac{1-e^{-s}}{s^2(1-e^{-2s})}$ (b) $F(s) = \dfrac{\coth s\pi/2}{s^2+1}$ (c) $F(s) = \dfrac{1-e^{-s}}{s(1-\frac{1}{2}e^{-2s})}$

8-22 A semiperiodic function $g(t)$ has a period of 2. Find its Laplace transform if the first pulse is defined as $f(t)$ in (a) Prob. 8-5a; (b) Prob. 8-5b; (c) Prob. 8-5c.

8-23 (a) Find $\mathcal{L}^{-1}[1/(s^2+1)]$ by use of convolution. (b) Solve, by Laplace transform, for $f(t)$:

$$f(t)u(t) + \int_0^t f(\tau)e^{(\tau-t)}\,d\tau = tu(t)$$

8-24 A transform driving-point admittance $V_{ab}(s)$ has a voltage source $Ve^{+\alpha t}$ (α real, positive) impressed. The initial conditions are adjusted so that the free component of the current response is identically zero. (a) Show that the power input to the terminal pair a-b from the source is $p_{ab}(t) = V^2Y(\alpha)e^{2\alpha t}$. (b) From the result of (a) prove that $Y(\alpha)$ is a positive number for positive α if the terminal pair is passive.

8-25 A transform admittance $Y(s)$ has a voltage source $v(t) = \frac{1}{2}e^{\mathrm{s}_g t} + \frac{1}{2}e^{\mathrm{s}_g^* t}$ impressed. The initial conditions are adjusted so that the free component of the response $i(t)$ is identically zero. (a) If $\mathrm{s}_g = -\alpha \pm j\omega$, show that the instantaneous power delivered to the admittance is

$$p(t) = \tfrac{1}{2}e^{-2\alpha t}\{\text{Re}\,[Y(\mathbf{s}_g)]\} + \tfrac{1}{2}e^{-2\alpha t}\,\text{Re}\,[Y(\mathbf{s}_g)e^{j2\omega t}]$$

(*b*) If $\mathbf{s}_g = j\omega$, show that the average power $P = (\omega/2\pi)\int_0^{2\pi/\omega} p(t)\,dt$ is given by $P = \frac{1}{2}\,\text{Re}\,[Y(j\omega)]$. Note that this proves that $\text{Re}\,[Y(j\omega)] > 0$ for passive terminal-pair networks. Compare this result with that of Prob. 8-24, and show how your proofs in Prob. 8-24 and in this problem illustrate the positive real property of driving-point immittances.

8-26 (*a*) It has been shown that a transform driving-point immittance has the form

$$Z(s) = H\frac{\prod_{z=0}^{z=Z}(s - s_z)}{\prod_{p=0}^{p=P}(s - s_p)}$$

where $P - Z = 0, +1$, or -1. Show that the angle of the complex factor $s - s_a$ cannot exceed 90° in magnitude if $\text{Re}\, s \geq 0$, provided that $\text{Re}\, s_a \leq 0$. (*b*) Using the positive real property of driving-point immittance functions and the result of (*a*), show that $P - Z = 0, +1$, or -1.

8-27 Prove the following statements: When a source $\phi(t) = Ae^{\alpha t}\cos(\omega t + \varphi)$ is impressed on a driving-point immittance, the component of the response at the source terminals due to the source cannot have the form $Kte^{\alpha t}\cos(\omega t + \delta)$ if $\alpha > 0$. The component of the response due to the source has the form $Be^{\alpha t}\cos(\omega t + \delta)$, and B cannot be zero.

8-28 Explain why each of the following functions cannot represent a physically realizable driving-point immittance. (*a*) $H(s) = s^2$; (*b*) $H(s) = (s - 1)/(s + 3)$; (*c*) $H(s) = s/(s^3 + 2s^2 + 2s + 1)$; (*d*) $H(s) = \sqrt{1 + s^2}$; (*e*) $H(s) = (s^2 + 2s + 2)/(s^3 + s + 1)$; (*f*) $(s^2 + 2s + 2)/(s + 10)$; (*g*) $s^2/(s^2 + 2)$.

8-29 (*a*) Show that the transfer function that relates v_2 to v_1 in the network of Fig. P8-29 has a zero in the right-half s plane. (*b*) Show that, in the sinusoidal steady state, the output amplitude is the same as the input amplitude and that the phase difference between the output sinusoid and the input sinusoid is $2\tan^{-1}\omega RC$.

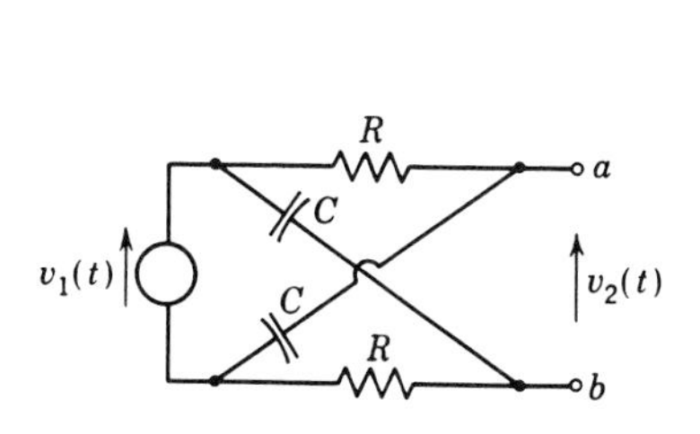

Fig. P8-29

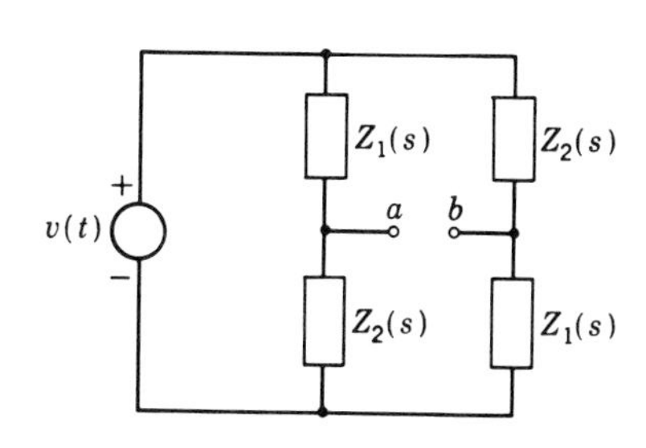

Fig. P8-30

8-30 (*a*) In Fig. P8-30, show that the transform network function which relates $v_{ab}(t)$ to $v(t)$ is $H(s) = [Z_2(s) - Z_1(s)]/[Z_1(s) + Z_2(s)]$. (*b*) Locate the zeros and poles of $H(s)$ if (1) $Z_1 = 1$, $Z_2 = s$; (2) $Z_1 = 1 + s$, $Z_2 = 1/(1 + s)$; (3) $Z_1 = s + s^{-1}$, $Z_2 = 1$.

8-31 (*a*) In the circuit shown in Fig. P8-31, prove that, if $Z^2 = Z_1^2 + 2Z_1Z_2$, then $Z_{ab} = Z$. (*b*) Show that the driving-point immittance Z is not in general physically realizable. (*c*) Is Z physically realizable if $Z_2 = kZ_1$, where k is a real constant? Explain and illustrate your answer.

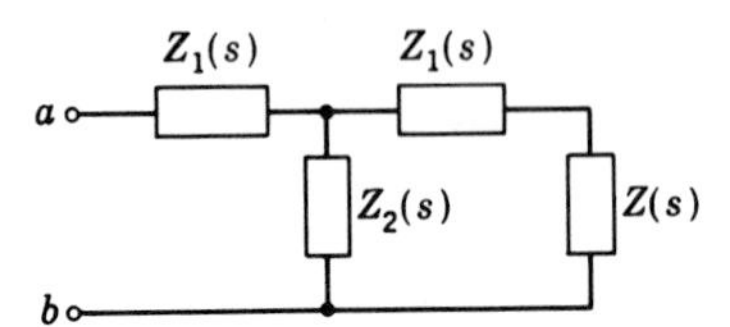

Fig. P8-31

8-32 (*a*) Show that the driving-point impedance of the terminal pair in Fig. P8-32*a* has a pole at $s = -1/RC$. (*b*) Show that the series connection of n different *R-C* circuits in Fig. P8-32*b* forms a driving-point impedance with n poles (at $-1/R_1C_1, -1/R_2C_2, \ldots, -1/R_nC_n$). (Assume $R_kC_k \neq R_jC_j$.)

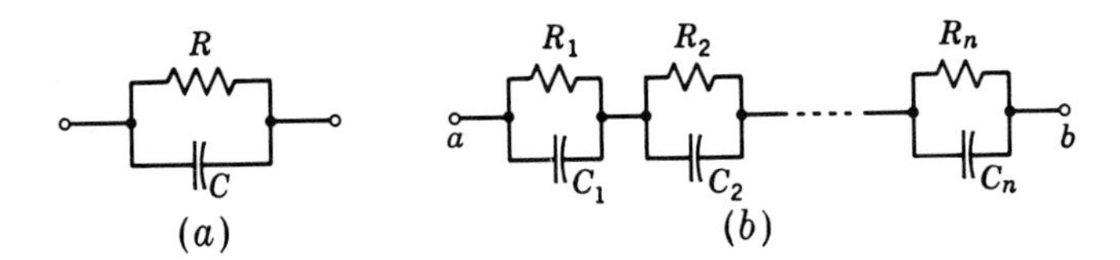

Fig. P8-32

8-33 A driving-point admittance function is given as $Y(s) = (s + 1)/s(s + 2)$. (*a*) Show that $Y(s) = Y_1 + Y_2$, where $Y_1(s) = 1/2s$; $Y_2 = 1/(2s + 4)$. (*b*) Identify $Z_1(s) = 2s$ and $Z_2(s) = 2s + 4$ as p elements, and draw a circuit which has the given driving-point admittance. (*c*) Show that $Z(s) = 1/Y(s) = Z_1 + Z_2$, where $Z_1 = s$; $Z_2 = 1/(1 + 1/s)$. (*d*) Use the result of (*c*) to draw a second circuit which has the prescribed driving-point admittance.

8-34 A driving-point impedance is given as

$$Z(s) = \frac{(s + 2)(s + 4)}{(s + 1)(s + 3)(s + 5)}$$

(*a*) Expand $Z(s)$ by the method of partial fractions into the form

$$Z(s) = \frac{A}{s + 1} + \frac{B}{s + 3} + \frac{C}{s + 5}$$

(*b*) Use the result of (*a*) to sketch a network with the prescribed $Z(s)$. Specify the numerical value of all elements.

Chapter 9 Circuits in the sinusoidal steady state

The steady-state response of networks to sinusoidal excitation is an important special case which merits further detailed treatment, for the following reasons.

1 The sinusoidal function is periodic. From the point of view of consumer of electric energy, the periodic waveform is practical, since its time variations are predictable.

2 Sinusoidal sources can be constructed to furnish large amounts of power with ease when compared with sources furnishing other waveforms.

3 The sinusoidal function is the only real periodic function which has an integral and a derivative of the same waveform as the function itself. In any linear circuit this means that the steady-state response to a sinusoidal source will be sinusoidal regardless of the types or number of circuit elements which are interconnected.

4 The analysis of circuits with nonsinusoidal periodic sources can often be carried out conveniently by representing the nonsinusoidal function as a sum of sinusoids and using superposition.

5 The steady-state response of a circuit to sinusoidal sources furnishes information about the response to other waveforms. This follows from the fact that sinusoidal functions can be represented as the sum of exponential functions; and the responses to exponential source functions are related to the pole-zero locations of the network function. Once the poles and zeros of a network function are established, both the free component of the response and the response to exponential source functions can be determined. Thus

the study of circuits with sinusoidal sources is important, not only because many practical systems use such sources (power stations, lights and appliances in homes, radio and television transmitters, electrical machinery, etc.), but also because many basic properties of circuits become evident through analysis or experimentation with sinusoidal sources *even if the circuit is not intended for use with such sources.*

In colloquial engineering terminology the steady-state analysis of circuits with a sinusoidal source is called *alternating-current* (a-c) circuit analysis.

9-1 General formula for sinusoidal steady-state response[1]

We recall that in a network a response $v_2(t)$ is related to a source $\phi(t)$ by the equilibrium equation

$$v_2(t) = [H(p)]\phi(t)$$

It is shown in Chap. 7 that when the source is $\phi(t) = \phi_m \cos(\omega t + \gamma)$,

$$\phi(t) = \operatorname{Re}(\boldsymbol{\phi}_m e^{j\omega t}) \qquad \boldsymbol{\phi}_m = \phi_m e^{j\gamma}$$

then, if $j\omega$ is not a pole of $H(s)$, the component of the response due to the source is given by

$$v_2(t)\Big]_s = \operatorname{Re}[H(j\omega)\boldsymbol{\phi}_m e^{j\omega t}] \tag{9-1}$$

If $H(s)$ has all its poles in the left-hand portion of the s plane then the free response component decays to zero as t increases so that $(v_2)_s$ is the steady-state response.

If the equation for $(v_2)_s$, Eq. (9-1), is rewritten

$$(v_2)_s = \operatorname{Re}(\mathbf{V}_m e^{j\omega t}) \tag{9-2}$$

then the amplitude phasor $\mathbf{V}_m$ is related to the source amplitude phasor $\boldsymbol{\phi}_m$ by

$$\mathbf{V}_m = H(j\omega)\boldsymbol{\phi}_m \tag{9-3}$$

For each value of ω, $H(j\omega)$ is a complex number whose magnitude and angle depend on ω. Expressing this fact in mathematical form,

$$H(j\omega) = h(\omega)\underline{/\theta(\omega)} \tag{9-4}$$

The steady-state-response amplitude will be $h(\omega)$ times the source amplitude, and the phase of the response sinusoid will lead that of the source sinusoid by θ; that is,

$$(v_2)_s = h(\omega)\phi_m \cos[\omega t + \gamma + \theta(\omega)] \tag{9-5}$$

RELATIONSHIP TO LAPLACE TRANSFORM The result for steady-state response, Eq. (9-1) or (9-5), can be obtained from Laplace transform if we recall that

$$V_1(s) = H(s)\Phi(s)$$

[1] Before proceeding it is suggested that the reader study the discussion of phasors in Appendix C.

and observe that, for

$$\phi(t) = \phi_m \cos \omega t$$

$$\mathcal{L}[\phi(t)] = \phi_m \frac{s}{s^2 + \omega^2}$$

Hence

$$V_2(s) = H(s) \frac{s}{s^2 + \omega^2} \phi_m \qquad (9\text{-}6)$$

If $H(s)$ has no pole at $s = j\omega$, the component of the response due to the source consists of the inverse Laplace transform of the two terms in the partial-fractions expansion of $V_2(s)$ that arise from the source poles. Equation (9-6) is therefore written

$$V_2(s) = \frac{\mathbf{K}_1}{s - j\omega} + \frac{\mathbf{K}_1^*}{s + j\omega} + R(s)$$

where $R(s)$ represents the remaining terms in the expansion. The Laplace transform of the component due to the source is

$$[V_2(s)]_s = \frac{\mathbf{K}_1}{s - j\omega} + \frac{\mathbf{K}_1^*}{s + j\omega}$$

where

$$\mathbf{K}_1 = H(j\omega) \frac{j\omega}{2j\omega} \phi_m$$

Hence

$$(v_2)_s = 2 \text{ Re } [\tfrac{1}{2}H(j\omega)\phi_m e^{j\omega t}] \qquad (9\text{-}7)$$

The only difference between Eqs. (9-7) and (9-1) is the angle γ, which has been taken as zero in the derivation of Eq. (9-7). It is left as an exercise for the reader to derive the result of Eq. (9-1) by use of Laplace transform when $\gamma \neq 0$.

Complex network function When a particular value of ω is substituted in $H(j\omega)$, then $H(j\omega)$ is a complex number, and the term *complex* network function is used rather than transform or operational network function. Below we denote such complex value by boldface type; for example, complex impedance **Z**, admittance **Y**, etc.

Solution of a-c circuit problems Although the solution of networks in the sinusoidal steady state is completely described by Eq. (9-1), problems involving such analysis occur so frequently in engineering that detailed procedures and a special terminology have been developed and are presented in the remainder of this chapter.

9-2 Single elements in the sinusoidal steady state

Table 9-1 shows the voltage-current relations for the three basic circuit elements in the sinusoidal steady state. We observe that for the energy-storing elements L and C voltage and current differ in phase by 90°. Such phase relations will be seen to be of some importance; they are conveniently expressed by phasor diagrams as shown in Table 9-1, and in the next section such diagrams are developed further.

Table 9-1 Sinusoidal response of the basic passive elements

Element	Equations	Amplitude	Phase relation	Phasor form $i = \text{Re}\,(\mathbf{I}_m e^{j\omega t})$ $v = \text{Re}\,(\mathbf{V}_m e^{j\omega t})$	Phasor diagram	Time relationship
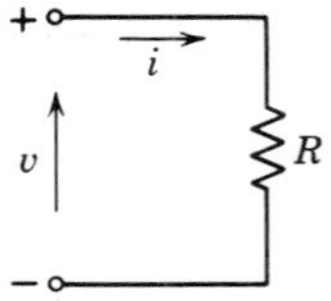 R	$i = I_m \cos(\omega t + \alpha)$ $v = V_m \cos(\omega t + \alpha)$	$\dfrac{V_m}{I_m} = R$	i and v are in phase	$\dfrac{\mathbf{V}_m}{\mathbf{I}_m} = R = \mathbf{Z}$ $\mathbf{Y} = G = 1/R$	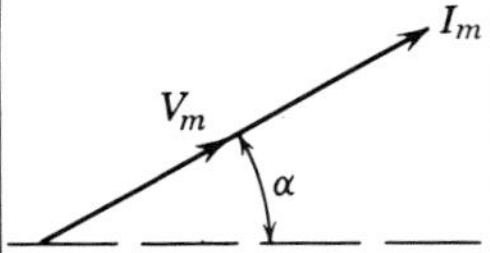 V_m, I_m, α	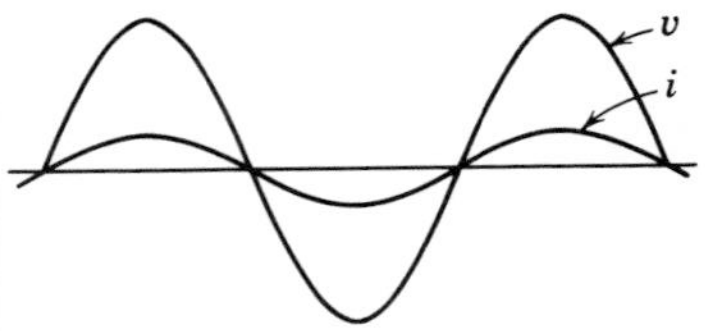 v, i
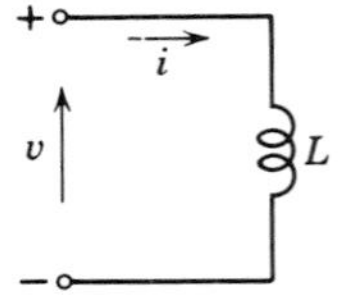 L	$i = I_m \cos(\omega t + \alpha)$ $v = V_m \cos\left(\omega t + \alpha + \dfrac{\pi}{2}\right)$	$\dfrac{V_m}{I_m} = \omega L = X_L$	i lags v by 90°	$\dfrac{\mathbf{V}_m}{\mathbf{I}_m} = j\omega L = \mathbf{Z} = jX_L$ $\mathbf{Y} = \dfrac{-j}{\omega L} = jB_L$ $\mathbf{V}_m = \omega L \mathbf{I}_m \angle \dfrac{\pi}{2}$	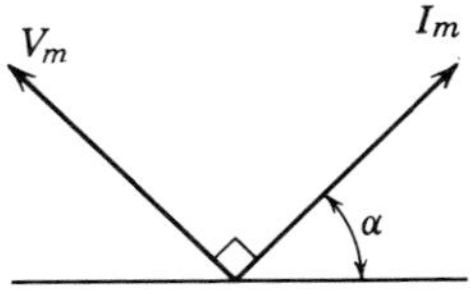 V_m, I_m, α	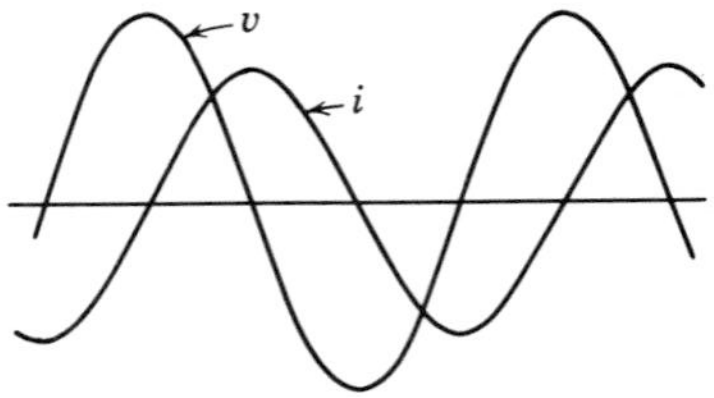 v, i
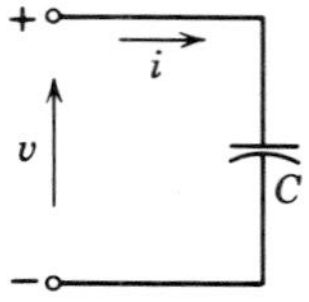 C	$i = I_m \cos(\omega t + \alpha)$ $v = V_m \cos\left(\omega t + \alpha - \dfrac{\pi}{2}\right)$	$\dfrac{V_m}{I_m} = \dfrac{1}{\omega C} = -X_C$	i leads v by 90°	$\dfrac{\mathbf{V}_m}{\mathbf{I}_m} = \dfrac{1}{j\omega C} = \mathbf{Z} = jX_C$ $\mathbf{Y} = j\omega C = jB_C$ $\mathbf{V}_m = \dfrac{1}{\omega C}\mathbf{I}_m \angle -\dfrac{\pi}{2}$	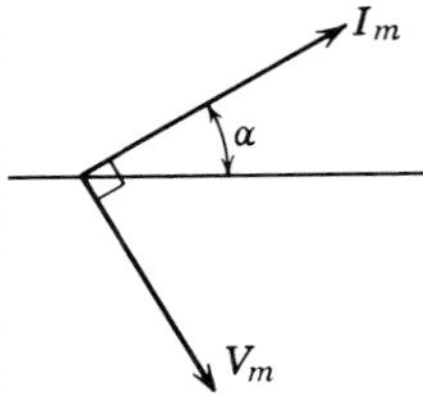 V_m, I_m, α	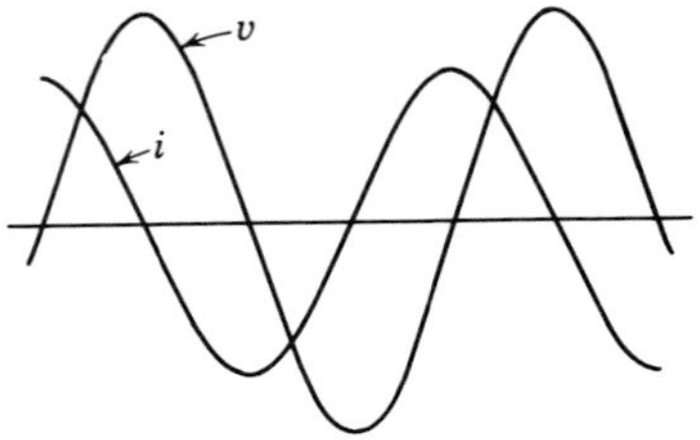 v, i

9-3 Reactance and susceptance

Since for any terminal pair *a-b* the (complex) impedance is always a complex number (as is any network function), it can be written in rectangular form,

$$\mathbf{Z}_{ab} = \text{Re } \mathbf{Z}_{ab} + j \text{ Im } \mathbf{Z}_{ab} \tag{9-8}$$

Similarly, for complex admittance,

$$\mathbf{Y}_{ab} = \text{Re } \mathbf{Y}_{ab} + j \text{ Im } \mathbf{Y}_{ab} \tag{9-9}$$

Since $\mathbf{Y} \equiv 1/\mathbf{Z}$,

$$\text{Re } \mathbf{Y}_{ab} + j \text{ Im } \mathbf{Y}_{ab} = \frac{1}{\text{Re } \mathbf{Z}_{ab} + j \text{ Im } \mathbf{Z}_{ab}} \tag{9-10}$$

Rationalizing the right side of Eq. (9-10),

$$\text{Re } \mathbf{Y}_{ab} + j \text{ Im } \mathbf{Y}_{ab} = \frac{\text{Re } \mathbf{Z}_{ab} - j \text{ Im } \mathbf{Z}_{ab}}{Z_{ab}^2} \tag{9-11}$$

$$\text{Re } \mathbf{Y}_{ab} = \text{Re } \frac{\mathbf{Z}_{ab}}{Z_{ab}^2} \tag{9-11a}$$

$$\text{Im } \mathbf{Y}_{ab} = - \text{ Im } \frac{\mathbf{Z}_{ab}}{Z_{ab}^2} \tag{9-11b}$$

The following terms are used in connection with complex immittance.

Reactance The imaginary part of a complex impedance is called the reactance of the impedance (or reactance for brevity) and is denoted by the symbol X.

Inductive reactance of an inductance The impedance of an inductance is $j\omega L$. Since $\text{Im } j\omega L = \omega L$, the product ωL is called the inductive reactance of L and is denoted by the symbol X_L. Thus, for an inductance, the complex impedance is given by $\mathbf{Z} = jX_L$.

Capacitive reactance of a capacitance The complex impedance of a capacitance is $1/j\omega C$. Since $\text{Im } (1/j\omega C) = -1/\omega C$, the quantity $-1/\omega C$ is called the capacitive reactance of C and is denoted by the symbol X_C. Thus, for a capacitance, the complex impedance is given by $\mathbf{Z} = jX_C$.

Susceptance The imaginary part of a complex admittance is called the susceptance (of the admittance) and is denoted by the symbol B.

Inductive susceptance of an inductance For an inductance,

$$\mathbf{Y} = \frac{1}{j\omega L}$$

The inductive susceptance of such an element is defined as $-1/\omega L$ and is denoted by the symbol B_L. Hence the admittance of an inductance can be written $\mathbf{Y} = jB_L$.

CAPACITIVE SUSCEPTANCE For a capacitance, $\mathbf{Y} = j\omega C$. The capacitive susceptance of such an element is ωC and is denoted by the symbol B_C. Hence the admittance of a capacitance is $\mathbf{Y} = jB_C$.

We observe that, for a single inductance or for a single capacitance, the susceptance is the negative reciprocal of the reactance. It must be emphasized that, in view of Eq. (9-11), this statement *does not* apply generally to any terminal pair, but applies only to the single elements. The definitions $X = \text{Im } \mathbf{Z}$ and $B = \text{Im } \mathbf{Y}$ *do* apply generally.

RESISTANCE AND CONDUCTANCE The resistance R of a complex impedance is defined as the real part of the impedance. The conductance G of an admittance is the real part of the admittance. Although, for the element R, $G = 1/R$, for a terminal pair a-b, $G_{ab} = \text{Re } \mathbf{Y}_{ab}$ is given by Eq. (9-11a) and is *not* the reciprocal of Re $\mathbf{Z}_{ab}$.

9-4 Power and energy relations in single elements

ENERGY-STORING ELEMENTS It was shown in the preceding sections that a sinusoidal voltage of the form

$$v_{ab}(t) = V_m \cos(\omega t + \alpha)$$

applied either to an inductance or to a capacitance produces a current

$$i_{ab}(t) = \frac{V_m}{X} \cos\left(\omega t + \alpha - \frac{\pi}{2}\right)$$

where $X = X_L = \omega L$ for an inductance, and $X = X_C = -1/\omega C$ for a capacitance. The power delivered *to* such an energy-storing element at every instant of time is given by the product $v_{ab}i_{ab}$. Thus the equation for the power at any instant of time is given by

$$p_{ab} = v_{ab}i_{ab} = \frac{V_m^2}{X} \cos(\omega t + \alpha) \cos\left(\omega t + \alpha - \frac{\pi}{2}\right) \qquad (9\text{-}12)$$

Since $\cos(\omega t + \alpha - \pi/2) = \sin(\omega t + \alpha)$ and since, for any angle x, $2 \sin x \cos x = \sin 2x$, Eq. (9-12) reduces to

$$p_{ab}(t) = \frac{V_m^2}{2X} \sin(2\omega t + 2\alpha) \qquad (9\text{-}13)$$

The result expressed in Eq. (9-13) is illustrated in Fig. 9-1a. In this figure it is assumed that X is a positive number; that is, $X = X_L$. We note, first, that the power oscillates at *twice* the frequency of the voltage or the current. Studying one complete cycle of the power oscillations is instructive: referring to Fig. 9-1a, in the interval $\omega t = A$ to $\omega t = B$, $p(t)$ is positive. This means that power is being furnished by the source to the inductance. The inductance is storing the energy (in its associated magnetic field), and at point B the energy which has been transferred to the inductance up to that time is equal to the area under the power curve. In the interval from B to C the

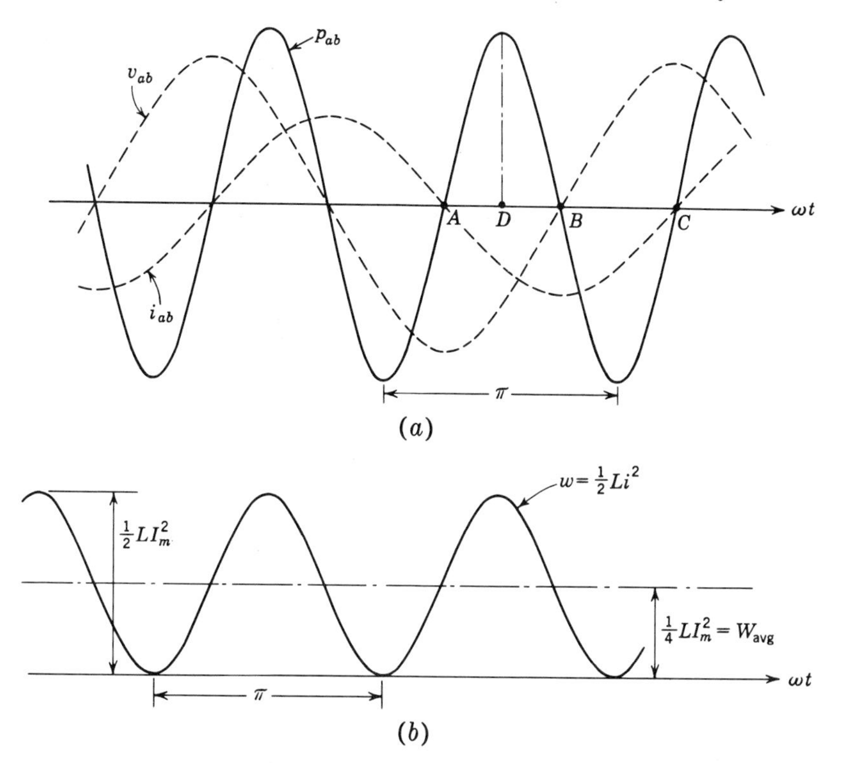

***Fig. 9-1** Power and energy as a function of time for an inductance in the sinusoidal steady state. (a) Voltage, current, and power as a function of time. (b) Stored energy as a function of time.*

power is negative. This means that, at point B, the power flow has reversed and the inductance is now furnishing energy to the source. At point C the total energy stored in the inductance is the same as at point A. We conclude, therefore, that the power supplied to an inductance averaged over one period of the voltage or current is zero, since at the end of every cycle no net energy has been transferred to the inductance. There exists, however, a periodic interchange of energy between the source and the inductance.

An interesting relationship is obtained by comparing the maximum stored energy with the maximum instantaneous power which flows. The energy stored in the inductance is given by

$$\begin{aligned} w_L &= \tfrac{1}{2}Li^2 \\ &= \tfrac{1}{4}LI_m{}^2[1 - \cos 2(\omega t + \alpha)] \end{aligned}$$

The waveform of w_L is shown in Fig. 9-1b. Since the energy is a maximum when the current is a maximum,

$$(w_L)_{\max} = \tfrac{1}{2}LI_m{}^2$$

We note from Fig. 9-1a that the value of the maximum power (point D) is

$$[p(t)]_{\max} = \tfrac{1}{2}V_mI_m = \tfrac{1}{2}\omega LI_m{}^2$$

so that the ratio of the maximum power to the maximum stored energy is the radian frequency ω (independent of voltage or current amplitude). Analogous relationships can be deduced for the capacitance.

RESISTANCE For a resistance R between terminals a-b a voltage v_{ab} will produce a current $i_{ab} = v_{ab}/R$. Hence, under sinusoidal conditions, if $v_{ab} = V_m \cos(\omega t + \alpha)$, the power delivered to the resistance is given by

$$p_{ab} = [V_m \cos(\omega t + \alpha)]\left[\frac{V_m}{R}\cos(\omega t + \alpha)\right]$$

or

$$p_{ab} = V_m I_m \cos^2(\omega t + \alpha) \qquad I_m = \frac{V_m}{R}$$

For any angle x, the identity $2\cos^2 x = 1 + \cos 2x$ holds. Hence

$$p_{ab}(t) = \frac{V_m I_m}{2}[1 + \cos(2\omega t + 2\alpha)]$$

Note that the power delivered to the resistance is either positive or zero at every instant of time. The equation for the power consists of a constant term and a sinusoidal term. Note that the frequency of the sinusoidal term is twice the frequency of the voltage or the current. The average value of the power over any complete cycle (or any number of complete cycles) is seen to be

$$P_{\text{avg}} = \frac{V_m I_m}{2} = P$$

[It is common usage, in discussing circuits in the sinusoidal steady state, to call the average power over one period simply the "power" and use the symbol P (capital, no special subscript).]

9-5 *Effective values*[1]

In sinusoidal steady-state calculations it is customary to specify voltages and currents in terms of their "effective," or "root-mean-square," values. For a periodic function $f(t)$ with period T, $[f(t) = f(t + T)]$, the rms value is defined through the relationship

$$F_{\text{rms}} = \sqrt{\frac{1}{T}\int_0^T f^2(t)\,dt} \tag{9-14}$$

Applying this definition to a sinusoidal function

$$f(t) = F_m \cos(\omega t + \alpha)$$

the rms value is $F_{\text{rms}} = F_m/\sqrt{2} = 0.707F_m$. It is customary *not* to use a subscript to denote effective (rms) value for voltages and currents. Thus the current $i(t) = I_m \cos(\omega t + \alpha)$ has the effective value $I = I_m/\sqrt{2}$. The utility of defining such an rms value is clear if we recall the formula for the power which is dissipated in a resistance R due to a current $i(t)$,

$$p(t) = [i(t)]^2 R$$

[1] A detailed discussion of rms values is given in Appendix A.

The average power which is dissipated in one period is

$$P = P_{\text{avg}} = \frac{1}{T} \int_0^T [i(t)]^2 R \, dt$$

If, instead of the current $i(t)$, a constant current I were to flow through the same resistance R, the power dissipated by the resistance would be

$$P = I^2 R \tag{9-15}$$

We note, therefore, that the *effective* value of a periodic current is that value which, when used in Eq. (9-15), will give the correct answer for the average power,

$$I_{eff} = \sqrt{\frac{P_{\text{avg}}}{R}} \tag{9-15a}$$

Using this definition for effective current in Eq. (9-14),

$$I_{eff} = \sqrt{\frac{1}{T} \int_0^T [i(t)]^2 \, dt}$$

so that the *effective* value of the current is the same as the rms value. This value of current is also referred to as the "heating" value. The reason for this term should be clear. A periodic current whose effective value is I will, when flowing through a resistance, convert electric energy into heat at an average rate equal to the rate at which constant current I converts electric energy into heat in the same resistance.

The relationship between amplitude and effective value for a sinusoid ($F = F_m/\sqrt{2}$), which is stated above, can be deduced without the formal integration indicated by Eq. (9-14). We start with the equation for power delivered to a resistance R_{ab}.

$$p_{ab}(t) = \frac{V_m I_m}{2} [1 + \cos (2\omega t + 2\alpha)]$$

Since the average value of a cosine wave is zero,

$$P_{\text{avg}} = \frac{V_m I_m}{2} = \frac{I_m{}^2}{2} R$$

Using the definition of effective value (9-15*a*), the effective value of a *sinusoidal* current is

$$I = \frac{1}{\sqrt{2}} I_m = 0.707 I_m \tag{9-15b}$$

This result may, of course, be verified by carrying out the integration, which is indicated above. It must be emphasized that the result expressed in Eq. (9-15*b*) applies *only* to *sinusoidal* currents (or voltages). For any other waveshape the definition [Eq. (9-14)] of rms values must be applied.[1]

[1] See Appendix A.

In this chapter we represent sinusoidal voltages and currents by complex numbers, i.e., phasors. Thus, if

$$v(t) = V_m \cos(\omega t + \alpha) \qquad \text{and} \qquad i(t) = I_m \cos(\omega t + \psi)$$

we represent $v(t)$ by the phasor $\mathbf{V}_m = V_m/\underline{\alpha}$ and $i(t)$ by the phasor $\mathbf{I}_m = I_m/\underline{\psi}$. The voltage and current phasors at the terminals of a terminal-pair network are then related by the complex immittance,

$$\mathbf{V}_m = \mathbf{Z}\mathbf{I}_m \qquad \text{or} \qquad \mathbf{I}_m = \mathbf{Y}\mathbf{V}_m \tag{9-16}$$

Since $V_m = \sqrt{2}\,V$ and $I_m = \sqrt{2}\,I$, where V and I are effective values, we can define the complex numbers $\mathbf{V}$ and $\mathbf{I}$ by the relations

$$\mathbf{V} = \frac{1}{\sqrt{2}}\mathbf{V}_m \qquad \text{and} \qquad \mathbf{I} = \frac{1}{\sqrt{2}}\mathbf{I}_m \tag{9-17}$$

Introducing (9-17) into (9-16), we note that $\mathbf{V}$ and $\mathbf{I}$ are related through the same complex immittances which relate $\mathbf{V}_m$ and $\mathbf{I}_m$,

$$\mathbf{V} = \mathbf{Z}\mathbf{I} \qquad \text{or} \qquad \mathbf{I} = \mathbf{Y}\mathbf{V} \tag{9-18}$$

We now observe that the numbers $\mathbf{V}$ and $\mathbf{I}$ are the phasors representing $(1/\sqrt{2})v(t)$ and $(1/\sqrt{2})i(t)$, respectively. Yet it is customary to say that $\mathbf{V}$ and $\mathbf{I}$ represent $v(t)$ and $i(t)$. To avoid confusion, the following terminology is used: The function

$$v(t) = V_m \cos(\omega t + \alpha) = \sqrt{2}\,V \cos(\omega t + \alpha)$$

may be represented by the *amplitude phasor* $\mathbf{V}_m = V_m/\underline{\alpha}$ or by the *effective phasor* $\mathbf{V} = V/\underline{\alpha}$. *In this book, from this point on, the term phasor is used synonymously with the term effective phasor.*

9-6 A series circuit

Before studying series and parallel combinations of elements, the series R-L circuit will be analyzed, because much can be learned about the essentials of a-c circuit analysis from a simple two-element combination. Consider, therefore, the circuit shown in Fig. 9-2*a*. We wish to find the component of $i(t)$ due to a sinusoidal source $v_{ab}(t)$ as indicated in Fig. 9-2*a*.

Before proceeding to the analytical treatment of this circuit, the following observations are appropriate (Fig. 9-2*a* and *b*):

1. When the inductance L is zero, the voltage v_{ab} and the current i are in phase and the ratio of their amplitudes (or of their effective values) is given by $V_m/I_m = V/I = R$.
2. When the resistance R is zero (and the inductance is not zero), the voltage v_{ab} leads the current i by 90° and the ratio of their effective values is given by $V/I = V_m/I_m = X_L$.

Because of these observations we expect that, when neither R nor L is zero, the voltage v_{ab} will lead the current i by an angle between 0 and 90° and that the ratio of their amplitudes will depend on both R and X_L.

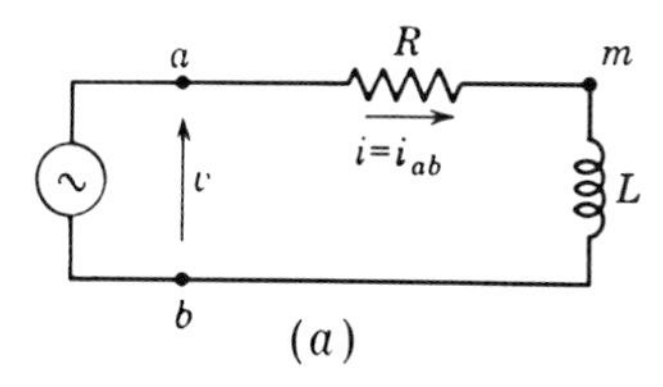

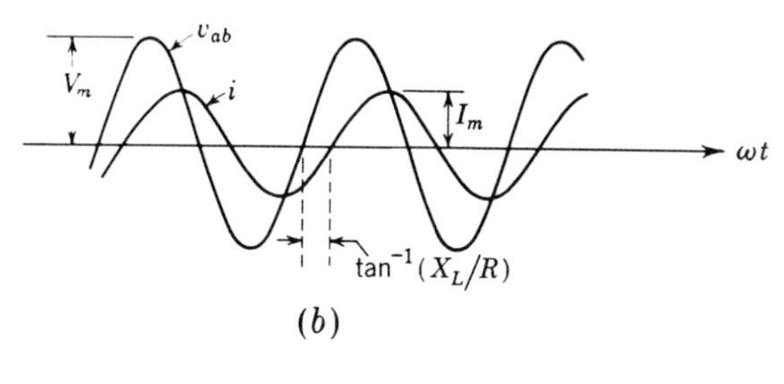

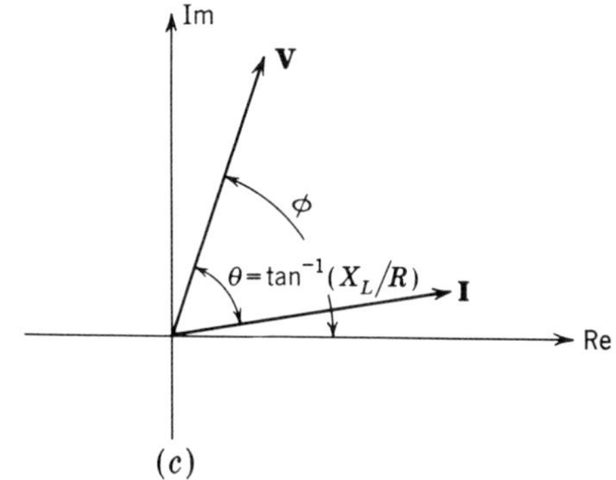

***Fig. 9-2** Series R-L circuit in the sinusoidal steady state. (a) Circuit diagram. (b) Time-domain relationship between v and i. (c) Phasor diagram.*

To deduce the analytical relationship between v_{ab} and $i(t)$ (Fig. 9-2a), we need recall only that the transform impedance of the series R-L circuit is $R + sL$, so that $Z(j\omega) = R + j\omega L = Z\underline{/\theta}$. Hence, if

$$v_{ab}(t) = \sqrt{2}\, V \cos(\omega t + \phi) = \operatorname{Re}(\sqrt{2}\,\mathbf{V}e^{j\omega t}) \qquad \mathbf{V} = Ve^{j\phi}$$

where $\mathbf{V}$ is the effective phasor representing v_{ab}, then

$$i(t) = \operatorname{Re}\left(\sqrt{2}\,\frac{\mathbf{V}}{R + j\omega L}\,e^{j\omega t}\right) \tag{9-19}$$

and the phasor representing $i(t)$ is $\mathbf{V}/(R + j\omega L) = \mathbf{I}$.

The complex impedance is, in polar form,

$$\mathbf{Z} = R + jX_L = Z\underline{/\theta}$$

where $$Z = \sqrt{R^2 + X_L{}^2} \qquad \theta = \tan^{-1}\frac{X_L}{R}$$

Hence the ratio of the effective values (or amplitudes) $V/I \ (= V_m/I_m) = Z$, and the voltage v_{ab} leads the current $i(t)$ by θ,

$$I\underline{/\phi - \theta} = \frac{V\underline{/\phi}}{Z\underline{/\theta}} \qquad i(t) = \frac{\sqrt{2}\,V}{Z}\cos(\omega t + \phi - \theta) \tag{9-20}$$

as illustrated in Fig. 9-2*c* and *b*. The same result can be obtained from a slightly different point of view. As we know, a sinusoidal voltage produces a sinusoidal current. If this current in the *R-L* circuit of Fig. 9-2*a* is represented by **I**, that is,

$$i(t) = \mathrm{Re}\,(\sqrt{2}\,\mathbf{I}e^{j\omega t})$$

then the phasor which represents the voltage v_{am} is $\mathbf{I}R$ and the phasor which represents v_{mb} is $j\mathbf{I}X_L$; that is,

$$v_{am} = \mathrm{Re}\,(\sqrt{2}\,\mathbf{I}Re^{j\omega t}) \qquad v_{mb} = \mathrm{Re}\,(\sqrt{2}\,\mathbf{I}jX_Le^{j\omega t})$$

The phasor which represents the sum of two sinusoids can be obtained by adding the phasors which represent the individual time functions,

$$\mathbf{V} = \mathbf{V}_{am} + \mathbf{V}_{mb}$$

or

$$\mathbf{V} = \mathbf{I}R + j\mathbf{I}X_L = (R + jX_L)\mathbf{I} \tag{9-21}$$

This phasor addition is illustrated in Fig. 9-3, where $\phi = \psi + \theta$.

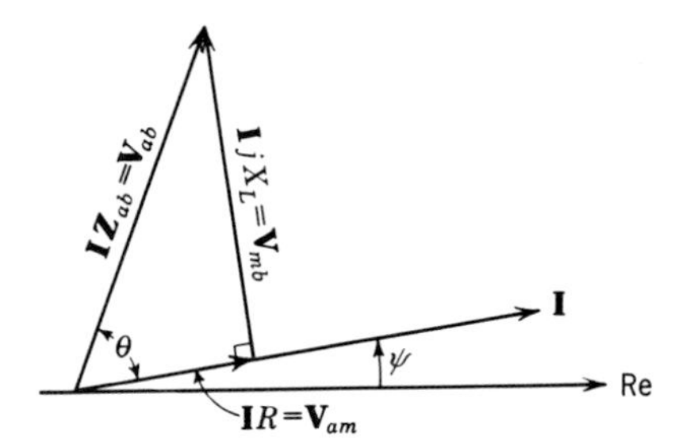

***Fig. 9-3** Phasor addition of voltages in the series R-L circuit of Fig. 9-2a.*

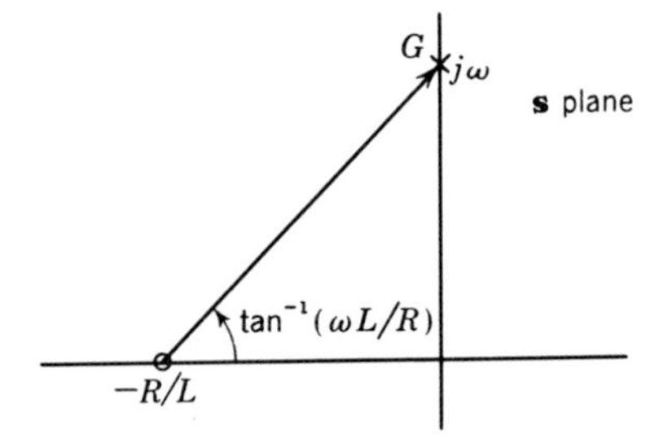

***Fig. 9-4** Relating the complex impedance of the series R-L circuit to the pole-zero diagram of the transform impedance.*

A diagram which shows the addition of voltage phasors in a circuit is called a *voltage phasor diagram*, or, because the addition of phasors resembles the addition of vectors, a *voltage vector diagram*.

As a final remark concerning the series *R-L* circuit, we observe that the transform impedance $Z(s)$ is

$$Z(s) = R + sL = L\left(s + \frac{R}{L}\right)$$

Hence the transform impedance has a scale factor L and a zero at $-R/L$ in the s plane. In Fig. 9-4 this zero is shown. For the sinusoidal case, $s + R/L$ is represented by the line joining $(-R/L)$ to G. Note that the angle which this line makes with the horizontal is the angle of $\mathbf{Z}$; that is,

$\tan^{-1}[\omega/(R/L)] = \tan^{-1}(\omega L/R)$. Thus, if the phase angle between $\mathbf{V}$ and $\mathbf{I}$ can be measured at a known frequency, the time constant L/R of the circuit, which characterizes the free response, can be inferred. It is this type of relationship between sinusoidal steady-state response and free response which extends the usefulness of a-c circuit analysis beyond the solution of specific steady-state problems.

9-7 *Series and parallel circuits*

If elements are connected in series to form a terminal-pair network, the complex impedance of the series combination is the sum of the individual impedances. This is evident, not only because complex impedance is a special case of transform impedance, but also because the addition of sinusoidal time functions can be carried out with phasors. Thus, if we consider the circuit of Fig. 9-5, the terminal pair marked $\mathbf{Z}_1$ implies that $\mathbf{V}_{am} = \mathbf{I}\mathbf{Z}_1$;

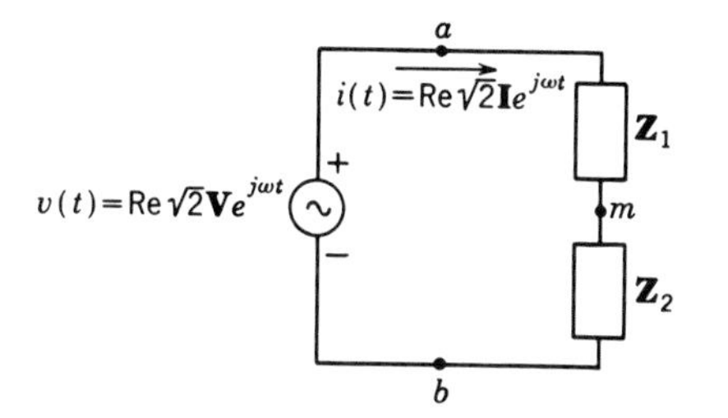

Fig. 9-5 Series combination of complex immittances.

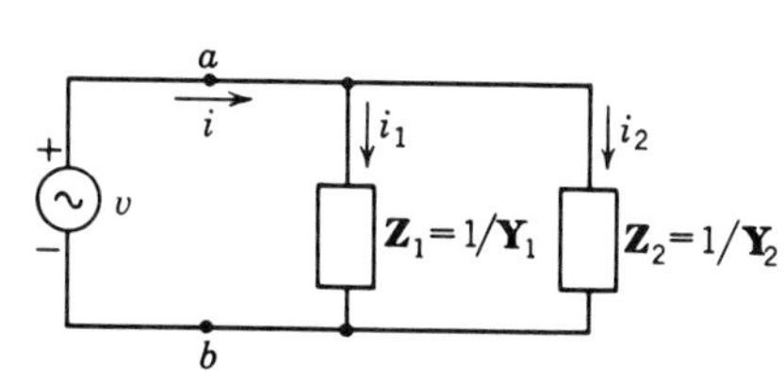

Fig. 9-6 Parallel connection of complex immittances.

similarly, $\mathbf{V}_{mb} = \mathbf{I}\mathbf{Z}_2$; hence, from Kirchhoff's voltage law, $\mathbf{V}_{ab} = \mathbf{I}(\mathbf{Z}_1 + \mathbf{Z}_2)$. Thus the phasor ratio $\mathbf{V}_{ab}/\mathbf{I}$, which defines the complex driving-point impedance $\mathbf{Z}_{ab}$, is given as the sum of $\mathbf{Z}_1$ and $\mathbf{Z}_2$; that is,

$$\mathbf{Z}_{ab} = \mathbf{Z}_1 + \mathbf{Z}_2$$

or in terms of admittances, $\mathbf{Z}_1 = 1/\mathbf{Y}_1$, $\mathbf{Z}_2 = 1/\mathbf{Y}_2$, $\mathbf{Z}_{ab} = 1/\mathbf{Y}_{ab}$, and

$$\mathbf{Y}_{ab} = \frac{1}{1/\mathbf{Y}_1 + 1/\mathbf{Y}_2} = \frac{\mathbf{Y}_1\mathbf{Y}_2}{\mathbf{Y}_1 + \mathbf{Y}_2}$$

Voltage-division formula Applying the operational or transform voltage-division formula to the a-c case, in Fig. 9-5,

$$\frac{\mathbf{V}_{mb}}{\mathbf{V}} = \frac{\mathbf{Z}_2}{\mathbf{Z}_1 + \mathbf{Z}_2} \qquad (9\text{-}22)$$

Thus the voltage phasors in a series circuit "divide" across the complex impedances in proportion to the values of these impedances.

Parallel elements Consider now the parallel branches shown in Fig. 9-6, where each terminal pair is specified through its complex admittance.

$$\mathbf{I}_1 = \mathbf{Y}_1\mathbf{V}_{ab} \qquad \mathbf{I}_2 = \mathbf{Y}_2\mathbf{V}_{ab}$$

Application of Kirchhoff's current law gives

$$\mathbf{I} = \mathbf{Y}_1\mathbf{V}_{ab} + \mathbf{Y}_2\mathbf{V}_{ab} = \mathbf{V}_{ab}(\mathbf{Y}_1 + \mathbf{Y}_2)$$

so that the driving-point admittance of parallel connected elements is the sum of the admittances. Thus

$$\mathbf{Y}_{ab} = \mathbf{Y}_1 + \mathbf{Y}_2$$

or in terms of impedances,

$$\mathbf{Z}_{ab} = \frac{\mathbf{Z}_1\mathbf{Z}_2}{\mathbf{Z}_1 + \mathbf{Z}_2}$$

We note further that

$$\frac{\mathbf{I}_1}{\mathbf{I}} = \frac{\mathbf{V}_{ab}\mathbf{Y}_1}{\mathbf{V}_{ab}(\mathbf{Y}_1 + \mathbf{Y}_2)} = \frac{\mathbf{Y}_1}{\mathbf{Y}_1 + \mathbf{Y}_2} \tag{9-23}$$

so that the complex *current-division formula* states: In a parallel circuit the current phasors divide through the admittances, in proportion to the values of the complex admittances. Comparison of Eq. (9-23) and Eq. (9-22) shows that the voltage-division and current-division formulas are duals of each other.

TERMINOLOGY In sinusoidal steady-state calculation the use of phasors and complex impedance is so common that the phrase "the phasor representing the current $i(t)$" is often omitted, and the phasor ($\mathbf{I}$) is referred to as "the current" in the circuit. Similarly, one speaks of a (complex) voltage $\mathbf{V}$, meaning the phasor representing the voltage

$$v(t) = \text{Re}\,(\mathbf{V}\,\sqrt{2}\,e^{j\omega t})$$

9-8 Examples, use of phasor diagram

Example 9-1 Series circuit In the series circuit shown in Fig. 9-7*a* calculate $i(t)$ in the steady state, and draw a voltage phasor diagram.

Solution Let the voltage $v_{ab}(t) = v(t)$ be represented by the phasor $\mathbf{V} = 100\underline{/0°}$; then $i(t)$ will be represented by the phasor $\mathbf{I} = I\underline{/-\theta}$, where $+\theta$ is the angle of the impedance $\mathbf{Z}_{ab}$ and $\mathbf{I} = \mathbf{V}/\mathbf{Z}_{ab}$. To calculate the complex impedance, we need the reactances of the energy-storing elements at the radian frequency of the source ($\omega = 1{,}000$).

$$X_{df} = -\frac{1}{1{,}000 \times 100 \times 10^{-6}} = -10 \text{ ohms}$$
$$X_{gh} = 1{,}000 \times 0.020 = +20 \text{ ohms}$$
$$X_{kb} = -\frac{1}{1{,}000 \times 40 \times 10^{-6}} = -25 \text{ ohms}$$

Between terminals *a-b* the complex impedance is now obtained as the sum of the impedances as follows:

$$\mathbf{Z}_{ab} = 6 - j10 + 10 + j20 + 4 - j25 = (20 - j15) \text{ ohms}$$

or $\quad \mathbf{Z}_{ab} = 25\underline{/-36.9°}$ ohms

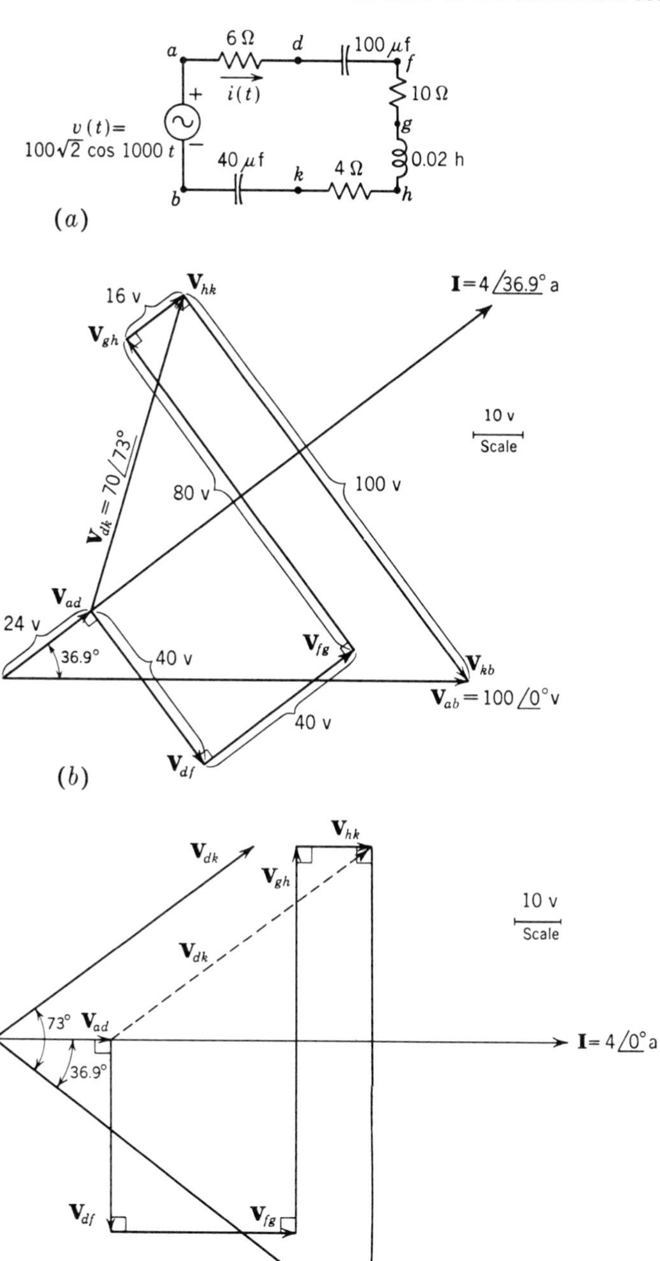

Fig. 9-7 Illustration for Example 9-1. ***(a) Circuit diagram. (b) Phasor diagram with V_{ab} as reference. (c) Phasor diagram with I as reference.***

The phasor **I** is given as the ratio of **V**/**Z**,

$$\mathbf{I} = \frac{100\underline{/0°}}{25\underline{/-36.9°}} = 4\underline{/36.9°}$$

and the solution for $i(t)$ is

$$i(t) = 4\sqrt{2} \cos (1{,}000t + 36.9°) \text{ amp}$$

The phasor diagram shown in Fig. 9-7*b* shows the addition of the voltage phasors across the individual elements. The "polygon" method of vector addition is used. Thus the tail of the next phasor in the sum is placed at the head of the preceding one. In Fig. 9-7*b* the voltage phasor $\mathbf{V}_{ab}$ is the reference phasor as stated in the problem. In Fig. 9-7*c* the phasor diagram is drawn with the current phasor as reference. This choice of reference is sometimes preferred in series connections of elements because the various voltage phasors for the elements are then drawn either horizontally or vertically; moreover, since the current in all parts of a series circuit is the same, it is natural to use the current phasor as the reference.

To illustrate the use of the phasor diagram, suppose that it is desired to find the voltage $\mathbf{V}_{dk}$. Since $\mathbf{V}_{dk} = \mathbf{V}_{df} + \mathbf{V}_{fg} + \mathbf{V}_{gh} + \mathbf{V}_{hk}$, the phasor $\mathbf{V}_{dk}$ can be located by drawing a phasor from the tail of $\mathbf{V}_{df}$ to the head of $\mathbf{V}_{hk}$. Performing this construction on Fig. 9-7*b*, we obtain the length of $\mathbf{V}_{dk}$ as 70 and the angle of $\mathbf{V}_{dk}$ with respect to the horizontal ($\mathbf{V}_{ab}$) as 73° leading. Hence

$$\mathbf{V}_{dk} = 70\underline{/73°} \qquad \text{or} \qquad v_{dk}(t) = 70\sqrt{2} \cos (1{,}000t + 73°)$$

Upon performing the same construction on Fig. 9-7*c*, the angle with the horizontal is 36°. But in this diagram **I** is reference, and the angle with $\mathbf{V}_{ab}$ is found by placing the phasor $\mathbf{V}_{dk}$ and $\mathbf{V}_{ab}$ tail to tail. Note that the resulting angle is 73° as before.

The same result can be obtained analytically. Since

$$\mathbf{V}_{dk} = \mathbf{I}\mathbf{Z}_{dk}$$

and

$$\mathbf{Z}_{dk} = -j10 + 10 + j20 + 4 = 14 + j10 = 17.2\underline{/35.6°}$$

Hence

$$\mathbf{V}_{dk} = (4\underline{/36.9°})(17.2\underline{/35.6°}) = 68.8\underline{/72.5°}$$

so that

$$v_{dk}(t) = 68.8\sqrt{2} \cos (1{,}000t + 72.5°)$$

This result is sufficiently close to the graphical answer to constitute a correct check.

Although in this particular example the phasor diagram is used merely as a check, we shall see in Example 9-2 that the phasor diagram can be a valuable tool in the solution of problems.

Example 9-2 Series circuit In the circuit shown in Fig. 9-8*a* the following *magnitudes* are known:

$$V_{ab} = 100 \qquad V_{a'b} = 200 \qquad I = 20 \text{ amp}$$

In addition, it is known that $X_L = 2R$. Solve for $\mathbf{Z}_{a'a}$ and $V_{a'a}$.

Solution We shall first show how the solution of this problem is obtained without reference to the phasor diagram. Since $\mathbf{Z}_{ab}$ is a resistance R_1, $\mathbf{V}_{ab}$, and **I** are in phase. If we let

$$\mathbf{I} = 20\underline{/0°}$$

then

$$\mathbf{V}_{ab} = 100\underline{/0°}$$

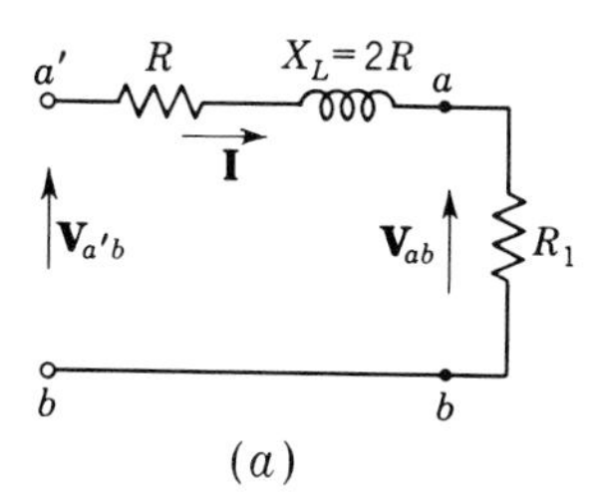

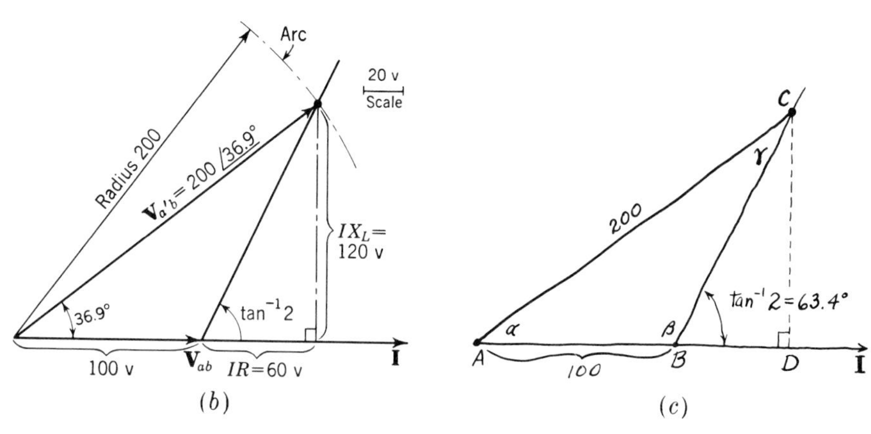

Fig. 9-8 Illustration for Example 9-2. (a) Circuit diagram. (b) Phasor diagram constructed to scale. (c) Free-hand sketch of phasor diagram.

Hence $R_1 = \frac{100}{20} = 5.0$ ohms. At terminals $a'b$ the magnitude of the impedance is $\frac{200}{20} = 10$ ohms. Hence

$$|\mathbf{Z}_{a'b}| = 10 = |5.0 + R + j2R|$$

or

$$[(5.0 + R)^2 + 4R^2]^{\frac{1}{2}} = 10$$

and squaring,

$$4R^2 + R^2 + 10R + 25 = 100$$

so that

$$5R^2 + 10R - 75 = 0$$

The solutions for R from this equation are $R = 3.0$ and $R = -5.0$. The negative value of R is discounted as extraneous. This value was introduced into the solution when $|\mathbf{Z}_{a'b}|$ was squared. Hence $R = 3.0$ ohms, $X_L = 6.0$ ohms, and $\mathbf{Z}_{a'a} = 3.0 + j6.0$ ohms. As a check we calculate $\mathbf{Z}_{a'b}$. $\mathbf{Z}_{a'b} = 3.0 + j6.0 + 5.0 = 8.0 + j6.0 = 10\angle +36.9°$, so that the magnitude is 10 as given. The voltage $V_{a'a}$ is given by $V_{a'a} = IZ_{a'a}$; hence

$$V_{a'a} = 20(|3 + j6|) = (20 \times 6.71) = 134.2 \text{ volts}$$

The solution of the quadratic equation can be avoided if a phasor diagram showing the voltage addition is drawn. In Fig. 9-8*b*, $\mathbf{V}_{ab}$ is drawn in phase with **I**. Since $\mathbf{V}_{a'a} + \mathbf{V}_{ab} = \mathbf{V}_{a'b}$, we need to add $\mathbf{V}_{a'a}$ to $\mathbf{V}_{ab}$. The magnitude of $\mathbf{V}_{a'a}$ is unknown, but we know that $\mathbf{V}_{a'a}$ will lead **I** by the angle of $\mathbf{Z}_{a'a}$. This angle is given and is equal to $\tan^{-1} 2$ (= 63.4°). We therefore draw a line from the end of the phasor $\mathbf{V}_{ab}$, making the angle 63.4° with the horizontal (**I**) and of indefinite length. Now the sum $\mathbf{V}_{a'a} + \mathbf{V}_{ab}$ must have a magnitude of 200; we swing an arc of radius 200 from the tail of the $\mathbf{V}_{ab}$ phasor. The intersection of this arc with the line at $\tan^{-1} 2$ must give the end point of both $\mathbf{V}_{a'b}$ and $\mathbf{V}_{a'a}$. We can now measure $\mathbf{V}_{a'a}$ and obtain its angle with respect to $\mathbf{I}$ to obtain $\mathbf{Z}_{a'a}$. Alternatively, IR and IX_L can be measured to obtain $\mathbf{Z}_{a'a}$.

We have explained the use of the phasor diagram on the assumption that the diagram has been drawn to scale. If this is inconvenient, one can draw the phasor diagram approximately to scale (i.e., freehand) and calculate the length and angles of the phasors by trigonometry. Such a sketch is shown in Fig. 9-8*c*. To avoid quadratic equations, the law of sines is used repeatedly.

Since β is the supplement of 63.4°, $\beta = 180° - 63.4° = 116.6°$. Then γ can be determined as

$$\frac{\sin \gamma}{\sin 116.6°} = \frac{100}{200}$$

and

$$\gamma = 26.5°$$

Hence

$$\alpha = 180 - 116.6 - 26.5 = 36.9°$$

Now BC can be calculated by using the law of sines again ($BC/AC = \sin \alpha / \sin \beta$), or CD is obtained: $200 \sin \alpha = 200 \times 0.6 = 120$. The solution is now virtually complete ($CD = 120 = IX_L$, $X_L = 6$, $R = 3$).

Example 9-3 A parallel circuit For the circuit shown in Fig. 9-9, calculate $i(t)$. The source voltage $v(t)$ is given as

$$v(t) = 156 \cos 377t$$

The element values are $R = 100$ ohms, $L = 0.20$ henry, $C = 20\ \mu$f.

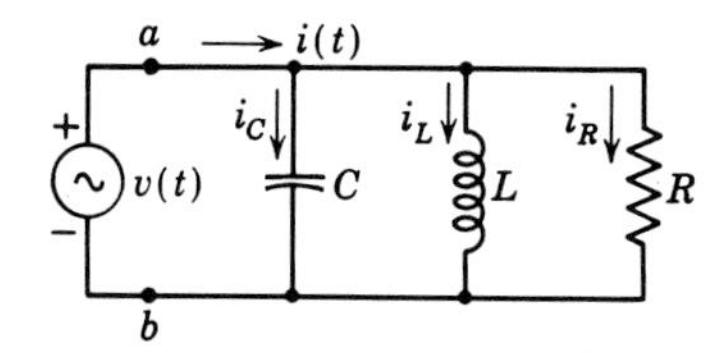

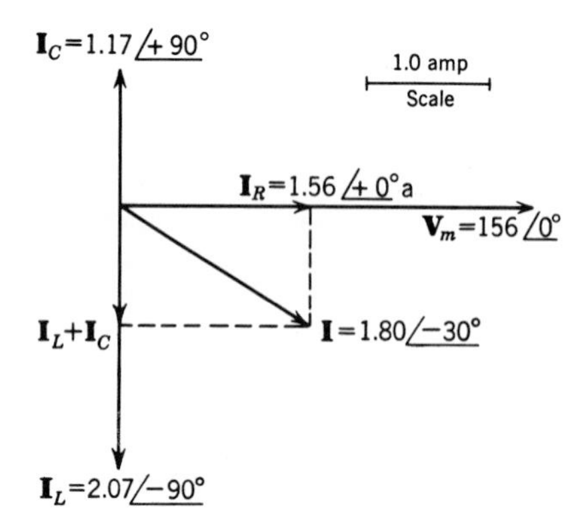

***Fig. 9-9** Parallel R-L-C circuit analyzed in Example 9-3 and the corresponding current phasor diagram.*

Solution Since $v(t) = 156 \cos 377t$, $V_m = 156$; eventually we shall need I_m, so that amplitude phasors will be used. Let

$$\mathbf{V}_m = 156\underline{/0}$$

To calculate $i(t)$, the complex admittance $\mathbf{Y}_{ab}$ is needed. We have

$$B_L = -\frac{1}{\omega L} \qquad B_L = -\frac{1}{377 \times 0.20} = -13.26 \times 10^{-3} \text{ mho}$$

and $$B_C = \omega C \qquad B_C = 377 \times 20 \times 10^{-6} = 7.54 \times 10^{-3} \text{ mho}$$

and $$G = \frac{1}{R} \qquad G = 0.01 = 10 \times 10^{-3} \text{ mho}$$

The complex admittance is

$$\mathbf{Y}_{ab} = (10 - j13.26 + j7.54) \times 10^{-3} \text{ mho}$$

or $$\mathbf{Y}_{ab} = 11.5 \times 10^{-3}\underline{/-30°} \text{ mho}$$

Hence

$$\mathbf{I}_m = 156\underline{/0°} \times 11.5 \times 10^{-3}\underline{/-30°} = 1.8\underline{/-30°} \text{ amp}$$

or $$i(t) = 1.8 \cos (377t - 30°) \text{ amp}$$

9-9 The series and the parallel R-L-C circuit The phase relations between voltage and current in the series and in the parallel R-L-C circuit are particularly interesting because of their various possibilities. The series R-L-C circuit shown in Fig. 9-10 has

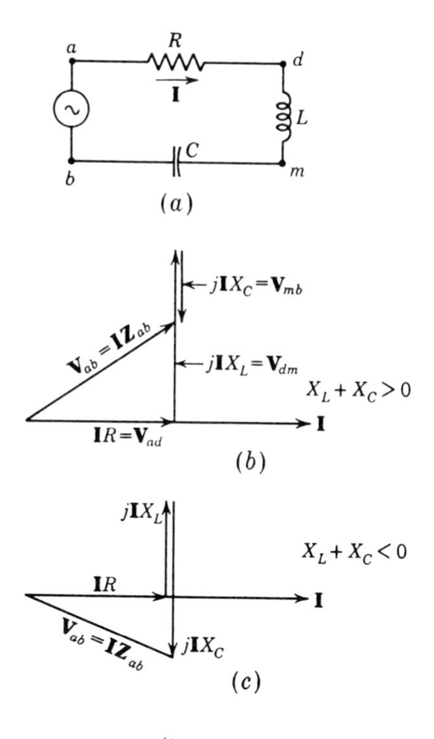

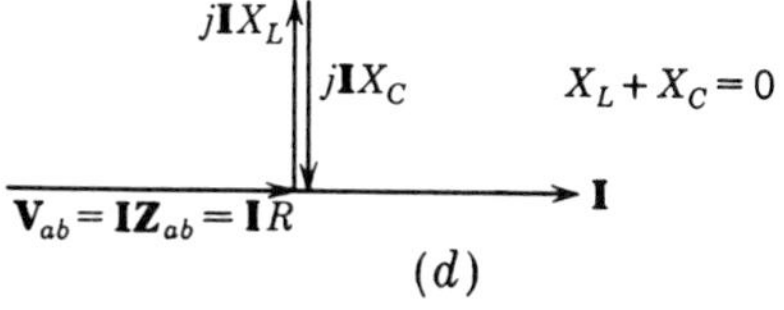

***Fig. 9-10** Phasor diagrams for a series R-L-C circuit. (a) Circuit diagram. (b) Voltage phasor diagram for $X > 0$. (c) Voltage phasor diagram for $X < 0$. (d) Voltage phasor diagram for $X = 0$.*

the complex impedance

$$\mathbf{Z}_{ab} = R + jX \qquad X = X_L + X_C$$

so that if

$$v_{ab}(t) = \sqrt{2}\, V \cos(\omega t + \phi)$$

then

$$i_{ab}(t) = \sqrt{2}\, \frac{V}{Z} \cos(\omega t + \phi - \theta)$$

where

$$\theta = \tan^{-1} \frac{X}{R}$$

We may now observe the following possible phase relations between v_{ab} and i:

(1) $\quad X_L + X_C > 0 \qquad \omega L > \dfrac{1}{\omega C} \qquad$ current i lags voltage v_{ab}

(2) $\quad X_L + X_C < 0 \qquad \omega L < \dfrac{1}{\omega C} \qquad$ current i leads voltage v_{ab}

(3) $\quad X_L + X_C = 0 \qquad \omega L = \dfrac{1}{\omega C} \qquad$ current i in phase with voltage v_{ab}

These possibilities are illustrated with the aid of voltage phasor diagrams in Fig. 9-10*b* to *d*. Condition (3), $X = 0$, is of particular interest and is discussed in detail in Chap. 10.

For the parallel *R-L-C* circuit shown in Fig. 9-11*a*, the admittance is

$$\mathbf{Y}_{ab} = G + jB \qquad B = B_L + B_C \qquad G = \frac{1}{R}$$

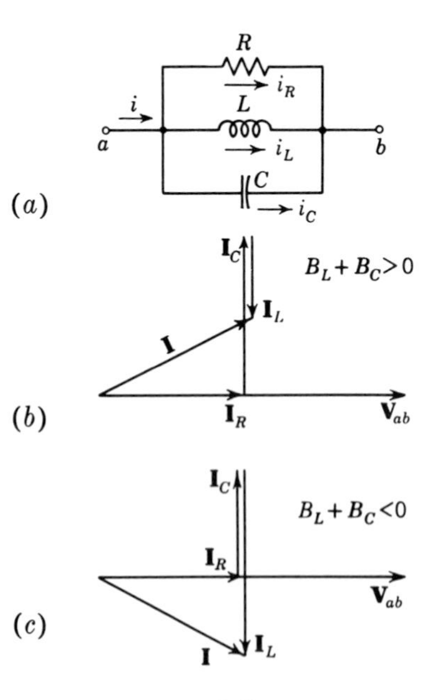

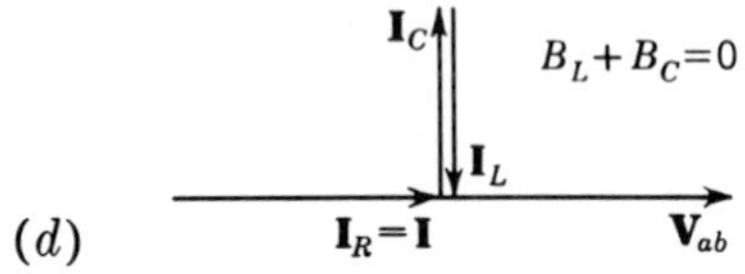

***Fig. 9-11** Phasor diagrams for a parallel R-L-C circuit. (a) Circuit diagram. (b) Current phasor diagram for $B > 0$. (c) Current phasor diagram for $B < 0$. (d) Current phasor diagram for $B = 0$.*

Hence, if the current $i(t)$ is given by

$$i(t) = \sqrt{2}\, I \cos(\omega t + \psi)$$

then

$$v_{ab}(t) = \sqrt{2}\,\frac{I}{Y} \cos\left(\omega t + \psi - \tan^{-1}\frac{B}{G}\right)$$

We again observe three possibilities:

(1) $B_L + B_C > 0 \qquad \dfrac{1}{\omega L} < \omega C \qquad v_{ab}$ lags i

(2) $B_L + B_C < 0 \qquad \dfrac{1}{\omega L} > \omega C \qquad v_{ab}$ leads i

(3) $B_L + B_C = 0 \qquad \dfrac{1}{\omega L} = \omega C \qquad v_{ab}$ in phase with i

These three possibilities are illustrated with the aid of current phasor diagrams in Fig. 9-11*b* to *d*.

9-10 Series-parallel circuits It should be clear now how an equivalent complex immittance can be found for any series-parallel connection of circuit elements by repeated replacement of series and parallel combinations with their equivalent immittances.

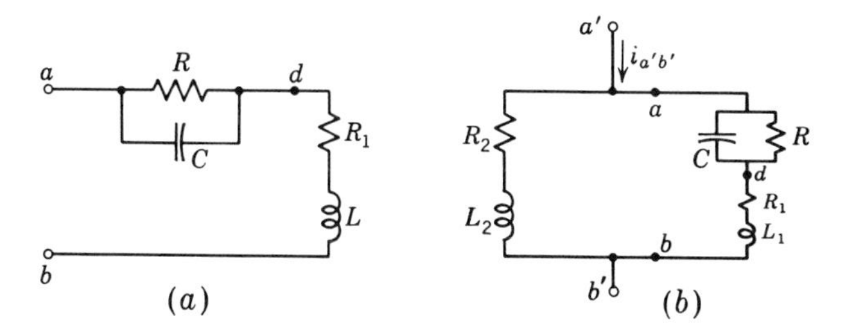

Fig. 9-12 Series-parallel circuits.

Consider, for example, the circuit shown in Fig. 9-12*a*. For this circuit

$$\mathbf{Z}_{ab} = \mathbf{Z}_{ad} + \mathbf{Z}_{db}$$

But

$$\mathbf{Z}_{db} = R_1 + j\omega L$$

and

$$\mathbf{Z}_{ad} = \frac{1}{\mathbf{Y}_{ad}} = \frac{1}{1/R + j\omega C}$$

Hence

$$\mathbf{Z}_{ab} = R_1 + j\omega L + \frac{1}{1/R + j\omega C} \tag{9-24}$$

For any particular value of ω the complex number $\mathbf{Z}_{ab}$ can be evaluated, and hence the relationship between v_{ab} and i_{ab} in the steady state is determined. If now the elements which form $\mathbf{Z}_{ab}$ are placed in parallel with a second set of elements as illustrated in Fig. 9-12*b*, the relationship between the sinusoids $v_{a'b'}$ and $i_{a'b'}$ is determined through the admittance $\mathbf{Y}_{a'b'}$,

$$\mathbf{Y}_{a'b'} = \mathbf{Y}_{ab} + \mathbf{Y}_2$$

$$\mathbf{Y}_2 = \frac{1}{R_2 + j\omega L_2}$$

$$\mathbf{Y}_{ab} = \frac{1}{\mathbf{Z}_{ab}}$$

and $\mathbf{Z}_{ab}$ is given by Eq. (9-24).

Example 9-4 Series-parallel ladder-type circuit Consider the circuit shown in Fig. 9-13. Assume that all the elements marked R are equal to 1.0 ohm and that

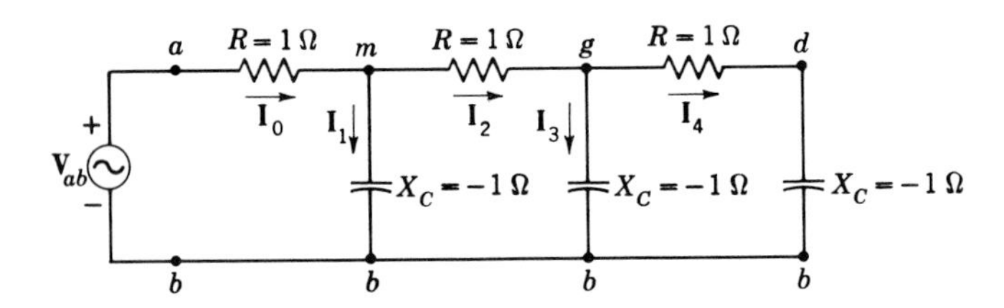

Fig. 9-13 Ladder circuit analyzed in Example 8-4.

all the elements marked X_C are equal to -1.0 ohm. Assume that the phasor $\mathbf{V}_{db}$ is known and that the expression for $\mathbf{V}_{ab}$ in terms of $\mathbf{V}_{db}$ is desired.

Solution $\mathbf{V}_{gb}$ is found by applying voltage division to the branch gdb:

$$\mathbf{V}_{gb} = \mathbf{V}_{db} \times \frac{1 - j}{-j} = (1 + j) \times \mathbf{V}_{db}$$

The phasor $\mathbf{I}_4$ is found in terms of V_{db} as

$$\mathbf{I}_4 = \frac{\mathbf{V}_{db}}{-j} = j\mathbf{V}_{db}$$

Similarly,

$$\mathbf{I}_3 = \frac{\mathbf{V}_{gb}}{-j} = \frac{1 + j}{-j}\mathbf{V}_{db}$$
$$= (-1 + j)\mathbf{V}_{db}$$

Hence $\mathbf{I}_2 = \mathbf{I}_3 + \mathbf{I}_4 = (-1 + j2)\mathbf{V}_{db}$. Since $\mathbf{V}_{mg} = 1 \times \mathbf{I}_2$, $\mathbf{V}_{mg} = (-1 + j2)\mathbf{V}_{db}$. Thus

$$\mathbf{V}_{mb} = \mathbf{V}_{mg} + \mathbf{V}_{gb}$$
$$= (-1 + j2)\mathbf{V}_{db} + (1 + j)\mathbf{V}_{db}$$
$$= (j3)\mathbf{V}_{db}$$

We can now write

$$\mathbf{I}_1 = \frac{\mathbf{V}_{mb}}{-j} = -3\mathbf{V}_{db}$$

and

$$\mathbf{I}_0 = \mathbf{I}_1 + \mathbf{I}_2 = (-4 + j2)\mathbf{V}_{db}$$

Since $\mathbf{V}_{am} = 1 \times \mathbf{I}_0$,

$$\mathbf{V}_{ab} = (-4 + j2)\mathbf{V}_{db} + \mathbf{V}_{mb}$$
$$= (-4 + j5)\mathbf{V}_{db}$$

so that

$$\mathbf{V}_{ab} = (6.4\underline{/128.6°}) \times \mathbf{V}_{db} \qquad \textit{Ans.}$$

Example 9-5 A bridge circuit In the circuit shown in Fig. 9-14 deduce the relationship between the parameters which will cause the current $i_{ab}(t)$ to be zero in the sinusoidal steady state.

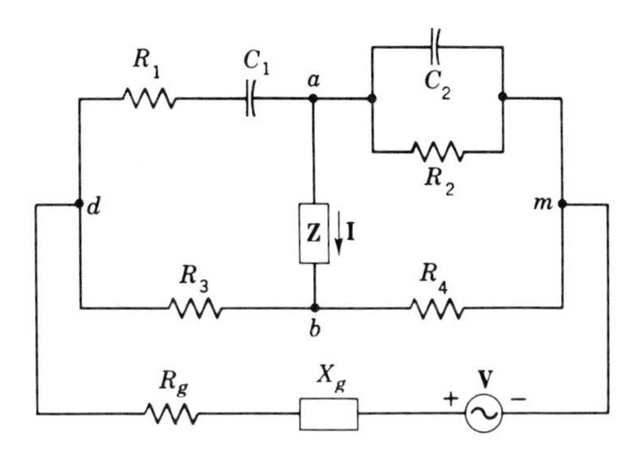

Fig. 9-14 An a-c bridge circuit.

Solution If $i_{ab}(t) = 0$, $\mathbf{I} = 0$ and $\mathbf{V}_{ab} = 0$ regardless of the value of $\mathbf{Z}$. Applying Kirchhoff's voltage law to loops *d-a-b-d* and *a-m-b-a*,

$$\mathbf{V}_{da} + \mathbf{V}_{ab} + \mathbf{V}_{bd} = 0$$

and

$$\mathbf{V}_{am} + \mathbf{V}_{mb} + \mathbf{V}_{ba} = 0$$

Since $\mathbf{V}_{ba}$ must be zero,

$$\begin{aligned} \mathbf{V}_{da} + \mathbf{V}_{bd} &= 0 \\ \text{and} \qquad \mathbf{V}_{am} + \mathbf{V}_{mb} &= 0 \end{aligned} \tag{9-25}$$

But

$$\begin{aligned} \mathbf{V}_{da} &= \mathbf{I}_{da}\mathbf{Z}_{da} \\ \mathbf{V}_{db} &= \mathbf{I}_{db}\mathbf{Z}_{db} \\ \mathbf{V}_{am} &= \mathbf{I}_{am}\mathbf{Z}_{am} \\ \mathbf{V}_{mb} &= \mathbf{I}_{mb}\mathbf{Z}_{mb} \end{aligned} \tag{9-26}$$

Since $\mathbf{I}_{ab}$ must be zero,

$$\begin{aligned} \mathbf{I}_{da} &= \mathbf{I}_{am} \\ \text{and} \qquad \mathbf{I}_{db} &= \mathbf{I}_{bm} \end{aligned} \tag{9-27}$$

Substituting Eqs. (9-25) in (9-26) and using (9-27),

$$\mathbf{I}_{da}\mathbf{Z}_{da} = -\mathbf{I}_{bd}\mathbf{Z}_{bd} \tag{9-28}$$

and

$$\mathbf{I}_{da}\mathbf{Z}_{am} = -\mathbf{I}_{bd}\mathbf{Z}_{mb} \tag{9-29}$$

Dividing (9-28) by (9-29),

$$\frac{\mathbf{Z}_{da}}{\mathbf{Z}_{am}} = \frac{\mathbf{Z}_{bd}}{\mathbf{Z}_{mb}} \tag{9-30}$$

This is the desired equation. Note that the result is independent of V_m, R_g, and X_g. For the impedances specified in this example, Eq. (9-30) becomes

$$\mathbf{Z}_{da}\mathbf{Y}_{am} = \frac{\mathbf{Z}_{bd}}{\mathbf{Z}_{mb}}$$

$$\left(R_1 + \frac{1}{j\omega C_1}\right)\left(\frac{1}{R_2} + j\omega C_2\right) = \frac{R_3}{R_4}$$

$$\frac{R_1}{R_2} + j\omega\left(C_2R_1 - \frac{1}{R_2C_1\omega^2}\right) + \frac{C_2}{C_1} = \frac{R_3}{R_4}$$

Hence $i_{ab}(t)$ is zero if

$$\frac{C_2}{C_1} = \frac{R_3}{R_4} - \frac{R_1}{R_2}$$

$$\omega^2 = \frac{1}{R_1 R_2 C_1 C_2}$$

Note that this final result involves ω, the radian frequency of the source.

Example 9-6 Use of Thévenin's theorem Use Thévenin's theorem to calculate the phasor $\mathbf{V}_{ab}$ in the circuit of Fig. 9-15*a*.

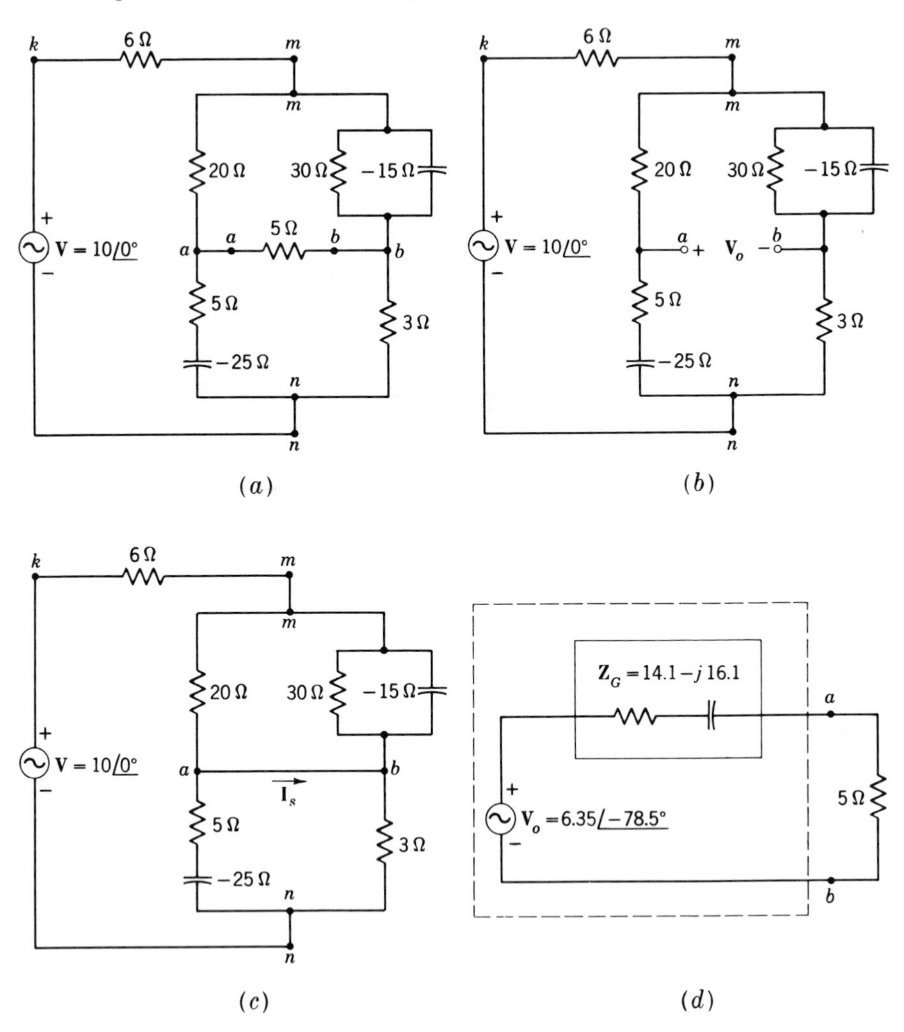

Fig. 9-15 Application of Thévenin's theorem for sinusoidal-steady-state calculations (Example 8-6). (a) Circuit diagram. (b) Circuit for calculation of open-circuit voltage. (c) Circuit for calculation of I_s. (d) Thévenin's equivalent at terminals a-b for the circuit of (a).

Solution Branch *a-b* is disconnected as shown in Fig. 9-15*b*, and $\mathbf{V}_{ab}$ is calculated. The phasor $\mathbf{V}_{ab}$ in Fig. 9-15*b* is $\mathbf{V}_o$, the open-circuit voltage phasor in Thévenin's equivalent circuit (Fig. 9-15*d*). To find $\mathbf{V}_o$, voltage division is applied three times as follows:

$$\mathbf{V}_{an} = \frac{\mathbf{Z}_{an}}{\mathbf{Z}_{man}}\mathbf{V}_{mn} \qquad \mathbf{V}_{bn} = \frac{\mathbf{Z}_{bn}}{\mathbf{Z}_{mbn}}\mathbf{V}_{mn} \qquad \mathbf{V}_{mn} = \frac{\mathbf{Z}_{mn}}{\mathbf{Z}_{kn}}\mathbf{V}_{kn}$$

so that

$$\mathbf{V}_o = \left(\frac{\mathbf{Z}_{an}}{\mathbf{Z}_{man}} - \frac{\mathbf{Z}_{bn}}{\mathbf{Z}_{mbn}}\right)\left(\frac{\mathbf{Z}_{mn}}{\mathbf{Z}_{kn}}\right) \times 10\underline{/0°} \tag{9-31}$$

The relevant impedance values and ratios are now calculated: $\mathbf{Z}_{an} = 5 - j25$, $\mathbf{Z}_{man} = 25 - j25$, so that $\mathbf{Z}_{an}/\mathbf{Z}_{man} = 0.60 - j0.40$; $\mathbf{Z}_{bn} = 3 + j0$, $\mathbf{Y}_{mb} = (1 + j2)/30$, $\mathbf{Z}_{mb} = 6(1 - j2) = 6 - j12$, $\mathbf{Z}_{mbn} = 9 - j12$, so that $\mathbf{Z}_{bn}/\mathbf{Z}_{mbn} = 3/(9 - j12) = 0.12 + j0.16$. To find $\mathbf{Z}_{mn}$ we first find $\mathbf{Y}_{mn} = \mathbf{Y}_{man} + \mathbf{Y}_{mbn} = 1/(25 - j25) + 1/(9 - j12) = (11 + j15)/450$, so that $\mathbf{Z}_{mn} = 450/(11 + j15) = 13 - j19.5 = 23.5\underline{/-56.4°}$ and $\mathbf{Z}_{kn} = 19 - j19.5 = 27.3\underline{/-45.8°}$. Substituting in Eq. (9-31) gives

$$\begin{aligned}\mathbf{V}_o &= (0.60 - j0.40 - 0.12 - j0.16)\left(\frac{23.5\underline{/-56.4°}}{27.3\underline{/-45.8°}}\right)(10\underline{/0°})\\ &= 6.35\underline{/-78.5°}\end{aligned}$$

To find the source impedance $\mathbf{Z}_G$ of Thévenin's equivalent circuit one could set $\mathbf{V} = 0$ in Fig. 9-15*b* and find $\mathbf{Z}_{ab}$. Since the driving-point immittance at terminals *a-b* is not a series-parallel combination of elements, it is more convenient to solve for $\mathbf{I}_s$, the short-circuit current phasor in the circuit of Fig. 9-15*c*. Since in that circuit $\mathbf{I}_s = \mathbf{I}_{ma} - \mathbf{I}_{an}$, we calculate from current division [Eq. (9-23)]

$$\mathbf{I}_{ma} = \frac{\frac{1}{20}}{\frac{1}{20} + \frac{1}{30} + j/15}\mathbf{I}_{km} \qquad \mathbf{I}_{an} = \frac{1/(5 - j25)}{1/(5 - j25) + \frac{1}{3}}\mathbf{I}_{km} \qquad \mathbf{I}_{km} = \mathbf{Y}_{km}\mathbf{V} \tag{9-32}$$

where the value $\mathbf{Y}_{km}$ must be calculated from Fig. 9-15*c*; the total admittance between terminals *m-a* (same as *m-b*) is $\mathbf{Y}_{ma} = \frac{1}{20} + \frac{1}{30} + j/15 = (5 + j24)/60$; $\mathbf{Z}_{ma} = 60/(5 + j4) = 7.32 - j5.86$; $\mathbf{Y}_{an} = 1/(5 - j25) + \frac{1}{3} = (8.88 + j)/26$; $\mathbf{Z}_{an} = 26/(8.88 + j) = 2.89 - j0.33$; $\mathbf{Z}_{km} = 6 + 7.32 - j5.86 + 2.89 - j0.33 = 17.4\underline{/-20.9°}$; and $\mathbf{Y}_{kn} = 0.0575\underline{/20.9°}$. Hence

$$\begin{aligned}\mathbf{I}_s &= \left[\frac{\frac{1}{20}}{\frac{1}{20} + \frac{1}{30} + j/15} - \frac{1/(5 - j25)}{1/(5 - j25) + \frac{1}{3}}\right](0.0575\underline{/20.9°})(10\underline{/0°})\\ &= 0.297\underline{/-29.6°} \text{ amp}\end{aligned}$$

Hence $\mathbf{Z}_G = \mathbf{V}_o/\mathbf{I}_s = (6.35\underline{/-78.5°})/(0.297\underline{/-29.6°}) = 21.4\underline{/-48.9°}$ as shown (in rectangular form, $21.4\underline{/-48.9°} = 14.1 - j16.1$) in Fig. 9-15*d*. In that diagram the 5-ohm load is in place, and from voltage division, $\mathbf{V}_{ab} = [(5/(19.1 - j16.1)](6.35\underline{/-78.5°}) = 1.27\underline{/-38.4°}$ volts.

9-11 Equivalent circuits

At this point it should be clear that the voltage-current relationship (driving-point immittance) for any passive terminal pair in the sinusoidal steady state at any one frequency is completely characterized by the complex immittance of the terminal pair. In any numerical case this means that the complex number

$$\mathbf{Z}_{ab} = R_{ab} + jX_{ab}$$

or

$$\mathbf{Y}_{ab} = G_{ab} + jB_{ab}$$

may represent a passive terminal pair at the input terminals *a-b* as indicated in Fig. 9-16.

As implied in Fig. 9-16*b* and *c*, this means that, *at any one frequency*, we can represent a terminal pair as a series or a parallel combination of only two elements. We shall now present several examples to illustrate this equivalent representation. In every case it must be remembered that we

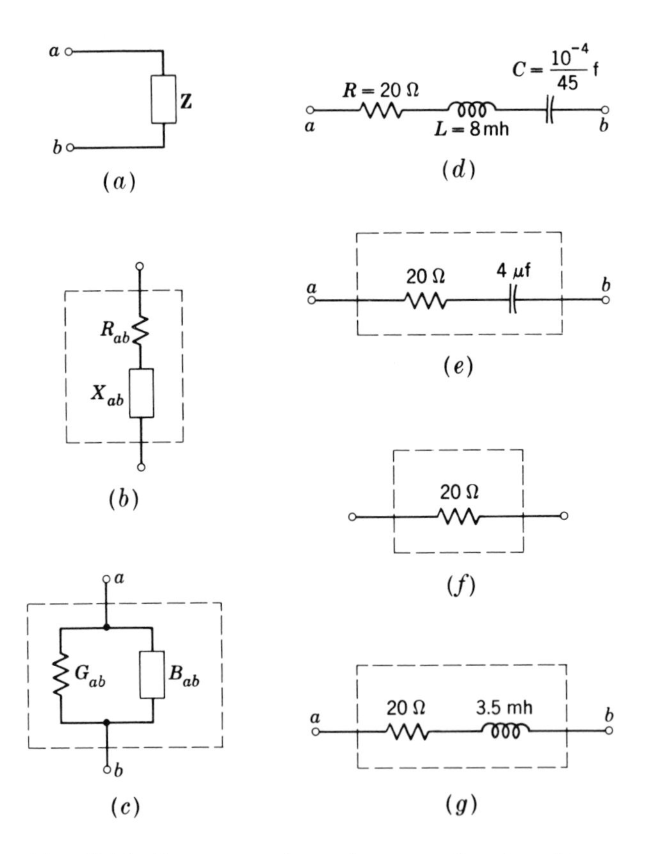

***Fig. 9-16** Representations for complex immittance. (a) Terminal pair. (b) Terminal pair as a series combination of resistance and reactance. (c) Terminal pair as a parallel combination of conductance and susceptance. (d) A series R-L-C circuit. (e) Representation of the terminal pair of (d) at a frequency when $X_C + X_L < 0$. (f) Representation of the terminal pair of (d) at a frequency where $X_C + X_L = 0$. (g) Representation of the terminal pair of (d) at a frequency where $X_C + X_L > 0$.*

are obtaining equivalent representations only with respect to the input terminals and only at one frequency in the sinusoidal steady state. The equivalent representations which are deduced on this basis will serve to obtain *not* the complete response of the original circuit, but *only* the component of the response (at the input terminals) to a sinusoidal source of the frequency at which the calculations are carried out.

Example 9-7 For the series R-L-C circuit shown in Fig. 9-16d deduce equivalent circuits with respect to terminals a-b consisting of two elements at the frequencies (a) $f_1 = 796$ hertz; (b) $f_2 = 1.5 \times 796$ hertz; (c) $f_3 = 2 \times 796$ hertz.

Solution (a) At the frequency $f_1 = 796$ hertz the radian frequency is

$$\omega_1 = 2\pi f_1 = 6.28 \times 796 = 5{,}000 \text{ rad/sec}$$

Hence

$$X_L = 5{,}000 \times 8 \times 10^{-3} = 40 \text{ ohms}$$

and

$$X_C = -\frac{1}{\frac{1}{45} \times 5{,}000 \times 10^{-4}} = -90 \text{ ohms}$$

so that

$$\mathbf{Z}_{ab} = 20 + j40 - j90$$
$$= 20 - j50 \qquad \text{at } \omega = \omega_1$$

This impedance can be interpreted as the series combination of a 20-ohm resistance and a -50-ohm capacitive reactance. Since the equivalent capacitive reactance is -50 ohms, the series capacitance $C = 1/(50 \times 5{,}000) = 4$ μf. This result is shown in Fig. 9-16e.

(b) At the frequency $\omega_2 = 1.5\omega_1$, $X_L = 60$ ohms, $X_C = -60$ ohms. Hence $\mathbf{Z}_{ab} = 20 + j60 - j60 = 20 + j0$ ohms at $\omega = \omega_2$. This result is illustrated in Fig. 9-16f.

(c) At the frequency $\omega = 2\omega_1$, $X_L = 2 \times 40 = 80$ ohms, $X_C = -\frac{1}{2} \times 90 = -45$ ohms. Hence $\mathbf{Z}_{ab} = 20 + j35$ at $\omega = \omega_3$.

This result can be interpreted as the series combination of a 20-ohm resistance and a 3.5×10^{-3} [$= 35/(2 \times 5{,}000)$] henry inductance as indicated in Fig. 9-16g.

Figure 9-16e to g shows three circuits which at the three frequencies given have the same complex value $\mathbf{Z}_{ab}$ as the original circuit of Fig. 9-16d. It must be emphasized that each of these equivalent circuits at the frequency at which they are equivalent has *the same sinusoidal steady-state response*, i.e., the same component of the response due to the source. These equivalent circuits have *neither* the same equilibrium equation *nor* the same complete response as the original circuit.

Example 9-8 For the circuit of Example 9-7 at the radian frequency $\omega = 5{,}000$ rad/sec, deduce a parallel equivalent circuit, and discuss other equivalent circuits.

Solution The impedance of the circuit at the frequency stated was found to be

$$\mathbf{Z}_{ab} = 20 - j50 \text{ ohms}$$

To obtain a parallel equivalent circuit, we shall find the complex admittance $\mathbf{Y}_{ab}$,

$$\mathbf{Y}_{ab} = \frac{1}{\mathbf{Z}_{ab}} = \frac{1}{20 - j50} = \frac{1}{10}\,\frac{1}{2 - j5}$$
$$= 0.1 \frac{1}{5.38\underline{/-68.2^\circ}} = 0.0185\underline{/+68.2^\circ}$$

or in rectangular form,

$$\mathbf{Y}_{ab} = 0.0069 + j0.0172 \text{ mho}$$

This admittance may be interpreted as the parallel combination of a 0.0069-mho conductance and a +0.0172-mho susceptance (capacitive). A 0.0172-mho susceptance corresponds at 5,000 rad/sec to a capacitance of 0.0172/5,000 = 3.45-μf capacitance. A 0.0069-mho conductance corresponds to a 1/0.0069 = 145-ohm resistance. These results are illustrated in Fig. 9-17. Again we must emphasize

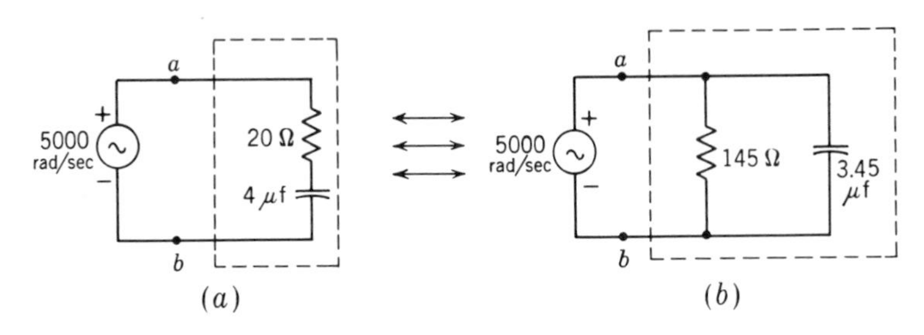

Fig. 9-17 Series and parallel two-element representations of a terminal pair at a particular frequency in the sinusoidal steady state.

that the equivalence which has been established applies only at the radian frequency of 5,000 in the sinusoidal steady state between terminals *a-b*. The current in the 145-ohm resistance or in the 3.45-μf capacitance of the equivalent parallel circuit has no meaning in the original circuit. The sum of these currents will, however, equal i_{ab} in the original circuit in the steady state if v_{ab} is the same sinusoidal voltage at the correct frequency.

Other equivalents are easily established. (An infinite variety is possible if more than two elements are allowed.) We may, for example, write

$$\mathbf{Z}_{ab} = 5 + 15 - j50$$

and define

$$\mathbf{Z}_1 = 15 - j50$$

Then

$$\mathbf{Y}_1 = \frac{1}{15 - j50} = 0.0192\underline{/73.3^\circ} = 0.00550 + j0.0183$$

This admittance may be interpreted as the parallel combination of a 181 (= 1/0.0055) ohm resistance and a 3.66 (= 0.0183/5,000) μf capacitance. Since $\mathbf{Z}_{ab} = 5 + \mathbf{Z}_1$, another equivalent circuit may be shown as indicated in Fig. 9-18.

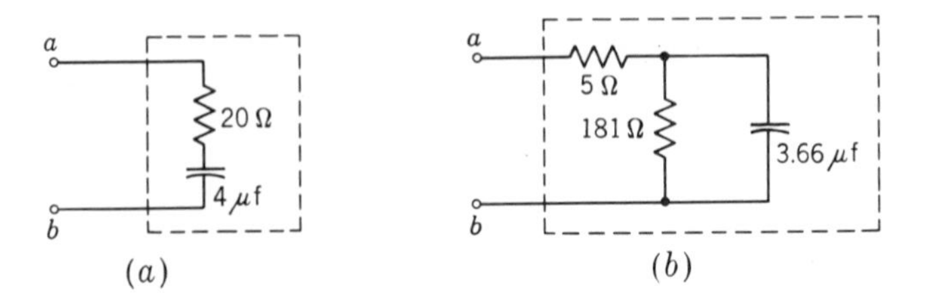

Fig. 9-18 Two terminal-pair networks that have the same complex admittance at a particular frequency in the sinusoidal steady state.

IMMITTANCE FOR TWO ELEMENTS Since a terminal pair, at a fixed frequency in the sinusoidal steady state, is completely specified by its complex immittance, it is possible to represent such a terminal pair either as a series or as a parallel combination of two elements.

A parallel G-B_C combination is shown in Fig. 9-19*a*. For this circuit the admittance is

$$\mathbf{Y}_{ab} = G + jB_C$$

and the complex impedance is

$$\mathbf{Z}_{ab} = \frac{1}{G + jB_C}$$

In rectangular form $\mathbf{Z}_{ab}$ is written

$$\mathbf{Z}_{ab} = \frac{1(G - jB_C)}{(G + jB_C)(G - jB_C)} = \frac{G}{G^2 + B_C{}^2} + j\frac{-B_C}{G^2 + B_C{}^2}$$

Consider now the series R-C circuit shown in Fig. 9-19*b*. For this circuit the impedance is

$$\mathbf{Z}_{a'b'} = R' + jX'_C$$

If we set

$$R' = \frac{G}{G^2 + B_C{}^2} \tag{9-33a}$$

and

$$X'_C = \frac{-B_C}{G^2 + B_C{}^2} \tag{9-33b}$$

then the circuits of Fig. 9-19*a* and *b* are equivalent (*in the sinusoidal steady state, at the frequency at which B_C was calculated with respect to terminals a-b and a'-b'*) if Eqs. (9-33) apply.

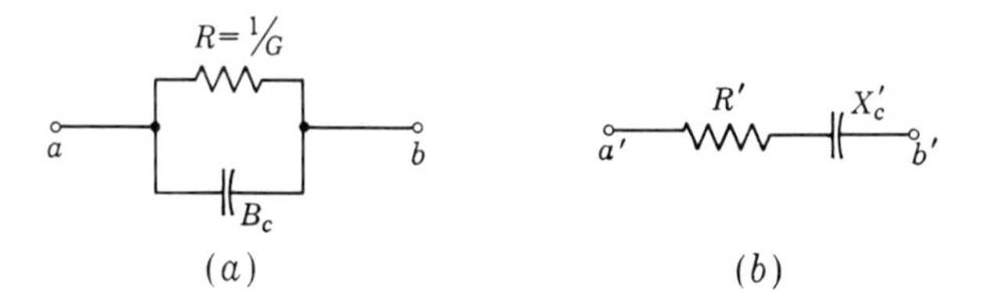

***Fig. 9-19** Parallel and series two-element combinations.*

Note that the equivalence was established by setting $R' = \text{Re}\,\mathbf{Z}_{ab}$ and $X' = \text{Im}\,\mathbf{Z}_{ab}$. In particular, note that R' is *not* the reciprocal of G; it is the real part of the complex impedance. We feel it necessary to warn the reader explicitly that the statement

$$\text{Re}\,\frac{1}{a + jb} = \frac{1}{a} \text{ is not true unless } b \text{ is zero}$$

If it is desired to "convert" the configuration of Fig. 9-19*b* into that of Fig. 9-19*a*, we should start with

$$\mathbf{Z}_{a'b'} = R' + jX'_C$$

Hence

$$\mathbf{Y}_{a'b'} = \frac{1}{R' + jX'_C} = \frac{R'}{R'^2 + X'^2_C} + j\frac{-X'_C}{R'^2 + X'^2_C}$$

If $\mathbf{Y}_{a'b'} = \mathbf{Y}_{ab}$, then

$$G = \frac{R'}{R'^2 + X_C{}^2}$$

and

$$B_C = \frac{-X_C'}{R'^2 + X_C'^2}$$

It must be emphasized again that the equivalence is limited to the steady-state terminal relations already cited.

9-12 Complete response with sinusoidal source

Example 9-9 ***Deenergizing an R-L circuit*** An R-L circuit as shown in Fig. 9-20 has been connected to a sinusoidal source

$$v(t) = V_m \cos(\omega t + \phi)$$

for a sufficiently long time so that the steady state is reached. At $t = 0$ the circuit is deenergized; i.e., the switch K is thrown from position 1 to position 2. Solve for $i(t)$ for all $t \geq 0^+$.

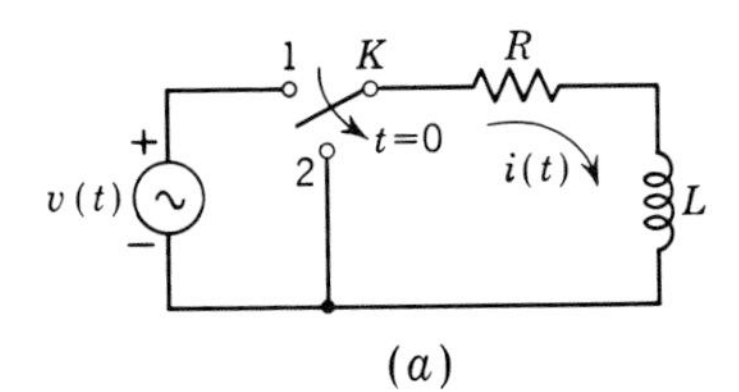

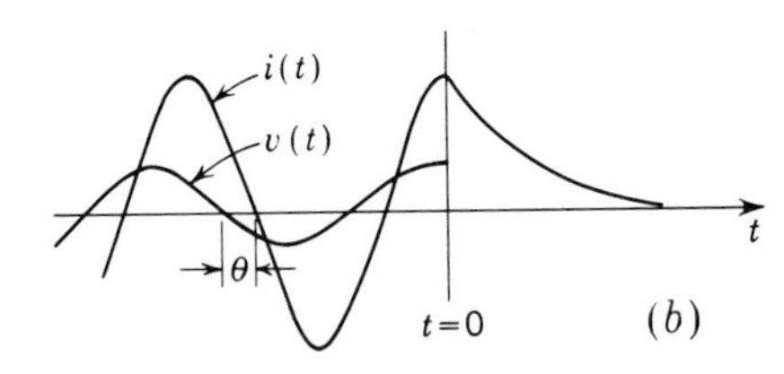

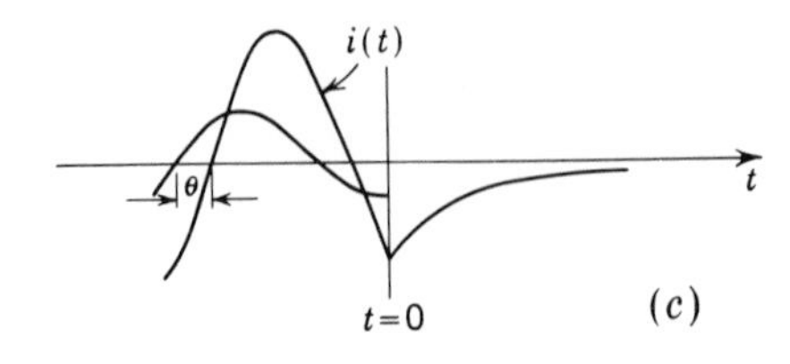

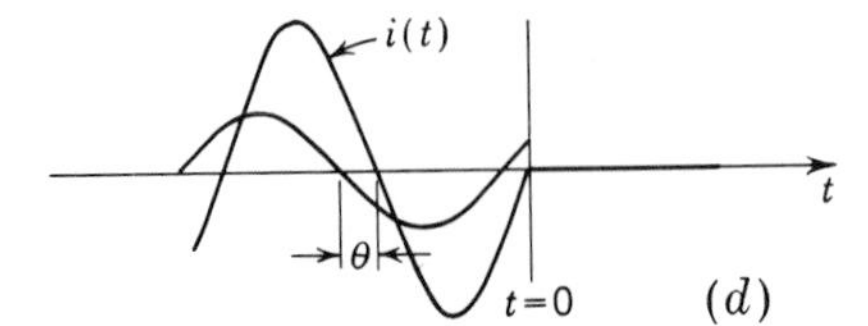

Fig. 9-20 ***Deenergizing an R-L circuit from a sinusoidal source. (a) Switching operation. (b) Response if $i(0^-) > 0$. (c) Response for $i(0^-) < 0$. (d) Response when $i(0^-) = 0$.***

Solution Since the circuit is source-free for $t \geq 0^+$, the solution for $i(t)$ is given by the familiar form

$$i(t) = i(0^+)e^{-t/T} \qquad T = \frac{L}{R},\ t \geq 0^+ \tag{9-34}$$

Because of the inductance we have the continuity condition

$$i(0^+) = i(0^-)$$

The value $i(0^-)$ is found by solving for the steady-state alternating current prior to $t = 0$ and setting $t = 0$ in the equation for $i(t)$.

Since

$$v(t) = \text{Re}\,(V_m e^{j(\omega t+\phi)})$$

$$i(t) = \text{Re}\left(\frac{V_m}{\mathbf{Z}}\, e^{j(\omega t+\phi)}\right)$$

where

$$\mathbf{Z} = R + j\omega L = Z\underline{/\theta}$$

Hence

$$i(t) = \frac{V_m}{Z}\,(\cos \omega t + \phi - \theta) \qquad t \leq 0^-$$

and

$$i(0^-) = \frac{V_m}{Z}\cos(\phi - \theta)$$

Substituting in (9-34), we have

$$i(t) = \left[\frac{V_m}{Z}\cos(\phi - \theta)\right] e^{-t/T} \qquad t \geq 0^+$$

We observe now that the response for $t \geq 0^+$ depends on $\phi - \theta$, *the phase of the steady-state current* at $t = 0$. Several possibilities are shown in Fig. 9-20. Note in particular Fig. 9-20*d*. In that diagram ϕ is chosen so that $\phi - \theta = \pi/2$ and $i(t) = 0$ for $t \geq 0^+$. If, in a practical problem, the angle ϕ is not known, one cannot determine the value $i(0^+)$ numerically; one can only delimit the value as at present, where $|i(0^+)|$ is at most V_m/Z and may be as small as zero.

Example 9-10 ***Use of time-domain method*** The circuit shown in Fig. 9-21 is in the steady state for $t < 0$ with the switch in position 1. At $t = 0$ the switch is thrown to position 2. Calculate $v_{ab}(t)$ for all $t \geq 0^+$.

Solution For $t < 0$, $v_{ab} = V_0$; hence $v_{ab}(0^+) = V_0$ (continuity condition). For $t > 0$, the component of the response v_{ab} due to the source is represented by the amplitude phasor $(\mathbf{V}_{ab})_m = [(1/j\omega C)/(R + j\omega C)](V_m\underline{/\phi})$, where $V_m\underline{/\phi}$ represents the source. Since

$$\frac{1/j\omega C}{R + 1/j\omega C} = \frac{1}{j\omega RC + 1} = \frac{1}{\sqrt{\omega^2R^2C^2 + 1}}\underline{/-\tan^{-1}\omega RC}$$

$$(v_{ab})_s = \frac{V_m}{\sqrt{\omega^2R^2C^2 + 1}}\cos(\omega t + \phi - \tan^{-1}\omega RC)$$

The source-free component of v_{ab} is exponential, $(v_{ab})_f = Ke^{-t/RC}$, since $v_{ab} = (v_{ab})_s + (v_{ab})_f$, to satisfy the initial condition

$$V_0 = \frac{V_m}{\sqrt{\omega^2R^2C^2 + 1}}\cos(\phi - \tan^{-1}\omega RC) + K$$

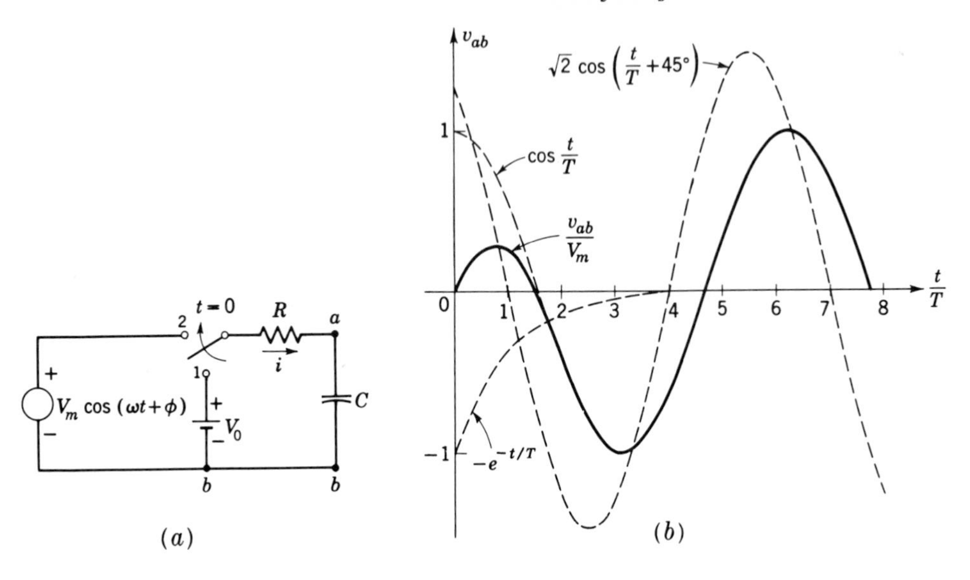

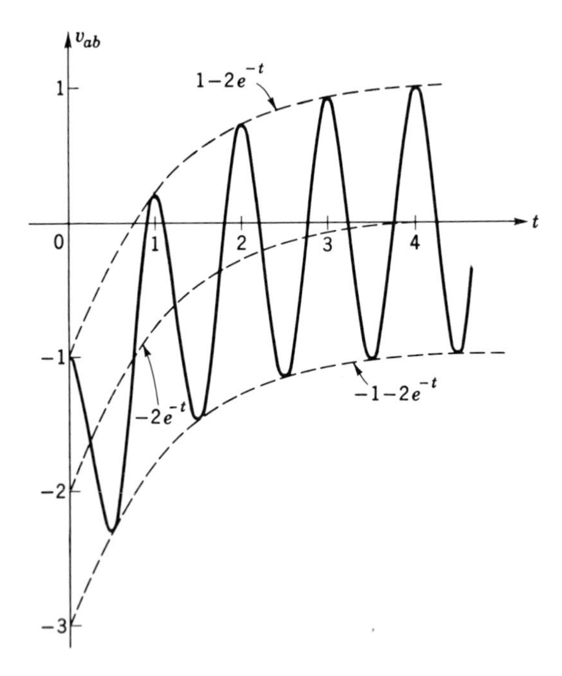

***Fig. 9-21** Energizing a series R-C circuit from an a-c source. (a) Switching operation. (b) Complete response for $v_{ab}(0^-) = V_0 = 0$, $\omega RC = 1$, $V_m = 1$, $\phi = 45°$. Note that, for $t > 3T$, only the steady-state component remains. The source voltage is shown for reference. (c) Complete response for $v_{ab}(0^-) = V_0 = -1 = -V_m/\sqrt{1 + 4\pi^2}$, $\omega T = 2\pi$, $\phi = \tan^{-1} 2\pi$.*

Solving for K and adding $(v_{ab})_f$ to $(v_{ab})_s$ gives the result

$$v_{ab} = \frac{V_m}{\sqrt{\omega^2R^2C^2+1}}\cos(\omega t + \phi - \tan^{-1}\omega RC) + \left[V_0 - \frac{V_m}{\sqrt{\omega^2R^2C^2+1}}\cos(\phi - \tan^{-1}\omega RC)\right]e^{-t/RC} \qquad t \geq 0^+ \qquad (9\text{-}35)$$

We note again that the transient component is zero if $V_0 = [v_{ab}(0)]_s$, that is, when the steady-state component satisfies the initial condition. To illustrate the waveform of the solution [Eq. (9-35)], two numerical cases are considered. Let $V_0 = 0$, $\omega RC = 1$, $V_m = \sqrt{2}$; in this case Eq. (9-35) becomes

$$v_{ab} = \cos(\omega t + \phi - 45°) - e^{-t/RC}\cos(\phi - 45°) \qquad t \geq 0^+$$

The waveform is the *sum* of an exponential and a sinusoidal term. Again the importance of ϕ, the phase of the source at the instant of switching, is apparent. Setting $\phi = 45°$, the transient component is maximized and $v_{ab} = \cos\omega t - e^{-t/RC}$, or since $\omega = 1/RC$,

$$v_{ab} = \cos\frac{t}{T} - e^{-t/T}$$

The construction for $T = 1$ is shown in Fig. 9-21*b*. As a second example we choose $V_0 = -V_m/\sqrt{\omega^2R^2C^2+1} = -1$, $\phi - \tan^{-1}\omega RC = 0$, $T = 1$ sec. Then

$$v_{ab}(t) = \cos 2\pi t - 2e^{-t}$$

The corresponding waveform is shown in Fig. 9-21*c*. Note that the waveforms of Fig. 9-21 are *sums* of exponentials and sinusoids, and should not be confused with the *products* constructed in Chap. 4.

Example 9-11 Use of Laplace transform Obtain the formula for the complete response $i(t)$ in a series R-L-C circuit where $R = 6$ ohms, $L = 1$ henry, $C = \frac{1}{25}$ farad, and the source is $v(t) = (20\cos 10t)u(t)$. Formulate the answer for the specific initial conditions $i(0^-) = 2$ amp, $(1/C)\int_{-\infty}^{0^-} i\,d\tau = 10$ volts.

Solution For $v(t) = (20\cos 10t)u(t)$, $V(s) = 20s/(s^2+100)$. Hence

$$\left(s + 6 + \frac{25}{s}\right)I(s) = \frac{20s}{s^2+100} + 2(1) - \frac{10}{s} = \frac{2s^3 + 10s^2 + 200s - 1{,}000}{s(s^2+100)}$$

or

$$I(s) = \frac{2s^3 + 10s^2 + 200s - 1{,}000}{(s+3-j4)(s+3+j4)(s-j10)(s+j10)} = \frac{\mathbf{K}_1}{s-j10} + \frac{\mathbf{K}_1^*}{s+j10} + \frac{\mathbf{K}_2}{s+3-j4} + \frac{\mathbf{K}_2^*}{s+3+j4}$$

where

$$\mathbf{K}_1 = \frac{2(j10)^3 + 10(j10)^2 + 200(j10) - 1{,}000}{[-100 + 6(j10) + 25](j20)} = 1.04\underline{/-51.4°}$$

$$\mathbf{K}_2 = \frac{2(-3+j4)^3 + 10(-3+j4)^2 + 200(-3+j4) - 1{,}000}{[(-3+j4)^2 + 100](j8)} = 2.05\underline{/80.5°}$$

Hence

$$i = 2.08\cos(10t - 51.4°) + 4.10e^{-3t}\cos(4t + 80.5°) \qquad t \geq 0^+ \qquad \textit{Ans.}$$

9-13 Power associated with a terminal pair

In the preceding it has been shown that, for the purpose of *sinusoidal steady-state calculations*, the relationship between voltage and current for a single-terminal-pair network is completely characterized by the complex

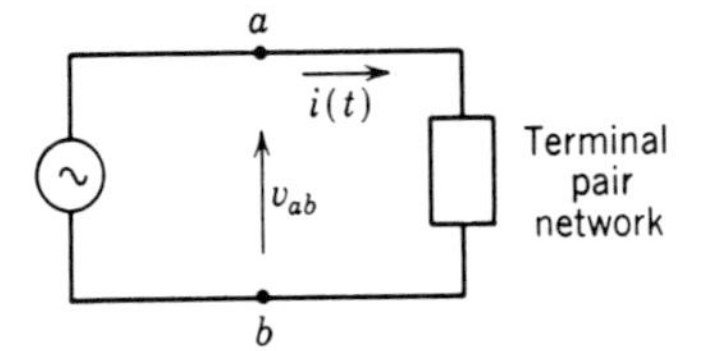

Fig. 9-22 Terminal-pair network with a sinusoidal source.

immittance. For such a terminal pair, with the reference directions for voltage and current as shown in Fig. 9-22, let

$$v_{ab} = \text{Re}\,(\sqrt{2}\,\mathbf{V}e^{j\omega t}) \qquad \mathbf{V} = Ve^{j\phi}$$

Then

$$i(t) = \text{Re}\,(\sqrt{2}\,\mathbf{I}e^{j\omega t}) \qquad \mathbf{I} = \mathbf{V}\mathbf{Y}_{ab} = \frac{\mathbf{V}}{\mathbf{Z}_{ab}} \qquad \mathbf{Z}_{ab} = Z\underline{/\theta}$$

Upon applying the definitions of voltage and current, the expression for the power which flows into the terminal pair at any instant of time is given by the product of v_{ab} and i,

$$p_{ab}(t) = v_{ab}(t)i(t)$$

or

$$p_{ab}(t) = [\text{Re}\,(\mathbf{V}\,\sqrt{2}\,e^{j\omega t})][\text{Re}\,(\mathbf{I}\,\sqrt{2}\,e^{j\omega t})] \tag{9-36}$$

The form of this expression is not convenient because the product of the real parts of two complex expressions is *not* the real part of the product. For this reason we introduce a result of the definition of the conjugate of a complex number, namely, if $\mathbf{z}$ is a complex number and $\mathbf{z}^*$ is its conjugate, then $\text{Re}\,\mathbf{z} = \frac{1}{2}(\mathbf{z} + \mathbf{z}^*)$; using this identity, we may write v_{ab} and i as follows:

$$v_{ab}(t) = \text{Re}\,(\sqrt{2}\,\mathbf{V}e^{j\omega t}) = \tfrac{1}{2}\sqrt{2}\,\mathbf{V}e^{j\omega t} + \tfrac{1}{2}\sqrt{2}\,(\mathbf{V}e^{j\omega t})^* \tag{9-37a}$$

and

$$i(t) = \text{Re}\,(\sqrt{2}\,\mathbf{I}e^{j\omega t}) = \tfrac{1}{2}\sqrt{2}\,\mathbf{I}e^{j\omega t} + \tfrac{1}{2}\sqrt{2}\,(\mathbf{I}e^{j\omega t})^* \tag{9-37b}$$

We now note that, for any two complex numbers $\mathbf{z}_1$ and $\mathbf{z}_2$, one can write

$$(\mathbf{z}_1\mathbf{z}_2)^* = \mathbf{z}_1^*\mathbf{z}_2^*$$

because, if $\mathbf{z}_1 = z_1\underline{/\alpha_1}$ and $\mathbf{z}_2 = z_2\underline{/\alpha_2}$, then

$$(z_1z_2\underline{/\alpha_1 + \alpha_2})^* = z_1z_2\underline{/-\alpha_1 - \alpha_2}$$

Hence

$$(\mathbf{V}e^{j\omega t})^* = \mathbf{V}^*(e^{j\omega t})^* = \mathbf{V}^*e^{-j\omega t}$$

and

$$(\mathbf{I}e^{j\omega t})^* = \mathbf{I}^*(e^{j\omega t})^* = \mathbf{I}^*e^{-j\omega t}$$

Equations (9-37a) and (9-37b) can now be rewritten in the form

$$v_{ab}(t) = \tfrac{1}{2}\sqrt{2}\,\mathbf{V}e^{j\omega t} + \tfrac{1}{2}\sqrt{2}\,\mathbf{V}^*e^{-j\omega t} \tag{9-38a}$$

$$i(t) = \tfrac{1}{2}\sqrt{2}\,\mathbf{I}e^{j\omega t} + \tfrac{1}{2}\sqrt{2}\,\mathbf{I}^*e^{-j\omega t} \tag{9-38b}$$

Upon introducing (9-38a) and (9-38b) into Eq. (9-36), the power flowing into the terminal pair may be written

$$p_{ab}(t) = \left(\frac{1}{\sqrt{2}}\mathbf{V}e^{j\omega t} + \frac{1}{\sqrt{2}}\mathbf{V}^*e^{-j\omega t}\right)\left(\frac{1}{\sqrt{2}}\mathbf{I}e^{j\omega t} + \frac{1}{\sqrt{2}}\mathbf{I}^*e^{-j\omega t}\right)$$

or multiplying out,

$$p_{ab}(t) = \tfrac{1}{2}\mathbf{V}\mathbf{I}^* + \tfrac{1}{2}\mathbf{V}^*\mathbf{I} + \tfrac{1}{2}\mathbf{V}\mathbf{I}e^{j2\omega t} + \tfrac{1}{2}\mathbf{V}^*\mathbf{I}^*e^{-j2\omega t} \tag{9-39}$$

Now if, as was stated at the outset,

$$\mathbf{V} = Ve^{j\phi} \qquad \text{then } \mathbf{V}^* = Ve^{-j\phi}$$

and

$$\mathbf{I} = Ie^{j(\phi-\theta)} \qquad \text{then } \mathbf{I}^* = Ie^{j(\theta-\phi)}$$

where θ is the angle of the impedance $\mathbf{Z}_{ab}$, then

$$\tfrac{1}{2}\mathbf{V}^*\mathbf{I} = \tfrac{1}{2}Ve^{-j\phi}Ie^{j(\phi-\theta)} = \tfrac{1}{2}VIe^{-j\theta}$$

and

$$\tfrac{1}{2}\mathbf{V}\mathbf{I}^* = \tfrac{1}{2}Ve^{j\phi}Ie^{j(\theta-\phi)} = \tfrac{1}{2}VIe^{j\theta} = \tfrac{1}{2}VI(e^{-j\theta})^*$$

Hence

$$\tfrac{1}{2}(\mathbf{V}^*\mathbf{I} + \mathbf{V}\mathbf{I}^*) = \tfrac{1}{2}\mathbf{V}^*\mathbf{I} + \tfrac{1}{2}(\mathbf{V}^*\mathbf{I})^* = \operatorname{Re}(\mathbf{V}^*\mathbf{I}) = \operatorname{Re}(\mathbf{V}\mathbf{I}^*)$$

We further note that

$$\tfrac{1}{2}\mathbf{V}\mathbf{I}e^{j2\omega t} = \tfrac{1}{2}(\mathbf{V}^*\mathbf{I}^*e^{-j2\omega t})^*$$

Hence the expression for $p_{ab}(t)$ reduces to

$$p_{ab}(t) = \operatorname{Re}(\mathbf{V}^*\mathbf{I}) + \operatorname{Re}(\mathbf{V}\mathbf{I}e^{j2\omega t}) \tag{9-40}$$

In trigonometric form this expression reads

$$p_{ab}(t) = VI\cos\theta + VI\cos(2\omega t + 2\phi - \theta) \tag{9-41}$$

In a section below we return to a study of Eq. (9-40); for the present we wish to examine the result in the form (9-41).

We note first that the value of the power delivered to a given terminal pair a-b in the sinusoidal steady state consists of two terms: a constant, $VI\cos\theta$, and a sinusoidal oscillation with radian frequency 2ω and amplitude VI. It is recalled here that in the case of single elements it was also observed that the fluctuating term in the expression for instantaneous power was a sinusoid of twice the frequency of the voltage or the current. Next we observe that Eq. (9-41) reduces to the expressions for single elements which were deduced in Sec. 9-4 if the appropriate conditions are imposed. If the terminal pair consists, for example, of an inductance L_{ab}, then $\theta = \pi/2$ and Eq. (9-41) reduces to Eq. (9-13).

A graph of the expression p_{ab} as a function of t or ωt is instructive. This time dependence is illustrated in Fig. 9-23. We observe that the oscillations of amplitude VI fluctuate about the average value $VI \cos \theta$ so that the maximum value of the instantaneous power is $VI(1 + \cos \theta)$. We note, moreover, that during part of each cycle (if $\cos \theta$ is not unity) negative power flows to terminals a-b. Since the condition $\cos \theta \neq 1$ means that the total susceptance or reactance between terminals a-b is not zero, this negative power flow is readily interpreted: during the time interval when $p_{ab}(t)$ is negative, the energy-storing elements in the "box" a-b (i.e., the reactances) are returning stored energy to the source.

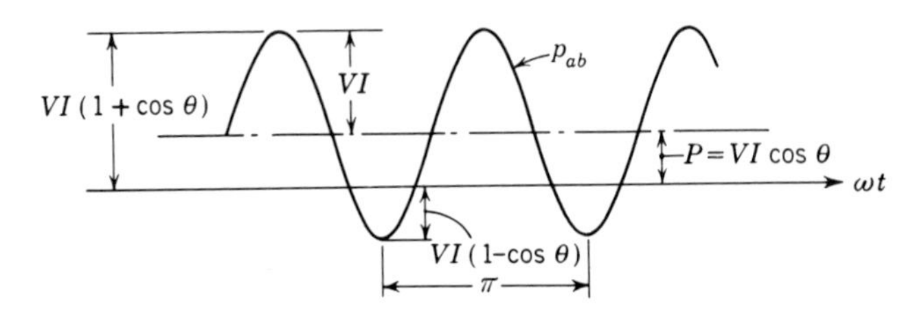

Fig. 9-23 Power as a function of time for a terminal pair in the sinusoidal steady state.

In the entire discussion so far it has been tacitly assumed that θ is in the first or fourth quadrant, i.e., that θ is an angle whose magnitude is between 0 and 90°. This assumption means also that $\text{Re}\,\mathbf{Z}_{ab} > 0$. We now conclude that this assumption is indeed necessary if the terminals pair a-b represents a combination of passive elements. If $\cos \theta$ were a negative number, then $VI \cos \theta$ would be negative and the *average* power flow *into* terminals a-b would be negative, so that the box a-b would represent a source of energy, i.e., an active terminal pair. The condition $\text{Re}\,\mathbf{Z}_{ab} \geq 0$ is a special case of the "positive real" property of driving-point immittance functions which is discussed in Sec. 8-13.

9-14 Average power

The average power delivered to the terminal pair a-b in the sinusoidal steady state is obtained from Eq. (9-41) by observing that the time-average value of the sinusoidal component is zero. Hence

$$P_{\text{avg}} = P_{ab} = VI \cos \theta \tag{9-42}$$

We recall here that V is the effective value of v_{ab}, I is the effective value of the current $i(t)$, and θ is the phase angle by which the voltage v_{ab} leads the current $i(t)$. The "average power" is used so commonly that this quantity is usually referred to simply as the "power." When the time function $p(t)$ is meant, a phrase like "instantaneous power" is used. The symbol for average power is the capital letter P, and no subscripts are employed except when it is necessary to identify the terminal pair in question.

Several formulas are equivalent to Eq. (9-42), and these will now be deduced. Since $\cos \theta$ is defined by the angle of the complex impedance (or admittance) between terminals *a-b*,

$$\cos\theta = \frac{R_{eq}}{|\mathbf{Z}_{ab}|} \tag{9-43}$$

It must be emphasized that the term R_{eq} in Eq. (9-43) refers to the *real part of the complex impedance* between terminals *a-b*. (We recall that Re $\mathbf{Z}_{ab}$ may depend on *all* the elements connected between terminals *a-b*.) Using Eq. (9-43) in (9-42),

$$P = VI\frac{R_{eq}}{Z_{ab}} \tag{9-44}$$

Since $V/Z_{ab} = I$, the expression (9-44) may be rewritten in the form

$$P = I^2R_{eq} \tag{9-45}$$

Alternatively, upon using the admittance and equivalent conductance between terminals *a-b*, $\cos\theta = G_{eq}/|\mathbf{Y}_{ab}|$, so that we may write

$$P = V^2G_{eq} \tag{9-46}$$

It must be emphasized that the equivalent conductance between terminals *a-b* is the real part of the complex admittance. For a terminal pair *a-b* the quantity G_{ab} is therefore *not*, in general, the reciprocal of R_{ab}. We rather note the relationship [setting (9-45) equal to (9-46)]

$$G_{eq}V^2 = R_{eq}I^2$$

so that

$$G_{eq}\frac{V^2}{I^2} = R_{eq}$$

Hence

$$G_{eq}|\mathbf{Z}|^2 = R_{eq} \qquad \text{or} \qquad G_{eq} = R_{eq}|\mathbf{Y}|^2 \tag{9-46a}$$

The reader is reminded that the relationship expressed in Eq. (9-46*a*) is discussed in Sec. 9-11, where equivalent impedances and admittances were discussed.

9-15 *Apparent power, power factor*

The expression for the average power in the sinusoidal steady state is given by any of the equivalent relations

$$P = VI\cos\theta = V^2 \operatorname{Re}\mathbf{Y} = I^2 \operatorname{Re}\mathbf{Z} \tag{9-47}$$

It is now recalled that *if* V and I were *not* the effective values of sinusoidal time functions but were a constant voltage across a terminal pair and a constant (direct) current flowing into the terminal pair, then the power which would be delivered to the terminal pair would be given by $P = VI$. For this reason the quantity VI is termed the "apparent power"; i.e., the result of multiplying the effective value of the voltage by the effective value of the current would give the correct answer for the power *if* direct current were flowing into a purely resistive terminal pair. In the present instance the quantity VI is not the power; we shall see that it can be interpreted physically and that it can also be an important auxiliary quantity in the solution of certain problems. For this reason we shall use the symbol P_a to denote the volt-ampere product, i.e., the apparent power,

$$P_a = VI = I^2|\mathbf{Z}| = V^2|\mathbf{Y}| \tag{9-48}$$

Dimensionally, the apparent power is volts $\times$ amperes = watts. To assure that this quantity is not confused with the average power, it is customary, in electrical engineering, not to refer to P_a as watts, but simply as "volt-amperes," abbreviated va. Thus

$$P_a = VI \qquad \text{va}$$

and

$$P = VI \cos \theta \qquad \text{watts}$$

The quantity $\cos \theta$, which, when multiplied by the apparent power, gives the correct answer for the average power, is referred to as the "power factor" (abbreviated as pf), and is sometimes multiplied by 100 and expressed as a percentage:

$$\cos \theta = \frac{P}{P_a} = \frac{R}{|\mathbf{Z}|} = \frac{G}{|\mathbf{Y}|} = \text{power factor} = \text{pf}$$
$$\text{pf} \times 100 = 100 \cos \theta = \text{power factor, \%} \tag{9-49}$$

Since the cosine of an angle θ is the same as the cosine of the negative angle θ, a statement of the power factor alone does not include information about the sign of the angle θ. It is customary to state that the power factor is *lagging* if the current i_{ab} *lags* the voltage v_{ab} so that θ (the angle of the impedance, the angle of voltage lead) is positive. In an "inductive" circuit, therefore, the power factor is lagging. In a capacitive circuit the current i_{ab} *leads* the voltage v_{ab}, and the power factor is *leading*. The angle of impedance in that case is negative.

We return now to the interpretation of $VI = P_a$. Note from Fig. 9-23 and from Eq. (9-41) that the instantaneous power is

$$p_{ab}(t) = P_a \cos (2\omega t + 2\phi - \theta) + P \tag{9-50}$$

It is therefore possible to interpret the term "apparent power" by stating that the apparent power is the amplitude of the oscillating component of the instantaneous power.

9-16 Complex power, reactive power

We now return to the expression for instantaneous power in complex form as given by Eq. (9-40).

$$p_{ab}(t) = \text{Re}\,(\mathbf{V}^*\mathbf{I}) + \text{Re}\,(\mathbf{V}\mathbf{I}e^{j2\omega t})$$

The quantity $\mathbf{V}^*\mathbf{I}$, written in exponential form, is

$$\mathbf{V}^*\mathbf{I} = VIe^{-j\theta} = P_a e^{-j\theta} = I^2\mathbf{Z}^* = V^2\mathbf{Y}$$

We shall call the complex quantity $P_a e^{-j\theta}$ the "complex power"[1] in the circuit and denote it by the symbol $\mathbf{P}_a$. We can now state that the absolute value of the complex power equals the apparent power, and the real part of the complex power is equal to the average power. The following question naturally arises: Does the imaginary part of the complex power have any physical interpretation?

The imaginary part of the complex power is given by the formula

$$\text{Im}\,(\mathbf{V}^*\mathbf{I}) = \text{Im}\,\mathbf{P}_a = VI \sin(-\theta)$$

This quantity is denoted by the symbol Q and called the reactive power,

$$Q = -VI \sin \theta \tag{9-51}$$

Although dimensionally the units of reactive power are watts, the use of the unit watt is reserved for the average of instantaneous power, and the units of reactive power are called "reactive volt-amperes," abbreviated var. To find physical meaning for reactive power, consider the terminal pair a-b

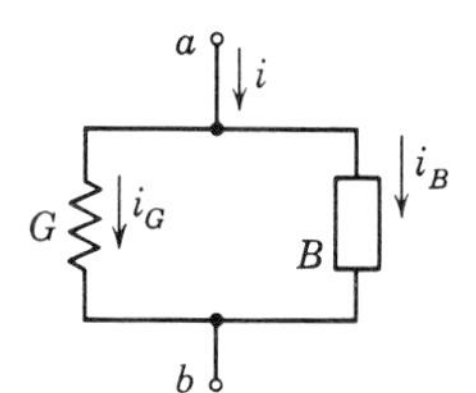

Fig. 9-24 Terminal-pair network represented by parallel conductance and susceptance for sinusoidal-steady-state analysis.

represented as the parallel combination of a conductance G and a susceptance B ($G = \text{Re}\,\mathbf{Y}_{ab}$ and $B = \text{Im}\,\mathbf{Y}_{ab}$) as shown in Fig. 9-24. If the voltage v_{ab} is given as before by

$$v_{ab} = \text{Re}\,(\mathbf{V}\sqrt{2}\,e^{j\omega t})$$

then the current in the susceptance B is given by

$$i_B(t) = \text{Re}\,(jB\mathbf{V}\sqrt{2}\,e^{j\omega t})$$

Hence the complex power which flows *to the susceptance B* is

$$(\mathbf{P}_a)_B = jV^2B$$

Since the instantaneous power to any element (or combination of elements) is given by

$$p(t) = \text{Re}\,\mathbf{P}_a + \text{Re}\,(\mathbf{VI}e^{j2\omega t})$$

[1] Some authors define complex power as $\mathbf{VI}^*$. In that case, since $\text{Re}\,(\mathbf{VI}^*) = \text{Re}\,(\mathbf{V}^*\mathbf{I})$, the real part of the complex power is also equal to the average power but $\text{Im}\,(\mathbf{VI}^*) = -\text{Im}\,(\mathbf{V}^*\mathbf{I})$, so that the sign of the imaginary part is opposite with this alternative definition. It will become apparent that the difference between the two definitions is trivial.

the instantaneous power flow to the susceptance B is

$$p_B(t) = V^2B \cos\left(2\omega t + \frac{\pi}{2} + 2\phi\right)$$

Now $V^2B = \text{Im}\,(V^2\mathbf{Y}) = Q$. Hence the reactive power Q is the maximum power which flows from the source to the element which represents all the energy-storing elements in the circuit during the course of a cycle. It is noted that this is an equivalent element whose value is a function of the value of all the circuit elements, as well as the frequency of the source. We can now write the complex power as follows,

$$\mathbf{P}_a = \mathbf{V}^*\mathbf{I} = P + jQ$$

Another interpretation of reactive power is possible. Since the power which flows to the energy-storing elements, represented by the susceptance B, is

$$p_B = Q \sin 2\omega t$$

the waveform of this power is the "double-frequency" sinusoid shown in Fig. 9-25*a*. The maximum *energy* which is delivered to the susceptance

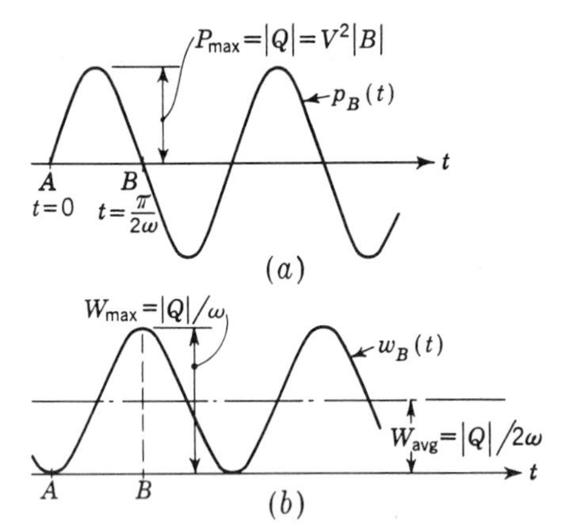

Fig. 9-25 Power and energy in a susceptance as a function of time.

B is given by the area under the curve from point A to point B. (In the time interval from $t = \pi/2\omega$ to $t = \pi/\omega$, this energy is returned to the source.) Since the frequency of the power fluctuations is twice the frequency of the voltage or current variations, a half cycle of power oscillations lasts from $\omega t = 0$ (point A) to $\omega t = \pi/2$ (point B). The maximum energy stored in the susceptance B is therefore given by

$$W_{\text{max}} = \int_0^{\pi/2\omega} Q \sin 2\omega t\, dt = -\frac{Q}{2\omega} \cos 2\omega t \Big|_0^{\pi/2\omega}$$

or

$$W_{\text{max}} = \frac{|Q|}{\omega} \tag{9-52}$$

The result of Eq. (9-52) may be stated in words:

$$|\text{Reactive power}| = \omega \times (\text{max value of stored energy})$$

The variation of the stored energy with time is shown in Fig. 9-25*b*. We observe that the reactive power is given by

$$|Q| = 2\omega \times (\text{avg stored energy in terminal pair})$$

A summary of the power relations is given in Table 9-2. Not all the formulas given here have been proved in the text; the reader is encouraged to prove them independently.

Table 9-2 Power relations in a terminal pair

Symbol or abbreviation	Name	Formulas	Comments
p_{ab}	Instantaneous power	$v \cdot i = \text{Re}(\mathbf{V}^*\mathbf{I}) + \text{Re}(\mathbf{VI}e^{j2\omega t})$	
P	Power	$P = VI\cos\theta = \text{Re}(\mathbf{V}^*\mathbf{I}) = I^2 \text{Re}\,\mathbf{Z} = V^2 \text{Re}\,\mathbf{Y}$	Time-average power
Q	Reactive power	$Q = -VI\sin\theta = \text{Im}(\mathbf{V}^*\mathbf{I}) = I^2 \text{Im}\,\mathbf{Z}^* = V^2 \text{Im}\,\mathbf{Y}$	Peak values of power to energy-storing elements
P_a	Apparent power	$P_a = VI = V^2Y = I^2Z = \|\mathbf{V}^*\mathbf{I}\|$	Peak value of oscillating component of $p(t)$
$\mathbf{P}_a$	Complex power	$\mathbf{P}_a = P + jQ = VI\cos\theta - jVI\sin\theta = V^2\mathbf{Y} = I^2\mathbf{Z}^*$	
pf	Power factor	$\text{pf} = \cos\theta = \frac{R}{Z} = \frac{G}{Y} = \frac{P}{P_a}$	Lagging for positive θ
rf	Reactive factor	$\text{rf} = -\sin\theta = \frac{-X}{Z} = \frac{Y}{B} = \frac{Q}{P_a}$	Lagging for positive θ

9-17 Use of complex power ADMITTANCES IN PARALLEL Consider two complex admittances, $\mathbf{Y}_1$ and $\mathbf{Y}_2$, connected in parallel across the terminals *a-b*. The total admittance $\mathbf{Y}_{ab}$ is

$$\mathbf{Y}_{ab} = \mathbf{Y}_1 + \mathbf{Y}_2$$

and the current $\mathbf{I}$ is given by

$$\mathbf{I} = \mathbf{V}_{ab}(\mathbf{Y}_1 + \mathbf{Y}_2) \tag{9-53}$$

With $\mathbf{V}_{ab}$ as reference, the phasor diagram which shows the addition of $\mathbf{I}_1$ to $\mathbf{I}_2$ to give $\mathbf{I}$ is drawn in Fig. 9-26*a*. In this diagram it is assumed that $\mathbf{Y}_1$ is inductive (negative imaginary part) and $\mathbf{Y}_2$ is capacitive (positive imaginary part).

Let us now multiply Eq. (9-53) by $\mathbf{V}_{ab}^*$. Since $\mathbf{V}_{ab}\mathbf{V}_{ab}^* = V_{ab}^2$, we obtain

$$\mathbf{V}^*\mathbf{I} = V_{ab}^2(\mathbf{Y}_1 + \mathbf{Y}_2)$$

or

$$\mathbf{V}^*\mathbf{I} = V_{ab}^2\mathbf{Y}_1 + V_{ab}^2\mathbf{Y}_2 \tag{9-54}$$

Equation (9-54) states that, *for parallel admittances, the complex power for the combination is the sum of the individual complex powers.* This relationship is particularly useful in problems in which the admittances are not explicitly specified but the power and power factor (and therefore the complex powers) of each parallel branch are known. The addition of the complex powers is shown in Fig. 9-26*b*. Note that the real axis represents power and the imaginary axis represents reactive power (since complex power is $P + jQ$). In Fig. 9-26*b* it is again assumed that $\mathbf{Y}_1$ is inductive and $\mathbf{Y}_2$ is capacitive.

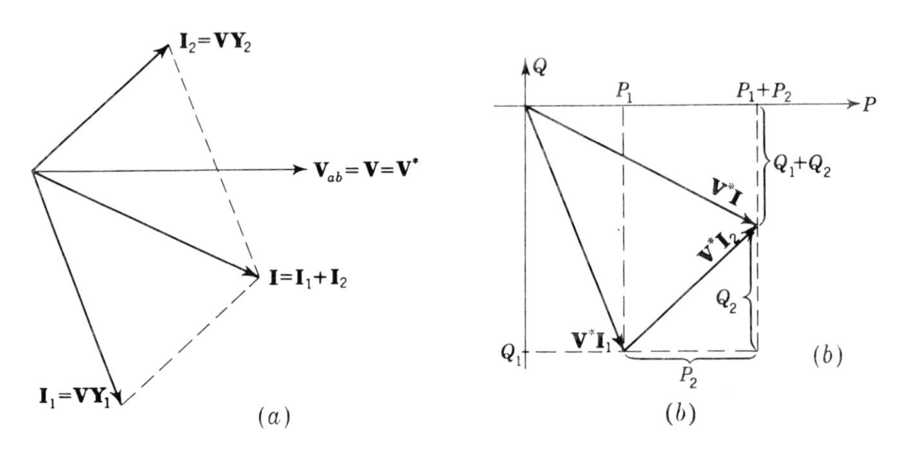

Fig. 9-26 Relationship between current phasor diagram (a) and (b) the addition of complex power.

IMPEDANCES IN SERIES We shall now show that the addition of complex powers gives the correct result for calculating the complex power of impedances in series.

The complex power for a circuit consisting of $\mathbf{Z}_1 + \mathbf{Z}_2 = \mathbf{Z}_{ab}$ is given by

$$\mathbf{V}_{ab}^*\mathbf{I}_{ab} = [(\mathbf{I}_{ab}\mathbf{Z}_1)^* + (\mathbf{I}_{ab}\mathbf{Z}_2)^*]\mathbf{I}_{ab}$$

Using the rules for manipulations of complex numbers,

$$\mathbf{V}_{ab}^*\mathbf{I}_{ab} = I_{ab}^2(\mathbf{Z}_1^* + \mathbf{Z}_2^*) = I_{ab}^2(\mathbf{Z}_1 + \mathbf{Z}_2)^*$$

so that the addition of the complex powers gives the complex power for the combination of series impedances.

Example 9-12 Three loads are connected in parallel across 660-volt a-c lines. These loads are specified as follows:

Load 1, 15 kw, 0.6 pf (lagging)
Load 2, 20 kw, 0.707 pf (leading)
Load 3, 10 kw, 0.4 pf (lagging)

The loads are connected to a substation by means of cables having a resistance of 1.1 ohms for the circuit and negligible reactance. Calculate the voltage at the substation.

Solution A schematic diagram is drawn as shown in Fig. 9-27*a*. Since for each load the power, power factor, and line voltage are given, the complex admittance can be calculated. For example, for load 1, $15{,}000 = 660 \times I_1 \times 0.6$. Then $\mathbf{Y}_1 = (I_1/660)\underline{/-\cos^{-1} 0.6}$ or $\mathbf{Z}_1 = (660/I_1)\underline{/\cos^{-1} 0.6}$.

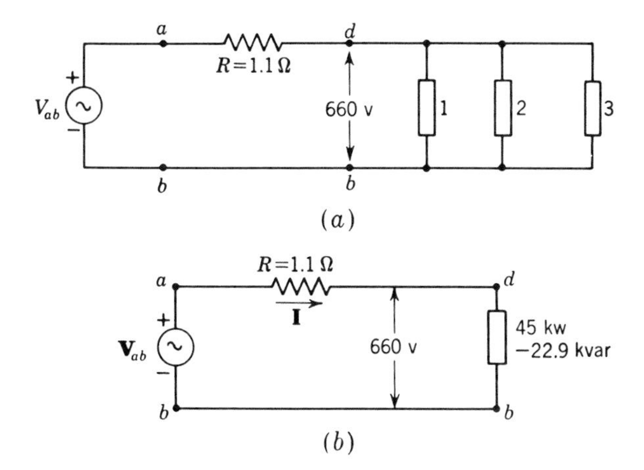

Fig. 9-27 Circuit for Example 9-12.

However, it is not necessary to calculate the impedances. The three loads can be combined into one equivalent load by the "complex-power method." The total power dissipated by the three loads is $P_{db} = P_1 + P_2 + P_3 = 15 + 20 + 10 = 45$ kw. The reactive powers are

$$\begin{aligned} Q_1 &= -15 \tan (\cos^{-1} 0.6) = -20 \text{ kvar} \\ Q_2 &= +20 \tan (\cos^{-1} 0.707) = +20 \text{ kvar} \\ Q_3 &= -10 \tan (\cos^{-1} 0.4) = -22.9 \text{ kvar} \end{aligned}$$

Hence the combination of the three loads, at terminals *d-b*, is equivalent to a single load which receives 45 kw and −22.9 kvar (Fig. 9-27*b*).

The apparent power for the load is

$$P_a = \sqrt{P^2 + Q^2} = 50.5 \text{ kva}$$

Hence the magnitude of the current I_{ad} is

$$I_{ad} = |\mathbf{I}_{ad}| = \frac{50.5 \times 10^3}{660} = 76.5 \text{ amp}$$

[Since $\tan^{-1} (22.9/45) = 27°$, $\mathbf{I}_{ad}$ lags $\mathbf{V}_{ab}$ by 27°.]

The voltage V_{ab} may be found in several ways. Continuing the use of power relations,

$$P_{ad} = 76.5^2 \times 1.1 = 6.46 \text{ kw}$$

Hence, since $Q_{ad} = 0$,

$$\begin{aligned} P_{ab} &= 45 + 6.46 \approx 51.5 \text{ kw} \\ Q_{ab} &= -22.9 \end{aligned}$$

The apparent power at terminals *a-b* is $\sqrt{51.5^2 + 22.9^2} = 56.3$ kva. Hence the voltage V_{ab} is, in magnitude,

$$V_{ab} = \frac{56{,}300}{76.5} = 736 \text{ volts} \qquad \textit{Ans.}$$

In this example the use of complex power was illustrated. The reader should check the results by other methods (calculation of admittances and impedances, solution by voltage phasor diagram, etc.).

PROBLEMS

9-1 In Fig. P9-1 the terminal pair *a-b* consists of a single element. Calculate the nature and the value of this element if the waveform marked v is (*a*) v_{ab}; (*b*) v_{ba}.

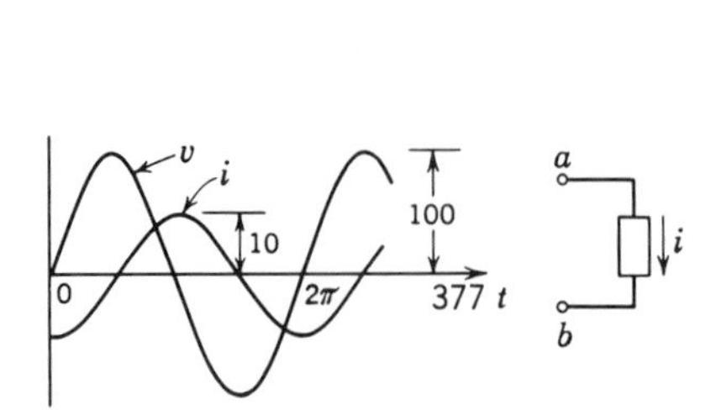

Fig. P9-1

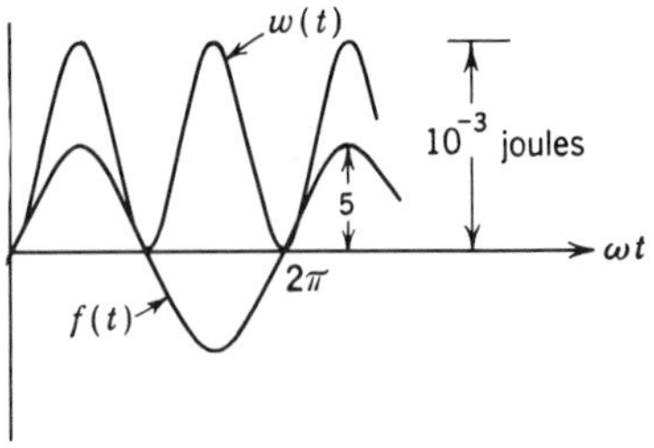

Fig. P9-2

9-2 Figure P9-2 shows a sinusoidal source function $f(t)$ and a curve of the energy which this source function delivers to an element. Calculate the nature and the value of this element if (*a*) $f(t) = v(t)$; (*b*) $f(t) = i(t)$.

9-3 A sinusoidal current $i_{ab} = \cos 4t$ flows through an element which is either an inductance ($L = 2$) or a capacitance ($C = \frac{1}{2}$). Calculate the value of the energy stored in this element in each case (L and C) at an instant when (*a*) $i = 0$; (*b*) $i = \frac{1}{2}$; (*c*) $i = 1$. The voltage across C is zero at $t = 0$.

9-4 The voltage across a capacitance has the form $v(t) = V_m \cos \omega t$. At $\omega t = \pi/3$ the energy stored in the capacitance is 5 joules. (*a*) What is the maximum value of the energy stored in the capacitance? (*b*) If $C = 10^{-3}$, calculate V_m.

9-5 In the circuit shown in Fig. P9-5 the source $i(t)$ is the sinusoid $i(t) = 5 \cos 10t$. In the steady state, $v_{ab}(t) = \cos (10t - 53.1°)$. (*a*) Calculate R and C. (*b*) Write the differential equation which relates $v_{ab}(t)$ to $i(t)$. (*c*) The source $i(t)$ is changed to be $i(t) = 5 \cos 5t$; calculate the steady-state response $[v_{ab}(t)]_s$.

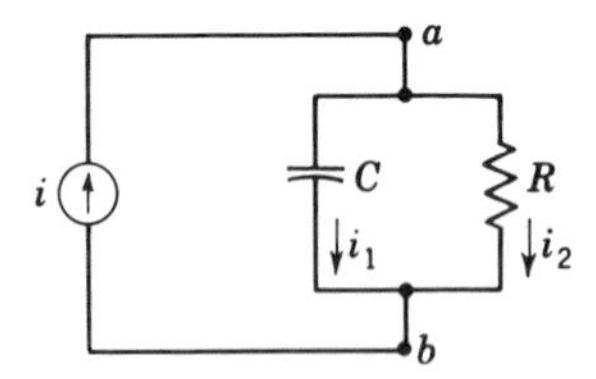

Fig. P9-5

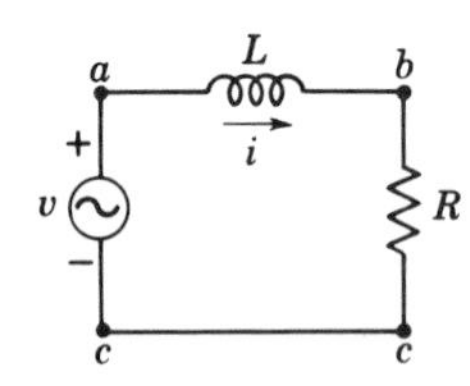

Fig. P9-6

9-6 The circuit shown in Fig. P9-6 is in the steady state. The rms value of v_{ab} is 100 volts, and the rms value of v_{ac} is 200 volts. Calculate (*a*) the rms value of v_{bc}; (*b*) the phase angle between v_{bc} and v_{ac}. (*c*) If v_{ac} is in phase with the function $\cos \omega t$, write the functions v_{ac}, v_{bc}, and v_{ab}.

9-7 The circuit of Fig. P9-5 is in the sinusoidal steady state. The rms value of i is 5 amp, and the rms value of i_1 is 3 amp. (*a*) Calculate the rms value of i_2. (*b*) Calculate the value of i_2 at an instant when (1) $i_1 = 3\sqrt{2}$ amp; (2) $i_1 = 3$ amp; (3) $i_1 = 0$; (4) $i = 0$; (5) $i = 5\sqrt{2}$ amp.

9-8 In the circuit shown in Fig. P9-8, $v_{db} = 100\sqrt{2}\cos 5{,}000t$. (*a*) Represent v_{db} as a phasor. (*b*) Calculate the phasor $\mathbf{V}_{ab}$. (*c*) Write $v_{ab}(t)$. (*d*) Draw a phasor diagram which shows $\mathbf{V}_{ad} + \mathbf{V}_{db} = \mathbf{V}_{ab}$.

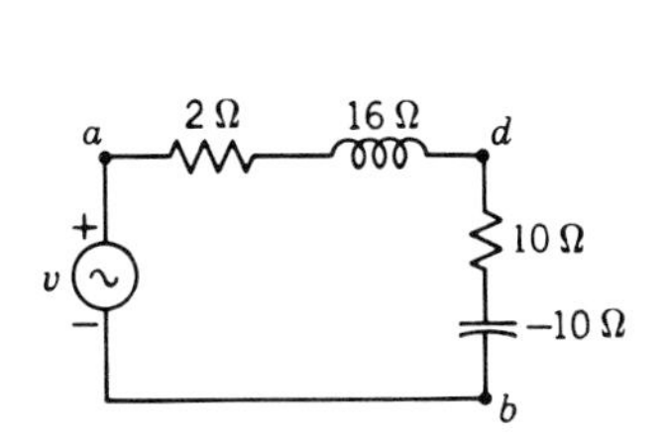

Fig. P9-8

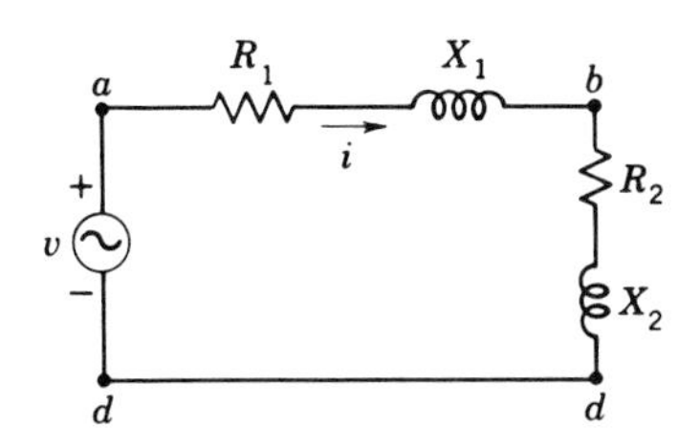

Fig. P9-9

9-9 In the circuit shown in Fig. P9-9, $R_1 = 25$, $X_1 = 10$ ohms, $X_2 = 26$ ohms. The rms value of v_{ab} is equal to the rms value of v_{bd}. (*a*) Calculate R_2. (*b*) Calculate the magnitude of the phasor ratio $\mathbf{V}_{ab}/\mathbf{V}_{ad}$. (*c*) If $V_{ad} = 100/0°$, find $\mathbf{V}_{ab}$ and $\mathbf{V}_{bd}$. (*d*) Calculate the value of v_{ab} when $v_{ad} = 100$ volts if $\mathbf{V}_{ad} = 100/0°$.

9-10 The circuit of Fig. P9-6 is in the steady state, $v(t) = 6 + 10\cos 2t$, $i(t) = 2 + A\cos(2t - 53.1°)$. Use the principle of superposition to calculate (*a*) R; (*b*) L; (*c*) A; (*d*) the steady-state equation for $i(t)$ if $v(t) = 10 + 5\cos t + 5\cos 2t$.

9-11 The circuit of Fig. P9-9 is in the steady state with $X_2 = 0$ and $i = 5\sqrt{2}\cos \omega t$. The rms value of v_{bd} is 50 volts, and the rms value of $v(t)$ is 105.8 volts. It is known that $3R_1 = X_1$. (*a*) Use a voltage phasor diagram drawn to scale to determine R_1 and X_1. (*b*) Calculate the phase angle between (1) $v(t)$ and v_{ab}; (2) $v(t)$ and i.

9-12 In the circuit of Fig. P9-9, $R_2 = 4$ ohms, $X_2 = 3$ ohms, $V_{ab} = 100$ volts, and $V_{ad} = 200$ volts. Calculate R_1 and X_1 if (*a*) $X_1/R_1 = 1$; (*b*) $X_1/R_1 = 10$.

9-13 In the circuit shown in Fig. P9-13, $V_{ab} = 100$ volts, and X_c is variable. Use a voltage phasor diagram to determine the smallest value of E for which the voltage $V_{ab} = 100$ volts.

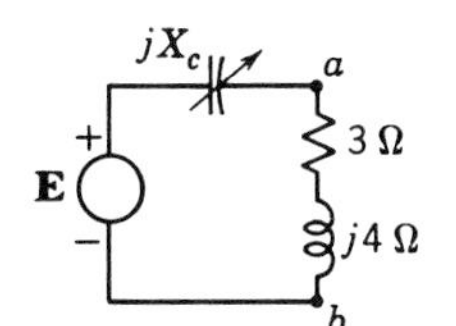

Fig. P9-13

9-14 In the circuit shown in Fig. P9-14, $V_{ab} = 100$ volts, $I = 5$ amp, $I_R = 2$ amp, $I_C = 10$ amp. Calculate R, L, and C if $\omega = 400$.

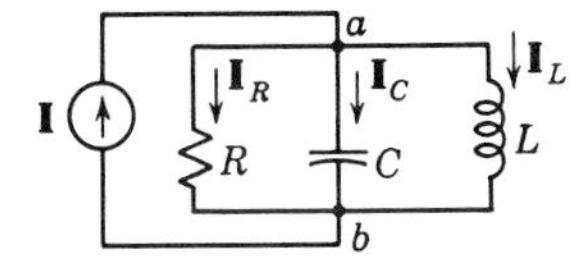

Fig. P9-14

9-15 In the circuit shown in Fig. P9-15, $R_1 = 6$ and $R_2 = 8$. It is known that $I_1 = I_2 = 10$ and $V_{ab} = 100$ volts. Calculate X_1, X_2, and I.

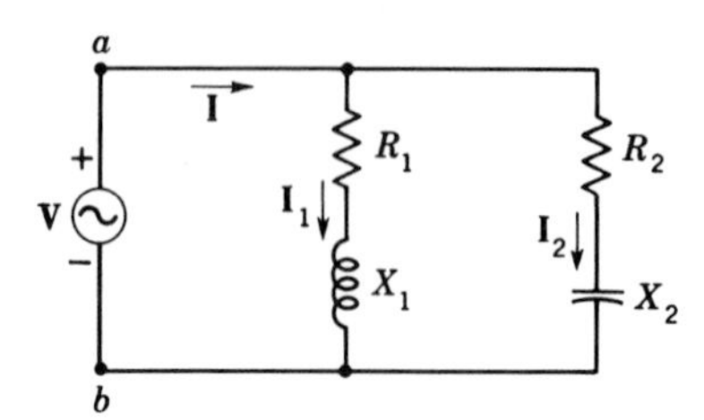

Fig. P9-15

9-16 In the circuit of Fig. P9-15, $R_2 = 0$, $R_1 = 5$ ohms, and $X_1 = 10$ ohms. (*a*) Calculate the value of X_2 if $\mathbf{V}$ and $\mathbf{I}$ are in phase. (*b*) If X_2 has the value of (*a*), calculate $\mathbf{I}_1$ and $\mathbf{I}_2$ if $\mathbf{I} = 1\underline{/0°}$. Illustrate your answer with the aid of a phasor diagram.

9-17 In Fig. P9-17, if $X = -R$, calculate the phasor ratio $\mathbf{V}_{ab}/\mathbf{V}$.

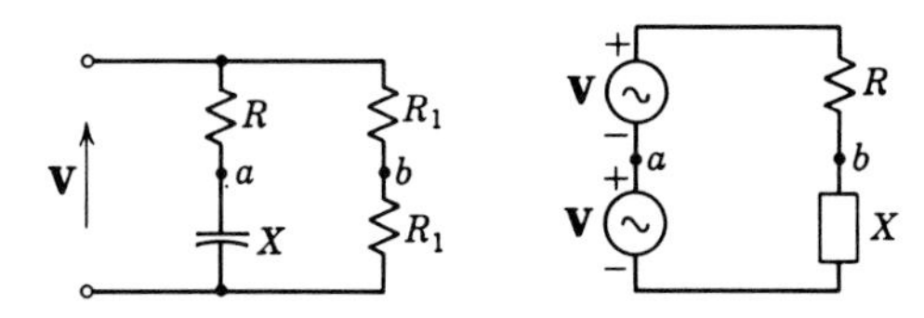

Fig. P9-17 *Fig. P9-18*

9-18 In the circuit shown in Fig. P9-18, the two sources have equal amplitudes and are in phase. Calculate the phasor ratio $\mathbf{V}_{ab}/\mathbf{V}$ if (*a*) $R = X$; (*b*) $3R = X$; (*c*) $3R = -X$; (*d*) $3X = -R$.

9-19 In the circuits of Figs. P9-17 and P9-18, show that the magnitude ratio V_{ab}/V is independent of the ratio X/R.

9-20 In the circuit shown in Fig. P9-20: (*a*) Calculate $\mathbf{I}_0$ and $\mathbf{V}_{ab}$ if $\mathbf{I} = 1\underline{/0°}$. (*b*) Use the result of (*a*) to calculate the driving-point impedance $\mathbf{Z}_{ab}$. (*c*) Use the result of (*a*) to determine the value of V_{ab} which gives $I = 2.5$ amp. (*d*) Use the result of (*a*) to determine the value of $\mathbf{I}$ when $\mathbf{V}_{ab} = 10\underline{/0°}$ volts.

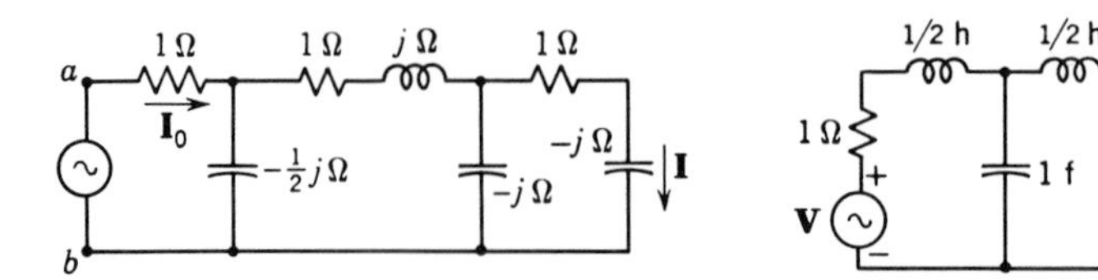

Fig. P9-20 *Fig. P9-21*

9-21 In Fig. P9-21: (*a*) Calculate the radian frequency at which V_2/V is $1/\sqrt{8}$. (*b*) Find the angle by which $\mathbf{V}_2$ lags $\mathbf{V}$ at the frequency of (*a*).

9-22 In Fig. P9-22, if $X_L = -X_c$, deduce Norton's equivalent circuit with respect to terminals *a-b*.

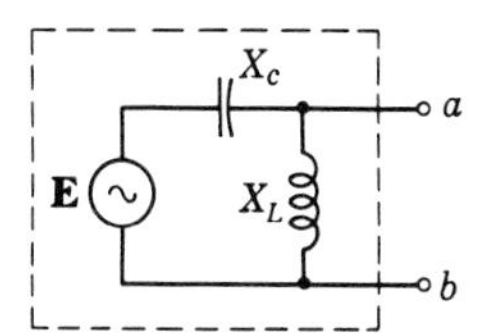

Fig. P9-22

9-23 The active terminal-pair network shown in Fig. P9-23 contains sinusoidal sources of the same frequency, in addition to R and L elements, and the steady-state conditions only are studied. The following data are obtained:

V_{ab}, rms volts	100	55	26.7
Z, ohms	Open circuit	$11 + j0$	$4 + j0$

(*a*) Deduce Thévenin's equivalent circuit for the active terminal pair in the sinusoidal steady state. (*b*) Calculate V_{ab} when (1) $\mathbf{Z} = \mathbf{Z}_s$, where $\mathbf{Z}_s$ is the internal impedance of the Thévenin source; (2) $\mathbf{Z} = |\mathbf{Z}_s|$; (3) $\mathbf{Z} = \mathbf{Z}_s^*$.

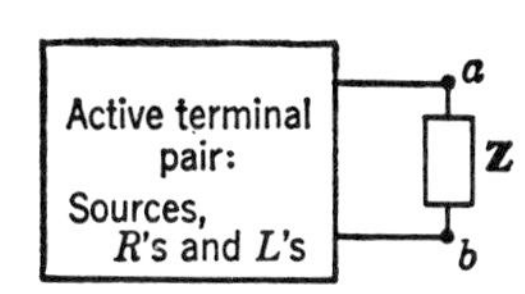

Fig. P9-23

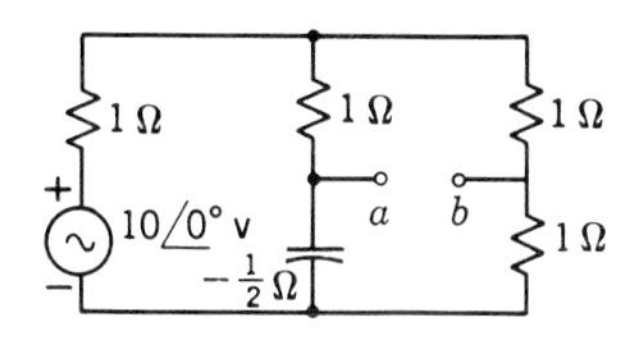

Fig. P9-24

9-24 In the circuit shown in Fig. P9-24: (*a*) Calculate $\mathbf{V}_{ab}$. (*b*) Calculate $\mathbf{I}_{ab}$ if terminals a and b are short-circuited. (*c*) Use Thévenin's theorem to calculate $\mathbf{V}_{ab}$ when an impedance $\mathbf{Z}_{ab}$ is connected between terminals *a-b* and (1) $\mathbf{Z}_{ab} = 1$; (2) $\mathbf{Z}_{ab} = 1 + j$.

9-25 Deduce the equation of balance for each of the bridge circuits shown in Fig. P9-25 (i.e., the relationship between the parameters so that $\mathbf{V}_{ab} = 0$).

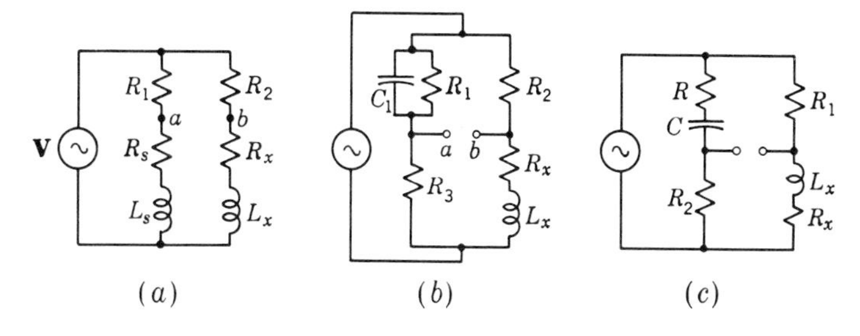

Fig. P9-25

9-26 For the terminal pair shown in Fig. P9-26: (*a*) Calculate $\mathbf{Y}_{ab}$ and $\mathbf{Z}_{ab}$. (*b*) Represent the terminal pair as the series combination of R' and C' if $\omega = 1$. (*c*) Show that if the series representation of (*b*) is to apply for all values of ω, then R' and C' must be functions of frequency. (*d*) Sketch $R'(\omega)$ and $C'(\omega)$ as obtained in (*c*) as functions of the radian frequency.

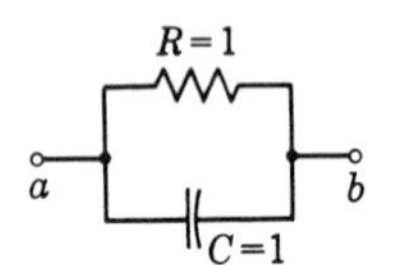

Fig. P9-26

9-27 Represent the complex driving-point immittance $\mathbf{Z}_{ab}$ of Fig. P9-20 as (*a*) the series connection of two elements; (*b*) the parallel connection of two elements. Use $\omega = 2$.

9-28 The series circuit shown in Fig. P9-28 is to be represented as the parallel combination of R' and X'_L. Plot R' as a function of X_L; and X'_L as a function of X_L.

1 Ω X_L
a b *Fig. P9-28*

9-29 Calculate the complete response $v_{ab}(t)$ in the circuit of Fig. P9-5 if $R = 6$, $C = 0.025$, and (*a*) $i(t) = 10 \cos 5t$, $v_{ab}(0^+) = 0$; (*b*) $i(t) = 10 \cos 5t$, $v_{ab}(0^-) = 30$; (*c*) $i(t) = 10 \cos 5t$, $v_{ab}(0^-) = -30$.

9-30 In Fig. P9-30 the circuit is in the steady state with the switch in position 2. At $t = 0$ the switch is thrown to position 1. If $v(t) = 5 \cos (3t + \phi)$ and $E = 6$, calculate $i(t)$ for all $t \geq 0^+$, and sketch the result for (*a*) $\phi = 0$; (*b*) $\phi = -\pi/2$.

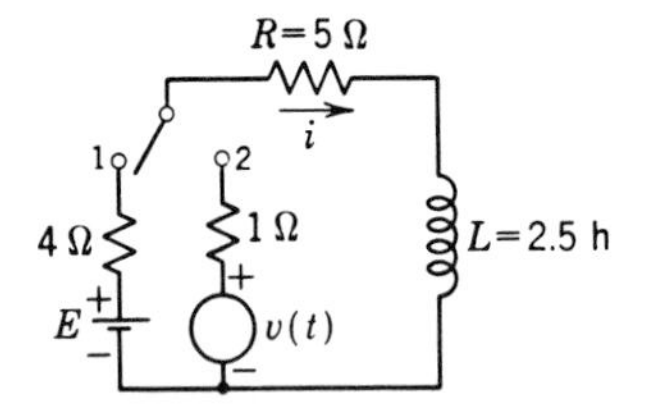

Fig. P9-30

9-31 In the circuit of Fig. P9-30, assume that steady state has been reached with the switch in position 1. At $t = 0$ the switch is thrown to position 2. Calculate and sketch $i(t)$ if $E = 18$, $v(t) = 14 \cos (3t + \phi)$, and sketch the result for (*a*) $\phi = 0$; (*b*) $\phi = -\pi/2$.

9-32 An initially deenergized parallel *R-C* circuit is connected across an ideal current source $i(t) = I_m \cos (\omega t + \phi)$ at $t = 0$. (*a*) Formulate the complete response for the voltage across the source $v(t)$. (*b*) From sketches of $v(t)$ estimate its maximum value and the instant after $t = 0$ at which it occurs if ϕ is chosen so that the *steady-state component of* v has its maximum at $t = 0$ in each of the following cases: (1) $\omega RC \ll 1$; (2) $\omega RC \gg 1$; (3) $\omega RC \approx 1$. (*c*) What is the complete response $v(t)$ if ϕ is chosen so that the steady-state component of v is zero at $t = 0$?

9-33 Sketch to scale each of the following functions: (*a*) $f(t) = e^{-t} - \cos 4\pi t$; (*b*) $f(t) = \cos 2\pi t - \frac{1}{2}e^{-t} \cos 10\pi t$.

9-34 In the circuit shown in Fig. P9-34, $R_1 = 10$ ohms, $R_2 = 0.2$ ohm, $L = 0.05$ henry, and $C = 0.05$ farad. The initial-energy storage is zero. The source function is

$v(t) = [5 \cos (18t + \phi)]u(t)$. (*a*) Find $v_2(0^+)$ and $(dv_2/dt)_{0^+}$ in terms of ϕ. (*b*) Find the complete response $v_2(t)$ if (1) $\phi = 0$; (2) $\phi = -\pi/2$.

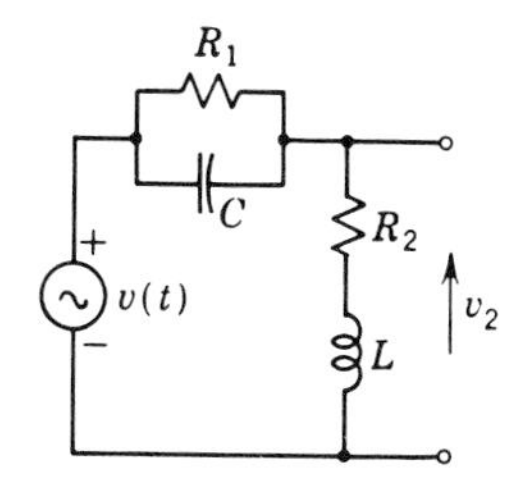

Fig. P9-34

9-35 In the circuit of Fig. P9-34, $R_1 = 1$ ohm, $R_2 = 3$ ohms, $L = 1$ henry, and $C = 1$ farad. The voltage across the capacitance at $t = 0^+$, $v(0^+) - v_2(0^+) = 8$ volts, and the current in the inductance at $t = 0^+$ is zero. The source function is given by $v(t) = [8 \cos (2t + \phi)]u(t)$. Find the complete response $v_2(t)$ for (*a*) $\phi = 0$; (*b*) $\phi = -\pi/2$.

9-36 In the series *R-L-C* circuit shown in Fig. P9-36, the source function has the form $v(t) = [V_m \cos (\omega t + \phi)]u(t)$ and the initial-energy storage is zero. The circuit is highly oscillatory ($R/2L = \alpha \ll \omega_0 = 1/\sqrt{LC}$), so that, for $\alpha t \ll 1$, the source-free component of the response $i(t)$ can be approximated by Re ($\mathbf{K}e^{j\omega_0 t}$) (see Prob.

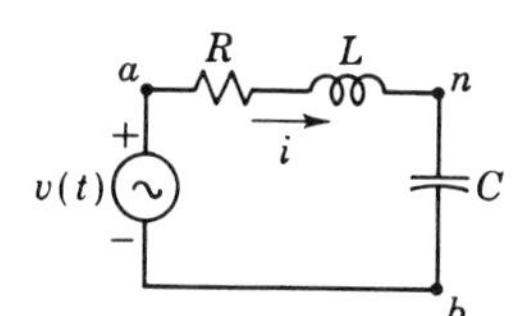

Fig. P9-36

4-43). The radian frequency of the source, ω, satisfies the inequality $\omega \leq 0.1\omega_0$. (*a*) Show that, because $\omega \leq 0.1\omega_0$, $\mathbf{Z}_{ab} \approx R + 1/j\omega C$, and because $\alpha \ll \omega_0$, $\mathbf{Z}_{ab}$ can be further approximated as $\mathbf{Z}_{ab} \approx 1/j\omega C$. (*b*) Show that the steady-state current is approximately given by $i_s(t) = I_m \cos (\omega t + \phi + \pi/2)$, $I_m = \omega C V_m$. (*d*) Show that the initial value of the derivative $(di/dt)_{0^+}$ is $(V_m \cos \phi)/L$. (*e*) Using the approximations outlined above, find the complete response (for $\alpha t \ll 1$) $i(t)$ for the cases $\phi = 0$ and $\phi = \pi/2$. (*f*) For each of the cases given in part *e*, estimate the largest value of $i(t)/I_m$ that occurs after $t = 0$. (*g*) Find the result required in part *f* if $\omega_0 \leq 0.1\omega$, so that $\mathbf{Z}_{ab} \approx j\omega L$ and $i_s(t) \approx I_m \cos (\omega t + \phi - \pi/2)$, $I_m = V_m/\omega L$.

9-37 In the circuit of Fig. P9-24, the capacitance stores zero energy at $t = 0$ and the source function is $v(t) = 10\sqrt{2} \cos 2t$. (*a*) Calculate the value of the capacitance. (*b*) Find the complete response $v_{ab}(t)$.

9-38 In Fig. P9-24, if $v(t) = 10 \cos 2t$, calculate in the steady state (*a*) the maximum value of the energy stored in the capacitance; (*b*) the values $v(t)$ can have when the condition of (*a*) occurs.

9-39 In Fig. P9-39, if $i(t) = \cos t$, calculate in the steady state (*a*) the average energy stored by the inductance; (*b*) the average energy stored by the capacitance; (*c*) the energy dissipated by the resistance in one period ($t = 0$ to $t = 2\pi$).

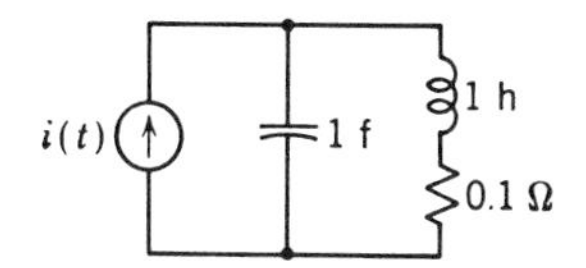

Fig. P9-39

9-40 The R-L circuit of Fig. P9-6 is initially deenergized. The source is adjusted so that $v(t) = V_m \cos(\omega t + \theta)$, $\theta = \tan^{-1}(\omega L/R)$. (*a*) Show that the complete response $i(t)$ is given by

$$i(t) = I_m(\cos \omega t - e^{-Rt/L}) \qquad I_m = \frac{V}{Z},\ t \geq 0^+$$

(*b*) Deduce an expression for the energy delivered to the inductance as a function of time. (*c*) Calculate the average energy stored in the inductance during the time interval $t = 0$ to $t = 2n\pi/\omega$, and show how the result is related to the reactive power as n increases to very large values.

9-41 In the circuit of Fig. P9-15, $R_1 = 6$ ohms, $X_1 = 10$ ohms, $R_2 = 24$ ohms, and $X_2 = -18$ ohms. (*a*) Calculate the effective value of v_{ab} if the average power delivered to the 6-ohm resistance is 500 watts. (*b*) For v_{ab} as in (*a*), calculate P_{ab}, Q_{ab}, the power, and the reactive power for each element.

9-42 The average power delivered to a load at 120 volts is 14.4 kw. The power is delivered through the 0.10-ohm resistance as shown in Fig. P9-42. Calculate and graph the efficiency of transmission $(P_{db}/P_{ab}) \times 100$ per cent as a function of load power factor and as a function of the angle which corresponds to the various load power factors.

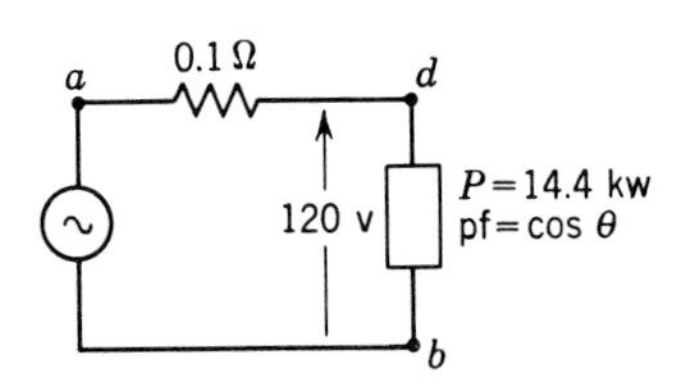

Fig. P9-42

9-43 A certain load draws 60 kw at a power factor of 0.5 lagging. A second load is connected in parallel. Calculate the complex power drawn by the second load if the parallel combination draws 60 kw at a power factor of (*a*) 0.6 lagging; (*b*) 0.8 lagging; (*c*) 0.9 lagging; (*d*) unity; (*e*) 0.9 leading.

9-44 The line voltage in Prob. 9-43 is 1,300 volts. Calculate the admittance of the "second" load for each of the cases enumerated in Prob. 9-43.

9-45 If load 1 draws 30 kw at unity power factor and load 2 draws 30 kw at a power factor of 0.45 (lagging), calculate the value of E and the power factor at terminals a-b in the circuit of Fig. P9-45.

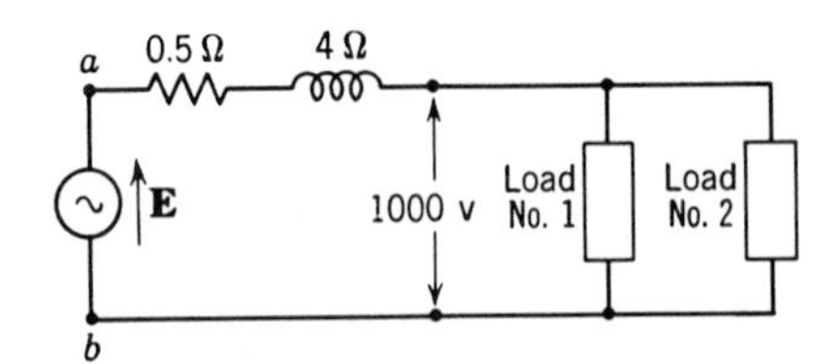

Fig. P9-45

9-46 The voltage phasor across a terminal-pair network has the form $\mathbf{V}_{ab} = \mathbf{V}_1(1 + \mathbf{\Gamma})$, where $\mathbf{\Gamma}$ is a complex number and the current entering the a terminal is given by the phasor $\mathbf{I}_{ab} = (\mathbf{V}_1/R)(1 - \mathbf{\Gamma})$. Show that the power and the reactive power delivered to the terminal pair are given by the expressions $P_{ab} = V_1^2(1 - \Gamma^2)/R$, and $Q_{ab} = -jV_1^2(\mathbf{\Gamma}^* - \mathbf{\Gamma})/R$.

Chapter 10 Adjustable circuits in the sinusoidal steady state

In Chap. 9 we discussed the techniques which are used to solve series-parallel a-c circuit problems. In each example we assumed that both the element values and the frequency of the source were uniquely specified. In this chapter we shall combine the techniques introduced in Chap. 9 with the pole-zero diagram to study circuits which are in the sinusoidal steady state but which are "adjustable." By an adjustable circuit we mean a circuit in which either the value of one (or more) element is changed *or* the frequency is changed *or* both are changed. It must be emphasized at the outset that we are not concerned with time-varying parameters. Rather, we consider circuits in which the change of the value of element (or frequency) is made in steps (possibly "small" steps) but after each change the steady state *only* is calculated. Thus we shall not be concerned with the transition (i.e., the transient) which occurs between the steady-state conditions as the element or the frequency is varied.

The importance of this subject matter in the analysis of electric circuits may be understood from the following observations:

1 In many circuits the elements actually have different values at various times; for example, the load on a motor or generator is usually not fixed.
2 The study of the adjustment of element values or frequency in the circuit can eventually lead us to the design of circuits which will have a desired response.

3 In many practical instances several sinusoidal sources of different frequency may be impressed on a circuit (either intentionally or unintentionally). A knowledge of circuit response over an appropriate range of frequencies makes it possible to analyze such a situation. As an example, consider the antenna circuit of a radio (or television) receiver. All stations (i.e., frequencies) are "picked up" by the antenna, but only the desired station is to be selected for listening.

4 The laboratory testing of circuits in the sinusoidal steady state at various frequencies can give information about the complete response of the circuit under various conditions. This follows from the fact that the pole-zero locations which characterize the complete response of a circuit to any source also determine the value of the complex network function. Consequently, it is possible to infer the pole-zero locations of a network function from the sinusoidal steady-state response at various frequencies.

10-1 Adjustable circuits

INTRODUCTION If we wish to study the effect of varying one element on the sinusoidal steady-state response of a circuit, we need to evaluate quantitatively the effect which the particular element has on the complex network function relating the source phasor to the response phasor. In two-element combinations this effect is easy to evaluate. Consider, for example, the series *R-L* circuit shown in Fig. 10-1*a*. In this circuit the resistance is

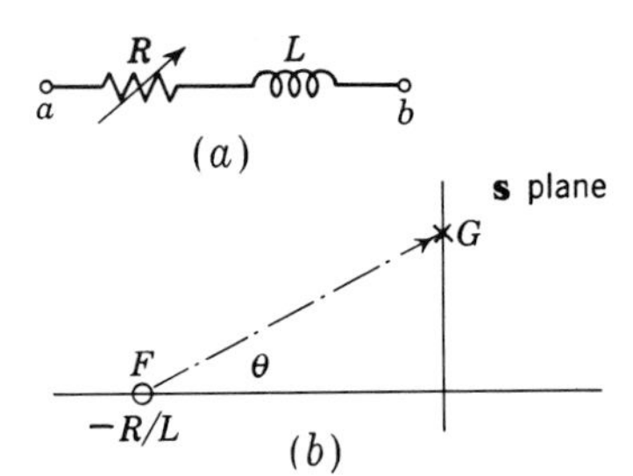

***Fig. 10-1** (a) Series R-L branch with adjustable resistance. (b) Pole-zero diagram for the impedance of (a).*

adjustable (in general, the adjustable elements are indicated on the circuit diagram by means of the arrow through the element symbol). In this circuit $Z_{ab}(s) = L(s + R/L)$; that is, the driving-point impedance has a zero at $s = -R/L$, as shown in Fig. 10-1*b*. Since a sinusoidal source is represented by a pole on the imaginary axis as at G, we observe that, as R is increased, the F point in Fig. 10-1*b* moves to the left, so that the distance FG increases and θ decreases. Thus we have determined the effect of varying R on the sinusoidal steady-state response, since the foregoing statement is equivalent to saying that the magnitude of the impedance increases and its angle decreases.

Considering circuits which consist of more than two elements, we recall that a change in the value of a single element generally influences all the pole-zero locations of the network function. For example, in the circuit of Fig. 10-2, the transform impedance at terminals a-b is

$$Z_{ab}(s) = 1 + \frac{1}{1 + sC} = \frac{s + 2/C}{s + 1/C}$$

Hence both the location of the zero (at $s = -2/C$) and the location of the pole (at $s = -1/C$) are changed if C is changed.

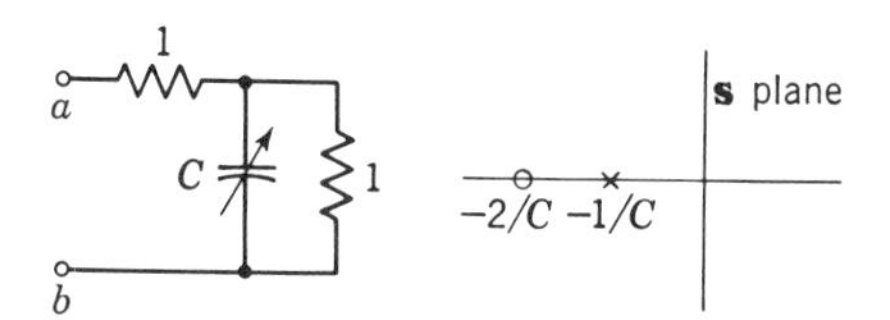

***Fig. 10-2** In the series-parallel circuit of (a) both the pole location and the zero location of $Z_{ab}(s)$ depend on the value of C.*

We conclude from this discussion that the "migration" of the poles and zeros in the complex-frequency plane which arises from a change in element value may be quite complex, so that the corresponding effect on the complex network function may require a considerable number of calculations. For this reason the effect of element variation is often studied by analyzing the complex network function itself rather than its pole-zero representation. This procedure is discussed in the later sections of this chapter.

If, on the other hand, the frequency of the *source* is adjusted, the s-plane diagram can be used conveniently because, for fixed element values, the pole-zero locations of the network function ("F points") remain fixed. As the source frequency varies, only the location of the pole at $s = j\omega$ varies. We have already illustrated the interpretation of G-point migration in Chap. 7 by studying the influence which this pole has on the factors $s_g - s_F$ of the network function. In the following sections a more detailed study of this problem is made.

10-2 Frequency response

AN R-C CIRCUIT As the first example of a circuit with fixed elements, consider the parallel R-C circuit with a sinusoidal source of adjustable frequency shown in Fig. 10-3a. The arrow through the source labeled f is intended to indicate that the frequency of the source is adjustable but that the amplitude of the source remains fixed. We shall study the admittance function $\mathbf{Y}_{ab}$, that is, the ratio of the phasors $\mathbf{I}/\mathbf{V}$ as a function of frequency. Again it is necessary to emphasize that we are concerned only with the steady-state response at the various frequencies—we shall not consider the (exceedingly difficult) problem of the transient which occurs while the frequency is changing.

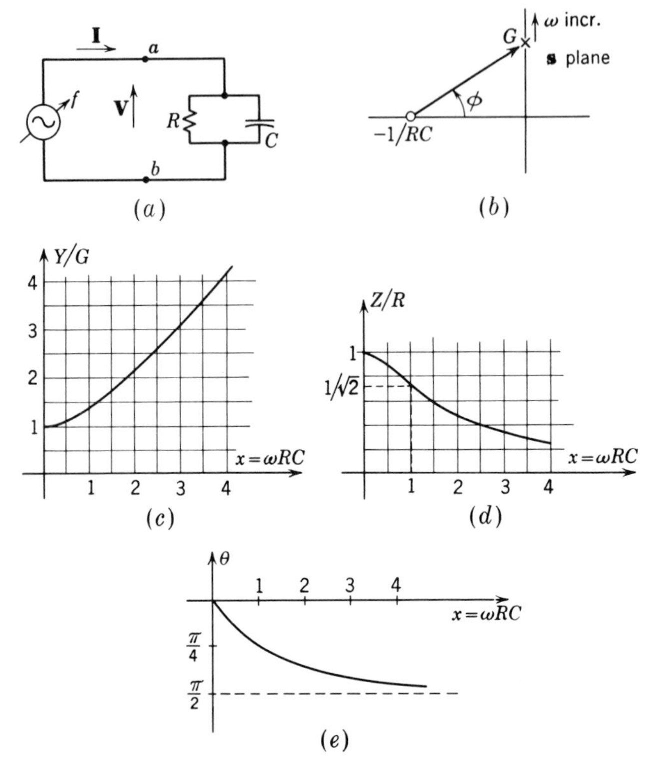

***Fig. 10-3** (a) Parallel R-C circuit with adjustable-frequency sinusoidal source. (b) Pole-zero diagram for calculation of Y(jω). (c) Normalized admittance as a function of normalized frequency. (d) Normalized impedance as a function of normalized frequency. (e) Angle of the complex impedance as a function of normalized frequency.*

The transform admittance of this circuit is

$$Y_{ab}(s) = \frac{1}{R} + sC = C\left(s + \frac{1}{RC}\right) \tag{10-1}$$

Hence we have a zero at $s = -1/RC$, as shown in Fig. 10-3b. As the radian frequency of the sources increases, the G point moves up as indicated by the arrow. Hence the admittance of the circuit increases in magnitude from $1/R$ at $s = 0$ to infinite value at infinite frequency. The magnitude of the complex impedance correspondingly decreases from R to 0. The angle of the admittance is ϕ, and is seen to increase from 0 to $\pi/2$ with increasing frequency.

It is customary to give the information concerning the frequency dependence of the network function by means of graphs which show how the absolute value of the network function and its angle depend on frequency. Such graphs are known as *frequency-response curves.* A curve which shows the magnitude of the network function as a function of frequency shows the amplitude (or effective value) of the response phasor when the amplitude (or effective value) of the source phasor is unity. If a current source of 1 amp amplitude (or rms value) is applied, the voltage amplitude (or rms value) across an impedance $\mathbf{Z}$ is Z volts. Curves that give the magnitude of the network function as a function of frequency are called *amplitude-response curves of the frequency response,* or *amplitude-response curves,* for brevity. Curves which give the angle of the network function as a function of frequency are called *phase-response curves* (of the frequency response). Although the frequency-response curves for complex networks are often obtained point by point from the pole-zero diagram, this is not necessary in the present example; for the parallel R-C circuit the admittance, in polar form, is

$$\mathbf{Y} = \sqrt{\frac{1}{R^2} + \omega^2 C^2}\ \underline{/\tan^{-1} \omega RC} \tag{10-2}$$

It is convenient to normalize the admittance with respect to the conductance $G = 1/R$,

$$\mathbf{Y}R = \frac{\mathbf{Y}}{G} = \sqrt{1 + \omega^2 R^2 C^2}\ \underline{/\tan^{-1} \omega RC} \tag{10-3}$$

Then the complex impedance, normalized with respect to R, is

$$\frac{1}{\mathbf{Y}R} = \frac{\mathbf{Z}}{R} = \frac{1}{\sqrt{1 + \omega^2 R^2 C^2}}\ \underline{/-\tan^{-1} \omega RC} \tag{10-4}$$

In these equations we notice that the radian frequency ω in both the magnitude and the angle is always multiplied by RC, the time constant of the R-C circuit. This fact again illustrates that the free response of the circuit (i.e., the time constant) is related to the sinusoidal steady-state response

For the purpose of drawing the frequency-response curves, it is convenient to introduce a new normalized frequency, say, x, where

$$x = \omega RC$$

(Since $\omega = 2\pi f$, $x = 2\pi \times$ ratio of time constant to period of source.) Using this variable,

$$\frac{\mathbf{Y}}{G} = \sqrt{1 + x^2}\ \underline{/\tan^{-1} x}$$

and

$$\frac{\mathbf{Z}}{R} = \frac{1}{\sqrt{1 + x^2}}\ \underline{/-\tan^{-1} x}$$

Figure 10-3*c* and *d* shows the amplitude curves for the admittance and the impedance function, respectively. The phase-response curve giving θ, the angle of the impedance, is shown in Fig. 10-3*e*.

The frequency-response curves suggest an interesting and frequently useful possibility. Suppose we were to apply a current source to the parallel R-C circuit of Fig. 10-3*a* and adjust the frequency until, in the steady state, v_{ab} lags $i(t)$ by 45°. At that frequency

$\omega RC = 1$. Since we presumably know the frequency, we could use this measurement to determine the time constant RC. Now notice that, if RC is known, the free component of the complete response is known to be of the form $Ke^{-t/RC}$. Consequently, at least for this circuit, we can obtain sufficient information for calculation of the complete response for any type of source applied by making measurements in the sinusoidal steady state.

Studying the graph of Z/R, we note that at the frequency

$$f_1 = \frac{1}{2\pi RC}$$

(which corresponds to $x = 1$), $Z = R/\sqrt{2}$. Hence, if an adjustable-frequency current source is applied to a parallel combination of R-C elements at the frequency f_1, the voltage V_{ab} will have an effective value which is 70.7 per cent of the value at direct current. If now a current source at fixed effective value, I, is applied to the R-C circuit, then at $\omega = 0$ (d-c) the power delivered to the circuit is I^2R. When $\omega_1 = 1/RC$, the power delivered to the circuit is $(V_{ab})^2/R = (0.707IR)^2/R = \frac{1}{2}I^2R$. Hence the frequency $f_1 = 1/2\pi RC$ is called the "half-power frequency" of the R-C circuit.

10-3 The parallel R-C circuit as a filter In this section we shall use the example of a parallel R-C circuit to illustrate the filtering property of circuits which contain energy-storing elements. In circuit analysis the term "filtering property of a circuit" means the characteristics of the circuit which result in different sinusoidal steady-state response amplitudes for a given fixed amplitude of the applied source and adjustable frequency.

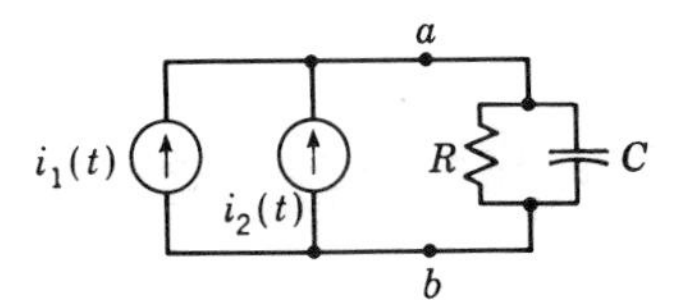

Fig. 10-4 Parallel R-C circuit with two ideal current sources.

Let us assume that the two current sources impressed on the parallel R-C circuit of Fig. 10-4 have the following forms: a d-c source

$$i_1(t) = I \tag{10-5}$$

and a sinusoid of amplitude I, radian frequency ω,

$$i_2(t) = I \cos \omega t \tag{10-6}$$

The response, in the steady state, may by superposition be written

$$v_{ab}(t) = (v_{ab})_1 + (v_{ab})_2$$

where $(v_{ab})_1$ is the steady-state response due to i_1 and $(v_{ab})_2$ is due to i_2. Evidently,

$$(v_{ab})_1 = IR \tag{10-7a}$$

Since the impedance of the parallel R-C circuit at the radian frequency ω is

$$\mathbf{Z}_{ab} = \frac{1}{\mathbf{Y}_{ab}} = \frac{1}{1/R + j\omega C} = \frac{R}{\sqrt{1 + R^2\omega^2C^2}} \underline{/-\tan^{-1}\omega RC}$$

we have

$$(v_{ab})_2 = IR\frac{1}{\sqrt{1 + R^2\omega^2C^2}} \cos(\omega t - \tan^{-1}\omega RC) \tag{10-7b}$$

Hence

$$v_{ab}(t) = IR\left[1 + \frac{1}{\sqrt{1 + R^2\omega^2C^2}}\cos(\omega t - \tan^{-1}\omega RC)\right] \tag{10-8}$$

Several numerical illustrations are useful. These are shown in Fig. 10-5, where the steady-state response waveforms are shown for four values of ωRC. [In these illustrations the time reference, and therefore the phase angles of $(v_{ab})_2$ with respect to i_2 are *not* shown.] Figure 10-5 shows that the sinusoidal component of the response waveform

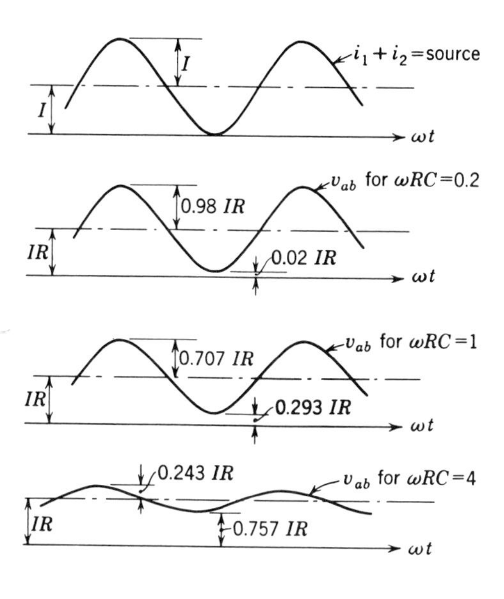

***Fig. 10-5** Source function and steady-state response of the circuit of Fig. 10-4 for $i_1 = I$ (d-c) and $i_2 = I\cos\omega t$ for various values of ωRC (normalized frequency).*

becomes less pronounced as the frequency is adjusted to larger values. This effect may be desirable (as it is in filters for rectifiers) if we do not want the steady-state response to be oscillatory; or it may be undesirable (as in a sound system) when we wish to transmit all frequencies with equal fidelity.

INTEGRATING AND DIFFERENTIATING PROPERTIES OF THE R-C CIRCUIT IN THE A-C STEADY STATE The frequency response of the R-C circuit can be related to the integrating and differentiating property of such circuits, as discussed in Chap. 5. Since in that chapter the series R-C circuit was used, we shall now discuss the frequency response of the series R-C circuit of Fig. 10-6. For this circuit, in the sinusoidal steady state the voltage

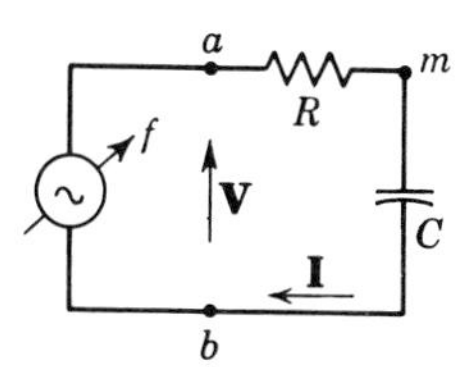

***Fig. 10-6** Series R-C circuit with adjustable-frequency constant-amplitude voltage source.*

across the capacitance is represented by the phasor

$$\mathbf{V}_{mb} = \mathbf{V}_{ab} \times \frac{1/j\omega C}{R + 1/j\omega C} = \frac{1}{j\omega RC + 1}\mathbf{V}_{ab} \tag{10-9}$$

If the radian frequency is sufficiently large so that $|j\omega RC| \gg 1$, then (approximately)

$$\mathbf{V}_{mb} \approx \frac{1}{j\omega RC}\mathbf{V}_{ab} \qquad \omega RC \gg 1 \tag{10-10}$$

Since division of a phasor by $j\omega$ represents integration with respect to time of the function represented by the phasor, we can write

$$v_{mb} \approx \frac{1}{RC}\int v_{ab}\,dt \qquad v_{ab} \text{ sinusoidal, } \omega RC \gg 1 \tag{10-11}$$

We observe that, when the product $\omega RC \gg 1$, the voltage across the capacitance is approximately proportional to the time integral of the source voltage. Similarly, the voltage across the resistance, v_{am}, represented by the phasor $\mathbf{V}_{am}$, is given by $\mathbf{V}_{am} = [(R/(R + 1/j\omega C)]\mathbf{V}_{ab}$, or

$$\mathbf{V}_{am} = \frac{j\omega RC}{1 + j\omega RC}\mathbf{V}_{ab}$$

so that, if $\omega RC \ll 1$, $\mathbf{V}_{am} \approx j\omega RC\mathbf{V}_{ab}$. Since multiplication of a phasor of $j\omega$ represents differentiation with respect to time of the function which is represented by the phasor, we can write

$$v_{am} \approx RC\frac{d}{dt}v_{ab} \qquad v_{ab} \text{ sinusoidal, } \omega RC \ll 1 \tag{10-12}$$

The conditions $\omega RC \ll 1$ (for the differentiating circuit) and $\omega RC \gg 1$ (for the integrating circuit) correspond to the inequalities (5-51) and (5-46) if the time functions are sinusoidal.

10-4 Frequency response of a series R-L-C circuit

The frequency response of *R-L-C* circuits is of particular importance because such circuits are the simplest form of "bandpass" filter. Referring to Fig. 10-7a, we observe that, if the source is a constant-amplitude voltage source, the current will have zero steady-state value at $\omega = 0$, because of the capacitance, and zero steady-state value as ω approaches infinity, because of the inductance. For this circuit the transform admittance is

$$Y(s) = \frac{1}{R + sL + 1/sC} = \frac{1}{L}\frac{s}{s^2 + (R/L)s + 1/L} \tag{10-13}$$

Setting $\alpha = R/2L$, $\omega_0{}^2 = 1/LC$,

$$Y(s) = \frac{1}{L}\frac{s}{s^2 + 2\alpha s + \omega_0{}^2}$$

or in factored form,

$$Y(s) = \frac{1}{L}\frac{s}{(s - s_1)(s - s_2)}$$

where $\qquad s_{1,2} = -\alpha \pm \sqrt{\alpha^2 - \omega_0{}^2}$

If we assume that the circuit is underdamped, $\alpha < \omega_0$, then we define $\omega_d = (\omega_0{}^2 - \alpha^2)^{\frac{1}{2}}$ and $\mathbf{s}_{1,2} = -\alpha \pm j\omega_d$. Hence $Y(s)$ has a zero at $s = 0$

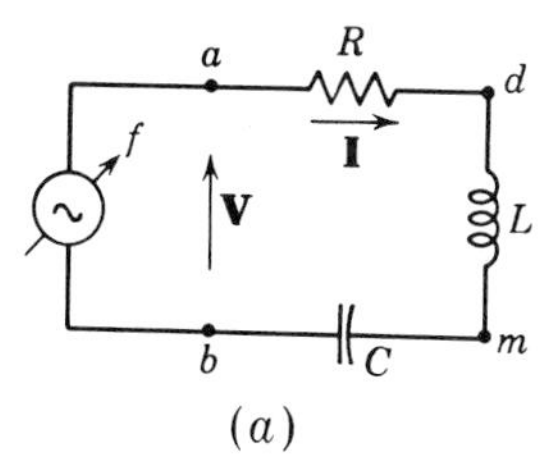

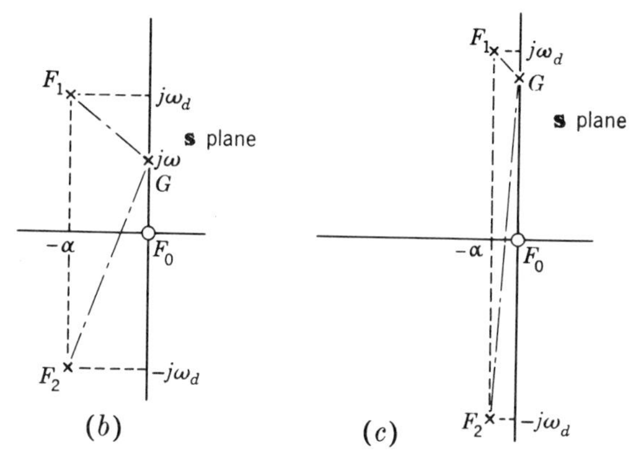

***Fig. 10-7** (a) Series R-L-C circuit with constant-amplitude adjustable-frequency sinusoidal voltage source. (b) Pole-zero diagram for the admittance of (a), $\mathbf{Y}_{ab}(j\omega)$, in a slightly oscillatory case. (c) Pole-zero diagram for $\mathbf{Y}_{ab}(j\omega)$ in a highly oscillatory case.*

and poles at $s = \mathbf{s}_1$ and $s = \mathbf{s}_2$. A sinusoidal source is represented by a pole at $s = j\omega$. The zeros and poles are shown in Fig. 10-7*b*. The magnitude of the complex admittance is given by the quotient of the products of the following distances:

$$Y = \frac{1}{L} \frac{F_0G}{(F_1G)(F_2G)} \tag{10-14}$$

Since $s = 0$ is a zero of $Y(s)$, $Y(0) = 0$. Since both F_1G and F_2G as well as F_0G approach infinity for infinite frequency, $Y(\infty) = 0$. Since $|Y(j\omega)|$ is a continuous function of ω, we expect a maximum value for some value of ω. It is seen that, if $\alpha \ll \omega_d$, then, as ω increases, in the neighborhood of $\omega = \omega_d$ the ratio of F_0G/F_2G changes very "slowly," but F_1G has a "large" percentage variation, having its minimum value at $\omega = \omega_d$. Thus we expect $Y(j\omega)$ to have a maximum *near* $\omega = \omega_d$ if $\alpha \ll \omega_0$, that is, in the highly oscillatory case. This can be verified by inspecting the pole-zero diagram for a highly oscillatory case as shown in Fig. 10-7*c*. The ratio of F_0G to F_2G changes very little as ω moves from G toward ω_d, whereas F_1G changes "rapidly."

From the study of the pole-zero diagram we conclude that the curve of $|Y(j\omega)|$ as a function of ω will have a maximum and that this maximum will be near $\omega = \omega_d$ in the highly oscillatory case. The exact frequency response of this circuit can be obtained either by constructing the graph of $|Y(j\omega)|$ by measuring the distances F_0G, F_1G, and F_2G for various ω and using Eq. (10-14) or by studying the complex admittance function analytically. The latter procedure is given in the next section.

10-5 The series R-L-C circuit

RESONANCE The exact analysis of the series R-L-C circuit is of interest for several reasons. First of all, this circuit finds considerable application inasmuch as many simple physical systems or devices can be represented by such a circuit. Second, we wish to compare the approximate answers of the preceding section with the exact answers. Finally, the manner in which the analysis is carried out is also applicable to other circuits.

For the series R-L-C circuit of Fig. 10-7*a* the complex impedance is

$$\mathbf{Z}_{ab}(j\omega) = R + j\omega L + \frac{1}{j\omega C} = R + jX$$

Since $X = \omega L - 1/\omega C$, we have $X = 0$ at $\omega^2 = 1/LC = \omega_0{}^2$. Hence, since $Z_{ab} = (R^2 + X^2)^{\frac{1}{2}}$, at $\omega = \omega_0$, Z_{ab} is a minimum. When the condition $\omega = \omega_0$ is satisfied, the circuit is said to be in "resonance." For the resonant condition, the following observations are made:

1 In resonance the complex impedance and the complex admittance of the series R-L-C circuit are real numbers. The complex impedance has the value R; the admittance has the value $1/R$.
2 In resonance the magnitude of the impedance of the series R-L-C circuit is a minimum, and the magnitude of its admittance is a maximum. Hence, if a constant-amplitude adjustable-frequency voltage source is impressed on this circuit, the amplitude of the current at resonance will be larger than at any other frequency.
3 The power factor at terminals *a-b* (Fig. 10-7*a*) is unity, since the voltage $v_{ab}(t)$ and the current $i_{ab}(t)$ will be in phase when $\omega = \omega_0$.
4 A study of the pole-zero diagram of the series R-L-C circuit (Fig. 10-7*c*) indicates that, when $\omega_d \approx \omega_0$, the impedance Z_{ab} has a minimum in the neighborhood of $\omega = \omega_d$. The results of the foregoing exact analysis show that this minimum occurs at $\omega = \omega_0$. Thus our study of the pole-zero diagram has enabled us to arrive at fairly good approximations.

The term "resonance" is used to describe three possible conditions in the sinusoidal steady state.

1 By analogy with the series R-L-C circuit, a terminal pair is in *series* resonance when the magnitude of the driving-point *admittance* is a *maximum.*

2 By analogy with the parallel *R-L-C* circuit (Sec. 10-9) *parallel* resonance is defined as that condition for which the magnitude of the *impedance* is a *maximum*. To contrast the series with the parallel *R-L-C* circuit the term *antiresonance* is also used as a synonym for parallel resonance.

3 *Unity-power-factor*, or *resistive*, resonance is that condition for which the driving-point immittance is real.

Although in the series (or parallel) *R-L-C* circuits the unity-power-factor condition occurs at the same frequency as the series (or parallel) resonant condition, for other terminal-pair networks these conditions generally occur at distinct frequencies.

Power and energy relations at resonance A study of the power and energy-storage conditions at resonant frequency is useful because the term "resonance" is thus clarified. Let us assume that a sinusoidal voltage source is applied to a series *R-L-C* circuit and that the frequency of the source is identical with the resonant frequency of the terminal pair.

Let this voltage be

$$v_{ab}(t) = \sqrt{2}\,V \cos \omega_0 t$$

represented by the phasor $\mathbf{V}_{ab} = V\underline{/0}$. The current in the circuit is then given by

$$i(t) = \sqrt{2}\,I_0 \cos \omega_0 t, \text{ represented by the phasor } \mathbf{I} = I_0\underline{/0^\circ}$$

where $I_0 = V/R$. The subscript zero refers to the resonant condition.

The average power in the circuit at resonance, denoted by the symbol P_0, is given by

$$P_0 = \frac{V^2}{R} = I_0{}^2R$$

Since the rms value (or the amplitude) of the current at resonance is larger than at any other frequency (because the magnitude of the admittance is a maximum), the average power dissipated in the resistance R is a maximum at resonance.

In Chap. 9 we saw that, at resonance, the steady-state equivalent circuit with respect to terminals *a-b* (of the series *R-L-C* circuit) is a resistance R. This equivalence is due to the fact that the voltage across L, v_{dm} in Fig. 10-7*a*, and the voltage across the capacitance, v_{mb}, are equal in amplitude and differ in phase by exactly 180°. To understand the role which the reactive elements play in resonance, the stored energy in each of these elements will be calculated. Using the current phasor as reference, $\mathbf{I} = I_0\underline{/0^\circ}$, we have $\mathbf{V}_{mb} = I_0/j\omega_0 C$; these phasors correspond to the time functions

$$i(t) = \sqrt{2}\,I_0 \cos \omega_0 t \qquad \text{and} \qquad v_{mb} = \sqrt{2}\,\frac{I}{\omega_0 C} \cos\left(\omega_0 t - \frac{\pi}{2}\right) = \frac{\sqrt{2}\,I}{\omega_0 C} \sin \omega_0 t$$

In a time interval t_1 to t_2 the energy which has been delivered to the inductance Δw_L is given by

$$\Delta w_L = \tfrac{1}{2}L[i(t_2)]^2 - \tfrac{1}{2}L[i(t_1)]^2 = LI_0{}^2 \cos^2 \omega_0 t_2 - LI_0{}^2 \cos^2 \omega_0 t_1 \tag{10-15}$$

In the same time interval the energy which has been delivered to the capacitance is given by

$$\Delta w_C = \tfrac{1}{2}C[v(t_2)]^2 - \tfrac{1}{2}C[v(t_1)]^2 = \frac{I^2}{\omega_0{}^2C} \sin^2 \omega_0 t_2 - \frac{I_0{}^2}{\omega_0{}^2C} \sin^2 \omega_0 t_1 \tag{10-16}$$

In Eq. (10-16) note that $\omega_0^2 = 1/LC$, so that

$$\frac{I_0^2}{\omega_0^2 C} = \frac{I_0^2 LC}{C} = LI_0^2 \tag{10-17}$$

Now we substitute Eq. (10-17) in Eq. (10-16) and add the result to Eq. (10-15) to obtain the total energy Δw which has been delivered to the energy-storing elements in the circuit in the time interval from t_1 to t_2.

$$\Delta w = LI_0^2[\cos^2 \omega_0 t_2 + \sin^2 \omega_0 t_2 - (\cos^2 \omega_0 t_1 + \sin^2 \omega_0 t_1)] \tag{10-18}$$

Since $\cos^2 x + \sin^2 x = 1$, Eq. (10-18) reads $\Delta w = 0$. Since t_2 and t_1 are arbitrary instants of time, we conclude that, with $\omega = \omega_0$, the *L-C combination* receives zero energy from the source at any time *in the steady state*.

Let us now examine again the individual expressions for Δw_L and Δw_C at certain time intervals. If we choose t_1 to be an instant of time at which $i(t)$ is zero, for example, $t_1 = -\pi/2\omega_0$, and t_2 an instant of time at which $i(t)$ is a maximum, for example, $t_2 = 0$, then for this time interval the inductance has received energy in the amount LI_0^2 and the capacitance has received energy in the amount $-LI_0^2$; the minus sign indicates that the capacitance has delivered energy to the other elements in the circuit. We conclude, therefore, that in the resonant condition there exists a continuous interchange of energy between inductance and capacitance. At any instant of time the *total* stored energy is $LI_0^2 = \frac{1}{2}L(\sqrt{2}\, I_0)^2$ or $\frac{1}{2}C(I_0 \sqrt{2}/\omega_0 C)^2$.

We must emphasize that the process of energy exchange which has been described is a steady-state phenomenon. The total energy stored by the energy-storing elements is supplied by the source during the transient time interval.

10-6 Selectivity, half-power points, and bandwidth

In the preceding section we concentrated on the resonant condition in the series *R-L-C* circuit, i.e., the sinusoidal steady state at the frequency $\omega = \omega_0 = 1/\sqrt{LC}$. In this section we shall discuss the frequency-response curves of the series *R-L-C* circuit. Although in any numerical case these curves can be obtained from a pole-zero diagram, it is instructive to deduce the shape of the admittance and impedance curves analytically. We start with the expression for the complex impedance $\mathbf{Z}_{ab}$,

$$\mathbf{Z}_{ab} = R + jX = R + j\left(\omega L - \frac{1}{\omega C}\right) \tag{10-19}$$

In order to make the results of this analysis more generally useful, it is convenient to normalize the complex impedance with respect to its minimum absolute value, that is, R, and to introduce a normalized frequency variable μ, which represents the ratio of any frequency to the series-resonant frequency. Thus

$$\mu = \frac{\omega}{\omega_0} = \frac{f}{f_0} = \omega \sqrt{LC} \tag{10-20}$$

Introducing μ into Eq. (10-19) and dividing by R, we have

$$\frac{\mathbf{Z}_{ab}}{R} = 1 + j\frac{1}{R}\left(\mu\omega_0 L - \frac{1}{\mu\omega_0 C}\right) \tag{10-21}$$

In Eq. (10-21) it is convenient to factor $\omega_0 L$ from the reactance term,

$$\frac{\mathbf{Z}_{ab}}{R} = 1 + j\frac{\omega_0 L}{R}\left(\mu - \frac{1}{\mu\omega_0^2 LC}\right) \tag{10-21a}$$

Using the definition of ω_0 in Eq. (10-21*a*), we obtain the form

$$\frac{\mathbf{Z}_{ab}}{R} = 1 + j\frac{\omega_0 L}{R}\left(\mu - \frac{1}{\mu}\right) \tag{10-21b}$$

At this point we recognize $\omega_0 L/R$ to be the quality factor Q_0 of the series *R-L-C* circuit discussed in Chap. 4. The symbol Q_0 should *not* be confused with the symbol for reactive power.

$$Q_0 = \frac{\omega_0}{2\alpha} = \frac{\sqrt{L/C}}{R} = \frac{\omega_0 L}{R} = \text{quality factor of } \textit{series } R\text{-}L\text{-}C \text{ circuit} \tag{10-22}$$

Introducing Q_0 in Eq. (10-21*b*), we have the following form for the complex impedance $\mathbf{Z}_{ab}$ of the series *R-L-C* circuit:

$$\frac{\mathbf{Z}_{ab}}{R} = 1 + jQ_0\left(\mu - \frac{1}{\mu}\right) \tag{10-23}$$

To obtain the expression for the absolute value of the impedance or admittance, we write Eq. (10-23) in polar form:

$$\frac{\mathbf{Z}_{ab}}{R} = \sqrt{1 + Q_0^2\left(\mu - \frac{1}{\mu}\right)^2} \Big/ \tan^{-1}\left[Q_0\left(\mu - \frac{1}{\mu}\right)\right] \tag{10-24}$$

Equation (10-24) embodies a very important conclusion: The shapes of the curves of the normalized impedance (absolute value) and the angle of the impedance θ, as a function of frequency, depend only on the value of Q_0. Thus, for the purpose of *comparing* the frequency responses of two *R-L-C* series circuits, we shall need to compare only the values of Q_0. We shall now discuss the curves of Z_{ab}/R, $Y_{ab}R$, and θ as a function of normalized frequency μ.

We have

Normalized absolute value of impedance:

$$\frac{Z_{ab}}{R} = \sqrt{1 + Q_0^2\left(\mu - \frac{1}{\mu}\right)^2} \tag{10-25}$$

Normalized absolute value of admittance:

$$Y_{ab}R = \frac{1}{\sqrt{1 + Q_0^2(\mu - 1/\mu)^2}} \tag{10-26}$$

Angle of impedance:

$$\theta = \tan^{-1} Q_0\left(\mu - \frac{1}{\mu}\right) \tag{10-27}$$

To plot Eqs. (10-25) and (10-26), we note that Z_{ab}/R is infinite at $\mu = 0$ and $\mu = \infty$. Similarly, $Y_{ab}R$ is zero at these extremes. At resonance

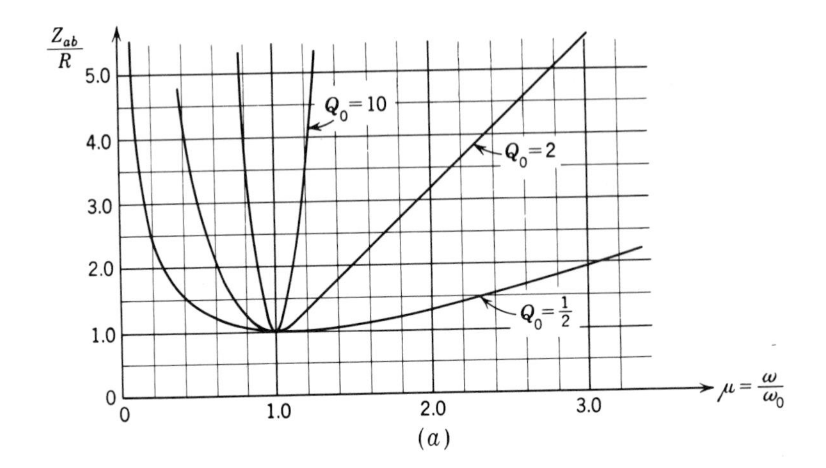

(a)

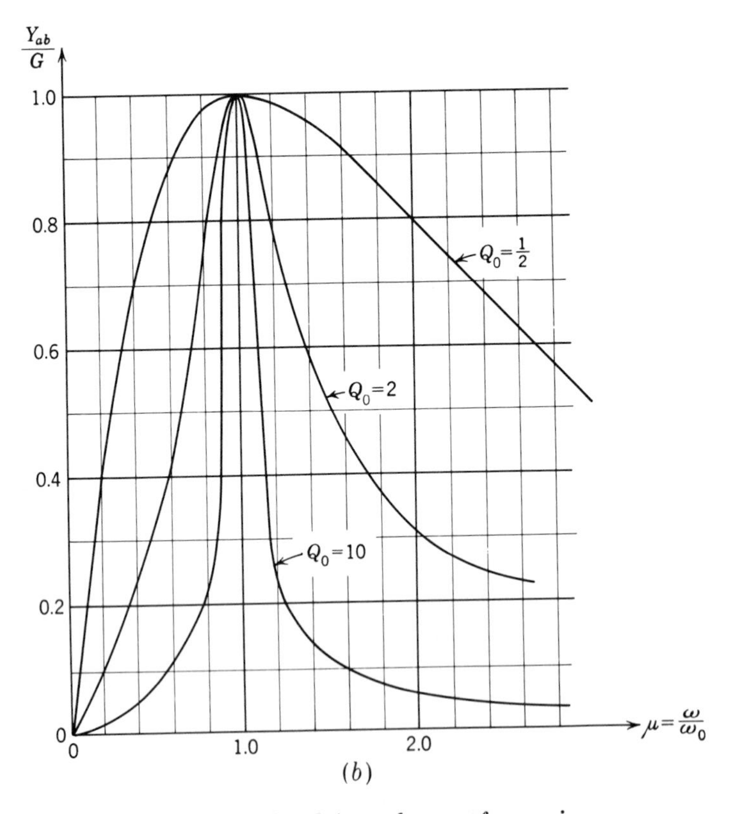

(b)

***Fig. 10-8** (a) Normalized impedance of a series R-L-C circuit as a function of normalized frequency with Q_0 as a parameter. (b) Normalized admittance of the series R-L-C circuit as a function of normalized frequency with Q_0 as a parameter.*

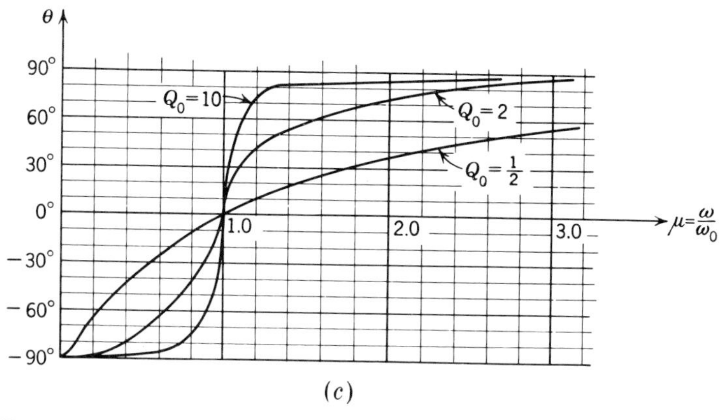

(c)

***Fig. 10-8** (c) Angle of the impedance of the series R-L-C circuit as a function of normalized frequency with Q_0 as a parameter.*

$\mu = 1$, and Z_{ab} and Y_{ab} have a minimum and a maximum, respectively. Curves for three values of Q_0 are shown in Fig. 10-8*a* and *b*. The curves of θ as a function of μ are shown in Fig. 10-8*c*. The curves of admittance or impedance as a function of frequency as shown in Fig. 10-8 are frequently used and are commonly referred to as *resonance curves.*

POWER AND ENERGY RELATIONS We shall now deduce energy and power relationships from an admittance curve such as that shown in Fig. 10-9.

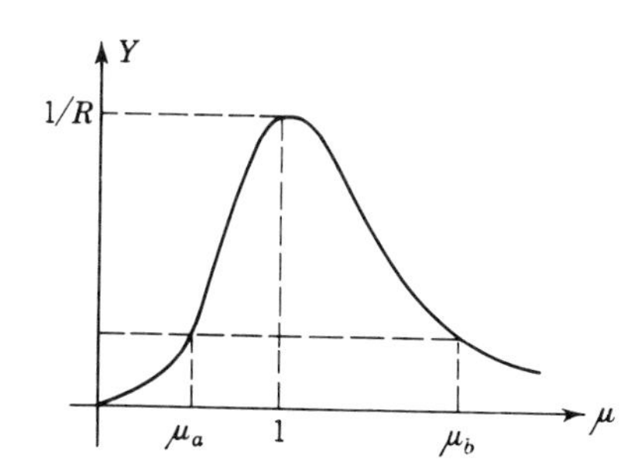

***Fig. 10-9** Magnitude of the admittance as a function of frequency for a series R-L-C circuit. Two frequencies at which the admittance has the same magnitude are indicated as μ_a and μ_b and $\mu_a\mu_b = 1$ [Eq. (10-28)].*

If a constant-amplitude adjustable-frequency voltage source is applied to the series *R-L-C* circuit, the effective value or the amplitude of the current at resonance ($\mu = 1$) is larger than at any other frequency because the circuit at resonance has maximum admittance. From the previous discussion we know that the power delivered to R at resonance is $P_0 = I_0{}^2R$, where $I_0 = V/R$. From Fig. 10-9 we observe that there are two values of frequency for which the admittance is a given fraction of $1/R$. We shall now show that these two frequencies have a geometric mean which is the resonant frequency, that is, $\mu_a\mu_b = 1$ or $\omega_a\omega_b = \omega_0{}^2$.

To prove this, we need recall only that, if Y is the same at the two frequencies, Z is also the same. Since $\mathbf{Z} = R + jX$, we conclude that the magnitude of the reactance is identical at the two frequencies. Now

$$\frac{X}{R} = Q_0\left(\mu - \frac{1}{\mu}\right)$$

Hence

$$\mu_a - \frac{1}{\mu_a} = -\left(\mu_b - \frac{1}{\mu_b}\right)$$

or

$$\mu_a{}^2\mu_b - \mu_b = -\mu_b{}^2\mu_a + \mu_a$$

or

$$\mu_a\mu_b(\mu_a + \mu_b) = \mu_a + \mu_b$$

so that

$$\mu_a\mu_b = 1 \tag{10-28}$$

Since, at the normalized frequencies μ_a and μ_b, the admittance has the same absolute value, if a constant-amplitude voltage source is impressed on the circuit, the current, and therefore the power at μ_a and μ_b, will be the same. We observe that the power at resonance is a maximum: the farther off resonance we adjust the frequency, the smaller the power transferred will be. Thus this circuit is a filter in the sense that a voltage of given amplitude far off resonance is unable to transfer as much power to the circuit as a voltage of the same amplitude near resonance. Since the admittance falls off as we depart from resonance in either direction ($\mu > 1$ or $\mu < 1$), the circuit is called a "bandpass filter" because the neighborhood near resonance forms a band of "favored" frequencies.

So far we have used vague terms like "near," "far," etc. In order to compare two resonant circuits, we need to set up some criterion which defines the "passed" frequencies. An arbitrary definition which is commonly accepted is the following: If a fixed-amplitude voltage source with adjustable frequency is impressed on the series *R-L-C* circuit, those frequencies for which the power transferred to R is more than *half the power transferred at resonance* form the *passband* of the circuit. The extreme frequencies at which the power transferred to R is *exactly half* the power transferred at resonance are referred to as the *half-power frequencies of the circuit.* Clearly, one of these half-power frequencies is above resonance. It is referred to as the "upper half-power frequency." The other is below resonance, and is called the "lower half-power frequency." The numerical difference between the upper and lower half-power frequencies is called the "bandwidth" of the circuit. We shall now calculate these significant values.

By definition, at the half-power frequencies, the power delivered to the circuit is $\frac{1}{2}P_0$. If the effective value of the current at these frequencies is I_1 and the effective value of the current at resonance is I_0, we have

$$I_1{}^2R = \tfrac{1}{2}I_0{}^2R$$

Hence

$$I_1 = \frac{1}{\sqrt{2}}I_0$$

Since the current at the half-power frequencies is $1/\sqrt{2}$ of the current at resonance,[1] the magnitude of the admittance at the half-power frequencies is $1/\sqrt{2}$ of the magnitude at resonance. Denoting by f_1 and f_2 the lower and upper half-power frequencies, respectively, we see that Y at f_1 and f_2 is $1/\sqrt{2}\,R$; the absolute value of the impedance at the half-power frequencies is $\sqrt{2}\,R$. Denoting the impedance at the half-power frequencies by $Z_{1,2}$, we have

$$|\mathbf{Z}_{1,2}| = \sqrt{2}\,R = |R + jX_{1,2}| \tag{10-29}$$

[1] The reader must be careful not to confuse this relationship, which involves $\sqrt{2}$, with the relationship between I_m and I at any one frequency.

where $X_{1,2}$ denotes the reactance at f_1 or f_2. We note that Eq. (10-29) reads

$$\sqrt{2}\, R = \sqrt{R^2 + X_{1,2}^2}$$

Solving,

$$X_{1,2} = \pm R$$

The half-power frequencies are therefore determined from the equations

$$\omega_1 L - \frac{1}{\omega_1 C} = -R \qquad (10\text{-}30a)$$

and

$$\omega_2 L - \frac{1}{\omega_2 C} = +R \qquad (10\text{-}30b)$$

The quadratic equation (10-30*a*) is now solved for ω_1. Using the relationship $\omega_0{}^2 = 1/LC$ and rejecting a negative solution for ω, we obtain

$$\omega_1 = -\frac{R}{2L} + \sqrt{\left(\frac{R}{2L}\right)^2 + \omega_0{}^2} = -\alpha + \sqrt{\omega_0{}^2 + \alpha^2} \qquad (10\text{-}31a)$$

Similarly, solving Eq. (10-30*b*) gives

$$\omega_2 = \frac{R}{2L} + \sqrt{\left(\frac{R}{2L}\right)^2 + \omega_0{}^2} = +\alpha + \sqrt{\omega_0{}^2 + \alpha^2} \qquad (10\text{-}31b)$$

Thus the bandwidth in radians per second is

$$\text{Bandwidth} = \omega_2 - \omega_1 = \frac{R}{L} = 2\alpha \qquad \text{rad/sec} \qquad (10\text{-}32)$$

If we form the ratio of resonant frequency to bandwidth, we obtain

$$\frac{\text{Resonant frequency}}{\text{Bandwidth}} = \frac{f_0}{f_2 - f_1} = \frac{\omega_0}{\omega_2 - \omega_1} = \frac{\omega_0 L}{R} = Q_0 \qquad (10\text{-}33)$$

We see now that the quantity Q_0 has the geometrical interpretation of ratio of resonant frequency to bandwidth. The number Q_0 therefore determines the "selectivity" of the *R-L-C* circuit. The term selectivity arises from the application of resonant circuits in radio and television receivers. In this application several stations broadcast simultaneously with different frequencies. The equivalent circuit of the antenna may be visualized as an *R-L-C* circuit to which several voltage sources of about equal amplitude, but with different frequencies, are applied simultaneously. It is the function of the resonant circuit to permit large current flow only for the desired station (frequency). The circuit is therefore adjusted so that the bandwidth of the circuit is sufficiently large for the signal of the station which is desired to be within the half-power frequencies but for the adjacent station to have a signal so far from resonance that it does not contribute materially to the signal which the receiver amplifies. The number Q_0 associated with the resonant circuit determines the suitability of a given circuit for this application. Hence Q_0 is called the "selectivity," or "quality ratio," of the circuit.

We mention here that, in the application cited above, reasonably large values of Q_0 (say, 50 or more) are often desirable, but the reader should not infer that there are no applications in which low Q_0 is desirable.

Although we have defined the Q of a series-resonant circuit as the ratio of resonant frequency to bandwidth, another interpretation which is associated with the ratio of stored to dissipated energy in the circuit is possible. We have shown that the stored energy in the circuit at resonance is constant and equal to LI_0^2. The power dissipated in the resistance is of course I_0^2R. The energy dissipated by the resistance in *one cycle* (Chap. 9) is given by

$$W_R = \frac{2\pi}{\omega_0} I_0^2 R \qquad \text{joules/cycle}$$

The ratio of the stored energy in the circuit to the energy dissipated per cycle is proportional to Q_0:

$$\frac{\text{Stored energy}}{\text{Energy dissipated/cycle}} = \frac{1}{2\pi}\frac{\omega_0 L I_0^2}{R I_0^2} = \frac{1}{2\pi} Q_0 \qquad (10\text{-}34)$$

In circuits other than series or parallel connection of R-L-C elements (whose Q's are $\omega_0 L/R$ and $R/\omega_0 L$, respectively), the quantity $\omega_0 L/R$ has no special significance. In such cases the Q of the circuit is defined by

$$Q_0 = 2\pi \frac{\text{peak stored energy}}{\text{energy dissipated/cycle}}$$

10-7 Approximate results obtained by the use of the pole-zero diagram

We shall now show how some useful approximate results can be obtained from the pole-zero diagram. Let us start by assuming a high-Q ($Q_0 > 5$) R-L-C circuit whose admittance has the pole-zero diagram shown in Fig. 10-10. As discussed in Sec. 10-4, in this diagram we assume that, as ω varies

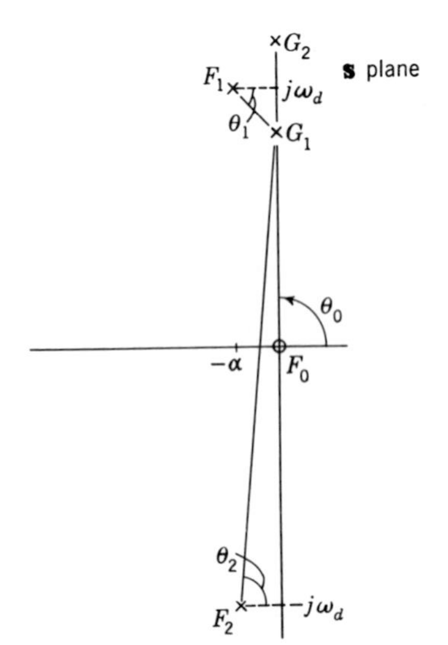

***Fig. 10-10** Pole-zero diagram for approximate admittance calculation in a highly oscillatory series R-L-C circuit.*

between G_1 and G_2, the distance ratio F_0G/F_2G remains approximately constant. This is justified if $\alpha \ll \omega_0$, because then

$$F_0G = \omega \qquad F_2G \approx \omega_d + \omega$$

And, near resonance, $\omega \approx \omega_d$, $F_0G/F_2G \approx \frac{1}{2}$. (The radian frequency ω_0 differs very little from ω_d if $\alpha/\omega_0 \ll 1$. For example, if $\omega_0 = 1$, $\alpha = 0.1$, $Q_0 = 5$; $\omega_d = \sqrt{0.99} \approx 1$.) Hence, near resonance, if $Q_0 > 5$,

$$Y \approx \frac{(1/L) \times \frac{1}{2}}{F_1G} \tag{10-35}$$

Now, at $\omega = \omega_d$, $F_1G = \alpha$. Hence the admittance is obtained at $\omega = \omega_d$ as $Y = 1/2L\alpha = 1/R$. (This is of course only approximate; we know that $Y = 1/R$ occurs at $\omega = \omega_0$, not at ω_d.) If we assume that the ratio F_0G/F_2G remains fixed near $\omega = \omega_d$, then Y is proportional to F_1G. But at $\omega = \omega_d - \alpha$ and $\omega = \omega_d + \alpha$, the distance F_1G is $\sqrt{2}\,\alpha$; that is, the admittance has an absolute value which is $1/\sqrt{2}\,R$. Hence $\omega_d \pm \alpha$ are the half-power points. Again the results, when compared with the exact analysis, are seen to be in good agreement. If, for example, $\omega_0 = 1$, $\alpha = 0.1$, the half-power frequencies are, from the exact equation (10-31),

$$\omega_{1,2} = \pm 0.1 + \sqrt{1.01} = 1.105, 0.905 \tag{10-36}$$

and the approximate analysis from the pole-zero diagram gives answers

$$\omega_{1,2} = \pm 0.1 + \sqrt{0.99} = 1.095, 0.895 \tag{10-37}$$

i.e., within about 1 per cent of the exact answers. Now we must be aware of the fact that this type of analysis is eventually applied to practical circuits, so that the component values will be known only with finite precision. Hence the small error which is introduced when the simpler and more convenient approximate analysis is used is often insignificant.

We note also that the bandwidth, as obtained from the approximate analysis, is $2\alpha = R/L$ rad. This is exactly the right answer. There is no particular significance in the fact that this answer is not approximate. It just happens that the errors in F_0G and F_2G compensate at these points.

If we continue the approximate calculation, then

$$Y \approx \frac{1}{2L(F_1G)} \tag{10-38}$$

or

$$Z \approx 2L(F_1G) \tag{10-39}$$

Now $F_1G = \sqrt{\alpha^2 + (\omega_d - \omega)^2}$; moreover, the angle of Y is

$$\text{Angle } \mathbf{Y} = \theta_0 - \theta_1 - \theta_2$$

Since $\theta_0 = \pi/2$ and $\theta_2 \approx \pi/2$, the angle of Y is, approximately, $-\theta_1$. This angle is given by

$$\theta_1 = -\tan^{-1}\frac{\omega_d - \omega}{\alpha} \tag{10-40}$$

Hence

$$\mathbf{Z} \approx 2L\sqrt{\alpha^2 + (\omega_d - \omega)^2}\ \underline{/-\theta_1} \tag{10-41}$$

near resonance. We recognize this complex quantity from Fig. 10-11 in rectangular form:

$$\mathbf{Z} \approx 2L\alpha + jL2(\omega_d - \omega)$$

Since $\alpha = R/2L$, $2L\alpha = R$; further, $L(\omega_d - \omega) \approx L(\omega_0 - \omega) = \omega_0 L[(\omega_0 - \omega)/\omega_0]$. Hence

$$\mathbf{Z} \approx R(1 + j2Q_0\delta) \tag{10-42}$$

where $\delta = 1 - (\omega/\omega_0) = 1 - \mu$. Equation (10-42) is a useful approximation to the exact formula for $\mathbf{Z}$. It applies when $Q_0 > 5$ and $|\delta| \ll 1$. (The requirement $\delta \ll 1$ corresponds to the assumption that the ratio of F_0G to F_2G remains constant; $Q_0 > 5$ corresponds to the assumption $\alpha/\omega_0 < 0.1$.)

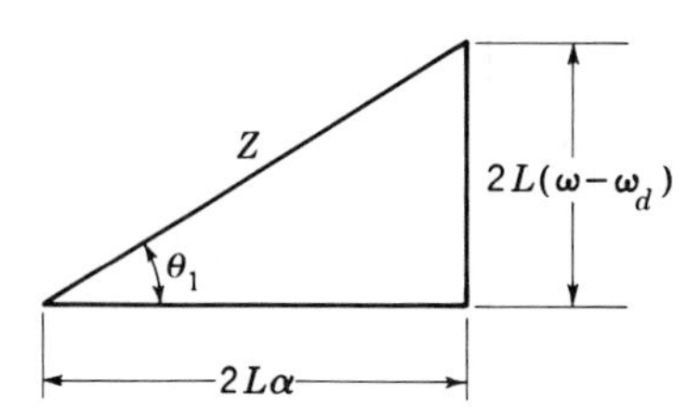

Fig. 10-11 ***Graphical interpretation of Eq. (10-41).***

From Eq. (10-42) we also deduce that, if $|\delta| \ll 1$, the resonance curve is symmetrical about the resonant frequency. This is also evident if we recall that the arithmetic mean and the geometric mean of two numbers which differ by a sufficiently small amount are approximately equal.

To summarize the application of the pole-zero diagram to frequency response, we point out that, in general, a network function has a *maximum* when the source frequency is "near" a *pole* and a *minimum* when the source frequency is "near" a *zero* of the transform-network function.

The significance of the pole-zero analysis lies not only in the fact that approximate results can be obtained for the frequency response, but also in the clear relationship between frequency response and free (transient) response which is inherent in the pole-zero diagram. The quantities ω_0, ω_1, and ω_2 are numbers which can be *measured* (and are derived) from a number of *sinusoidal steady-state conditions* at different frequencies. Thus, if we apply a constant-amplitude variable-frequency source to the series *R-L-C* circuit and measure the radian frequency at which the current (or the voltage across *R*) has the largest amplitude, we have measured ω_0. If we adjust the frequency to those values at which the current has 70.7 per cent of the resonant value, ω_1 and ω_2 can be determined. Using these values, we can now calculate

$$2\alpha = \omega_2 - \omega_1 \tag{10-43}$$

Knowing α and ω_0, we now have all the information needed to predict (i.e., calculate) the free component of the response, and hence the complete response of the circuit! This possibility is one of the significant accomplishments which sinusoidal-steady-state calculations and experiments make possible. This indirect method for predicting transient components from sinusoidal-steady-state measurements is important because measurements of transient response are frequently more difficult to perform than sinusoidal-steady-state measurements.

10-8 ***Low-Q circuit*** Much of the discussion has revolved around highly oscillatory (high-*Q*) circuits. It should not be inferred that low-*Q* circuits are without significance.

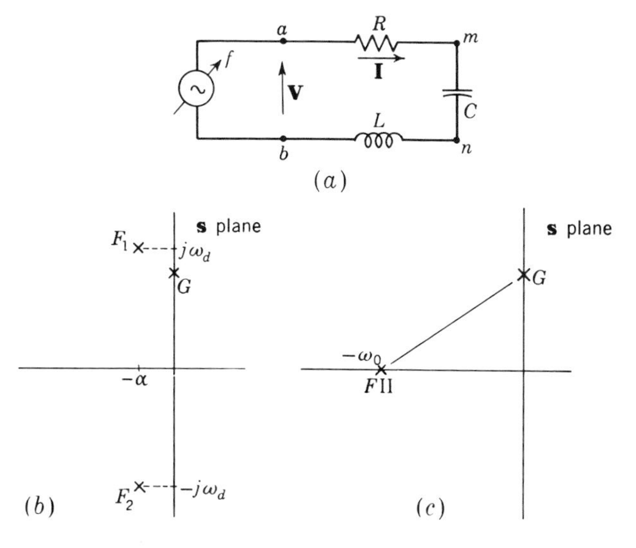

***Fig. 10-12** (a) Series R-L-C circuit with adjustable-frequency constant-amplitude sinusoidal source and pole-zero diagram of voltage gain function that relates v_{mn} to v in (b) a highly oscillatory case and (c) the critically damped case.*

Let us therefore consider the *R-L-C* series circuit shown in Fig. 10-12*a*. We know that, for this circuit, a constant-amplitude adjustable-frequency voltage source will produce maximum current amplitude at $\omega = \omega_0$. Since the voltage across the resistance is proportional to the current, it is also a maximum at resonance. In many applications, however, the voltage across the capacitance, v_{mn}, or across the inductance, v_{nb}, is the desired response. We shall now demonstrate that these voltages do not have their maximum amplitude at the resonant frequency ω_0. This is immediately evident from the pole-zero diagram. Consider the phasor ratio $\mathbf{V}_{mn}/\mathbf{V}$:

$$\frac{\mathbf{V}_{mn}}{\mathbf{V}} = \frac{1/j\omega C}{R + j\omega L + 1/j\omega C} = \frac{1}{(j\omega)^2 LC + j\omega RC + 1}$$

The transform network function (voltage gain function) of interest is

$$H(s) = \frac{1}{s^2LC + sRC + 1} = \frac{1}{LC}\frac{1}{s^2 + (R/L)s + 1/LC}$$

Hence

$$H(s) = \frac{\omega_0^2}{(s - s_1)(s - s_2)}$$

where $s_{1,2} = -\alpha \pm \sqrt{\alpha^2 - \omega_0^2}$, $\alpha = R/2L$, $\omega_0^2 = 1/LC$. This network function, in contrast to the driving-point admittance, has no zero at $s = 0$. The poles are as before. Now, in a high-Q circuit, the poles are close to the imaginary axis as in Fig. 10-12*b*. Hence we still have F_1G varying more rapidly than F_2G near $\omega = \omega_d$, so that v_{mn} will have maximum amplitude near $\omega = \omega_d \approx \omega_0$. Let us now turn to a low-Q circuit, e.g., a critically damped circuit, $Q_0 = \frac{1}{2}$. In this case

$$s_1 = s_2 = -\omega_0 = -\alpha$$

and we have a double pole at $s = -\omega_0$ on the real axis as shown in Fig. 10-12*c*. In this case the denominator $(FG)^2$ increases with increasing ω, and the only maximum amplitude of v_{mn} occurs at $s = 0$. In general it can be shown that, if $Q_0 > 1/\sqrt{2}$, the voltage v_{mn} will have one maximum at $\omega = 0$, the other at

$$\omega = \omega_p = \omega_0 \sqrt{1 - \frac{1}{2Q_0^2}} \tag{10-44}$$

The radian frequency ω_p is less than ω_0. The reason for this is easy to understand from the following exposition.

The absolute value of the voltage across the capacitance is written

$$|\mathbf{V}_c| = |\mathbf{I}|\,|X_c| \tag{10-45}$$

or

$$V_c = I\frac{1}{\omega C} \tag{10-46}$$

To locate the maximum, we differentiate (10-46) with respect to ω.

$$\frac{\partial V_c}{\partial \omega} = \frac{\partial I}{\partial \omega}\frac{1}{\omega C} + I\left(-\frac{1}{\omega^2 C}\right)$$

Now, at $\omega = 0$, $V_c = V$ since $I = 0$, and at $\omega = \infty$, $V_c = 0$ since $X_c = 0$. Hence we expect that, under certain conditions of $\partial V_c/\partial\omega = 0$, between zero and infinite frequencies V_c has a maximum. We know already that, *at resonance*, I has a maximum, that is, $\partial I/\partial\omega\big|_{\omega=\omega_0} = 0$. Hence, at resonance, $\partial V_c/\partial\omega\big|_{\omega=\omega_0} = -I/\omega_0^2 C$, so that the maximum voltage across the capacitance does *not* occur at resonance. Since at resonance $\partial V_c/\partial\omega$ is negative, the point $\omega = \omega_0$ must be on the decreasing portion of the V_c versus ω curve (Fig. 10-13). Therefore we expect that the maximum capacitance voltage occurs at a frequency below resonance. Four typical curves for V_c/V as a function of μ are shown in Fig. 10-13.

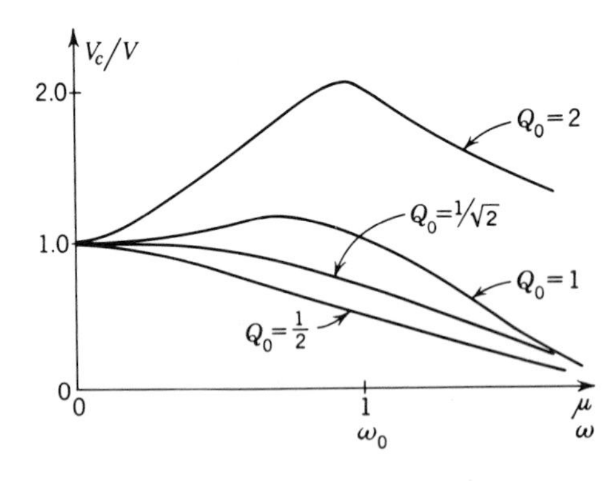

***Fig. 10-13** Amplitude-response ratio for the phasor ratio $\mathbf{V}_{mn}/\mathbf{V}$ (Fig. 10-12) illustrating that maxima occur below the series resonant frequency. For $Q_0^2 < \frac{1}{2}$, the maxima occur at $\omega = 0$.*

10-9 The parallel R-L-C circuit The parallel *R-L-C* circuit is the dual of the series circuit, and we shall see that, by using this fact *carefully*, many results of the series circuit can be applied.

The *admittance* of the three elements *R*, *L*, and *C* in parallel has the form

$$\mathbf{Y}_{ab} = G + j\omega C + \frac{1}{j\omega L}$$

The magnitude of $\mathbf{Y}_{ab}$ is a minimum at the radian frequency $\omega_0 = 1/\sqrt{LC}$. This frequency defines the parallel or antiresonant frequency of the circuit. Since

$$\mathbf{Y}_{ab} = G + j\omega C\left(1 - \frac{1}{\omega^2 LC}\right)$$

we use the symbol ω_0 to denote $1/\sqrt{LC}$ and obtain

$$\mathbf{Y}_{ab} = G + j\omega C\left(1 - \frac{\omega_0^2}{\omega^2}\right)$$

If we normalize the admittance with respect to G and the frequency with respect to ω_0 (as before, let $\mu = \omega/\omega_0$), we write

$$\frac{\mathbf{Y}_{ab}}{G} = 1 + j\frac{\omega_0\mu C}{G}\left(1 - \frac{1}{\mu^2}\right)$$

or

$$\frac{\mathbf{Y}_{ab}}{G} = 1 + j\frac{\omega_0 C}{G}\left(\mu - \frac{1}{\mu}\right)$$

We now *define* the number Q_0 *of a parallel-resonant circuit* as the ratio $\omega C/G$,

$$Q_0 \equiv \frac{\omega_0 C}{G} = \frac{R}{\omega_0 L}$$

so that the equation for the normalized admittance reads

$$\frac{\mathbf{Y}_{ab}}{G} = 1 + jQ_0\left(\mu - \frac{1}{\mu}\right) \tag{10-47}$$

Equation (10-47) shows that the *admittance of the parallel R-L-C circuit* has a form which is identical to the *impedance of the series-resonant circuit.* All the results of the previous analysis are therefore applicable, as is seen from the tabulation below.

Series R-L-C circuit	*Parallel R-L-C circuit*
Resonance at $\omega = \omega_0 = \dfrac{1}{\sqrt{LC}}$	Resonance at $\omega = \omega_0 = \dfrac{1}{\sqrt{LC}}$
$Q_0 = \dfrac{\omega_0 L}{R}$	$Q_0 = \dfrac{\omega_0 C}{G} = \dfrac{R}{\omega_0 L}$
At resonance: $\lvert\mathbf{Z}\rvert$ is minimum $\lvert\mathbf{Y}\rvert$ is maximum	At resonance: $\lvert\mathbf{Y}\rvert$ is minimum $\lvert\mathbf{Z}\rvert$ is maximum
High Q if $\dfrac{\omega_0 L}{R} \geq 5$	High Q if $\dfrac{\omega_0 C}{G} \geq 5 \left(\dfrac{R}{\omega_0 L} \geq 5\right)$

In every instance where the results of the series circuit are to apply, it must be remembered that an observation for the series circuit which describes the current if a voltage source is impressed corresponds to a result in the parallel case concerning the voltage if a current source is applied.

10-10 Comparison of lossy and lossless circuits

For an oscillatory R-L-C series circuit the admittance is

$$Y = \frac{1}{L}\frac{s}{(s - \mathbf{s}_1)(s - \mathbf{s}_2)}$$

where $\mathbf{s}_{1,2} = -\alpha \pm j\omega_d$.

For a series *L-C* circuit the admittance is

$$Y = \frac{1}{sL + 1/sC} = \frac{1}{L}\frac{s}{s^2 + \omega_0^2}$$

where $\omega_0^2 = 1/LC$. In Fig. 10-14 the pole-zero diagrams for an *R-L-C* and an *L-C* circuit are shown. It is assumed that in both these circuits $LC = 1$. The difference between the two pole-zero diagrams is in the location of the

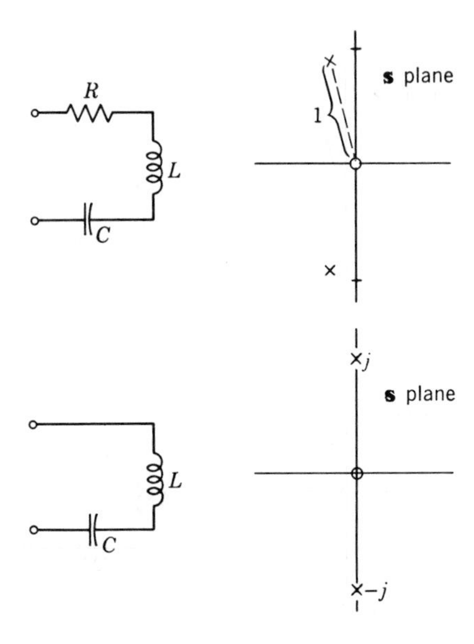

Fig. 10-14 ***Comparison of pole-zero diagrams for a highly oscillatory and a lossless circuit.***

poles. In the lossless (*L-C*) case the poles are on the imaginary axis; hence a sinusoidal source, at the frequency of the pole ($s = j$), will produce a double pole so that the component of the response due to the source has the form $t \cos (t + \psi)$. This component approaches infinity as t approaches infinity. (This is verified by observing that $X = 0$ at $\omega = 1$.) Comparing the *R-L-C* and the *L-C* circuits, we observe the following: At the frequency $\omega = \omega_0$ the lossless circuit results in "infinite" response, but the lossy circuit has maximum response. Furthermore, the effect of adding small damping to the *L-C* circuit moves the poles into the left half plane, off the $j\omega$ axis, but the resonant frequency remains approximately equal to the frequency of the damped oscillations ($\omega_0 \approx \omega_d$).

The relationship between the *L-C* and the *R-L-C* circuits can be generalized as follows: If a circuit is slightly damped so that all the poles and zeros of its network functions are "near" the imaginary axis, they can be located approximately by first ignoring the resistance. (The term "near" means that, for every pole and zero at $s = \mathbf{s}_1$, we require $|\text{Re } \mathbf{s}_1| \ll |\text{Im } \mathbf{s}_1|$.) Thus, in the *slightly damped* circuit, the maxima and minima of response will occur near the poles and zeros of the corresponding lossless circuit.

It is noted that the circuit is slightly damped if the resistance in series with an L or C element is small compared with the element reactance at the zero-pole frequencies and if the resistance in parallel with an L or C element is large compared with the reactance at these frequencies.

Example 10-1 Determine the frequency response of the circuit shown in Fig. 10-15a. The source is an ideal current source. The response is v_{ab}. Use and justify approximate methods.

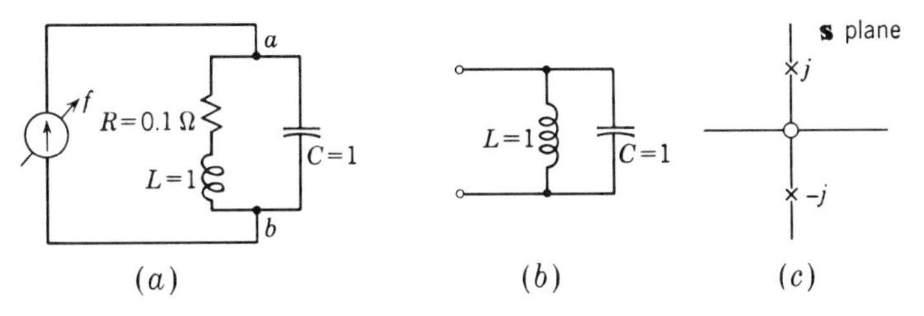

Fig. 10-15 Illustration of how the low-loss circuit (a) can be analyzed approximately by use of a lossless circuit (b) and the pole-zero diagram (c). See Example 10-1.

Solution Setting the series resistance to zero, we obtain the circuit of Fig. 10-15b. For this circuit $Y(s) = s + 1/s$ or $Y(s) = (s^2 + 1)/s$. Hence $Z(s) = s/(s^2 + 1)$, which corresponds to a zero at $s = 0$ and a pole pair at $s = \pm j$, as shown in Fig. 10-15c. We therefore expect maximum response at $s = j$, $\omega = 1$. At $\omega = 1$, $X_L = 1$, since $R = 0.1$, $R/X_L = 0.1$; hence the circuit of Fig. 10-15a is slightly damped, and we expect its Z_{ab} to have a maximum in the neighborhood of $\omega = 1$. This impedance at $s = j$ is

$$Z(j) = \frac{(0.1 + j)/j}{0.1} = 10 - j$$

Thus, from the pole-zero diagram, we conclude that Z has a maximum value $\sqrt{101}$ ohms at $\omega = 1$. This is an approximate result. To illustrate the value of the approximate method, the exact maximum of Z_{ab} is calculated. Consider the circuit of Fig. 10-15a.

$$Y(s) = s + \frac{1}{0.1 + s} = \frac{s^2 + 0.1s + 1}{s + 0.1}$$

or

$$Z(s) = \frac{s + 0.1}{s^2 + 0.1s + 1}$$

so that the poles of $Z(s)$ are at $s = 0.05 \pm j\sqrt{1 - 0.05^2}$ and the zero is at $s = -0.1$. We observe that the poles are very close to the values obtained by using $R = 0$ and that the zero is at $s = -0.1$ rather than zero. If we study $Z(j\omega)$,

$$Z(j\omega) = \frac{j\omega + 0.1}{1 - \omega^2 + 0.1j}$$

its absolute value is

$$Z(\omega) = \frac{\sqrt{0.01 + \omega^2}}{\sqrt{(1 - \omega^2) + 0.01}}$$

We find after elaborate computation that $Z(\omega)$ is a maximum at $\omega = 1.00453$, within 0.5 per cent of 1.

Example 10-2 In a certain low-loss network the response $v_2(t)$ is related to the source $v_1(t)$ by the equation

$$v_2(t) = \frac{1}{3} \frac{p(p^2 + 2p + 200)}{(p^2 + p + 50)(p^2 + 3p + 500)} v_1(t)$$

Use the approximate pole-zero-diagram method to sketch the amplitude response curve $|H(j\omega)| = h(\omega)$.

Solution The transform network function is given as

$$H(s) = \frac{s(s^2 + 2s + 200)}{3(s^2 + s + 50)(s^2 + 3s + 500)}$$

We calculate (approximately) the following pole-zero locations:

Zeros:	$s = 0$	$s \approx -1 \pm j14$
Poles:	$s \approx -0.5 \pm j7$	$s \approx -1.5 \pm j22$

The corresponding pole-zero diagram is shown in Fig. 10-16a. We observe that the low-loss approximations apply because, for each pole and zero, $|\text{Re } s| \ll |\text{Im } s|$ and the distances between poles and zeros are all large compared with their distance from the axis.

We note that $h(0) = 0$ (a minimum) and that $h(14)$ will be an approximate minimum. To calculate the value $h(14)$ we note that

$$h(14) = \frac{(Z_0G_1)(Z_1G_1)(Z_2G_1)}{3(P_1G_1)(P_2G_1)(P_3G_1)(P_4G_1)}$$

where the distances Z_1G_1, etc., can all be approximated by their vertical components. Hence

$$h(14) \approx \frac{(14) \times (1) \times (28)}{3(7) \times (21) \times (8) \times (36)}$$
$$\approx 0.0031$$

Similarly, maxima occur at $\omega \approx 7$ and $\omega = 22$. The values of these maxima are approximated by

$$h(7) = \frac{(Z_0G_2)(Z_1G_2)(Z_2G_2)}{3(P_1G_2)(P_2G_2)(P_3G_2)(P_4G_2)}$$
$$\approx \frac{(7)(7)(21)}{3(\frac{1}{2}) \times (14) \times (15) \times (29)}$$
$$\approx 0.11$$

$$h(22) = \frac{(Z_0G_3)(Z_1G_3)(Z_2G_3)}{3(P_1G_3)(P_2G_3)(P_3G_3)(P_4G_3)}$$
$$\approx \frac{(22) \times (8) \times (36)}{3(15) \times (29) \times (\frac{3}{2}) \times (44)}$$
$$\approx 0.074$$

To sketch the response curve $h(\omega)$, we calculate the following approximate additional values: In the neighborhood of G_1, the distance Z_1G controls the magnitude of $h(\omega)$. If all other distances are assumed constant near $\omega = 14$, we note (Fig. 10-16b) that

$$h(15) \approx h(13) \approx h(14) \times \sqrt{2}$$

Similarly,

$$h(6.5) \approx h(7.5) = \frac{h(7)}{\sqrt{2}}$$

and

$$h(20.5) \approx h(23.5) = \frac{h(22)}{\sqrt{2}}$$

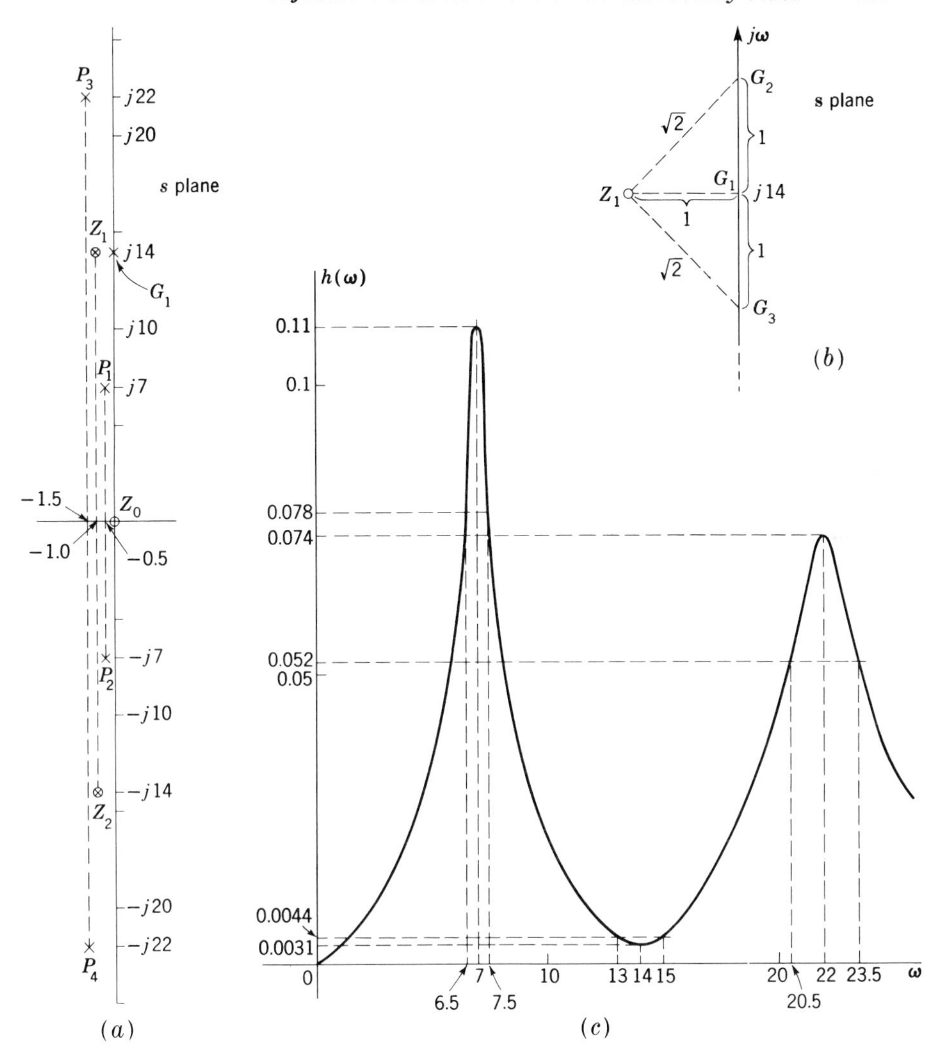

***Fig. 10-16** (a) Pole-zero diagram for the low-loss transform network function analyzed in Example 10-2. In this diagram G_1 is at j14 for the calculation of $h(14) = |H(j14)|$. (b) Portion of the s plane for calculation of $h(13)$ and $h(15)$. In this calculation the values of distances from Z_0, P_1, and P_2 and Z_2, P_3, and P_4 to G_2 and G_3 approximated by those to G_1. (c) Frequency response $h(\omega)$ as obtained by approximate analysis in Example 10-2.*

We now have the following table of values:

ω	0	6.5	7	7.5	13	14	15	20.5	22	23.5
$h(\omega)$	0	0.078	0.11	0.078	0.0044	0.0031	0.0044	0.052	0.074	0.052

The corresponding graph is shown in Fig. 10-16*c*.

FURTHER DISCUSSION OF RESONANCE In the foregoing discussion we associate the term "resonance" with special response conditions in the sinusoidal steady state such as maximum or minimum response-source amplitude, unity power factor, etc. These definitions are derived and related to the special conditions which prevail when a network is excited by a source function which is either a natural mode of the network or which results in zero response due to source.

We recall that a source

$$\phi(t) = \text{Re}\,(\boldsymbol{\phi} e^{\mathbf{s}_g t})$$

applied to a network which is specified through $H(p)$ produces y_s, the response component due to the source given by

$$y_s(t) = \text{Re}\,(\boldsymbol{\phi} t e^{\mathbf{s}_g t})$$

when $\mathbf{s}_g$ coincides with a simple pole of $H(s)$. We observe that, for low-loss networks, maxima in the sinusoidal-steady-state response occur when $\mathbf{s}_g = j\omega = \text{Im}\,\mathbf{s}_F$, where $\mathbf{s}_F$ is a characteristic root of the network (i.e., a pole of H). We observe that such a resonance will generally[1] occur for *all* responses of the network since the poles of the network characterize the (same) source-free response component for all responses.

When in the source function $\phi(t) = \text{Re}\,(\boldsymbol{\phi} e^{\mathbf{s}_g t})$ the value of $\mathbf{s}_g$ corresponds to a zero of $H(s)$, the component of the response due to the source is zero. This type of resonance is associated with minima in the sinusoidal steady state, and must also be associated with the particular response for which the numerator polynomial of $H(p)$ operates on the source waveform to produce a forcing function that is identically zero.

From the viewpoint of the $\mathbf{s}$ plane, minima in response waveforms are associated with the approximate coincidence of a source pole with a network zero, whereas maxima are related to the (approximate) coincidence of a source pole with a network pole.

10-11 Pure-reactance networks

In the preceding sections we saw how a lossless network can be used to give approximate but sufficiently precise answers about slightly lossy circuits. For this reason the study of networks which contain no resistance is useful. Such circuits are called "pure-reactance" or "pure-susceptance" networks. In this section we shall discuss the properties of pure-reactance driving-point immittances.

In Chap. 8 it was shown that driving-point immittances have the form

[1] For discussion of exceptions, see Sec. 12-3.

$H(s) = N(s)/D(s)$, where N and D are polynomials in s. Further, it was shown that neither the zeros nor the poles of driving-point immittance functions can be in the right-half s plane. If a pole or zero of a driving-point immittance is in the left-half plane, the corresponding free response will be exponentially damped. But in a pure-reactance network, any energy which is initially stored in the source-free network cannot decay, since there are no dissipating elements. Hence, for a purely reactive network, the poles and zeros must be on the imaginary axis, and the factors of N and D have the form

$$(s - j\omega_A)(s + j\omega_A) = s^2 + \omega_A{}^2$$

In addition, we observe that the complex driving-point immittance must be imaginary because the average power input in the sinusoidal steady state must be zero. Hence, for pure-reactance networks,

$$Z(s)\Big|_{s=j\omega} = Z(j\omega) = \frac{1}{Y(j\omega)} = jX(\omega)$$

where X is a real function of the (radian) frequency. The factors $s^2 + \omega_A{}^2$ are real for $s = j\omega$, but $Z(s)$ itself is imaginary. It follows that $Z(s)$ must have a factor of s either in N or in D. Thus, for pure-reactance networks, the degree of the numerator polynomial *must* differ from the degree of the denominator polynomial by unity. This fact implies that such functions must have either a zero or a pole at $s = 0$. The same conclusion can be reached in another way: If there is a path from one input terminal of the terminal pair to the second terminal which does not pass through a capacitance, the driving-point impedance at $s = 0$ is zero; if every path passes through a capacitance, the driving-point impedance at $s = 0$ is infinity. Since these are the only two possibilities, s must be a factor of the numerator or denominator of the driving-point impedance.

The general form of the driving-point impedance for a purely reactive network is therefore

$$Z(s) = Hs^{\pm 1}\frac{(s^2 + \omega_A{}^2)(s^2 + \omega_B{}^2)\cdots}{(s^2 + \omega_1{}^2)(s^2 + \omega_2{}^2)\cdots}$$

Another condition must be satisfied by the function $Z(s)$. This condition states that zeros and poles must alternate. Thus, if

$$Z(s) = s\frac{(s^2 + \omega_A{}^2)(s^2 + \omega_B{}^2)\cdots}{(s^2 + \omega_1{}^2)(s^2 + \omega_2{}^2)\cdots}$$

we have a zero at $s = 0$ and

$$\omega_1 < \omega_A < \omega_2 < \omega_B < \cdots$$

whereas, if $Z(s)$ has the form

$$Z(s) = \frac{1}{s}\frac{(s^2 + \omega_A{}^2)\cdots}{(s^2 + \omega_1{}^2)\cdots}$$

it is necessary that

$$\omega_A < \omega_1 < \omega_B < \cdots$$

This property is called the *separation* property of zeros and poles, and is part of a theorem which deals with the synthesis of pure-reactance driving-point immittances. This theorem is called Foster's reactance theorem, after its discoverer, R. M. Foster. The proof of the separation property is beyond the scope of this book.[1]

10-12 Reactance and susceptance—frequency curves

Since, for a pure-reactance terminal pair a-b, $Z_{ab} = jX_{ab}$ and $Y_{ab} = jB_{ab}$, it follows that the driving-point reactance X_{ab} is always the negative reciprocal of the susceptance B_{ab}. Moreover, if X_{ab} is positive, this means that in the sinusoidal steady state v_{ab} leads i_{ab} by 90°; if X_{ab} is negative, v_{ab} lags i_{ab} by 90°. Hence a graph of X_{ab} or B_{ab} as a function of frequency shows not only the magnitude ratio V_{ab}/I_{ab}, but also the phase relation between v_{ab} and i_{ab}. The latter is obtained from the sign of X_{ab} or B_{ab}. In this section we shall first study the curves of X and B for the two basic energy-storing elements and show how these curves can be applied to more elaborate circuits.

For an inductance L_{ab} the reactance is $X_{ab} = \omega L_{ab}$ and the susceptance is $-1/\omega L_{ab}$. A curve of reactance as a function of frequency is therefore a straight line, as shown in Fig. 10-17a. The curve of susceptance as a function of frequency for an inductance is the hyperbola shown in Fig. 10-17b.

[1] See, for example, M. Javid and E. Brenner, "Analysis, Transmission, and Filtering of Signals," chap. 13, McGraw-Hill Book Company, New York, 1963.

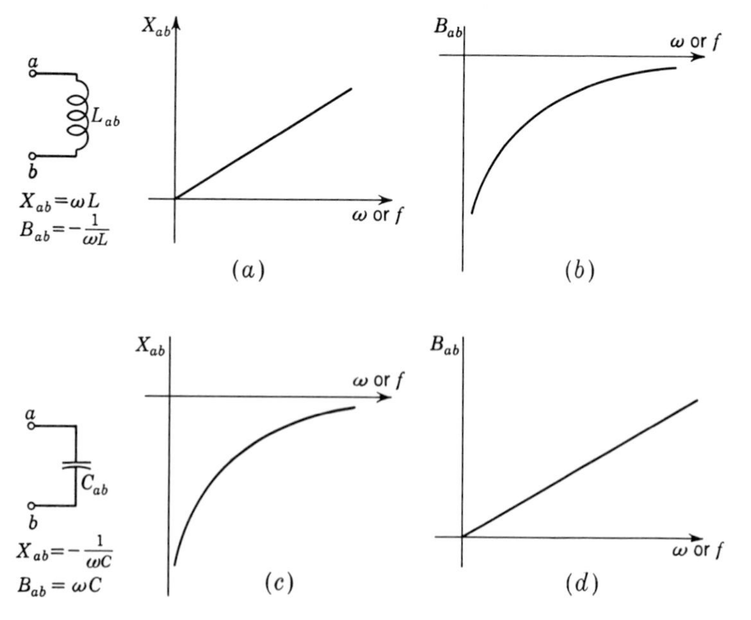

***Fig. 10-17** Reactance- and susceptance-frequency curves for a single inductance and a single capacitance.*

For a capacitance C_{ab} we observe that the reactance curve is a hyperbola and the susceptance curve is a straight line as shown in Fig. 10-17*c* and *d*.

If we now desire to find the reactance-frequency curve of two series elements, we need only add the individual reactance curves, and to obtain the susceptance-frequency curve of two parallel elements, we need only add the individual susceptance curves, because series reactances and parallel susceptances add. In this procedure, as well as in the subsequent examples, we shall not be concerned with the values of the reactances at the various frequencies, although the graphical method which will be described could be used in any numerical case. We shall rather concentrate on determining the sign of X or B and the shape of the curves.

Let us consider first the reactance-frequency curve of a series *L-C* circuit, shown in Fig. 10-18*a*. The reactance-frequency curves for these two elements are drawn in Fig. 10-18*b* and *c*. Since $X_{ab} = X_L + X_C$, by adding the two curves of Fig. 10-18*b* and *c* we note that, at $\omega = 0$, $X = -\infty$; at $\omega = \infty$, $X = +\infty$; and at some point marked ω_0, $X = 0$. At that point $X_L + X_C = 0$ or $\omega_0 L = +1/\omega_0 C$. Hence the radian frequency at which the reactance changes sign is at $\omega_0 = 1/\sqrt{LC}$. The curve for X_{ab} is shown in Fig. 10-18*d*.

In Fig. 10-18 we observe that, at $\omega = \omega_0$, $X_{ab} = 0$. Hence, at that frequency, the impedance $Z_{ab} = 0$. It follows that $s = \pm j\omega_0$ are zeros of the driving-point impedance and poles of the admittance. Thus, if an ideal sinusoidal *current* source of radian fre-

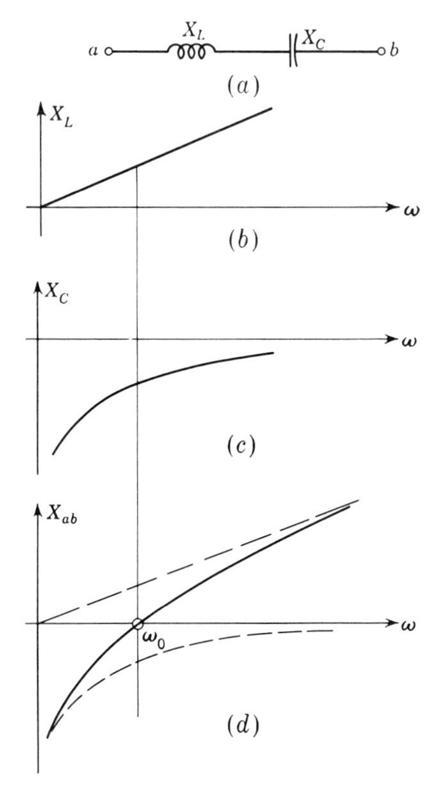

***Fig. 10-18** Reactance-frequency curve of a series L-C circuit obtained by addition of the reactance-frequency curves of an inductance and of a capacitance.*

quency ω_0 is impressed on the circuit, the voltage response $v_{ab}(t)$ due to this source is zero. If, on the other hand, a voltage source of radian frequency ω_0 is impressed, the response component due to the source is $t \cos(\omega_0 t + \psi)$, since a pole of the admittance function corresponds to the pole of the source function. The reactance functions discussed above are special cases of network functions, and the terms zeros and poles which are applied to transform network functions are also applied to reactance functions. Thus ω_0 is a zero of the reactance of the series L-C circuit (Fig. 10-18).

The definitions that follow are used in the discussion of pure-reactance circuits:

Series-resonant frequency: A series-resonant frequency is a frequency at which the driving-point reactance has a zero.

Parallel- (or anti-) resonant frequency: A parallel-resonant frequency is a frequency at which the driving-point reactance has a pole.

The reason for the definition of parallel-resonant frequency becomes clear if we study the parallel L-C circuit of Fig. 10-19. We note that, at $\omega = \omega_0$, the reactance is infinite. It is also evident that the susceptance curve of the series L-C circuit is identical in shape

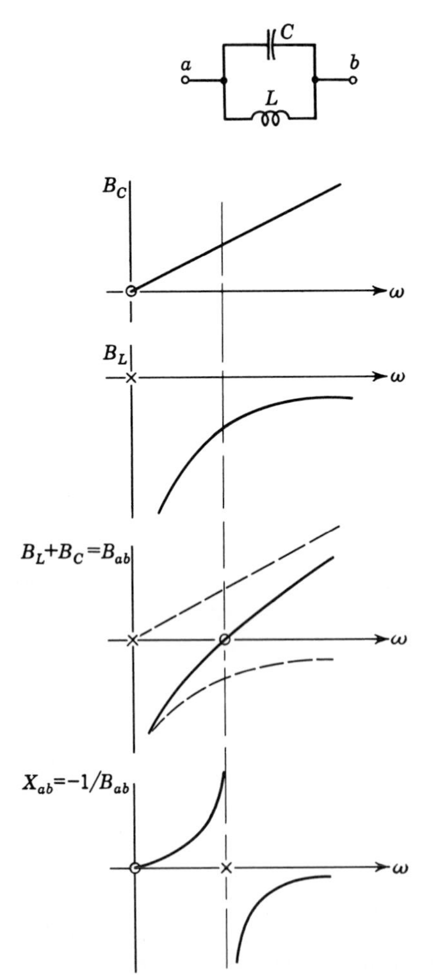

***Fig. 10-19** Susceptance- and reactance-frequency curves constructed for a parallel L-C circuit.*

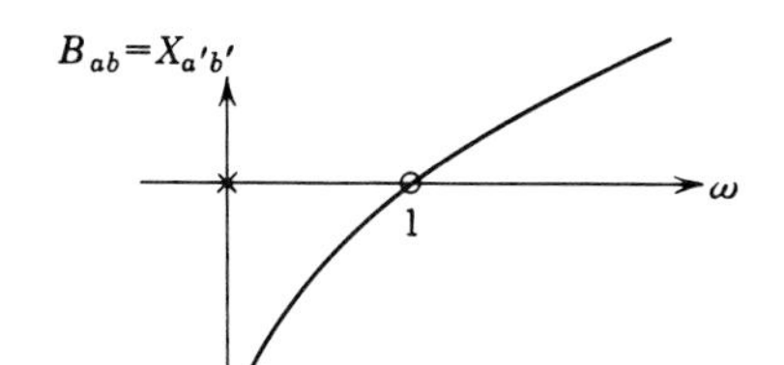

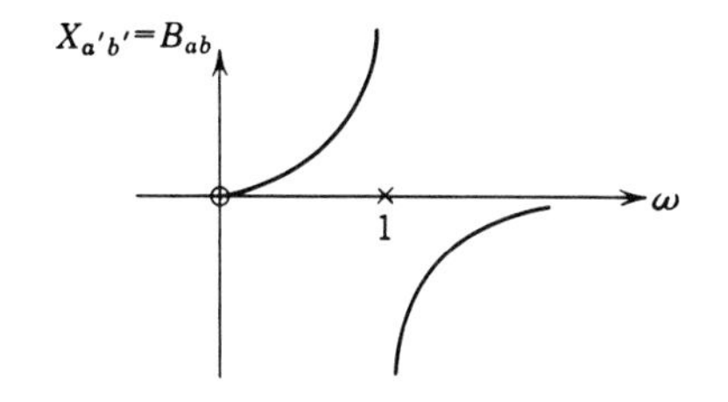

***Fig. 10-20** Illustration of duality. In this example $\mathbf{L} = \mathbf{C}'$ and $\mathbf{C} = \mathbf{L}'$; hence the same curve applies to $\mathbf{B}_{ab}$ and $\mathbf{X}_{a'b'}$. In general, the shape of the susceptance-frequency curve of the parallel circuit is the same as that of the reactance-frequency curve of the series circuit, but the scales need to be changed.*

with the reactance curve of the parallel L-C circuit, and vice versa. This is again a manifestation of the duality of the two circuits (Fig. 10-20).

As an example of a circuit which consists of more than two elements, consider the pure-reactance circuit shown in Fig. 10-21*a*. Let it be required to sketch the shape of X_{ab} as a function of the radian frequency ω for the circuit of Fig. 10-21*a*. To construct these curves, we may start with branch mb. If we add the susceptance of branch mb to the susceptance of C_2, we obtain B_{db}. If we then find X_{db} and add X_{ad}, the desired result, X_{ab}, is obtained. These steps are shown in Fig. 10-21*b* to *g*.

10-13 Circuits with adjustable elements

Introduction Since changing a single element in a network generally changes all pole-zero locations, the pole-zero diagram is not a convenient method for studying the corresponding changes in network functions. Instead, the complex-network-function *locus* can be constructed in the complex plane. (A locus is a collection of points that satisfy a specified condition; for instance, the locus of all points in a plane equidistant from a given point is a circle.)

Series impedances and parallel admittances To introduce the subject of complex-network-function loci, consider the series combination shown in Fig. 10-22*a*. As indicated on the circuit diagram, the resistance R is adjustable and the capacitive reactance is fixed at -3 ohms. Hence the imaginary part of $\mathbf{Z}_{ab}$ is constant and the real part varies. Defining the $\mathbf{Z}_{ab}$ plane (complex impedance plane) of Fig. 10-22*b* as a set of rectangular coordinates where the abscissa is the real part of $\mathbf{Z}_{ab}$ and the ordinate the imaginary part of $\mathbf{Z}_{ab}$, the horizontal line from X_C to the right represents the locus of $\mathbf{Z}_{ab}$ as R varies between zero and infinity. For any particular value

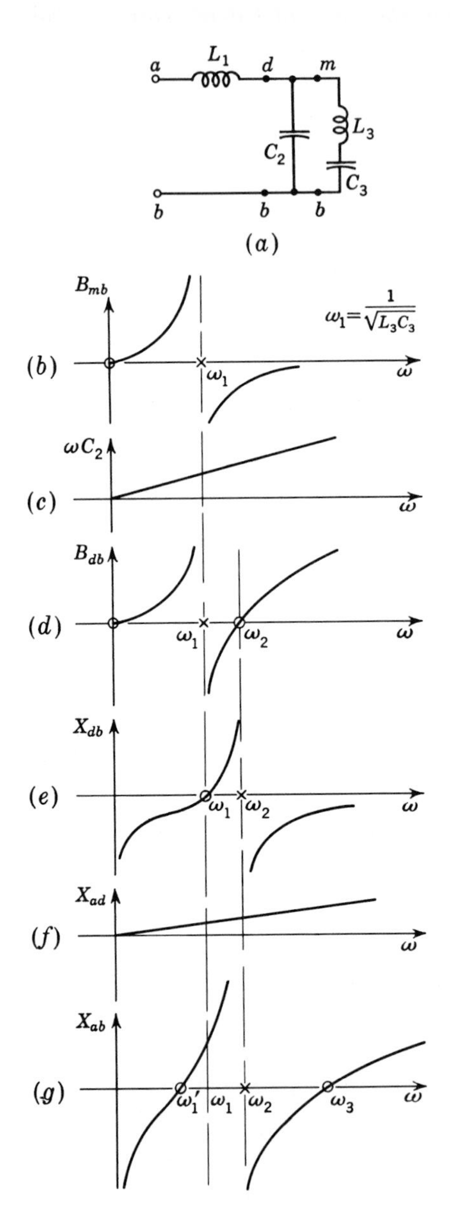

Fig. 10-21 Construction of the reactance-frequency curve for the four-element terminal pair shown in (a).

of R, say, $R = R_1 = 4$, the vector from the origin to the linear locus (dashed line) represents the corresponding complex impedance $\mathbf{Z}_{ab}$. A second example is shown in Fig. 10-22c, where $\mathbf{Y}_{ab} = 1/(4 + j2) + \frac{1}{20} + jB_C$, which works out to $\mathbf{Y}_{ab} = 0.15 - j0.2 + jB_C$. Hence, if $B_C = 0$,

$$\mathbf{Y}_{ab} = \mathbf{Y}_0 = 0.15 - j0.2$$

and the imaginary part of $\mathbf{Y}_{ab}$ increases as B_C increases as shown in Fig. 10-22d. If, for example, $B_C = 0.5$ ($X_C = -2$), $\mathbf{Y}_{ab} = \mathbf{Y}_1 = 0.15 + j0.3$, as shown in Fig. 10-22$d$. Note that in each of the two examples given, an

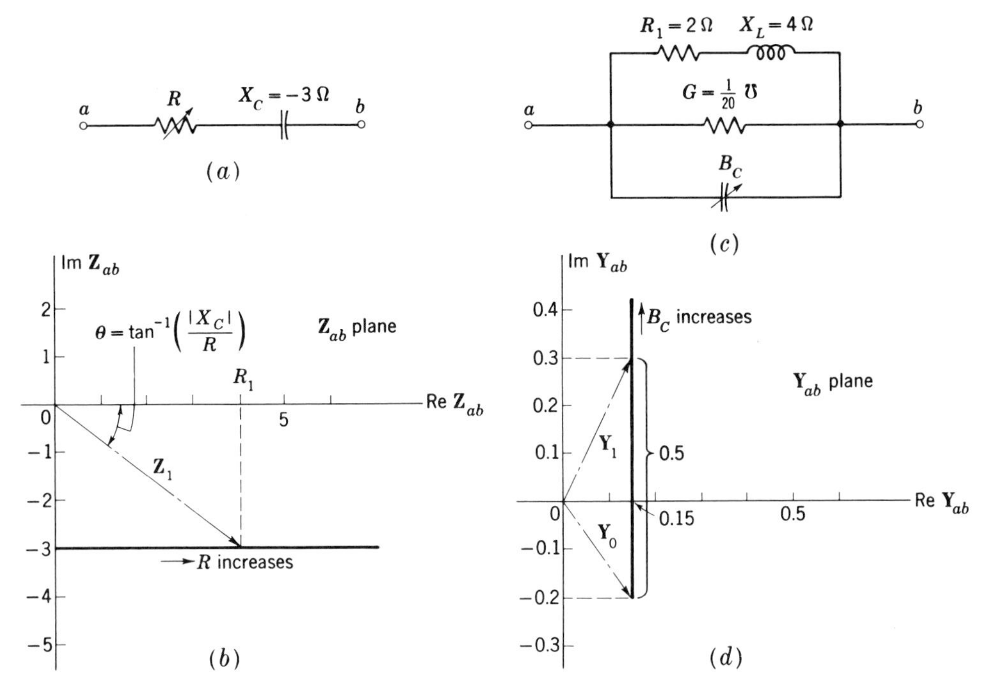

Fig. 10-22 ***(a) A two-element series circuit with an adjustable element. (b) Impedance locus for the circuit of (a). (c) Parallel circuit with an adjustable element. (d) Admittance locus for the circuit of (c).***

arrow has been associated with each locus to show how an increase in the variable parameter value affects the value of **Z** or **Y**.

Inverse locus Since series circuits are generally excited by voltage sources, and parallel circuits by current sources, it is often necessary to find the impedance locus of a parallel combination or the admittance locus of a series circuit. Given a driving-point impedance locus $\mathbf{Z} = Z\underline{/\theta}$, the admittance locus can of course be constructed point by point. Using $\mathbf{Y} = 1/\mathbf{Z} = (1/Z)\underline{/-\theta}$, one can measure (or calculate) Z and θ for several values of the adjustable element, calculate the corresponding values of **Y**, and plot them in the **Y** plane as shown in Fig. 10-23. In Fig. 10-23*a* the locus $\text{Re}\,\mathbf{Z} = 4$ is given so that $\mathbf{Z}_1 = 5\underline{/37°}$, $\mathbf{Y}_1 = \frac{1}{5}\underline{/-37°}$, $\mathbf{Z}_2 = \sqrt{32}\,\underline{/-45°}$, $\mathbf{Y}_2 = (1/\sqrt{32})\underline{/45°}$, $\mathbf{Z}_3 = 4\underline{/0°}$, $\mathbf{Y}_3 = \frac{1}{4}\underline{/0°}$. The three values of **Y** ($\mathbf{Y}_1$, $\mathbf{Y}_2$, and $\mathbf{Y}_3$), are shown in Fig. 10-23*b*, and the corresponding locus is suggested by the curve that connects them.

Before discussing the details of an inverse locus, we make the following observations and definitions:

1 A point **Z** in the first quadrant of the **Z** plane will be in the fourth quadrant of the **Y** plane, and vice versa.

2 If in the **Z**-plane locus the point $\mathbf{Z}_1$ is closer to the origin than the point $\mathbf{Z}_2$, then in the **Y** plane, $\mathbf{Y}_1 = 1/\mathbf{Z}_1$ will be farther from the origin than $\mathbf{Y}_2 = 1/\mathbf{Z}_2$. If $\mathbf{Z}_1 = 0$, then $\mathbf{Y}_2 = \infty$, and vice versa.

3 The "ohms" scale on an impedance locus is determined independently of the "mhos" scale on its inverse locus; for example, the length OZ_3 (ohms) in Fig. 10-23*a* and its inverse, the length OY_3 (mhos) in Fig. 10-23*b*, were chosen independently, for convenience.

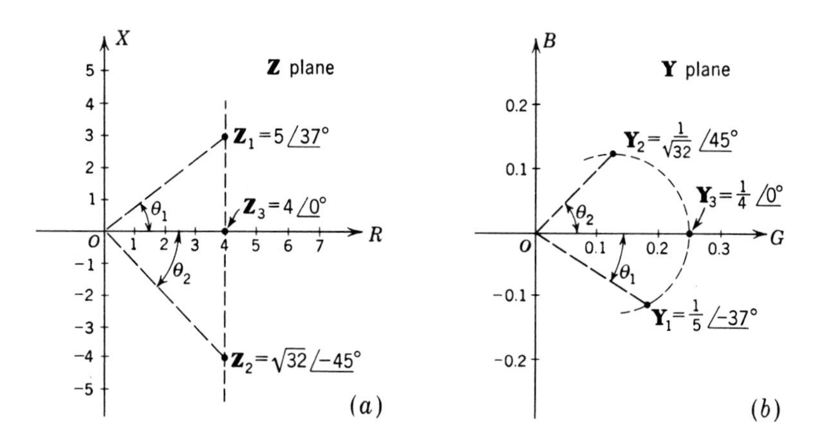

Fig. 10-23 Example of "point-by-point" inversion.

4 The function $\mathbf{Y} = 1/\mathbf{Z}$ is said to *map* a locus in the **Z** plane into its *inverse* in the **Y** plane. Each pair of corresponding points are said to be inverses of each other, $\mathbf{Z}_1 = 1/\mathbf{Y}_1$, $\mathbf{Y}_1 = 1/\mathbf{Z}_1$, etc.

5 If two points $\mathbf{Z}_1$ and $\mathbf{Z}_2$ in the **Z** plane are separated by a certain distance, their inverses $\mathbf{Y}_1$ and $\mathbf{Y}_2$ are generally *not* separated by the same distance. However, the points $\mathbf{Z}_1$ and $\mathbf{Z}_2$ are seen from the origin through an angle Z_1OZ_2, and *this angle is identical* with the angle Y_2OY_1, presented by $\mathbf{Y}_2$ and $\mathbf{Y}_1$ to the origin in the **Y** plane. In other words, in mapping $\mathbf{Z} = 1/\mathbf{Y}$, distances are *not* preserved, but angles as seen from the respective origins remain the same (although quadrants change).

10-14 Inversion of straight lines and circles

Since the application of locus inversion will be to convert impedance to admittance loci, as well as admittance to impedance loci, we temporarily change the notation and discuss the relationship between a circle or a straight line in the **z** plane with rectangular coordinates x and y and its inverse in the **w** plane with coordinates u and v. (If in an application $\mathbf{z} = \mathbf{Z}$, then $x = R$, $y = X$, $\mathbf{w} = \mathbf{Y}$, $u = G$, $v = B$. If $\mathbf{z} = \mathbf{Y}$, then $x = G$, $y = B$, $\mathbf{w} = \mathbf{Z}$, $u = R$, and $v = X$.) *We shall now show that the inverse of a* *straight line or a circle in either plane* (**z** *or* **w**) *is also a straight line or a circle in the other plane* (**w** *or* **z**).

In the **z** plane a circle of radius r with center at a and b is given by

$$(x - a)^2 + (y - b)^2 = r^2$$

or expanding and collecting terms,

$$A(x^2 + y^2) + Bx + Cy + D = 0 \tag{10-48}$$

where $A = 1$, $B = -2a$, $C = -2b$, and $D = a^2 + b^2 - r^2$. Equation (10-48) represents all circles and (if $A = 0$) represents also all straight lines in the **z** plane. We now let $\mathbf{z} = x + jy$ and $\mathbf{w} = u + jv$; then, if $\mathbf{z} = 1/\mathbf{w}$,

$$\mathbf{z} = x + jy = \frac{1}{u + jv} = \frac{u - jv}{u^2 + v^2}$$

$$x = \frac{u}{u^2 + v^2} = \frac{u}{w^2} \qquad y = \frac{-v}{u^2 + v^2} = \frac{-v}{w^2} \tag{10-49}$$

Since x and y are on the locus defined by Eq. (10-48), we substitute from Eqs. (10-49) into Eq. (10-48) and obtain

$$A\left(\frac{u^2}{w^4} + \frac{v^2}{w^4}\right) + \frac{Bu}{w^2} - \frac{Cv}{w^2} + D = 0 \tag{10-50}$$

Multiplying both sides of Eq. (10-50) by $w^2 = u^2 + v^2$, we have

$$D(u^2 + v^2) - Cv + Bu + A = 0 \tag{10-51}$$

Equation (10-51) describes a circle (or straight line if $D = 0$) in the **w** plane,

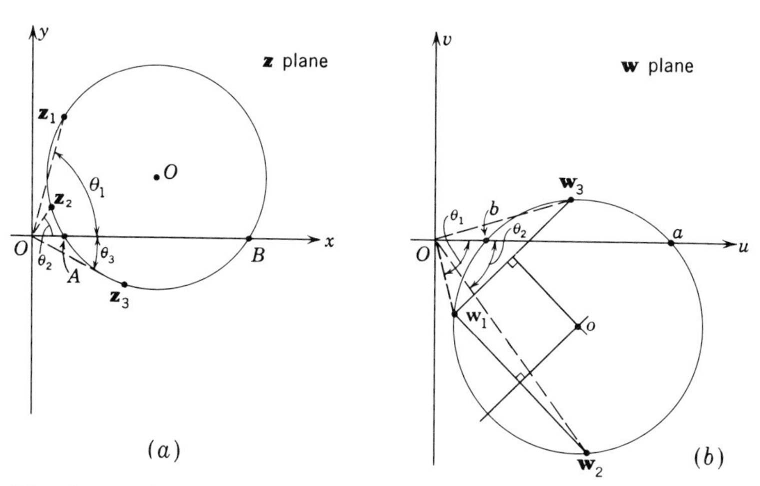

***Fig. 10-24** Graphical method for inversion of a circular locus: The inverse of three points is located; then the perpendicular bisector of two chords determines the location of the center in the inverted locus.*

thus showing that inverse loci of circles are circles, a straight line being considered a circle of infinite radius. It is left as an exercise for the reader to prove that all straight lines and circles *that pass through the origin* have inverse loci that are straight lines.

The derivation above does not give explicit formulas for the determination of the radius and center of the inverted circle. Such formulas are not necessary since, if we *know* the inverse locus to be a circle, we could find it by locating three points on the inverse locus as shown in Fig. 10-24. This method is not always the best, since it is difficult to determine how to choose three points in the **z** plane to obtain three points in the **w** plane that are sufficiently far apart to permit accurate construction. Further, the method developed below allows us to find easily the value of the variable element in a resultant locus.

10-15 Inverse-diameter method

Let the circle of Fig. 10-25a with center P be in the **z** plane, and let it be required to find its inversion in the **w** plane. If the origin O of the **z** plane

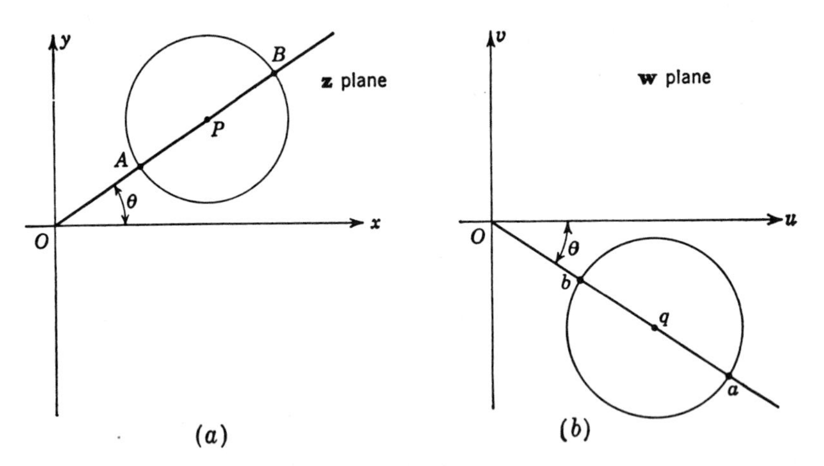

***Fig. 10-25** Inversion of circular locus by the "inverse diameter method."*

is joined to the center P and extended to cut the circle at A and B, these will be the points on the circle nearest and farthest from the origin, respectively. Therefore the inversion of A in the **w** plane will be farthest away from the origin, and that of B will be closest to the origin of the **w** plane.

If these points in the **w** plane are designated as a and b, respectively, they will both lie on a line abO such that the angle $\underline{/aOu}$ in the **w** plane will be equal (in magnitude) to the angle $\underline{/AOx}$ in the **z** plane. Once the points a and b are found, a little thought will show that the center of the circle in the **w** plane will be midway between a and b on the abO line. The reader is warned that this center, *the point q, is not the inversion of the P, the center of the circle in the* **z** *plane, nor is the length ab equal to the reciprocal of the length AB.* It is recalled that in the process of inversion only the angles subtended to the origin are maintained and that the distances will *not* be preserved.

Example 10-3 Determine the admittance locus of the circuit of Fig. 10-26*a*, which is identical with Fig. 10-22*a*.

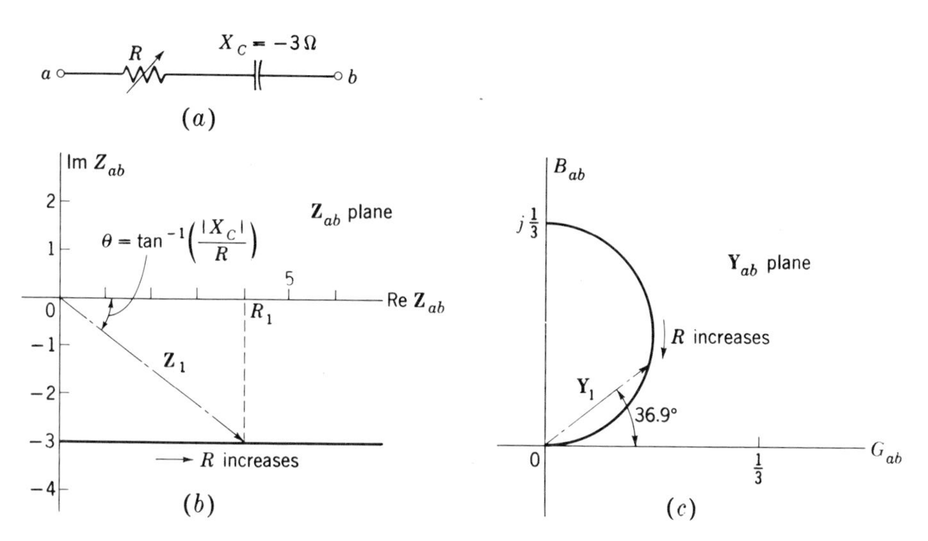

Fig. 10-26 *(a)* ***Adjustable R-C series circuit.*** *(b)* ***Its impedance locus.*** *(c)* ***The corresponding admittance locus.***

Solution The impedance locus is shown in Fig. 10-26*b*; it is identical with that of Fig. 10-22*b*. Since the straight line in the $\mathbf{Z}_{ab}$ plane goes to infinity, the locus in the Y_{ab} plane passes through the origin. In the $\mathbf{Z}_{ab}$ plane the point $-j3$ is closest to the origin; hence the point $\mathbf{Y}_{ab} = +j\frac{1}{3}$ is farthest from the origin, and the diameter of the circle in the **Y** plane extends on the imaginary axis from zero to $j\frac{1}{3}$. The relevant portion of the locus of $\mathbf{Y}_{ab}$ (Fig. 10-26*c*) is the semicircle in the first quadrant (because the $\mathbf{Z}_{ab}$ locus is in the fourth quadrant). To determine how a point on the $\mathbf{Y}_{ab}$ locus corresponds to a desired point on the $\mathbf{Z}_{ab}$ locus we choose, for example, $R = R_1 = 4$. For this value, $|\theta| = \tan^{-\frac{3}{4}} = 36.9°$. Hence $\mathbf{Y}_1$ is the vector from the origin to the circular locus in Fig. 10-26*c* at the angle $+36.9°$, and measuring $\mathbf{Y}_1 = 0.2\underline{/36.9°} = 0.16 + j0.12$ and $G_{ab} = 0.16$, $R_{ab} = 0.12$ mho. To find which value of R is needed to obtain a given point on the $\mathbf{Y}_{ab}$ locus, we again use the angle correspondence. For example, if we desire that Re $\mathbf{Y}_{ab}$ be a maximum, then from Fig. 10-26*c*, Re $\mathbf{Y}_{ab} = \frac{1}{6}$ and the angle is 45° up from the horizontal. Hence the angle in the $\mathbf{Z}_{ab}$ is 45° down from the horizontal and $R_1 = 3$ ohms.

Example 10-4 Repeated locus inversion Construct the admittance locus $\mathbf{Y}_{ab}$ for the circuit shown in Fig. 10-27*a*, and determine from the locus the value of X_C so that Im $\mathbf{Y}_{ab}$ = maximum.

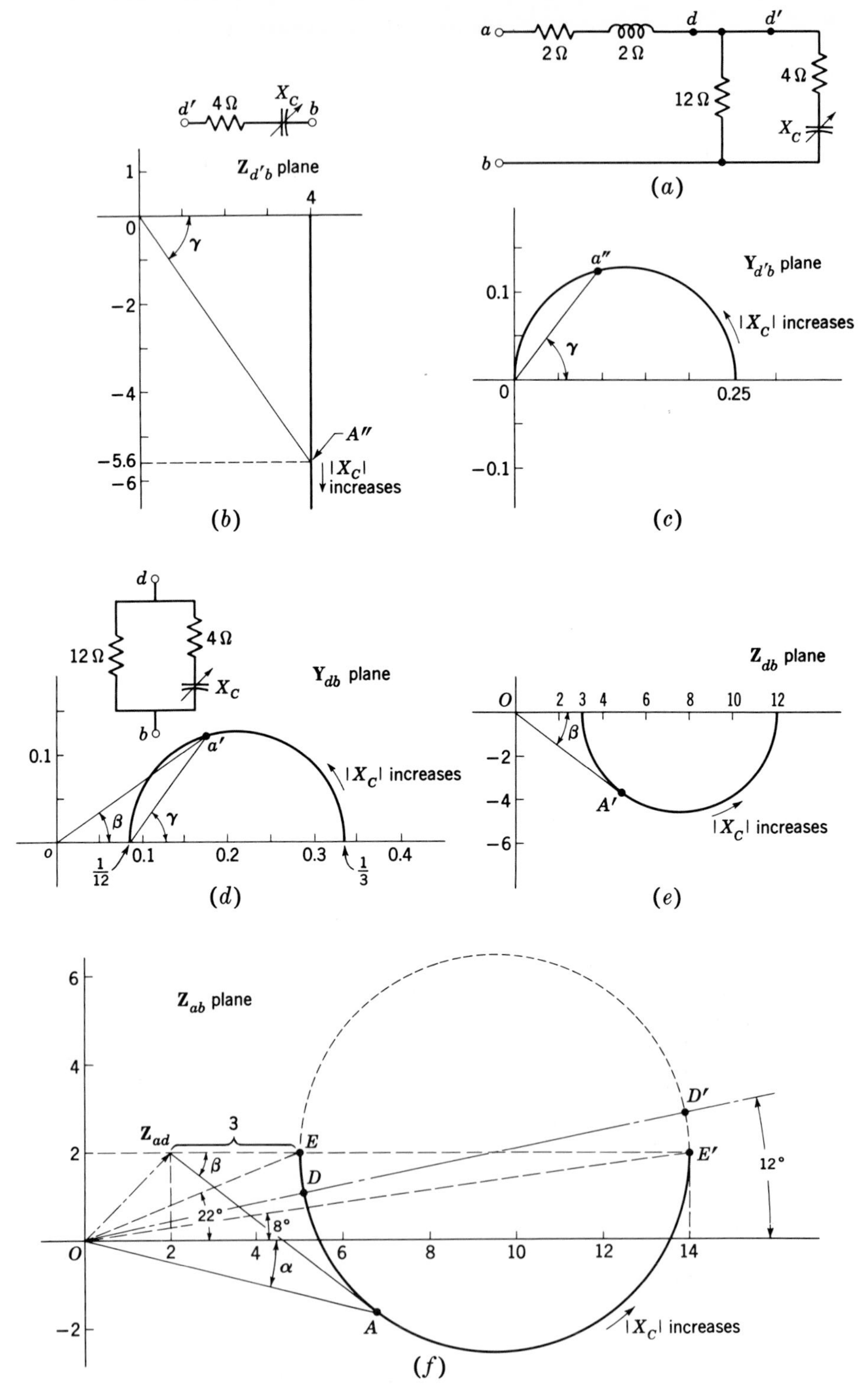
a
2 Ω
2 Ω
d
d′
4 Ω
12 Ω
X_C
b
(a)
d′
4 Ω
X_C
b
Z_d′b plane
1
0
γ
4
−2
−4
A″
−5.6
−6
|X_C| increases
(b)
Y_d′b plane
a″
0.1
|X_C| increases
γ
0
0.25
−0.1
(c)
d
12 Ω
4 Ω
X_C
b
Y_db plane
0.1
a′
|X_C| increases
β
γ
o
0.1
0.2
0.3
0.4
1/12
1/3
(d)
Z_db plane
O
2
3
4
6
8
10
12
β
−2
−4
A′
|X_C| increases
−6
(e)
6
Z_ab plane
4
D′
3
Z_ad
E
2
β
E′
D
12°
22°
8°
O
2
4
6
8
10
12
14
α
−2
A
|X_C| increases
(f)

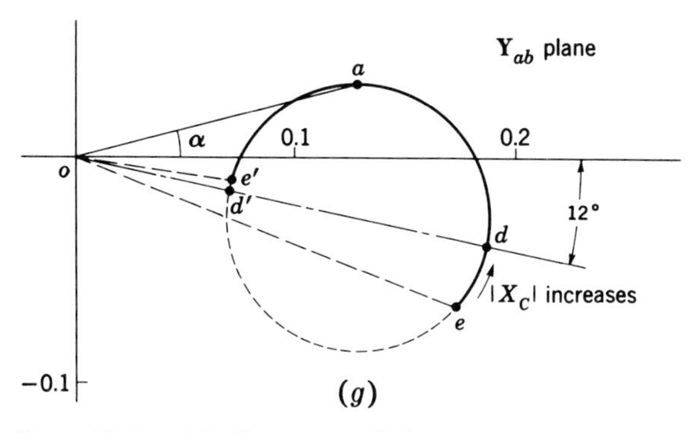

***Fig. 10-27** (a) Series-parallel circuit with an adjustable element. (b) Locus of $4 + jX_C$. (c) Inverse locus of (b). (d) Locus of (c) shifted by $\frac{1}{12}$. (e) Inverse of locus of (d). (f) Locus of (e) shifted by $2 + j2$. (g) Inverse of locus of (f).*

Solution We begin with the locus for the impedance of the series R-X_C branch, $\mathbf{Z}_{d'b}$, shown in Fig. 10-27*b*. Next the locus of $\mathbf{Z}_{d'b}$ is inverted to give $\mathbf{Y}_{d'b}$ as shown in Fig. 10-27*c*. To find the locus of $\mathbf{Y}_{db}$, the constant admittance $\frac{1}{12}$ mho is added to the locus of $\mathbf{Y}_{d'b}$, shifting the semicircle of Fig. 10-27*c* to the right by $\frac{1}{12}$ as shown in Fig. 10-27*d*. Since $\mathbf{Y}_{db}$ is in series with $Z_{ad} = 2 + j2$, we construct $\mathbf{Z}_{db}$, the inverse locus of $\mathbf{Y}_{db}$ as shown in Fig. 10-27*e*. The semicircular locus of $\mathbf{Z}_{db}$ is seen to have its diameter on the real axis since $\mathbf{Z}_{db}$ is real when $|X_C| = 0$ and infinite, intersecting the real axis at $\mathbf{Z}_{db} = 12$ ($|X_C| =$ infinite) and $\mathbf{Z}_{db} = (12)(4)/(12 + 4) = 3$ ($X_C = 0$).

Next the constant impedance $2 + j2$ is added to the locus of $\mathbf{Z}_{db}$ to give the locus of $\mathbf{Z}_{ab}$ as shown in Fig. 10-27*f*. Since $\mathbf{Y}_{ab}$ is desired, the circle has been completed, so that the inverse diameter can be located; however, only the inverse of the solid-line portion of the circle will constitute the $\mathbf{Y}_{ab}$ locus. The diameter extends from D to D'; its inverse is shown in Fig. 10-27*g* from d' to d in the fourth quadrant at the angle $\approx 12°$. Measuring distances, we obtain $OD = 5.2$ and $OD' = 14.1$, so that $od \approx 0.19$ and $od' = 0.07$; the circle is now constructed. To find e and e', the inverse of E and E', respectively, the angle between the horizontal and OE is found; it is 22°, locating e at 22° in the fourth quadrant, "far" from the origin on the circle. Similarly, the angle for e' is 8°. As the locus in Fig. 10-27*f* passes through D, the circular arc in Fig. 10-27*g* must pass through d; hence the portion of the circle drawn with a solid line constitutes the locus of $\mathbf{Y}_{ab}$. To find the value of X_C for which Im $\mathbf{Y}_{ab}$ is a maximum, we locate the point a on the $\mathbf{Y}_{ab}$ locus, measure α, and refer the point a to the $\mathbf{Z}_{ab}$ plane, where it is marked A. To find which point in Fig. 10-27*e* corresponds to A, we measure angle $EZ_{ad}A$ in Fig. 10-27*f* and transfer it to the origin in Fig. 10-27*e*, giving point A'. Measuring β and transferring it to the first quadrant in Fig. 10-27*d* gives point a' as shown. We now measure γ in Fig. 10-27*d* and transfer it to the origin of Fig. 10-27*c*, giving point a''. The angle γ is now transferred to Fig. 10-27*b*, giving A'', which corresponds to $X_C = -5.6$ ohms. Although the graphical method for solution may appear tedious, the reader can convince himself of its value by setting up the problem of maximizing Im $\mathbf{Y}_{ab}$ analytically.

10-16 Transfer-function loci

Consider now a source (phasor) $\mathbf{V}$ energizing a network that contains a variable element as shown in Fig. 10-28*a*, and let us discuss the locus of the

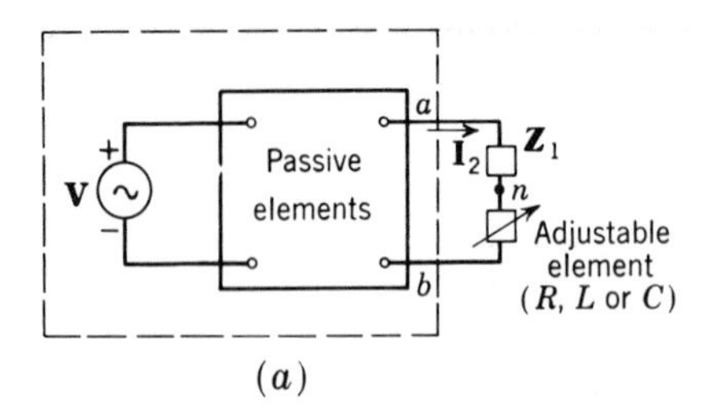

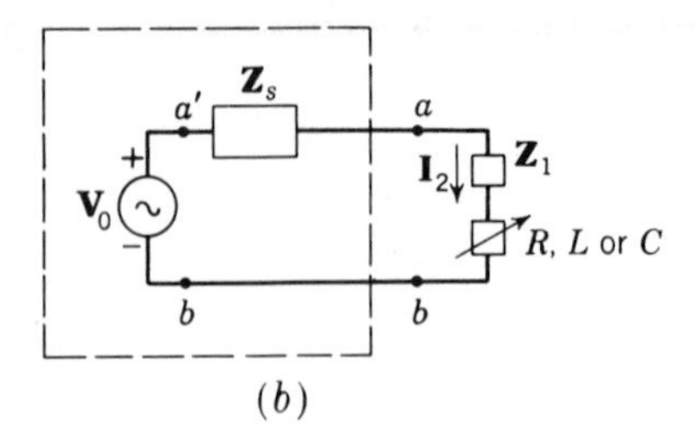

***Fig. 10-28** (a) Definition of transfer-function problem. (b) Thévenin's equivalent to (a) at terminals a-b.*

voltage gain $\mathbf{V}_{ab}/\mathbf{V}$ or the transfer admittance $\mathbf{I}_2/\mathbf{V}$. In that circuit both $\mathbf{V}_{ab}$ and $\mathbf{V}$ are associated with the variable branch.

The desired locus may be found by use of Thévenin's theorem at terminals a-b. The elements to the left of terminals a-b in Fig. 10-28a are represented by Thévenin's equivalent circuit, $\mathbf{V}_0$ in series with $\mathbf{Z}_s$, as shown in Fig. 10-28b. In that circuit $\mathbf{V}_0 = \mathbf{H}_0\mathbf{V}$, where $\mathbf{H}_0$ is the complex open-circuit voltage gain. Hence the locus of $\mathbf{I}_2/\mathbf{V}_0$ is the admittance locus of the series combination of $\mathbf{Z}_s$, $\mathbf{Z}_1$, and the variable element, that is,

$$\frac{\mathbf{I}_2}{\mathbf{V}_0} = \mathbf{Y}_{a'b}$$

Since $\mathbf{V}_0 = \mathbf{H}_0\mathbf{V}$, writing $\mathbf{H}_0$ in polar form, $\mathbf{H}_0 = h_0/\underline{\phi}$,

$$\frac{\mathbf{I}_2}{\mathbf{V}} = \mathbf{Y}_{a'b}(h_0/\underline{\phi}) \tag{10-52}$$

It follows that the locus $\mathbf{I}_2/\mathbf{V}$ is obtained by multiplying the scale of the locus $\mathbf{Y}_{ab}$ by h_0 and rotating it through the angle ϕ. Similarly,

$$\mathbf{V}_{ab} = \mathbf{V}_0 - \mathbf{I}_2\mathbf{Z}_s = \mathbf{V}_0(1 - \mathbf{Y}_{a'b}\mathbf{Z}_s)$$

or

$$\frac{\mathbf{V}_{ab}}{\mathbf{V}} = \mathbf{H}_0(1 - \mathbf{Z}_s\mathbf{Y}_{a'b}) \tag{10-53}$$

so that the locus of $\mathbf{V}_{ab}/\mathbf{V}$ is obtained by multiplication of $\mathbf{Y}_{a'b}$ by $\mathbf{Z}_s$, subtraction of the product from 1, and multiplication by $\mathbf{H}_0$. It must be remembered that all graphical multiplications by complex numbers are scale changes and rotations. Since the locus of $\mathbf{Y}_{a'b}$ is a circular arc, the transfer-function loci will also be circular arcs; hence, if desired, the three-point method of locating a circle can be used.

If the variable element is not in the desired output branch, the voltage across the variable element or the current through it can be calculated first and then related to the desired variable, as illustrated in the following example.

Example 10-5 In the circuit of Fig. 10-29a, obtain the transfer-function locus $\mathbf{V}_2/\mathbf{V}$.

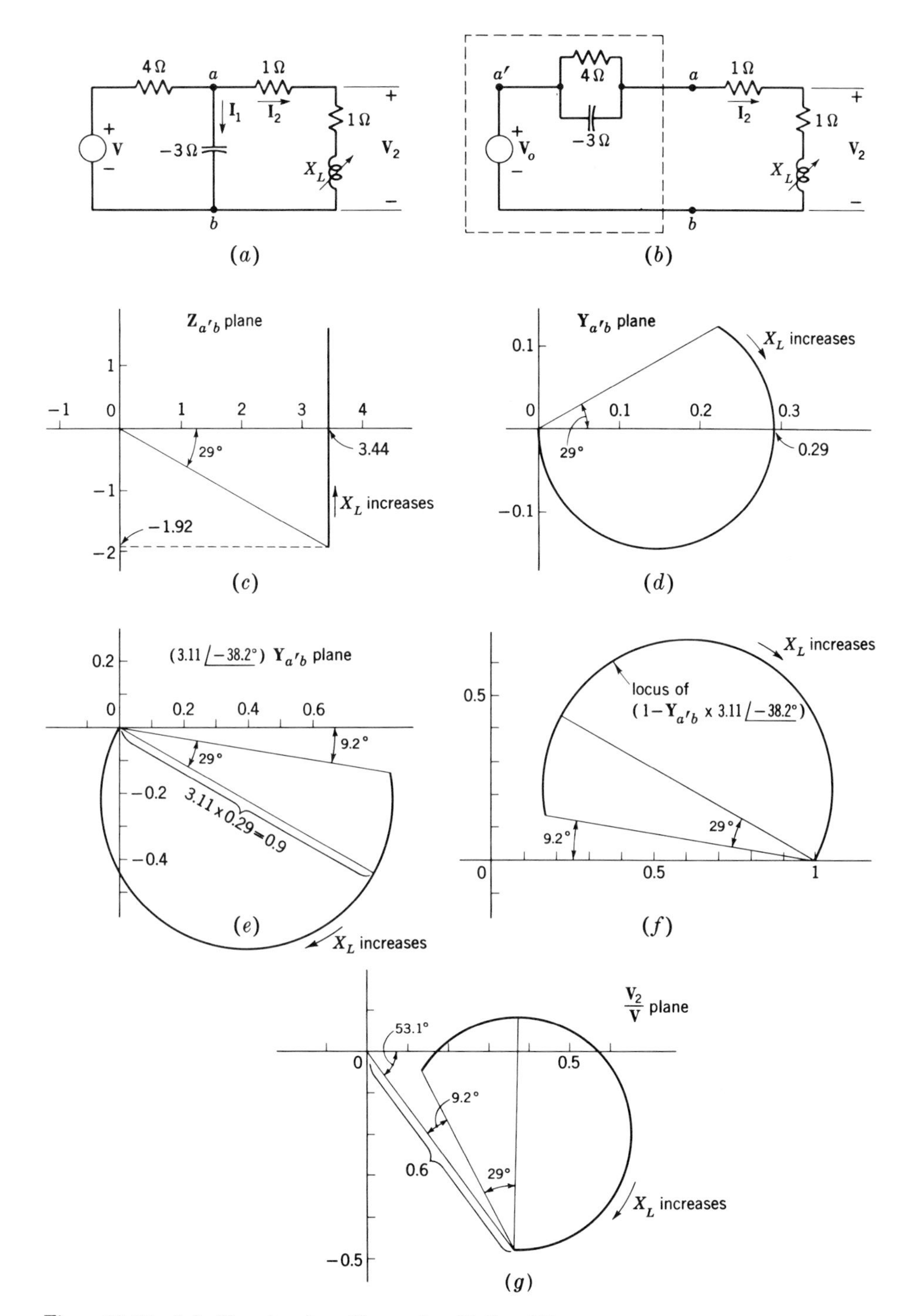

***Fig. 10-29** (a) Circuit for Example 10-5. (b) Thévenin's equivalent to (a) at terminals a-b. (c) Locus of $\mathbf{Z}_{a'b}$. (d) Locus of $\mathbf{Y}_{a'b}$. (e) Locus (d) rotated by −38.2° and scaled. (f) Locus of 1 − [locus of (e)]. (g) Locus of $\mathbf{V}_2/\mathbf{V}$.*

Solution Applying Thévenin's theorem to the left of terminals *a-b*, the circuit of Fig. 10-29*b* is obtained; in that circuit $\mathbf{V}_0 = \mathbf{V}[-j3/(4 - j3)] = (0.6\underline{/-53.1°})\mathbf{V}$. In Fig. 10-29*b*, $\mathbf{Z}_{a'b} = 3.44 - j1.92 + jX_L$, and has the locus shown in Fig. 10-29*c*. The corresponding admittance locus is shown in Fig. 10-29*d*. We observe that

$$\mathbf{V}_2 = \mathbf{V}_0 - \mathbf{I}_2(2.44 - j1.92)$$
$$= \mathbf{V}_0[1 - (2.44 - j1.92)\mathbf{Y}_{a'b}]$$

or

$$\frac{\mathbf{V}_2}{\mathbf{V}} = (0.6\underline{/-53.1°})[1 - (3.11\underline{/-38.2°})\mathbf{Y}_{a'b}]$$

The construction of the locus is shown in Fig. 10-29*e* to *g*.

If we want to find the locus of $\mathbf{I}_1/\mathbf{V}$, we write $\mathbf{I}_1 = \mathbf{V}_{ab}/-j3 = \mathbf{V}_0(1 - \mathbf{Z}_{a'a}\mathbf{Y}_{a'b})/(-j3)$; hence

$$\frac{\mathbf{I}_1}{\mathbf{V}} = \frac{0.6\underline{/-53.1°}}{3\underline{/90°}}[1 - (1.44 - j1.92)\mathbf{Y}_{a'b}]$$

Construction of this locus is left as an exercise for the reader.

Discussion of examples From Examples 10-3 to 10-5, we observe that locus diagrams can be used to study network-function *sensitivity*. By this is meant the effect on a network function produced by a change in an element value. For example, Fig. 10-29*g* shows that, as X_L increases from zero, the phase of $\mathbf{V}_2/\mathbf{V}$ of Fig. 10-29*a* varies slowly near zero but reaches $-53.1°$ as X_L becomes infinite. The value of X_L corresponding to the point where the locus has a horizontal tangent is the value where the phase changes least rapidly, whereas the points near vertical tangency represent most rapid variation of imaginary part with little change in real part. The example of Fig. 10-27 shows, similarly, limits of admittance variation caused by changing one element. It is clear that such locus diagrams can be used to determine whether or not precise adjustment of an element is necessary for achieving a value of network function within a prescribed tolerance: if, for instance, a locus is a short arc far from the origin, the element value is not critical.

10-17 Examples Two parallel admittances As illustration of the application of the locus diagrams in circuits in which one element is adjusted, we consider a parallel circuit such as is shown in Fig. 10-30. In this example we are assuming that X_c is adjustable. The reader is urged to construct the locus if one of the other elements is adjusted (Prob. 10-40). The problem is formulated as follows: Find the locus of the end point

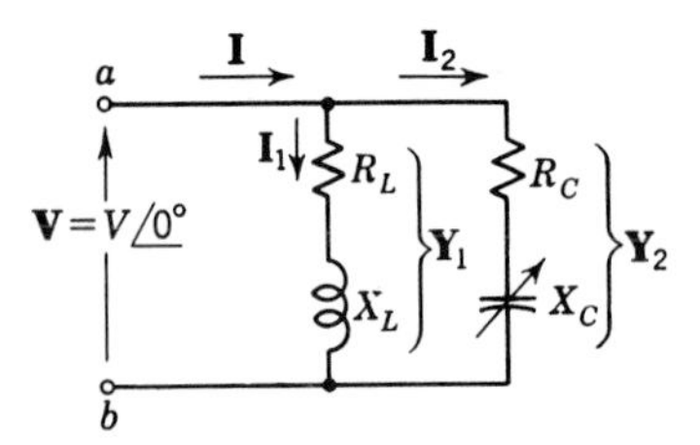

Fig. 10-30 Two parallel branches with one adjustable element.

of the phasor **I** if $|X_c|$ is adjusted from zero (infinite capacitance) to infinity (zero capacitance). It should be clear that the case $\mathbf{V} = 1\underline{/0}$ will give the admittance locus. The problem will be done for this case since there will be no loss in generality, only a scale change being involved.

Since $\mathbf{Y}_{ab} = \mathbf{Y}_1 + \mathbf{Y}_2$ the graphical representation of $\mathbf{Y}_{ab}$ will be the sum of the fixed value $\mathbf{Y}_1$ and the adjustable value $\mathbf{Y}_2$. Now

$$\mathbf{Y}_1 = \frac{1}{R_L + jX_L} = \frac{1}{Z_1}\underline{/-\theta_1}$$

where $\theta_1 = \tan^{-1}(X_L/R)$. The complex number $\mathbf{Y}_1$ is represented by the line OA in Fig. 10-31a. To this line we must now add $\mathbf{Y}_2$. If $X_c = 0$, then $\mathbf{Y}_2 = 1/R_c$. This

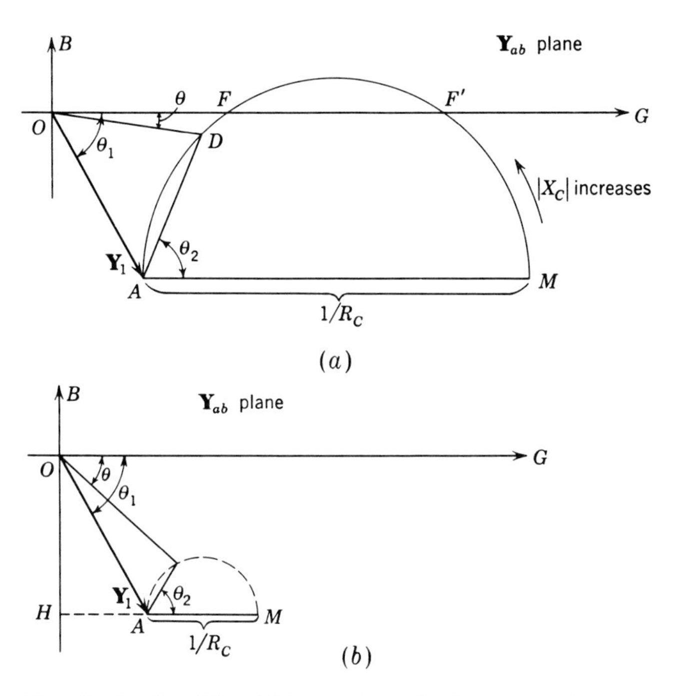

***Fig. 10-31** Possible driving-point admittance loci for the circuit of Fig. 10-30. (a) The resistance R_C is sufficiently small so that unity power factor can be achieved by adjustment of X_C. (b) The resistance of R_C is so large that no unity-power-factor condition can be achieved by adjustment of X_C.*

condition is shown in Fig. 10-31a, where the line AM has been constructed equal in length to $1/R_c$ and parallel to the G axis. Line OM therefore represents the admittance $\mathbf{Y}_{ab}$ if $X_c = 0$. If X_c is infinite, $\mathbf{Y}_2 = 0$ and $\mathbf{Y}_{ab} = \mathbf{Y}_1$, represented by OA. The locus of the admittance $\mathbf{Y}_2$ we know to be a semicircle, with AM as diameter. This semicircle has been constructed in Fig. 10-31a. If, for some value of X_c, line AD is drawn such that $\tan\theta_2 = |X_c|/R_c$, then AD represents $\mathbf{Y}_2$ and the line OD represents $\mathbf{Y}_{ab}$. We conclude that a line from the origin to a point on the semicircle will represent the admittance $\mathbf{Y}_{ab}$

for a value of X_c corresponding to θ_2. Of particular interest are lines OF and OF' because, for these points, $\mathbf{V}_{ab}$ is real, the susceptances of $\mathbf{Y}_1$ and $\mathbf{Y}_2$ having canceled each other's effect. In this case the voltage $\mathbf{V}_{ab}$ and the current $\mathbf{I}$ will be in phase. This condition is the unity-power-factor resonance ($\cos\theta_{ab} = 0$), and is discussed further, below. For the present we wish to point out that this unity-power-factor condition can be obtained only if $1/R_c$ is sufficiently large so that the semicircle will intersect the real axis.

Figure 10-31*b* shows a condition where R_c is so large ($1/R_c$ is so small) that unity power factor cannot be produced by adjustment of X_c. It is of interest to calculate the limits of R_c for which unity power factor can occur. The susceptance of $\mathbf{Y}_1$ is the projection of OA on the imaginary (B) axis. If unity power factor can be produced by adjustment of X_c, the radius of the circle must at least be equal to OH (Fig. 10-31*b*). In this case the requirement would be

$$Y_1 \sin|\theta_1| \leq \frac{1}{2R_c}$$

or, since $\sin\theta_1 = X_L/Z_1$ and $Y_1 = 1/\sqrt{R_L{}^2 + X_L{}^2}$,

$$\frac{X_L}{R_L{}^2 + X_L{}^2} \leq \frac{1}{2R_c}$$

Hence we conclude that

One unity-power-factor condition is possible if $\quad R_c = \dfrac{R_L{}^2 + X_L{}^2}{2X_L}$

No unity-power-factor condition is possible if $\quad R_c > \dfrac{R_L{}^2 + X_L{}^2}{2X_L}$

Two unity-power-factor conditions are possible if $R_c < \dfrac{R_L{}^2 + X_L{}^2}{2X_L}$

It must be emphasized that the last condition does not ensure that unity power factor will occur in the circuit. The condition states that unity power factor is *possible* by adjustment of X_c.

The above results can also be obtained analytically. The condition $\cos\theta_{ab} = 1$ corresponds to $\text{Im}\,\mathbf{Y}_1 + \text{Im}\,\mathbf{Y}_2 = 0$. (See Prob. 10-39.)

Unity power factor and minimum current for parallel connection of impedances
In Fig. 10-30, if R_c is sufficiently small so that unity power factor (resonance) is possible, the values of the capacitive reactance for these conditions may be determined graphically in any particular numerical case. In the locus diagram of Fig. 10-32, line AF or AF' is

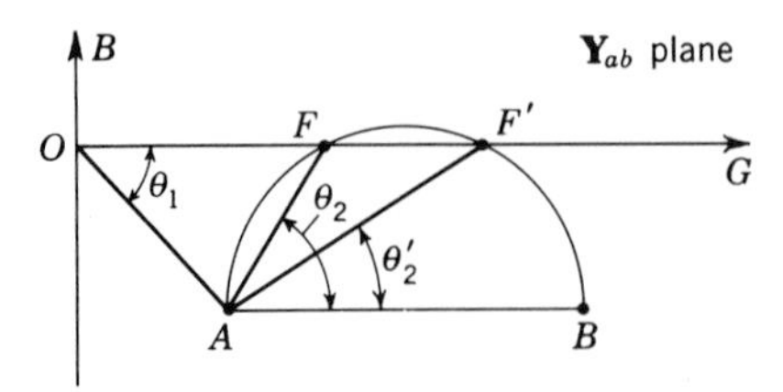

***Fig. 10-32** Graphical determination of the unity-power-factor conditions for the circuit of Fig. 10-30.*

drawn, and θ_2 or θ_2' is measured. Since θ_2 is the angle of Y_2, $\tan\theta_2 = |X_c|/R_c$ and $\tan\theta_2' = |X_c'|/R_c$. The two values of X_c can now be calculated. It is, of course, possible to obtain the same result analytically.

A second interesting possibility is the adjustment of the circuit to have minimum admittance. Graphically, this situation is analyzed with ease; we ask the question: What is the shortest distance from point O to a point on the semicircle? (See Fig. 10-33.) The answer is, of course, a line from O to point H so that, if OH is prolonged, it will intersect AB at K, the center of the semicircle. The absolute value of the minimum admittance $Y_{ab} = Y_{\min}$ is easily calculated. Since $OH = OK - HK$, we calculate OK first from the right triangle OPK.

$$OK = \sqrt{(Y_1 \sin \theta_1)^2 + \left(Y_1 \cos \theta_1 + \frac{1}{2R_c}\right)^2}$$

Since $HK = 1/2R_c$,

$$|\mathbf{Y}_{\min}| = \sqrt{Y_1^2 \sin^2 \theta_1 + \left(Y_1 \cos \theta_1 + \frac{1}{2R_c}\right)^2} - \frac{1}{2R_c} \tag{10-54}$$

Although Eq. (10-54) can be simplified, we shall not waste space doing so since, in any numerical case, the result is obtained from the construction immediately. This method of obtaining the minimum admittance (and therefore the maximum impedance) is a good

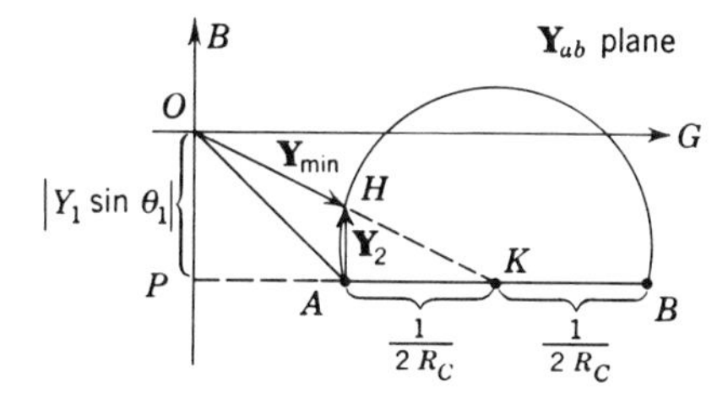

***Fig. 10-33** Graphical determination of the condition for minimum admittance in the circuit of Fig. 10-30.*

example of the usefulness of the locus method of solution. Analytically, the proof of Eq. (10-54) is quite involved. The reader should convince himself of this statement by setting up the problem analytically.

10-18 Power-factor correction

Another application of the locus diagram is to the subject of power-factor adjustment, or power-factor correction. To understand the nature of the problem, consider a load receiving power from a source through cables which have resistance as shown in Fig. 10-34. We shall assume that the power dissipated

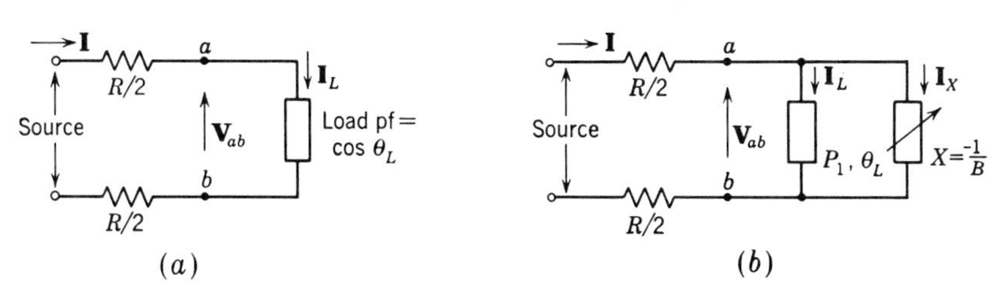

***Fig. 10-34** Load receiving energy from a resistive source. (a) Circuit without additional parallel admittance. (b) Circuit with power-factor-correcting admittance.*

by the load, P_{ab}, and the magnitude of the voltage across the load, V_{ab}, must be a certain value. Since the circuit is a series circuit, $I = I_L$, the value of the current I is

$$|\mathbf{I}| = \frac{P_{ab}}{V_{ab}\cos\theta_L}$$

The power factor of the load, $\cos\theta_L$, has not yet been specified. Let us now calculate the power dissipated in the cable. This power is $I^2(R/2 + R/2)$. Hence

$$\text{Power dissipated in cable} = \left(\frac{P_{ab}}{V_{ab}\cos\theta_L}\right)^2 R$$

We note that, for $\cos\theta_L = 1$, this quantity is less than for any other power factor. At a power factor of ($\cos\theta_L =$) 0.5, for example, four times the power loss occurs in the cables than would occur at unity power factor. If the power factor of the load is small, this power loss may be excessive, and often is accompanied by overheating of the cables (or of the generator if some part of R represents internal resistance of the generator). For these reasons it is often practical to connect a second admittance in parallel with the given load for the purpose of adjusting the power factor. If this second admittance is used to adjust the power factor only, it should be a pure susceptance; otherwise it will dissipate energy. Such a connection is shown in Fig. 10-34b; the added admittance has been labeled $X = -1/B$, to indicate an ideal power-factor-adjusting element.

If the load is inductive, it can be represented as an admittance with a negative imaginary part such as the admittance $\mathbf{Y}_1$ in the circuit of Fig. 10-30. In order to increase the power factor, the element X must be a capacitive reactance. Figure 10-35 shows

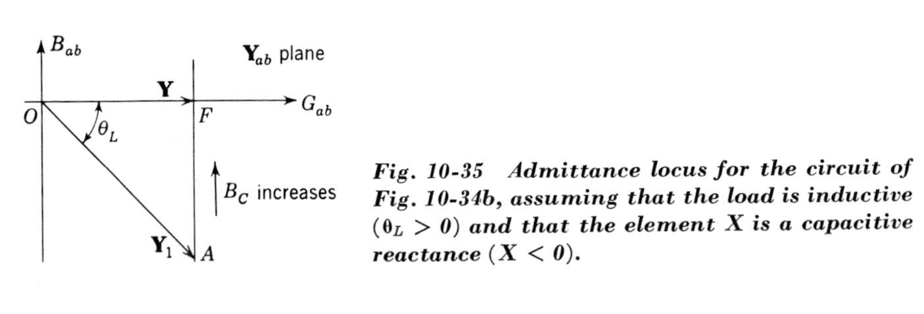

***Fig. 10-35** Admittance locus for the circuit of Fig. 10-34b, assuming that the load is inductive ($\theta_L > 0$) and that the element X is a capacitive reactance ($X < 0$).*

the locus of the total admittance $\mathbf{Y}_{ab}$ for adjustable capacitive reactance. Note that this locus is a special case of the diagram shown in Fig. 10-31a, when R_c is zero, so that the circle has become a straight line from A upward. One finite value of X_C only is possible for unity power factor. This value corresponds to the admittance represented by line AF (Fig. 10-35). Note that the minimum admittance and unity-power-factor conditions coincide. Quantitatively, unity power factor is obtained when

$$\text{Im } \mathbf{Y}_1 = -Y_1 \sin\theta_L + B_c = 0$$

or

$$\omega C = \frac{X_L}{R^2 + X_L^2}$$

From another point of view (which is very popular in power-system study), we recall from Chap. 9 that the complex powers in circuits add. Adjusting the capacitance in the present example can therefore be regarded as adjustment of the reactive power. If the complex power for the load is $P_L + jQ_L$, then, for unity power factor, the reactive power of the parallel branch must be $-Q_L$. Figure 10-36 shows the complex-power diagram for two changed power-factor conditions. Note that, to raise the power factor from cos θ to cos θ', the capacitive reactive power needed is $Q' = P_L\,(\tan\theta - \tan\theta')$.

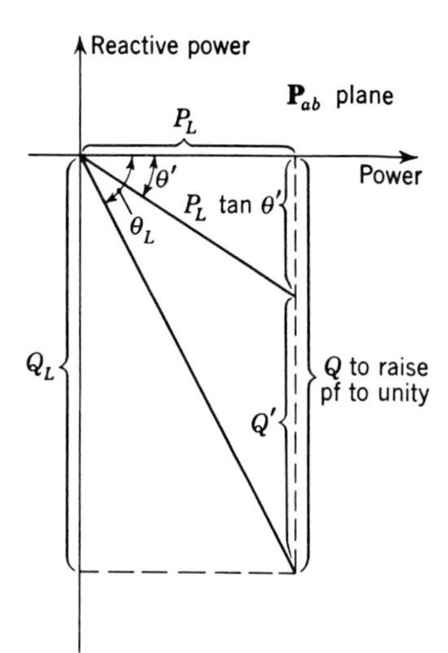

Fig. 10-36 Complex power locus corresponding to the admittance locus of Fig. 10-35.

10-19 Amplitude and phase curves A study of the locus of the complex driving-point immittance or transfer function in the complex plane gives information about the ratio of the response phasors to the source phasor. It is sometimes convenient to present the dependence of the complex network function by means of curves in which the magnitude of the phasor ratio is plotted as a function of the adjustable element, as well as curves which show the phase difference between source and response. Such curves are called amplitude and phase curves, respectively.

If the locus has been constructed to scale, the amplitude and phase curves can be obtained quite quickly. For each value of the adjustable element for which data are desired, we locate the corresponding point on the locus. The distance from the origin to the point in question gives the magnitude of the network function, and the angle of the line joining the origin to the point in question gives its angle. A compass, scale, and protractor give the information which is needed quite quickly and accurately if a reasonably careful drawing of the locus has been made.

10-20 Maximum-power-tranfer theorem

In Chap. 6 we discussed the power transfer from a resistive voltage source to a resistive load. As the final example of circuits with adjustable elements, we shall now consider the question of the power-transfer capabilities in the sinusoidal steady state, for sources with complex internal impedance.

Let us consider an ideal sinusoidal current source in parallel with admittance $\mathbf{Y}_S$, representing a nonideal current source (this source could be Norton's equivalent for an active terminal pair). Figure 10-37*a* shows such a source connected to an admittance $\mathbf{Y}_T$. Since each admittance can be repre-

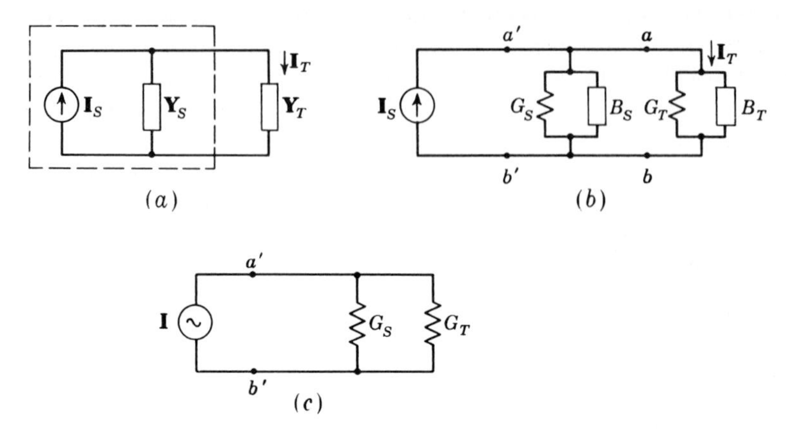

Fig. 10-37 ***Current source with internal admittance $\mathbf{Y}_S$ connected to a load $\mathbf{Y}_T$. (b) Circuit of (a), with each admittance represented as the parallel combination of conductance and susceptance. (c) Equivalent circuit if $B_S + B_T = 0$.***

sented as the parallel combination of a conductance and susceptance, an equivalent circuit is shown in Fig. 10-37*b*.

To calculate the power delivered to the load, we note that this power is the power delivered to the conductance G_T. Let P_T be the symbol for this power,

$$P_T = V_{ab}^2 G_T$$

and calculate $\mathbf{V}_{ab}$ from

$$\mathbf{V}_{ab}(\mathbf{Y}_S + \mathbf{Y}_T) = \mathbf{I}_s \tag{10-55}$$

Hence

$$\mathbf{V}_{ab} = \frac{\mathbf{I}_s}{\mathbf{Y}_S + \mathbf{Y}_T}$$

Since

$$\mathbf{Y}_S = G_S + jB_S \qquad \text{and} \qquad \mathbf{Y}_T = G_T + jB_T$$

$$Y_S = \sqrt{(G_T + G_S)^2 + (B_T + B_S)^2}$$

the magnitude of $\mathbf{V}_{ab}$ is

$$V_{ab} = \frac{I}{\sqrt{(G_T + G_S)^2 + (B_T + B_S)^2}}$$

and the expression for P_T becomes

$$P_T = \frac{I_s^2 G_T}{(G_T + G_S)^2 + (B_T + B_S)^2}$$

The maximum power which is transferred to G_T can be deduced with ease from the above equation if we recall that susceptance may be either a positive or a negative number. If we set the total susceptance to the right of a' and b' to zero,

$$B_T + B_S = 0 \qquad \text{or} \qquad B_T = -B_S \tag{10-56}$$

then

$$P_T = \frac{I_s^2 G_T}{(G_T + G_S)^2}$$

This equation gives the expression for maximum power transfer if the susceptances only are adjustable. For this condition the equivalent circuit is shown in Fig. 10-37*c*. But for this circuit we have shown, in Chap. 6, that the condition for maximum power transfer for a resistive current source *with any waveform* is $G_T = G_S$. Hence we conclude that maximum power transfer is achieved in the circuit of Fig. 10-37 if $\mathbf{Y}_S = \mathbf{Y}_T^*$. In words: If a sinusoidal current source is represented as the parallel combination of an ideal source $\mathbf{I}_s$ and a parallel (source) admittance $\mathbf{Y}_S$, then, in the sinusoidal steady state, the maximum power is transferred to a load admittance $\mathbf{Y}_T$ if the source admittance and the load admittance are *conjugate* complex numbers.

An interesting theorem concerning power transfer deals with the case in which the magnitude, but not the angle of $\mathbf{Y}_T$, is adjustable. To visualize such a case, we may imagine that $\mathbf{Y}_S$ is complex but $\mathbf{Y}_T$ is real, that is, a resistance. When the magnitude but not the angle of $\mathbf{Y}_T$ is adjustable, the condition at which maximum power transfer takes place is expressed by

$$|\mathbf{Y}_T| = |\mathbf{Y}_S| \tag{10-57}$$

i.e., the magnitudes are adjusted to be equal. To prove this, we write

$$\mathbf{Y}_T = Y_T(\cos\theta_T + j\sin\theta_T) = Y_T\underline{/\theta_T}$$
$$\mathbf{Y}_S = Y_S\underline{/\theta_S}$$

and substituting this form in Eq. (10-57),

$$\mathbf{V}_{ab} = \frac{\mathbf{I}_s}{Y_S\underline{/\theta_S} + Y_T\underline{/\theta_T}}$$

Then the power delivered to G_T is given by

$$P_T = \frac{I_s^2 Y_T \cos\theta_T}{(Y_S\cos\theta_S + Y_T\cos\theta_T)^2 + (Y_S\sin\theta_S + Y_T\sin\theta_T)^2} \tag{10-58}$$

Since all parameters except Y_T are fixed, Eq. (10-57) can be maximized by setting $\partial P_T/\partial Y_T = 0$. Carrying out the operation, we find that $Y_T = Y_S$ corresponds to this condition (Prob. 10-45).

For a series circuit containing voltage source with internal impedance $\mathbf{Z}_S$ (that is, Thévenin's equivalent representation of an active terminal pair) (Fig. 10-38), the condition for maximum power transfer to a load impedance $\mathbf{Z}_T$ is

$$\mathbf{Z}_T = \mathbf{Z}_S^*$$

The proof of this, as well as the condition $|\mathbf{Z}_T| = |\mathbf{Z}_S|$ when the angle of $\mathbf{Z}_T$ is fixed, is left as an exercise for the reader (Prob. 10-44).

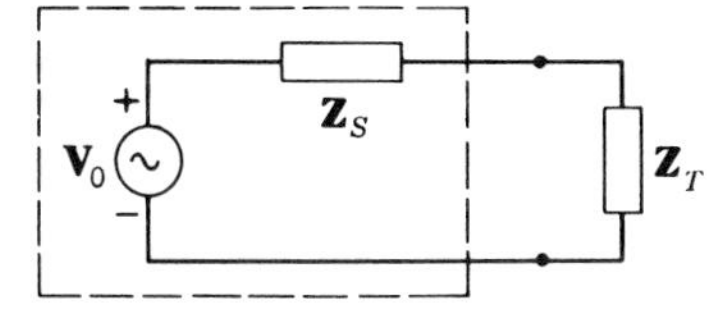

***Fig. 10-38** Series circuit. For maximum power transfer to $\mathbf{Z}_T$, $\mathbf{Z}_T$ must be the complex conjugate of $\mathbf{Z}_S$.*

PROBLEMS

10-1 For the series circuit shown in Fig. P10-1, $v(t) = \sqrt{2}\, V \cos \omega t$, and in the steady state, $i_{ab}(t) = \sqrt{2}\, I \cos (\omega t - \theta)$; calculate and sketch to scale the following frequency-response curves: (*a*) V/I; (*b*) I/V; (*c*) θ; (*d*) V_{ab}/V; (*e*) angle of $\mathbf{V}_{ab}$.

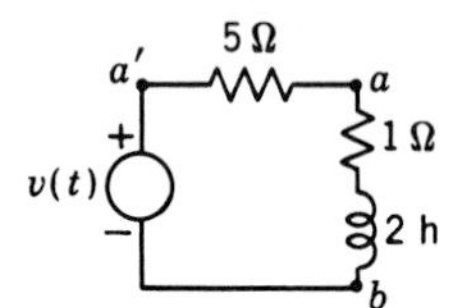

Fig. P10-1

10-2 In the circuit of Fig. P10-1, $v(t) = 12 + 12 \cos 3t$. Calculate and sketch as a function of time the steady-state functions (*a*) $i_{a'a}(t)$; (*b*) $v_{ab}(t)$.

10-3 Calculate the complete response $i_{a'a}(t)$ in the circuit of Fig. P10-1 if $v(t) = 12 + 12 \cos 3t$ and if $i(0^+) = 0$.

10-4 Discuss the integrating and differentiating properties of a series *R-L* circuit from the sinusoidal-steady-state viewpoint; in particular, deduce the range of frequencies for which the above-named properties occur if it is assumed that any phase shift above 89.43° corresponds to 90°.

10-5 Use a pole-zero diagram to deduce the amplitude- and phase-response curves of the frequency response for I_1/I and for V_{ab}/I (Fig. P10-5).

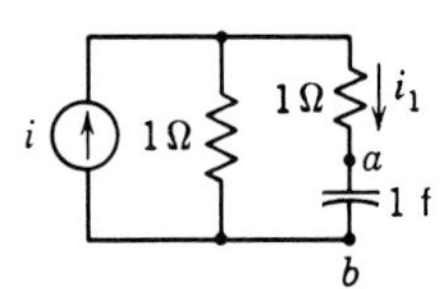

Fig. P10-5

10-6 In the circuit of Fig. P10-5, $i(t) = 2 + 2 \cos \frac{1}{2}t$; calculate and sketch the steady-state functions $i_1(t)$ and $v_{ab}(t)$.

10-7 A "discriminator" is a circuit in which the output-voltage amplitude is proportional to the frequency of the source. Show that, in a series *R-C* circuit supplied by a voltage source, the voltage amplitude across *R* is proportional to frequency if $\omega RC \ll 1$.

10-8 Show that, in a series *R-L-C* circuit, the value of Q_0 can be defined as the ratio of the rms value of the voltage across the capacitance of the rms value of the source voltage at resonance.

10-9 (*a*) Sketch to scale a curve of Y_{ab} as a function of frequency if between terminals *a-b* there is the series connection of 40 ohms, 10^{-3} henry, and 10^{-7} farad. Choose unequal frequency increments so that the Y_{ab} increments are approximately equal. *Hint:* Use normalized axes. (*b*) Sketch the angle of Y_{ab} for the circuit of (*a*) as a function of frequency.

10-10 The admittance of a series *R-L-C* circuit is a maximum at a frequency of 5 kc and 50 per cent of its maximum at a frequency of 4 kc. (*a*) At which other frequency is the admittance 50 per cent of its maximum value? (*b*) What is the angle of the complex admittance at 4 kc and at the frequency of (*a*)? (*c*) Calculate the upper and lower half-power frequencies of the circuit. (*d*) Calculate Q_0.

10-11 The lower and upper half-power frequencies of a series *R-L-C* circuit are 5 and 20 kc, respectively. (*a*) Calculate Q_0. (*b*) Calculate the frequencies at which the angle of the complex admittance is $\pm 10°$.

10-12 In the theory of oscillators, *R-L-C* circuits are important; in particular, the rate of change of the angle of the immittance with frequency is of significance in connection with stability. Start with the equation $\theta(\mu) = \tan^{-1} Q_0(\mu - 1/\mu)$ and show that $(d\theta/d\mu)_{\mu=1} = 2Q_0$.

10-13 The circuit shown in Fig. P10-13 is named, after its discoverer, the Wien bridge. (*a*) If $R_1 = 2$ and $R_2 = 1$, obtain the transform network function which relates

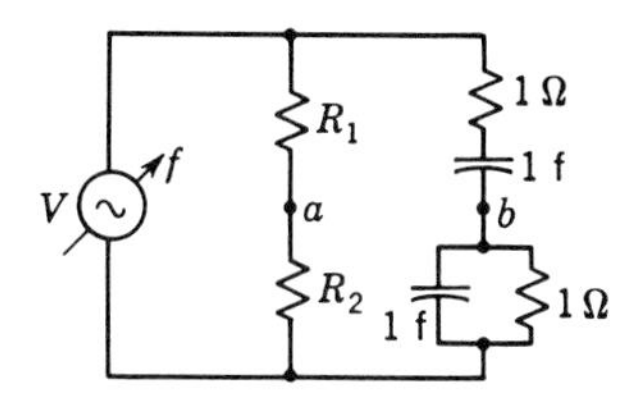

Fig. P10-13

v_{ab} to v. Locate the zeros and poles. (*b*) Plot the frequency-response curve for V_{ab}/V, using the values of R_1 and R_2 given in (*a*). (*c*) Show that, when $\omega = 1$, V_{ab} is in phase with V regardless of the values of R_2 and R_1. (*d*) Locate the zeros and poles for $R_2/(R_1 + R_2) = \frac{1}{2}$, and plot the amplitude and phase response for V_{ab}/V point by point. Normalize with respect to $(V_{ab}/V)_{\min}$. (*e*) Repeat part *d* for $R_2/(R_1 + R_2) = \frac{1}{4}$, and plot the results on the same set of axes as are used in (*d*).

10-14 It is known from experiment that a certain *R-L-C* series circuit has $Q_0 = 12$, $\omega_0 = 8{,}000$, and $R = 7$ ohms. (*a*) Calculate *L* and *C*. (*b*) Use the values of *R*, *L*, and *C* to calculate the driving-point impedance of this circuit at a radian frequency $\omega = 8{,}370$. Use the slide rule. (*c*) Use the pole-zero diagram to calculate the approximate driving-point impedance at $\omega = 8{,}370$ as explained in Sec. 10-7. Use the slide rule. (*d*) Comment on the precision of the results obtained in (*c*) as compared with (*b*).

10-15 Prove Eq. (10-44).

10-16 In attempting to determine the parameters of a series-resonant circuit (Fig. P10-16), the voltage across *R* cannot always be obtained because the terminals

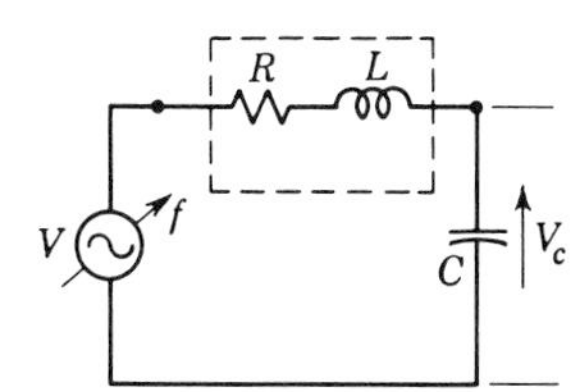

Fig. P10-16

of *R* may not be accessible. The frequency response V_c/V is therefore obtained. (*a*) Show that the maximum value of V_c/V is given by

$$\left(\frac{V_c}{V}\right)_{\max} = Q_0 \frac{1}{\sqrt{1 - 1/4Q_0^2}} \qquad Q_0^2 > \frac{1}{2}$$

(*b*) Calculate the percentage errors in the half-power frequencies if the half-power frequencies are taken at $V_c = 0.707(V_c)_{\max}$ rather than at $I = 0.707 I_{\max}$ when $Q_0 = 2$.

10-17 A constant-amplitude variable-frequency sinusoidal current source is impressed on a parallel *R-L-C* circuit. Let $\mu = \omega/\omega_0$, $Q_0 = \omega_0 C/G$, $\omega_0^2 = 1/LC$, and calculate the value of μ at which (*a*) the current through the inductance is a maximum; (*b*) the current through the capacitance is a maximum; (*c*) the current through the resistance is a maximum.

10-18 In the circuit of Fig. 10-15*a* the rms value of the source current is 1; calculate (*a*) the average value of energy stored by the capacitance if $\omega = 1$; (*b*) the energy dissipated in the 0.1-ohm resistance during one cycle if $\omega = 1$.

10-19 In the circuit of Fig. 10-15*a* change R from 0.1 to 1. (*a*) Is the approximate method used in Sec. 10-10 now justified? (*b*) Calculate $Z(j)$ for $R = 1$. (*c*) Calculate the radian frequency at which $|Z(j\omega)|$ is a maximum.

10-20 In the circuit of Fig. 10-15*a* replace the 0.1-ohm resistance by an adjustable R. Calculate the maximum value of R so that the maximum impedance occurs at a value of ω which is within 5 per cent of 1.00.

10-21 A transform network function has poles at $s = -0.4 \pm j5$ and $s = -2 \pm j20$ and zeros at $s = 0$ and $s = -1 \pm j10$. When the source function is $10 \cos 5t$, the response amplitude is 5. (*a*) Calculate H. (*b*) Calculate *approximately* the maxima and minima of the complex network function as the frequency is varied, and sketch the amplitude-response curves.

10-22 A transform network function has simple poles at $s = -\alpha_1 \pm j9$ and $s = -\alpha_2 \pm j11$ and a double zero at $s = 0$. Sketch approximately the amplitude response using a point-by-point method from $\omega = 7$ to $\omega = 13$ if (*a*) $\alpha_1 = \alpha_2 = \frac{1}{2}$; (*b*) $\alpha_1 = \alpha_2 = 2$. Use $H = 100$.

10-23 A network is known to be "low-loss." The frequency-response curve that relates input to output, $h(\omega) = |H(j\omega)|$, is shown in Fig. P10-23. Use the approximation technique described in Sec. 10-10 to find $H(s)$.

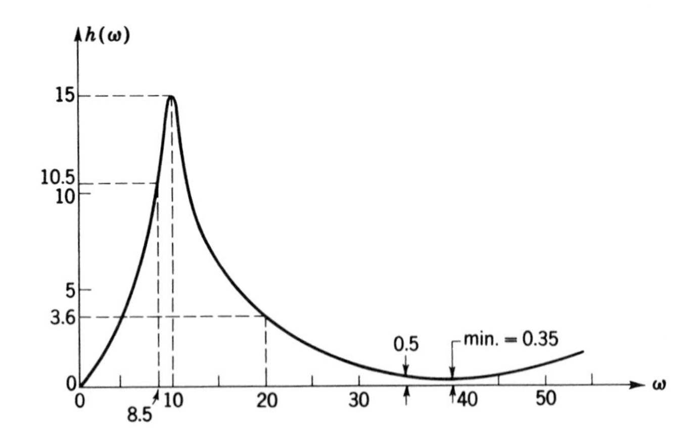

Fig. P10-23

10-24 The frequency-response curve of a certain "low-loss" transfer function approaches infinity as $\omega \to 0$; it has a minimum near $\omega = 10$ and a maximum near $\omega = 30$. (*a*) Explain why the network has poles at $s = 0$ and $s_p \approx -\alpha_p \pm j\omega_p$ and zeros $s_0 \approx -\alpha_0 \pm j\omega_0$, and state the values ω_p and ω_0. (*b*) Use the approximate pole-zero diagram to find α_p and α_0 in each of the following: (1) $h(\omega_p + 1) \approx 0.7h(\omega_p)$, $h(\omega_0 + 1) \approx 1.4h(\omega_0)$; (2) $h(\omega_p + 0.2) \approx 0.7h(\omega_p)$, $h(\omega_0 + 0.2) \approx 1.4h(\omega_0)$.

10-25 For the driving-point impedance shown in Fig. P10-25: (*a*) Sketch the reactance frequency curve if the circuit is made lossless. (*b*) Locate the zeros and poles of the lossless $Z_{ab}(j\omega)$. (*c*) Calculate the actual value $\mathbf{Z}_{ab}$ of the given lossy circuit at the frequencies of (*b*).

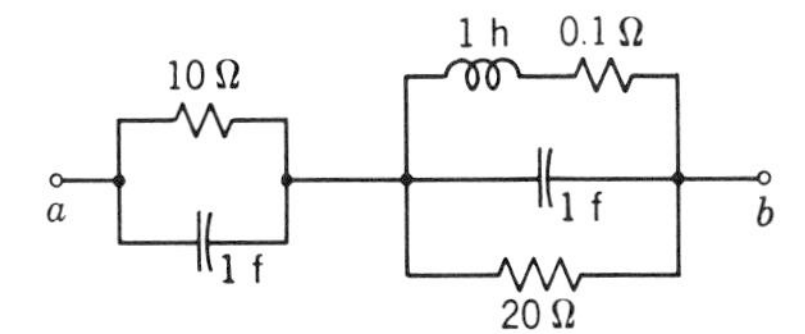

Fig. P10-25

10-26 Sketch the reactance and susceptance frequency curves for the reactive terminal-pair networks shown in Fig. P10-26. Assume that no special relationship between the parameters exists.

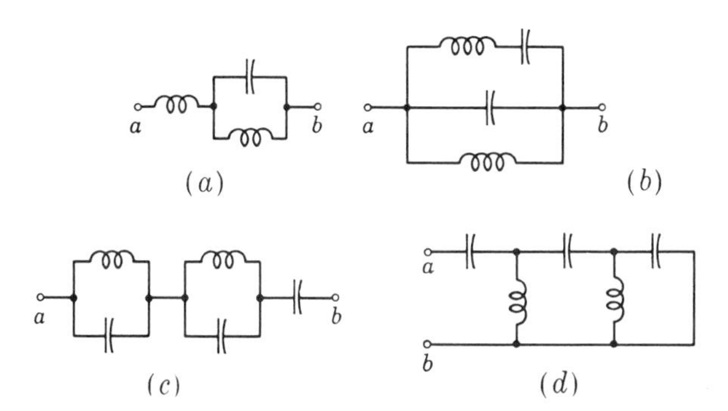

Fig. P10-26

10-27 In the circuit shown in Fig. P10-27: (*a*) Assume that $R = 0$ and locate the zeros and poles of $Z_{ab}(s)$ with L as a parameter if $LC = 1$ and the element marked X is a unit capacitance. (*b*) Repeat (*a*) if the element marked X is a unit inductance.

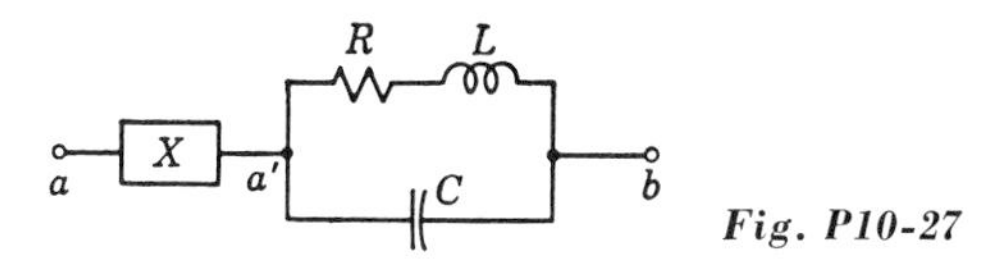

Fig. P10-27

10-28 In the circuit of Fig. P10-27, it is desired that Z_{ab} have a maximum at $\omega = 10^6$ and a minimum at $\omega = \omega' = \frac{1}{2} \times 10^{+6}$. It is also known that $R = 100$ and $L = 10^{-2}$ henry. (*a*) Calculate C, and show that the result, within slide-rule accuracy, is not affected by R. (*b*) Calculate $\mathbf{Z}_{a'b}$ at $\omega = 10^6$. (*c*) Calculate $\mathbf{Z}_{a'b}$ at $\omega = \omega'$. (*d*) Determine the element value at X (L or C) which results in minimum Z_{ab} at ω'. (*e*) Calculate $\mathbf{Z}_{ab}$ at $\omega = \omega'$, using the value of X from (*d*). (*f*) Repeat parts *a* to *e*, using $\omega' = 2 \times 10^6$.

10-29 Foster's reactance theorem states that zeros and poles for *driving*-point reactances and susceptances must alternate. Locate the zeros and poles of the transform *transfer* functions v_2/v_1 (Fig. P10-29), and observe that the separation property does not apply to transfer functions.

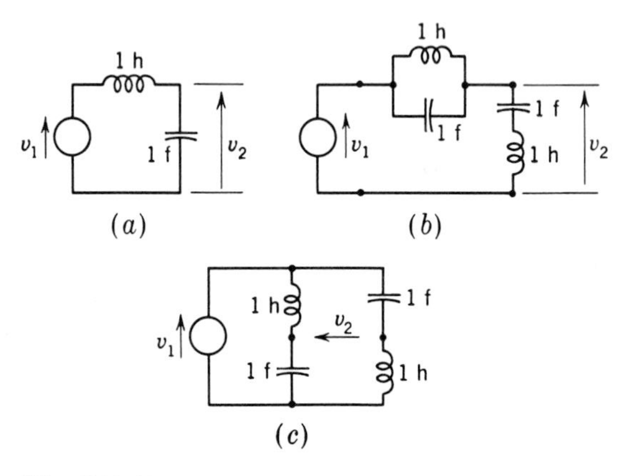

Fig. P10-29

10-30 A transform driving-point impedance function has the form $Z_{ab}(s) = 12[(s^2 + 1)/s(s^2 + 4)]$. (*a*) Show that this function is physically realizable as a pure-reactance terminal pair. (*b*) Show that $Z_{ab}(s) = Z_1(s) + Z_2(s)$, where $Z_1(s) = 3/s$ and $Z_2(s) = 9s/(s^2 + 4)$. (*c*) Sketch the network consisting of the series connection of Z_1 and Z_2. Give the values of all elements. *Hint:* $Z_2(s) = 1/Y_2(s)$.

10-31 (*a*) Show that the locus of the immittance of a circuit which contains only one energy-storing element as the *frequency* is adjusted is the same as the locus obtained when the element is adjustable. (*b*) Sketch the impedance and admittance loci for a series *R-L-C* circuit as the *frequency* is adjusted between zero and infinity.

10-32 Sketch the driving-point impedance and admittance loci for the terminal-pair networks shown in Fig. P10-32.

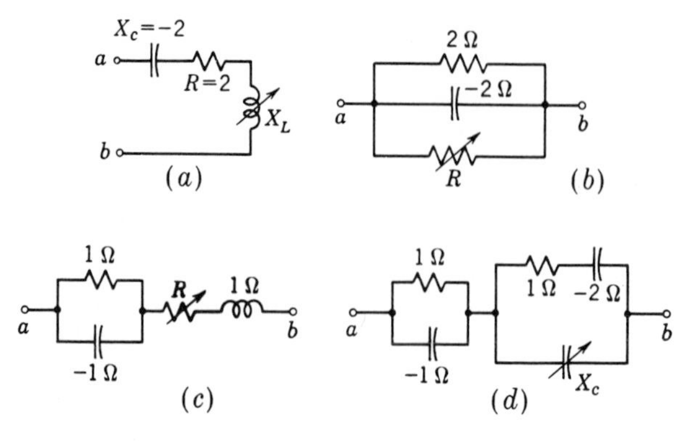

Fig. P10-32

10-33 The admittance locus $\mathbf{Y}_{ab}$ is given in Fig. P10-33; the adjustable element is R. Show that, when $OB = \mathbf{Y}_{ab}$, the power ratio P_{db}/P_{ab} is the distance ratio DB/AB.

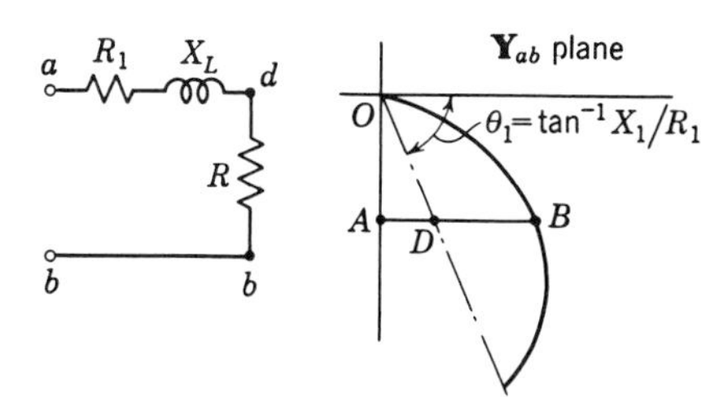

Fig. P10-33

10-34 From the admittance locus of Fig. P10-32*d* obtain the value of $\mathbf{Y}_{ab}$ in polar form if $X_c = -\frac{1}{2}, -\frac{3}{4}, -1$, and -2.

10-35 In Fig. P10-35: (*a*) Sketch the transfer-function locus $\mathbf{V}_2/\mathbf{V}_1$. (*b*) From the locus of (*a*) find the value of X_c for which the phase difference between $\mathbf{V}_2$ and $\mathbf{V}_1$ is a maximum.

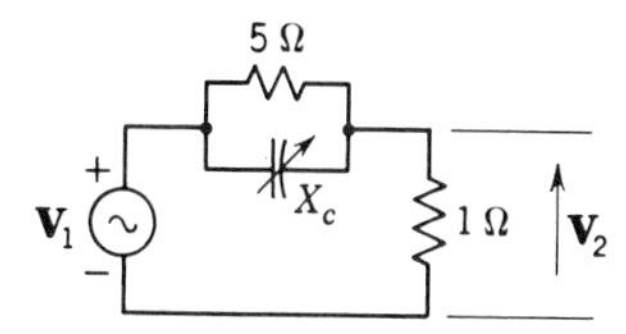

Fig. P10-35

10-36 In Fig. P10-36: (*a*) Sketch the transfer-function locus $\mathbf{V}_2/\mathbf{V}_1$. (*b*) Assume that $C = 1$, and use the result of (*a*) to find the frequency for which the phase difference between $\mathbf{V}_1$ and $\mathbf{V}_2$ is a maximum.

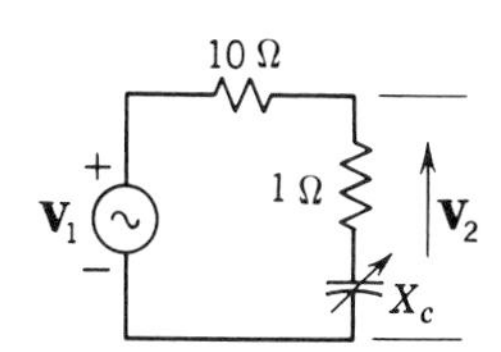

Fig. P10-36

10-37 In the circuit of Fig. P10-13 assume that $R_1 = 1$ and R_2 is adjustable. (*a*) Sketch the transfer-function locus $\mathbf{V}_{ab}/\mathbf{V}$ for $\omega = 1$. (*b*) Repeat (*a*) if $\omega = 1.5$.

10-38 From the locus of Prob. 10-36, construct to scale a curve which gives the phase angle between $\mathbf{V}_2$ and $\mathbf{V}_1$ as a function of frequency.

10-39 In the circuit shown in Fig. P10-39, X_c is the adjustable element. Deduce analytically the inequalities which determine whether or not unity power factor can be achieved by adjustment of X_c.

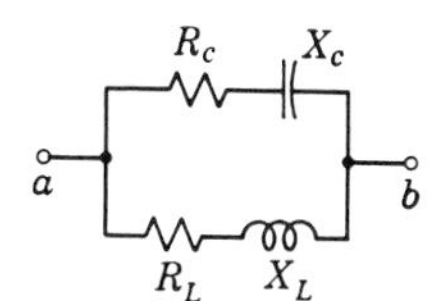

Fig. P10-39

10-40 In the circuit of Fig. P10-39 use an admittance-locus diagram to deduce the conditions for unity power factor at terminals *a-b* if all elements are fixed except (*a*) R_c; (*b*) R_L; (*c*) X_L.

10-41 In the circuit of Fig. P10-39, $R_L = 1$, $X_L = 2$. Use a locus diagram to calculate the minimum value of Y_{ab} and the corresponding $\mathbf{Y}_{ab}$ and X_c if (*a*) $R_c = 0$; (*b*) $R_c = 0.2$; (*c*) $R_c = 1$; (*d*) $R_c = 4$.

10-42 Two loads are connected in parallel. One load draws 100 kva at a power factor of 0.4 lagging; the second load draws 70 kva at an adjustable power factor ($-90° \leq \theta_2 \leq 90°$). Calculate (*a*) the power factor of the parallel combination if $\theta_2 = -90°$ (leading); (*b*) the maximum and minimum power factor of the parallel combination as θ_2 is adjusted. *Hint:* Construct the locus of the complex power of the combination.

10-43 A load draws 50 kw at a power factor of 0.5 lagging. The voltage across the load is 1,000 volts. The cables joining the load to the source have a total resistance R_0. A capacitive element is connected in parallel with the 50-kw load. (*a*) Plot carefully to scale the power dissipated in R_0, normalized with respect to the 0.5-pf condition as a function of the power factor of the parallel combination as the capacitive element is adjusted. (*b*) Plot the reactive power drawn by the capacitive element as a function of the power factor of the parallel combination.

10-44 Prove the maximum-power-transfer theorem for a (Thévenin) voltage source.

10-45 Show that, when a load $\mathbf{Z}_T$ is connected to a source whose internal impedance is $\mathbf{Z}_s$, then, if the magnitude but not the angle of $\mathbf{Z}_T$ is adjustable, maximum power transfer occurs for $Z_T = Z_s$.

10-46 The Thévenin equivalent of an active terminal-pair network in the sinusoidal steady state is a voltage source $V_o = 24$ volts and an internal impedance $\mathbf{Z}_s = 3 + j4$. Calculate the power delivered by this source to a load $\mathbf{Z}_T$ if (*a*) $\mathbf{Z}_T = 3 - j4$; (*b*) $\mathbf{Z}_T = 3 + j4$; (*c*) $\mathbf{Z}_T = 5$; (*d*) $\mathbf{Z}_T = 10$.

10-47 In Fig. P10-47: (*a*) Calculate the value $\mathbf{Z}_T$ for which the active terminal pair delivers maximum power of $\mathbf{Z}_T$; (*b*) for $\mathbf{Z}_T$ as in (*a*), calculate the power delivered by the ideal source **V**.

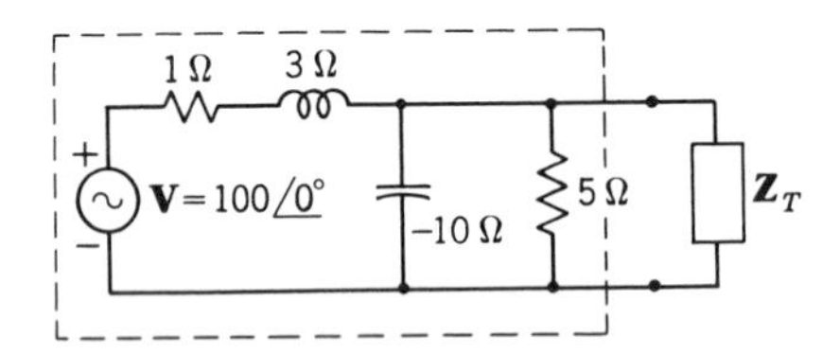

Fig. P10-47

10-48 In the circuit shown in Fig. P10-48 the value of C is adjusted so that Im $\mathbf{Y}_{ab} = 0$ at $\omega = 10^6$. (*a*) Calculate Re $\mathbf{Y}_{ab}$ at $\omega = 10^6$. (*b*) Obtain an expression for the power delivered to the 1,000-ohm resistance when $\omega = 10^6$. (*c*) Obtain an expression for the power delivered to the 1,000-ohm resistance when $\omega = 0$ (d-c).

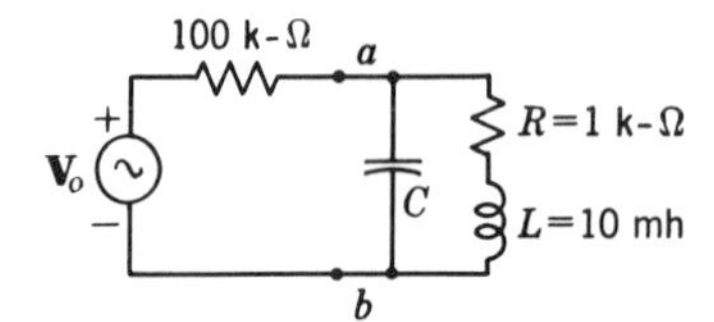

Fig. P10-48

Chapter 11 General analysis techniques I—simultaneous-equations methods

In the preceding chapters the formulation of equilibrium equations (network functions) for series-parallel circuits was discussed in detail. In this chapter we consider the formulation of such equations for circuits consisting of elements that are arbitrarily interconnected. To do this, we shall first treat briefly bridge circuits (delta-wye conversion) and then discuss the general simultaneous-equations techniques. These general techniques not only are useful in the solution of specific examples, but they can also be employed to deduce certain general properties of networks (Chap. 12). It will be seen that the matrix methods for network description enable us to treat networks without explicitly specifying how many elements there are (as long as the number is finite) or how they are interconnected.

It is necessary to point out that the general analysis techniques make it possible to reduce the formulation and solution of network equations to a standard routine. Such routines are useful when the network is complex or when computer solutions are employed. If, however, one is confronted with a specific problem, involving a series-parallel circuit such as has been treated in the preceding chapters, then the simple techniques (for example, voltage-division or current-division formulas) are most efficient for solution.

11-1 Passive three-terminal networks

Before discussing the general analysis techniques of networks that involve simultaneous equations, we extend the series-parallel branch analysis technique by use of *delta-wye* and *wye-delta* conversion. A passive "delta" is a portion of a network consisting of three passive driving-point immittances (initially not energized) that form a loop with *three* accessible terminals as shown in Fig. 11-1*a*. A "wye" is a three-immittance combination where

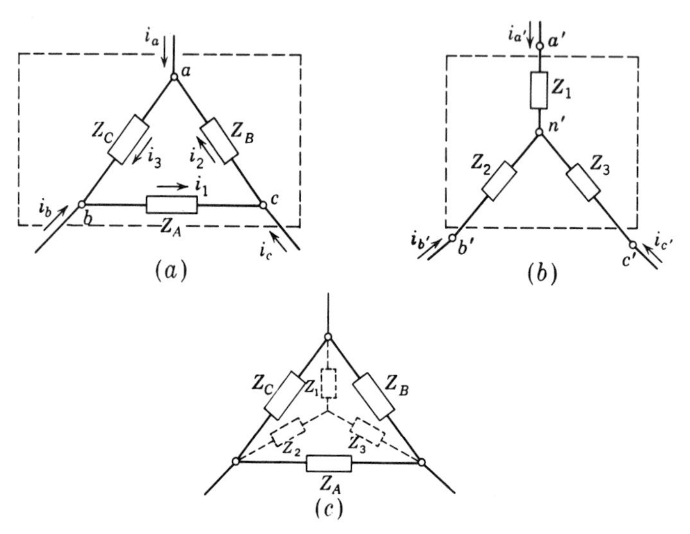

Fig. 11-1 ***(a) Three elements connected to form a delta. (b) Three elements connected to form a wye. (c) Diagram used in delta-wye conversion.***

each of the three p elements have one terminal connected to a common point and the remaining three terminals accessible as shown in Fig. 11-1*b*. The term delta is used because the structure of Fig. 11-1*a* resembles the Greek letter capital delta (Δ); the term wye is used because the structure of Fig. 11-1*b* resembles the letter wye (Y). We show below that, *with respect to the three accessible terminals a-b-c*, either network can be made equivalent by relating the immittances Z_1, Z_2, and Z_3 properly to Z_A, Z_B, and Z_C. Such delta-to-wye or wye-to-delta conversion is often useful if it results in a series-parallel circuit.

The formulas for wye-to-delta conversion (representing the three-terminal network of Fig. 11-1*b* in the form of Fig. 11-1*a*) are shown below to be

$$Z_A = \frac{Z_1Z_2 + Z_2Z_3 + Z_3Z_1}{Z_1} = \frac{Y_1 + Y_2 + Y_3}{Y_2Y_3} \tag{11-1a}$$

$$Z_B = \frac{Z_1Z_2 + Z_2Z_3 + Z_3Z_1}{Z_2} = \frac{Y_1 + Y_2 + Y_3}{Y_3Y_1} \tag{11-1b}$$

$$Z_C = \frac{Z_1Z_2 + Z_2Z_3 + Z_3Z_1}{Z_3} = \frac{Y_1 + Y_2 + Y_3}{Y_1Y_2} \tag{11-1c}$$

The formulas for delta-to-wye conversion (representing the three-terminal network of Fig. 11-1*a* in the form of Fig. 11-1*b*) are shown to be

$$Z_1 = \frac{Z_B Z_C}{Z_A + Z_B + Z_C} = \frac{Y_A}{Y_A Y_B + Y_B Y_C + Y_C Y_A} \tag{11-2a}$$

$$Z_2 = \frac{Z_C Z_A}{Z_A + Z_B + Z_C} = \frac{Y_B}{Y_A Y_B + Y_B Y_C + Y_C Y_A} \tag{11-2b}$$

$$Z_3 = \frac{Z_A Z_B}{Z_A + Z_B + Z_C} = \frac{Y_C}{Y_A Y_B + Y_B Y_C + Y_C Y_A} \tag{11-2c}$$

MNEMONIC AID FOR DELTA-WYE AND WYE-DELTA CONVERSION To remember Eqs. (11-1) and (11-2) independently of the element labeling, the wye is drawn inside the delta as shown in Fig. 11-1*c*. We refer to each impedance of the delta as a "side" of the triangle formed by the delta. For delta-to-wye conversion [Eqs. (11-2)] each impedance of the wye is given as the product of the two *adjacent* sides of the delta divided by the sum of the delta impedances. For wye-to-delta conversion one forms the sum of products $N = Z_1Z_2 + Z_2Z_3 + Z_3Z_1$; and then each impedance of a side of the delta is N divided by the *opposite* impedance of the wye. Similar rules can be formulated in terms of admittances.

SYMMETRICAL DELTA AND WYE CONFIGURATIONS If $Z_1 = Z_2 = Z_3 = Z_Y$, then $Z_A = Z_B = Z_C = Z_\Delta$, and the delta and wye are termed symmetrical. For symmetrical deltas and wyes the conversion formulas (11-1) and (11-2) reduce to

$$Z_\Delta = 3Z_Y \tag{11-3}$$

PROOF OF EQS. (11-1) AND (11-2)[1] If, in Figs. 11-1*a* and 11-1*b*, $i_{a'} = i_a$, $i_{b'} = i_b$, and $i_{c'} = i_c$ when $v_{ab} = v_{a'b'}$ and $v_{bc} = v_{b'c'}$ and $v_{ca} = v_{c'a'}$, then the two three-terminal pairs are equivalent at their terminals.

From Fig. 11-1*a* we observe that

$$i_a = Y_C v_{ab} + Y_B v_{ac} \tag{11-4}$$

In Fig. 11-1*b* we note that $i_{c'} = -i_{a'} - i_{b'}$ and

$$v_{a'b'} = Z_1 i_{a'} - Z_2 i_{b'} \tag{11-5a}$$

$$v_{a'c'} = Z_1 i_{a'} - Z_3(-i_{a'} - i_{b'}) \tag{11-5b}$$

[1] In an alternative approach to this problem, the equivalence can be established by equating the driving-point impedances in the two circuits as follows:

$$Z_{ab} = Z_{a'b'} \qquad Z_{bc} = Z_{b'c'} \qquad Z_{ca} = Z_{c'a'}$$

In each calculation of such a driving-point impedance, the third terminal is ignored. Thus $Z_{ab} = (Z_A + Z_B)Z_C/(Z_A + Z_B + Z_C)$, $Z_{a'b'} = Z_1 + Z_2$, etc. From the three simultaneous equations which are obtained in this manner, the relationship between the numbered and lettered impedance is deduced. This method is left as an exercise for the reader. It is pointed out that this approach shows that the results arrived at are necessary. The proof of necessity and sufficiency is as given in the text.

Multiplying Eq. (11-5*a*) by Z_3 and (11-5*b*) by Z_2 and adding, we obtain

$$(Z_1Z_2 + Z_2Z_3 + Z_3Z_1)i_{a'} = Z_3v_{a'b} + Z_2v_{a'c}$$

or

$$i_{a'} = \frac{Z_3}{Z_1Z_2 + Z_2Z_3 + Z_3Z_1} v_{a'b'} + \frac{Z_2}{Z_1Z_2 + Z_2Z_3 + Z_3Z_1} v_{b'c'} \qquad (11\text{-}6)$$

Setting $i_{a'} = i_a$, $v_{a'b} = v_{ab'}$, $v_{a'c'} = v_{ac}$, we obtain by comparison with Eq. (11-4)

$$Y_C = \frac{Z_3}{Z_1Z_2 + Z_2Z_3 + Z_3Z_1} \qquad Z_C = \frac{Z_1Z_2 + Z_2Z_3 + Z_3Z_1}{Z_3} \qquad (11\text{-}7)$$

$$Y_B = \frac{Z_2}{Z_1Z_2 + Z_2Z_3 + Z_3Z_1} \qquad Z_B = \frac{Z_1Z_2 + Z_2Z_3 + Z_3Z_1}{Z_2} \qquad (11\text{-}8)$$

Equation (11-7) is seen to be identical with Eq. (11-1*c*), and (11-8) is the same as (11-1*b*). From symmetry, Eq. (11-1*a*) can be obtained by changing 3 to 1, 1 to 2, 2 to 3, and C to A.

The formulas for delta-to-wye conversion are obtained, observing that, in the delta of Fig. 11-1*a*, $v_{bc} = v_{ac} - v_{ab}$ and

$$i_b = -Y_Cv_{ab} + Y_A(v_{ac} - v_{ab}) \qquad (11\text{-}9)$$

Solving Eqs. (11-9) and (11-4) simultaneously for v_{ab} gives the result

$$v_{ab} = \frac{Y_A}{Y_AY_B + Y_BY_C + Y_CY_A} i_a - \frac{Y_B}{Y_AY_B + Y_BY_C + Y_CY_A} i_b \qquad (11\text{-}10)$$

Setting $v_{ab} = v_{a'b'}$, $i_a = i_{a'}$ and $i_b = i_{b'}$ we obtain Eqs. (11-2*a*) and (11-2*b*) by comparison of Eq. (11-10) with Eq. (11-5*a*). Equation (11-2*c*) is then obtained from symmetry.

Example 11-1 In the circuit shown in Fig. 11-2*a* use delta-wye conversion to obtain the network function relating v_{bn} to v.

Solution We shall change the delta *a-b-c* consisting of the 4-, 4-, and 8-ohm loop to a wye. The circuit which results is shown in Fig. 11-2*b*. This circuit is a series-parallel circuit with

$$R_1 = \frac{4 \times 4}{16} = 1 \text{ ohm} \qquad R_2 = R_3 = \frac{4 \times 8}{16} = 2 \text{ ohms}$$

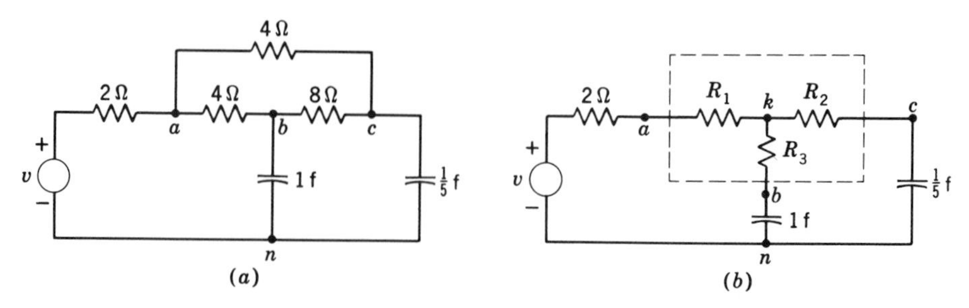

***Fig. 11-2** Use of delta-wye conversion; see Example 11-1.*

We now apply voltage division twice,

$$v_{bn} = \frac{\dfrac{(2 + 5/p)(2 + 1/p)}{4 + 6/p}}{\dfrac{(2 + 5/p)(2 + 1/p)}{4 + 6/p} + 1 + 2} \frac{1/p}{2 + 1/p} v$$

which reduces to

$$v_{bn} = \frac{2p + 5}{16p^2 + 30p + 5} v$$

The foregoing example is fairly simple because no energy-storing elements are involved in the conversion. The algebraic complexity introduced by energy-storing elements is quite considerable, so that delta-wye and wye-delta conversions do not find extensive application.

It must also be noted that the wye obtained from a delta or the delta obtained from a wye, which are equivalent to each other at the three terminals, may not consist of physically realizable elements. If, for instance, the three elements of a delta are $Z_A = 1/pC = Z_B$, $Z_C = R$, then

$$Z_1 = Z_3 = \frac{R/pC}{R + 2/pC} = \frac{R}{pRC + 2}$$

$$Z_2 = \frac{(1/pC)^2}{R + 2/pC} = \frac{1}{p^2R^2C^2 + 2pC}$$

and Z_2 is not physically realizable because the degree of the numerator polynomial differs from that of the denominator by more than one and $Z_2(s)$ is not a positive real function. The converted circuit can, *with respect to its three terminals,* nevertheless be used for calculations.

11-2 Solution of networks using simultaneous equations. Introduction

In this section we shall introduce the notion of working with several circuit variables simultaneously with the purpose of eventually obtaining an equilibrium equation which relates one of these variables to the source functions. This method requires that a circuit variable (i.e., voltage or current) be assigned symbolically to each element (either primitive element or p element, depending on how the problem is formulated) and then that Kirchhoff's laws be applied until a sufficient number of independent simultaneous equations has been obtained (i.e., as many equations as there are variables) so that the problem is (uniquely) specified. This procedure raises the following questions: (1) In any given circuit how many variables should be chosen as unknowns? (2) How can we be sure that the formulated equations are independent? The remainder of this section will be devoted to an illustration which is intended to show how these questions arise.

Consider the circuit of Fig. 11-3 as an example in which it is required to relate the current $i_3(t)$ to the source functions $v(t)$ and $i(t)$. It is clear that if we knew the current in each element, we should also know the voltage across each element, because we know the voltage-current relationships for the elements. Similarly, knowledge of the voltages across each element would give us the currents in the elements by means of the same basic equations. Now the waveforms of the ideal source $v(t)$ and the ideal current source are assumed to be known. There are four other elements, R_1, R, R_3, and C, all passive. We shall now symbolically associate a variable with each of these elements. The circuit variables chosen as the unknowns are $i_1(t)$, $i_2(t)$, $i_3(t)$, and $v_{ab}(t)$. To find the waveform of these variables, we need four independent equations. In this circuit we have three junctions, a, b, and d, so that Kirchhoff's current law can be applied three times, as follows:

At junction a: $$i_1 - pCv_{ab} - i_3 = 0 \tag{11-11}$$
At junction d: $$i_3 - i_2 + i(t) = 0 \tag{11-12}$$
At junction b: $$-i_1 + i_2 - i(t) + pCv_{ab} = 0 \tag{11-13}$$

We observe that Eq. (11-13) can be obtained by adding (11-11) to (11-12) and multiplying the resulting equation by -1. It is therefore clear that, although we have three junctions, Kirchhoff's current law can be applied fruitfully only twice. Any two of the three equations can be chosen for further use. Since we have four circuit variables, we need two more equations. Clearly, these equations must be obtained by application of Kirchhoff's voltage law, since the current law has been "exhausted." Kirchhoff's voltage law states that the sum of the voltages in a closed path (loop) of the circuit is zero. A study of Fig. 11-3 shows that the following are some of the loops of the network:

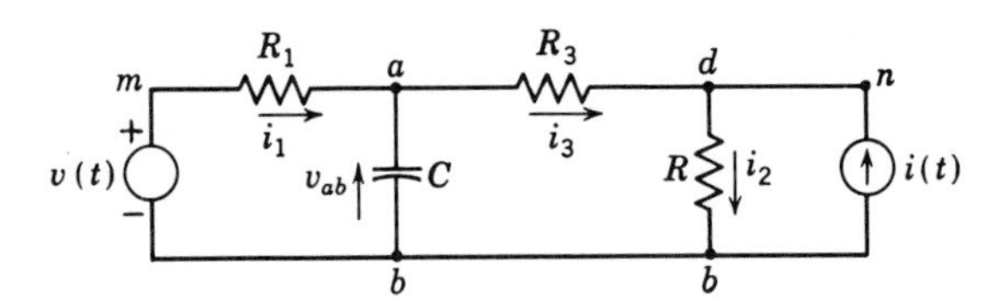

***Fig. 11-3** A network and its branch variables.*

m-a-b-m; *m-b-d-a-m*; *a-b-d-a*; *b-n-d-b*; etc. Clearly, we need information which will tell us which loops to choose, so that application of Kirchhoff's voltage law will result in independent equations, without trial and error. An elementary introduction to network geometry answers the questions concerning the number of independent equations that can be written for a given network, and the next section deals with this subject.

11-3 Network terminology and geometry

TERMINOLOGY In discussing how the elements forming a network are interconnected, certain terms were introduced in Chap. 3. It is convenient to restate the definitions of these terms here.

Node: A point in the network which is common to two or more elements.
Junction: A node formed by three or more elements.
Branch: An element or a series connection of elements between junctions. A branch is *active* if it includes one or more ideal sources; otherwise it is *passive*.
Loop: A closed path through the circuit elements of a network.
Mesh: In a given "planar" network diagram, a loop which does not enclose or encircle another loop and which cannot be divided into other loops.

It is noted here that, by redrawing a network diagram, a mesh in the old diagram can be made to become a loop in the new diagram, and vice versa. We therefore talk about a mesh or a loop in connection with a specific drawing of the network diagrams. We point out again that the term *element* may apply either to a primitive element such as R, L or C or to an operational element. The definitions given above apply in any case.

NETWORK GEOMETRY (TOPOLOGY)[1] Whether a network is shown with its primitive circuit elements or in the form of p elements, its geometry can be studied, regardless of the nature of the circuit elements or p-element branches. Since all the elements dealt with so far, except mutual inductance, have two terminals, and the latter can be represented by a combination of such elements (as is shown in Chap. 15), then between these two terminals all circuit elements, as well as the p-element branches, can be geometrically represented by lines between two terminals.

Line elements: A line which represents a circuit element or a p element is called a line element.

Graph of a network: When the elements of a network are replaced by line elements, the resulting structure is called the graph of the network.

Planar network: A network whose graph can be drawn on a plane (e.g., a sheet of paper) so that no line elements cross is called a planar network. In a nonplanar network the definition of meshes cannot be applied.

GEOMETRICAL REPRESENTATION The representation of a network by its graph is called the "geometrical" representation, or *linear graph,* of the network. In Fig. 11-4*a* and *b*, a passive network and its linear graph are illustrated. Exactly as in the network diagram, the small circles shown in the graph are called *nodes;* a node common to more than two elements is also a *junction.* The definitions of loop and mesh are similarly extended to apply to the network graphs of networks.

TREES AND LINKS The graph of a network such as that shown in Fig. 11-4*b* includes a number of loops in which current may flow when the network is energized. It is possible to remove some of the line elements from the graph in such a manner as to destroy (open) all the loops and *leave all the nodes connected to each other.* In Fig. 11-4*c* examples of several such arrangements are shown in that illustration; the dashed lines indicate the line elements which have been removed, and all the nodes are joined by the solid-line elements.

The reader should sketch more of these arrangements. Removal of appropriate line elements in each case causes the network to lose all its

[1] For a detailed study of topology the reader is referred to E. A. Guillemin, "Introductory Circuit Theory," chaps. 1 and 2, John Wiley & Sons, Inc., New York, 1953.

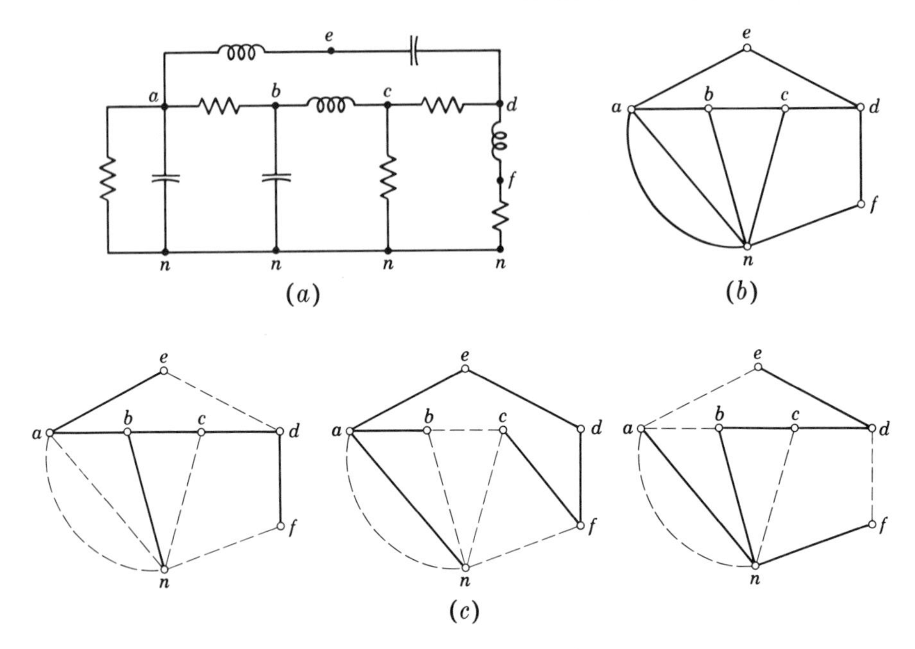

***Fig. 11-4** (a) A passive network. (b) Linear graph for the network of (a). (c) Trees and links of the graph of (b). The solid lines are the elements of the respective trees, and the dashed lines are the links.*

loops, so that the network no longer possesses any closed paths for the flow of current.

Tree: A tree of a network is the graph made by a certain number of its line elements such that all the nodes of the network are connected but no closed path is formed. There are many trees[1] possible for a given network. In Fig. 11-4*c* some of the trees of the graph of network of Fig. 11-4*a* are shown in solid lines.

Link: For a given tree all the lines in the graph of a network which do not belong to the tree are called the links of the tree. In Fig. 11-4*c* the links of the graph are indicated by dashed lines.

Basic properties of trees and links Whereas the tree has the property of connecting all the nodes without forming a closed path, from Fig. 11-4 it is seen that the links have the following important property: Starting with a tree, every time a link is added to the graph, a new loop can be traced which is different from previous loops inasmuch as it includes the new link, not included in the previous loops. In this sense, to each link there corresponds a loop which is "independent" from other loops. It can

[1] See, for example, L. Weinberg, Number of Trees in a Graph, *Proc. IRE*, vol. 46, no. 12, pp. 1954–1955, December, 1958.

be shown that the number of meshes in a network is identical with the number of independent loops of the network, which, as stated above, is equal to the number of the links of the network. Furthermore, it can be shown[1] that the application of Kirchhoff's voltage law to independent loops of the network[2] will result in the maximum number of independent equations, which is equal to the number of the links of the network.

NUMBER OF LINE ELEMENTS IN A TREE It is seen that the number of line elements in all possible trees of a given network are equal (six in Fig. 11-4). This can be proved by observing that, in tracing a tree, starting with unconnected nodes, the first line element connects two nodes together, the second line element will connect three nodes, and so on. Thus, if

$$N = \text{total number of nodes}$$

then

$$N - 1 = \text{number of line elements in the tree} \tag{11-14}$$

Similarly, if

$$E = \text{total number of line elements in the graph}$$

and

$$l = \text{number of links} = \text{number of meshes}$$

then

$$E = (N - 1) + l$$

or

$$l = E - N + 1 \tag{11-15}$$

Equations (11-14) and (11-15) are used in the determination of the minimum (necessary and sufficient) number of Kirchhoff-law equations which are required for obtaining the response of a network.

11-4 Networks with sources—entries[3]

There are two ways in which ideal sources can be introduced into a passive network, namely:

1 An ideal source can be connected between two nodes of a source-free network; this is called a *soldering-iron* entry. This term is used because, in this type of connection, the entry does not disturb the existing branches of the network; we visualize the source introduced by soldering leads to a node pair of the network and connecting the source to those leads.

2 An ideal source can be introduced in a branch of the source-free network; this is called a *pliers entry.* This term is used because insertion of the source requires cutting an existing lead in the network (e.g., with a pair of pliers) and inserting the ideal source.

We shall now study the relationship between source-free networks and the corresponding networks which result when ideal sources are introduced by means of the entries described.

[1] Guillemin, *op. cit.*

[2] We have so far illustrated network geometry only for source-free networks. Networks with sources are treated in Sec. 11-4.

[3] The terminology used in this section follows Guillemin, *op. cit.*

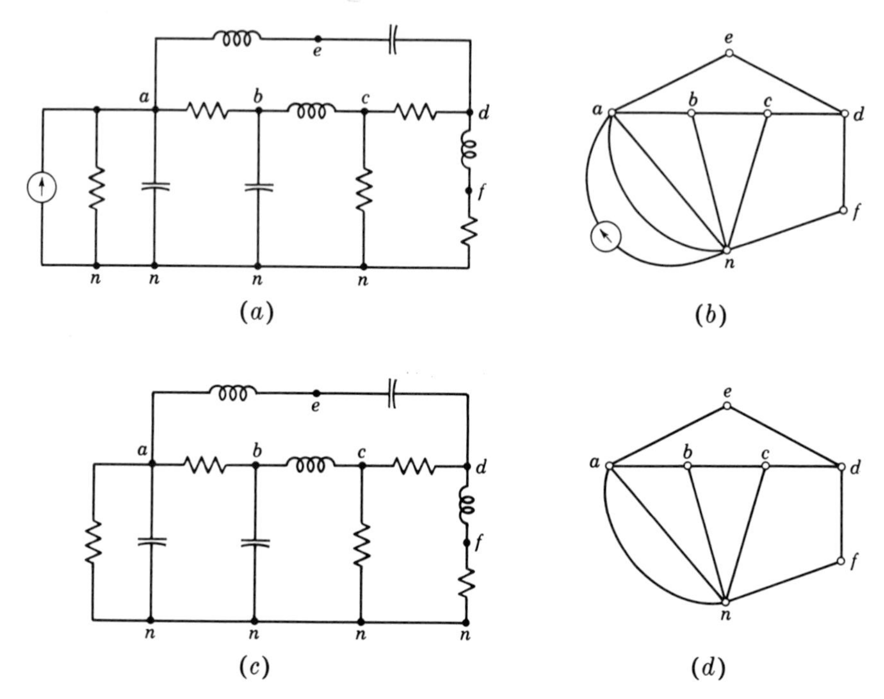

***Fig. 11-5** (a) Circuit of Fig. 11-4a, with ideal current source added at terminals a-n (soldering-iron entry). (b) Linear graph of (a), including the active element. (c) Source-free circuit corresponding to (a). (d) Linear graph of (c).*

SOLDERING-IRON CURRENT-SOURCE ENTRY In Fig. 11-5*a* is shown the network of Fig. 11-4*a* with a current source added by means of soldering-iron entry at nodes *a* and *n*. In Fig. 11-5*b* we have drawn the graph of this network, including a line element for the ideal source. To show that this line element represents an ideal source, the current source's symbol (circle with arrow) is used for the active branch. Now recall that any complete response of the network will include the "source-free" response component. Further, this component is derived by obtaining the general solution of the source-free network (equations). The source-free network is obtained by setting the sources to zero. When a current source is set to zero, it represents a "branch" carrying zero current, that is, an open circuit. If, in Fig. 11-5*a* and the corresponding graph, Fig. 11-5*b*, we set $i \equiv 0$, the source-free network of Fig. 11-5*c*, and its graph, Fig. 11-5*d*, result. We draw the following conclusion (which will be seen to be significant):

The introduction of an ideal current source into a network by soldering-iron entry does not change the structure of the corresponding source-free network. In other words, if we have given a certain number of passive elements connected in a particular fashion and we introduce a *current* source by *soldering-iron* entry, and if subsequently we set the source to zero, the original source-free network results. Thus certain properties (i.e., the free modes) of the source-free network remain unaltered with the source introduced as described.

PLIERS CURRENT-SOURCE ENTRY In Fig. 11-6a is shown the network of Fig. 11-4a with a current source added by means of a pliers entry into the resistive branch connecting nodes a and n, and the corresponding graph is shown in Fig. 11-6b. We now inquire about the corresponding source-free

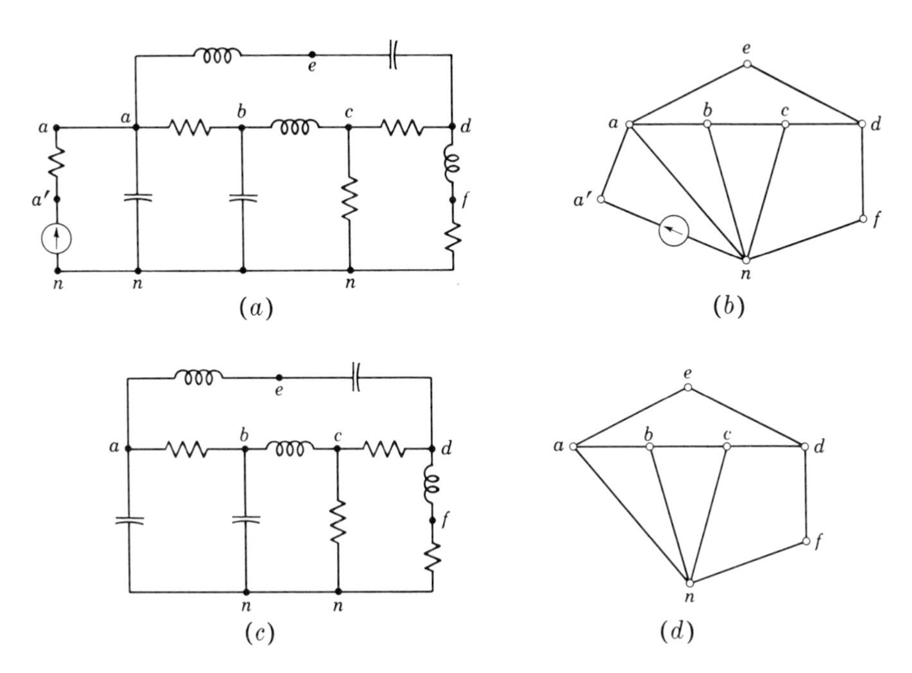

***Fig. 11-6** (a) Circuit of Fig. 11-4a with an ideal current source introduced into resistive branch a-n. (b) Linear graph of network of (a). (c) Source-free network corresponding to (a). (d) Linear graph of (c). Note that this graph has only one link between a and n.*

network. We observe that if $i_{nb'}$ is set to zero, the current in branch a'-n, and therefore *also* in the passive branch a'-a will be zero. As a result the source-free network corresponding to that of Fig. 11-6a will not include the resistance between a and a' since no current can be in it. The source-free network corresponding to Fig. 11-6a is shown in Fig. 11-6c, and its graph in Fig. 11-6d. We draw the following conclusion:

The introduction of an ideal current source into a network by pliers entry changes the structure of the corresponding source-free network. Hence, if we have given a source-free network whose characteristic roots we know and introduce current source by pliers entry, the source-free component of the response will have different characteristic roots than the original source-free network.

Example 11-2 (*a*) Find the characteristic roots of the source-free network given in Fig. 11-7*a*. (*b*) Introduce an ideal constant current source I_0 into the network shown in Fig. 11-7*b* by means of soldering-iron entry at nodes *a-n*, and find the form of the complete response for v_{an}. (*c*) Introduce a current source into the 3-ohm branch by means of a pliers entry and find the complete response for v_{an}.

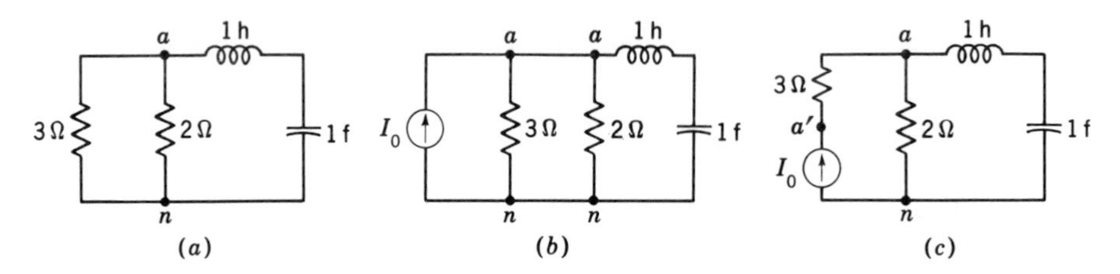

Fig. 11-7 ***(a) A source-free network. (b) Network of (a) with ideal current source introduced by soldering-iron entry. (c) Network of (a) with ideal current source introduced by pliers entry.***

Solution (*a*) The source-free circuit is an *R-L-C* series circuit with $L = 1$, $C = 1$, $R = 3 \times 2/5 = 1.2$, so that $\alpha = R/2L = 0.6$, $\omega_0^2 = 1$, and the characteristic roots are given by $-\alpha \pm (\alpha^2 - \omega_0^2)^{\frac{1}{2}}$, which gives

$$\mathbf{s}_{1,2} = -0.6 \pm j0.8$$

(*b*) A constant source I_0 (with reference arrow chosen up) has been introduced between nodes *a* and *n* in Fig. 11-7*b*. The operational driving-point admittance at the source terminals is

$$Y_{an}(p) = \frac{1}{3} + \frac{1}{2} + \frac{1}{p + 1/p} = \frac{p^2 + 1.2p + 1}{1.2(p^2 + 1)}$$

Hence

$$Z_{an}(p) = \frac{1.2(p^2 + 1)}{p^2 + 1.2p + 1}$$

and

$$v_{an} = \frac{1.2(p^2 + 1)}{p^2 + 1.2p + 1} I_0$$

$$(p^2 + 1.2p + 1)v_{an} = 1.2(p^2 + 1)I_0 = 1.2I_0$$

The characteristic equation is

$$s^2 + 1.2s + 1 = 0$$

so that the characteristic roots are

$$\mathbf{s}_{1,2} = -0.6 \pm j0.8$$

The component of the response due to the source is $v_{an} = V_{an} = 1.2I_0$, and the complete response has the form

$$v_{an} = 1.2I_0 + \text{Re}\,(\mathbf{K}e^{(-0.6+j0.8)t})$$

We observe that the source-free component of v_{an} is characterized by the characteristic roots of the given source-free circuit, Fig. 11-7*a*.

(*c*) In Fig. 11-7*c* the current source I_0 (with arbitrarily chosen reference direction) has been introduced in the 3-ohm branch. We observe that $i_{a'a} = I_0$ *independently* of the 3-ohm resistance, so that, applying Kirchhoff's current law at a gives

$$I_0 = \tfrac{1}{2}v_{an} + \frac{1}{p + 1/p}v_{an}$$
$$= \frac{p^2 + 2p + 1}{2(p^2 + 1)}v_{an}$$

or
$$(p^2 + 2p + 1)v_{an} = 2(p + 1)I_0 = 2I_0$$

The characteristic equation is

$$s^2 + 2s + 1 = 0$$
$$s_{1,2} = -1$$

and the component of v_{an} due to the source is $v_{an} = V_{an} = 2I_0$. Hence the form of the complete response is

$$v_{an} = 2I_0 + (K_1 + K_2t)e^{-t}$$

We note that the characteristic roots of the circuit of Fig. 11-7c are *not* the same as those of Fig. 11-7a.

VOLTAGE SOURCE, PLIERS ENTRY A study of the examples shown in Fig. 11-8a to d shows that the *introduction of an ideal voltage source into a network*

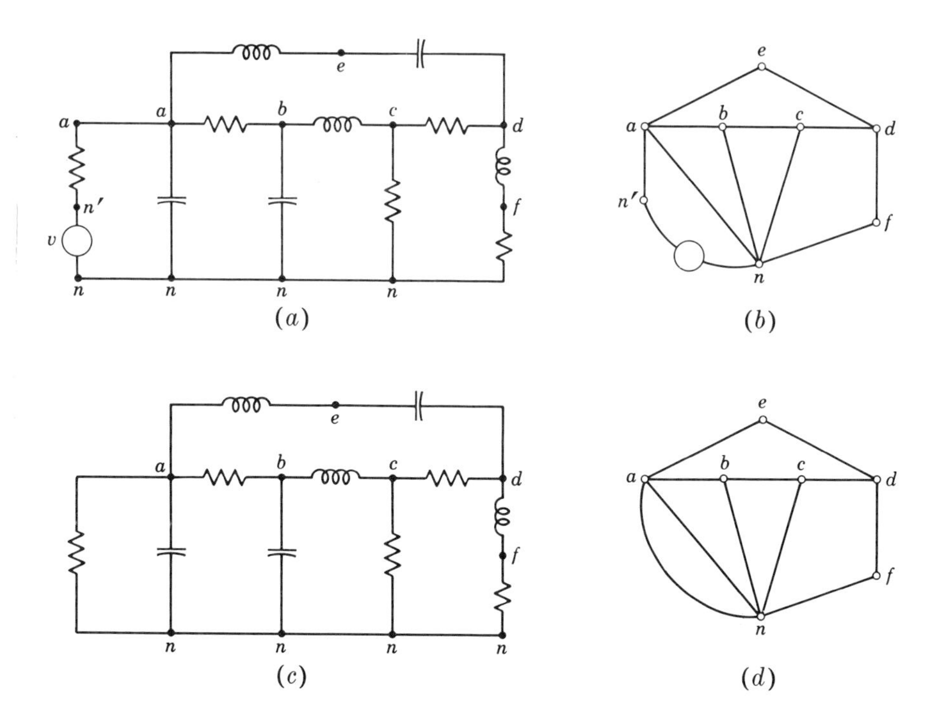

***Fig. 11-8** (a) Circuit of Fig. 11-4a with ideal voltage source introduced by pliers entry in the resistive branch a-n. (b) Linear graph corresponding to (a). (c) Source-free network corresponding to (a). Since the voltage source is set to zero, node n′ has become the same as n. (d) Linear graph corresponding to (c).*

by means of a pliers entry does not change the structure of the corresponding source-free network. Consequently, given the characteristic roots of a source-free network, they can be applied to a network if voltage sources have been introduced by pliers entry.

VOLTAGE SOURCE, SOLDERING-IRON ENTRY It is left as an exercise for the reader to demonstrate that *introduction of an ideal voltage source by soldering-iron entry changes the structure of the source-free network.* (See Prob. 11-12.)

SUMMARY *Introduction of ideal voltage sources by pliers entry and of ideal current sources by soldering-iron entry does not change the structure of the corresponding network with ideal sources set to zero.* Thus the study of network graphs and equations for source-free networks has application to the networks with ideal sources only when the ideal sources have been introduced in the fashion described above.

11-5 Network reduction Sketching the linear graph of a network is sometimes useful in analysis because the graph shows clearly the structure of the network. Elements in series and in parallel can then be clearly identified and combined into p elements. Using p elements for passive series and parallel combinations, the network can be structurally reduced to eliminate nodes that are not junctions. Using, in addition, source conversion, all nodes that are not junctions can be removed *provided* that all ideal voltage sources are inserted by pliers entry and all ideal current sources by soldering-iron entry. The following example illustrates this technique.

Example 11-3 In the circuit of Fig. 11-9*a*, use network reduction to formulate the equilibrium equation for v_{mn}.

Solution The linear graph for the circuit of Fig. 11-9*a* is shown in Fig. 11-9*b*. We observe immediately that C_2 and C_3 are in parallel, as are R_3, C_4, and R_6. The series-parallel combinations L-R_5-R_4 and R_0-R_1-C_1 are also identified. Representing these combinations by p elements, all nodes except a that are not also junctions are eliminated as shown in Fig. 11-9*c*. To eliminate node a, source conversion is used as shown in Fig. 11-9*d*. Additional source conversion gives the circuit of Fig. 11-9*e*, where

$$v_{mn} = \frac{1}{Y_E + Y_D}(i + i_A) \tag{11-16}$$

where $Z_A = R_0 + 1/(G_1 + pC_1)$, $Y_B = p(C_2 + C_3)$, $Z_C = 1/Y_C$, $Y_C = G_4 + 1/(R_5 + pL)$, $Y_D = G_3 + G_6 + pG_4$, and

$$Y_E = \frac{1}{Z_C + 1/(Y_A + Y_B)} \tag{11-17}$$

$$i_A = \frac{Y_A}{Z_C(Y_A + Y_B) + 1} \tag{11-18}$$

Proof of the relations given in Eqs. (11-17) and (11-18) is left as an exercise for the reader.

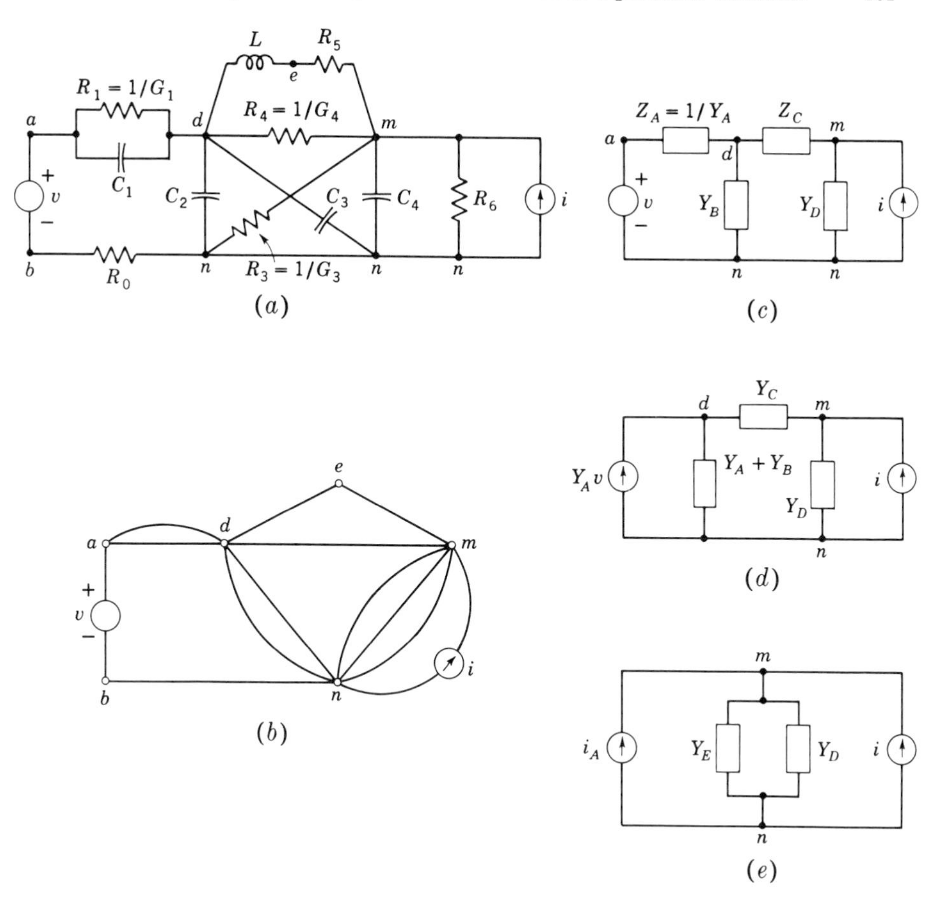

***Fig. 11-9** (a) A complex network. (b) Linear graph for the network of (a). (c) Simplified structure of (a) obtained by use of p elements. (d) Network of (c) after conversion of the voltage source $v - Z_A$. (e) A network equivalent to (a) at terminals m-n obtained by repeated source conversion from (d).*

We can now point out several considerations which can be used as guides to determine whether the reduction of complexity of the network structure by use of p elements, conversion of sources, or both are desirable:

1 What is the circuit problem? If the reduction of the network eliminates terminals at which responses are to be found, the procedure may not be desirable.
2 How much algebra is involved in the reduction? The combination of elements either in operational form or in complex form may be very time-consuming, and unless the desired terminals are kept accessible, a considerable amount of algebra may be necessary before the final solution is obtained.

In addition to these two questions, another problem concerned with use of p elements in network reduction arises. This problem arises from the fact that certain aspects of network behavior are hidden when combinations of elements are treated as a unit. This is the problem of mode coupling, and is discussed in Chap. 12. Nevertheless, network reduction can be of value in the solution of some problems.

11-6 The general problem of network analysis

We shall formulate the general problem of network analysis as follows: (1) Given a network which contains sources and passive elements, relate the variables associated with the passive elements to the source functions. (It is a matter of indifference here whether p elements or primitive elements are used, because we deal with whatever variables are associated with the given elements.) (2) Find the complete response.

Note that we have not stated what is known and what is unknown in the circuit. For the purpose of explaining the analytical procedure, we shall assume that the source functions as well as the element values or operators are known. As we progress, it will become evident that the analytical procedure can sometimes be reversed so that the source functions for a required response or the element values can be the unknowns.

Assuming that the source functions and the network elements are known, then, if there are E passive elements, S_i current sources, and S_v voltage sources in the network, it appears that there are $2E + S_i + S_v$ unknowns. We arrive at this result as follows: We associate with each passive element two variables, namely, the voltage across the terminals of the element and the current which enters (or leaves) a terminal of the element. Hence there are $2E$ unknowns associated with the passive elements. Since, for each ideal voltage source, the voltage across the terminals is known, the current through the voltage source is the only unknown associated with each such source, and there are S_v of these. For each ideal current source the voltage across its terminals is the only unknown, and this adds S_i unknowns. Hence we have $2E + S_i + S_v$ unknowns.

Now it is immediately evident that knowledge of the circuit elements means that we know the relationship between the current which enters the terminal of an element and the voltage across the terminals of the element because it is exactly this relationship which is used to define the circuit elements or p elements in the first place. We say, therefore, that, starting with $2E$ unknowns, E of them, either the voltages or the currents associated with each passive element, can be found in terms of the other E, and we shall take the number of these unknowns to be E. Next we shall show that the $S_v + S_i$ unknowns associated with the ideal sources are not independent unknowns, but are known if the E unknowns associated with the passive elements are found. In the network there are several possible ways in which a source can be connected. We shall consider each of these in turn.

1 An ideal voltage source may be in series with another ideal voltage source as shown in Fig. 11-10a. If that is the case, then, by application of the definition of a voltage source, the two sources can be represented as

a single source. In Fig. 11-10*a*,

$$v_{ab} = v_{am} + v_{mb} \tag{11-19}$$

2 An ideal current source, in parallel with another ideal current source, can be represented as a single current source as shown in Fig. 11-10*b*, where

$$i(t) = i_1(t) + i_2(t) \tag{11-20}$$

3 Two ideal voltage sources $v_1(t)$ and $v_2(t)$ cannot be in parallel unless $v_1 \equiv v_2$. Similarly, two ideal current sources $i_1(t)$ and $i_2(t)$ cannot be in series unless $i_1 \equiv i_2$. This follows from the definition of ideal sources.

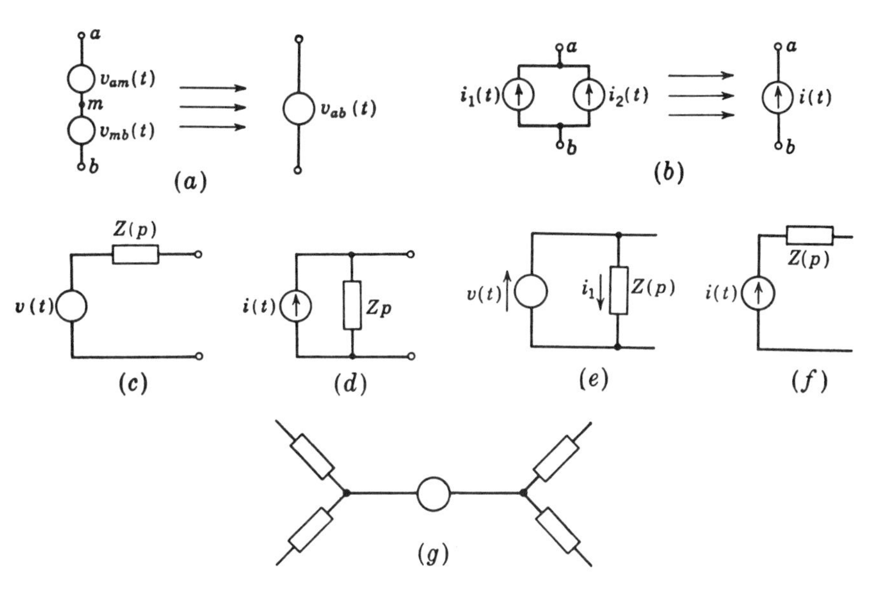

Fig. 11-10 *An enumeration of the manner in which sources can appear in a network.*

4 An ideal voltage source can be in series with a passive element, that is, inserted by pliers entry, as shown in Fig. 11-10*c*. In this case it is evident that, once the current in the passive element is known, the current in the source is known as well.

5 An ideal current source may be in parallel with a passive element by soldering-iron entry, as in Fig. 11-10*d*. In this case the voltage across the element is identical with the voltage across the current source.

6 An ideal voltage source may be in parallel with a passive element by soldering-iron entry, as in Fig. 11-10*e*. In this case the current in the element is immediately known, $i_1 = Y(p)v(t)$, so that one connection of this type in a network gives $E - 1$ unknowns associated with passive elements plus one additional unknown, the current in the voltage source, or altogether, E unknowns. Although this case may occur "on paper," it does not represent a practical situation. (Why?)

7 An ideal current source may be in series with a passive element by soldering-iron entry, as in Fig. 11-10*f*. In this case the current in $Z(p)$ is known, and the voltage across the current source is not known. Again such an arrangement, which may occur on paper, does not represent a practical situation.

8 An ideal source may be connected between two junctions, as in Fig. 11-10*g*. If the currents in all the passive elements are known, the voltage across such a source is found by application of Kirchhoff's voltage law around any loop which includes the source, and the current in such a source is found by application of Kirchhoff's current law at either terminal of the source.

We conclude, therefore, that *the circuit is completely specified through E variables, where E is the number of passive elements in the circuit, and only one variable, either voltage or current, is chosen for each passive element.*

We now recall that, for a source-free circuit with E (line) elements and N nodes, the number of geometrically independent loops is $E - N + 1$, and this number corresponds to the number of independent voltage-law equations which can be written for such a circuit. If we now consider a circuit in which ideal voltage sources have been introduced by pliers entry, then, as the reader can easily prove, the number of links in a tree are the same if the tree is drawn with sources or if sources are not included. Hence the number of independent equations which can be obtained in such circuits as a result of application of the voltage law is $E - N + 1$. Since these are unknowns, the remaining equations are obtained by application of the current law $N - 1$ times. A little thought will show that the minimum number of such equations results when all nodes are junctions.

In the ensuing sections we discuss the formulation of such equations in detail. We shall see that the choice of variables can reduce the total number of simultaneous equations below E. By this we mean that the E unknowns can be found by solving fewer than E *simultaneous* equations and subsequently applying a sufficient number of Kirchhoff-law relationships to find all E unknowns.

11-7 Formulation of network equations using branch variables[1]

We know that, in a circuit in which all element values and source functions are known, there are as many unknowns as there are passive elements. To obtain the Kirchhoff-law equations from which these unknowns can be determined, one may begin by assigning to each passive element, symbolically, a voltage or current variable. The voltage law is then applied to each mesh or geometrically independent loop, and the current law at each independent junction, to give the required number of simultaneous equations to solve for the unknowns. This method will be illustrated by means of an example.

[1] The reader may wish to study Appendix D, which deals with matrices and determinants, before reading the ensuing sections on simultaneous-equations methods in network analysis.

Example 11-4 In the resistance network of Fig. 11-11*a* find the currents in all resistances as a function of the source functions $v_A(t)$ and $i_B(t)$.

Solution Since only one type of passive element (resistance) is involved, the relationship between the currents in the resistances and the source functions will be algebraic. We shall first illustrate the use of branch currents as variables. The circuit contains seven resistances; we have accordingly assigned the variables i_1 to i_7 to the resistances. (The reference directions chosen are arbitrarily assigned.) We now seek seven independent relationships for finding the seven unknown currents. The network graph shows five meshes and four junctions. One of these meshes, *d-d'-n-d* consists of a resistive current source which can be changed to a voltage source and can therefore be eliminated. Hence we shall have four Kirchhoff-

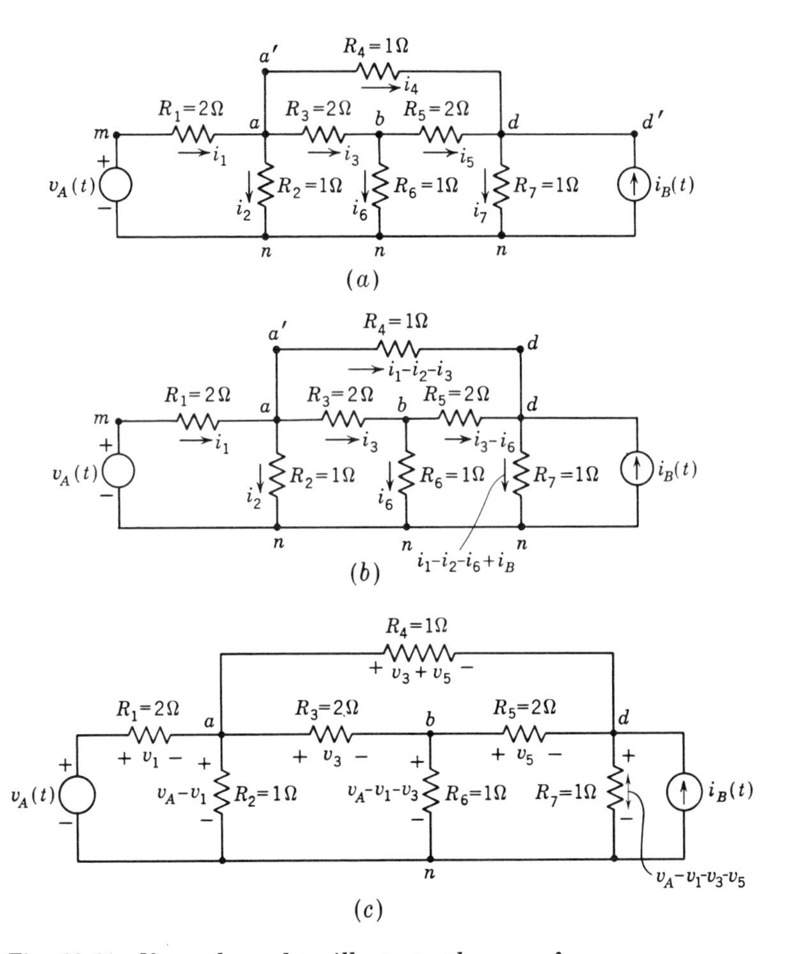

Fig. 11-11 ***Network used to illustrate the use of branch variables.*** ***(a) A branch current has been assigned to each element. (b) Assignment of branch currents so that the Kirchhoff current law is satisfied at every junction. (c) Assignment of element voltages so that the Kirchhoff voltage law is satisfied for every loop.***

voltage-law equations for the remaining four meshes and three Kirchhoff-current-law equations. From the voltage law, using the numerical values of the elements, we have for

Mesh *m-a-n-m*:	$2i_1 + i_2 = v_A(t)$	(11-21*a*)
Mesh *a-b-n-a*:	$2i_3 + i_6 - i_2 = 0$	(11-21*b*)
Mesh *b-d-n-b*:	$2i_5 + i_7 - i_6 = 0$	(11-21*c*)
Mesh *a-b-d-a′-a*:	$2i_3 + 2i_5 - i_4 = 0$	(11-21*d*)

From the current law at

Junction *a*:	$i_1 - i_2 - i_3 - i_4 = 0$	(11-21*e*)
Junction *b*:	$i_3 - i_6 - i_5 = 0$	(11-21*f*)
Junction *d*:	$i_5 - i_7 + i_B(t) + i_4 = 0$	(11-21*g*)

[Note that addition of Eqs. (11-21*e*), (11-21*f*), and (11-21*g*) gives the same result as application of Kirchhoff's current law at junction *n*.] Equations (11-21) can now be solved simultaneously for the seven unknowns i_1 to i_7.

It is possible to reduce the labor involved in solution and in setting up equations if, in the assignment of current symbols to the elements, Kirchhoff's current law is satisfied at every junction. Thus, at each junction, we start assigning symbols to each branch current except one, where the symbol assigned to this branch will depend on the symbols assigned to other branches, such that Kirchhoff's current law is satisfied. Referring to Fig. 11-11*b* (this circuit is identical with the circuit of Fig. 11-11*a*), we have assigned i_1 to R_1, i_2 to R_2, i_3 to R_3, but the current in R_4 is chosen to be $i_1 - i_2 - i_3$, so that the manner in which symbols are assigned to the branch variables assures that Kirchhoff's current law is satisfied at junction *a*. Proceeding to junction *b*, we assign i_6 to R_6, but call the current in R_5, $i_3 - i_6$, so that the current law is satisfied at junction *b*. At junction *d* Kirchhoff's current law is satisfied by assigning the current $i_1 - i_2 - i_6 + i_B(t)$ to R_7. We now have four unknowns, i_1, i_2, i_3, and i_6. Note, however, that when these currents are found, the currents in R_4, R_5, and R_7 must be found through further algebraic operations.

Since Kirchhoff's current law is already satisfied at every junction, the four mesh equations are now written

Mesh *m-a-n-m*:	$2i_1 + i_2 = v_A(t)$	(11-22*a*)
Mesh *a-b-n-a*:	$2i_3 + i_6 - i_2 = 0$	(11-22*b*)
Mesh *b-d-n-b*:	$2(i_3 - i_6) + 1[i_1 - i_2 - i_6 + i_B(t)] - i_6 = 0$	(11-22*c*)
Mesh *a-b-d-a′* = *a*:	$2i_3 + 2(i_3 - i_6) - 1(i_1 - i_2 - i_3) = 0$	(11-22*d*)

Rewriting (11-22*a*) to (11-22*d*) and collecting terms, we have

$$2i_1 + i_2 = v_A(t) \tag{11-23a}$$
$$-i_2 + 2i_3 + i_6 = 0 \tag{11-23b}$$
$$i_1 - i_2 + 2i_3 - 4i_6 = -i_B(t) \tag{11-23c}$$
$$-i_1 + i_2 + 5i_3 - 2i_6 = 0 \tag{11-23d}$$

Solving for $i_1(t)$, by Cramer's rule we have

$$i_1 = \frac{\begin{vmatrix} v_A & +1 & 0 & 0 \\ 0 & -1 & +2 & +1 \\ -i_B & -1 & +2 & -4 \\ 0 & +1 & +5 & -2 \end{vmatrix}}{\begin{vmatrix} +2 & +1 & 0 & 0 \\ 0 & -1 & +2 & +1 \\ +1 & -1 & +2 & -4 \\ -1 & +1 & +5 & -2 \end{vmatrix}} = \frac{-35v_A + 9i_B}{-89} = \tfrac{35}{89}v_A - \tfrac{9}{89}i_B \tag{11-24}$$

Similarly,

$$i_2 = \frac{-19v_A - 18i_B}{-89} = \tfrac{19}{89}v_A + \tfrac{18}{89}i_B \tag{11-25}$$

$$i_3 = \tfrac{6}{89}v_A + \tfrac{1}{89}i_B \tag{11-26}$$

$$i_6 = \tfrac{7}{89}v_A + \tfrac{16}{89}i_B \tag{11-27}$$

Using Kirchhoff's current law, at junctions a, b, and n in Fig. 11-11a we have the branch currents i_4, i_5, and i_7,

$$i_4 = i_1 - i_2 - i_3 = \tfrac{10}{89}v_A - \tfrac{28}{89}i_B$$
$$i_5 = i_3 - i_6 = -\tfrac{1}{89}v_A - \tfrac{15}{89}i_B$$
$$i_7 = i_1 - i_2 - i_6 + i_B = \tfrac{9}{89}v_A + \tfrac{46}{89}i_B$$

USE OF VOLTAGE VARIABLES If we symbolically assign a voltage to each resistance in Fig. 11-11a, we can write the seven equations (11-21) as before. Exactly as in the case of current variables, the number of equations can be reduced below seven. This is achieved if we assign voltage variables so that Kirchhoff's *voltage* law is satisfied by the manner in which the variables are chosen. Starting with R_1, the voltage v_1 is assigned to R_1 as in Fig. 11-11c. Now we note that the voltage across R_2 (with reference directions as chosen) must be $v_A(t) - v_1$ because Kirchhoff's voltage law around mesh m-a-n-m must be satisfied. Proceeding now to another mesh, say, mesh a-b-n-a, we assign v_3 to R_3 as shown. The voltage across R_6 is seen to be $v_A - v_1 - v_3$. Assigning v_5 to R_5, we have $v_A - v_1 - v_3 - v_5$ across R_7 and $v_3 + v_5$ across R_4 (all with arbitrarily chosen reference directions as indicated in the figure). Now Kirchhoff's voltage law around every mesh is satisfied by the manner in which the voltage variables have been assigned. We now proceed to satisfy Kirchhoff's current law.

At junction a:
$$\frac{v_1}{2} - \frac{v_A - v_1}{1} - \frac{v_3}{2} - \frac{v_3 + v_5}{1} = 0 \tag{11-28}$$

At junction b:
$$\frac{v_3}{2} - \frac{v_A - v_1 - v_3}{1} - \frac{v_5}{2} = 0 \tag{11-29}$$

At junction d:
$$\frac{v_5}{2} - \frac{v_A - v_1 - v_3 - v_5}{1} + i_B + \frac{v_3 + v_5}{1} = 0 \tag{11-30}$$

Collecting terms in each of these equations, we have

$$+\tfrac{3}{2}v_1 - \tfrac{3}{2}v_3 - v_5 = v_A$$
$$+v_1 + \tfrac{3}{2}v_3 - \tfrac{1}{2}v_5 = v_A$$
$$+v_1 + 2v_3 + \tfrac{5}{2}v_5 = v_A - i_B$$

Solving, we have

$$v_1 = \frac{70v_A}{89} - \frac{18i_B}{89} \tag{11-31}$$

$$v_3 = \frac{12v_A}{89} + \frac{2i_B}{89} \tag{11-32}$$

$$v_5 = -\frac{2v_A}{89} - \frac{30i_B}{89} \tag{11-33}$$

The remaining voltages are found by application of the voltage law.

$$v_{an} = v_A(t) - v_1 \qquad v_{bn} = v_A(t) - v_1 - v_3$$
$$v_{ad} = v_3 + v_5 \qquad v_{dn} = v_A - v_1 - v_3 - v_5$$

The numerical result may be obtained by the reader by using direct substitution, and the results for the branch currents may then be checked.

In this example we have shown how the number of simultaneous equations can be reduced below E to either $E - (N - 1)$ or $N - 1$ by choice of current or voltage symbols such that either the current or the voltage law is satisfied "automatically." Two systematic procedures which accomplish the same aim (i.e., reducing the number of simultaneous equations below E) are the "node-voltage" and the "mesh-current" methods, which are treated in the next two sections.

11-8 The method of node voltages

In this section we shall show how, by the choice of certain voltages as the variables of a network, simultaneous equations can be obtained by the application of Kirchhoff's current law alone. Furthermore, it will be seen that, when these voltages are found, all other variables of the network can be found. Since we plan to obtain the simultaneous equations of the network by application of Kirchhoff's current law, the voltage variables must be chosen such that Kirchhoff's voltage law will be satisfied for every loop. For the present it will be *convenient* to assume that all voltage sources in the network have internal immittances and that the network has been drawn with all sources shown as (i.e., converted to) current sources. We shall further represent the network by use of operational elements, but remember that the results hold when each operational element is a primitive element. The use of operational elements in the circuit reduces the network, as we have shown, so that all nodes are junctions.

To develop the node method, we assign a number or letter to all nodes in the network. In Fig. 11-12*a* a typical node, which is assigned the number 1, is shown. This node has the following connections to it:

1. The admittance $Y_A(p)$ connects node 1 to node 2.
2. The current source consisting of $i_B(t)$ and $Y_B(p)$ connects node 1 to node 3.
3. The current source consisting of $i_C(t)$ and $Y_C(p)$ connects node 1 to node 4.
4. The admittance $Y_D(p)$ connects node 1 to node R.
5. The admittance $Y_E(p)$ connects node 1 to node 5.
6. The admittance zero connects node 1 to node 6 (this means that there is no connection between node 1 and node 6).
7. The ideal current source $i_E(t)$ connects node 1 to node 7.

In Fig. 11-12*a* only those circuit elements which are connected to node 1 are shown. Nodes 2, 3, etc., are also interconnected, but we are now concerned only with the conditions at the "typical node," which we have numbered node 1. We further assume that the network contains no mutual inductance. (Circuits with mutual inductance will be treated in Chap. 15.)

Let us apply Kirchhoff's current law at node 1. The sum of the currents flowing into node 1 is given by the following equations (for v_{21} read "v two one," not "v twenty-one," etc.):

$$Y_A(p)v_{21} + Y_B(p)v_{31} + i_B(t) + Y_C(p)v_{41} - i_C(t) + Y_D(p)v_{R1} + Y_E(p)v_{51} - i_E(t) = 0$$

or

$$Y_A(p)v_{21} + Y_B(p)v_{31} + Y_C(p)v_{41} - Y_D(p)v_{R1} + Y_E(p)v_{51} = -i_B + i_C + i_E \quad (11\text{-}34)$$

Multiplying (11-34) by -1, we have

$$Y_A(p)v_{12} + Y_B(p)v_{13} + Y_C(p)v_{14} + Y_D(p)v_{1R} + Y_E(p)v_{15} = i_B - i_C - i_E \quad (11\text{-}35)$$

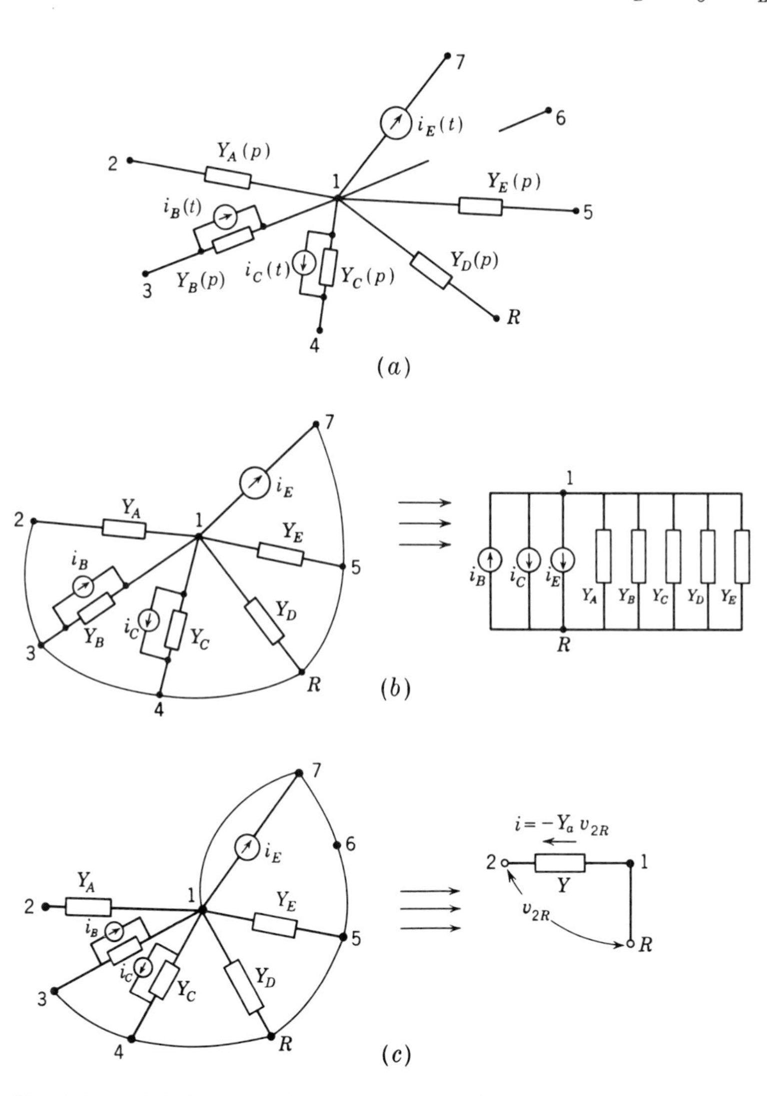

***Fig. 11-12** (a) Portion of a network showing how node 1 can be connected to other nodes. (b) All nodes except node 1 are connected to the reference node when Y_{11} is determined. (c) All nodes except node 2 are connected to the reference node when Y_{12} is determined.*

REFERENCE NODE AND NODE VOLTAGES In Eq. (11-35) the currents which flow through passive elements into node 1 are expressed in terms of the voltage between node 1 and the other nodes. But these voltages can all be expressed by applying Kirchhoff's voltage law around loops in the circuit, all of which include an arbitrarily chosen node, called the reference node. We have chosen node R to be the reference node, thus:

$$\begin{aligned} v_{12} &= v_{1R} + v_{R2} = v_{1R} - v_{2R} \\ v_{13} &= v_{1R} + v_{R3} = v_{1R} - v_{3R} \\ v_{14} &= v_{1R} + v_{R4} = v_{1R} - v_{4R} \\ v_{15} &= v_{1R} + v_{R5} = v_{1R} - v_{5R} \\ v_{1R} &= v_{1R} + v_{RR} = v_{1R} \end{aligned} \tag{11-36}$$

In this equation the voltage associated with a branch (such as v_{12} associated with branch joining junctions 1 and 2) is determined in terms of the difference between the voltage between the nodes of that branch and a given reference node: $v_{12} = v_{1R} - v_{2R}$. The voltage of a given node with respect to a specified reference node is called the node voltage.

Substituting Eqs. (11-36) into (11-35), we may write

$$Y_A(v_{1R} - v_{2R}) + Y_B(v_{1R} - v_{3R}) + Y_C(v_{1R} - v_{4R}) + Y_D v_{1R} + Y_E(v_{1R} - v_{5R}) = i_B - i_C - i_E$$

or

$$[Y_A(p) + Y_B(p) + Y_C(p) + Y_D(p) + Y_E(p)]v_{1R} + [-Y_A(p)]v_{2R} + [-Y_B(p)]v_{3R} + [-Y_C(p)]v_{4R} + [-Y_E(p)]v_{5R} = i_B - i_C - i_E \tag{11-37}$$

Now in Eq. (11-37) we note that the terms are arranged so that the right-hand side of the equation contains only the source functions. We observe that $i_B - i_C - i_E$ is equal to the sum of the currents from ideal sources *flowing into node* 1. If the reference arrow of the current source is away from the node, the source is assigned a negative sign in the summation. We denote this sum as $i_1(t)$.

$$i_B - i_C - i_E = i_1(t) \tag{11-38}$$

NODAL CURRENT SUPPLY In general, we denote the sum of the currents which flow from ideal current sources into node k as i_k and call it the current supply of node k. Note that i_k is the sum of the currents from the *ideal* sources and does not include currents associated with the node k through passive elements.

NODAL SELF-ADMITTANCE Next we study the terms on the left-hand side of Eq. (11-37). We observe that, if we set

$$v_{2R} = v_{3R} = v_{4R} = v_{5R} \equiv 0$$

but $\quad v_{1R} \neq 0 \quad$ i.e., if $v_{kR} = 0$, for $k \neq 1$,

then

$$[Y_A(p) + Y_B(p) + Y_C(p) + Y_D(p) + Y_E(p)]v_{1R} = i_1(t) \tag{11-39}$$

Now v_{kR} can be made zero by connecting node k to node R. In Fig. 11-12*b* the portion of the network shown in Fig. 11-12*a* is redrawn, but all nodes except node 1 are connected to node R. We now define the *self-admittance of node k*, denoted by the symbol $Y_{kk}(p)$, as follows:

> If all nodes except node k are connected to the reference node, then the application of Kirchhoff's current law at node k will result in an equation of the form $Y_{kk}(p)v_{kR}(t) = i_k(t)$. $Y_{kk}(p)$ is called the operational self-admittance of node k.

From Eq. (11-39) it is seen that in a network (which does not contain mutual inductance) the self-admittance of a node is given by the sum of the admittances between the node in question and the reference node if all nodes except the node in question are connected to the reference node. Hence, in the above example, $Y_{11}(p) = Y_A + Y_B + Y_C + Y_D + Y_E$.

MUTUAL ADMITTANCE BETWEEN TWO NODES Observe that, in Fig. 11-12*a*, if all nodes except node 2 but including node 1 are connected to the reference node, as shown in Fig. 11-12*c*, then the current which flows *out of* node 1 owing to a voltage v_{2R} is $(-Y_A)v_{2R}$. Based on this observation, we define the *mutual admittance between two nodes, k* and n, as follows:

> If all nodes except node k are connected to the reference node, then the current which flows *out of* node n due to a voltage v_{kR} is $Y_{nk}v_{kR}$, where Y_{nk} is the operational mutual admittance between nodes n and k.

We observe from Eq. (11-38) that (in the absence of mutual inductance in the network) the mutual admittance between two nodes is given by the *negative* of the admittance of the element which directly connects these two nodes.

Setting $Y_{11} = Y_A + Y_B + Y_C + Y_D + Y_E$, $Y_{12} = -Y_A$, $Y_{13} = -Y_B$, $Y_{14} = -Y_C$, $Y_{15} = -Y_E$, Kirchhoff's current law at node 1 is expressed through the equation

$$Y_{11}(p)v_{1R} + Y_{12}(p)v_{2R} + Y_{13}(p)v_{3R} + Y_{14}(p)v_{4R} + Y_{15}(p)v_{5R} = i_1(t)$$

In a network which contains N nodes, we have for node 1

$$Y_{11}(p)v_{1R} + Y_{12}(p)v_{2R} + \cdots + Y_{1n}v_{nR} + Y_{1N}v_{NR} = i_1(t)$$

where $n = N - 1$. If the node numbered N is the reference node,

$$v_{NR} = v_{RR} = 0$$

The term $Y_{1N}v_{NR}$ in the equation for node 1 can then be omitted. Moreover, from the definition of self-admittance, it is seen that the self-admittance of the reference node is zero. Using the symbols defined above, we now write the equations which express Kirchhoff's current law at each node except at the reference node. This will result in $n = N - 1$ equations, where N is the total number of the nodes.

$$\begin{aligned}
Y_{11}v_{1R} + Y_{12}v_{2R} + Y_{13}v_{3R} + \cdots + Y_{1n}v_{nR} &= i_1(t) \\
Y_{21}v_{1R} + Y_{22}v_{2R} + Y_{23}v_{3R} + \cdots + Y_{2n}v_{nR} &= i_2(t) \\
\vdots \qquad\qquad\qquad\qquad & \quad \vdots \\
Y_{n1}v_{1R} + Y_{n2}v_{2n} + Y_{n3}v_{3R} + \cdots + Y_{nn}v_{nR} &= i_n(t)
\end{aligned} \tag{11-40}$$

The node equation for the reference node reads

$$Y_{R1}v_{1R} + Y_{R2}v_{2R} + Y_{R3}v_{3R} + \cdots + Y_{nR}v_{nR} = i_R \tag{11-41}$$

This equation is superfluous because we already have the n simultaneous equations (11-40) for the n variables v_{1R} to v_{nR}, which completely specify the network. To show explicitly that Eq. (11-41) is not independent of the system of Eqs. (11-40), we observe that

$$Y_{R1} = -(Y_{11} + Y_{21} + \cdots + Y_{n1}) \tag{11-42}$$

since Y_{R1} is the negative of the admittances connected between node R and node 1. In Fig. 11-12*a*,

$$\begin{aligned}
Y_{R1} &= -Y_D \\
&= -(Y_A + Y_B + Y_C + Y_D + Y_E - Y_A - Y_B - Y_C - Y_E)
\end{aligned}$$

Hence

$$Y_{Rk} = -(Y_{kk} + Y_{k1} + Y_{k2} + \cdots + Y_{kn}) \tag{11-43}$$

Hence the sum of Eqs. (11-40) gives Eq. (11-41), and the latter equation is not independent of the others.

SINGLE-INDEX NOTATION FOR NODE VOLTAGES We have seen that the voltage of any node with respect to a specified reference node of the network is called the node voltage of that node. Since the node voltages in the network are always with respect to a specified reference node, these voltages can be referred to by means of a single index. Thus, in place of v_{1R}, we write v_1; in place of v_{2R}, v_2; etc. It is understood that the reference node will be marked on the circuit diagram so that the single-index notation is meaningful. Commonly, the "ground symbol" ⏚ is used to mark the reference node.

FURTHER DISCUSSION OF THE NODE EQUATIONS Using the single-index notation, the node equations have the form

$$\begin{aligned}
Y_{11}v_1 + Y_{12}v_2 + \cdots + Y_{1n}v_n &= i_1 \\
\vdots \qquad\qquad\qquad & \quad \vdots \\
Y_{n1}v_1 + Y_{n2}v_2 + \cdots + Y_{nn}v_n &= i_n
\end{aligned} \tag{11-44a}$$

To represent this set of simultaneous equations, the matrix form is particularly convenient. We define the node admittance matrix $\|Y_{jk}\|$

$$\|Y_{jk}\| = \begin{Vmatrix} Y_{11} & Y_{12} & \cdots & Y_{1n} \\ Y_{21} & Y_{22} & \cdots & Y_{2n} \\ \cdot & \cdot & \cdots & \cdot \\ Y_{n1} & Y_{n2} & \cdots & Y_{nn} \end{Vmatrix} \tag{11-44b}$$

so that the node equations are written as the matrix product

$$\|Y_{jk}\| \times \|v_j\| = \|i_j\| \tag{11-44c}$$

where $\|v_j\|$ is the node-voltage column matrix, and $\|i_j\|$ is the nodal supply-current column matrix. The solution for the node voltages is obtained by solution of these equations.

THE SUMMATION NOTATION FOR THE SIMULTANEOUS EQUATIONS OF NETWORK The set of equations given in Eq. (11-44*a*) can be written

$$\sum_{k=1}^{k=n} Y_{jk}v_k = i_j \qquad j = 1, 2, 3, \ldots, n \tag{11-44d}$$

where the subscript k is a subscript of summation, and subscript j is called the parametric subscript. For every value of j from 1 to n, Eq. (11-44*d*) will be the result of the application of Kirchhoff's laws to the node corresponding to the chosen j. Thus Eq. (11-44*d*) represents a set of $n = N - 1$ simultaneous equations, where N is the total number of the nodes in the network.

PROPERTIES OF THE NODE ADMITTANCE MATRIX The matrix (11-44*b*) has a principal diagonal which consists of the self-admittances at the nodes. We now note that $Y_{12}v_2$ is the current which flows out of node 1 (with node 1 and all nodes except 2 connected to the reference node) due to a voltage v_2 at node 2, and $Y_{21}v_1$ is the current out of node 2 due to a voltage v_1 at node 1 if all nodes except node 1 are connected to the reference node. Hence, for admittances composed of the elements R, L, and C, $Y_{12} = Y_{21}$. Similarly, $Y_{jk} = Y_{kj}$. (This property of the elements is called the "bilateral" property.) Consequently, the node admittance matrix is symmetrical with respect to the principal diagonal.

Example 11-5 Find the node equations for the circuit of Fig. 11-13.

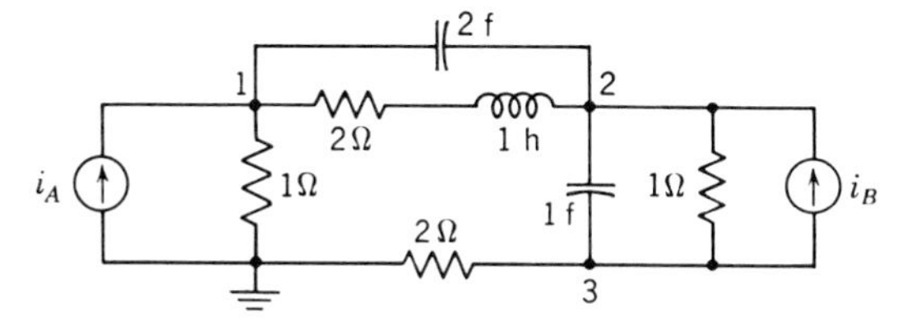

Fig. 11-13 Circuit used to illustrate formulation of node equations; see Example 11-5.

Solution The reference node is given, and the junctions are numbered. If we are not interested in the voltage across the inductance, the node between the 2-ohm resistance and the 1-henry inductance may be eliminated by using a p element for that branch. Since there are three other nodes (in addition to the reference node), the equations are

$$\begin{aligned} Y_{11}(p)v_1 + Y_{12}(p)v_2 + Y_{13}(p)v_3 &= i_1 \\ Y_{21}(p)v_1 + Y_{22}(p)v_2 + Y_{23}(p)v_3 &= i_2 \\ Y_{31}(p)v_1 + Y_{32}(p)v_2 + Y_{33}(p)v_3 &= i_3 \end{aligned}$$

where the nodal current supplies are $i_1(t) = i_A(t)$, $i_2(t) = i_B(t)$, $i_3(t) = -i_B(t)$. The node admittances are

$$\begin{aligned} Y_{11}(p) &= 1 + 2p + \frac{1}{2+p} = \frac{2p^2 + 5p + 3}{2+p} \\ Y_{22}(p) &= 1 + 3p + \frac{1}{2+p} = \frac{3p^2 + 7p + 3}{2+p} \\ Y_{33}(p) &= \left(\frac{3}{2} + p\right) = \frac{3 + 2p}{2} \\ Y_{12} = Y_{21} &= -\left(2p + \frac{1}{2+p}\right) = -\frac{2p^2 + 4p + 1}{2+p} \\ Y_{13} = Y_{31} &= 0 \\ Y_{23} = Y_{32} &= -(1+p) \end{aligned}$$

In matrix form the node equations read

$$\left\| \begin{matrix} \dfrac{2p^2+5p+3}{2+p} & -\dfrac{2p^2+4p+1}{2+p} & 0 \\ -\dfrac{2p^2+4p+1}{2+p} & \dfrac{3p^2+7p+3}{2+p} & -(1+p) \\ 0 & -(1+p) & \dfrac{3+2p}{2} \end{matrix} \right\| \times \left\| \begin{matrix} v_1 \\ v_2 \\ v_3 \end{matrix} \right\| = \left\| \begin{matrix} i_A \\ i_B \\ -i_B \end{matrix} \right\|$$

THE NUMBER OF NECESSARY AND SUFFICIENT NODE EQUATIONS In Sec. 11-6 dealing with network topology we noted that the number of the elements in the tree of a network is equal to $n = N - 1$, where N is the number of the nodes of the network. In this section we have seen that the node method uniquely describes the network by obtaining $n = N - 1$ nodal equations. This fact shows that the number of elements in the tree of a network determines the number of independent nodal equations required for unique description of the network. It is seen that this number is $n = N - 1$.

11-9 The method of mesh currents[1]

In this section we shall show how the choice of certain currents as the variables in a network can be used in connection with Kirchhoff's voltage law alone to obtain a set of equations which uniquely describe the network. In other words, we shall assign current variables in such a manner that Kirchhoff's current law at every junction in the network is satisfied by the choice of variables. For the present it will be convenient to assume that all

[1] As noted above, in this book the terms "mesh" and "independent loop" are used synonymously, although, strictly speaking, the term mesh applies to planar networks only.

the current sources in the network have internal immittances and that the network has been drawn with such sources converted to voltage sources. We shall further assume that there is no mutual inductance in the network and that the elements are represented as operational elements. (The case with mutual inductance is discussed in Chap. 15.)

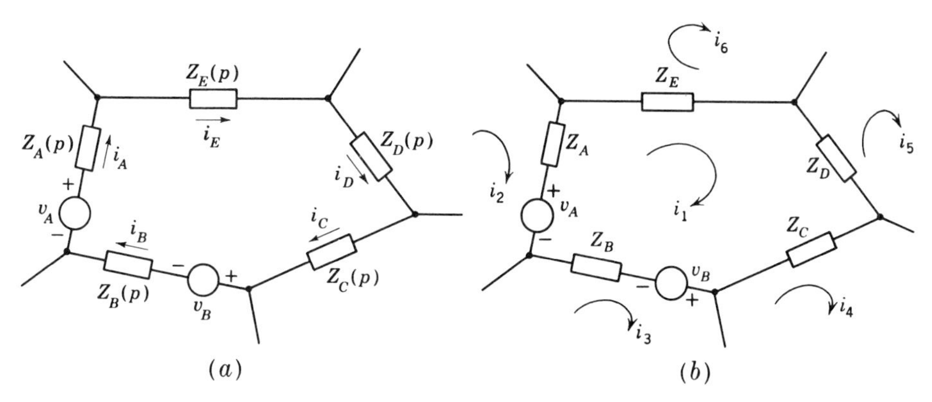

Fig. 11-14 A typical mesh in a network.

In Fig. 11-14*a* a typical mesh of the network is shown. The application of Kirchhoff's voltage law around this mesh (clockwise) gives the equation

$$-v_A + Z_A(p)i_A + Z_E(p)i_E + Z_D(p)i_D + Z_C(p)i_C + v_B + Z_B(p)i_B = 0$$

or

$$Z_A(p)i_A + Z_B(p)i_B + Z_C(p)i_C + Z_D(p)i_D + Z_E(p)i_E = v_A(t) - v_B(t) \quad (11\text{-}45)$$

We observe that Eq. (11-45) is written with the branch currents i_A, i_B, i_C, i_D, and i_E as the variables.

MESH CURRENTS We define six new variables which are dimensionally currents (Fig. 11-14*b*) such that

$$\begin{aligned} i_A &= i_1 - i_2 \\ i_B &= i_1 - i_3 \\ i_C &= i_1 - i_4 \\ i_D &= i_1 - i_5 \\ i_E &= i_1 - i_6 \end{aligned} \quad (11\text{-}46)$$

In other words, each branch current has been taken as the *difference* between i_1 and another of the new numbered current variables. Equation (11-45), using the new variables, reads

$$(Z_A + Z_B + Z_C + Z_D + Z_E)i_1 + (-Z_Ai_2) + (-Z_Bi_3) + (-Z_Ci_4) + (-Z_Di_5) + (-Z_Ei_6) = v_A - v_B \quad (11\text{-}46a)$$

If we now associate each of the currents i_1 to i_6 with a *mesh* as indicated in Fig. 11-14*b*, the relationships (11-46) are satisfied, and since each of the numbered currents *circulates* in a mesh, it enters *and* leaves each node so that Kirchhoff's current law at every node or junction is satisfied. Hence, if we choose these "new" variables, called *mesh* currents, as the circuit variables, we need only write the Kirchhoff-voltage-law equations for each mesh in the circuit. A typical mesh equation is Eq. (11-46*a*), the equation for mesh 1 in Fig. 11-14*b*. We note that all mesh currents are assigned clockwise; and once a choice has been made, we associate that circulation not only with the mesh current, but also with the geometric entity, namely, the mesh itself. The mesh equation (11-46*a*) has the form

$$Z_{11}(p)i_1 + Z_{12}(p)i_2 + Z_{13}(p)i_3 + Z_{14}(p)i_4 + Z_{15}(p)i_5 + Z_{16}(p)i_6 = v_1$$

where the Z_{ij} operators and v_1 are identified by comparison with Eq. (11-46*a*). We now introduce the following definitions.

MESH VOLTAGE SUPPLY The sum of voltage sources in the branches of a mesh, summed up in a direction *opposite* to the reference direction of the mesh current, is called the mesh voltage supply. The mesh voltage supply is denoted by letter v with a single subscript referring to its respective mesh. In Fig. 11-14*b* the mesh voltage supply of mesh number 1 is obtained by starting from a node and adding up voltage sources in a counterclockwise direction. This gives

$$v_1 = v_A - v_B$$

SELF-IMPEDANCE OF A MESH If all meshes except mesh k are *opened* and a mesh current i_k flows in mesh k, the sum of the voltages across the passive elements in mesh k summed up in the direction of the reference arrow of i_k is given by $Z_{kk}i_k$, where $Z_{kk}(p)$ is the self-impedance of mesh k. If the network contains no mutual inductance, the sum of the impedances of the elements in the mesh is called the self-impedance of that mesh. In mesh 1 (Fig. 11-14*b*)

$$Z_{11} = Z_A + Z_B + Z_C + Z_D + Z_E$$

MUTUAL IMPEDANCE OF TWO MESHES If all currents except i_k are zero and if we add the voltages across the passive elements in mesh m due to i_k, the addition being done in the direction of the reference arrow of i_k, then the sum of these voltages is $Z_{mk}i_k$, where Z_{mk} is called the mutual impedance between the two meshes k and m. From this definition it is seen that, if two meshes in a network have a common branch (which does not include a mutual inductance), the negative of impedance of the branch is the mutual impedance of the two meshes when the corresponding mesh currents have opposite directions in that branch. This is denoted by Z, with two subscripts corresponding to the two meshes sharing the common branch. Thus, in Fig. 11-14*b*, $Z_{12} = -Z_A$, $Z_{13} = -Z_B$, etc. For bilateral elements, $Z_{mn} = Z_{nm}$.

The mesh equations for an m-mesh network can now be written in the form

$$\begin{aligned} Z_{11}(p)i_1 + Z_{12}(p)i_2 + \cdots + Z_{1m}i_m &= v_1 \\ Z_{21}(p)i_1 + Z_{22}(p)i_2 + \cdots + Z_{2m}i_m &= v_2 \\ &\vdots \\ Z_{m1}(p)i_1 + Z_{m2}(p)i_2 + \cdots + Z_{mm}i_m &= v_m \end{aligned} \tag{11-47}$$

or in matrix form

$$\|Z_{jk}\| \times \|i_j\| = \|v_j\|$$

where $\|Z_{jk}\|$ is the mesh impedance matrix,

$$\|Z_{jk}\| = \begin{Vmatrix} Z_{11} & Z_{12} & \cdots & Z_{1m} \\ Z_{21} & Z_{22} & \cdots & Z_{2m} \\ \cdot & \cdot & \cdots & \cdot \\ Z_{m1} & Z_{m2} & \cdots & Z_{mm} \end{Vmatrix} \tag{11-48}$$

We observe that the mesh impedance matrix is symmetrical with respect to its principal diagonal if the elements in the network are bilateral.

Summation notation for mesh equations By analogy with the use of summation notation for node equations, the mesh equations of Eq. (11-47) can be written

$$\sum_{k=1}^{k=m} Z_{jk}i_k = v_j \qquad j = 1, 2, \ldots, m$$

where m is the number of meshes in the network, and the i_k and v_j are the mesh currents and mesh voltage supplies, respectively.

Example 11-6 Find the mesh equations for the network of Fig. 11-15.

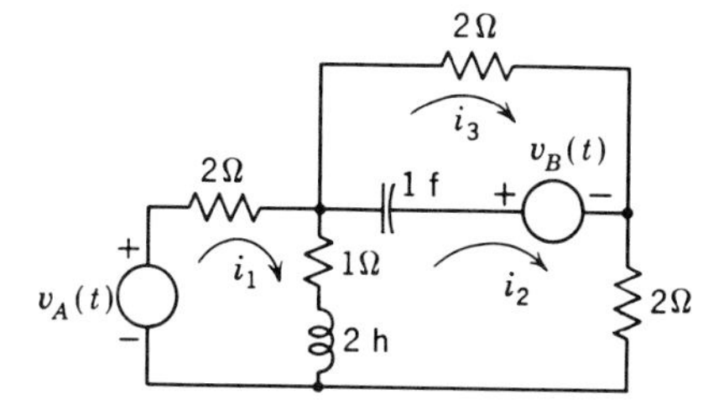

Fig. 11-15 Circuit used to illustrate formulation of mesh equations; see Example 11-6.

Solution For the three-mesh network shown, the mesh equations in symbolic form are

$$\begin{aligned} Z_{11}(p)i_1 + Z_{12}(p)i_2 + Z_{13}(p)i_3 &= v_1 \\ Z_{21}(p)i_1 + Z_{22}(p)i_2 + Z_{23}(p)i_3 &= v_2 \\ Z_{31}(p)i_1 + Z_{32}(p)i_2 + Z_{33}(p)i_3 &= v_3 \end{aligned}$$

where $\qquad v_1 = v_A \qquad v_2 = -v_B \qquad v_3 = v_B$

and $Z_{11} = 3 + 2p$, $Z_{22} = 3 + 2p + 1/p$, $Z_{33} = 2 + 1/p$, $Z_{12} = Z_{21} = -1 - 2p$, $Z_{23} = Z_{32} = -1/p$, $Z_{13} = Z_{31} = 0$.

THE NUMBER OF NECESSARY AND SUFFICIENT MESH EQUATIONS In Sec. 11-6 dealing with network topology, we noted that the number of the links in the graph of a network is identical with the number of independent loops of that graph. Since meshes are independent loops, their number is equal to the number of links given by $E - N + 1$, where E and N are the number of passive elements and nodes of the network, respectively. Thus the number of mesh equations required for the unique description of the network is $E - N + 1$.

11-10 Network equations using state variables

In Chaps. 5 and 6 we showed how to describe a network by state equations. Before we deal with the results of such formulation in the ensuing discussion, we briefly review in this section the salient features of such a formulation. We recall that the state variables of a network focus attention on the energy-storing elements in the network rather than on the network geometry. In this method the variables are the voltages across capacitances and the currents in inductances. The resulting simultaneous equations have the property that they contain a minimum number of derivatives since the state equations consist of *first-order differential equations equal in number to the independent energy-storing elements.* In contrast, the node or mesh equations may include a derivative (or integral) associated with the voltage-current relationship of an energy-storing element twice (or more often) if the element is common to two (or more) nodes or loops.

The simplest method for setting up state equations is to replace each inductance by a dependent source $i = \pm(1/pL)v$ and each capacitance by a dependent source $v = \pm(1/pC)i$ (where the choice of sign depends on the reference directions chosen for v and i) and then applying the principle of superposition. The method was explained in detail in Sec. 6-9. An example will serve to refresh the reader's memory.

Example 11-7 Formulate the state equations for the network shown in Fig. 11-16*a*. Reference directions have been assigned for the voltages and currents associated with each energy-storing element.

Solution Figure 11-16*b* shows the circuit of Fig. 11-16*a* with the energy-storing elements represented by dependent sources. In Fig. 11-16*c* to *g* the circuit is shown with each source save one, in turn, set to zero. We define the following state variables:

$$y_1 \equiv v_1 \qquad y_2 \equiv v_2 \qquad y_3 \equiv i_L$$

In the table below we show the relationship between state variables when each source is acting alone.

	Fig. 11-16*c*	Fig. 11-16*d*	Fig. 11-16*e*	Fig. 11-16*f*	Fig. 11-16*g*
$i_1 = \frac{1}{3}py_1 =$	$\frac{1}{2}v_A$	0	$-\frac{1}{2}y_1$	0	$-y_3$
$i_2 = \frac{1}{2}py_2 =$	0	i_B	0	0	y_3
$v_L = \frac{3}{2}py_3 =$	0	0	y_1	$-y_2$	$-5y_3$

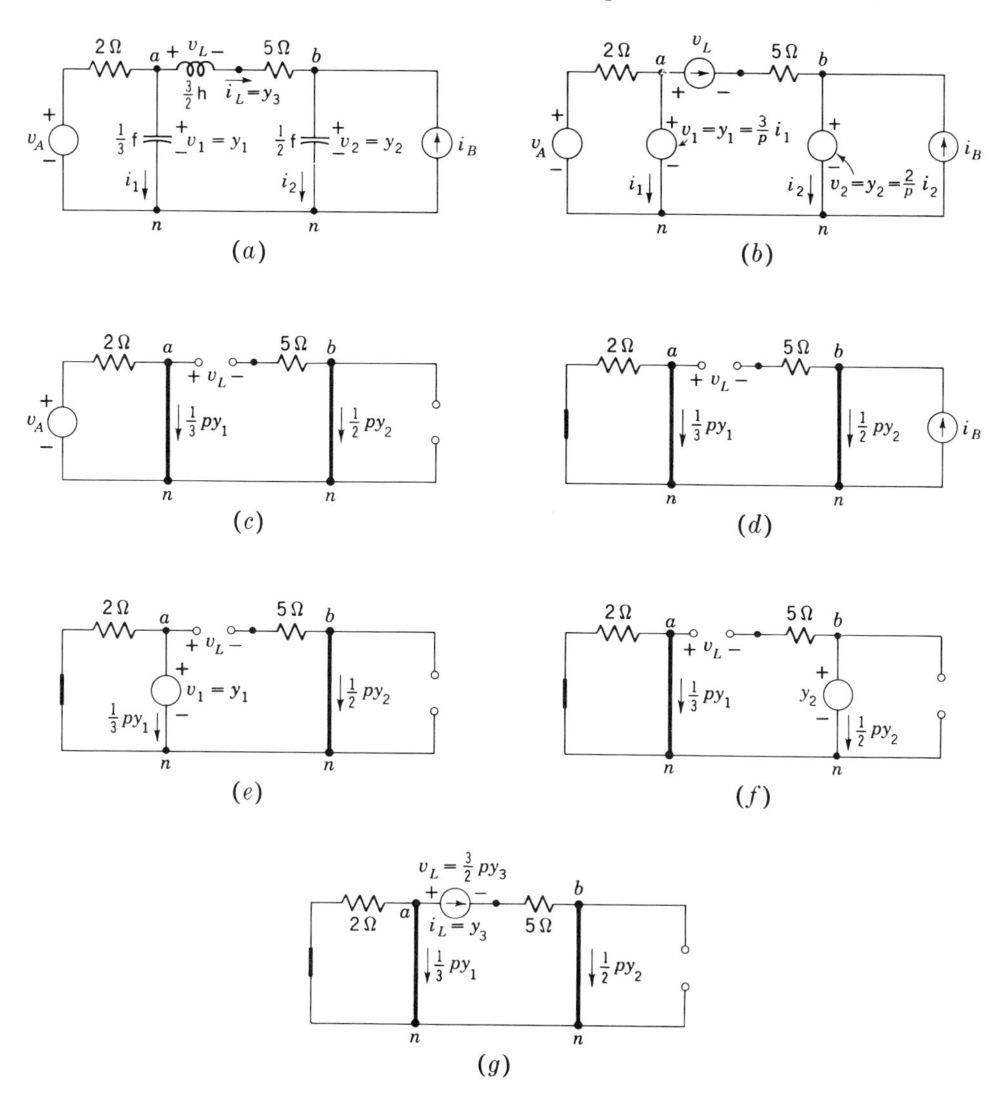

Fig. 11-16 (*a*) *Network used to illustrate formulation of state equations.* (*b*) *Circuit of* (*a*) *with dependent sources substituted for the energy-storing elements.* (*c*) *Circuit of* (*b*) *with all sources except* v_A *set to zero.* (*d*) *Circuit of* (*b*) *with all sources except* i_B *set to zero.* (*e*) *Circuit of* (*b*) *with all sources except* v_1 *set to zero.* (*f*) *Circuit of* (*b*) *with all sources except* v_2 *set to zero.* (*g*) *Circuit of* (*b*) *with all sources except* i_L *set to zero.*

Using superposition, we obtain the equations

$$\begin{aligned} \tfrac{1}{3}py_1 &= -\tfrac{1}{2}y_1 & & - y_3 + \tfrac{1}{2}v_A \\ \tfrac{1}{2}py_2 &= & & y_3 + \ i_B \\ \tfrac{3}{2}py_3 &= \quad y_1 & - y_2 & - 5y_3 \end{aligned}$$

In standard form these equations are

$$\begin{aligned} \frac{dy_1}{dt} &= py_1 = -\tfrac{3}{2}y_1 & & - 3y_3 + \tfrac{3}{2}v_A \\ \frac{dy_2}{dt} &= py_2 = & & 2y_3 + 2i_B \\ \frac{dy_3}{dt} &= py_3 = \tfrac{2}{3}y_1 - \tfrac{2}{3}y_2 & & - \tfrac{10}{3}y_3 \end{aligned} \tag{11-49}$$

These equations have the desired standard form

$$\begin{aligned} py_1 &= a_{11}y_1 + a_{12}y_1 + a_{13}y_2 + g_1 \\ py_2 &= a_{21}y_1 + a_{22}y_2 + a_{23}y_3 + g_2 \\ py_3 &= a_{31}y_1 + a_{32}y_2 + a_{33}y_3 + g_3 \end{aligned}$$

with $a_{11} = -\frac{3}{2}$, $a_{12} = 0$, $a_{13} = -3$, $g_1 = \frac{3}{2}v_A$, $a_{21} = 0$, $a_{22} = 0$, $a_{23} = 2$, $g_2 = 2i_B$, $a_{31} = \frac{2}{3}$, $a_{32} = -\frac{2}{3}$, $a_{33} = -\frac{10}{3}$, and $g_3 = 0$.

MATRIX FORM OF STATE EQUATIONS The state equations have the form

$$\|py_i\| = \|A\| \times \|y_i\| + \|g\|$$

where the $\|A\|$ matrix is the square matrix, termed *state matrix*,

$$\|A\| = \begin{Vmatrix} a_{11} & a_{12} & \cdots & a_{1n} \\ a_{21} & a_{22} & \cdots & a_{2n} \\ \cdot & \cdot & \cdots & \cdot \\ a_{n1} & a_{n2} & \cdots & a_{nn} \end{Vmatrix}$$

in which all the elements a_{ik} are algebraic expressions; that is, they include no differential operators.

Another form of the state equations is obtained if we combine the column matrix

$$\|py_i\| = \begin{Vmatrix} py_1 \\ py_2 \\ py_3 \\ \cdot \\ \cdot \\ \cdot \\ py_n \end{Vmatrix}$$

with the product $\|A\| \times \|y_i\|$. To do this we use the unit matrix, defined as

$$\|1\| = \begin{Vmatrix} 1 & 0 & 0 & \cdots & 0 \\ 0 & 1 & 0 & \cdots & 0 \\ 0 & 0 & 1 & \cdots & 0 \\ \cdot & \cdot & \cdot & \cdots & \cdot \\ 0 & 0 & 0 & \cdots & 1 \end{Vmatrix}$$

and note that

$$\|py_i\| = p\|1\| \times \|y_i\|$$

Hence the state equations have the form

$$[p\|1\| - \|A\|] \times \|y_i\| = \|g_i\| \tag{11-50}$$

where $[p\|1\| - \|A\|]$ is the matrix

$$[p\|1\| - \|A\|] \equiv \|B\| = \begin{Vmatrix} p - a_{11} & -a_{12} & -a_{13} & \cdots & -a_{1n} \\ -a_{21} & p - a_{22} & -a_{23} & \cdots & -a_{2n} \\ -a_{31} & -a_{32} & p - a_{33} & \cdots & -a_{3n} \\ \cdot & \cdot & \cdot & \cdots & \cdot \\ -a_{n1} & -a_{n2} & -a_{n3} & \cdots & p - a_{nn} \end{Vmatrix}$$

We note that the $\|B\|$ matrix has only one differential operator per row, always on the principal diagonal.

The g functions in the state equations are all linear combinations of source functions, and are termed *generalized source functions.*

11-11 Equilibrium equation for a single circuit variable

In this section we show how simultaneous equations are manipulated to obtain the equilibrium equation for a single variable.

Suppose that we have an n-node network which is specified through the node equations

$$\|Y(p)\| \times \|v\| = \|i\|$$

or in detail,

$$\begin{aligned} Y_{11}(p)v_1 + Y_{12}(p)v_2 + \cdots + Y_{1n}(p)v_n &= i_1(t) \\ Y_{21}(p)v_1 + Y_{22}(p)v_2 + \cdots + Y_{2n}(p)v_n &= i_2(t) \\ \cdot \qquad\qquad\qquad\qquad & \quad \cdot \\ \cdot \qquad\qquad\qquad\qquad & \quad \cdot \\ \cdot \qquad\qquad\qquad\qquad & \quad \cdot \\ Y_{n1}(p)v_1 + Y_{n2}(p)v_2 + \cdots + Y_{nn}(p)v_n &= i_n(t) \end{aligned}$$

We now make use of the quasi-algebraic character of the operators $Y_{jk}(p)$ and solve for a desired variable by Cramer's rule. If, for example, the equilibrium equation for $v_1(t)$ is desired, then

$$v_1 = \frac{\begin{vmatrix} i_1 & Y_{12} & \cdots & Y_{1n} \\ i_2 & Y_{22} & \cdots & Y_{2n} \\ \cdot & \cdot & \cdots & \cdot \\ i_n & Y_{n2} & \cdots & Y_{nn} \end{vmatrix}}{\begin{vmatrix} Y_{11} & Y_{12} & \cdots & Y_{1n} \\ Y_{21} & Y_{22} & \cdots & Y_{2n} \\ \cdot & \cdot & \cdots & \cdot \\ Y_{n1} & Y_{n2} & \cdots & Y_{nn} \end{vmatrix}} \tag{11-51}$$

Now the determinant in the denominator of Eq. (11-51) is the determinant of the node admittance matrix, and this determinant is denoted by the symbol $D_y(p)$. Applying Laplace's expansion of determinants to the numerator of Eq. (11-51) and expanding about the first column, we obtain

$$v_1 = \frac{\begin{vmatrix} Y_{22} & \cdots & Y_{2n} \\ \cdot & \cdots & \cdot \\ Y_{n2} & \cdots & Y_{nn} \end{vmatrix}}{D_y(p)} i_1 + (-1) \frac{\begin{vmatrix} Y_{12} & \cdots & Y_{1n} \\ \cdot & \cdots & \cdot \\ Y_{n2} & \cdots & Y_{nn} \end{vmatrix}}{D_y(p)} i_2 + \cdots + (-1)^{n+1} \frac{\begin{vmatrix} Y_{12} & \cdots & Y_{1n} \\ \cdot & \cdots & \cdot \\ Y_{(n-1)2} & \cdots & Y_{(n-1)n} \end{vmatrix}}{D_y(p)} i_n$$

or, using the definition of cofactors and letting $(F_y)_{k1}$ be the cofactor of row k, column 1, of D_y,

$$v_1 = \frac{1}{D_y(p)} [(F_y)_{11}i_1 + (F_y)_{21}i_2 + (F_y)_{31}i_3 + \cdots + (F_y)_{n1}i_n]$$

This expression can be abbreviated to read

$$v_1 = \frac{1}{D_y(p)} \sum_{k=1}^{n} (F_y)_{k1}i_k \tag{11-52a}$$

For any node voltage v_a,

$$v_a = \frac{1}{D_y(p)} \sum_{k=1}^{n} (F_y)_{ka}i_k \tag{11-52b}$$

Equation (11-52*b*) is interpreted as

$$D_y(p)v_a = \sum_{k=1}^{n} (F_y)_{ka}i_k \tag{11-52c}$$

Equation (11-52*c*) is the operational form of the equilibrium equation which relates a node voltage v_a to the source functions i_k.

If we start with the mesh equations of a circuit, the result for the relation between a mesh current i_a and the source functions v_k is

$$D_z(p)i_a = \sum_{k=1}^{m} (F_z)_{ka}v_k \tag{11-53}$$

where $D_z(p)$ is the determinant of the mesh-impedance matrix and $(F_z)_{ka}$ is its cofactor of row k, column a. Proof of Eq. (11-53) is left as an exercise for the reader.

In the case of state equations, we begin with the form

$$\begin{aligned}(p - a_{11})y_1 + \quad (-a_{12})y_2 + \cdots + (-a_{1n})y_n &= g_1(t)\\ (-a_{21})y_1 + (p - a_{22})y_2 + \cdots + (-a_{2n})y_n &= g_2(t)\\ &\vdots\\ (-a_{n1})y_1 + \quad (-a_{n2})y_2 + \cdots + (-a_{nn})y_n &= g_n(t)\end{aligned} \tag{11-54}$$

where we have defined the following matrices (Sec. 11-10),

$$\|A\| = \begin{Vmatrix} a_{11} & \cdots & a_{1n} \\ a_{21} & \cdots & a_{2n} \\ \cdot & \cdots & \cdot \\ a_{n1} & \cdots & a_{nn} \end{Vmatrix} \tag{11-55}$$

and

$$\|B\| = p\|1\| - \|A\| \tag{11-56}$$

that is,

$$\|B\| = \begin{Vmatrix} p - a_{11} & -a_{12} & \cdots & -a_{1n} \\ -a_{21} & p - a_{22} & \cdots & -a_{2n} \\ \cdot & \cdot & \cdots & \cdot \\ -a_{n1} & -a_{n2} & \cdots & p - a_{nn} \end{Vmatrix} \tag{11-56a}$$

so that

$$\|B\| \times \|y\| = \|g\| \tag{11-57}$$

Hence the state variable y_a is given by

$$y_a = \frac{1}{D_B(p)} \sum_{k=1}^{n} (F_B)_{ka} g_k \tag{11-57a}$$

where $D_B(p)$ is the determinant of the B matrix and $(F_B)_{ka}$ is its cofactor of row k, column a.

11-12 Network functions

Study of Eq. (11-52*c*) shows that the response (node voltage) v_a is related to a nodal supply current i_k through the network function (*transfer impedance*)

$$H(p) = \frac{(F_y)_{ka}}{D_y(p)} \tag{11-58}$$

Similarly, from Eq. (11-53), we observe that the response (mesh current) i_k is related to the mesh voltage supply v_a through the network function (*transfer admittance*)

$$H(p) = \frac{(F_z)_{ka}}{D_z(p)} \tag{11-59}$$

and from Eq. (11-57) we note that the state variable y_k is related to the generalized source function g_a through the network function

$$H(p) = \frac{(F_B)_{ka}}{D_B(p)} \tag{11-60}$$

Hence, no matter which formulation is used, the relationship between a response variable $r_a(t)$ and the source functions has the form

$$r_a(t) = H_{1a}(p)s_1(t) + H_{2a}(p)s_2(t) + \cdots + H_{sa}(p)s_s(t) \tag{11-61}$$

where each network function is the ratio of a cofactor to a determinant, that is,

$$H_{ka}(p) = \frac{F_{ka}}{D}$$

and $D = D_z$ when $r_a = i_a$ (mesh current) and $s_1, \ldots, s_s$ are $v_1, \ldots, v_m$ (the mesh supply voltages)
$D = D_y$ when $r_a = v_a$ (node voltage) and $s_1, \ldots, s_n$ are $i_1, \ldots, i_n$ (the nodal supply currents)
$D = D_B$ when $r_a = y_a$ (state variable) and $s_1, \ldots, s_a$ are $g_1, \ldots, g_a$ (the generalized source functions)

It follows that the equilibrium equation for a response $r_a(t)$ may be written in the form

$$D(p)r_a(t) = F_{1a}(p)s_1(t) + F_{2a}(p)s_2(t) + \cdots + F_{sa}(p)s_s(t) \tag{11-62}$$

STRUCTURE OF EQUILIBRIUM EQUATIONS, STATE VARIABLES If state equations are used to arrive at the equilibrium equation (11-62), the determinant $D(p)$ is the determinant of the B matrix.

$$\|B\| = \begin{Vmatrix} p - a_{11} & -a_{12} & \cdots & -a_{1n} \\ -a_{21} & p - a_{22} & \cdots & -a_{2n} \\ \cdot & \cdot & \cdots & \cdot \\ -a_{n1} & -a_{n2} & \cdots & p - a_{nn} \end{Vmatrix} \tag{11-63}$$

From the structure of this matrix, Eq. (11-63), we conclude that $D(p)$ is a polynomial in p of degree n, where n is equal to the number of state equations that describe the system. Similarly, we observe that any cofactor of D, $F_{ka}(p)$ is a polynomial of degree $n - 1$. Thus the equilibrium equation for a state variable y_a has the form

$$\begin{aligned}(a_0 + a_1p + a_2p^2 + \cdots + a_np^n)y_a &= (b_{11} + \cdots + b_{1(n-1)}p^{n-1})g_1 \\ &+ (b_{21} + \cdots + b_{2(n-1)}p^{n-1})g_2 + (b_{n1} + \cdots + b_{n(n-1)}p^{n-1})g_n\end{aligned}$$

where a_k and b_{ik} are constants whose values depend on the values of the passive elements. In words, the equilibrium equation shows that a linear combination of a response y_a with certain of its derivations is balanced by a linear combination of the source functions with certain of their derivatives.

Example 11-8 Obtain the equilibrium equations for all the state variables in the circuit of Fig. 11-17. The state variables y_1, y_2, and y_3 are defined as i_1, i_2, and v, respectively.

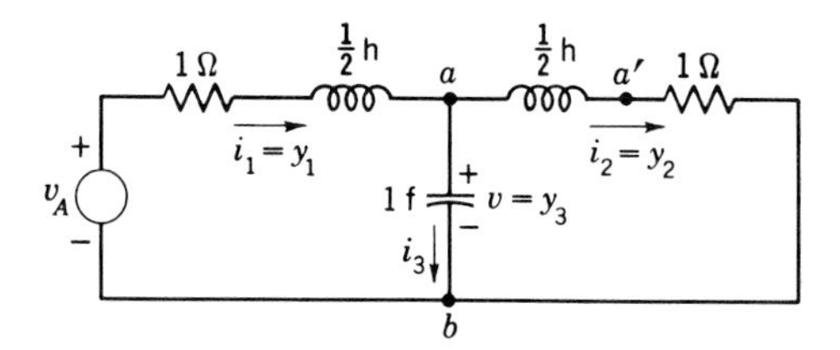

***Fig. 11-17** Circuit for Example 11-8.*

Solution It is left as an exercise for the reader to show that the state equations have the form

$$\left\| \begin{matrix} p+2 & 0 & 2 \\ 0 & p+2 & -2 \\ -1 & 1 & p \end{matrix} \right\| \times \left\| \begin{matrix} y_1 \\ y_2 \\ y_3 \end{matrix} \right\| = \left\| \begin{matrix} 2v_A \\ 0 \\ 0 \end{matrix} \right\| \tag{11-64}$$

The determinant of the B matrix is given by

$$D_B(p) = \begin{vmatrix} p+2 & 0 & +2 \\ 0 & p+2 & -2 \\ -1 & 1 & p \end{vmatrix}$$

which results in the polynomial

$$D_B(p) = p^3 + 4p^2 + 8p + 8$$

The response y_1 satisfies the equilibrium equation obtained by Cramer's rule.

$$y_1 = \frac{\begin{vmatrix} 2v_A & 0 & 2 \\ 0 & p+2 & -2 \\ 0 & 1 & p \end{vmatrix}}{D_B(p)}$$

or

$$(p^3 + 4p^2 + 8p + 8)y_1 = (2p^2 + 4p + 4)v_A \tag{11-65}$$

Similarly, we obtain for y_2 and y_3, respectively,

$$(p^3 + 4p^2 + 8p + 8)y_2 = \begin{vmatrix} p+2 & 2v_A & 2 \\ 0 & 0 & -2 \\ -1 & 0 & p \end{vmatrix}$$

or

$$(p^3 + 4p^2 + 8p + 8)y_2 = 4v_A \tag{11-66}$$

and

$$(p^3 + 4p^2 + 8p + 8)y_3 = \begin{vmatrix} p+2 & 0 & 2v_A \\ 0 & p+2 & 0 \\ -1 & 1 & 0 \end{vmatrix}$$

or

$$(p^3 + 4p^2 + 8p + 8)y_3 = (2p + 4)v_A$$

NETWORK FUNCTIONS From the preceding discussion we observe that a state variable of a network is related to a source function through an operational network function which always has the form of a ratio of polynomials in p.

We shall now show that this form is also arrived at by use of mesh, node, or branch equations. We begin by assuming that the node equations of a network have been set up and that these *node equations were written using primitive elements* when all the nodes are *not* junctions. Several typical nodes of a network are shown in Fig. 11-18. In this example $Y_{11} = G_1 + G_2 + 1/R + pC$, $Y_{22} = 1/R + 1/pL$, $Y_{13} = -G_2 - pC$, etc. In

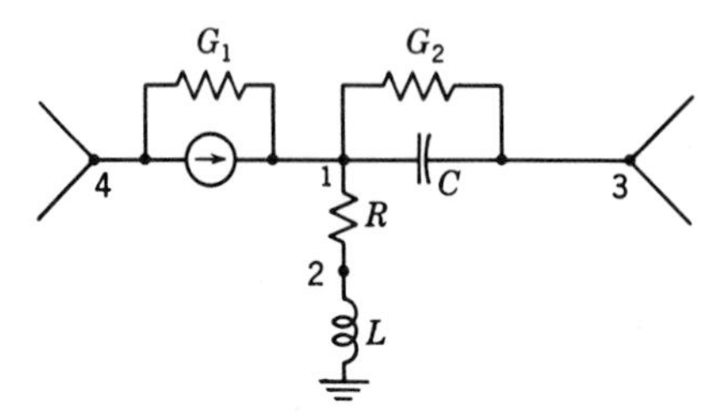

Fig. 11-18 Portion of a network showing four nodes and the reference node.

general, we conclude that, if primitive elements are used, each node admittance has the general form

$$Y_{jk} = G_{jk} + pC_{jk} + \frac{1}{pL_{jk}}$$

or

$$Y_{jk}(p) = \frac{p^2C_{jk} + pG_{jk} + 1/L_{jk}}{p} \tag{11-67}$$

where C_{jk}, G_{jk}, L_{jk} are negative if $j \neq k$ and positive if $j = k$. The determinant $D_y(p)$ is obtained by adding certain products of node admittances in accordance with the rules of determinant manipulation. It therefore follows that $D_y(p)$ has the form

$$D_y(p) = \frac{b_0 + b_1p + b_2p^2 + \cdots + b_kp^k}{p^r} \tag{11-68}$$

i.e., it is the ratio of a polynomial in p to p raised to a power r, where the value of k and r will depend on the "anatomy" of the network. Now any cofactor of the determinant $D_y(p)$ will also contain elements whose form is as shown in Eq. (11-67). Therefore such a cofactor will also be the ratio of a polynomial in p to a power of p,

$$F_y = \frac{a_0 + a_1p + a_2p^2 + \cdots + a_lp^l}{p^q} \tag{11-68a}$$

where the value of l and q will again depend on the structure of the network. From Eqs. (11-68) and (11-68*a*) it is seen that the general form of a network function is as expected,

$$H(p) = \frac{F_y}{D_y} = \frac{\sum_{n=0}^{N} a_np^n}{\sum_{m=0}^{M} b_mp^m} \tag{11-69}$$

where N, M, a_n, and b_m are determined by the structure of the network. A little thought will show that the above derivation applies to mesh equations by application of duality. If it is recalled how mesh and node equations are deduced from branch relations, the structure of a network function as given by Eq. (11-69) is seen to apply to branch formulation as well.

If p elements, instead of primitive elements, are used to set up the node, mesh, or branch equations, we note that each term in the determinant of the node or mesh matrix may itself be a ratio of polynomials in p. A little thought will show that, after all fractions are cleared, the ratio of a cofactor to its determinant will be a ratio of polynomials in p.

11-13 Free modes of response; coupling

We recall from Eq. (11-61) that a response of a network $r(t)$ is related to a source function $s(t)$ by the operational network function $H(p)$ which is a ratio of polynomials in p; that is,

$$r(t) = [H(p)]s(t)$$

and
$$H(p) = \frac{N(p)}{D(p)} = \frac{b_0 + b_1p + \cdots + b_np^m}{a_0 + a_1p + \cdots + a_np^n} \tag{11-70}$$

If $N(p)$ and $D(p)$ have common factors, they can be canceled (the significance of such common factors is discussed further below). Let us assume that such common factors have been canceled, so that the equilibrium equation is

$$[D(p)]r(t) = [N(p)]s(t)$$

When the source function is set to zero $[s(t) \equiv 0]$, one obtains the source-free equilibrium equation

$$D(p)r_f(t) = 0$$

which has a solution of the form $r_f = Ke^{st}$, so that

$$D(p)Ke^{st} = 0$$

We recall that

$$\begin{aligned} D(p)Ke^{st} &= (a_0 + a_1p + \cdots + a_np^n)Ke^{st} \\ &= (a_0 + a_1s + \cdots + a_ns^n)Ke^{st} \\ &= D(s)Ke^{st} \end{aligned}$$

Hence the value or values of s which allow Ke^{st} to be a solution of the source-free equilibrium equation can be found by solution of the *characteristic equation*

$$D(s) = 0 \tag{11-71}$$

or
$$a_0 + a_1s + \cdots + a_ns^n = 0$$

Equation (11-71) has n roots (*characteristic roots*) $s_1, s_2, \ldots, s_n$. If no two of these roots are equal, the source-free response component can be written as the sum of n linearly independent modes as

$$r_f(t) = K_1e^{s_1t} + K_2e^{s_2t} + \cdots + K_ne^{s_nt} \tag{11-72}$$

If the characteristic equation has multiple roots the solution is modified accordingly as explained in Chaps. 7 and 8; thus, for example, if

$$D(s) = (s + 2)^2(s + 3)^3(s + 4)$$

then
$$r_f = (K_1 + K_2t)e^{-2t} + (K_3 + K_4t + K_5t^2)e^{-3t} + K_6e^{-4t}$$

If any of the roots are complex, its conjugate is also a root, and the two exponentials can be combined into real form. For example, the characteristic equation

$$(s + 2)(s + 3 - j4)(s + 3 + j4) = 0$$

corresponds to the source-free response component

$$r_f(t) = \mathbf{K}_1e^{-2t} + \text{Re}\,(\mathbf{K}e^{(-3+j4)t})$$

COUPLING BETWEEN MODES AND A SOURCE When a single external ideal source is applied to an initially deenergized network[1] and different responses are calculated, it is possible that some or all free modes which appear in the free component of the response for one response function do not appear in the corresponding component in another response function. Thus, for instance, if y_1, y_2, and y_3 are three different responses of an initially deenergized network excited by a source, it may happen that

$$y_1 = (y_1)_s + K_1 e^{s_1 t} + K_2 e^{s_2 t}$$

but

$$y_2 = (y_2)_s + K_3 e^{s_1 t}$$

and

$$y_3 = (y_3)_s$$

That is, the mode $e^{s_2 t}$ does not appear in the complete response for y_2, although it appears in y_1; y_3 does not contain any free modes. Another way of saying this is to say that the mode $K_4 e^{s_2 t}$ in y_2 has the mode amplitude $K_4 = 0$. When this occurs one says that the input is *not coupled* to the mode $e^{s_2 t}$ in the output y_2.

In this connection it is pointed out that the output of a circuit may not be a state variable, but may be a linear combination of state variables; e.g., if the output is $v_2(t)$, it may be that

$$v_2(t) = A_1 y_1(t) + A_2 y_2(t)$$

If, for zero-initial-energy storage, the output of a network does not include a free mode (which is present in the general solution of the source-free network), that mode is said to be uncoupled from the input.

Example 11-9 Find the form of the free response components for the state variables y_1, y_2, and y_3 in Example 11-8.

Solution The equilibrium for the state variables are given in Example 11-8 as

$$(p^3 + 4p^2 + 8p + 8)y_1 = (2p^2 + 4p + 4)v_A$$
$$(p^3 + 4p^2 + 8p + 8)y_2 = 4v_A$$
$$(p^3 + 4p^2 + 8p + 8)y_3 = (2p + 4)v_A$$

The polynomial $p^3 + 4p^2 + 8p + 8$ is factored:

$$(p^3 + 4p^2 + 8p + 8) = (p + 2)(p + 1 - j\sqrt{3})(p + 1 + j\sqrt{3})$$

To obtain the free response component for y_1, we factor also

$$2p^2 + 4p + 4 = 2(p + 1 - j)(p + 1 + j)$$

Hence

$$y_1 = \frac{N(p)}{D(p)} v_A$$

where N and D have no common factors. The characteristic equation is

$$(s + 2)(s + 1 - j\sqrt{3})(s + 1 + j\sqrt{3}) = 0$$

so that

$$y_{1f} = K_1 e^{-2t} + \text{Re}\,(\mathbf{K} e^{(-1+j\sqrt{3})t}) \tag{11-73a}$$

[1] In the terminology of state theory, a network that stores zero energy at $t = 0^-$ is said to be in the *zero state*.

For y_2, $N(p) = 4$, so that it has no common factor with D, and y_{2f} has the same form as y_{1f}.

In the case of the state variable y_3, we have the relationship

$$y_3 = \frac{2(p+2)}{(p+2)(p+1-j\sqrt{3})(p+1+j\sqrt{3})} v_A(t)$$

and canceling the common factor $p + 2$,

$$y_3 = \frac{2}{(p+1-j\sqrt{3})(p+1+j\sqrt{3})} v_A(t)$$

Thus, for y_3, the source-free response component is seen to be

$$y_{3f} = \text{Re}\,(\mathbf{K}e^{(-1+j\sqrt{3})t}) \tag{11-73b}$$

Example 11-9 clearly shows that all the responses of a network do not include the same source-free response component when a single source excites an initially deenergized network. Specifically, the mode e^{-2t} is not coupled to the source when the output is $y_3(t)$ and the input is $v_A(t)$.

In the discussion preceding Example 11-9, it is stated that a mode which occurs in one response function of a network may not be present in another response function when the initial-energy storage is zero. We shall now show why the condition of zero-initial-energy storage is necessary. First of all, we observe that the circuit used in Fig. 11-17 includes three independent energy-storing elements—three (independent) states. Hence one can, in general, specify three arbitrary initial conditions. It is therefore clear that, with *arbitrary* initial-energy-storage conditions, one cannot find the response $y_3(t)$ using as free component the solution (11-73b) since this component includes only two constants, Re **K** and Im **K**. We now recall that initial-energy storage can be represented by initial-condition generators. Such sources can be placed in series or in parallel with the energy-storing elements. Now, to find the complete response of a network, one may use superposition, that is, find the complete response due to each source as is explicitly shown by the Laplace transform,[1] Chap. 8. Since each initial-condition generator is in a different branch (or between different nodes) of the network, the response produced by it is related to it by a different network function. Review of Example 11-9 shows why the mode e^{-2t} is not coupled to the input for the output y_3 because the network function relating y_3 to v_A has the factor $p + 2$ in the numerator, canceling the corresponding factor in the denominator. (In other words, the pole at $s = -2$ is canceled by a zero at $s = -2$.) If now a source is placed in a different branch, the numerator of the new network function may not have a common factor with the denominator, and all possible modes may appear at the output. Hence, in a selected output, a mode may be coupled to an initial-condition generator, and thus appear in that output, although the mode in question is not coupled to the external source.

[1] In the terminology of state theory the response when the initial-energy storage is zero is termed *zero-state response*.

POSSIBLE FREE MODES Since all possible free modes of a network do not necessarily appear in every response for every input, we now ask the following question: Given a network, how can one determine *all possible* free modes?

This question is answered by recalling how the state equations are set up: Each energy-storing element is, in turn, considered an ideal source. Hence the number of state equations corresponds exactly to the number of *independent* energy-storing elements. Thus, if in state equations

$$\|B\| \times \|y\| = \|g\|$$

all sources are set to zero, the resulting equations

$$\|B\| \times \|y_f\| = 0$$

all have the form

$$D_B(p)y_f = 0$$

where $D_B(p)$ is the determinant of the B matrix. This determinant is called the *characteristic determinant of the state matrix.* When it is reduced to the polynomial $D_B(p)$, this polynomial is termed the characteristic polynomial of the network. The characteristic equation

$$D_B(s) = 0$$

yields *all* characteristic roots and corresponding source-free modes of the network that can occur.[1]

MESH AND NODE EQUATIONS If a network is described by a set of node equations, i.e.,

$$\|Y\| \times \|v\| = \|i\|$$

then, when the source currents are set to zero, the characteristic determinant associated with the node admittance matrix is

$$D_y(p) = \det \|Y\|$$

The equation $D_y(s) = 0$ will then yield all modes of the node voltage which can be excited by current sources connected between the nodes. If primitive elements are used, this characteristic equation will include all possible modes because each state variable will be a linear function of node voltages. If p elements are used, it is possible that common factors have been canceled in setting up the p elements; hence all modes may not be included in $D_y(p)$.

The same statement applies to mesh equations: All possible network modes are obtained only when primitive elements are used.

Example 11-10 Find the free modes in the circuit of Fig. 11-19 from the node equations using (*a*) primitive elements, (*b*) p elements.

Solution (*a*) Using primitive elements, the node admittance matrix is

$$\|Y\| = \left\| \begin{matrix} 3 & -1 & -1 \\ -1 & 1 + \dfrac{1}{p} & 0 \\ -1 & 0 & 1 + \dfrac{1}{p} \end{matrix} \right\|$$

[1] In this statement it is assumed that additional sources will not be inserted with entries that change the structure of the source-free network; see Sec. 11-4.

and its determinant is

$$D(p) = \frac{(p+1)(p+3)}{p^2}$$

and $$D(s) = 0 \qquad \text{for } s = -1,\ s = -3$$

Hence all possible free modes are K_1e^{-t} and K_2e^{-3t}.

(b) Using p elements, we have the single term

$$\|Y_{ik}\| = Y(p) = 1 + \frac{2}{p+1} = \frac{p+3}{p+1}$$

and $D(s) = 0$ gives $s = -3$, and the characteristic root $s = -1$ is not found by this analysis.

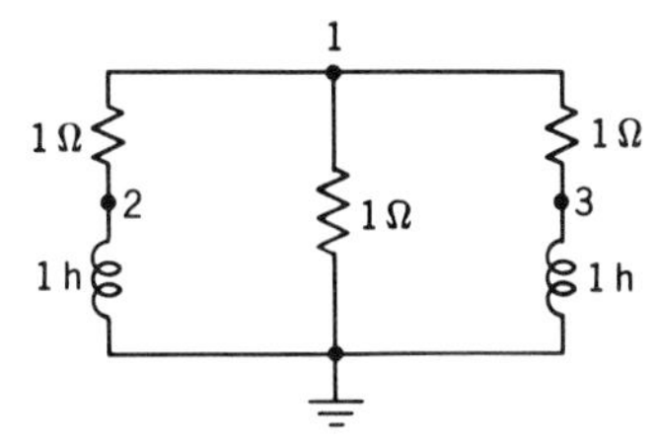

Fig. 11-19 Circuit for Example 11-10.

11-14 Solution of examples. General discussion

Although, generally speaking, the determination of the complete response of a network involves finding the functions which represent *all* its variables, it frequently happens that one need not solve for all voltages and currents in a network, but only for selected variables. Thus, for example, in a five-node network, only the node voltages $v_1(t)$ and $v_5(t)$ may be of interest. The procedure for obtaining the complete response function desired can be divided into the following steps:

1 Formulate the transform or operational network equations. If Laplace transform is used, initial-energy storage will be accounted for by sources.
2 Manipulate the network equations (for example, by Cramer's rule) to obtain an equilibrium equation that relates the desired response function to the sources.
3 Formulate the complete response. If Laplace transform is used, the complete response is obtained by solving the equation obtained in step 2 for the transform of the desired response and inverting, using, for example, the partial-fractions expansion. If the time-domain method is to be used, the free and forced response components are found, and the complete response is formulated with arbitrary constants. These constants are then evaluated by use of continuity and/or derived initial conditions.

11-15 The component of the response due to the source

Once the equilibrium equations for the desired response have been formulated the component of the response due to a source is found by the methods already discussed. If Laplace transform has been used, those terms in the partial-fractions expansion that represent source poles are evaluated as explained in Chap. 8. If sinusoidal steady-state response is desired, phasors and complex immittances can be used from the outset, as explained in connection with series-parallel circuits in Chap. 9. If d-c steady-state response only is desired, the d-c equivalent circuit can be analyzed (by simultaneous-equation methods if desired) directly, and the algebra involved in formulation $H(p)$ or $H(s)$ can be avoided. In the d-c case, the mesh impedances or node admittances are not functions of p or s, and the terms "mesh resistance" and "node conductance" are used.

Example 11-11 Use the result of Example 11-8 to find the steady-state solution for v_{ab} if $v_A(t) = 15 \cos (2t - 30°)$.

Solution The state equations are given in Example 11-8; the state variable y_3 corresponds to v_{ab}. From Eq. (11-66),

$$y_3 = \frac{2p + 4}{p^3 + 4p^2 + 8p + 8} v_A$$

For sinusoidal steady-state calculation we *replace* p with $j\omega = j2$ and v_A with $\mathbf{V}_{A_m} = 15\underline{/-30°}$ and define the amplitude phasor $\mathbf{V}_m$ so that $v_{ab}(t) = \text{Re}\,(\mathbf{V}_m e^{j2t})$. Thus

$$\mathbf{V}_m = \frac{(2)(j2) + 4}{(j2)^3 + 4(j2)^2 + 8(j2) + 8} \times 15\underline{/-30°} = 7.5\underline{/-120°}$$

so that

$$[v_{ab}(t)]_s = 7.5 \cos (2t - 120°) \qquad \textit{Ans.}$$

Example 11-12 For the network shown in Fig. 11-20*a*, find the steady-state solutions for v_3 and i_3, using node equations if $i_A = I_A = 2$ amp and $i_B = I_B = 3$ amp.

Solution The d-c equivalent circuit is drawn; see Fig. 11-20*b*. Using capital letters for the steady-state node voltages, the node equations are

$$\begin{aligned}
(\tfrac{1}{6} + \tfrac{1}{3} + \tfrac{1}{2})V_1 + (-\tfrac{1}{3})V_2 + (-\tfrac{1}{2})V_3 &= 2 - 3 = -1 \\
(-\tfrac{1}{3})V_1 + (\tfrac{1}{3} + \tfrac{2}{3} + 1)V_2 + (-1)V_3 &= 0 \\
(-\tfrac{1}{2})V_1 + (-1)V_2 + (1 + \tfrac{1}{2} + \tfrac{1}{2})V_3 &= 3
\end{aligned}$$

Solving for V_1 and V_3 gives the result

$$V_1 = \frac{1}{\frac{35}{18}} = 0.514 \text{ volt}$$

$$V_3 = \frac{\frac{13}{3}}{\frac{35}{18}} = 2.23 \text{ volts}$$

so that $I_3 = (V_1 - V_3)/2 = -0.859$ amp.

Example 11-13 The circuit of Fig. 11-20a is in the sinusoidal steady state with $i_A = I_{A_m} \cos 4t$ and $i_B = I_{B_m} \cos (4t + \phi)$. Find I_{B_m}/I_{A_m} and ϕ, so that $(v_3)_s \equiv 0$.

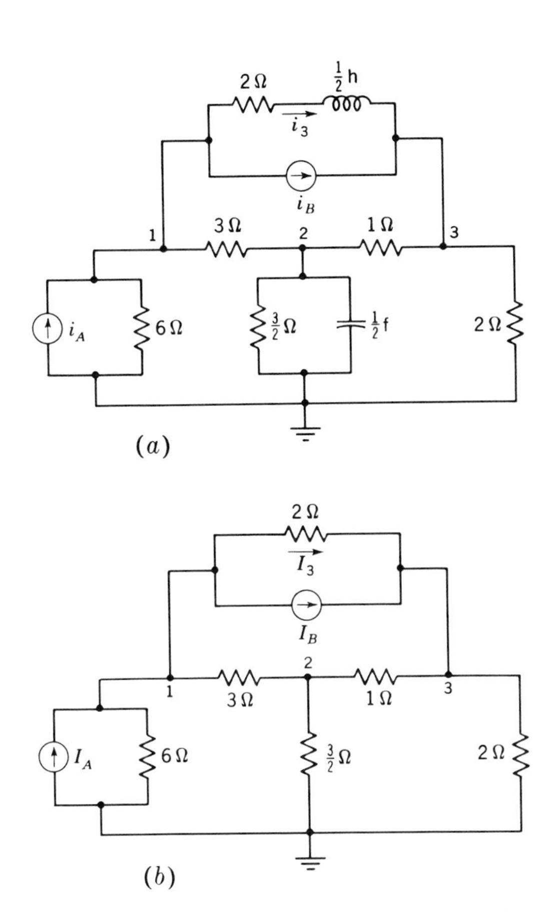

Fig. 11-20 *(a) A network and (b) its d-c equivalent.*

Solution The complex node admittances are

$$\mathbf{Y}_{11} = \tfrac{1}{6} + \tfrac{1}{3} + \frac{1}{2 + j2} = \frac{3 - j}{4}$$

$$\mathbf{Y}_{22} = \tfrac{1}{3} + \tfrac{2}{3} + 1 + j2 = 2 + j2$$

$$\mathbf{Y}_{33} = 1 + \tfrac{1}{2} + \frac{1}{2 + j2} = \frac{7 - j}{4}$$

$$\mathbf{Y}_{12} = \mathbf{Y}_{21} = -\tfrac{1}{3}$$

$$\mathbf{Y}_{13} = \mathbf{Y}_{31} = \frac{-1}{2 + j2} = \frac{-1 + j}{4}$$

$$\mathbf{Y}_{23} = \mathbf{Y}_{32} = -1$$

The amplitude phasor representing i_A is $I_{A_m}\underline{/0}$; i_B is represented by $I_{B_m}\underline{/\phi}$. The amplitude phasor $\mathbf{V}_{3m}$ is zero if the numerator determinant for $\mathbf{V}_{3m}$ is zero, that is, if

$$\begin{vmatrix} \mathbf{Y}_{11} & \mathbf{Y}_{12} & \mathbf{I}_{A_m} - \mathbf{I}_{B_m} \\ \mathbf{Y}_{12} & \mathbf{Y}_{22} & 0 \\ \mathbf{Y}_{13} & \mathbf{Y}_{23} & \mathbf{I}_{B_m} \end{vmatrix} = 0$$

or

$$\begin{vmatrix} \dfrac{3-j}{4} & -\frac{1}{3} & \mathbf{I}_{A_m} - \mathbf{I}_{B_m} \\ -\frac{1}{3} & 2+j2 & 0 \\ \dfrac{-1+j}{4} & -1 & \mathbf{I}_{B_m} \end{vmatrix} = \tfrac{4}{3}\mathbf{I}_{Am} + \frac{5+j9}{3}\mathbf{I}_{B_m} = 0$$

Hence

$$\frac{\mathbf{I}_{B_m}}{\mathbf{I}_{A_m}} = \frac{I_{B_m}}{I_{A_m}}\underline{/\phi} = 0.39\underline{/119^\circ}$$

so that $\phi = 119^\circ$ and the amplitude ratio $I_{B_m}/I_{A_m} = 0.39$.

Example 11-14 Find the expression for the node voltage phasor $\mathbf{V}_2$ in the circuit of Fig. 11-21. Note that $G_m\mathbf{V}_1$ is a controlled source.

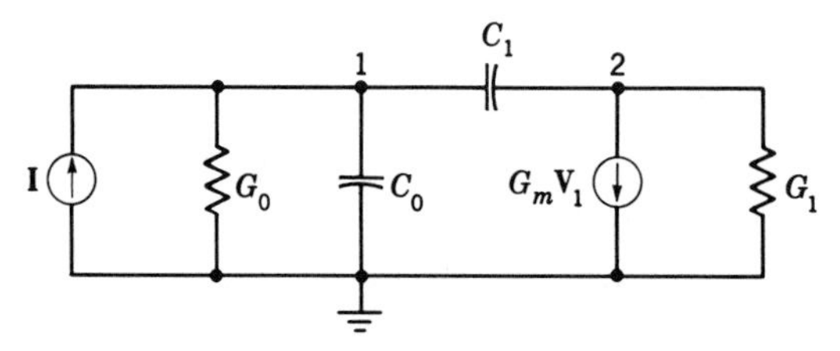

Fig. 11-21 Network in the sinusoidal steady state with controlled source.

Solution The current-law equation at node 1 is in phasor form,

$$[G_0 + j\omega(C_0 + C_1)]\mathbf{V}_1 + (-j\omega C_1)\mathbf{V}_2 = \mathbf{I}$$

and at node 2,

$$(-j\omega C_1 + G_m)\mathbf{V}_1 + (G_1 + j\omega C_1)\mathbf{V}_2 = 0$$

Note that in this network $\mathbf{Y}_{12} = -j\omega C_1 \neq \mathbf{Y}_{21} = -j\omega C_1 + G_m$. This is a characteristic of nonreciprocal network (Chap. 12).

Solving for $\mathbf{V}_2$ gives the result

$$\mathbf{V}_2 = \frac{(+j\omega C_1 - G_m)\mathbf{I}_1}{G_0G_1 - \omega^2 C_0C_1 + j\omega[C_1G_0 + (C_0 + C_1)G_1 + C_1G_m]}$$

Example 11-15 Solve the circuit shown in Fig. 11-22 for v_2 using mesh currents.

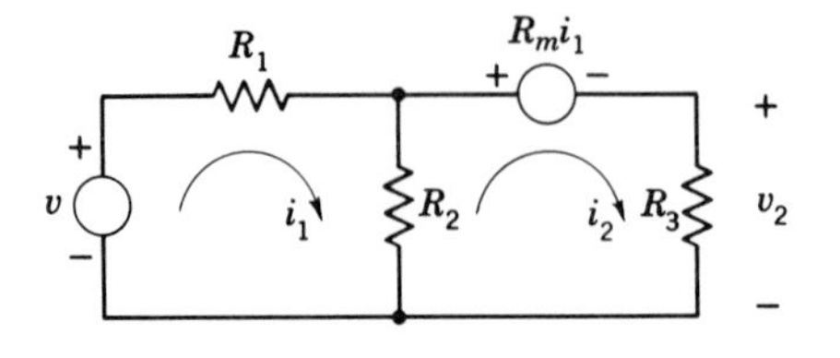

Fig. 11-22 Resistive network with controlled source.

Solution The mesh equations are written applying the voltage law in each loop:

$$(R_1 + R_2)i_1 + (-R_2)i_2 = v$$
$$(-R_2 + R_m)i_1 + (R_2 + R_3)i_2 = 0$$

Note that in this circuit the mesh-impedance matrix is not symmetrical because of the presence of the controlled source. Solving,

$$i_2 = \frac{(R_2 - R_m)v}{R_1R_2 + R_2R_3 + R_3R_1 + R_2R_m}$$

and $v_2 = R_3 i_2$.

11-16 Complete response calculations using Laplace transform

If in a network the value of the voltage across each capacitance and the current in each inductance is known at $t = 0^-$, the use of initial-condition generators (which are ideal sources) allows the formulation of the Laplace transform of the complete response by use of transform immittances. If the required values at $t = 0^-$ are not given explicitly, they must be found or the solution expressed in terms of symbols representing these values. The following examples illustrate the technique.

Example 11-16 The circuit shown in Fig. 11-23*a* is in the steady state with the switch in position *a*. At $t = 0$, the switch is moved to position *b*. The current source $i(t)$ is given by $i(t) = -10 + 2t$. Find $v_2(t)$ for all $t \geq 0^+$.

Solution To find $v_2(0^-)$ and $v_{13}(0^-)$, the d-c steady state with the switch in position *a* is analyzed. The circuit is a series circuit; from voltage division, $v_1(0^-) = 10$ volts, $v_2(0^-) = 3$ volts, $v_3(0^-) = 1$ volt, and $v_{13}(0^-) = 10 - 1 = 9$ volts. The circuit showing transform initial-condition sources is shown in Fig. 11-23*b*. Noting that $Y_{11} = 1 + s$, $Y_{22} = \frac{3}{2} + s$, $Y_{33} = 3 + s$, $Y_{12} = -\frac{1}{2}$, $Y_{13} = Y_{31} = -s = Y_{21}$, and $Y_{23} = Y_{32} = -1$, and letting $\mathcal{L}[i] = I(s)$,

$$V_3(s) = \frac{\begin{vmatrix} 1+s & -\frac{1}{2} & I(s)+9 \\ -\frac{1}{2} & \frac{3}{2}+s & 3 \\ -s & -1 & -9 \end{vmatrix}}{\begin{vmatrix} 1+s & -\frac{1}{2} & -s \\ -\frac{1}{2} & \frac{3}{2}+s & -1 \\ -s & -1 & \frac{3}{2}+s \end{vmatrix}} = \frac{-\frac{9}{2}s - \frac{15}{4} + (s^2 + \frac{3}{2}s + \frac{1}{2})I(s)}{\frac{5}{2}s^2 + 3s + \frac{7}{8}}$$

We now introduce $I(s) = \mathcal{L}[-10 + 2t] = (2 - 10s)/s^2$ and factor the denominator. The result is

$$V_3(s) = \frac{-5.8s^3 - 6.7s^2 - 0.8s + 0.4}{s^2(s + 0.5)(s + 0.7)}$$

Expanding in partial fractions gives

$$V_3(s) = \frac{1.14}{s^2} + \frac{-6.2}{s} + \frac{-3}{s + 0.5} + \frac{3.4}{s + 0.7}$$

so that

$$v_3(t) = (1.14t - 6.2 - 3e^{-0.5t} + 3.4e^{-0.7t})u(t)$$

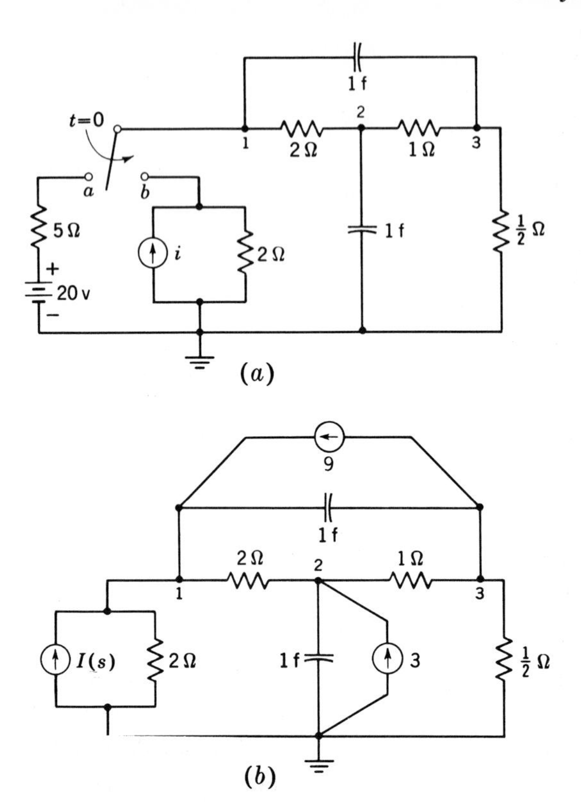

***Fig. 11-23** (a) A three-node circuit, analyzed in Example 11-16. (b) Circuit of (a) with initial-condition transform sources.*

Example 11-17 In the circuit of Fig. 11-24a, $i_1(0^-) = 4$ amp, $i_2(0^-) = 2$ amp, and $v_{ab}(0^-) = 10$ volts. The source is given by $v(t) = 10 \sin 4t$. Solve for $i_2(t)$ for $t \geq 0^+$, using mesh analysis.

Solution The circuit showing transform initial-condition sources is shown in Fig. 11-24b. Note that the source in the 8-henry branch is $\mathcal{L}\{1[i_1(0^-) - i_2(0^-)]\}$. The solution for $\mathcal{L}[i_2(t)] = I_2(s)$ can be written directly in determinant form {using $V(s) = \mathcal{L}[v(t)] = 40/(s^2 + 16)$} as

$$I_2(s) = \frac{\begin{vmatrix} 5 + \frac{3}{2}s & 4 + V(s) & -\frac{1}{2}s \\ -s & -1 & -\frac{1}{2}s \\ -\frac{1}{2}s & -3 - \dfrac{10}{s} & s + \dfrac{2}{s} \end{vmatrix}}{\begin{vmatrix} 5 + \frac{3}{2}s & -s & -\frac{1}{2}s \\ -s & 5 + \frac{3}{2}s & -\frac{1}{2}s \\ -\frac{1}{2}s & -\frac{1}{2}s & s + \dfrac{2}{s} \end{vmatrix}}$$

$$= \frac{-25s - 20 - \dfrac{10}{s} + (s^2 + 2)V(s)}{\frac{25}{2}s^2 + \frac{55}{2}s + 30 + \dfrac{50}{s}}$$

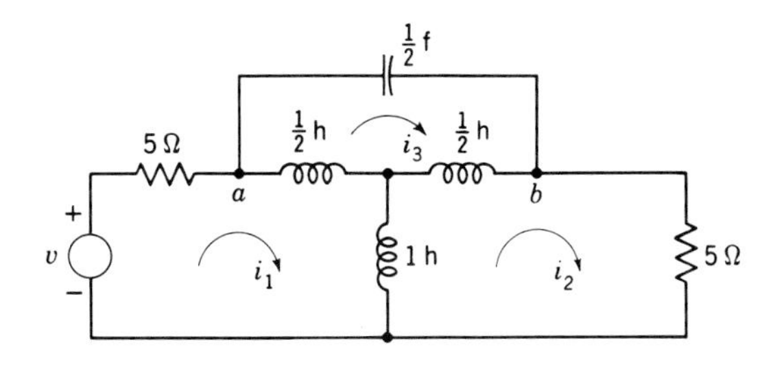

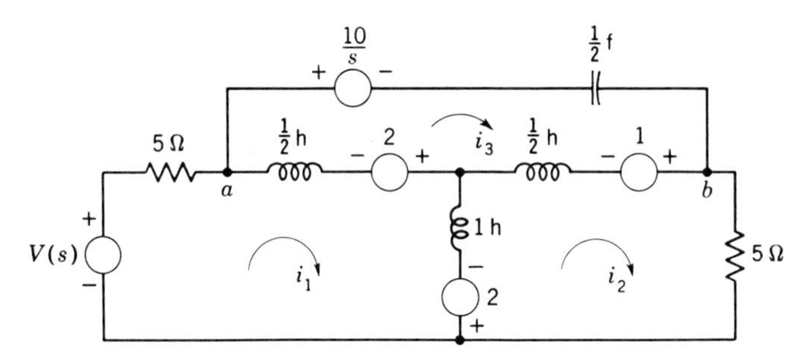

***Fig. 11-24** (a) Circuit with three meshes. (b) Circuit of (a) with transform initial-condition sources shown.*

Substituting $V(s) = 40/(s^2 + 16)$, clearing fractions, and factoring the denominator,

$$I_2(s) = \frac{-2s^4 + 1.6s^3 - 32.8s^2 - 19.2s - 1.28}{(s^2 + 16)(s + 2)(s^2 + 0.2s + 2)}$$

Expanding in partial fractions,

$$\mathbf{I}_2(s) = \frac{\mathbf{K}_1}{s - j4} + \frac{\mathbf{K}_1^*}{s + j4} + \frac{K_2}{s + 2} + \frac{\mathbf{K}_3}{2 + 0.1 - j1.4} + \frac{\mathbf{K}_3^*}{s + 0.1 + j1.4}$$

where $\mathbf{K}_1 = 0.36\underline{/-60^\circ}$

$K_2 = 1.12$

$\mathbf{K}_3 = 0.93\underline{/-152^\circ}$

so that

$$i_2(t) = 0.72 \cos (4t - 60^\circ) + 1.12e^{-2t} + 1.86e^{-0.1t} \cos (1.4t - 152^\circ)$$

11-17 Complete-response calculation using time-domain method

Exactly as in the case of series-parallel circuits, the complete response of a network can be found by superposition of the forced and free components of response and choosing the constants in the free component so that the complete response satisfies the initial conditions. An example is sufficient to illustrate this technique.

Example 11-18 In the circuit of Fig. 11-25, $v_1(0^+) = 4$, $v_2(0^+) = 10$, and $i(t) = 8u(t)$. Solve for $v_1(t)$ using time-domain analysis.

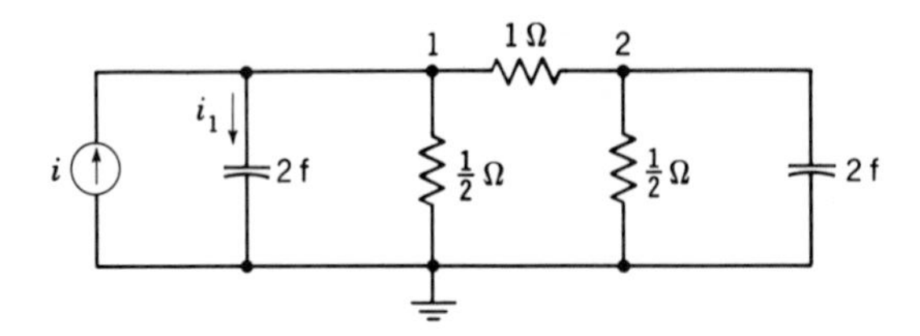

Fig. 11-25 Circuit diagrams for Example 11-18.

Solution From a d-c equivalent circuit, $v_{1s} = V_1 = 8 \times (\frac{3}{2} \times \frac{1}{2})/(\frac{3}{2} + \frac{1}{2}) = 3$ volts. The determinant of the node admittance matrix is

$$\begin{vmatrix} Y_{11} & Y_{12} \\ Y_{21} & Y_{22} \end{vmatrix}$$

where $Y_{11} = Y_{22} = 3 + 2s$, $Y_{12} = Y_{21} = -1$. The characteristic equation is

$$(3 + 2s)^2 - 1 = 0 \qquad s_1 = -1,\ s_2 = -2$$

and the complete response has the form

$$v_1 = 3 + K_1e^{-t} + K_2e^{-2t} \tag{11-74}$$

To find K_1 and K_2, we must use the initial conditions $v_1(0^+) = 4$ and $v_2(0^+) = 10$. Since we do not wish to solve for $v_2(t)$, it is convenient to derive the initial condition for dv_1/dt. (Simultaneous solution for v_1 and v_2 is discussed in Chap. 12.) To find dv_1/dt, we note that v_1 may be a state variable and that the node equation for node 1 is also a state equation, that is,

$$2pv_1 + 3v_1 - v_2 = i$$

so that

$$\frac{dv_1}{dt} = -\tfrac{3}{2}v_1 + \tfrac{1}{2}v_2 + \tfrac{1}{2}i$$

Setting $t = 0^+$, $(dv_1/dt)_{0^+} = -\frac{3}{2}(4) + \frac{1}{2}(10) + 4 = 3$ volts/sec. We now substitute $t = 0$ in Eq. (11-74), differentiate Eq. (11-74) term by term with respect to t, and again set $t = 0^+$. The result is

$$3 + K_1 + K_2 = 4$$
$$0 - K_1 - 2K_2 = 3$$

Solving, $K_1 = -\frac{1}{3}$, $K_2 = \frac{4}{3}$, so that

$$v_1(t) = 3 - \tfrac{1}{3}e^{-t} + \tfrac{4}{3}e^{-2t} \qquad t \geq 0^+ \qquad \textit{Ans.}$$

11-18 The compensated attenuator—an application In many practical applications a "resistance voltage divider" as shown in Fig. 11-26*a* is placed into a circuit with the intention of obtaining a voltage $v_2(t)$ which is a predetermined fraction of $v_1(t)$ (that is, v_2 is "attenuated" with respect to v_1) regardless of the waveform of $v_1(t)$. Now, if all elements are ideal, then $v_2(t) = R_2v_1/(R_1 + R_2)$ and the output waveform will be identical with the input waveform for all time. Unfortunately, it often happens that a capacitance C_2 appears across R_2, unavoidably, because of the nature of the device which is placed across the terminals of R_2. This situation is represented by the circuit of Fig. 11-26*b*. Under these conditions the effect of the capacitance introduces a free response component. Assuming for the present that v_1 is generated by an ideal source, the effect of this capacitance is to introduce a time delay because the voltage across it cannot change abruptly even if the input voltage v_1 changes abruptly. To analyze the situation,

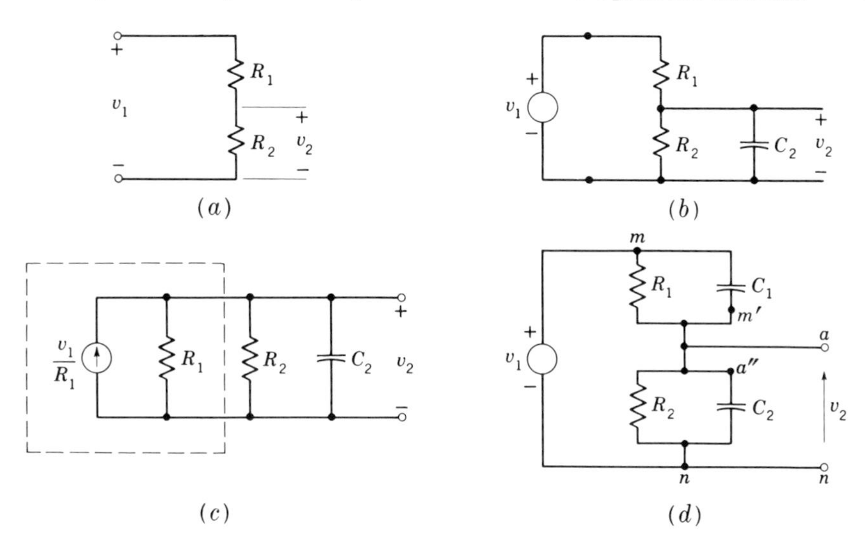

***Fig. 11-26** Analysis of attenuator circuit, compensation.*

we treat the series combination of v_1 and R_1 as a resistive voltage source, convert to a current source as shown in Fig. 11-26*c*, and observe that the free component of the response will be an exponential decay with a time constant $R_1R_2C_2/(R_1 + R_2)$. We recall that the time delays introduced by *R-C* circuits were discussed in detail in Chap. 5.

This circuit can be modified so that $v_2(t)$ is proportional to $v_1(t)$ [as long as $v_1(t)$ remains an ideal source] by placing a capacitance C_1 across R_1 as shown in Fig. 11-26*d*. In this new circuit the loop consisting of $v_1(t)$, C_1, and C_2 is lossless, so that impulses of current can take place. Let us now assume that $v_1(t)$ is a step function, $v_1(t) = Vu(t)$. Further, assume that, at $t = 0^-$, each capacitance is uncharged: $v_{ma}(0^-) = 0$, $v_{an}(0^-) = 0$. Now, at $t = 0^+$, $v_{ma}(0^+) + v_{an}(0^+) = V$. Consequently, the values of the voltages v_{ma} and v_{an} must change in zero time, so that an impulse of current must flow through C_1 and C_2. Hence, at $t = 0$, infinite current flows in the capacitances, so that the current in the resistances (which must be finite) is at that instant neglected. Consequently, at $t = 0^+$, the voltage across the capacitances will be divided inversely as the capacitances.

$$v_{ma}(0^+) = V\frac{C_2}{C_1 + C_2} \qquad v_{an}(0^+) = V\frac{C_1}{C_1 + C_2}$$

The same initial values can be deduced from the initial-value theorem of Laplace transform (Sec. 8-6). In Fig. 11-26*d*, $v_{am}(0^-) = 0$, $v_2(0^-) = 0$, and $v_1(t) = Vu(t)$, so that $V_1(s) = V/s$. Noting that $Z_{am} = R_1/(1 + sR_1C_1)$, $Z_{an} = R_2/(1 + sR_2C_2)$, and $V_2(s) = Z_{am}V_1(s)/Z_{mb}$, we obtain, after some algebra,

$$V_2(s) = \frac{R_2}{R_1 + R_2}\frac{1 + sR_1C_1}{1 + s(C_1 + C_2)[R_1R_2/(R_1 + R_2)]} \times \frac{V}{s}$$

and $\lim_{s\to\infty} sV_2(s) = VC_1/(C_1 + C_2)$ as before. Now the steady-state value of v_2 is evidently $VR_2/(R_1 + R_2)$. If we set

$$\frac{R_2}{R_1 + R_2} = \frac{C_1}{C_1 + C_2}$$

or

$$R_2C_1 + R_2C_2 = R_1C_1 + R_2C_1 \qquad \text{or} \qquad R_1C_1 = R_2C_2$$

then the initial value will be the same as the steady-state value and no transient occurs. The circuit with $R_1C_1 = R_2C_2$ is called a "compensated" attenuator.

We can reach the same conclusion from the solution of the equilibrium equation:

$$v_2 = \frac{R_2}{R_1 + R_2} V + Ke^{-t/T} \qquad T = R_pC_p,\ t \geq 0^+$$

Using the initial condition,

$$\frac{C_1}{C_1 + C_2} V = \frac{R_2}{R_1 + R_2} V + K$$

Thus

$$K = \frac{R_1C_1 - R_2C_2}{(C_1 + C_2)(R_2 + R_2)} V$$

so that $K = 0$ for $R_1C_1 = R_2C_2$.

We have based this example on the assumption that $v_1(t)$ is generated by an ideal source.

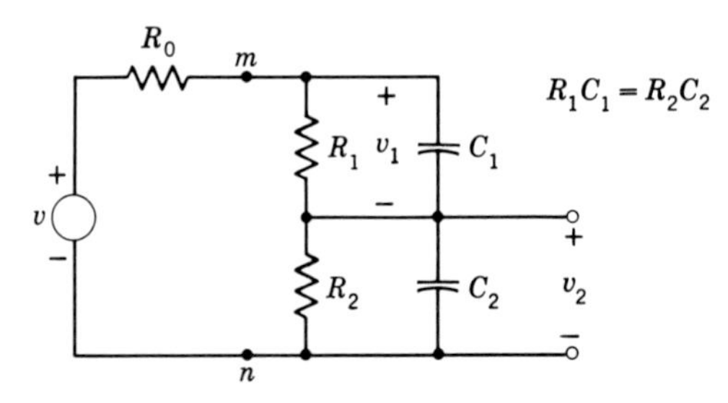

Fig. 11-27 Compensated attenuator with input from a resistive source.

Analysis of the compensated attenuator with source resistance is an excellent example of the problem of mode coupling. In Fig. 11-27 such a circuit is shown. In this circuit we obtain, by voltage division,

$$v_2 = \frac{\dfrac{1}{1/R_2 + pC_2}}{\dfrac{1}{1/R_2 + pC_2} + \dfrac{1}{1/R_1 + pC_1} + R_0} v$$

Simplifying and defining, $T = R_1C_1 = R_2C_2$, gives the equation

$$v_2 = \frac{R_2}{R_1 + R_2 + R_0 + pTR_0} v$$

so that the only mode in v_2 coupled to the input v is the mode

$$(v_2)_f = Ke^{-t/T'}$$

where

$$T' = T \frac{R_0}{R_0 + R_1 + R_2}$$

The state equations can be shown to be (using $T = R_1C_1 = R_2C_2$),

$$\frac{dv_1}{dt} = -\left(\frac{1}{T} + \frac{1}{R_0C_1}\right) v_1 - \frac{1}{R_0C_1} v_2 + \frac{1}{R_0C_1} v$$

$$\frac{dv_2}{dt} = -\frac{1}{R_0C_2} v_1 - \left(\frac{1}{T} + \frac{1}{R_0C_2}\right) v_2 + \frac{1}{R_0C_2} v$$

so that the characteristic determinant is

$$D(p) = \begin{vmatrix} p + \dfrac{1}{T} + \dfrac{1}{R_0C_1} & \dfrac{1}{R_0C_1} \\ \dfrac{1}{R_0C_1} & p + \dfrac{1}{T} + \dfrac{1}{R_0C_2} \end{vmatrix}$$

It is now convenient to eliminate C_2 by use of the relationship $C_2 = R_1C_1/R_2$. Hence

$$D(p) = \begin{vmatrix} p + \dfrac{1}{T} + \dfrac{1}{R_0C_1} & \dfrac{1}{R_0C_1} \\ \dfrac{R_2/R_1}{R_0C_1} & p + \dfrac{1}{T} + \dfrac{R_2/R_1}{R_0C_1} \end{vmatrix}$$

Expanding, we get the polynomial

$$p^2 + p\left(\frac{2}{T} + \frac{1 + R_2/R_1}{R_0C_1}\right) + \frac{1}{T^2} + \frac{1}{T}\frac{1 + R_2/R_1}{R_0C_1}$$

We now write

$$\begin{aligned} \frac{2}{T} + \frac{1 + R_2/R_1}{R_0C_1} &= \frac{1}{T} + \frac{1}{T} + \frac{1 + R_2/R_1}{R_0C_1} \\ &= \frac{1}{T} + \frac{R_0C_1 + T + R_2T/R_1}{TR_0C_1} \\ &= \frac{1}{T} + \frac{1}{T}\frac{R_0R_1C_1 + R_1{}^2C_1 + R_2R_1C_1}{T_0R_0R_1C_1} \\ &= \frac{1}{T} + \frac{1}{T}\frac{R_0 + R_1 + R_2}{R_0} \\ &= \frac{1}{T} + \frac{1}{T'} \end{aligned}$$

The characteristic equation is seen to be

$$s^2 + s\left(\frac{1}{T} + \frac{1}{T'}\right) + \frac{1}{TT'} = 0$$

which has the roots

$$s_1 = -\frac{1}{T} \qquad s_2 = -\frac{1}{T'}$$

The mode $e^{-t/T}$ is not coupled to the output v_1 or v_2 when the source is v. To illuminate the concept of mode coupling further, we consider the circuit with the two sources as shown in Fig. 11-28. Note that v_A may represent an initial-condition generator.

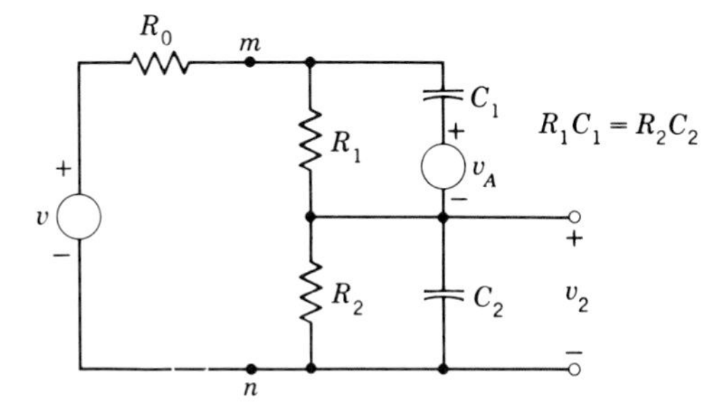

***Fig. 11-28** Compensated attenuator with initial-condition generator v_A.*

For this circuit the state equation can be shown to be

$$\left(p + \frac{1}{T} + \frac{1}{R_0C_1}\right)v_1 + \frac{1}{R_0C_1}v_2 = \frac{v}{R_0C_1} + \left(\frac{1}{T} - \frac{1}{R_1C_1}\right)v_A$$

$$\frac{1}{R_0C_2}v_1 + \left(p + \frac{1}{T} + \frac{1}{R_0C_2}\right)v_2 = \frac{v}{R_0C_2} - \frac{1}{R_0C_2}v_A$$

Solving for v_2, we obtain, by application of Cramer's rule,

$$v_2(t) = \frac{1}{D(p)}\begin{vmatrix} p + \dfrac{1}{T} + \dfrac{1}{R_0C_1} & \dfrac{v}{R_0C_1} + \left(-\dfrac{1}{T} - \dfrac{1}{R_1C_1}\right)v_A \\ \dfrac{1}{R_0C_2} & \dfrac{v}{R_0C_2} - \dfrac{1}{R_0C_2}v_A \end{vmatrix}$$

After some algebra this expression becomes

$$v_2(t) = \frac{\dfrac{1}{R_0C_2}\left(p + \dfrac{1}{T}\right)v - \dfrac{p}{R_0C_2}v_A}{D(p)}$$

$$= \frac{1}{R_0C_2}\left[\frac{p + \dfrac{1}{T}}{\left(p + \dfrac{1}{T}\right)\left(p + \dfrac{1}{T'}\right)}v + \frac{-p}{\left(p + \dfrac{1}{T}\right)\left(p + \dfrac{1}{T'}\right)}v_A\right]$$

We observe that, when the circuit is excited at the terminals of v, then at the output v_2, the mode $e^{-t/T}$ is not coupled to v, while excitation with the source v_A will cause this mode to appear at the output v_2. As stated before, the source v_A may represent an initial-condition generator; hence we have illustrated the fact that the phenomenon of modes uncoupled to the input is associated only with networks that store zero initial energy.

PROBLEMS

11-1 In the circuit of Fig. P11-1: (*a*) Use Thévenin's theorem to show that $(39p + 20)i_{ab} = 3pv$. (*Hint:* Replace the capacitance by an open circuit to calculate v_0 and by a short circuit to calculate i_s.) (*b*) Use delta-wye conversion to obtain the result of (*a*).

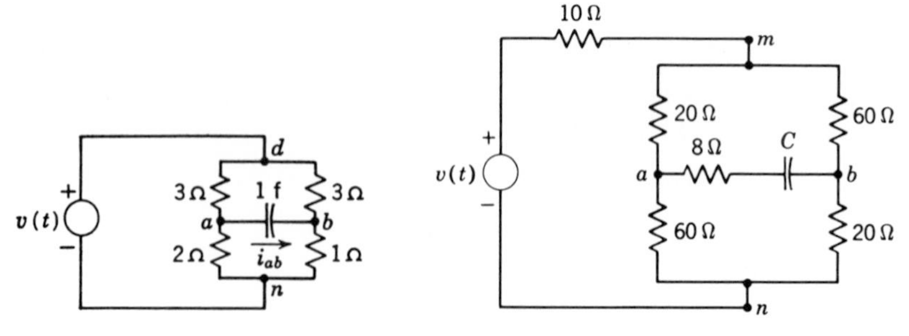

Fig. P11-1 *Fig. P11-2*

11-2 The circuit shown in Fig. P11-2 is operating in the sinusoidal steady state with $v(t) = 80 \cos \omega t$, and ω is chosen so that $X_C = -40$ ohms. (*a*) Calculate the steady-state voltage $[v_{ab}(t)]_s$ by use of Thévenin's theorem, using the open-circuit voltage–short-circuit current method (replacing the $8 + jX_c$ branch by open and short circuits to find $\mathbf{V}_0$ and $\mathbf{I}_s$, respectively). (*b*) Remove the $8 + jX_c$ branch, set the source to zero, and find the resistance in Thévenin's equivalent circuit by delta-wye conversion. (*c*) Apply delta-wye or wye-delta conversion to the circuit as given to find $[v_{ab}(t)]_s$.

11-3 In the circuit shown in Fig. P11-3 the three *p* elements labeled Z_1 are identical with each other, and the three elements labeled Z_0 are identical with each other. (*a*) Draw a circuit equivalent to the given circuit at terminals *a-b*, converting the three Z_0 elements to a wye. (*b*) If $Z_0 = 30$ ohms and $Z_1 = 10$ ohms (i.e., all resistances), what is the numerical value of Z_{ab}? (*c*) If the elements marked Z_0 are 3-farad capacitances and the elements labeled Z_1 are 1-farad capacitances, what is $Z_{ab}(p)$? (*d*) $v_{ab} = 60 \cos 10{,}000t$. Z_0 is the series combination of a 3-ohm resistance and a 0.0003-henry inductance. Z_1 is the series combination of a 1-ohm resistance and a 100-μf capacitance. Calculate (1) $i(t)$ in the steady state; (2) $v_{db}(t)$ in the steady state.

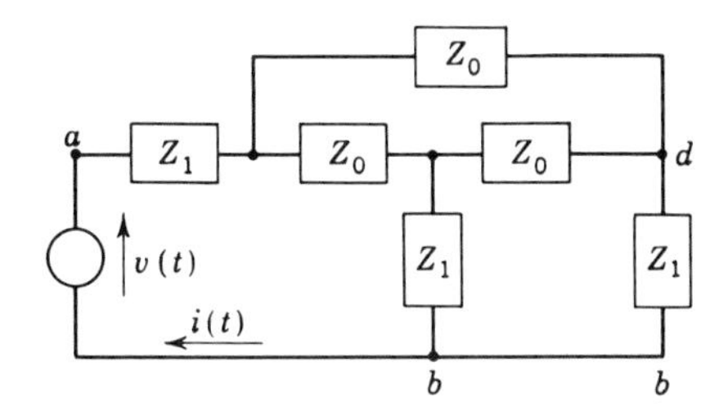

Fig. P11-3

11-4 Represent the three-terminal network given in Fig. P11-4 as (*a*) a wye; (*b*) a delta.

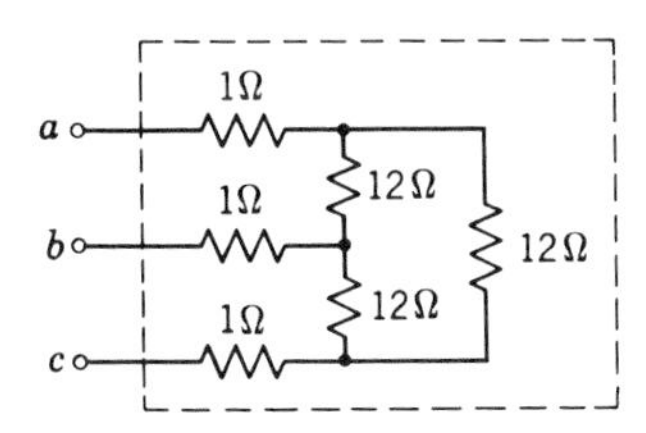

Fig. P11-4

11-5 (*a*) Represent the three-terminal network given in Fig. P11-5 as a wye. (*b*) Represent the same three-terminal network as a delta. (*c*) An ideal constant voltage source $V_{ab} = 120$ volts is connected across terminals *a-b*. No connection is made to terminal *c*. Calculate the power delivered by this source.

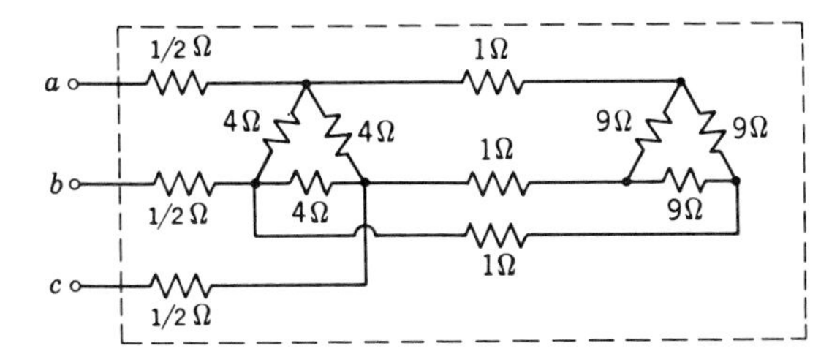

Fig. P11-5

11-6 In Fig. P11-6, at a given frequency, $\mathbf{Z}_\Delta = 3.6 + j4.8$. If $\mathbf{Z}_y$ is purely capacitive, find its value at this frequency such that the equivalent three-terminal network is a purely resistive delta or wye.

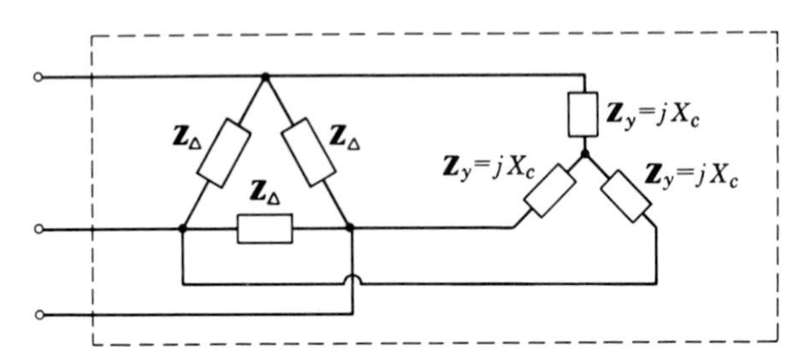

Fig. P11-6

11-7 Calculate the transform driving-point impedance of the terminal pair shown in Fig. P11-7 if (*a*) $Z_1 = Z_4 = Z = 1$, $Z_2 = Z_3 = 1/s$; (*b*) $Z = 1$, $Z_1 = Z_4 = s$, $Z_2 = Z_3 = 1/s$.

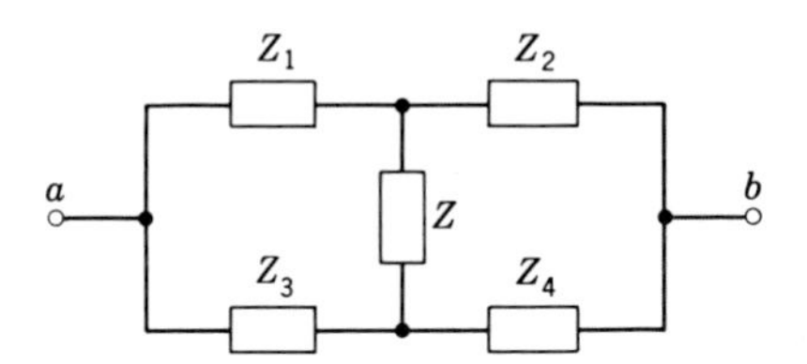

Fig. P11-7

11-8 Deduce the formulas for delta-wye and wye-delta conversion by the method suggested in the footnote on page 389.

11-9 The linear graph of a network is shown in Fig. P11-9. (*a*) What is the number of (1) nodes, (2) junctions, (3) meshes, (4) line elements in a tree? (*b*) Draw five trees for the given graph.

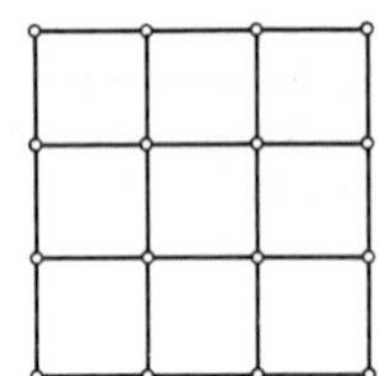

Fig. P11-9

11-10 Draw the graph of a network which has five nodes, all junctions, and each junction connected to all other junctions. Is this network planar?

11-11 Why should a designer of printed circuits be concerned with the planar or nonplanar character of a network?

11-12 (*a*) Show that the introduction of an ideal voltage source by soldering-iron entry changes the structure of the corresponding source-free network. (*b*) Calculate the characteristic roots of the source-free network shown in Fig. P11-12. (*c*) An ideal voltage source is introduced by soldering-iron entry at terminals *a*-*b*. Find the form of the source-free response component for v_{db}.

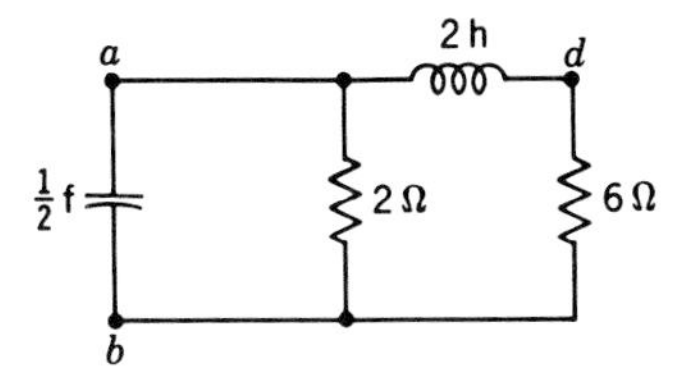

Fig. P11-12

11-13 The circuit of Fig. P11-12 stores zero energy at $t = 0$. Find the complete response $v_{ab}(t)$ in the following cases: (*a*) An ideal current source $i(t) = 10u(t)$ is introduced between terminals *a*-*b* by soldering-iron entry (reference arrow points toward terminal *a*). (*b*) An ideal voltage source $v_{db}(t) = 10u(t)$ is introduced (by soldering-iron entry) between terminals *d*-*b*. (*c*) An ideal voltage source $20u(t)$ is introduced by pliers entry in series with the 2-ohm resistance (the minus sign of the reference polarity marks is at *b*). (*d*) An ideal current source $10u(t)$ is introduced in series with the 2-ohm resistance by pliers entry (reference arrow points to *a*).

11-14 In the circuit of Fig. P11-2 the branch *a*-*b* is removed (open circuit), and the source $v(t)$ is replaced by a current source $i(t)$ (with reference arrow pointing toward *m*). Find Thévenin's equivalent circuit with respect to terminals *a*-*b*.

11-15 In the circuit of Fig. P11-15 find $v_{ab}(t)$ if $i(0^-) = 0$ for the following cases: (*a*) An ideal current source, $i(t) = (24 \cos 3t)u(t)$, is introduced by soldering-iron entry at terminals *a*-*b* (reference arrow toward *a*). (*b*) The current source of (*a*) is replaced by a voltage source, $v_{ab}(t) = (24 \cos 3t)u(t)$. (*c*) A voltage source, $v_{ab}(t) = (24 \cos 3t)u(t)$, is introduced in series with the 4-ohm resistance. (*d*) An ideal current source (with reference arrow pointing to *a*), $i(t) = (24 \cos 3t)u(t)$, replaces the voltage source of part *c*.

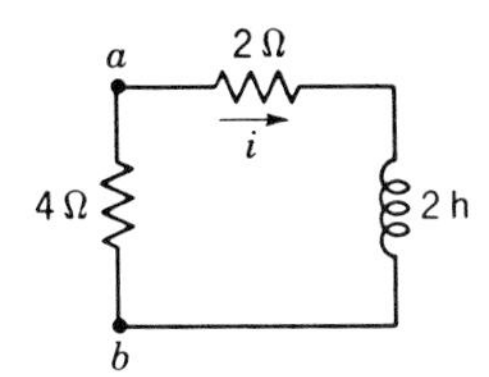

Fig. P11-15

11-16 (*a*) In the circuit of Fig. P11-16 obtain the equilibrium equation that relates $v_{ab}(t)$ to the source functions by reduction of the given circuit to a series circuit

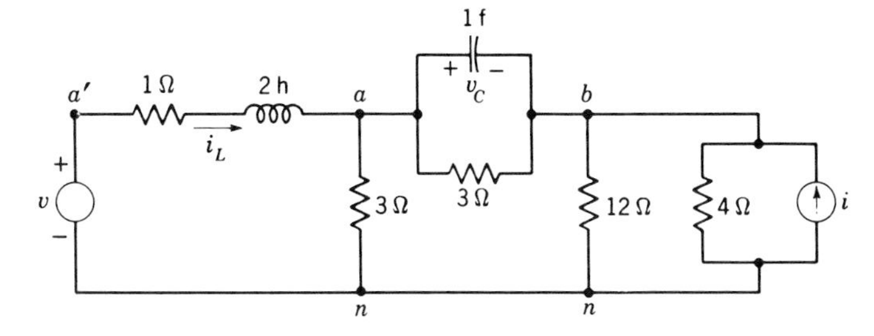

Fig. P11-16

and application of voltage division. (b) Verify the result of part a by use of Thévenin's theorem. (c) Find the component of the response v_{ab} due to each of the following sets of source functions: (1) $v(t) = 10e^{-2t}$, $i(t) = 5e^{-2t}$; (2) $v(t) = 20 \cos t$, $i(t) = 20 \cos 2t$.

11-17 (a) In the circuit of Fig. P11-16 apply Kirchhoff's current law at junctions a and b to obtain the two simultaneous equations for v_{an} and v_{bn}. (b) Use the result of part a to obtain an equilibrium equation for v_{ab}.

11-18 (a) In the circuit of Fig. P11-16 represent the parallel combination of the source $i(t)$ with the 4- and 12-ohm resistances as a resistive voltage source, and write the two simultaneous equations with i_L and v_C as independent variables to describe the network. (b) Eliminate i_L from the equation of part a to obtain the equilibrium equation for $v_C = v_{ab}$.

11-19 In Fig. P11-19: (a) Obtain two simultaneous equations for I_1 and V_{ab}. (b) Calculate I_1 and V_{ab}. (c) Calculate the power delivered by each source and the power dissipated by each resistance.

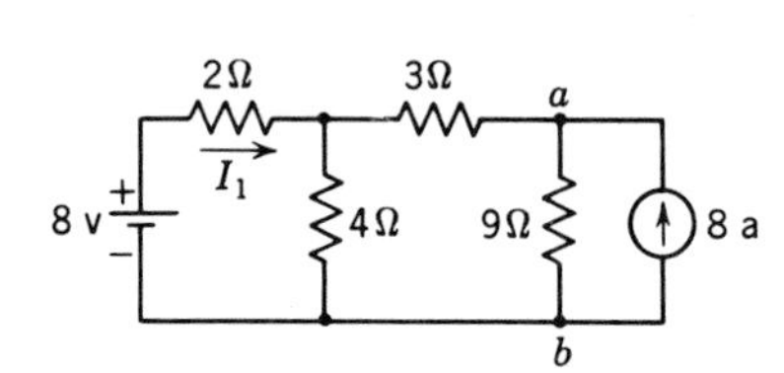

Fig. P11-19

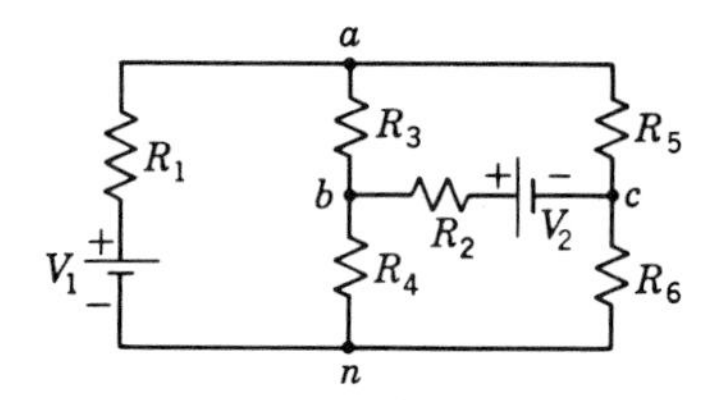

Fig. P11-20

11-20 If $R_1 = R_2 = 1$, $R_3 = R_4 = 2$, $R_5 = 4$, $R_6 = 3$ (all in ohms), $V_1 = 10$ volts and $V_2 = 10$ volts, calculate the power delivered to each resistance and the power delivered by each source in Fig. P11-20, using (a) branch currents as variables; (b) voltage across elements as variables; (c) mesh currents as variables; (d) junction voltages as variables.

11-21 In Fig. P16-21: (a) Represent the combination of Z_2 and $v(t)$ as a current source, and write the node equations for v_1 and v_2. (b) Write the node equations for v_1 and v_2 if each of the p elements (Z_1 to Z_5) is a 1-ohm resistance. (c) Write the node equations if $Z_1 = Z_4 = 1$, $Z_2 = 1/2p$, $Z_3 = 3$, $Z_5 = p$, $i_a = u(t)$, $i_b = 3u(t)$, $v = 2tu(t)$. (d) Draw the circuit diagram which corresponds to the information given in (c). (e) If $v(t) = 0$ and $i_a(t) = 3 \cos \frac{1}{2}t$, calculate $i_b(t)$ so that $v_1(t) \equiv 0$ in the steady state; use the numerical values of the passive elements given in (c). (f) If the circuit elements i_a and i_b are as in (e), calculate $v(t)$ so that $v_1(t) = 5 \cos \frac{1}{2}t$ in the steady state.

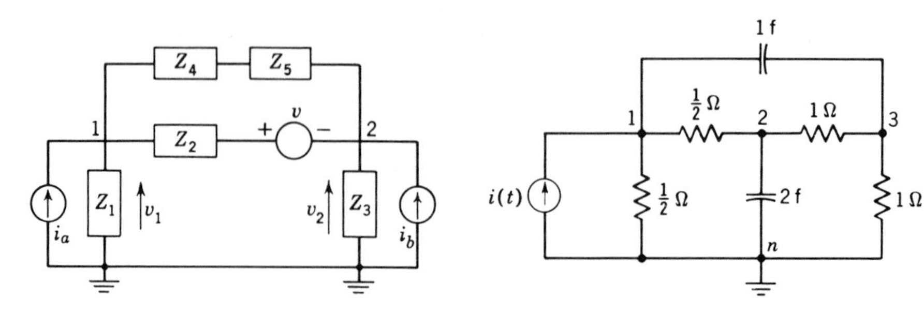

Fig. P11-21 *Fig. P11-22*

11-22 Write the node equations for the circuit shown in Fig. P11-22.

11-23 Write the mesh equations for the circuit of Fig. P11-23.

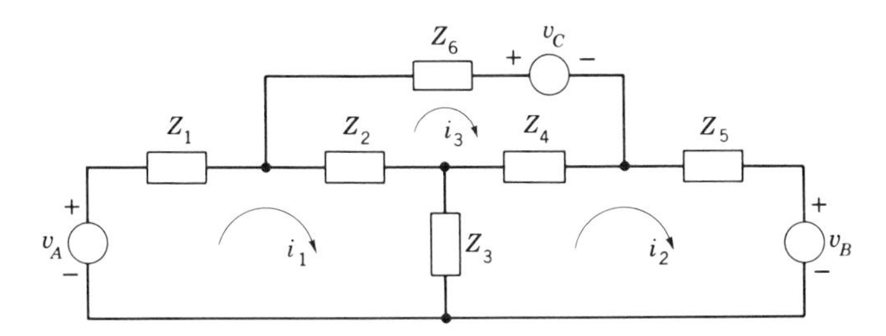

Fig. P11-23

11-24 In the circuit of Fig. P11-23, all passive elements are resistive; their values are $Z_1 = Z_2 = 2$ ohms, $Z_3 = 1$ ohm, $Z_4 = Z_5 = 2$ ohms, and $Z_6 = 4$ ohms. All sources are constant, and $v_A = V_A = 10$ volts. Find the relationship between v_B $(= V_B)$ and v_C $(= V_C)$ so that $i_1 = I_1 = 0$.

11-25 (*a*) Obtain the state equations for the circuit of Fig. P11-12, using v_{ab} and i_{ad} (in the inductance) as variables. (*b*) Modify the equation of part *a* if an ideal current source $i_S(t)$ is introduced by soldering-iron entry between terminals *a-d* with reference arrow pointing to (*a*).

11-26 (*a*) Obtain the state equations for the circuit of Fig. P11-16, using i_L and v_C as variables. (*b*) Obtain the equilibrium equation for v_C by eliminating i_L and pi_L from the equations of (*a*).

11-27 Write the state equations for the circuit of Fig. P11-22, using as variables $y_1 = v_{2n}$ and $y_2 = v_{13}$.

11-28 Show that in an *R-C* ladder circuit of the type shown in Fig. P11-28*a*, each state equation can be obtained from a node equation by multiplication with a constant. (*b*) Show that in the *R-L* ladder circuit shown in Fig. P11-28*b*, each state equation is related to a mesh equation by multiplication with a constant.

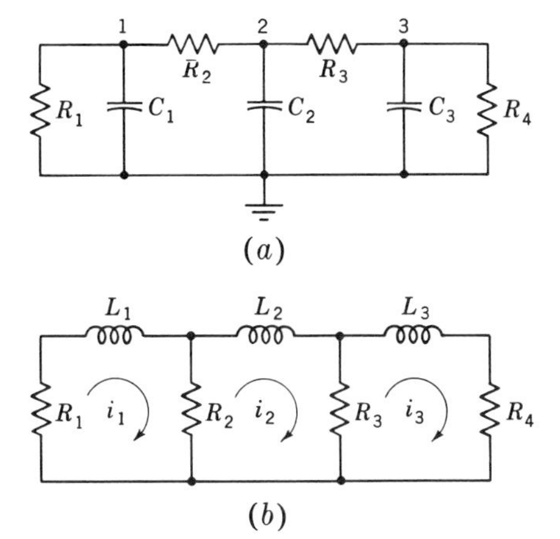

Fig. P11-28

11-29 In the circuit of Fig. P11-29 let $L = \frac{1}{2}$ henry, $C_1 = C_2 = \frac{1}{2}$ farad, $R_1 = R_2 =$ 1 ohm, and obtain the operational transfer function that relates the response v_2 to the source by using (*a*) source conversion and voltage division; (*b*) node equations; (*c*) state equations.

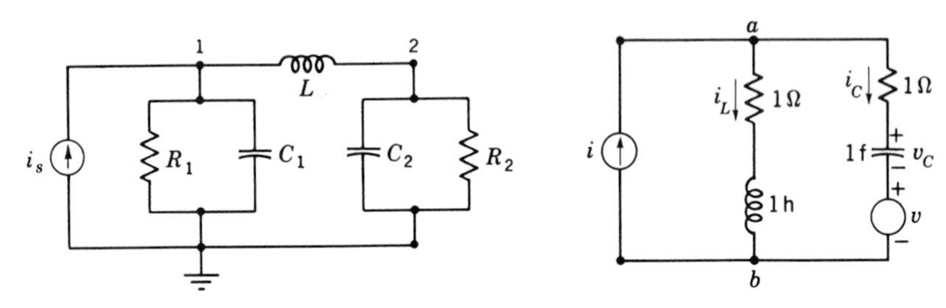

Fig. P11-29

Fig. P11-30

11-30 The element values in the circuit of Fig. P11-30 have been selected so that mode coupling may be illustrated. (*a*) Set all sources to zero and show that the state equations are $pi_L = -2i_L + v_C$, $pv_C = -i_L$. (*b*) Show that, in general, a source-free response component is given by $y_f = (K_1 + K_2t)e^{-t}$. (*c*) Show that (for $v \equiv 0$) v_C is related to i by the equation $(p + 1)v_C = i$. Which mode in v_C is not coupled to i? (*d*) Set $i \equiv 0$, and find the transfer function that relates v_C to v. Are all possible modes in v_C coupled to v? (*e*) Show, for the output v_{ab}, that the mode e^{-t} is not coupled to i but is coupled to v. (*f*) Show that, independently of the initial conditions, the mode te^{-t} does not exist in v_{ab}. (This is not a case of mode decoupling, but is a problem of observability; see Sec. 12-3.)

11-31 The element values in the circuit shown in Fig. P11-31 have been selected to illustrate mode coupling. (*a*) Find all possible free modes of the circuit. (*b*) Show that, when the circuit is initially not energized, one of the possible modes does not appear in i_1. (*c*) Show from the transfer admittance that relates i_1 to an initial-condition generator for one of the inductances that all possible modes occur in i_1 when arbitrary initial conditions are assigned.

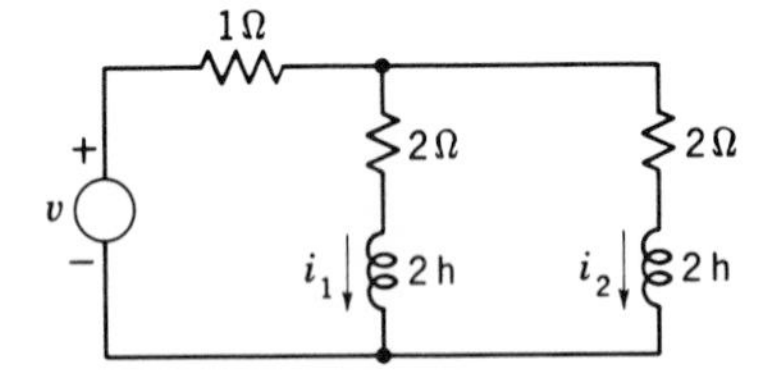

Fig. P11-31

11-32 In the circuit of Fig. P11-31 the source $v(t)$ is given by $v(t) = 40u(t)$. Find the complete response $i_1(t)$ if the initial conditions are (1) $i_1(0^+) = i_2(0^+) = 0$; (2) $i_1(0^+) = -i_2(0^+) = 6$ amp; (3) $i_1(0^+) = 6$ amp, $i_2(0^+) = 0$.

11-33 In the circuit of Fig. P11-29, $R_1 = R_2 = R$, $C_1 = C_2 = 1$ farad, and $L = 1$ henry. (*a*) Find all possible free modes (in terms of R) of the circuit. (*b*) Find the range of values for R that give nonoscillatory source-free response (all characteristic roots real).

11-34 In a certain two-node circuit, $Y_{11} = Y_{22} = (p^2 + 4p/R + 1)/p$ and $Y_{12} = Y_{21} = -(p + 2/R)$. (*a*) Deduce the range of values for R so that the circuit has no oscillatory source-free modes (i.e., all characteristic roots are real). (*b*) Sketch such a circuit, giving numerical values for all elements if $R = 12$.

11-35 In a certain two-mesh circuit, $Z_{11} = Z_{22} = (p^2 + 3Rp + 1)/p$ and $Z_{12} = Z_{21} = -(R + 1/p)$. Deduce the range of values for R so that all characteristic roots are real.

11-36 In the circuit of Fig. P11-36, find I_3 so that $V_1 = 0$ when $I_1 = 1$ amp and $I_2 = 2$ amp.

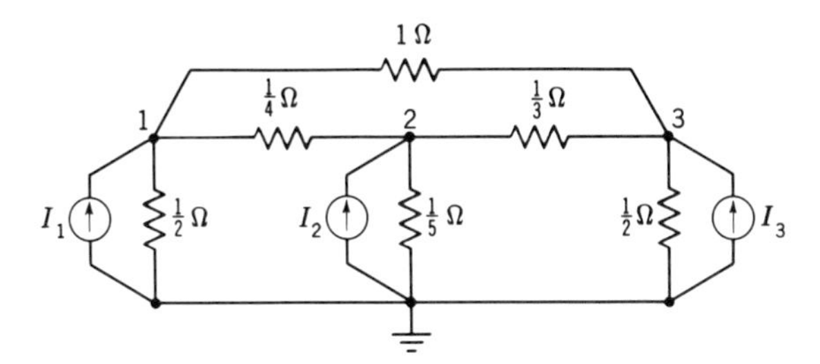

Fig. P11-36

11-37 In the circuit of Fig. P11-36, $I_2 = I_3 = 0$. Calculate the ratio V_3/V_1.

11-38 In the circuit of Fig. P11-36 calculate (*a*) V_3 when $I_1 = 5$ amp, $I_2 = I_3 = 0$; (*b*) V_1 when $I_3 = 5$ amp, $I_1 = I_2 = 0$.

11-39 In the circuit of Fig. P11-29, $i_s(t) = 8 \cos 4t$. Calculate $v_2(t)$ in the steady state if $C_1 = C_2 = \frac{1}{2}$ farad, $R_1 = R_2 = 1$ ohm, and $L = \frac{1}{2}$ henry.

11-40 In the circuit of Fig. P11-29, $i_s(t) = 10e^{-t} \cos 2t$. Calculate the component of the response v_2 due to the source if $C_1 = C_2 = 1$ farad, $L = 1$ henry, $R_1 = R_2 = 1$ ohm.

11-41 In the circuit of Fig. P11-28*a* a current source $i_1(t)$ is connected between node 1 and the reference node. All elements have unit value. (*a*) Calculate the ratio $(v_3)_s/(v_1)_s$ (where the subscript s indicates component due to the source) if $i(t) = 30u(t)$. (*b*) Calculate the phasor ratio $\mathbf{V}_3/\mathbf{V}_1$ if (1) $i(t) = 60 \cos 3t$; (2) $60 \cos 4t$.

11-42 In the circuit of Fig. P11-28*b*, let all elements have unit value. A voltage source is inserted in series with R_1 so that the mesh supply voltage is $v_1(t)$. Calculate the ratio of power delivered to R_4 to power delivered to R_1 in the steady state if (*a*) $v_1(t) = 10$; (*b*) $v_1(t) = 60 \cos 3t$.

11-43 In the circuit of Fig. P11-43, find the steady-state solution for $v_C(t)$ if $i_1(t) = 16 \cos 5t$ and $i_2(t) = 4 \cos (5t - 90°)$ by each of the following methods: (*a*) source conversion; (*b*) node equations; (*c*) state equation.

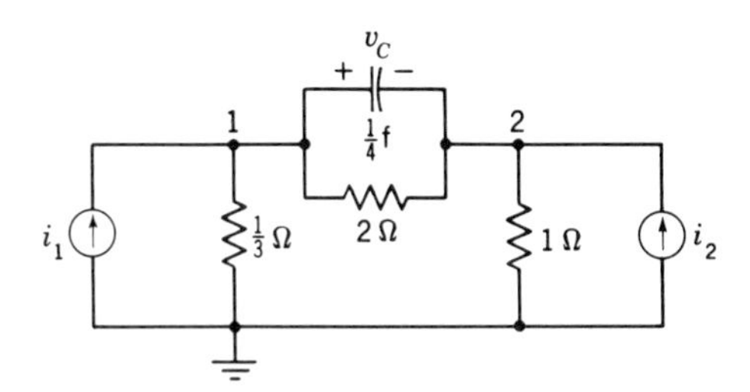

Fig. P11-43

11-44 Formulate the state equation for v_C in the circuit of Fig. P11-43 and solve for $(v_C)_s$ if $i_1 = 10e^{-3t}$ and $i_2(t) = 5e^{-t} \cos 5t$.

11-45 (*a*) Write the mesh equations for Fig. P11-23. (*b*) The p elements in the circuit are defined as follows: Z_1 and Z_5, 4-ohm resistances; Z_2, parallel combination of 4-ohm resistance and $\frac{1}{16}$-farad capacitance; Z_3, 2-henry inductance; Z_4, series combination of 5-ohm resistance and 3-henry inductance; Z_6, $\frac{1}{3}$-farad capacitance. The sources are sinusoidal, and all have the same radian frequency $\omega = 3$. Write the complex mesh equations. (*c*) In (*b*) calculate the phasor ratio $\mathbf{I}_3/\mathbf{I}_1$ if $\mathbf{V}_B = \mathbf{V}_C = 0$ and $\mathbf{V}_A \neq 0$.

11-46 In the circuit of Fig. P11-23, $v_B = v_C \equiv 0$, and $v_A = 141.4 \cos 1{,}000t$. The elements are defined as follows: $Z_1 = R_1$, $Z_2 = R_2 + pL_2$, $Z_3 = 1/pC_3$, $Z_4 = R_4 + pL_4$, $Z_5 = R_5$, $Z_6 = 1/pC_6$. The complex mesh equations are

$$
\begin{aligned}
(20 - j10)\mathbf{I}_1 + (j20)\mathbf{I}_2 + (-15 - j10)\mathbf{I}_3 &= 100 + j0 \\
(j20)\mathbf{I}_1 + (20 - j10)\mathbf{I}_2 + (-10 - j10)\mathbf{I}_3 &= 0 \\
(-15 - j10)\mathbf{I}_1 + (-10 - j10)\mathbf{I}_2 + (25 + j10)\mathbf{I}_3 &= 0
\end{aligned}
$$

(*a*) Calculate the values of R_1, R_2, R_4, R_5, L_2, L_4, C_3, and C_6. (*b*) Calculate $i_1(t)$ in the steady state. (*c*) Calculate the average power delivered by the source.

11-47 The simultaneous equations

$$
\begin{aligned}
3\frac{d^2x}{dt^2} + 12\frac{dx}{dt} + 20x \qquad + 8\frac{dy}{dt} + 10y &= 50 \cos 2t \\
8\frac{dx}{dt} + 10x + 3\frac{d^2y}{dt^2} + 6\frac{dy}{dt} + 16y &= 0
\end{aligned}
$$

have a solution of the form $x = A \cos (2t + \theta)$. Calculate A and θ.

11-48 The state equations of a certain linear system are given by $py_1 = -3y_1 + y_2 + 12 \cos 2t$; $py_2 = 2y_1 - 4y_2$. (*a*) Find the steady-state solution for y_1. (*b*) Find the complete response $y_1(t)$ if $y_1(0^-) = 0$ and $y_2(0^-) = 0$.

11-49 In the circuit of Fig. P11-49 use state equations to find the component of the response v_C due to the source function in the following cases: (*a*) $v(t) = 5 \cos 3t$, $i(t) = 2 \sin 3t$; (*b*) $v(t) = 10e^{-2t}$, $i(t) = 4 \cos 3t$.

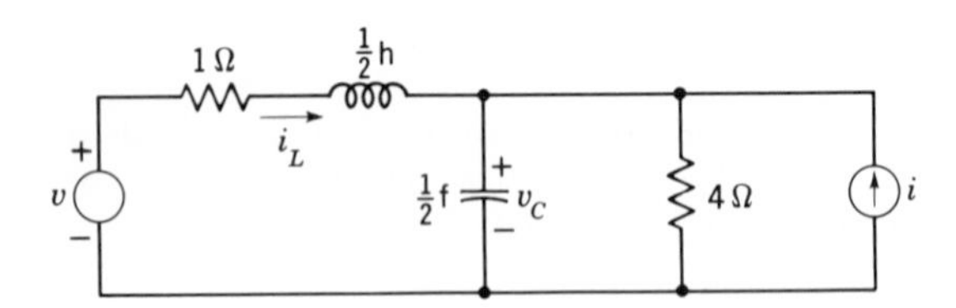

Fig. P11-49

11-50 In the circuit of Fig. P11-29, let $L = 1$ henry, $C_1 = C_2 = \frac{1}{2}$ farad, $R_1 = R_2 = 1$ ohm, $i_s(t) = 2u(t)$. The circuit stores zero initial energy. (*a*) Find the initial values $v_1(0^+)$, $v_2(0^+)$, $(dv_1/dt)_{0^+}$, $(dv_2/dt)_{0^+}$, $(d^2v_1/dt^2)_{0^+}$, and $(d^2v_2/dt^2)_{0^+}$. (*b*) Find the characteristic roots. (*c*) Find the complete response $v_2(t)$ without and with use of Laplace transform.

11-51 In the circuit of Fig. P11-16 find the complete response $v_C(t)$ if $i_L(0^+) = i_L(0^-) = 1$ amp, $v_C(0^+) = v_C(0^-) = 0$, $i(t) = 2tu(t)$, and $v(t) = 2 \cos 3tu(t)$.

11-52 The circuit shown in Fig. P11-52 is in the steady state with the switch open. The switch is closed at $t = 0$. Solve for the voltage across the capacitance.

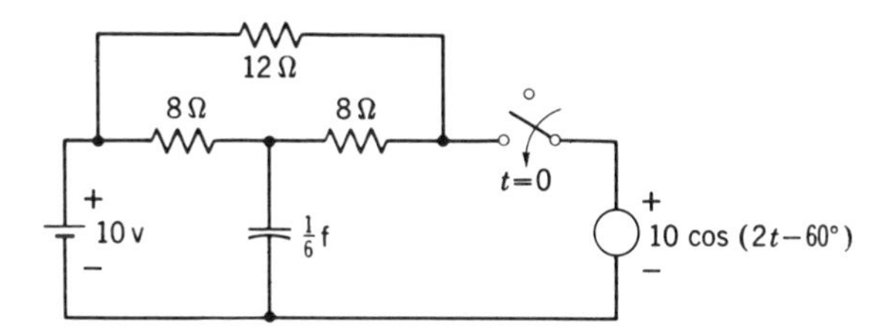

Fig. P11-52

11-53 Find the complete response $i(t)$ in the circuit of Fig. P11-53 if the elements store zero initial energy for the following source functions: (a) $v(t) = (10 \cos t)u(t)$; (b) $v(t) = (10 \sin t)u(t)$; (c) $v(t) = (10 + 10t)u(t)$; (d) $v(t) = 10e^{-t} \cos t$; (e) $v(t) = 10e^{-4t}$; (f) $v(t) = 10e^{-t}$.

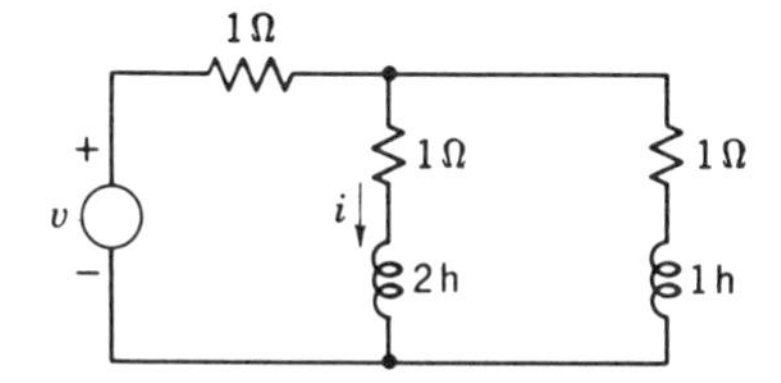

Fig. P11-53

11-54 In a certain network problem the component of the response $v_2(t)$ due to the source is given by $(v_2)_s = 20 \cos 3t$. The characteristic roots are $s_1 = -3$, $s_{2,3} = -3 \pm j5$. It is known that $v_2(0^+) = 0$, $(dv_2/dt)_{0^+} = -90$, $(d^2v_2/dt^2)_{0^+} = -1{,}260$. Find the complete response $v_2(t)$.

11-55 In the circuit of Fig. P11-43 replace the $\frac{1}{4}$-farad capacitance by an initial unenergized 1-henry inductance, and solve for v_{12}, using a state equation with $i_1(t) = 12u(t)$ and $i_2(t) = 3tu(t)$.

11-56 It is the purpose of this problem to show the effect on the output waveform of an imperfectly compensated attenuator with zero source resistance. Such a circuit is shown in Fig. P11-56; when $\Delta C = 0$ the attenuator is (perfectly) compensated. The initial-energy storage is zero. Let $v(t) = 10u(t)$ and find $v_2(t)$ for all $t \geq 0^+$ in the following cases: (a) $\Delta C = 0$; (b) $\Delta C = 0.2$; (c) $\Delta C = -0.2$.

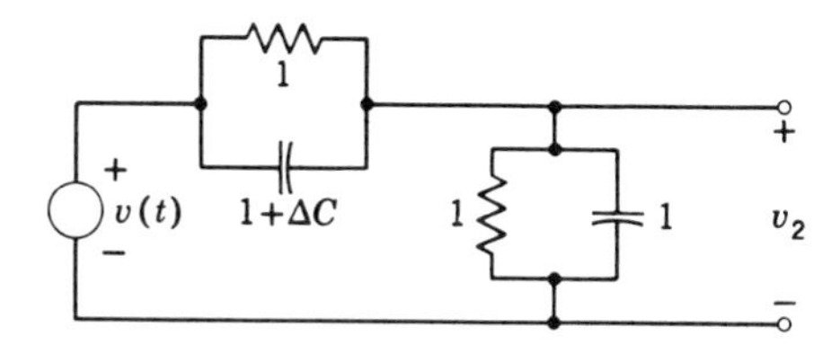

Fig. P11-56

11-57 It is the purpose of this problem to illustrate numerically the effect of source resistance on the output waveform of an attenuator. In Fig. P11-57 a resistive attenuator is shown. (*a*) Calculate $v_2(t)$. (*b*) A unit capacitance C_1 is placed (unavoidably) across the output terminals *b-n*. For zero-initial-energy storage find $v_2(t)$ for all $t \geq 0^+$. (*c*) With the capacitance C_1 of part *b* in place, the attenuator is now compensated by addition of a capacitance $C_2 = \frac{1}{9}$ across terminals *a-b*. For zero-initial-energy storage, formulate $v_2(t)$ for this circuit, and compare the result with that of part *b*.

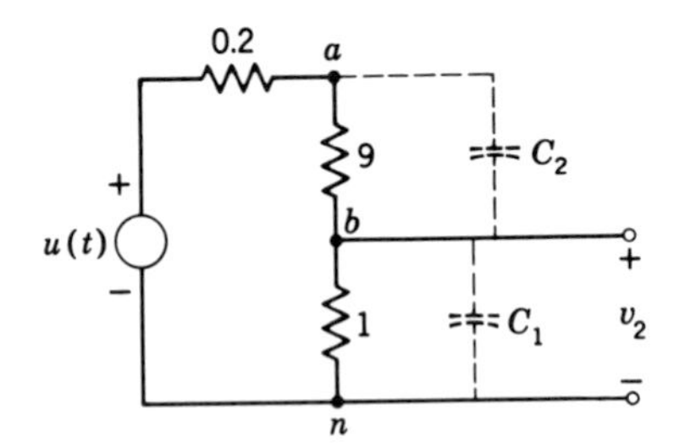

Fig. P11-57

11-58 Although the discussion in the text deals with passive networks, the analysis techniques described may be applied to active linear systems. In Fig. P11-58*a*, an element of an active system is shown by the "block" marked A. This symbol represents an ideal voltage amplifier; that is, if a voltage v_1 appears across terminals *a-b*, then $v_{cd} = Av_1$. The element is, however, unilateral; a voltage applied at terminals *c-d* produces no response at terminals *a-b*. Further, the driving-point *admittance* at terminals *a-b* is zero. The block can also be represented by the controlled source of Fig. P11-58*b*. Note that there is no path from *a* to *b* unless external elements are connected. (*a*) In the circuit of Fig. P11-58*c* show that (1) $v_C = v_1 = [1/(p+1)]v_2 = [1/(p+1)]Av_1$; (2) $[p + (1 - A)]v_1 = 0$; (3) the circuit is unstable for $A > 1$. (*b*) In the circuit of Fig. P11-58*d* show that (1) $[p/(p+1)](v_2 - v_1) = (1+p)v_1$; (2) $p^2v_1 + p(3-A)v_1 + v_1 = 0$; (3) the network has a free response characterized by a pole pair in the right-half *s* plane if $A > 3$ (this fact makes it possible to use active networks as "oscillators").

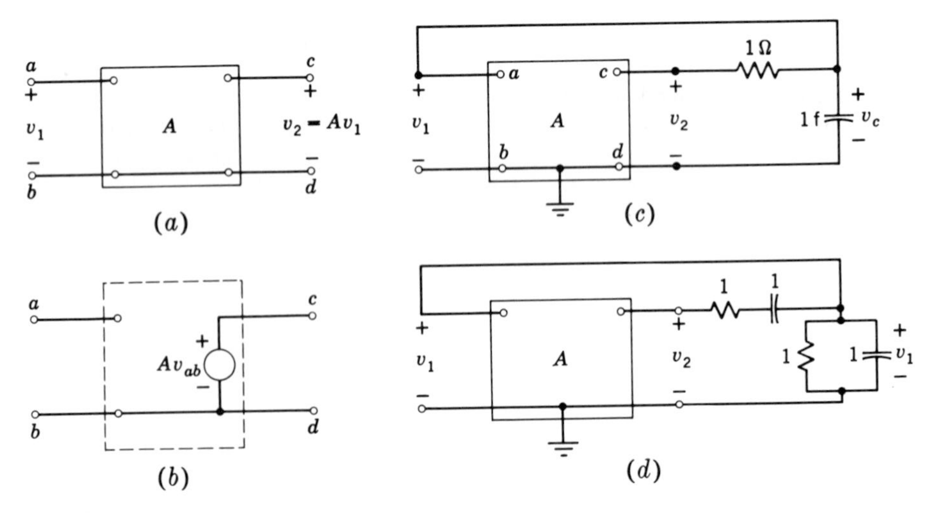

Fig. P11-58

Chapter 12 General analysis techniques II—network theorems

In Chap. 11 certain general properties of networks were discussed so as to develop techniques for the solution of network problems. In this chapter we deal with those of the general network properties that are derived from the state, mesh, and node equations.

12-1 Relationship between free-mode amplitudes—introduction

In this section we begin discussion of the relationship between the mode amplitudes of a given mode in the various network variables. For the purpose of this discussion we shall use state variables, but the results will be seen to be applicable to all network descriptions that involve simultaneous linear equations with constant coefficients.

To pinpoint the specific problem under discussion, consider a system described by two state variables y_1 and y_2, and assume that the characteristic roots are real and distinct, i.e., that the characteristic equation is

$$D(s) = (s - s_1)(s - s_2) \qquad s_1 \neq s_2;\ s_1,\ s_2 \text{ real}$$

For such a system the solution for the state variables y_1 and y_2 is

$$\begin{aligned} y_1 &= (y_1)_s + K_{11}e^{s_1t} + K_{12}e^{s_2t} \\ y_2 &= (y_2)_s + K_{21}e^{s_1t} + K_{22}e^{s_2t} \end{aligned} \tag{12-1}$$

We shall show that, independently of the initial conditions, there is a fixed relationship between the mode amplitudes of each mode; that is, there is a fixed relationship between K_{11} and K_{21}, the mode amplitudes of e^{s_1t} in y_1 and y_2, respectively. A similar relationship exists between K_{12} and K_{22}, the mode amplitudes e^{s_2t}. In Sec. 12-2 we shall discuss the corresponding relationship in the general case where there are n free modes.

To deduce the desired relationship, we write the state equations of a two-state system in general form, as follows:

$$\begin{aligned}(p - a_{11})y_1 + \quad (-a_{12})y_2 &= g_1(t) \\ (-a_{21})y_1 + (p - a_{22})y_2 &= g_2(t)\end{aligned} \tag{12-2}$$

and substitute the solution

$$\begin{aligned}y_1 &= (y_1)_s + (y_1)_f \\ y_2 &= (y_2)_s + (y_2)_f\end{aligned} \tag{12-3}$$

where

$$\begin{aligned}y_{1f} &= K_{11}e^{s_1t} + K_{12}e^{s_2t} \\ y_{2f} &= K_{21}e^{s_1t} + K_{22}e^{s_2t}\end{aligned} \tag{12-4}$$

Substitution from Eqs. (12-3) in (12-2) gives the result

$$\begin{aligned}(p - a_{11})y_{1s} + (-a_{12})y_{2s} + (p - a_{11})(y_1)_f + (-a_{12})(y_2)_f &= g_1(t) \\ (-a_{21})y_{2s} + (p - a_{22})y_{2s} + (-a_{21})(y_1)_f + (p - a_{12})(y_2)_f &= g_2(t)\end{aligned}$$

Since y_{1s} and y_{2s} are chosen to satisfy the inhomogeneous state equations, we recognize that

$$\begin{aligned}(p - a_{11})(y_1)_s + (-a_{12})y_{2s} &= g_1(t) \\ (-a_{21})(y_2)_s + (p - a_{22})y_{2s} &= g_2(t)\end{aligned}$$

so that, as expected, the source-free component of the response must satisfy the homogeneous state equations

$$\begin{aligned}(p - a_{11})(y_1)_f + \quad (-a_{12})(y_2)_f &= 0 \\ (-a_{21})(y_1)_f + (p - a_{22})(y_2)_f &= 0\end{aligned} \tag{12-5}$$

We now substitute the form of the free response components as given by Eq. (12-4) in Eq. (12-5), recalling that $pe^{st} = se^{st}$. The result is

$$[(s_1 - a_{11})K_{11} - a_{12}K_{21}]e^{s_1t} + \quad [(s_2 - a_{12})K_{12} - a_{12}K_{22}]e^{s_2t} = 0 \tag{12-6}$$

$$[(-a_{21})K_{11} + (s_1 - a_{22})K_{21}]e^{s_1t} + [(-a_{21})K_{12} + (s_2 - a_{22})K_{22}]e^{s_2t} = 0 \tag{12-7}$$

In Eqs. (12-6) and (12-7), note that, since $s_1 \neq s_2$ (assumed), e^{s_1t} and e^{s_2t} are linearly independent. Hence the coefficients of each exponential term in each of these two equations must be zero. This results in the following two sets of simultaneous equations:

$$(s_1 - a_{11})K_{11} + \quad (-a_{12})K_{21} = 0 \tag{12-8a}$$
$$(-a_{21})K_{11} + (s_1 - a_{22})K_{21} = 0 \tag{12-8b}$$

$$(s_2 - a_{11})K_{12} + \quad (-a_{12})K_{22} = 0 \tag{12-9a}$$
$$(-a_{21})K_{12} + (s_2 - a_{22})K_{22} = 0 \tag{12-9b}$$

Equations (12-8) and (12-9) appear to be homogeneous simultaneous equations for the constants K_{ij}, and may therefore have only the trivial solutions $K_{ij} = 0$. We note, however, that the determinants

$$D(s_k) = \begin{vmatrix} s_k - a_{11} & -a_{12} \\ -a_{21} & s_k - a_{22} \end{vmatrix} \qquad k = 1, 2$$

are zero because $s_k = s_1$ or s_2 have been chosen as the characteristic roots so that $D(s_1) = D(s_2) = 0$. Hence Eqs. (12-8*a*) and (12-8*b*) are not independent; neither are Eqs. (12-9*a*) and (12-9*b*). Therefore (see Appendix *D*), one equation in each set can be deleted as superfluous. Deleting Eqs. (12-8*b*) and (12-9*b*) gives the result

$$(s_1 - a_{11})K_{11} + (-a_{12})K_{21} = 0 \qquad (12\text{-}10)$$
$$(s_2 - a_{11})K_{12} + (-a_{12})K_{22} = 0 \qquad (12\text{-}11)$$

and we have the result

$$\frac{K_{11}}{K_{21}} = -\frac{-a_{12}}{s_1 - a_{11}} \qquad (12\text{-}12)$$
$$\frac{K_{12}}{K_{22}} = -\frac{-a_{12}}{s_2 - a_{11}} \qquad (12\text{-}13)$$

It is left as an exercise for the reader to show that retaining Eqs. (12-8*b*) and (12-9*b*) while deleting (12-8*a*) and (12-9*a*) leads to the same result.

Equations (12-12) and (12-13) show the relationship between the mode amplitudes of each free mode in the two responses y_1 and y_2. An immediate consequence of this relationship is apparent: If the *initial conditions* are chosen so that a given mode has zero amplitude in the response y_1, it will also have zero amplitude in y_2. Note that this *mode suppression* by choice of *particular numerical initial conditions* differs from the phenomenon of uncoupled modes. We know that a mode may not be coupled to the input in y_2 but may occur in y_1, for initially deenergized networks; in the present discussion we note that, if a mode is suppressed by choice of *initial conditions*, it does not appear in an output. This general question of whether or not all possible modes appear in an output is of importance because it cautions us to exercise care in the analysis of experimental data or in drawing conclusions from analysis of a single transfer function.

Example 12-1 (*a*) In the circuit of Fig. 12-1, find the complete response for $y_1 = v_1$ and $y_2 = v_2$ and show how the modes are related. (*b*) Show how free modes may be suppressed if initial conditions are properly chosen for the following three cases: $v \equiv 0$; $v = 10$; $v = 10 \cos 2t$.

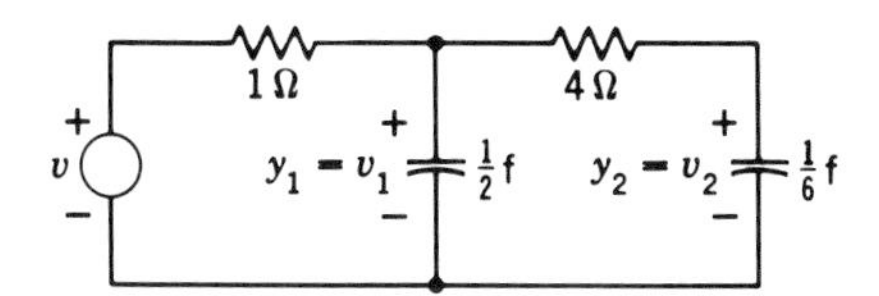

Fig. 12-1 Circuit analyzed in Example 12-1.

Solution (*a*) For the given circuit the state equations are

$$\frac{dv_1}{dt} = -\tfrac{5}{2}v_1 + \tfrac{1}{2}v_2 + 2v$$
$$\frac{dv_2}{dt} = \tfrac{3}{2}v_1 - \tfrac{3}{2}v_2 \qquad (12\text{-}14)$$

The characteristic equation is

$$D(s) = \begin{vmatrix} s + \frac{5}{2} & -\frac{1}{2} \\ -\frac{3}{2} & s + \frac{3}{2} \end{vmatrix} = 0$$

or

$$s^2 + 4s + 3 = 0 \qquad (12\text{-}15)$$

so that the characteristic roots are

$$s_{1,2} = -1, \ -3$$

Hence the free response components are

$$v_{1f} = K_{11}e^{-t} + K_{12}e^{-3t}$$
$$v_{2f} = K_{21}e^{-t} + K_{22}e^{-3t}$$

Using the first state equation with $v = 0$, we obtain the results corresponding to Eqs. (12-12) and (12-13) as follows:

$$(-1 + \tfrac{5}{2})K_{11} - \tfrac{1}{2}K_{21} = 0 \qquad \text{or} \qquad 3K_{11} = K_{21}$$
$$(-3 + \tfrac{5}{2})K_{12} - \tfrac{1}{2}K_{22} = 0 \qquad \text{or} \qquad K_{12} = -K_{22}$$

Thus the complete response for v_1 and v_2 can be written in terms of two constants, for example, K_{11} and K_{12}.

$$v_1 = (v_1)_s + K_{11}e^{-t} + K_{12}e^{-3t}$$
$$v_2 = (v_2)_s + 3K_{11}e^{-t} - K_{12}e^{-3t}$$

(*b*) *Source-free circuit* ($v \equiv 0$). In this case the solution consists of the free modes only:

$$v_1 = K_{11}e^{-t} + K_{12}e^{-3t}$$
$$v_2 = 3K_{11}e^{-t} - K_{12}e^{-3t} \qquad (12\text{-}16)$$

If we choose $v_1(0^+) = V_{10}$ (arbitrary) but $v_2(0^+) = 3V_{10}$, then

$$K_{11} + K_{12} = V_{10}$$
$$3K_{11} - K_{12} = 3V_{10}$$

so that $K_{11} = V_{10}$ and $K_{12} = 0$. Thus, choosing the initial conditions for v_1 and v_2 in the ratio of K_{11}/K_{21} suppresses the mode e^{-3t} in *both* v_1 and v_2.

The source is constant with $v = 10$. In this case the components of the response due to the source are

$$(v_1)_s = (v_2)_s = 10$$

and the complete response has the form

$$v_1 = 10 + K_{11}e^{-t} + K_{12}e^{-3t}$$
$$v_2 = 10 + 3K_{11}e^{-t} - K_{12}e^{-3t}$$

If we choose $v_1(0^+) = V_{10}$ (arbitrary) and $v_2(0^+) = 3V_{10} - 20$, then

$$K_{11} + K_{12} = V_{10} - 10$$
$$3K_{11} - K_{12} = 3V_{10} - 30$$

and $K_{11} = V_{10} - 10$, $K_{12} = 0$, so that the mode e^{-3t} has been suppressed.

The source is sinusoidal with $v = 10 \cos 2t$. In this case the components of the response due to the source are found readily, using amplitude phasors. Let

$$(v_1)_s = \text{Re}\,(\mathbf{V}_{1m}e^{j2t}) \qquad (v_2)_s = \text{Re}\,(\mathbf{V}_{2m}e^{j2t})$$

and use the state equations (12-14) as follows:

$$j2\mathbf{V}_{1m} = -\tfrac{5}{2}\mathbf{V}_{1m} + \tfrac{1}{2}\mathbf{V}_{2m} + 2(10\underline{/0^\circ})$$
$$j2\mathbf{V}_{2m} = \tfrac{3}{2}V_{1m} - \tfrac{3}{2}\mathbf{V}_{2m}$$

or

$$(\tfrac{5}{2} + j2)\mathbf{V}_{1m} - \tfrac{1}{2}\mathbf{V}_{2m} = 20 + j0$$
$$(-\tfrac{3}{2})\mathbf{V}_{1m} + (\tfrac{3}{2} + j2)\mathbf{V}_{2m} = 0$$

Solving,

$$\mathbf{V}_{1m} = -4.46 + j4.31 = 6.21\underline{/136^\circ}$$
$$\mathbf{V}_{2m} = -0.462 + j3.69 = 3.73\underline{/97.1^\circ}$$

Hence the complete response is

$$v_1(t) = 6.21 \cos (2t + 136^\circ) + K_{11}e^{-t} + K_{12}e^{-3t}$$
$$v_2(t) = 3.73 \cos (2t + 97.1^\circ) + 3K_{11}e^{-t} - K_{12}e^{-3t}$$

Since

$$K_{11} + K_{12} = v_1(0^+) - 6.21 \cos 136^\circ = v_1(0^+) + 4.46$$

and

$$3K_{11} - K_{12} = v_2(0^+) - 3.73 \cos 97.1^\circ = v_2(0^+) + 0.462$$

the constant K_{12} can be made zero by choosing

$$v_2(0^+) + 0.462 = 3[v_1(0^+) + 4.46]$$

We note also that if we choose

$$v_1(0^+) = -4.46 \qquad v_2(0^+) = -0.462$$

then both free modes are suppressed.

12-2 General relationship between mode amplitudes

The state equations of a source-free network with n independent energy-storing elements have the form

$$\begin{Vmatrix} p - a_{11} & -a_{12} & \cdots & -a_{1n} \\ -a_{21} & p - a_{22} & \cdots & -a_{2n} \\ \cdot & \cdot & \cdots & \cdot \\ -a_{n1} & -a_{n2} & \cdots & p - a_{nn} \end{Vmatrix} \times \begin{Vmatrix} y_{1f} \\ y_{2f} \\ \cdot \\ y_{nf} \end{Vmatrix} = \begin{Vmatrix} 0 \\ 0 \\ \cdot \\ 0 \end{Vmatrix} \tag{12-17}$$

or

$$\|B(p)\| \times \|y_f\| = \|0\| \tag{12-17a}$$

where

$$\|B(p)\| = \begin{Vmatrix} p - a_{11} & \cdots & -a_{nn} \\ \cdot & \cdots & \cdot \\ -a_{1n} & \cdots & p - a_{nn} \end{Vmatrix}$$

The general solution of the set of linear homogeneous first-order simultaneous equations (12-17) is found through the characteristic equation

$$D_B(s) = 0 \tag{12-18}$$

where $D_B(s)$ is the characteristic determinant:

$$D_B(s) = \begin{vmatrix} s - a_{11} & -a_{12} & \cdots & -a_{1n} \\ -a_{21} & s - a_{22} & \cdots & -a_{2n} \\ \cdot & \cdot & \cdots & \cdot \\ -a_{n1} & -a_{n2} & \cdots & s - a_{nn} \end{vmatrix} \tag{12-19}$$

In the ensuing discussion it is assumed that *all the characteristic modes are distinct.* For this case the general solution for the state variables is

$$\begin{aligned} y_{1f} &= K_{11}e^{s_1t} + K_{12}e^{s_2t} + \cdots + K_{1n}e^{s_nt} \\ y_{2f} &= K_{21}e^{s_1t} + K_{22}e^{s_2t} + \cdots + K_{2n}e^{s_nt} \\ &\vdots \\ y_{nf} &= K_{n1}e^{s_1t} + K_{n2}e^{s_2t} + \cdots + K_{nn}e^{s_nt} \end{aligned} \tag{12-20}$$

where the n different characteristic roots $s_1, \ldots, s_n$ are the solutions of the characteristic equation (12-18). Of the n^2 constants (mode amplitudes) $K_{11}, \ldots, K_{nn}$, n are determined by the initial conditions and the remaining $n^2 - n$ constants found through the general relationship between mode amplitudes, which is discussed below.

These relationships are of the form

$$K_{ik} = \frac{(F_D)_{1i}}{(F_D)_{11}} K_{1k} \qquad k = 1, 2, 3, \ldots, n; i = 1, 2, 3, \ldots, n$$

where $(F_D)_{1i}$ is the cofactor of determinant $D_B(s_k)$, obtained by deleting the ith column and first row. Thus $n - 1$ of the n mode amplitudes of the kth mode e^{s_kt} are directly obtainable from the mode amplitude K_{1k}.

To show how the above relationship is obtained, we observe that the solutions (12-17) can be written as the sum of column vectors:

$$\begin{Vmatrix} y_{1f} \\ y_{2f} \\ \cdot \\ \cdot \\ \cdot \\ y_{nf} \end{Vmatrix} = \begin{Vmatrix} K_{11}e^{s_1t} \\ K_{21}e^{s_1t} \\ \cdot \\ \cdot \\ \cdot \\ K_{n1}e^{s_1t} \end{Vmatrix} + \begin{Vmatrix} K_{12}e^{s_2t} \\ K_{22}e^{s_2t} \\ \cdot \\ \cdot \\ \cdot \\ K_{n2}e^{s_2t} \end{Vmatrix} + \cdots + \begin{Vmatrix} K_{1n}e^{s_nt} \\ K_{2n}e^{s_nt} \\ \cdot \\ \cdot \\ \cdot \\ K_{nn}e^{s_nt} \end{Vmatrix} \tag{12-21}$$

or

$$\begin{Vmatrix} y_{1f} \\ y_{2f} \\ \cdot \\ \cdot \\ \cdot \\ y_{nf} \end{Vmatrix} = \begin{Vmatrix} K_{11} \\ K_{21} \\ \cdot \\ \cdot \\ \cdot \\ K_{n1} \end{Vmatrix} e^{s_1t} + \begin{Vmatrix} K_{12} \\ K_{22} \\ \cdot \\ \cdot \\ \cdot \\ K_{n2} \end{Vmatrix} e^{s_2t} + \cdots + \begin{Vmatrix} K_{1n} \\ K_{2n} \\ \cdot \\ \cdot \\ \cdot \\ K_{nn} \end{Vmatrix} e^{s_nt} \tag{12-21a}$$

Substitution of the solutions as given by Eq. (12-21) into Eq. (12-17*a*) gives the matrix equation

$$\|B(p)\| \times \left(\begin{Vmatrix} K_{11} \\ K_{21} \\ \cdot \\ \cdot \\ \cdot \\ K_{n1} \end{Vmatrix} e^{s_1 t} + \begin{Vmatrix} K_{12} \\ K_{22} \\ \cdot \\ \cdot \\ \cdot \\ K_{n2} \end{Vmatrix} e^{s_2 t} + \cdots + \begin{Vmatrix} K_{1n} \\ K_{2n} \\ \cdot \\ \cdot \\ \cdot \\ K_{nn} \end{Vmatrix} e^{s_n t} \right) = \|0\|$$

or since the distributive law applies to matrix operations,

$$\sum_{k=1}^{n} \|B(p)\| \times \begin{Vmatrix} K_{1k} \\ K_{2k} \\ \cdot \\ \cdot \\ \cdot \\ K_{nk} \end{Vmatrix} e^{s_k t} = \|0\| \tag{12-22}$$

Since the solutions $e^{s_k t}$ are linearly independent, each of the terms in the sum (12-22) must be a null matrix; that is,

$$\|B(p)\| \times \begin{Vmatrix} K_{1k} \\ K_{2k} \\ \cdot \\ \cdot \\ \cdot \\ K_{nk} \end{Vmatrix} e^{s_k t} = \|0\| \qquad k = 1, 2, \ldots, n$$

Recalling that $pe^{s_k t} = s_k e^{s_k t}$, we obtain the result

$$\|B(p)\| \times \begin{Vmatrix} K_{1k} \\ K_{2k} \\ \cdot \\ \cdot \\ \cdot \\ K_{nk} \end{Vmatrix} e^{s_k t} = \|B(s_k)\| \times \begin{Vmatrix} K_{1k} \\ K_{2k} \\ \cdot \\ \cdot \\ \cdot \\ K_{nk} \end{Vmatrix} e^{s_k t}$$

Hence

$$\|B(s_k)\| \times \begin{Vmatrix} K_{1k} \\ K_{2k} \\ \cdot \\ \cdot \\ \cdot \\ K_{nk} \end{Vmatrix} = \|0\| \qquad k = 1, 2, \ldots, n \tag{12-23}$$

Equations (12-23) represent n sets of equations, each of which is a set of linear homogeneous algebraic equations of the form

$$\begin{aligned}
(s_k - a_{11})K_{1k} + (-a_{12})K_{2k} + \cdots + (-a_{1n})K_{nk} &= 0\\
(-a_{21})K_{1k} + (s_k - a_{22})K_{2k} + \cdots + (-a_{2n})K_{nk} &= 0\\
&\vdots\\
(-a_{n1})K_{1k} + (-a_{n2})K_{2k} + \cdots + (s_k - a_{nn})K_{nn} &= 0
\end{aligned} \tag{12-24}$$

These equations are linearly dependent because $D_B(s_k) = 0$; in fact, this relationship is used to determine the roots s_k. We now add the initial conditions; that is, we require that the mode amplitudes satisfy not only Eqs. (12-24), but also the n conditions

$$(y_m)_f \Big|_{t=0^+} = \sum_{k=1}^{n} K_{mk} \tag{12-25}$$

We observe that Eqs. (12-24) are n^2 conditions and Eq. (12-25) imposes an additional n condition on the mode amplitudes. Hence these constants are overspecified unless some of the equations are redundant. We know, however, that since $D_B(s_k) = 0$, at least one equation in each of the set represented by (12-24) is superfluous.

Deleting one of the equations in (12-24), we can solve for $n - 1$ constants in each set in terms of the nth constant; that is, deleting the first equation as superfluous,

$$\begin{aligned}
(s_k - a_{22})K_{2k} + (-a_{23})K_{3k} + \cdots + (-a_{2n})K_{nk} &= a_{21}K_{1k}\\
(-a_{32})K_{2k} + (s_k - a_{33})K_{3k} + \cdots + (-a_{3n})K_{nk} &= a_{31}K_{1k}\\
&\vdots\\
(-a_{n2})K_{2k} + (-a_{n3})K_{3k} + \cdots + (s_k - a_{nn})K_{nk} &= a_{n1}K_{1k}
\end{aligned} \tag{12-26}$$

We note that the matrix of the coefficients for $K_{2k}, \ldots, K_{nk}$ is obtained by deletion of row 1 and column 1 in the B matrix. It follows that the coefficients in Eqs. (12-26) have a determinant that is the cofactor of the first column and first row of the $D_B(s_k)$. Denoting this cofactor by $(F_D)_{11}$ and assuming[1] that $(F_D)_{11} \neq 0$, we can solve for K_{2k} in the form

$$K_{2k} = \frac{\begin{vmatrix} a_{21} & -a_{23} & \cdots & -a_{2n}\\ a_{31} & s_k - a_{33} & \cdots & -a_{3n}\\ \cdot & \cdot & \cdots & \cdot\\ a_{n1} & -a_{n3} & \cdots & s_k - a_{nn} \end{vmatrix}}{(F_D)_{11}} K_{1k} \tag{12-27}$$

Comparing the numerator of Eq. (12-27) with the determinant $D_B(s_k)$ [Eq. (12-19)], we observe that it is obtained from $D_B(s_k)$ by deletion of row 1, column 2, and that the elements of the first column in the remaining rows

[1] If $F_{11} = 0$, another row may be deleted (see also Sec. 12-3).

are the negatives of those in $D_B(s_k)$. A little thought will show that the numerator of Eq. (12-27) is the cofactor of $D_B(s_k)$ associated with row 1 and column 2.

$$K_{2k} = \frac{(F_D)_{12}}{(F_D)_{11}} K_{1k}$$

Similarly, it can be shown that

$$K_{ik} = \frac{(F_D)_{1i}}{(F_D)_{11}} K_{1k}$$

Another way in which this relationship can be expressed is

$$\frac{K_{1k}}{(F_D)_{11}} = \frac{K_{2k}}{(F_D)_{12}} = \frac{K_{3k}}{(F_D)_{13}} = \cdots = \frac{K_{nk}}{(F_D)_{1n}} \tag{12-28a}$$

Equation (12-28*a*) is the desired relationship between the mode amplitudes. The reader familiar with linear algebra will identify the column vectors

$$\begin{Vmatrix} K_{1k} \\ K_{2k} \\ \cdot \\ \cdot \\ \cdot \\ K_{nk} \end{Vmatrix} = K_{1k} \begin{Vmatrix} 1 \\ \dfrac{(F_D)_{12}}{(F_D)_{11}} \\ \cdot \\ \cdot \\ \cdot \\ \dfrac{(F_D)_{1n}}{(F_D)_{11}} \end{Vmatrix} \tag{12-28b}$$

as the characteristic vectors of the state matrix. Among its properties is that every one of the n characteristic vectors [obtained by substituting $k = 1, 2, 3, \ldots, n$ in the left side of (12-28*b*)] is orthogonal to every row vector of the state matrix as indicated by Eq. (12-23).

Example 12-2 In the circuit of Fig. 12-2, obtain the relationship between mode amplitudes.

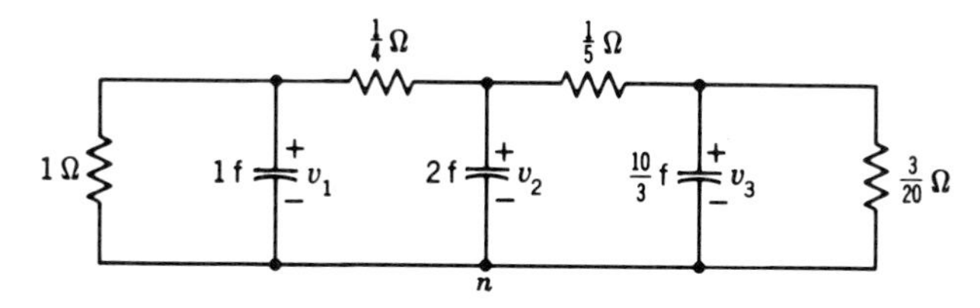

Fig. 12-2 Circuit analyzed in Example 12-2.

Solution We observe that the state variables v_1, v_2, and v_3 are identical with the node voltages as indicated on the diagram. The node equations are

$$\begin{Vmatrix} p+5 & -4 & 0 \\ -4 & 2p+9 & -5 \\ 0 & -5 & \frac{10}{3}p + \frac{35}{3} \end{Vmatrix} \times \begin{Vmatrix} v_1 \\ v_2 \\ v_3 \end{Vmatrix} = \begin{Vmatrix} 0 \\ 0 \\ 0 \end{Vmatrix}$$

Hence the characteristic determinant is given by

$$D_B(s) = \begin{vmatrix} s+5 & -4 & 0 \\ -4 & 2s+9 & -5 \\ 0 & -5 & \frac{10}{3}s + \frac{35}{3} \end{vmatrix}$$
$$= \tfrac{20}{3}(s^3 + 13s^2 + 44s + 32)$$
$$= \tfrac{20}{3}(s+1)(s+4)(s+8)$$

so that the characteristic roots are $s_1 = -1$, $s_2 = -4$, $s_3 = -8$.

We now calculate the cofactors used in Eqs. (12-28a):

$$(F_D)_{11} = \begin{vmatrix} 2s+9 & -5 \\ -5 & \frac{10}{3}s + \frac{35}{3} \end{vmatrix} = \begin{cases} \frac{100}{3} & \text{for } s = -1 \\ -\frac{80}{3} & \text{for } s = -4 \\ 80 & \text{for } s = -8 \end{cases}$$

$$(F_D)_{12} = -\begin{vmatrix} -4 & -5 \\ 0 & \dfrac{10s+35}{3} \end{vmatrix} = \begin{cases} \frac{100}{3} & \text{for } s = -1 \\ -\frac{20}{3} & \text{for } s = -4 \\ -60 & \text{for } s = -8 \end{cases}$$

$$(F_D)_{13} = \begin{vmatrix} -4 & 2s+9 \\ 0 & -5 \end{vmatrix} = 20$$

The free response components are written

$$v_1 = K_{11}e^{-t} + K_{12}e^{-4t} + K_{13}e^{-8t}$$
$$v_2 = K_{21}e^{-t} + K_{22}e^{-4t} + K_{23}e^{-8t}$$
$$v_3 = K_{31}e^{-t} + K_{32}e^{-4t} + K_{33}e^{-8t}$$

The mode amplitudes of $(v_2)_f$ and $(v_3)_f$ are expressed in terms of K_{11}, K_{12}, and K_{13} by use of Eq. (12-28) as follows:

$$K_{21} = K_{11} \left.\frac{(F_D)_{21}}{(F_D)_{11}}\right]_{s=-1} = K_{11}\frac{\frac{100}{3}}{\frac{100}{3}} = K_{11} \qquad K_{31} = K_{11}\left.\frac{(F_D)_{31}}{(F_D)_{11}}\right]_{s=-1}$$
$$= \frac{20}{\frac{100}{3}}K_{11} = \tfrac{3}{5}K_{11}$$

Similarly,

$$K_{22} = K_{12}\frac{-\frac{20}{3}}{-\frac{80}{3}} = \tfrac{1}{4}K_{12} \qquad K_{32} = K_{12}\frac{20}{-\frac{80}{3}} = -\tfrac{3}{4}K_{12}$$
$$K_{23} = -K_{13}\tfrac{60}{80} = -\tfrac{3}{4}K_{13} \qquad K_{33} = K_{13}\tfrac{20}{80} = \tfrac{1}{4}K_{13}$$

Hence the free response components are

$$v_1 = K_{11}e^{-t} + K_{12}e^{-4t} + K_{13}e^{-8t}$$
$$v_2 = K_{11}e^{-t} + \tfrac{1}{4}K_{12}e^{-4t} - \tfrac{3}{4}K_{13}e^{-8t}$$
$$v_3 = \tfrac{3}{5}K_{11}e^{-t} - \tfrac{3}{4}K_{12}e^{-4t} + \tfrac{1}{4}K_{13}e^{-8t}$$

or in matrix form,

$$\begin{Vmatrix} v_1 \\ v_2 \\ v_3 \end{Vmatrix} = \begin{Vmatrix} 1 & 1 & 1 \\ 1 & \frac{1}{4} & -\frac{3}{4} \\ \frac{3}{5} & -\frac{3}{4} & \frac{1}{4} \end{Vmatrix} \times \begin{Vmatrix} K_{11}e^{-t} \\ K_{12}e^{-4t} \\ K_{13}e^{-8t} \end{Vmatrix} \tag{12-29}$$

MODE SUPPRESSION If the initial conditions[1] are given as $y_{kf}(0^+)$, then if

$$\begin{Vmatrix} y_{1f}(0^+) \\ y_{2f}(0^+) \\ \cdot \\ \cdot \\ \cdot \\ y_{nf}(0^+) \end{Vmatrix} = \begin{Vmatrix} K_{1k} \\ K_{2k} \\ \cdot \\ \cdot \\ \cdot \\ K_{nk} \end{Vmatrix} \tag{12-30}$$

all modes except s_k are suppressed.

COMPLEX ROOTS The derivation of the relationship between mode amplitudes applies in the case of complex roots as well as real roots. We recall that the mode amplitudes for a pair of complex conjugate roots are complex conjugate themselves; hence the relationship for such a pair of roots can be found by considering only one complex root.

MULTIPLE ROOTS Since the proof above assumes that the free modes are all of the form $e^{s_k t}$, the case of multiple roots when the form is $t^n e^{s_k t}$ is not included. The reader is referred to the literature on linear algebra for study of the characteristic vectors in such cases.

12-3 Observability

We have already shown that, when a circuit is excited by a single source and is *initially deenergized*, certain outputs may not include all modes. In such cases the modes not appearing in that output are said to be uncoupled from the input (Sec. 11-13). In Secs. 12-1 and 12-2 it is shown that the initial conditions can be chosen so that certain modes are suppressed. A third phenomenon may occur; namely, in a circuit excited by a single source, a mode or modes may not appear in an output *independently of the initial conditions.* In such cases one says that the mode or modes that do not occur in the output are *not observable.* An example of nonobservability can be seen in Example 12-2, where the following result was obtained:

$$v_1 = K_{11}e^{-t} + K_{12}e^{-4t} + K_{13}e^{-8t}$$
$$v_2 = K_{11}e^{-t} + \tfrac{1}{4}K_{12}e^{-4t} - \tfrac{3}{4}K_{13}e^{-8t}$$

If the output is defined as the voltage across the $\frac{1}{4}$-ohm resistance,

$$v_{\text{out}} = v_1 - v_2$$

then

$$v_{\text{out}} = \tfrac{3}{4}K_{12}e^{-4t} + \tfrac{7}{4}K_{13}e^{-8t} \tag{12-31}$$

Hence, in the output, the mode e^{-t} is not observable; that is, no matter which initial values are chosen for v_1, v_2, and v_3, the output [Eq. (12-31)] will include only two modes.

[1] Note that, in the presence of sources, the above theory applies provided that the difference $y_{kf}(0^+) - y_{ks}(0^+)$ is used instead of $y_{kf}(0^+)$.

Recognition that certain modes may not be observable in a given output and input independently of the initial conditions is of great importance in linear-system theory because a system may be unstable and yet the selected output may appear stable.

In Fig. 12-3, a network with negative resistances is shown.[1] Such a network with negative resistances cannot be built with passive elements but may be the equivalent circuit of certain active feedback systems.

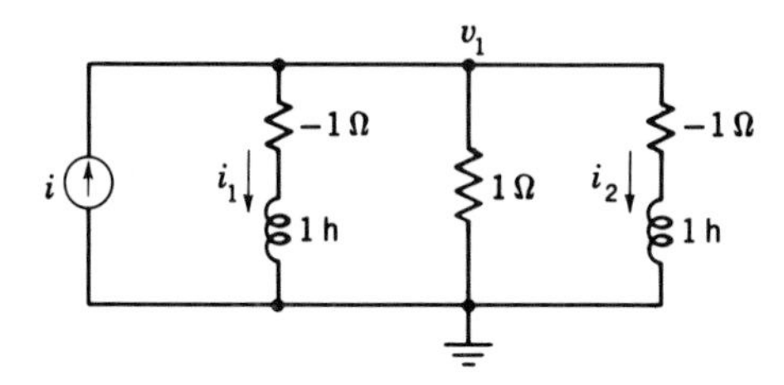

***Fig. 12-3** Network excited by a single source containing negative resistance.*

For the network of Fig. 12-3, the driving-point impedance is

$$Z(p) = \frac{1}{[2/(-1 + p)] + 1} = \frac{p - 1}{p + 1}$$

so that the differential equation relating v_1 to i is

$$(p + 1)v_1 = (p - 1)i$$

We observe, first, that only one characteristic root appears in the equation for v_1, $s = -1$, which indicates a decaying transient. Since there are two independent energy-storing elements, we may believe that we are dealing with a problem of a mode that is not coupled to the input for the selected output. In this case, however, we shall show that a second mode does appear in other outputs (such as v_2), but does *not appear in v_1 independently of the initial conditions; that is, it is not observable in v_1.* To show this, we analyze the state equations, which are

$$\begin{aligned} pi_1 &= 0 - i_2 + i \\ pi_2 &= -i_1 + 0 + i \end{aligned}$$

The characteristic determinant is

$$D_B(s) = \begin{vmatrix} s & 1 \\ 1 & s \end{vmatrix}$$

or

$$D_B(s) = s^2 - 1 = 0$$

which has the solutions

$$s_1 = -1 \qquad s_2 = +1$$

The free response components for i_1 and i_2 are

$$\begin{aligned} i_{1f} &= K_{11}e^{-t} + K_{12}e^{t} \\ i_{2f} &= K_{21}e^{-t} + K_{22}e^{t} \end{aligned}$$

[1] The example cited is similar to that used by L. A. Zadeh and C. A. Desoer, "Linear System Theory," p. 417, McGraw Hill Book Company, New York, 1963.

Substitution into one of the source-free equations ($i \equiv 0$) shows that

$$pi_1 + i_2 = 0$$

or

$$(-K_{11} + K_{21})e^{-t} + (K_{12} + K_{22})e^{t} = 0$$

so that

$$K_{21} = K_{11}$$
$$K_{22} = -K_{12}$$

Hence

$$i_1 = K_{11}e^{-t} + K_{12}e^{t} + i_{1s}$$
$$i_2 = K_{11}e^{-t} - K_{12}e^{t} + i_{2s}$$

The output v_1 is given by

$$v_1 = -(i_1 + i_2) + i$$

Hence

$$v_1 = -(2K_{11}e^{-t} + i_{1s} + i_{2s}) + i$$

Thus, independently of K_{11} and K_{12}, that is, independently of the initial conditions, the mode e^t does not appear in v_1. We note, however, that the voltage across the inductance which carries i_2 is

$$pi_2 = \frac{di_{2s}}{dt} - K_{11}e^{-t} - K_{12}e^{t}$$

and that this response increases indefinitely with time, indicating that the network is unstable.

Normal coordinates The subject of mode coupling, observability, and related system behavior can be generalized by the introduction of *normal coordinates.* If the characteristic roots s_n of a network are all distinct then a normal coordinate is defined as the function $Ke^{s_n t}$. We shall illustrate the use of normal coordinates with a two-state network; the general theory is developed in the state theory literature.[1]

We start with the state equations

$$py_1 = a_{11}y_1 + a_{21}y_2 + g_1$$
$$py_2 = a_{21}y_1 + a_{22}y_2 + g_2$$

and define

$$y_1 = q_1 + q_2$$
$$y_2 = \mu_1 q_1 + \mu_2 q_2$$

where μ_1 and μ_2 are ratios K_{21}/K_{11} and K_{22}/K_{12} respectively, i.e.,

$$\mu_1 = \frac{s_1 - a_{11}}{a_{12}} \qquad \mu_2 = \frac{s_2 - a_{11}}{a_{12}}$$

If $y_1 = q_1 + q_2$ and $y_2 = \mu_1 q_1 + \mu_2 q_2$ are introduced into the state equations, we obtain after much algebra the differential equations for q_1 and q_2 as

$$pq_1 = s_1 q_1 + B_1 g_1$$
$$pq_2 = s_2 q_2 + B_2 g_2 \qquad (12\text{-}32)$$

where

$$B_1 = \frac{-\mu_2 g_1 + g_2}{\mu_1 - \mu_2} \qquad B_2 = \frac{g_1 - \mu_1 g_2}{\mu_1 - \mu_2}$$

We observe that the equations for q_1 and q_2 are not coupled, i.e., they can be solved independently of each other. Moreover, the free component of each

[1] See, for example, Zadeh and Desoer, *op. cit.*

q is an exponential, i.e.,

$$q_1 = q_{1s} + K_{11}e^{s_1t} \qquad q_2 = q_{2s} + K_{12}e^{s_2t}$$

The functions we have denoted by q are the normal coordinates of the system described by the original state equations.

We now recall that any output $z(t)$ is a linear combination of the states y_n and the sources. Since the states and the normal coordinates are also linearly related, we can write for any output

$$z(t) = C_1q_1 + C_2q_2 + D_1g_1 + D_2g_2$$

where C_1, C_2, D_1, D_2 are constants. If it turns out that for a chosen output z the value C_1 is zero then the mode q_1 is not observable in that output.

CONTROLLABILITY If in Eqs. (12-32) $B_1 = 0$ then q_1 is independent of the source waveforms and is said to be *noncontrollable.*

In the example of Fig. 12-3 we let

$$i_1 = q_1 + q_2 \qquad i_2 = q_1 - q_2$$

so that the state equations

$$pi_1 = -i_2 + i \qquad pi_2 = -i_1 + i$$

become

$$pq_1 + pq_2 = -q_1 + q_2 + i$$
$$pq_1 - pq_2 = -q_1 - q_2 + i$$

or

$$pq_1 = -q_1 + 2i$$
$$pq_2 = q_2$$

We note that q_2 is of the form Ke^t and independent of the source. Hence this mode is *not controllable.* For the output v_1 we have

$$v_1 = i_1 + i_2 = q_1$$

Hence v_1 is independent of q_2 so that the mode q_2 is *not observable* in the output v_1 when the source i is used. In contrast, the output voltage across the inductance carrying i_2 is $pi_2 \equiv v_2 \doteq -i_1 + i = -q_1 - q_2 + i$ and both modes are observable in it. If however the initial conditions are chosen so that $q_2(0^+) = 0$ then $q_2 = 0$ for all $t \geq 0^+$, and this mode is *not coupled* to the input v with the output v_2.

12-4 Superposition and driving-point immittance

In this section we discuss certain network properties which are most conveniently studied through node or mesh equations.

> *Principle of superposition* In a network which is composed of linear elements the component of the response due to several ideal sources acting simultaneously is equal to the sum of the components of the response due to each source acting individually.

This statement has already been discussed several times. The superposition principle is the direct consequence of the linearity of the elements, i.e., a consequence of linear equations. Thus the assumption of linearity constitutes the proof of the superposition principle. Nevertheless, we wish to duplicate some of the discussion concerning superposition before the discussion of other network theorems, because all these theorems are a consequence of the superposition property. As we have seen earlier in this chapter, the node and mesh formulation of network equations exhibits the superposition principle in a particularly useful and lucid fashion.

The node equations for an n-node network are

$$\begin{aligned}
Y_{11}(p)v_1 + Y_{12}(p)v_2 + Y_{13}(p)v_3 + \cdots + Y_{1n}(p)v_n &= i_1 \\
Y_{21}(p)v_1 + Y_{22}(p)v_2 + Y_{23}(p)v_3 + \cdots + Y_{2n}(p)v_n &= i_2 \\
\cdot \qquad\qquad\qquad\qquad & \quad \cdot \\
\cdot \qquad\qquad\qquad\qquad & \quad \cdot \\
\cdot \qquad\qquad\qquad\qquad & \quad \cdot \\
Y_{n1}(p)v_1 + Y_{n2}(p)v_2 + Y_{n3}(p)v_3 + \cdots + Y_{nn}(p)v_n &= i_n
\end{aligned}$$

so that

$$v_1 = \frac{(F_y)_{11}}{D_y} i_1 + \frac{(F_y)_{21}}{D_y} i_2 + \cdots + \frac{(F_y)_{n1}}{D_y} i_n \tag{12-33}$$

If in Eq. (12-33) all current sources are set to zero except i_1, then

$$v_1 \text{ due to } i_1 = \frac{(F_y)_{11}}{D_y} i_1$$

Similarly, if all sources except i_2 are set to zero, then

$$v_1 \text{ due to } i_2 = \frac{(F_y)_{21}}{D_y} i_2$$

Hence Eq. (12-33) is an expression of the principle of superposition since it reads

$$v_1 = (v_1 \text{ due to } i_1) + (v_1 \text{ due to } i_2) + \cdots + (v_1 \text{ due to } i_n)$$

If we had written the mesh equations of an m-mesh network, we should have solved for the mesh current i_k and obtained the result

$$i_k = \sum_{a=1}^{m} \frac{(F_z)_{ak}}{D_z} v_a$$

so that $i_k = (i_k \text{ due to } v_1) + (i_k \text{ due to } v_2) + \cdots + (i_k \text{ due to } v_m)$.

We have already presented many examples in which the response of particular networks was determined by applications of the superposition principle. At this point it is important to recognize a general result of superposition. If we wish to study networks in general form, it is quite sufficient that we study the network behavior under the influence of a single ideal source; moreover, we may assume that the network is initially deenergized. This follows first of all from the fact that the initial-energy storage in networks can be represented by ideal sources (initial-condition generators); hence assuming nonzero initial-energy storage corresponds to assuming that the network has several sources. Now, if we prove a statement with the assumption that the network contains only a single ideal source, then, by using the superposition principle, we can prove that statement for any number of sources acting simultaneously, since the general procedure of proof will be the same for all sources.

DRIVING-POINT IMMITTANCE If we wish to impress an ideal source on a source-free network without affecting its natural modes, there are two ways in which this can be done. Either we can connect an ideal current source to a node pair (soldering-iron entry) or we can insert an ideal voltage source (pliers entry). An operational driving-point immittance is then defined in connection with the terminals of the source as that operator which relates the voltage across the source to the current which the source furnishes. Thus, if a current source is impressed between two nodes a and b with reference direction as shown in Fig. 12-4a, the driving-point immittance at ter-

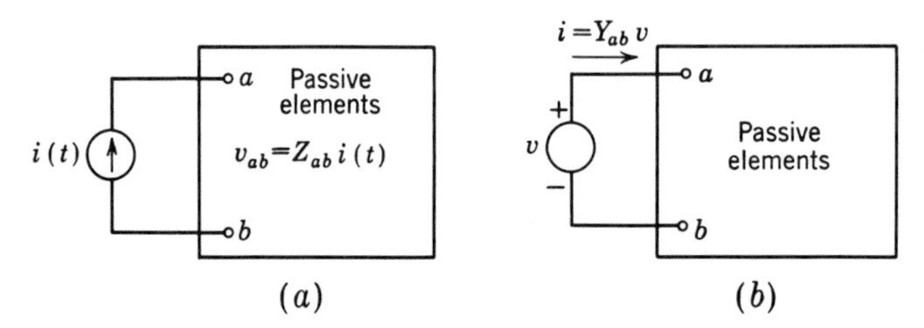

Fig. 12-4 Driving-point immittance functions.

minals a-b, $Z_{ab}(p)$, is given by the operational relationship $v_{ab}(t) = Z_{ab}(p)i(t)$. If a voltage source is inserted into a network between terminals a-b as shown in Fig. 12-4b, the driving-point admittance $Y_{ab}(p)$ is defined through the relationship $i(t) = Y_{ab}(p)v_{ab}(t)$. It should be evident that the driving-point immittance of a given network depends on the terminals used to insert the source.

The use of node and mesh equations makes it possible to deduce a general formula for driving-point immittance. Suppose that a current source $i(t)$ is inserted between node 1 and the reference node such that $i(t)$ flows into node 1 and that this is the only source in the network. Then, from Eq. (12-33),

$$v_1(t) = \frac{(F_y)_{11}}{D_y} i(t)$$

so that the driving-point impedance between node 1 and the reference node, $Z_{1R}(p)$, is

$$Z_{1R}(p) = \frac{(F_y)_{11}}{D_y(p)} \tag{12-34}$$

If a voltage source $v(t)$ is inserted in mesh 1 such that only $i_1(t)$ flows in the source, then, if the reference direction for $v(t)$ is in the direction of mesh 1,

$$i_1(t) = \frac{(F_z)_{11}}{D_z} v(t)$$

Hence the driving-point admittance at the terminals of the source is $Y(p) = (F_z)_{11}/D_z(p)$.

12-5 The reciprocity theorem

All the theorems discussed so far deal with the behavior of a network at a terminal pair. The reciprocity theorem describes how certain voltages and currents in two different branches of a network are related. Thus it will deal with *two terminal pairs* of a network. The reciprocity theorem is most simply stated as follows:

> *Reciprocity theorem* If an ideal current source $i(t)$ is impressed between node 1 and the reference node [such that $i(t)$ enters node 1] of a passive network and if the resulting component of the node voltage $v_2(t)$ due to this source is $v_a(t)$, then impressing the same source current $i(t)$ between node 2 and the reference node [such that $i(t)$ enters node 2] will result in a contribution to the node voltage at node 1 which will be $v_a(t)$.

To understand this statement of the reciprocity theorem, consider the circuit indicated in Fig. 12-5*a*. In this figure the component of $v_2(t)$ due to

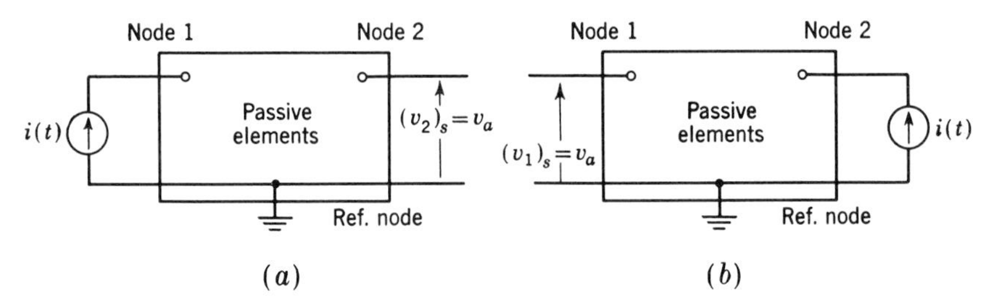

Fig. 12-5 Illustration of the reciprocity theorem for node voltages.

$i(t)$ is $v_a(t)$. Now in Fig. 12-5*b* we have the same passive elements, but $i(t)$ is applied between node 2 and the reference node. The reciprocity theorem tells us that now $[v_1(t)]_s = v_a(t)$.

To prove the reciprocity theorem in the above form, we refer to the node equations (12-32). If $i_1(t) = i(t)$ but all other node current sources are zero,

$$v_2(t) = \frac{(F_y)_{12}}{D_y} i(t) \equiv v_a(t)$$

If now $i_2(t) = i(t)$ and all other current sources are zero,

$$v_1(t) = \frac{(F_y)_{21}}{D_y} i(t) \equiv v_b(t)$$

Now we recall that our circuit elements are bilateral $[Z_{ab}(p) = Z_{ba}(p)]$, so that the node-admittance matrix is symmetrical with respect to the principal diagonal. Hence $(F_y)_{21} = (F_y)_{12}$, and $v_b(t) \equiv v_a(t)$.

In the statement of the reciprocity theorem, we have discussed the relationship between node voltages if a current source is connected between the reference node and either one of two other nodes. We can extend the reciprocity theorem so that it deals with any two node pairs. Consider

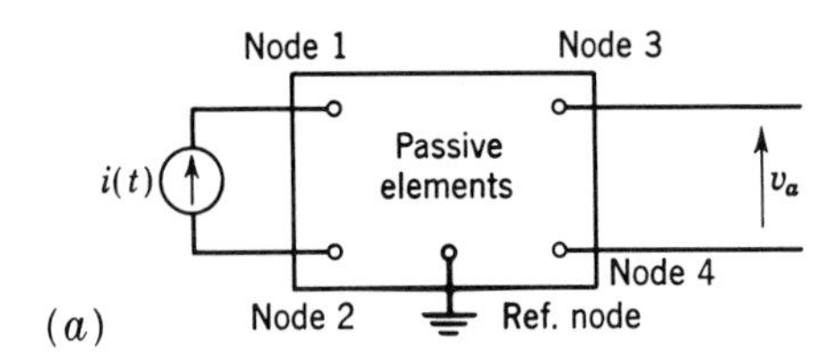

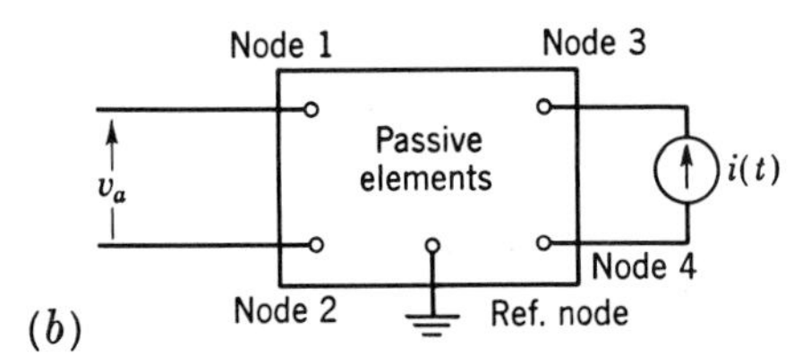

Fig. 12-6 Extension of reciprocity theorem to node-pair voltages.

the network represented by Fig. 12-6a, with a current source impressed between nodes 1 and 2. The node equations will be

$$\begin{aligned}
Y_{11}v_1 + Y_{12}v_2 + Y_{13}v_3 + Y_{14}v_4 + \cdots + Y_{1n}v_n &= i(t) \\
Y_{21}v_1 + Y_{22}v_2 + Y_{23}v_3 + Y_{24}v_4 + \cdots + Y_{2n}v_n &= -i(t) \\
Y_{31}v_1 + Y_{32}v_2 + \cdots\cdots\cdots + Y_{3n}v_n &= 0 \\
\vdots \qquad\qquad & \quad \vdots \\
Y_{n1}v_1 + \cdots\cdots\cdots\cdots\cdots + Y_{nn}v_n &= 0
\end{aligned}$$

Hence

$$\begin{aligned}
v_3(t) &= \frac{(F_y)_{13}}{D_y} i + \frac{(F_y)_{23}}{D_y}(-i) \\
v_4(t) &= \frac{(F_y)_{14}}{D_y} i + \frac{(F_y)_{24}}{D_y}(-i)
\end{aligned} \tag{12-35}$$

and

$$v_a = v_3 - v_4$$

Now if we connect the ideal source to the node pair 3-4 as shown in Fig. 12-6b, we obtain from the new node equations

$$\begin{aligned}
v_1 &= \frac{(F_y)_{31}}{D_y} i + \frac{(F_y)_{41}}{D_y}(-i) \\
v_2 &= \frac{(F_y)_{32}}{D_y} i + \frac{(F_y)_{42}}{D_y}(-i)
\end{aligned} \tag{12-36}$$

and since $F_{kj} = F_{jk}$ as before, by comparing Eqs. (12-35) and (12-36), we have $v_1 - v_2 = v_a$.

The reciprocity theorem can also be used when the network is active provided that we interpret the result as dealing only with the component of the response due to the external source $i(t)$.

It is also possible to state the reciprocity theorem in connection with a voltage source:

> If an ideal voltage source $v(t)$ introduced in branch b_1 produces a component of current $i_a(t)$ due to $v(t)$ in branch b_2, then the same component of current $i_a(t)$ due to $v(t)$ will be produced in branch b_1 when the voltage source $v(t)$ is introduced in branch b_2.

(The proper reference directions for voltages and currents are illustrated in Fig. 12-7.)

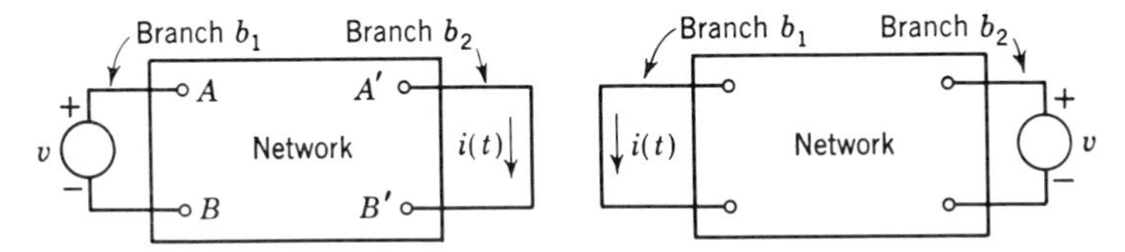

***Fig. 12-7** Reference directions for statement of the reciprocity theorem with voltage sources.*

In this ("voltage") form of the theorem, unless each of the two branches referred to belongs to one mesh only (but not the same mesh), the proof of this form is somewhat cumbersome.

Example 12-3 The "box" in Fig. 12-8 contains only resistances. It is known that, when $v_1(t) = 30t$, $v_2(t) \equiv 0$, $i_1(t) = 5t$, and $i_2(t) = 2t$. Calculate $i_1(t)$ when $v_1(t) = 30t + 60$ and $v_2(t) = 60t + 15$.

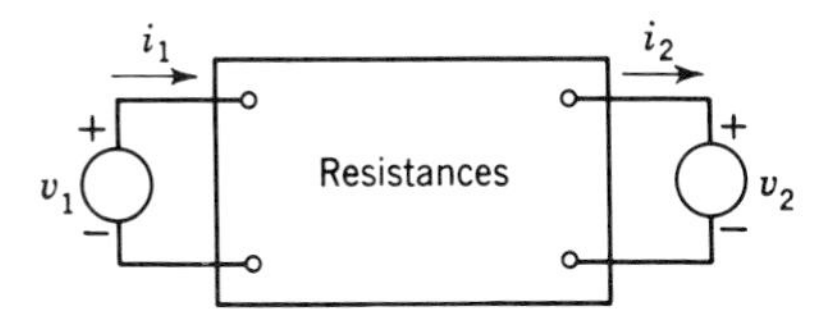

***Fig. 12-8** Resistive network with two accessible branches.*

Solution We shall tabulate as follows:

				Reason
$v_1 = 30t$	$v_2 \equiv 0$	$i_1 = 5t$	$i_2 = 2t$	Given
$v_1 \equiv 0$	$v_2 = 30t$	$i_1 = -2t$	i_2 not known	Reciprocity
$v_1 \equiv 0$	$v_2 = 60t + 15$	$i_1 = -4t - 1$	i_2 not known	Superposition
$v_1 = 30t + 60$	$v_2 = 0$	$i_1 = 5t + 10$	$i_2 = 2t + 4$	Superposition
$v_1 = 30t + 60$	$v_2 = 60t + 15$	$i_1 = t + 9$	i_2 not known	Superposition

12-6 Introduction to two-terminal-pair networks

We have discussed the relationship of driving-point or transfer immittances to the mesh or node formulation of network equations. In this section we shall study the analytical description of two-terminal-pair networks (also called "two-ports"). By a two-terminal-pair network we mean a passive interconnection of initially deenergized circuit elements such that two pairs of terminals are accessible. Referring to Fig. 12-9, the interconnection of elements within the rectangular box is a two-terminal-pair

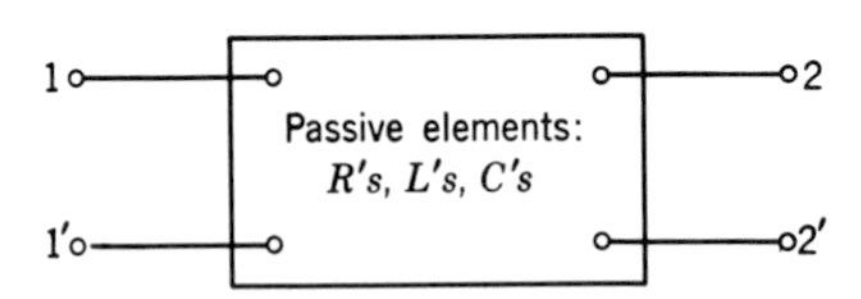

Fig. 12-9 A passive two-terminal-pair (two-port) network.

network. The two accessible terminal pairs are 1-1′ and 2-2′, usually called the input and output terminals, respectively. In discussing such networks we deal with the relationship between v_1, i_1, v_2, i_2, that is, with the variables that are associated with the two terminal pairs; we do *not* consider such functions as v_{12}. In Fig. 12-10 we have indicated schematically a general

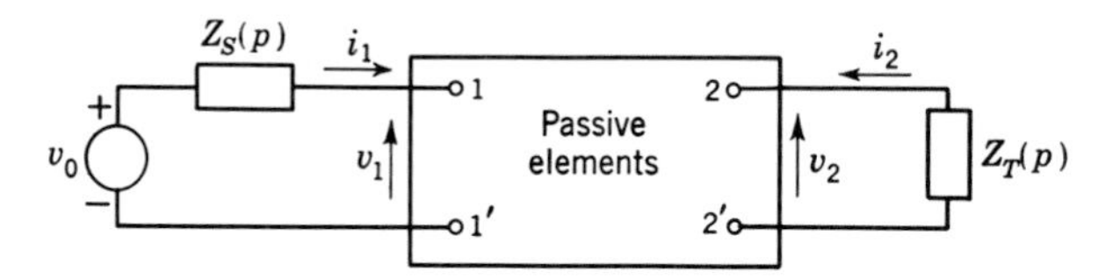

Fig. 12-10 A two-port supplied by a source with internal immittance terminated into an impedance Z_T.

problem of signal transmission. A two-terminal-pair network is interposed between a source and a load. How can the input voltage and current (v_1 and i_1) be related to load voltage and current (v_2 and i_2)? We know immediately that $v_2 = -Z_T i_2$. On the other hand, the relationship between v_1 and i_2 depends on the elements of the two-terminal pair as well as on the load. A simple way of arriving at a useful result for the schematic of Fig. 12-10 is to apply mesh[1] analysis (had the current-source equivalent for v_0 and Z_s been chosen, node analysis would be used). As indicated in Fig. 12-11, we assume that v_0 and Z_s are part of mesh 1 only, and no other mesh, and that Z_T is part of mesh 2 only, and no other mesh.

If we calculate the mesh impedances, all these except Z_{11} and Z_{22} are independent of source and load. For Z_{11} we may write $Z_{11} = Z_s + Z'_{11}$

[1] We use the term *mesh* synonymously with *independent loop*.

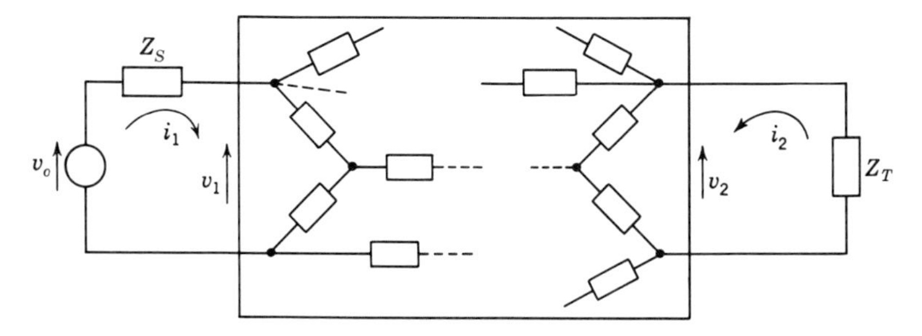

Fig. 12-11 Choice of mesh currents for analysis of a two-port.

and, for Z_{22}, $Z_{22} = Z_T + Z'_{22}$, where Z'_{11} and Z'_{22} are the contributions of the two-terminal pair to the mesh impedances of meshes 1 and 2, respectively. Hence the mesh equations for meshes 1 and 2, assuming m meshes, read

$$\begin{aligned} Z'_{11}i_1 + Z_s i_1 + Z_{12}i_2 + \cdots + Z_{1m}i_m &= v_0 \\ Z_{21}i_1 + Z'_{22}i_2 + Z_T i_2 + \cdots\cdots\cdots &= 0 \end{aligned}$$

Since $v_o - Z_s i_1 = v_1$ and $v_2 = -Z_T i_2$, the m-mesh equations read

$$\begin{aligned} Z'_{11}i_1 + Z_{12}i_2 + Z_{13}i_3 + \cdots + Z_{1m}i_m &= v_1 \\ Z_{21}i_1 + Z'_{22}i_2 + Z_{23}i_3 + \cdots + Z_{2m}i_m &= v_2 \\ Z_{31}i_1 + Z_{32}i_2 + Z_{33}i_3 + \cdots + Z_{3m}i_m &= 0 \\ \vdots \qquad\qquad\qquad &\quad \vdots \\ Z_{m}i_1 + Z_{m2}i_2 + Z_{m3}i_3 + \cdots + Z_{mm}i_m &= 0 \end{aligned}$$

The mesh equations are written in the above form because of the fact that the impedance functions which appear in this set of equations depend *only* on the two-terminal pair and do not depend on source or load. To obtain a relationship between the input and output variables which does not depend on Z_s and Z_T, we may solve the above equations for i_1 and i_2 only since none of the other currents are accessible.

Using Cramer's rule,

$$i_1 = \frac{\begin{vmatrix} v_1 & Z_{12} & \cdots & Z_{1m} \\ v_2 & Z'_{22} & \cdots & Z_{2m} \\ 0 & \cdot & \cdots & \cdot \\ 0 & \cdot & \cdots & \cdot \\ 0 & \cdot & \cdots & Z_{mm} \end{vmatrix}}{\begin{vmatrix} Z'_{11} & Z_{12} & Z_{13} & \cdots & Z_{1m} \\ Z_{21} & Z'_{22} & Z_{23} & \cdots & \cdot \\ Z_{31} & Z_{32} & Z_{33} & \cdots & Z_{mm} \end{vmatrix}} = \frac{\begin{vmatrix} v_1 & Z_{12} & \cdots & Z_{1m} \\ v_2 & Z'_{22} & \cdots & \cdot \\ 0 & \cdot & \cdots & \cdot \\ 0 & \cdot & \cdots & \cdot \\ 0 & \cdot & \cdots & Z_{mm} \end{vmatrix}}{D'_z}$$

where the prime (that is, D_z') indicates that the mesh impedances for meshes 1 and 2 do not include source or load. By Laplace's expansion,

$$\begin{aligned} i_1 &= \frac{(F_z')_{11}}{D_z'} v_1 + \frac{(F_z')_{21}}{D_z'} v_2 \\ i_2 &= \frac{(F_z')_{12}}{D_z'} v_1 + \frac{(F_z')_{22}}{D_z'} v_2 \end{aligned} \tag{12-37}$$

It follows that the relationship between the four variables associated with the two-terminal pairs is completely defined through the four operators $(F_z')_{11}/D_z'$, $(F_z')_{22}/D_z'$, $(F_z')_{21}/D_z'$, and $(F_z')_{12}/D_z'$. If the network is bilateral and obeys the reciprocity theorem, then $F_{21} = F_{12}$. Hence three operators are sufficient to specify the terminal relationships for a two-terminal-pair network!

Dimensionally, each of the terms F_{ij}'/D_z' is an admittance function. We use lower-case letters to abbreviate,

$$\begin{aligned} y_{11}(p) &= \frac{(F_z')_{11}}{D_z'} \\ y_{22}(p) &= \frac{(F_z')_{22}}{D_z'} \\ y_{12}(p) &= y_{21}(p) = \frac{(F_z')_{12}}{D_z'} \end{aligned}$$

Hence we can write

$$\begin{aligned} i_1 &= y_{11}(p)v_1 + y_{12}(p)v_2 \\ i_2 &= y_{12}(p)v_1 + y_{22}(p)v_2 \end{aligned} \tag{12-38}$$

Now our result so far can be stated in words as follows: The terminal relations of a two-terminal-pair network are specified through the three operators y_{11}, y_{22}, and y_{12}. These operators can be obtained from the mesh equations or the node equations as above. Even if this statement were the only conclusion from the foregoing discussion, the results would be important. In addition, however, we shall now show that these operators can be determined, without study of the mesh or node equations, simply by calculation (or "measurement") of the terminal relationships under special conditions.

12-7 The short-circuit admittance matrix

If we inspect Eqs. (12-38), we observe that, when $v_2 \equiv 0$,

$$\begin{aligned} i_1 &= y_{11}v_1 \qquad v_2 \equiv 0 \\ i_2 &= y_{12}v_1 \qquad v_2 \equiv 0 \end{aligned}$$

Now setting $v_2 \equiv 0$ means short-circuiting terminal pair 2-2′. Hence y_{11} is the driving-point admittance at terminal pair 1-1′, with terminal pair 2-2′ short-circuited. Similarly, y_{12} is the transfer admittance from terminal pair 1-1′ to 2-2′ with v_2 short-circuited, and y_{22} is the driving-point admittance at terminal pair 2-2′ with pair 1-1′ short-circuited! Since, in matrix form, Eqs. (12-38) are written

$$\begin{Vmatrix} y_{11} & y_{12} \\ y_{12} & y_{22} \end{Vmatrix} \times \begin{Vmatrix} v_1 \\ v_2 \end{Vmatrix} = \begin{Vmatrix} i_1 \\ i_2 \end{Vmatrix}$$

the matrix

$$\begin{Vmatrix} y_{11}(p) & y_{12}(p) \\ y_{21}(p) & y_{22}(p) \end{Vmatrix} \equiv \|y\|$$

is referred to as the *short-circuit admittance matrix* of the two-terminal-pair network. We again point out that the elements of the matrix need not be obtained from the mesh equations but are determined from transfer and driving-point admittances. From this point of view, Eqs. (12-38) *define* the short-circuit admittance matrix.

12-8 The open-circuit impedance matrix

If Eqs. (12-38) are solved for v_1 and v_2, we obtain

$$v_1 = \frac{y_{22}}{y_{11}y_{22} - y_{12}^2} i_1 + \frac{-y_{12}}{y_{11}y_{22} - y_{12}^2} i_2$$
$$v_2 = \frac{-y_{12}}{y_{11}y_{22} - y_{12}^2} i_1 + \frac{y_{11}}{y_{11}y_{22} - y_{12}^2} i_2$$

The terms $y_{ij}/(y_{11}y_{22} - y_{12}^2)$ are dimensionally impedance functions. Hence we can write

$$\begin{aligned} v_1 &= z_{11}i_1 + z_{12}(i_2) \\ v_2 &= z_{12}i_1 + z_{22}(i_2) \end{aligned} \tag{12-39}$$

Exactly as before, the operators z_{11}, z_{22}, and z_{12} need *not* be considered as defined either through the short-circuit admittance matrix or through the mesh or node equations. We rather observe that

$$v_1 = z_{11}i_1 \text{ if } i_2 \equiv 0 \qquad v_2 = z_{12}i_1 \text{ if } i_2 \equiv 0$$
$$v_1 = z_{12}i_2 \text{ if } i_1 \equiv 0 \qquad v_2 = z_{22}i_2 \text{ if } i_1 \equiv 0$$

Hence these operators are determined as driving-point or transfer impedances under open-circuit conditions. The parameters in matrix form

$$\begin{Vmatrix} z_{11} & z_{12} \\ z_{12} & z_{22} \end{Vmatrix}$$

are referred to as the *open-circuit impedance matrix* of the two-terminal pair.

12-9 Circuit interpretation of short-circuit admittance and open-circuit impedance matrices

Each of Eqs. (12-38),

$$\begin{aligned} i_1 &= y_{11}v_1 + y_{12}v_2 \\ i_2 &= y_{21}v_1 + y_{22}v_2 \end{aligned} \tag{12-38}$$

can be interpreted as expressing Kirchhoff's current law at a junction. This is done in Fig. 12-12 when the current i_1 enters the junction a and

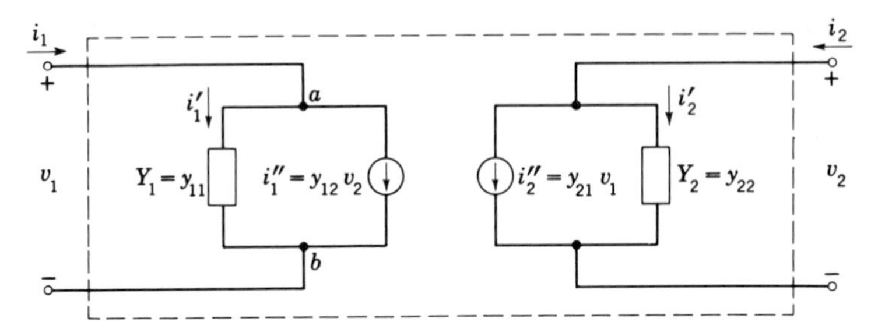

Fig. 12-12 Interpretation of a short-circuit admittance matrix as a network with two dependent sources and two driving-point admittances.

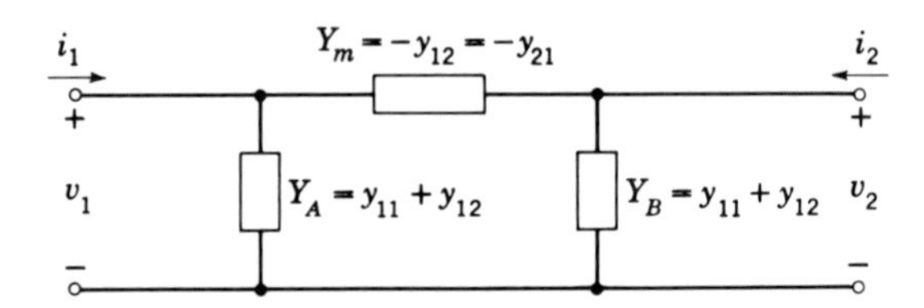

Fig. 12-13 Interpretation of a short-circuit admittance matrix as a three-element network when $y_{12} = y_{21}$.

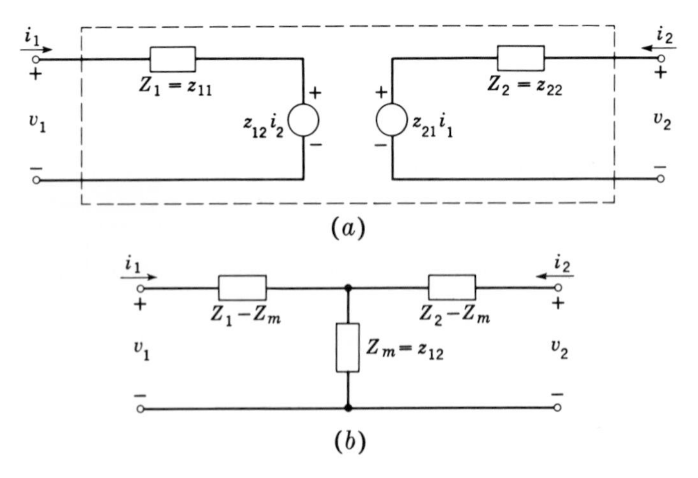

Fig. 12-14 (a) General circuit interpretation of the open-circuit impedance matrix. (b) Network to represent a two-port in terms of open-circuit impedance functions when $z_{12} = z_{21}$.

divides into i_1' and i_1''. In that illustration i_1' is identified as the current through a driving-point immittance $Y_1(p) = y_{11}(p)$, and i_1'' is represented as a dependent source. Similarly, i_2 is represented as the sum i_2' and i_2''.

When the two-port obeys the reciprocity theorem one can replace the dependent sources with bilateral element as shown in Fig. 12-13. It is left as an exercise for the reader to show that the open-circuit impedance representation can be visualized in general through the network of Fig. 12-14*a* and reduces to Fig. 12-14*b* when the reciprocity theorem applies.

12-10 The hybrid matrix

A third method of representing two-ports is by expressing v_1 and i_2 in terms of v_1 and i_1. The equations in this case are written in the form

$$v_1 = h_{11}(p)i_1 + h_{12}(p)v_2 \qquad (12\text{-}40a)$$

$$i_2 = h_{21}(p)i_1 + h_{22}(p)v_2 \qquad (12\text{-}40b)$$

Equations (12-40) define the h matrix as follows:

$$\begin{aligned} h_{11} &= \frac{v_1}{i_1} && \text{with } v_2 \equiv 0 \\ h_{12} &= \frac{v_1}{v_2} && \text{with } i_1 \equiv 0 \\ h_{21} &= \frac{i_2}{i_1} && \text{with } v_2 \equiv 0 \\ h_{22} &= \frac{i_2}{v_2} && \text{with } i_1 \equiv 0 \end{aligned} \qquad (12\text{-}40c)$$

From the relationships (12-40*a*) we observe that

h_{11} = driving-point impedance at terminal pair 1 with terminal pair 2 short-circuited

h_{12} = voltage gain v_1/v_2 with terminal pair 1 open; that is, $h_{12}v_2$ is the voltage produced at terminal pair 1 when it is open-circuited and an ideal source v_2 is placed across terminal pair 2

h_{21} = current gain i_2/i_1 with terminal pair 2 short-circuited; that is, $h_{21}i_1$ is the current produced in a short circuit placed across terminal pair 2 with an ideal current source i_1 connected to terminal pair 1

h_{22} = driving-point admittance at terminal pair 2 with terminal pair 1 open-circuited

These interpretations make clear the term "hybrid" as used to describe the h functions: each h function is obtained from a different type of driving-point or transfer-function measurement.

The relationship between the h matrix and the short-circuit admittance matrix is easily shown. Thus, if the first of Eqs. (12-38) is written

$$v_1 = \frac{1}{y_{11}}\, i_1 - \frac{y_{12}}{y_{11}}\, v_2 \qquad (12\text{-}41)$$

we have the relationships $h_{11} = 1/y_{11}$, $h_{12} = -y_{12}/y_{11}$. Introducing (12-41) into the second of Eqs. (12-37) gives

$$i_2 = y_{21}\left(\frac{1}{y_{11}}\,i_1\right) + \left(-\frac{y_{21}y_{12}}{y_{11}} + y_{22}\right)v_2 \tag{12-42}$$

so that $$h_{21} = \frac{y_{21}}{y_{11}} \qquad \text{and} \qquad h_{22} = \frac{y_{11}y_{22} - y_{12}y_{21}}{y_{11}}$$

Similar relations can be derived to relate the h matrix to the z matrix. The circuit interpretation of the h matrix is obtained by interpreting Eq. (12-40*a*) as a series connection of h_{11} ohms and a dependent voltage source $h_{21}v_2$ and identifying Eq. (12-40*b*) as a junction where the current i_2 splits into an admittance h_{22} and a controlled current source $h_{21}i_1$.

12-11 Impedance level

The reader has undoubtedly noticed that the numerical values assigned to the circuit parameters in the numerical examples and in many of the problems at the ends of the chapters were not "practical." The magnitudes of the values which were chosen for the R, L, and C elements were between 0.1 and 10 (ohms, henrys, and farads, respectively). Now, we recall from elementary physics that such values (for example, a 1-farad capacitance) may not correspond to practical devices. We have chosen these simple numbers for two reasons: First, the principles connected with the solutions of problems can be exhibited just as clearly with simple numbers as with annoying decimals, and the operations are easier to follow with small integral numbers. Second, as we shall show in this section, a change of the scale can often be used either to simplify the numbers in a problem or, more important, to *reinterpret* a problem in such a manner that the numbers are simple and still correspond to a problem with "practical" numbers.

Circuit problems can be formulated in a variety of ways. For example, mesh, node, or other equations may be used, but in every case we make use of the operational immittance. Thus the equations with which we deal have terms of the form

$$Z(p)i = v \qquad \text{or} \qquad Y(p)v = i$$

Now, if an operational impedance is multiplied by a constant, say, k_L (a constant for change of level), then

$$v(t) = k_L Z(p)\,\frac{1}{k_L}\,i(t) \tag{12-43}$$

Now we can interpret Eq. (12-43) in two ways. The obvious way is to state that one side was multiplied and divided by a constant and nothing has changed. Alternatively, we may say that Eq. (12-43) states: If the operational impedance is multiplied by k_L, then $1/k_L$ times as much current is needed to produce the same voltage as before the multiplication. This interpretation is used to change the impedance level of a network.

Operational impedances are composed of sums, products, and quotients of terms which have the form

$$Z(p) = R + pL + \frac{1}{pC}$$

Hence
$$k_L Z(p) = k_L R + pk_L L + \frac{1}{pC/k_L} \tag{12-44}$$

It follows that the impedance level of a network can be changed if every resistance is multiplied by a constant, every inductance is multiplied by the same constant, and every capacitance is divided by that constant. (For convenience we often choose $k_L = 10^n$, n = positive or negative integer.)

Example 12-4 In the circuit of Fig. 12-15*a* it is desired that the value of no capacitance should exceed 1 μf. Draw a terminal pair whose immittance has the same poles and zeros (the same frequency response) as the terminal pair shown but which satisfies the limitation on capacitance values.

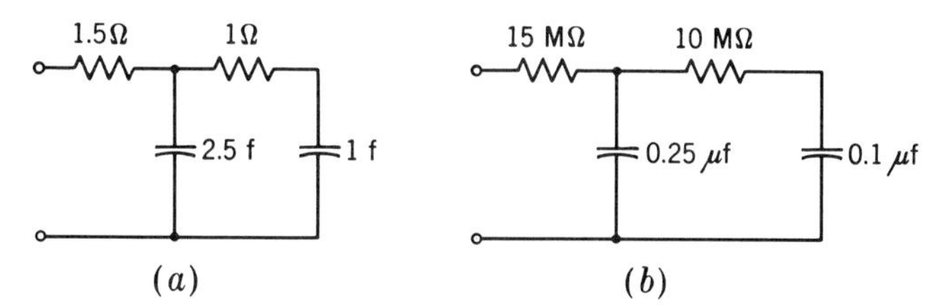

Fig. 12-15 ***Circuits to illustrate change in impedance level.***

Solution Since the specification deals with reducing capacitance values, $k_L > 1$. The larger capacitance controls. Choosing $k_L = 10^7$, a network that meets the requirement is shown in Fig. 12-15*b*.

The results of the above discussion may be formally stated as follows:

Let $Z_G(j\omega)$ be a driving-point or transfer impedance of a *given* network G, and let it be required to design a *new* network N with the same topological configuration as G but with the requirement that the impedance $Z_N(j\omega)$ and $Z_G(j\omega)$ satisfy the condition

$$Z_N(j\omega) = k_L Z_G(j\omega)$$

where k_L is a positive real constant. The new network is obtained from the given network by multiplying every resistance and every inductance in the given network by k_L and dividing every capacitance by k_L.

Figure 12-16 shows the magnitude of admittance of a series R-L-C circuit, as well as the magnitude of admittance when the *impedance* level is changed by a factor of $k_L = 2$, as a function of radian frequency.

From Fig. 12-16 it is seen that at every frequency the admittance of one network is twice the admittance of the other network. We also note that

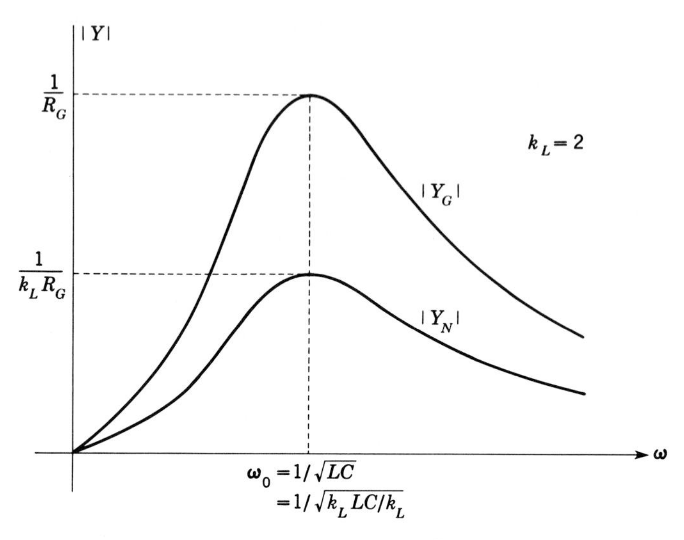

Fig. 12-16 Illustration of impedance-level change in a series R-L-C circuit. The diagram is drawn for $k_L = 2$.

the two graphs have identical features; for example, they have the same half-power frequencies, as well as frequency at which they reach their peak.

We observe, further, that the new network N and the given network G will have the same pole-zero distribution for any network function. This can be shown by observing that network functions may be obtained as the ratio of certain cofactors to the determinant of the mesh-impedance matrix. Now the determinant of the mesh-impedance matrix of N will be $(k_L)^m$ times that of G; hence the pole-zero locations are unaffected by impedance level scaling.

12-12 Change of time and frequency scale

We now consider the problem of time scaling. Suppose that, for a given network, where we have used simple numbers, we obtain poles and zeros of the undesired order of magnitude. Thus, for example, we may have in the network a free mode with a time constant of 1 sec when we really desire a microsecond time constant. It is also possible that the frequency-response curve of a network function has the correct shape but not the desired frequency scale. For example, a series *R-L-C* circuit in which $R = 0.1$ ohm, $C = 1$ farad, $L = 1$ henry may have the (desired) $Q_0 = 10$ but a radian resonant frequency of 1. We may be satisfied with the value Q_0, but may desire a radian resonant frequency of 100,000. By choice of proper scale factors the results obtained by the use of analysis of circuits with simple numbers can be used.

The problem is then to modify the values of the elements in a network so that the shapes of its amplitude- and phase-frequency curves are unaltered but every aspect of these new curves occurs at a frequency k_F times the old frequency. To satisfy this requirement, we note that the frequency response of a network may be computed from the ratios of length of lines joining the zeros and poles of the network function to the pole $j\omega$ representing the source function. If the shape of the frequency response is to be unaltered, this ratio must be maintained when ω is changed to $k_F\omega$.

We formulate the question as follows: In the given network each inductance will have a reactance $X_G = \omega_1 L_G$ at ω_1. What shall the value of L be so that each inductance will have the value $\omega_1 L_G$ at $k_F\omega_1$? The answer is, evidently, to change L_G to $L_N = L_G/k_F$. Similarly, if we replace each C_G by $C_N = C_G/k_F$, all reactances of the new network will give, at $k_F\omega$, the reactance that the given network has at ω. Since the resistances are independent of frequency, they remain unchanged.

The results of the foregoing discussion may be formally stated as follows: Let $H_G(j\omega)$ denote a network function of a given network G, and let it be required to design a new network with the same topological configuration as G such that the network function of the new network, $H_N(j\omega)$, satisfies the condition

$$H_N(j\omega) = H_G\left(j\frac{\omega}{k_F}\right) \tag{12-45}$$

where k_F is a real positive number. The new network is derived from the given network by dividing the value of every L and C in the given network by k_F and leaving the value of R's unaltered.

As an example, suppose that we wish to change the resonant frequency of a series R, L, C circuit by a factor of 2 and that the network function of interest is $H_G(j\omega) = \dfrac{1}{R + j\omega L + 1/j\omega C}$. According to this rule,

$$H_N(j\omega) = \frac{1}{R + j\omega(L/2) + 2/j\omega C}$$

The magnitudes of $H_G(j\omega)$ and $H_N(j\omega)$ are plotted in Fig. 12-17. In this figure any feature of $H_G(j\omega)$ occurring at a radian frequency ω occurs in

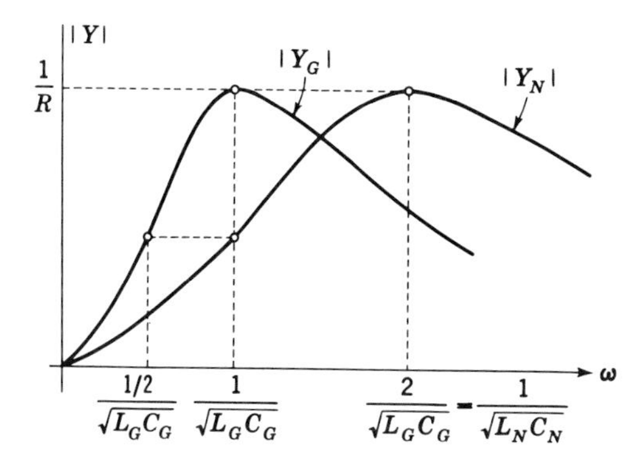

Fig. 12-17 Illustration of frequency scaling, $k_F = 2$, in a series R-L-C circuit.

$H_N(j\omega)$ at 2ω. For example, the peaks occur at frequencies $1/\sqrt{L_G C_G}$, and $2/\sqrt{L_G C_G}$, the lower half-power frequencies, occur at frequencies in the ratio 2:1, etc. Note that, although the bandwidth of $H_N(j\omega)$ is twice the bandwidth of $H_G(j\omega)$, the two *relative bandwidths* (Q_0) are the same. We also note that, at some frequencies, the magnitude of $H_N(j\omega)$ is smaller than the magnitude of $H_G(j\omega)$, and at other frequencies, the reverse is true.

The significance of the term scaling becomes clear when we observe that, if in Fig. 12-17 we had plotted $|Y|$ as a function of ω/ω_0, where ω_0 is the radian frequency at which $|Y|$ is a maximum, the same graph could apply to both $|Y_G|$ and $|Y_N|$ but the value of ω_0 would be different for the two networks G and N; $(\omega_0)_G = \frac{1}{2}(\omega_0)_N$, where, for the given network, $\alpha = \alpha_G = R/2L_G$, and for the new network, $\alpha = \alpha_N = R/2L_N$. Since $L_N = L_G/2$, $\alpha_N = 2\alpha_G$. We have already shown that $(\omega_0)_N = 2(\omega_0)_G$; hence each characteristic root of the new network is twice the corresponding characteristic root of the given network, $(s_{1,2})_N = 2(s_{1,2})_G$. A little thought shows that this relationship indicates a speeding-up of the time scale; the new time constant $1/\alpha_N = T_N$ is half that of the given network.

The above result is general; that is, if in a given network G, having poles and zeros at s_G, the frequency scale is changed by k_F to give the new network N, then all poles and zeros of N will be at $s_N = k_F s_G$ and the time scale of the free component will be divided by k_F.

To prove the foregoing statement we begin with a typical term in the transform mesh-impedance matrix

$$[Z_{ij}(s)]_G = sL_G + R + \frac{1}{sC_G} \tag{12-46}$$

In the new network the corresponding term will be

$$[Z_{ij}(s)]_N = s\frac{L_G}{k_F} + R + \frac{k_F}{sC_G} \tag{12-47}$$

In Eq. (12-47) we substitute S for s/k_F. This gives

$$[Z_{ij}]_N = SL_G + R + \frac{1}{SC_G} \tag{12-48}$$

Comparing Eq. (12-46) with (12-48), we observe that the forms are identical, but S in (12-48) appears for s in (12-46). Thus, if $D_{Z_G}(s) = 0$ at $s = s_G$, then $D_{Z_n}(S) = 0$ for $S = s_G$. Since $S = s/k_F$, $s_N = k_F s_G$. Hence the above reasoning applies to the cofactor of the determinant D_Z. We conclude that multiplication of the frequency scale by k_F multiplies all poles and zeros by k_F and divides the time scale by k_F.

Example 12-5 In a series *R-L-C* circuit, $L = 1$, $C = 1$, and $R = \frac{1}{2}$. Change the impedance level so that the maximum magnitude of the admittance is 10^{-3} mho, and change the time scale so that $\omega_d = 10^6$.

Solution We begin by changing the impedance level. Since $R = \frac{1}{2}$ ohm, $G = 2$ mhos. For a series *R-L-C* circuit the maximum admittance occurs at resonance and is equal to G. Hence the impedance level must be increased by $k_L = 1{,}000/\frac{1}{2} = 2{,}000$. This requires $L = 2{,}000$ henrys, $C = 5 \times 10^{-4}$ farad, and $R = 1{,}000$ ohms.

To find the required frequency scaling factor we calculate ω_d in the given network. Since $\alpha = R/2L = 0.25$, $\omega_0 = 1/\sqrt{LC} = 1$, $\omega_d = 1 - (0.75)^2 = 0.97$, and $k_F = 10^6/0.97 = 1.03 \times 10^6$. Hence, in the final network, $R = 1{,}000$ ohms, $L = 2{,}000/(1.03 \times 10^6) = 1.94$ mh, and $C = 5 \times 10^{-4}/(1.03 \times 10^6) = 485$ pf.

In this example we first changed the impedance level and then the frequency scale, but since the two procedures are independent, the reverse order could just as well have been used.

PROBLEMS

12-1 In the *R-L-C* circuit shown in Fig. P12-1, let the source function be zero. (*a*) Find the characteristic roots s_1 and s_2 of the circuit. (*b*) Express $i(t)$ in terms of $v_C = K_{11}e^{s_1t} + K_{12}e^{s_2t}$. (*c*) Find the relationship between the initial conditions $v_C(0^+)$ and $i(0^+)$ so that (1) the mode with the larger time constant is suppressed; (2) the mode with the smaller time constant is suppressed. (*d*) For arbitrary initial conditions, what is the output (1) v_{ad}; (2) v_{bn}?

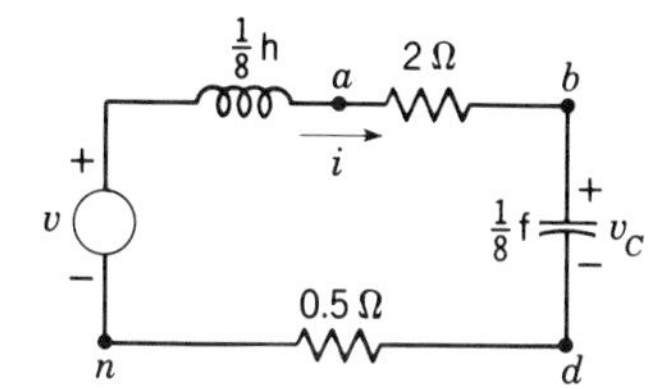

Fig. P12-1

12-2 In the circuit of Fig. P12-1, $v_C(0^+) = 5$ volts. Find the initial value $i(0^+)$ for each of the following cases: (*a*) The source function is $v(t) = (37 + 32t)u(t)$, and the free mode with the larger time constant is suppressed. (*b*) The source function is as in part *a*, but the free mode with the smaller time constant is suppressed. (*c*) The source is 16 cos 8*t*, and the mode with the larger time constant is suppressed.

12-3 In the circuit of Fig. P12-1, find $v_{ad}(t)$ for all $t \geq 0^+$ if $v(t) = (37 + 32t)u(t)$ if $v_{bd}(0^+) = 3$ volts and $i(0^+) = 4$ amp.

12-4 In a critically damped *R-L-C* circuit with a voltage source $v(t)$, the free modes are $K_1e^{-\omega_0t}$ and $K_2te^{-\omega_0t}$. Show how the initial values of the voltage across the capacitance $v_C(0^+)$ and the current $i(0^+)$ can be chosen so that either or both modes are suppressed. Use the reference directions for v, i, and v_C (but not the numerical values) shown in Fig. P12-1.

12-5 For the circuit of Fig. P12-5: (*a*) Show that all possible free modes are given by $(K_1 + K_2t)e^{-t}$. (*b*) Show that if the initial-energy storage is zero, neither mode appears in the output v_{an}. (*c*) Insert initial-condition generators in series with L and C, and show that the reason neither mode occurred in part *b* is that the mode te^{-t} is not observable at the terminals *a-n* and the mode e^{-t} is not coupled to the source v. (*d*) Show that, for arbitrary initial conditions, both modes occur in i.

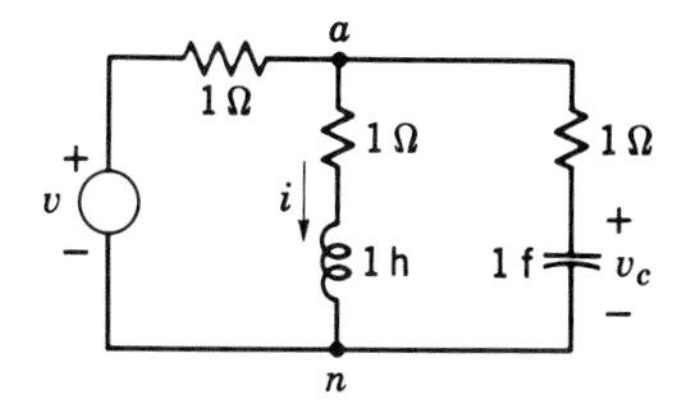

Fig. P12-5

12-6 It is the purpose of this problem to show that certain modes can never have their mode amplitudes influenced by the source. Such modes are said to be *"not controllable."* In the circuit of Fig. P12-6: (*a*) Show that, if $i_1(0^+) = I_{10}$ and $i_2(0^+) = I_{20}$, then $v_2(t)$ is given by $v_2(t) = 0.2V + (0.8V - 8I_{10})e^{-5t} + R_2 I_{20} e^{-(R_1+R_2)t/L}$. (*b*) Show that, in the output v_2, the mode Ke^{-5t} can be suppressed by choice of I_{10}. (*c*) What is the output v_2 if $0.8V - 8I_{10} = 0$ and $(R_1 + R_2)/L = 5$?

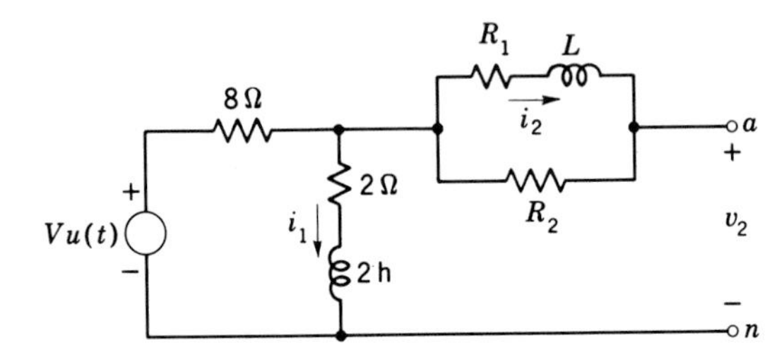

Fig. P12-6

12-7 A certain active linear system has the input $f(t)$ and two independent energy-storing elements, and is described completely by the two state equations $py_1 = -y_1 + 3y_2 + f(t)$, $py_2 = 4y_1 - 2y_2 - \frac{4}{3}f(t)$. (*a*) Find and solve the characteristic equation and show that the system is unstable. (*b*) Show that, if the output y_1 is selected, the unstable mode (increasing exponential) is not coupled to the input $f(t)$. (*c*) Show that, when the output is defined as $x = y_1 - y_2$, then, independently of the initial conditions, the unstable mode does not appear in x, that is, it is not observable in x.

12-8 In a certain source-free system the state variables have the response given by

$$\begin{Vmatrix} y_{1f} \\ y_{2f} \\ y_{3f} \end{Vmatrix} = \begin{Vmatrix} 1 & 1 & 1 \\ \frac{1}{2} & 2 & -1 \\ -\frac{1}{4} & 1 & 2 \end{Vmatrix} \times \begin{Vmatrix} K_1 e^{-t} \\ K_2 e^{-3t} \\ K_3 e^{-5t} \end{Vmatrix}$$

An output x is defined by $x = y_1 + ay_2 + by_3$. Find values for a and b so that (1) the mode e^{-t} is not observable in x; (2) the mode e^{-5t} is not observable in x.

12-9 Use mesh equations to obtain the transform driving-point admittance of the circuit shown (*a*) in Fig. P12-9*a*; (*b*) in Fig. P12-9*b*. Comment on the number of modes evident from the poles of $Y(s)$.

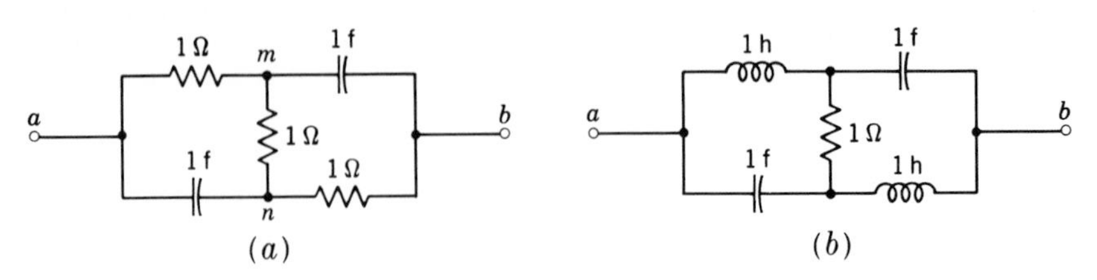

Fig. P12-9

12-10 Use node equations to calculate the transform driving-point impedances for the two circuits of Fig. P12-9*a* and *b*. Comment on the number of modes that are evident from the poles of $Z(s)$.

12-11 (*a*) In the network given in Fig. P12-11, $Z_{11} = Z_{22}$ if the parallel combination of 1 farad and $\frac{1}{4}$ henry is treated as a p element. Use the method of mesh currents

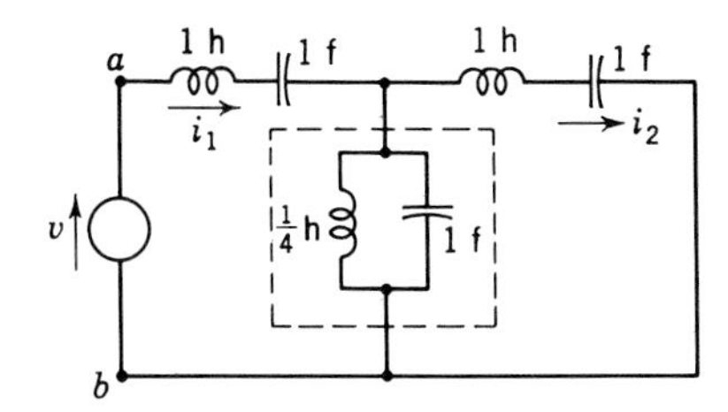

Fig. P12-11

with p-element mesh impedances to locate the zeros and poles of the driving-point admittance $Y_{ab}(s)$ and of the transform transfer admittance which relates i_2 to v. (*b*) Assume that the circuit is built of capacitors with negligible losses and inductors which are represented as series *R-L* branches. Assume further that, at radian frequencies which exceed $\frac{1}{2}$, the inductive reactance of each inductor is more than five times its resistance. Use these assumptions in conjunction with the pole-zero location of (*a*) to sketch the shape of the frequency-response curves for I_1/V and I_2/V in the sinusoidal steady state.

12-12 In a certain resistive network the node admittances are $Y_{11} = 16$, $Y_{22} = 10$, $Y_{33} = 9$, $Y_{12} = Y_{21} = -2$, $Y_{13} = Y_{31} = -4$, $Y_{23} = Y_{32} = -3$. Calculate (*a*) the driving-point resistance between nodes 1 and the reference node; (*b*) the transfer impedance relating v_2 to i_1 if a current source i_1 is connected between node 1 and the reference node.

12-13 When ideal source $v(t) = v_{11'}(t)$ is impressed on the terminal pair 1-1′ in Fig. P12-13, then $v_{22'} = \frac{1}{2}v(t)$ if terminals 2-2′ are left open and $v_{22'} = \frac{1}{3}v(t)$ when a 5-ohm resistance is connected between terminals 2-2′. (*a*) Calculate $v_{22'}$ when a 3-ohm resistance is connected to terminals 2-2′. (*b*) If $v_{11'}(t) = 3u(t)$, calculate the complete response $v_{22'}$ if the parallel combination of a 10-ohm resistance and a 0.15-farad, initially deenergized capacitance, is connected between 2-2′.

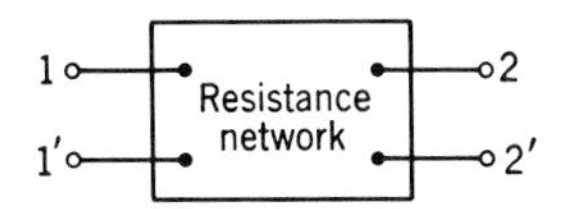

Fig. P12-13

12-14 In the network of Fig. P12-13 an ideal current source $i_A(t)$ is connected between 1 and 1′ (so that the reference arrow of i_A points toward terminal 1), and an ideal source $i_B(t)$ is connected between terminals 2 and 2′ (so that its reference arrow points toward terminal 2). When $i_A \equiv 5$, $i_B \equiv 0$, $v_{11'} \equiv 100$ and $v_{22'} \equiv 30$. Calculate i_B so that $v_{11'} = 0$.

12-15 When $v_1(t) = 10 \cos \omega t$ and $v_2 = 0$, then, in the steady state, $i_1(t) = 2 \cos (\omega t - 36.9°)$ and $i_2(t) = 2.82 \sin (\omega t + 8.1°)$. Calculate the steady-state function $i_1(t)$ if (*a*) $v_2(t) = 10 \cos \omega t$, $v_1(t) \equiv 0$; (*b*) $v_2(t) = 10 \cos \omega t$, $v_1(t) = 10 \sin \omega t$. The circuit is shown in Fig. P12-15.

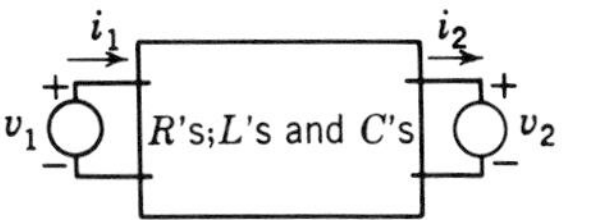

Fig. P12-15

12-16 Two two-ports are connected in series as shown in Fig. P12-16. In making the connection, no elements in either network are short-circuited. Show that the open-circuit impedance matrix which relates v_1, i_1, v_2, and i_2 is the sum of the open-circuit impedance matrices of N_A and N_B.

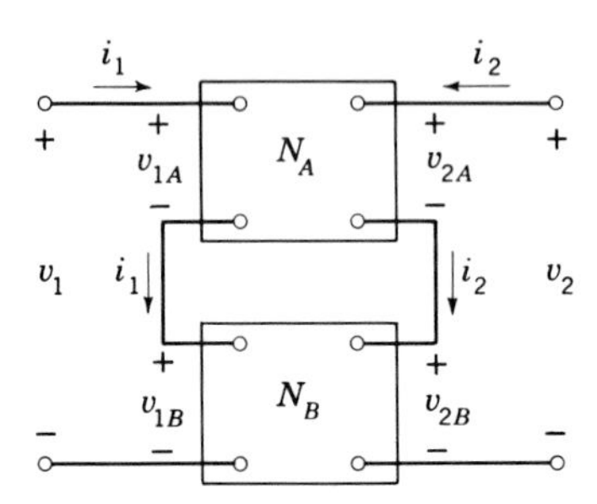

Fig. P12-16

12-17 Figure P12-17 shows a parallel connection of two ports. In making this connection, no elements in either network are short-circuited. Show that the short-circuit admittance matrix of the combination (relating v_1, i_1, v_2, i_2) is the sum of the short-circuit admittance matrices of N_A and N_B.

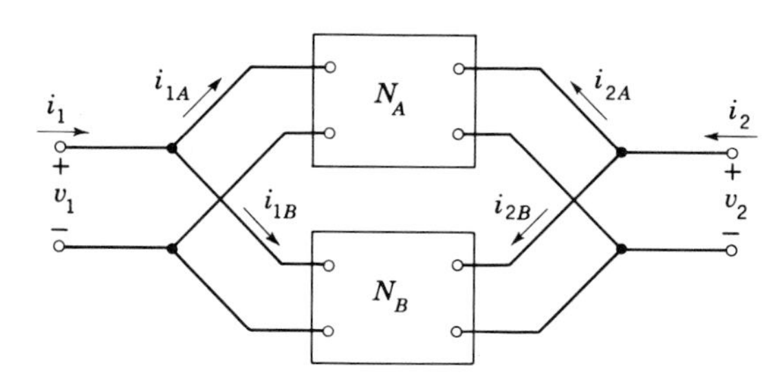

Fig. P12-17

12-18 (*a*) With reference directions as shown in Fig. P12-18*a*, the cascade matrix is defined by

$$\begin{Vmatrix} v_1 \\ i_1 \end{Vmatrix} = \begin{Vmatrix} A(p) & B(p) \\ C(p) & D(p) \end{Vmatrix} \times \begin{Vmatrix} v_2 \\ i_2 \end{Vmatrix}$$

Show that the cascade matrix that relates v_1, i_1 to v_2, i_2 in the cascade connection

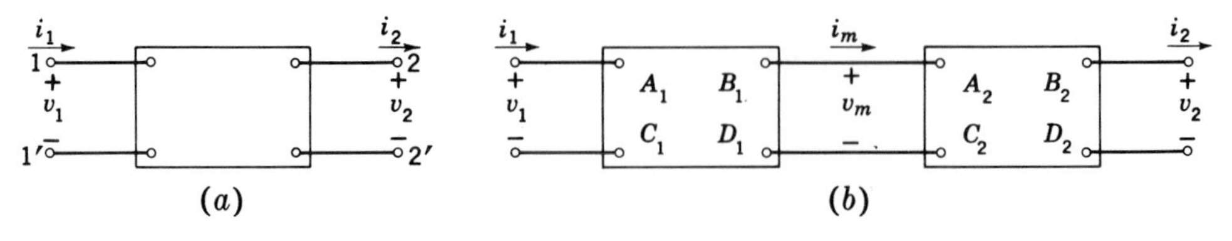

Fig. P12-18

of Fig. P12-18*b* is the matrix product

$$\begin{Vmatrix} A_1 & B_1 \\ C_1 & D_1 \end{Vmatrix} \times \begin{Vmatrix} A_2 & B_2 \\ C_2 & D_2 \end{Vmatrix}$$

(*b*) In Fig. P12-18*a*, an impedance $Z_T(p)$ is placed across terminals 2-2′. Show that the driving-point impedance at terminals 1-1′ is given by $[A(p)Z_T(p) + B(p)]/[C(p)Z_T(p) + D(p)]$. (*c*) Show that the transfer admittance that relates i_2 to a source v_1 is $1/[A(p)Z_T(p) + B(p)]$. (*d*) An ideal voltage source $v_1(t)$ is placed across terminals 1-1′ (Fig. P12-18*a*), and terminals 2-2′ are left open-circuited. Obtain (in terms of A, B, C, D, v_1) Thévenin's equivalent circuit with respect to terminals 2-2′.

12-19 In a two-port, the following driving-point immittances are measured or calculated: Driving-point impedance at terminal pair 1 with terminal pair 2 short-circuited, Z_{s1}; driving-point impedance at terminal pair 1 with terminal pair 2 open, Z_{01}; driving-point impedance at terminal pair 2 with terminal pair 1 short-circuited, Z_{s2}; driving-point impedance at terminal pair 2 with terminal pair 1 open, Z_{02}. Show that $Z_{s1}Z_{02} = Z_{01}Z_{s2}$.

12-20 In a certain resistive two-port (Fig. P12-18*a*), $v_1 = 2v_2 + 20i_2$, $i_1 = v_2/12 + 4i_2/3$. Calculate (*a*) the driving-point impedance if an 8-ohm resistance is connected across terminals 2-2′ so that $v_2 = 8i_2$; (*b*) the time constant of the circuit if $v_1 \equiv 0$ and if a $\frac{1}{2}$-farad capacitance replaces the 8-ohm resistance of part *a*.

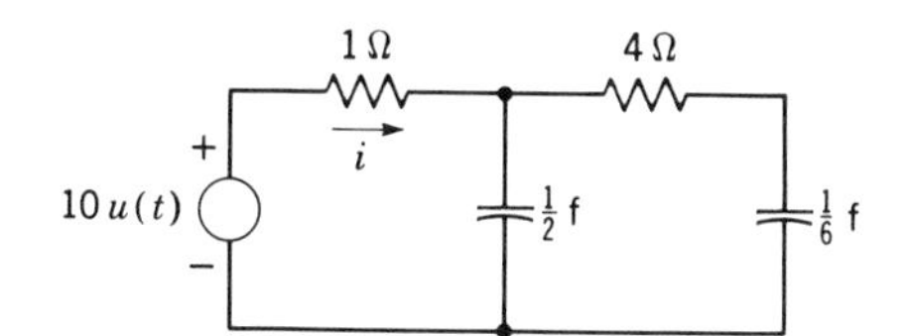

Fig. P12-21

12-21 In the network shown in Fig. P12-21, the capacitances are uncharged at $t = 0^-$. (*a*) Find $i(0^+)$. (*b*) Find the characteristic roots. (*c*) Change the impedance level of the network so that $i(0^+) = 10^{-4}$ amp, and change the frequency scale so that the longest time constant of the transient component is 5 μsec. Draw the new network, giving numerical values for all the elements.

12-22 A series *R-L-C* circuit has the characteristic roots $s_{1,2} = -5 \pm j12$ when $R = 10$ ohms. (*a*) Calculate the resonant frequency. (*b*) Describe how the impedance level and the frequency scale should be changed so that the circuit resistance is 600 ohms and the resonant frequency is 15 k-hertz. (*c*) What are the characteristic roots of the network obtained by the scale changes of part *b*?

12-23 The circuit shown in Fig. P12-23 is in the sinusoidal steady state. (*a*) Determine the frequency at which $v_{an}(t)$ and $v(t)$ differ in phase by 180°. (*b*) Change the impedance level so that the driving-point impedance approaches 600 ohms as ω approaches infinity and then change the frequency scale so that the phase condition of part *a* is fulfilled at 550 hertz.

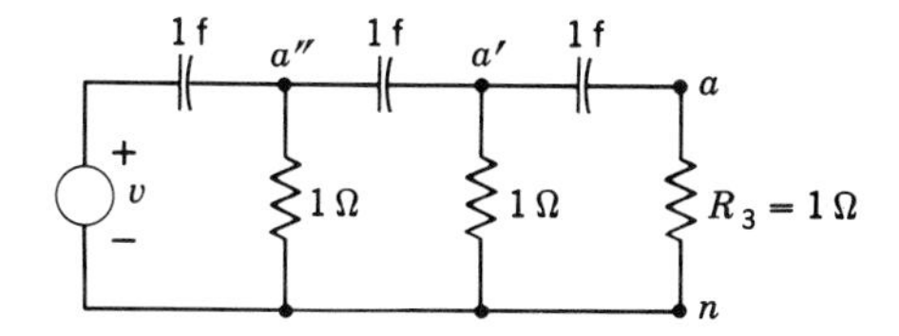

Fig. P12-23

12-24 Given the circuit shown in Fig. P12-24: (*a*) If $R = 1, L = 10, C = 0.1$, locate the poles and zeros of $Z_{ab}(s)$, and compare the results with the pole-zero location for $R = 0, L = 10, C = 0.1$. (*b*) Use the pole-zero diagram of (*a*) to calculate the maximum value of $Z_{ab}(j\omega)$. (*c*) Use scale changes to calculate the values of R, L, and C if the maximum value of Z_{ab} has a real part which is 40,000 ohms at a frequency of 100 k-hertz.

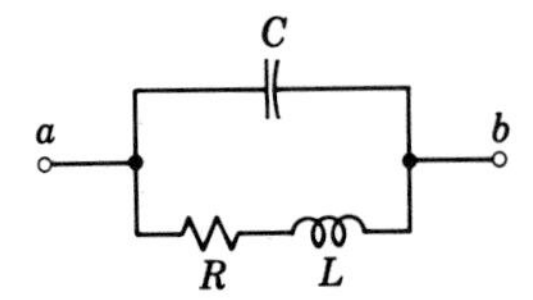

Fig. P12-24

12-25 (*a*) In the circuit of Fig. P12-25, $R_L = 1$, $L = \frac{1}{2}$, $C = 1$, $R_C = 1$. The source is sinusoidal. Calculate and sketch the angle by which v_{ab} lags v in the sinusoidal steady state as a function of frequency. (*b*) If the frequency scale is changed so that v_{ab} lags v by 135° at 400 k-hertz, and if the level is changed so that $R_L = R_C = 600$, calculate L and C.

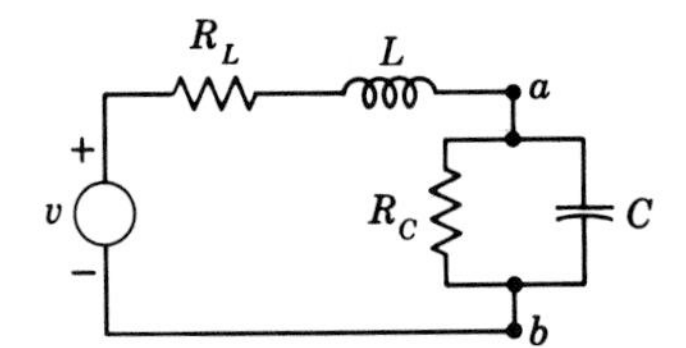

Fig. P12-25

12-26 In the circuit of Fig. P12-26, $R_1 = 1$ megohm, $R_2 = 250$ k-ohms, $C_1 = 167$ pf, and $C_2 = 500$ pf. It is desired to locate the zeros and poles of $Z_{ab}(s)$. Do this by first changing the impedance level so that $R_1 = 1$ ohm and then changing the time scale so that $C_2 = 1$ farad. Find the zeros and poles of the driving-point impedance of the network that has undergone scale change, and then apply the result to the given network.

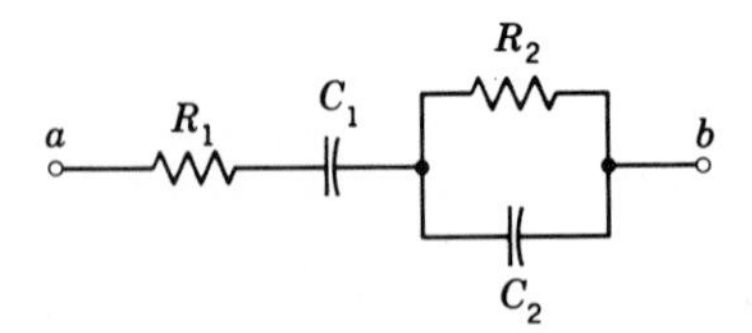

Fig. P12-26

Chapter 13 General analysis techniques III—flow graphs

In addition to the matrix methods (state, node, and mesh variables) for network analysis discussed in Chap. 11, there are other methods for formulating network functions. This chapter deals with the technique termed "topological method" and the closely related "signal-flow method."[1]

13-1 Topological method

Topology deals with the properties of geometrical figures that remain unaltered when the figure is deformed. Its relevance to network analysis stems from the observation that the performance of an electric network is unaltered when the "wires" are bent.[2] In Sec. 10-4 the terms tree and link are defined and the linear graph of networks is used to demonstrate certain properties of networks (entries, independent loops, etc.). To present the basic ideas of the topological method concisely, the following definitions are introduced.

TREE VALUE This is the product of the admittances of the branches forming the tree in the source-free network. Using p elements for the

[1] A brief introduction is presented here. Detailed discussion is given in S. J. Mason and H. J. Zimmerman, "Electronic Circuits, Signals and Systems," John Wiley & Sons, Inc., New York, 1960.

[2] This statement is true only at low frequencies.

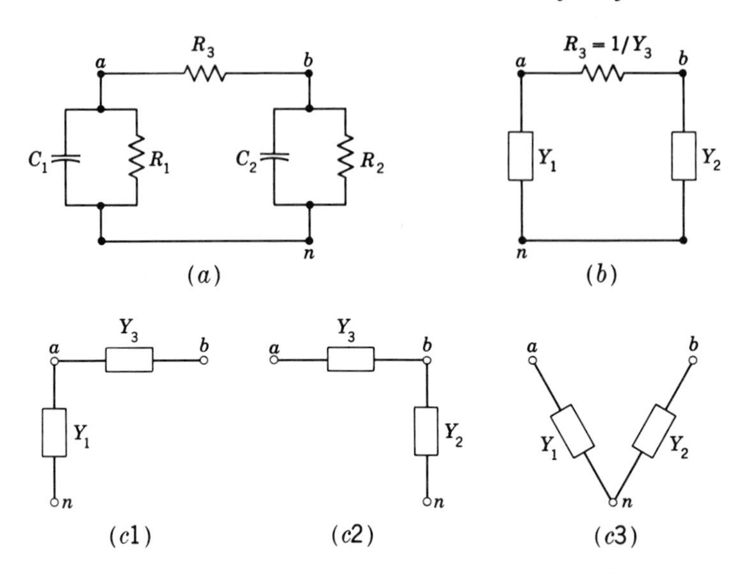

***Fig. 13-1** (a) A source-free network. (b) Network of (a) using p elements for the R_1-C_1 and R_2-C_2 combinations. (c) The three trees of the network of (b).*

network of Fig. 13-1*a* as indicated in Fig. 13-1*b*, the tree values are, for the three trees shown in Fig. 13-1*c*, Y_1Y_3, Y_3Y_2, Y_1Y_2, respectively, where $Y_1 = 1/R_1 + pC_1$, $Y_2 = 1/R_2 + pC_2$, $Y_3 = 1/R_3$.

NETWORK DETERMINANT This is the sum of the tree values of all the distinct trees in the graph of the source-free network; the network determinant is designated by the symbol Δ (Greek capital delta). With reference to Fig. 13-1*b*, $\Delta = Y_1Y_2 + Y_2Y_3 + Y_3Y_1$.

PATH BETWEEN TWO TERMINALS This is a continuous succession of branches terminating at two terminals, chosen so that no node is encountered more than once. In the network of Fig. 13-1*b*, Y_2 is a path between terminals *b* and *n*, as are the branches $Y_3 - Y_1$.

PATH VALUE This is the product of the branch admittances in a path between two terminals. Its value will be denoted by P_k for the *k*th path. For example, in Fig. 13-1*b*, the paths between terminals *b*-*n* have the path values $P_1 = Y_2$, $P_2 = Y_1Y_3$.

PATH COFACTOR When all branches in a given path between two terminals are short-circuited, the determinant of the remaining network is termed *path cofactor* of the given path. If the given path happens to include all nodes of the original network, its cofactor is taken as unity. The cofactor of path P_k is denoted by Δ_k. With reference to path $P_1 = Y_2$ in Fig. 13-1*b*, $\Delta_1 = Y_1 + Y_3$, and for P_2, $\Delta_2 = 1$.

SOURCE AND METER BRANCH When a source is added to a network either by pliers or soldering-iron entry, this source forms a *source branch* (S). The response of a network is either a current or a voltage; if the response is a current, we visualize the response as being "measured" by an ammeter inserted by pliers entry into the branch whose current is the response, and refer to the added branch as the *ammeter branch*. When the response is a voltage, we add a "voltmeter" between the desired terminals, and use the term *voltmeter branch* for the branch so added. Meter branches are identified by the letter M on diagrams and are assigned unit admittance.

TRANSMISSION PATH This is a path terminating on the terminals of the source and containing the meter branch; we deal with only one source at one time and use superposition when the network includes several sources.

TRANSMISSION-PATH VALUE P'_k This is the product of the branch admittances in the kth transmission path, with the meter-branch admittance taken as unity. A + (or −) sign is assigned to P'_k if a positive current in the direction of the path will cause a positive (negative) meter deflection. Figure 13-2 shows the network of Fig. 13-1*b* with a voltage source v inserted by pliers entry into the C_1 branch. In this figure v_{bn} has been chosen as the response, and the meter branch M inserted. For the transmission path a'-a-b-M-n, $P'_1 = pC_1 \times (1/R_3) \times 1 = pC_1/R_3$.

TRANSMISSION-PATH COFACTOR The determinant of the network which remains when all branches of the kth transmission path, including the meter branch, are short-circuited is termed the kth-transmission-path cofactor and denoted by Δ'_k. In Fig. 13-1*a* for transmission path a-b-n, $\Delta'_1 = 1$ because this path includes all nodes.

NETWORK TRANSMISSION The term network transmission is identical with the term network function; that is, it is the operator that operates on a source function to produce a response. We denote the network transmission by T [or $T(p)$].

NETWORK-TRANSMISSION FORMULA We state without proof[1] that

$$T = \frac{\sum\limits_k P'_k \Delta'_k}{\Delta} \tag{13-1}$$

that is, the network transmission is the quotient of the sum of the products of all transmission-path values with their cofactors to the network deter-

[1] Mason and Zimmerman, *op. cit.*, chap. 4.

minant. We state also without proof the formula

$$\Delta = \sum_j P_j \Delta_j \tag{13-2}$$

that is, one may find Δ as the sum of the products of all path values between *any* two convenient terminals (of the source-free network) with their cofactors. Note that, if the determinant of a source-free network is known, it can be used in Eq. (13-1) if voltage sources are inserted by pliers entry and current sources by soldering-iron entry (Sec. 11-3).

In the example of the source-free network (Fig. 13-1) and the network with the voltage source v inserted by pliers entry (Fig. 13-2), we calculate the network transmission relating v_{bn} to v as follows: For the enumeration of trees (Fig. 13-1*c*), using

$$G_1 = \frac{1}{R_1} \qquad G_2 = \frac{1}{R_2} \qquad G_3 = \frac{1}{R_3} \tag{13-3}$$

$$\begin{aligned} \Delta &= Y_1Y_2 + Y_2Y_3 + Y_3Y_1 \\ &= (G_1 + pC_1)(G_2 + pC_2) + (G_2 + pC_2)G_3 + G_3(G_1 + pC_1) \\ &= p^2C_1C_2 + p(C_1G_2 + C_2G_1 + C_2G_3 + C_1G_3) + G_1G_2 + G_2G_3 + G_3G_1 \end{aligned} \tag{13-4}$$

or Δ may be computed from Eq. (13-2), using terminals *b*-*n*.

$$\begin{aligned} \Delta &= P_1\Delta_1 + P_2\Delta_2 \\ &= Y_2(Y_1 + Y_3) + Y_1Y_3(1) \end{aligned} \tag{13-5}$$

Equation (13-5) is seen to be identical with Eq. (13-4). We now have the denominator Δ for use in Eq. (13-1). To obtain the numerator, we observe

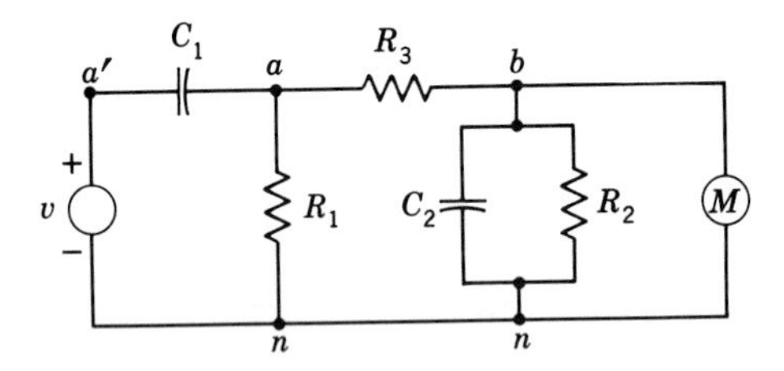

***Fig. 13-2** Network of Fig. 13-1 with voltage-source branch introduced by pliers entry in the R_1 branch and with a meter branch introduced for the voltage v_{bn}.*

that there is only one transmission path from v to M in Fig. 13-2. As stated before, its value is pC_1G_3, and its cofactor is 1. Hence

$$v_{ab} = Tv$$

where

$$T = \frac{pC_1G_3}{p^2(C_1C_2) + p(C_1G_2 + C_2G_1 + C_1G_3 + C_2G_3) + G_1G_2 + G_2G_3 + G_3G_1}$$

13-2 Signal flow graphs

We recall that the response of a network may always be found as the solution to a set of simultaneous (differential) equations. *A signal flow graph is a graphical representation of a set of such equations.* In this representation we use lines to represent operations and circles with symbols within them to represent variables. For example, if the signal y is related to the signal x by $y = ax$, this means that, given x, we operate on it by multiplying by a to get y. In Fig. 13-3a this operation is shown in the form

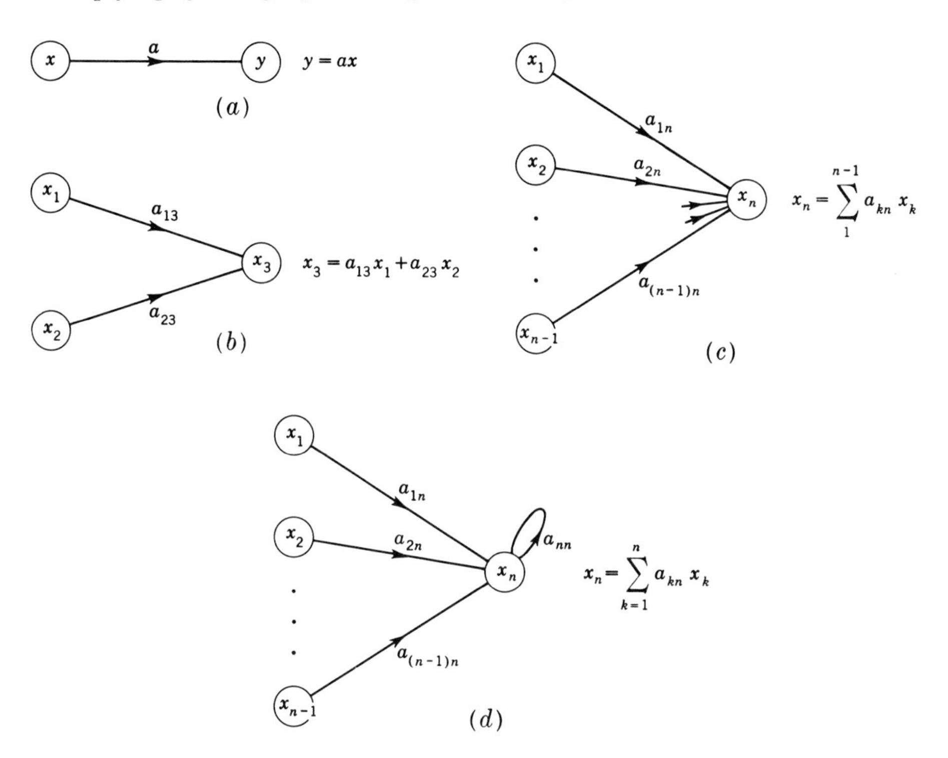

Fig. 13-3 Signal-flow-graph symbolism.

of a signal flow graph. To get y (the arrow of a points to y), we begin with x and multiply by a (the line is marked a). If several lines end at one circle and point to it, the variable is given by the sum of the operations implied by each line. Thus, in Fig. 13-3b, we have illustrated $x_3 = a_{13}x_1 + a_{23}x_2$, and the graph of Fig. 13-3c represents the equation $x_n = \sum_{1}^{n-1} a_{kn}x_k$.

One may think of the circles as "reservoirs" whose contents are x_1, x_2, etc., and of the lines as the "pipes" which take the x's from the reservoirs to the "output" (in the direction of the arrow), modifying the contents in the process through the operation (multiplication by a_{kn} in the example of Fig. 13-3c).

In Fig. 13-3d we show the equation

$$x_n = \sum_{k=1}^{n} a_{kn}x_k \tag{13-6}$$

and we observe that the term $a_{nn}x_n$ is represented by a "self-loop." This loop takes x_n from the reservoir, operates on it by a_{nn}, and adds the result $a_{nn}x_n$ to the "mix" in the reservoir containing x_n.

Suppose that we have a network with a source $s(t)$ and two independent energy-storing elements. The state equations, using y_1 and y_2 as state variables, are

$$a_{11}y_1 + a_{12}y_2 + b_1s = py_1 \equiv y_3 \tag{13-7a}$$

$$a_{21}y_1 + a_{22}y_2 + b_2s = py_2 \equiv y_4 \tag{13-7b}$$

$$y_1 = \frac{1}{p}\,y_3 \tag{13-7c}$$

$$y_2 = \frac{1}{p}\,y_4 \tag{13-7d}$$

Note that, since each "reservoir" is to be a variable, we have defined $py_1 \equiv y_3$ and $py_2 \equiv y_4$ through Eqs. (13-7*c*) and (13-7*d*); these relationships will appear on the signal flow graph. To draw the signal flow graph we begin by drawing five reservoirs as in Fig. 13-4*a*. We now add the lines to show how y_3 is obtained in Eqs. (13-7); this is shown in Fig. 13-4*b*. We then add to Fig. 13-4*b* the lines representing the operations described by Eqs. (13-7*b*) to (13-7*d*), as is shown in the final result, Fig. 13-4*c*; this graph conveys exactly the same information as Eqs. (13-7).

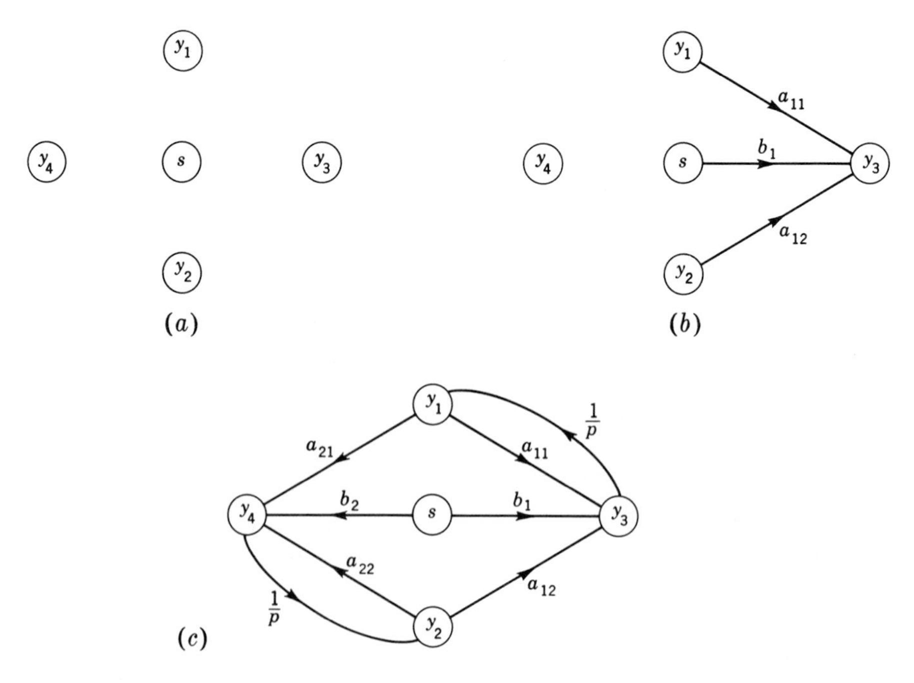

***Fig. 13-4** Signal-flow development for Eqs. (13-7). (a) Establishments of variables. (b) Representation for the first of Eqs. (13-7). (c) Complete signal flow diagram.*

Corresponding to each form in which a set of equations can be written, a signal flow graph can be drawn. If Eqs. (13-7) are written in the form

$$\begin{aligned} \frac{1}{p - a_{11}} (a_{12}y_2 + b_1 s) &= y_1 \\ \frac{1}{p - a_{22}} (a_{21}y_1 + b_2 s) &= y_2 \end{aligned} \tag{13-8}$$

the signal flow graph of Fig. 13-5 applies.

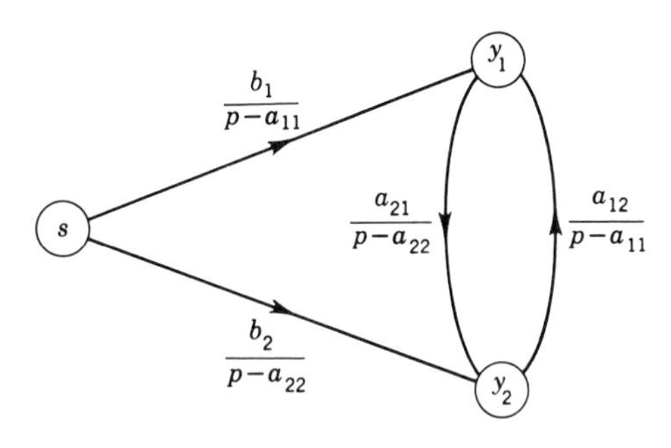

***Fig. 13-5** Signal flow diagram for Eqs. (13-8). This graph represents the same system as that shown in Fig. 13-4c, but the variables y_3 and y_4 have been eliminated.*

The form of the signal flow graph used in a specific problem depends on the nature of the problem. If, in the example cited, s is known and y_1 is desired, we seek the signal flow graph that represents the network equation $y_1 = H_1 s$, where H_1 is to be expressed in terms of a_{11}, a_{12}, a_{21}, a_{22}, b_1, b_2, and p; in other words, we seek the signal flow graph of Fig. 13-6.

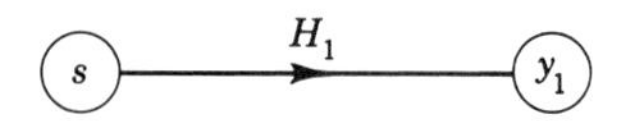

***Fig. 13-6** Transmission signal flow graph in the desired final form.*

13-3 Signal-flow-graph reduction

In the ensuing discussion we use the conventional term "node" to denote what we have described as a reservoir and identify it by the variable it represents as shown in Fig. 13-7. The flow lines are referred to as (directed) branches. A signal flow graph may be reduced in complexity through step-by-step elimination of nodes or branches. In Fig. 13-7 are shown four basic equivalences that can be used for this purpose. These equivalents are either self-evident or can easily be proved. Other, more complicated equivalents that are used for node elimination are shown in Fig. 13-8. To prove these, the equations relating the variables are written, and the variable y_m is eliminated; this is left as an exercise for the reader. Another type of equivalence is shown in Fig. 13-9, where a new node is defined in order to eliminate a self-loop e. Proof of this follows from observing that $y_5 = ay_1 + by_2 + ey_5$ or $y_5(1 - e) = ay_1 + by_2 \equiv y$. Exactly as elimination of a node does not eliminate the *variable* that it represents in the original problem, addition of

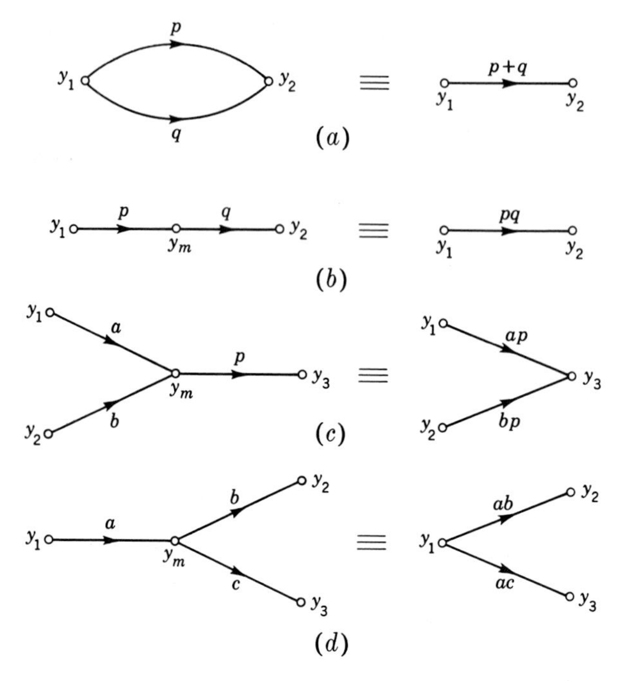

Fig. 13-7 Basic equivalences. (a) Parallel operation (addition). (b) Cascade operation (multiplication). (c, d) Distributive laws.

a node, such as y, does not generate a new physical variable, it only generates an auxiliary variable for use in analysis.

Example 13-1 Use the signal-flow-graph reduction to solve the state equations given by Eqs. (13-7) in the form of Fig. 13-6.

Solution The set of equations (13-7) is shown in Fig. 13-10a on page 499, which is similar to Fig. 13-4a. Solution of the problem consists of elimination of all nodes except y_1 and s. In Fig. 13-10b is shown the portion of the signal flow diagram which shows all branches leading to node y_3. Node y_3 is now eliminated by use of the identity shown in Fig. 13-8c; the result is shown in Fig. 13-10(b2). The self-loop is now eliminated by use of the identity of Fig. 13-8e; the result is shown in Fig. 13-10(b3). In Fig. 13-10c the elimination of node y_4 is shown by the same two-step procedure. The new signal flow diagram, with nodes y_3 and y_4 eliminated, is obtained by combining Figs. 13-10(b3) and 13-10(c3); the result is shown in Fig. 13-10d. We observe that Fig. 13-10d is identical with Fig. 13-5. We now eliminate node y_2 from Fig. 13-10d by use of the identity of Fig. 13-10b [note that the branch $b_1/(p - a_{11})$ is unchanged]; the result is shown in Fig. 13-10c. The parallel branches $b_1/(p - a_{11})$ and $b_2a_{12}/[(p - a_{11})(p - a_{22})]$ are now combined as shown in Fig. 13-10f. In Fig. 13-10f an additional branch with unit multiplier has been added, so that the identity of Fig. 13-8d can be used to obtain the final result shown in Fig. 13-10g. The result is $y_1 = Hs$, where

$$H = \frac{b_1(p - a_{22}) + b_2a_{12}}{(p - a_{11})(p - a_{22}) - a_{12}a_{21}} \tag{13-9}$$

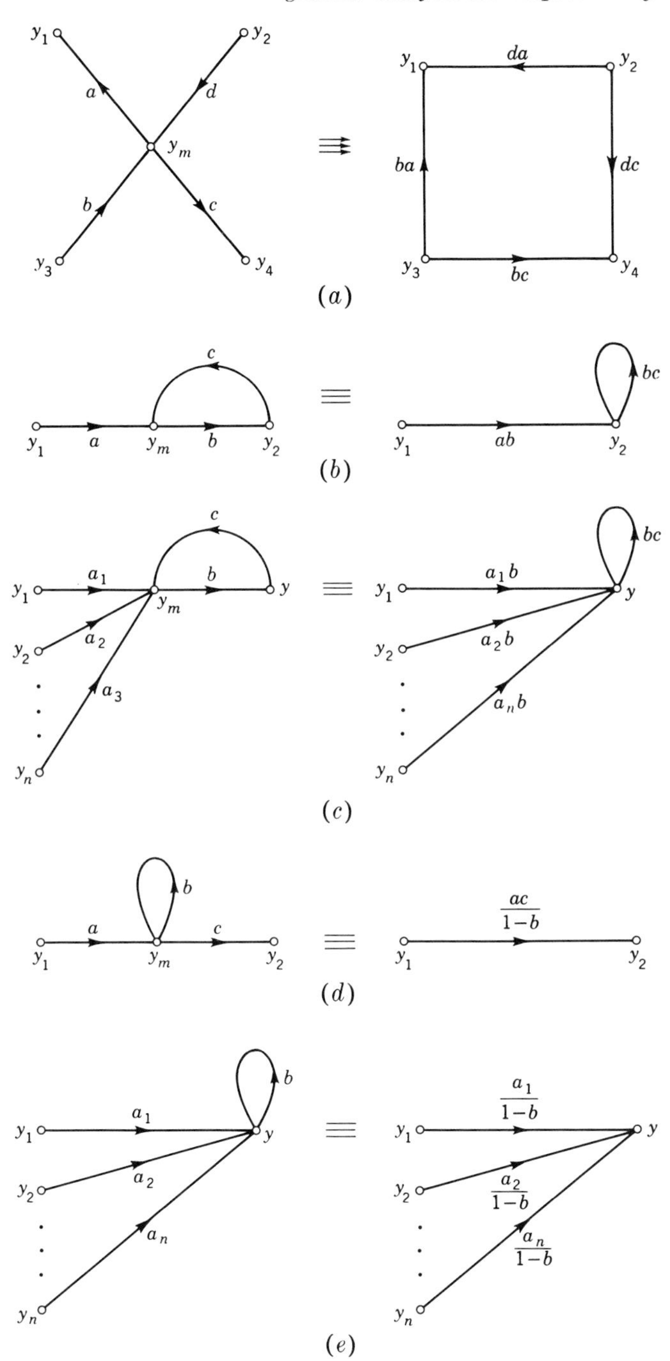

Fig. 13-8 Examples of node elimination.

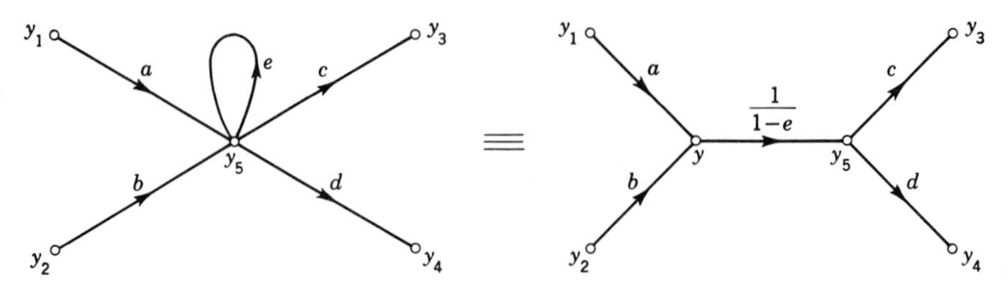

Fig. 13-9 Replacement of a self-loop addition of a node.

The reader may verify the result given by Eq. (13-9) by solution of the simultaneous equations (13-7) or (13-8).

13-4 Mason's network transmission formula[1]

In Example 13-1 we have already indicated how signal flow reduction can be used to solve for an input-output relationship. To clarify the problem generally, in a signal flow diagram, one defines *source* and *sink* nodes. A source node is one into which *no branch enters* (no branch arrow points to it), and a *sink* is a node from which *no branch leaves* (no branch arrow points away from it). In the signal flow graph of Fig. 13-11*a*, a source and a sink are easily identified and marked. In the case of the graph shown in Fig. 13-11*b*, no source or sink is clearly identified. Where sources or sinks are to be introduced into such a graph [introducing a source into the graph corresponds to applying a source in the network or system represented by the graph by forcing a variable corresponding to a node to be $s(t)$, whereas introducing a sink corresponds to choosing the "output" variable], branches with unit transmission are introduced. Thus, if in the graph of Fig. 13-11*b* we fix y_1 to be $s(t)$ and choose y_3 as the output, the source and sink are identified as shown in Fig. 13-11*c*.

The operator that relates a chosen sink variable (output or response) to a single source variable of a flow graph is termed the transmission T (or transfer function) of the graph. As in the case of the topological method (Sec. 13-1), a formula for finding T has been developed. To apply this formula correctly, the following definitions are needed; some of these definitions are similar to those used in the topological method.

NODE SPLITTING By this term is meant the replacement of a node by two nodes, one of which is a source and the other a sink, in the manner described as follows: Given a node k with the associated variable y_k as shown in Fig. 13-12*a*, it is split into the nodes k', which is a sink, and k'', which is a source, as shown in Fig. 13-12*b*; associated with nodes k' and k'' are the new variables y_k' and y_k''.

[1] Mason and Zimmerman, *op. cit.*

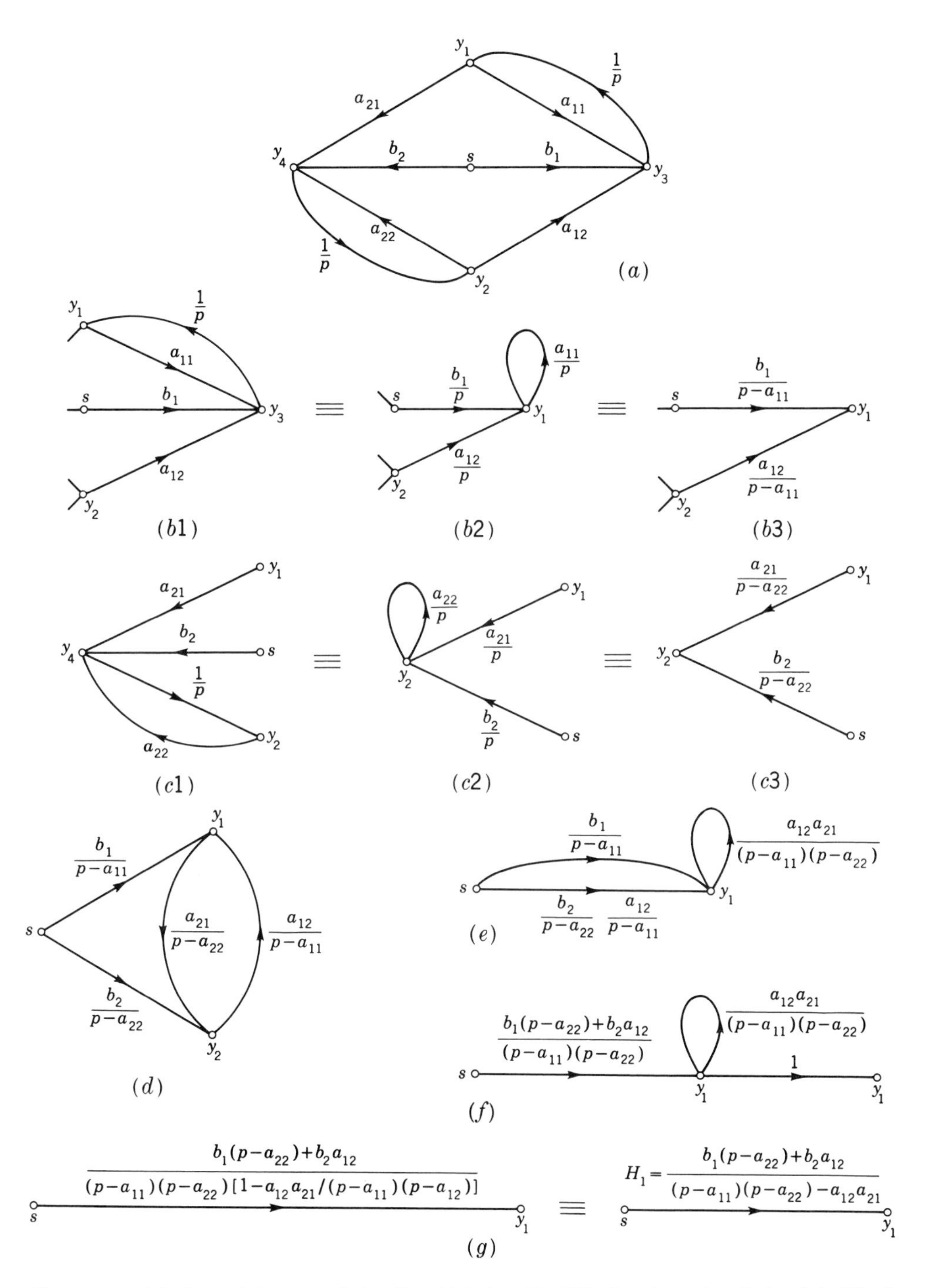

Fig. 13-10 *(**a**) **Signal flow graph for Eqs. (13-7).** (**b**) **Elimination of node 3.** (**c**) **Elimination of node 4.** (**d**) **Signal flow graph of (a) after elimination of nodes 3 and 4. This graph is also shown in Fig. 13-8.** (**e**) **Elimination of node 2.** (**f**) **Same as (e) with unit branch added.** (**g**) **Final result in the form of Fig. 13-6.***

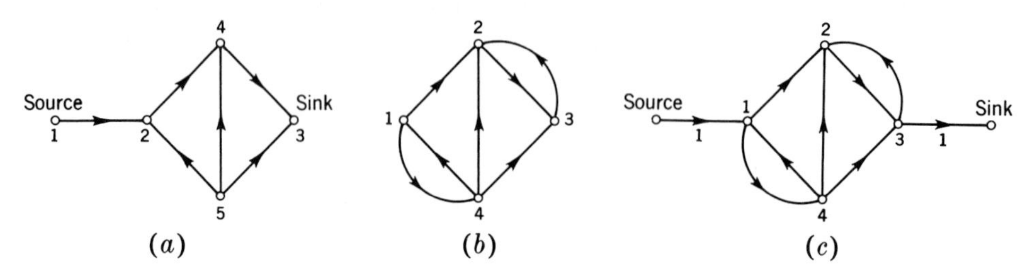

Fig. 13-11 (*a*) *Illustration of source and sink.* (*b*) *Signal flow diagram with no defined source and sink.* (*c*) *Introduction of unit branches to make node 1 the source and node 3 the sink.*

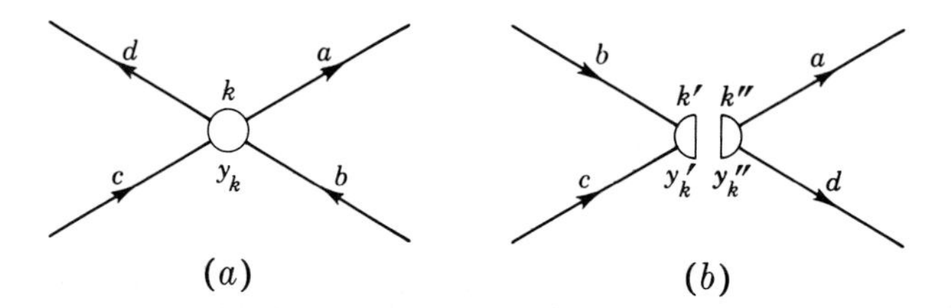

Fig. 13-12 *Node splitting.*

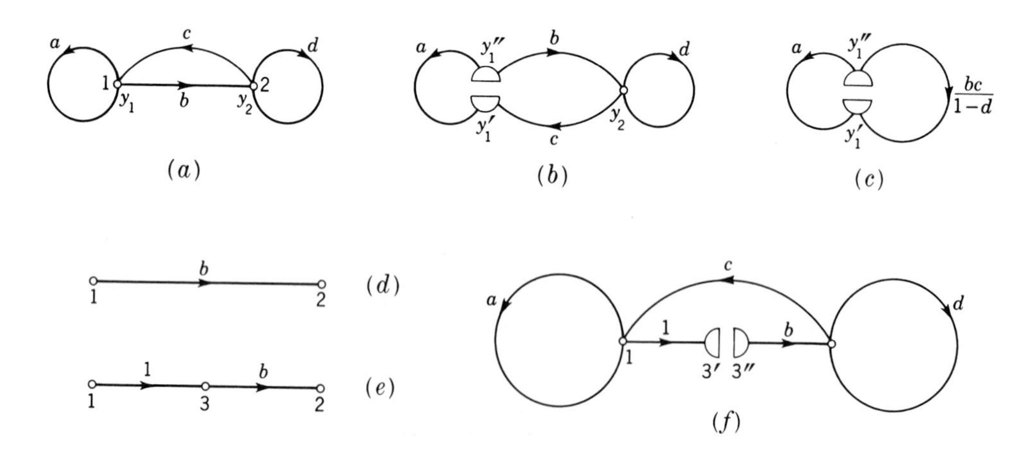

Fig. 13-13 (*a*) *A signal flow graph.* (*b*) *Graph with node 1 split for calculation of* τ_1. (*c*) *Equivalent to* (*b*) *with node 2 eliminated.* (*d*) *A branch in a signal flow graph.* (*e*) *Interior node 3 introduced in branch b.* (*f*) *Graph of* (*a*) *with interior node 3 split for calculation of branch loop transmission.*

LOOP TRANSMISSION OF A NODE The loop transmission of node k is defined as the transfer function τ_k which relates y'_k to y''_k by $y'_k = \tau_k y''_k$ when the node k is split.

For example, given the signal flow graph shown in Fig. 13-13*a*, the loop transmission of node 1 is calculated as follows: Node 1 is split as shown in Fig. 13-13*b*, and τ_1 is calculated by considering y''_1 as the input and y'_1 as the output. In Fig. 13-13*b* signal-flow-graph reduction is used to eliminate node 2 (by use of the identity of Fig. 13-8*d*) to give Fig. 13-13*c*. We observe that we now have two parallel branches, so that $\tau_1 = a + bc/(1 - d)$. The loop transmission of a node is zero when there is no path from k'' to k' *in the direction of the branches;* that is, when there is no *feedback path* to the kth node (a feedback path is one that begins and ends at the same node).

INTERIOR NODE OF A BRANCH A branch b may be split by introducing a new node as shown in Fig. 13-13*d* and *e*, where node 3 has been introduced as an *interior node of branch b*. Note that, in Fig. 13-13*e*, a unit branch enters node 3, and that the branch leaving node 3 has the value b, the same as the original branch in Fig. 13-13*d*.

LOOP TRANSMISSION OF A BRANCH The loop transmission of a branch is defined to be the loop transmission of an interior node of that branch. For branch b of Fig. 13-13*a*, the loop transmission is calculated from Fig. 13-13*f*; the result, which may be verified by the reader as an exercise, is $\tau_3 = bc/[(1 - a)(1 - d)]$.

PARTIAL LOOP TRANSMISSION OF A NODE The partial loop transmission of the node k of a graph with n nodes is defined as the loop transmission of node k with nodes $k + 1, k + 2, \ldots, n$ split. It is denoted by τ'_k.

DETERMINANT OF A FLOW GRAPH The determinant of a flow graph is denoted by Δ and defined as

$$\Delta = (1 - \tau'_1)(1 - \tau'_2) \cdots (1 - \tau'_n) = \prod_1^n (1 - \tau'_k) \qquad (13\text{-}10)$$

It can be shown that Δ is a property of the *graph* and is independent of how the nodes are numbered.

SOURCE-TO-SINK TRANSMISSION PATH A source-to-sink transmission path is a continuous string of branches oriented from the source node to the sink node along which no node is encountered more than once. Its value, denoted by P_k for the kth path, is the product of the values of the branches in that path.

COFACTOR OF THE kTH TRANSMISSION PATH This is denoted by Δ_k and is defined as the determinant of the graph which results when *all* branches connected to nodes of the kth path are removed, that is, when the (transmission) values of these branches are set to zero.

Using the foregoing definitions, we state without proof Mason's transmission formula as follows:

The transmission T of a network (system) is given by

$$T = \frac{\sum_{k=1}^{n} P_k \Delta_k}{\Delta} \tag{13-11}$$

Example 13-2 Find the transmission T_{14} for the flow graph shown in Fig. 13-14a.

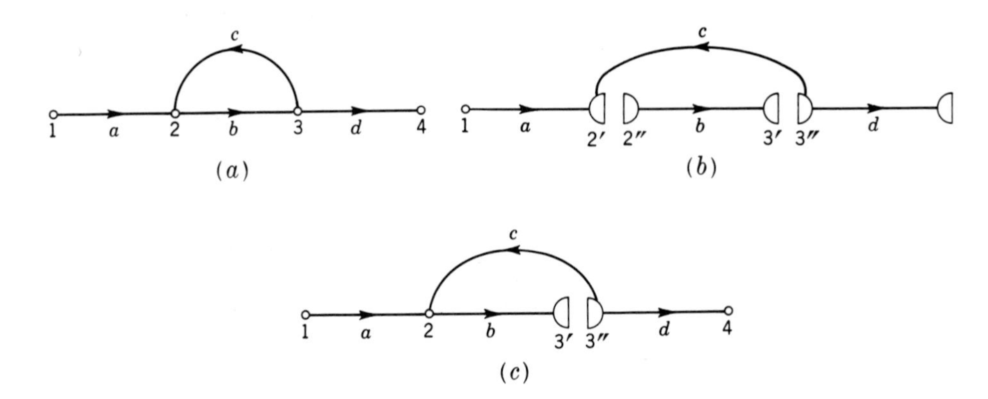

***Fig. 13-14** (a) A signal flow graph. (b) Graph for calculation of τ_2'. (c) Graph for calculation of τ_3'.*

Solution To find Δ, we write Eq. (13-10) for $n = 4$.

$$\Delta = (1 - \tau_1')(1 - \tau_2')(1 - \tau_3')(1 - \tau_4')$$

To find τ_1', all nodes are split, and since there is no feedback to node 1, $\tau_1' = 0$. To find τ_2', nodes 2, 3, and 4 are split as shown in Fig. 13-14b. Since starting at 2″ there is no path to 2′, it follows that $\tau_2' = 0$. To find τ_3', nodes 3 and 4 are split as shown in Fig. 13-14c. Starting at 3″, the transmission to 3′ is seen to be cb; hence $\tau_3' = cb$. Finally, we note that $\tau_4' = 0$ since there is no feedback path to it. Hence, in this example, $\Delta = 1 - cb$. There is only one transmission path from 1 to 4, and its value is $P_1 = abd$. If one path is removed, no loop remains, so that its cofactor is 1 (all τ_k's are zero). Hence, using Eq. (13-11), we have

$$T_{14} = \frac{abd}{1 - cb}$$

This result can of course be obtained immediately by flow-graph reduction, using the identities of Fig. 13-8b and d. The purpose of the example was, however, to illustrate the use of Eq. (13-11).

ALTERNATIVE METHOD FOR EVALUATION OF Δ AND Δ_k Mason[1] has shown that the determinants used in the transmission formula (13-11) can be conveniently calculated by focusing attention on the closed loops in the flow graph. A (closed) loop is defined as a path beginning and terminating at the same node, traced in the direction of the branch arrows without passing through any node more than once. To state the formula for the evaluation of the determinants, the following definitions are needed.

LOOP VALUE The loop value of a loop is the product of the branch transmission values in the loop. For example, in the flow graphs of Fig. 13-13*a*, there are three loops; their values are a, bc, and d. In the graph of Fig. 13-10*d* there is only one loop; its value is $[a_{21}/(p - a_{22})][a_{12}/(p - a_{11})]$. Note that, in that graph, the branches s-y_1-y_2-s do not form loops. The value of the kth loop in a graph is denoted by L_k.

TOUCHING LOOPS If two loops share one or more nodes, they are termed "touching loops." For example, of the three loops in the flow graph of Fig. 13-13, the loops with values a and bc are touching loops, whereas the loops a and d are not touching; the two loops with values a_{11}/p and a_{22}/p in the graph of Fig. 13-10*a* are not touching loops.

PRODUCT VALUE OF ORDER k In the flow graph of Fig. 13-15, there are five loops with loop values $L_1 = a_1b_1$, $L_2 = a_2b_2$, $L_3 = a_3b_3$, $L_4 = a_4b_4$, and

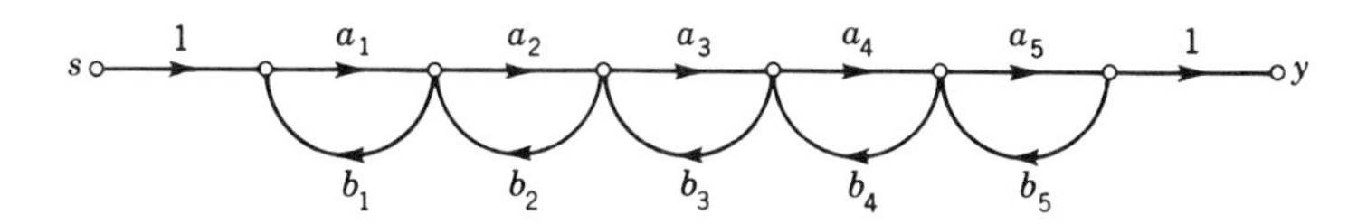

Fig. 13-15 A signal flow graph with five loops.

$L_5 = a_5b_5$. In the evaluation of a network determinant certain products of these loop values are used. The *product value of order 1* is defined identically with loop values. In the flow graph of Fig. 13-15, the product values of order 1 are therefore L_1, L_2, L_3, L_4, and L_5. A *product value of order 2* is defined as the product of the loop values of two *nontouching* loops. In the graph of Fig. 13-15, the product values of order 2 are L_1L_3, L_1L_4, L_1L_5, L_2L_4, L_2L_5, and L_3L_5. In general, a product value of order k is defined as the product of k loop values of *nontouching* loops. In Fig. 13-15 the only product value of order 3 is $L_1L_3L_5$, and there are none of higher order.

[1] *Ibid.*

Sum of order k The sum of all loop products of order k will be referred to as the "sum of order k," denoted by S_k. In Fig. 13-15,

$$\begin{aligned}
S_1 &= L_1 + L_2 + L_3 + L_4 + L_5 \\
S_2 &= L_1L_3 + L_1L_4 + L_1L_5 + L_2L_4 + L_2L_5 + L_3L_5 \\
S_3 &= L_1L_3L_5 \\
S_k &= 0 \qquad k > 3
\end{aligned}$$

Mason's rule With the foregoing definitions it can be shown that the determinant of a flow graph, Δ (which is used in Eq. 13-11) is given by

$$\Delta = 1 - S_1 + S_2 - \cdots + (-1)^k S_k + \cdots + (-1)^n S_n \qquad (13\text{-}12)$$

where S_n is the highest-order nonzero sum. The value n is, of course, the same as the largest number of nontouching loops in the graph. Each cofactor Δ_k [for use in the numerator of Eq. (13-11)] can be obtained from Δ by setting to zero in Δ all those branch values that are connected to nodes of the kth transmission path. In other words, those loop values that include loops touching one or more nodes of the kth transmission path are set to zero in Δ to obtain Δ_k.

These procedures are illustrated in the following example.

Example 13-3 Figure 13-16 shows the signal flow diagram of Fig. 13-10a [representing Eqs. (13-7)] modified in two ways: First, the s has been identified as source node,

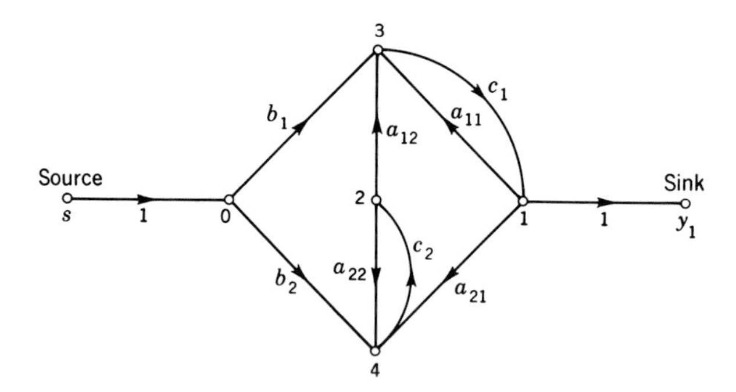

Fig. 13-16 Signal flow graph similar to Fig. 13-10a analyzed in Example 13-3.

and y_1 as sink node, and second, the two branches whose value is $1/p$ in Fig. 13-10a have been identified by the symbols c_1 and c_2 so that they can be distinguished when the values Δ_k are found from Δ. The transmission from s to y_1 is to be determined by use of Eq. (13-11) with the aid of Eq. (13-12).

Solution To identify all the loops in the graph, we examine each node in turn. There are no loops including node 0. For node 1 we observe the two loops $L_1 = a_{11}c_1$, $L_2 = a_{21}c_2a_{12}c_1$. For node 2 we have $L_3 = a_{22}c_2$ as the only new loop. Nodes 3 and 4 yield no new loops. Hence

$$S_1 = a_{11}c_1 + a_{12}a_{21}c_1c_2 + a_{22}c_2$$

Loops L_1 and L_2 are touching (they share node 1); L_1 and L_3 are nontouching; L_2 and L_3 touch. Hence

$$S_2 = a_{11}c_1a_{22}c_2$$

All higher-order products are zero, so that, using Eq. (13-12),

$$\Delta = 1 - (a_{11}c_1 + a_{12}a_{21}c_1c_2 + a_{22}c_2) + a_{11}a_{22}c_1c_2 \tag{13-13}$$

There are two transmission paths: $P_1 = b_1c_1$ and $P_2 = b_2c_2a_{12}c_1$. The cofactor Δ_1 is obtained by setting c_1 to zero in Eq. (13-13) as

$$\Delta_1 = 1 - a_{22}c_2 \tag{13-14}$$

The cofactor Δ_2 is obtained by setting $c_1 = 0$, $c_2 = 0$, and $a_{12} = 0$ in Eq. (13-13):

$$\Delta_2 = 1 \tag{13-15}$$

Using Eqs. (13-13) to (13-15), together with P_1 and P_2 in Eq. (13-11), the desired transmission is

$$T = \frac{(b_1c_1)(1 - a_{22}c_2) + (b_2c_2a_{12}c_1)(1)}{1 - (a_{11}c_1 + a_{22}c_2 + a_{12}a_{21}c_1c_2) + a_{11}a_{22}c_1c_2} \tag{13-16}$$

To relate this result to the flow graph of Fig. 13-10a, we let $c_1 = c_2 = 1/p$, so that

$$\begin{aligned} T &= \frac{(b_1/p)(1 - a_{22}/p) + (a_{12}b_2/p^2)}{1 - (a_{11}/p + a_{22}/p + a_{12}a_{21}/p^2) + a_{11}a_{22}/p^2} \\ &= \frac{b_1(p - a_{22}) + b_2a_{12}}{p^2 - p(a_{11} + a_{22}) + a_{11}a_{22} - a_{12}a_{21}} \\ &= \frac{b(p - a_{22}) + b_2a_{12}}{(p - a_{11})(p - a_{22}) - a_{12}a_{21}} \end{aligned}$$

13-5 Topological procedure for finding the flow graph of a network

So far we have considered only the problem of obtaining a transmission (network) function from a system described by its equations. We have illustrated the procedure with state equations; the reader can practice the procedure using node and mesh equations. It is possible to obtain the signal flow graph of a network directly from the network diagram without first writing network equations. The rule for deducing this signal flow graph, based on the topological properties of networks, is the following:

> Select one of the trees of the network so as to include *all* its voltage sources and *none* of its current sources. Next draw the flow graph which gives the link voltages of this tree in terms of the tree-branch voltages. Then draw the flow graph that gives the tree-branch currents in terms of the link currents. Finally, connect the nodes of the two signal flow graphs with branches consisting of the tree-branch *impedance* of the selected tree (in accordance with the voltage-current relationship of the branch) and the link admittances of the links selected similarly. The number of nodes in the graph will be

equal to twice the number of passive elements plus the number of sources.

Example 13-4 The signal flow graph for the circuit of Fig. 13-17a is to be constructed.

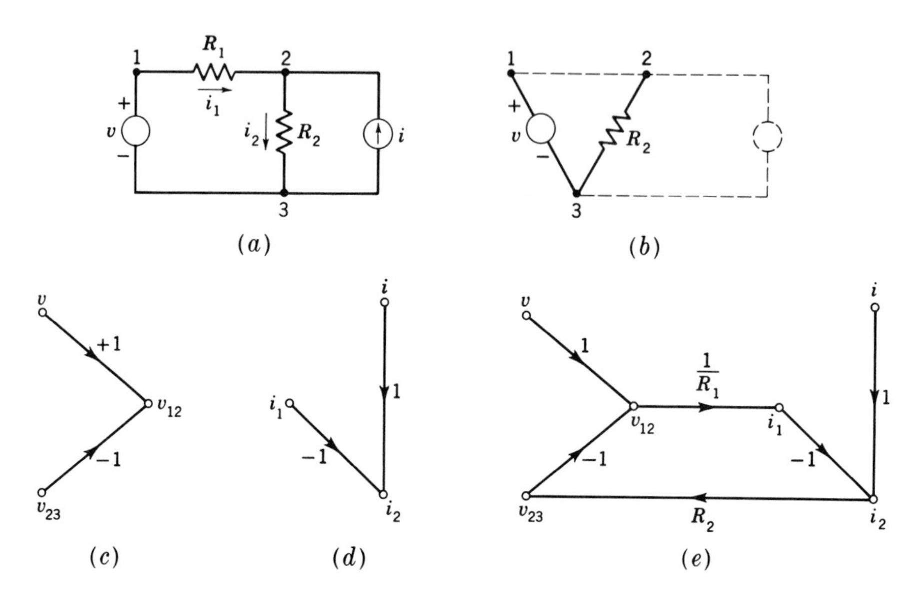

Fig. 13-17 ***(a) A network. (b) A tree of (a), including all voltage sources but no current sources. (c) Relationship between tree-branch and link voltages. (d) Relationship between tree-branch and link currents. (e) Signal flow graph for the network of (a).***

Solution A tree not including i is selected as in Fig. 13-17b. Only one link voltage must be expressed in terms of the tree-branch voltages: $v_{12} = v - v_{23}$; this relationship is shown in Fig. 13-17c. The link currents are i and i_1; we note that $i_2 = i_1 - i$ and express this in Fig. 13-17d. In Fig. 13-17e the two parts of the flow graph are shown. For the selected tree $v_{23} = R_2 i_2$; hence nodes v_{23} and i_2 are connected with a branch of value R_2. (The direction of the branch arrow is the direction of the operation "multiply.") Since $i_1 = v_{12}/R_1$, we have connected i_1 and v_{12} with the branch $1/R_1$. The reader may wish to construct the diagram, using $v - R_1$ as the starting tree.

Example 13-5 Draw the signal flow graph for the network of Fig. 13-18a.

Solution The tree shown in Fig. 13-18b has been selected, and we have for the voltages $v_{12} = v - v_{23} - v_{34}$, $v_{25} = v_{23} + v_{35}$, and $v_{45} = -v_{34} + v_{35}$, as shown on the left side of Fig. 13-18c. For the tree-branch currents we have $i_L = i_0 - i_c$, $i_1 = i_0 - i_2$, $i_3 = i_2 - i_c$, as shown on the right side of Fig. 13-18c; the connections shown are the impedances for the tree-branch voltage-current connections, and the admittances for the link voltage-current connection, with proper sign and direction used in each case.

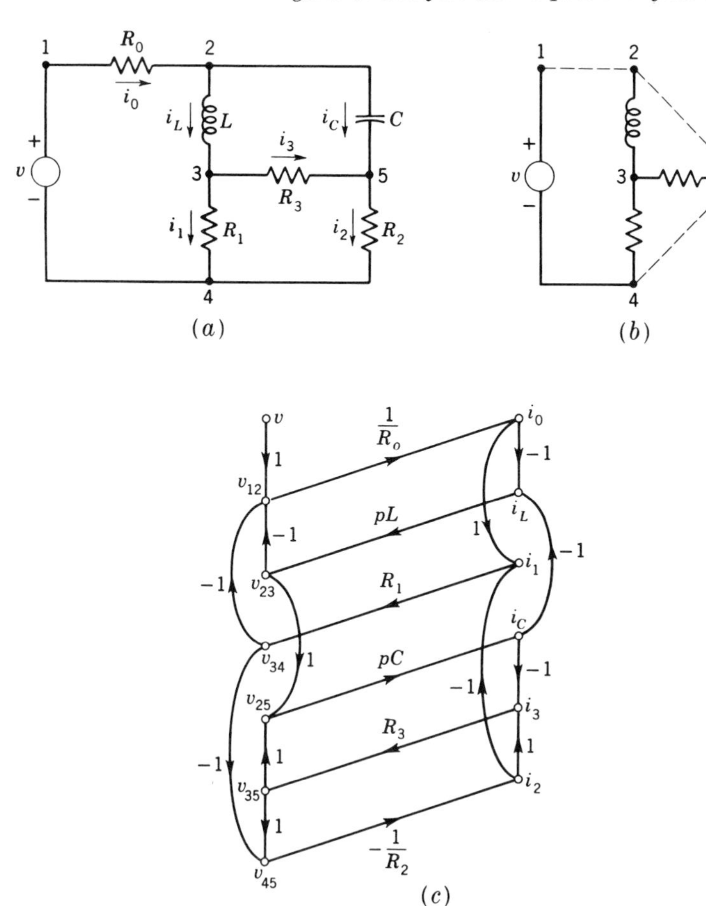

Fig. 13-18 A network and the development of its signal flow graph.

PROBLEMS

13-1 In the source-free network of Fig. P13-1: (*a*) Find the network determinant by an enumeration of distinct trees. (*b*) Verify the answer of part *a* by use of Eq. (13-2). (*c*) Insert a voltage source v by pliers entry in the Y_1 branch, and obtain the transmission relating v_{bn} to v by use of Eq. (13-1). (*d*) Find the driving-point impedance (in terms of Y_1, Y_2, Y_3, and Y_4) at the terminals of the source of part *c* by use of Eq. (13-1).

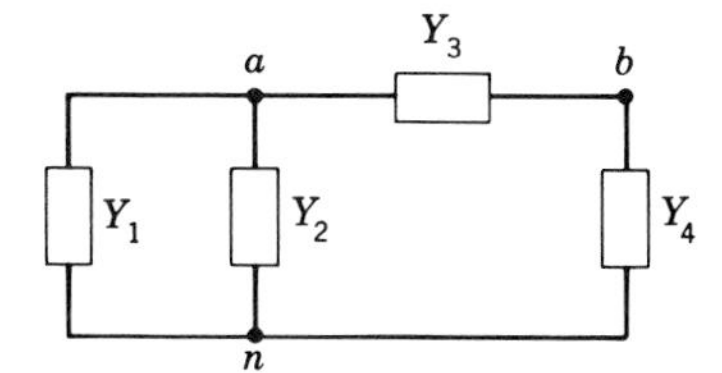

Fig. P13-1

13-2 In the network of Fig. P13-1, an ideal current source i entering the node a is inserted by soldering-iron entry at terminals a-b. Use Eq. (13-1) to express v_{an} and v_{ab} in terms of i.

13-3 In the network of Fig. 13-2 set $v \equiv 0$ and find the differential equations that relate i_{ab} (in R_3) and v_{bn} to an ideal current source $i_S(t)$ inserted by soldering-iron entry at terminals a-n.

13-4 In a certain linear system the output v_2 is related to the input v_1 by the equations

$$\begin{aligned} v_1 + v_f &= v_i \\ v_2 &= A(p)v_i \\ v_f &= B(p)v_2 \end{aligned}$$

Draw the signal flow diagram, and find the operational relationship between v_1 and v_2.

13-5 In a certain resistive circuit the mesh currents i_1, i_2, and i_3 are related to a source v by the equations

$$\begin{aligned} i_1 &= \tfrac{1}{5}v_1 + \tfrac{4}{5}i_2 \\ i_2 &= \tfrac{1}{3}i_1 + \tfrac{1}{2}i_3 \\ i_3 &= \tfrac{3}{8}i_2 \end{aligned}$$

(*a*) Draw the signal flow graph that represents these equations. (*b*) Find the relationship between i_1 and v_1. (*c*) Rewrite the given equations in the standard form of mesh equations, and solve these equations for i_1 in terms of v. (*d*) Draw a circuit diagram that corresponds to the given equations.

13-6 Prove the equivalences indicated in Figs. 13-7 and 13-8.

13-7 In Fig. P13-7 is shown the most general one-loop flow graph. Determine the transmission T_{16} by (*a*) signal-flow-graph reduction; (*b*) Mason's gain formula [Eq. (13-11)], using (1) node splitting to evaluate the determinants, (2) Eq. (13-12) to evaluate the determinants.

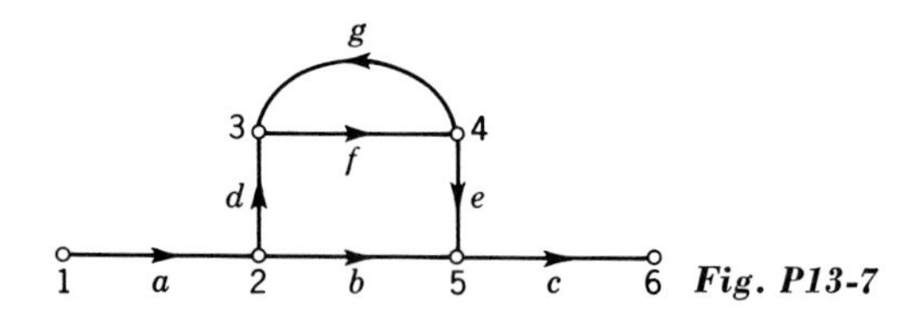

Fig. P13-7

13-8 In the flow graph of Fig. P13-8: (*a*) Enumerate all the loops and find their loop values. (*b*) Find the product values of all orders. (*c*) Find the cofactors Δ_1 and Δ_2 if $P_1 = ab_1cd$.

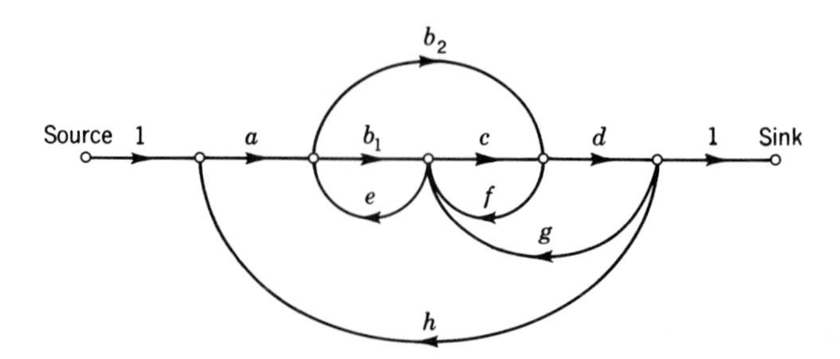

Fig. P13-8

13-9 Signal-flow-graph analysis is particularly useful if the system being analyzed is described by many variables but its graph has only a few loops. This problem is intended to illustrate this situation. Let x be a source and y_1, y_2, y_3, y_4, and y_5 the system variables, related by

$$\begin{aligned} y_1 &= x + ay_2 + by_5 \\ y_2 &= cy_3 + dy_5 \\ y_3 &= ey_1 \\ y_4 &= fy_3 + gy_5 \\ y_5 &= hy_4 \end{aligned}$$

(*a*) Show that Fig. P13-9 represents the signal flow graph of this system. (*b*) Enumerate the loops. (*c*) State the transmission-path values and their cofactors (in terms of $a, b, \ldots, h$) if the output is in turn $y_1, y_2, \ldots, y_5$. (*d*) For the numerical case $x = 20$, $a = 3$, $b = c = d = 1$, $e = 2$, $f = 5$, $g = 2$, and $h = -\frac{1}{2}$, find all the y's. Compare this technique with the direct solution of the simultaneous equations.

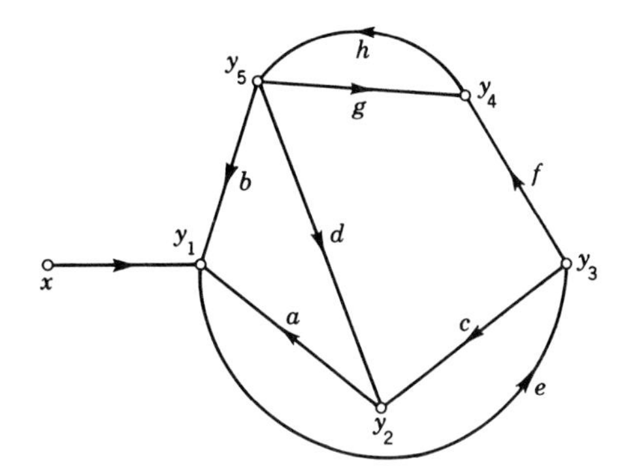

Fig. P13-9

13-10 When a flow graph can be identified as consisting of two sections with only one feedback branch as shown in Fig. P13-10*a*, the analysis can be simplified by the

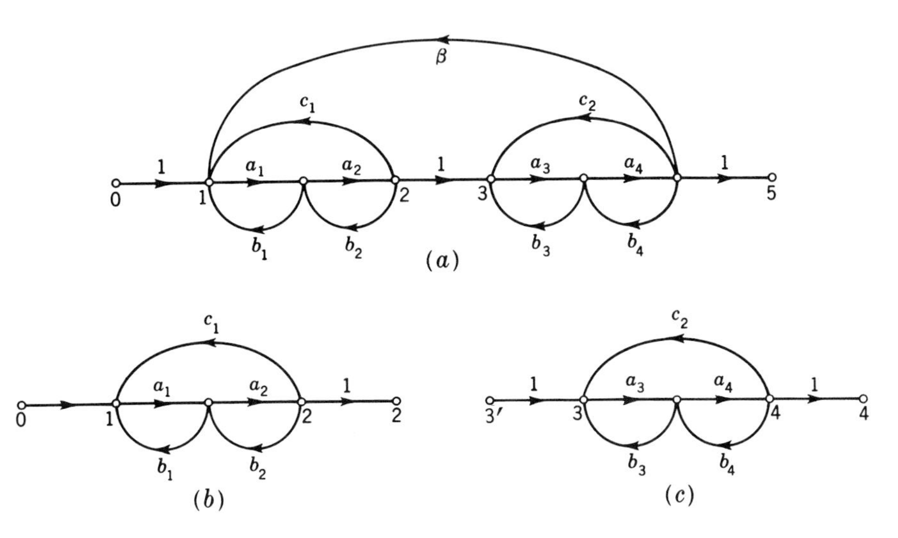

Fig. P13-10

procedure illustrated in this example. (*a*) Find T_{12} in Fig. P13-10*b*. (*b*) Find T_{34} in Fig. P13-10*c*. (*c*) Show that T_{05} in Fig. P13-10*a* is given by $T_{12}T_{34}/(1 - \beta T_{02}T_{3'4})$.

13-11 Figure P13-11 shows a bilateral ladder network. (*a*) Show that the node equations are

$$v_1 = \frac{1}{Y_{11}}(i_s - Y_{12}v_2)$$

$$v_k = \frac{1}{Y_{kk}}(-Y_{k-1,k}v_{k-1} - Y_{k+1,k}v_{k+1}) \qquad k = 2, \ldots, 6$$

$$v_7 = \frac{1}{Y_n}(-Y_{67}v_6)$$

where Y_{ij} are the node admittances (for example, $Y_{11} = Y_0 + Y_1$, $Y_{12} = -Y_1$, etc.). (*b*) Draw the signal flow graph for the network, compare your result to the type of graph shown in Fig. 13-15, and generalize the result to an n-node ladder.

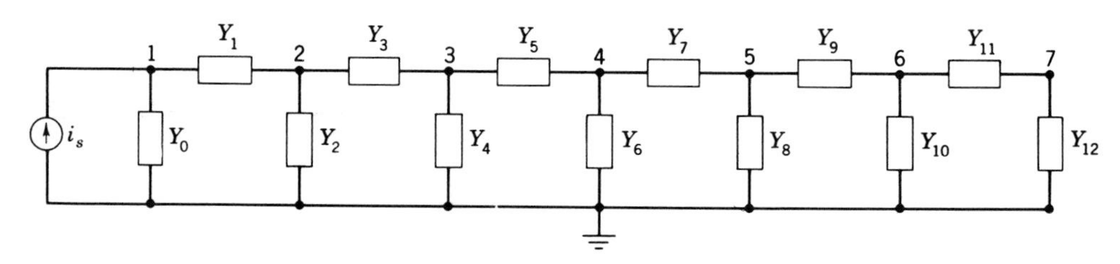

Fig. P13-11

13-12 The network of Fig. P13-1 has a voltage source inserted (by pliers entry) in series with Y_1. Use the method of Sec. 13-5 to find the flow graph of the network, and use this result to obtain the results required in Probs. 13-1*c* and *d*.

Chapter 14 Three-phase circuits

This chapter deals with circuits which are supplied by certain special combinations of sinusoidal sources. We shall be exclusively concerned with the sinusoidal steady-state conditions in such circuits. The discussion presented here will serve to illustrate the application of general principles of circuit analysis described in the preceding chapters. However, the circuits discussed are themselves of intrinsic importance, especially in power systems. We shall show that certain symmetrical connections of sources (and loads) are advantageous from the economical point of view, especially when large amounts of power are involved. All the circuits which we shall discuss contain several sources. We emphasize at the outset that all these sinusoidal sources will be assumed to operate at a single frequency. This assumption is true of the interconnected power systems. In the United States the power system operates at the frequency of 60 hertz (with very few exceptions[1]).

14-1 Sources

A practical sinusoidal source of electric energy can be represented as the series combination of an ideal voltage source and a complex impedance or as the parallel combination of an ideal current source and a complex admittance, as indicated in Fig. 14-1*a* and *b*.

If a passive combination of elements, forming a single-terminal-pair

[1] Air-borne equipment and other *isolated* systems are often operated at other frequencies.

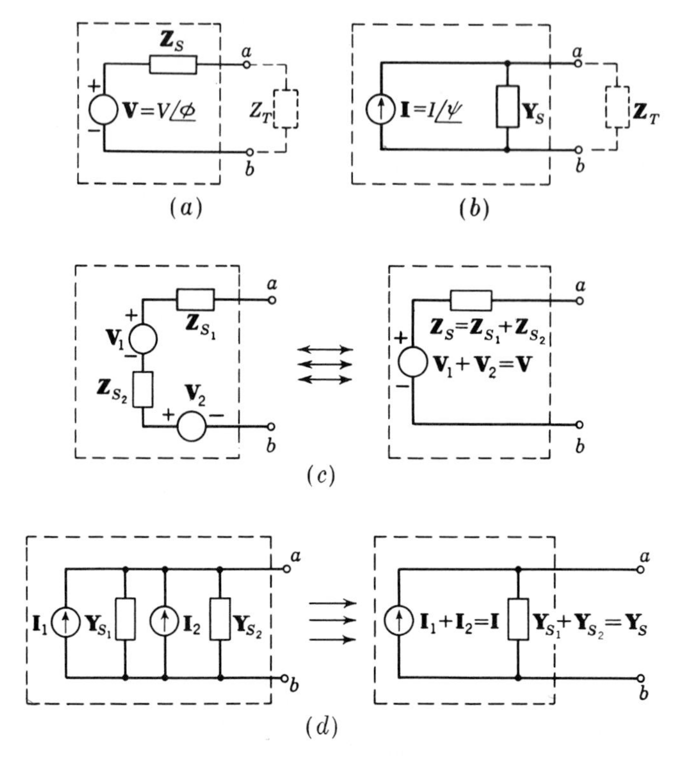

Fig. 14-1 Examples of single-phase sources.

network and represented by impedance $\mathbf{Z}_T$, is connected to such a source (as indicated by the dashed portion of Fig. 14-1*a* and *b*), one may, for purposes of calculating the circuit response in the steady state, choose the phase angle of the ideal source (ϕ or ψ) arbitrarily. A circuit which contains only one such source is distinguished by the fact that all phase relations between voltages and currents in every pair of points of the circuit can be related to the phase angle of this single source. For this reason a practical two-terminal sinusoidal source represented by an internal impedance (or admittance) and an ideal sinusoidal source is called a *single-phase source*.

If a circuit contains several sources (of like frequency), three possibilities may occur:

1 The combination of sources may be equivalent to a single-phase source.
2 Special relationships may exist between the amplitude and/or phase of various sources.
3 Neither (1) nor (2) occurs.

In the next section we shall consider briefly circuits with a single-phase source. We recall here that all the circuits discussed in Chap. 9 have sources of this type. In subsequent sections we shall discuss those circuits with special combinations of sources which are called three-phase sources. We shall see that combinations of single-phase sources which do not involve special phase relationships are also of interest in this discussion. The circuit in which these sources occur receives its name from the type of source used. Thus, unless a special relationship between the sources exists, the circuit is called a "single-phase" circuit with one or several sources (as the case may be).

Although circuits with several sources can always be analyzed by using the principle of superposition, so that the effect of each source can be accounted for separately, we shall see that the application of the network procedures developed in the previous chapters makes such a cumbersome scheme unnecessary.

14-2 Single-phase circuits SOURCES IN SERIES If two single-phase sources are connected in series as shown in Fig. 14-1*c*, the combination of these sources is equivalent to the single-phase source shown in Fig. 14-1*c*. This statement follows immediately from application of Kirchhoff's voltage law between terminals *a-b*. In either circuit, if a current represented by the phasor $\mathbf{I}_{ab}$ flows through an external load connected to terminals *a-b*, the phasor $\mathbf{V}_{ab}$ is $\mathbf{V}_1 + \mathbf{V}_2 - \mathbf{I}_{ab}(\mathbf{Z}_{s1} + \mathbf{Z}_{s2})$, and therefore the two circuits are equivalent at the terminals *a-b*.

SOURCES IN PARALLEL If two practical single-phase sources are in parallel, we may represent them as current sources as shown in Fig. 14-1*d*. The combination of these two sources, with respect to terminals *a-b*, is equivalent to the single source shown in Fig. 14-1*d*. This follows from the application of Kirchhoff's current law at terminal *a* or *b*. For any voltage $\mathbf{V}_{ab}$ the current $\mathbf{I}_{ab}$ flowing through an external admittance is given by $\mathbf{I}_{ab} = \mathbf{I}_1 + \mathbf{I}_2 - \mathbf{V}_{ab}(\mathbf{Y}_{s1} + \mathbf{Y}_{s2})$, and therefore the two circuits are equivalent at the terminals *a-b*.

THREE-WIRE SINGLE-PHASE CIRCUIT The three-wire single-phase circuit is discussed here, not only because of its intrinsic importance, but also because this connection is often confused with the three-phase connection, which will be discussed later. A three-wire single-phase source consists of two single-phase sources in which each source has its terminals accessibly connected, as shown in Fig. 14-2*a*. Such a source is a three-terminal or three-wire source.

In practically all instances an additional restriction is imposed on the three-terminal connection of Fig. 14-2*a*. In a *three-wire single-phase* source the two source voltages $\mathbf{V}_1$ and $\mathbf{V}_2$ are usually understood to be identical; i.e., they are to have the same amplitude, and they are to be in phase with each other. Moreover, the complex internal impedances of the sources must be identical. This situation is indicated in Fig. 14-2*b*. The internal common point of the two sources is marked n' and is referred to as the "neutral" point of the source.

A typical application of a three-wire system is shown in Fig. 14-2*c*. In this circuit two identical complex loads $\mathbf{Z}_1$ are connected together and joined through an impedance $\mathbf{Z}_n$ to the neutral point of the source. In addition, an impedance $\mathbf{Z}_2$ is connected across the terminals *a-b* (the impedance of the wires joining the loads to the *a* and *b* points of the source may be included in $\mathbf{Z}_s$). With this special symmetry of the load impedances and the source, the circuit forms a balanced three-wire *single-phase* system. By "balanced system" we mean that both the two sources and the loading of the sources are identical. We shall now show that with this balanced connection $\mathbf{V}_{nn'} = 0$.

The voltage sources are converted to current sources, and the impedances are labeled with the values of the corresponding admittances so that the circuit shown in Fig. 14-3 is obtained. This circuit is equivalent to the original circuit with respect to terminals

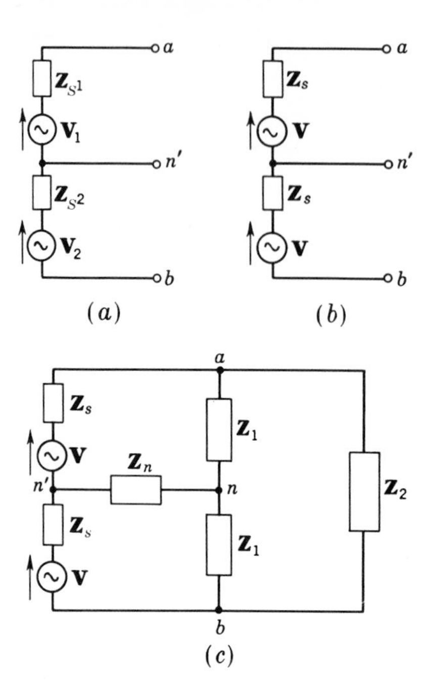

Fig. 14-2 Three-wire systems. (a) Three-wire source. (b) Three-wire single-phase source. (c) Balanced three-wire single-phase system.

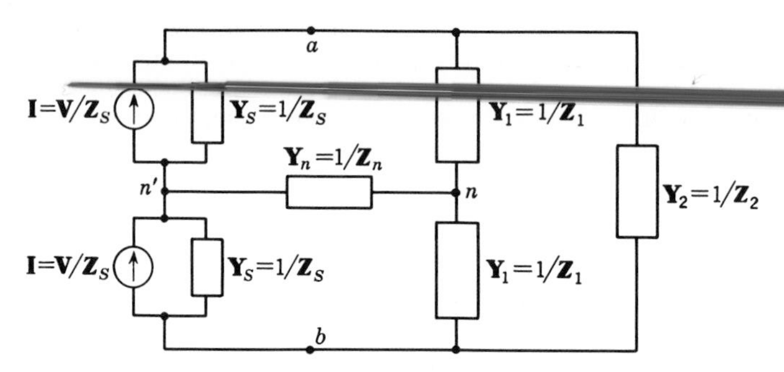

Fig. 14-3 Balanced three-wire single-phase system using current sources.

a, n', and b. Using junction n' as the reference node, we have from the node equations

$$\mathbf{V}_n = \frac{\begin{vmatrix} \mathbf{Y}_s + \mathbf{Y}_1 + \mathbf{Y}_2 & -\mathbf{Y}_2 & \mathbf{I} \\ -\mathbf{Y}_2 & \mathbf{Y}_s + \mathbf{Y}_1 + \mathbf{Y}_2 & -\mathbf{I} \\ -\mathbf{Y}_1 & -\mathbf{Y}_1 & 0 \end{vmatrix}}{\mathbf{D}_y} \equiv \frac{\mathbf{N}}{\mathbf{D}_y} \tag{14-1}$$

where $\mathbf{D}_y$ is the complex determinant of the node-admittance matrix.

In the numerator determinant of Eq. (14-1) we add row 3 to row 2 and then add the new row 2 to row 1.

$$\mathbf{N} = \begin{vmatrix} \mathbf{Y}_s + \mathbf{Y}_1 + \mathbf{Y}_2 & -\mathbf{Y}_2 & \mathbf{I} \\ -\mathbf{Y}_2 & \mathbf{Y}_s + \mathbf{Y}_1 + \mathbf{Y}_2 & -\mathbf{I} \\ -\mathbf{Y}_1 & -\mathbf{Y}_1 & 0 \end{vmatrix} = \begin{vmatrix} \mathbf{Y}_s & \mathbf{Y}_s & 0 \\ -\mathbf{Y}_2 & \mathbf{Y}_s + \mathbf{Y}_2 & -\mathbf{I} \\ -\mathbf{Y}_1 & -\mathbf{Y}_1 & 0 \end{vmatrix} \tag{14-2}$$

The determinant on the right-hand side of Eq. (14-2) is evaluated by expansion about column 3, row 2.

$$\mathbf{N} = -(-\mathbf{I})[-\mathbf{Y}_s\mathbf{Y}_1 - (\mathbf{Y}_s)(-\mathbf{Y}_1)] \equiv 0$$

Hence $\mathbf{V}_n = \mathbf{V}_{nn'} \equiv 0$.

For a balanced three-wire single-phase circuit we conclude as follows:

1 Since $\mathbf{V}_{nn'} = 0$, the current in the neutral $\mathbf{I}_{nn'} = 0$. Hence the functioning of the circuit is independent of the admittance $\mathbf{Y}_n$. This admittance may therefore be removed without affecting the circuit as long as the circuit remains balanced.

2 Since $\mathbf{I}_{nn'} = 0$, $\mathbf{V}_{an} = \mathbf{V}_{nb}$; this is seen from the fact that the current in the two equal impedances $\mathbf{Z}_1$ is the same current. Hence $\mathbf{V}_{ab} = 2\mathbf{V}_{an}$. This states that the voltage across the impedance $\mathbf{Z}_2$ has twice the value of the voltage across $\mathbf{Z}_1$. We have, therefore, a source which furnishes two voltages, one twice the other. Frequently, two such voltages are needed. In a home, if $|\mathbf{V}_{an}|$ is, for example, 110 volts and $|\mathbf{V}_{ab}| = 220$ volts, then terminal pairs *a-n* and *b-n* can be used for lighting, and terminal pair *a-b*, where the higher voltage is available, may be used for a major appliance such as a clothes drier or an electric stove.

Although the discussion has revolved around the balanced case, where the connection from n to n' is not needed, this connection is not omitted in practice because of the effect on the various voltages if the load becomes unbalanced ($\mathbf{Z}_{an} \neq \mathbf{Z}_{nb}$) (see Prob. 14-1).

14-3 Three-phase sources

A source which can be represented as three single-phase voltage sources connected to form a wye or a delta as shown in Fig. 14-4 forms a balanced three-phase source provided that the following special relationships exist:

1 The three complex impedances are equal to each other.

$$\mathbf{Z}_{s1} = \mathbf{Z}_{s2} = \mathbf{Z}_{s3}$$

2 The magnitudes of the source voltages are equal to each other.

In the wye: $\quad |\mathbf{V}_1| = |\mathbf{V}_2| = |\mathbf{V}_3| = V_y$

In the delta: $\quad |\mathbf{V}_a| = |\mathbf{V}_b| = |\mathbf{V}_c| = V_\Delta$

3 The source voltages add up to zero at every instant of time so that their phasor sum is zero.

In the wye: $\quad \mathbf{V}_1 + \mathbf{V}_2 + \mathbf{V}_3 = 0$

or

In the delta: $\quad \mathbf{V}_a + \mathbf{V}_b + \mathbf{V}_c = 0$

If three phasors have equal absolute values and add up to zero, they must differ in phase by exactly 120° because the graphical addition of these phasors by the polygon method must result in an equilateral triangle. In Fig. 14-4, if in the wye connection we choose $\mathbf{V}_1$ as the reference, we have

$$\mathbf{V}_1 = V_y\underline{/0^\circ}$$

and either $\quad \mathbf{V}_2 = V_y\underline{/+120^\circ} \quad$ and $\quad \mathbf{V}_3 = V_y\underline{/-120^\circ}$

or $\quad \mathbf{V}_2 = V_y\underline{/-120^\circ} \quad$ and $\quad \mathbf{V}_3 = V_y\underline{/+120^\circ}$

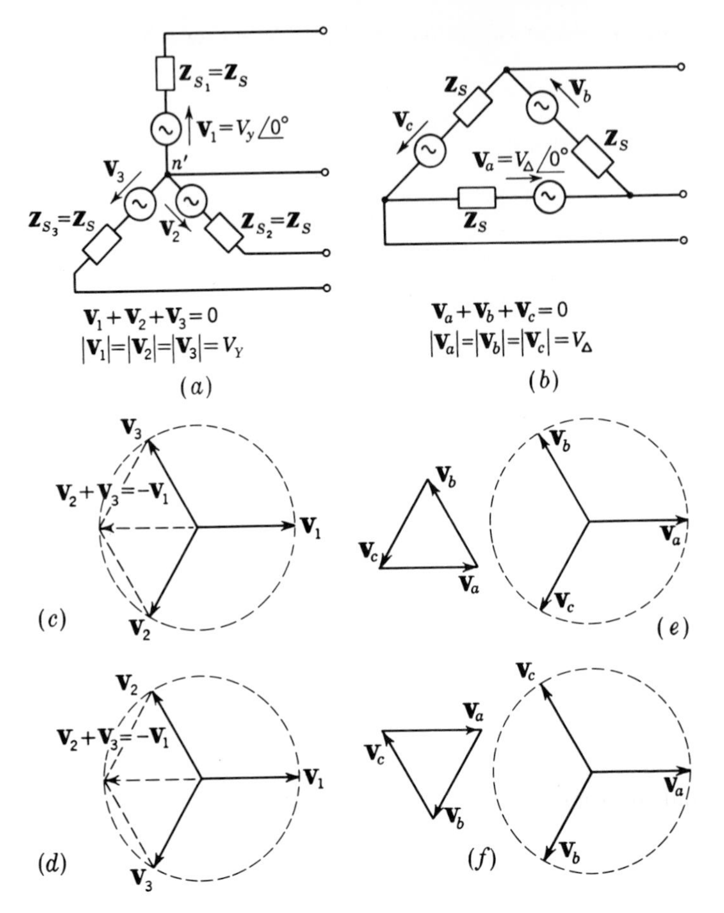

Fig. 14-4 Balanced three-phase sources. ***(a) Balanced three-phase wye-connected source. (b) Balanced three-phase delta-connected source. (c) Phasor diagram for 1-2-3 sequence in (a). (d) Phasor diagram for 1-3-2 sequence in (a). (e) Phasor diagram for a-c-b sequence in (b). (f) Phasor diagram for a-b-c sequence in (b).***

Similarly, if in the delta connection we choose $\mathbf{V}_a$ as reference, we have

$$\mathbf{V}_a = V_\Delta \underline{/0°}$$

and either $\mathbf{V}_b = V_\Delta \underline{/+120°}$ and $\mathbf{V}_c = V_\Delta \underline{/-120°}$

or $\mathbf{V}_b = V_\Delta \underline{/-120°}$ and $\mathbf{V}_c = V_\Delta \underline{/+120°}$

The possibilities are shown in Fig. 14-4*c* to *f*. For each source we have shown the two possible phase relations. The term used to describe the difference between the phase relations of Fig. 14-4*c* and *d* (or between Fig. 14-4*e* and *f*) is the *phase sequence*, or *phase order*. By phase sequence we mean the order in which the three sources reach their maxima. For the relationship shown in Fig. 14-4c, if we observe the voltages $\mathbf{V}_1$, $\mathbf{V}_2$, and $\mathbf{V}_3$, we find the following:

$$v_1(t) \text{ is maximum at } t = 0$$

$$v_2(t) \text{ is maximum at } \omega t = \frac{2\pi}{3} \text{ or } 120°$$

$$v_3(t) \text{ is maximum at } \omega t = \frac{4\pi}{3} \text{ or } 240°$$

$$v_1(t) \text{ is maximum at } \omega t = 2\pi \text{ or } 360°$$

$$\cdots\cdots\cdots\cdots\cdots\cdots$$

Hence the phase sequence is $\overline{1\text{-}2\text{-}3}\text{-}\overline{1\text{-}2\text{-}3} \cdots$. Starting with the source labeled 1, the phase sequence is expressed as 1-2-3.

Similarly, if we start when v_a is a maximum in Fig. 14-4e, the phase order is a-c-b.

On *paper*, the phase sequence appears to be a trivial matter because we can always renumber the sources to obtain either phase order. Practically, we must point out that the three sources shown in Fig. 14-4 may represent a single physical device, i.e., a three-phase generator. Once we label the actual wires, the phase sequence is fixed. The direction of rotation of a three-phase motor will depend on the phase sequence of the three-phase source. Knowledge of the actual phase order in the physical generator may therefore be a matter of importance. We shall illustrate in a later section how phase sequence may be determined experimentally.

In the definition of a balanced three-phase source we have stated three requirements: equal internal impedances for the three sources, equal voltage amplitudes, and a phase difference of exactly 120° between the voltages. Lest the reader should wonder about the rigidity of these requirements, we point out that in practice these conditions can be approximated very closely, for example, by construction of a single rotating machine in which three separate windings are made identically (to ensure equal impedances and equal voltage magnitudes) and are placed symmetrically in the mechanical structure of the machine (to ensure the correct phase relationships).

When *any* of the conditions for balance is not satisfied, the source is called an *unbalanced* (or unsymmetrical) three-phase source. We understand that, for purposes of analysis, an unbalanced three-phase source is merely an interconnection of single-phase sources.

14-4 Delta- and wye-connected sources

EQUIVALENT IDEAL SOURCES In order to show how delta- and wye-connected sources can replace each other, we shall first consider the ideal, balanced sources shown in Fig. 14-5a. These sources are equivalent with respect to terminals a', b', c', provided that the voltages $\mathbf{V}_a$, $\mathbf{V}_b$, and $\mathbf{V}_c$ have a special relationship to $\mathbf{V}_1$, $\mathbf{V}_2$, and $\mathbf{V}_3$. Note that the junction point of the sources in the equivalent wye connection (point n') is not accessible

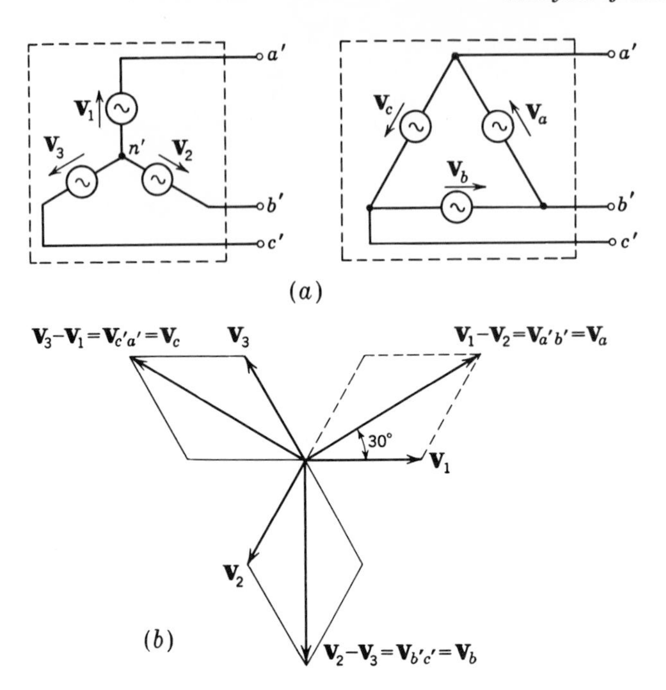

***Fig. 14-5** Ideal three-phase voltage sources. (**a**) Wye- and delta-connected sources. (**b**) Phasor diagram used to establish equivalence.*

to the circuit external to the source. To establish the conditions for equivalence, we note first that the voltages between a', b', and c' are independent of the currents which may be drawn by an external circuit connected to these points, because all sources are ideal. Hence the two sources are equivalent if the voltages between any pairs of terminals correspond exactly as in the tabulation below.

$$\begin{array}{llll} & \textit{Wye} & \textit{Delta} & \\ \mathbf{V}_{a'b'} = & \mathbf{V}_1 - \mathbf{V}_2 = & \mathbf{V}_a & \quad (14\text{-}3) \\ \mathbf{V}_{b'c'} = & \mathbf{V}_2 - \mathbf{V}_3 = & \mathbf{V}_b & \quad (14\text{-}4) \\ \mathbf{V}_{c'a'} = & \mathbf{V}_3 - \mathbf{V}_1 = & \mathbf{V}_c & \quad (14\text{-}5) \end{array}$$

Equations (14-3), (14-4), and (14-5) show how the equivalent delta is obtained, given the voltages in the wye. Note that, since the balanced case was assumed,

$$|\mathbf{V}_1 - \mathbf{V}_2| = |\mathbf{V}_2 - \mathbf{V}_3| = |\mathbf{V}_3 - \mathbf{V}_1| = \sqrt{3}\,|\mathbf{V}_1| \qquad (14\text{-}6)$$

as can be seen from the phasor diagram shown in Fig. 14-5*b*. We conclude from this diagram that the magnitude of each equivalent source in the delta is $\sqrt{3}$ times the magnitude of a source in the wye. Both sources form a balanced system of voltages: in the delta each source has the same magnitude, and the sum of the phasors is zero. Hence no current flows in any of the delta-connected sources in the absence of external loading. Note that the condition for equivalence at the external terminals a', b', and c' causes the voltages $\mathbf{V}_a$, $\mathbf{V}_b$, $\mathbf{V}_c$ to be 30° displaced from $\mathbf{V}_1$, $\mathbf{V}_2$, and $\mathbf{V}_3$, respectively. Since in any particular example the phase differences between the various voltages can be obtained from a phasor diagram with ease, the principal conclusion here is the magnitude relationship. If V_y is the magnitude of a source voltage in the wye and V_Δ is the magnitude of a source voltage in its equivalent delta connection, these magnitudes are related by the number $\sqrt{3}$; $V_\Delta = \sqrt{3}\,V_y$.

EQUIVALENT PRACTICAL SOURCES Figure 14-6*a* and *b* shows sources with internal impedances. For the balanced case these are equivalent at their terminals if

$$|\mathbf{V}_a| = |\mathbf{V}_b| = |\mathbf{V}_c| \qquad \mathbf{Z}_a = \mathbf{Z}_b = \mathbf{Z}_c \tag{14-7}$$

and

$$\mathbf{V}_1 = \mathbf{V}_a - \mathbf{V}_c \qquad \mathbf{V}_2 = \mathbf{V}_b - \mathbf{V}_a \qquad \mathbf{V}_3 = \mathbf{V}_c - \mathbf{V}_b \tag{14-8}$$

$$\mathbf{Z}_1 = \mathbf{Z}_2 = \mathbf{Z}_3 = \tfrac{1}{3}\mathbf{Z}_a \tag{14-9}$$

or

$$\mathbf{V}_a = \mathbf{V}_1 - \mathbf{V}_2 \qquad \mathbf{V}_b = \mathbf{V}_2 - \mathbf{V}_3 \qquad \mathbf{V}_c = \mathbf{V}_3 - \mathbf{V}_1 \tag{14-10}$$

Example 14-1 In the balanced delta-connected generator shown in Fig. 14-6*c*

$$|\mathbf{V}_a| = |\mathbf{V}_b| = |\mathbf{V}_c| = 173 \text{ volts} \tag{14-11}$$

and

$$\mathbf{V}_a + \mathbf{V}_b + \mathbf{V}_c = 0$$

Calculate the absolute value of the current phasor $\mathbf{I}$, in the 6-ohm resistance.

To illustrate the usefulness of conversion of the delta source to a wye source, this problem is first done without conversion of the source.

Solution Method 1 (*with original source*) Using $\mathbf{V}_a$ as reference and assuming *a-b-c* phase order, we write

$$\mathbf{V}_a = 173\underline{/0°}$$
$$\mathbf{V}_b = 173\underline{/-120°}$$
$$\mathbf{V}_c = 173\underline{/+120°}$$

Since terminal *c* is not connected to any external elements, we may combine the two branches containing $\mathbf{V}_c$ and $\mathbf{V}_b$ into a single branch as shown in Fig. 14-6*d*.

Since $\mathbf{V}_d = \mathbf{V}_b + \mathbf{V}_c$, $\mathbf{V}_d = 173\underline{/-120°} + 173\underline{/+120°} = 173\underline{/180°}$. Converting the voltage sources to current sources, we obtain the circuit of Fig. 14-6*e*. Hence we may calculate

$$\mathbf{V}_{ab} = \left(\frac{173\underline{/0°}}{3 + j9} - \frac{173\underline{/180°}}{6 + j18}\right)\mathbf{Z}_{ab} \tag{14-12}$$

where $\mathbf{Z}_{ab}$ is the impedance of the three parallel branches shown in Fig. 14-6*e*;

$$\mathbf{Y}_{ab} = \frac{1}{6} + \frac{1}{6 + j18} + \frac{1}{3 + j9} = \frac{1}{6} + \frac{1}{2 + j6}$$

and

$$\mathbf{Z}_{ab} = \frac{(6)(2 + j6)}{8 + j6}$$

Since

$$\frac{173/0^\circ}{3+j9} - \frac{173/180^\circ}{6+j18} = \frac{173}{2+j6}$$

$$\mathbf{V}_{ab} = \frac{173/0}{2+j6}\,\frac{(6)(2+j6)}{8+j6} \qquad |\mathbf{V}_{ab}| = \left|\frac{(173)(6)}{8+j6}\right| = \frac{(173)(6)}{10}$$

Hence

$$I = 17.3 \text{ amp} \qquad\qquad \textit{Ans.}$$

Method 2 (*by conversion of the delta source*) The delta source can be replaced by a balanced wye source in which each single-phase source generates an amplitude of $173/\sqrt{3} = 100$ volts and has an internal impedance of $(3+j9)/3 = 1 + j3$ ohms as shown in Fig. 14-6*f*.

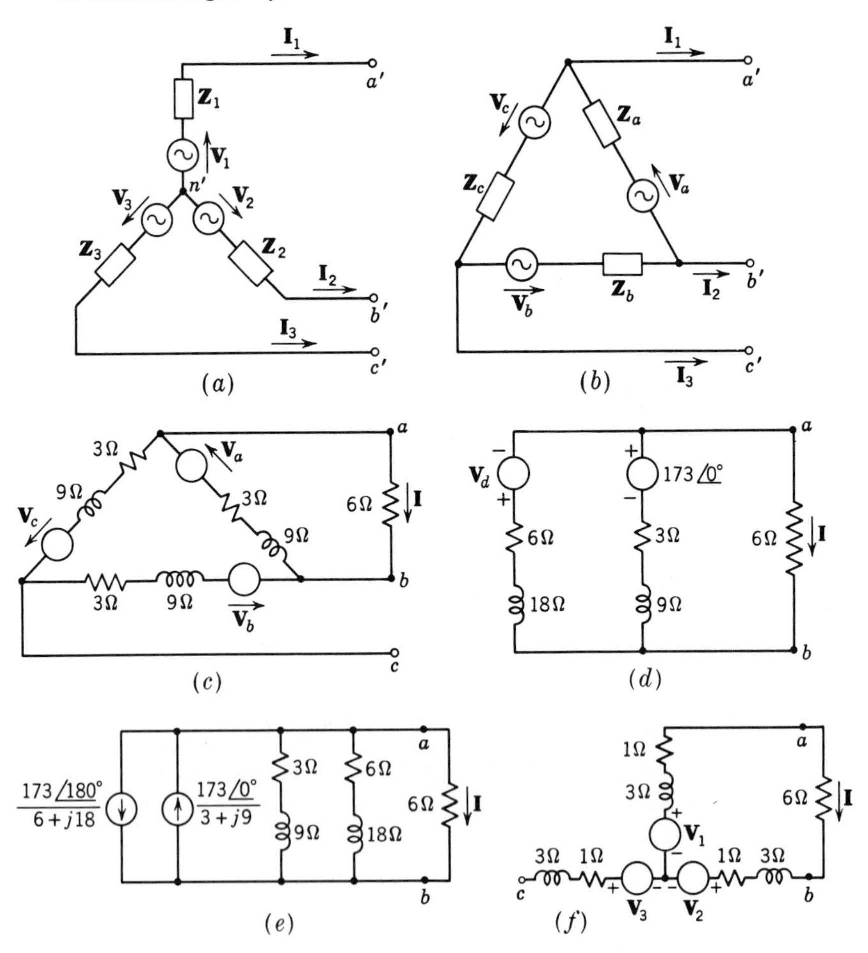

Fig. 14-6 (a) Wye-connected source with internal impedances. (b) Delta-connected source with internal impedances. (c) Circuit for Example 14-1. (d) Equivalent to (c). (e) Current-source form of (d). (f) Circuit of (e) after (source) delta-wye conversion.

The magnitude of the current $\mathbf{I}$ is given by

$$|\mathbf{I}| = \frac{|\mathbf{V}_1 - \mathbf{V}_2|}{|2 + j6 + 6|} = \left| \frac{\mathbf{V}_1 - \mathbf{V}_2}{8 + j6} \right| = \frac{|\mathbf{V}_1 - \mathbf{V}_2|}{10}$$

Since the source is balanced, the difference between $\mathbf{V}_1$ and $\mathbf{V}_2$ has the absolute value $V_1\sqrt{3}$. Hence we have the solution immediately:

$$I = \frac{100\sqrt{3}}{10} = 17.3 \text{ amp} \qquad \textit{Ans.}$$

It is seen that conversion from delta into wye reduces the amount of work entailed in this particular problem.

Reconnection of sources A problem quite different from equivalent three-phase sources is posed by the relationship between a delta- and a wye-connected source, which may be formed by connecting three single-phase sources. We shall term this problem the "reconnection" of sources. Consider the three single-phase sources shown in Fig. 14-7*a*. Since the internal impedances of the three sources are equal and the source

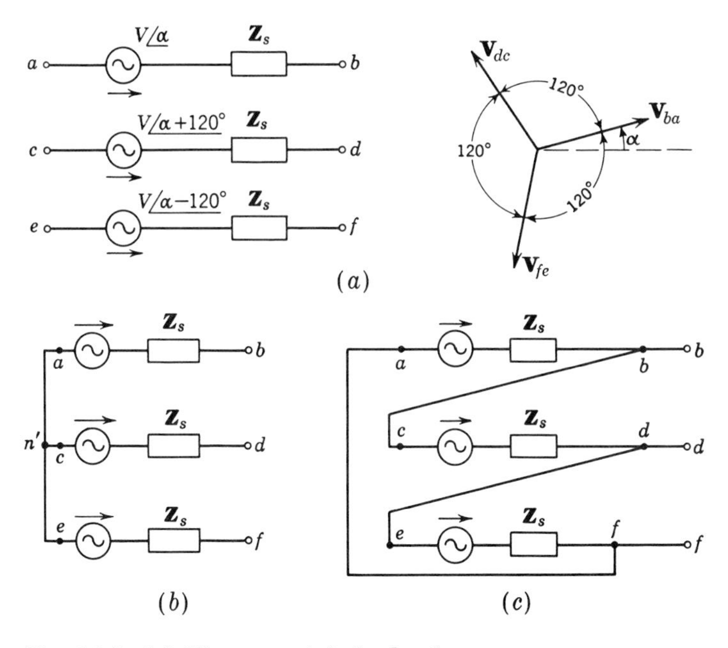

Fig. 14-7 ***(a) Three special single-phase sources. (b) Balanced three-phase wye-connected source formed with the sources of (a). (c) Balanced three-phase delta-connected source formed from (a).***

voltages are equal and add up to zero, balanced three-phase sources may be formed by connecting these single-phase sources properly.

A study of the phasor diagram shows that a wye-connected source may be formed by joining points *a*, *c*, and *e* to form the neutral. This connection is shown in Fig. 14-7*b*.

Since $\mathbf{V}_{ab} + \mathbf{V}_{cd} + \mathbf{V}_{ef} = 0$, a balanced delta-connected source may be formed by joining points *bc*, *de*, and *fa* as shown in Fig. 14-7*c*.

We note now that in the wye-connected source the voltage between any two of the three terminals of the source (*b-d*, *d-f* and *f-b*) is in magnitude $\sqrt{3}\,V$, and the corresponding terminals in the delta connection make available the open-circuit voltage V. In return for the higher voltage which the wye connection furnishes, we shall show later, the current "capacity" of this source is less. Qualitatively, this can be seen by noting that, in the wye, any current which flows from one of the terminals, say, terminal *b*, flows directly through one of the single-phase sources, whereas the corresponding current in the delta connection may flow through either of two paths (that is, *a-b* and *c-d*).

PHASE GENERATORS, LINE AND PHASE VOLTAGES Each of the three single-phase sources used to form a three-phase source is called a phase generator. The magnitude of voltage associated with each phase generator in a balanced source is called the phase voltage of the source. The magnitude of the voltage which is available between any two of the three line terminals is called a "line-to-line" voltage, or for brevity, a "line" voltage. For an ideal wye-connected balanced three-phase source, the ratio of line to phase voltage is the *square root of* 3.

14-5 Balanced three-phase circuits

BALANCED LOAD Three *equal complex impedances*, connected in delta or in wye, form a balanced three-phase load. Exactly as in the case of three-phase sources, the representation of a three-phase load as three separate terminal pairs does not mean that the physical device which is so represented actually consists of three separate units. In the case of loads, it is possible for a single unit (such as a "three-phase motor") to have an equivalent circuit which is represented by a delta or a wye; it is also possible to form a wye or delta load from separate units.

Three-phase balanced loads are shown in Fig. 14-8*a* and *b*. Since every delta can be converted to a wye, any three-phase load can be represented as either a wye or a delta. We have seen that, in the balanced case, the transformation between delta and wye reduces to

$$\mathbf{Z}_Y = \tfrac{1}{3}\mathbf{Z}_\Delta \tag{14-13}$$

In the three-phase loads of Fig. 14-8, terminals *a*, *b*, and *c* are referred to as the "line terminals" of the load. In the wye-connected load of Fig. 14-8*a*

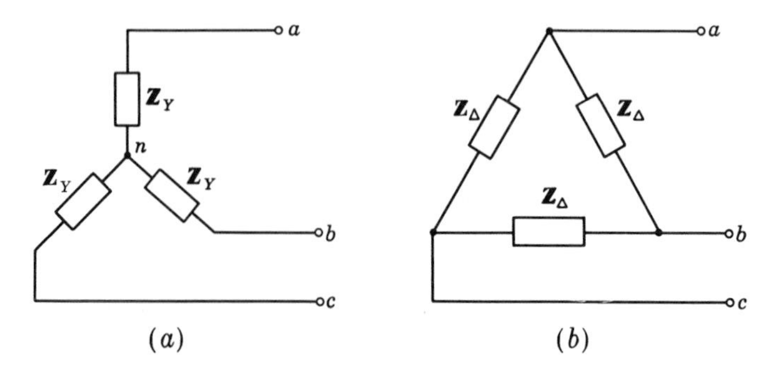

Fig. 14-8 Balanced three-phase loads.

the junction of the three impedances (point n) is referred to as the "neutral point" of the load. In each load each of the impedances $\mathbf{Z}_Y$ or $\mathbf{Z}_\Delta$ is called a "phase impedance."

BALANCED CIRCUIT If a balanced load is supplied by a balanced three-phase source, the combination is referred to as a *balanced three-phase circuit.* If either source or load is unbalanced, the circuit is unbalanced. In practice, the generator is usually balanced, but the load may be unbalanced. We shall first study, in detail, balanced circuits.

In Fig. 14-9a, a balanced three-phase wye-connected source is shown connected to a balanced wye load by means of three cables connecting the line

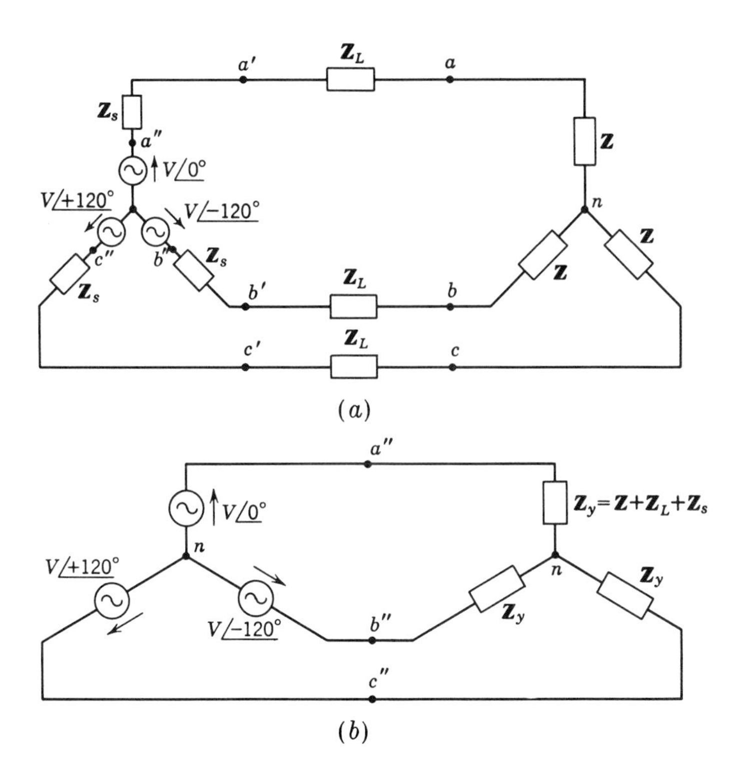

Fig. 14-9 Balanced three-phase circuits.

terminals of the load to those of the source. Since the cables have equal impedances $\mathbf{Z}_L$, an equivalent circuit is shown in Fig. 14-9b. In this circuit the internal impedances of the source and the line impedances have been added to the impedances of the load so that the resulting circuit consists of an ideal three-phase wye-connected source connected to a balanced wye load. We shall see that every balanced circuit can be reduced to a circuit as shown in Fig. 14-9b by appropriate use of delta-wye conversion.

We shall now show how every balanced three-phase circuit can be solved by solving an equivalent single-phase circuit. The problem we set ourselves is the calculation of the relationship between the currents which flow from the source to the load, the line voltages, and the power transferred.

14-6 The representative single-phase problem

Since every balanced circuit can be represented as the combination of a balanced wye-connected load and an ideal wye-connected three-phase source, we shall first study this circuit in detail. Examples of how the results which are obtained in this case apply to other balanced circuits will be cited in subsequent sections.

Consider, therefore, the balanced three-phase circuit shown in Fig. 14-10. Let it be required to solve for the line currents $\mathbf{I}_a$, $\mathbf{I}_b$, and $\mathbf{I}_c$ in terms of the

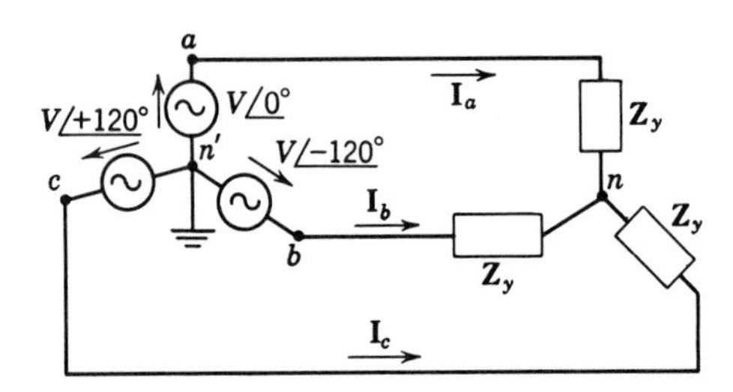

Fig. 14-10 Balanced three-phase circuit showing reference directions.

phase voltages of the source, $\mathbf{V}_{an'}$, $\mathbf{V}_{bn'}$, $\mathbf{V}_{cn'}$. We note that the circuit has only two junction points; hence it can be solved by means of a single node equation: If the voltage $V_{nn'}$ is known, the currents can be calculated by the application of Kirchhoff's voltage law. Although each voltage source in series with $\mathbf{Z}_y$ can be converted to a current source, we need not apply this formalism. At junction n Kirchhoff's current law reads

$$\mathbf{I}_a + \mathbf{I}_b + \mathbf{I}_c = 0 \tag{14-14}$$

but

$$\mathbf{I}_a = \frac{\mathbf{V}_{an}}{\mathbf{Z}_y} \qquad \mathbf{I}_b = \frac{\mathbf{V}_{bn}}{\mathbf{Z}_y} \qquad \mathbf{I}_c = \frac{\mathbf{V}_{cn}}{\mathbf{Z}_y} \tag{14-15}$$

and

$$\mathbf{V}_{an} = \mathbf{V}_{an'} - \mathbf{V}_{nn'} \tag{14-16a}$$

$$\mathbf{V}_{bn} = \mathbf{V}_{bn'} - \mathbf{V}_{nn'} \tag{14-16b}$$

$$\mathbf{V}_{cn} = \mathbf{V}_{cn'} - \mathbf{V}_{nn'} \tag{14-16c}$$

Substituting (14-16) in (14-15) and applying (14-14), we obtain

$$\frac{\mathbf{V}_{an'}}{\mathbf{Z}_y} + \frac{\mathbf{V}_{bn'}}{\mathbf{Z}_y} + \frac{\mathbf{V}_{cn'}}{\mathbf{Z}_y} = \frac{3\mathbf{V}_{nn'}}{\mathbf{Z}_y} \tag{14-17}$$

or

$$\mathbf{V}_{an'} + \mathbf{V}_{bn'} + \mathbf{V}_{cn'} = 3\mathbf{V}_{nn'} \tag{14-17a}$$

But in a balanced three-phase source the sum of the phase voltages (with respect to n') is zero; hence

$$\mathbf{V}_{an'} + \mathbf{V}_{bn'} + \mathbf{V}_{cn'} = 3\mathbf{V}_{nn'} = 0 \tag{14-18}$$

and

$$\mathbf{V}_{nn'} = 0$$

We conclude that, in a balanced three-phase circuit, the voltage between the neutral point of the source and the neutral point of the load is zero.

Since the voltage between points n and n' is always zero, no current will flow between these points if they are connected together through any impedance. The circuit shown in Fig. 14-11a is therefore equivalent to the circuit shown in Fig. 14-10. It follows that the two neutral points may be joined by a wire of zero impedance, as shown in Fig. 14-11b. We now observe that the current $\mathbf{I}_a$ may be calculated by application of Kirchhoff's voltage law in loop a-n-n'-a: $\mathbf{I}_a = \mathbf{V}_{an'}/\mathbf{Z}_y$. Similarly, $\mathbf{I}_b = \mathbf{V}_{bn'}/\mathbf{Z}_y$, and $\mathbf{I}_c = \mathbf{V}_{cn'}/\mathbf{Z}_y$. If we now separate the three parts of the circuit, as shown

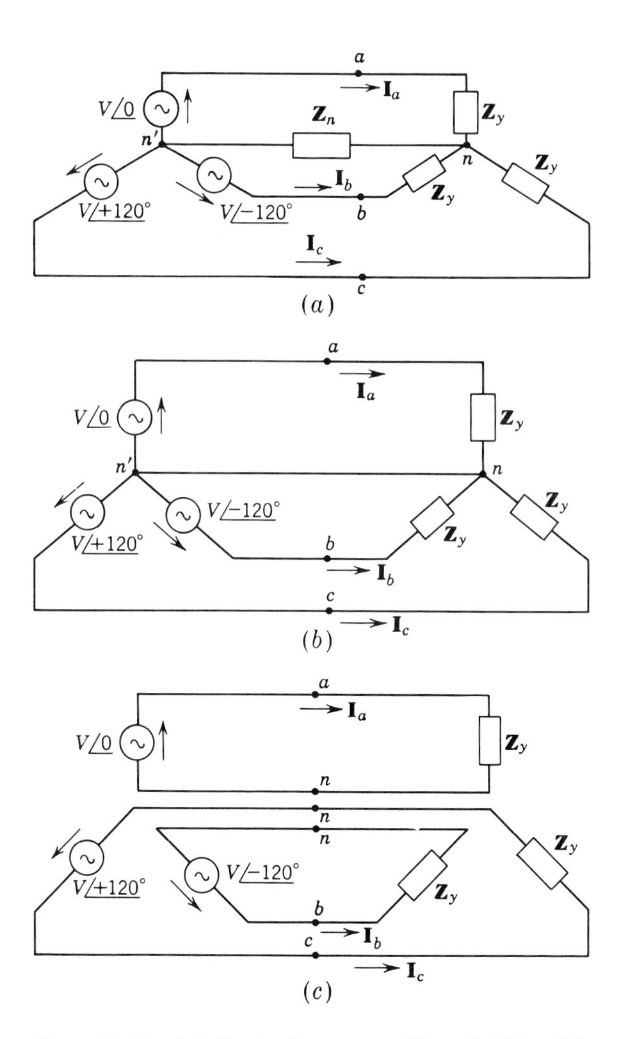

***Fig. 14-11** (a) Equivalent to Fig. 14-10. (b) Equivalent to (a). (c) The three single-phase circuits shown are equivalent to the balanced three-phase circuit.*

in Fig. 14-11*c*, each current in each of the three single-phase circuits will be identical with the current in one of each of the three line currents in the three-phase circuit. To solve the three-phase circuit, it is necessary only to solve any *one* of these single-phase circuits shown in Fig. 14-11*c*. For example, if $\mathbf{Z}_y = Z_y\underline{/\theta}$, then

$$\mathbf{I}_a = \frac{V\underline{/0}}{Z\underline{/\theta}} = \frac{V\underline{/-\theta}}{Z}$$

Since the voltages are symmetrical, the currents will also differ in phase by 120°, so that, for the circuits of Fig. 14-11,

$$\mathbf{I}_a = \frac{\mathbf{V}}{\mathbf{Z}_y}$$
$$\mathbf{I}_b = \mathbf{I}_a(1\underline{/-120°})$$
$$\mathbf{I}_c = \mathbf{I}_a(1\underline{/+120°})$$

Example 14-2 Calculate the magnitude of a line current in the circuit shown in Fig. 14-12*a*.

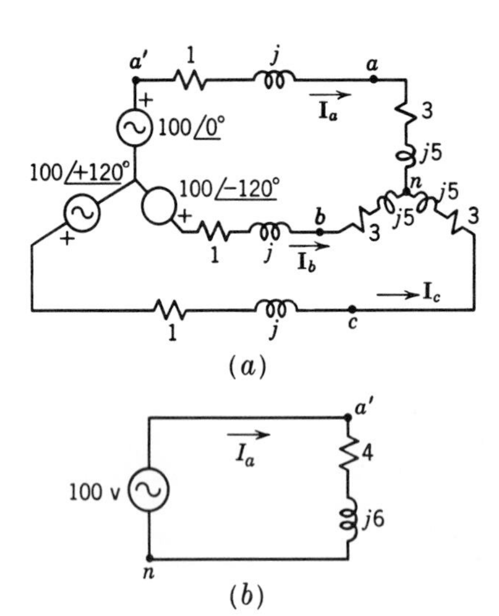

Fig. 14-12 ***(a) A balanced three-phase circuit. (b) A representative phase of the circuit shown in (a).***

Solution We immediately recognize the circuit as balanced and separate one-phase, as shown in Fig. 14-12*b*. Hence

$$|\mathbf{I}_a| = \left|\frac{100}{4 + j6}\right| = 13.85 \text{ amp}$$

Since

$$|\mathbf{I}_a| = |\mathbf{I}_b| = |\mathbf{I}_c|$$

the solution is complete.

14-7 Line and phase quantities

THE SQUARE ROOT OF 3 If a three-phase load is represented by a three-terminal box marked "load" as shown in Fig. 14-13a, then, in view of the

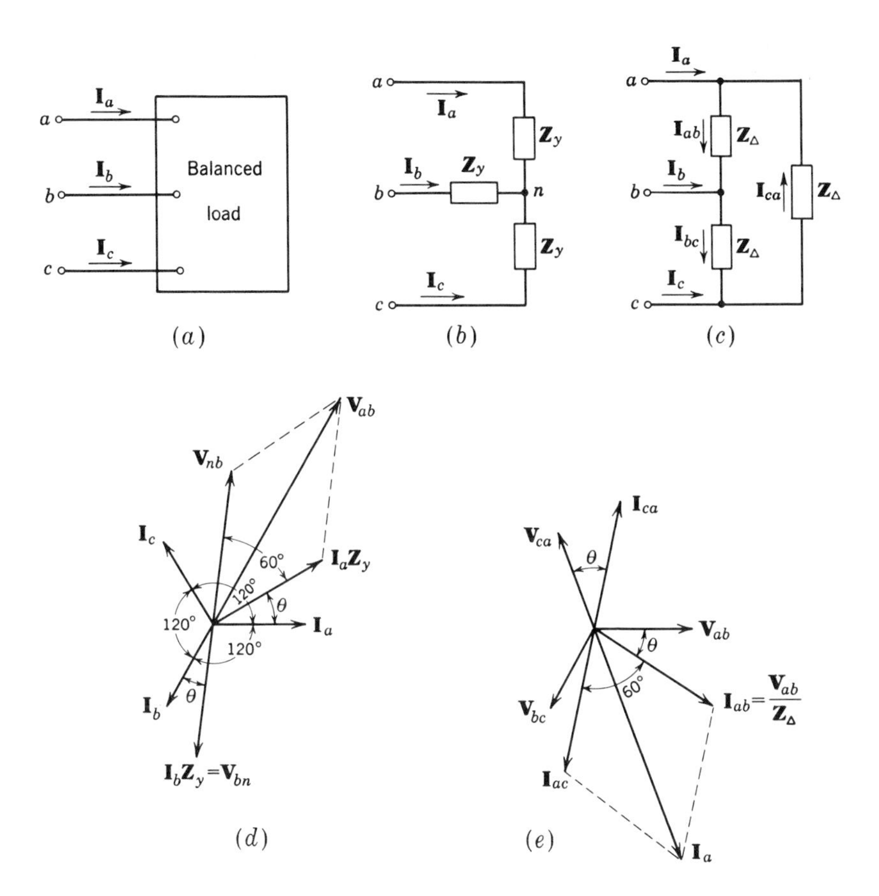

Fig. 14-13 (a) Symbol for a three-terminal load. (b) Balanced wye-connected load. (c) Balanced delta-connected load. (d) Phasor addition of phase voltages in the wye connection to give a line voltage. (e) Phasor addition of phase currents in the delta to give a line current.

delta-wye equivalence, the internal connections of the impedances in the box are immaterial with respect to the three accessible terminals. For purposes of calculating the currents entering the terminals, we may, if we choose, assume either delta or wye connection internally. The voltages across the individual impedances and the current through them are, however, of interest. We shall now relate the internal voltages and currents in the balanced delta and wye to the quantities at the line terminals. The voltages $\mathbf{V}_{ab}$, $\mathbf{V}_{bc}$, and $\mathbf{V}_{ca}$ at the terminals of the load are referred to as the line voltages. Since, in the balanced *circuit*, they are all equal in magnitude, we shall designate their magnitude by the symbol V_L,

$$V_L = |\mathbf{V}_{ab}| = |\mathbf{V}_{bc}| = |\mathbf{V}_{ca}|$$

The currents entering the terminals of the load are equal in magnitude, and this magnitude is called the line current, denoted by the symbol I_L,

$$I_L = |\mathbf{I}_a| = |\mathbf{I}_b| = |\mathbf{I}_c|$$

In a balanced load the magnitudes of the voltages across each impedance are equal, and the magnitude of the current through each impedance is the same. These magnitudes are called the phase voltage and phase current, respectively, and will be identified by the subscript p. Thus, in the balanced *wye*,

$$V_p = |\mathbf{V}_{an}| = |\mathbf{V}_{bn}| = |\mathbf{V}_{cn}| \qquad I_p = |\mathbf{I}_{an}| = |\mathbf{I}_{bn}| = |\mathbf{I}_{cn}|$$

and in the balanced *delta*,

$$V_p = |\mathbf{V}_{ab}| = |\mathbf{V}_{bc}| = |\mathbf{V}_{ca}| \qquad I_p = |\mathbf{I}_{ab}| = |\mathbf{I}_{bc}| = |\mathbf{I}_{ca}|$$

Evidently, a line current in a wye is identical with a phase current,

In the wye: $$I_L = I_p$$

whereas in a delta, line and phase voltages are identical,

In the delta: $$V_L = V_p$$

To relate the line and phase voltages (*magnitudes*) in a balanced wye, with a-b-c sequence, let $\mathbf{I}_a = I_p\underline{/0^\circ}$, $\mathbf{I}_b = I_p\underline{/-120^\circ}$, and $\mathbf{I}_c = I_p\underline{/+120^\circ}$. A line voltage, such as $\mathbf{V}_{cb}$, may be obtained by the application of Kirchhoff's voltage law.

$$\mathbf{V}_{ab} = \mathbf{V}_{an} + \mathbf{V}_{nb} \qquad \mathbf{V}_{an} = \mathbf{I}_a\mathbf{Z}_y \qquad \mathbf{V}_{nb} = -\mathbf{I}_b\mathbf{Z}_y$$

(Evidently, $|\mathbf{V}_{an}| = |\mathbf{V}_{bn}| = V_p$.) Hence

$$\mathbf{V}_{ab} = (\mathbf{I}_a - \mathbf{I}_b)\mathbf{Z}_y$$

But

$$\mathbf{I}_a - \mathbf{I}_b = I_p(1\underline{/0^\circ} - 1\underline{/-120^\circ}) = I_p\sqrt{3}\,\underline{/30^\circ}$$

thus

$$\mathbf{V}_{ab} = \sqrt{3}\,I_p\mathbf{Z}_y\underline{/30^\circ}$$

or, using only absolute values,

In the balanced wye: $$|\mathbf{V}_{ab}| = \sqrt{3}\,V_p = V_L$$

This result, illustrated by the phasor diagram shown in Fig. 14-13*d*, could also be obtained by assuming a balanced wye-connected source, joining the neutral points, and noting that, for the source, $V_p\sqrt{3} = V_L$, as was shown in Sec. 14-4.

To find the relationship between line and phase currents in a balanced delta, as shown in Fig. 14-13*c*, let

$$\mathbf{V}_{ab} = V_L\underline{/0}$$

and

$$\mathbf{V}_{ca} = V_L\underline{/+120^\circ}$$

Then

$$\mathbf{I}_{ab} = \frac{V_L\underline{/0}}{\mathbf{Z}_\Delta} \qquad \mathbf{I}_{ac} = -\frac{V_L\underline{/+120^\circ}}{\mathbf{Z}_\Delta}$$

Now, the line current $\mathbf{I}_a$ is given by Kirchhoff's current law at point a,

$$\mathbf{I}_a = \mathbf{I}_{ab} + \mathbf{I}_{ac}$$

or

$$\mathbf{I}_a = \frac{V_L}{\mathbf{Z}_\Delta}(1 - 1\underline{/+120^\circ}) = \sqrt{3}\,\frac{V_L}{\mathbf{Z}_\Delta}\underline{/-30^\circ}$$

But since $|\mathbf{V}_L/\mathbf{Z}_\Delta| = I_p$, we have the result

In a balanced delta: $\qquad \sqrt{3}\, I_p = I_L$

This result is illustrated by the phasor diagram shown in Fig. 14-13*e*.

It must be emphasized that the relationships

For a balanced wye: $\qquad V_L = \sqrt{3}\, V_p$

For a balanced delta: $\qquad I_L = \sqrt{3}\, I_p$

are magnitude relationships. If the phase relationships between the various currents and voltages are required, they are easily obtained from a phasor diagram or by analytical application of Kirchhoff's laws.

14-8 Power relationships

Since every balanced three-phase circuit can be represented as three individual single-phase circuits as in Fig. 14-11*c*, the total power delivered to the balanced three-phase load equals three times the power delivered to each phase. Denoting the power per phase by P_p, we have

$$P_p = V_p I_p \cos\theta$$

where θ *is the angle of a phase impedance.* Hence the total power delivered to the balanced three-phase load is given by

$$P = 3V_p I_p \cos\theta \tag{14-19}$$

In three-phase circuit applications it is preferable to express, where possible, all final results in terms of *line* quantities. In the case of Eq. (14-19), we note that, for both delta- and wye-connected loads,

$$V_p I_p = \frac{1}{\sqrt{3}} V_L I_L \tag{14-20}$$

Because in a delta

$$V_p = V_L \qquad \text{and} \qquad I_p = \frac{1}{\sqrt{3}} I_L$$

and in a wye

$$V_p = \frac{1}{\sqrt{3}} V_L \qquad \text{and} \qquad I_p = I_L$$

the expression for the power delivered to a three-phase load may be written

$$P = \sqrt{3}\, V_L I_L \cos\theta \qquad \text{watts} \tag{14-21a}$$

Note that θ is the angle between a *phase voltage* and the corresponding *phase current*, i.e., the angle of an (equivalent) phase impedance.

Similarly, the total reactive power delivered to the three-phase load is given by

$$Q = -\sqrt{3}\, V_L I_L \sin\theta \qquad \text{var} \tag{14-21b}$$

Since the complex power per phase is $P_p + jQ_p$, we define the complex power for a three-phase load as

$$\mathbf{P}_a = \text{complex power} = 3P_p + j3Q_p$$
$$\mathbf{P}_a = \sqrt{3}\, V_L I_L(\cos\theta - j\sin\theta) \tag{14-21c}$$

In a *single-phase* circuit the absolute value of the complex power is the apparent power for the circuit. We now extend this definition to a balanced three-phase circuit. For a balanced three-phase circuit the apparent power is

$$P_a = \sqrt{3}\, V_L I_L \qquad \text{va} \tag{14-21d}$$

The definition of power factor is also retained,

$$\text{pf} = \frac{P}{P_a} = \cos\theta \tag{14-21e}$$

Thus the power factor for a balanced three-phase load is equal to the power factor of one of the phases, i.e., the cosine of the angle of an (equivalent) phase impedance.

We recall here that the use of complex power simplified many single-phase problems (such as combinations of parallel loads and power-factor correction). Since we can always reduce a balanced three-phase problem to an equivalent single-phase problem by using a representative phase, the method of complex-power addition (discussed in Chaps. 9 and 10) for single-phase circuits can be applied to balanced three-phase circuits.

14-9 Balanced three-phase compared with single-phase We shall now discuss some advantages which make three-phase circuits preferable to single-phase circuits. This discussion will also serve to illustrate how some of the results of the previous articles are applied. Additional examples will be presented in the next section.

POWER TRANSMISSION Suppose that a certain load requires a power P, at a power factor $\cos\theta$, and at a line voltage V_L. This load is to be connected to a generator by means of cables. Let us assume that we have the option of connecting the load as a single-phase load as shown in Fig. 14-14*a* or as a balanced three-phase load as shown in Fig. 14-14*b*. We shall compare the power loss in the cables for the two cases. For the single-phase connection (Fig. 14-14*a*) we have called the resistance of each of the

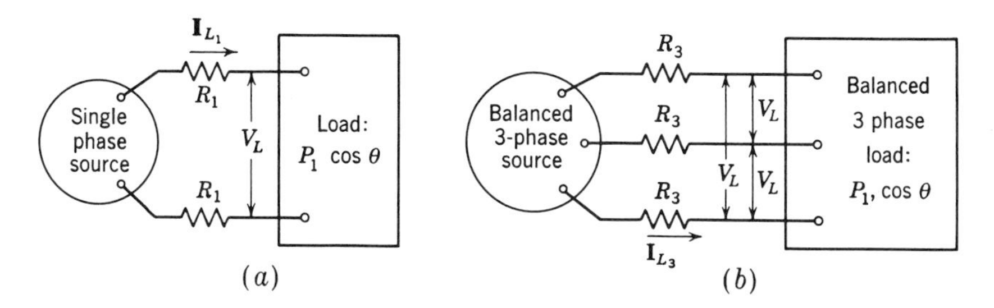

***Fig. 14-14** A single-phase and a balanced three-phase circuit.*

two cables R_1, and for the balanced three-phase load (Fig. 14-14*b*), each of the three cables has a resistance R_3.

In the *single-phase* connection the line current I_{L1} is given by

$$I_{L1} = \frac{P}{V_L \cos\theta}$$

Since the power loss in the two cables is $2(I_{L1}^2 R_1)$, we have

$$\text{Power loss in single phase} = \frac{P^2}{V_L^2 \cos^2\theta} 2R_1$$

In the balanced *three-phase* connection the magnitude of a line current is given by

$$I_{L3} = \frac{P}{\sqrt{3}\, V_L \cos\theta}$$

Since the power loss in the three cables is $3(I_{L3}^2 R_3)$, we have

$$\text{Power loss in three-phase} = \frac{P^2}{3V_L^2 \cos^2\theta} \times 3R_3$$

Hence, if we divide the two results for comparison,

$$\frac{\text{Power loss in three-phase transmission}}{\text{Power loss in single-phase transmission}} = \frac{R_3}{2R_1} \tag{14-22}$$

If we use identical cables, the power loss in three-phase transmission is only half the loss in the cables for the single-phase connection. If we *allow* the same power loss in either case, each of the three cables in the three-phase connection may have twice the resistance of a cable in the single-phase case. Now, the resistance of a cable varies inversely with the cross-sectional area, and the length of the cables is the same in either case. The volume of the conducting material used therefore compares as follows: If $R_3 = 2R_1$ (equal loss in transmission), the volume $(\text{Vol})_3$ of a three-phase cable is half of $(\text{Vol})_1$, the volume of a single-phase cable. Since the total volume of conducting material in three-phase is $3(\text{Vol})_3$ and in single-phase is $2(\text{Vol})_1$, we have $(\text{Vol})_3/(\text{Vol})_1 = \frac{3}{4}$. This saving of conducting material and weight is of importance. Although this example has dealt with the advantage of three-phase transmission, similar savings occur in weight and size of three-phase machines.

INSTANTANEOUS POWER We shall now demonstrate that the *sum* of the *instantaneous* powers which is delivered to a balanced three-phase load in a balanced circuit is *constant* and equal to the average power. We recall that, in a single-phase circuit, the instantaneous power consists of a constant component and a sinusoid (which oscillates at twice the frequency of the source voltage). Since the instantaneous power in a three-phase circuit is constant, if the load is a three-phase motor, this motor will develop a constant torque, whereas a single-phase motor develops a pulsating torque.

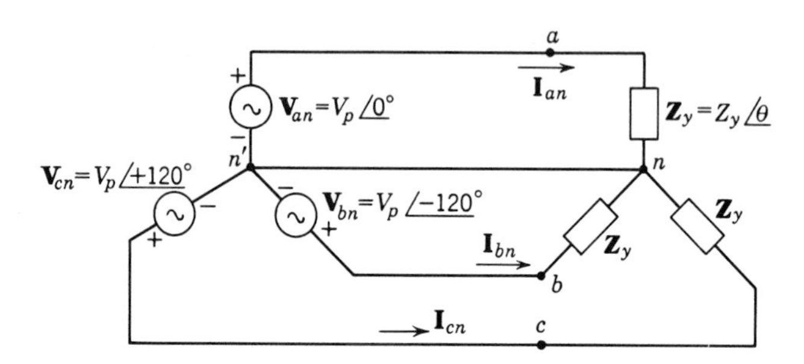

Fig. 14-15 A balanced three-phase circuit.

To obtain the expression for the instantaneous power delivered to all three phases of a balanced three-phase load from a balanced source, consider the circuit shown in Fig. 14-15. The total power delivered to all three phases at any instant of time is

$$p(t) = p_{an}(t) + p_{bn}(t) + p_{cn}(t)$$

where

$$p_{an}(t) = v_{an}i_{an} = \text{power to phase } a$$
$$p_{bn}(t) = v_{bn}i_{bn} = \text{power to phase } b$$
$$p_{cn}(t) = v_{cn}i_{cn} = \text{power to phase } c$$

Since $\mathbf{V}_{an} = V_p\underline{/0}$, we may write

$$v_{an}(t) = \sqrt{2}\,V_p \cos \omega t \qquad v_{bn} = \sqrt{2}\,V_p \cos(\omega t - 120°) \qquad v_{cn} = \sqrt{2}\,V_p \cos(\omega t + 120°)$$

Hence

$$i_{an}(t) = \sqrt{2}\,I_p \cos(\omega t - \theta) \qquad \text{where } I_p = \frac{V_p}{Z_y}$$
$$i_{bn}(t) = \sqrt{2}\,I_p \cos(\omega t - \theta - 120°)$$
$$i_{cn} = \sqrt{2}\,I_p \cos(\omega t - \theta + 120°)$$

Then

$$p_{an} = 2V_pI_p \cos \omega t \cos(\omega t - \theta) = V_pI_p \cos\theta + V_pI_p \cos(2\omega t - \theta)$$
$$p_{bn} = 2V_pI_p \cos(\omega t - 120°) \cos(\omega t - \theta - 120°)$$
$$= V_pI_p \cos\theta + V_pI_p \cos(2\omega t - 240° - \theta)$$
$$p_{cn} = 2V_pI_p \cos(\omega t + 120°) \cos(\omega t - \theta + 120°)$$
$$= V_pI_p \cos\theta + V_pI_p \cos(2\omega t + 240° - \theta)$$

Since

$$\cos(2\omega t - \theta) + \cos(2\omega t - 240° - \theta) + \cos(2\omega t + 240° - \theta) \equiv 0$$

we have

$$p(t) = p_{an} + p_{bn} + p_{cn} = 3V_pI_p \cos\theta = P \tag{14-23}$$

as was stated at the beginning of the discussion.

14-10 Examples of balanced three-phase calculations ***Example 14-3*** Two balanced three-phase loads are connected in parallel (Fig. 14-16*a*). Load 1 draws

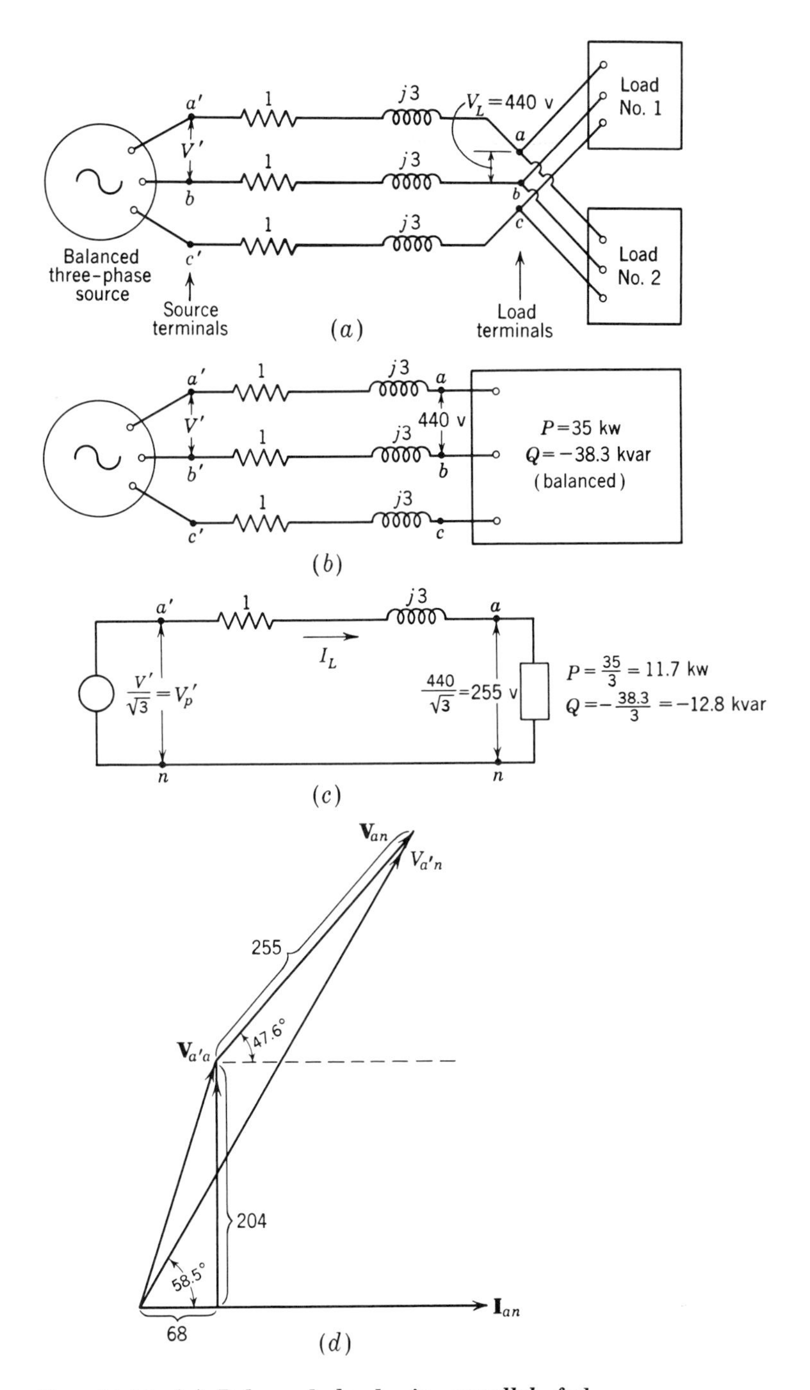

***Fig. 14-16** (a) Balanced loads in parallel fed through cables from a balanced source. (b) Representing the two balanced loads as a single equivalent load. (c) The representative single-phase problem for (b). (d) Voltage-phasor addition in the circuit of (c).*

20 kw at a lagging power factor of 0.85; load 2 draws 15 kw at a lagging power factor of 0.5. The line voltage at the terminals of the load is 440 volts. The load terminals are connected to the source terminals by means of cables which have an impedance of $1 + j3$ ohms each. Calculate (*a*) the line current drawn by the combination of the two loads; (*b*) the line voltage at the terminals of the source; (*c*) the power factor at the load terminals and at the source terminals.

Solution A variety of methods for solving this problem is possible. We shall illustrate one very convenient method in which complex power is used. The two loads in parallel may be replaced by an equivalent load with respect to the load terminals by addition of complex power.

For load 1: $\quad P_1 = 20 \text{ kw} \qquad \cos\theta_1 = 0.85 \qquad \theta_1 = 31.7°$

Hence

$$Q_1 = -P_1 \tan\theta_1 = (-20)(0.614) = -12.3 \text{ kvar}$$

For load 2: $\quad P_2 = 15 \text{ kw} \qquad \cos\theta_2 = 0.5 \qquad \theta_2 = 60°$

Hence

$$Q_2 = -P_2 \tan\theta_2 = (-15)(1.73) = -26 \text{ kvar}$$

The combination draws

$$P = P_1 + P_2 = 20 + 15 = 35 \text{ kw}$$

and

$$Q = Q_1 + Q_2 = -12.3 - 26 = -38.3 \text{ kvar}$$

The equivalent balanced three-phase circuit is shown in Fig. 14-16*b*.

We now formulate the representative single-phase problem, assuming a wye connection for source and load. Each phase draws $P/3$ and $Q/3$. The phase voltage for a wye connection is $1/\sqrt{3}$ times the line voltage. Hence the representative single-phase circuit is as shown in Fig. 14-16*c*. In this circuit the apparent power for the load is

$$P_a = \sqrt{11.7^2 + 12.8^2} = 17.4 \text{ kva}$$

The magnitude of the line current is now calculated.

$$I_L = \frac{17.4 \times 10^3}{255} = 68 \text{ amp} \qquad \textit{Ans. (a)}$$

This result can also be obtained without referring to the equivalent single-phase problem. For the three-phase load $P = 35$ kw, $Q = -38.3$ kvar. Hence

$$\sqrt{3}\, V_L I_L = \sqrt{35^2 + 38.3^2} = 52.5 \text{ kvar}$$

Thus

$$I_L = \frac{52.2 \times 10^3}{1.73 \times 440} = 68 \text{ amp}$$

We have chosen to show the single-phase method first, in view of the following parts of the problem: To calculate the line voltage V' at the terminals of the source, we calculate first the phase voltage V_p' in the representative single-phase problem. Since $V_p' = |\mathbf{V}_{a'n}|$, we apply Kirchhoff's voltage law.

$$\mathbf{V}_{a'n} = \mathbf{V}_{a'a} + \mathbf{V}_{an}$$

To carry out this phasor addition, we may choose any reference. For this series circuit it is convenient to write

$$\mathbf{I}_{an} = I_L\underline{/0} = 68\underline{/0°}$$

Hence $\mathbf{V}_{an}$ will be given as

$$\mathbf{V}_{an} = 255\underline{/\theta}$$

where θ is the power-factor angle of the single-phase load. Because $\tan\theta = -Q/P$, we have

$$\tan\theta = \frac{12.8}{11.7} \qquad \theta = 47.6°$$

The voltage $\mathbf{V}_{a'a}$ is given by $\mathbf{I}_{an}\mathbf{Z}_{a'a}$.

$$\mathbf{V}_{a'a} = (68\underline{/0°})(1 + j3) = 68 + j204$$

Hence

$$\mathbf{V}_{a'n} = 68 + j204 + 255\underline{/47.6°}$$

or

$$\mathbf{V}_{a'n} = 68 + j204 + 172 + j188 = 240 + j392 = 460\underline{/58.5°}$$

This phasor addition is illustrated by Fig. 14-16*d*. We need the magnitude of $\mathbf{V}_{a'n}$; since this value is V'_p,

$$V'_p = 460 \text{ volts}$$

Thus

$$V' = 460\sqrt{3} = 796 \text{ volts} \qquad \textit{Ans. (b)}$$

Note that we do not know the angle of V' (with respect to I_{an}) since line and phase voltages are not in phase. This angle, if required, could be obtained from a phasor diagram of line and phase voltages if the phase sequence were specified.

The power factor at the load terminals is immediately obtained either as the cosine of the angle between $\mathbf{V}_{an}$ and $\mathbf{I}_{an}$,

$$\begin{aligned}\text{pf at load terminals} &= \cos 47.6° \\ &= 0.674 \text{ lagging} \qquad \textit{Ans. (c)}\end{aligned}$$

or as the ratio of the power to the apparent power,

$$\text{pf}\Big)_{\text{load}} = \frac{P}{P_a} = \frac{11.7}{17.4} = \frac{35}{52.5} = 0.67 \text{ lagging} \qquad \textit{Check}$$

At the generator terminals, the power factor is given by

$$\text{pf}\Big)_{gen} = \cos \angle\begin{matrix}\mathbf{V}_{a'n} \\ \mathbf{I}_{a'n}\end{matrix} = \cos 58.5° = 0.523 \text{ lagging} \qquad \textit{Ans. (c)}$$

The results can be checked by the use of complex power.

$$P_{a'n} = P_{a'a} + P_{an} = I_L^2 \times 1 + 11.7 \times 10^3$$
$$P_{a'n} = 4.6 + 11.7 = 16.3 \text{ kw}$$

$$-Q_{a'n} = \frac{(I_L)^2 \times 3}{1{,}000} + 12.8 = 26.6 \text{ kvar}$$

We now write

$$|\mathbf{V}_{a'n}\mathbf{I}_{a'n}| = \sqrt{16.3^2 + 26.6^2} = 31.2 \text{ kva}$$

Hence, at the terminals of the source,

$$\text{pf}\Big)_{gen} = \frac{16.3}{31.2} = 0.522 \qquad \textit{Check}$$

and

$$V'_p = \frac{31{,}200}{68} = 460 \qquad \textit{Check}$$

The reader should not be concerned by the unrealistically large energy dissipation and energy storage in the cables. The numbers were purposely chosen so that these results are unrealistically large and none of the quantities are negligible.

Example 14-4 The balanced three-phase load shown in Fig. 14-17 draws 20 kw at a lagging power factor of 0.4. The frequency is 60 hertz. (*a*) Calculate the total

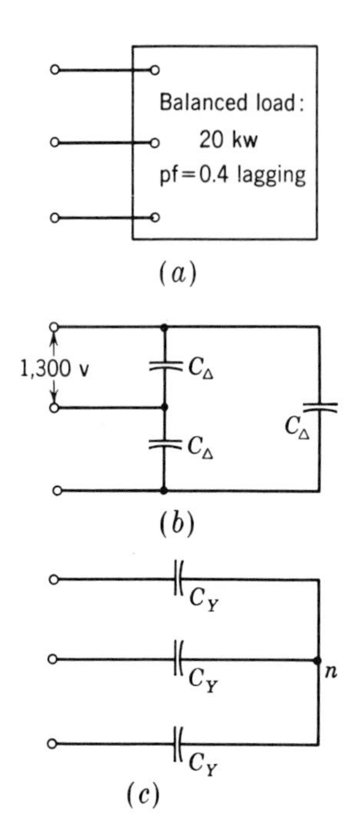

***Fig. 14-17** (a) A balanced load. (b) Delta-connected capacitances. (c) Wye-connected capacitances.*

kva rating of three capacitances which, when connected in delta or in wye in parallel with the load, will raise the power factor of the combination to 0.9 lagging. (*b*) If the line voltage at the load terminals is 1,300 volts, calculate the value of each capacitance of (*a*) for (1) delta connection; (2) wye connection.

Solution The reactive power drawn by the load alone is

$$Q = -20 \tan (\cos^{-1} 0.4) = -45.8 \text{ kvar}$$

If the power factor were 0.9 lagging, the reactive power drawn by the load would be Q'.

$$Q' = -20 \tan (\cos^{-1} 0.9) = -9.8 \text{ kvar}$$

Hence the three capacitances combined must draw 45.8 − 9.8 kvar or, calling this quantity Q_c,

$$Q_c = +36.0 \text{ kvar} \qquad \textit{Ans.}$$

Each capacitance draws

$$Q_p = +\frac{36.0}{3} = 12.0 \text{ kvar} \qquad \textit{Ans.}$$

If the capacitances are delta-connected, as shown in Fig. 14-17*b*, the voltage across each is a line voltage V_L given as 1,300 volts.

Since the capacitances draw no average power, the apparent power in each equals, numerically, the magnitude of the reactive power. The magnitude of a phase current in a capacitance is given by

$$V_L I_c = Q_p = 12{,}000$$

or

$$I_c = \frac{12{,}000}{1{,}300} = 9.23 \text{ amp}$$

Hence

$$\omega C_\Delta = B_C\Big)_\Delta = \frac{9.23}{1{,}300} = 7.10 \times 10^{-3} \text{ mho}$$

and

$$C_\Delta = \tfrac{1}{377} \times 7.10 \times 10^{-3} = 18.8\ \mu\text{f} \qquad \textit{Ans. } (b_1)$$

If the capacitances are wye-connected, as shown in Fig. 14-17*c*, then, by delta-wye conversion,

$$(B_C)_y = 3(B_C)_\Delta$$

Thus

$$C_y = 56.4\ \mu\text{f} \qquad \textit{Ans. } (b_2)$$

This result can be obtained independently. The voltage across each capacitance in the wye is $1{,}300/\sqrt{3} = 752$ volts. Hence a phase current I'_c in a wye-connected capacitance is given by

$$I'_c = \frac{12{,}000}{752} = 16 \text{ amp}$$

and

$$(B_C)_y = \tfrac{16}{752} = 21.3 \times 10^{-3} \text{ mho}$$

$$C_y = \frac{21.3}{0.377} \times 10^{-6} = 56.4\ \mu\text{f}$$

14-11 Unbalanced three-phase circuits

As was stated earlier, the calculation of an unbalanced three-phase circuit is carried out by node (or mesh) analysis because the special symmetry, which allows us to replace a balanced three-phase problem by a representative single phase, no longer exists. It should also be evident that the advantages of three-phase over single-phase are lost if the circuit becomes severely unbalanced. In this section we shall illustrate the calculation of several unbalanced circuits.

THE OPEN-DELTA (V) CONNECTION Consider the unbalanced three-phase load shown in Fig. 14-18. The two impedances are equal, but a third impedance which, if equal to the others, would form a balanced delta when

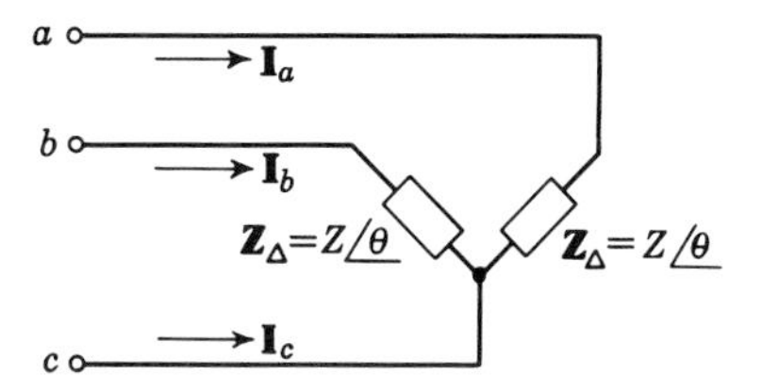

Fig. 14-18 The open-delta (V) connection.

connected across terminals a-b is omitted. The line voltages at the load terminals are assumed balanced and of a-b-c sequence. Hence

$$\mathbf{V}_{ab} = V_L\underline{/0^\circ}$$
$$\mathbf{V}_{cb} = V_L\underline{/-120^\circ}$$
$$\mathbf{V}_{ca} = V_L\underline{/+120^\circ}$$

We shall calculate the line currents $\mathbf{I}_a$, $\mathbf{I}_b$, and $\mathbf{I}_c$. Since the voltage across each impedance is known, we have

$$\mathbf{I}_{bc} = \frac{\mathbf{V}_{bc}}{\mathbf{Z}_\Delta} = \frac{V_L}{Z}\underline{/-120^\circ - \theta}$$
$$\mathbf{I}_{ca} = \frac{\mathbf{V}_{ca}}{\mathbf{Z}_\Delta} = \frac{V_L\underline{/+120^\circ - \theta}}{Z}$$

Hence

$$\mathbf{I}_a = -\mathbf{I}_{ca} = \frac{V_L}{Z}\underline{/-60^\circ - \theta}$$

and

$$\mathbf{I}_b = \mathbf{I}_{bc} = \frac{V_L}{Z}\underline{/-120^\circ - \theta}$$

Since

$$\mathbf{I}_c = -\mathbf{I}_b - \mathbf{I}_a$$

the line current $\mathbf{I}_c$ is given by

$$\mathbf{I}_c = \frac{V_L}{Z}(-1\underline{/-60^\circ - \theta} - 1\underline{/-120^\circ - \theta})$$

The phasor addition is illustrated in Fig. 14-19 for an acute positive angle θ. The result is

$$\mathbf{I}_c = \sqrt{3}\,\frac{V_L}{Z}\underline{/90^\circ - \theta}$$

Since the three line currents are unequal, if this load is connected to a source by means of cables, the power loss in the c cable would be three times the loss in the a or b cable. In addition, the voltage across the cable impedances would be unequal and unbalanced. Consequently, if the source

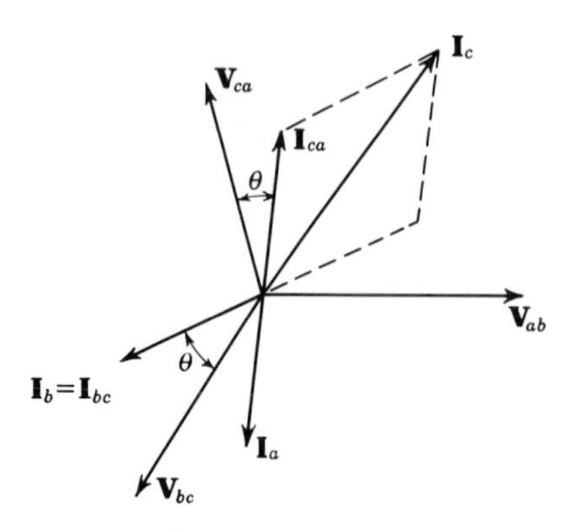

Fig. 14-19 Phasor diagram for Fig. 14-18.

furnishes balanced voltages, the assumption of balanced load voltage may be unjustified.

UNBALANCED WYE Consider next the unbalanced circuit shown in Fig. 14-20a. Assuming the source voltages to be known, the line current can

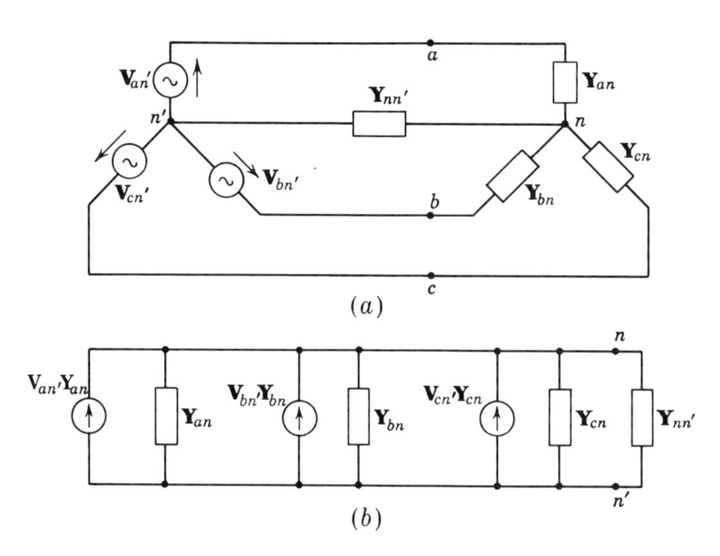

***Fig. 14-20** (a) An unbalanced three-phase circuit. (b) Current-source equivalent with respect to terminals n-n′ for the circuit of (a).*

be calculated if the voltage of a, b, and c with respect to the neutral point of the load is known. The voltage $\mathbf{V}_{nn'}$ is easily calculated by using the node method. An equivalent circuit with respect to terminals n and n' can be obtained by converting each voltage source to a current source. This circuit is shown in Fig. 14-20b. Application of Kirchhoff's current law at junction n gives

$$\mathbf{V}_{an'}\mathbf{Y}_{an} + \mathbf{V}_{bn'}\mathbf{Y}_{bn} + \mathbf{V}_{cn'}\mathbf{Y}_{cn} = \mathbf{V}_{nn'}(\mathbf{Y}_{an} + \mathbf{Y}_{bn} + \mathbf{Y}_{cn} + \mathbf{Y}_{n'n})$$

or

$$\mathbf{V}_{nn'} = \frac{\mathbf{V}_{an'}\mathbf{Y}_{an} + \mathbf{V}_{bn'}\mathbf{Y}_{bn} + \mathbf{V}_{cn'}\mathbf{Y}_{cn}}{\mathbf{Y}_{an} + \mathbf{Y}_{bn} + \mathbf{Y}_{cn} + \mathbf{Y}_{n'n}}$$

If the neutrals n and n' are joined through a zero impedance (infinite admittance), then $\mathbf{V}_{nn'}$ is zero and the voltage across any phase impedance ($\mathbf{Z}_{an}$, $\mathbf{Z}_{bn}$, or $\mathbf{Z}_{cn}$) is not dependent on the other impedance. If, however, $\mathbf{Z}_{n'n}$ is appreciable, the voltages across the individual phase impedances will influence each other.

As a numerical example, consider the balanced voltage with a-b-c phase order,

$$\mathbf{V}_{an'} = V\underline{/0} \qquad \mathbf{V}_{bn'} = V\underline{/-120} \qquad \mathbf{V}_{cn'} = V\underline{/+120^\circ}$$

and the unbalanced load admittances

$$\mathbf{Y}_{an} = 1 + j0 \text{ mho} \qquad \mathbf{Y}_{bn} = -j1 \text{ mho} \qquad \mathbf{Y}_{cn} = 0.5 + j0 \text{ mho}$$

and let the admittance of the neutral line be $0 - j3$ mhos. Then

$$\begin{aligned} \mathbf{V}_{an'}\mathbf{Y}_{an} &= V(1 + j0) \\ \mathbf{V}_{bn'}\mathbf{Y}_{bn} &= V(1\underline{/-120^\circ})(-j) = V(-0.866 + j0.5) \\ \mathbf{V}_{cn'}\mathbf{Y}_{cn} &= V(1\underline{/+120^\circ})(0.5) = V(-0.25 + j0.433) \end{aligned}$$

Hence

$$\begin{aligned} \mathbf{V}_{an'}\mathbf{Y}_{an} + \mathbf{V}_{bn'}\mathbf{Y}_{bn} + \mathbf{V}_{cn'}\mathbf{Y}_{cn} &= V(0.116 + j0.933) \\ &= 0.93V\underline{/82.9^\circ} \end{aligned}$$

Also

$$\mathbf{Y}_{an} + \mathbf{Y}_{bn} + \mathbf{Y}_{cn} + \mathbf{Y}_{n'n} = 1.5 - j4 = 4.28\underline{/-69.5^\circ}$$

Thus

$$\begin{aligned} \mathbf{V}_{nn'} &= \frac{0.937V}{4.28}\underline{/82.9^\circ + 69.5^\circ} \\ &= 0.219V\underline{/152.4^\circ} \end{aligned}$$

Therefore

$$\begin{aligned} \mathbf{V}_{an} &= \mathbf{V}_{an'} - \mathbf{V}_{nn'} = V\underline{/0} - 0.219V\underline{/152.4^\circ} \\ &= V\underline{/0} + V(0.194 - j0.103) \end{aligned}$$

Similarly,

$$\begin{aligned} |\mathbf{V}_{an}| &= 1.2V \\ |\mathbf{V}_{bn}| &= 1.02V \\ |\mathbf{V}_{cn}| &= 0.788V \end{aligned}$$

We conclude that the load voltages are unbalanced. The line currents will be given by

$$\begin{aligned} |\mathbf{I}_{an}| &= |\mathbf{V}_{an}|\,|\mathbf{Y}_{an}| = 1.2V \qquad \text{amp} \\ |\mathbf{I}_{bn}| &= |\mathbf{V}_{bn}|\,|\mathbf{Y}_{bn}| = 1.02V \qquad \text{amp} \\ |\mathbf{I}_{cn}| &= |\mathbf{V}_{cn}|\,|\mathbf{Y}_{cn}| = 0.394V \qquad \text{amp} \end{aligned}$$

The current in the neutral is

$$|\mathbf{I}_{n'n}| = |\mathbf{V}_{n'n}|\,|\mathbf{Y}_{nn}| = 0.657V \qquad \text{amp}$$

so that the currents are unbalanced.

If we now assume the *opposite* phase order,

$$\mathbf{V}_{an'} = V\underline{/0} \qquad \mathbf{V}_{bn'} = V\underline{/+120^\circ} \qquad \mathbf{V}_{cn'} = V\underline{/-120^\circ}$$

then

$$\mathbf{V}_{an'}\mathbf{Y}_{an} = V(1 + j0)$$

But

$$\begin{aligned} \mathbf{V}_{bn'}\mathbf{Y}_{bn} &= (V\underline{/+120^\circ})(-j) = V(0.866 + j0.5) \\ \mathbf{V}_{cn'}\mathbf{Y}_{cn} &= (V\underline{/-120^\circ})(0.5) = V(-0.25 - j0.433) \end{aligned}$$

Hence, in this case,

$$\mathbf{V}_{an'}\mathbf{Y}_{an} + \mathbf{V}_{bn'}\mathbf{Y}_{bn} + \mathbf{V}_{cn'}\mathbf{Y}_{cn} = V(1.61 + j0.07)$$

and

$$|\mathbf{V}_{n'n}| = 0.377V$$

This result differs from the previous result.

We conclude that, in an unbalanced circuit, the magnitudes of the line currents depend on the phase order.

14-12 An application of unbalanced circuits. Phase sequence We have just seen that line currents (and therefore phase voltage in a wye) depend on the phase sequence in unbalanced three-phase circuits. This result may be used to build a circuit which has the function of indicating the phase sequence for a balanced source. Such a phase-sequence indicator is shown in Fig. 14-21. We shall show that the magnitude

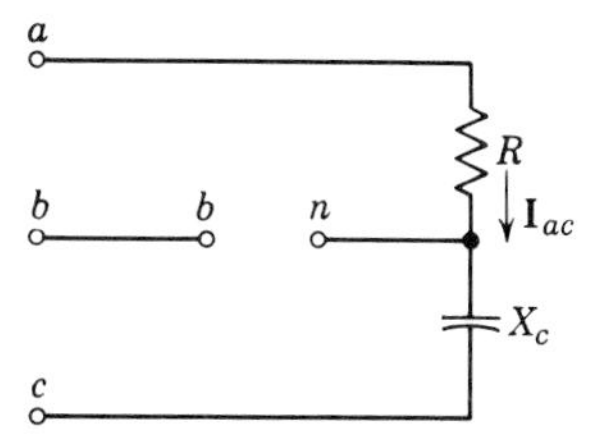

Fig. 14-21 A phase-sequence indicator.

V_{bn} is either larger or smaller than the line voltage V_L, depending on the phase sequence. The magnitude of V_{bn} for the two possible phase orders is easily obtained from a phasor diagram drawn for each case. Since $\mathbf{V}_{ab} + \mathbf{V}_{bc} + \mathbf{V}_{ca} = 0$, the three phasors which represent these voltages form an equilateral triangle as in Fig. 14-22*a* for the *ab-bc-ca* sequence and in Fig. 14-22*b* for the *ab-ca-bc* sequence. In either case $\mathbf{I}_{ac}$ leads $\mathbf{V}_{ac}$ by $\theta = \tan^{-1} |X_c|/R$ as shown in the phasor diagrams in Fig. 14-22.

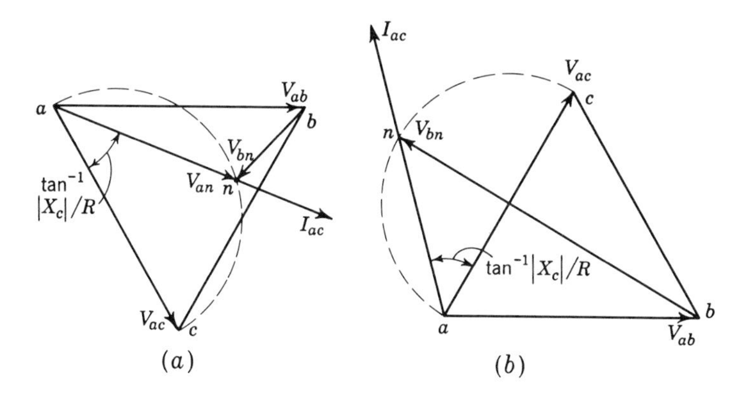

Fig. 14-22 Phasor diagrams for the circuit of Fig. 14-21, showing conditions for the two possible phase orders.

Since $\mathbf{V}_{an}$ is in phase with $\mathbf{I}_{ac}$ and $\mathbf{V}_{nb}$ lags $\mathbf{I}_{ac}$ by 90°, point n is located on the phasor diagrams, so that angle $\angle anc$ is a right angle. For any ratio of X_c to R the locus of point n is on the dashed semicircle. Since in Fig. 14-22*a* the distance from b to n, which represents the magnitude $|V_{bn}|$, is less than V_L, and in Fig. 14-22*b* $|V_{bn}| > V_L$, this device can be used to indicate phase sequence.

Other unbalanced circuits can also be used to indicate phase sequence (Prob. 14-16).

14-13 Wattmeters We assume that the reader is familiar with the principles involved in the operation of a wattmeter. In this section we shall give an expression for the reading of an a-c wattmeter in terms of the voltages and currents applied to its voltage and current coils, respectively. A wattmeter is symbolically shown as in

Fig. 14-23*a*. In this representation the symbol of inductance is used to represent the current coil, and the symbol of resistance represents the voltage coil of the meter. For an ideal meter the inductance (impedance) of the current coil is assumed to be zero ($v_{cc'} = 0$ for any $i_{cc'}$), and the resistance (impedance) of the voltage coil is assumed infinite ($i_{pp'} = 0$ for any $v_{pp'}$).

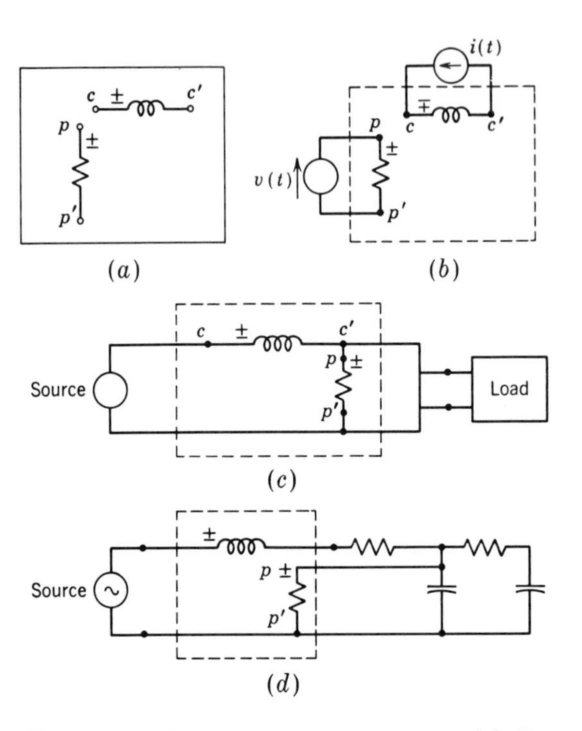

Fig. 14-23 Wattmeter connections.* (*a*) *Symbolic representation of a wattmeter.* (*b*) *Separate sources excite voltage and current coils.* (*c*) *A wattmeter connected to read single-phase power.* (*d*) *A wattmeter connection in a single-phase circuit. The wattmeter reading does not correspond to the power delivered to the circuit.

The mechanism of an a-c wattmeter is such that if a current source $i(t)$ is applied to the current coil with the reference arrow of the source pointing to the terminal marked $\pm$, as shown in Fig. 14-23*b*, and if a voltage source $v(t)$ is applied to the voltage coil with its reference arrow pointing to the terminal marked $\pm$, and if the functions $v(t)$ and $i(t)$ are periodic with period T, then the wattmeter reading will be

$$W = \frac{1}{T} \int_0^T v(t)i(t)\, dt$$

Two points are to be noted here:

1 Although the meter is called an a-c wattmeter, the voltages and currents associated with it need not be sinusoidal and the above expression is valid so long as $v(t)$ and $i(t)$ are periodic and have the same period.

2 The reading of the meter is not necessarily the average power delivered to a load. For example, in Fig. 14-23*b*, for given $v(t)$ and $i(t)$, there is no power involved, but the meter will still read $W = (1/T) \int_0^T vi\, dt$. The only time that the reading of the meter is to be interpreted as power (delivered to a terminal pair) is when $i(t)$ and $v(t)$ applied to the current and voltage coils of the meter are those associated with a (load) terminal pair. In Fig. 14-23*c* the meter is connected so that the current through its current coil is the same as current in the terminal pair (current through voltage coil of the meter is assumed to be zero), and the voltage across the voltage coil of the meter is the same in the voltage across the terminal pair. Therefore, if the source is periodic and the circuit is in steady state, the reading of the wattmeter is identical with the average power delivered to the terminal pair.

On the other hand, even with one source, it is possible to connect a wattmeter so that the needle will deflect but its reading is not associated with power. In Fig. 14-23*d* an example of such a connection is given. We shall distinguish between average power and the reading of a wattmeter by using the symbol W for wattmeter reading and P for power. In any case the reading of a wattmeter is given by

$$W = (i_{\text{current coil}} v_{\text{volt coil}})_{\text{avg}}$$

In this sense the wattmeter is an integrating and averaging device, defined by the expression

$$W = (v_{pp'} i_{cc'})_{\text{avg}} = \frac{1}{T} \int_0^T v_{pp'} i_{cc'}\, dt$$

If the current through the current coil and the voltage across the voltage coil are sinusoidal with the same frequency, then

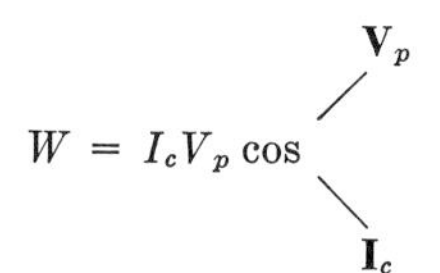

In words: Under sinusoidal conditions, the reading of the wattmeter is the product of the rms values of the current through its current coil and the voltage across its voltage coil multiplied by the cosine of the angle between those two phasors. In this interpretation the reference arrows of the voltage and current must be pointing to the polarity signs ± of their respective coils.

14-14 Measurement of three-phase power It should be evident that the connections shown in Fig. 14-24 can be used to measure the power in either a delta- or a wye-connected load. We have shown a wattmeter for each leg of the delta or wye; hence each wattmeter reads the power delivered to the load which is correspondingly numbered. Each of these connections employs three wattmeters and represents what is referred to as the "three-wattmeter method" and for the following reasons is generally not useful. In the case of a delta load, the current coils of the wattmeters must be inserted in series with each phase impedance, and these may not be accessible. In the case of the wye-connected load, each voltage coil is connected to the neutral point, and this point may not be accessible. In the case of parallel connections of delta- and wye-connected loads, the number of wattmeters which this method requires makes this scheme hopelessly inadequate.

In Fig. 14-25 the two-wattmeter method used in connection with three-wire systems is shown. We shall now demonstrate that the (algebraic) sum of the readings of the two meters is the average power delivered to the three-terminal load. This is true so long as the sources in the network are periodic, with the same period, and the network

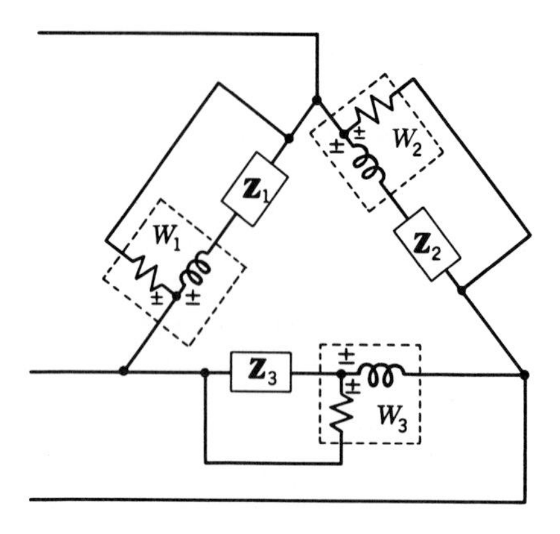

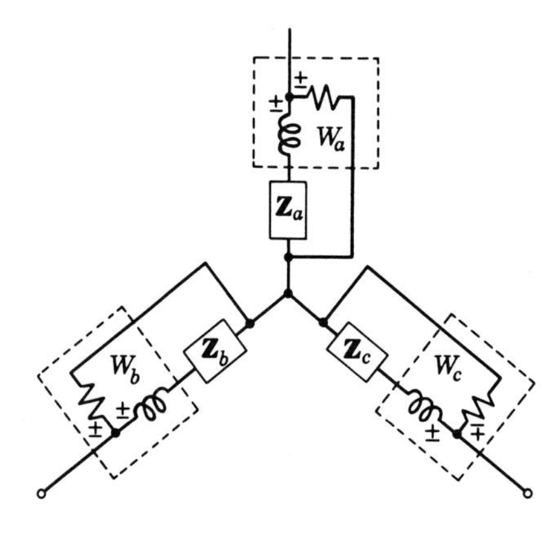

Fig. 14-24 *The three-wattmeter method for measuring three-phase power.*

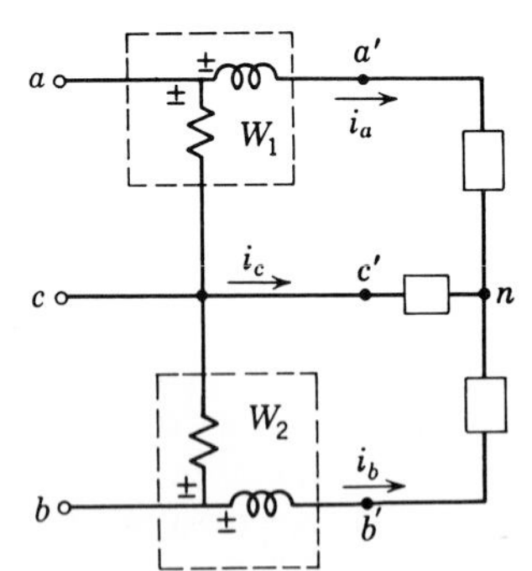

Fig. 14-25 *The two-wattmeter method for measuring three-phase power.*

is in steady state. In Fig. 14-25 the load is shown as wye-connected. This is for convenience and does not affect the generality of the following arguments.

In accordance with the definition of the wattmeter readings and the polarities shown in Fig. 14-25,

$$W_1 + W_2 = \frac{1}{T}\int_0^T (v_{ac}i_a + v_{bc}i_b)\,dt$$

$$v_{ac} = v_{an} + v_{nc} \qquad v_{bc} = v_{bn} + v_{nc}$$

$$W_1 + W_2 = \frac{1}{T}\int_0^T [(v_{an} + v_{nc})i_a + (v_{bn} + v_{nc})i_b]\,dt$$

Adding and subtracting $v_{cn}i_c$ inside the integral sign and noting that $v_{cn} = -v_{nc}$, we have

$$W_1 + W_2 = \frac{1}{T}\int_0^T [v_{an}i_a + v_{bn}i_b + v_{cn}i_c + v_{nc}(i_c + i_b + i_a)]\,dt$$

In this expression, $i_a + i_b + i_c$ is the sum of the currents entering n and is identically zero. It is noted that this sum is zero regardless of the waveform of the currents. We have assumed that there is no voltage across the current coils and therefore $v_{an} = v_{a'n}$, $v_{bn} = v_{b'n}$, and $v_{cn} = v_{c'n}$, but $1/T \int_0^T v_{a'n}i_a\,dt$ is the average power delivered to the load $a'n$, with similar interpretation for other terms in the integral. This shows that $W_1 + W_2$ is the total power delivered to the three-terminal load.

Although the two-wattmeter method requires neither that voltages and currents be sinusoidal nor that the load be balanced, the following discussion shows that, when the system given in Fig. 14-25 is a three-phase balanced system, then, from the readings of W_1 and W_2, the reactive power as well as the average power can be computed. In Fig.

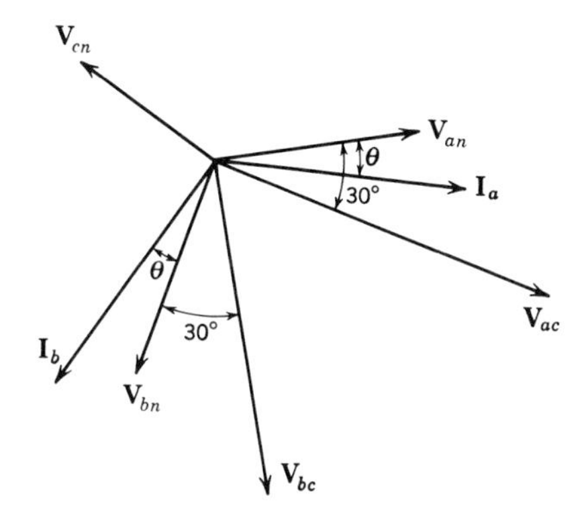

***Fig. 14-26** Phasor diagram for the two-wattmeter method in a balanced three-phase circuit.*

14-26 the phasor diagram of a three-phase balanced system, with the load angle θ, is shown, such that $\mathbf{I}_a$ lags $\mathbf{V}_{an}$ by θ and $\mathbf{I}_b$ lags $\mathbf{V}_{bn}$ by the same angle. From this phasor diagram it is seen that

$$W_1 = V_{ac}I_a \cos \underline{/V_{ac}, I_a} = V_L I_L \cos (30° - \theta) \tag{14-24}$$

$$W_2 = V_{bc}I_b \cos \underline{/V_{bc}, I_b} = V_L I_L \cos (30° + \theta) \tag{14-25}$$

where V_L and I_L are line voltage and line currents, respectively.

$$W_1 + W_2 = V_L I_L[\cos (30° - \theta) + \cos (30° + \theta)] = V_L I_L(2 \cos \theta \cos 30°)$$

$$= \sqrt{3}\, V_L I_L \cos \theta = P \tag{14-26}$$

$$W_1 - W_2 = V_L I_L[\cos (30° - \theta) - \cos (30° + \theta)] = V_L I_L(2 \sin \theta \sin 30°)$$

$$= V_L I_L \sin \theta = -\frac{1}{\sqrt{3}} Q \tag{14-27}$$

$$\tan \theta = -\sqrt{3}\,\frac{W_1 - W_2}{W_1 + W_2} \tag{14-28}$$

Equations (14-26) to (14-28) show that, from the values of W_1 and W_2, the average power, reactive power, and power factor of the load can be computed. From Eq. (14-25) it is seen that, if $\theta < 60°$, then (with the polarities shown in Fig. 14-25) the value of $\cos (30 + \theta)$, and therefore W_2, will be positive. At $\theta = 60°$, $W_2 = V_L I_L \cos 90° = 0$, and for $\theta > 60°$, $30° + \theta > 90°$ and W_2 will be negative. In such cases the meter polarity must be reversed (to read upscale) and then $P = |W_1| - |W_2|$.

PROBLEMS

14-1 In Fig. P14-1 calculate v_{an}/v_{nb} if $R_s = \frac{1}{2}$, $R_1 = 5$, $R_3 = 10$, $v = V = 100$, and (*a*) $R_2 = 5$; (*b*) $R_2 = 10$, $R_n = 20$; (*c*) $R_2 = 10$, $R_n = 0$; (*d*) $R_2 = 10$, $1/R_n = 0$.

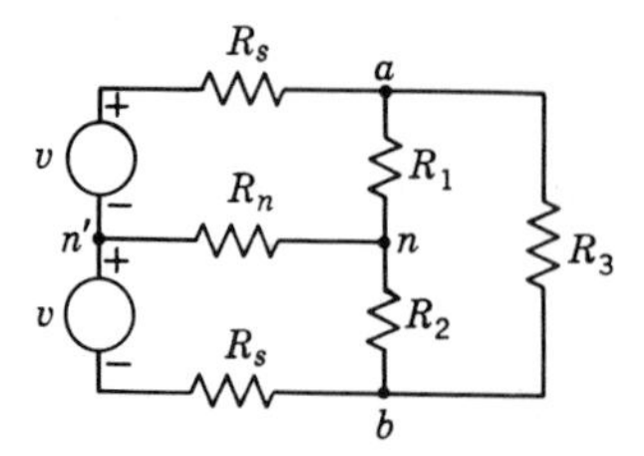

Fig. P14-1

14-2 In Prob. 14-1, $v(t) = \text{Re}\,\sqrt{2}\,100e^{j\omega t}$. Calculate the phasor ratio $\mathbf{V}_{an}/\mathbf{V}_{nb}$ if $R_s = \frac{1}{2}$ and R_1 is replaced by $\mathbf{Z}_1 = 3 + j4$, R_3 is replaced by $6 + j8$, and (*a*) R_2 is replaced by $\mathbf{Z}_2 = 3 + j4$; (*b*) $R_n = 0$ and $R_2 = 5$; (*c*) $1/R_n = 0$ and $R_2 = 6$.

14-3 In Fig. P14-3 use a voltage phasor diagram drawn to scale of V volts = 2 in. to show that the three-terminal source a'-b'-c' is a balanced three-phase source. (Is m' the "neutral" of the three-phase source?)

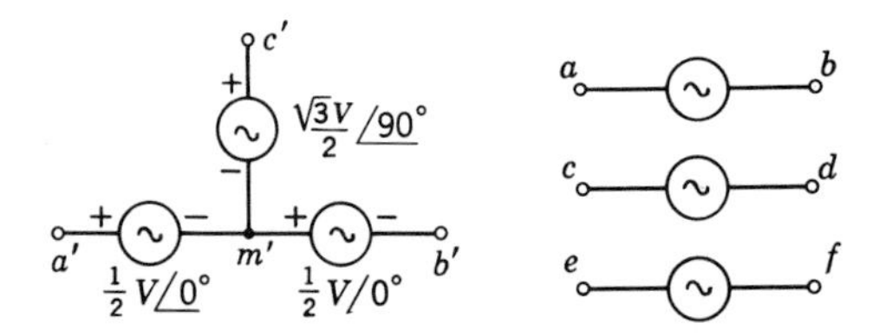

Fig. P14-3 ***Fig. P14-4***

14-4 Three ideal sources are given in Fig. P14-4 whose phasors are $\mathbf{V}_{ab} = V\underline{/0°}$, $\mathbf{V}_{cd} = V\underline{/60°}$, $\mathbf{V}_{ef} = V\underline{/-60°}$. How should these sources be connected to form a balanced three-phase source which is (*a*) wye-connected; (*b*) delta-connected? In each case state the phase sequence.

14-5 A balanced three-phase load consists of three equal impedances. When these impedances are connected in wye across balanced three-phase lines, line voltage V_L, the line current is 1. Complete the following table for the line and phase quantities indicated, assuming that the line voltage remains V_L as the connections are changed.

Connection		Generator		Line	Load	
Generator	Load	V_p	I_p	I_L	V_p	I_p
Wye	Wye	$V_L/\sqrt{3}$	1	1	$V_L/\sqrt{3}$	1
Wye	Delta					
Delta	Delta					
Delta	Wye					

14-6 A balanced three-phase circuit consists of a delta-connected ideal source furnishing line voltages of 173 volts and three load impedances of $5 + j5$ ohms each. Calculate the phase voltage and phase current in the load and generator if the three impedances form a load which is (*a*) delta-connected; (*b*) wye-connected.

14-7 In the balanced three-phase circuit shown in Fig. P14-7, $R = 12$ ohms, $X_L = 18$ ohms, $X_C = -10$ ohms, $R' = 0.5$ ohm, $X' = 1.25$ ohms, $V_{ab} = 220$ volts. Calculate the magnitude of the line-to-line voltage at the generator terminals.

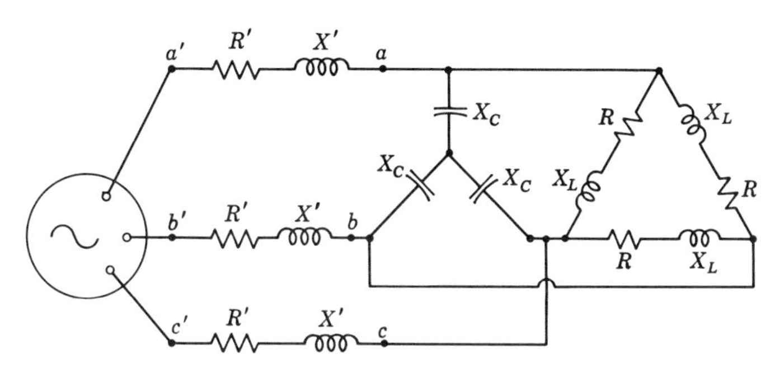

Fig. P14-7

14-8 Two balanced three-phase loads are connected in parallel across balanced three-phase lines. One load draws 50 kw at a lagging power factor of 0.4; the second load draws 50 kw at a lagging power factor of 0.707. (*a*) Calculate the power factor of the combination. (*b*) Calculate the magnitude of a line current if $V_L = 220$ volts.

14-9 (*a*) Calculate the kva size of a three-phase capacitance bank which when connected in parallel with the loads of Prob. 14-8 results in a power factor of 0.8 lagging for the combination. (*b*) If $V_L = 220$ volts, calculate the susceptance of each capacitance of the three-phase bank in (*a*) if this bank is (1) delta-connected; (2) wye-connected.

14-10 The balance three-phase load of Fig. P14-10 is connected to a balanced source. It is known that $\mathbf{V}_{ab} = 173\underline{/0°}$, $\mathbf{I}_{an} = 5\underline{/-45°}$. (*a*) Calculate the complex

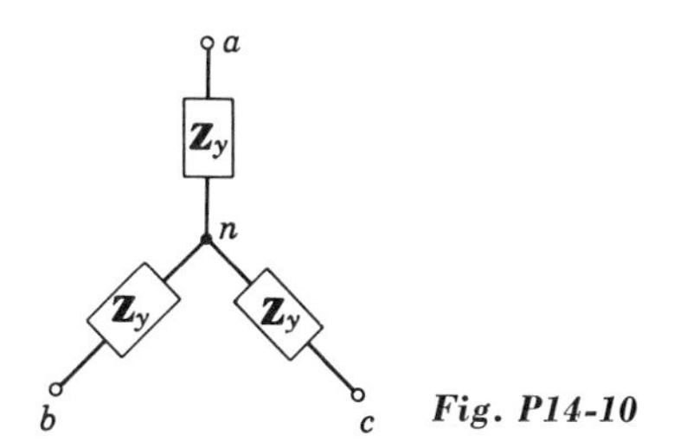

Fig. P14-10

value $\mathbf{Z}_y$ if $\mathbf{V}_{bc} = 173\underline{/-120°}$ (*a-b-c* phase order). (*b*) Calculate the complex value $\mathbf{Z}_y$ for the phase order *a-c-b*.

(*c*) For (*a*) and (*b*) draw a phasor diagram showing the nine phasors $\mathbf{V}_{ab}$, $\mathbf{V}_{bc}$, $\mathbf{V}_{ca}$, $\mathbf{V}_{an}$, $\mathbf{V}_{bn}$, $\mathbf{V}_{cn}$, $\mathbf{I}_{an}$, $\mathbf{I}_{bn}$, and $\mathbf{I}_{cn}$.

14-11 Assume that the open-delta (V) load of Fig. P14-11 is connected to a balanced three-phase source. (*a*) If the maximum line current allowed in any one line is I_L, obtain an expression for maximum power delivered to the two impedances. (*b*) If $I_{L\max}$ is the same as in (*a*), compare the result of (*a*) to the power delivered to three impedances $\mathbf{Z}$ connected in delta.

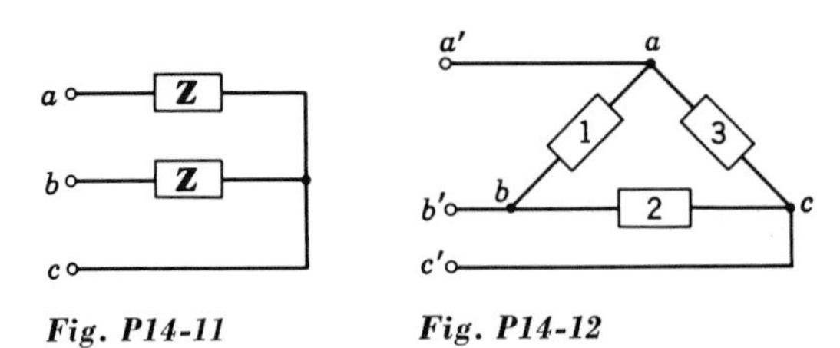

Fig. P14-11 *Fig. P14-12*

14-12 If $\mathbf{V}_{ab} = 200\underline{/0}$, $\mathbf{V}_{bc} = 200\underline{/-120°}$, $\mathbf{V}_{ca} = 200\underline{/+120°}$, and if the three loads are specified as follows, calculate $\mathbf{I}_{a'a}$, $\mathbf{I}_{b'b}$, and $\mathbf{I}_{c'c}$: load 1, 20 kw, unity power factor; load 2, 12 kw, 0.6 power factor lagging; load 3, 16 kw, 0.8 power factor leading. The circuit is shown in Fig. P14-12.

14-13 Calculate $\mathbf{I}_{a'a}$, $\mathbf{I}_{b'b}$, and $\mathbf{I}_{c'c}$ if $\mathbf{V}_{a'b'} = 200\underline{/0°}$, $\mathbf{V}_{b'c'} = 200\underline{/+120°}$, and $V_{c'a'} = 200\underline{/-120°}$. The circuit is shown in Fig. P14-13.

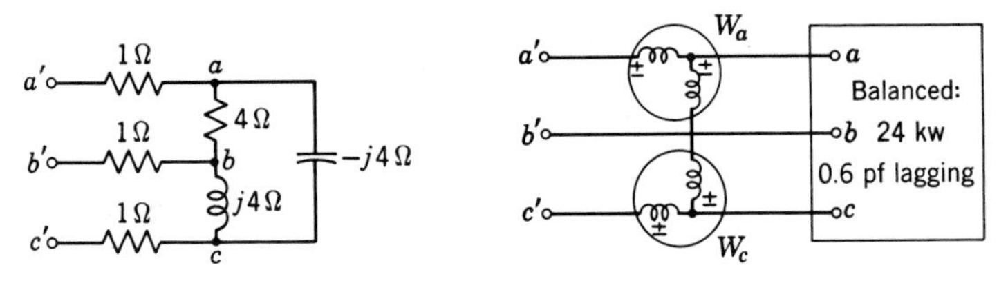

Fig. P14-13 *Fig. P14-14*

14-14 Calculate the reading of W_a and W_c if the phase sequence is (*a*) *a-b-c*; (*b*) *a-c-b*. The circuit is shown in Fig. P14-14.

14-15 In the circuit of Prob. 14-14, a wattmeter is connected so that the current in the current coil is $\mathbf{I}_{a'a}$ and the voltage across the voltage coil is $\mathbf{V}_{bc}$. Show that the magnitude of the reading of this wattmeter is proportional to reactive power.

14-16 Terminals *a-b-c* are connected to a balanced three-phase source. Show that the magnitude of the ratio $\mathbf{V}_{an}/\mathbf{V}_{bn}$ is more or less than unity, depending only on the phase order. The circuit is shown in Fig. P14-16.

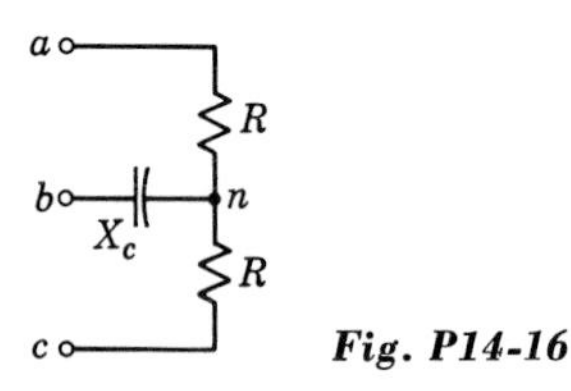

Fig. P14-16

14-17 The combination of the three-phase and single-phase loads in Fig. P14-17 forms an unbalanced load. The line voltages are balanced and are given by

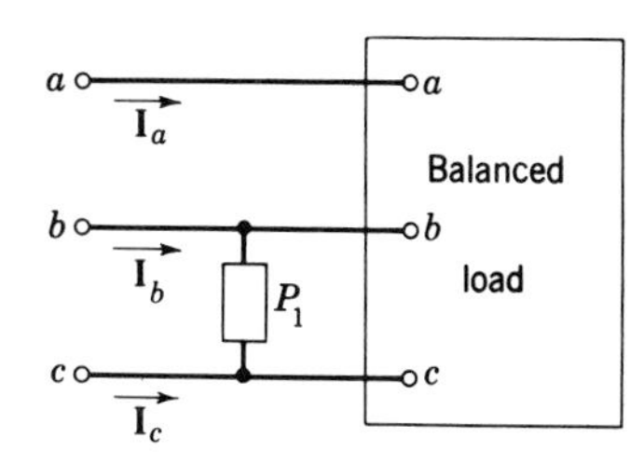

Fig. P14-17

$V_{ab} = 220\underline{/0}$, $V_{bc} = 220\underline{/-120°}$, $V_{ca} = 220\underline{/+120°}$. The balanced load draws 50 kw at a lagging power factor of 0.500; the single-phase load P_1 draws 30 kw at unity power factor. (*a*) Calculate $\mathbf{I}_a$, $\mathbf{I}_b$, and $\mathbf{I}_c$. (*b*) Two wattmeters are connected to read total power. One of these wattmeters is in line *a*, the other in line *c*. Calculate the reading of each wattmeter.

Chapter 15 Circuits with magnetic coupling

In our study of networks we have made extensive use of the concept of two-terminal elements called terminal pairs. All the networks so far studied have been distinguished by one common feature, namely, that their elements are connected to each other in such a manner that, between any two nodes of the network, there exists an uninterrupted path made of network elements. The nodes and meshes of such networks are said to be coupled together by "conduction." The equilibrium equations of the coupled networks are in the form of "simultaneous," or "coupled," equations. When we say that the coupling between two meshes is through conduction, we mean that the mutual impedance between meshes consists of passive elements which are shared by those two meshes. Thus, in Fig. 15-1, meshes 1 and 2 are coupled by means of the resistance R, and are said to be resistively coupled. In Fig. 15-2, meshes 1 and 2 are coupled inductively through

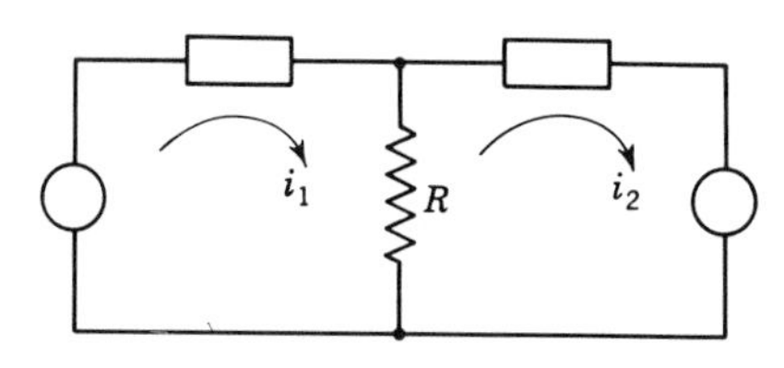

***Fig. 15-1** Two meshes coupled conductively through a resistance.*

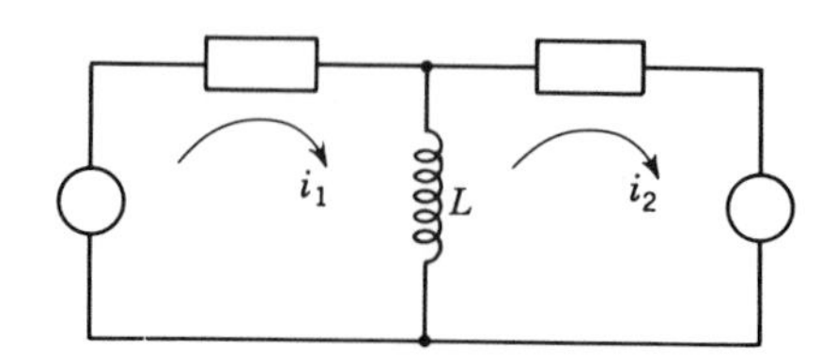

***Fig. 15-2** Two meshes coupled conductively through an inductance.*

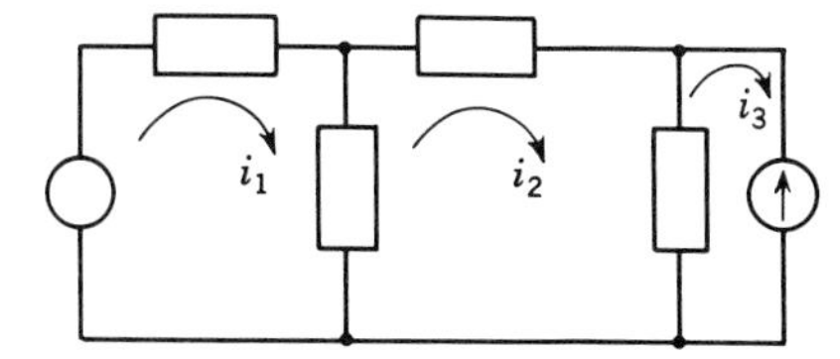

Fig. 15-3 Meshes 1 and 3 are coupled "indirectly" through conduction.

conduction. In Fig. 15-3, meshes 1 and 3 are coupled (indirectly) through conduction because each of these two meshes shares an element with mesh 2 (although not the same element).

Now, at the outset of this book, we stated that the purpose of circuit analysis is to provide "models" of the basic electromagnetic-field problems. In this chapter we shall study the circuit representation of the transfer of energy from one region of space to another through a magnetic field, a problem we have so far ignored. In circuit analysis this phenomenon is represented by a two-terminal-pair network called "mutual inductance." Two loops coupled by a mutual inductance are said to be coupled by induction. In contrast to coupling through conduction (where two loops share a branch), with coupling through induction there is no common element between the two coupled loops (either directly or indirectly).

15-1 Mutual inductance

In Chap. 1 we mentioned that, if a charge is accelerated in one region of space, a force will be exerted on charges located in other regions. From this we concluded that, when a charge moves with nonuniform velocity, an electric field will be established all over space, because of the acceleration of this charge. To say that a charge moves with nonuniform velocity in space is analogous in circuit language to saying that $d^2q/dt^2 = di/dt$ is not zero in a terminal pair. Therefore, when a current i flows in a terminal pair such that di/dt is not zero, in addition to the magnetic energy stored in the space and given by $\frac{1}{2}Li^2$, we have to account for the electric field intensity established in other regions of space. This phenomenon is accounted for by introducing a dependent source in one terminal pair to represent the effect of the variation of the current in another terminal pair.

In saying that a current $i_1(t)$ flows in a terminal pair we have implicitly assumed that there exists a closed path (circuit) through which the current flows. The magnetic energy associated with the flow of current in this path is accounted for by an inductance L_1. The reader is reminded that this inductance accounts for the magnetic energy associated with the "entire closed path" in which $i_1(t)$ flows, and therefore L_1 is a property of a closed path, although it is symbolically shown in a "lump" by a terminal pair. Any path of current will also have a resistance R_1. We may assume that the current i_1 is caused by a voltage source $v(t)$ as shown in Fig. 15-4a. For convenience of reference, we shall call the closed path of Fig. 15-4a, which includes the source, the "primary." Now consider another closed path, which need not, to begin with, contain any material medium and may be thought of as a closed line drawn through space. We shall call this closed path the "secondary."

Because of the variations of i_1 in the primary, an electric field will exist at every point in the secondary. In conformity with the circuit concept we shall lump the whole effect of this electric field around the secondary (the line integral of the electric-field intensity around the closed path) into a dependent voltage source. Theoretical considerations, as well as experimental evidence, show that the waveform of this dependent voltage source is proportional to di_1/dt and that the constant of proportionality depends on the "geometry" of the primary and secondary paths. The proportionality constant is called the *mutual inductance* between the two paths and is designated by letter M. Thus the effect of variation of i_1 in a primary as observed in the secondary will be accounted for by a dependent source $v_{i_2} = M(di_1/dt)$ as shown in Fig. 15-4*b*. This voltage is called the *voltage induced in the secondary due to* i_1 *in the primary*. Although the voltages pL_1i_1 or R_1i_1 can also be represented as dependent sources, we do not choose to do so at this time.

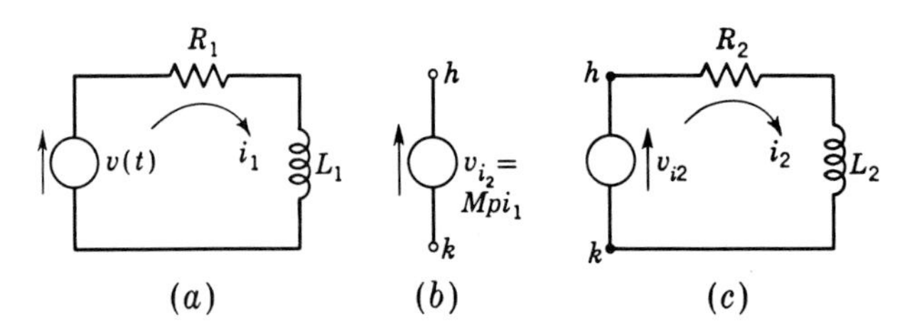

***Fig. 15-4** Circuit interpretation of mutual inductance. (a) Current in a mesh. (b) Dependent source between two terminals not in the mesh of* i_1*, induced by* i_1*. (c) The energy transferred to the circuit elements in mesh 2 represents the energy transferred through the magnetic field.*

In Fig. 15-4*b* we have not as yet shown a closed circuit, but only the induced voltage in the form of a dependent source. Suppose that the secondary path was such that a current could flow in it. Then, because of the induced voltage v_{i_2}, a current i_2 will flow in the secondary path. The magnetic energy associated with i_2 flowing in the secondary is accounted for by associating an inductance L_2 with this path. Similarly, a resistance R_2 must account for the dissipation of electromagnetic energy into heat.

Thus, to a good degree of approximation, the secondary path can be represented by a series connection of the dependent source v_{i_2}, R_2, and L_2, as shown in Fig. 15-4*c*. In this figure the reference arrow of i_2 is drawn arbitrarily from h to k through the R_2-L_2 combination. (We could have chosen the opposite direction if we wished.) Because a time-varying current i_2 flows in the secondary, an electric field will be created at every point of the primary. The effect of this field must be represented in the primary by a dependent voltage source whose waveform will be proportional to di_2/dt. The constant of proportionality will depend on the geometry of the two paths. Theoretical and experimental considerations show that this constant is the same M as discussed above. Therefore, to account for the effect of variation of i_2, we include a dependent source $v_{i_1} = M(di_2/dt)$ in the primary.

COEFFICIENT OF COUPLING, k We have said that the value of the mutual inductance between two paths (circuits) depends on the geometry of these paths. We remember that the inductance of each path, L_1 and L_2, also depended on the geometry of the individual paths. Analytical as well as experimental studies show that the value of M between two circuits is related to the inductances of those circuits, L_1 and L_2, through the relation

$$M = k\sqrt{L_1L_2} \qquad (15\text{-}1)$$

where k is dimensionless and depends again on the geometry of the paths. In the following paragraphs we shall show that $|k| \leq 1$. The factor k in Eq. (15-1) is called the coupling coefficient of the two paths. The unit of mutual inductance is the same as the unit of inductance (the henry in the mks system). From Eq. (15-1) it is seen that the mutual inductance presupposes the inductive property of the two paths.

The energy-storing characteristic of a circuit in a magnetic field is designated by the symbol associated with L. Since mutual inductance represents transfer of energy through a magnetic field, the symbol chosen for mutual inductance is, by convention, made up of symbols for the inductances of the primary and the secondary.

SYMBOL AND POLARITY OF MUTUAL INDUCTANCE Once we have chosen reference directions for the currents i_1 and i_2, the reference directions for the compensating sources pMi_1 and pMi_2 must conform to the geometry of the paths of the current.

In circuit diagrams the information about the geometry of the two paths is shown by "polarity dots" (or other marks such as $\pm$) on the inductances associated with the paths. The conventional symbol of mutual inductance between two paths of current whose inductances are L_1 and L_2 is shown in Fig. 15-5*a*. In this figure the dots imply that dependent sources are in series with L_1 and L_2 in such a manner that, with no current in L_2, the voltage of its dotted point with respect to its undotted point is equal to Mpi_1, where i_1 is the current whose reference arrow goes from the dotted point to the undotted point through L_1. A similar relationship is understood for the voltage across L_1 when there is no current flowing through it. This interpretation is shown in Fig. 15-5*b*. It is seen that, in the explicit representation of Fig. 15-5*b*, the reference directions for the dependent sources are obtained exactly in the manner described above. In Fig. 15-5*b* the depend-

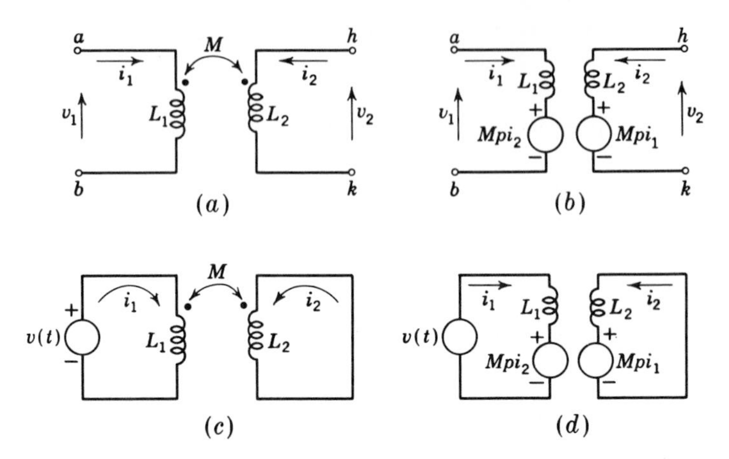

***Fig. 15-5** (a) Circuit symbol for mutual inductance between two paths which have self-inductances L_1 and L_2. (b) The assumed current directions, together with the reference dots in (a), result in the indicated polarities for the two compensating sources which represent the effect of mutual inductance. (c) Circuit used to study the range of values of k. (d) Circuit of (c) with compensating sources used to represent the effect of mutual inductance.*

ent source Mpi_2 represents contribution to the voltage v_{ab} due to the magnetic coupling between the paths represented by L_1 and L_2. Similarly, Mpi_1 represents the contribution to v_{hk} due to this magnetic coupling.

PROOF THAT COUPLING COEFFICIENT CANNOT EXCEED UNITY The following analysis shows that the magnitude of k can never exceed unity. In Fig. 15-5c, $v(t)$ is a voltage source with any arbitrary waveform. At $t = 0$ this source is applied to the primary of a (deenergized) mutual inductance whose secondary is short-circuited. Since there is no resistance in this network, any energy supplied by the source must be stored in the magnetic field, and over any period of time 0 to t_1, the amount of energy delivered by the source to the network must be nonnegative. This is true since the network itself is passive and, having started with no stored energy, cannot deliver an amount of energy to the source exceeding the value it has received from the source. With reference to Fig. 15-5c the energy delivered by the source to the network, in time zero to t_1, is

$$w(t_1) = \int_0^{t_1} vi_1\, dt \tag{15-2}$$

From Fig. 15-5d,

$$v = L_1 \frac{di_1}{dt} + M \frac{di_2}{dt} \tag{15-2a}$$

$$M \frac{di_1}{dt} = -L_2 \frac{di_2}{dt} \tag{15-2b}$$

Substituting for di_2/dt from Eq. (15-2b) in Eq. (15-2a), we have

$$v = L_1 \frac{d}{dt} i_1 - \frac{M^2}{L_2} \frac{d}{dt} i_1 = \frac{1}{L_2} (L_1L_2 - M^2) \frac{d}{dt} i_1 \tag{15-2c}$$

Substituting for v from Eq. (15-2c) in Eq. (15-2), we have

$$\begin{aligned} w(t_1) &= \int_0^{t_1} \left[\frac{1}{L_2} (L_1L_2 - M^2) \frac{d}{dt} i_1 \right] i_1 \, dt \\ &= \frac{L_1L_2 - M^2}{L_2} \int_{i_1=0}^{i_1=i_1(t_1)} i_1 \, di_1 \\ &= \frac{L_1L_2 - M^2}{2L_2} [i_1(t_1)]^2 \end{aligned}$$

Since $w(t_1)$ cannot be negative and $[i_1(t_1)]^2$ is always positive,

$$L_1L_2 - M^2 = L_1L_2(1 - k^2) \geq 0 \qquad k^2 \leq 1$$

which proves that the magnitude of k is less than or equal to unity.

Mutual inductance as a two-port A study of Fig. 15-5b shows that the two equations representing the mutual inductance of Fig. 15-5a are

$$\begin{aligned} v_1 &= pL_1i_1 + pMi_2 = z_{11}i_1 + z_{12}i_2 \\ v_2 &= pMi_1 + pL_2i_2 = z_{21}i_1 + z_{22}i_2 \end{aligned}$$

Recalling the definition of a two-port (two-terminal pair) from Chap. 12, we observe that a mutual inductance can be considered a two-port defined by the above equations.

15-2 Series connection of mutual inductance

The terminals of the two-port representing a mutual inductance can be connected in different manners, changing the two-port into a terminal pair. Such connections are referred to as series or parallel connections of mutual inductance, and will be discussed in this and the following sections.

If one of the output terminals of the two-port representing a mutual inductance is connected to a terminal in the input side, as shown in Fig. 15-6a and b, then the resulting circuit is a terminal pair. This terminal pair behaves as an inductance, whose value will now be found. In Fig. 15-6c the effect of mutual inductance is shown explicitly by compensating sources for the connection of Fig. 15-6a. Since both the input and output currents are the same $i(t)$, the waveforms of the two compensating sources will be identical and given by $pMi(t)$. The polarity of the compensating sources is determined by observing that, in L_1, the reference arrow of current goes *from* the dotted point *through* L_1; therefore, with no current in L_2, pMi is the voltage of the dotted point of L_2 (point b) with respect to its undotted point (point b'). But, with no current in L_2, $v_{bd} = 0$, and $v_{bb'} = v_{db'}$. Therefore $pMi = v_{db}$, as shown in Fig. 15-6c.

In L_2 the reference arrow of the current points *through* L_2 *to* the dotted point; therefore, with no current in L_1, $-pMi$ is the voltage of the dotted point of L_1 (point a) with respect to its undotted point (point a'). With no current in L_1, $-pMi = v_{aa'} = v_{na'}$ or $v_{a'n} = pMi$, as shown in Fig. 15-6c.

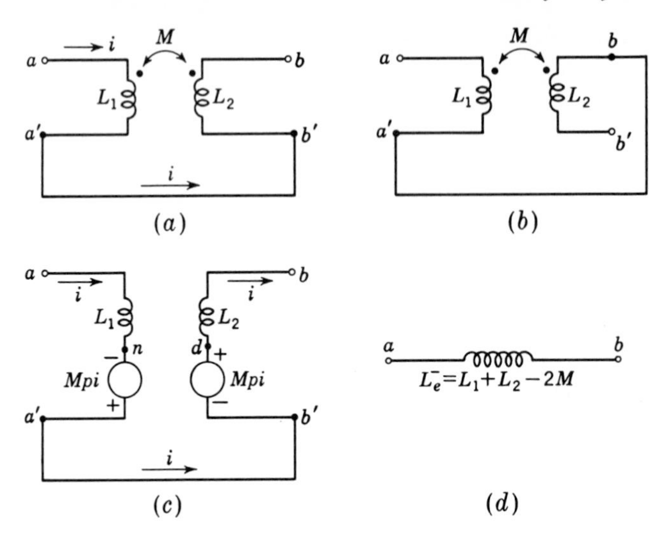

***Fig. 15-6** (a,b) Series connection of mutual inductance. (c) Circuit of (a) with compensating sources representing the effect of mutual inductance. (d) Equivalent inductance between terminals a-b for the circuit of (a).*

Now if, in Fig. 15-6c, v_{ab} is expressed in terms of i, we have

$$\begin{aligned} v_{ab} &= pL_1i - pMi - pMi + pL_2i \\ &= p(L_1 + L_2 - 2M)i = pL_e^-i \end{aligned}$$

where L_e^- is an equivalent inductance between a and b as shown in Fig. 15-6d, such that

$$L_e^- = L_1 + L_2 - 2M$$

In this case the two undotted terminals of the mutual inductance were connected together. If a dotted point and undotted point were connected as shown in Fig. 15-6b, the resulting equivalent inductance would be

$$L_e^+ = L_1 + L_2 + 2M$$

The reader should verify this relationship.

The two series connections are useful because they furnish a means of measuring mutual inductance between coils. We need measure only the equivalent inductances for each connection; then

$$\tfrac{1}{4}(L_e^+ - L_e^-) = M$$

or if the measurements are carried out in the sinusoidal steady state,

$$\tfrac{1}{4}(\omega L_e^+ - \omega L_e^-) = \omega M \equiv X_m$$

where X_m is called the *mutual reactance* of M at the radian frequency ω (Prob. 15-1).

In another application, two coils are made so that the mutual inductance between them can be adjusted. The series connection is then used to make a variable inductor.

15-3 Parallel connection of mutual inductance

When the output-input terminals of the mutual inductance are connected as shown in Fig. 15-7*a*, the connection is referred to as a parallel connection.

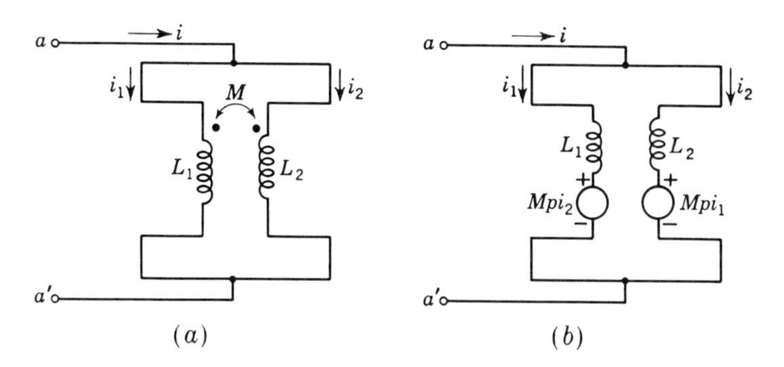

Fig. 15-7 Parallel connection of mutual inductance.

In this case, from the explicit representation of Fig. 15-7*b*, we have

$$v_{aa'} = pL_1i_1 + pMi_2$$
$$v_{aa'} = pMi_1 + pL_2i_2$$

Solving for i_1 and i_2 and adding, we have

$$i_1 + i_2 = \frac{L_1 + L_2 - 2M}{p(L_1L_2 - M_2)} v_{aa'} = i$$
$$v_{aa'} = p\,\frac{L_1L_2 - M^2}{L_1 + L_2 - 2M}\,i = pL_\pi^+ i$$

where L_π^+ is the equivalent parallel inductance and is given by

$$L_\pi^+ = \frac{L_1L_2 - M^2}{L_1 + L_2 - 2M}$$

If a dotted point were connected to an undotted point, the resulting equivalent inductance would be

$$L_\pi^- = \frac{L_1L_2 - M^2}{L_1 + L_2 + 2M}$$

The reader should verify this expression.

15-4 Mutual inductance in multielement networks

Since the compensating sources used to represent the effect of mutual inductance are dependent on the currents, the natural method for the formulation of the equilibrium equation of circuits with mutual inductance is the method of mesh currents. If meshes 1 and 2 are coupled through the mutual inductance M_{12}, the induced voltages are

$$v_{i_2} = \pm pM_{12}i_1 \qquad \text{and} \qquad v_{i_1} = \pm pM_{12}i_2$$

where the choice of sign depends on the relative location of the dots. Consequently, mutual inductance between meshes 1 and 2 will introduce the term $\pm pM_{12}$ into the mesh impedances Z_{12} and Z_{21}. Mutual inductance within a mesh will also affect the mesh impedance of such a mesh. An illustrative example will serve to clarify these statements.

Example 15-1 Let it be required to formulate the mesh equations for the circuit shown in Fig. 15-8. Note that, in this circuit, mesh 1 is magnetically coupled to mesh 2 through both M_1 and M_2 and that there is magnetic coupling within mesh 1 due to M_1. The mesh equations for this two-mesh circuit read

$$\begin{aligned} Z_{11}(p)i_1 + Z_{12}(p)i_2 &= v_1(t) \\ Z_{21}(p)i_1 + Z_{22}(p)i_2 &= -v_2(t) \end{aligned}$$

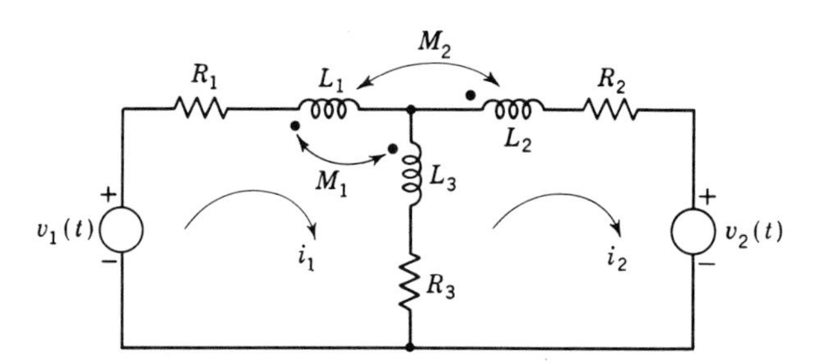

Fig. 15-8 A two-mesh circuit in which the coupling between the two meshes is partially by conduction and partially by induction.

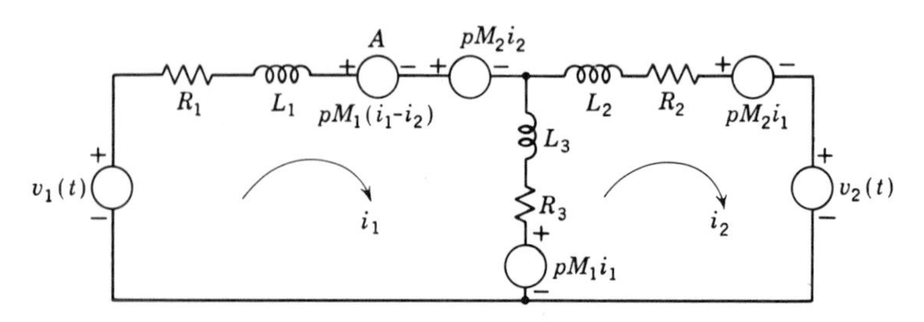

Fig. 15-9 Equivalent to Fig. 15-8. The effect of the mutual inductances is represented by compensating sources.

To find the mesh impedances, we use the definitions for mesh impedance given in Chap. 11, so that the effect of mutual inductance can be taken into account correctly.

$Z_{11}(p)i_1$ = sum of voltages in mesh 1 due to i_1 ($i_2 \equiv 0$), added in the direction of mesh 1
$Z_{12}(p)i_2$ = sum of voltages in mesh 1 due to i_2 ($i_1 \equiv 0$), added in the direction of mesh 1
$Z_{21}(p)i_1$ = sum of voltages in mesh 2 due to i_1 ($i_2 \equiv 0$), added in the direction of mesh 2
$Z_{22}(p)i_2$ = sum of voltages in mesh 2 due to i_2 ($i_1 \equiv 0$), added in the direction of mesh 2

To apply these definitions, it is convenient to redraw the circuit, replacing the effect of the mutual inductances by compensating sources. This circuit is shown in Fig. 15-9. Note that the source A depends on the branch current $i_1 - i_2$ which flows in L_3. To obtain the mesh impedances, we now apply the definitions given above for the self- and mutual impedances. If $i_2 \equiv 0$, then in mesh 1,

$$R_1i_1 + pL_1i_1 + pM_1i_1 + pL_3i_1 + pM_1i_1 + R_3i_1 = \Sigma \text{ voltages in mesh 1}$$

Hence

$$Z_{11}(p) = R_1 + R_3 + p(L_1 + L_3 + 2M_1)$$

This result could also have been obtained by noting that the equivalent series inductance in mesh 1 is $L_1 + L_3 + 2M_1$ (M_2 has no effect, since we set $i_2 \equiv 0$).

If $i_1 \equiv 0$, then in mesh 1,

$$-pL_3i_2 + pM_2i_2 - R_3i_2 - pM_1i_2 = \Sigma \text{ voltages in mesh 1}$$

so that

$$Z_{12}(p) = -R_3 - p(L_3 - M_2 + M_1)$$

If $i_2 \equiv 0$, then in mesh 2,

$$-R_3i_1 - pL_3i_1 + pM_2i_1 - pM_1i_1 = \Sigma \text{ voltages in mesh 2}$$

so that

$$Z_{21}(p) = Z_{12}(p)$$

Finally, in mesh 2, if $i_1 \equiv 0$,

$$(R_2 + R_3)i_2 + p(L_2 + L_3)i_2 = \Sigma \text{ voltages in mesh 2}$$

so that

$$Z_{22}(p) = R_2 + R_3 + (pL_2 + pL_3)$$

The mesh equations for the circuit of Fig. 15-8 therefore read

$$\begin{aligned}
[(R_1 + R_3) + p(L_1 + L_3 + 2M_1)]i_1 + [-R_3 - p(L_3 + M_1 - M_2)]i_2 &= v_1(t) \\
[-R_3 - p(L_3 + M_1 - M_2)]i_1 + \quad [R_2 + R_3 + p(L_2 + L_3)]i_2 &= -v_2(t)
\end{aligned}$$

SIGN OF M In the case where only one mutual inductance exists, we recognize that the effect of reversing the position of *one* of the reference dots (but *not* both) is equivalent to assigning a negative number to the mutual inductance (or to the corresponding coefficient of coupling). Hence, in circuits with only one mutual inductance, it is always possible to choose the reference dots so that M (or k) is a positive number. In circuits where there are several mutual inductances associated with each loop (see, for example, Prob. 15-8), a separate set of dots or other marks must generally be used if all mutual inductances are to be given by positive numbers. In order to avoid such multiplicity of symbols it is generally convenient to allow M (and k) to be either positive or negative and use a single set of dots for all mutual inductances as in Prob. 15-8.

15-5 The linear transformer

A two-mesh circuit in which the coupling between meshes is entirely magnetic is shown in Fig. 15-10. Such a circuit is called a linear-transformer

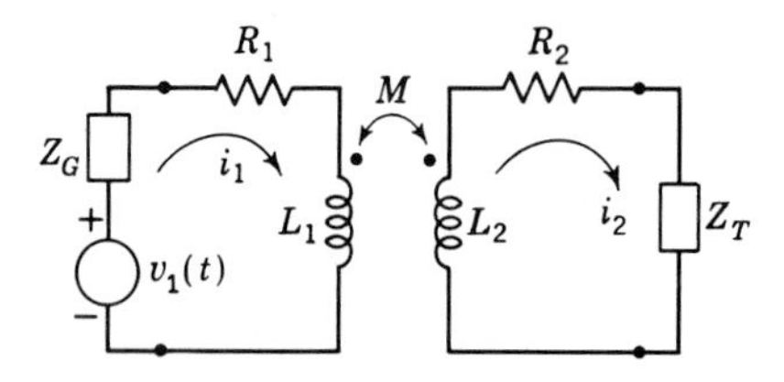

Fig. 15-10 Linear-transformer circuit.

circuit. We shall use the term *linear* because, in practice, such circuits are often representations of coils coupled together through an iron core. In such cases the equivalent representation is an approximation, since the inductance of iron-cored coils is a function of the current through the coils. In Fig. 15-10 it is assumed that the elements are linear (such circuits are sometimes called "air-core" transformers). In the illustration, R_1 and L_1 can be interpreted as the equivalent circuit of the coil which is magnetically coupled to a second coil represented by R_2 and L_2. Z_T is a load impedance, and the series combination of Z_G and $v_1(t)$ forms the (Thévenin) equivalent circuit of a nonideal source. For the assumed current direction relative to the dots, the mesh equations are

$$Z_{11}(p)i_1 + Z_{12}(p)i_2 = v_1(t)$$
$$Z_{21}(p)i_1 + Z_{22}(p)i_2 = 0$$

where $\quad Z_{11}(p) = Z_G + R_1 + pL_1 \qquad Z_{22}(p) = Z_T + R_2 + pL_2$

$$Z_{12}(p) = Z_{21}(p) = -pM$$

In the sinusoidal steady state, if $v_1(t)$ is represented by the phasor $\mathbf{V}_1$, the complex mesh equations are

$$(R_1 + j\omega L_1 + \mathbf{Z}_G)\mathbf{I}_1 \qquad + (-j\omega M)\mathbf{I}_2 = \mathbf{V}_1$$
$$(-j\omega M)\mathbf{I}_1 + (R_2 + j\omega L_2 + \mathbf{Z}_T)\mathbf{I}_2 = 0$$

To study some relationships in the linear transformer, it is convenient to consider the circuit shown in Fig. 15-11, where a source $v_1(t)$ is applied

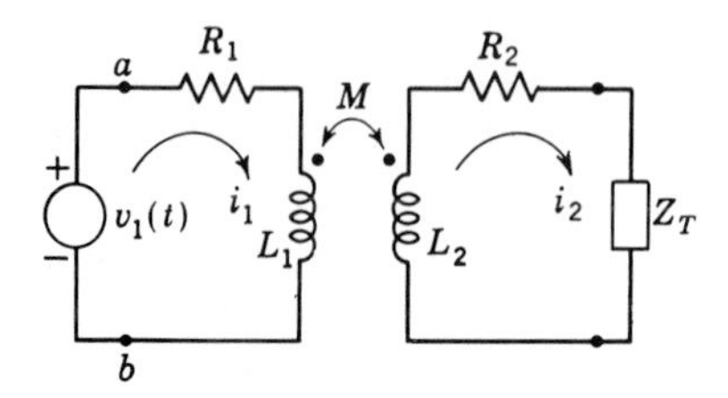

Fig. 15-11 Linear transformer with ideal source and arbitrary loading.

directly to the terminals of the coil represented by R_1 and L_1. The results which we shall obtain can be applied when a nonideal source is impressed at these terminals, because the function $v_1(t)$ can then be reinterpreted as the sum of the ideal source and the compensating source which accounts for the internal impedance of the nonideal source. At present, however, we are concerned with the voltage-current relationship of the transformer considered as a two-port, so that $v_1(t)$ will be treated as an ideal source.

For the circuit of Fig. 15-11, the mesh equations are

$$\begin{aligned}(R_1 + pL_1)i_1 \qquad\qquad\qquad - pMi_2 &= v_1(t)\\ -pMi_1 + [R_2 + pL_2 + Z_T(p)]i_2 &= 0\end{aligned}$$

so that in operational form,

$$i_1(t) = \frac{R_2 + pL_2 + Z_T}{(R_1 + pL_1)[R_2 + pL_2 + Z_T(p)] - p^2M^2}\, v_1(t)$$

and

$$i_2(t) = \frac{pM}{(R_1 + pL_1)[R_2 + pL_2 + Z_T(p)] - p^2M^2}\, v_1(t)$$

Several relationships are of interest and will now be deduced.

Driving-point impedance The operational driving-point impedance at terminals *a-b* is defined by $Z_{ab}i_1 = v_1$. For the circuit under discussion the result is

$$Z_{ab}(p) = R_1 + pL_1 + \frac{-p^2M^2}{R_2 + pL_2 + Z_T(p)}$$

We note that $Z_{ab} = Z_{11} - Z_{12}{}^2/Z_{22}$; if $i_2 \equiv 0$, $Z_{ab} = Z_{11}$. The term $-Z_{12}{}^2/Z_{22}$ which adds to Z_{11} is referred to as the *impedance of the secondary reflected into the primary*. For the sinusoidal steady state, the complex driving-point impedance is

$$\mathbf{Z}_{ab} = R_1 + j\omega L_1 + \frac{\omega^2M^2}{R_2 + j\omega L_2 + \mathbf{Z}_T}$$

We note in passing that, for "small" values of ωM, this quantity depends only very slightly on $\mathbf{Z}_T$, the load impedance.

Current ratio From the equation of mesh 2, the ratio of the currents in operational form is

$$\frac{i_2}{i_1} = \frac{pM}{R_2 + pL_2 + Z_T(p)}$$

Voltage ratio The ratio of the voltage across the load impedance to the input voltage is, in operational form,

$$\frac{v_2}{v_1} = \frac{i_2Z_T}{v_1} = \frac{pMZ_T(p)}{(R_1 + pL_1)(R_2 + pL_2 + Z_T) - p^2M^2}$$

15-6 The ideal transformer

A two-port shown in Fig. 15-12*a*, such that

$$v_2(t) = nv_1(t)$$
$$i_2(t) = \frac{1}{n} i_1(t)$$

where n is a positive number, is called an ideal transformer. If n is larger than unity, the transformer is called an ideal step-up transformer; otherwise it is called a step-down transformer.

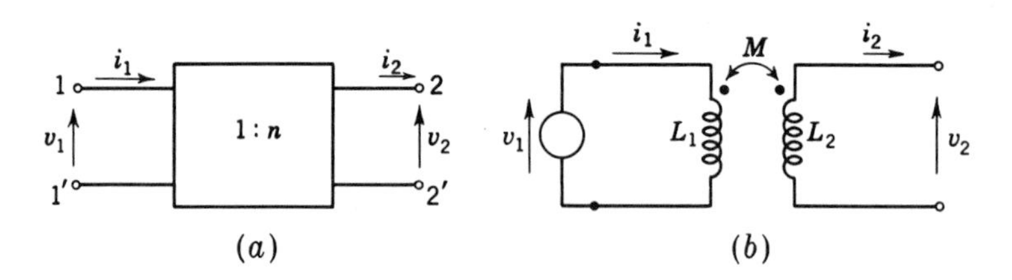

***Fig. 15-12** (a) Symbol for an ideal transformer. (b) Pertinent to the representation of an ideal transformer by mutual inductance.*

If we wish to represent an ideal transformer by a mutual inductance, as shown in Fig. 15-12*b*, then we begin with the defining equations of mutual inductance,

$$v_1 = L_1pi_1 - Mpi_2$$
$$-v_2 = -Mpi_1 + L_2pi_2$$

Imposing the conditions

$$v_2 = nv_1 \; i_2 = \frac{1}{n} i_1$$

it is seen that, with a source v_1 applied to the input of the two-port, then with open-circuited output, i_1 must be zero since i_1 is made proportional to i_2 (and for open circuit at the output port $i_2 \equiv 0$). But from the defining equations for mutual inductance, we have $v_1 = L_1pi_1$ when i_2 is zero. It follows that L_1 must be infinite so that i_1 will be zero with nonzero v_1.

By a similar argument it can be shown that L_2 must also be infinite and the coefficient of coupling must be unity.

Those readers who are familiar with transformer theory will note that, if L_1 is infinite, the magnetizing current will be zero, and since we have not included any resistance in the circuit representation of the ideal transformer, the losses must also be zero, resulting in a fixed ratio between the currents and voltages at the input and output sides of the ideal transformer.

LOSSLESS TRANSFORMER The transformer of Fig. 15-11 is lossless if $R_1 = R_2 = 0$, but is not ideal. For the lossless transformer the driving-point impedance, and the voltage and current ratios, by the results of the preceding sections, are

$$Z_{ab}(p) = pL_1 + \frac{-p^2M^2}{pL_2 + Z_T(p)}$$

Current ratio: $$\frac{i_2}{i_1} = \frac{pM}{pL_2 + Z_T(p)}$$

Voltage ratio: $$\frac{v_2}{v_1} = \frac{pMZ_T(p)}{p^2(L_1L_2 - M^2) + pL_1Z_T(p)}$$

PERFECTLY COUPLED AND LOSSLESS TRANSFORMER A transformer is perfectly coupled and lossless (but not ideal) if

$$R_1 = R_2 = 0$$

and $$L_1L_2 - M^2 = 0$$

Such an arrangement is not "ideal" because the current ratio, as well as the input impedance, depends on the values of inductances and not on their ratio. A lossless, perfectly coupled transformer which is not ideal is sometimes referred to as a "pair of perfectly coupled lossless coils."

The three significant expressions for this case are

$$Z_{ab}(p) = \frac{pL_1Z_T}{pL_2 + Z_T}$$

$$\frac{i_2}{i_1} = \frac{p\sqrt{L_1L_2}}{pL_2 + Z_T}$$

$$\frac{v_2}{v_1} = \frac{p\sqrt{L_1L_2}}{pL_1} = \sqrt{\frac{L_2}{L_1}}$$

Note that, for the perfectly coupled lossless transformer, the voltage ratio is independent of the waveform.

IDEAL TRANSFORMER We have stated in effect that an ideal transformer is a lossless, perfectly coupled transformer whose inductances are infinite. To obtain its properties, we therefore take the following limits:

$$L_2 \to \infty \qquad L_1 \to \infty \qquad \text{but} \qquad \frac{L_1}{L_2} \to \text{finite}$$

If we examine the operational expressions which were deduced for the preceding cases, the limits stated above will give $Z_T(p)$ negligible when compared with pL_2. In the sinusoidal steady state this means that the condition $|\mathbf{Z}_T| \ll |j\omega L_2|$ will approximate the conditions for an ideal transformer if the resistance and coupling conditions are approximately fulfilled. Theoretically, however, the two-port "ideal transformer" is defined with infinite inductance because then the expressions which are to be deduced will be independent of the impressed waveforms.

For the ideal transformer, the driving-point impedance and the voltage and current ratios are seen to be

Current ratio: $$\frac{i_2}{i_1} = \sqrt{\frac{L_1}{L_2}} = \frac{1}{n}$$

Voltage ratio: $$\frac{v_2}{v_1} = \sqrt{\frac{L_2}{L_1}} = n$$

Driving-point impedance:

$$Z_{ab}(p) = \frac{L_1}{L_2} Z_T(p) = \frac{1}{n^2} Z_T(p)$$

(Note that $v_2 i_2 = v_1 i_1$.) The symbol shown in Fig. 15-13 will be used to

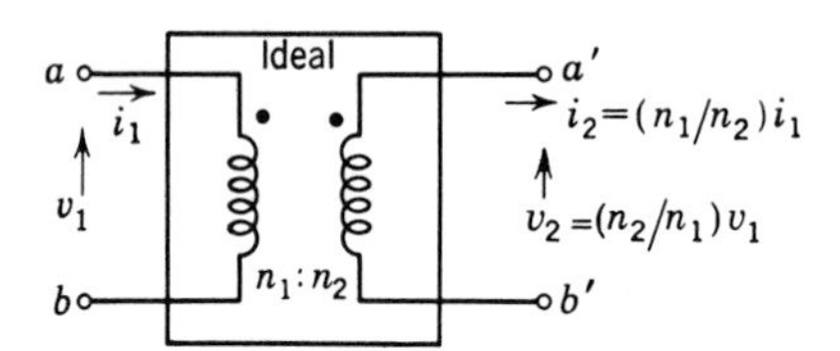

***Fig. 15-13** Another symbol for an ideal transformer. This symbol is intended to be suggestive of the use of mutual inductance to achieve the properties of an ideal transformer.*

represent an ideal transformer. The transformer in this illustration is marked with the symbol $n_1{:}n_2(=1/n)$. From the above it is seen that

$$\frac{n_1}{n_2} = \sqrt{\frac{L_1}{L_2}} = \frac{1}{n}$$

To have the polarity dots on the ideal transformer symbol of Fig. 15-13 consistent with the definition of Fig. 15-12a, it is necessary that the reference arrows for both voltages and both currents be as shown in the two figures. However, if in Fig. 15-13 the reference arrow for v_1 is chosen in the opposite direction to that shown, that is, if $v_1 = v_{ba}$ instead of v_{ab}, then

$$v_2 = v_{a'b'} = -\left(\frac{n_2}{n_1}\right) v_1$$

Analogous statements apply to the reference arrows for currents.

Example 15-2 In the circuit shown in Fig. 15-14 determine $\mathbf{Z}_{ab}$, $\mathbf{I}_1$, $\mathbf{I}_2$.

Solution The driving-point impedance at the input terminals of the ideal transformer is

$$\mathbf{Z}_{a'b} = (R_2 + jX_c)\left(\frac{n_1}{n_2}\right)^2$$

Hence

$$\mathbf{Z}_{ab} = R_1 + \left(\frac{n_1}{n_2}\right)^2 R_2 + jX_L + jX_c\left(\frac{n_1}{n_2}\right)^2$$

and

$$\mathbf{I}_1 = \frac{\mathbf{V}}{\mathbf{Z}_{ab}} \qquad \mathbf{I}_2 = -\frac{n_1}{n_2}\mathbf{I}_1$$

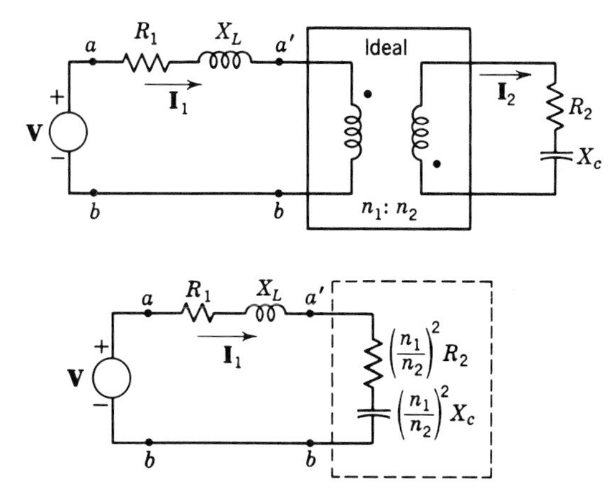

Fig. 15-14 Circuit to illustrate calculation with ideal transformer.

15-7 Equivalent circuit for the nonideal transformer

In this section the mesh equations of a nonideal transformer will be studied for the purpose of arriving at several equivalent circuits. In Fig. 15-15

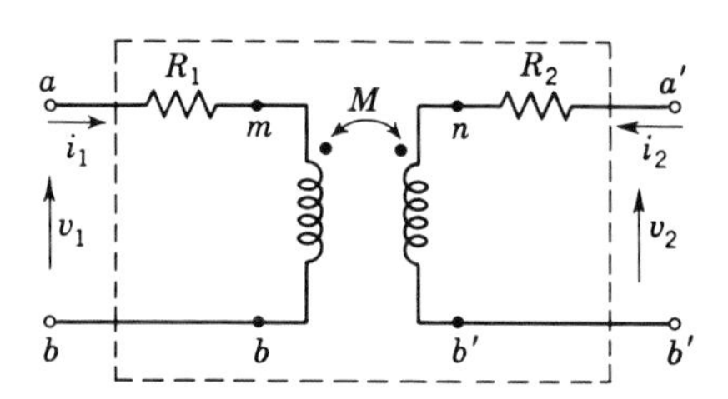

Fig. 15-15 A (nonideal) transformer as a two-port.

a transformer circuit is shown. Since we are interested in the relationships between v_1, i_1, v_2, and i_2, the linear transformer is shown as a two-port represented by the dashed box. For this circuit the mesh equations are

$$(R_1 + pL_1)i_1 + (+pM)i_2 = v_1 \tag{15-3a}$$
$$(+pM)i_1 + (R_2 + pL_2)i_2 = v_2 \tag{15-3b}$$

If we have Eqs. (15-3*a*) and (15-3*b*), *the circuit shown in Fig.* 15-16 *can be drawn to fit these equations* without regard to the original circuit since we

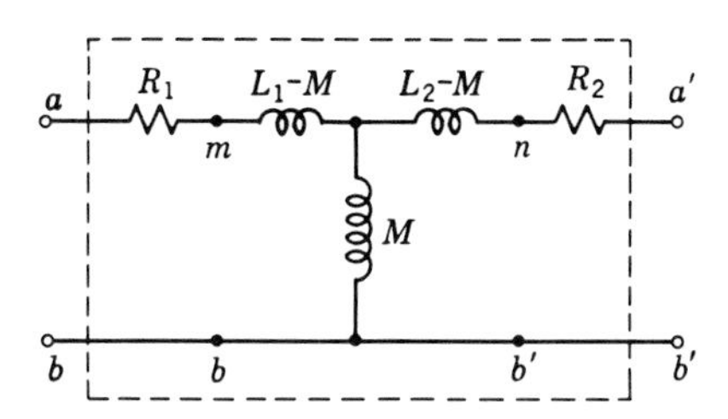

Fig. 15-16 T-section representation of mutual inductance and of the transformer of Fig. 15-15.

are concerned only with voltages and currents on the outside of the box. The circuit of Fig. 15-16 is therefore equivalent to the circuit of Fig. 15-15 with respect to the terminal pairs *a-b* and *a'-b'*.

T-SECTION REPRESENTATION OF MUTUAL INDUCTANCE From Figs. 15-15 and 15-16 it is seen that the mutual inductance M between the terminals *m-b*, *n-b'* has been replaced by a two-port composed of three inductances connected in the form of a T. This representation of a mutual inductance is called its T-section representation.

Since the value of mutual inductance is given by $M = k(L_1L_2)^{\frac{1}{2}}$, in the T-section representation it is possible for either $L_1 - M$ or $L_2 - M$ (Fig. 15-16) to be a negative number and therefore represent a nonphysical (i.e., nonrealizable) element. Although this "objection" to the circuit of Fig. 15-16 can be overcome, it must be understood that the circuit in question is *on paper* equivalent to the original circuit, with respect to the two pairs of terminals, because it satisfies the same mesh equations as the circuit of Fig. 15-15.

A second remark concerning the equivalent circuit of Fig. 15-16 is necessary. In the original circuit of the transformer, the input terminals *a-b* are isolated from the output terminals *a'-b'* because the coupling between the meshes is entirely by induction. In the equivalent circuit of Fig. 15-16, the coupling between the two meshes is through the inductance of value M, so that the isolating property of the transformer has not been preserved.

The two objections mentioned above can be overcome (artificially) by manipulating the mesh equations (15-3*a*) and (15-3*b*) and using an ideal transformer.

Let us start by supposing that $L_1 - M$ is negative. Then $L_2 - M$ is positive because M is at most the geometric mean between L_1 and L_2. Let us choose a number a, less than unity, so that $L_1 - aM$ is positive. The mesh equations (15-3*a*) and (15-3*b*) may now be written as follows:

$$(R_1 + pL_1)i_1 + paM\frac{i_2}{a} = v_1 \tag{15-4a}$$

$$(paM)i_1 + (R_2a^2 + pL_2a^2)\frac{i_2}{a} = av_2 \tag{15-4b}$$

Equation (15-4*a*) is identical with Eq. (15-3*a*), and Eq. (15-4*b*) is a times Eq. (15-3*b*). Note now that the following four quantities are related in Eqs. (15-4): i_1, v_1, i_2/a, and av_2. We can therefore interpret the mesh equations (15-4) as representing the circuit shown in Fig. 15-17*a*. If we wish to make an equivalent circuit in which the current i_2 and the voltage v_2 appear, an ideal transformer whose turns ratio is a can be introduced. Hence, with respect to the two terminal pairs *a-b* and *a'-b'*, the circuit shown in Fig. 15-17*b* is equivalent for all values of the "number" a. Note that the "number" a can be chosen in such a way that each element within the dashed box in Fig. 15-17*b* is physically realizable and that the use of the ideal transformer preserves the isolating property of the actual transformer.

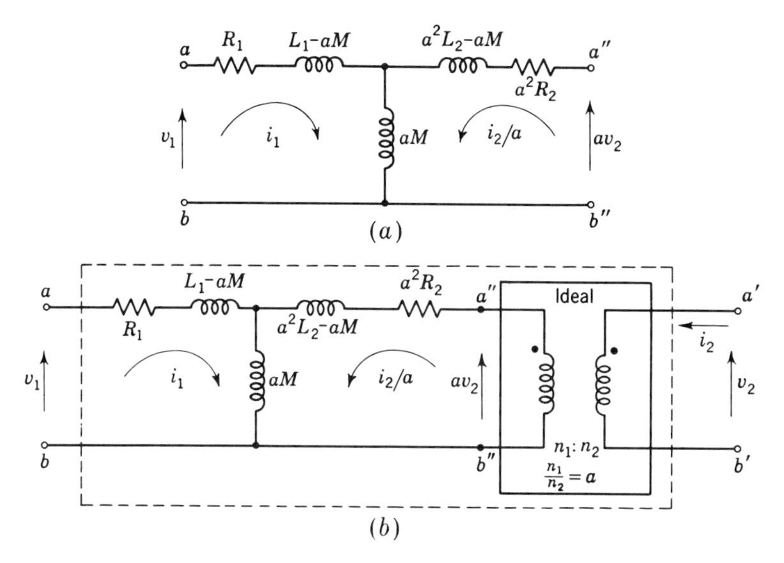

Fig. 15-17 ***(a) The mesh equations for this two-port are the same as the equation for the two-port of Fig. 15-15, but the current and voltage in the second mesh do not correspond. (b) The two-port shown is equivalent to the two-port of Fig. 15-15 at the accessible terminal pairs a-b and a′-b′.***

The most popular value chosen for the "number" a is the turns ratio which the actual transformer has. Then, if

$$a = \left(\frac{L_1}{L_2}\right)^{\frac{1}{2}}$$

we have $aM = (L_1/L_2)^{\frac{1}{2}}k(L_1L_2)^{\frac{1}{2}} = kL_1 \leq L_1$, and the equivalent circuit contains the elements with values as shown in Fig. 15-18. Note that, for $k = 1$, $R_1 = R_2 = 0$, and L_1 infinite, only the ideal transformer is left. The

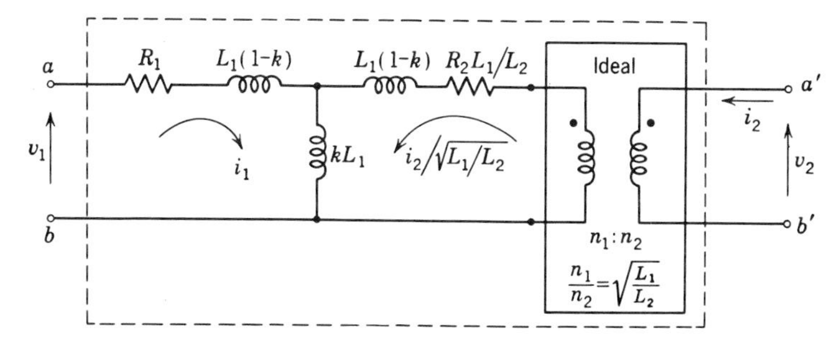

Fig. 15-18 ***In this equivalent circuit of a (nonideal) transformer the elements are physically realizable, and the isolating property of the transformer has been retained.***

arrangements of the elements R_1, $L_1(1-k)$, R_2L_1/L_2, and $k\,L_1$ in Fig. 15-18 conveniently represent the "imperfections" of the actual transformer.

15-8 Adjustable circuits with mutual inductance. Introduction

In Chap. 7 it was shown how transform network functions are related to sinusoidal steady-state response and complete response through the pole-zero diagram. In Chap. 10 this relationship was explored for simple series, parallel, and series-parallel *R-L-C* circuits. In the remaining parts of this chapter the frequency response of circuits with mutual inductance is discussed. In addition, the influence of adjusting the coefficient of coupling is studied. The problem is formulated as follows: In a two-mesh network the meshes are coupled through mutual inductance *only;* how does variation of the coefficient of coupling affect the responses of the circuit, and what is the frequency response of certain commonly used networks? As the various circuits are studied, the possible applications will be pointed out. In general, we observe that each voltage induced by magnetic coupling is proportional to the *rate of change* of a current. Hence, if a source which is the sum of a constant and a time function is impressed in mesh 1 of the circuit, the constant source, in the steady state, has no effect in mesh 2 since constant sources produce only constant steady-state currents.

To fix the basic idea, a pair of coupled inductors, represented by the circuit of Fig. 15-19*a*, are first studied. (Each inductor is represented as a series *R-L* circuit.) An ideal source is placed in series with R_1 and L_1 to form mesh 1; mesh 2 is left open. Since mesh 1 includes the only source in the circuit, it is referred to as the primary. Mesh 2 is the secondary. We shall study the transform voltage-gain function, which relates $v_2(t)$ to $v_1(t)$, and the transform transfer impedance, which relates $v_2(t)$ to $i_1(t)$. In the former case the source in the primary is an ideal voltage source; in the latter case it is an ideal current source.

Since the secondary is open, $i_2 \equiv 0$; hence $v_{i1} \equiv 0$. The effect of the mutual inductance is therefore represented by the single source $v_{i_2} = pMi_1$ as shown in Fig. 15-19*b*. From Fig. 15-19*b* we observe that $(pL_1 + R_1)i_1 = v_1$ so that

$$v_2 = \frac{pM}{pL_1 + R_1}v_1 = \frac{pk\sqrt{L_1L_2}}{pL_1 + R_1}v_1 \tag{15-5}$$

or

$$v_2 = \sqrt{\frac{L_2}{L_1}}\frac{pk}{p + R_1/L_1}v_1 \tag{15-6}$$

The transform voltage gain is

$$H(s) = \sqrt{\frac{L_2}{L_1}}\,k\,\frac{s}{s + R_1/L_1}$$

It is convenient to study the ratio $H(s)/\sqrt{L_2/L_1}$ since $\sqrt{L_2/L_1}$ is the voltage gain of an ideal transformer. Thus

$$K(s) \equiv \frac{H(s)}{\sqrt{L_2/L_1}} = k\,\frac{s}{s + R_1/L_1}$$

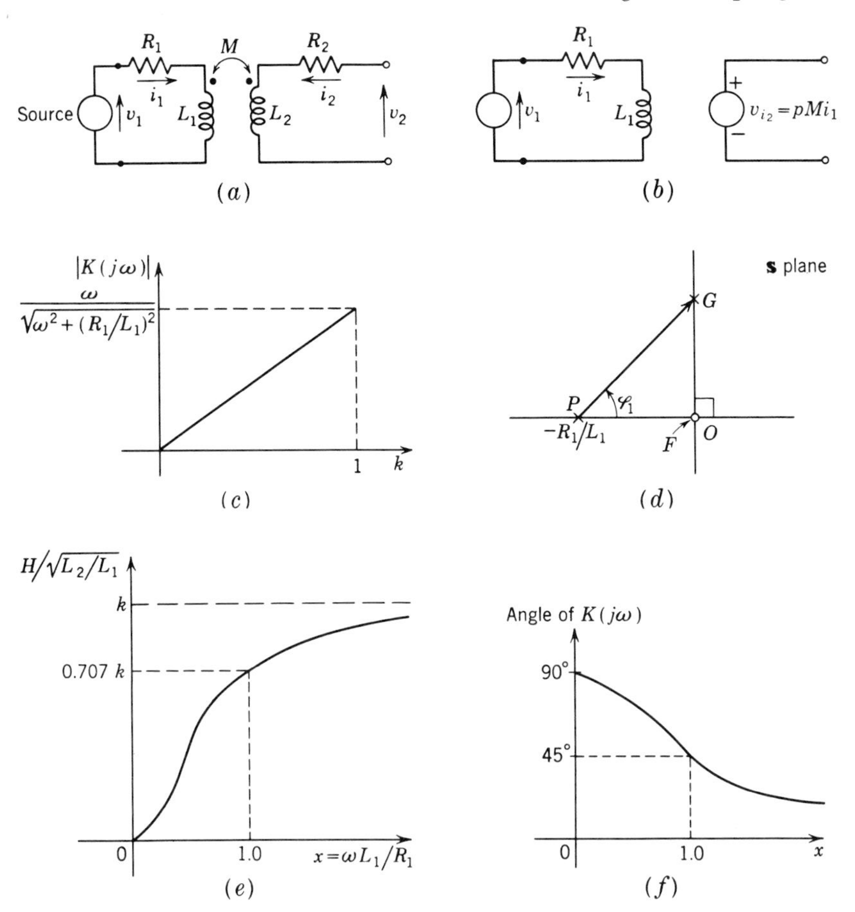

***Fig. 15-19** (a) Circuit representation of a pair of magnetically coupled coils. (b) Equivalent circuit of (a) with secondary open-circuited. (c) Normalized open-circuit voltage gain as a function of coupling for the circuit of (a) in the sinusoidal steady state. (d) Pole-zero diagram for the transform voltage gain. (e) Amplitude response as a function of frequency. (f) Phase response as a function of frequency.*

Hence the coefficient of coupling influences only the scale factor; i.e., for any value $s = s_g$, $K(s)$ is proportional to k. This result is shown for $s_g = j\omega$ in Fig. 15-19*c*. The frequency-response curve is obtained from the pole-zero diagram of Fig. 15-19*d*: $K(s)$ has a zero at $s = 0$ and a pole at $s = -R_1/L_1$. Hence, in terms of the indicated distances,

$$K(j\omega) = k \frac{OG}{PG} \underline{/90° - \varphi_1}$$

The amplitude-response curve is seen to be zero at $\omega = 0$ and approaches k as $\omega \to \infty$. Analytically, let $x = \omega L_1/R_1$; then

$$\frac{H}{\sqrt{L_2/L_1}} = k \frac{x}{\sqrt{1 + x^2}}$$

This curve is shown in Fig. 15-19*e*. The corresponding phase response is shown in Fig. 15-19*f*. We observe that this network may find application as a "high-pass filter." If the voltage source $v_1(t)$ is nonideal but is represented as the series combination of R_s–L_s–ideal voltage source, the results obtained above are applicable, provided that R_1 includes R_s and L_1 includes L_s. The value of k must then be chosen (reduced) to give the correct value of M (see Prob. 15-21).

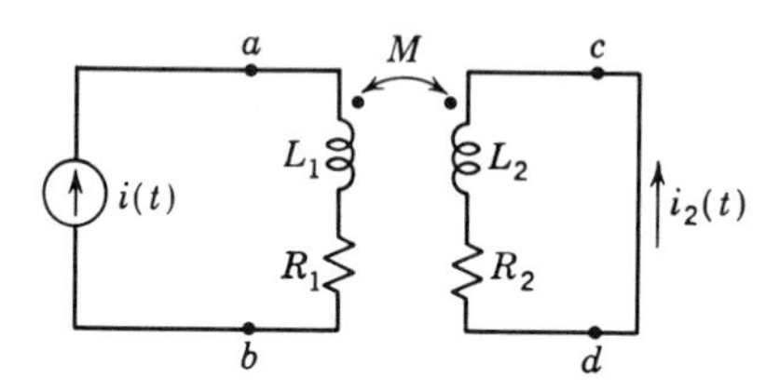

***Fig. 15-20** Circuit of Fig. 15-19a with short-circuited secondary.*

If terminals *c-d* are short-circuited and if the source is an ideal current source $i(t) = i_1(t)$, as shown in Fig. 15-20, then, from the mesh equation for mesh 2,

$$(R_2 + pL_2)i_2 = -pMi_1 \qquad (15\text{-}7)$$

Equation (15-7) is of the same form as (15-6); we observe that the frequency response for the current ratio I_2/I_1 in the short-circuit case is identical with the frequency response of the voltage ratio V_2/V_1 in the open-circuit case. Thus the result shown in Fig. 15-19*e* is applicable to the current ratio if x is redefined as $\omega L_2/R_2$ and the current gain is normalized with respect to $\sqrt{L_1/L_2}$.

15-9 Introduction to tuned transformer circuits

In many applications of coupled inductors, capacitances appear across the input and output terminals as shown in Fig. 15-21*c*. Either or both of these capacitances may appear in the circuit diagram to represent elements which have been placed into the actual circuit intentionally or to represent the effect of "stray capacitances" (e.g., due to wiring). The frequency responses of the transfer functions which can be associated with such circuits (called "tuned transformer circuits") are of interest because these circuits find wide application as bandpass filters.

In the transmission of signals it is often necessary to design a two-port which has the following characteristics:

1 In a certain band of radian frequencies ω_1 to ω_2 all sinusoidal signals are to be passed with the *same* gain (or attenuation) and must suffer the

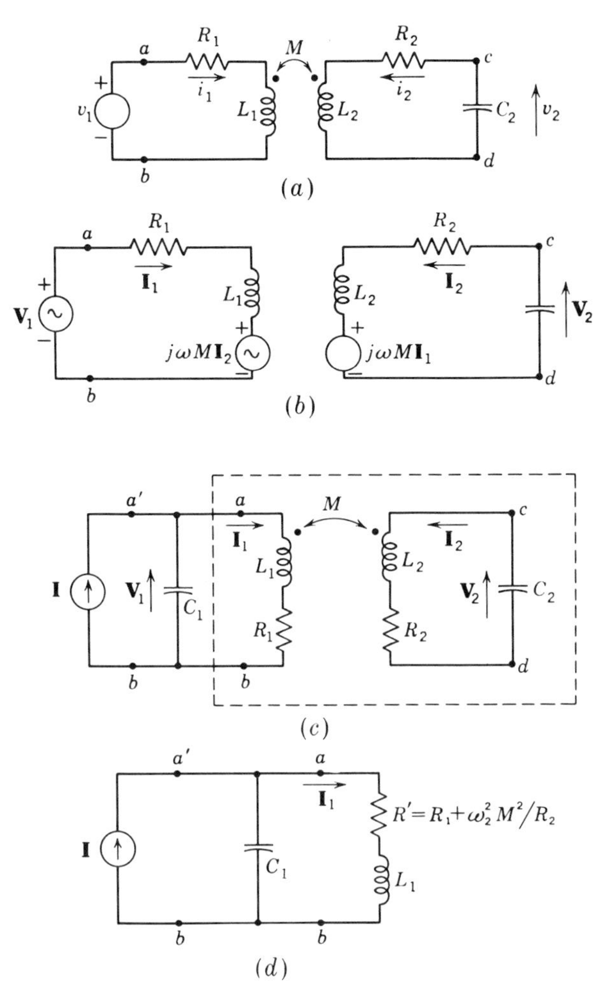

***Fig. 15-21** (a) Singly tuned transformer circuit. (b) Circuit of (a) in the sinusoidal steady state. The effect of the mutual inductance is represented by the dependent sources. (c) Doubly tuned transformer circuit. (d) Equivalent circuit of the primary mesh of (c) at the resonant frequency of the secondary.*

same *time* delay. Thus, if an input is $v_1(t) = V_m \cos \omega t$, the output v_2 in that band is to be $v_2(t) = HV_m \cos [\omega(t - t_d)]$, where H and t_d are constants. In terms of a complex transfer function this means that

$$\frac{\mathbf{V}_2(j\omega)}{\mathbf{V}_1(j\omega)} = H\underline{/-\omega t_d} \qquad \omega_1 < \omega < \omega_2$$

2 For frequencies outside of the band ω_1 to ω_2, it is desired that all sinusoids be attenuated (reduced in amplitude) as much as possible.

The reasons for the above requirements are discussed in Sec. 16-17. In Fig. 15-20*c* an "ideal" bandpass-filter characteristic is shown. In Sec. 10-6 we discussed an example of a bandpass filter, namely, the series *R-L-C* circuit. Comparison of Fig. 10-8*a* with Fig. 15-20*c* shows that, in the *R-L-C* circuit, the rate at which the amplitude response diminishes with frequency outside the passband (half-power points), depends on the parameter Q_0. In the *R-L-C* circuit, therefore, if the radian resonant frequency ω_0 as well as the half-power frequencies ω_1 and ω_2 are specified, then, since $Q_0 = \omega_0/(\omega_2 - \omega_1)$, we have no element which can be adjusted to "control" the attenuation outside the half-power frequencies.

A tuned transformer circuit, such as the circuit shown in Fig. 15-21*c*, is one example of bandpass filters in which the required flexibility of adjustment is available and frequency-response curves of the shape shown in Fig. 15-27*c* can be obtained.

The principles used to study the frequency response of circuits were discussed in Chaps. 7 and 10. We recall that the use of the pole-zero diagram is of great utility in this connection. Now we observe that the circuit of Fig. 15-21*c* contains four energy-storing elements. Consequently, we expect four poles to characterize its transfer functions. Since these poles are obtained by factoring a polynomial, the amount of labor involved in a general study of the circuit of Fig. 15-21*c* is quite large and very tedious. In this, as in subsequent sections, an introduction to tuned transformer circuits is presented; in this introduction certain simplifying assumptions concerning the relationship between the parameters is made. The use of these simplifying assumptions reduces the labor involved in the analysis and yet gives practical results because the assumed conditions are often (nearly) fulfilled.

The following analysis shows the necessity for optimizing certain network parameters to fulfill the above specifications. We start with the circuit shown in Fig. 15-21*a*, where a capacitance appears only across the "output" terminals. This circuit is called the "singly" tuned circuit. The sinusoidal steady-state conditions, using compensating sources, are shown in Fig. 15-21*b*. From mesh 2, we have

$$\frac{\mathbf{I}_2}{\mathbf{I}_1} = \frac{-j\omega M}{R_2 + j(\omega L_2 - 1/\omega C_2)}$$

and

$$\frac{\mathbf{V}_2}{\mathbf{I}_1} = \frac{-\mathbf{I}_2/j\omega C_2}{\mathbf{I}_1} = \frac{M/C_2}{R_2 + j(\omega L_2 - 1/\omega C_2)} \tag{15-8}$$

From Eq. (15-8) we conclude that, if an ideal current source is inserted in the primary, V_2 is a maximum at the frequency at which

$$\omega L_2 - \frac{1}{\omega C_2} = 0$$

independently of M. Thus $\mathbf{V}_2/\mathbf{I}_1 = M/R_2C_2$ when $\omega^2 = 1/L_2C_2$. At this frequency $V_2/I_1 = M/C_2R_2$. Hence, if I_1 is kept constant (i.e., if the source is an ideal current source), the smallest possible value of R_2 and the largest possible value of M are "favorable." Once these have been obtained, V_2 is proportional to I_1. Now, if an ideal current source furnishes the current whose phasor is $\mathbf{I}$, it is possible to use a second capacitance C_1 in the primary to increase I_1 above I, as shown in Fig. 15-21c. We shall now discuss the requirements for M and R_2 which, with C_1 in the circuit, would maximize V_2. We recall, from Sec. 15-5, that

$$\mathbf{Z}_{ab} = R_1 + j\omega L_1 + \frac{\omega^2 M^2}{\mathbf{Z}_{22}}$$

and in the present discussion ($\omega = 1/\sqrt{L_2C_2} \equiv \omega_2$), $\mathbf{Z}_{22} = R_2$. Hence

$$\mathbf{Z}_{ab} = R' + j\omega_2 L_1 \qquad R' = R_1 + \frac{\omega_2{}^2 M^2}{R_2}$$

as shown in Fig. 15-21d. In that circuit we use the current-division formula

$$\mathbf{I}_1 = \mathbf{I}\,\frac{1/j\omega_2 C_1}{R' + j\omega_2 L_1 + 1/j\omega_2 C_1}$$

Hence, if we choose $L_1C_1 = L_2C_2 = 1/\omega_2{}^2$,

$$\frac{\mathbf{I}_1}{\mathbf{I}} = \frac{1}{j\omega_2 R' C_1} = \frac{\sqrt{L_1/C_1}}{jR'}$$

so that, if $\sqrt{L_1/C_1}/R' > 1$, $I_1 > I$. Now we observe that this condition requires the smallest possible value of R'. Since $R' = R_1 + (\omega_2 M)^2/R_2$, this implies that a small M/R_2 is favorable. This is precisely the opposite of the "conclusion" reached at the outset!

This discussion shows that the choice of M (that is, k) as well as of the other circuit parameters for maximum V_2 is not obvious. In the following sections an introduction to quantitative treatment of this problem is given.

15-10 The singly tuned circuit

In this section we discuss the singly tuned circuit of Fig. 15-22a. In this circuit the mesh impedances are

$$Z_{11}(p) = R_1 + pL_1 \qquad Z_{22} = R_2 + pL_2 + \frac{1}{pC_2} \qquad Z_{12} = pk\sqrt{L_1L_2}$$

The mesh equation for mesh 2 gives

$$Z_{12}(p)i_1 + Z_{22}(p)i_2 = 0$$

If the source in the primary is an ideal current source $i_1(t)$, then

$$i_2(t) = -\frac{Z_{12}(p)}{Z_{22}(p)}\,i_1(t)$$

or

$$i_2(t) = -\frac{pk\sqrt{L_1L_2}}{R_2 + pL_2 + 1/pC_2}\,i_1(t)$$

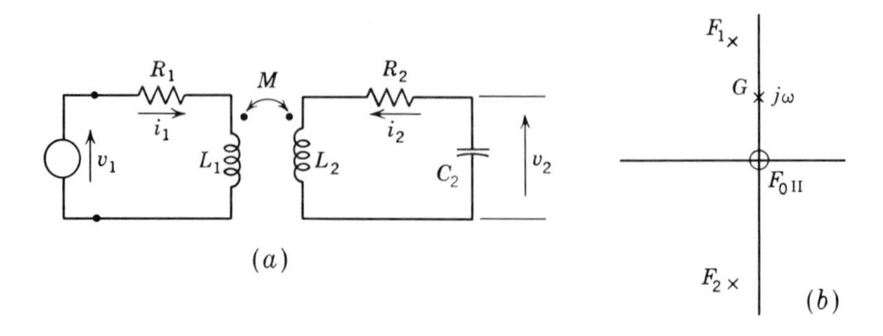

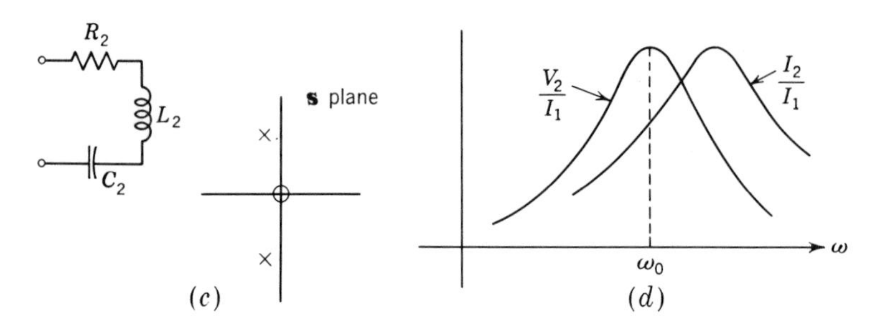

***Fig. 15-22** (a) Singly tuned circuit. (b) Pole-zero diagram for the transform current gain, i_2/i_1, assuming $Q_2 > \frac{1}{2}$. (c) Series R-L-C circuit and the pole-zero diagram of its transform driving-point admittance. (d) Frequency-response curves of the singly tuned circuit.*

The transform current gain function which relates i_2 to i_1 is denoted by $H_i(s)$.

$$H_i(s) = -\sqrt{\frac{L_1}{L_2}} \frac{s^2 k}{s^2 + (R_2/L_2)s + 1/L_2C_2} \tag{15-9}$$

We now define the radian resonant frequency of mesh 2 as $\omega_0 = 1/(L_2C_2)^{\frac{1}{2}}$ and its quality factor (its Q) as $Q_2 = \omega_0 L_2/R_2 = \sqrt{L_2/C_2}/R_2$ and write

$$H_i(s) = -\sqrt{\frac{L_1}{L_2}} \frac{s^2 k}{s^2 + (\omega_0/Q_2)s + \omega_0^2}$$

We observe that the current gain is proportional to the coefficient of coupling. The frequency response for any value of k is found from the pole-zero pattern. The function $H_i(s)$ has a double zero at $s = 0$ and poles at $s = s_{1,2}$, where s_1 and s_2 are the roots of the equation

$$s^2 + \frac{\omega_0 s}{Q_2} + \omega_0^2 = 0$$

Assuming the oscillatory case ($Q_2 > \frac{1}{2}$), a typical pole-zero diagram is shown in Fig. 15-22b. We now recall that, for a series R-L-C circuit, the transform driving-point admittance is (Fig. 15-22*c*)

$$Y_{ab}(s) = \frac{(1/L_2)s}{s^2 + (R_2/L_2)s + 1/L_2C_2}$$

The pole pattern which corresponds to this function is shown in Fig. 15-22*c*. In Chap. 10 it was shown that $Y_{ab}(j\omega)$ has a maximum at $\omega = \omega_0 = 1/\sqrt{LC}$. Since $H_i(j\omega)$ is proportional to the product of $j\omega$ and $Y_{ab}(j\omega)$, the maximum for $H_i(j\omega)$ will occur, not at $\omega = \omega_0$, but at a frequency above this value. This follows by observing that

$$|H_i(j\omega)| = A\omega|Y_{ab}|$$

where A is a proportionality constant. Hence

$$\frac{\partial|H_i|}{\partial\omega} = A|Y_{ab}| + \omega A\,\frac{\partial|Y_{ab}|}{\partial\omega}$$

Since $\partial|Y_{ab}|/\partial\omega$ at $\omega = \omega_0$ is zero, $\partial|H_i|/\partial\omega$ is positive at that frequency. Hence the derivative $\partial H_i/\partial\omega$ is zero at a frequency at which $\partial|Y_{ab}|/\partial\omega$ is negative; this point will be after the peak of Y_{ab}, that is, for $\omega > \omega_0$. Detailed analysis shows that $|H_i(j\omega)|$ has a maximum when

$$\frac{\omega^2}{\omega_0^2} = \frac{2Q_2^2}{2Q_2^2 - 1} \qquad Q_2^2 > \tfrac{1}{2}$$

If we study the transfer impedance v_2/i_1, we have $v_2 = (-1/pC_2)i_2$, or

$$v_2 = \frac{k\sqrt{L_1L_2}\,p\omega_0^2}{p^2 + (\omega_0/Q_2)p + \omega_0^2}\,i_1 \tag{15-10}$$

Note that the transfer impedance which relates v_2 to i_1 has its maximum at $\omega = \omega_0$. These results are illustrated in the frequency-response curves of Fig. 15-22*d*.

The study of the voltage ratio v_2/v_1 is more complex than the discussion above because the transform voltage-gain function has three poles which must be located as the solutions of a cubic equation. We can approach this problem fruitfully by considering, first, the lossless case shown in Fig. 15-23*a*. For this circuit,

$$\begin{aligned} pL_1i_1 \qquad\qquad &+ pMi_2 = v_1 \\ pMi_1 + \frac{L_2}{p}\,&(p^2 + \omega_0^2)i_2 = 0 \end{aligned}$$

where $\omega_0^2 = 1/L_2C_2$. We also have $i_2 = -pC_2v_2$. Hence

$$v_2 = \frac{k\sqrt{L_1L_2}/C_2}{L_1L_2(p^2 + \omega_0^2) - k^2p^2L_1L_2}\,v_1$$

or

$$v_2 = \sqrt{\frac{L_2}{L_1}}\,\frac{k\omega_0^2}{p^2(1 - k^2) + \omega_0^2}\,v_1 \tag{15-11}$$

The corresponding transform network function has a pole pair at

$$s = \frac{\pm j\omega_0}{\sqrt{1 - k^2}}$$

as shown in Fig. 15-23*b*. Hence, in the low-loss case ($R_{1,2} \ll \omega_0 L_{1,2}$), we expect maximum response near the frequencies $\omega_0/\sqrt{1 - k^2}$. Thus the frequency at which V_2/V_1 has a maximum value in general depends on the coefficient of coupling k.

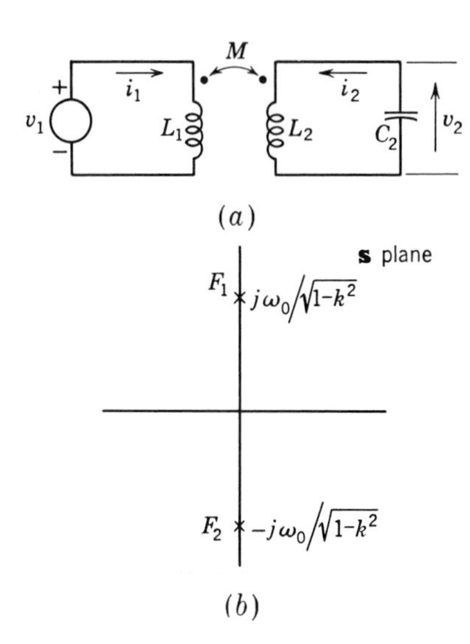

***Fig. 15-23** (a) Lossless singly tuned circuit. (b) Pole-zero diagram of the transform voltage gain v_2/v_1.*

Let us now consider the phasor ratio $\mathbf{V}_2/\mathbf{V}_1$ at any radian frequency $\omega = \mu\omega_0$. We have, from Eq. (15-11),

$$\frac{\mathbf{V}_2}{\mathbf{V}_1} = \sqrt{\frac{L_2}{L_1}} \frac{k}{(1 - \mu^2) + \mu^2 k^2} \tag{15-12}$$

Hence, for every value $\mu > 1$, there exists a special value of k which gives maximum response ratio (the "maximum" is infinite in the lossless case). In particular, if we choose $k = \sqrt{1 - 1/\mu^2}$, the expression (15-12) is infinite.

We now apply the ideas of the above discussion to the lossy case, Fig. 15-22*a*. For convenience in analysis we choose to find the value of k for which the voltage gain is a maximum at the resonant frequency of the secondary mesh. When $\omega = \omega_0 = 1/\sqrt{L_2 C_2}$, $\mathbf{Z}_{11} = R_1 + j\omega_0 L_1$, $\mathbf{Z}_{22} = R_2$, $\mathbf{Z}_{12} = j\omega_0 k \sqrt{L_1 L_2}$. Hence

$$\frac{\mathbf{V}_2}{\mathbf{V}_1} = \frac{-\mathbf{I}_2/j\omega_0 C_2}{\mathbf{V}_1} = \frac{k\sqrt{L_1 L_2}/C_2}{R_1 R_2 + j\omega_0 L_1 R_2 + \omega_0^2 k^2 L_1 L_2}$$

or

$$\frac{\mathbf{V}_2}{\mathbf{V}_1} = \sqrt{\frac{L_2}{L_1}} \frac{k}{1/Q_1 Q_2 + k^2 + j/Q_2} \tag{15-13}$$

where $Q_2 = \omega_0 L_2/R_2 = \sqrt{L_2/C_2}/R_2$ is the quality factor of the secondary mesh, and Q_1 is the reactance-resistance ratio of the primary at $\omega = \omega_0$. Differentiating V_2/V_1 with respect to k and setting the result equal to zero, we obtain $k = k_c$, where

$$k_c = \left(\frac{1}{Q_2}\sqrt{1 + \frac{1}{Q_1^2}}\right)^{\frac{1}{2}} \tag{15-14}$$

The value of k_c given by Eq. (15-14) is called the "critical coupling of the singly tuned circuit." It is that value of coupling which gives maximum voltage gain *at the resonant frequency of the secondary.*

The value given in Eq. (15-14) applies only when Q_2 and Q_1 are sufficiently large so that $k_c < 1$. If Eq. (15-14) gives $k_c > 1$, maximum voltage gain at ω_0 can be obtained with (as close to) unity coupling (as possible). A curve for $k_c < 1$ is shown in Fig. 15-24.

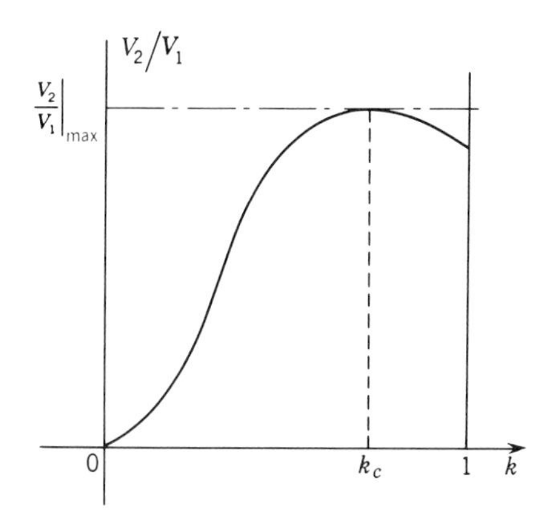

***Fig. 15-24** Voltage gain of the singly tuned circuit at the resonant frequency of the secondary as a function of coefficient of coupling.*

15-11 The doubly tuned circuit

We now turn to the doubly tuned circuit, excited by an ideal current source as shown in Fig. 15-25a. By representing this current source in par-

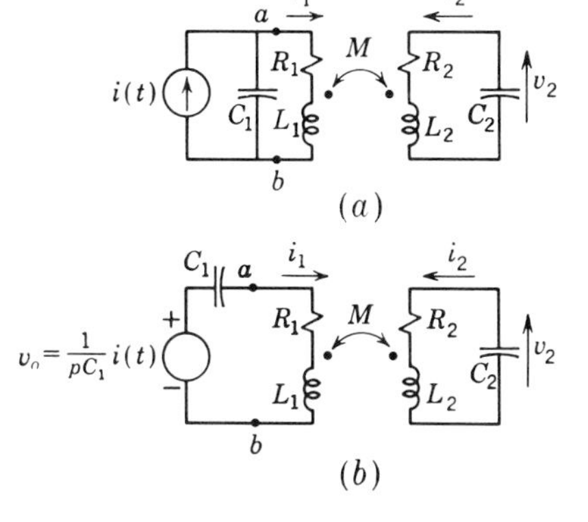

***Fig. 15-25** (a) Doubly tuned circuit excited by an ideal current source. (b) Circuit of (a) using voltage-source representation for the $i(t)$-C_1 combination.*

allel with C_1 as a voltage source, the circuit shown in Fig. 15-25*b* results. In order to simplify the analysis, we assume that $1/L_1C_1 = 1/L_2C_2 = \omega_0^2$; that is, the resonant frequency of both primary and secondary is the same frequency. (This assumption is fulfilled in many practical circuits.) The mesh equations are

$$Z_{11}i_1 + Z_{12}i_2 = v_o(t)$$
$$Z_{12}i_1 + Z_{22}i_2 = 0$$

where
$$Z_{11}(p) = \frac{L_1}{p}\left(p^2 + \frac{R_1}{L_1}p + \omega_0^2\right) \qquad Z_{12}(p) = pk\sqrt{L_1L_2}$$
$$Z_{22}(p) = \frac{L_2}{p}\left(p^2 + \frac{R_2}{L_2}p + \omega_0^2\right) \qquad v_o(t) = \frac{1}{pC_1}i(t)$$

Conditions at the resonant frequency $f_0 = \omega_0/2\pi$ To study the relationships between the phasors at the resonant frequency, we replace p with $j\omega_0$ and define $Q_1 = \omega_0L_1/R_1$, $Q_2 = \omega_0L_2/R_2$. Then

$$\mathbf{Z}_{11} = R_1 \qquad \mathbf{Z}_{22} = R_2 \qquad \mathbf{Z}_{12} = jk\sqrt{R_1R_2Q_1Q_2} \qquad \mathbf{V}_o = \frac{1}{j\omega_0C_1}\mathbf{I}$$

and
$$\mathbf{Z}_{11}\mathbf{Z}_{22} - \mathbf{Z}_{12}^2 = R_1R_2 + k^2R_1R_2Q_1Q_2 = R_1R_2(1 + k^2Q_1Q_2)$$

Hence

$$\frac{\mathbf{I}_2}{\mathbf{I}} = -\sqrt{\frac{L_1}{L_2}}\frac{k}{1/Q_1Q_2 + k^2} \tag{15-15}$$

Differentiating Eq. (15-15) with respect to k and setting the result to zero gives $k = k_c$, where

$$k_c = \frac{1}{\sqrt{Q_1Q_2}} \tag{15-16}$$

The value k_c given by Eq. (15-16) is that value of coupling which produces maximum current gain at the resonant frequency and which is called "critical coupling for the doubly tuned circuit." Since $\mathbf{V}_2 = -\mathbf{I}_2/j\omega_0C_2$, this value of coupling also gives maximum transfer impedance V_2/I at $\omega = \omega_0$.

Frequency response Before considering the general case, the lossless circuit of Fig. 15-26*a* is analyzed. We again assume $L_1C_1 = L_2C_2$; then

$$Z_{11}(p) = (L_1/p)(p^2 + \omega_0^2) \qquad Z_{22}(p) = (L_2/p)(p^2 + \omega_0^2)$$
$$Z_{12}(p) = pk\sqrt{L_1L_2}$$

Solving the mesh equations,

$$i_2 = \frac{-p^2k\sqrt{L_1L_2}/C_1}{L_1L_2(p^2 + \omega_0^2)^2 - p^4k^2L_1L_2}\,i$$

or
$$i_2 = -k\sqrt{\frac{L_1}{L_2}}\,\omega_0^2\,\frac{p^2}{(p^2 + \omega_0^2)^2 - p^4k^2}\,i$$

Thus

$$i_2 = -k\sqrt{\frac{L_1}{L_2}}\,\omega_0{}^2 \frac{p^2}{(p^2 + \omega_0{}^2 + p^2k)(p^2 + \omega_0{}^2 - p^2k)}\,i \qquad (15\text{-}17)$$

and $\quad v_2 = -\dfrac{1}{pC_2}\,i_2$

Both these network functions have poles at

$$s^2(1 \pm k) + \omega_0{}^2 = 0$$

or, if $k \neq 1$,

$$s = \pm \frac{j\omega_0}{\sqrt{1+k}} \qquad s = \pm \frac{j\omega_0}{\sqrt{1-k}}$$

The transform current gain which relates i_2 to i has a double zero at $s = 0$, whereas the transfer impedance v_2/i has a simple zero. The pole-zero dia-

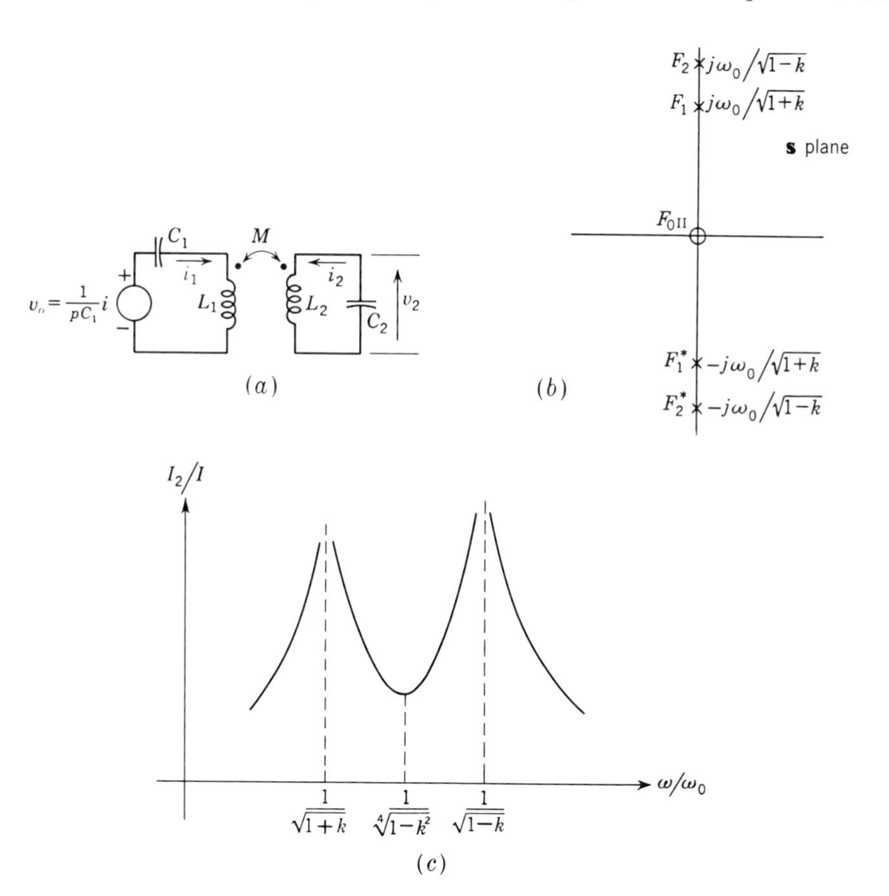

***Fig. 15-26** (a) Lossless version of the circuit of Fig. 15-25b. (b) Pole-zero diagram for the transform current gain, i_2/i_1, of the lossless doubly tuned circuit. (c) Current gain as a function of frequency for the pole-zero pattern of (b).*

gram for the current gain is shown in Fig. 15-26*b*. We observe that the response of the lossless circuit is maximum (infinite) at the pole frequencies. In the slightly (high-Q) lossy case, we expect that the poles move off the $j\omega$ axis, and maxima occur near the pole frequencies of the lossless circuit. In addition, we note that the current ratio will have its maxima at frequencies above those of the transfer impedance because of the double zero at the origin in the former case. The results for the lossless case are illustrated in Fig. 15-26*c*. We note that, as $k \to 0$, a double pole forms at $s = \pm j\omega_0$. Thus critical coupling for the lossless case ($Q_1Q_2 = \infty$) is zero coupling.

Although at first glance it may seem strange that maximum response does not occur at ω_0, the following physical reasoning explains this phenomenon. At resonance, $Z_{22} = 0$ in the lossless case. Hence the impedance of the secondary, reflected into the primary, $\omega_0^2M^2/Z_{22}$, is infinite. Thus i_1 can only be infinitesimal, so that an infinitesimal voltage pMi_1 is induced in the secondary. This produces finite current in the zero impedance.

We now apply this result to the lossy case. The expected pole-zero diagram is shown in Fig. 15-27*a* for the current-gain function. For any frequency the source is represented by a pole as at G. We now assume that, near ω_0, F_1^*G, F_0G, and F_2^*G are constant. Hence, near ω_0, the current gain function H_i $[i_2 = H_i(p)i]$ is proportional to $1/(F_1G)(F_2G)$. In Fig. 15-27*b* an enlarged portion of the pole-zero diagram, showing the region near $s = j\omega_0$, is given. Now, as ω moves from ω_A to ω_B, F_1G_1 increases and F_2G_1 decreases. It follows that α_1 and α_2, as well as k (which determines $\omega_B - \omega_A$), will determine whether the product $(F_1G_1)(F_2G_1)$ decreases or increases between ω_A and ω_B. We therefore conclude that the frequency-response curve for I_2/I and V_2/I may have a maximum or a minimum between ω_A and ω_B. If a maximum occurs, it is the only maximum. It can be shown that, in the lossy case (finite Q), the frequency-response curve for I_2/I has only one peak and that peak is at ω_0 if:

1. The coefficient of coupling is less than or equal to the value of k_c given in Eq. (15-16), provided that $Q_1 = Q_2$; that is, $k_c \leq 1/Q_1 = 1/Q_2$.
2. The coefficient of coupling is less than (or equal to) a value k_t (called "transitional coupling") if $Q_1 \neq Q_2$. The value of k_t is given by

$$k_t = \frac{\sqrt{Q_1^2 + Q_2^2}}{\sqrt{2Q_1Q_2}}\,k_c$$

The curves of Fig. 15-27*c* illustrate these results.

If $Q_1 = Q_2$, the poles of the transfer functions can easily be located exactly. Referring to Fig. 15-25*b*, we have, if $L_1C_1 = L_2C_2 = 1/\omega_0^2$,

$$Z_{11}(p) = pL_1 + R_1 + \frac{1}{pC_1} = \frac{L_1}{p}\left(p^2 + \frac{\omega_0}{Q_1}p + \omega_0^2\right)$$

$$Z_{22}(p) = pL_2 + R_2 + \frac{1}{pC_2} = \frac{L_2}{p}\left(p^2 + \frac{\omega_0}{Q_2}p + \omega_0^2\right)$$

$$Z_{12}(p) = pk\sqrt{L_1L_2}$$

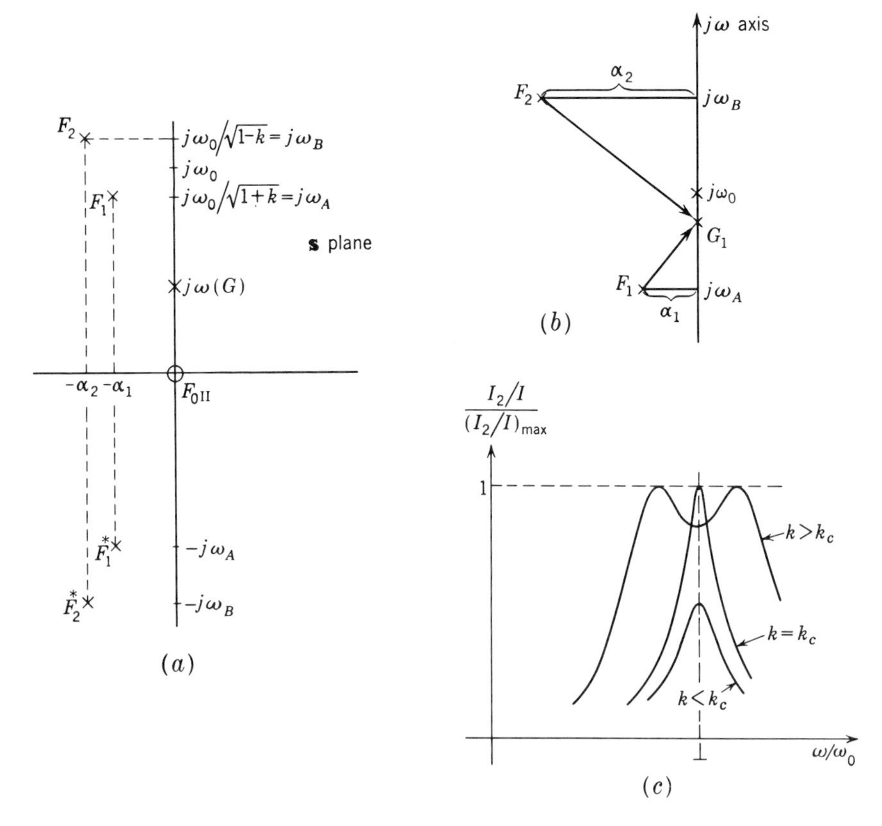

***Fig. 15-27** (a) Pole-zero diagram for the transform current gain in a low-loss (high-Q) doubly tuned circuit: $\alpha_1/\omega_0 \ll 1$, and $\alpha_2/\omega_0 \ll 1$. (b) Region of the s plane near $s = j\omega_0$. (c) Current gain as a function of frequency with coefficient of coupling as a parameter for the high-Q case, $Q_1 = Q_2 \gg 1$.*

Hence, if $Q_1 = Q_2 = Q_0$, we have

$$p^2(Z_{11}Z_{22} - Z_{12}^2) = L_1L_2\left[\left(p^2 + \frac{\omega_0}{Q_0}p + \omega_0^2\right)^2 - k^2p^4\right]$$

and
$$i_2 = -k\omega_0^2\sqrt{\frac{L_1}{L_2}}\frac{p^2}{[p^2 + (\omega_0/Q_0)p + \omega_0^2]^2 - k^2p^4}\,i(t)$$

Thus the poles are located as the roots of

$$\left[\left(s^2 + \frac{\omega_0}{Q_1}s + \omega_0^2\right) + ks^2\right]\left[\left(s^2 + \frac{\omega_0}{Q_0}s + \omega_0^2\right) - ks^2\right] = 0$$

at the four locations:

$$s_p = \omega_0\left[-\frac{1}{2Q_0(1 \pm k)} \pm \sqrt{\frac{1}{4Q_0^2(1 \pm k)^2} - \frac{1}{1 \pm k}}\right]$$

Or, if Q_0 is sufficiently large so that the discriminant is negative,

$$\mathbf{s}_p = \frac{-\omega_0}{2Q_0(1 \pm k)} \pm j \frac{\omega_0}{\sqrt{1 \pm k}} \sqrt{1 - \frac{1}{4Q_0^2(1 \pm k)}}$$

Hence, for this case in Fig. 15-27*b*,

$$\alpha_1 = \frac{-\omega_0}{2Q_0(1 + k)} \qquad \omega_A = \frac{\omega_0}{\sqrt{1 + k}} \sqrt{1 - \frac{1}{4Q_0^2(1 + k)}}$$

$$\alpha_2 = \frac{-\omega_0}{2Q_0(1 - k)} \qquad \omega_B = \frac{\omega_0}{\sqrt{1 - k}} \sqrt{1 - \frac{1}{4Q_0^2(1 - k)}}$$

We observe again that the values ω_A and ω_B correspond very closely to the values obtained in the lossless case (Fig. 15-26*b*) if $Q_0 \gg 1$. The frequency-response curves can now be obtained by the method illustrated in Chap. 10, that is, by solving the geometry of Fig. 15-27*b*. This matter and related topics such as calculation of bandwidth is not pursued here because the main object of the preceding discussion has been to show how the principles of the preceding chapters can be applied to explain the behavior (e.g., double peaks) of the tuned circuits. Further details (relative height of the two peaks, bandwidth, etc.) are found in the applications literature.[1] Example E-2 in Appendix E illustrates the use of a digital computer for problems of this algebraic complexity.

PROBLEMS

15-1 In the circuit shown in Fig. P15-1 a 100-volt (rms) sinusoidal source is applied between terminals as indicated in the tabulation below. Connections are made

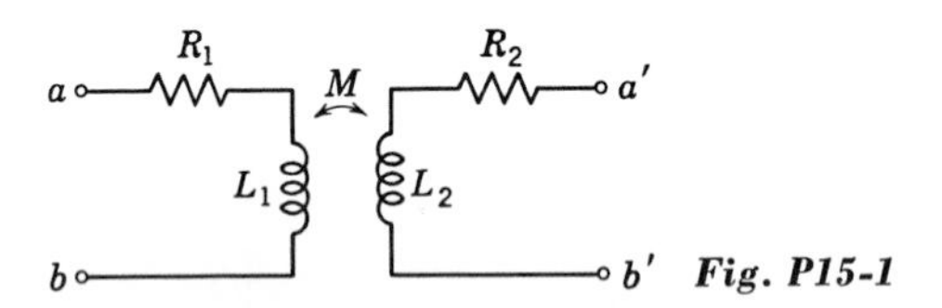

Fig. P15-1

as indicated, and the results of several measurements at the input terminals are also given. The polarity dots are omitted, and their location may be deduced from the data below. (*a*) Calculate R_1, R_2, and the coefficient of coupling k. (*b*) Calculate L_1 and L_2 if the frequency is 60 hertz.

[1] See, for example, F. E. Terman, "Radio Engineering," chap. 3, particularly secs. 3–4, McGraw-Hill Book Company, New York, 1947; E. J. Angelo, Jr., "Electronic Circuits," pp. 350–375, McGraw-Hill Book Company, New York, 1958; or G. E. Valley, Jr., and H. Wallman (eds.), "Vacuum Tube Amplifiers," vol. 18, Massachusetts Institute of Technology Radiation Laboratory Series, chap. 5, pp. 201–221, McGraw-Hill Book Company, New York, 1948.

Apply source between	Connect	Rms, amp	Avg power, watts
a-a'	b to b'	7.07	500
a-b'	b to a'	0.90	Not measured
a-b	a' and b not connected	4.90	96

15-2 For the terminal pair shown in Fig. P15-2: (*a*) Calculate the equivalent inductance L_{ab} as a function of the coefficient of coupling k. (*b*) Let L_0 be the value of L_{ab} when $k = 0$. Plot the ratio L_{ab}/L_0 as a function of k for (1) $L_2 = L_1$; (2) $L_2 = 2L_1$; (3) $L_2 = 10L_1$ for $-1 \leq k \leq +1$.

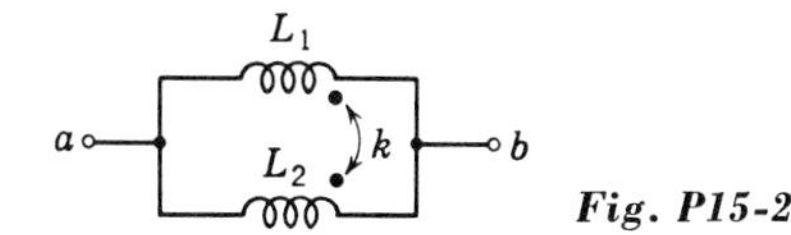

Fig. P15-2

15-3 Sketch the reactance- and susceptance-frequency curves for the pure reactance network given in Fig. P15-3.

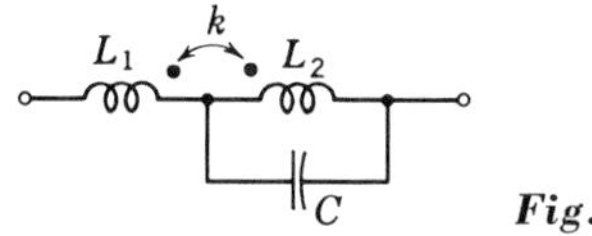

Fig. P15-3

15-4 If $v(t)$ is the ramp function $v(t) = Vt/T_1u(t)$: (*a*) Calculate v_{ab} for all $t \geq 0^+$, assuming no initial-energy storage. (*b*) Discuss the usefulness of this circuit as a differentiator. The circuit is shown in Fig. P15-4. (*c*) Calculate $v_{ab}(t)$ in the steady state if $v(t) = V_m \cos \omega t$.

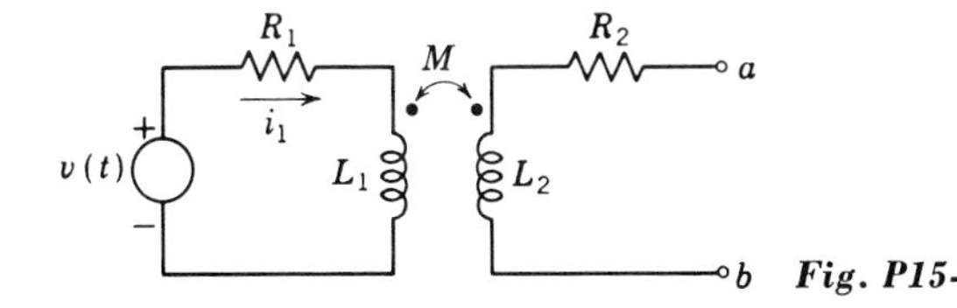

Fig. P15-4

15-5 Write state equations for (*a*) the circuit of Fig. P15-4 with terminals *a*-*b* short-circuited; (*b*) the circuit of Fig. P15-9.

15-6 In the circuit of Fig. P15-4 terminals *a*-*b* are short-circuited. It is known that $L_1/R_1 = L_2/R_2 = 1$. Solve for the complete response $i_{ab}(t)$ (current through the short circuit) for $t \geq 0^+$ if $|k| < 1$ and if the initial-energy storage is zero. The source is a unit step function.

15-7 In the circuit of Fig. P15-4 terminals *a*-*b* are short-circuited. The current in each mesh at $t = 0^-$ is zero. Let $R_1 = 1$, $R_2 = 2$, $L_1 = L_2 = 2$, $k = 1$. The source is $u(t)$. (*a*) Show that $i_{ab}(0^+) \neq 0$ and calculate its value. *Hint:* Use the initial-value theorem of Laplace transform. (*b*) Calculate $i_1(0^+)$. (*c*) Justify your answers to parts *a* and *b* using continuity of stored energy. (*d*) Obtain the complete response for $i_1(t)$ and $i_{ab}(t)$ for $t \geq 0^+$.

15-8 In the circuit of Fig. P15-8: (*a*) Calculate the mesh impedances $Z_{11}(p)$, $Z_{12}(p)$, $Z_{22}(p)$. (*b*) If $L_1 = L_2 = 0.1$ henry, $L_3 = 0.2$ henry, $M_{12} = 0.05$ henry, $M_{23} = 0.1$ henry, $M_{13} = -0.10$ henry, and $R_1 = R_2 = R_3 = 50$ ohms, calculate the complex mesh impedances at a radian frequency of 1,000/sec.

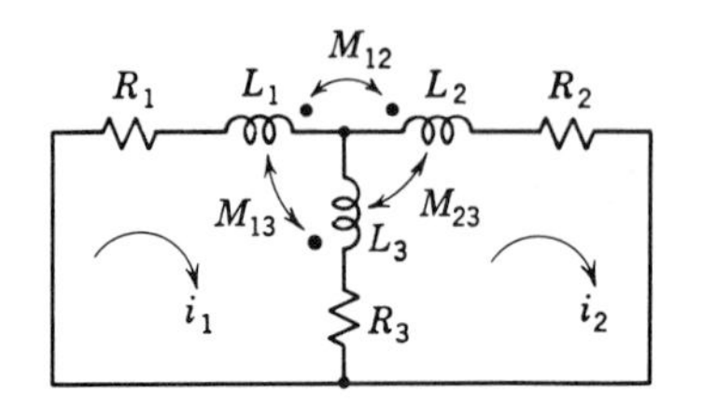

Fig. P15-8

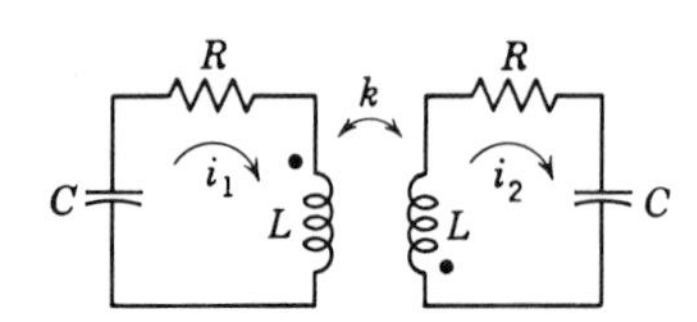

Fig. P15-9

15-9 Write the characteristic equation for the mesh currents, and deduce the condition(s) for nonoscillatory transient components in the circuit of Fig. P15-9. ($M = kL$, $0 < k < 1$.)

15-10 In the circuit of Fig. P15-10, $R_1 = 1.0$ ohm, $R_2 = 10$ ohms, $X_{L1} = 9.0$ ohms, $X_{L2} = 90$ ohms, $k = 0.8$, $X_c = -50$ ohms, $R_T = 50$ ohms. Calculate $\mathbf{V}_{ab}$.

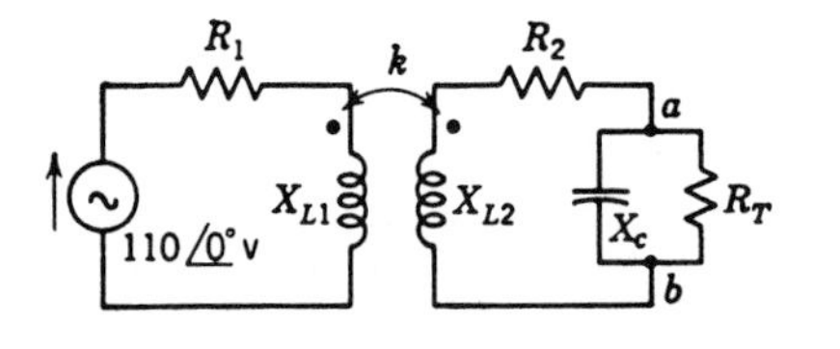

Fig. P15-10

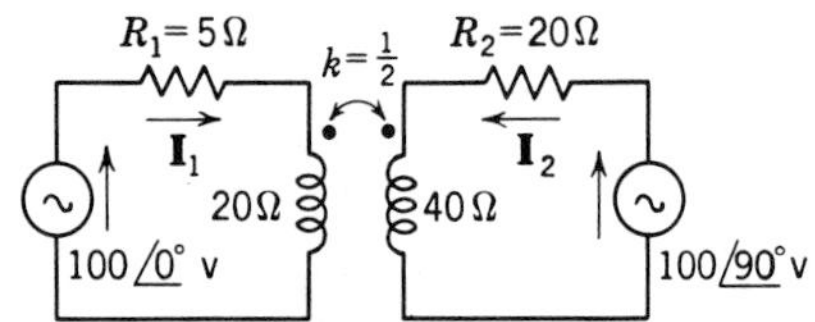

Fig. P15-11

15-11 The circuit shown in Fig. P15-11 is in the sinusoidal steady state. Calculate (*a*) $\mathbf{I}_1$, $\mathbf{I}_2$; (*b*) the power delivered to R_1 and the power delivered to R_2; (*c*) the power delivered by each ideal source.

15-12 In the circuit of Fig. P15-4, $R_1 = 2$ ohms, $L_1 = 0.01$ henry, $R_2 = 20$ ohms, $L_2 = 0.1$ henry, $M = 0.03$ henry. A load whose equivalent circuit is the parallel combination of a 100-ohm resistance and a 12.5-μf capacitance is connected across terminals *a-b*. If $v_{ab} = 100\sqrt{2}\cos 400t$, calculate the source function $v(t)$.

15-13 In the circuit of Fig. P15-4, $R_1 = 2$ ohms, $L_1 = 0.01$ henry, $R_2 = 20$ ohms, $L_2 = 0.1$ henry, $M = 0.03$ henry. If $v(t) = 100\sqrt{2}\cos 400t$, deduce Thévenin's equivalent circuit for sinusoidal steady-state analysis ($\mathbf{V}_o$ and $\mathbf{Z}_s$) with respect to terminals *a-b*.

15-14 If $L_1 = 16L_2$ and if $k = 1$, obtain Thévenin's equivalent circuit with respect to terminals *c-d* in the "autotransformer" circuit of Fig. P15-14.

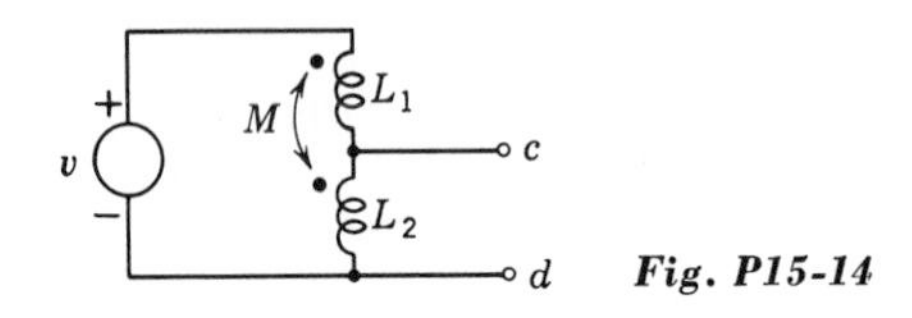

Fig. P15-14

15-15 In the (induction regulator) shown in Fig. P15-15, k is adjustable. Obtain Thévenin's equivalent circuit with respect to terminals a-b.

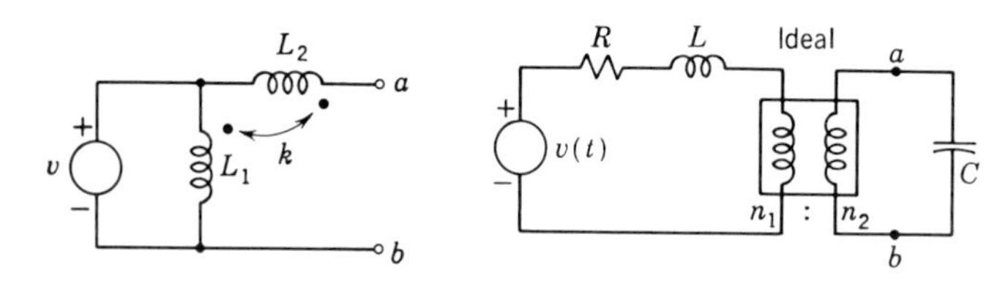

Fig. P15-15 ***Fig. P15-16***

15-16 In the circuit shown in Fig. P15-16, $R = 10$ ohms, $L = 0.001$ henry, $C = 10^{-6}$ farad, $n_1/n_2 = a$. Calculate the value of a so that the circuit is (*a*) critically damped; (*b*) oscillatory, with two cycles of decaying oscillations per time constant of envelope decay; (*c*) overdamped, with one root of the characteristic equation equal to twice the undamped (natural) radian frequency.

15-17 A voltage source has an internal resistance of 10,000 ohms. It is to supply a load resistance, value 10 ohms. Calculate (*a*) the turns ratio of an ideal transformer which will ensure that maximum power is transferred to the 10-ohm resistance; (*b*) the ratio of the power transferred in (*a*) to the power which would be transferred to the 10-ohm resistance if this resistance were connected directly across the terminals of the resistive generator.

15-18 What is a mechanical analogue of an ideal transformer?

15-19 (*a*) In the circuit of Fig. P15-16 use the numerical values $R = 0.2$, $L = 0.1$, $C = 1.25$, $n_1/n_2 = \frac{1}{2}$, and calculate the complete response for $v_{ab}(t)$ if no energy is stored at $t = 0^-$ and if the source is $v(t) = u(t)$. (*b*) Calculate the value of a resistance to be placed in parallel with C across terminals a-b so that the circuit is critically damped.

15-20 Calculate the power delivered by each source and the power dissipated by each resistance in the circuit of Fig. P15-20.

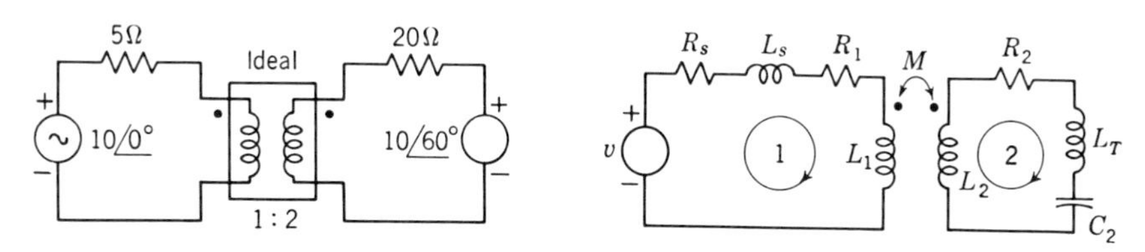

Fig. P15-20 ***Fig. P15-21***

15-21 Although the coefficient of coupling is usually defined as the ratio of mutual inductance to the geometric mean of two inductances ("coil self-inductances"), that is, $k = M/\sqrt{L_1L_2}$, one can also define k as the ratio of mutual inductance to the geometric mean of two mesh inductances, that is, $k = M/\sqrt{L_{11}L_{22}}$. If the latter definition is used in the circuit shown in Fig. P15-21, (*a*) obtain an expression for k; (*b*) calculate k numerically if $4L_1 = L_2$, $M = 1.5L_1$, and (1) $L_s = L_T = 0$, (2) $L_s = L_T = L_1$, (3) $L_s = L_1$, $L_T = L_2$.

15-22 In the two networks shown in Fig. P15-22: (*a*) Show that the transform current gain i_2/i is the same for both. (*b*) Calculate the values of the elements for each network if $R = 600$ ohms and if the pole is at $s = -5{,}000$. (*c*) Sketch the amplitude and phase response for $\mathbf{I}_2/\mathbf{I}$ in the sinusoidal steady state.

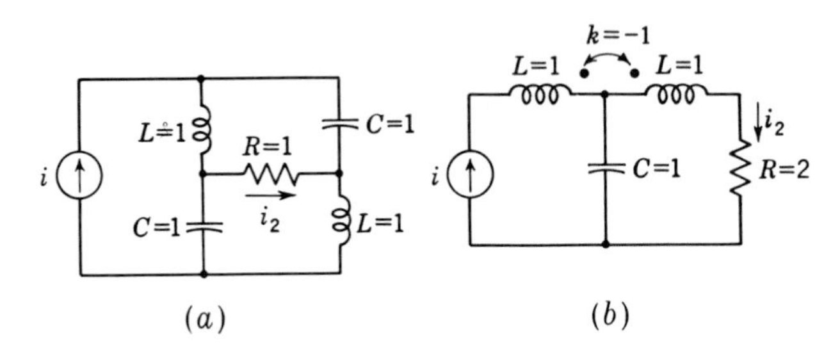

Fig. P15-22

15-23 Calculate the transform driving-point impedance for each of the networks of Fig. P15-22.

15-24 In the text, critical coupling is defined as that condition which maximizes the absolute value of a transfer function at resonance. It is the purpose of this problem to show that this definition, although useful, is arbitrary because, in general, the value of k which maximizes a transfer function depends on the frequency. (*a*) Consider a complex transfer function of the form $\mathbf{H} = \mathbf{A}[k/(\mathbf{B} + k^2)]$, where k is adjustable, and show that $H_{\max}$ is obtained when $k^2 = B$. (*b*) Show that the form given in (*a*) applies to a singly or doubly tuned circuit if an ideal voltage source of radian frequency ω, whose phasor is $\mathbf{V}$, is impressed and if the response is either $\mathbf{V}_2$ or $\mathbf{I}_2$. Find $\mathbf{A}$ and $\mathbf{B}$ for these cases. (*c*) Show that the form given in (*a*) applies to a doubly tuned circuit with an ideal current source impressed. Find $\mathbf{A}$ and $\mathbf{B}$ for the transfer functions $\mathbf{V}_2/\mathbf{I}$ and $\mathbf{I}_2/\mathbf{I}$.

15-25 To explain why the coupling for maximum gain is a function of frequency (Prob. 15-24), consider the lossless singly tuned circuit shown in Fig. P15-25. (*a*) Use

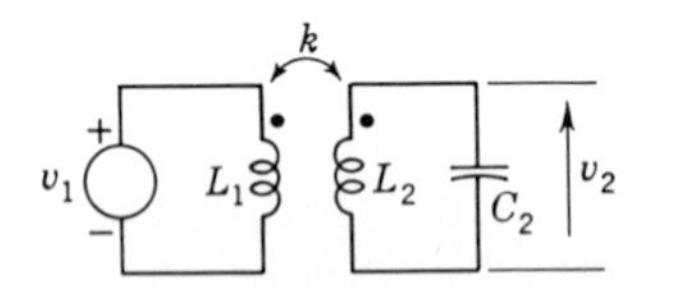

Fig. P15-25

the numerical values $L_1 = 1 = L_2$, $C_2 = 1$, and locate the zeros and poles of the transform transfer function which relates v_2 to v. (*b*) If the circuit is low-loss, how is the coefficient of coupling related to the frequency at which V_2/V is a maximum? (*c*) Calculate the location of the poles and the value of C_2 if the resonant frequency of the secondary is 455 k-hertz, if $L_1 = L_2 = 1.0$ mh, and if (1) $k = 0.8$, (2) $k = 0.1$.

15-26 In the singly tuned circuit of Fig. P15-25, $L_1 = L_2 = 1$, $C_2 = 1$, and a resistance $R_2 = 0.02$ is in series with L_2. The voltage source is replaced by a current source i in parallel with a 50-ohm resistance. The circuit is in the sinusoidal steady state. (*a*) Use the result of Prob. 15-24 to determine the relationship between k and ω so that at each frequency V_2/I is a maximum (assume $\mathbf{Z}_{11} = 50$).

(*b*) Calculate approximately the range of frequencies for which a value of k which is less than unity gives maximum transfer impedance V_2/I.

15-27 Consider the geometry given in Fig. P15-27. It is shown in Prob. 15-28 that this geometry can be used to deduce the frequency response of a high-Q doubly

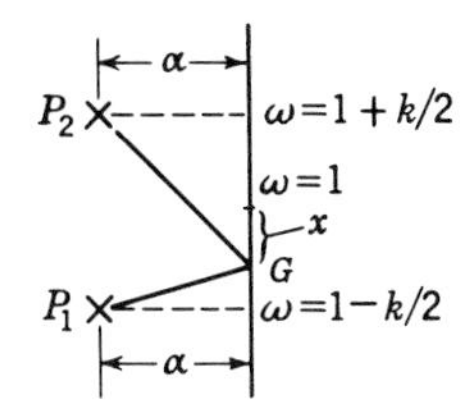

Fig. P15-27

tuned circuit. (*a*) Show that the square of the product of the distances $(P_1G)^2(P_2G)^2$ is given by $(P_1G)^2(P_2G)^2 = \alpha^4 + 2\alpha^2(k^2/4 + x^2) + (k^2/4 - x^2)^2$. (*b*) Show that, if $\alpha > k/2$, the product $(P_1G)(P_2G)$ is a minimum at $x = 0$. (*c*) Show that, if $\alpha < k/2$, the product $(P_1G)(P_2G)$ is a maximum at $x = 0$ and has two minima at $x = \pm(\frac{1}{4}k^2 - \alpha^2)^{\frac{1}{2}}$.

15-28 (*a*) In the symmetrical, doubly tuned circuit of Fig. P15-28, show that, if $L = 1$, $C = 1$, $R < 1$, the poles of the transform transfer functions, i_2/i and v_2/i, are at

$$s = -\frac{R/2}{1 \pm k} \pm \sqrt{\frac{R^2/4}{(1 \pm k)^2} - \frac{1}{1 \pm k}}$$

(*b*) Show that, for $R \ll 1$ and $k \ll 1$, the poles are approximately at $s = -R/2 \pm j(1 \mp k/2)$. (*c*) Use the approximate pole-zero diagram and the result of Prob. 15-27 to show that the frequency-response curves for V_2/I and I_2/I have two peaks at approximately $\omega = 1 \pm \frac{1}{2}k$ if $k > R$ and only one peak at $\omega = 1$ for $k < R$. (*d*) Show that the condition $k = R$ gives the value of critical coupling $k = k_c = 1/\sqrt{Q_1Q_2}$. (*e*) Calculate the two radian frequencies at which the I_2/I has the same magnitude as at $\omega = 1$ if $k = 2R = 0.1$.

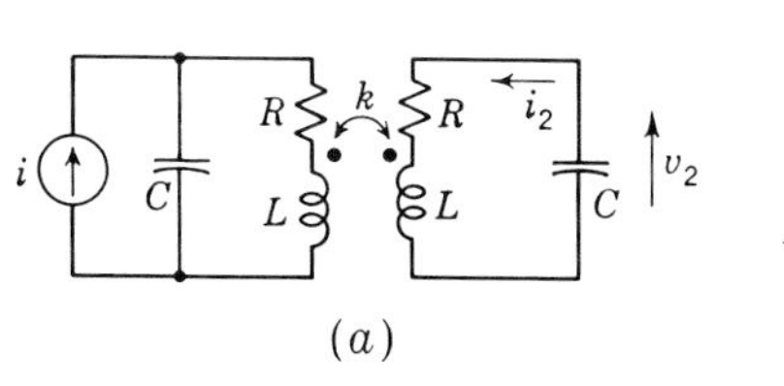

ω
$1-\frac{k}{\sqrt{2}}$ $1-\frac{k}{2\sqrt{2}}$ 1 $1+\frac{k}{2\sqrt{2}}$ $1+\frac{k}{\sqrt{2}}$
(b)

Fig. P15-28

15-29 In the doubly tuned circuit of Fig. P15-28, the coefficient of coupling is 0.03, and $R = 0.02$. (*a*) Calculate the location of the poles of the transform network function v_2/i. (*b*) Calculate the two radian frequencies at which V_2/I has the same value as at $\omega = 1$. (*c*) Change the frequency scale and the impedance level so that the resonant frequency is 455 k-hertz and $L = 1.0$ mh (calculate C and R). (*d*) Calculate for the circuit of (*c*) the two frequencies at which the response ratios are the same as at 455 k-hertz.

Chapter 16 Harmonic analysis and Fourier series

In this chapter we shall show that a function which is the sum of sinusoidal functions whose frequencies are integral multiples of each other is a periodic function and that (under certain conditions) a periodic function can be represented as the sum of a finite or infinite number of sinusoidal functions. By such formulation we shall be able to represent periodic source functions as the sum of sinusoidal source functions. Hence the component of the response due to periodic source functions can be obtained as the sum (superposition) of sinusoidal functions. In addition, we shall find that the concept of representing a waveform as the sum of a number of sinusoidal functions can be developed into a powerful mathematical tool which will shift the emphasis from the time variation of waveforms (time-domain analysis) to its frequency components (frequency-domain analysis).

16-1 Periodic functions

If two sinusoidal functions $f_1(t) = B_1 \sin \omega t$ and $f_2(t) = B_2 \sin 2\omega t$ are added to give $f(t)$,

$$f(t) = B_1 \sin \omega t + B_2 \sin 2\omega t$$

then $f(t)$ is periodic with period $T = 2\pi/\omega$, because

$$f\left(t + \frac{2\pi}{\omega}\right) = B_1 \sin (\omega t + 2\pi) + B_2 \sin (2\omega t + 4\pi) = f(t)$$

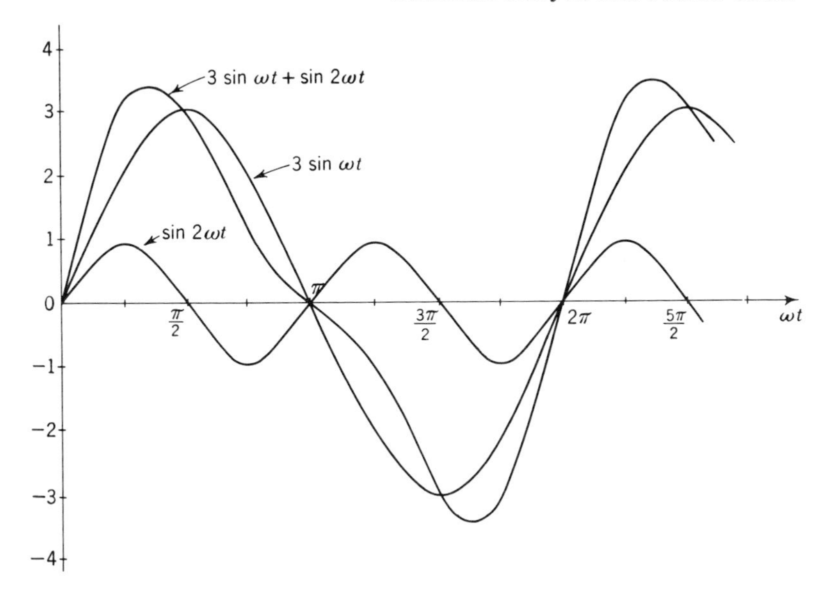

***Fig. 16-1** The addition of the functions 3 sin ωt and sin 2ωt. The resultant has the period 2π/ω.*

In Fig. 16-1 this addition is illustrated for the case $B_1 = 3$, $B_2 = 1$. The function $f(t)$ is seen to be periodic with period $2\pi/\omega$, but not sinusoidal. A little thought will show that the function

$$f(t) = C_0 + C_1 \cos(\omega t + \varphi_1) + C_2 \cos(2\omega t + \varphi_2) + \cdots + C_n \cos(n\omega t + \varphi_n) \quad (16\text{-}1)$$

is also periodic with period $2\pi/\omega$ if n is an integer. In the following sections we shall show that periodic functions can be represented as the sum of sinusoidal functions in the form of Eq. (16-1); this sum can have an infinite or a finite number of terms. For practical purposes a finite number of terms can be used to approximate a given periodic function to any desired degree of accuracy. Thus, for example, the waveform of Fig. 16-2*a* can be approximated by the sum of a constant term and the four sinusoidal terms given in Eq. (16-2).

Before discussing the procedure which is followed when it is desired to represent a periodic function in the form of Eq. (16-1), it is important to realize the purpose of such a procedure. Let us therefore assume that the waveform of Fig. 16-2*a* is the waveform of an ideal voltage source, and is approximated by

$$v(t) = \frac{3}{2} - \frac{3}{\pi}\left(\sin\frac{2\pi}{3}t + \frac{1}{2}\sin\frac{4\pi}{3}t + \frac{1}{3}\sin\frac{6\pi}{3}t + \frac{1}{4}\sin\frac{8\pi}{3}t\right) \quad (16\text{-}2)$$

where the period of the waveform is 3. Suppose that this waveform is applied to the series combination of a 1-ohm resistance and a $3/\pi$-henry inductance. Let it be required to find the steady-state current in this circuit.

From the superposition theorem we know that the response to several sources is the

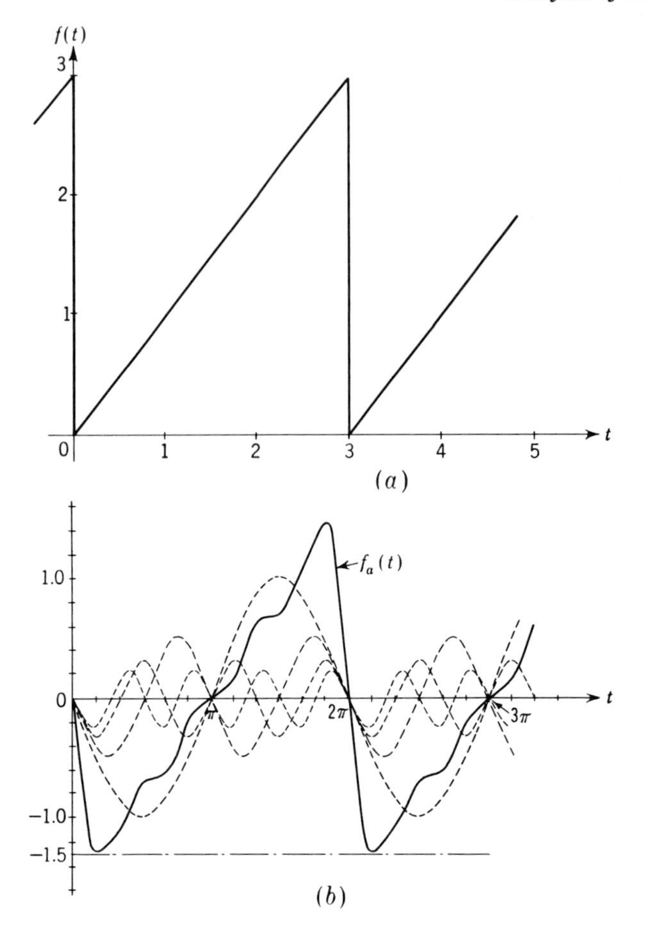

***Fig. 16-2** (a) A periodic triangular waveform. (b) The waveform $f_a(t)$ can be used to approximate the triangular waveform of (a) if the $f = 0$ line is shifted to -1.5 and if the time scales are matched. The equation for $f_a(t)$ is $f_a(t) = -(\sin t + \frac{1}{2}\sin 2t + \frac{1}{3}\sin 3t + \frac{1}{4}\sin 4t)$.*

sum of the responses to the individual sources. Now we write Eq. (16-2) in the form

$$v(t) = V_0 + \sum_{k=1}^{4} v_k(t) \qquad v_k = -\frac{3}{k\pi}\sin k\frac{2\pi}{3}t \qquad V_0 = \frac{3}{2} \tag{16-3}$$

and calculate the steady-state response due to each source.

The response to $V_0 = \frac{3}{2}$ volts is $I_0 = \frac{3}{2}$ amp.

At the radian frequency $\omega_1 = 2\pi/3$,

$$R + j\omega_1 L = 1 + j2 = 2.24\underline{/63.4°} \text{ ohms}$$

Hence the component of the steady-state response due to the source $v_1(t)$, denoted by

$i_1(t)$, is

$$i_1(t) = -\frac{3/\pi}{2.24}\sin\left(\frac{2\pi}{3}t - 63.4°\right) = -\frac{3}{\pi}\left[0.446\sin\left(\frac{2\pi}{3}t - 63.4°\right)\right]$$

At the radian frequency $\omega_2 = 4\pi/3$, the complex impedance of the circuit is

$$R + j\omega_2 L = 1 + j4 = 4.13\underline{/76°}$$

so that the response due to v_2 is

$$i_2(t) = -\frac{3/2\pi}{4.13}\sin\left(\frac{4\pi}{3}t - 76°\right) = -\frac{3}{\pi}\left[0.121\sin\left(\frac{4\pi}{3}t - 76°\right)\right]$$

At the radian frequency $6\pi/3$ the complex impedance of the circuit is

$$1 + j6 = 6.08\underline{/80.5°}$$

so that the steady-state response to v_3 is

$$i_3(t) = -\frac{3/3\pi}{6.08}\sin\left(\frac{6\pi}{3}t - 80.5°\right) = -\frac{3}{\pi}\left[0.0548\sin\left(\frac{6\pi}{3}t - 80.5°\right)\right]$$

Similarly, the steady-state response to v_4 is

$$i_4(t) = -\frac{3/4\pi}{8.07}\sin\left(\frac{8\pi}{3}t - 82.9°\right) = -\frac{3}{\pi}\left[0.031\sin\left(\frac{8\pi}{3}t - 82.9°\right)\right]$$

By using superposition, the steady-state response of the R-L circuit to $v(t)$ is

$$i(t) = \frac{3}{2} - \frac{3}{\pi}\left[0.446\sin\left(\frac{2\pi}{3}t - 63.4°\right) + 0.121\sin\left(\frac{4\pi}{3}t - 76°\right)\right.$$
$$\left. + 0.0548\sin\left(\frac{6\pi}{3}t - 80.5°\right) + 0.031\sin\left(\frac{8\pi}{3}t - 82.9°\right)\right]$$

The steady-state response waveform can now be constructed point by point by superposition of the different terms which form $i(t)$. From the expressions for $v(t)$ and $i(t)$, we observe that *the source function and the steady-state response function do not have the same waveform.*

16-2 Fundamental and harmonics

Corresponding to a specified radian frequency ω, the sinusoidal function $C_1 \cos(\omega t + \varphi_1)$ is called a fundamental function, or fundamental. For any integral value n the sinusoidal function $C_n \cos(n\omega t + \varphi_n)$ is said to be a harmonic (function) of the fundamental $C_1 \cos(\omega t + \varphi_1)$.

For a fundamental and its harmonics, there is no restriction placed on the values or relative values of C_1, φ_1, C_n, and φ_n. For example, the functions $1.8 \cos(2\omega t - 23°)$ and $15.1 \cos(2\omega t + 57°)$ are both second harmonics of the fundamental $2.3 \cos(\omega t + 11°)$. For this fundamental, $2.1 \cos(3\omega t + 25°)$ is a third harmonic, but $\cos(3.1\omega t + 36°)$ is not a harmonic of $\cos(\omega t + 11°)$ since the ratio of the two frequencies is not an integer. It is seen that $\cos \omega t$ and $\cos 3.1\omega t$ are both harmonics of $\cos 0.1\omega t$, the former being its tenth harmonic and the latter its 31st. We observe that, whenever the ratio of frequencies of two sinusoids is a rational number, the two functions are both harmonics of a common fundamental. If the ratio of the two frequencies is not a rational number, the two functions are not harmonically related and may be said to be *incommensurate.* For example, the functions $\cos t$ and $\cos \sqrt{5}\, t$ are incommensurate since the ratio of their frequencies is $\sqrt{5}$, which is an irrational number.

Sum of a fundamental and its harmonics In Sec. 16-1 we saw that the sum of a fundamental and its second harmonic is a periodic function whose period is that of the fundamental. We now generalize that statement by observing that, if the infinite sum

$$f(t) = \sum_{k=0}^{k=\infty} C_k \cos(k\omega t + \varphi_k) \tag{16-4}$$

exists (i.e., converges), this sum represents a periodic function with period $2\pi/\omega$. This is seen from the fact that every term in the sum has the period $2\pi/\omega$, and therefore their sum has the same period, namely, $2\pi/\omega$, and therefore $f(t) = f(t + 2\pi/\omega)$.

In Fig. 16-2*b* the sum of four terms

$$f_a(t) = -(\sin t + \tfrac{1}{2}\sin 2t + \tfrac{1}{3}\sin 3t + \tfrac{1}{4}\sin 4t)$$

or

$$f_a(t) = -\sum_{n=1}^{n=4} \frac{1}{n}\sin nt \tag{16-5}$$

is shown. In Eq. (16-5) the fundamental is $\sin t$ with radian frequency $\omega = 1$ and the period $T = 2\pi/\omega = 2\pi$. The summation contains the fundamental $\sin t$ and the first three harmonics. From Fig. 16-2*b* it is seen that the result of the summation is a periodic function with period $T = 2\pi$.

Two points are noted in connection with the sum of a number of sinusoidal functions:

1. If a summation includes a number of harmonics but no fundamental, the sum is still periodic with the period of the fundamental. This is seen from the fact that, in the summation of Eq. (16-4), no restriction is placed on the C_n, and C_1 can be equal to zero.
2. If a sum of sinusoidal functions contains an incommensurate term, the sum is not periodic.

We have now established the fact that the sum of a number of harmonic functions is periodic. It is natural to ask whether *any* periodic function can be written as the sum of a fundamental of the same period as the periodic function and its harmonics. This question has been answered by mathematicians, and the answer is presented below under the name of the Fourier-series theorem.[1]

[1] Named after its discoverer, the French mathematician Jean Baptiste Joseph Fourier (1758–1830).

16-3 Fourier-series theorem

Any periodic function $f(t)$ with period T, such that, in an interval T, $f(t)$ is always finite but not necessarily continuous (i.e., "piecewise continuous"), can be represented by the sum of a constant, a fundamental of period T, and its harmonics. This sum is called a Fourier series, and has the form

$$f(t) = f(t + T) = \sum_{k=0}^{k=\infty} C_k \cos\left(k\frac{2\pi}{T}t + \varphi_k\right) \tag{16-6}$$

If, at a point $t = t_1$, $f(t)$ is discontinuous, the summation results in the value

$$\sum_{k=0}^{\infty} C_k \cos\left(k\frac{2\pi}{T}t_1 + \varphi_k\right) = \frac{1}{2}[f(t_1^+) + f(t_1^-)] \tag{16-6a}$$

The proof of this theorem will not be given here. We note, however, that this theorem is by no means self-evident and requires proof. The fact that the sum of a fundamental and its harmonics is a periodic function does not prove that a periodic function can be written as the sum of a fundamental and its harmonics. The conditions regarding the finiteness of the function and the requirement of piecewise continuity are met by all functions of interest in circuit analysis, i.e., all periodic source and response functions.[1]

Assuming the theorem to be true, we now have to study the evaluation of the amplitudes C_n and the phase angles φ_n of the harmonics which correspond to any periodic function $f(t)$. The calculation of these quantities is based on a certain property of sinusoidal functions, referred to as their *orthogonality* property. This property is most easily discussed when the sinusoidal functions are written in exponential form. For this reason we discuss the various forms of the Fourier series before discussing the evaluation of the coefficients C_k and the phase angles φ_k.

16-4 Trigonometric and exponential form of the Fourier series

The general term in the Fourier series, Eq. (16-6), can be written in exponential form,

$$C_k \cos\left(k\frac{2\pi}{T}t + \varphi_k\right) = \tfrac{1}{2}C_k e^{j[k(2\pi/T)t+\varphi_k]} + \tfrac{1}{2}C_k e^{-j[k(2\pi/T)t+\varphi_k]}$$

or

$$C_k \cos\left(k\frac{2\pi}{T}t + \varphi_k\right) = (\tfrac{1}{2}C_k e^{j\varphi_k})e^{jk(2\pi/T)t} + (\tfrac{1}{2}C_k e^{-j\varphi_k})e^{-jk(2\pi/T)t} \tag{16-7}$$

[1] If it is desired to express a function which is not periodic but which is piecewise continuous in an interval t_1 to t_2 by a Fourier series, it is possible to do so by assuming that the function has a period $T = t_2 - t_1$ and expanding. The result, which has the form of Eq. (16-6), is then used only in the interval t_1 to t_2. This technique is commonly used in boundary-value problems (where, incidentally, space coordinates, not time, are the variables).

Hence the series (16-6) can be written as follows:

$$f(t) = C_0 + \sum_{k=1}^{\infty} (\tfrac{1}{2}C_k e^{j\varphi_k}) e^{jk(2\pi/T)t} + \sum_{k=1}^{\infty} (\tfrac{1}{2}C_k e^{-j\varphi_k}) e^{-jk(2\pi/T)t} \qquad (16\text{-}8)$$

Now we define the complex quantities $\mathbf{D}_k$,

$$\mathbf{D}_k \equiv \tfrac{1}{2}C_k e^{j\varphi_k} \qquad k = 1, 2, 3, \ldots$$

and the real quantity

$$D_0 \equiv C_0$$

Then Eq. (16-8) can be written in the form

$$f(t) = D_0 + \sum_{k=1}^{\infty} \mathbf{D}_k e^{jk(2\pi/T)t} + \sum_{k=1}^{\infty} \mathbf{D}_k^* e^{-jk(2\pi/T)t}$$

We now define

$$\mathbf{D}_{-k} \equiv \mathbf{D}_k^* = \tfrac{1}{2}C_k e^{-j\varphi_k}$$

Then

$$\sum_{k=1}^{\infty} \mathbf{D}_k^* e^{-jk(2\pi/T)t} = \sum_{k=1}^{\infty} \mathbf{D}_{-k} e^{-jk(2\pi/T)t} = \sum_{k=-1}^{-\infty} \mathbf{D}_k e^{jk(2\pi/T)t}$$

Hence $f(t)$ can be written as the sum

$$f(t) = \sum_{k=-\infty}^{k=+\infty} \mathbf{D}_k e^{jk(2\pi/T)t} \qquad (16\text{-}9)$$

Note that the summation in Eq. (16-9) extends from $-\infty$ to $+\infty$; that is,

$$f(t) = D_0 + \mathbf{D}_1 e^{j(2\pi/T)t} + \mathbf{D}_2 e^{j(4\pi/T)t} + \cdots + \mathbf{D}_{-1} e^{-j(2\pi/T)t} + \mathbf{D}_{-2} e^{-j(4\pi/T)t} + \cdots$$

The form (16-9) is called the *exponential*, or *complex*, form of the Fourier series, and is particularly useful because one (complex) quantity, $\mathbf{D}_k$, contains both the amplitude and the phase angle of each harmonic.

Another (trigonometric) form of the Fourier series is obtained by expanding

$$\cos\left(k\frac{2\pi}{T}t + \varphi_k\right) = \cos k\frac{2\pi}{T}t \cos\varphi_k - \sin k\frac{2\pi}{T}t \sin\varphi_k$$

We define

$$\begin{aligned} A_0 &\equiv C_0 \\ A_k &\equiv C_k \cos\varphi_k = 2 \operatorname{Re} \mathbf{D}_k \qquad & k = 1, 2, 3, \ldots \\ B_k &\equiv -C_k \sin\varphi_k = -2 \operatorname{Im} \mathbf{D}_k \qquad & k = 1, 2, 3, \ldots \end{aligned}$$

and rewrite Eq. (16-6) in the form

$$f(t) = A_0 + \sum_{k=1}^{\infty} \left(A_k \cos k\frac{2\pi}{T}t + B_k \sin k\frac{2\pi}{T}t\right) \qquad (16\text{-}10)$$

Table 16-1 The three forms of a Fourier series

Complex form	Trigonometric form 1	Trigonometric form 2
$f(t) = \sum_{k=-\infty}^{+\infty} \mathbf{D}_k e^{jk(2\pi/T)t}$	$f(t) = C_0 + \sum_{k=1}^{\infty} C_k \cos\left(k \frac{2\pi}{T} t + \varphi_k\right)$ $= D_0 + \sum_{k=1}^{\infty} 2 \operatorname{Re}\left(\mathbf{D}_k e^{jk(2\pi/T)t}\right)$	$f(t) = A_0 + \sum_{k=1}^{\infty} \left(A_k \cos k \frac{2\pi}{T} t + B_k \sin k \frac{2\pi}{T} t\right)$

Conversion formulas for $k = 1, 2, 3, \ldots$

$$\mathbf{D}_k = \tfrac{1}{2} C_k \underline{/\varphi_k} \qquad \mathbf{D}_{-k} = \mathbf{D}_k^* \qquad \mathbf{D}_k = \tfrac{1}{2} A_k - j\tfrac{1}{2} B_k$$

$$C_k = 2D_k \qquad \varphi_k = \text{angle } \mathbf{D}_k \qquad C_k = \sqrt{A_k^2 + B_k^2} \qquad \varphi_k = -\tan^{-1} \frac{B_k}{A_k}$$

$$A_k = 2 \operatorname{Re} \mathbf{D}_k \qquad B_k = -2 \operatorname{Im} \mathbf{D}_k \qquad A_k = C_k \cos \varphi_k \qquad B_k = -C_k \sin \varphi_k$$

For $k = 0$,

$$D_0 = C_0 = A_0 \qquad B_0 \equiv 0$$

CONVERSION OF SERIES FROM ONE FORM TO ANOTHER The three forms in which the Fourier series can be expressed are summarized in Table 16-1. The formulas used in conversion from one form to another are also listed in that table.

Example 16-1 Convert the Fourier series

$$f(t) = \sum_{k=-\infty}^{+\infty} \frac{1}{1 + jk} e^{jkt}$$

to both trigonometric forms.

Solution $D_0 = 1$; hence $C_0 = 1$, $A_0 = 1$. Since $\mathbf{D}_k = 1/(1 + jk)$,

$$\mathbf{D}_k = \left(\frac{1}{\sqrt{1 + k^2}}\right) \underline{/-\tan^{-1} k}$$

$$f(t) = 1 + \sum_{k=1}^{\infty} \frac{2}{\sqrt{1 + k^2}} \cos(kt - \tan^{-1} k)$$

Also

$$\mathbf{D}_k = \frac{1}{1 + jk} = \frac{1}{1 + k^2} - \frac{jk}{1 + k^2}$$

Hence

$$A_k = \frac{2}{1 + k^2} \qquad B_k = \frac{2k}{1 + k^2}$$

and

$$f(t) = 1 + 2 \sum_{k=1}^{\infty} \frac{\cos kt + k \sin kt}{1 + k^2}$$

16-5 The orthogonality property of sinusoidal functions[1]

A number of functions with one or more common characteristics which distinguish them from other functions is called a set of functions. As an example, periodic functions form a set of functions whose distinguishing characteristic is that they are periodic. A particular set of functions which is of special interest in analysis is called an *orthogonal* set. To define the property of orthogonality, we first define a useful symbol, δ_{mn}, called Kronecker's delta.[2] This symbol is zero or unity, depending on whether the subscripts m and n are unequal or equal.

$$\delta_{mn} = \begin{cases} 0 & m \neq n \\ 1 & m = n \end{cases}$$

For example, $\delta_{12} = 0$, and $\delta_{44} = 1$.

[1] A somewhat more detailed discussion is given in M. Javid and E. Brenner, "Analysis, Transmission, and Filtering of Signals," chaps. 2 and 3, McGraw-Hill Book Company, New York, 1963.

[2] Named after the mathematician Leopold Kronecker (1823–1891).

Definition of orthogonal functions If the members of a set of complex functions[1] are designated by $\mathbf{f}_1(t), \mathbf{f}_2(t), \ldots, \mathbf{f}_m(t), \ldots, \mathbf{f}_n(t), \ldots,$ and if there exists an interval $a < t < b$ and a function $w(t)$ called the "weighting function," such that

$$\int_a^b \mathbf{f}_m(t)\mathbf{f}_n^*(t)w(t)\,dt = A_{mn}\delta_{mn}$$

where A_{mn} are constants, then in the interval $a < t < b$, the set of functions is said to be orthogonal with respect to the chosen weighting function.

We shall not discuss orthogonal functions in general, but shall state that the harmonic functions are orthogonal over a period of the fundamental with respect to the weighting function $w(t) = 1$. To illustrate this property, consider

$$\mathbf{f}_m(t) = e^{jm(2\pi/T)t}$$
$$\mathbf{f}_n^*(t) = e^{-jn(2\pi/T)t}$$

We have

$$\int_0^T \mathbf{f}_m(t)\mathbf{f}_n^*(t)\,dt = \int_0^T e^{j(m-n)(2\pi/T)t}\,dt \tag{16-11}$$

Now in the integral (16-11), if $m = n$, we have

$$\int_0^T dt = T = \text{const}$$

and if $m \neq n$,

$$\int_0^T e^{j(m-n)(2\pi/T)t} = \frac{e^{j(m-n)(2\pi/T)T} - 1}{j(m-n)(2\pi/T)} = 0$$

Since $e^{j(m-n)2\pi} = 1$ for integral values of m and n,

$$\int_0^T e^{jm(2\pi/T)t}e^{-jn(2\pi/T)t}\,dt = T\delta_{mn}$$

Hence the set of functions $e^{jm(2\pi/T)t}$ is *orthogonal* in an interval $0 < t < T$.

Similarly, it can be shown that sinusoidal functions which are harmonically related are orthogonal because

$$\int_0^{2\pi} \sin mx \sin nx\,dx = \pi\delta_{mn}$$
$$\int_0^{2\pi} \sin mx \cos nx\,dx = 0$$
$$\int_0^{2\pi} \cos mx \cos nx\,dx = \pi\delta_{mn}$$

16-6 Evaluation of Fourier coefficients

The set of complex values $\mathbf{D}_k$ or the set A_k, B_k is referred to as the Fourier coefficients. We shall now show how these coefficients are related to the periodic function $f(t)$ which the Fourier series represents. If we start with

[1] Real functions are considered special cases of complex functions whose imaginary part is zero.

the complex form

$$f(t) = \sum_{k=-\infty}^{+\infty} \mathbf{D}_k e^{jk(2\pi/T)t} \tag{16-12}$$

and multiply both sides of Eq. (16-12) by $e^{-jn(2\pi/T)t}$, then

$$f(t)e^{-jn(2\pi/T)t} = \sum_{k=-\infty}^{+\infty} \mathbf{D}_k e^{j(k-n)(2\pi/T)t} \tag{16-13}$$

Now, if both sides of Eq. (16-13) are integrated between the limits $t = 0$ to $t = T$ with respect to t, we have

$$\int_0^T f(t)e^{-jn(2\pi/T)t}\, dt = \int_0^T \left(\sum_{k=-\infty}^{+\infty} \mathbf{D}_k e^{j(k-n)(2\pi/T)t} \right) dt$$

Interchanging summation and integration,[1]

$$\sum_{k=-\infty}^{+\infty} \int_0^T \mathbf{D}_k e^{j(k-n)(2\pi/T)t}\, dt = \sum_{k=-\infty}^{\infty} \mathbf{D}_k T \delta_{kn} = \mathbf{D}_n T$$

since in the above summation all the δ_{kn} except for $k = n$ are zero. Hence

$$\mathbf{D}_k = \frac{1}{T} \int_0^T f(t)e^{-jk(2\pi/T)t}\, dt \tag{16-14}$$

Using the conversion formulas from Table 16-1,

$$A_0 = D_0 = \frac{1}{T} \int_0^T f(t)\, dt = [f(t)]_{\text{avg}} \tag{16-15a}$$

$$A_k = 2 \operatorname{Re} \mathbf{D}_k = \frac{2}{T} \int_0^T f(t) \cos k \frac{2\pi}{T} t\, dt \tag{16-15b}$$

$$B_k = -2 \operatorname{Im} \mathbf{D}_k = \frac{2}{T} \int_0^T f(t) \sin k \frac{2\pi}{T} t\, dt \tag{16-15c}$$

Thus the problem of representing a given waveform as a Fourier series is reduced to the "routine" problem of evaluating the integral (16-14) or the equivalent forms (16-15).

Before presenting examples of the evaluation of Fourier coefficients, we shall discuss some interesting aspects of this procedure. First, we observe that, because of the orthogonality property of the sinusoidal functions, the formulas which we obtained allow us to find any one term in the Fourier series without finding all the other terms. Moreover, it can be shown (as we stated above) that the infinite Fourier series will converge to the value of the function at any point of continuity. To understand fully what this implies, we shall consider another method of using trigonometric series to approximate a periodic function. Suppose that it is known that a function

[1] Such interchanges are not always possible. The procedure can be justified in the present instance.

$f(t)$ whose period is 2π can be approximated by the series

$$f_a(t) = A_1 \cos t + A_2 \cos 2t + A_3 \cos 3t$$

[i.e., suppose that we have evaluated A_0, B_1, B_2, B_3 from (16-15*a*) and (16-15*c*) and obtained zero as the answer].

For this approximation how shall we choose A_1, A_2, and A_3? One method is, of course, to apply Eq. (16-15*b*) and find the Fourier coefficients. An alternate method is the following: Choose three points in the interval $0 < t < T$, the points t_1, t_2, and t_3, and then choose A_1, A_2, and A_3 so that $f_a(t) = f(t)$ at the three chosen points; i.e., let

$$\begin{aligned} f_a(t_1) &= f(t_1) = A_1 \cos t_1 + A_2 \cos 2t_1 + A_3 \cos 3t_1 \\ f_a(t_2) &= f(t_2) = A_1 \cos t_2 + A_2 \cos 2t_2 + A_3 \cos 3t_2 \\ f_a(t_3) &= f(t_3) = A_1 \cos t_3 + A_2 \cos 2t_3 + A_3 \cos 3t_3 \end{aligned}$$

and now solve these three simultaneous equations for A_1, A_2, and A_3. The disadvantages of this method compared with Fourier's method are clear. First, we need to solve simultaneous equations, which means that the coefficients cannot be determined independently of each other. Second, if we have found A_1, A_2, and A_3 and sketched $f_a(t)$ [which is supposed to approximate $f(t)$], if the approximation is not accurate enough and we desire another harmonic, this other harmonic (for example, A_4) cannot be determined without recalculating A_1, A_2, and A_3. Finally, this alternative method assures us only that f_a corresponds to $f(t)$ at the three chosen instants t_1, t_2, and t_3. We have no information about the nature of the approximation at other instants during the cycle. In the next section we shall discuss the nature of the approximation which is implied when the Fourier coefficients are used and when a finite number of terms in the series is used to approximate a given periodic function.

16-7 Squared-error property of Fourier series

If a periodic function $f(t)$ with period[1] 2π is expressed as the Fourier series

$$f(t) = C_0 + \sum_{k=1}^{\infty} C_k \cos (kt + \varphi_k)$$

then, in practice, we use a finite number of terms, i.e., the first N terms of the Fourier series, to approximate $f(t)$.

$$f(t) \approx f_N(t) = \sum_{k=0}^{N} C_k \cos (kt + \varphi_k)$$

Now the actual value of $f(t)$ differs from the values obtained by use of the

[1] Choosing the period as 2π does not detract from the generality of the discussion because a time-scale change can always be used to apply the result to arbitrary period T.

first N terms in the sum $f_N(t)$ by a function $\epsilon(t)$, which is called the error.

$$\epsilon(t) = f(t) - f_N(t)$$

In one period the average error is

$$E_{\text{avg}} = \frac{1}{2\pi} \int_0^{2\pi} \epsilon(t)\, dt$$

The average error is not a useful number in determining the "quality" of the approximation because positive errors will tend to cancel negative errors. Thus, if in the interval 0 to 2π the function $f(t) = t$ is approximated by $f_N(t) = \pi$, $\epsilon = t - \pi$ and

$$\int_0^{2\pi} (t - \pi)\, dt = \left[\frac{t^2}{2} - \pi t\right]_0^{2\pi} = 2\pi(\pi - \pi) = 0$$

so that the average error is zero for this (gross) approximation.

To avoid this difficulty, Gauss[1] introduced the idea of mean squared error M,

$$M = \frac{1}{2\pi} \int_0^{2\pi} \epsilon^2(t)\, dt \tag{16-16}$$

It can be shown[2] that choice of the coefficients in the Fourier series is such that the mean squared error $(1/2\pi) \int_0^{2\pi} [f(t) - f_N(t)]^2\, dt$ is minimized for any given N. In other words, if we wish to approximate a periodic function by means of, say, five harmonics, then using the Fourier coefficients will result in the smallest possible mean squared error; any other choice of coefficients results in a larger mean squared error. Moreover, as we have already stated, the mean squared error will become zero if the Fourier series is allowed to become infinite.

Example 16-2 The function $v_a(t) = B_1 \sin t$ is to be used to approximate the periodic function $v(t)$ shown in Fig. 16-3 so that the mean squared error is a minimum. Show that the value of B_1 for this condition is the Fourier coefficient given by Eq. (16-15*c*).

[1] Karl Friedrich Gauss (1777–1855).
[2] See, for example, Javid and Brenner, *loc. cit.*

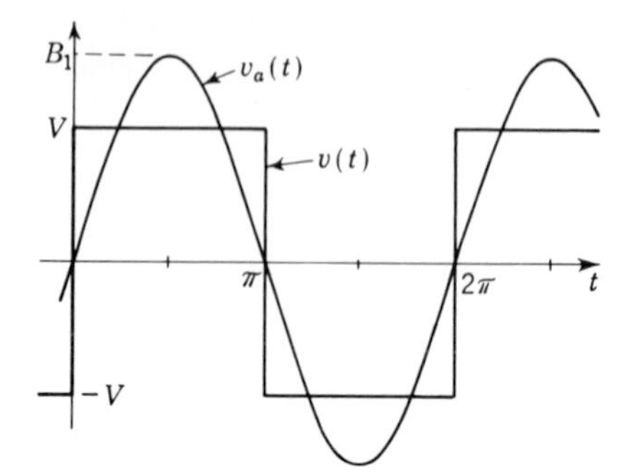

Fig. 16-3 The square wave $v(t)$ is to be approximated by the sine wave $v_a(t)$.

Solution Since

$$v(t) = +V \qquad 0 < t < \pi$$
$$v(t) = -V \qquad \pi < t < 2\pi$$

the error ϵ is given by

$$\epsilon = V - B_1 \sin t \qquad 0 < t < \pi$$
$$\epsilon = -V - B_1 \sin t \qquad \pi < t < 2\pi$$

Hence the mean squared error M is given as

$$M = \frac{1}{2\pi}\left[\int_0^{\pi} (V - B_1 \sin t)^2\, dt + \int_0^{2\pi} (-V - B_1 \sin t)^2\, dt\right]$$

Integrating and substituting the limits,

$$M = V^2 - \frac{4}{\pi} VB_1 + \frac{B_1{}^2}{2}$$

In order to make M a minimum, $\partial M/\partial B_1$ is formed and set to zero.

$$\frac{\partial M}{\partial B_1} = -\frac{4V}{\pi} + B_1 \qquad \frac{\partial M}{\partial B_1} = 0 \text{ gives } B_1 = \frac{4V}{\pi}$$

Since $\partial^2 M/\partial B_1{}^2 = +1$, a positive number, the value $B_1 = 4V/\pi$ gives minimum mean squared error. The Fourier coefficient B_1 is found by use of Eq. (16-15*c*), by using $2\pi/T = 1$.

$$B_1 = \frac{1}{\pi}\int_0^{2\pi} v(t) \sin t\, dt$$

or

$$B_1 = \frac{1}{\pi}\left[\int_0^{\pi} V \sin t\, dt + \int_0^{2\pi} (-V) \sin t\, dt\right] = 4V/\pi$$

exactly as was obtained by minimizing the mean squared error.

16-8 *Evaluation of Fourier coefficients, examples*

In Sec. 16-6 the complex and real Fourier coefficients were given as definite integrals. If it is desired to find the Fourier series for a given function, the function can be specified either graphically by its waveform or analytically over an interval such as $t = 0$ to $t = T$ by means of equations.

ANALYTICAL EVALUATION OF FOURIER COEFFICIENTS In order to use the integral expressions for the Fourier coefficients correctly, it is necessary that a sufficient number of equations be written to specify $f(t)$ over at least one complete period. In Fig. 16-4 three periodic functions are shown. These functions are specified analytically as follows:

For $f_a(t)$ the period is 3; that is, $f_a(t) = f_a(t + 3)$. The equations which define this function are

$$f_a(t) = t + 1 \qquad -1 < t < 2$$
$$f_a(t) = t - 2 \qquad 2 < t < 5$$
$$f_a(t) = t - 5 \qquad 5 < t < 8$$

We observe that the equation $f_a(t) = t - 2$ is valid only in the stated interval $2 < t < 5$; a different analytical expression is required to specify the function in each 3-sec interval.

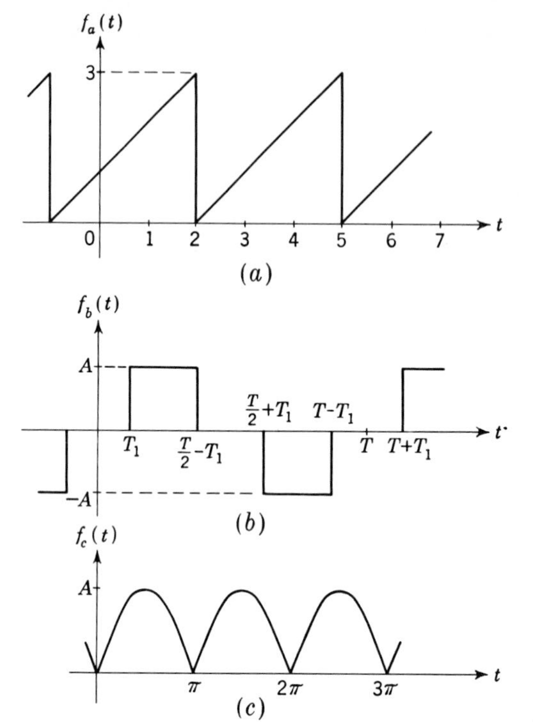

Fig. 16-4 ***Three periodic waveforms.***

For the function $f_b(t)$, the following five equations specify the function in the interval from $t = 0$ to $t = T$:

$$f_b(t) = 0 \qquad 0 < t < T_1$$
$$f_b(t) = A \qquad T_1 < t < \left(\frac{T}{2} - T_1\right)$$
$$f_b(t) = 0 \qquad \left(\frac{T}{2} - T_1\right) < t < \left(\frac{T}{2} + T_1\right)$$
$$f_b(t) = -A \qquad \left(\frac{T}{2} + T_1\right) < t < (T - T_1)$$
$$f_b(t) = 0 \qquad (T - T_1) < t < T$$

The function $f_c(t)$, for $A = 1$, is given as

$$f_c(t) = |\sin t|$$

or

$$f_c(t) = \sin t \qquad 0 < t < \pi$$
$$f_c(t) = \sin (t - \pi) \qquad \pi < t < 2\pi$$
$$f_c(t) = \sin (t - 2\pi) \qquad 2\pi < t < 3\pi$$

Clearly, the expression $\sin t$ is not valid in the interval $\pi < t < 2\pi$; we observe that $f_c(t)$ has only positive values, whereas the function $\sin t$ in the interval $\pi < t < 2\pi$ has negative values. Note, however, that $\sin (t - \pi)$ does give the correct answer for $f_c(t)$ in that interval.

When a function is specified by means of several equations over one period, the integral which defines the Fourier coefficients must be expressed as the sum of several integrals. For example, the complex coefficients $\mathbf{D}_k$ are given by

$$\mathbf{D}_k = \frac{1}{T}\int_0^T f(t)e^{-jk(2\pi/T)t}\,dt$$

To evaluate D_k for the waveform $f_a(t)$, for example, we write

$$D_k = \frac{1}{T}\left[\int_0^2 (t+1)e^{-jk(2\pi/T)t}\,dt + \int_2^3 (t-2)e^{-jk(2\pi/T)t}\,dt\right]$$

Example 16-3 Find the Fourier series for the sawtooth waveform of Fig. 16-5.

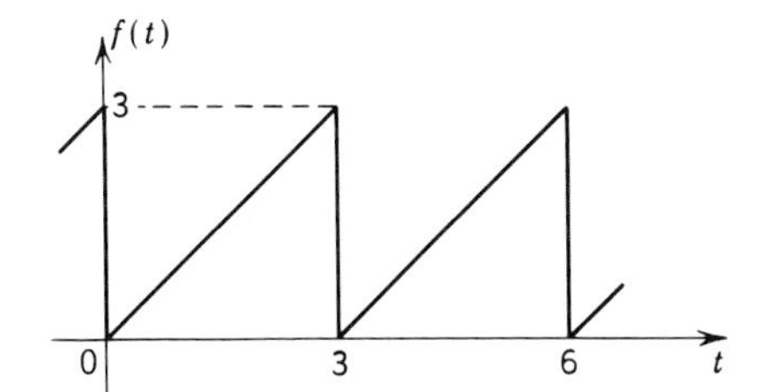

Fig. 16-5 Periodic sawtooth waveform.

Solution This waveform has a period of 3. The equation of the waveform in the interval 0 to T can be expressed by the single equation $f(t) = t$, $0 < t < 3$. Using the complex form of the Fourier series, we have

$$\mathbf{D}_k = \tfrac{1}{3}\int_0^3 te^{-jk(2\pi/3)t}\,dt$$

Integrating by parts,

$$3\mathbf{D}_k = \frac{te^{-jk(2\pi/3)t}}{-jk(2\pi/3)}\Bigg]_0^3 + \frac{1}{jk(2\pi/3)}\int_0^3 e^{-jk(2\pi/3)t}\,dt$$

or

$$3\mathbf{D}_k = \frac{3e^{-jk2\pi} - 0}{-jk(2\pi/3)} + \frac{1}{k^2(4\pi^2/9)}(e^{-jk2\pi} - 1)$$

Since $e^{-jk2\pi} = 1$,

$$\mathbf{D}_k = j\frac{3}{2\pi k} = j\frac{0.478}{k} \qquad k \neq 0 \tag{16-17}$$

We observe, first, that this result does not apply for $k = 0$, and the value D_0 will have to be found separately. Second, the coefficients $\mathbf{D}_k$, which are generally complex, are, for this example, purely imaginary. This result can be predicted by certain symmetry considerations of $f(t)$ which are discussed in Sec. 16-11.

To calculate the value D_0, we observe that

$$D_0 = \frac{1}{T}\int_0^T f(t)\,dt = \text{avg value of } f(t)$$

In this example we can find D_0 by calculating the area of the triangular pulse between $t = 0$ and $t = 3$ and dividing by the period.

$$D_0 = (\tfrac{1}{2})(3)(3)(\tfrac{1}{3}) = \tfrac{3}{2} = 1.5$$

Hence we can write $f(t)$ in complex form.

$$f(t) = 1.5 + j0.478\left(e^{j(2\pi/3)t} + \tfrac{1}{2}e^{j(4\pi/3)t} + \tfrac{1}{3}e^{j(6\pi/3)t} + \cdots \right.$$
$$\left. + -e^{-j(2\pi/3)t} + \frac{1}{-2}e^{-j(4\pi/3)t} + \frac{1}{-3}e^{-j(6\pi/3)t} + \cdots\right)$$

This series can be converted to real form either by use of the formulas in Table 16-1 or by "pairing" the terms $e^{j[k(2\pi/3)]t}$ with the terms $e^{-jk(2\pi/3)t}$; using the latter method, we write

$$f(t) = 1.5 + 0.478\left(\frac{e^{j(2\pi/3)t} - e^{-j(2\pi/3)t}}{-j} + \frac{e^{j(4\pi/3)t} - e^{-j(4\pi/3)t}}{-j2}\right.$$
$$\left. + \frac{e^{j(6\pi/3)t} - e^{-j(6\pi/3)t}}{-j3} + \cdots\right)$$

or

$$f(t) = 1.5 - 0.956\left(\sin\frac{2\pi}{3}t + \frac{1}{2}\sin\frac{4\pi}{3}t + \frac{1}{3}\sin\frac{6\pi}{3}t + \cdots + \frac{1}{k}\sin\frac{2\pi k}{3}t + \cdots\right)$$

This result shows that, for this function, the real coefficients are

$$A_0 = 1.5 \qquad A_k = 0,\ k = 1, 2, 3, \ldots \qquad B_k = -\frac{0.956}{k}$$

The reader should verify this result by application of the formulas (16-15) which define A_k and B_k.

With reference to this example, the following observations can be made:

1 A periodic waveform has been written as the sum of a constant value and an infinite number of sinusoidal functions. The constant value is the average value of the function and is referred to as the "d-c" component of the function.

2 In the real (trigonometric) form of the Fourier series there are only *sine* terms; the cosine terms are zero (except for A_0). This is not a general result, and applies only to certain types of functions, which are studied below.

3 The amplitude of the higher harmonics, B_n, decreases with n, the order of the harmonic. This is not always so. For example, it is possible that, for a given periodic function, the amplitude of the fifth harmonic is greater than the amplitude of the fourth harmonic. It is, however, always true that the amplitude of the nth harmonic approaches zero as n approaches infinity. In other words, although the amplitude of successive harmonics may "go up and down" as n increases, they will all tend to zero as n increases. This property of the Fourier coefficients (amplitudes of harmonics) can be proved by use of the conditions stated in the Fourier-series theorem.

4 Since, in this example, the amplitude of the kth harmonic decreases as $1/k$, for some practical purposes it will be sufficient to represent such a function as the sum of a finite number of terms of the Fourier series. In Fig. 16-2*b* the sum of a fundamental and the first four harmonics is shown for a series in which the harmonic amplitudes decrease as $1/k$. A comparison of Fig. 16-2*a* and *b* shows that, except for the difference in their average

values and a scale factor $2\pi/3$ in their period, the two functions are similar in waveform, f_a being a fairly good approximation to $f(t)$. If more harmonics are added to $f_a(t)$ (with proper amplitudes, of course), the resulting waveshape will resemble the sawtooth waveform more and more closely.

GRAPHICAL DETERMINATION OF FOURIER COEFFICIENTS It frequently happens that a periodic function is specified graphically (e.g., by means of an oscillogram). To find the Fourier series for such a function, it may not be convenient to use "curve fitting" for the purpose of obtaining equations which describe the function over one period. It is, however, possible to evaluate the real Fourier coefficients A_k and B_k graphically by evaluating the corresponding integrals graphically.

Example 16-4 Evaluate graphically the Fourier coefficient B_1 for the waveform of Fig. 16-5. (The waveform of Fig. 16-5 is identical with that of Fig. 16-7*a*.)

Solution From Eq. (16-15*c*), we have

$$B_1 = \frac{2}{3}\int_0^3 t \sin \frac{2\pi}{3} t \, dt \tag{16-18}$$

To evaluate B_1 graphically, we plot the integrand $t \sin (2\pi/3)t$ and evaluate the area under this graph between $t = 0$ and $t = 3$. The graph of $t \sin (2\pi/3)t$ is shown in Fig. 16-6.

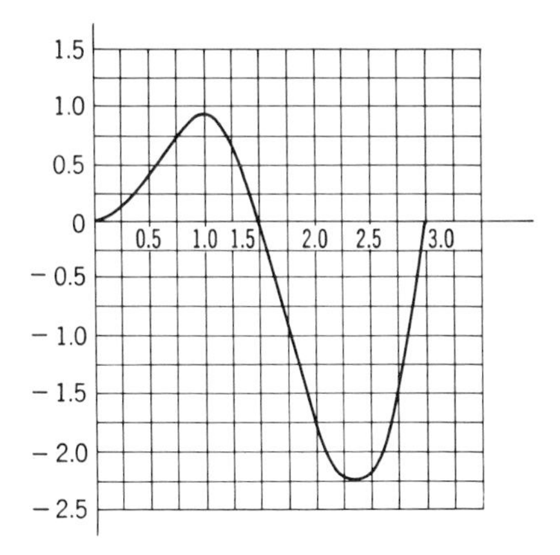

Fig. 16-6 Graph of the function t sin $(2\pi/3)t$ used in the graphical evaluation of the coefficient B_1 for the waveform of Fig. 16-5.

The area under the curve $t \sin (2\pi/3)t$ in the interval $t = 0$ to $t = 3$ may be found by counting the number of "positive" and "negative" squares in Fig. 16-6. In Fig. 16-6 each square is $(\frac{1}{4})(\frac{1}{4}) = \frac{1}{16}$. There are approximately 11.4 squares above the line $f(t) = 0$ and about 34.5 squares below the $f(t) = 0$ line; therefore the area under the curve is $(11.4 - 34.5)/16$, and from Eq. (16-18),

$$B_1 = \frac{\frac{2}{3}(11.4 - 34.5)}{16} = -0.96$$

The value of B_1 obtained by analytical means was $B_1 = -3/\pi = -0.955$, which agrees closely with the above value. To find B_2, a plot of the function $t \sin (4\pi/3)t$ is made, and from the graph of this function the area under the curve is found. The value of B_2 is two-thirds of this area. The other Fourier coefficients are similarly found.

In order to evaluate the integrals $f(t) \sin k(2\pi/3)t$ or $f(t) \cos k(2\pi/3)t$ graphically, systematic tabulations which introduce short cuts into the calculations can be devised (Probs. 16-11 and 16-12).

16-9 Time displacement

We have seen that the Fourier series for the sawtooth function shown in Fig. 16-7*a* is

$$f_a(t) = 1.5 - \frac{3}{\pi} \sum_{k=1}^{\infty} \frac{1}{k} \sin \left(k \frac{2\pi}{3} t\right) = 1.5 - \frac{3}{\pi} \sum_{k=1}^{\infty} \frac{1}{k} \cos \left(k \frac{2\pi}{3} t - \frac{\pi}{2}\right) \tag{16-19}$$

The function $f_b(t)$ shown in Fig. 16-7*b* has the same waveform as $f_a(t)$ but

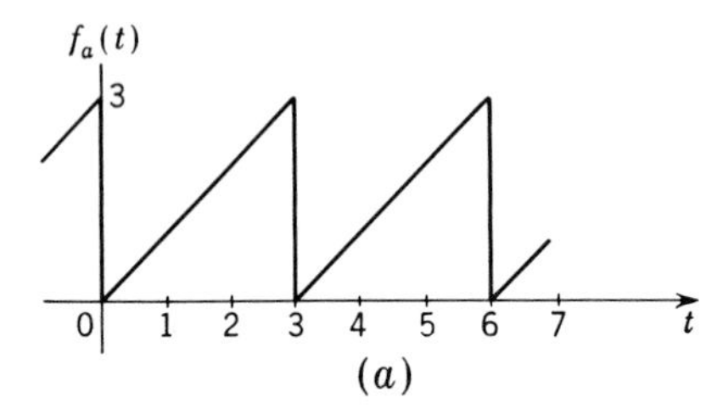

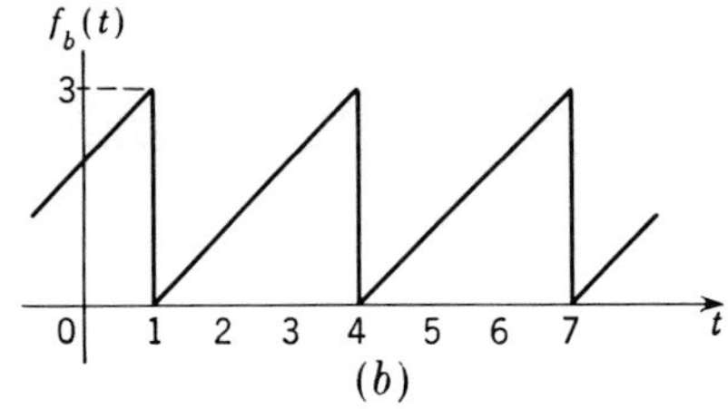

***Fig. 16-7** (a) Sawtooth waveform. (b) Waveform of (a) delayed by one unit of time.*

is delayed in time by one unit with respect to $f_a(t)$; that is,

$$f_b(t) = f_a(t - 1)$$

Thus the Fourier series for $f_b(t)$ can be obtained from the Fourier series for $f_a(t)$ [Eq. (16-19)] by replacing t with $t - 1$ in Eq. (16-19).

$$f_b(t) = 1.5 - \frac{3}{\pi} \sum_{k=1}^{\infty} \frac{1}{k} \cos \left[k \frac{2\pi}{3} (t - 1) - \frac{\pi}{2}\right] \tag{16-20a}$$

or

$$f_b(t) = 1.5 - \frac{3}{\pi} \sum_{k=1}^{\infty} \frac{1}{k} \cos \left[k \frac{2\pi}{3} t + \left(-\frac{2\pi k}{3} - \frac{\pi}{2}\right)\right] \tag{16-20b}$$

Now Eq. (16-20*b*) is the Fourier series for the function $f_b(t)$ in the form

$$f_b(t) = C_0 + \sum_{k=1}^{\infty} C_k \cos \left(k \frac{2\pi}{T} t + \varphi_k\right) \tag{16-21}$$

and Eq. (16-19) is the corresponding trigonometric form for the function $f_a(t)$. A comparison of Eqs. (16-19) and (16-20*b*) shows that the time displacement leaves the value of the coefficients C_k unchanged but changes the angle φ_k of each harmonic. This result, illustrated for a particular example, can be proved in general in the form of the following theorem.

TIME-DISPLACEMENT THEOREM If a function $f_a(t)$ has the Fourier series

$$f_a(t) = \sum_{k=-\infty}^{+\infty} \mathbf{D}_k e^{jk(2\pi/T)t} = C_0 + \sum_{k=1}^{\infty} C_k \cos\left(k \frac{2\pi}{T} t + \varphi_k\right) \qquad (16\text{-}22)$$

then this function delayed in time by t_1 has the Fourier series

$$f_a(t - t_1) = \sum_{k=-\infty}^{+\infty} (\mathbf{D}_k e^{-jk(2\pi/T)t_1}) e^{jk(2\pi/T)t}$$

$$= C_0 + \sum_{k=1}^{\infty} C_k \cos\left(k \frac{2\pi}{T} t + \varphi_k - \frac{2\pi t_1}{T} k\right) \qquad (16\text{-}23)$$

Since (16-23) was obtained from Eq. (16-22) by replacing t with $t - t_1$, the proof of the statement is evident. We observe that the effect of a time delay t_1 on the original complex coefficients D_k is the change in phase angle by the amount $k2\pi t_1/T$. Thus

$$(\mathbf{D}_k)_{\text{delayed function}} = (\mathbf{D}_k)_{\text{original function}} e^{-jk(2\pi t_1/T)} \qquad (16\text{-}24)$$

We summarize this discussion as follows: If the function $f_b(t)$ has the same waveform as $f_a(t)$ but is delayed in time by t_1, the harmonic amplitudes of the two functions are identical and the phase angle of the kth harmonic of f_b is $k2\pi t_1/T$ rad less than the phase angle of the corresponding harmonic of $f_a(t)$.

The most important application of this theorem deals with the converse: If two functions have the same harmonic amplitudes and if the phase angle differs so that the phase shift between corresponding harmonics is proportional to the frequency of the harmonic, the two functions have the same waveform.

To illustrate the importance of this statement, we recall that networks which contain energy-storing elements have their own characteristic frequency response, as discussed in Chaps. 7, 10, and 15. Suppose, now, that the source function which is impressed on a network is periodic, so that it can be expressed as a Fourier series. The steady-state response to such a source function can be calculated by superposition, i.e., by using phasor representation for each harmonic. Thus the response for each harmonic has an amplitude which is determined by the absolute value of the complex network function at the frequency of the harmonic. The phase angle of a harmonic of the response waveform will differ from the phase angle of the corresponding harmonic of the source waveform by the angle of the network function at the frequency of the harmonic. A little thought will show that the steady-state response waveform has the same shape as the source waveform only if all response harmonics have the same relative amplitudes as the source harmonics *and* if the phase angles of the response harmonics differ from those of the source harmonics proportionally to the frequency of the harmonics. For this condition it is necessary that the *network* function have the *same* absolute value for all frequencies and a *phase angle which is proportional to frequency.*[1] This matter was mentioned in Sec. 15-9, and is discussed further in the section on ideal filters (Sec. 16-18). When generalized, it becomes a cornerstone in the theory of transmission of signals.

16-10 Symmetry properties of certain waveforms

For the waveform of Fig. 16-7a, the Fourier series has a constant and sine terms only, and, for Fig. 16-7b,

$$f_b(t) = 1.5 - \frac{3}{\pi}\sum_{k=1}^{\infty}\frac{1}{k}\cos\left[k\frac{2\pi}{3}t - \left(\frac{2\pi k}{3} + \frac{\pi}{2}\right)\right]$$

or
$$f_b(t) = 1.5 - \frac{3}{\pi}\sum_{k=1}^{\infty}\left[\frac{1}{k}\cos\left(\frac{2\pi k}{3} + \frac{\pi}{2}\right)\cos k\frac{2\pi t}{3} - \frac{1}{k}\sin\left(\frac{2\pi k}{3} + \frac{\pi}{2}\right)\sin k\frac{2\pi}{3}t\right]$$

i.e., for $f_a(t)$, $A_0 = 1.5$, $A_k = 0\ (k > 1)$, $B_k = -3/\pi k$; and for $f_b(t)$, $A_0 = 1.5$, $A_k = -(3/\pi k)\cos(2\pi k/3 + \pi/2)$ if $k > 1$,

$$B_k = -\left(\frac{3}{\pi k}\right)\sin\left(\frac{2\pi k}{3} + \frac{\pi}{2}\right)$$

The fact that $f_a(t)$ has a simpler form is the result of certain symmetry properties of $f_a(t)$. Such symmetry properties will now be considered in general.

EVEN FUNCTION A function $f(t)$ is said to be "even" with respect to the axis $t = a$ if $f(a + t) = f(a - t)$. In particular, if the axis of symmetry is $t = 0$ so that $f(t) = f(-t)$, the function is said to be an *even function.* The term even function is used because the Maclaurin series for such a function contains only even powers of the independent variable. As an example of an even function, we cite $f(t) = at^2$ or $f(t) = A\cos t$. Since $(-t)^2 = t^2$ and $\cos t = \cos(-t)$, these two functions are even functions. Inspection of Fig. 16-8a shows that the function shown, $f_1(t)$, is an even function. In Fig. 16-8b, $f_2(t)$ is symmetrical with respect to $t = 2$. In general, if $f(t)$ is an even function, then $f(t - a)$ is even with respect to $t = a$. It is left to the reader to show that sums and products of even functions are even functions. An even function can be identified by observing that the graph of the function for positive t is the mirror image of the graph for negative t. If a mirror is placed at $t = 0$ perpendicular to the t axis, the

[1] Such networks are called "linear-phase networks."

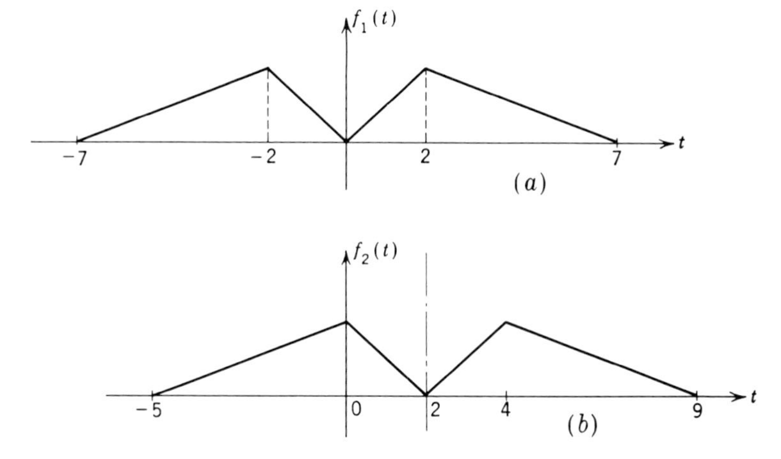

***Fig. 16-8** (a) The function $f_1(t)$ is an even function. (b) The function $f_2(t)$ is even with respect to (the axis) $t = 2$.*

waveform of $f(t)$ for negative values of t will be the image of its waveform for positive values of t. For this reason even functions are said to have "zero-axis" symmetry.

ODD FUNCTION A function is said to be odd with respect to an axis $t = b$ if $f(b + t) = -f(b - t)$. In particular, if $b = 0$ so that

$$f(t) = -f(-t)$$

then the function is said to be an *odd function*. The functions t^3 and $\sin t$ are examples of odd functions. In Fig. 16-9a an odd function is shown.

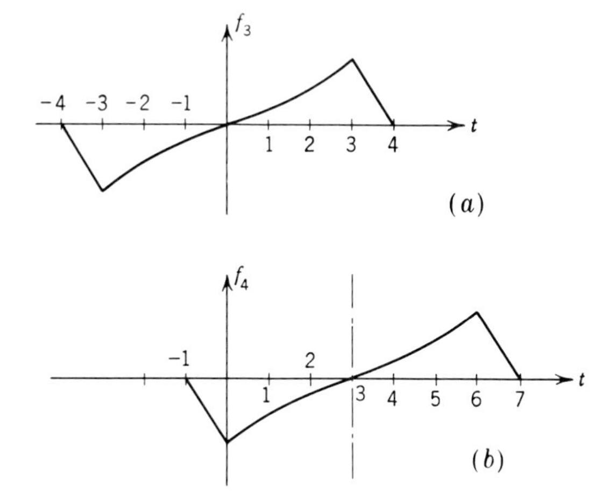

***Fig. 16-9** (a) The function $f_3(t)$ is an odd function. (b) The function $f_4(t)$ is odd with respect to (the axis) $t = 3$.*

The function $f_4(t)$ shown in Fig. 16-9*b* is odd with respect to $t = 3$ because $f_4(t+3) = -f_4(-t+3)$.

The sum of two odd functions is an odd function, but the product of two odd functions is an even function. To prove the latter, let $f_5(t)$ and $f_6(t)$ be two odd functions. Let $f(t) = f_5(t)f_6(t)$. Then $f(-t) = f_5(-t)f_6(-t)$. Since f_5 and f_6 are odd, $f_5(-t)f_6(-t) = (-)f_5(t)(-)f_6(t) = f_5(t)f_6(t) = f(t)$. Hence the product f_5f_6 is even.

EXPRESSION OF ARBITRARY FUNCTION IN TERMS OF ITS EVEN AND ODD COMPONENTS Let $f(t)$ be an arbitrary function. We can write

$$2f(t) = f(t) - f(-t) + f(t) + f(-t)$$

or

$$f(t) = \tfrac{1}{2}[f(t) + f(-t)] + \tfrac{1}{2}[f(t) - f(-t)] = \phi_1(t) + \phi_2(t)$$

In this equation $\phi_1(t) = \frac{1}{2}[f(t) + f(-t)]$ is an even function since

$$\phi_1(-t) = \tfrac{1}{2}[f(-t) + f(t)] = \phi_1(t)$$

Similarly, it can be shown that $\phi_2(t)$ is an odd function. Thus an arbitrary function is expressed as the sum of an even and an odd function.

INTEGRATION OF EVEN AND ODD FUNCTIONS In evaluating Fourier coefficients, integrals of the form

$$\int_{-a}^{+a} f(x)\sin x\,dx = \int_{-a}^{+a} F(x)\,dx$$

occur. We shall now show that, if $F(x)$ is an odd function, then

$$\int_{-a}^{+a} F(x)\,dx \equiv 0$$

independently of the value of a. To show this, we write

$$\int_{-a}^{+a} F(x)\,dx = \int_{-a}^{0} F(x)\,dx + \int_{0}^{a} F(x)\,dx$$

Now in the integral $\int_{-a}^{0} F(x)\,dx$, let $y = -x$; then

$$\int_{-a}^{0} F(x)\,dx = -\int_{a}^{0} F(-y)\,dy = \int_{0}^{a} F(-y)\,dy$$

Therefore

$$\int_{-a}^{+a} F(x)\,dx = \int_{0}^{a} F(-y)\,dy + \int_{0}^{a} F(x)\,dx$$

Since x and y are ("dummy") variables of integration,

$$\int_{0}^{a} F(-y)\,dy = \int_{0}^{a} F(-x)\,dx$$

Hence

$$\int_{-a}^{a} F(x)\,dx = \int_{0}^{a} [F(x) + F(-x)]\,dx \tag{16-25}$$

If $F(x)$ is an odd function, then $F(x) + F(-x) = 0$ and the integral (16-25) is zero. If $F(x)$ is an even function, then $F(x) + F(-x) = 2F(x)$. Hence

$$\int_{-a}^{+a} F(x)\,dx = \begin{cases} 0 & \text{if } F(x) \text{ is odd} \\ 2\int_0^a F(x)\,dx & \text{if } F(x) \text{ is even} \end{cases} \tag{16-26}$$

HALF-WAVE SYMMETRY If $f(t)$ is periodic with period T so that

$$f(t) = f(t + T)$$

and if, in addition,

$$f(t) = -f\left(t + \frac{T}{2}\right)$$

then the periodic function is said to have *half-wave* symmetry. In Fig. 16-10*a*, a waveform with half-wave symmetry is shown.

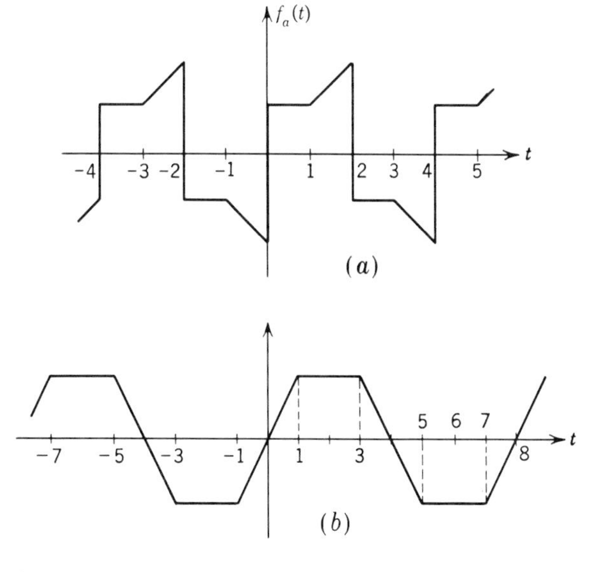

***Fig. 16-10** (a) A waveform with half-wave symmetry. (b) A waveform with odd quarter-wave symmetry.*

QUARTER-WAVE SYMMETRY If a periodic function has half-wave symmetry and in addition is either an even or an odd function, it is said to have even or odd quarter-wave symmetry. In Fig. 16-10*b* a waveform with odd quarter-wave symmetry is shown.

It is indicated in the next section that the calculation of the real Fourier coefficients is simplified if the periodic function has any of the symmetry properties discussed above.

16-11 Fourier coefficient of symmetrical waveforms

If $f(t)$ is an even function then $f(t)\cos(k2\pi t/T)$ is also even and $f(t)\sin\omega t$ is odd. In such cases it follows from Eq. (16-26) that $B_k = 0$ and A_k is given by

$$A_k = \frac{4}{T}\int_0^{T/2} f(t)\cos\left(k\frac{2\pi}{T}t\right)dt \tag{16-27}$$

Similar relations apply in cases of other symmetry; the pertinent formulas are given in Table 16-2. Proof of the formulas in Table 16-2 is left as an exercise for the reader.

Example 16-5 Find the Fourier series for the periodic odd function shown in Fig. 16-11.

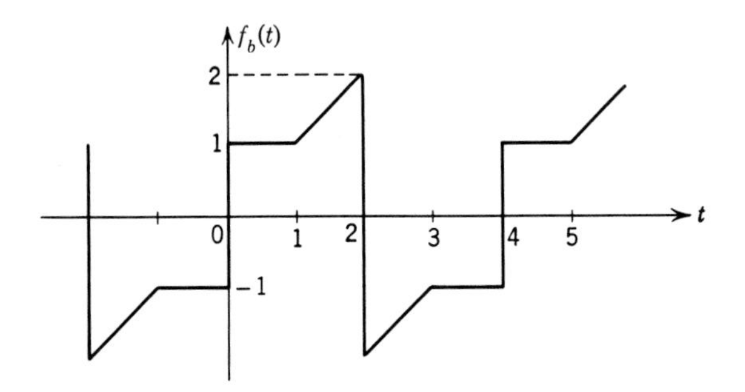

Fig. 16-11 ***An odd function.***

Solution The period T is 4. The function $f(t)$ is defined by the four equations

$$\begin{aligned} f(t) &= 1 && 0 < t < 1 \\ f(t) &= t && 1 < t < 2 \\ f(t) &= -4 + t && 2 < t < 3 \\ f(t) &= -1 && 3 < t < 4 \end{aligned}$$

If no use of the symmetry properties is made, four integrals must be evaluated to determine $\mathbf{D}_k$ or eight integrals to evaluate A_k and B_k. Since $f(t)$ is odd, we know immediately that $A_k = 0$. The general term B_k is then given by

$$B_k = \frac{4}{4}\int_0^1 (1)\sin k\frac{2\pi}{4}t\,dt + \frac{4}{4}\int_1^2 t\sin k\frac{2\pi}{4}t\,dt$$

Integrating and substituting the limits, we have

$$B_k = \frac{2}{\pi}\left[\frac{1-2(-1)^k}{k} - \frac{2}{\pi k^2}\sin k\frac{\pi}{2}\right] \tag{16-28}$$

Enumerating,

$$B_1 = \frac{2}{\pi}\left[1 - 2(-1) - \frac{2}{\pi}\sin\frac{\pi}{2}\right] = \frac{2}{\pi}\left[3 - \frac{2}{\pi}\right]$$

$$B_2 = \frac{2}{\pi}\left[\frac{1-2(-1)^2}{2} - \frac{2}{4\pi}\sin\pi\right] = -\frac{1}{\pi}$$

. . . .

Table 16-2 Fourier coefficients of symmetrical periodic waveforms

Type of symmetry	Conditions	Form of the Fourier series	Formulas for the coefficients
Odd function	$f(t) = -f(-t)$	$f(t) = \sum_{k=1}^{\infty} B_k \sin k \frac{2\pi}{T} t$	$A_k = 0;\ B_k = \frac{4}{T}\int_0^{T/2} f(t) \sin k \frac{2\pi}{T} t\, dt$
Even function	$f(t) = f(-t)$	$f(t) = \sum_{k=0}^{\infty} A_k \cos k \frac{2\pi}{T} t$	$B_k = 0;\ A_k = \frac{4}{T}\int_0^{T/2} f(t) \cos k \frac{2\pi}{T} t\, dt;\ A_0 = \frac{2}{T}\int_0^{T/2} f(t)\, dt$
Half-wave	$f(t) = -f\left(t + \frac{T}{2}\right)$	$f(t) = \sum_{k=1}^{\infty} \left[A_{2k-1} \cos (2k-1) \frac{2\pi}{T} t + B_{2k-1} \sin (2k-1) \frac{2\pi}{T} t \right]$	$\left.\begin{matrix} A_{2k-1} \\ B_{2k-1} \end{matrix}\right\} = \frac{4}{T}\int_0^{T/2} f(t) \begin{matrix}\cos \\ \sin\end{matrix} (2k-1) \frac{2\pi}{T} t\, dt$
Odd quarter-wave	$f(t) = -f(-t)$ and $f(t) = -f\left(t + \frac{T}{2}\right)$	$f(t) = \sum_{k=1}^{\infty} B_{2k-1} \sin (2k-1) \frac{2\pi}{T} t$	$A_k = 0;\ B_{2k-1} = \frac{8}{T}\int_0^{T/4} f(t) \sin (2k-1) \frac{2\pi}{T} t\, dt$
Even quarter-wave	$f(t) = f(-t)$ and $f(t) = -f\left(t + \frac{T}{2}\right)$	$f(t) = \sum_{k=1}^{\infty} A_{2k-1} \cos (2k-1) \frac{2\pi}{T} t$	$B_k = 0;\ A_{2k-1} = \frac{8}{T}\int_0^{T/4} f(t) \cos (2k-1) \frac{2\pi}{T} t\, dt$

Thus

$$f(t) = \frac{2}{\pi} \sum_{k=1}^{\infty} \left[\frac{1 - 2(-1)^k}{k} - \frac{2}{\pi k^2} \sin k \frac{\pi}{2} \right] \sin k \frac{\pi}{2} t$$

Example 16-6 Find the Fourier series for the function $f_a(t)$ shown in Fig. 16-12*a*, and use the result to obtain the Fourier series for $f_b(t)$ shown in Fig. 16-12*b*.

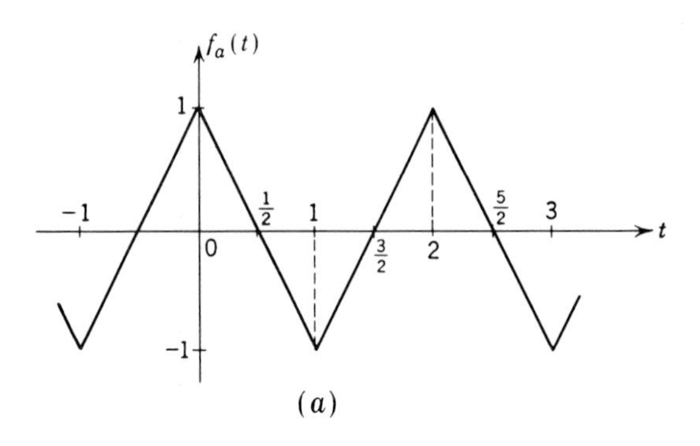

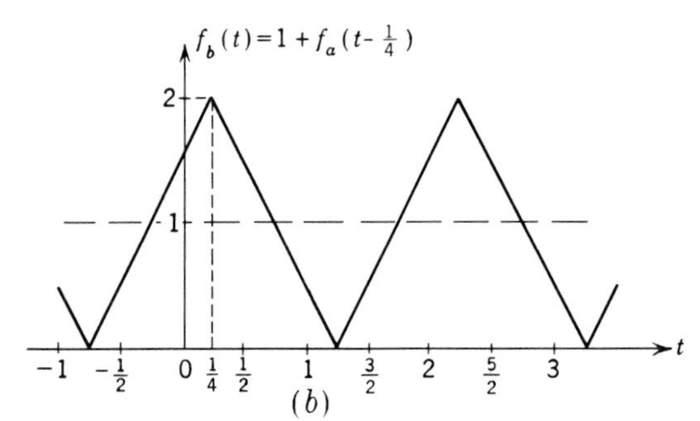

***Fig. 16-12** (a) An even function. (b) The function $f_b(t)$, as obtained from $f_a(t)$ by translation of the origin of coordinates, is neither even nor odd.*

Solution Since $f_a(t)$ has even quarter-wave symmetry, we need the formulas for f_a only in the interval $t = 0$ to $t = T/4 = 0.5$. In this interval

$$f_a(t) = 1 - 2t \qquad 0 < t < 0.5$$

Since f_a is an even function, $B_k = 0$. Since f_a has half-wave symmetry, $A_0 = 0$, $A_k = 0$, and k is even. Hence

$$A_{2k-1} = \frac{8}{2} \int_0^{0.5} (1 - 2t) \cos (2k - 1)\pi t \, dt$$

or

$$A_{2k-1} = \frac{8}{(2k - 1)^2 \pi^2} \tag{16-29}$$

Hence

$$f_a(t) = \frac{8}{\pi^2} \sum_{k=1}^{\infty} \frac{1}{(2k - 1)^2} \cos (2k - 1)\pi t$$

Since $f_b(t) = f_a(t - \frac{1}{4}) + 1$,

$$f_b(t) = 1 + \frac{8}{\pi^2} \sum_{k=1}^{\infty} \frac{1}{(2k-1)^2} \cos \left[(2k-1)\pi \left(t - \frac{1}{4} \right) \right]$$

or

$$f_b(t) = 1 + \frac{8}{\pi^2} \sum_{k=1}^{\infty} \frac{1}{(2k-1)^2} \cos \left[(2k-1)\pi t - (2k-1)\frac{\pi}{4} \right]$$

16-12 Differentiation and integration of Fourier series

The Fourier-series theorem states that, for any value of the independent variable t, the Fourier series converges to the value of its corresponding periodic function $f(t)$, except at points of discontinuity of $f(t)$, where the series converges to $\frac{1}{2}[f(t^-) + f(t^+)]$. When the periodic function has no discontinuities, the series converges everywhere to the value of the periodic function, and the derivative of this function will be equal to the sum of the derivatives of its harmonic components.

$$f(t) = \sum_{k=0}^{\infty} C_k \cos (k\omega t + \varphi_k)$$

$$\frac{d}{dt} f(t) = \sum_{k=1}^{\infty} -k\omega C_k \sin (k\omega t + \varphi_k)$$

or

$$\frac{d}{dt} f(t) = \sum_{k=1}^{\infty} k\omega C_k \cos \left(k\omega t + \varphi_k + \frac{\pi}{2} \right) \tag{16-30}$$

In this term-by-term differentiation the derivative of the average value of $f(t)$ is zero; thus we observe that the average value of the waveform of the derivative of a continuous periodic function is always zero.

If $f(t)$ has discontinuities, then, at the points of discontinuity, the derivative will be an impulse function whose strength is equal to the jump of the function at the point of discontinuity.

From the above discussion it is seen that if C_k, the Fourier coefficients of continuous periodic function, are known, the Fourier coefficients of the derivative of that waveform are known and are equal to $k\omega C_k$.

INTEGRATION The integral of a periodic waveform whose average value is nonzero will not be a periodic function. This is seen from

$$\int f(t)\, dt = \int \left[A_0 + \sum_{k=1}^{\infty} C_k(k\omega t + \varphi_k) \right] dt$$

$$= A_0 t + \int \sum_{k=1}^{\infty} C_k \cos (k\omega t + \varphi_k)\, dt \tag{16-31}$$

The term $A_0 t$ is not periodic, and therefore the integral is not periodic. However, if the average value of $f(t)$ is zero, the integral $\int f(t)\, dt$ will be periodic and its Fourier coefficients are related to those of $f(t)$ by

$$\int f(t)\,dt = \sum_{k=1}^{\infty} \int C_k \cos(k\omega t + \varphi_k)\,dt = K + \sum_{k=1}^{\infty} \frac{C_k}{k\omega} \cos\left(k\omega t + \varphi_k - \frac{\pi}{2}\right)$$

In Fig. 16-13*a*, a periodic waveform is shown whose Fourier series is

$$f(t) = -\frac{3}{\pi} \sum_{k=1}^{\infty} \frac{1}{k} \sin k \frac{2\pi}{3} t \tag{16-32}$$

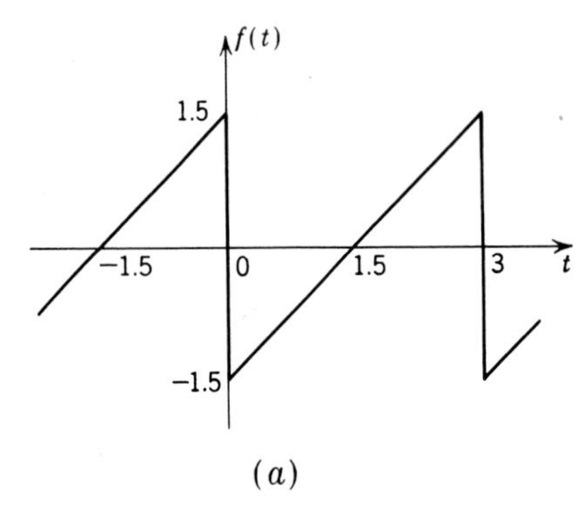

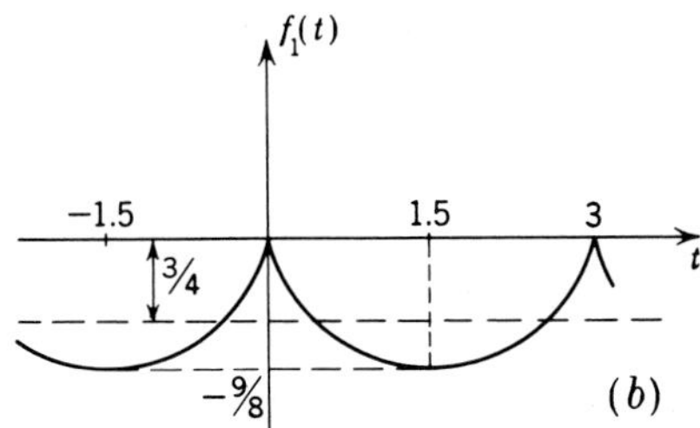

***Fig. 16-13** (a) A periodic waveform $f(t)$. (b) The function $f_1(t)$ is an integral of $f(t)$, chosen so that $f_1(0) = 0$.*

In Fig. 16-13*b* an integral of this waveform, $f_1(t)$, is shown. The Fourier series of $f_1(t)$ is given by

$$f_1(t) = \int_{-\infty}^{t} f(\tau)\,d\tau = K - \frac{3}{\pi} \sum_{n=1}^{\infty} \int \frac{1}{n} \sin n \frac{2\pi}{3} t = K + \frac{9}{2\pi^2} \sum_{n=1}^{\infty} \frac{1}{n^2} \cos n \frac{2\pi}{3} t$$

The value of K for the particular function $f_1(t)$ is obtained by observing that

$$f_1(0) = 0 = K + \frac{9}{2\pi^2} \sum_{n=1}^{\infty} \frac{1}{n^2}$$

$$K = -\frac{9}{2\pi^2} \sum_{n=1}^{\infty} \frac{1}{n^2}$$

The value of K can also be found from the waveform of $f(t)$. In the range $0 < t < 3$, $f(t) = t - 1.5$,

$$f_1(t) = 0.5t^2 - 1.5t \qquad 0 < t < 3$$

which checks at $t = 0$; that is, $f_1(0) = 0$.

$$K = \frac{1}{1.5}\int_0^{1.5} (0.5t^2 - 1.5t)\,dt = -0.75$$

From the above discussion it is seen that the effect of integration of a waveform is to reduce the relative magnitude of the higher harmonics. In the case of the waveform of Fig. 16-13*a*, the ratio of the amplitude of adjacent terms is

$$\frac{C_n}{C_{n+1}} = \frac{n+1}{n} = 1 + \frac{1}{n}$$

whereas for the integral of this waveform the ratio of the amplitude of the nth harmonic to that of the $(n+1)$st is

$$\frac{(n+1)^2}{n^2} = 1 + \frac{2}{n} + \frac{1}{n^2} > 1 + \frac{1}{n}$$

In general, the Fourier coefficients of integral of a waveform decrease at a faster rate than those of the function itself. In Chap. 2 we noted that the process of integration "smooths" a waveform, and this smoothing effect is the cause of the reduction of the amplitude of the higher harmonics. Similarly, differentiating a waveform will increase its "sharpnesses," and this will cause larger amplitudes for the higher harmonics.

In general, if we have two waveforms of the same period, but one changing its value more "rapidly" than the other, then, if we compare the Fourier series of the two waveforms, we shall find that the one corresponding to the waveform with rapid variation will have "larger" relative amplitudes for its higher harmonics.

USE OF DIFFERENTIATION FOR EVALUATION OF FOURIER COEFFICIENTS[1] For certain types of waveforms it is possible to compute the Fourier coefficients with little or no integration. In Fig. 16-14*a*, a trapezoidal waveform is shown. To find the complex Fourier coefficients of this waveform from the formula

$$\mathbf{D}_k = \frac{1}{T}\int_0^T f(t)e^{-jk\omega t}\,dt$$

three separate integrals must be evaluated in the regions $0 < t < 1$, $1 < t < 2$, $2 < t < 4$. This is a somewhat long and tedious procedure. However, if the waveform of Fig. 16-14*a* is differentiated twice, the waveforms of Fig. 16-14*b* and *c* are obtained. Let

$$f(t) = \sum_{n=-\infty}^{+\infty} \mathbf{D}_n e^{jn\omega t}$$

Then

$$\frac{d}{dt}f(t) = f'(t) = \sum_{n=-\infty}^{+\infty} jn\omega\mathbf{D}_n e^{jn\omega t} = \sum_{n=-\infty}^{+\infty} \mathbf{D}'_n e^{jn\omega t}$$

$$\frac{d^2}{dt^2}f(t) = f''(t) = \sum_{n=-\infty}^{+\infty} -\omega^2 n^2\mathbf{D}_n e^{jn\omega t} = \sum_{n=-\infty}^{+\infty} \mathbf{D}''_n e^{jn\omega t}$$

[1] See J. F. Gibbons, A Simplified Procedure for Finding Fourier Coefficients, *Proc. IRE*, vol. 45, no. 2, p. 243.

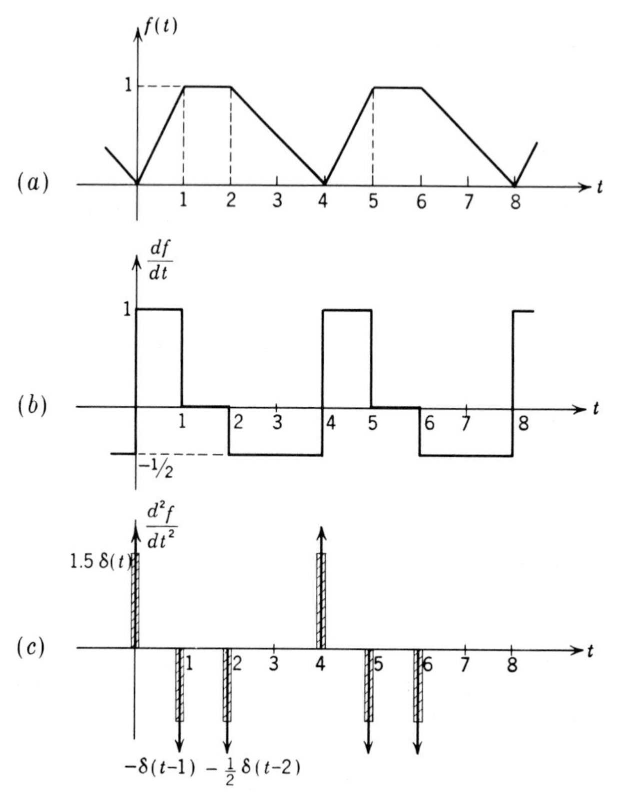

Fig. 16-14 A periodic waveform and the waveforms of its first two derivatives.

where the $\mathbf{D}_n''$ are the Fourier coefficients of the function $f''(t)\,dt$ and

$$\mathbf{D}_n = -\frac{1}{\omega^2 n^2}\mathbf{D}_n''$$

In the interval of time $0 \leq t < 4^-$, the function $f''(t)$ can be written

$$f''(t) = 1.5\delta(t) - \delta(t-1) - \tfrac{1}{2}\delta(t-2)$$

Now, if we find $\mathbf{D}_n''$ from $f''(t)$, we have

$$\mathbf{D}_n'' = \frac{1}{T}\int_{0^-}^{T^-} f''(t)e^{-jn\omega t} = \frac{1}{4}\int_{0^-}^{4^-} [1.5\delta(t) - \delta(t-1) - \tfrac{1}{2}\delta(t-2)]e^{-jn\omega t}\,dt$$

But we know that

$$\int_{t=x^-}^{t=x^+} \delta(t-x) = 1 \qquad \text{and} \qquad \int_{t=x^-}^{x^+} e^{at}\delta(t-x)\,dt = e^{ax}$$

Therefore

$$\mathbf{D}_n'' = \tfrac{1}{4}(1.5 - e^{-jn\omega} - \tfrac{1}{2}e^{-j2n\omega})$$

In this case

$$\omega = 2\pi/4 = \pi/2,$$
$$\mathbf{D}_n'' = \tfrac{1}{4}(1.5 - e^{-jn\pi/2} - \tfrac{1}{2}e^{-jn\pi}) = \tfrac{1}{4}[1.5 - (-j)^n - \tfrac{1}{2}(-j)^{2n}]$$

and
$$\mathbf{D}_n = -\frac{1}{\omega^2 n^2}\mathbf{D}_n'' = -\frac{1}{\pi^2 n^2}[1.5 - (-j)^n - \tfrac{1}{2}(-j)^{2n}]$$

Thus, if the waveform lends itself to differentiation in such a manner that the result of successive differentiation is a number of delta functions, the evaluation of the Fourier coefficients does not entail integration of the form $\mathbf{D}_n = (1/T)\int_0^T f(t)e^{-jn\omega t}\,dt$, but may be computed from expressions such as $\int_{0^-}^{t_1^+} \delta(t - t_1)e^{-jn\omega t}\,dt = e^{-j\omega n t_1}$.

In addition to simplifying the computation, this method clearly shows the relationship between Fourier coefficients of a waveform and its derivatives and also introduces the concept of representation of the derivative of a function as a sum of a number of delta functions.

16-13 Discrete frequency spectra

Since a periodic function can be represented as the sum of terms of the form $C_k \cos(2\pi kt/T + \varphi_k)$ (that is, as the sum of its harmonics), the specification of the amplitudes C_k and the phase angles φ_k of these harmonics uniquely determines the periodic function. These specifications can be presented in graphical form by a plot of the values C_k against integral (discrete) values of k, together with a similar plot for φ_k. The Fourier series for the periodic function shown in Fig. 16-15*a* can be written in the form

$$f(t) = \frac{1}{6}\left[1 + 2\sum_{k=1}^{\infty} \frac{\sin(k\pi/6)}{k\pi/6} \cos\left(\frac{k\pi t}{3} - \frac{k\pi}{6}\right)\right] \tag{16-33}$$

Hence

$$C_0 = \frac{1}{6} \qquad C_1 = \frac{1}{3}\frac{\sin(\pi/6)}{\pi/6} \quad \cdots \qquad \varphi_1 = \frac{-\pi}{6},\ \varphi_2 = \frac{-\pi}{3},\ \cdots$$

In Fig. 16-15*b* the values of C_k are shown as a function of k, and in Fig. 16-15*c* the corresponding values of φ_k are presented. (Note that C_k is negative for k between 6 and 12; it is often customary to plot the absolute values, $|C_k|$, which are of course positive. Since $-\cos x = \cos(x \pm \pi)$, a phase angle of π is then added or subtracted from φ_k when C_k is negative.) We observe that C_k and φ_k have values only for integral values of k, that is, at those frequencies which are integral multiples of the fundamental frequency. Hence the spectra are referred to as "discrete" frequency spectra.

For the waveform of Fig. 16-16*a*, we obtain

$$C_k = 2\frac{\tau}{T}\frac{\sin(k\pi\tau/T)}{k\pi\tau/T} \qquad \varphi_k = -\frac{k\pi\tau}{T} \tag{16-34}$$

Proof of these relations is left as an exercise for the reader.

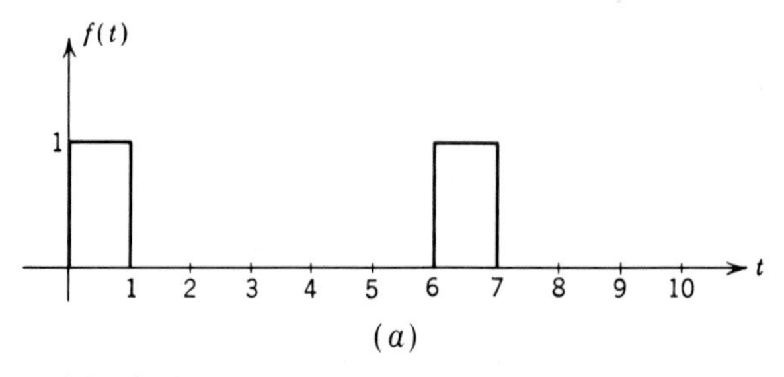

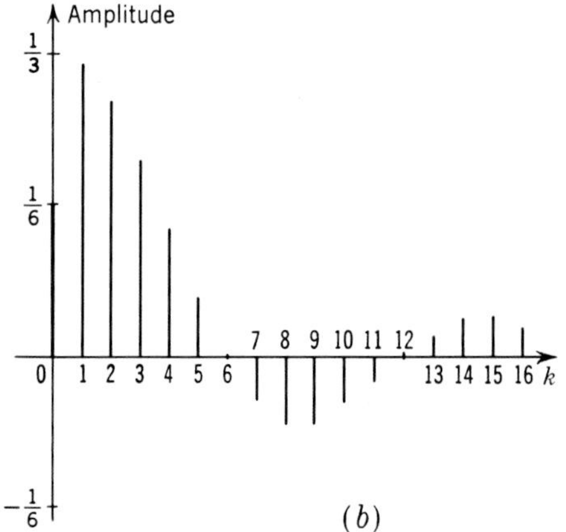

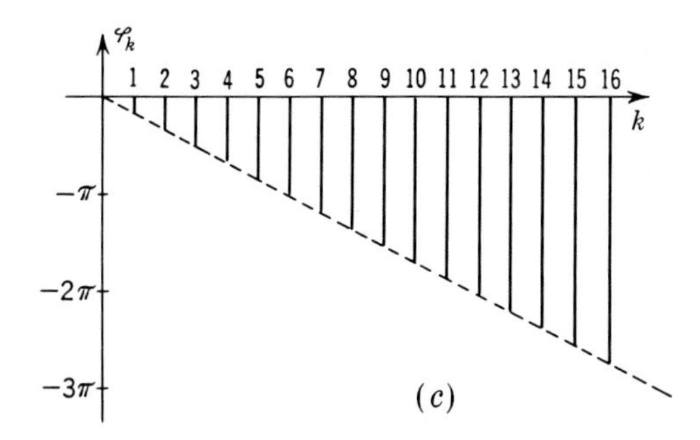

***Fig. 16-15** (a) Periodic rectangular pulses. (b) Plot of C_k as a function of k for the pulses of (a). (c) Plot of ϕ_k as a function of k for the pulses of (a).*

In Eq. (16-34), if we let $k\pi\tau/T = x_k$, then we have

$$C_k = 2 \frac{\tau}{T} \frac{\sin x_k}{x_k} \tag{16-35}$$

In Eq. (16-35) we have changed the (discrete) variable k to the (discrete) variable x_k. In frequency analysis, terms of the form $(\sin x)/x$ appear so frequently that it is useful to become acquainted with its graph. Because of the usage of this function in sampling problems, it is called the sampling function, and the symbol Sa x is assigned to it; Sa $x \equiv (\sin x)/x$.

Figure 16-16*b* shows Sa x plotted against x. In plotting this waveform x is dimensionless (radians). This plot is made for continuous variations of x, but in our present applications we are interested only in *selected* (*discrete*) values of x, such as $x_k = k\pi\tau/T$. Since $-1 \le \sin x \le +1$, it is seen that Sa x goes to zero as x goes to infinity. However, as x increases, $\sin x$ oscillates between positive and negative values, and this is reflected in the variation of Sa x. It is noted that $(\sin x)/x$ is unity for $x = 0$.

In the discrete frequency spectrum of the waveform shown in Fig. 16-16a, given by Eq. (16-34), the values of $TC_k/2\tau$ vary in such a manner that the line representing them always ends on Sa $(k\pi\tau/T)$. The value of C_0 is exempt from this rule in this instance.

Discrete complex frequency spectra It is often convenient to use the complex form of the Fourier series to obtain a frequency spectrum. Since $C_k = 2D_k$, one can plot D_k as a function of k. Since $D_k = D_{-k}$, the amplitude spectrum obtained in this manner always has even symmetry. Similarly, since $\varphi_k = -\varphi_{-k}$, the phase spectrum has odd symmetry. For the waveform of Fig. 16-16a,

$$\mathbf{D}_k = \frac{\tau}{T} \frac{\sin (k\pi\tau/T)}{(k\pi\tau/T)} e^{-j(k\pi\tau/T)}$$

Hence $D_k/D_0 = |\text{Sa}\ x_k|$, where $x_k = k\pi\tau/T$ and $\varphi_k = -x_k$, except that where Sa x_k is negative, an odd multiple of π is added to φ_k, so that no information is lost. The spectra obtained in this manner for the waveform of Fig. 16-16a in the case $\tau/T = \frac{1}{6}$ are shown in Fig. 16-17a and b.

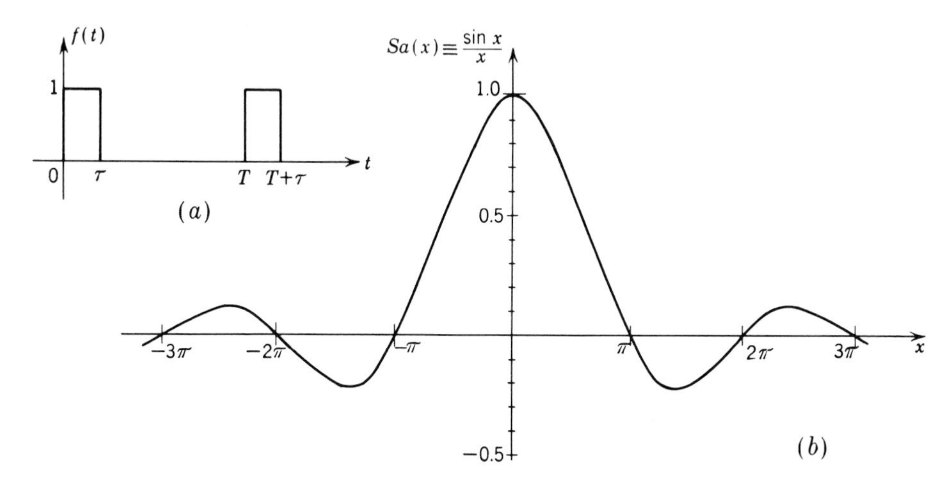

***Fig. 16-16** (a) Periodic rectangular pulses of duration τ and repetition rate 1/T. (b) Graph of the function Sa x ≡ (sin x)/x.*

The reader recognizes that each of the lines in the spectrum of Fig. 16-15b corresponds to the amplitude of a harmonic. For example, in Fig. 16-15b, the line corresponding to $k = 3$ is $C_3 = 0.287$, the amplitude of the third harmonic in the Fourier series of the waveform of Fig. 16-15a. In Fig. 16-17a we have "amplitudes" corresponding to both positive and negative

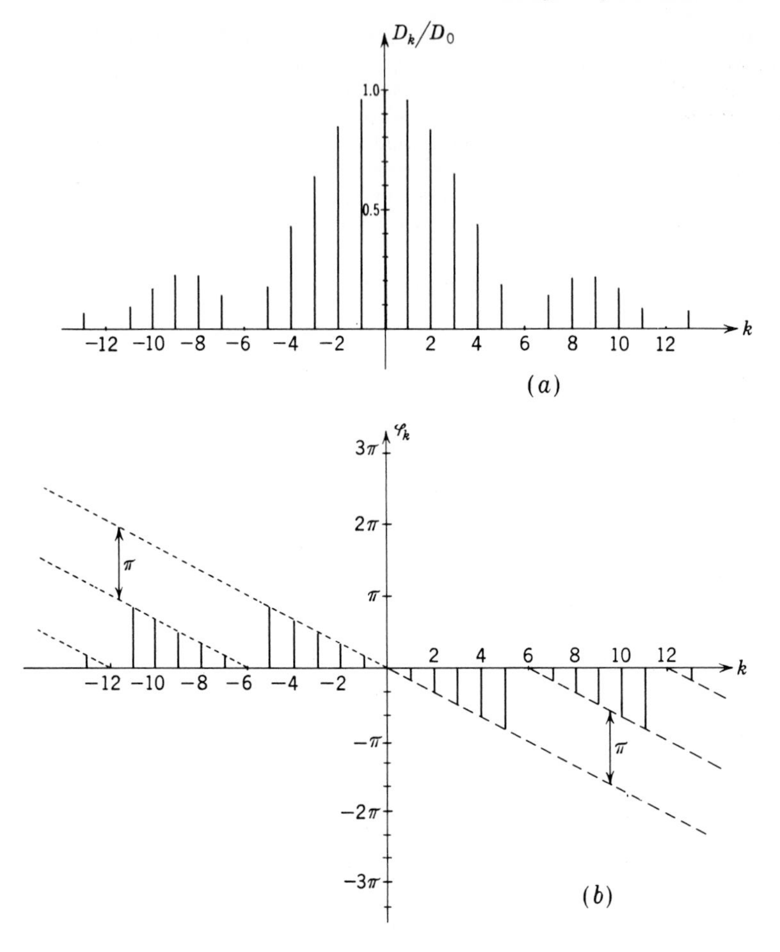

***Fig. 16-17** (a) Values of $\mathbf{D}_k/\mathbf{D}_0$ for the waveform of Fig. 16-16a, where $\tau/T = \frac{1}{6}$. This waveform corresponds to that shown in Fig. 16-15a. (b) Phase spectrum corresponding to the amplitude spectrum shown in (a). Note that an odd multiple of π has been introduced for $6 < k < 12$ because absolute values are shown in (a).*

values of k or $k\omega$. For example, $D_{-3} = D_3 = 0.143$. This is caused by the fact that, in obtaining the complex Fourier series, the identity

$$C_k \cos (k\omega t + \varphi_k) = \tfrac{1}{2}\mathbf{D}_k e^{jk\omega t} + \tfrac{1}{2}\mathbf{D}_{-k} e^{-jk\omega t}$$

was used, and both $\mathbf{D}_k$ and $\mathbf{D}_{-k}$ refer to the same harmonic, namely, $C_k \cos (k\omega t + \varphi_k)$.

It is noted that a periodic function can be specified by giving its waveform in the time domain or its complex spectrum (i.e., both the amplitude and the phase) in the frequency domain. In this book we have so far been chiefly concerned with the time-domain analysis of network variables. In this chapter we have already shown the advantage of analysis in the frequency domain. This subject is fully studied in more advanced courses in analysis, but in the remainder of this chapter we shall become acquainted with the basic concepts of analysis in the frequency domain.

16-14 Power content of periodic waveforms

In dealing with functions which represent voltages and currents, the square of these functions occurs whenever we deal with power or energy problems. For this reason a quantity which is associated with the square of the function is defined; this quantity, which is closely related to the rms value, is called the *power content* of the function, defined as the mean squared value.

$$\text{Power content} = F^2 = \frac{1}{T}\int_0^T [f(t)]^2\, dt \tag{16-36}$$

We now recall that the rms value of a periodic function is defined by

$$F = F_{\text{rms}} = \sqrt{\frac{1}{T}\int_0^T [f(t)]^2\, dt}$$

It follows that the power content of a periodic waveform is the mean squared value of the waveform. The term "power content" is used because a voltage with mean squared value V^2, or a current with mean squared value I^2, in a unit resistance results in power of value V^2 or I^2, respectively. More precisely, the term "power content referred to unit resistance" may be used in place of "power content."

From the definition of rms values it is clear that the power content of a periodic waveform can be calculated without reference to Fourier series. It is, however, the object of this section to relate the value of F^2 to the Fourier coefficients. In particular, we shall show that

$$F^2 = \sum_{-\infty}^{+\infty} \mathbf{D}_k\mathbf{D}_k^* \tag{16-37}$$

Proof Let the periodic function $f(t)$ be written as the Fourier series

$$\begin{aligned} f(t) = D_0 &+ \mathbf{D}_1 e^{j\omega t} + \mathbf{D}_2 e^{j2\omega t} + \cdots + \mathbf{D}_k e^{jk\omega t} + \cdots \\ &+ \mathbf{D}_{-1} e^{-j\omega t} + \mathbf{D}_{-2} e^{-j2\omega t} + \cdots + \mathbf{D}_{-k} e^{-jk\omega t} + \cdots \end{aligned}$$

Then

$$[f(t)]^2 = D_0 \sum_{-\infty}^{+\infty} \mathbf{D}_k e^{jk\omega t} + \mathbf{D}_1 e^{j\omega t} \sum_{-\infty}^{+\infty} \mathbf{D}_k e^{jk\omega t} + \cdots$$

or

$$[f(t)]^2 = \sum_{k=-\infty}^{+\infty} \sum_{m=-\infty}^{+\infty} \mathbf{D}_k \mathbf{D}_m e^{j(k+m)\omega t} \tag{16-38}$$

Integrating both sides of Eq. (16-38) with respect to t between the limits 0 and $T = 2\pi/\omega$, and using the orthogonality property

$$\int_0^T e^{j(k+m)\omega t}\, dt \begin{cases} = 0 & k + m \neq 0 \\ = T & k + m = 0 \end{cases}$$

we have the result

$$\frac{1}{T}\int_0^T [f(t)]^2\,dt = F^2 = \sum_{k=-\infty}^{+\infty} \mathbf{D}_k\mathbf{D}_k^*$$

Since

$$\sum_{k=-\infty}^{+\infty} \mathbf{D}_k\mathbf{D}_k^* = \mathbf{D}_0^2 + \mathbf{D}_1\mathbf{D}_{-1} + \mathbf{D}_{-1}\mathbf{D}_1 + \cdots$$

we can also write

$$F^2 = D_0^2 + 2\sum_{k=1}^{\infty} D_k^2 \tag{16-39}$$

Using the relationships between $\mathbf{D}_k$, C_k, A_k, and B_k, we have the alternative forms

$$F^2 = C_0^2 + \sum_{k=1}^{\infty}\frac{C_k^2}{2} = A_0^2 + \tfrac{1}{2}\sum_{k=1}^{\infty}(A_k^2 + B_k^2) \tag{16-40}$$

We now observe that

$$F^2 = C_0^2 + \sum_{k=1}^{\infty}\left(\frac{C_k}{\sqrt{2}}\right)^2 = \sum_{k=0}^{\infty}(\text{rms value of } k\text{th harmonic})^2$$

Thus the mean squared value of a waveform is given as the sum of the mean squared values of its harmonics.

Example 16-7 The current in a 5-ohm resistance is given by

$$i(t) = 5 + 14.14\cos t + 7.07\cos 2t$$

Calculate the average power delivered to the resistance.

Solution For the given waveform, the amplitudes of the harmonics and the corresponding rms values are

$$I_0 = 5 \qquad I_{1m} = 14.14 \qquad I_1 = 10 \qquad I_{2m} = 7.07 \qquad I_2 = 5$$

Hence

$$I_{\text{rms}} = I = \sqrt{5^2 + 10^2 + 5^2} = \sqrt{150}$$

and since $P = I^2R$, $P = (150)(5) = 750$ watts.

Application to steady-state-power calculations If, at the terminals a-b of a network, the voltage v_{ab} is defined by the Fourier series

$$v_{ab}(t) = \sum_{k=0}^{\infty} V_{k_m}\cos(k\omega t + \varphi_k)$$

and the current entering terminal a is $i(t)$ defined similarly as

$$i(t) = \sum_{k=0}^{\infty} I_{k_m}\cos(k\omega t + \psi_k)$$

then the power input as a function of time is given by the double sum

$$p_{ab}(t) = \sum_{n=0}^{\infty}\sum_{k=0}^{\infty} V_{k_m}I_{m_n}\cos(k\omega t + \varphi_k)\cos(n\omega t + \psi_k)$$

Using the orthogonality property of sinusoidal functions in the integral $\int_0^T p_{ab}\,dt$, we obtain the following expression for the *average* power, P_{ab}:

$$\frac{1}{T}\int_0^T p_{ab}\,dt = P_{ab} = V_0 I_0 + \tfrac{1}{2}[V_{1m}I_{1m}\cos(\varphi_1 - \psi_1) + V_{2m}I_{2m}\cos(\varphi_2 - \psi_2) + \cdots] \quad (16\text{-}41)$$

or if we let $\varphi_k - \psi_k = \theta_k$, that is, the angle by which the kth-harmonic voltage leads the kth-harmonic current, and if we express the amplitudes V_{km} and I_{km} by means of their rms values (for example, $I_{km} = \sqrt{2}\,I_k$), then we have the result

$$P = P_0 + P_1 + P_2 \cdots = \sum_{k=0}^{\infty} P_k \quad (16\text{-}42)$$

where $P_k = V_k I_k \cos\theta_k$

In words: The average power delivered to a terminal pair is the sum of the average powers delivered by the individual harmonics.

This remarkable "superposition of average power" property is an exceedingly attractive feature of Fourier analysis as applied to circuit problems. We note that, in general, power is not a linear function of voltage or current and cannot be superposed; yet the orthogonality property of sinusoids which are harmonically related permits the superposition of the average powers delivered by the harmonics to give total average power.

Example 16-8 The voltage at terminals a-b of a terminal pair is

$$v_{ab} = 100 + 100\cos t + 50\cos 2t + 30\cos 3t$$

and the current entering terminal a is

$$i(t) = 10\cos(t - 60°) + 2\cos(3t - 135°)$$

Calculate the average power delivered to the terminal pair a-b.

Solution $V_0 = 100$, $I_0 = 0$, and hence $P_0 = 0$. Since $V_1 = (100)(0.707)$ and $I_1 = (10)(0.707)$, $V_1 I_1 = 500$. For the fundamental the phase difference between voltage and current is 60°, cos 60° = 0.5, and hence $P_1 = 250$ watts. Because $i_2 = 0$, $P_2 = 0$. For the third harmonic $V_3 I_3 = 30$, and $\cos\theta_3 = -0.707$; hence $P_3 = -21.2$, $P = 250 - 21.2 = 228.8$ watts.

16-15 Power content in a given frequency range

We have seen that the power content of a periodic waveform is associated with the sum of the squares of its Fourier coefficients. The systems or devices studied in communication problems respond differently to different frequencies. In Chaps. 7, 10, and 15, this phenomenon was discussed in terms of the bandwidth of the circuit representing the device. For this reason it is often of interest to know the power content of a waveform in a specified range (band) of frequencies.

The power content of a periodic waveform whose fundamental frequency is f_0 in the band of frequencies from f_1 to f_2 is

$$\sum_{k=m}^{k=n} \tfrac{1}{2}C_k^2$$

where m is the next largest integer to the number f_1/f_0, and n is the next smallest integer to the number f_2/f_0. Thus, for example, if a periodic waveform has a fundamental frequency of 5, then the power content in the band $f_1 = 7$ to $f_2 = 21$ is

$$\frac{1}{2}\sum_{k=2}^{k=4} C_k^2 = \frac{C_2^2 + C_3^2 + C_4^2}{2}$$

For the sawtooth waveform of Fig. 16-13a,

$$f(t) = -\frac{3}{\pi}\sum_{k=1}^{\infty}\frac{1}{k}\sin k\frac{2\pi t}{3} \tag{16-43}$$

the power content is

$$F_q = \frac{1}{3}\int_0^3 (-1.5 + t)^2\,dt = 0.75 = \frac{1}{2}\sum_{k=1}^{\infty}\left(-\frac{3}{k\pi}\right)^2$$

and the power content in the band $f = 1.1$ to $f = 2$ is

$$\frac{1}{2}\sum_{k=4}^{k=6}\left(-\frac{3}{k\pi}\right)^2 = \frac{9}{2\pi^2}\left(\frac{1}{16} + \frac{1}{25} + \frac{1}{36}\right) = 0.0297$$

Hence, in the band $1.1 < f < 2$, the energy content is 7.91 per cent of the total (since $7.91 = 0.0297/0.75$).

16-16 Frequency-selective networks

If a network contains energy-storing elements, then in general, the steady-state response waveform due to a nonsinusoidal periodic source will not be the same as the source waveform. We may explain this by observing that the network responds differently to the individual harmonics of the input waveform.

To illustrate this approach, consider the R-C network shown in Fig. 16-18a as a two-port. Let the input waveform be given by the finite Fourier series

$$v_1(t) = 100(\cos t + \cos 3t + \cos 5t)$$

as shown in Fig. 16-18b.

For the R-C two-port of Fig. 16-18a, the phasor ratio $\mathbf{V}_2/\mathbf{V}_1$ at any radian frequency ω is

$$\frac{\mathbf{V}_2}{\mathbf{V}_1} = \frac{1/j\omega C}{R + 1/j\omega C} = \frac{1}{j\omega RC + 1}$$

$$= \frac{1}{\sqrt{\omega^2R^2C^2 + 1}}\underline{/-\tan^{-1}\omega RC} \tag{16-44}$$

Since the fundamental radian frequency of the input $v_1(t)$ is unity, the phasor ratio for the kth harmonic is

$$\left.\frac{\mathbf{V}_2}{\mathbf{V}_1}\right|_k = \frac{1}{\sqrt{k^2R^2C^2 + 1}}\underline{/-\tan^{-1}kRC} \tag{16-45}$$

It follows from the principle of superposition that the output waveform in the steady

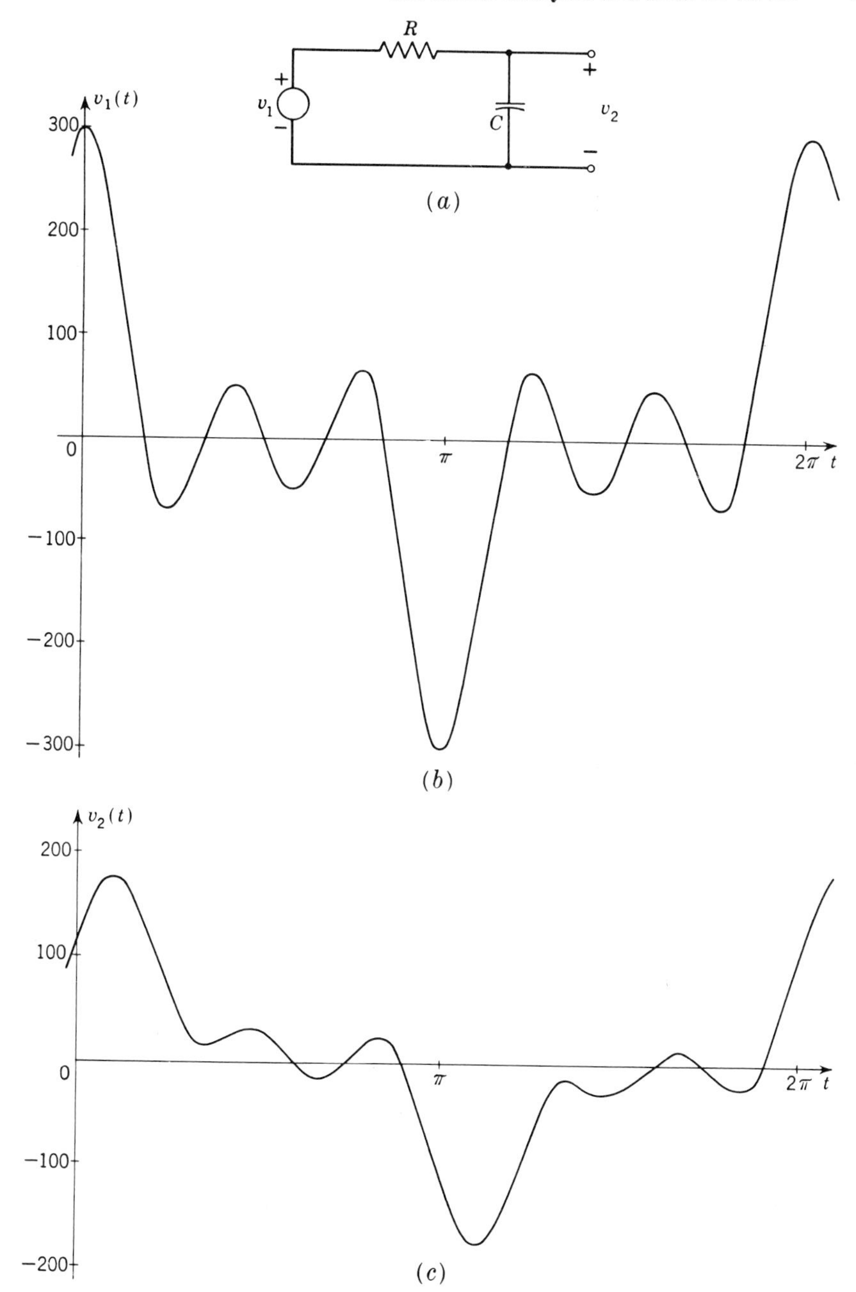

***Fig. 16-18** (a) An R-C two-port. (b) Graph of $v_1(t) = 100(\cos t + \cos 3t + \cos 5t)$. (c) Graph of the steady-state response $v_2(t)$ when $RC = \frac{1}{2}$.*

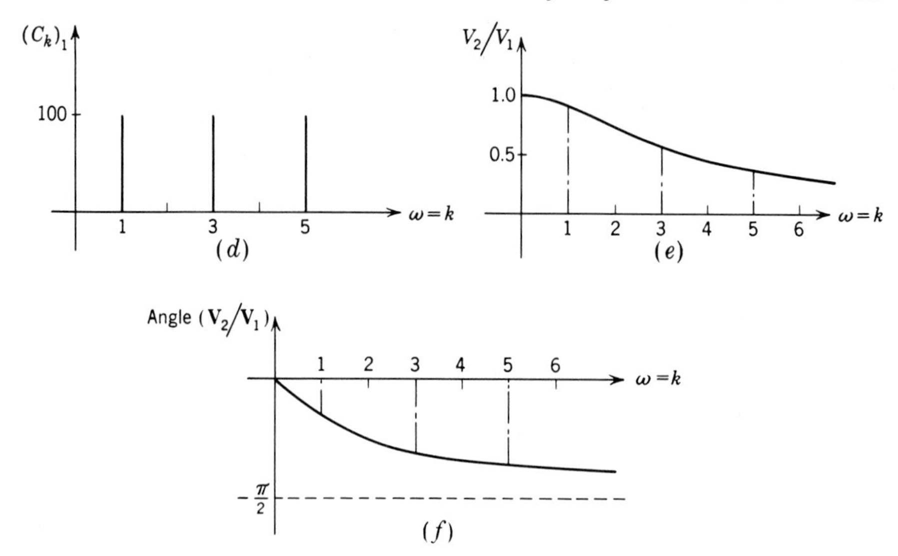

***Fig. 16-18** (d) Spectrum of $v_1(t)$. (e) Amplitude response of the two-port. (f) Phase response of the two-port.*

state is given by the equation

$$v_2(t) = 100\left[\frac{1}{\sqrt{R^2C^2+1}}\cos(t - \tan^{-1} RC) + \frac{1}{\sqrt{9R^2C^2+1}}\cos(3t - \tan^{-1} 3RC) + \frac{1}{\sqrt{25R^2C^2+1}}\cos(2t - \tan^{-1} 5RC)\right]$$

The waveform of $v_2(t)$ is shown in Fig. 16-18*c* for the case $RC = \frac{1}{2}$.

Another way of looking at the result of Fig. 16-18*c* is to study how the spectrum of the input waveform is modified by the network to produce the response spectrum. For the input waveform $v_1(t)$, the amplitude spectrum is shown in Fig. 16-18*d*. The amplitude and phase responses of the network for $RC = \frac{1}{2}$ are shown in Fig. 16-18*e* and *f*. From Eq. (16-45), it follows that each line in the amplitude spectrum of the steady-state response is found by multiplying the corresponding spectral amplitude of the input by the magnitude of the network (transfer) function. Each line in the phase spectrum is found by adding the phase angle of the input signal at each harmonic to the corresponding angle of the network function at that frequency. The results, in the example under discussion, are shown dotted in Fig. 16-18*e* and *f*. Thus, in general,

$$\begin{pmatrix}\text{Complex spectrum}\\ \text{of response}\end{pmatrix} = \begin{pmatrix}\text{complex network}\\ \text{function}\end{pmatrix}\begin{pmatrix}\text{complex spectrum}\\ \text{of input}\end{pmatrix}$$

16-17 Distortionless transmission

If the relationship between the source function $v_1(t)$ and a response function $v_2(t)$ is of the form

$$v_2(t) = h_0 v_1(t - T_0)$$

then the network transmits the input signal without distortion. In the above expression it is seen that the waveform of the response is delayed in time by T_0 with respect to the input and that the magnitude of the output is h_0 times the magnitude of a corresponding phase of the input. If h_0 exceeds 1, it is referred to as the "gain" of the network; if h_0 is less than 1, it is called the "attenuation."

If $v_1(t)$ is a periodic signal whose fundamental radian frequency is ω_0 and if its spectrum is given by $\mathbf{D}_k = \frac{1}{2}C_k e^{j\varphi k}$, from the definition of distortionless transmission it follows that the amplitude coefficients in the response v_2 must be $h_0 D_k$ and the phase angles in the response must be $\varphi_k - k\omega_0 T_0$. We conclude that, for distortionless transmission, the transfer function of the network, $H(j\omega)$, must have the form

$$H(j\omega) = \frac{\mathbf{V}_2}{\mathbf{V}_1} = h(\omega)e^{j\theta(\omega)} = h_0 e^{-j\omega T_0}$$

i.e., the amplitude response $h(\omega)$ (which in general varies with frequency) must be constant h_0, and the phase response must be proportional to frequency. If this condition is met, then

$$\text{Complex Fourier coefficient in the response} = h_0 \mathbf{D}_k e^{-jk\omega_0 T_0}$$

where $\mathbf{D}_k$ is the corresponding Fourier coefficient in the input. The required frequency response of a network for distortionless transmission is shown in Fig. 16-19.

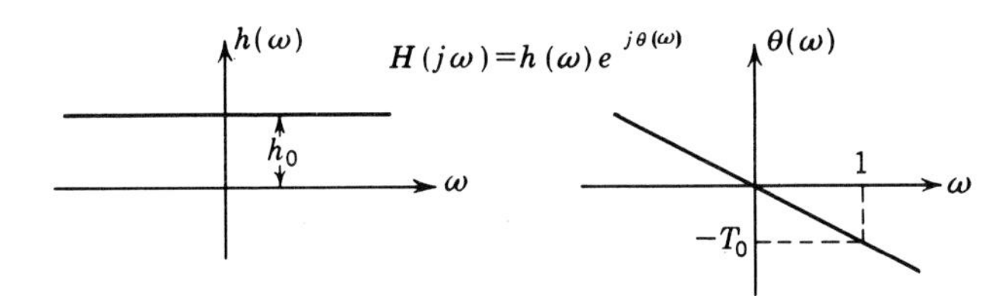

***Fig. 16-19** Amplitude and phase characteristics of a network which transmits signals without distortion.*

16-18 Ideal filters

From the frequency-response curve of the R-C network of Fig. 16-18*a* shown in Fig. 16-18*e*, we observe that the network passes the low frequencies with little change in amplitude and with almost linear phase shift ($\tan^{-1} \omega RC \approx \omega RC$ if $\omega RC \ll 1$), and the high frequencies ($\omega RC > 1$) are attenuated to a greater degree. We recall that this discrimination is called "filtering action" and that in this sense the R-C two-port of Fig. 16-18*a* is called a "low-pass filter." Similarly, an R-L-C circuit or the tuned transformer circuits are referred to as a "bandpass filters."

Now, in practice, no network can satisfy the requirements for distortionless transmission at all frequencies. Because of ever-present "stray" capacitance between input terminals and output terminals, and because of lead inductance, the high frequencies will eventually be attenuated. To discuss distortionless transmission, therefore, we restrict the discussion to a *band of frequencies.* In this connection it is useful to define an "ideal" filter as a filter which passes a band of frequencies with uniform attenuation (gain) and with linear phase characteristics within this band; moreover, an ideal filter must furnish infinite attenuation outside the passband. The frequency characteristics of some ideal filters are shown in Fig. 16-20. (Note

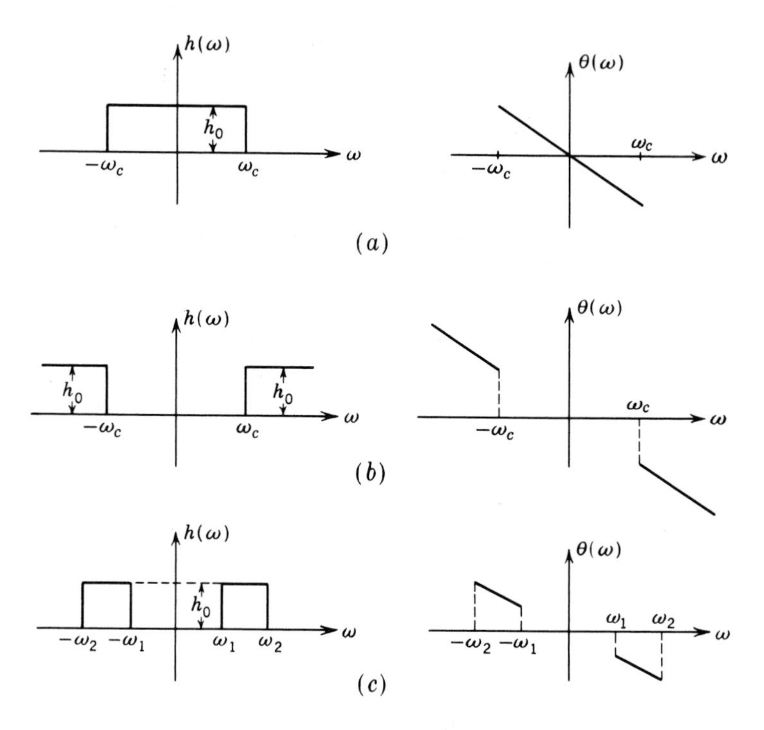

***Fig. 16-20** Frequency-domain characteristics of ideal filters.* $H(j\omega) = h(\omega)e^{j\theta(\omega)}$. *(a) Ideal low-pass filter. (b) Ideal high-pass filter. (c) Ideal band-pass filter.*

that both "negative" and "positive" frequencies are shown; this is done as a convenience when the complex Fourier coefficients are used.)

Examining the transfer characteristics of the ideal filters, we observe that the "cutoff" is with infinite slope, so that, for example, in the case of the low-pass filter (Fig. 16-20*a*), $h(\omega) = h_0$ at $\omega = \omega_c^-$ and $h(\omega) = 0$ at $\omega = \omega_c^+$. It can be shown that, in any network, the frequency-response curve must be a continuous function of ω with continuous derivatives and that the infinite slope shown cannot be attained with a finite number of *R-L-C* elements.[1]

[1] It can also be shown that the use of an infinite number of elements can result in infinite $dh/d\omega$ provided that infinite time delay is associated with the filter.

Although ideal filters are not physically attainable, it is frequently possible to obtain useful results by approximating a frequency-response curve with "sharp" cutoff, with ideal filter characteristics such as are shown in Fig. 16-20.

16-19 Sampling theorem[1] BAND-LIMITED SIGNALS A periodic function is a *band-limited* function if in its (discrete) spectrum the values of $\mathbf{D}_n$ are identically zero for $|n| > N$, where N is a finite integer. For example, the function

$$v_1(t) = 100(\cos t + \cos 3t + \cos 5t)$$

shown in Fig. 16-18*b* is band-limited because the harmonics for $n > 5$ are identically zero. From the above definition it follows that an ideal low-pass or bandpass filter whose input is an arbitrary periodic function has an output which is a band-limited periodic function. For many practical purposes the output of nonideal filters (transmission systems) may be considered band-limited.

According to the above definition, a band-limited periodic function has the form

$$f(t) = \sum_{k=-N}^{+N} \mathbf{D}_k e^{jk(2\pi/T)t} = C_0 + \sum_{k=1}^{N} C_k \cos\left(k\frac{2\pi}{T}t + \varphi_k\right) \tag{16-46}$$

We observe from Eq. (16-46) that if T is known, we need only the $2N + 1$ *numbers*

$$C_0, C_1, \ldots, C_N; \qquad \varphi_1, \varphi_2, \ldots, \varphi_N$$

to specify the *function* $f(t)$.

> *Sampling theorem* A band-limited periodic function with no harmonics of order higher than N is uniquely specified by its value at $2N + 1$ instants within one period.

The proof of the theorem follows from Eq. (16-46). If we substitute the $2N + 1$ values $f(t_1), f(t_2), \ldots, f(t_{2N+1})$, referred to as samples of f, in Eq. (16-46) we obtain $2N + 1$ simultaneous equations for the unknowns

$$C_0, C_1, \ldots, C_N \qquad \varphi_1, \ldots, \varphi_N$$

or

$$D_0 \qquad \text{Re}\,\mathbf{D}_1, \ldots, \text{Re}\,\mathbf{D}_N \qquad \text{Im}\,\mathbf{D}_1, \ldots, \text{Im}\,\mathbf{D}_N$$

The solution of these equations uniquely specifies the Fourier series and therefore specifies $f(t)$.

UNIFORM SAMPLING We observe that the instants $t_1, \ldots, t_{2N+1}$ can in principle be chosen arbitrarily. Most often one chooses these instants at equal intervals of $T/(2N + 1)$. This type of sampling is called *uniform* sampling. This theorem may be used to evaluate Fourier coefficients of a band-limited function by sampling. However, the sampling procedure is not merely used for evaluation of Fourier coefficients, but has the following practical applications.

PULSE AMPLITUDE MODULATION In Fig. 16-21 is illustrated a function $f(t)$ whose period is 2π and is known to be band-limited with $N = 2$. Five uniform samples are marked at $t = 0, 2\pi/5, 4\pi/5, 6\pi/5, 8\pi/5$. Below $f(t)$ a pulse train is shown. This pulse train is periodic with the same period as $f(t)$, each period consisting of five pulses whose amplitude and spacing correspond to the sampled values of $f(t)$. Fourier analysis of the

[1] A more detailed discussion is given in Javid and Brenner, *op. cit.*

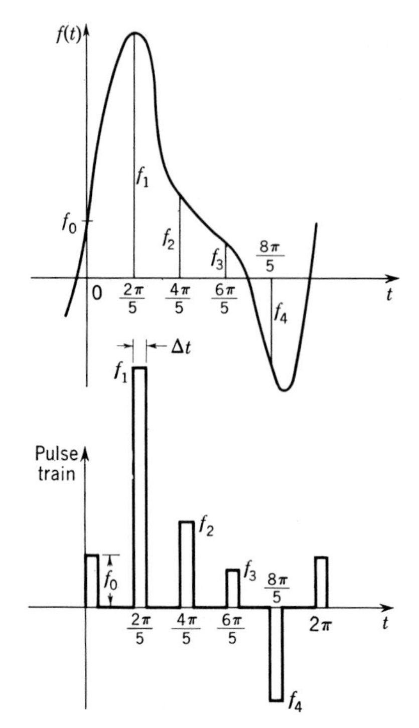

***Fig. 16-21** The periodic function $f(t)$ is known to have no harmonics of higher order than the second. Shown are five uniform samples $f_0, \ldots, f_4$. The pulse train has five pulses per period, and the height of the pulses corresponds to the sampled values of f.*

pulse train shows (Prob. 16-32) that it is not band-limited but that, for the pulse train D_0, $\mathbf{D}_1$ and $\mathbf{D}_2$ are proportional to the corresponding coefficients of $f(t)$.

If the pulse train is passed through an "ideal" low-pass filter which passes its second but not third and higher harmonics, the output of the low-pass filter will be proportional to $f(t)$. We conclude that, to transmit the band-limited signal, it is sufficient to transmit the pulse train. The comparative economy in power required to transmit the train of pulses, although evident, may not be an important factor. More important is the fact that, in the interval between the sampled pulses, the transmission channel may be used for transmitting other signals. The term used to describe the use of one "communication channel" for several "messages" is *multiplexing*. Special filter networks and techniques are used to interlace and separate signals for multiplex transmission.

16-20 Applications to nonlinear circuits

In the entire book so far we have discussed only linear circuits, i.e., circuits composed of linear elements which are described by linear equilibrium equations. As a consequence, the treatment of circuit analysis has in fact been a study of the superposition principle and its consequences. Indeed, the idea of Fourier series stems directly from the idea of superposition, i.e., the representation of source and response function as the sum of sinusoids so that each response component can be calculated individually.

When a circuit includes one (or more) nonlinear elements, the equilibrium equation is not linear and, as can easily be demonstrated, the superposition principle does not apply. In spite of this, it is possible to use the concept of Fourier series to discuss certain properties of nonlinear circuits. It is the object of this section to illustrate in a very elementary and introductory fashion some of these properties.

In a *linear* circuit the component of the response due to the source will include a harmonic for every harmonic of the input. For a given signal the amplitude and phase of these harmonics depend only on the network; they are unique and independent of the initial conditions.

In a *nonlinear* circuit the following possibilities exist:

1 *Harmonics which are not present in the input signal occur in the steady-state response.* Thus, for example, in a rectifier, the input may be sinusoidal, and the output will contain a d-c component and an infinite number of harmonics. As a second example, consider an electronic oscillator. The input is a constant voltage, and the output may be sinusoidal.
2 *Subharmonics may be present in the output.* In certain nonlinear circuits, the fundamental frequency in the response may be a proper fraction of the fundamental frequency of the source function.
3 *The response amplitudes may not be a unique function of the network parameters.* In certain nonlinear circuits, the response depends on how the steady state is approached. Thus, for a given input signal, the network may have a different steady-state response, depending on the initial conditions. This property also illustrates that no free and forced response components can be superposed.

Although the analysis (and design) of linear circuits (i.e., circuits which are to be realized by "sufficiently" linear elements) can be systematized and standardized, as indeed we have shown in the preceding chapters, no such standardization is possible in the case of nonlinear circuits. Each problem must be examined individually, and solutions, usually approximate, must be worked out for special numerical cases. In fact, many times it is necessary to use experimental evidence as the starting point in an analytical solution.

PROBLEMS

16-1 Make careful freehand sketches of the following functions: (*a*) $f_a(x) = 10 \cos x + 5 \cos 2x$; (*b*) $f_b(x) = 10 \sin x + 5 \cos 2x$; (*c*) $f_c(x) = 10 \cos (x - \pi/4) + 5 \cos 2x$; (*d*) $f_d(x) = 10 \cos x + 5 \cos 3x$; (*e*) $f_e(x) = 10 \cos x - 5 \sin 3x$.

16-2 A voltage source $v(t) = 10 + 10 \cos \omega t + 5 \cos 2\omega t$ is impressed on the series *R-C* circuit as shown in Fig. P16-2. It is known that $R = 1/\omega C$. Calculate

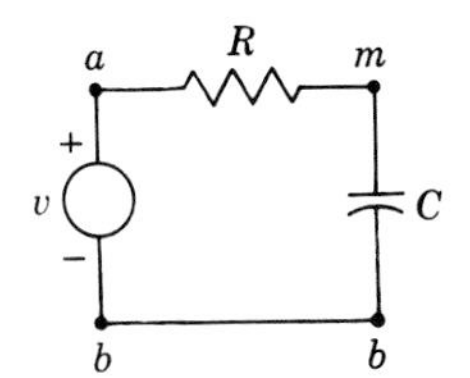

Fig. P16-2

(*a*) the steady-state functions v_{am} and v_{mb}; (*b*) the complete responses v_{am} and v_{mb} if $\omega = 1/RC = 10$ and if (1) $v_{mb}(0^+) = 0$, (2) $v_{mb}(0^+) = 25$.

16-3 In the circuit of Fig. P16-2, the source is given by $v(t) = 10 \cos \omega t + 10 \cos n\omega t$. If $\omega RC = 1$, calculate the smallest integral value of n so that, in the steady-state component of v_{mb}, the amplitude of the nth harmonic is less than 5 per cent of the amplitude of the fundamental.

16-4 A voltage source $v(t) = 10 + 10 \cos \omega t + 10 \cos 2\omega t + 10 \cos 3\omega t$ is impressed on the series *R-L-C* circuit shown in Fig. P16-4. Calculate the steady-state functions v_{am}, v_{mn}, and v_{nb} if $\omega L = 1.5$, $1/\omega C = 6.0$, and (*a*) $R = 3.0$; (*b*) $R = 0.60$. (*c*) For each case calculate Q_0 of the circuit, and use the result of this example to explain the bandpass-filter property of this circuit.

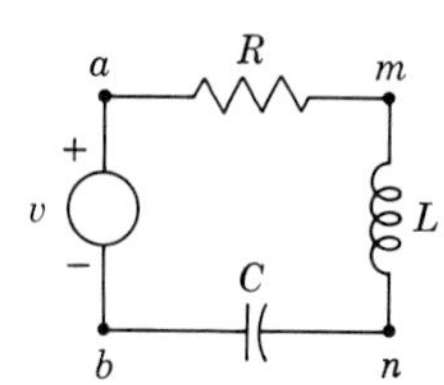

Fig. P16-4

16-5 The polynomial functions $P_0(x) = 1$, $P_1(x) = x$, $P_2(x) = \frac{1}{2}(3x^2 - 1)$, $P_3(x) = \frac{1}{2}(5x^3 - 3x)$, $P_4(x) = \frac{1}{8}(35x^4 - 30x^2 + 3)$ are the first five members of a set of functions (termed Legendre polynomials) which are orthogonal in the interval $-1 \le x \le 1$ with respect to the weighting function $w(x) = 1$. Verify this statement by application of the definition of orthogonality.

16-6 The set of functions $T_n(x) = (1/2^{n-1}) \cos (n \cos^{-1} x)$, called Tchebycheff polynomials, are orthogonal in the interval $-1 \le x \le +1$ with respect to the weighting function $w(x) = (1 - x^2)^{-\frac{1}{2}}$. (*a*) Prove the orthogonality with the aid of the substitution $\cos^{-1} x = \theta$. (*b*) Expand $T_n(x)$ for $n = 1, 2, 3$, and 4, and show that the results are polynomials in x.

16-7 (*a*) Show that $\int_0^{2\pi} \left(\sum_{-n}^{+n} \mathbf{D}_k e^{jkx} \right)^2 dx = 2\pi \sum_{-n}^{+n} \mathbf{D}_k \mathbf{D}_{-k}$. (*b*) If a periodic function $f(x)$ with period 2π, $f(x) = f(x + 2\pi)$, is approximated by $f_a(x)$ such that $f_a(x) = \sum_{-n}^{+n} \mathbf{D}_k e^{jkx}$, then the error in using the approximation is $\epsilon = f(x) - f_a(x)$. Use the result of (*a*) to show that the mean squared error M is given by

$$M = \frac{1}{2\pi} \int_0^{2\pi} [f(x)]^2 \, dx - \frac{1}{\pi} \left[f(x) \sum_{-n}^{+n} \mathbf{D}_k e^{jkx} \right] dx + \sum_{-n}^{+n} \mathbf{D}_k \mathbf{D}_{-k}$$

(*c*) Show that the mean squared error is a minimum if $\mathbf{D}_k$ is calculated as the complex Fourier coefficient according to Eq. (16-14).

16-8 The periodic function shown in Fig. 16-3, which is defined by the formula $v(t) = V$, $0 < t < \pi$, $v(t) = -V$, $\pi < t < 2\pi$, and $v(t) = v(t + 2\pi)$, is to be approximated by the finite Fourier series $v_a(t)$ as given below. Calculate the mean squared error if (*a*) $v_a(t) = B_1 \sin t$; (*b*) $v_a(t) = B_1 \sin t + B_3 \sin 3t$.

16-9 Obtain the Fourier series in complex form for the waveforms given in Fig. P16-9, and express the answer in the form of Eq. (16-6).

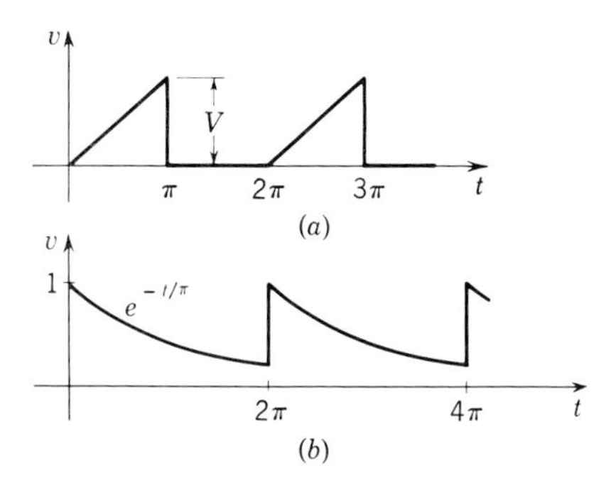

Fig. P16-9

16-10 (*a*) Show that the Fourier series for the waveform given in Fig. P16-10 is

$$v(t) = \frac{1}{\pi} \sum_{k=1}^{\infty} \left[\frac{\sin k\pi/2}{k} \cos kt + \frac{(1 - \cos k\pi/2)}{k} \sin kt \right]$$

(*b*) Use the result of (*a*) to calculate the value of $v(t)$ *as given by the Fourier series* at the points of discontinuity $t = 0$ and $t = \pi/2$. *Hint:* Recall that the Maclaurin series for $\tan^{-1} x$ is $\tan^{-1} x = \sum_{1}^{\infty} \frac{(-1)^{k+1} x^{2k-1}}{2k-1}$.

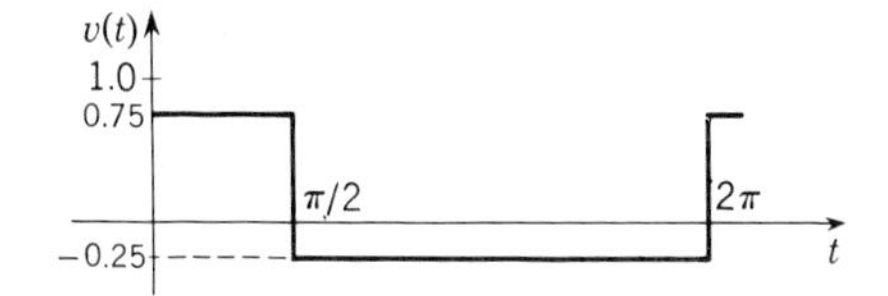

Fig. P16-10

16-11 (*a*) Assume that the time scale for a periodic function $f(t)$ has been normalized so that the period is 2π. Show, by use of the trapezoidal rule, dividing the period into n equal intervals, that the coefficients of the Fourier series in trigonometric form can be obtained approximately by use of the formulas

$$A_0 = \frac{1}{n} \sum_{m=0}^{n-1} f\left(m \frac{2\pi}{n}\right)$$

$$A_k = \frac{2}{n} \sum_{m=0}^{n-1} f\left(m \frac{2\pi}{n}\right) \cos\left(km \frac{2\pi}{n}\right)$$

$$B_k = \frac{2}{n} \sum_{m=0}^{n-1} f\left(m \frac{2\pi}{n}\right) \sin\left(km \frac{2\pi}{n}\right)$$

(*b*) Show that the formulas given for A_k and B_k are ambiguous unless they are used only to calculate the harmonic coefficients for $0 < k < n/2$ for waveforms which are known to have negligible harmonics for $k > n/2$. To show this, first let $n = 6$, and expand the formula for A_1; then use the same formula for A_5, and compare the result. Do the same for A_2 and A_4. Finally, show in general that the use of the formula for A_k and A_{n-k} gives $A_k = A_{n-k}$, and discuss the resulting ambiguity. Repeat for B_k. (*c*) Use the formulas of (*a*) to calculate B_1 and B_3 for the waveform of Fig. 16-3. Use $n = 8$. (When f is discontinuous at $m2\pi/n$, use its average value.) Compare your result with the exact values. (*d*) Calculate A_0 for the waveforms of Fig. P16-9, using (1) $n = 4$, (2) $n = 6$, (3) $n = 8$, and compare the result with the exact values.

16-12 To expedite the computational work in connection with the approximate formulas of Prob. 16-11, the work can be arranged in tabular form, called, after its discoverer, the Runge schedule. In particular, it is observed that $f(m2\pi/n)$ is multiplied by the various factors such as $\cos(km2\pi/n)$. A typical Runge chart is shown below for $n = 6$. For use, tracing paper is placed over such a chart, and the numbers for a particular example are entered in the spaces. Note that in the chart use of the identity $\cos(km2\pi/6) = \cos[k(6 - m)2\pi/6]$ has been made. The abbreviation $f_m = f(m2\pi/n)$ is used. Note that the A_3 column is omitted since no B_3 column can be used for $n = 6$. Use this chart to evaluate the first three terms of the Fourier series for the waveform in Fig. P16-9*a*.

	Multiplying factors, cosine coefficients		
	A_0	A_1	A_2
f_0	1.00	1.00	1.00
$f_1 + f_5$	1.00	0.50	−0.50
$f_2 + f_4$	1.00	−0.50	−0.50
f_3	1.00	−1.00	1.00
×	Add column above	Add column above	Add column above
Ans.:	Divide by 6	Divide by 3	Divide by 3

	Multiplying factors, sine coefficients	
	B_1	B_2
$f_1 - f_5$	0.87	0.87
$f_2 - f_4$	0.87	−0.87
×	Add column above	Add column above
Ans.:	Divide by 3	Divide by 3

16-13 Construct a chart similar to that given in Prob. 16-12 for $n = 12$. Use it to calculate the first three terms of the Fourier series for the waveform of Fig. P16-9*a*, and compare the result with that obtained in Prob. 16-12.

16-14 (*a*) Prove that $\sum_{k=0}^{n-1} e^{jk2\pi/n} = 0$ where n is an integer. *Hint:* $e^{jk2\pi/n} = (e^{j2\pi/n})^k$ so that the series given is a finite geometric series. (*b*) Prove that

$$\sum_{k=0}^{n-1} \cos\frac{k2\pi}{n} = 0 \qquad \text{and} \qquad \sum_{k=0}^{n-1} \sin\frac{k2\pi}{n} = 0$$

16-15 (*a*) Use the result of Prob. 16-14 to prove that

$$\sum_{m=0}^{n-1} \cos m\frac{k2\pi}{n} = \begin{cases} 0 & \text{if } \dfrac{k}{n} \text{ is not an integer} \\ n & \text{if } \dfrac{k}{n} \text{ is an integer} \end{cases}$$

(*b*) Given an even function whose Fourier series has no constant term, that is, $f(x) = \sum_{k=1}^{\infty} A_k \cos kx$, show that

$$\sum_{k=1}^{\infty} A_{nk} = \frac{1}{n} \sum_{m=0}^{n-1} f\left(m \frac{2\pi}{n}\right)$$

(*c*) Show that the result of (*b*) applies when the function is neither even nor odd.
(*d*) Use this result to evaluate A_3, A_2, and A_1 for the waveform of Fig. P16-10, assuming that $A_k = 0$ for $k \geq 4$. *Note:* This problem is an introduction to the "method of selected ordinates." Details may be found in R. H. Frazier, "Elementary Electric Circuit Theory," pp. 279–286, McGraw-Hill Book Company, New York, 1945.

16-16 For the waveform shown in Fig. P16-16: (*a*) Obtain the Fourier series analytically if the point $t = 0$ is at A. (*b*) Use the result of (*a*) to write the Fourier series if $t = 0$ is at (1) B; (2) C.

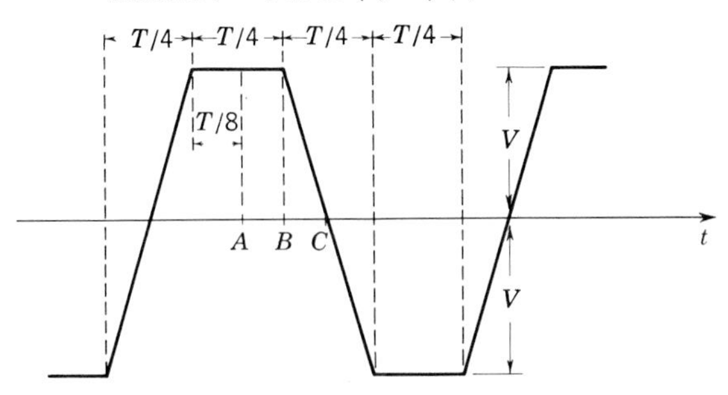

Fig. P16-16

16-17 Referring to Fig. P16-17: (*a*) Obtain analytically the Fourier series for $f_a(x)$. (*b*) Use the result of (*a*) to obtain the Fourier series for $f_b(x)$ by superposition.

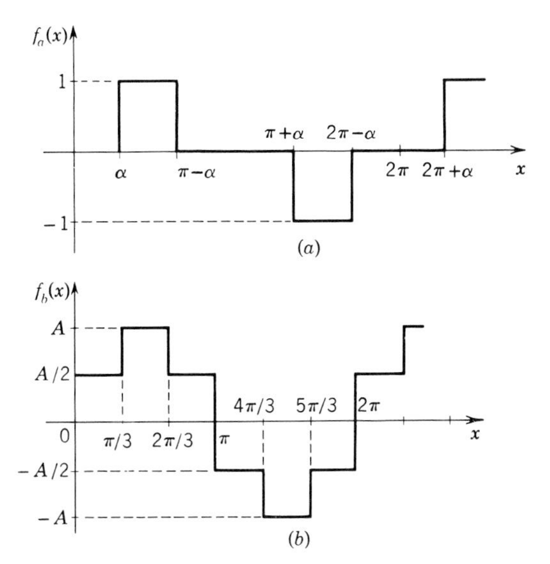

Fig. P16-17

16-18 (*a*) A function is defined through $f_a(x) = V_m \sin(x/2)$, $0 \le x \le 2\pi$, $f_a(x) = f_a(x + 2\pi)$ as shown in Fig. P16-18. Find, analytically, the Fourier series for $f_a(x)$. (*b*) Sketch the function $(V_m/2) \sin(x/2) + \frac{1}{2}f_a(x)$, and write its Fourier series. (*c*) Write the Fourier series for the waveforms of (*a*) and (*b*) if the variable is changed to $\omega t = x/2$.

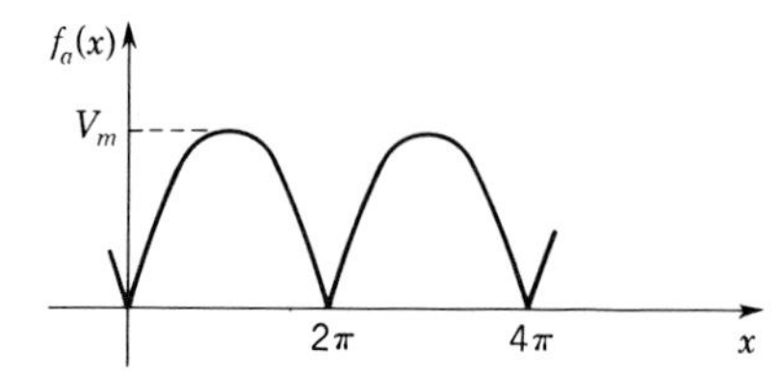

Fig. P16-18

16-19 Find the Fourier series for the derivative of the waveform of Prob. 16-16*a*, and correlate the answer with the result of Prob. 16-17*a*.

16-20 Sketch the amplitude and phase spectrum of the waveforms of (*a*) Prob. 16-16*a* (*b*) Fig. P16-17*a* for $\alpha = \pi/4$; (*c*) Fig. P16-17*a* for $\alpha = 9\pi/20$.

16-21 Calculate the rms value of the following waveforms, both exactly by applying the definition of rms value to the waveform directly, and approximately by use of the Fourier coefficients: (*a*) waveforms of Probs. 16-18*a* and 16-18*b*; (*b*) waveform of Fig. P16-17*a* with $\alpha = \pi/4$; (*c*) waveform of Fig. P16-16.

16-22 The voltage across a terminal pair *a-b* is $v_{ab} = 100 + 100 \cos \omega t + 30 \cos 3\omega t$. The current entering the *a* terminal is

$$i_{ab} = 50 \cos(\omega t - 45°) + 30 \sin(3\omega t - 60°) + 30 \cos 5\omega t$$

Calculate (*a*) the rms value of v_{ab}; (*b*) the rms value of i_{ab}; (*c*) the average power input P_{ab}.

16-23 (*a*) Calculate the average power delivered by the source in Probs. 16-4*a* and 16-4*b* (Fig. P16-4). (*b*) In Prob. 16-4*a* calculate the average value of the energy stored in L and in C if $\omega = 100$ rad/sec.

16-24 An amplitude-modulated signal is defined through the equation $f(t) = [1 + m(t)] \cos \omega_c t$. Obtain the frequency spectrum if (*a*) $m(t) = m_0 \cos \omega_s t$; (*b*) $m(t) = m_0 \cos \omega_s t + \frac{1}{2} m_0 \cos 2\omega_s t$. For each case sketch the amplitude spectrum, using $\omega_c = 2{,}000\omega_s$. *Hint:* Use trigonometric identities.

16-25 For the waveforms of Prob. 16-24 assume that $\omega_c = 1{,}000\omega_s$, and calculate the power content of the waveforms with m_0 as a parameter. What fraction of the power content is contributed by the harmonic at ω_c?

16-26 In Prob. 16-24*a* let $\omega_s = 0.001$, $\omega_c = 1$. Calculate the steady-state voltage across the resistance in an *R-L-C* circuit if $f(t)$ is a voltage source, $LC = 1$, and Q_0 is (*a*) 10; (*b*) 50; (*c*) 200. Use $m_0 = 1$.

16-27 A certain active nonlinear device has an output voltage v_o which is related to the input v_i through the equation $v_o = v_i + v_i^2$. Obtain the amplitude spectrum of the output voltage if (*a*) $v_i = \sin \omega t$; (*b*) $v_i = \sin \omega_1 t + \sin \omega_2 t$; (*c*) $v_i = 1 + \sin \omega t$.

16-28 Show that the function in Fig. P16-28—$f(t) = \sum_{n=1}^{\infty} u(t - 2n\pi)$—can be written in the form

$$f(t) = \left(\frac{t}{2\pi} - \frac{1}{2} + \frac{1}{\pi} \sum_{k=1}^{\infty} \frac{\sin kt}{k}\right) u(t)$$

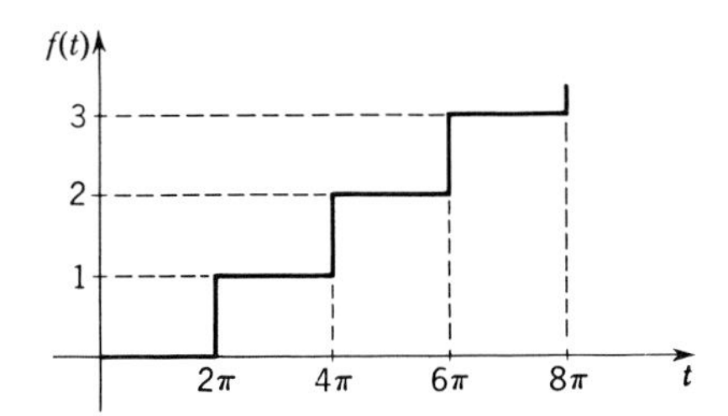

Fig. P16-28

16-29 If the function of Prob. 16-28 is a source function, what are the corresponding pole locations in the s plane?

16-30 The periodic signal $v(t)$ shown in Fig. P16-30 is applied to the input of an ideal low-pass filter whose cutoff frequency is 2.1 cps. Sketch the amplitude spectrum of the output voltage if (*a*) $T = 2$ sec; (*b*) $T = 4$ sec; (*c*) $T = 8$ sec.

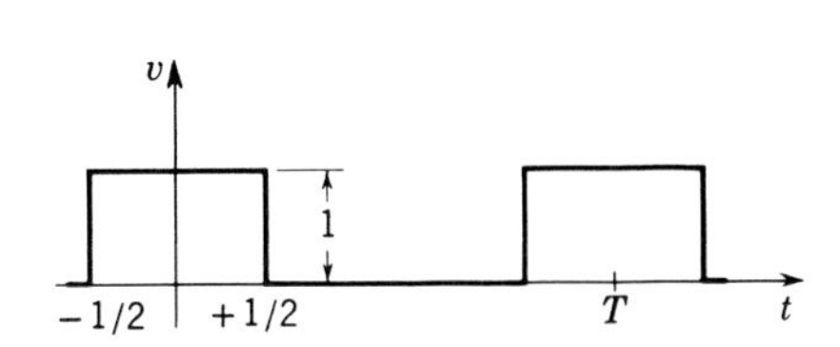

Fig. P16-30

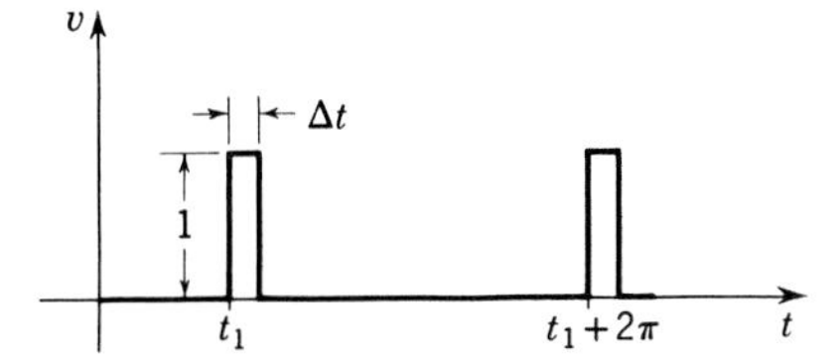

Fig. P16-31

16-31 (*a*) The periodic pulses shown in Fig. P16-31 form the input to an ideal low-pass filter whose cutoff frequency f_c is such that the highest harmonic passed is the Nth. The time delay caused by the filter is negligible. Show that the Fourier series of the output voltage v_0 is

$$v_0 = \frac{\Delta t}{2\pi} \sum_{k=-N}^{k=+N} \frac{\sin\,(k\,\Delta t/2)}{k\,\Delta t/2}\, e^{jk(t-t_1-\Delta t/2)}$$

(*b*) Show that, if Δt is sufficiently small so that $N\,\Delta t/2 \ll 1$, the result of (*a*) can be approximated by

$$v_0 \approx \frac{\Delta t}{2\pi} \sum_{k=-N}^{k=+N} e^{jk(t-t_1)}$$

16-32 The signal shown in Fig. 16-21 is known to consist of a d-c component and the first two harmonics only, that is, $f(t) = D_0 + \mathbf{D}_1 e^{jt} + \mathbf{D}_{-1} e^{-jt} + \mathbf{D}_2 e^{j2t} + \mathbf{D}_{-2} e^{-j2t}$. (*a*) Show that

$$5D_0 = \sum_{m=0}^{4} f\left(m\frac{2\pi}{5}\right)$$

$$5\mathbf{D}_1 = \sum_{m=0}^{4} f\left(m\frac{2\pi}{5}\right) e^{-jm2\pi/5}$$

$$5\mathbf{D}_k = \sum_{m=0}^{4} f\left(m\frac{2\pi}{5}\right) e^{-jk2\pi m/5} \qquad k = 0,\ \pm 1,\ \pm 2$$

Hint: Use the result of Prob. 16-14. (*b*) The signal $f(t)$ has the values $f(m2\pi/5)$ at the instants indicated. Show that the periodic train of narrow pulses shown in Fig. 16-21 is represented by a Fourier series whose coefficients are proportional to the coefficients of $f(t)$ for $k = 0$, 1, and 2. (Use the result of Prob. 16-31.) (*c*) Generalize the result of this problem by proving the following theorem: If a periodic function $f(t)$ whose period is 2π and that has no harmonics above $k = N$ is represented by $2N + 1$ equally spaced rectangular pulses, so that each pulse at $t = 2\pi m/(2N + 1)$ has the amplitude $f[m2\pi/(2N + 1)]$, then these pulses, when passed through a low-pass filter which passes the Nth but not the $(N + 1)$st harmonic, will result in a filter output which has the same waveform as $f(t)$.

16-33 A periodic signal $m(t)$ has no harmonics above the frequency f_m. Hence, if $m(t)$ is passed through an ideal low-pass filter with cutoff of f_m, of unit gain and no time delay, the output is $m(t)$. Show that the signal $[1 + m(t)] \cos 2\pi f_c t$, $f_c \gg f_m$, will pass through an ideal bandpass filter with passband from $f_c - f_m$ to $f_c + f_m$ without distortion.

Appendix A   Averages and rms values

In the study of response of networks, emphasis is placed on the relationship between source and response waveforms. In many practical problems the feature of interest may not be the details of the shape of waveforms, since this may be known in advance; for example, response to any exponential or sinusoidal source is exponential or sinusoidal, respectively. Alternatively, the shape of the waveform may be too complicated and not conveniently expressible in terms of known functions. For example, if the source has the waveform of speech signal (audio signal), both the source and response waveforms have irregular shapes, and a discussion of their waveforms may not be very informative. Consequently, other methods must be found for the description of "irregular" waveforms. The Fourier-series approach discussed in Chap. 16 and Fourier transform[1] are two such methods of signal analysis. In this Appendix we consider two basic values, average and mean squared value, which are used to characterize waveforms.

A-1 Average and mean squared value of a waveform in an interval

In Fig. A-1, a waveform $f(t)$ is shown. The average of $f(t)$ *in the interval* $T_{ij} = t_j - t_i$ is defined as

$$(F_{\text{avg}})_{T_{ij}} = \frac{1}{t_j - t_i} \int_{t_i}^{t_j} f(t)\, dt$$

[1] See, for example, M. Javid and E. Brenner, "Analysis, Transmission, and Filtering of Signals," McGraw Hill Book Company, New York, 1963.

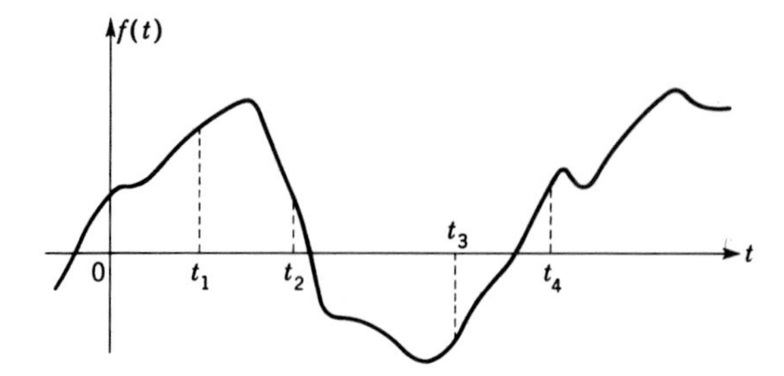

Fig. A-1 A waveform $f(t)$.

This average value is interpreted as the algebraic value of the *area under the $f(t)$ in interval T_{ij} divided by T_{ij}.* It is clear that this average depends on the choice of the interval (t_i,t_j). For example, in Fig. A-1,

$$(F_{avg})_{T_{12}} > 0 \qquad \text{and} \qquad (F_{avg})_{T_{23}} < 0 \qquad (F_{avg})_{T_{14}} \simeq 0$$

Thus, in general, when we discuss average value of a waveform, we must either explicitly state the interval, or the interval of averaging must be implicitly understood.

Very often it may be sufficient to know the average value of a waveform over a stated interval. Under other conditions the average value of the *square* of a waveform is of importance. For example, let $i(t)$ be a current in a load resistance R. The average power delivered to this resistance in the interval (t_1,t_2) is

$$(P_{avg})_{T_{12}} = \frac{1}{t_2 - t_1} \int_{t_1}^{t_2} Ri^2(t)\, dt \qquad \text{watts}$$

If $R = 1$ ohm, the average power is

$$(P_{avg})_{T_{12}} \text{ referred to 1 ohm} = \frac{1}{t_2 - t_1} \int_{t_1}^{t_2} i^2(t)\, dt \qquad \text{watts/ohm}$$

The quantity $[1/(t_2 - t_1)] \int_{t_1}^{t_2} i^2(t)\, dt$ is the average (or mean) square of $i(t)$, and is referred to by the abbreviation "ms value of i." The square root of this quantity has the same units as i, and is denoted by I_{rms} (or often by capital I without subscript) and referred to as the "rms value of i."

$$I_{rms} = \sqrt{\frac{1}{t_2 - t_1} \int_{t_1}^{t_2} [i(t)]^2\, dt} \qquad \text{amp} \tag{A-1}$$

As stated above, an interval (t_2,t_1) is implicitly assumed in discussing I_{rms}. The same definition for rms value is used not only for current, but also for voltage and other waveforms. Thus, for a function $f(t)$, the rms value in an interval t_1 to t_2 is

$$F_{rms} = \sqrt{\frac{1}{t_2 - t_1} \int_{t_1}^{t_2} [f(t)]^2\, dt} \tag{A-2}$$

A-2 Periodic waveforms

Many waveforms of interest in system analysis have repetitive patterns. Among these are the sinusoidal waveform of the voltage and currents in electric power systems, the rectangular pulses of the master "clock" of computers, and the sinusoidal bursts of radar transmitters. Such waveforms have a fixed interval called period in which the patterns repeat, and their averages and rms values are considered over the interval of a period. They are referred to as periodic waveforms defined as follows: Let P be a constant value. Then a function (waveform) $f(t)$ which satisfies the condition

$$f(t) = f(t + nP) \qquad n = 1, 2, 3, 4, \ldots$$

is called a periodic function (waveform with period P). In Fig. A-2 the waveform of a periodic function is shown. Besides the term period, the following terms are used in discussing periodic waveforms.

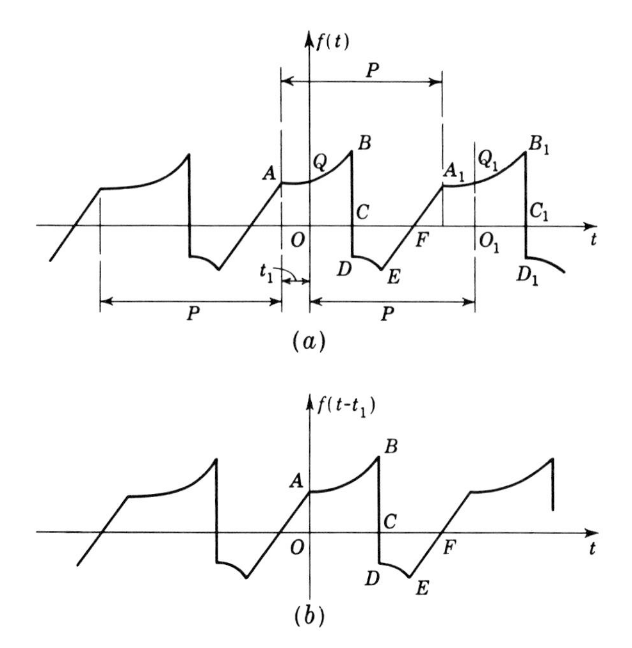

Fig. A-2 (a) A periodic waveform. (b) Waveform of (a) retarded by t_1.

CYCLE The portion of the waveform in an interval t to $t + P$ is called one cycle of that waveform. In Fig. A-2a the portion of the waveform between A and A_1 or between B and B_1 constitutes one cycle of the waveform.

REPETITION FREQUENCY The reciprocal of the period is called repetition frequency and is designated by the letter f; $f = 1/P$. It is seen that the repetition frequency of a waveform is equal to the number of cycles of a waveform which occur in one unit of time.

PHASE Each point on a cycle of a periodic waveform is a *phase* of that waveform. In Fig. A-2a, the point A is a phase of the waveform, and it is seen that each phase repeats itself at intervals of one period. Most often we deal with particular phases (points) of a waveform where something special happens to the waveform. For example, at point A the waveform of Fig. A-2a changes its slope discontinuously, and this is a special phase (point) of the waveform. At B the waveform has its largest value, at C it has a negative-going zero, and at F it has a positive-going zero. All these phases repeat at intervals of one period.

PHASE DIFFERENCE If $f(t)$ is a periodic function and t_1 is a constant value, then $f(t - t_1)$ will also be a periodic function. Each phase of $f(t)$ will occur in $f(t - t_1)$ at a later time, the difference of occurrence of each phase being t_1. This time is called the phase difference between the functions $f(t)$ and $f(t - t_1)$. If t_1 is a positive value, then $f(t - t_1)$ is said to lag $f(t)$ by t_1 and $f(t)$ is said to lead $f(t - t_1)$ by t_1. It is clear that $f(t + t_1)$ leads $f(t)$ by t_1

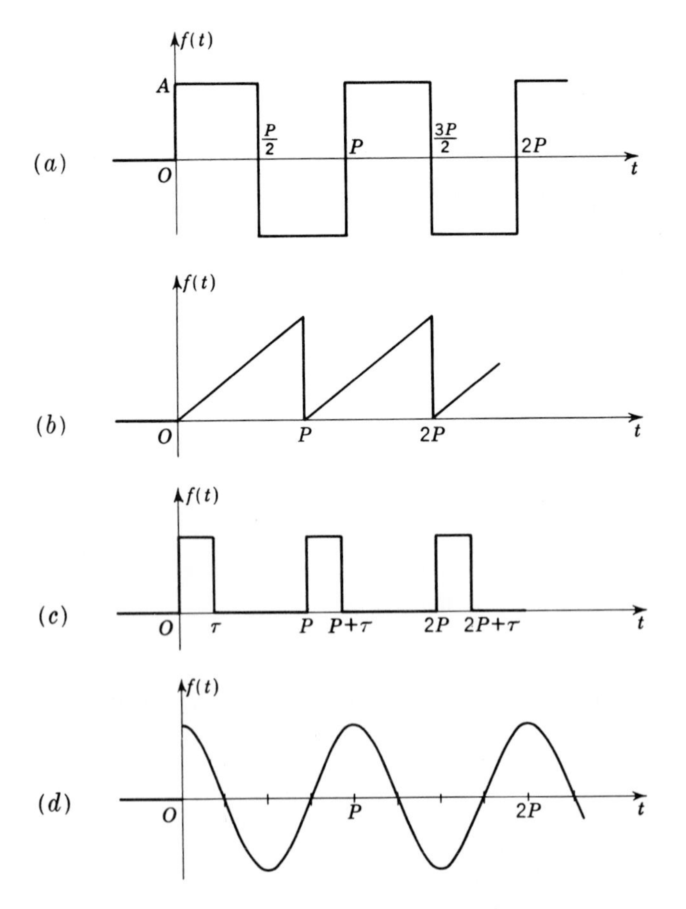

Fig. A-3 Examples of semiperiodic waveforms.

and $f(t)$ lags $f(t + t_1)$ by t_1. We also say that $f(t - t_1)$ is obtained from $f(t)$ by retarding it by t_1, and $f(t + t_1)$ is obtained from $f(t)$ by advancing it by t_1. Figure A-2*b* shows the waveform of Fig. A-2*a* retarded by an amount t_1. It is seen that each phase of the waveform of Fig. A-2*b* (such as A or B, etc.) occurs at a time t_1 later than its occurrence in the waveform of Fig. A-2*a*.

SEMIPERIODIC WAVEFORMS If $f(t)$ is a periodic waveform, then $f(t)u(t)$ is called a semiperiodic waveform. It is seen that the value of the semiperiodic waveform $f(t)u(t)$ is zero for all $t < 0$ and is periodic for all $t > 0$. Figure A-3 shows some basic semiperiodic waveforms encountered in circuit analysis.

A-3 Quantities associated with periodic waveforms

The following values associated with a waveform are of practical interest:

1 POSITIVE (OR NEGATIVE) PEAK OF A WAVEFORM This is the largest positive (or negative) value of the function. It is not necessarily a mathematical maximum (or minimum); the derivative may or may not be zero. Point B in Fig. A-2 is a positive peak; point E is a negative peak.

2 PEAK-TO-PEAK VALUE OF A WAVEFORM This is the algebraic difference between the positive and negative peaks. In Fig. A-2 the vertical distance between B and E corresponds to the peak-to-peak value.

3 AVERAGE VALUE OF THE WAVEFORM OVER ONE CYCLE By definition the average value of the periodic function $f(t)$ whose period is P is given by

$$F_{\text{avg}} = \frac{1}{P} \int_0^P f(t)\, dt = \frac{1}{P} \int_{t_1}^{P+t_1} f(t)\, dt \tag{A-3}$$

With reference to the waveform in Fig. A-2, it is seen that the integral $\int_0^P f(t)\, dt$ is the algebraic sum of the areas $OQBC$, $CDEF$, and $FA_1Q_1O_1$, where the area $CDEF$ has a negative value. Thus the average value of a periodic waveform is the area under one cycle of the waveform divided by its period.

4 ROOT-MEAN-SQUARED VALUE OF A PERIODIC WAVEFORM The quantity $(1/P) \int_0^P f^2(t)\, dt$ of a periodic function $f(t)$ with period P is termed the mean of the square of that function. We note that, although the average value of some periodic functions may be zero, the rms value of no periodic function can be zero. This is due to the fact that, in the integral $\int_0^P f(t)\, dt$, $f(t)$ may have positive or negative values in the interval $0 \le t \le P$, but in the integral $\int_0^P f^2(t)\, dt$, $f^2(t)$ is always positive.

From the above definition it is seen that

$$F_{\text{rms}} = \sqrt{\frac{1}{P}\int_0^P [f(t)]^2\,dt} \tag{A-4}$$

If a current with a periodic waveform whose rms value is I_{rms} flows through a resistance R, the time average of power delivered to R is given by $I_{\text{rms}}{}^2R$. If a constant current I flows through R, the power is I^2R. Hence the rms value of a periodic current is also referred to as the "effective," or "heating," value of the current.

Example A-1 For the current waveform shown in Fig. A-4*a* determine the average and the effective value.

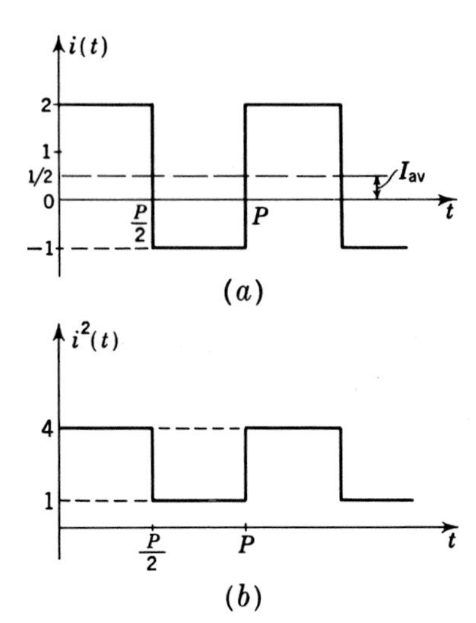

Fig. A-4 ***(a) A semiperiodic waveform and its average value. (b) Waveform for calculation of the rms value for (a).***

Solution For the average value we need the area under the waveform.

$$\text{Area} = 2 \times \frac{P}{2} - 1 \times \frac{P}{2} = \frac{P}{2}$$

The average value is

$$I_{\text{avg}} = \frac{\text{area}}{\text{base}} = \frac{P/2}{P} = \tfrac{1}{2}\text{ amp}$$

To obtain the effective value, we need the area under the curve $[i(t)]^2$. This curve is shown in Fig. A-4*b*. By inspection the area is

$$\text{Area} = 4 \times \frac{P}{2} + 1 \times \frac{P}{2} = \tfrac{5}{2}P$$

Hence the effective value is

$$I = \sqrt{\frac{\text{area}}{\text{base}}} = \sqrt{2.5\,P/P} = \sqrt{2.5} = 1.58\text{ amp}$$

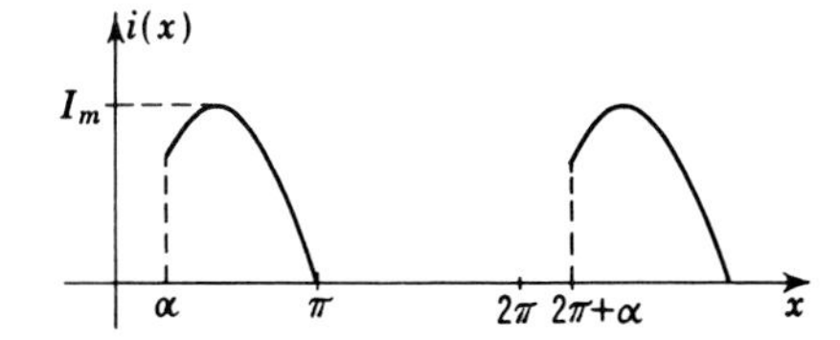

Fig. A-5 Sinusoidal pulses analyzed in Example A-2.

Example A-2 The equations of the waveform shown in Fig. A-5 are

$$\begin{aligned} i(x) &= 0 & 0 < x < \alpha \\ i(x) &= I_m \sin x & \alpha < x < \pi \\ i(x) &= 0 & \pi < x < 2\pi \\ i(x) &= i(x + 2\pi) & \end{aligned}$$

The average value of $i(x)$ is the area under the curve divided by the period.

$$I_{\text{avg}} = \frac{1}{2\pi} \int_\alpha^\pi I_m \sin x \, dx = \frac{I_m}{2\pi} \Big[- \cos x \Big]_\alpha^\pi = \frac{I_m}{2\pi} (1 + \cos \alpha)$$

The effective value is found by applying Eq. (A-3):

$$I_{\text{rms}} = I_{eff} = \sqrt{\frac{I_m{}^2}{2\pi} \int_\alpha^\pi \sin^2 x \, dx} = I_m \sqrt{\frac{1}{2\pi} \int_\alpha^\pi \left(\frac{1}{2} - \frac{1}{2} \cos 2x \right) dx}$$

$$I = I_m \sqrt{\frac{1}{4\pi} \left[x - \frac{1}{2} \sin 2x \right]_\alpha^\pi}$$

or

$$I = \frac{I_m}{2\sqrt{\pi}} \sqrt{(\pi - \alpha) + \frac{1}{2} \sin 2\alpha}$$

PROBLEMS

A-1 Find the average and rms values for the waveforms shown in Fig. A-3. In each case assume that the positive peak value of the waveform is A.

A-2 (*a*) Show that the rms value of the triangular periodic waveform shown in Fig. PA-2 is $A/\sqrt{3}$ independently of the value t_1. (*b*) Sketch the waveform $f_1(t) = f(t) - A/2$, where $f(t)$ is given in Fig. PA-2. (*c*) Show that the rms value of $f_1(t)$ is $A/(2\sqrt{3})$ independently of t_1.

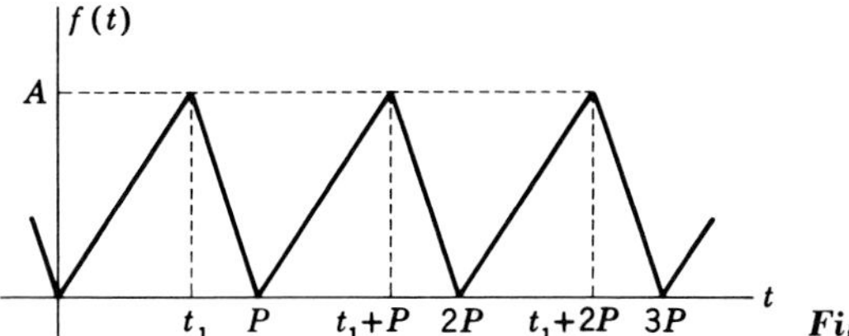

Fig. PA-2

A-3 (*a*) Show that the rms value of the periodic, full-wave-rectified sine wave, $f_1(t) = V_m|\sin \omega t|$, is the same as for the unrectified sine wave, $f_2(t) = V_m \sin \omega t$. (*b*) Sketch the periodic waveform $f_3(t) = (V_m \sin \omega t)u(\sin \omega t)$. (*c*) Show that the rms value of $f_3(t)$ is $V_m/2$. (*d*) Show that the average value of $f_1(t)$ is $2V_m/\pi$. (*e*) Show that the average value of $f_3(t)$ is V_m/π.

A-4 The periodic waveform shown in Fig. PA-4 represents the voltage applied to a 10-ohm resistance. (*a*) Calculate the average power delivered to the resistance. (*b*) Calculate the peak power delivered to the resistance.

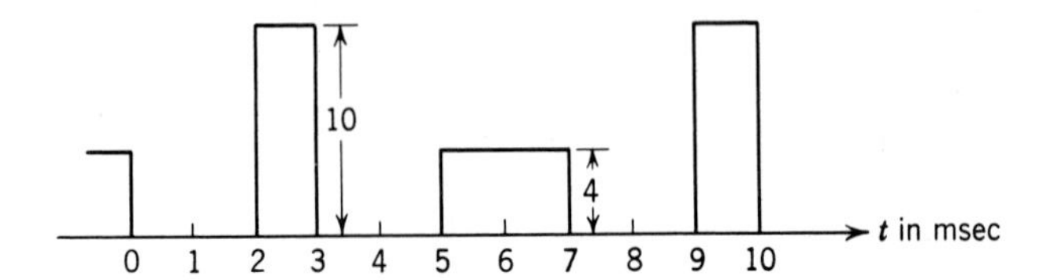

Fig. PA-4

Appendix B Complex numbers

The class of real numbers is defined by certain rules of operation such as addition, multiplication, division, extracting square roots, etc. When we apply the rule for extracting square roots to negative real numbers, we arrive at a result which cannot be a real number because the square of a real number is always a positive real number. In this sense the class of real numbers is not complete. In other words, we are faced with the situation where some of the operations which were defined in connection with the real numbers cannot be performed. To overcome this difficulty, mathematicians have introduced the class of numbers called "imaginary numbers," which obey the same rules of operation as real numbers, but whose squares are always negative real numbers. The terms "real" and "imaginary" are not used in their colloquial sense, but in the mathematical sense. The choice of the term imaginary is unfortunate because of its colloquial implication.

The concept of real integral numbers is formed through association with objects: the words "two eyes" or "ten fingers" are intuitively meaningful. The concept of negative numbers, on the other hand, arises from the insistence that the algebraic operations of addition and subtraction always can be performed. Thus in the statement $x + 5 = 3$, the symbol x cannot represent a number unless we admit negative numbers into the system of numbers. In terms of such algebraic operations, imaginary numbers are introduced in connection with relationships such as

$$s^2 = -\omega_0^2$$

If we relate the imaginary number s with the real number ω_0 through the relation

$$s = j\omega_0 \tag{B-1}$$

where j is the unit imaginary number and ω_0 is the real number indicating how many imaginary units there are in the imaginary number s, then, squaring both sides of (B-1), we have

$$s^2 = -\omega_0^2 = j^2\omega_0^2$$

where $$j^2 \equiv -1$$

and $$j \equiv +\sqrt{-1} \tag{B-2}$$

Equation (B-2) may be taken to be the defining equation of the unit imaginary number[1] j.

B-1 Complex numbers

We become acquainted with imaginary numbers in the process of extracting the square root of a negative real number. Suppose now we attempt to extract the square root of an imaginary number. Here we are faced with an operation whose result is neither a real nor an imaginary number. This difficulty is overcome by introducing another class of numbers called *complex numbers.* We shall see that both real and imaginary numbers are special cases of complex numbers. It can be shown that the result of application of all algebraic operations on a complex number is a complex number. By contrast, this statement could not be made about either real or imaginary numbers.

Let lower-case letters such as $a, b, c, d, \ldots$ denote real numbers. Then the combination $a + jb$ is called a complex number, denoted by a boldface letter, often capital $\mathbf{A} = a + jb$. a is called the real part and b the imaginary part of the complex number $\mathbf{A}$. Note that, by this definition, the imaginary part of a complex number is a real number. Symbolically, the real and imaginary parts of complex numbers are indicated by the abbreviations Re and Im. Thus, for the complex number $\mathbf{A} = a + jb$, we write

$$\text{Re}\,\mathbf{A} = a \qquad \text{and} \qquad \text{Im}\,\mathbf{A} = b$$

In a numerical case, if $\mathbf{A} = -3 - j6$, $\text{Re}\,\mathbf{A} = -3$, $\text{Im}\,\mathbf{A} = -6$.

B-2 Addition of complex numbers

Let

$$\mathbf{A} = a + jb \tag{B-3}$$

$$\mathbf{B} = c + jd \tag{B-4}$$

Then, by definition,

$$\mathbf{A} + \mathbf{B} = \mathbf{B} + \mathbf{A} \equiv \mathbf{D}$$

[1] The reader will recall that mathematicians denote the unit imaginary number by the letter i; engineers use j because the symbol i is used for current.

where **D** is a complex number such that

$$\mathbf{D} = (a + c) + j(b + d)$$

The sum of two complex numbers is another complex number whose real part is the sum of the real parts and whose imaginary part is the sum of the imaginary parts of the components.

SUBTRACTION OF COMPLEX NUMBERS Subtraction can be taken to be a special case of addition resulting in

$$\mathbf{A} - \mathbf{B} \equiv \mathbf{E}$$
$$\mathbf{E} = (a - c) + j(b - d)$$

where **A** and **B** are given by (B-3) and (B-4).

B-3 Multiplication of complex numbers

In multiplying two complex numbers, each number is treated as a binomial. When in the product the term j^2 appears, it can be replaced by the number -1.

$$\mathbf{B} \cdot \mathbf{A} = \mathbf{A} \cdot \mathbf{B} = (a + jb)(c + jd) = ac + jad + jbc + j^2bd$$
$$= (ac - bd) + j(ad + bc)$$

Thus the product of two complex numbers results in another complex number.

B-4 Conjugate of a complex number

The conjugate of a complex number **A**, denoted by the symbol **A***, is a complex number whose real part is identical with the real part of **A** and whose imaginary part is the negative of the imaginary part of **A**. Thus, if

$$\mathbf{A} = a + jb$$

then

$$\mathbf{A}^* = a - jb$$

The product of a complex number and its conjugate is a real number:

$$\mathbf{A} \cdot \mathbf{A}^* = (a + jb)(a - jb) = a^2 - (jb)^2 = a^2 + b^2$$

Let

$$A = +\sqrt{a^2 + b^2}$$

Then A is called the magnitude of the complex number **A**. It is clear that the magnitude of **A*** is also A.

B-5 Division of complex numbers

The result of dividing two complex numbers is another complex number. To find the ratio **A**/**B**, where **A** and **B** are complex numbers given by (B-3) and (B-4), the method called "rationalization" is used. This is illustrated in

$$\frac{\mathbf{A}}{\mathbf{B}} = \frac{\mathbf{A}}{\mathbf{B}} \cdot \frac{\mathbf{B}^*}{\mathbf{B}^*} = \frac{\mathbf{A} \cdot \mathbf{B}^*}{B^2}$$

The operation of division of two complex numbers is reduced to multiplication of two complex numbers by multiplying the numerator and denominator of $\mathbf{A}/\mathbf{B}$ by the conjugate of the denominator. Since $\mathbf{BB}^* = B^2$ is a real number, then $\mathbf{D} = \mathbf{A}/\mathbf{B}$ is a complex number such that

$$\mathbf{D} = \frac{\mathbf{A}}{\mathbf{B}} = \frac{\mathbf{AB}^*}{B^2} = \frac{(a + jb)(c - jd)}{c^2 + d^2} = \frac{ac + bd}{c^2 + d^2} + j\frac{bc - ad}{c^2 + d^2}$$

B-6 The complex plane

The concept of abstract integral numbers such as 3 or 5 without reference to any object is difficult to explain. Even more difficult is the concept of abstract negative real numbers or irrational real numbers or transcendental numbers. In the case of integers we form our understanding of abstract numbers through their association with objects. For other types of numbers this means is not available. For example, the number $e = 2.7183 \cdots$, the base of natural logarithms, cannot be associated with objects visually or otherwise. One way of describing such numbers is through the artifice of an axis, called the real-number axis. If a straight line starts from point O and continues indefinitely in a given direction, as in Fig. B-1, and on this

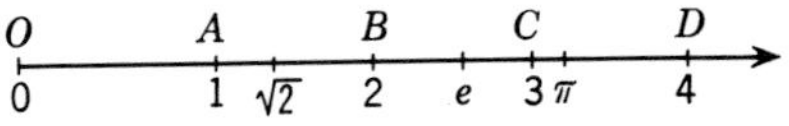

O A B C D
0 1 √2 2 e 3 π 4

Fig. B-1 The axis of positive real numbers.

line a given length OA is taken to represent unity, then each and every point on this line will correspond to a positive real number. If $OB = 2(OA)$, the point B corresponds to the number 2. Irrational numbers such as $\sqrt{2}$ or transcendental numbers such as π are also located on this axis.

When the axis of the real numbers is extended on both sides of point O as in Fig. B-2, then if the part OA and its continuation are taken to contain

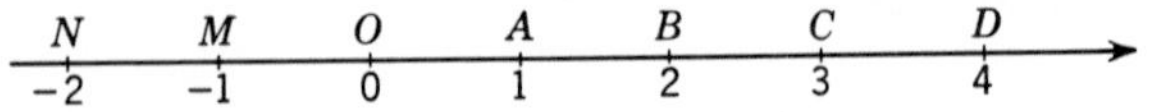

N M O A B C D
−2 −1 0 1 2 3 4

Fig. B-2 The axis of real numbers.

the set of all positive real numbers, the part OM and its continuation will contain the set of all negative real numbers.

We see, therefore, that every real number can be associated with a point on an axis, and vice versa. We now recall that any complex number is defined through two real numbers, namely, the real and the imaginary part of the complex number. A geometrical interpretation of complex numbers can therefore be achieved if we recall that each point in a plane is also defined through two real numbers. Starting with a set of mutually perpendicular axes, if one of these axes is taken to be the axis of the real numbers and the other the axis of the imaginary numbers, the plane of the two axes is called the *complex plane*, and with each point on this plane a complex number can be associated. The coordinate of a point along the real axis will be the real part of the complex number, and its coordinate along the imaginary axis will be the imaginary part.

Often the real and imaginary parts of a complex number are designated by x and y, respectively, and the complex number itself by $\mathbf{z}$. Thus $\mathbf{z} = x + jy$. If x and y represent the coordinates of a point in a set of rectangular cartesian coordinates (Fig. B-3), then to each point (x,y) in the

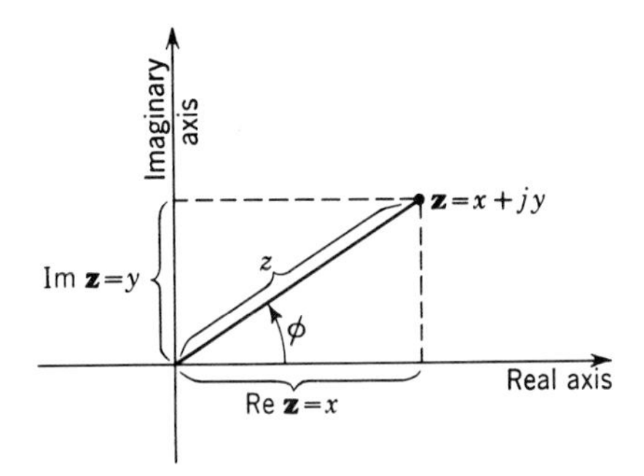

Fig. B-3 The complex plane.

cartesian plane there may therefore be made to correspond exactly one complex number, and the cartesian plane may serve as a representation of all the complex numbers. In this representation all the real numbers ($y = 0$) will be found on the x axis, or "real axis"; the y axis will be the locus for all the purely imaginary numbers and will be called the "imaginary axis." Instead of considering the point (x,y) as the representative of the complex number $\mathbf{z} = x + jy$, we may as well consider the corresponding radius vector as such. And instead of characterizing the complex number by its cartesian coordinates (x,y), we may as well characterize it by its polar coordinates. In that case it is customary among electrical engineers to write

$$\mathbf{z} = z\underline{/\phi} \qquad \text{(read "}z\text{ at an angle }\phi\text{")}$$

where z is called the "magnitude," or "absolute value," and ϕ the "argument," or "angle," of the complex number $\mathbf{z}$. The magnitude z is also indicated symbolically by $z = |\mathbf{z}|$.

The following equations relate the polar and rectangular form of a complex number

$$z = \sqrt{x^2 + y^2} \qquad \phi = \tan^{-1} \frac{y}{x}$$

and

$$\operatorname{Re} \mathbf{z} = x = z \cos \phi$$
$$\operatorname{Im} \mathbf{z} = y = z \sin \phi$$

So far we have seen that a complex number may be written in three different forms:

$$\mathbf{z} = z\underline{/\phi} = x + jy = z(\cos\phi + j\sin\phi)$$

where $z\underline{/\phi}$ is called the *polar form* of $\mathbf{z}$, $x + jy$ is called the *rectangular form* of $\mathbf{z}$, and $z(\cos\phi + j\sin\phi)$ is called the *trigonometric form* of $\mathbf{z}$.

B-7 The exponential form

Factoring out z in the trigonometric expression for $\mathbf{z}$, we may write

$$\mathbf{z} = z \cdot f(\phi)$$

where
$$f(\phi) = \cos\phi + j\sin\phi$$

We see that in the complex plane (Fig. B-4) $f(\phi)$ represents a point on a circle whose center is at the origin and whose radius is 1. The abscissa of

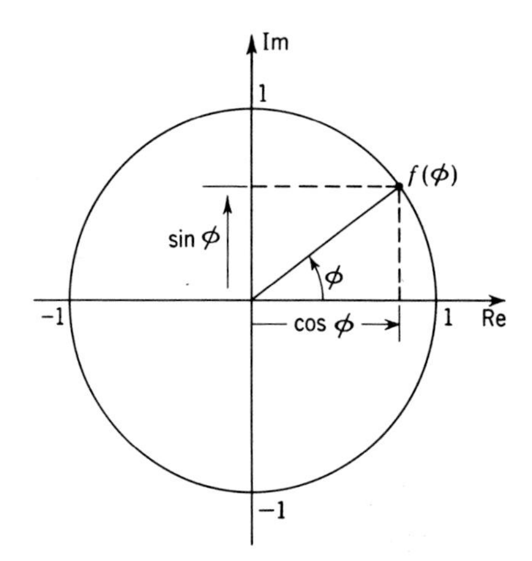

***Fig. B-4** The unit circle in the complex plane represents the function $f(\phi) = \cos\phi + j\sin\phi$.*

the point is $\cos\phi$; the ordinate has the length $\sin\phi$. Or we may also say that $f(\phi)$ represents a complex number of magnitude 1 at an angle ϕ: $f(\phi) = 1\underline{/\phi}$.

We note the values of $f(\phi)$ for some particular values of ϕ:

$$f(0) = +1 \qquad f(\tfrac{1}{2}\pi) = +j$$
$$f(\pi) = -1 \qquad f(\tfrac{3}{2}\pi) = -j$$
$$f(2\pi) = f(0) = 1$$

In order to find out whether or not we can identify $f(\phi)$ with one of the functions we know from elementary analysis, we differentiate it with respect to ϕ:

$$\frac{df}{d\phi} = -\sin\phi + j\cos\phi = j(\cos\phi + j\sin\phi)$$

We see that f obeys the differential equation

$$\frac{df}{d\phi} = jf$$

or in words, f is a function whose derivative is proportional to itself. We know that functions that have this property are the exponential functions and therefore

$$f(\phi) = Ke^{j\phi}$$

Because

$$f(0) = 1$$

if follows that $K = 1$, and therefore

$$f(\phi) = \cos\phi + j\sin\phi = e^{j\phi}$$

This equation shows the close relationship which exists between the trigonometric and the exponential functions. It was discovered by the mathematician Leonhard Euler (1707–1783).

It follows from the previous discussion that $e^{j\phi}$ may be represented in the complex plane by a radius vector of magnitude 1 at an angle ϕ (Fig. B-5).

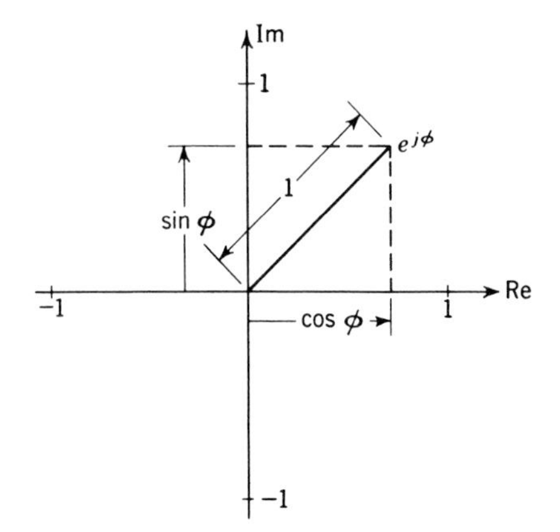

***Fig. B-5** Representation of $e^{j\phi}$ in the complex plane.*

Its real part is $\cos\phi$, its imaginary part $\sin\phi$. Some special values of $e^{j\phi}$ are

$$e^{j\pi/2} = j \qquad e^{j\pi} = -1 \qquad e^{j3\pi/2} = -j \qquad e^{j2\pi} = 1$$

We note that

$$e^{j(\phi \pm 2k\pi)} = e^{j\phi} \qquad k = \text{integer}$$

that is, the exponential function with imaginary argument is a periodic function having the period 2π.

We may now write any complex number of magnitude z and argument ϕ in the form

$$\mathbf{z} = ze^{j\phi}$$

which is called the *exponential form*. Strictly speaking, in the exponential form the angle of a complex number must be specified in radians. Although this is true, the use of radians is not convenient in numerical work because the slide rules which we all use are marked in degrees. It is therefore permissible to give the angle in exponential and polar forms in degrees, provided that the degree symbol is shown. This symbol ° is then interpreted to mean the factor $\pi/180$, which is used to convert degrees to radians. Thus, for example,

$$6e^{j\pi/4} = 6e^{j45°} = 6e^{j45\pi/180}$$

The reader is cautioned not to omit the degree symbol if degrees are meant since we must interpret a "dimensionless" angle as an angle in radians.

B-8 Graphical manipulation of complex numbers

In Fig. B-6 two complex numbers $\mathbf{z}_1 = x_1 + jy_1$ and $\mathbf{z}_2 = x_2 + jy_2$ are shown. We may consider either the point z_1 or the radius vector Oz_1 to "represent" the complex number $\mathbf{z}_1$, with a similar representation for $\mathbf{z}_2$. Let Oz be the resultant of Oz_1 and Oz_2 as obtained by the "parallelogram" rule. Let point z represent the complex number $\mathbf{z}$. A study of Fig. B-6 shows that the real part of $\mathbf{z}$ is $x_1 + x_2$ and its imaginary part is $y_1 + y_2$. Therefore

$$\mathbf{z} = (x_1 + x_2) + j(y_1 + y_2)$$

By definition of complex numbers,

$$\mathbf{z}_1 + \mathbf{z}_2 = x_1 + jy_1 + x_2 + jy_2 = (x_1 + x_2) + j(y_1 + y_2) = \mathbf{z}$$

Thus the sum of two complex numbers $\mathbf{z}_1$ and $\mathbf{z}_2$ can be obtained by the graphical method illustrated in Fig. B-6. This method is often the most convenient way of finding the sum of two complex numbers, especially if these numbers are given in their polar or exponential form. The subtraction of two complex numbers, $\mathbf{z}_1 - \mathbf{z}_2$, can be treated as the addition of $\mathbf{z}_1 + (-\mathbf{z}_2)$. In Fig. B-7 the relation between $\mathbf{z}_2$ and $-\mathbf{z}_2$ is shown. In this figure, if Oz_2 represents $\mathbf{z}_2$, then Oa, which is collinear with Oz_2, with

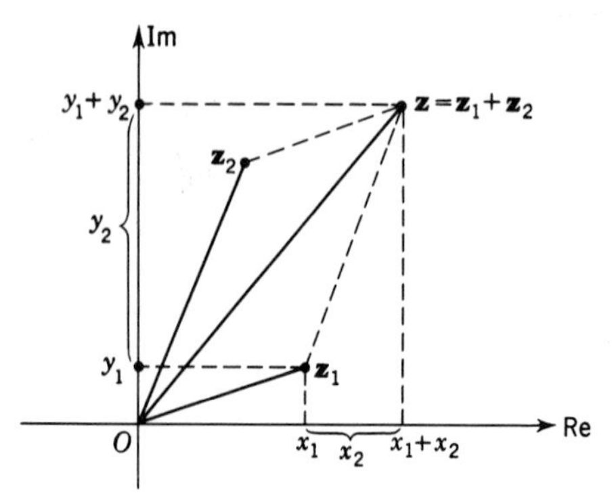

Fig. B-6 Addition of complex numbers.

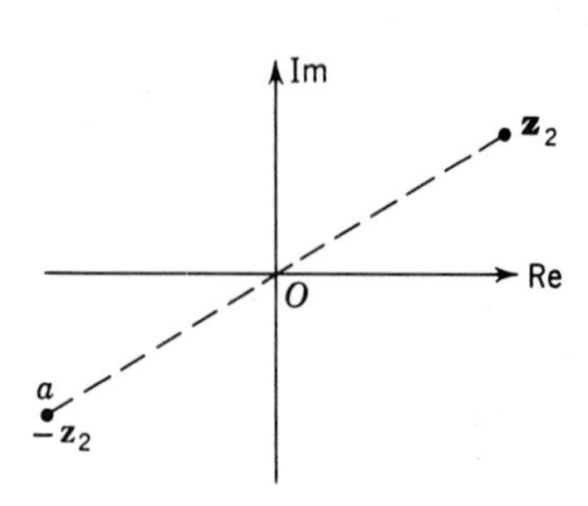

Fig. B-7 A complex number and its negative.

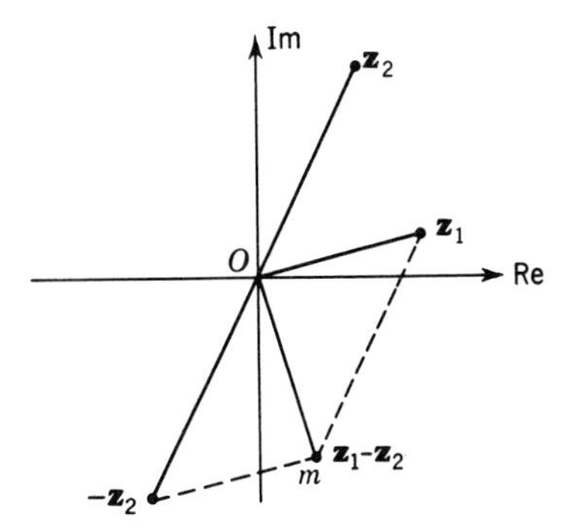

Fig. B-8 Subtraction of complex numbers.

$Oz_2 = Oa$, will represent $-\mathbf{z}_2$. In Fig. B-8 the sum of $\mathbf{z}_1 + (-\mathbf{z}_2)$ is shown. In this figure Om represents $\mathbf{z}_1 - \mathbf{z}_2$.

Figures B-6 and B-8 show that the rule for adding and subtracting complex numbers is identical with the rule for adding two-dimensional vectors in the xy space. This resemblance between vector addition and the addition of complex numbers is responsible for the practice of calling a diagram showing the addition of complex numbers, as in Fig. B-6, a "vector" diagram. As is evident from the succeeding operations (multiplication, division), the resemblance between vector algebra and ordinary algebra (which applies to complex numbers) *is restricted to the rule for addition.*

B-9 Use of exponential form in multiplication and division

Let

$$\mathbf{z}_1 = z_1 e^{j\phi_1} = z_1 \underline{/\phi_1} \qquad \mathbf{z}_2 = z_2 e^{j\phi_2} = z_2 \underline{/\phi_2}$$

Then

$$\mathbf{z}_1\mathbf{z}_2 = z_1 z_2 e^{j\phi_1} e^{j\phi_2} = z_1 z_2 e^{j(\phi_1+\phi_2)} = z_1 z_2 \underline{/\phi_1 + \phi_2}$$

and

$$\frac{\mathbf{z}_1}{\mathbf{z}_2} = \frac{z_1 e^{j\phi_1}}{z_2 e^{j\phi_2}} = \frac{z_1}{z_2} e^{j(\phi_1-\phi_2)} = \frac{z_1}{z_2} \underline{/\phi_1 - \phi_2}$$

In words, the product (quotient) of two complex numbers is a complex number whose magnitude equals the product (quotient) of the magnitudes and whose angle equals the sum (difference) of the angles of the two original complex numbers.

B-10 Powers and roots

Let $\mathbf{z} = ze^{j\phi}$; then

$$\mathbf{z}^n = z^n e^{jn\phi} = z^n \underline{/n\phi} = z^n \cos n\phi + jz^n \sin n\phi \tag{B-5}$$

In words, the nth power of a complex number is a complex number whose magnitude is the nth power of the magnitude and whose angle is n times the angle of the original complex number. Equation (B-5) is also valid for fractional powers. It must be remembered, however, that any integral multiple of 2π may be added or subtracted to ϕ without changing the values of $e^{j\phi}$:

$$e^{j\phi} = e^{j(\phi+2k\pi)} \qquad k \text{ an integer}$$

Therefore, if we wish to find the *square root* of a complex number $\mathbf{z}$, we shall find *two* values:

$$(\sqrt{\mathbf{z}})_1 = z^{\frac{1}{2}}e^{j\phi/2} = \sqrt{z}\,\underline{/\tfrac{\phi}{2}} = \sqrt{z}\cos\frac{\phi}{2} + j\sqrt{z}\sin\frac{\phi}{2}$$

$$(\sqrt{\mathbf{z}})_2 = z^{\frac{1}{2}}e^{j(\phi+2\pi)/2} = \sqrt{z}\,\underline{/\tfrac{\phi}{2}+\pi} = -\sqrt{z}\cos\frac{\phi}{2} - j\sqrt{z}\sin\frac{\phi}{2}$$

Similarly, we shall always find *three* different values for the *cube root* of a complex number:

$$(\sqrt[3]{\mathbf{z}})_1 = z^{\frac{1}{3}}e^{j\phi/3} = \sqrt[3]{z}\,\underline{/\tfrac{\phi}{3}}$$

$$(\sqrt[3]{\mathbf{z}})_2 = z^{\frac{1}{3}}e^{j(\phi+2\pi)/3} = \sqrt[3]{z}\,\underline{/\tfrac{\phi}{3}+\tfrac{2\pi}{3}}$$

$$(\sqrt[3]{\mathbf{z}})_3 = z^{\frac{1}{3}}e^{j(\phi+4\pi)/3} = \sqrt[3]{z}\,\underline{/\tfrac{\phi}{3}+\tfrac{4\pi}{3}}$$

As an example, consider the three values of $\sqrt[3]{1}$ illustrated in Fig. B-9. In

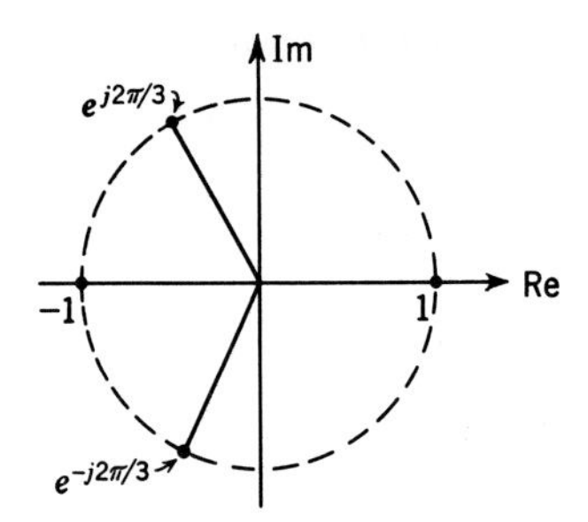

***Fig. B-9** The three cube roots of unity.*

this figure it is seen that

$$(\sqrt[3]{1})_1 = 1\underline{/0} = 1 + j0$$

$$(\sqrt[3]{1})_2 = 1\,\underline{/\tfrac{2\pi}{3}} = -\frac{1}{2} + j\frac{\sqrt{3}}{2}$$

$$(\sqrt[3]{1})_3 = 1\,\underline{/\tfrac{4\pi}{3}} = -\frac{1}{2} - j\frac{\sqrt{3}}{2}$$

B-11 The logarithm

If $\mathbf{z} = ze^{j\phi}$, we have

$$\ln \mathbf{z} = \ln (ze^{j\phi}) = \ln z + \ln e^{j\phi} = \ln z + j\phi$$

Again it must be remembered that any integer multiple of 2π may be added to or subtracted from ϕ without changing the value of $e^{j\phi}$. Therefore the logarithm of a complex number is not single-valued since

$$\ln \mathbf{z} = \ln z + j(\phi + 2\pi n) \qquad n = 0, \pm 1, \pm 2, \pm 3, \ldots$$

Sometimes a particular value of n such as $n = 0$ is chosen to define a "principal value" of $\ln \mathbf{z}$.

B-12 Exponential form of the conjugate

The conjugate of

$$\mathbf{z} = x + jy$$

is given by

$$\mathbf{z}^* = x - jy = ze^{-j\phi} = z\underline{/-\phi}$$

The complex number $\mathbf{z}$ and its conjugate $\mathbf{z}^*$ are shown in Fig. B-10. From this figure it is seen that $\mathbf{z}^*$ is the "image" of $\mathbf{z}$ with respect to the real axis.

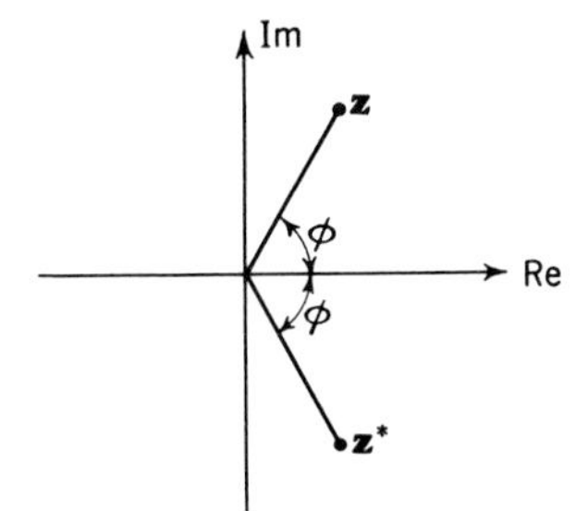

Fig. B-10 A complex number and its conjugate.

From the foregoing definition of the conjugate it follows that

$$\operatorname{Re} \mathbf{z} = \tfrac{1}{2}(\mathbf{z} + \mathbf{z}^*) \tag{B-6}$$

$$\operatorname{Im} \mathbf{z} = \frac{1}{2j}(\mathbf{z} - \mathbf{z}^*) \tag{B-7}$$

$$z^2 = |\mathbf{z}|^2 = \mathbf{z}\mathbf{z}^* \tag{B-8}$$

In the special case where $\mathbf{z} = e^{j\phi}$, $\mathbf{z}^* = e^{-j\phi}$, Eqs. (B-6) and (B-7) yield the important relations

$$\cos \phi = \tfrac{1}{2}(e^{j\phi} + e^{-j\phi})$$

$$\sin \phi = \frac{1}{2j}(e^{j\phi} - e^{-j\phi})$$

These are illustrated by the diagrams in Fig. B-11.

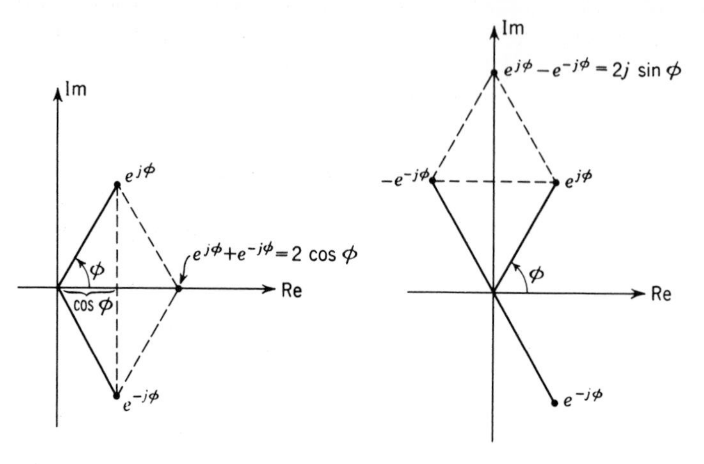

Fig. B-11 Representation of cos φ and sin φ by use of complex numbers.

B-13 Graphical representation of multiplication and division

The complex number $e^{j\phi}$ has a magnitude of unity and an angle ϕ. Multiplication of the complex number $\mathbf{A} = Ae^{j\phi_1}$ by $e^{j\phi_2}$ results in

$$\mathbf{B} = Ae^{j\phi_1}e^{j\phi_2} = Ae^{j(\phi_1+\phi_2)}$$

The complex number $\mathbf{B}$ is obtained by rotating the radius vector OA counterclockwise through an angle ϕ_2 (Fig. B-12).

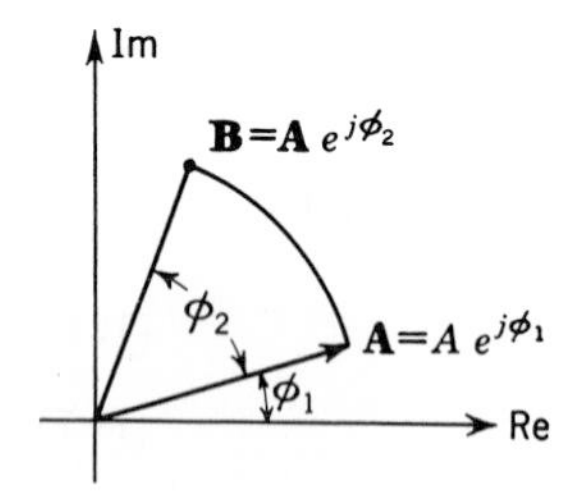

Fig. B-12 Multiplication of $Ae^{j\phi_1}$ by $e^{j\phi_2}$.

It is also noted that

$$\begin{aligned}
e^{j0} &= \cos 0 + j\sin 0 = 1 = j^0 \\
e^{j\pi/2} &= \cos\frac{\pi}{2} + j\sin\frac{\pi}{2} = j = j^1 \\
e^{j\pi} &= \cos\pi + j\sin\pi = -1 = j^2 \\
e^{j3\pi/2} &= \cos\frac{3\pi}{2} + j\sin\frac{3\pi}{2} = -j = j^3 \\
e^{j2\pi} &= \cos 2\pi + j\sin 2\pi = 1 = j^4
\end{aligned}$$

Since $j = e^{j\pi/2}$, the multiplication of a complex number **A** by j is graphically represented by a rotation of the radius vector OA by $\pi/2$ rad. Multiplication by j^2 or (-1) is equivalent to rotation of the radius vector representing the multiplicand by π rad, and so on.

B-14 Application of complex numbers

Real numbers are used in quantitative statements about one particular measurable attribute of a phenomenon. For example, the magnitude of a current at a given instant is indicated by means of a real number. Certain phenomena have measurable attributes which require simultaneous indication by two real numbers. For example, the position of a point on a cartesian plane is indicated by two real numbers. Also, a sinusoidal function of a known frequency can be uniquely described by two real numbers, its amplitude and its phase angle. In this way, a complex number composed of the two measurable attributes of the phenomenon can be used to describe it. Thus we can indicate a point on the cartesian plane by the complex number $x + jy$ or a sinusoidal function of amplitude A and phase angle θ by the complex number $A\underline{/\theta}$. The latter representation is discussed in Appendix C.

PROBLEMS

B-1 If **A** and **B** are two complex numbers, prove the following identities: (*a*) $\text{Re}\,\mathbf{A} + \text{Re}\,\mathbf{B} = \text{Re}\,(\mathbf{A} + \mathbf{B})$; (*b*) $\text{Re}\,\mathbf{A} = \text{Im}\,(j\mathbf{A})$; (*c*) $\text{Im}\,\mathbf{A} = \text{Re}\,(-j\mathbf{A})$; (*d*) $\text{Im}\,\mathbf{A} = \text{Re}\,(j\mathbf{A}^*)$; (*e*) $(\mathbf{AB})^* = \mathbf{A}^*\mathbf{B}^*$.

B-2 Show that, if **A** and **B** are two complex numbers, then (*a*) $(\text{Re}\,\mathbf{A})(\text{Re}\,\mathbf{B}) \neq \text{Re}\,(\mathbf{AB})$; (*b*) $|\mathbf{A}| + |\mathbf{B}| \neq |\mathbf{A} + \mathbf{B}|$.

B-3 Prove that $e^{j\phi} = \cos\phi + j\sin\phi$ by expanding the function into Maclaurin's series.

B-4 If A is any real positive number, prove that $|A^j| = 1$.

B-5 Prove that $(\cos\theta + j\sin\theta)^n = \cos n\theta + j\sin n\theta$ (De Moivre's theorem).

B-6 Prove each of the following identities: (*a*) $\sin(x + jy) = \frac{1}{2}(e^y + e^{-y})\sin x + \frac{1}{2}j(e^y - e^{-y})\cos x$; (*b*) $\cos(x + jy) = \frac{1}{2}(e^y + e^{-y})\cos x - \frac{1}{2}j(e^y - e^{-y})\sin x$; (*c*) $\text{Re}\,e^{x+jy} = e^x\cos y$.

B-7 In the polynomial equation $a_0 + a_1x + a_2x^2 + \cdots + a_nx^n = 0$, the coefficients $a_0, a_1, \ldots, a_n$ are real. Prove that, if **Z** is a solution of this equation, then $\mathbf{Z}^*$ is also a solution.

B-8 Find the five distinct solutions of $x^5 = 1$, and express the answers in exponential, polar, and rectangular form.

B-9 If $\mathbf{A} = 3.53 + j2.48$, $\mathbf{B} = -0.125 + j0.436$, $\mathbf{C} = 126 + j19.0$, express in polar and rectangular form the numbers (*a*) $\mathbf{A} \cdot \mathbf{B}$; (*b*) $\mathbf{A} \cdot \mathbf{B}/\mathbf{C}^*$; (*c*) $\mathbf{A}^2$; (*d*) $\mathbf{A} \cdot \mathbf{C}/\mathbf{B}$; (*e*) $\sqrt{\mathbf{A}}$.

B-10 Convert to polar form (*a*) $5.60 + j3.23$; (*b*) $0.560 + j3.23$; (*c*) $0.0560 + j3.23$; (*d*) $-5.60 + j2.32$; (*e*) $5.60 - j2.32$; (*f*) $2.32 - j5.60$.

B-11 Convert to rectangular form (*a*) $40.8\underline{/110°}$; (*b*) $40.8\underline{/70°}$; (*c*) $40.8\underline{/-70°}$; (*d*) $(40.8\underline{/70°})(0.656\underline{/-30°})$; (*e*) $8.56\underline{/-87°}$.

B-12 Evaluate Re **A** if (*a*) $\mathbf{A} = 10\underline{/89.2^\circ}$; (*b*) $\mathbf{A} = 10\underline{/92^\circ}$; (*c*) $\mathbf{A} = 10e^{3+j4}$; (*d*) $\mathbf{A} = 10e^{-1+j0.77}$; (*e*) $\mathbf{A} = j5e^{-2+j0.5}$; (*f*) $\mathbf{A} = (4.56 + j6.38)/(6.20 - j3.60)$.

B-13 Evaluate **A** and express the answer in polar and rectangular form:

(*a*) $$\mathbf{A} = \frac{(4 - j2)(1 + j3) - 5\underline{/45^\circ}}{\sqrt{4.5}\,\underline{/15^\circ}(1 + j) - 3\underline{/-60^\circ}}$$

(*b*) $$\mathbf{A} = \frac{(7.86 + j18.4)\ \mathrm{Re}\ 78\underline{/120^\circ} + \mathrm{Im}\ 214\underline{/-120^\circ}}{(4.78 + j6.14)^2}$$

(*c*) $\mathbf{A} = (2.46 + j37.2)(4.24 + j41.3) - (j38.0)^2$

B-14 Find **K** and express the answer in polar form if (*a*) Re $\mathbf{K} = 7$ and Re $[(-5 + j14)\mathbf{K}] = 21$; (*b*) Re $\mathbf{K} = 8$ and Im $[(-2 + j10)\mathbf{K}] = 64$.

B-15 Solve the following equations for the real positive value A and the angle ϕ in degrees: (*a*) $100\underline{/0^\circ} + A\underline{/73.8^\circ} = 200\underline{/\phi}$; (*b*) $100\underline{/-60^\circ} + A(2 + j3) = 200\underline{/\phi}$.

B-16 If $\mathbf{A} = (10 + 5\underline{/\phi})/(10 - 5\underline{/\phi})$, find the maximum value of A and the corresponding angle ϕ (in degrees).

B-17 Evaluate the following terms and express the answer in polar form: (*a*) j^j; (*b*) $\sqrt{j}$; (*c*) $(5 + j2)^j$; (*d*) $(5 - j5)^{1+j}$.

B-18 Defining the principal value of the angle of **A** to be between 0 and 2π, evaluate and express in rectangular and polar form ln **A** if (*a*) $\mathbf{A} = 14.1\underline{/45^\circ}$; (*b*) $\mathbf{A} = j$; (*c*) $\mathbf{A} = -20$.

Appendix C Phasor representation of sinusoidal functions

Sinusoidal functions are a special type of periodic function which satisfy the differential equations[1] of the form

$$\frac{d^2}{dt^2} f(t) + \omega^2 f(t) = 0$$

where ω is a constant and the solution of the equation is in the form

$$f(t) = A \cos \omega t + B \sin \omega t$$

ω is called the radian frequency of the sinusoidal function.

Frequently, voltages and currents in a circuit are sinusoidal. In applying Kirchhoff's laws to such circuits, it will therefore be necessary to add sinusoidal functions. In this Appendix we show that the addition of sinusoids which have the same frequency results in another sinusoidal function. Moreover, the use of complex-number representation for such functions transforms the summation of such transcendental functions to algebraic operations.

C-1 Basic properties of sinusoids

The function

$$f_1(t) = A \cos \omega t$$

[1] One way to define functions is to specify them as the solutions of differential equations. For example, the exponential function $e^{-\alpha t}$ is defined as a solution of the equation $df/dt + \alpha f = 0$.

may be visualized, from elementary trigonometry, as having at every instant of time t the value given by the ratio of the adjacent leg to the hypotenuse of a right triangle in which the angle formed by these two legs is increasing with time at the rate ω rad/sec. A similar visualization can be made for

$$f_2(t') = B \sin \omega t'$$

In this case the opposite leg and the hypotenuse are involved. Figure C-1a shows the waveform of $f_1(t) = A \cos \omega t$, and Fig. C-1b shows $f_2(t') = B \sin \omega t'$.

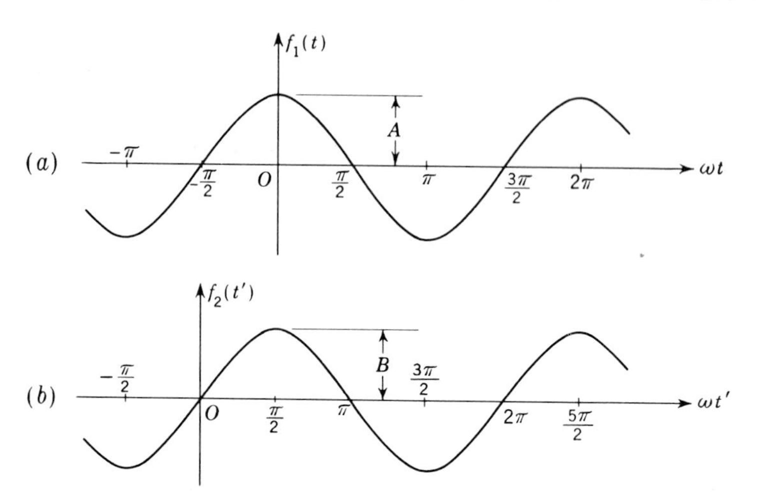

***Fig. C-1** Waveform of sinusoids.*

With sinsuoidal functions it is often convenient to normalize the time scale and choose as independent variable ωt instead of t. In such a case the period P of the function corresponds to $\omega P = 2\pi$. The waveforms of Fig. C-1 are shown with ωt as the independent variable. Examining the waveshapes of Fig. C-1a and b, we note that they are identical, the only difference being the choice of location of the f axis, $\omega t = 0$. For this reason, distinction between sine and cosine functions is often unnecessary, and the term *sinusoidal function* is used to describe either function or a linear combination of both.

Referring to Fig. C-1, we note that, when $t = 0$, $t' = \pi/2\omega$. This is an expression of the trigonometric identity concerning complementary angles, $\cos \omega t = \sin (\omega t + \pi/2)$. Pursuing this, consider Fig. C-2, where a sinusoid

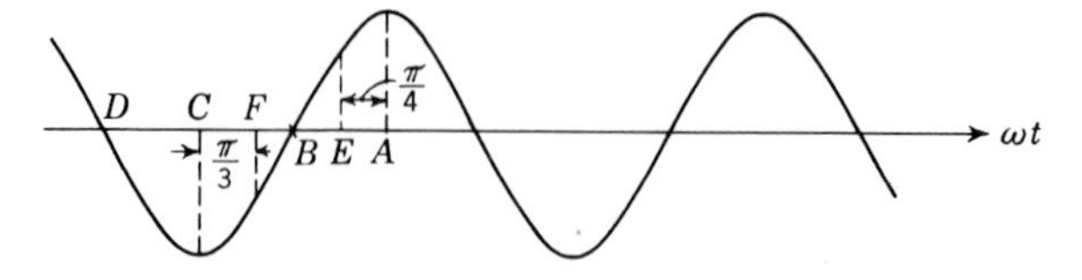

***Fig. C-2** Choice of time reference determines the functional form of the sinusoid as in Table C-1.*

is sketched without the ordinate axis. The equation which describes this waveform depends on the location of the point where t (or ωt) is zero. Table C-1 indicates several such choices and the resulting equation. The con-

Table C-1 Data pertinent to Fig. C-2

$\omega t = 0$ at point	Time function
A	$\cos \omega t$
B	$\sin \omega t = \cos\left(\omega t - \frac{\pi}{2}\right)$
C	$-\cos \omega t = \cos(\omega t + \pi)$
D	$-\sin \omega t = \sin(\omega t + \pi) = \cos\left(\omega t + \frac{\pi}{2}\right)$
E	$\cos\left(\omega t - \frac{\pi}{4}\right) = \sin\left(\omega t + \frac{\pi}{4}\right)$
F	$-\cos\left(\omega t + \frac{\pi}{3}\right) = \cos\left(\omega t - \frac{2\pi}{3}\right) = \sin\left(\omega t - \frac{\pi}{6}\right)$

clusion we wish to draw at present is that a general form of the sinusoid is

$$f(t) = A\cos(\omega t + \alpha) = A\sin\left(\omega t + \alpha + \frac{\pi}{2}\right)$$

PERIODICITY The sinusoidal time function is periodic, as expressed by the identity

$$\cos(\omega t + \alpha) = \cos(\omega t + 2n\pi + \alpha) \qquad n = \pm 1, \pm 2, \ldots$$

The relation between its period P and its radian frequency ω is

$$\omega P = 2\pi \qquad \text{or} \qquad P = \frac{2\pi}{\omega}$$

The reciprocal of the period is called the frequency (f), which becomes

$$f = \frac{1}{P} = \frac{\omega}{2\pi}$$

WAVESHAPE PROPERTY From the time-derivative relationships

$$\frac{d}{dt}\sin \omega t = \omega\cos\omega t \qquad \frac{d}{dt}\cos\omega t = -\omega\sin\omega t$$

we observe that *the sinusoid is the only periodic real function which has a derivative of the same shape as the function itself.* It is this particular property of sinusoidal functions which makes them so useful in analysis.

AMPLITUDE The function

$$f(t) = A \cos(\omega t + \alpha)$$

reaches the maximum value A when $\omega t + \alpha = 0$ or $2n\pi$; the minimum is $-A$ and occurs at $\omega t + \alpha = (2n - 1)\pi$ (n = integer). The positive maximum value of the function is called the *amplitude* of the sinusoid. Note that the relationship

$$\cos \omega t = -\cos(\omega t \pm \pi)$$

makes it possible to choose A always as a positive number, e.g.,

$$f(t) = -10 \cos\left(\omega t + \frac{\pi}{3}\right) = 10 \cos\left(\omega t - \frac{2\pi}{3}\right)$$

PHASE ANGLE We have already defined phase difference in connection with the periodic functions (Appendix B). Two sinusoidal waveforms of the same frequency can differ only in their amplitude and phase. Figure C-3 shows the waveforms of $f_1(t) = A \cos \omega t$ and $f_2(t) = B \sin \omega t$. The two

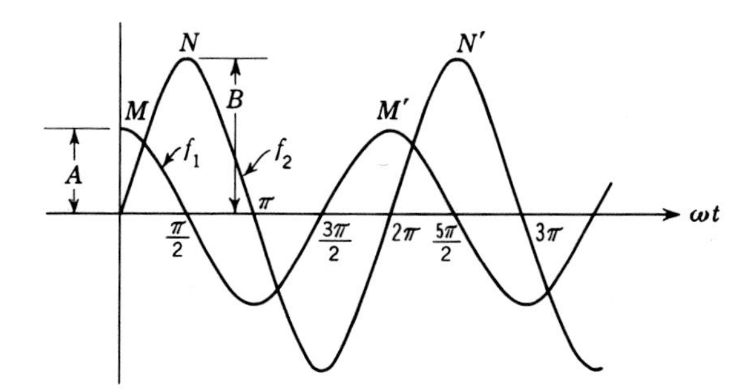

Fig. C-3 Illustration of phase difference between two sinusoidal functions. The function f_2 lags the function f_1 by $\pi/2$.

waveforms have the same frequency but differ in amplitude and phase. For example, $f_1(t)$ reaches its maximum at $\omega t = 0$, whereas $f_2(t)$ has its maximum at $\omega t = \pi/2$. Inspection of Fig. C-3 shows that every phase of $f_1(t)$ occurs in $f_2(t)$ at exactly $\pi/2$ rad later. Thus we say that a sine function *lags* a cosine function by $\pi/2$ rad (90°). It is of course possible to say that the maximum N of $f_2(t)$ occurs earlier than maximum M' of $f_1(t)$ by $3\pi/2$. We may therefore say, $f_2(t) = B \sin \omega t$ leads $f_1(t) = A \cos \omega t$ by $3\pi/2$ rad (270°). It is seen that lagging by $\pi/2$ and leading by $3\pi/2$ are identical conditions. This point is evident from the fact that $\omega t - \pi/2$ differs from $\omega t + 3\pi/2$ by 2π, which is the period of the sinusoid on the ωt scale.

The ideas from the special example of Fig. C-3 may be generalized to compare any two sinusoids which have the *same frequency*. Thus, in Fig. C-4, f_1 leads f_2 by α rad (or f_2 leads f_1 by $2\pi - \alpha$) or f_2 lags f_1 by α. This is

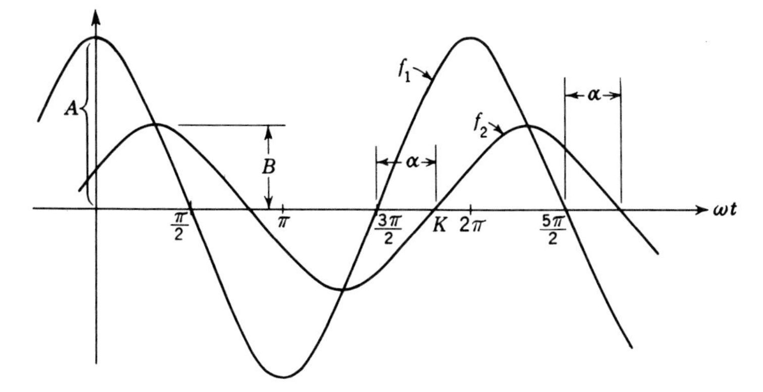

Fig. C-4 Illustration of phase difference between two sinusoids.

expressed analytically

$$f_1(t) = A \cos \omega t$$
$$f_2(t) = B \cos (\omega t - \alpha)$$

The angle α is called the phase (angle) difference of the two sinusoids. Note that, if the time reference ($\omega t = 0$) is chosen at K, then

$$f_1(t) = A \cos \left(\omega t - \frac{\pi}{2} + \alpha\right) = A \sin (\omega t + \alpha)$$

$$f_2(t) = B \cos \left(\omega t - \frac{\pi}{2}\right) = B \sin \omega t$$

Thus, regardless of the choice of origin, the phase difference remains the same.

To conclude, it needs to be pointed out that *a phase difference on the "angle scale" corresponds to a time difference on the time scale.* Thus, if f_2 lags f_1 by α, f_2 is delayed in time by α/ω sec with respect to f_1.

C-2 Addition of sinusoids

We have already seen (in connection with the *L-C* circuit) that the sum of two sinusoids may occur in circuit problems. The addition of two or more sinusoidal time functions of the same frequency but with different amplitude and phase is common in circuit analysis; moreover, it is almost always desired to express the sum as a single sinusoidal term.

Consider as an example

$$f(t) = A \cos \omega t + B \sin \omega t \tag{C-1}$$

This function may be reduced to the form

$$f(t) = C \cos (\omega t + \gamma) \tag{C-2}$$

by the method of identities in trigonometry. Since

$$\cos(\omega t + \gamma) = \cos \omega t \cos \gamma - \sin \omega t \sin \gamma$$

we get

$$f(t) = A \cos \omega t + B \sin \omega t = C \cos \gamma \cos \omega t - C \sin \gamma \sin \omega t \tag{C-3}$$

If Eq. (C-3) is to hold for all t, by comparison of the coefficient of $\sin \omega t$ and $\cos \omega t$ on both sides of the equation it is seen that

$$\begin{aligned} A &= C \cos \gamma \\ B &= -C \sin \gamma \end{aligned} \tag{C-4}$$

or

$$\begin{aligned} C &= +\sqrt{A^2 + B^2} \\ \gamma &= \tan^{-1} \frac{-B}{A} \end{aligned} \tag{C-5}$$

Hence Eq. (C-1) may be rewritten in the desired form,

$$f(t) = \sqrt{A^2 + B^2} \cos \left(\omega t + \tan^{-1} \frac{-B}{A} \right) \tag{C-6}$$

This result, as well as the addition indicated by Eq. (C-1), is illustrated in Fig. C-5 for the special case $A = B = 1$. Although the conversion of Eq. (C-1) to the form

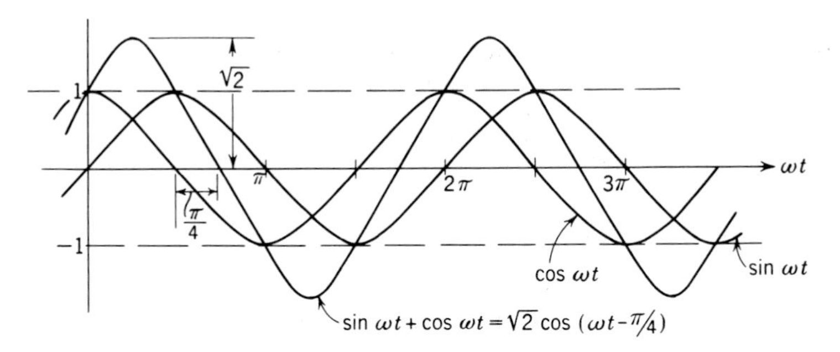

Fig. C-5 Addition of two sinusoids.

(C-6) is done with little difficulty by trigonometric expansion, the addition of two terms of the form

$$f(t) = A \cos(\omega t + \alpha) + B \cos(\omega t + \beta)$$

or of three terms, e.g.,

$$f(t) = A \cos(\omega t + \alpha) + B \sin(\omega t + \beta) + D \cos(\omega t + \delta)$$

is very cumbersome by this method. In Sec. C-4 we study a method of adding any number of sinusoidal functions with like frequency. This method is understood through study of the complex function $e^{j\omega t}$.

C-3 The complex function $e^{j\omega t}$. Phasors

In Appendix B we discussed Euler's equation, that is,

$$e^{j\phi} = \cos \phi + j \sin \phi \tag{C-7}$$

where ϕ is a real number (in radians). We may extend this equation to the case when ϕ is a real function of t such as

$$\phi = \omega t$$

where ω is in radians per second. Thus we have

$$e^{j\omega t} = \cos \omega t + j \sin \omega t \tag{C-8}$$

The transition from Eq. (C-7), which is a representation of a complex *number* in two forms (exponential and trigonometric), to Eq. (C-8), relating a complex exponential *function* of the real variable t to two sinusoidal functions of t, is an important step. Recalling the relationship between the real parts of a complex number and its conjugates, we have

$$\begin{aligned} \cos \omega t &= \operatorname{Re} e^{j\omega t} = \tfrac{1}{2}(e^{j\omega t} + e^{-j\omega t}) \\ \sin \omega t &= \operatorname{Im} e^{j\omega t} = \frac{1}{2j}(e^{j\omega t} - e^{-j\omega t}) \end{aligned} \tag{C-9}$$

The above relationships are very useful. From Eq. (C-8) it is seen that the function $e^{j\omega t}$ is a complex function whose real and imaginary parts are the cosine and the sine functions, respectively. This relationship may be illustrated graphically as in Fig. C-6. The circle with the center at the origin and unit radius (unit circle) represents all possible values of the function $e^{j\omega t}$.

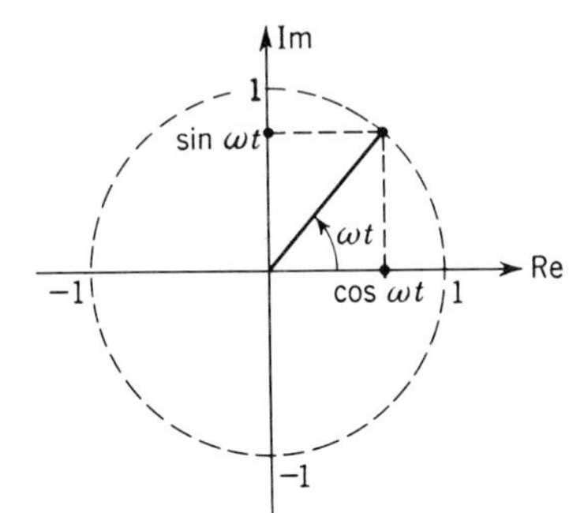

***Fig. C-6** Representation of the complex function $e^{j\omega t}$.*

At any instant of time the value of $e^{j\omega t}$ is found by drawing the radius of the circle at the angle ωt rad with the horizontal. The real part of $e^{j\omega t}$ is the projection of this line on the real axis (i.e., it is $\cos \omega t$); the imaginary part is the projection on the imaginary axis.

Consider now the two functions

$$\begin{aligned} f_1(t) &= \cos \omega t = \operatorname{Re} e^{j\omega t} \\ f_2(t) &= \cos (\omega t + \alpha) = \operatorname{Re} e^{j(\omega t + \alpha)} \end{aligned}$$

Both these functions have unit amplitude, the same frequency, but f_2 leads f_1 by the angle α. (In the discussion of this problem, it is assumed that α is an acute positive angle.) Figure C-7a shows how the values of f_1 and f_2 are obtained at $t = t_1$; they are the projections of lines OA_1 and OB_1 on the real axis. Figure C-7*b* shows how they may be obtained at $t = t_2 > t_1$. Figure C-7*c* shows how these values are obtained at $t = 0$. Note that $f_2(t)$ at $t = t_2$ as shown in Fig. C-7*b* is the projection of OB_2 on the negative real axis; hence $f_2(t_2)$ is a negative number.

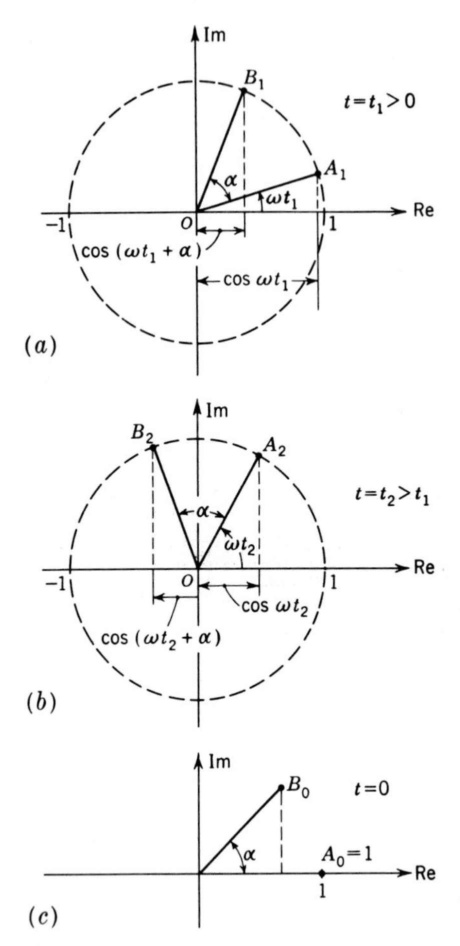

***Fig. C-7** Values of two complex functions at different instants of time.*

Let us again examine the method used to find the value of

$$f(t) = \cos(\omega t + \alpha) = \text{Re } e^{j(\omega t+\alpha)}$$

at any instant of time. We draw the unit radius as in Fig. C-8*a* at an angle $\omega t + \alpha$ with the real axis, then take the projection of that radius on the real axis. Figure C-8*a* shows three such instants of time. Note that, as t increases, the radius whose projection is generating $f(t)$ rotates *counterclockwise*.

Alternatively, the value of $f(t)$ at any instant of time may be found by drawing the radius at the angle α (in the positive, i.e., counterclockwise, direction) with the real axis, rotating the axes clockwise through the angle

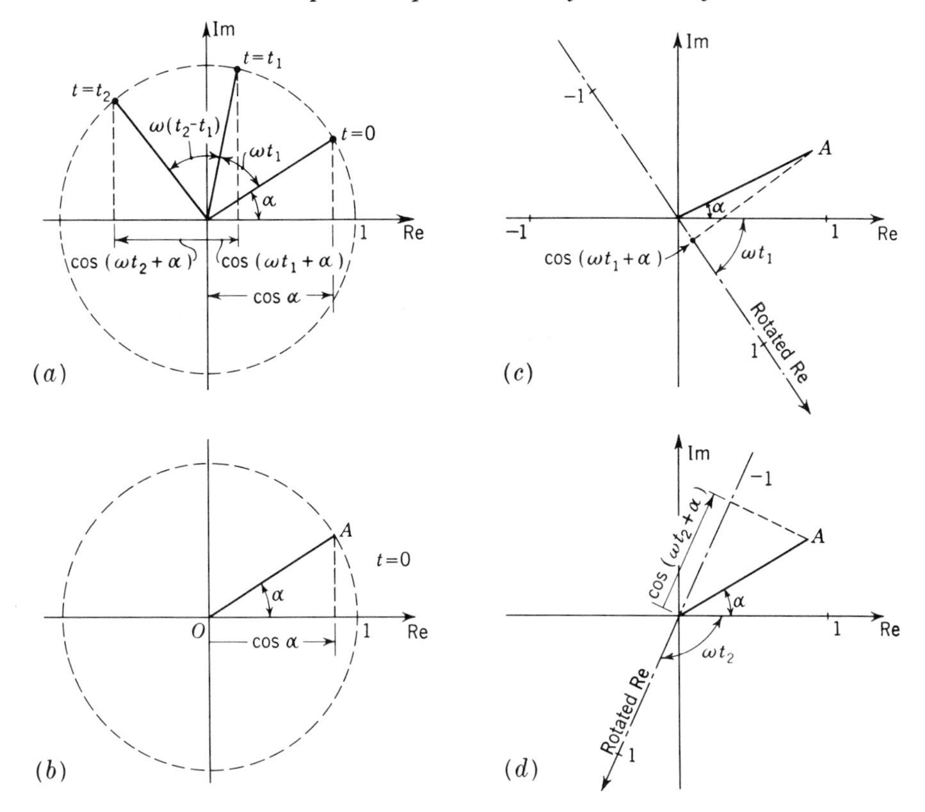

Fig. C-8 Determination of instantaneous values (a) by rotation of the radius vector; (b–d) by rotation of the real axis.

ωt, and then taking the projection on the rotated real axis as illustrated in Fig. C-8*b* to *d* for the same three instants used in Fig. C-8*a*. Since by this method the axes rotate and the radius representing the exponentials remains the *same*, the radius OA in Fig. C-8*b* may be taken to *represent* the sinusoid $f(t)$. We may therefore associate with every sinusoidal function a *complex number* which represents the amplitude and phase of the sinusoid with respect to a time (phase) reference. When a complex number is used to represent a sinusoidal time function, the graphical representation of this complex number is called a *phasor*.[1]

[1] To avoid ambiguity, it is customary to place an arrowhead at the point where the complex number is located. For this reason, and because the addition of complex numbers resembles vector addition, the term "vector" is often used to mean phasor. The term "sinor" is also used as a synonym for phasor. It must be noted that we have defined a phasor to be a complex *number*. There is some tendency to confuse the complex *function* $\mathbf{A}e^{j\omega t}$ with the phasor $\mathbf{A}$ which *represents* (*but is not equal to*) the real part of that function. The authors, following a suggestion of B. J. Starkey, "Laplace Transforms for Electrical Engineers," Philosophical Library, Inc., New York, 1955, believe that the term "animated phasor," or "activated phasor," may appropriately describe the function $\mathbf{A}e^{j\omega t}$.

Thus the complex number $1\underline{/0^\circ}$ may be used to represent the sinusoidal time function $f(t) = \cos \omega t$. The number $A\underline{/\alpha}$ would then represent the time function $A \cos (\omega t + \alpha)$.

Since in the addition of any two complex numbers the identity

$$\text{Re } \mathbf{z}_1 + \text{Re } \mathbf{z}_2 = \text{Re } (\mathbf{z}_1 + \mathbf{z}_2)$$

applies, the resultant of two sinusoids may be obtained by adding the phasors which represent the two sinusoids. To obtain the value of the resulting time function at any instant of time, the resulting phasor is projected on a properly rotated real axis. Before supplementing the graphical proof which is outlined above by an analytical proof, a simple graphical example will be done.

Consider

$$f(t) = \cos \omega t + \sin \omega t$$

or

$$f(t) = \cos \omega t + \cos (\omega t - 90^\circ)$$

Since $\cos \omega t = \text{Re } e^{j\omega t}$ and $\sin \omega t = \text{Re } e^{j(\omega t - \pi/2)}$, the value of $f(t)$ at $t = 0$ is given by the projection of the sum $1\underline{/0} + 1\underline{/-90^\circ}$ on the real axis as in Fig. C-9a. At the instant

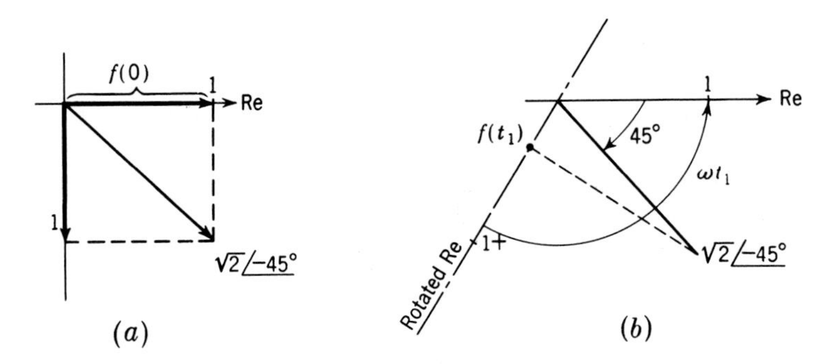

Fig. C-9 Illustrating addition of two phasors.

of time $t = t_1$ the value $f(t_1)$ is illustrated in Fig. C-9b. It is emphasized that the meaning of the resultant phasor

$$1\underline{/0} + 1\underline{/-90^\circ} = \sqrt{2}\,\underline{/-45^\circ}$$

is that

$$f(t) = \sqrt{2} \cos (\omega t - 45^\circ)$$

C-4 Analytical proof of phasor addition

Consider

$$f(t) = A_1 \cos (\omega t + \alpha_1) + A_2 \cos (\omega t + \alpha_2)$$

We may write

$$f(t) = \text{Re } (A_1 e^{j\alpha_1} e^{j\omega t}) + \text{Re } (A_2 e^{j\alpha_2} e^{j\omega t})$$

Again recalling the identity for any two complex quantities,

$$\text{Re}\,\mathbf{z}_1 + \text{Re}\,\mathbf{z}_2 = \text{Re}\,(\mathbf{z}_1 + \mathbf{z}_2)$$
$$f(t) = \text{Re}\,[(A_1 e^{j\alpha_1} + A_2 e^{j\alpha_2})e^{j\omega t}]$$

Defining

$$A_1 e^{j\alpha_1} + A_2 e^{j\alpha_2} = Ce^{j\gamma}$$
$$f(t) = C \cos(\omega t + \gamma)$$

Thus the sum $f(t)$ is obtained by the following representations:

The phasor $A_1\underline{/\alpha_1}$ represents $A_1 \cos(\omega t + \alpha_1)$
The phasor $A_2\underline{/\alpha_2}$ represents $A_2 \cos(\omega t + \alpha_2)$
The phasor $A_1\underline{/\alpha_1} + A_2\underline{/\alpha_2} = C\underline{/\gamma}$ represents $f(t)$

Because $\sin \omega t = \cos(\omega t - \pi/2)$, phasors representing a function $B \sin(\omega t + \beta)$ may be drawn immediately by observing that the phasor $1\underline{/0}$ represents $\cos \omega t$, the phasor $1\underline{/-90°}$ represents $\sin \omega t$, and $1\underline{/-90° + \beta}$ represents $\sin(\omega t + \beta)$. Several examples are shown in Fig. C-10.

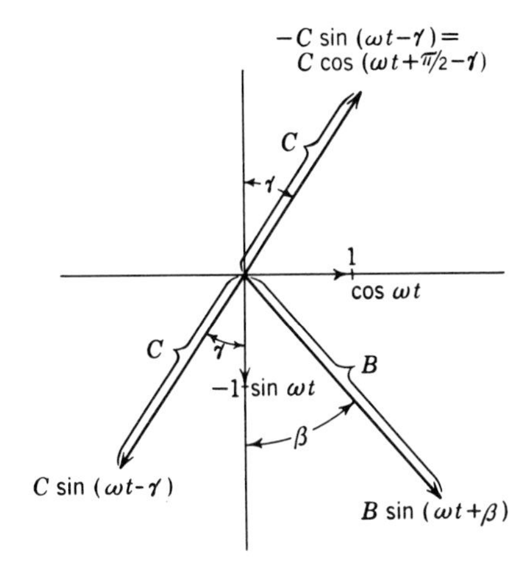

Fig. C-10 Representations of sinusoidal functions by phasors.

It is noted that, when we say that the phasor $A\underline{/\alpha}$ represents a sinusoidal function, by this we mean that this sinusoidal function is given by

$$\text{Re}\,(Ae^{j\alpha}e^{j\omega t}) = A \cos(\omega t + \alpha)$$

We could have just as well used the convention of taking the imaginary part to give

$$\text{Im}\,(Ae^{j\alpha}e^{j\omega t}) = A \sin(\omega t + \alpha)$$

This would be equivalent to rotating the imaginary axis by ωt and projecting $A\underline{/\alpha}$ on the rotated imaginary axis. In this book we choose the convention that the phasor $1\underline{/\alpha}$ represents $\text{Re}\,e^{j(\omega t+\alpha)}$ rather than $\text{Im}\,e^{j(\omega t+\alpha)}$.

Example C-1 Express

$$f(t) = 100 \cos(\omega t + 30°) - 60 \sin(\omega t + 45°) - 40 \cos(\omega t - 45°)$$

as a single sinusoid by use of phasor addition.

Solution Let

$$100 \cos (\omega t + 30°) \text{ be represented by } 100\underline{/30°}$$

Then

$$-60 \sin (\omega t + 45°) \text{ is represented by } 60\underline{/+45° - 90° \pm 180°}$$

and $$-40 \cos (\omega t - 45°) \text{ is represented by } 40\underline{/-45° \pm 180°}$$

These representations are illustrated in Fig. C-11. We therefore add

$$\begin{aligned} \mathbf{C} &= 100\underline{/30°} + 60\underline{/135°} + 40\underline{/135°} \\ &= 86.6 + j50 - 70.7 + j70.7 = 15.9 + j120.7 \\ &= 122\underline{/82.5°} \end{aligned}$$

or $$f(t) = 122 \cos (\omega t + 82.5°)$$

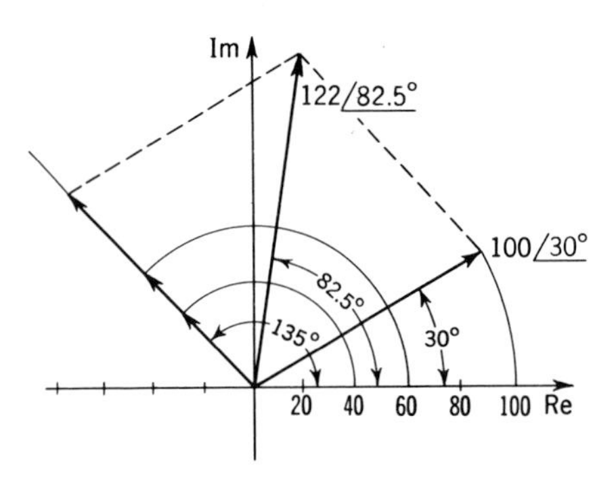

***Fig. C-11** Graphical addition of three phasors for Example C-1.*

C-5 Summary of representation of sinusoidal functions by phasors

We recall that in Appendix B we mentioned that complex numbers are suitable for representing phenomena with two measurable attributes. If the radian frequency of a sinusoidal function is known, the only other two factors which would completely specify the function are its amplitude and the phase angle. The phasor (complex number) $A\underline{/\alpha}$ gives these two factors in one complex number. It must be remembered that the complex number $A\underline{/\alpha}$ *represents* the sinusoidal function $A \cos (\omega t + \alpha)$ but is *not* equal to it. This is shown by writing

$$A \cos (\omega t + \alpha) \rightarrow A\underline{/\alpha}$$

where the arrow means "represents." The second statement is shown by

$$A\underline{/\alpha} \neq A \cos (\omega t + \alpha)$$

to indicate that the representation is not mathematical equality. Similarly, whereas $A\underline{/\alpha} + B\underline{/\beta}$ represents the sum of $A \cos (\omega t + \alpha) + B \cos (\omega t + \beta)$, it is *not* correct to write $A\underline{/\alpha} + B \cos (\omega t + \beta)$ to represent the sum of the two sinusoidal functions. This is a mixture of two different notations which does not convey the required information.

The justification for adding phasors to represent the sum of their respective sinusoidal functions lies in the fact that

$$\begin{aligned} A \cos(\omega t + \alpha) &= \text{Re}\,(Ae^{j(\omega t+\alpha)}) = \text{Re}\,(Ae^{j\alpha}e^{j\omega t}) \\ B \cos(\omega t + \beta) &= \text{Re}\,(Be^{j(\omega t+\beta)}) = \text{Re}\,(Be^{j\beta}e^{j\omega t}) \end{aligned}$$

Now, since the phasor method is used to add only sinusoidal functions of the same frequency, the term $e^{j\omega t}$ and the symbol (operator) Re will occur in all linear operations such as additions, subtraction, differentiation, and integration. We may implicitly assume these recurrences and by $Ae^{j\alpha}$ may understand that it represents $\text{Re}\,(Ae^{j\alpha}e^{j\omega t})$ and by $Be^{j\beta}$, that it represents $\text{Re}\,(Be^{j\beta}e^{j\omega t})$. Therefore, when we write $Ae^{j\alpha} + Be^{j\beta}$, we imply $\text{Re}\,(Ae^{j\alpha}e^{j\omega t}) + \text{Re}\,(Be^{j\beta}e^{j\omega t})$, which is $A \cos(\omega t + \alpha) + B \cos(\omega t + \beta)$. In this sense the phasor is a shorthand notation with specified rules of operation.

Finally, when the phasor $A\underline{/\alpha}$ is shown graphically by an arrow of length A at an angle α with the reference (measured counterclockwise from the horizontal), this means that this line represents the sinusoidal function $A \cos(\omega t + \alpha)$, inasmuch as the value of this function at any time t_1 is found by projecting the phasor $A\underline{/\alpha}$ on an axis which makes the angle ωt_1 with reference in the clockwise direction. If the projection of the arrow is along the positive direction of the horizontal axis rotated through ωt_1, then $f(t_1)$ has a positive value; otherwise $f(t_1)$ will be negative.

A word of caution: Whereas any sinusoidal function *can* be represented by a complex number (dealing with one frequency only), not all complex numbers are to be taken to represent sinusoidal functions. For example, $3\underline{/\pi/4}$ *may* represent $3 \cos(\omega t + \pi/4)$. On the other hand, it may be just a complex number and have nothing to do with sinusoidal functions.

C-6 Phasor representation of derivatives of sinusoids

We now show that when a sinusoidal function $f(t)$ with radian frequency ω is represented by the phasor $\mathbf{F}_m$, its derivative with respect to time is represented by the phasor $j\omega\mathbf{F}_m$. To prove this we observe that if

$$f(t) = F_m \cos(\omega t + \phi) = \text{Re}\,(\mathbf{F}_m e^{j\omega t})$$

then

$$\frac{df}{dt} = \frac{d}{dt}[\text{Re}\,(\mathbf{F}_m e^{j\omega t})] \tag{C-10}$$

We now use the important identity[1]

$$\frac{d}{dt}(\text{Re}\,e^{j\omega t}) = \text{Re}\left[\frac{d}{dt}(e^{j\omega t})\right] \tag{C-11}$$

Hence the operation of differentiation and taking the real part can be exchanged in Eq. (C-10) so that

$$\begin{aligned} \frac{df}{dt} &= \text{Re}\left[\frac{d}{dt}(\mathbf{F}_m e^{j\omega t})\right] \\ &= \text{Re}\,[(j\omega\mathbf{F}_m)e^{j\omega t}] \end{aligned} \tag{C-12}$$

[1] Proof of this identity is left as an exercise for the reader; see Prob C-7.

Letting $\mathbf{D}_m = j\omega\mathbf{F}_m$ in Eq. (C-12), we have

$$\frac{df}{dt} = \text{Re}\,(\mathbf{D}_m e^{j\omega t}) \qquad \mathbf{D}_m = j\omega\mathbf{F}_m \tag{C-13}$$

Equation (C-13) shows that the phasor representing df/dt, $\mathbf{D}_m$, is $j\omega\mathbf{F}_m$, that is, $j\omega$ times the phasor representing $f(t)$.

Example C-2 Find the sinusoid $f(t)$ if

$$v(t) = 30 \cos (3t - 20°)$$

and

$$\frac{d^3v}{dt^3} + 4\frac{d^2v}{dt^2} + 19\frac{dv}{dt} + 76v = f(t) \tag{C-14}$$

Solution Since $v(t) = 30 \cos (3t - 20°) = \text{Re}\,[(30\underline{/-20°})e^{j3t}]$, it is represented by the phasor $\mathbf{V}_m = 30\underline{/-20°}$. Hence d^nv/dt^n is represented by $(j3)^n\mathbf{V}_m$. Letting $f(t)$ be given by $f(t) = F_m \cos (3t + \phi)$, it is represented by the phasor $F_m\underline{/\phi}$. Substituting the phasors for the corresponding time functions in Eq. (C-14) gives

$$[(j3)^3 + 4(j3)^2 + 19(j3) + 76](30\underline{/-20°}) = F_m\underline{/\phi}$$

or evaluating the complex number in the brackets,

$$\begin{aligned}(j3)^3 + 4(j3)^2 + 19(j3) + 76 &= 40 + j30 \\ &= 50\underline{/36.9°}\end{aligned}$$

Hence

$$\begin{aligned}F_m\underline{/\phi} &= (50\underline{/36.9°})(30\underline{/-20°}) \\ &= 1{,}500\underline{/16.9°}\end{aligned}$$

so that

$$f(t) = 1{,}500 \cos (3t + 16.9°) \qquad \textit{Ans.}$$

C-7 Generalization of the phasor method for complex exponential functions

In this section we deal with the addition and differentiation of functions that have the form $f(t) = Fe^{\sigma t} \cos (\omega t + \phi)$. The function

$$f(t) = Fe^{\sigma t} \cos (\omega t + \phi)$$

can be written as the real part of a complex exponential as follows:

$$\begin{aligned}f(t) &= Fe^{\sigma t} \cos (\omega t + \phi) \\ &= \text{Re}\,(Fe^{\sigma t}e^{j(\omega t+\phi)}) \\ &= \text{Re}\,(Fe^{j\phi}e^{(\sigma+j\omega)t})\end{aligned}$$

Defining the generalized phasor $\mathbf{F}$ as

$$\mathbf{F} = Fe^{j\phi}$$

and the complex number

$$\mathbf{s} = \sigma + j\omega$$

we can write

$$f(t) = \text{Re}\,(\mathbf{F}e^{\mathbf{s}t}) \tag{C-15}$$

where we shall refer to $\mathbf{F}$ as the generalized phasor representing $f(t)$.

ADDITION The sum consisting of two (or more similar) terms

$$K_1 e^{\sigma t} \cos (\omega t + \gamma_1) + K_2 e^{\sigma t} \cos (\omega t + \gamma_2) = f(t)$$

can be combined into

$$f(t) = K e^{\sigma t} \cos (\omega t + \gamma) \tag{C-16}$$

where

$$K\underline{/\gamma} = K_1\underline{/\gamma_1} + K_1\underline{/\gamma_2} \tag{C-17}$$

Proof of addition We let, in accordance with Eq. (C-14),

$$K_1 e^{\sigma t} \cos (\omega t + \gamma_1) = \text{Re}\,(\mathbf{K}_1 e^{\mathbf{s}t})$$
$$K_2 e^{\sigma t} \cos (\omega t + \gamma_2) = \text{Re}\,(\mathbf{K}_2 e^{\mathbf{s}t})$$

where $\mathbf{K}_1 = K_1\underline{/\gamma_1} \qquad \mathbf{K}_2 = K_2\underline{/\gamma_2} \qquad \mathbf{s} = \sigma + j\omega$

Thus

$$f(t) = \text{Re}\,(\mathbf{K}_1 e^{\mathbf{s}t}) + \text{Re}\,(\mathbf{K}_2 e^{\mathbf{s}t})$$
$$= \text{Re}\,[(\mathbf{K}_1 + \mathbf{K}_2) e^{\mathbf{s}t}]$$

so that, letting

$$\mathbf{K}_3 = \mathbf{K}_1 + \mathbf{K}_2$$

Eqs. (C-15) and (C-16) are seen to be valid.

Example C-3 Express $f(t)$ in the form $Ke^{-3t} \cos (5t + \gamma)$ if

$$f(t) = 10e^{-3t} \cos (5t - 30°) + 15e^{-3t} \cos (5t + 75°)$$

Solution We write

$$f(t) = \text{Re}\,[(10\underline{/-30°}) e^{(-3+j5)t} + (15\underline{/75°}) e^{(-3+j5)t}]$$
$$= \text{Re}\,[(10\underline{/-30°} + 15\underline{/75°}) e^{(-3+j5)t}]$$

Since

$$[10\underline{/-30°} + 15\underline{/75°}] = 15.8\underline{/37°}$$
$$f(t) = 15.8e^{-3t} \cos (5t + 37°) \qquad \textit{Ans.}$$

DIFFERENTIATION If $\mathbf{s}$ is any complex number, then if the function $f(t) = \text{Re}\,(\mathbf{F}e^{\mathbf{s}t})$, its derivative with respect to time is $\text{Re}\,(\mathbf{sF}e^{\mathbf{s}t})$. (See Prob. C-8.) Thus, if $f(t)$ is represented by the generalized phasor $\mathbf{F}$, then df/dt is represented by the generalized phasor $\mathbf{sF}$.

Example C-4 Find the function $f(t) = Ke^{-3t} \cos (5t + \gamma)$ if $v = 6e^{-3t} \cos (5t + 15°)$ and if

$$\frac{d^3v}{dt^3} + 4\frac{d^2v}{dt^2} + 10\frac{dv}{dt} + 20v = f \tag{C-18}$$

Solution Let $\mathbf{s} = -3 + j5$ and $\mathbf{V} = 6\underline{/15°}$; then $v = \text{Re}\,(\mathbf{V}e^{\mathbf{s}t})$ and $d^nv/dt^n = \text{Re}\,(\mathbf{s}^n\mathbf{V}e^{\mathbf{s}t})$. We let $f(t) = \text{Re}\,(\mathbf{K}e^{\mathbf{s}t})$, where $\mathbf{K} = K\underline{/\gamma}$, and replace the time functions in Eq. (C-17) by their generalized phasors:

$$[(-3 + j5)^3 + 4(-3 + j5)^2 + 10(-3 + j5) + 20](6\underline{/15°}) = K\underline{/\gamma}$$

The quantity in the brackets is evaluated

$$[(-3 + j5)^3 + 4(-3 + j5)^2 + 10(-3 + j5) + 20] = 124 - j60 = 138\underline{/-25.8°}$$

Hence

$$K\underline{/\gamma} = (138\underline{/-25.8°})(6\underline{/15°})$$
$$= 828\underline{/-10.8°}$$

and

$$f(t) = 828e^{-3t} \cos (5t - 10.8°) \qquad \textit{Ans.}$$

PROBLEMS

C-1 (*a*) Sketch as a function of t the function $f(t) = 23 \cos (5{,}000t - 30°)$. (*b*) For the function of (*a*) what are (1) its amplitude; (2) its radian frequency; (3) its frequency? (*c*) What is the phase relation of $f(t)$ to the function (1) $\cos 5{,}000t$; (2) $\sin 5{,}000t$; (3) $\sin (5{,}000t + 60°)$; (4) $\sin (5{,}000t - 60°)$; (5) $-\sin (5{,}000t - 60°)$?

C-2 Use phasor representation to perform the following additions: (*a*) $f_a(t) = 34 \cos (\omega t - 14°) + 40 \sin (\omega t + 13°) - 20 \sin \omega t$; (*b*) $f_b(t) = 150 \cos (\omega t - 120°) + 200 \cos \omega t - 75 \sin \omega t$.

C-3 The following functions are to be represented by phasors: (*a*) $10 \sin \omega t$; (*b*) $10 \sin (\omega t + 30°)$; (*c*) $-5 \cos (\omega t - 17°)$; (*d*) $-4 \sin (\omega t - 143°)$. What are the phasors that represent each of these functions if (1) $\cos \omega t$ is represented by $1\underline{/0}$; (2) $\sin \omega t$ is represented by $1\underline{/0}$?

C-4 Solve the following equation for the positive real number A and the angle ϕ in degrees: $100 \cos \omega t + A \cos (\omega t + 60°) = 173 \cos (\omega t + \phi)$.

C-5 In a certain loop *a-b-c-d-a* of a circuit, $v_{ab} = 30 \cos \omega t$, $v_{bc} = 40 \sin \omega t$, $v_{cd} = 5 \cos (\omega t + 37°)$. Calculate v_{da}.

C-6 The phasor $1\underline{/0°}$ represents $\cos \omega t$. The function $f_1(t)$ is represented by $10\underline{/45°}$, and $f_2(t)$ is represented by $12\underline{/-30°}$. At the instants (*a*) $\omega t_a = 0$, (*b*) $\omega t_b = \pi/6$, (*c*) $\omega t_c = 3\pi/4$, (*d*) $\omega t_d = 3\pi/2$, find (1) $f_1(t)$; (2) $f_2(t)$; (3) $f_1(t) + f_2(t)$; (4) $f_1(t) - f_2(t)$.

C-7 Prove the following identities:

$$(a)\ \text{Re}\left[\frac{d}{dt}(e^{j\omega t})\right] = \frac{d}{dt}(\text{Re}\ e^{j\omega t})$$

$$(b)\ \text{Im}\left[\frac{d}{dt}(e^{j\omega t})\right] = \frac{d}{dt}(\text{Im}\ e^{j\omega t})$$

$$(c)\ \frac{d}{dt}(\text{Re}\ e^{j\omega t}) = \text{Im}\left[\frac{d}{dt}(je^{j\omega t})\right]$$

C-8 (*a*) Prove that, for any complex function of a real variable, $\mathbf{w}(t) = u(t) + jv(t)$, $d/dt\ (\text{Re}\ \mathbf{w}) = \text{Re}\ [d/dt(\mathbf{w})]$, and $d/dt\ (\text{Im}\ \mathbf{w}) = \text{Im}\ (d\mathbf{w}/dt)$. (*b*) Use the result of part *a* to show that, if $\mathbf{s}$ is any complex number, then $d/dt\ (\text{Re}\ e^{\mathbf{s}t}) = \text{Re}\ [d/dt(e^{\mathbf{s}t})]$, and $d/dt\ (\text{Im}\ e^{\mathbf{s}t}) = \text{Im}\ [d/dt(e^{\mathbf{s}t})]$. (*c*) Prove the identities of part *b* by performing the operations indicated in each sequence.

C-9 Find the real positive number A and the angle ϕ in degrees in each of the following cases; in each instant $v = 25 \cos (4t - 17°)$:

$$(a)\ \frac{d^3v}{dt^3} + 5\frac{d^2v}{dt^2} + 30\frac{dv}{dt} + 80v = A \cos (4t + \phi)$$

$$(b)\ \frac{d^3v}{dt^3} + 10\frac{d^2v}{dt^2} + 11\frac{dv}{dt} + 25v = A \cos (4t + \phi)$$

C-10 (*a*) If $f(t) = \text{Re}\ (30e^{(-2+j4)t})$, find the value of d^3f/dt^3 at the instant $t = 0$. (*b*) For $f(t)$ as given in part *a*, find $g(t) = f + df/dt$. (*c*) For $f(t)$ as given in part *a*, find A and ϕ if

$$\frac{d^2f}{dt^2} + \frac{df}{dt} + f = Ae^{-2t} \cos (4t + \phi)$$

Appendix D *Matrices and determinants*

D-1 Matrices

In the study of systems, such as electric power systems, communication systems, and other systems characterized by sets of simultaneous equations, we deal with a set of numbers (coefficients), operators (time or space and other differential operators), and functions whose proper combinations determine the behavior of the system (the solution of the problem). In a systematic analysis of such problems these numbers, operators, or functions, which we shall refer to as "elements," often occur in the form of an "array" of "rows" and "columns," which can be manipulated in accordance with certain specified rules. Such an array of elements is called a matrix. With this in mind we define a matrix to be a "table" of elements arranged in a *rectangular* array of rows and columns, subject to certain rules of manipulation.

In the following sections we define certain terms associated with matrices. These definitions may seem arbitrary, and the purpose of a definition may not be immediately clear, but later experience will justify them. Some definitions are made to facilitate precision in statements about the properties of matrices.

D-2 Subscript notation

We may designate a "typical" element of a matrix by a letter such as a and show its position in the array by two subscripts indicating its row and its column, respectively. For example, a_{25} denotes the element whose position is at the "intersection" of the second row and fifth column, and a_{ij} denotes the element whose position is at the "intersection" of ith row and jth column. Consider the matrix given in the tabular form

$$\begin{array}{c|c|c} 2 & R & L\dfrac{d}{dt} \\ \hline -1 & -M\dfrac{d}{dt} & 4 \end{array} \qquad \text{(D-1)}$$

This matrix has two rows and three columns; its elements are $a_{11} = 2$, $a_{12} = R$, $a_{13} = L(d/dt)$, $a_{21} = -1$, $a_{22} = -M(d/dt)$, $a_{23} = 4$. A matrix is designated by a capital letter such as A and is written

$$A = \|a_{ij}\| \qquad \text{(D-2)}$$

where i takes up all the row numbers and j the column numbers of the matrix. The matrix of Eq. (D-1) can be written

$$A = \begin{Vmatrix} a_{11} & a_{12} & a_{13} \\ a_{21} & a_{22} & a_{23} \end{Vmatrix} = \|a_{ij}\| \qquad \begin{array}{l} i = 1, 2 \\ j = 1, 2, 3 \end{array}$$

D-3 Order of a matrix

A matrix is called an m-by-n matrix where m is the number of its rows and n the number of its columns. The matrix of Eq. (D-1) is a two-by-three (2×3) matrix.

D-4 Matrix algebra

A matrix is not a "number," and it does not have a "value." It is simply an arrangement of "elements" ordered in rows and columns. Furthermore, no restriction is placed on the nature (dimensions) of the elements; they can have varied dimensions. Although we have said that matrices are not numbers in the ordinary sense of the word, we may define certain rules for manipulations of matrices, analogous to those defined for algebraic numbers. Such rules may be stated quite arbitrarily, but it is doubtful that arbitrary rules would result in manipulations which are useful (in the study of physical systems). Therefore the rules for manipulation of matrices are drawn up in such a manner that their use simplifies the study and analysis of practical problems.

D-5 Equal matrices

If two matrices have the same order and each element in one matrix is identical with its corresponding element (in the same row and column) in the other matrix, the two matrices are said to be equal.

D-6 Addition of matrices

If the matrices $A = \|a_{ij}\|$ and $B = \|b_{kl}\|$ are of the same order, the matrix $C = \|c_{mn}\|$ is said to be the sum of the matrices A and B if $c_{mn} = a_{mn} + b_{mn}$.

Example D-1 Let

$$A = \begin{Vmatrix} 2 & \frac{d}{dt} & 0 & \frac{d}{dx} \\ \frac{d}{dy} & 3 & 1 & 0 \end{Vmatrix}$$

$$B = \begin{Vmatrix} 1 & 1 & 1 & 1 \\ 2 & 2 & 2 & 2 \end{Vmatrix}$$

Then

$$A + B = \begin{Vmatrix} 2+1 & \frac{d}{dt}+1 & 0+1 & \frac{d}{dx}+1 \\ \frac{d}{dy}+2 & 3+2 & 1+2 & 0+2 \end{Vmatrix}$$

The addition is defined only for matrices of the same order. Subtraction can be considered as a special case of addition. It is noted that $A + B = B + A$.

The rules of addition and subtraction of matrices are easily understood. To define other matrix operations, we give an example from common experience; and based on the manipulations in this example, generalizations and definitions will be made.

D-7 Item, price, and payment matrices

Consider two customers making purchases from a supermarket. Customer X purchases 2 loaves of bread, 6 cans of orange juice, and $\frac{1}{2}$ lb of butter, and customer Y buys 3 loaves of bread, 0 cans of orange juice, and $\frac{3}{4}$ lb of butter. The purchases of the two customers can be set up as a matrix of 2×3 (2 customers and 3 items) as follows:

Customer	Loaves of bread	Cans of juice	Pounds of butter
X	2	6	$\frac{1}{2}$
Y	3	0	$\frac{3}{4}$

We could add items or customers to this matrix and change the order of the matrix, but the above example suffices for our purpose. We call this matrix the "items" matrix. Now, at the cash desk, the clerk has a price list as follows:

Bread	23 cents per loaf
Orange juice	11 cents per can
Butter	72 cents per lb

This can be considered as a 3×1 unit price matrix. Customer X pays

$$2 \times 23 + 6 \times 11 + \tfrac{1}{2} \times 72 = 148 \text{ cents}$$

Notice that 148 cents is arrived at from a summation whose first term is obtained by taking the first element of the first row in the items matrix

$$\begin{Vmatrix} 2 & 6 & \frac{1}{2} \\ 3 & 0 & \frac{3}{4} \end{Vmatrix} \tag{D-3}$$

and multiplying it by the first element of the first column in the price matrix

$$\begin{Vmatrix} 23 \\ 11 \\ 72 \end{Vmatrix} \tag{D-4}$$

thus getting 2×23. The second term in the summation is obtained by taking the second element in the first row of the items matrix and multiplying by the second element of the price matrix, which gives 6×11. The third term in the summation is the product of the third term in the first row of the items matrix multiplied by the third term of the price matrix, giving $\frac{1}{2} \times 72$.

Customer Y must pay $3 \times 23 + 0 \times 11 + \frac{3}{4} \times 72 = 123$ cents. The value 123 cents is arrived at in a similar manner to that of 148 cents, except that the terms of the summation are obtained by multiplying the proper elements of the second row of the items matrix by those of the price matrix. Thus the payments made to the cash clerk are

Customer	Payment, cents
X	148
Y	123

This is a 2×1 matrix, which we call the payment matrix.

$$\begin{Vmatrix} 148 \\ 123 \end{Vmatrix} \tag{D-5}$$

The elements of the payment matrix are obtained from elements of the items and price matrices through a definite procedure, which is now generalized in the following rules for matrix multiplication.

D-8 Compatible matrices

If the matrix A is of the order $m \times n$ (for example, $m = 5$, $n = 3$) and matrix B is of the order $n \times k$ ($n = 3$, and, for example, $k = 7$), the matrix B is said to be compatible with matrix A. Thus, for B to be compatible with A, the number of rows of B must be identical with the number of columns of A. In the above numerical examples, B is compatible with A, since the number of rows of B is the same (three) as that of columns of A. But A is not compatible with B, since the number of rows of A is 5 and the number of columns of B is 7. In the example of purchases from the supermarket, the price matrix is compatible with the items matrix, since the number of rows in the price matrix is the same (three) as the number of columns of the items matrix. But the items matrix is not compatible (two rows) with the price matrix (one column).

D-9 Matrix multiplication

If the matrix $B = \|b_{ij}\|$ with n rows and q columns is compatible with matrix $A = \|a_{ij}\|$ of p rows and n columns, the matrix $C = \|c_{ij}\|$ is said to be the product $A \times B$, if

$$c_{ij} = \sum_{k=1}^{n} a_{ik}b_{kj}$$

$C = A \times B$, where the parametric subscript i assumes the values 1, 2, . . . , p and the parametric subscript j assumes the values 1, 2, . . . , q.

From this definition the following points are to be noted:

1 It is possible for the product $A \times B$ to exist without $B \times A$ existing. This happens when B is compatible with A but A is not compatible with B.

2 Even when both $A \times B$ and $B \times A$ exist, in general,

$$A \times B \neq B \times A$$

3 If A is of the order $p \times n$ and B is of the order $n \times q$, then

$$A \times B = C$$

is a matrix of the order $p \times q$; that is, C has as many rows as A and as many columns as B.

4 The element c_{11} of $C = A \times B$ is obtained by taking the sum of a number of terms obtained from the product of "proper" elements of A and B.

$$c_{11} = \sum_{k=1}^{n} a_{1k}b_{k1} = a_{11}b_{11} + a_{12}b_{21} + a_{13}b_{31} + \cdots + a_{1n}b_{n1}$$

$$c_{21} = \sum_{k=1}^{n} a_{2k}b_{k1} = a_{21}b_{11} + a_{22}b_{21} + a_{23}b_{31} + \cdots + a_{2n}b_{n1}$$

$$\vdots \qquad \vdots \qquad \vdots$$

$$c_{12} = \sum_{k=1}^{n} a_{1k}b_{k2} = a_{11}b_{12} + a_{12}b_{22} + a_{13}b_{32} + \cdots + a_{1n}b_{n2}$$

Note that the inner subscripts such as 2 in $a_{12}b_{21}$ or 3 in $a_{13}b_{32}$ are always equal.

With the above definition it is seen that the payment matrix of the supermarket example is the product of the items matrix and the unit price matrix,

$$\|\text{items matrix}\| \times \|\text{unit price matrix}\| = \|\text{payment matrix}\|$$

$$\begin{Vmatrix} 2 & 6 & \frac{1}{2} \\ 3 & 0 & \frac{3}{4} \end{Vmatrix} \times \begin{Vmatrix} 23 \\ 11 \\ 72 \end{Vmatrix} = \begin{Vmatrix} (2 \times 23) + (6 \times 11) + (\frac{1}{2} \times 72) \\ (3 \times 23) + (0 \times 11) + (\frac{3}{4} \times 72) \end{Vmatrix} = \begin{Vmatrix} 148 \\ 123 \end{Vmatrix}$$

We may extend the example of the supermarket purchase and say that the prices are given as 23 cents per loaf, 11 cents per can, and 72 cents per pound if paid for in cash and 24 cents per loaf, 12 cents per can, and 76 cents per pound if charged to account. In this case the unit price matrix will be a 3 × 2 matrix.

Item	Cash price, cents	Credit price, cents
Loaf of bread	23	24
Can of juice	11	12
Pound of butter	72	76

Now the product of the items matrix and the new price matrix is

$$\begin{Vmatrix} 2 & 6 & \frac{1}{2} \\ 3 & 0 & \frac{3}{4} \end{Vmatrix} \times \begin{Vmatrix} 23 & 24 \\ 11 & 12 \\ 72 & 76 \end{Vmatrix}$$

$$= \begin{Vmatrix} (2 \times 23 + 6 \times 11 + \frac{1}{2} \times 72) & (2 \times 24 + 6 \times 12 + \frac{1}{2} \times 76) \\ (3 \times 23 + 0 \times 11 + \frac{3}{4} \times 72) & (3 \times 24 + 0 \times 12 + \frac{3}{4} \times 76) \end{Vmatrix}$$

$$= \begin{Vmatrix} 148 & 158 \\ 123 & 129 \end{Vmatrix}$$

The new payment matrix is composed of elements which are the prices that X must pay, cash or credit (148 or 158), and Y must pay, cash or credit (123 or 129). The relationship $c_{ij} = \sum_{k=1}^{n} a_{ik}b_{kj}$ indicates the operation of taking the kth element of the ith row in A (items matrix) and multiplying it by the kth element of the jth column in B (price matrix) and adding the results for all k (all items) to give the c_{ij} element (the payment customer $i = X$ or $i = Y$ must make if he pays j = cash or j = charge account).

D-10 Square matrix

If the number of the rows and columns of a matrix are identical, the matrix is called a square matrix.

PRINCIPAL ELEMENTS OF A SQUARE MATRIX In the square matrix

$$A = \|a_{jk}\|$$

the terms a_{jj} are called the principal elements of the matrix. Thus, in a square matrix of order 4, the terms a_{11}, a_{22}, a_{33}, and a_{44} are the principal elements.

PRINCIPAL DIAGONAL OF A SQUARE MATRIX The "diagonal" formed by the principal elements of a square matrix is called the principal diagonal of the matrix.

SYMMETRICAL MATRIX If in a square matrix $a_{jk} = a_{kj}$, that matrix is called symmetrical. Network matrices (mesh or node) of reciprocal networks are symmetrical, since $Y_{21} = Y_{12}$, etc.

D-11 Fundamental products of elements of a square matrix

At the beginning of this Appendix we noted that, for the purpose of precision of statement, we must define certain terms. In the following paragraphs some of these terms are defined. A square matrix A of order n (n rows and n columns) has n^2 elements. Consider a term obtained by multiplication of n of such elements; a variety of such terms is possible, and the number increases as n increases. Among such products a particular set is of particular interest. The members of this set are chosen in such a manner that, for example, if a_{23} is one of the elements in the product, none of the other elements are chosen from row 2 or column 3 of the matrix. Or in general, if a_{ij} is one of the elements in the product, none of the other elements are chosen from the ith row or jth column. We call such products the "fundamental" products of the square matrix.

As a specific example, we consider a fourth-order square matrix (four rows and four columns, with $n = 4$), $A = \|a_{ij}\|$. We may ask ourselves: How many fundamental products exist for such a matrix? The answer to this question is found if we systematize the problem by noting that, since not more than one element can be taken from one row, each of the elements in a fundamental product must be taken from a different row. Therefore a typical product can be written as $a_{1x}a_{2y}a_{3z}a_{4r}$, where x, y, z, and r refer to the column position of the various elements. Now, if $x = 2$, then y, z, and r can no longer be 2. We may start by giving x the choice of all the possible number of columns (one to four for a fourth-order matrix). Once the value of x is selected, y can assume (in this example) three values, z two values, and r one value. Thus there will be $4 \times 3 \times 2 \times 1 = 4! = 24 = n!$ different products of the form $a_{1x}a_{2y}a_{3z}a_{4r}$, where no two elements belong to the same column or the same row. These products can be obtained by ordering the elements in accordance with their rows, that is, $a_{1-}a_{2-}a_{3-}a_{4-}$, and then filling the blanks by the order chosen for the columns. These orders are given in Table D-1.

Table D-1

x	y	z	r	x	y	z	r	x	y	z	r	x	y	z	r
1	2	3	4	2	1	3	4	3	1	2	4	4	1	2	3
1	2	4	3	2	1	4	3	3	1	4	2	4	1	3	2
1	3	2	4	2	3	1	4	3	2	1	4	4	2	1	3
1	3	4	2	2	3	4	1	3	2	4	1	4	2	3	1
1	4	2	3	2	4	1	3	3	4	1	2	4	3	1	2
1	4	3	2	2	4	3	1	3	4	2	1	4	3	2	1

The corresponding products are $a_{11}a_{22}a_{33}a_{44}$, $a_{12}a_{21}a_{33}a_{44}$, $a_{13}a_{21}a_{32}a_{44}$, $a_{14}a_{21}a_{32}a_{43}$, with the remaining rows found from Table D-1.

D-12 Assignment of positive or negative signs to fundamental products

A fundamental product of a (numerical) matrix may be either a positive or a negative number. Besides this sign, which will depend on the value of the elements in the product, we associate a positive or negative sign with the product (we polarize the product) in accordance with a rule illustrated in the following example.

If, in the example of a fourth-order matrix, the elements of the fundamental products are ordered in accordance with their row numbers and the sequence of the column numbers of the elements examined, one of the sequences given in Table D-1 will result. Any of these (column numbers) sequences can be changed into a 1 2 3 4 (ordered) sequence by successive interchange of *adjacent* numbers. For example, 2 1 4 3 can be changed to 2 1 3 4 and then to 1 2 3 4, a total of two changes of *adjacent* numbers. The sequence 1 2 4 3 can be changed into 1 2 3 4 in one step. If for a product the number of such changes is odd, the negative of that fundamental product will be called a "polarized product." If the number of changes is even, the polarized product is identical with the fundamental product. Thus the polarized product corresponding to $a_{11}a_{22}a_{34}a_{43}$ is $-a_{11}a_{22}a_{34}a_{43}$, and the polarized product corresponding to $a_{12}a_{21}a_{34}a_{43}$ is this product itself.

D-13 Determinant of a square matrix

The sum of the polarized products of a square matrix is called the determinant of that matrix. If the square matrix is $A = \|a_{ij}\|$, the determinant is shown as $D_A = |a_{ij}|$.

The determinant of a matrix is defined in the above manner because of its use in the solution of a set of simultaneous equations. This will become apparent as we treat such equations.

Example D-2 Find the determinant of $A = \begin{Vmatrix} 1 & 2 \\ 3 & 4 \end{Vmatrix}$.

Solution The fundamental products of A are $1 \times 4 = 4$ and $3 \times 2 = 6$. The polarized products are 4 and -6. The sum of these two gives

$$D_A = \begin{vmatrix} 1 & 2 \\ 3 & 4 \end{vmatrix} = 4 - 6 = -2$$

D-14 Properties of determinants

RULE 1 INTERCHANGE OF ROWS AND COLUMNS A determinant is a "quantity" (number, operator, function) associated with a square array of elements. If the rows and columns of a determinant are interchanged, the value of the new determinant is identical with the value of the original determinant.

Example D-3

$$\begin{vmatrix} 1 & 2 \\ 3 & 4 \end{vmatrix} = \begin{vmatrix} 1 & 3 \\ 2 & 4 \end{vmatrix}$$

The reason for this is that changing the rows and columns of determinants does not change the value of the polarized products. In Example D-3 the polarized products of the original determinant are 4 and -6, and those of the new determinant are also 4 and -6. A detailed study of any determinant will show the generality of the argument.

Because of this property, any statement made about rows also applies to columns. For the sake of brevity, in such statements we shall not repeat "rows or columns"; rather, we shall simply make the statement about rows with the implicit understanding that the statement is equally applicable to columns.

RULE 2 INTERCHANGE OF ROWS An interchange of adjacent rows of a determinant forms a new determinant. The fundamental products of the new determinant are identical with those of the original determinant but will have the opposite polarity. For example, the fundamental products of each of the two determinants

$$\begin{vmatrix} 1 & 2 \\ 3 & 4 \end{vmatrix} \quad \text{and} \quad \begin{vmatrix} 3 & 4 \\ 1 & 2 \end{vmatrix}$$

are $1 \times 4 = 4$ and $3 \times 2 = 6$. The polarized products of the first determinant are 4 and -6. The polarized products of the second determinant are 6 and -4.

A detailed study of the general form of a determinant will show that this change of polarity takes place whenever adjacent rows of a determinant are interchanged. Since the value of a determinant is the sum of its polarized products, an interchange of adjacent rows of a determinant reverses the sign of the determinant. Thus

$$\begin{vmatrix} 1 & 2 \\ 3 & 4 \end{vmatrix} = -2 \quad \text{and} \quad \begin{vmatrix} 3 & 4 \\ 1 & 2 \end{vmatrix} = +2$$

RULE 3 IDENTICAL ROWS If two rows of a determinant are identical (their corresponding elements are identical), the value of the determinant is zero.

Proof We first consider the case when the two adjacent rows are identical. Then an interchange of the adjacent rows must reverse the sign of the determinant. But such an interchange does not change the determinant because the two rows are identical. A quantity whose negative is equal to itself must be zero.

Example D-4 From property 2,

$$\begin{vmatrix} 1 & 2 & 3 \\ 1 & 2 & 3 \\ 4 & 5 & 6 \end{vmatrix} = - \begin{vmatrix} 1 & 2 & 3 \\ 1 & 2 & 3 \\ 4 & 5 & 6 \end{vmatrix}$$

The second determinant is obtained by interchanging the position of the first and second rows in the first determinant, but no change in the array results. The above determinant is equal to the negative of itself and must be zero.

The proof, for the case when the rows are not adjacent, is left to the reader as an exercise.

RULE 4 MULTIPLICATION OF ELEMENTS OF A ROW BY A CONSTANT If all the elements in a row of a determinant are multiplied by a constant, the value of the determinant is multiplied by the constant.

Proof Each fundamental product contains one and only one element from the row which is multiplied by the constant. Therefore all the products are multiplied by the constant, and thus the resulting sum of polarized products is multiplied by the constant.

Before studying other properties of determinants we shall define a few terms.

D-15 Minors, cofactors, and complements

If in a determinant $D_A = |a_{mn}|$ the rth row and cth column are deleted, the remaining determinant is called the *minor* of the element a_{rc} and is denoted by $(M_A)_{rc}$. In the determinant

$$D_B = \begin{vmatrix} b_{11} & b_{12} & b_{13} \\ b_{21} & b_{22} & b_{23} \\ b_{31} & b_{32} & b_{33} \end{vmatrix}$$

there are nine possible minors, among which are

$$(M_B)_{11} = \begin{vmatrix} b_{22} & b_{23} \\ b_{32} & b_{33} \end{vmatrix} \qquad (M_B)_{23} = \begin{vmatrix} b_{11} & b_{12} \\ b_{31} & b_{32} \end{vmatrix}$$

If the minor M_{rc} is multiplied by $(-1)^{r+c}$, the result is called the *cofactor* of row r and column c. Thus, if $r + c$ is an even number, the cofactor is identical with the minor; and if $r + c$ is an odd number, the cofactor is the negative of the minor. We denote the cofactor associated with row r and column c of the determinant D_B by $(F_B)_{rc}$. Thus

$$(F_B)_{rc} = (-1)^{r+c}(M_B)_{rc} \tag{D-6}$$

The element a_{rc} is called the *complement* of the cofactor F_{rc}.

D-16 Evaluation of determinant by Laplace's development

A systematic and convenient method for evaluating determinants is by Laplace's development. According to this method the value of a determinant may be found as follows.

Select either a row or a column of the determinant. Obtain all the cofactors for that row or column. The value of the determinant is the sum of all the products of the cofactors multiplied by their complements. Thus the third-order determinant

$$D_A = \begin{vmatrix} a_{11} & a_{12} & a_{13} \\ a_{21} & a_{22} & a_{23} \\ a_{31} & a_{32} & a_{33} \end{vmatrix}$$

may be written

$$D_A = a_{11}\begin{vmatrix} a_{22} & a_{23} \\ a_{32} & a_{33} \end{vmatrix} + (-1)a_{21}\begin{vmatrix} a_{12} & a_{13} \\ a_{32} & a_{33} \end{vmatrix} + a_{31}\begin{vmatrix} a_{12} & a_{13} \\ a_{22} & a_{23} \end{vmatrix}$$

or $$D_A = (-1)a_{21}\begin{vmatrix} a_{12} & a_{13} \\ a_{32} & a_{33} \end{vmatrix} + a_{22}\begin{vmatrix} a_{11} & a_{13} \\ a_{31} & a_{33} \end{vmatrix} + (-1)a_{23}\begin{vmatrix} a_{11} & a_{12} \\ a_{31} & a_{32} \end{vmatrix}$$

or in general, if a column c is used,

$$D_A = a_{1c}(F_A)_{1c} + a_{2c}(F_A)_{2c} + a_{3c}(F_A)_{3c} \qquad c = 1 \text{ or } 2 \text{ or } 3$$

and if a row r is used,

$$D_A = a_{r1}(F_A)_{r1} + a_{r2}(F_A)_{r2} + a_{r3}(F_A)_{r3} \qquad r = 1 \text{ or } 2 \text{ or } 3$$

The general form for Laplace's development gives

$$D_A = \sum_{j=1}^{n} a_{jc}(F_A)_{jc} = \sum_{j=1}^{n} a_{rj}(F_A)_{rj}$$

where n is the order of the determinant.

Using column 1 in the following numerical example,

$$D = \begin{vmatrix} 8 & 2 & -4 \\ 3 & 6 & 2 \\ 1 & 2 & 3 \end{vmatrix} = 8\begin{vmatrix} 6 & 2 \\ 2 & 3 \end{vmatrix} - 3\begin{vmatrix} 2 & -4 \\ 2 & 3 \end{vmatrix} + 1\begin{vmatrix} 2 & -4 \\ 6 & 2 \end{vmatrix}$$

Applying Laplace's development to a two-column two-row determinant,

$$\begin{vmatrix} a_{11} & a_{12} \\ a_{21} & a_{22} \end{vmatrix} = a_{11}a_{22} - a_{21}a_{12}$$

We observe, therefore, that repeated application of Laplace's development to the evaluation of the cofactors eventually results in a determinant of order 2, which is evaluated as above. It is evidently advantageous to choose the column or row for Laplace's development such that (where possible) as many complements are zero as is possible.

Example D-5

$$D_1 = \begin{vmatrix} 8 & 2 & -4 \\ 3 & 6 & 2 \\ 0 & 2 & 3 \end{vmatrix} = 8(18 - 4) - 3(6 + 8) + 0(4 + 24) = 70$$

This value is identical with the sum of the polarized products of the matrix. The proof of Laplace's development is not given here. It may be proved by tabulation of the polarized products in groups where the first elements of the products are the same. The sum of such a group is the product of a complement with its cofactor.

D-17 Properties of cofactors

From Laplace's development we have seen that

$$D_A = |a_{ij}| = \sum_{m=1,2,3,\ldots,n} a_{mc}(F_A)_{mc} = \sum_{m=1,2,3,\ldots,n} a_{rm}(F_A)_{rm} \qquad \text{(D-7)}$$

where n is the order of the determinants. In this summation each cofactor $(F_A)_{mc}$ is multiplied by its own complement a_{mc}. We now show that, if the respective cofactor of row 1 [that is, $(F_A)_{1m}$] is multiplied by complements taken from other rows (such as a_{2m}) and the results added, zero results; i.e.,

$$\sum_{m=1,2,3,\ldots,n} a_{2m}(F_A)_{1m} = \sum_{m=1,2,3,\ldots,n} a_{m2}(F_A)_{m1} = \cdots = 0$$

To prove this, consider the determinant D_A' obtained from D_A by replacing its first row with its second row but keeping the second row unchanged.

$$D_A' = \begin{vmatrix} a_{21} & a_{22} & a_{23} & \cdots & a_{2n} \\ a_{21} & a_{22} & a_{23} & \cdots & a_{2n} \\ a_{31} & a_{32} & \cdot & \cdots & \cdot \\ \cdot & \cdot & \cdot & \cdots & \cdot \\ a_{n1} & a_{n2} & a_{n3} & \cdots & a_{nn} \end{vmatrix}$$

Since two rows of D_A' are identical, $D_A' \equiv 0$. Laplace's development of D_A' gives

$$D_A' = a_{21}(F_A)_{11} + a_{22}(F_A)_{12} + a_{23}(F_A)_{13} + \cdots + a_{2n}(F_A)_{1n} = 0$$

This can be written

$$\sum_{m=1,2,\ldots,n} a_{2m}(F_A)_{1m} = 0$$

which proves the above statement. If the first row is replaced by the third row and the third row is left unchanged, we arrive at

$$\sum_{m=1,2,\ldots,n} a_{3m}(F_A)_{1m} = 0 = \sum_{\substack{m=1,2,\ldots,n \\ i \neq j}} a_{im}(F_A)_{jm}$$

By a similar procedure we can show that

$$\sum_{m=1,2,\ldots,n} a_{mi}(F_A)_{mj} = 0 \qquad \text{for } i \neq j \tag{D-8}$$

D-18 Properties of determinants (continued)

In Sec. D-14 four of the properties of determinants were studied. By the use of Laplace's development and the properties of the cofactors, the following properties of determinants can be examined:

RULE 5 Any row of a determinant may be added to another row without changing the value of determinant.

Proof Consider the determinant $D_A = |a_{ij}| = \sum_{j=1}^{n} a_{ij}(F_A)_{ij}$. Let D'_A be obtained from D_A by adding the second row of D_A to its first row. Then

$$D'_A = \sum_{j=1}^{n} (a_{1j} + a_{2j})(F_A)_{1j} = \sum_{j=1}^{n} a_{1j}(F_A)_{1j} + \sum_{j=1}^{n} a_{2j}(F_A)_{1j}$$

In Sec. D-17 it is seen that $\sum_{j=1}^{n} a_{2j}(F_A)_{1j} = 0$, which shows that

$$D_A = D'_A$$

RULE 6 If all terms except those on the principal diagonal of a determinant are zero, the value of the determinant is equal to the product of the terms on the principal diagonal.

Proof The statement that the terms of the principal diagonal are zero can be expressed as

$$a_{ij} = \delta_{ij}a_{ij}$$

where δ_{ij} is Kronecker's delta, defined as

$$\delta_{ij} = 1 \qquad i = j$$
$$\delta_{ij} = 0 \qquad i \neq j$$

$$D_A = \sum_{j=1}^{n} a_{1j}(F_A)_{1j} = \sum_{j=1}^{n} \delta_{1j}a_{1j}(F_A)_{1j} = a_{11}(F_A)_{11}$$

A similar argument shows that $(F_A)_{11}$ is equal to the product of a_{22} by another factor which is a_{33} times a_{44}, and so on. Thus

$$D_A = a_{11}a_{22}a_{33} \cdots a_{nn}$$

To illustrate some of these properties, consider the three-column three-row determinant D.

$$D = \begin{vmatrix} 8 & 2 & -4 \\ 3 & 6 & 2 \\ 1 & 2 & 3 \end{vmatrix}$$

This determinant is evaluated by application of various properties of determinants given in Rules 1 to 6 in Secs. D-14 and D-18.

Apply Rule 4, multiplying row 1 by $\frac{1}{2}$.

$$\tfrac{1}{2}D = \begin{vmatrix} 4 & 1 & -2 \\ 3 & 6 & 2 \\ 1 & 2 & 3 \end{vmatrix}$$

Apply Rule 5, adding row 1 to row 2.

$$\tfrac{1}{2}D = \begin{vmatrix} 4 & 1 & -2 \\ 7 & 7 & 0 \\ 1 & 2 & 3 \end{vmatrix}$$

Now multiply column 2 by -1, and add to column 1.

$$\tfrac{1}{2}D = \begin{vmatrix} 3 & 1 & -2 \\ 0 & 7 & 0 \\ -1 & 2 & 3 \end{vmatrix}$$

Now multiply column 1 by 2, and add to column 2.

$$\tfrac{1}{2}D = \begin{vmatrix} 3 & 7 & -2 \\ 0 & 7 & 0 \\ -1 & 0 & 3 \end{vmatrix}$$

Now subtract row 2 from row 1.

$$\tfrac{1}{2}D = \begin{vmatrix} 3 & 0 & -2 \\ 0 & 7 & 0 \\ -1 & 0 & 3 \end{vmatrix}$$

Now multiply row 1 by $\frac{1}{3}$, and add to row 3.

$$\tfrac{1}{2}D = \begin{vmatrix} 3 & 0 & -2 \\ 0 & 7 & 0 \\ 0 & 0 & \frac{7}{3} \end{vmatrix}$$

Finally, multiply row 3 by $\frac{6}{7}$, and add to row 1.

$$\tfrac{1}{2}D = \begin{vmatrix} 3 & 0 & 0 \\ 0 & 7 & 0 \\ 0 & 0 & \frac{7}{3} \end{vmatrix}$$

Hence, using Rule 6,

$$\tfrac{1}{2}D = 3 \times 7 \times \tfrac{7}{3} = 49$$

or

$$D = 98$$

D-19 Application to linear equations

The system of n simultaneous linear equations

$$\begin{aligned} a_{11}x_1 + a_{12}x_2 + \cdots + a_{1n}x_n &= y_1 \\ a_{21}x_1 + a_{22}x_2 + \cdots + a_{2n}x_n &= y_2 \\ \cdot \qquad\qquad\qquad & \quad \cdot \\ \cdot \qquad\qquad\qquad & \quad \cdot \\ \cdot \qquad\qquad\qquad & \quad \cdot \\ a_{n1}x_1 + a_{n2}x_2 + \cdots + a_{nn}x_n &= y_n \end{aligned} \tag{D-9}$$

can be written as the product of the matrices

$$\begin{Vmatrix} a_{11} & a_{12} & a_{13} & \cdots & a_{1n} \\ a_{21} & a_{22} & a_{23} & \cdots & a_{2n} \\ \cdot & \cdot & \cdot & \cdots & \cdot \\ a_{n1} & a_{n2} & a_{n3} & \cdots & a_{nn} \end{Vmatrix} \times \begin{Vmatrix} x_1 \\ x_2 \\ \cdot \\ x_n \end{Vmatrix} = \begin{Vmatrix} y_1 \\ y_2 \\ \cdot \\ y_n \end{Vmatrix} \tag{D-9a}$$

If we denote the matrix of the coefficients by $\|a_{rc}\|$, then Eqs. (D-9) can be expressed as

$$\|a_{rc}\| \times \|x_r\| = \|y_r\|$$

where $\|a_{rc}\|$ is always a square matrix, and $\|x_r\|$ and $\|y_r\|$ have one column only and are called column matrices.

D-20 Cramer's rule

Cramer's rule concerns the solution of linear equations by determinants. This rule states that the solution for any variable x_m in the system of equations

$$\begin{aligned} a_{11}x_1 + a_{12}x_2 + \cdots + a_{1m}x_m + \cdots + a_{1n}x_n &= y_1 \\ a_{21}x_1 + a_{22}x_2 + \cdots + a_{2m}x_m + \cdots + a_{2n}x_n &= y_2 \\ a_{31}x_1 + a_{32}x_2 + \cdots + a_{3m}x_m + \cdots + a_{3n}x_n &= y_3 \\ \cdot \qquad & \quad \cdot \\ \cdot \qquad & \quad \cdot \\ \cdot \qquad & \quad \cdot \\ a_{n1}x_1 + a_{n2}x_2 + \cdots + a_{nm}x_m + \cdots + a_{nn}x_n &= y_n \end{aligned}$$

is given by

$$x_m = \frac{\overset{mth\ column}{\begin{vmatrix} a_{11} & a_{12} & \cdots & y_1 & \cdots & a_{1n} \\ a_{21} & a_{22} & \cdots & y_2 & \cdots & a_{2n} \\ \cdot & \cdot & \cdots & \cdot & \cdots & \cdot \\ a_{n1} & a_{n2} & \cdots & y_n & \cdots & a_{nn} \end{vmatrix}}}{D_A}$$

that is,

$$x_m = \frac{\text{determinant of the } a\text{'s with column } m \text{ replaced by the } y\text{'s}}{\text{determinant of the } a\text{'s}}$$

Thus, for example, the set of equations

$$\begin{aligned} 5i_1 - 2i_2 - i_3 &= 4 \\ -2i_1 + 10i_2 - 3i_3 &= 2 \\ -i_1 - 3i_2 + 4i_3 &= 0 \end{aligned}$$

has the solutions

$$i_1 = \frac{\begin{vmatrix} 4 & -2 & -1 \\ 2 & 10 & -3 \\ 0 & -3 & 4 \end{vmatrix}}{D} \qquad i_2 = \frac{\begin{vmatrix} 5 & 4 & -1 \\ -2 & 2 & -3 \\ -1 & 0 & 4 \end{vmatrix}}{D} \qquad i_3 = \frac{\begin{vmatrix} 5 & -2 & 4 \\ -2 & 10 & 2 \\ -1 & -3 & 0 \end{vmatrix}}{D}$$

where

$$D = \begin{vmatrix} 5 & -2 & -1 \\ -2 & 10 & -3 \\ -1 & -3 & 4 \end{vmatrix}$$

The proof of Cramer's rule is quite straightforward. We show that Cramer's rule applies for the solution x_1. Since we can rearrange the column in Eq. (D-9) to call any unknown x_1, this proof is general. Now let us multiply the first of the equations by the cofactor $(F_A)_{11}$, the second by the cofactor $(F_A)_{21}$, the third by $(F_A)_{31}$, etc. The equations now read

$$\begin{aligned} a_{11}(F_A)_{11}x_1 + a_{12}(F_A)_{11}x_2 + a_{13}(F_A)_{11}x_3 + \cdots + a_{1n}(F_A)_{11}x_n &= (F_A)_{11}y_1 \\ a_{21}(F_A)_{21}x_1 + a_{22}(F_A)_{21}x_2 + a_{23}(F_A)_{21}x_3 + \cdots + a_{2n}(F_A)_{21}x_n &= (F_A)_{21}y_2 \\ \vdots \qquad\qquad & \qquad \vdots \\ a_{n1}(F_A)_{n1}x_1 + a_{n2}(F_A)_{n1}x_2 + a_{n3}(F_A)_{n1}x_3 + \cdots + a_{nn}(F_A)_{n1}x_n &= (F_A)_{n1}y_n \end{aligned}$$

Now we add all the equations, and the result reads

$$\begin{aligned} &[a_{11}(F_A)_{11} + a_{21}(F_A)_{21} + a_{31}(F_A)_{31} + \cdots + a_{n1}(F_A)_{n1}]x_1 \\ &\quad + [a_{12}(F_A)_{11} + a_{22}(F_A)_{21} + a_{32}(F_A)_{31} + \cdots + a_{n2}(F_A)_{n1}]x_2 + \cdots \\ &\qquad = (F_A)_{11}y_1 + (F_A)_{21}y_2 + (F_A)_{31}y_3 + \cdots + (F_A)_{n1}y_n \end{aligned}$$

But from the properties of cofactors given in Eq. (D-8) and the above equation, we see that the coefficient of x_1 is D_A and the coefficients of x_2, $x_3, \ldots, x_n$ are identically zero. Therefore

$$D_A x_1 = y_1(F_A)_{11} + y_2(F_A)_{21} + \cdots + (F_A)_{n1}y_n$$

$$x_1 = \frac{\sum_{m=1,2,\ldots,n} y_m(F_A)_{m1}}{D_A} \tag{D-10}$$

We can show that, for other x such as x_b,

$$x_b = \frac{\sum_{m=1,2,\ldots,n} y_m(F_A)_{mb}}{D_A}$$

which proves Cramer's rule.

D-21 Inverse matrix

We can now restate the problem posed by n simultaneous linear equations in a slightly different fashion. Equation (D-10) states how the $\|y_r\|$ matrix is obtained if the $\|a_{rc}\|$ and the $\|x_r\|$ matrix is given. In other words, the multiplication $\|a_{rc}\| \times \|x_r\|$ transforms the x's into the y's through linear combinations. The problem of finding the x's, given the y's (that is, solving the simultaneous equations), is then the problem of finding the transformation which transforms the y's into x's; or given

$$\|a_{rc}\| \times \|x_r\| = \|y_r\|$$

we wish to find a matrix $\|b_{rc}\|$ such that

$$\|b_{rc}\| \times \|y_r\| = \|x_r\|$$

But we know, from Cramer's rule and from Laplace's development, that

$$x_e = y_1 \frac{(F_A)_{1e}}{D_A} + y_2 \frac{(F_A)_{2e}}{D_A} + \cdots + y_n \frac{(F_A)_{ne}}{D_A}$$

Hence the $\|b\|$ matrix has the elements $b_{rc} = (F_A)_{cr}/D_A$.

We now define the $\|b_{rc}\|$ matrix as the *inverse* of the $\|a_{rc}\|$ matrix. Note that $D_A \neq 0$ is a necessary condition for the existence of the $\|b\|$ matrix. In general, then,

$$\text{Inverse of } \|a_{rc}\| = \|a_{rc}\|^{-1} = \left\| \frac{(F_A)_{cr}}{D_A} \right\|$$

For example,

$$\begin{Vmatrix} a_{11} & a_{12} \\ a_{21} & a_{22} \end{Vmatrix}^{-1} = \begin{Vmatrix} \dfrac{a_{22}}{a_{11}a_{22} - a_{12}a_{21}} & \dfrac{-a_{12}}{a_{11}a_{22} - a_{12}a_{21}} \\ \dfrac{-a_{21}}{a_{11}a_{22} - a_{21}a_{12}} & \dfrac{a_{11}}{a_{11}a_{22} - a_{12}a_{21}} \end{Vmatrix} \qquad \text{if } a_{11}a_{22} - a_{12}a_{21} \neq 0$$

$$\text{For } A = \begin{Vmatrix} 1 & 2 \\ 3 & 4 \end{Vmatrix}$$

$$A^{-1} = \begin{Vmatrix} \dfrac{4}{4-6} & \dfrac{-2}{4-6} \\ \dfrac{-3}{4-6} & \dfrac{1}{4-6} \end{Vmatrix} = \begin{Vmatrix} \left(-\dfrac{1}{2}\right)(4) & \left(+\dfrac{1}{2}\right)(2) \\ \left(+\dfrac{1}{2}\right)(3) & \left(-\dfrac{1}{2}\right)(1) \end{Vmatrix}$$

D-22 Column and row vectors, transpose

Very often a matrix consists of a single column, as, for example, the matrix of x's or of y's in Eq. (D-9a). Such a matrix is termed column matrix, or column vector.[1] We shall use a small boldface letter with an arrow over it to denote such a matrix. Thus, for example,

$$\vec{\mathbf{x}} = \begin{Vmatrix} x_1 \\ x_2 \\ \cdot \\ \cdot \\ \cdot \\ x_n \end{Vmatrix} \qquad \vec{\mathbf{y}} = \begin{Vmatrix} y_1 \\ y_2 \\ \cdot \\ \cdot \\ \cdot \\ y_n \end{Vmatrix} \tag{D-11}$$

TRANSPOSE The transpose of the matrix $\|A\| = \|a_{ij}\|$ denoted by $\|A\|_T = \|B\| = \|b_{ij}\|$ is such that $b_{ij} = a_{ji}$. Thus the rows of $\|A\|_T$ are the

[1] The term "vector" is used by analogy to space vectors where a vector has three components. In the case of column vectors in matrix analysis, the matrix has one column but an arbitrary though finite number of elements.

columns of $\|A\|$, and columns of $\|A\|_T$ are the rows of $\|A\|$. For example, if

$$\|A\| = \begin{Vmatrix} 5 & 6 & 7 \\ 2 & -4 & 3 \end{Vmatrix}$$

then its transpose is

$$\|A\|_T = \begin{Vmatrix} 5 & 2 \\ 6 & -4 \\ 7 & 3 \end{Vmatrix}$$

Row vector The transpose of a column vector is termed a row vector. If $\vec{\mathbf{x}}$ is the column vector defined in Eq. (D-11), its transpose consists of a single row,

$$\vec{\mathbf{x}}_T = \|x_1 \quad x_2 \quad \cdots \quad x_n\| \tag{D-12}$$

Using the above definition of column vectors, the simultaneous linear equations (D-9*a*) can be written

$$\|A_{ij}\|\vec{\mathbf{x}} = \vec{\mathbf{y}} \tag{D-13}$$

and we say that solving these equations amounts to finding the solution column vector $\vec{\mathbf{x}}$. The square matrix $\|A_{ij}\|$ can be written as the row matrix

$$\|A\| = \|\vec{\mathbf{a}}_1 \quad \vec{\mathbf{a}}_2 \quad \cdots \quad \vec{\mathbf{a}}_n\| \tag{D-14}$$

when each of the elements of the matrix A is a column vector.

$$\vec{\mathbf{a}}_1 = \begin{Vmatrix} a_{11} \\ a_{12} \\ \cdot \\ \cdot \\ \cdot \\ a_{1n} \end{Vmatrix} \qquad \vec{\mathbf{a}}_2 = \begin{Vmatrix} a_{21} \\ a_{22} \\ \cdot \\ \cdot \\ \cdot \\ a_{2n} \end{Vmatrix} \tag{D-15}$$

Using these definitions, we may write Eq. (D-13) as

$$\|\vec{\mathbf{a}}_1 \quad \vec{\mathbf{a}}_2 \quad \cdots \quad \vec{\mathbf{a}}_n\| \times \vec{\mathbf{x}} = \vec{\mathbf{y}}$$

D-23 Homogeneous equations

If all the elements in the $\vec{\mathbf{y}}$ matrix [Eq. (D-16)] are zero, the equations are *homogeneous* linear algebraic equations. If the A matrix is a square matrix with nonzero determinant, then, from Cramers' rule, the only solution vector $\vec{\mathbf{x}}$ is one in which all elements are zero; this is said to be the trivial case. If, on the other hand, the determinant of the A matrix is zero, it follows that (by Rule 3, Sec. D-14) the equations are not independent, since two or more of the given simultaneous equations are not linearly independent. For example, in the homogeneous linear equations

$$\begin{aligned} x_1 + x_2 + x_3 &= 0 \\ 3x_1 + 2x_2 - 4x_3 &= 0 \\ 2x_1 + x_2 - 5x_3 &= 0 \end{aligned} \tag{D-16}$$

we observe that the third can be derived from the first two by subtraction. It can therefore be deleted as superfluous. The two remaining equations are

$$\begin{aligned} x_1 + x_2 + x_3 &= 0 \\ 3x_1 + 2x_2 - 4x_3 &= 0 \end{aligned} \tag{D-17}$$

These cannot be solved for x_1, x_2, and x_3, but they do imply a fixed relationship between x_1, x_2, and x_3. To determine this relationship we write Eq. (D-17) in the form

$$\begin{aligned} x_1 + x_2 &= -x_3 \\ 3x_1 + 2x_2 &= 4x_3 \end{aligned} \tag{D-18}$$

and solve for x_1 and x_2 in terms of x_3, using Cramers' rule.

$$x_1 = -6x_3 \qquad x_2 = -7x_3$$

Hence we conclude that $x_1{:}x_2{:}x_3 = 6{:}-7{:}1$; that is, x_1, x_2, and x_3 must be in a certain proportion. The solution vector can be expressed in terms of x_3 as

$$\vec{\mathbf{x}} = x_3 \begin{Vmatrix} 6 \\ -7 \\ 1 \end{Vmatrix}$$

and the value of x_3 is arbitrary. Thus we can write

$$\vec{\mathbf{x}} = K_1 \begin{Vmatrix} 6 \\ -7 \\ 1 \end{Vmatrix} = K_2 \begin{Vmatrix} 1 \\ -\frac{7}{6} \\ \frac{1}{6} \end{Vmatrix} = \cdots$$

In the above example we had three equations, two of which were linearly independent. If we have more than one dependent equation, more solution vectors exist. Consider, for example, the set of linear homogeneous equations

$$\begin{aligned} x_1 + x_2 + x_3 - x_4 &= 0 \\ 4x_1 + 2x_2 + x_3 + 2x_4 &= 0 \\ 5x_1 + 3x_2 + 2x_3 + x_4 &= 0 \\ 9x_1 + 5x_2 + 3x_3 + 2x_4 &= 0 \end{aligned} \tag{D-19}$$

In Eqs. (D-19) the third was obtained by adding the first and the second, and the fourth is the sum of the second and third. Hence the last two are superfluous, and we write

$$\begin{aligned} x_1 + x_2 &= -x_3 + x_4 \\ 4x_1 + 2x_2 &= -x_3 - 2x_4 \end{aligned} \tag{D-20}$$

Solving in terms of x_3 and x_4, we obtain

$$x_1 = \tfrac{1}{2}x_3 + 2x_4 \qquad x_2 = -\tfrac{3}{2}x_3 + 3x_4$$

or

$$\begin{Vmatrix} x_1 \\ x_2 \\ x_3 \\ x_4 \end{Vmatrix} = x_3 \begin{Vmatrix} \frac{1}{2} \\ -\frac{3}{2} \\ 1 \\ 0 \end{Vmatrix} + x_4 \begin{Vmatrix} 2 \\ 3 \\ 0 \\ 1 \end{Vmatrix} \tag{D-21}$$

Since no information is given about x_3 and x_4, they are arbitrary constants; that is, Eqs. (D-19) are satisfied for all x_3 and x_4 provided that x_1 and x_2 satisfy (D-21).

RANK AND NULLITY Given a square matrix of order n, its rank r is defined as the number of independent rows it has and its *nullity* is defined as $n - r$. In the matrix of coefficients of Eqs. (D-16), the order is 3, the rank is 2, and the nullity is 1. In the case of Eqs. (D-19), the order of the matrix of coefficients is 4, the rank is 2, and the nullity is 2. From the above discussion it follows that, in the equation

$$\|A_{ij}\|\vec{\mathbf{x}} = 0$$

the nullity of the A matrix is equal to the number of superfluous equations, which is equal to the number of arbitrary constants in the solution.

D-24 The characteristic-value problem

In problems involving linear equations (see, for example, Secs. 12-1 and 12-2), one often desires to find the number or numbers s which satisfy the equation

$$\|A_{ij}\|\vec{\mathbf{x}} = s\vec{\mathbf{x}}$$

that is,

$$\begin{Vmatrix} a_{11} & a_{12} & \cdots & a_{1n} \\ \cdot & \cdot & \cdots & \cdot \\ a_{n1} & a_{n2} & \cdots & a_{nn} \end{Vmatrix} \times \begin{Vmatrix} x_1 \\ x_2 \\ \cdot \\ \cdot \\ \cdot \\ x_n \end{Vmatrix} = \begin{Vmatrix} sx_1 \\ sx_2 \\ \cdot \\ \cdot \\ \cdot \\ sx_n \end{Vmatrix} \tag{D-22}$$

The values of s which satisfy Eq. (D-22) are termed the *characteristic values* of the A matrix. Corresponding to s_k, the kth value of s satisfying (D-22), there is a column vector $\vec{\mathbf{x}}_{ck}$ which satisfies (D-22). This column vector is called a *characteristic vector* of the matrix A [Eqs. (D-27) to (D-29)].

From the rules of matrix addition and subtraction (Sec. D-6), it follows that Eq. (D-22) requires that $\vec{\mathbf{x}}$ satisfy the homogeneous linear equations

$$\begin{Vmatrix} a_{11} - s & a_{12} & \cdots & a_{1n} \\ a_{21} & a_{22} - s & \cdots & a_{2n} \\ \cdot & \cdot & \cdots & \cdot \\ a_{n1} & a_{n2} & \cdots & a_{nn} - s \end{Vmatrix} \times \begin{Vmatrix} x_1 \\ x_2 \\ \cdot \\ \cdot \\ \cdot \\ x_n \end{Vmatrix} = \begin{Vmatrix} 0 \\ 0 \\ \cdot \\ \cdot \\ \cdot \\ 0 \end{Vmatrix} \qquad \text{(D-23)}$$

We define the matrix $\|B\|$

$$\|B\| = \begin{Vmatrix} a_{11} - s & \cdots & a_{1n} \\ \cdot & \cdots & \cdot \\ a_{n1} & \cdots & a_{nn} - s \end{Vmatrix}$$

which is *termed the characteristic matrix of the A matrix,* and observe that nontrivial solutions of Eqs. (D-23) are possible only if the determinant of the B matrix is zero:

$$\det \|B\| = 0$$

that is, the so-called characteristic equation

$$D_B(s) = \begin{vmatrix} a_{11} - s & a_{12} & \cdots & a_{1n} \\ a_{21} & a_{22} - s & \cdots & a_{2n} \\ \cdot & \cdot & \cdots & \cdot \\ a_{n1} & a_{n2} & \cdots & a_{nn} - s \end{vmatrix} = 0 \qquad \text{(D-24)}$$

Note that the values of s found from the nth-degree algebraic equation (D-24) are the *characteristic values* of the A matrix. We observe that Eq. (D-24) has n solutions, which may all be distinct or not, and that the characteristic values may be real, imaginary, or complex.

We observe, further, that for each characteristic value, the rank of the B matrix is less than its order. If there are n distinct characteristic values, $s_1, \ldots, s_n$, it can be shown that the nullity of the B matrix is one and therefore one of the x's can be chosen arbitrarily for each characteristic value. [In the case of multiple characteristic roots of Eq. (D-24), the situation is more complicated, and the reader is referred to the mathematics literature.]

D-25 Characteristic vectors

We consider now only the case where the characteristic equation (D-24) has n distinct solutions. For this case the set of equations (D-23) is satisfied for n values of s; that is,

$$\begin{Vmatrix} a_{11} - s_k & a_{12} & \cdots & a_{1n} \\ \cdot & \cdot & \cdots & \cdot \\ a_{1n} & a_{2n} & \cdots & a_{nn} - s_k \end{Vmatrix} \times \begin{Vmatrix} x_1 \\ x_2 \\ \cdot \\ \cdot \\ \cdot \\ x_n \end{Vmatrix} = \begin{Vmatrix} 0 \\ 0 \\ \cdot \\ \cdot \\ \cdot \\ 0 \end{Vmatrix} \qquad k = 1, \ldots, n \qquad \text{(D-25)}$$

The set of equations (D-25) contains one superfluous equation; let us assume that it is the first one, so that we delete the first equation. We now choose x_1 as an arbitrary constant (which may be conveniently taken to be -1), and we solve for $x_2, \ldots, x_n$ in terms of x_1, for each characteristic value of s. To do this we must solve the equation

$$\begin{Vmatrix} a_{22} - s_k & a_{23} & \cdots & a_{2n} \\ \cdot & \cdot & \cdots & \cdot \\ a_{n2} & a_{n3} & \cdots & a_{nn} - s_k \end{Vmatrix} \times \begin{Vmatrix} x_2 \\ x_3 \\ \cdot \\ \cdot \\ \cdot \\ x_n \end{Vmatrix} = \begin{Vmatrix} a_{21}x_1 \\ a_{31}x_1 \\ \cdot \\ \cdot \\ \cdot \\ a_{n1}x_1 \end{Vmatrix} = \begin{Vmatrix} a_{21} \\ a_{31} \\ \cdot \\ \cdot \\ \cdot \\ a_{n1} \end{Vmatrix} \qquad \begin{matrix} k = 1, \ldots, n \\ x_1 = -1 \end{matrix}$$

Solving by Cramers' rule, we observe that, for example,

$$x_{2k} = \frac{\begin{vmatrix} a_{21} & a_{23} & \cdots & a_{2n} \\ \cdot & \cdot & \cdots & \cdot \\ a_{n1} & a_{n2} & \cdots & a_{nn} - s_k \end{vmatrix}}{\begin{vmatrix} a_{22} - s_k & \cdots & a_{2n} \\ \cdot & \cdots & \cdot \\ a_{n2} & \cdots & a_{nn} - s_k \end{vmatrix}} \qquad k = 1, \ldots, n$$

where, in x_{2k}, the subscript k indicates that the characteristic value s_k is used. We observe that the denominator of the above equation is the cofactor of $D(s_k)$ for the first column and row and that the numerator is the cofactor of the first row and second column; i.e.,

$$x_{2k} = \frac{(F_{D_k})_{12}}{(F_{D_k})_{11}}$$

where the subscript k indicates that the characteristic value s_k is used. Similarly, it can be shown that

$$x_{3k} = \frac{(F_{D_k})_{13}}{(F_{D_k})_{11}}$$

$$x_{jk} = \frac{(F_{D_k})_{1j}}{(F_{D_k})_{11}} \qquad \text{(D-26)}$$

It is to be understood that, for each value of s_k, Eq. (D-26) yields a solution vector of the form given by Eq. (D-23); these solution vectors are termed *characteristic vectors of the A matrix.*

Appendix E Computer solution of network problems

At present there exist[1] computer programs and computer-input facilities by means of which one may "sketch" a circuit diagram, assign values to circuit elements and sources and command the computer to compute and display the waveform of any of the desired responses. Using these facilities the impulsive or step response as well as the frequency response of the network may be displayed. These facilities may also be used for the purpose of design, inasmuch as the computer may vary the value assigned to various elements of the network to achieve a prespecified response. The details of such facilities and programs are, of course, very lengthy and cannot be discussed here. References to system-analysis programs may be found in literature.[2] However, a large part of such programs is concerned with input-output manipulations and evaluation of the network functions, and the response calculations may occupy only a small portion of the program. In this appendix we give two examples to show how a program may be written to evaluate circuit response if the network function or the impulsive response is known. The algorithms used in these programs are not necessarily the most efficient or the most accurate. They are chosen for their simplicity of illustrating the use of computer programs in calculating circuit

[1] W. H. Ninke, Graphic 1—A Remote Graphical Display Console System, *Proc. of Fall Joint Computer Conf.*, vol. 27, pp. 834–846, 1965.

[2] A detailed description of such programs and many references may be found in the book "System Analysis by Digital Computers," edited by F. F. Kuo and J. F. Kaiser, John Wiley & Sons, Inc., New York, 1966.

response. In one example the program evaluates the response of a network to a source at instants $t_k = k\,\Delta t$, where $k = 1, 2, \ldots, N$ and N and Δt are prespecified values. In the other example the program evaluates the frequency response of a doubly tuned circuit at radian frequencies $\omega_k = k\,\Delta\omega$ where $k = 1, 2, \ldots, N$ and N and $\Delta\omega$ are prespecified values.

E-1 Computer usage of superposition principle[1]

We have shown in Chap. 2 [Eq. (2-13)] that the response of a linear system to any input can be approximated as the sum of the responses to step-function increments of input. Thus if $a(t)$ is a response of a linear system that stores zero energy[2] at $t = 0$ to the source $u(t)$, then the response v_2 to a source $v_1(t)u(t)$ at the instant $t = n\,\Delta t$ is given approximately by

$$v_2(n\,\Delta t) = v_1(0^+)a(t) + \sum_{m=1}^{n} (\Delta v_1)_m a(n\,\Delta t - m\,\Delta t) \tag{E-1}$$

where $(\Delta v_1)_m$ is the increment of v_1 occurring at the instant $m\,\Delta t$. Equation (E-1) can be made exact by making the increments infinitesimal, thus changing the summation to an integration [see the discussion in connection with Eq. (E-5) below]. One method of computing network responses numerically consists of using the summation form (E-1), with sufficiently small increments. This requires that $a(t)$ be known either as a formula or as an experimental response record.

If the response of a system to the unit-impulse source $\delta(t)$ rather than to the step-function source $u(t)$ is known, a summation similar to Eq. (E-1) can be used to relate (approximately) the response to a source $v_1(t)$ to response of the network to $\delta(t)$. To deduce this formula we denote by $h(t)$ the response v_2 when $v_1 = \delta(t)$. [The reader may recall that $h(t) = \mathcal{L}^{-1}[H(s)]$, see Eq. (8-32$a$) with $V_1(s) = 1$.] From the related sources theorem we have

$$h(t) = \frac{da(t)}{dt} \qquad h(t - t_1) = \frac{d}{dt}a(t - t_1)$$

or

$$h(t - t_1)\,\Delta t \approx \Delta a\,(t - t_1) \tag{E-2}$$

We now consider an input $v_1(t)$ approximated by rectangular pulses as shown in Fig. E-1. From Fig. E-1 we observe that

$$v_1(n\,\Delta t) \approx \sum_{m=0}^{n} v_1(m\,\Delta t)[u(n\,\Delta t - m\,\Delta t) - u(n\,\Delta t - m\,\Delta t - \Delta t)]$$

Now the response to $v_1(m\,\Delta t)u(n\,\Delta t - m\,\Delta t)$ is $v_1(m\,\Delta t)a(n\,\Delta t - m\,\Delta t)$ so

[1] The formulas discussed in this section can also be interpreted as convolution formulas, see Sec. 8-12.

[2] When using superposition formulas, initial stored energy must be represented by initial-condition generators.

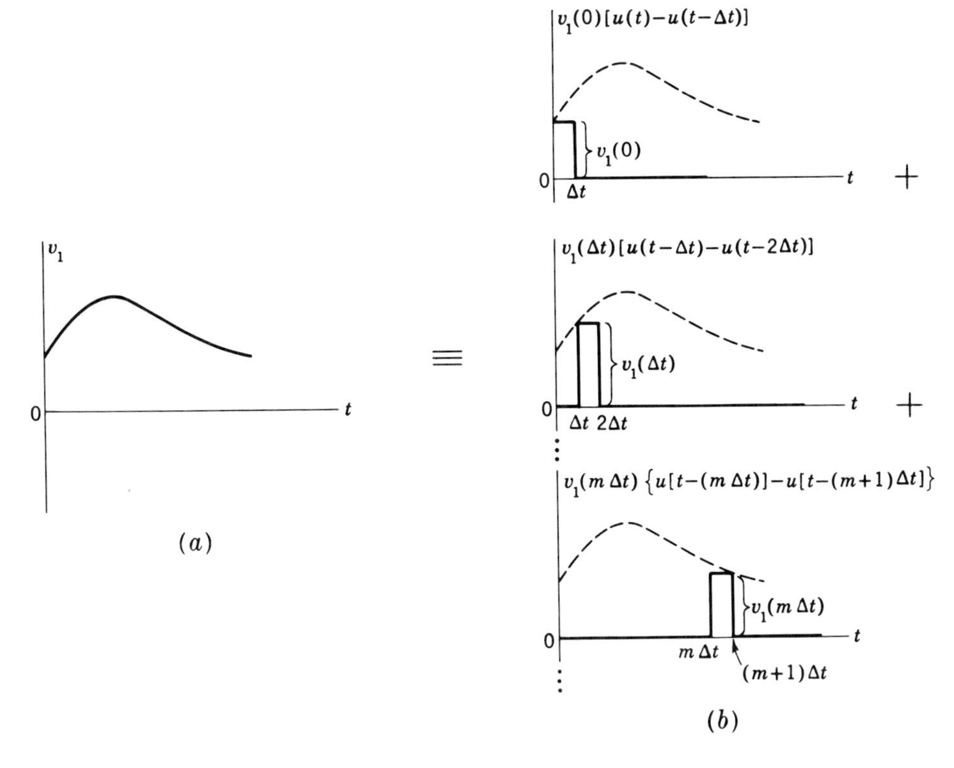

***Fig. E-1** The waveform shown in (a) can be represented approximately as the sum of the rectangular pulses shown in (b).*

that by superposition

$$v_2(n\,\Delta t) \approx \sum_0^n v_1(m\,\Delta t)[a(n\,\Delta t - m\,\Delta t) - a(n\,\Delta t - m\,\Delta t - \Delta t)]$$

The quantity in the brackets above is seen to be the increment Δa in the interval $m\,\Delta t$ to $(m+1)\,\Delta t$; from Eq. (E-2) this is $h(n\,\Delta t - m\,\Delta t)\,\Delta t$. Hence

$$v_2(n\,\Delta t) \approx \sum_0^n [v_1(m\,\Delta t)\,\Delta t\,h(n\,\Delta t - m\,\Delta t)] \tag{E-3}$$

Equation (E-3) is the desired equation relating approximately the response to a source $v_1(t)$ to the response when $v_1(t) = \delta(t)$.

Superposition integrals If in Eq. (E-1) or (E-3) we let $\Delta t \to d\tau$ $m\,\Delta t \to \tau$, $n\,\Delta t \to t$ then we obtain the superposition integrals.[1] In the

[1] See also Sec. 8-12.

case of Eq. (E-3) the result is

$$v_2(t) = \int_0^t v_1(\tau)h(t - \tau)\, d\tau \tag{E-4}$$

where it has been assumed implicitly that $v_1 \equiv 0$ for $\tau < 0$ and $h(t) \equiv 0$ for $t < 0$. The first of these assumptions is a matter of convenience (we begin the problem at $t = 0$) while the second states that no output can occur before the input is applied. In the case of Eq. (E-1) we recognize that $(\Delta v_1)_m \approx (dv/dt)_{t=m\,\Delta t} \times \Delta t$ and obtain

$$v_2(t) = v_1(0^+)a(t) + \int_0^t \frac{dv_1(\tau)}{d\tau}\, a(t - \tau)\, d\tau \tag{E-5}$$

Equations (E-4) and (E-5) are the "exact" version of the summations (E-3) and (E-1) respectively.

Example E-1 In the RC circuit of Fig. E-2 the response to $v_1(t) = \delta(t)$ is $v_2(t) = h(t) = (1/RC)e^{-t/RC}u(t)$. Find the response to $v_1(t) = V \sin \omega t u(t)$ for specified values of R, C, ω and draw N samples of $v_2(t)$ at intervals of Δt, where N and Δt are also specified.

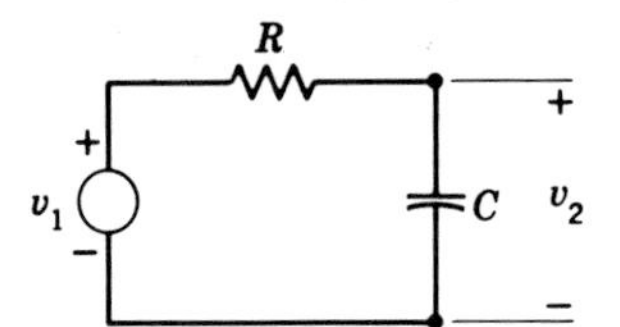

Fig. E-2 Circuit for Example E-1.

Solution Using Eq. (E-3) we have

$$v_2(n\,\Delta t) = \sum_{m=0}^{n} V \sin\,(\omega m\,\Delta t)\,\Delta t\,\frac{1}{RC}\,e^{-(n-m)\,\Delta t/RC}$$

Defining the program variables

WDT → $\omega\,\Delta T$
FM → m where F refers to "floating-point" value
FNM → $n - m$
TDRC → $\Delta t/RC$
V2(N) → $v_2(n\,\Delta t)$

we have

$$\text{V2(N)} = \text{TDRC} \sum_{\text{FNM}=\text{N}}^{\text{FNM}=\text{O}} \sin\,(\text{WDT} * \text{FM})/\text{EXP}(\text{FNM} * \text{TDRC})$$

Table E-1 Fortran program for output of the R-C circuit of Fig. E-2

```
                         COMPUTATION CENTER

              SCHOOL OF ENGINEERING AND ARCHITECTURE
                         THE CITY COLLEGE

                                              FORTRAN SOURCE LIST
ISN          SOURCE STATEMENT

  0 $IBFTC GREEN
  1          CALL NOHD
  2           DIMENSION V2(1000)
  3           COMMON INN (4100)
  4           READ 6,NR
  6 6         FORMAT(I4)
  7           DO 100 IR=1,NR
 10           READ 7,R,C,W,NS,DT
 12 7         FORMAT(F9.3,F9.6,F9.3,I5,F6.3)
 13           PRINT 8
 14 8         FORMAT(1H1,////)
 15           VHI=0.0
 16           WDT=W*DT
 17           TDRC=DT/(R*C)
 20           DO 1 N=1, NS
 21           V2(N)=0.0
 22           DO3M=1,N
 23           FM=FLOAT(M)
 24           FNM=FLOAT(N-M)
 25           V2(N)=V2(N)+SIN(WDT*FM)/EXP(FNM*TDRC)
 26           IF(ABS(V2(N)).GT.VHI) VHI=ABS(V2(N))
 31 3         CONTINUE
 33 1         CONTINUE
 35           DO2 L=1,NS
 36           INN(L)=(V2(L)*50.0)/VHI
 37 2         CONTINUE
 41           I1=1
 42           I2=NS
 43           CALL GRAFPN (I1,I2)
 44 100       CONTINUE
 46           CALL EXIT
 47           END
```

The instructions with internal numbers (ISN)16 to 25 of the program of Table E-1 with the DO loop with ISN number 22 implements this summation. This program has additional instructions for drawing the graph of the output V2(N). A subroutine GRAFPN (I1, I2) is used for plotting samples I1 through I2 stored in locations INN. The calling program of Table E-1 defines I1 and I2 as I1 = 1 and I2 = NS. The program parameters R, C, ω, Δt and number of samples, NS, are read from an input card. The number NR read from a separate card indicates how many data cards are to be read. Typical results are shown in Figs. E-3 and E-4.

E-2 Additional example

Since computers can perform a large number of arithmetic operations rapidly they can be used to advantage when functions of many parameters are to be analyzed. The following example illustrates such a case involving an algebraic function.

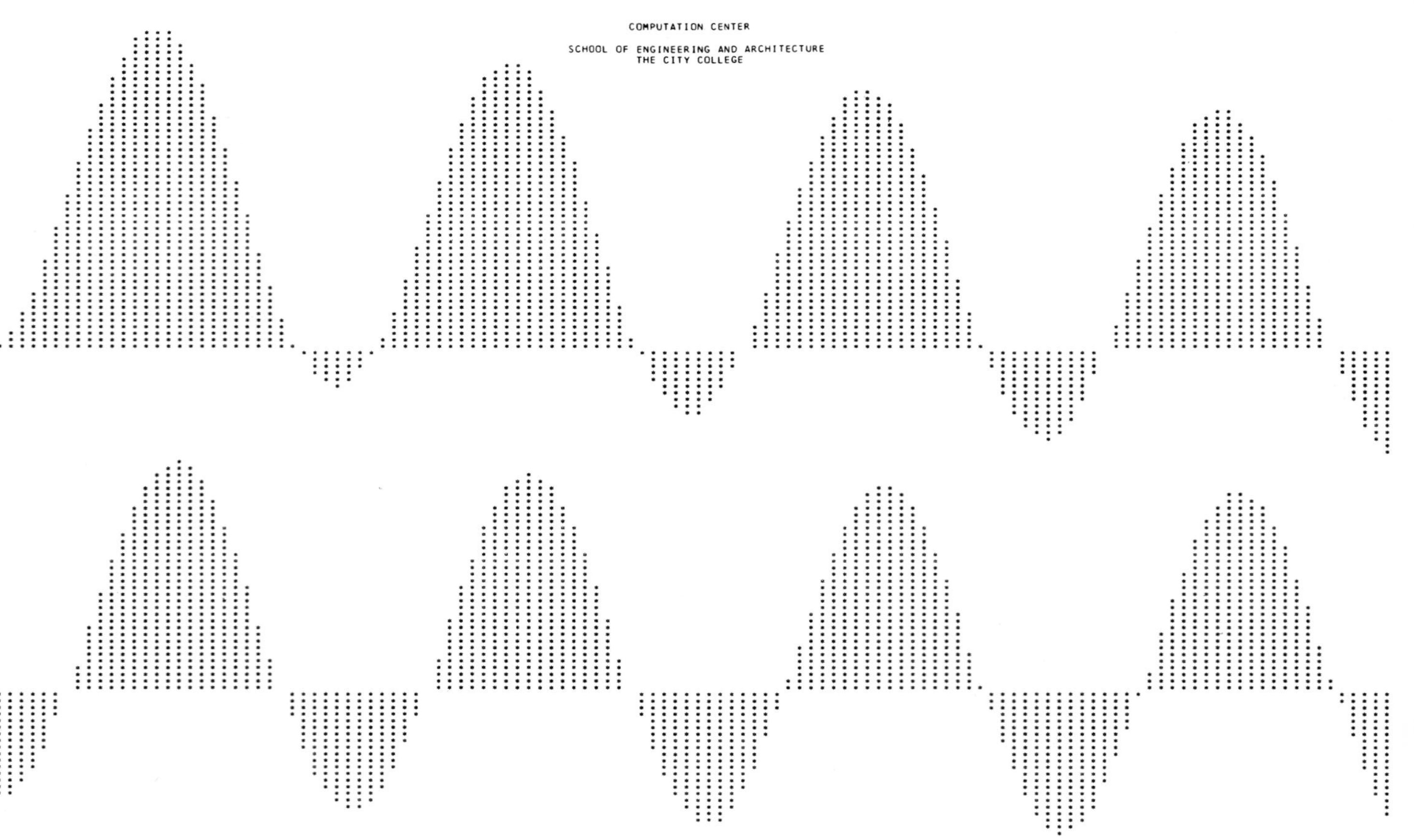

Fig. E-3 Computed response for the output $v_2(t)$ of the R-C circuit shown in Fig. E-2 for $R = 25$, $C = 1$ and $v_1(t) = \sin t$. The output samples are shown with normalized ordinates; they are computed at intervals $\Delta t = 0.2$ sec. The lower portion of the illustration immediately follows the upper portion. Note that the time constant corresponds to about 4 cycles ($25/2\pi$) of the source sinusoid so that the transient is quite pronounced.

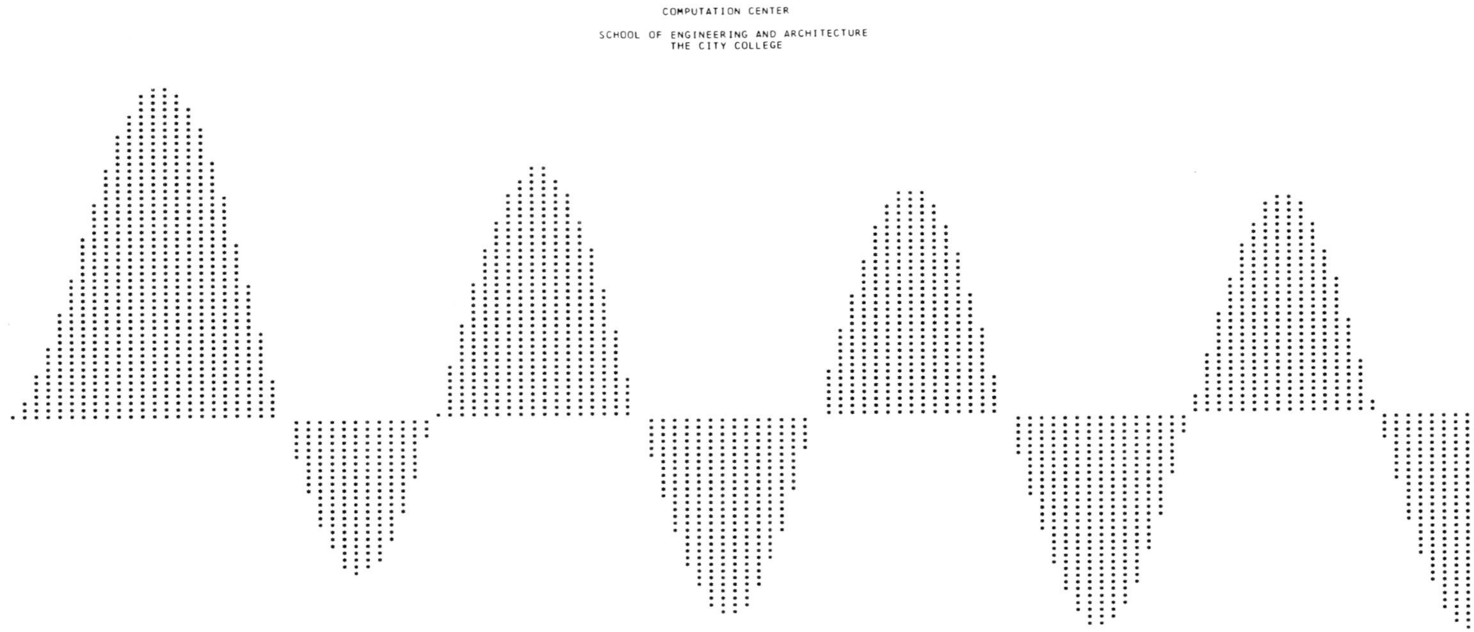

COMPUTATION CENTER

SCHOOL OF ENGINEERING AND ARCHITECTURE
THE CITY COLLEGE

Fig. E-4 **Computed response $v_2(t)$ of the R-C circuit of Fig. E-2 for $R = 5$, $C = 1$ and $v_1(t) = \sin t$. The output samples are computed and shown with normalized ordinates at intervals $\Delta t = 0.2$ sec. Note that in this case the time constant $RC = 5$, differing from a period of sin t by less than 25 per cent, so that the transient is less noticeable than in Fig. E-3.**

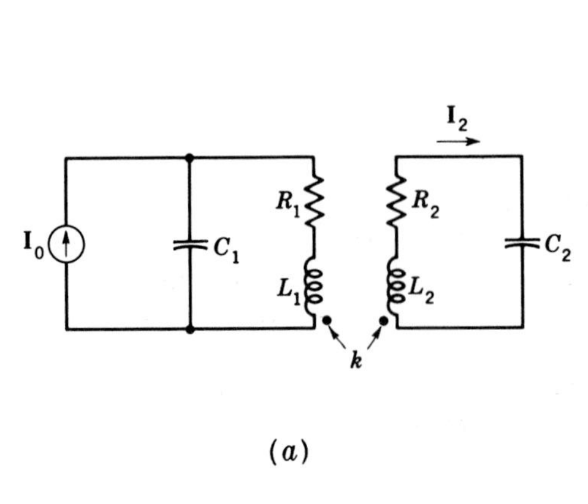

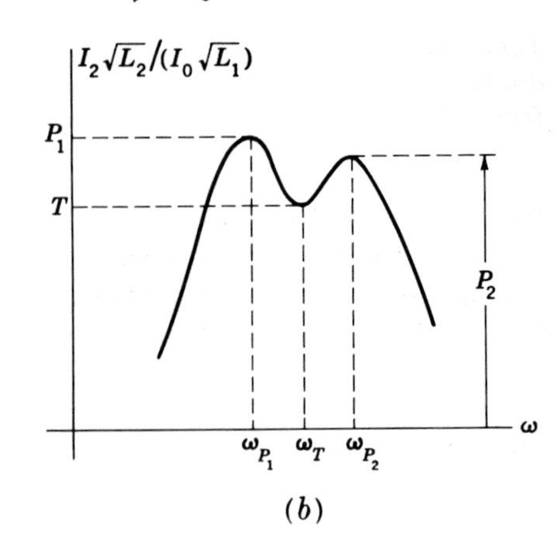

***Fig. E-5** (a) Doubly tuned circuit energized by ideal-current source. (b) Typical magnitude response ratio $\mathbf{I}_2/\mathbf{I}_0$ showing the quantities calculated in Example E-2.*

Example E-2 In the doubly tuned circuit of Fig. E-5*a* calculate the magnitude of the phasor ratio I_2/I_0 as a function of frequency for various resonant frequencies, quality ratios, Q, and coefficients of coupling.

Solution The phasor ratio $\mathbf{I}_2/\mathbf{I}_0$ can be shown to be given by

$$\frac{\mathbf{I}_2}{\mathbf{I}_0} = \sqrt{\frac{L_1}{L_2}} \frac{-k\omega^2\omega_1^2}{\omega^4(1 - k^2) - \omega^2(\omega_1^2 + \omega_2^2 + \omega_1\omega_2/Q_1Q_2) + \omega_1^2\omega_2^2 + j[\omega\omega_1\omega_2(\omega_1/Q_2 + \omega_2/Q_1) - \omega^3(\omega_1/Q_1 + \omega_2/Q_2)]}$$

where

$$\omega_1 = \frac{1}{\sqrt{L_1C_1}} \qquad \omega_2 = \frac{1}{\sqrt{L_2C_2}}$$

$$Q_1 = \frac{\omega_1 L_1}{R_1} \qquad Q_2 = \frac{\omega_2 L_2}{R_2}$$

For convenience we normalize with respect to $\sqrt{L_1/L_2}$ and study the normalized magnitude ratio $y = \sqrt{L_2}\, I_2/\sqrt{L_1}\, I_0$.

We recall that a typical frequency-response curve of a doubly tuned circuit has the shape shown in Fig. E-5*b*. Of interest are the peaks P_1 and P_2 and trough T, and the frequencies at which they occur ω_{P_1}, ω_{P_2}, and ω_T. These vary as circuit characteristics, such as coupling coefficient k, Q_1 and Q_2, and ω_1, ω_2 the resonant frequencies of the primary and secondary, are adjusted.

The values of k, ω_1, ω_2, Q_1, Q_2, the initial and final frequencies ω_I and ω_F for which the magnitude of I_2/I_0 is plotted at the increment of frequencies $\Delta\omega$, are punched on input cards. The value of N read from a separate card indicates the number of data cards which should be read. The program (see Table E-2) reads a data card and evaluates and plots a value y, proportional to the magnitude of I_2/I_0 for the values of circuit characteristics indicated on data cards. Some of these graphs are shown in Fig. E-6. In addition the program finds the values of P_1 and P_2 as well as the trough value T and the corresponding frequencies and tabulates these as shown in Table E-3. In the program of Table E-2 the follow-

Table E-2 Fortran program for plotting the frequency response of a doubly tuned circuit. The values shown in Table E-3 are also obtained from this program.

```
                              COMPUTATION CENTER

                    SCHOOL OF ENGINEERING AND ARCHITECTURE
                              THE CITY COLLEGE

                                             FORTRAN SOURCE LIST
ISN        SOURCE STATEMENT

  0 $IBFTC RESON
  1        DIMENSION S(1000)
  2        COMMON INN(4100)
  3         DIMENSION P(30,30)
  4        CALL NOHD
  5         READ 6,N
  7 6       FORMAT(I4)
 10         DO 100 IR=1,N
 11         LOJ=0
 12         LOJ2=1
 13        READ 2, W1, W2, Q1, Q2, DW, WI, WF,AKAY
 14 2       FORMAT(8F7.3)
 15        M = (WF-WI)/DW
 16        Q1Q2=Q1*Q2
 17        W12= W1**2
 20        W22 = W2**2
 21        W1W2=W1*W2
 22        W= WI-DW
 23        SHI=0.
 24        DO 99 I=1,M
 25        W = W+DW
 26        V2 = W**2
 27        V3= V2*W
 30        V4= V3*W
 31        A=(V4*(1.-AKAY**2)-V2*(W12+W22+(W1W2/Q1Q2))+W1W2**2)**2
 32        B=(W*W1*W2*((W1/Q2)+(W2/Q1))-V3*((W1/Q1)+(W2/Q2)))**2
 33        S(I)=(AKAY*V2*W12)/SQRT(A+B)
 34        IF (S(I) .GT. SHI) SHI=S(I)
 37         IF(I-1) 99,99,201
 40 201     IF(LOJ2-1)101,101,99
 41 101     IM=I-1
 42         IF(LOJ-1) 70,71,72
 43 70      IF(S(I)-S(IM)) 50,99,99
 44 50      P1=S(IM)
 45         WP1=W-DW
 46         LOJ=1
 47         GO TO 99
 50 71      IF(S(I)-S(IM)) 99,99,81
 51 81      T=S(IM)
 52         WT=W-DW
 53         LOJ=2
 54         GO TO 99
 55 72      IF(S(I)-S(IM)) 60,99,99
 56 60      P2=S(IM)
 57         WP2=W-DW
 60         LOJ2=2
 61 99     CONTINUE
 63         P(IR,1)=AKAY
 64         P(IR,2)=W1
 65         P(IR,3)=W2
 66         P(IR,4)=Q1
 67         P(IR,5)=Q2
 70         P(IR,6)=P1
 71         P(IR,9)=WP1
 72         IF(LOJ2-1)901,901,902
```

Table E-2 (Continued)

```
ISN          SOURCE STATEMENT

 73 901        P(IR,12)=1.0
 74            GO TO 77
 75 902        P(IR,7)=T
 76            P(IR,8)=P2
 77            P(IR,10)=WT
100            P(IR,11)=WP2
101            P(IR,12)=2.0
102 77        DO 93 L=1,M
103           INN(L)=(S(L)*50.0)/SHI
104 93        CONTINUE
106           PRINT 80, W1, W2, Q1, Q2, DW, AKAY, M, SHI
107 80        FORMAT (8H1FOR W1=,F6.3,4H W2=,F6.3,4H Q1=.F7.3,4H Q2=,F7.3,
             14H DW=,F6.3,3H K=,F7.4, 3H M=,I6,5H SHI=,F7.3,
             2 22H FREQUENCY RESPONSE IS)
110           PRINT 500
111 500       FORMAT(1H )
112           PRINT 500
113           I1=1
114           I2=M
115           CALL GRAFPN(I1,I2)
116 100       CONTINUE
120            PRINT 800
121 800        FORMAT(1H1,////)
122            PRINT 600
123 600        FORMAT(1H ,30X,69HCOMPUTED FREQUENCY RESPONSE CHARACTERISTICS OF
             1DOUBLY TUNED CIRCUITS.)
124            PRINT 801
125 801        FORMAT(1H ,/)
126            PRINT 700
127 700        FORMAT(1H ,20X,1HK,8X,2HW1,7X,2HW2,7X,2HQ1,7X,2HQ2,7X,2HP1,7X,
             1 1HT,8X,2HP2,7X,3HWP1,6X,2HWT,7X,3HWP2)
130            PRINT 801
131            DO 400 L=1,N
132            IF(P(L,12)-1.0) 991,991,990
133 990        PRINT 992,(P(L,NI),NI=1,11)
140 992        FORMAT(1H ,15X,11F9.3)
141            GO TO 400
142 991        PRINT 993,(P(L,1),P(L,2),P(L,3),P(L,4),P(L,5),P(L,6),P(L,9))
147 993        FORMAT(1H ,15X,6F9.3,4X,4HNONE,5X,4HNONE,1X,F9.3,4X,4HNONE,5X,
             1 4HNONE)
150 400        CONTINUE
152           CALL EXIT
153           END
```

ing symbols are used

$\omega_1 \to$ W1	= resonant frequency of primary
$\omega_2 \to$ W2	= resonant frequency of secondary
$Q_1 \to$ Q1	= quality factor of primary
$Q_2 \to$ Q2	= quality factor of secondary
$\Delta\omega \to$ DW	= increment of frequency at which computation is made
$\omega_I \to$ WI	= the initial frequency at which computation is made
$\omega_F \to$ WF	= the final frequency at which computation is made
$k \to$ AKAY	= coupling coefficient

In this program instructions ISN 15 through 22 define the variables; samples are calculated from ISN 24 through 33. The remainder of the program deals with the location of the peaks and the troughs, plotting graphs such as shown in Fig. E-6 and producing Table E-3.

COMPUTATION CENTER

SCHOOL OF ENGINEERING AND ARCHITECTURE
THE CITY COLLEGE

COMPUTED FREQUENCY RESPONSE CHARACTERISTICS OF DOUBLY TUNED CIRCUITS.

K	W1	W2	Q1	Q2	P1	T	P2	WP1	WT	WP2
0.010	1.000	1.000	50.000	50.000	20.002	NONE	NONE	1.000	NONE	NONE
0.020	1.000	1.000	50.000	50.000	25.001	NONE	NONE	1.000	NONE	NONE
0.040	1.000	1.000	50.000	50.000	25.095	20.001	24.095	0.985	1.000	1.020
0.040	1.000	1.200	50.000	50.000	4.424	0.943	4.334	1.000	1.095	1.200
0.040	1.000	1.200	50.000	50.000	4.424	0.943	4.334	1.000	1.095	1.200
0.040	1.000	1.400	50.000	50.000	2.070	0.246	2.046	1.000	1.185	1.400
0.010	1.000	1.000	50.000	20.000	9.091	NONE	NONE	1.000	NONE	NONE
0.040	1.000	1.000	50.000	20.000	15.499	NONE	NONE	0.995	NONE	NONE
0.080	1.000	1.000	50.000	20.000	15.005	10.811	13.939	0.965	1.000	1.035
0.020	1.000	1.100	50.000	20.000	4.484	1.654	1.929	1.000	1.055	1.095
0.020	1.100	1.000	50.000	20.000	2.341	2.005	5.391	1.005	1.040	1.100
0.010	1.100	1.000	50.000	50.000	2.857	1.147	2.856	1.000	1.050	1.100
0.020	1.100	1.000	50.000	50.000	5.650	2.225	5.627	1.000	1.050	1.100
0.040	1.100	1.000	50.000	50.000	10.417	3.968	10.483	0.995	1.050	1.105
0.040	1.200	1.000	50.000	50.000	6.370	1.359	6.241	1.000	1.095	1.200
0.040	1.400	1.000	50.000	50.000	4.058	0.482	4.010	1.000	1.185	1.400
0.100	1.000	1.000	10.000	10.000	5.075	NONE	NONE	0.980	NONE	NONE
0.200	1.000	1.000	10.000	10.000	5.405	3.947	4.541	0.925	1.015	1.100
0.400	1.000	1.000	10.000	10.000	5.882	2.248	3.905	0.850	1.050	1.275

Table E-3 Peak and trough values and their radian frequencies as defined in Fig. E-5b and in the text for various values of coupling and Q. The frequency increment used is $DW = 0.005$ rad/sec. To correlate the results with those derived in Chap. 15, note that for $Q_1 = Q_2 = 50$ and with $\omega_1 = \omega_2$ ($W1 = W2 = 1$), critical coupling corresponds to $k = 0.02$ and only one peak occurs for this and smaller values of coupling. When the coupling is increased to twice critical ($k = 0.04$), two peaks are observed. The peak at the higher frequency is smaller than that at the lower frequency; this can be seen through the pole-zero diagram (see Prob. 15-28). We observe that most of the other results shown in this table can be deduced from the formula with elaborate, time-consuming calculations, while a computer program produces results in equally short time for all combinations of parameters.

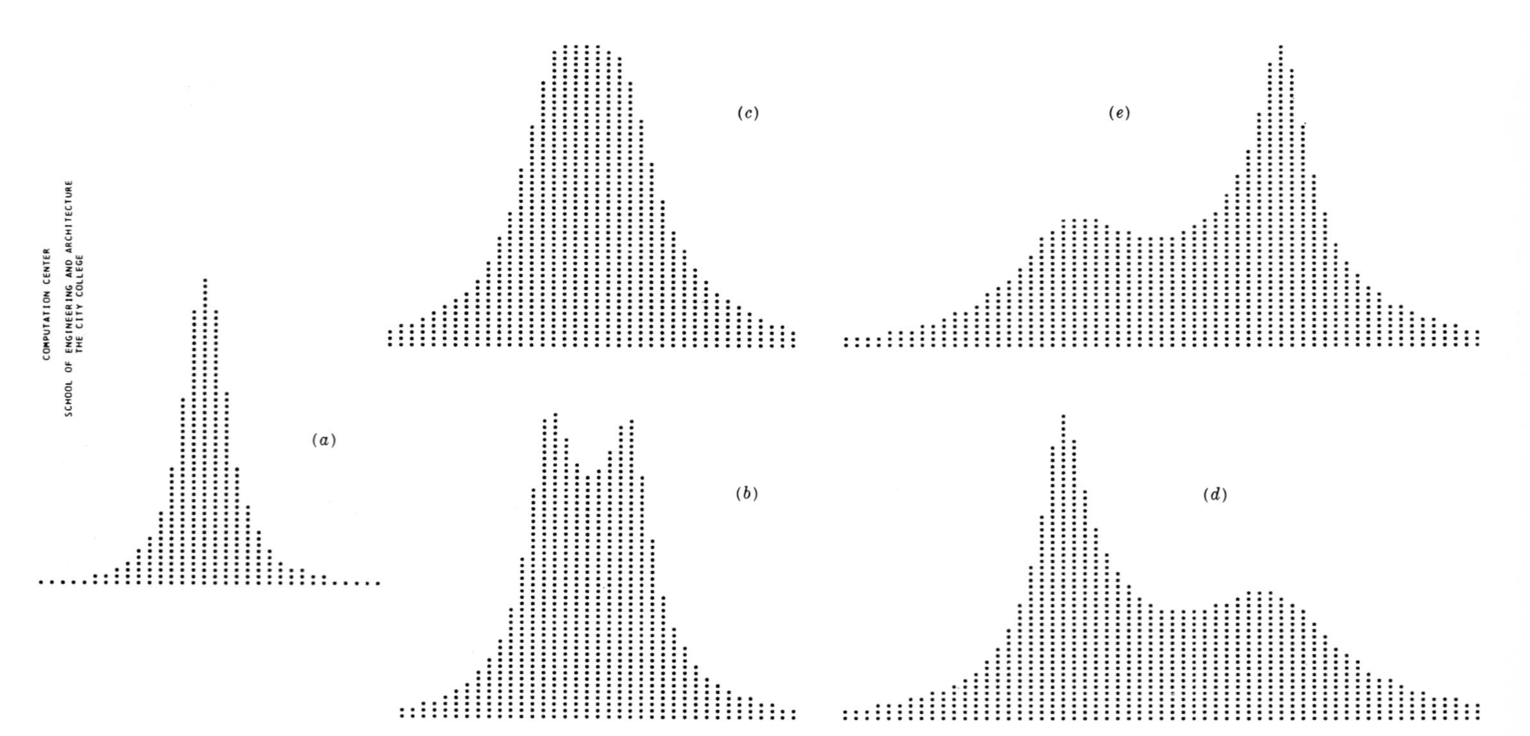

Fig. E-6 Histogram of the frequency response for the circuit of Fig. E-5a. The values of the parameters for the histograms of (a), (b), (c), (d) and (e) correspond respectively to lines 1, 3, 8, 10, and 11 in Table E-3. The frequency scale in each case can be inferred by noting WP1 from Table E-3 and using the fact that the increment in radian frequency between samples is DW = 0.005 rad/sec. The ordinates have been normalized in the histograms; actual values can be found from the values given for the peaks in Table E-3.

Index

Index